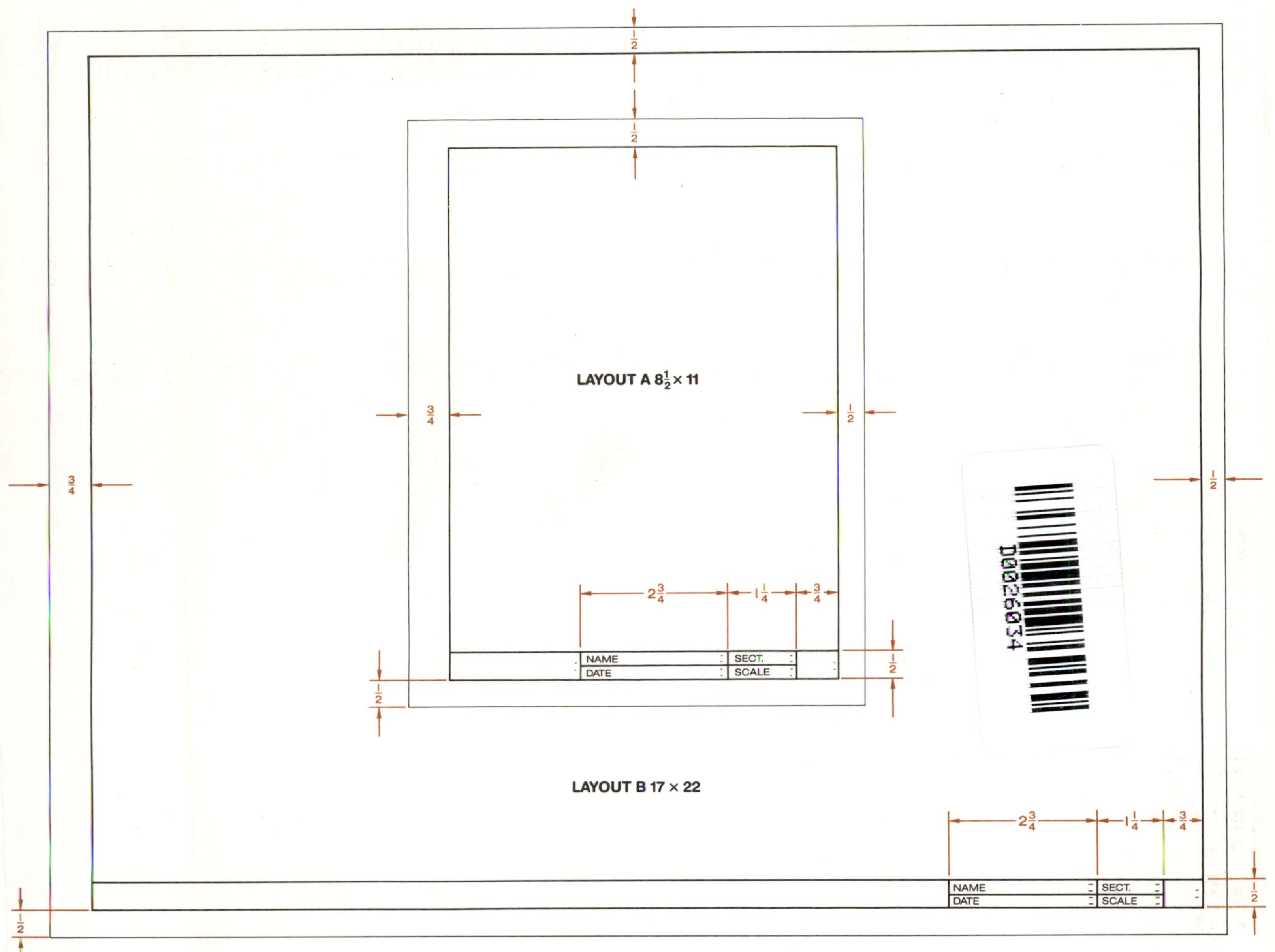

LAYOUT A $8\frac{1}{2} \times 11$
NAME
DATE
SECT.
SCALE
LAYOUT B 17×22
NAME
DATE
SECT.
SCALE

Engineering Graphics

Engineering Graphics

Herbert W. Yankee
Professor Emeritus, Mechanical Engineering
Worcester Polytechnic Institute

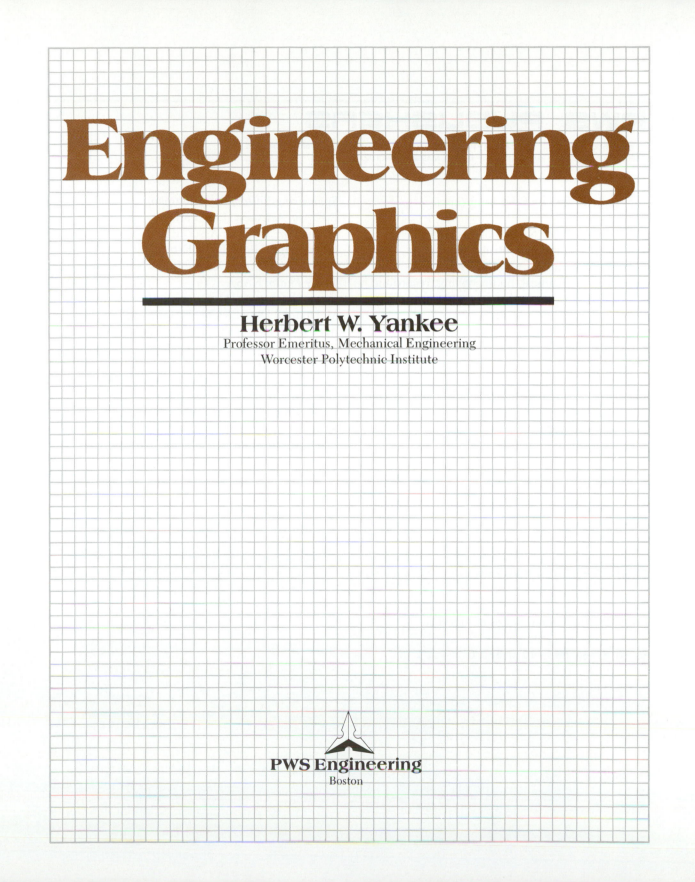

PWS Engineering
Boston

PWS PUBLISHERS

Prindle, Weber & Schmidt • Duxbury Press • PWS Engineering • Breton Publishers •
Statler Office Building • 20 Park Plaza • Boston, Massachusetts 02116

PWS Publishers is a division of Wadsworth, Inc.

Library of Congress Cataloging in Publication Data

Yankee, Herbert W.
Engineering Graphics.

1. Engineering graphics. 2. Computer graphics.
I. Title.
T353.Y365 1985 604.2 84-28513
ISBN 0-534-04167-1

ISBN 0-534-04167-1

Printed in the United States of America

85 86 87 88 89—10 9 8 7 6 5 4 3 2

Sponsoring Editor: Ray Kingman
Editorial Assistant: Jane Parker
Production: Del Mar Associates
Manuscript Editor: Dave Estrada
Interior and Cover Design: Louis Neiheisel
Cover Photo: NASA
Cover Illustration: McDonnell Aircraft, division McDonnell Douglas Corporation
Art Coordinators: Kim Fraley, Louis Neiheisel
Interior Illustration: Kim Fraley, Kristy Paulson, Florence Fujimoto, Kevin Canty, Gregory Miller, Mary Envall, Shirley Collier, Fiona King, Cher Threinen, Louis Neiheisel
Typesetting: Thompson Type
Cover Printing: New England Book Components
Printing and Binding: Halliday Lithograph

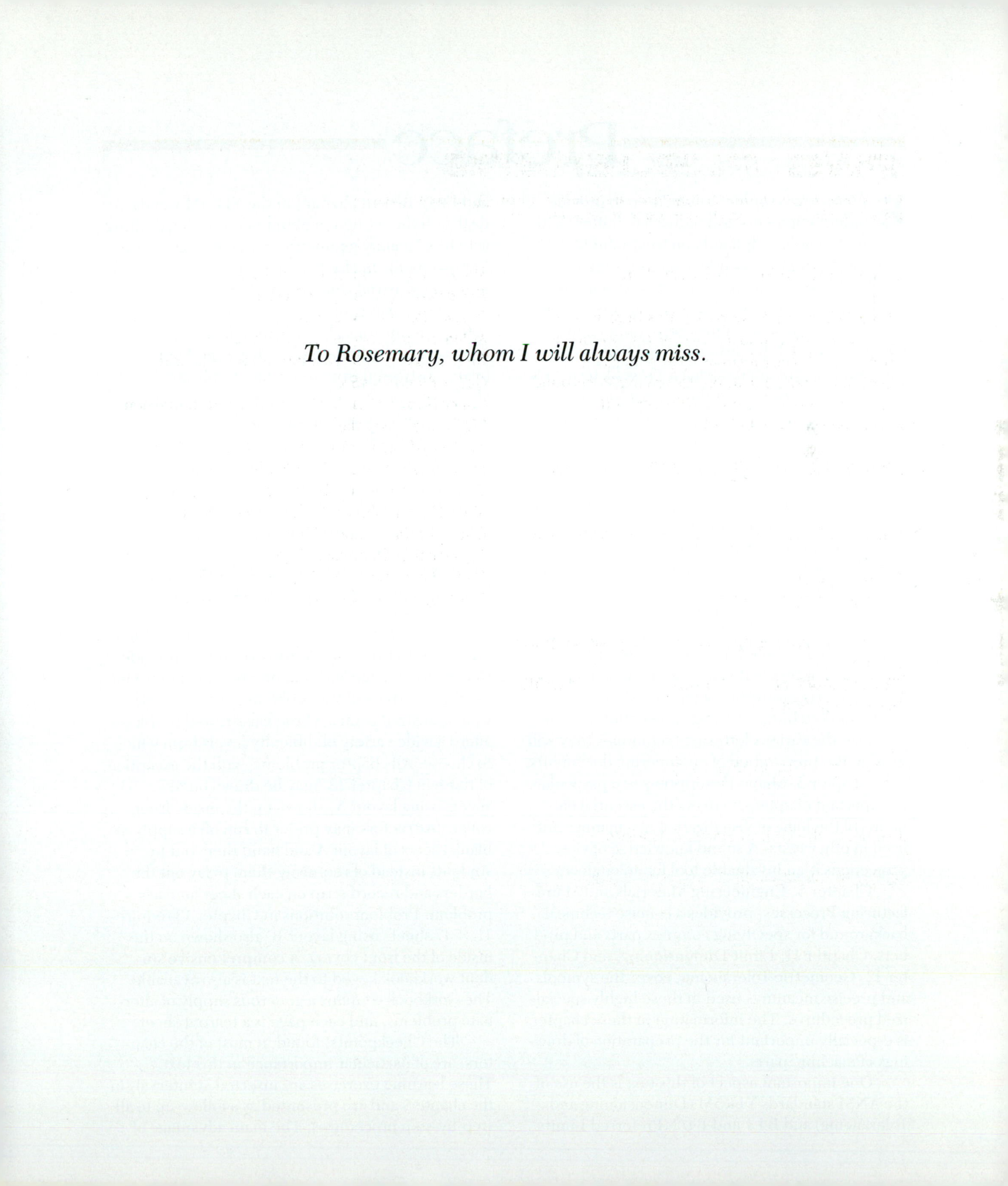

To Rosemary, whom I will always miss.

Preface

Engineering Graphics was written to provide student engineers and industrial drafters with a competency-based, hands-on treatment of engineering graphics and basic computer graphics. The book assumes no prerequisites. It contains a practical sequence with emphasis on current topics in graphics. The text is purposely confined to core topics—material that is needed and that can be reasonably assimilated during a 14-week term. The format is well suited for today's classroom needs. Its use will help students develop the professional literacy in graphics that every engineer needs when making sketches and when using and interpreting drawings.

Engineering graphics is the basis of all design and has an important place in all types of engineering practice. Concise graphical documents are required before virtually any product can be manufactured. Graphics also serves the engineer as a foundation for problem analysis and research.

Techniques for freehand sketching, a basic skill required by all engineers, are presented in the early chapters of this text, as are techniques for engineering lettering, a logical extension of freehand sketching. The text shows students how to utilize the various lettering techniques they will need in the preparation of engineering documents.

Chapter 5, Shape Description, is a particularly important chapter. It covers the essential elements of the logic of visualization in a manner not used in other texts. A sound knowledge of view projections is an invaluable tool for an engineer.

Chapter 8, Engineering Materials and Manufacturing Processes, provides a concise technical background for specifying complex parts and products. Chapter 11, Limit Dimensioning, and Chapter 12, Geometric Tolerancing, cover the symbols and precise meanings used in these highly specialized procedures. The information in these chapters is especially important for the preparation of drawings of machine parts.

One important aspect of this text is the use of the ANSI standards Y14.5M (Dimensioning and Tolerancing) and B4.1 and B4.2 (Preferred Limits and Fits). Recent changes in the Y14.5M standard deal with the reduction of time-consuming drafting activities, mainly geometric symbology, as well as improvements in clarity and increasing commonality with industrial companies all over the world.

The book is profusely illustrated with nearly 500 up-to-date, clearly understandable line drawings, sketches, photographs, charts, and tables (many abstracted from important publications), all of which are keyed to the text material. The artwork has been prepared to professional artist-quality standards, and many of the drawings have been reproduced from actual industrial prints in current use. Where appropriate, to aid in self-study, the artwork illustrates incorrect as well as correct examples. Each caption includes a reference to a specific section in the book in which the corresponding illustration is explained. In this way the student is carefully guided in the technique being illustrated.

Engineering Graphics contains over 600 high-quality end-of-the-chapter problems that provide thorough coverage and reinforcement of important principles. Many of the problems were adapted from industrial prints. These class-tested problems afford a wide variety of difficulty levels from which to choose. All chapter problems, with the exception of those in Chapter 13, may be drawn on $8\frac{1}{2} \times 11$ sheets using layout A, shown on the inside front cover. Instructors may prefer to run off a supply of blank sheets of layout A and hand them out to students instead of requiring them to lay out the borders and record strip on each sheet for each problem. Problem solutions in Chapter 13 require 11×17 sheets using layout B (also shown on the inside of the front cover). A comprehensive student workbook keyed to the text is also available. The workbook contains a generous supply of alternate problems, and each page is a tearout sheet.

The Checkpoints, found in most of the chapters, are of particular importance in this text. These learning exercises are inserted at intervals in the chapters and are presented as a follow-up to all step-by-step procedures. The main advantage of the Checkpoints is that students can make an immediate check on their progress instead of waiting to work the problems at the end of the chapter.

A procedural approach is used throughout the book that includes clearly stated step-by-step explanations, which in most cases, accompany corresponding illustrations placed on the same page. Students will find this procedural method complete and easy to understand and use. Some of the material in this book has served as the basis for a series of self-paced individual modules that were successfully used by hundreds of engineering students in graphics courses over a ten-year period at Worcester Polytechnic Institute.

This text also utilizes a two-color format, which promotes ease in reading and understanding and emphasizes important points in each illustration. Technical terms are defined as they are used and key words are given in italics. The proper use of the SI system is stressed throughout the book. The metric system and various drafting scales are carefully explained in Chapter 3, and differences in applying the English and metric systems of tolerancing are treated in Chapter 11.

The final chapter, Computer Graphics, pulls together the various computer applications referred to throughout the book. It explains the use of the computer—a standard tool of the engineer. The basic principles are carefully outlined, and modern computer hardware is carefully explained and illustrated. Several straightforward computer programs are given together with complete explanations. Chapter 15 was written by John T. Demel and Michael J. Miller of The Ohio State University; they are authors of *Introduction to Computer Graphics*, published in 1984 by the Brooks/Cole Engineering Division.

I wish to acknowledge the many valuable suggestions received from my former colleagues at Worcester Polytechnic Institute, notably those of Professors Ladislav H. Berka, Robert L. Norton, and Kenneth E. Scott. I am especially indebted to Greg Santini, WPI '85, for his careful work on the problem material. To the countless industrial plants across the nation who provided many of the engineering documents used in this book and to the many competent reviewers for their useful input, I express my grateful appreciation.

Herbert W. Yankee
February 1985
New London, New Hampshire

Contents

1 Single-View Sketching and Engineering Lettering

Page 1

1.1 MATERIALS 2
1.2 SKETCHING TECHNIQUES 3
1.3 LINE QUALITY 3
1.4 STRAIGHT LINES 4
- Horizontal Lines 4
- Vertical Lines 5
- Inclined Lines 6
- Sketching Border Lines 6
- Dividing a Line into Equal Parts 7

1.5 CIRCLES AND ARCS 7
- Sizes up to 1 Inch Diameter 7
- Sizes Greater Than 1 Inch Diameter 8

1.6 LAYING OUT ANGLES 10
1.7 LAYING OUT A SINGLE-VIEW SKETCH 11
- Outside-in Method 11
- Inside-out Method 12

1.8 COMPUTER-GENERATED SINGLE-VIEW DRAWINGS 13
1.9 ENGINEERING LETTERING 13
1.10 LETTER STYLES 14
1.11 LETTERING PRACTICE 15
1.12 GUIDE LINES FOR LETTERING 16
- Drawing Horizontal Guide Lines 17
- Drawing Vertical Guide Lines 17
- Drawing Inclined Guide Lines 17

1.13 SPACING LETTERS AND WORDS 18
1.14 MECHANICAL LETTERING AIDS 19
1.15 COMPUTER-GENERATED LETTERING 20

2 Pictorial Sketching

Page 34

2.1 OBLIQUE SKETCHING 35
2.2 OBLIQUE SKETCHING AXES 36
2.3 TYPES OF OBLIQUE SKETCHES 36
2.4 MAKING AN OBLIQUE SKETCH 37
2.5 HIDDEN LINES 37
2.6 RECEDING LINE DIRECTIONS 38
2.7 CIRCLES IN OBLIQUE SKETCHES 39
2.8 CIRCLES NOT PARALLEL TO THE FRONT, TOP, OR RIGHT-SIDE PLANES 40
2.9 CHOICE OF POSITION 41
2.10 AXONOMETRIC POSITIONS 42
2.11 ISOMETRIC SKETCHING 43
2.12 MAKING AN ISOMETRIC SKETCH 43
2.13 ISOMETRIC AND NONISOMETRIC LINES 44
2.14 INCLINED PLANES 44
2.15 OBLIQUE PLANES 45
2.16 CHOICE OF POSITION 45
2.17 ARCS AND CIRCLES IN ISOMETRIC SKETCHES 46
2.18 RIGHT CIRCULAR CYLINDER 47
2.19 ISOMETRIC CIRCLE POSITIONS 48
2.20 CENTERING A PICTORIAL DRAWING 49
2.21 PICTORIAL SECTIONS 50
2.22 COMPUTER-GENERATED PICTORIALS 51

3 Use of Instruments

Page 58

3.1 TYPICAL EQUIPMENT AND MATERIALS 59
Work Stations 59
T Squares 60
Drafting Machines 60
Parallel-Ruling Straightedges 60

3.2 PENCILS 61
3.3 POINTING THE PENCIL 62
3.4 TRIANGLES 62
3.5 DRAWING HORIZONTAL AND VERTICAL LINES 63
3.6 DRAWING INCLINED LINES 64
3.7 DRAWING PARALLEL LINES 66
3.8 DRAWING PERPENDICULAR LINES 66
3.9 SCALES 68
Civil Engineer Scales 70
Mechanical Engineer Scales 71
Architect Scales 72
Metric Scales 73

3.10 THE PROTRACTOR 77
3.11 THE COMPASS 79
3.12 TEMPLATES 80
3.13 IRREGULAR CURVES 80
3.14 DIVIDERS 82
3.15 ERASING SHIELD 83

4 Engineering Geometry

Page 96

4.1 GEOMETRIC FIGURES 97
Angles 97
Triangles 98
Quadrilaterals 98
Polygons 99
Circles and Arcs 99
Solids 100

4.2 DIVIDING A LINE 101
Two Equal Parts 101
Any Number of Equal Parts 102
Proportional Parts 102

4.3 DIVIDING THE SPACE BETWEEN TWO LINES 103
4.4 DRAWING PERPENDICULAR LINES 104
4.5 DRAWING PARALLEL LINES 104
4.6 LAYING OUT AN ANGLE 106
4.7 BISECTING AN ANGLE 106
4.8 CONSTRUCTING A TRIANGLE 106
4.9 CONSTRUCTING A SQUARE 106
4.10 CONSTRUCTING A PENTAGON 109
4.11 CONSTRUCTING A HEXAGON 110
4.12 CONSTRUCTING AN OCTAGON 111
4.13 DRAWING A CIRCLE THROUGH THREE POINTS 113
4.14 FINDING THE CENTER OF A CIRCLE 113
4.15 DRAWING A CIRCLE TANGENT TO A LINE AT A GIVEN POINT 113
4.16 DRAWING A LINE TANGENT TO A CIRCLE THROUGH A POINT ON A CIRCLE 114
4.17 DRAWING A LINE TANGENT TO A CIRCLE THROUGH A POINT OUTSIDE A CIRCLE 114
4.18 DRAWING TANGENT ARCS 114
4.19 DRAWING A REVERSE, OR OGEE, CURVE 119
4.20 CONIC SECTIONS 122
4.21 CONSTRUCTING AN ELLIPSE 123
4.22 ELLIPSE TEMPLATES 127
4.23 CONSTRUCTING A PARABOLA 128
4.24 CONSTRUCTING A HYPERBOLA 130

Shape Description

Page 139

5.1 ORTHOGRAPHIC PROJECTION OR MULTIVIEW DRAWING 140
5.2 THE SIX PRINCIPAL VIEWS 142
5.3 THE RELATIONSHIP BETWEEN THE VIEWS 143
5.4 LINE TECHNIQUES 143
Visible Lines 143
Hidden Lines 143
Center Lines 144
Precedence of Lines 144
5.5 CHOICE OF VIEWS 145
5.6 ONE-VIEW DRAWINGS 146
5.7 TWO-VIEW DRAWINGS 146
5.8 THREE-VIEW DRAWINGS 150
5.9 ALTERNATE VIEW POSITION 152
5.10 PLANE SURFACES 153
Normal Surfaces 153
Inclined Surfaces 154
Oblique Surfaces 155
5.11 ADJACENT SURFACES 157
5.12 SURFACE IDENTIFICATION 157
5.13 SIMILARITY OF SURFACE CONFIGURATION 160
5.14 THE USE OF PICTORIAL SKETCHES 160
5.15 CONSTRUCTING A THIRD PRINCIPAL VIEW 160
5.16 CYLINDRICAL SURFACES 166
5.17 INTERSECTING CYLINDERS 168
5.18 INTERSECTIONS OF CYLINDERS AND PRISMS 169
5.19 FILLETS, ROUNDS, AND RUNOUTS 169
5.20 PHANTOM LINES 172

6 Auxiliary Views

Page 188

6.1 AUXILIARY VIEWS 189
6.2 PRIMARY AUXILIARY VIEWS 190
6.3 PARTIAL AUXILIARY VIEWS 191
6.4 TYPES OF PRIMARY AUXILIARY VIEWS 192
6.5 AUXILIARY VIEW CONSTRUCTION 192
Front-Adjacent Auxiliary Views 192
Top-Adjacent Auxiliary Views 193
Side-Adjacent Auxiliary Views 194
6.6 PLOTTING CURVES IN AUXILIARY VIEWS 195
6.7 SECONDARY AUXILIARY VIEWS 198

Sectional Views and Conventional Representation

Page 213

7.1 SECTIONAL VIEWS 214
The Cutting Plane 216
Section Lining 216
7.2 VISUALIZING A SECTIONAL VIEW 218
Visible Lines 218
Hidden Lines 219
7.3 FULL SECTIONS 219
7.4 HALF SECTIONS 221
7.5 OFFSET SECTIONS 224
7.6 ALIGNED SECTIONS 227
7.7 REVOLVED SECTIONS 230
7.8 REMOVED SECTIONS 231
7.9 AUXILIARY SECTIONS 233
7.10 BROKEN-OUT SECTIONS 234
7.11 MEANING OF CONVENTIONAL REPRESENTATION 234
7.12 CONVENTIONAL REPRESENTATION SECTIONS 235
Section Lining 235
Spokes in Sectional Views 236
Partial Views 236
Rotated Features 237
Breaks 238

8 Engineering Materials and Manufacturing Processes

Page 251

8.1 GENERAL GUIDELINES FOR DESIGN 252
8.2 ENGINEERING MATERIALS 252
8.3 METALS AND ALLOYS 252
8.4 FERROUS METALS 253
8.5 NUMBERING SYSTEMS 253
- AISI-SAE System for Steel 253
- Unified System for Metals and Alloys 254

8.6 NONFERROUS METALS 255
- Aluminum 255
- Magnesium 256
- Copper 256
- Copper-Base Alloys 256
- Zinc-Base Alloys 256
- Nickel-Base Alloys 256

8.7 PROPERTIES OF METALS AND ALLOYS 256
8.8 PRIMARY MANUFACTURING PROCESSES 257
8.9 CASTING METHODS 258
- Sand Casting 258
- Plaster Mold Casting 259
- Shell Molding 259
- Investment Casting 259
- Permanent Mold Casting 260
- Die Casting 260

8.10 FORGING 262
8.11 ASSEMBLY METHODS 264
- Welds 264
- Adhesives 264
- Threaded Fasteners 264

8.12 PRESSWORKING METHODS 265
8.13 POWDER METALLURGY 266
8.14 SECONDARY MANUFACTURING PROCESSES 266
8.15 TYPES OF MACHINE TOOLS 267
8.16 MACHINING OPERATIONS 267
- Sawing 267
- Milling 267
- Broaching 270
- Shaping 271
- Turning and Boring 271
- Grinding 274
- Hole Making Operations 275

8.17 TOLERANCES FOR VARIOUS PROCESSES 276
8.18 SURFACE FINISHING PROCESSES 276
8.19 MICROFINISHING 278
8.20 SURFACE FINISH CONTROL 279
8.21 NUMERICAL CONTROL (NC) 282
- Principles 282
- Advantages 283
- Applications 283

8.22 COMPUTER-AIDED MANUFACTURING 283
- Computer Numerical Control (CNC) 283
- Direct Numerical Control (DNC) 284

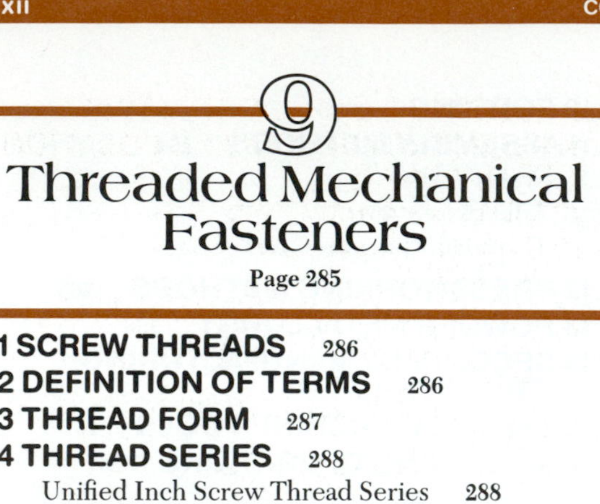

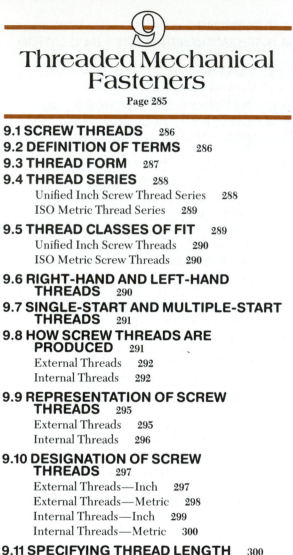

Threaded Mechanical Fasteners

Page 285

9.1 SCREW THREADS 286
9.2 DEFINITION OF TERMS 286
9.3 THREAD FORM 287
9.4 THREAD SERIES 288
- Unified Inch Screw Thread Series 288
- ISO Metric Thread Series 289

9.5 THREAD CLASSES OF FIT 289
- Unified Inch Screw Threads 290
- ISO Metric Screw Threads 290

9.6 RIGHT-HAND AND LEFT-HAND THREADS 290
9.7 SINGLE-START AND MULTIPLE-START THREADS 291
9.8 HOW SCREW THREADS ARE PRODUCED 291
- External Threads 292
- Internal Threads 292

9.9 REPRESENTATION OF SCREW THREADS 295
- External Threads 295
- Internal Threads 296

9.10 DESIGNATION OF SCREW THREADS 297
- External Threads—Inch 297
- External Threads—Metric 298
- Internal Threads—Inch 299
- Internal Threads—Metric 300

9.11 SPECIFYING THREAD LENGTH 300
9.12 THREAD ENGAGEMENT 302
9.13 SCREW FASTENERS 303
9.14 BOLTS AND NUTS 304
- Bolt Types 304
- Nut Types 304
- Metric Bolts and Nuts 304

9.15 DRAWING OR SKETCHING A HEXAGON BOLT OR NUT 304
- Drawing or Sketching a Square Bolt or Nut 305
- Preferred Position for Bolts and Nuts 305
- Chamfer Angle 305

9.16 BOLT LENGTHS 306
9.17 SPECIFYING BOLTS AND NUTS 306
- American National Standard Bolts 306
- Metric Hexagon Bolts 306

9.18 STUDS 307
9.19 SPECIFYING STUDS 307
9.20 CAP SCREWS 307
9.21 CAP SCREW LENGTHS 308
9.22 CAP SCREW THREAD LENGTH 308
9.23 SPECIFYING CAP SCREWS 308
9.24 MACHINE SCREWS 309
9.25 MACHINE SCREW THREAD LENGTHS 309
9.26 SPECIFYING MACHINE SCREWS 309
9.27 MACHINE SCREW NUTS 309
9.28 SPECIFYING MACHINE SCREW NUTS 309
9.29 SET SCREWS 310
9.30 SPECIFYING SET SCREWS 310
9.31 ADDITIONAL THREADED FASTENERS 310

10 Size Description

Page 314

10.1 SCALE OF DRAWING 316
10.2 DIMENSIONING TECHNIQUES 316
- Dimension Lines 316
- Extension Lines 317
- Arrowheads 317
- Center Lines 317
- Leaders 318

10.3 ARRANGEMENT OF DIMENSION AND EXTENSION LINES 318
10.4 DIMENSION FIGURES 319
10.5 PLACEMENT OF DIMENSION FIGURES 319
- Reading Directions 319
- On or Off the Views 320

10.6 DIMENSIONING SYSTEMS 320
- Fractional Inch 320
- Fractional and Decimal Inch Combinations 321
- Decimal Inch 321
- Rounding Off a Decimal Inch 322
- The Metric System 323

10.7 DUAL DIMENSIONING 323
10.8 DESIGNATION OF UNITS OF MEASUREMENT 324
10.9 FINISHED SURFACES 324
10.10 DEGREE OF ACCURACY 326
10.11 DIMENSIONING IN LIMITED SPACES 327
10.12 OVERALL DIMENSIONS 328
10.13 DIMENSIONING REPETITIVE FEATURES 328
10.14 DIMENSIONING ANGLES 329
10.15 DIMENSIONING ARCS 330
10.16 ROUNDED CORNERS 331
10.17 CURVED OUTLINES 331
10.18 PARTS WITH ROUNDED ENDS 332
- Low Precision 333
- High Precision 333
- Partially Rounded Ends 333

10.19 DIMENSIONING CYLINDERS 334
- Solid Cylinders 334
- Hollow Cylinders 335

10.20 GEOMETRIC SHAPE ANALYSIS 336
10.21 SELECTION OF SIZE AND LOCATION DIMENSIONS 337
10.22 CONTOUR DIMENSIONING 339
10.23 SYMMETRICAL CURVED OUTLINES 340
10.24 IRREGULAR CURVED OUTLINES 340
10.25 LOCATION DIMENSIONING OF HOLES 341
10.26 SIZE DIMENSIONING OF MACHINED HOLES 344
- Format for Specification 345
- Repetitive Holes 345
- Drilled Holes 346
- Counterdrilled Holes 346
- Reamed Holes 347
- Bored Holes 348
- Counterbored Holes 348
- Spotfaced Holes 348
- Countersunk Holes 350
- Tapped Holes 350

10.27 DIMENSIONING CHAMFERS 351
10.28 DIMENSIONING TAPERS 352
10.29 DIMENSIONING MACHINING CENTERS 353
10.30 DIMENSIONING SQUARE SHAPES 353
10.31 DIMENSIONING KEYSEATS AND KEYWAYS 354
10.32 DIMENSIONING KNURLING 355
10.33 DIMENSIONING NECKS AND UNDERCUTS 356
10.34 DIMENSIONING THREAD RELIEFS 356
10.35 DIMENSIONING A HALF SECTION 357
10.36 SUPERFLUOUS DIMENSIONS 358
10.37 OUT-OF-SCALE DIMENSIONS 358
10.38 TABULAR DRAWINGS 359
10.39 DIMENSIONING FOR CAD AND CAM 360

11 Limit Dimensioning
Page 381

11.1 SYSTEMS OF LIMITS AND FITS 382
11.2 TOLERANCING 382
11.3 ENGLISH SYSTEM—DEFINITION OF BASIC TERMS 382
11.4 FIT 385
11.5 ALLOWANCE 385
11.6 DESIGN CONSIDERATIONS 385
11.7 THE BASIC HOLE SYSTEM 386
11.8 THE BASIC SHAFT SYSTEM 387
11.9 STANDARD FITS—ENGLISH SYSTEM 388
Running and Sliding Fits 388
Locational Fits 389
Force Fits 389
11.10 METHOD OF COMPUTING THE ENGLISH SYSTEM OF FITS AND LIMITS 390
11.11 SI METRIC SYSTEM—DEFINITION OF BASIC TERMS 391
11.12 FIT 392
11.13 TOLERANCE SYMBOLS 392
Individual Part Limits 392
Mating Part Limits 392
11.14 PREFERRED BASIC SIZES 393
11.15 PREFERRED FITS 394
11.16 BASIS OF PREFERRED FITS 395
11.17 METHOD OF COMPUTING THE SI METRIC SYSTEM OF FITS AND LIMITS 396
11.18 ACCUMULATION OF TOLERANCES 398
Chain Dimensioning 398
Baseline Dimensioning 398
Direct Dimensioning 398

12 Geometric Tolerancing
Page 401

12.1 TOLERANCES OF POSITION 402
12.2 METHOD OF EXPRESSING POSITIONAL TOLERANCES 403
Feature Control Frame 403
Geometric Characteristic Symbols 403
Material Condition Symbol 403
Datum Reference Letters 404
Feature Control Placement 404
Datum Feature Symbol 404
12.3 APPLICATION OF A POSITIONAL TOLERANCE 405
12.4 THE MMC PRINCIPLE 405
12.5 RFS AS RELATED TO POSITIONAL TOLERANCING 407
12.6 LMC AS RELATED TO POSITIONAL TOLERANCING 408
12.7 CONCENTRICITY 409
12.8 SYMMETRY 410
12.9 TOLERANCES OF FORM 410
12.10 STRAIGHTNESS TOLERANCE 410
12.11 FLATNESS TOLERANCE 411
12.12 CIRCULARITY OR ROUNDNESS TOLERANCE 411
12.13 CYLINDRICITY TOLERANCE 412
12.14 PROFILE TOLERANCE 412
12.15 ANGULARITY TOLERANCE 413
12.16 PARALLELISM TOLERANCE 413
12.17 PERPENDICULARITY TOLERANCE 414
12.18 RUNOUT TOLERANCES 415
12.19 SPECIFYING DATUMS 415
12.20 THE THREE DATUM PLANE SYSTEM 416
Flat Datum Features 416
Cylindrical Datum Features 418

13 Production Drawings

Page 420

13.1 DETAIL DRAWINGS 421
13.2 FUNCTIONAL DESIGN LAYOUT 423
13.3 PRODUCTION DESIGN ASSEMBLY DRAWING 424
13.4 WORKING ASSEMBLY DRAWINGS 426
13.5 EXPLODED PICTORIAL ASSEMBLY DRAWINGS 426
13.6 BILL OF MATERIAL OR PARTS LIST 428
13.7 SECTIONING TECHNIQUES 429
13.8 TITLE BLOCKS 430
13.9 REVISION BLOCK 432
13.10 STANDARD DRAWING SHEET SIZES 432
13.11 CHECKING DRAWINGS 433

14 Principles of Basic Descriptive Geometry

Page 465

14.1 PROJECTION OF A POINT 466
14.2 POSITION OF A POINT 466
14.3 PROJECTION OF A LINE 467
14.4 DIRECTION AND POSITION OF A LINE 467
14.5 KINDS OF LINES 468
14.6 TRUE LENGTH OF A LINE 470
14.7 POINT VIEW OF A LINE 471
14.8 PROJECTION OF A POINT ON A LINE 473
14.9 BEARING AND AZIMUTH OF A LINE 473
14.10 SLOPE OF A LINE 474
14.11 GRADE OF A LINE 475
14.12 PLANE SURFACES 475
14.13 TYPES OF PLANES 476
14.14 LOCATING POINTS IN A PLANE 476
14.15 TRUE-LENGTH LINES IN PLANES 478
14.16 EDGE VIEW OF A PLANE 478
14.17 TRUE SIZE OF A PLANE 479
14.18 STRIKE OF A PLANE 480
14.19 SLOPE OF A PLANE 480
14.20 DIP OF A PLANE 481
14.21 BASIC CONSTRUCTIONS 481
14.22 FINDING THE SHORTEST DISTANCE FROM A POINT TO A PLANE 481
14.23 FINDING THE ANGLE BETWEEN A LINE AND A PLANE 482
14.24 FINDING THE INTERSECTION OF A LINE AND A PLANE 482
14.25 FINDING THE SHORTEST DISTANCE BETWEEN TWO PARALLEL PLANES 484
14.26 FINDING THE DIHEDRAL ANGLE BETWEEN TWO PLANES 484
14.27 FINDING THE SHORTEST DISTANCE BETWEEN TWO SKEW LINES 485
14.28 FINDING THE LINE OF INTERSECTION BETWEEN TWO PLANES 486
Cutting Plane Method 486
Edge View Method 487

Computer Graphics

Page 515

15.1 COMPUTER GRAPHICS IN ENGINEERING 516

15.2 COMPUTER-AIDED DESIGN/COMPUTER-AIDED MANUFACTURING (CAD/CAM) 518

Mechanical Engineering 518
Civil Enginnering 519
Chemical Engineering 520
Structural Engineering 521
Electrical Engineering 522

15.3 GENERAL COMPUTER-GRAPHICS SYSTEM HARDWARE 522

A Typical Workstation 522
Hardcopy Output Devices and File Storage Devices 529

15.4 GRAPHICS SOFTWARE 532

The Menu 532
Mode Settings 533
Prompts and Other Messages 534
Drawing Routines 534
Drawing Aids 535
Storing and Retrieving the Drawing 536
Hardcopy 537
Program Interaction 538

15.5 DRAWING WITH A COMPUTER GRAPHICS SYSTEM 539

Using Grid Paper 540
Establishing a Scale 540
Blocking In the Views (Outside-In) 540
Sketching the Details 541
Laying Out the Drawing Area on the Screen 541
Drawing the Entities 542

15.6 MODIFYING AN EXISTING DRAWING 543

15.7 SOME DRAWING EXERCISES USING A SYSTEM 544

Creating a Border and Title Block 544
Drawing the Three Principal Views of a Cut Block from the Isometric Drawing 545
Drawing an Isometric of a Cut Block from the Three Principal Views 546

15.8 CREATING AND SAVING A DRAWING 548

A Hex Nut 548
A Simple Half-Section 550
Retrieving, Dimensioning, and Resaving Each of the Drawings 550

Appendix

Page 553

Table A-1 Screw Threads, American National Standard Unified and American National **553**

Table A-2 Screw Threads, Metric **555**

Table A-3 Inch Twist Drill Sizes, Numbered and Lettered **556**

Table A-4 Metric Twist Drill Sizes **557**

Table A-5 Screw Threads, Square and Acme **558**

Table A-6 Acme Threads, General Purpose **558**

Table A-7 Hex Bolts, American National Standard **559**

Table A-8 Square Bolts, American National Standard **560**

Table A-9 Hex Nuts, American National Standard **561**

Table A-10 Square Nuts, American National Standard **562**

Table A-11 Square and Hex Machine Screw Nuts, American National Standard **563**

Table A-12 Metric Hex Bolts, Hex Cap Screws, and Socket Cap Screws **564**

Table A-13 Metric Hex Nuts (Style 1 and Style 2) **565**

Table A-14 Cap Screws, Slotted and Socket Head, American National Standard **566**

Table A-15 Machine Screws, American National Standard **567**

Table A-16 Metric Machine Screws **569**

Table A-17 Metric Machine Screw Lengths **569**

Table A-18 Set Screws, Hexagon Socket, Slotted Headless, and Square Head, American National Standard **570**

Table A-19 Socket Head Shoulder Screws, American National Standard **571**

Table A-20 Shaft Center Sizes **571**

Table A-21 Keys—Square, Flat, Plain Taper, and Gib Head **572**

Table A-22 Inch Small Rivets, American National Standard **573**

Table A-23 Inch Large Rivets, American National Standard **574**

Table A-24 Plain Washers, American National Standard **575**

Table A-25 Lock Washers, American National Standard **577**

Table A-26 Cotter Pins, American National Standard **578**

Table A-27 Taper Pins, American National Standard **579**

Table A-28 Straight Pins, American National Standard **580**

Table A-29 Clevis Pins, American National Standard **581**

Table A-30 Wire Gage Standard **582**

Table A-31 Woodruff Keys, American National Standard **584**

Table A-32 Woodruff Key Sizes for Different Shaft Diameters **584**

Table A-33 Woodruff Key-Seat Dimensions **585**

Table A-34 Pratt and Whitney Round-End Keys **586**

Table A-35 Running and Sliding Fits, American National Standard **587**

Table A-36 Clearance Locational Fits, American National Standard **590**

Table A-37 Transition Locational Fits, American National Standard **593**

Table A-38 Interference Locational Fits, American National Standard **595**

Table A-39 Force and Shrink Fits, American National Standard **596**

Table A-40 Preferred Hole Basis Clearance Fits—Cylindrical Fits, American National Standard **598**

Table A-41 Preferred Hole Basis Transition and Interference Fits—Cylindrical Fits, American National Standard **601**

Table A-42 Preferred Shaft Basis Clearance Fits—Cylindrical Fits, American National Standard **603**

Table A-43 Preferred Shaft Basis Transition and Interference Fits—Cylindrical Fits, American National Standard **606**

INDEX **609**

Engineering Graphics

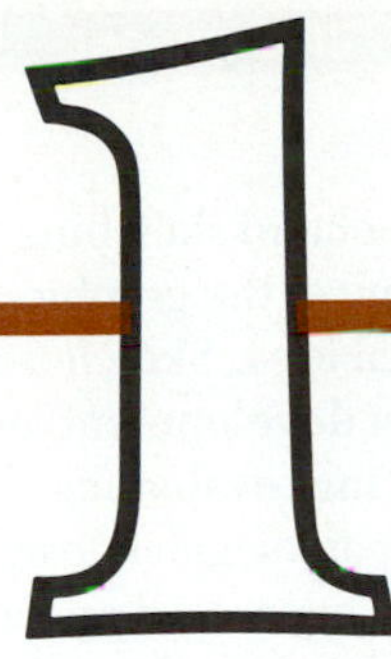

Single-View Sketching and Engineering Lettering

OBJECTIVES

After completing this chapter you should have gained the following abilities:

1. To make a single-view sketch quickly and clearly using simple, straightforward techniques.

2. To recognize and properly use the following three conventional types of lines: construction lines, center lines, and visible-object lines.

3. To distinguish between and use recommended methods for sketching horizontal lines, vertical lines, inclined lines, circles, and arcs.

4. To use the procedure called "blocking-in" as a method of constructing the shape of objects.

5. To maintain proper proportions.

6. To divide a line into equal parts.

7. To lay out lines at approximate angles.

8. To maintain acceptable line quality.

9. To be aware of how a well-executed single-view sketch helps the engineer convey an idea.

10. To know the proportion and strokes involved in making letters and numerals in the single-stroke Gothic alphabet.

11. To develop the necessary level of competence in the style of lettering (vertical or inclined) that seems best for you.

12. To maintain a clear mental image of the various letter forms of this plain, legible style of engineering lettering. You should understand that the stroke directions suggested for each letter and numeral may properly vary with the individual.

13. To recognize that the most important requirements in lettering are legibility and speed of execution.

14. To be aware that good lettering greatly adds to a neat and pleasing appearance on a drawing or a sketch and certifies pride at a professional level as well as assures clear communication.

15. To discover that lettering is not related to writing skills. Rather, lettering entails essentially the same techniques used in single-view sketching. In fact, a letter form is actually a two-dimensional diagram.

16. To recognize the need for guide lines. Use the T square and drafting triangles or a drafting machine scale to draw horizontal, vertical, and inclined guide lines.

17. To properly space the individual letters, words, and lines of lettering.

18. To rapidly letter notes and specifications using words that are ⅛ inch high and fractions that are ¼ inch high.

19. To know the recommended applications for the use of uppercase and lowercase letters.

20. To place notes only in a horizontal position on a drawing or a sketch.

21. To use an F or H grade pencil to make dark, sharp letters and numerals that clearly stand out on a drawing or a sketch.

22. To be conscious of the need for continuous, earnest effort in order to improve your lettering skills.

As an engineer you must be proficient at freehand sketching because it is often important to record some of the graphic details of a verbal statement or of a hazy mental idea. Sketches are usually produced in the early stages of product development and are considered a basic tool by engineers for clarifying, evaluating, and recording preliminary ideas or concepts. After an original design idea has been released from your thoughts and expressed on paper, your mind is free for further analytical and creative thinking. You need considerable practice to be able to sketch without concentrating on the mechanics of drawing. Sketching should be done as easily and freely as writing, and your mind should be free to concentrate upon an idea, not upon the technique of sketching the idea.

Before you make an accurate drawing you will often prepare freehand sketches that show several possible solutions to a given problem so that a choice may be made from among them. While your preliminary design sketches may show only a partial solution to a more complex and complete system, sufficient detail should be shown so that you can compare the relative merits of one idea to another.

To prepare final working drawings, detail drafters often work from freehand design sketches made by engineers. In addition, engineers often use freehand sketches to illustrate a written report. A sketch is usually far more effective than words. In some cases, neat and orderly sketches are as effective as drawings made with instruments.

A freehand sketch may properly consist of only a few hastily sketched lines and a few scribbled notes. Rough or incomplete sketches are sometimes used to express and record quickly and graphically the essential meaning of what might have been a vague idea, perhaps originally presented orally to the sketcher. Figure 1-1 shows a somewhat complete single-view sketch, which is used to graphically convey precise technical information. This type of drawing is typical of an engineer's freehand sketch. It is an example of effective line work, neat lettering, and careful layout. You must follow an orderly system and have a strong desire to excel in order to make such neat, well-executed sketches.

Single-view sketches are useful for graphically portraying the shape of flat objects, such as templates, gaskets, simple gears, cams and plates, or sheet metal parts. Civil engineers frequently prepare sketches of plot plans, septic systems, or drainage fields. Electrical engineers make freehand sketches as an aid in selecting and arranging the positions for the necessary components in the design of a circuit or a transmission path for a proposed system. Chemical engineers often use single-view sketches during the preliminary stages of planning piping layouts or in the analysis of flow diagrams, fluid power circuits, and so forth.

1.1 MATERIALS

The materials required for sketching—paper, pencil, and eraser—are easily available. Engineers often use pads of plain white paper or cross-section (coordinate) paper. The printed lines on cross-section paper are helpful in sketching. A size value may be assigned to the squares, which generally results in a sketch with more uniform spacing of lines and with nearly correct proportional distances. Figure 1-2 illustrates a typical example of a simple sketch prepared by an engineer on cross-section paper. Soft grade leads such as F or H are best for freehand sketching.

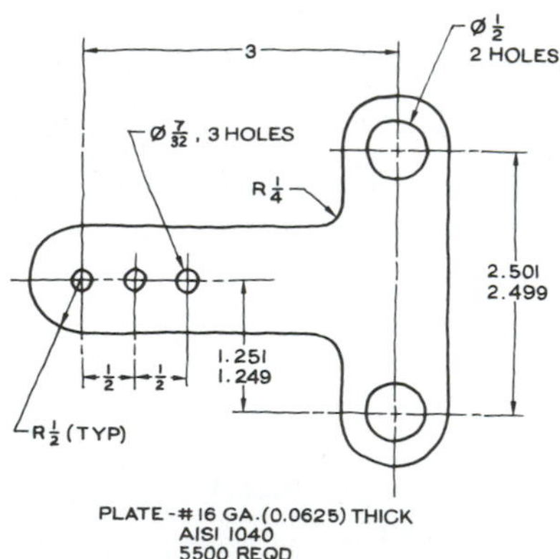

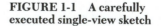

FIGURE 1-1 A carefully executed single-view sketch

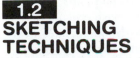

1.2 SKETCHING TECHNIQUES

To produce acceptable sketches you must follow an orderly procedure consisting of simple, fundamental techniques. To begin with, when sketching, hold the pencil loosely and at a greater distance from the point than in normal writing. This distance may range from about 1½ to 2 inches. Eye-hand coordination and muscle control are important and come with practice. Most successful sketchers use finger movement but hold the wrist stationary, particularly when sketching short lines. For longer lines, less finger action with a stiffer arm movement proves to be more effective. Sketching lines, particularly long lines, in a direction away from the body produces best results. As you sketch, rotate the pencil slightly between the thumb and fingers. This practice helps to maintain a uniformly sharp point.

Sketch lines by pulling the pencil lightly over the paper through a series of short strokes. These light construction-weight dashed lines can later be connected and darkened with a heavier line. Construction lines should be sketched so lightly that they will not reproduce; therefore, they need not be erased. Guard against pushing the pencil because of the possibility of tearing or puncturing the paper. Because more freedom of movement is necessary when sketching, avoid fastening the paper to the working surface. The paper should be free to move across the working surface as the sketch is developed.

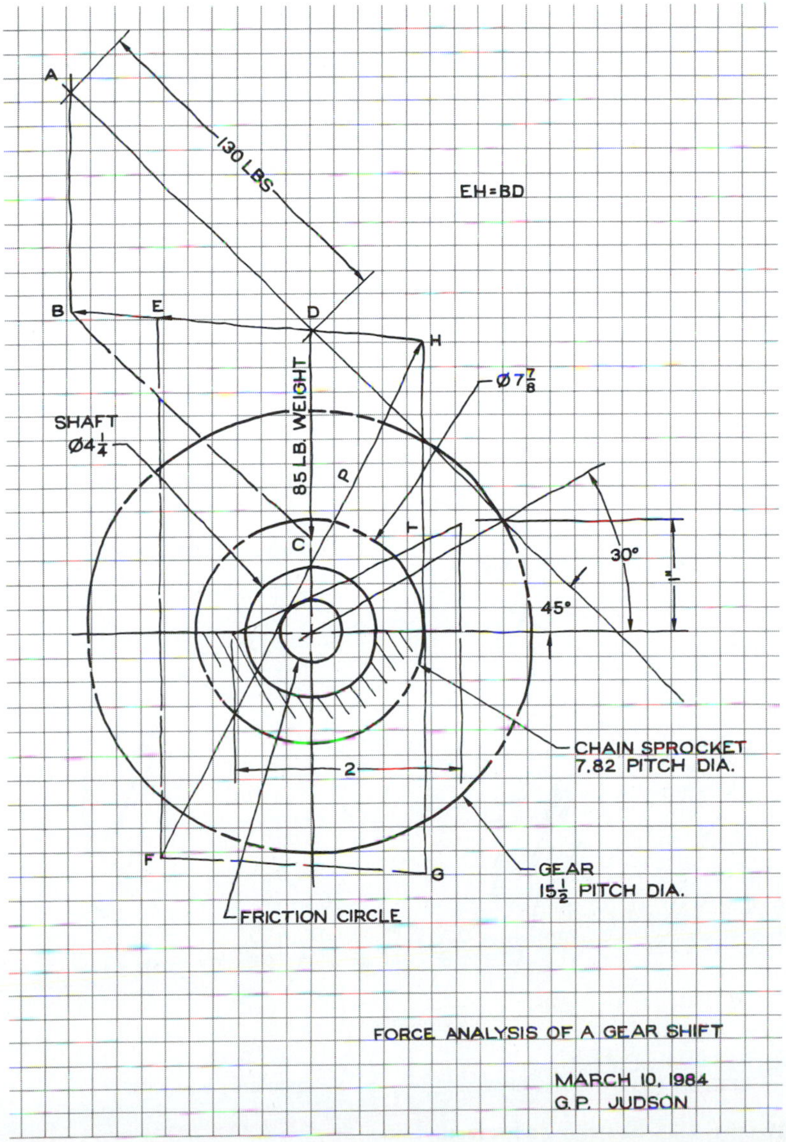

FIGURE 1-2 A single-view sketch on cross-section paper *(Section 1.1)*

1.3 LINE QUALITY

Final lines on a sketch should be bold, dark, and clean-cut. Such lines are called *visible* (or *object*) lines. Straight lines should follow a reasonably straight direction with a minimum of waviness. Try to avoid using ragged, overlapping strokes as you sketch the lines. Freehand circles should be as round as possible. As you develop more competence your eye will become more sensitive to corners that are not square, lines that are not parallel, or circles that are lopsided. Also, with a reasonable amount of practice you will note your freehand lines will become less ragged, wavy, and fuzzy.

1.4 STRAIGHT LINES

Since most sketches consist mainly of straight lines (horizontal, vertical, or inclined), it is important for you to learn the simple techniques involved in sketching these lines. Start by holding the pencil loosely and at a comfortable distance back from the point. Holding the pencil too close to the point may obscure your vision of the area around the point.

The following paragraphs describe some recommended methods for sketching various kinds of straight lines used in two-dimensional sketching. The directions apply to right-handed sketchers. From time to time left-handers must adjust directions that obviously were intended for right-handed people. If it is any consolation, your left-handed author shares your frustrations in coping with a right-handed world.

Horizontal Lines

Assume you want to sketch a horizontal line AB about 7 inches long. Position the paper so that the line to be sketched will be approximately at a right angle to your forearm. As shown in Step 1 of Figure 1-3, establish the desired end points A and B with a short dash. Do not attempt to draw long lines with one continuous stroke. Instead, working from left to right, or in a direction away from your body, sketch a series of non-overlapping short construction lines about ½ inch long starting at point A and ending at point B

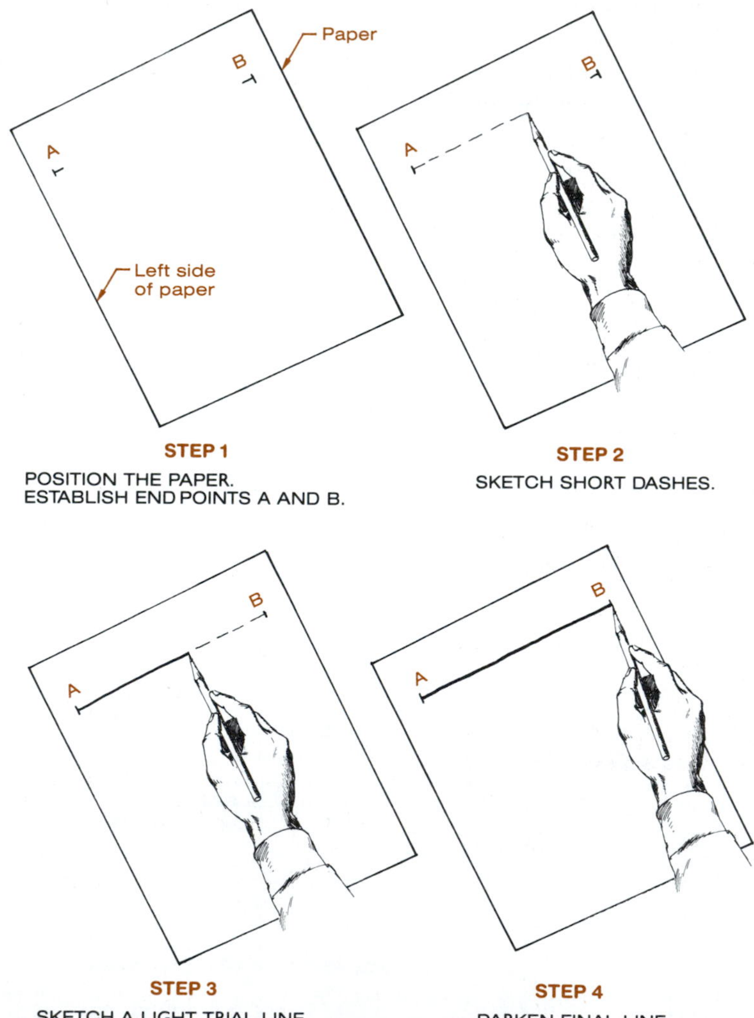

FIGURE 1-3 **Sketching a horizontal line** ***(Section 1.4)***

(Step 2). Aim for point B as you sketch the light dashed line, keeping it as nearly parallel as possible to the top edge of the paper. Now, using a relaxed finger movement while holding your wrist fairly rigid, connect the short dashes with a light trial line (Step 3). Test for straightness by holding the paper away from you at arm's length, examining the trial line. Straighten it if necessary. Finally, apply a bit more pressure to the pencil and carefully improve the straightness as you darken in the trial line AB (Step 4). Retain this drawing for use in the following sections.

CHECKPOINT 1
SKETCHING HORIZONTAL LINES

Review Steps

1. Position the paper and establish the end points of the desired line.
2. Sketch a series of short lines. Aim for the end point.
3. Sketch a light trial line.
4. Darken in the final line.

Practice sketching line AB, shown in Figure 1-3, in the same relative position on a plain sheet of 8½ × 11 inch paper. Sketch about a dozen more horizontal lines spaced by eye about ½ inch apart. Try to keep the lines reasonably parallel.

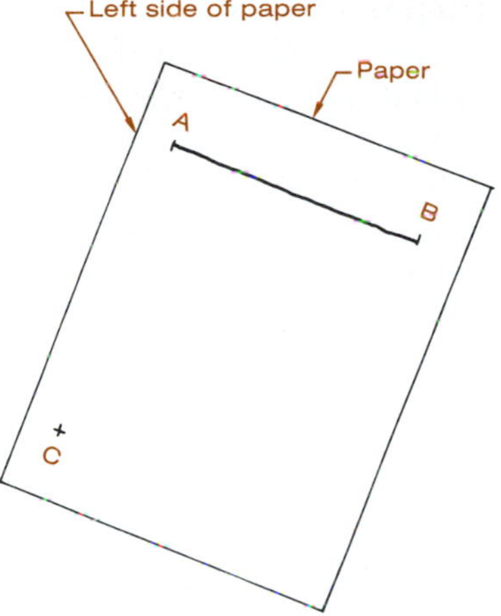

ESTABLISH POINT C.

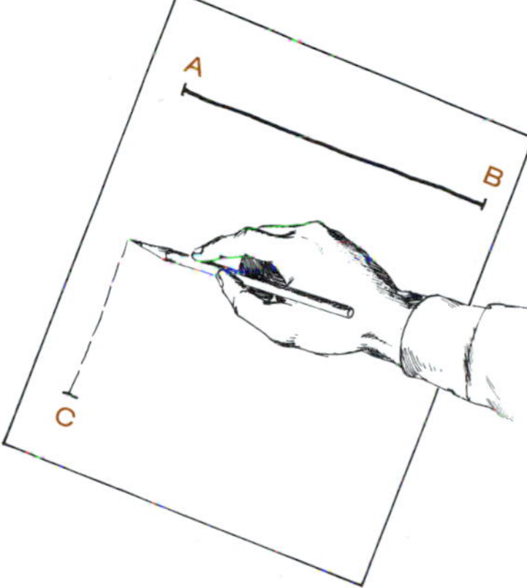

STEP 2
SKETCH SHORT DASHES.

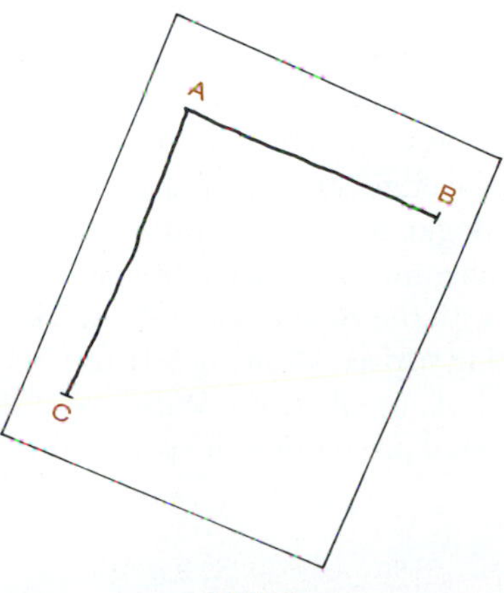

STEP 3
DARKEN FINAL LINE.

Vertical Lines

For this example you will sketch a vertical line CA about 7 inches long. Vertical lines are sketched in the identical manner as horizontal lines. As you did before, when sketching the horizontal line, shift the paper so that the new line to be sketched will be approximately at right angles to your forearm. As shown in Step 1 of Figure 1-4, establish the position of point C with a short dash located approximately the same distance from the left side of the paper as point A. Sketch a series of short dashes while keeping your aim toward point A (Step 2). The dashed line CA should be as parallel as possible to the left edge of the paper and at right angles to line AB. Finally, improve the straightness while gradually darkening in the trial line CA (Step 3).

CHECKPOINT 2
SKETCHING VERTICAL LINES

Vertical lines are sketched in the same manner as horizontal lines. Sketch a vertical line CA using the method shown in Figure 1-4. Practice by slowly and carefully sketching a dozen or so additional vertical lines spaced by eye about ½ inch apart and as parallel as possible.

FIGURE 1-4 Sketching a vertical line ***(Section 1.4)***

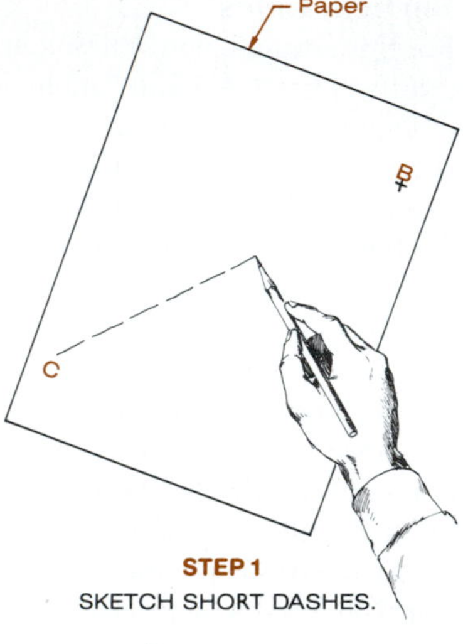

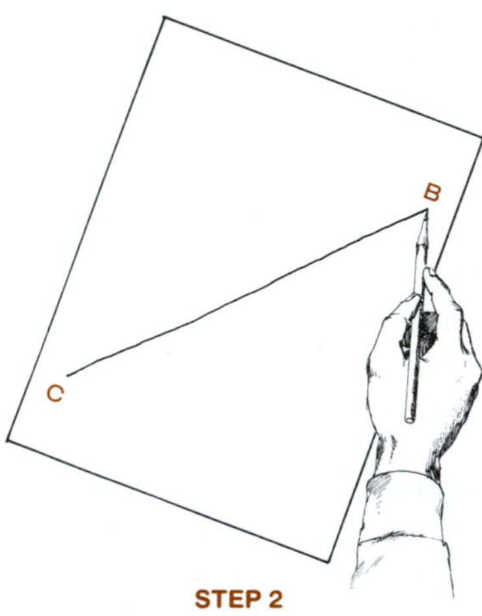

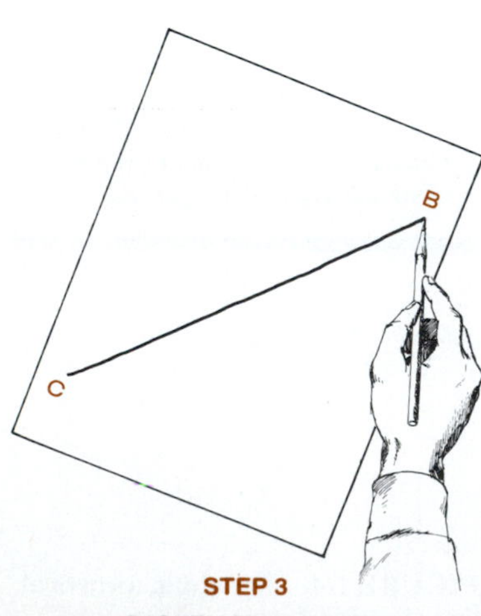

FIGURE 1-5 **Sketching an inclined line** ***(Section 1.4)***

Inclined Lines

Assume you want to sketch an inclined line. Using the same techniques outlined previously for sketching horizontal and vertical lines, once again shift the paper so that the desired line CB will be about 90° to your hand. As shown in Figure 1-5, start at point C and sketch a series of short dashes while directing your aim toward point B (Step 1). Connect the short dashes with a trial line (Step 2). Straighten the line if necessary and darken the final line CB (Step 3).

CHECKPOINT 3
SKETCHING INCLINED LINES

Inclined lines are sketched in the same manner as horizontal and vertical lines. Practice sketching line CB together with a number of additional inclined lines on another plain sheet of 8½ × 11 inch paper. Space the inclined lines about ½ inch apart and as parallel as possible. Continue to strive for clean-cut, dark, uniform lines.

Sketching Border Lines

Figure 1-6 illustrates a useful method for sketching straight lines such as border lines around the edges of a sheet of paper. Begin, as shown in Step 1, by aligning one edge of the paper with the right edge of the working surface (drawing board, desk, or tabletop). With one hand preventing the paper from slipping, hold the pencil firmly in the other hand. Extend one finger as a guide against the edge of the working surface. The distance the line will be spaced in from the edge is regulated by the distance from the guide finger to the pencil point (Step 2). Slowly sketch the line by pulling the pencil toward you (Step 3). Unlike when sketching

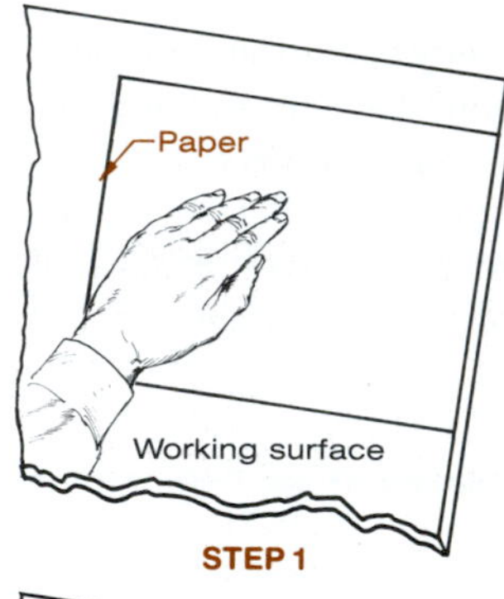

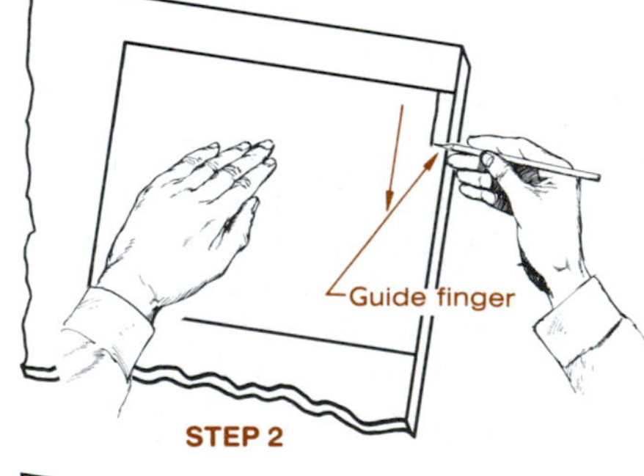

FIGURE 1-6 **Sketching a border line** ***(Section 1.4)***

other kinds of lines, it is not necessary for you to first produce a light trial line. Instead, you should produce a bold, crisp line at the first stroke.

You can use this method to produce freehand lines spaced as much as 1½ inches in from the edge of the paper. Beyond this distance, positioning the pencil becomes awkward.

> **CHECKPOINT 4**
> **SKETCHING BORDER LINES**
>
> Review Steps
> 1. Align one edge of the paper with the right edge of the working surface.
> 2. Use your left hand to prevent the paper from slipping.
> 3. Extend one guide finger against the working edge.
> 4. Pull the pencil toward you.
> 5. Produce a dense, final line.
>
> Practice this method by sketching several differently spaced border lines around the edges of a plain sheet of 8½ × 11 inch paper. Try to produce neat square corners where the lines intersect. Avoid extending one line beyond another.

Dividing a Line into Equal Parts

Figure 1-7 shows the steps involved in dividing a line by eye into four approximately equal parts. Start by carefully estimating the center point of the line (Step 1) and then the center point of each half (Step 2).

This basic technique occurs again and again in sketching, where you must learn to develop an eye for good proportion. Equal distances and various spacings—one-sixth, one-fifth, one-fourth, one-third, or one-half, for example—often must be established by eye when sketching. Practice is required before you will develop an acceptable level of competence in estimating distances.

> **CHECKPOINT 5**
> **DIVIDING A LINE INTO A REQUIRED NUMBER OF PARTS**
>
> Review the steps given in Figure 1-7 and begin by sketching four horizontal lines on a plain sheet of 8½ × 11 inch paper. Carefully divide the first line into three equal divisions, the second line into five equal divisions, the third line into seven equal divisions, and the final line into eight equal divisions

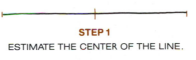

GIVEN LINE
STEP 1
ESTIMATE THE CENTER OF THE LINE.
STEP 2
ESTIMATE THE CENTER OF EACH HALF.

FIGURE 1-7 Dividing a line into four parts ***(Section 1.4)***

1.5 CIRCLES AND ARCS

Sizes up to 1 Inch Diameter

Before sketching a circle or an arc, sketch a horizontal and a vertical line that intersect at the midpoint of the desired circle, as shown in Step 1 of Figure 1-8. These lines will be changed later to center lines. Mark off the radius distance (estimated by eye) on each line (also in Step 1). As shown in Step 2, begin by lightly sketching one-quarter of the circle and then gradually sketch one-half of the circle (Step 3) followed by the other half (Step 4). It may be more natural to first sketch the *left* half of the circle and then to complete the other half after rotating the paper. Make corrections until the circle looks smoothly circular and then create the visible outline of the circle by darkening it in. Now change the light horizontal and vertical construction lines intersecting at the center of the

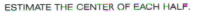

FIGURE 1-8 Sketching a circle: sizes up to 1 inch in diameter ***(Section 1.5)***

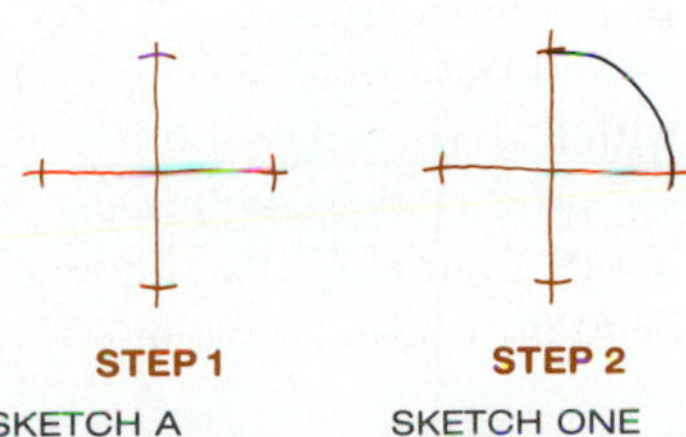
STEP 1
SKETCH A HORIZONTAL AND A VERTICAL LINE AND MARK OFF RADIUS DISTANCE.
STEP 2
SKETCH ONE QUARTER OF THE CIRCLE.

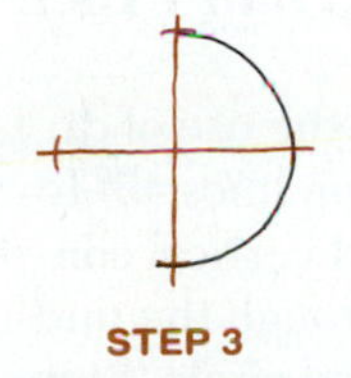
STEP 3
SKETCH ONE HALF OF THE CIRCLE.

STEP 4
SKETCH THE OTHER HALF, DARKEN THE CIRCLE, AND SKETCH THE CENTER LINES.

circle into center lines, as shown in Step 4 of Figure 1-8.

A center line, used to indicate the center of a circle or an arc, is dark but thin enough to contrast well with the other lines on the drawing. It should be drawn as a long line interrupted periodically by single short dashes. The short dashes should cross at the center of the circle. Make each short dash about 1/8 inch long with about a 1/16 inch space on either side. The long dashes will vary in length depending upon the total length of the center line. When working with small circles, showing a short dash at the center is not practicable. In such a case a continuous line is drawn and the dash is omitted. The three holes shown on the left side of Figure 1-1 illustrate the convention for center lines on small holes.

You will very likely find that, when sketching relatively small circles (1/4 inch in diameter and less, for example), construction lines can be entirely omitted. Good quality small circles can be easily sketched by using only the center point as a guide and sketching the circle in one continuous movement.

Sizes Greater Than 1 Inch Diameter

Figure 1-9 shows the use of diagonal construction lines 45° to the horizontal and vertical construction lines through the midpoint of the desired circle (Step 1). Diagonal construction lines help to obtain a truer circle, particularly for the larger sizes. The estimated (equal) radial distances are marked off on each line, which yields an additional four points, making a total of eight points through which the desired circle may be sketched.

Figure 1-10 shows another method that is particularly useful when sketching large circles. Begin as before by sketching light horizontal and vertical lines that intersect at right angles at the midpoint of the

CHECKPOINT 6
SKETCHING CIRCLES

Review Steps

1. Sketch two construction lines that intersect at right angles. Add diagonals if desired.
2. Mark off the radius distance on each line.
3. Lightly sketch one-quarter of the circle.
4. Sketch the other quadrants.
5. Darken the final circle.
6. Sketch the center lines.

Using a plain sheet of 8½ × 11 inch paper, sketch circles of several different sizes. On the same sheet, sketch a dozen or so circles 1/4 inch in diameter or less using only a center point as a guide.

FIGURE 1-9 Sketching a circle: sizes greater than 1 inch in diameter ***(Section 1.5)***

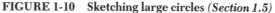

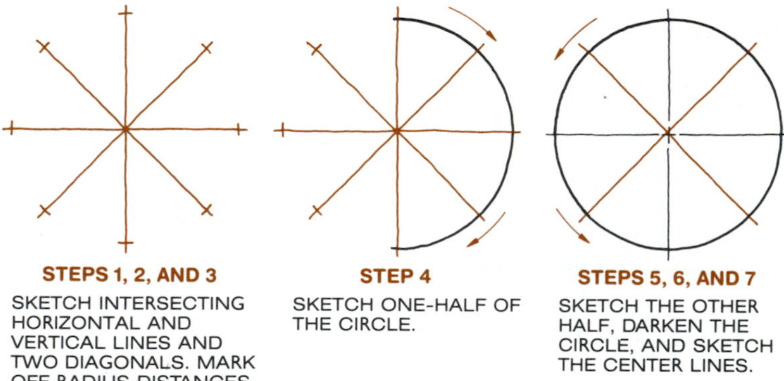

STEPS 1, 2, AND 3
SKETCH INTERSECTING HORIZONTAL AND VERTICAL LINES AND TWO DIAGONALS. MARK OFF RADIUS DISTANCES.
STEP 4
SKETCH ONE-HALF OF THE CIRCLE.
STEPS 5, 6, AND 7
SKETCH THE OTHER HALF, DARKEN THE CIRCLE, AND SKETCH THE CENTER LINES.

FIGURE 1-10 Sketching large circles ***(Section 1.5)***

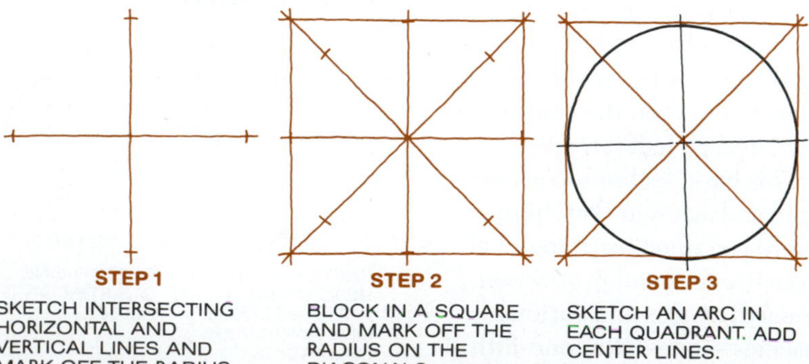

desired circle. Mark off the radius distance on each line (Step 1). Next, block in an enclosing square construction box (Step 2), adding diagonals if you so desire. Finally, lightly sketch an arc in each quadrant of the square, gradually correcting the shape of the circle until it is acceptable. Add center lines to your sketch and darken the final circle (Step 3).

Figure 1-11 illustrates several methods of sketching arcs. Notice that the center points of the arcs and the points of tangency at the midpoints of the sides—as at (c), for example—are carefully approximated on the constructions. Small arcs, like small circles, may be satisfactorily sketched without the aid of construction lines.

Examples of recommended constructions for sketching composite figures consisting of both circles and arcs are shown in Figure 1-12. Note the consistent use of construction lines for blocking in the main outlines of the figures and the circles and arcs. Only the desired final portion of the arc is darkened.

CHECKPOINT 7
SKETCHING ARCS

Review the steps for sketching circles. Using a plain sheet of 8½ × 11 inch paper, copy the arc constructions shown in Figure 1-11 to a convenient size. Avoid shortcuts.

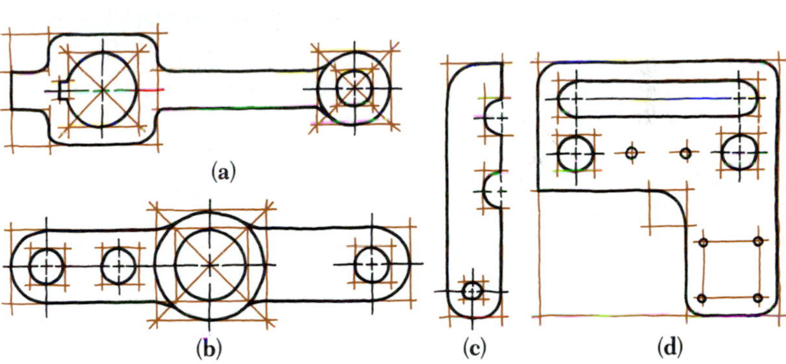

FIGURE 1-12 Sketching arcs and circles *(Section 1.5)*

FIGURE 1-11 Sketching arcs *(Section 1.5)*

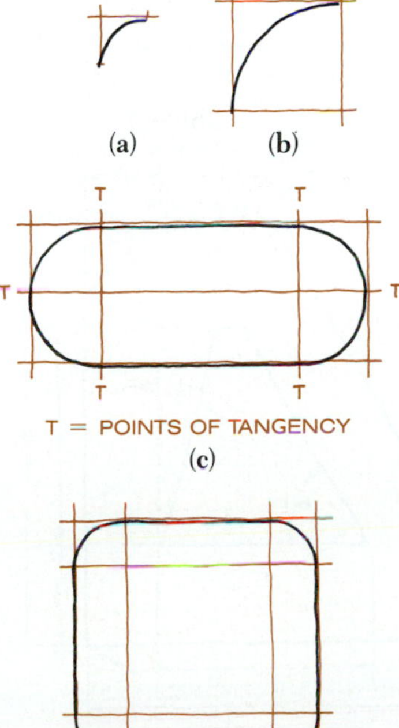

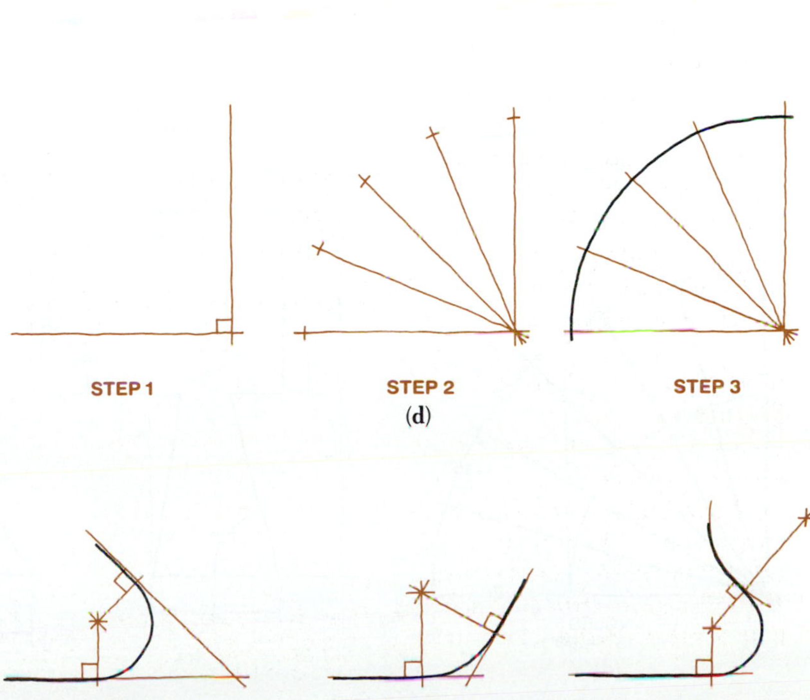

FIGURE 1-13 Sketching approximate angles *(Section 1.6)*

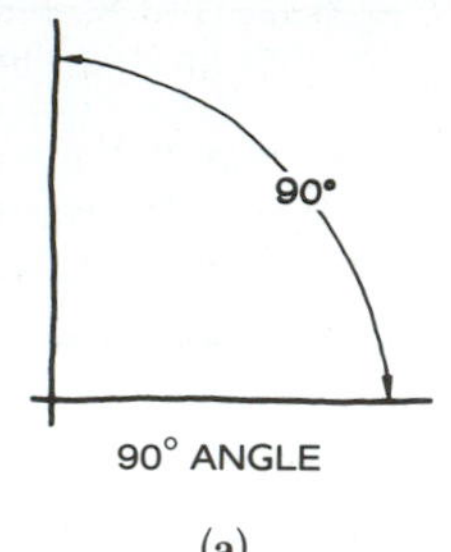

(a)

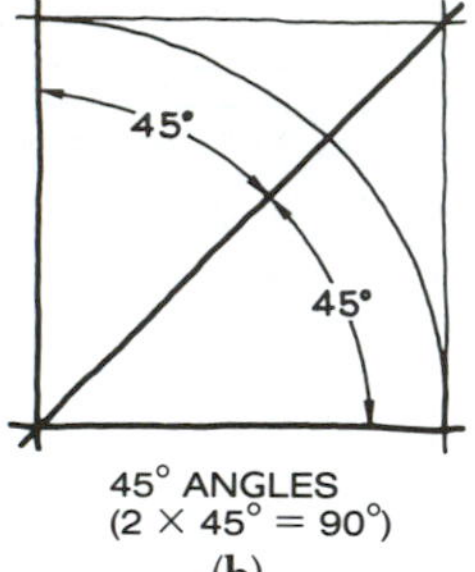

(b)

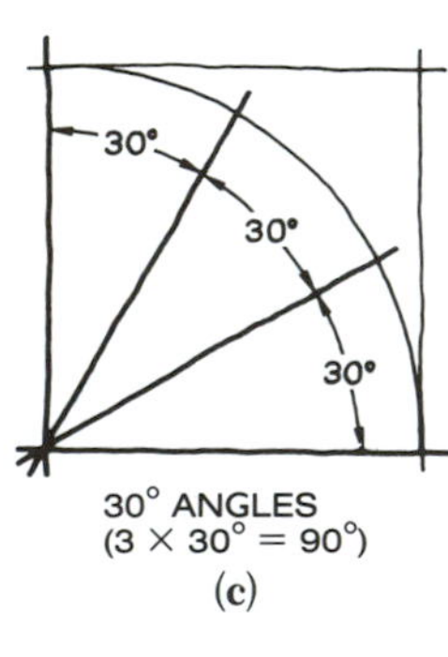

(c)

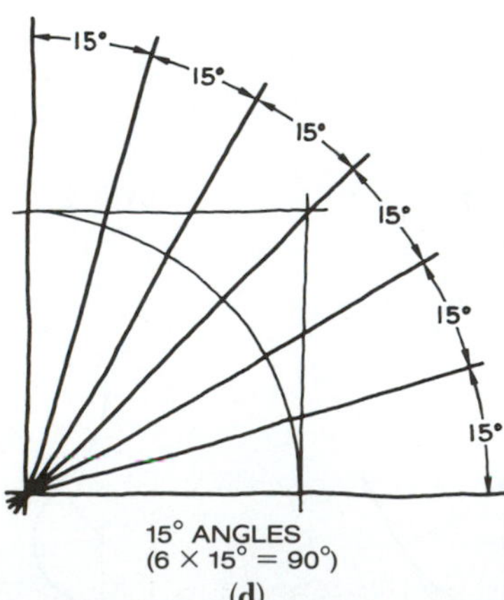

(d)

1.6 LAYING OUT ANGLES

Another sketching skill is the ability to approximate a desired angle. Figure 1-13 illustrates some useful methods for estimating angles, starting with an angle of 90° and working down in 15° increments. In each case the desired angle is sketched by bisecting or trisecting a previously constructed angle. An angle of approximately 7°, for example, can be obtained by bisecting a 15° angle. Note how the arc used in the construction aids in estimating the desired angles.

CHECKPOINT 8
SKETCHING ANGLES USING PROPORTIONAL TECHNIQUES

Using a plain sheet of 8½ × 11 inch paper, copy Figures 1-13(a) through 1-13(d). On a second sheet of paper, sketch the objects shown in Figure 1-14. Select a suitable size so that your four solutions will fit neatly on a sheet of 8½ × 11 inch paper. The dimensions on each problem should help you to estimate the proper proportions. (*Hint*: For each sketch, start by blocking in the main outline and then work out the constructions for each angle.)

FIGURE 1-14 Checkpoint 8 problems

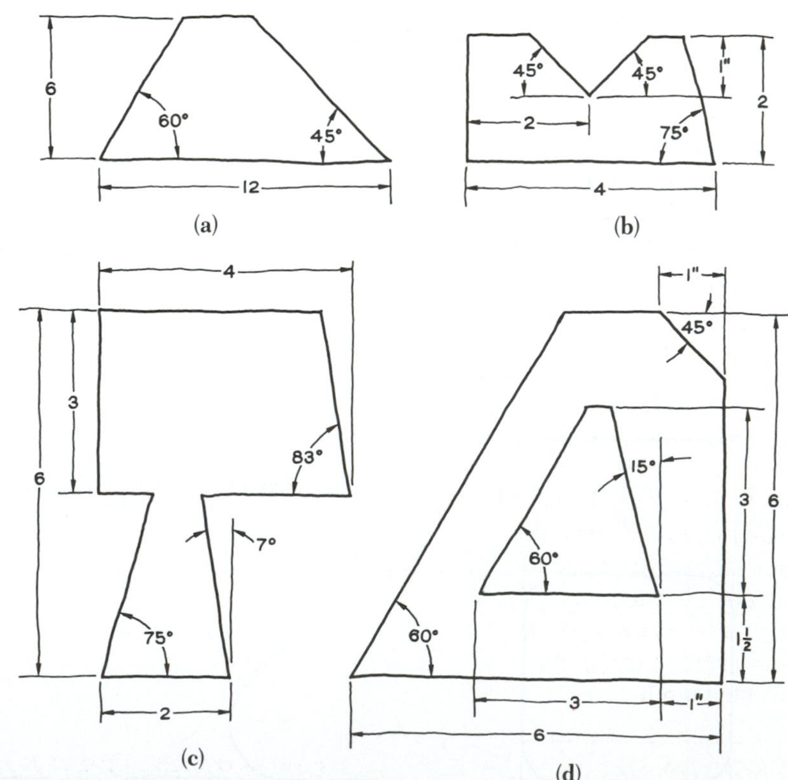

1.7 LAYING OUT A SINGLE-VIEW SKETCH

A well-executed sketch requires some careful preliminary planning. In many cases it is helpful to begin the sketch by blocking in the main outline using construction lines. Next, add the main center lines and gradually sketch the internal details of the object. The use of these basic steps in laying out a sketch are necessary to obtain good results. Sketches are not made to a precise size, but close adherence to the correct *proportions* is important. Proper proportion is achieved by carefully estimating by eye the relationship of each of the various distances on the object to be sketched. Decide upon a convenient size for your sketch depending upon the space available on the paper and the complexity of the object to be sketched. Avoid "mini-sketches"; important detail is difficult to display at a reduced size.

The *outside-in* (or basic-form) method and the *inside-out* (or axes) method are two recommended procedures for laying out a single-view sketch. Both methods require the use of a construction-box technique in a manner similar to that for laying out circles. The procedure you should select depends upon the configuration of the object you wish to sketch.

Outside-in Method

If the outline of the object is basically made up of straight lines, using the outside-in method is generally easier. In this case, begin by blocking in the figure with a framework of light construction lines. The proportion of the framework is determined by studying the main outlines—that is, the width and height of the object you wish to sketch.

In Figure 1-15 the various steps involved in making an outside-in, single-view sketch of an object are illustrated. For this example assume that you have a drawing of the object and that you wish to sketch it at a reduced size. Begin the sketch, as shown in Step 1, by blocking in the main outline of the figure using light construction lines to aid in estimating the correct proportions. Observe in this step how the relative proportion of the width versus the height is maintained at a reduced size. Now block in the proportions of the other cuts and features of

FIGURE 1-15 Sketching an object at a reduced size using the outside-in method *(Section 1.7)*

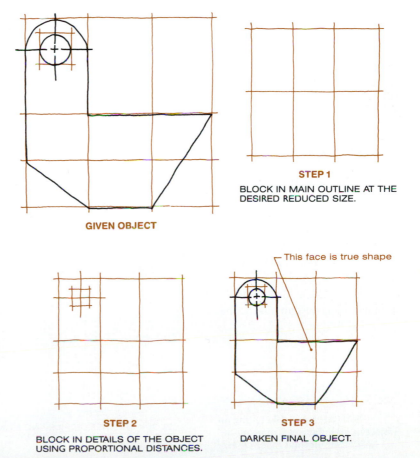

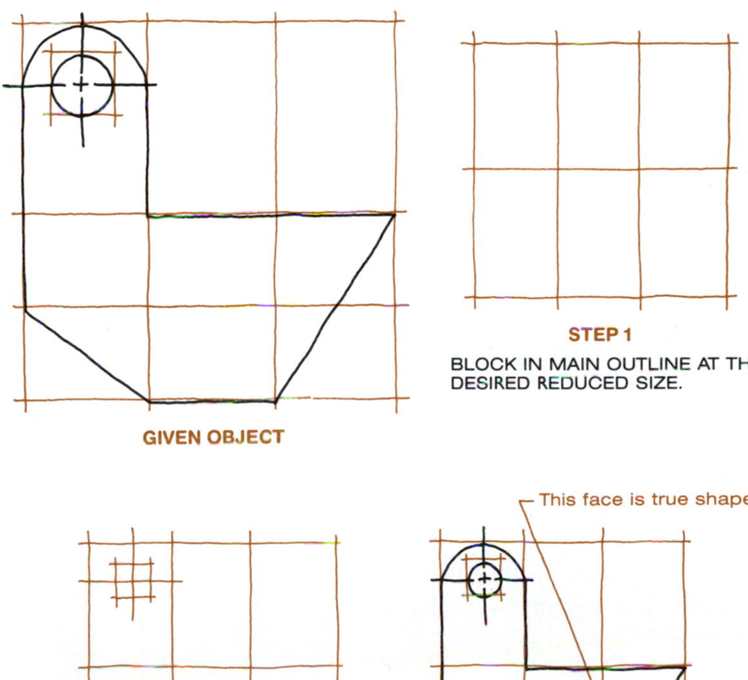
GIVEN OBJECT
STEP 1
BLOCK IN MAIN OUTLINE AT THE DESIRED REDUCED SIZE.
This face is true shape
STEP 2
BLOCK IN DETAILS OF THE OBJECT USING PROPORTIONAL DISTANCES.
STEP 3
DARKEN FINAL OBJECT.

the object (Step 2). Finally, darken all of the visible edges of the object (Step 3).

Single-view sketches always show the front plane or "face" of the object in *true shape*, but the scale may be represented at any convenient size. A surface appears true shape when the observer views it perpendicularly.

> **CHECKPOINT 9**
> **MAKING A SINGLE-VIEW SKETCH**
>
> Review Steps
>
> 1. Carefully study the proportions of the figure to be sketched.
> 2. Lightly construct a box for the main outline.
> 3. Gradually sketch construction lines representing the other cuts and details of the object.
> 4. Darken the final figure.
>
> Practice by copying the figure shown in Figure 1-15, Step 4. Use a plain sheet of 8½ × 11 inch paper, and fit your sketch neatly on the *top half* of the sheet. Do not erase your construction lines.

Inside-out Method

Figure 1-16 illustrates the procedure for laying out a single-view sketch using the inside-out method. This method is particularly useful for sketching symmetrical objects, cylinders, cones, and splines. Unlike the object shown in Figure 1-15, the object shown in Figure 1-16 is composed of a number of arcs, circles, and straight lines. For this example assume you want a sketch that is larger than the original drawing. After studying the general proportions of the original figure, begin the sketch by working inside-out. Sketch the main center lines or axes that form the central skeleton of the object and then sketch the diagonals, as shown (Step 1). In Step 2 carefully estimate the relative proportions of the various features of the object and block in these details. Mark off the radii of the arcs and circles

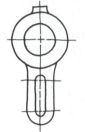

GIVEN OBJECT

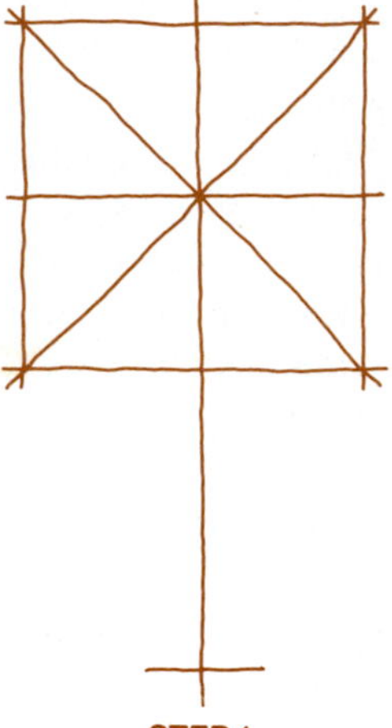

STEP 1
SKETCH CENTER LINES AND MAIN CONSTRUCTION BOX USING PROPORTIONAL DISTANCES AT THE DESIRED ENLARGED SIZE.

STEP 2
BLOCK IN OTHER DETAILS AND MARK OFF RADII USING PROPORTIONAL DISTANCES.

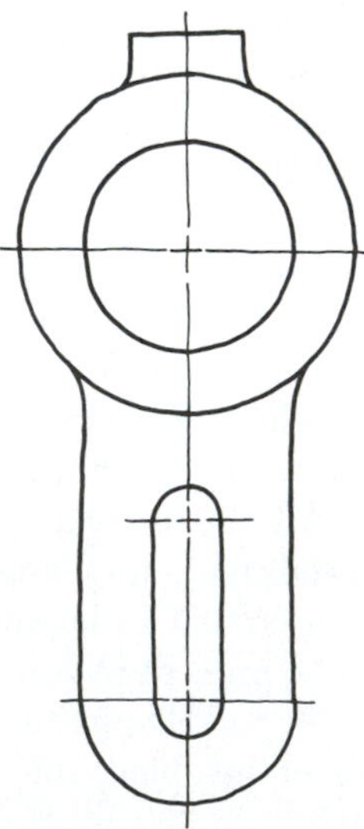

STEP 3
DARKEN FINAL OBJECT.

FIGURE 1-16 Sketching an object at an enlarged size using the inside-out method *(Section 1.7)*

and lightly sketch these features. Finally, darken the visible lines and center lines to complete the sketch (Step 3).

CHECKPOINT 10
MAKING A SINGLE-VIEW SKETCH

Review Steps

1. Carefully study the proportions of the figure to be sketched.
2. Lightly sketch construction lines representing the main center lines of the object.
3. Gradually sketch other lines representing the cuts and features of the object.
4. Darken the final figure.

Practice by copying Figure 1-16 in a horizontal position on the lower half of the sheet that you used to copy the example in Checkpoint 9. Do not erase your construction lines.

1.8 COMPUTER-GENERATED SINGLE-VIEW DRAWINGS

Computer-aided drafting is a designer's method of creating and storing accurate geometrical data. It can also contribute to product improvement because it affords an opportunity for better analysis of geometry, consistant application of standards, and greater accuracy and control, resulting in fewer errors.

Single-view drawings can be displayed on the graphics screen at the press of a keyboard button, if the data have been previously programmed, or by the flick of a light pen on the tablet. Depending upon the instructions given, the data will be accepted by the computer as direct input or as commands to "find" or recall previously stored input. A computer system can generate pictures with precise right angles, center lines, parallels, intersections, circles, fillets (inside corners), or rounds (outside corners). When working with computers, a designer who is not fully satisfied with his or her individual creations can try many other variants. The computer lets the designer try out many versions before commitment to a final shape, much in the same way an original sketch of a design concept is first prepared on paper and then is gradually revised until the desired result is obtained.

Electronic drafting systems are not intended to compete with sketch pads. In many cases it takes longer to write a program for a drawing using automated drafting techniques than it does to manually prepare a sketch or drawing on paper. Automated drafting systems are not generally justified unless a company has high volume or complex drafting or design work. Approximately 20 percent of industrial drafting work is currently being accomplished on automated drafting equipment and this percentage will rapidly increase.

1.9 ENGINEERING LETTERING

Many graphical documents used in engineering, whether prepared freehand or with instruments, require rapidly executed lettering. Examples are explanatory notes, specifications, detail titles, data in bills of materials and revision notes, and title blocks. The ability to letter quickly and legibly is important for drafters and engineers. All that is really required of the beginner is an earnest desire to learn how to use a straightforward and almost universally adopted lettering style. By practicing a few simple rules, you can easily master the art of lettering. The distinctive style of engineering lettering is so well established that previous engineering training is usually assumed by the general public whenever a sample of an engineer's plain and legible lettering is observed on a mailing envelope, on bank checks, or on various other forms.

The technique that is employed in making letters and numerals is a logical continuation of the methods previously discussed in this chapter for single-view sketching. When sketching an object, you first block in its outline with construction lines. In a similar fashion, when lettering, you will use pairs of light, horizontal guide lines and randomly spaced vertical or inclined guide lines. As in sketching, use a fairly soft pencil lead to produce dark, opaque letters and numerals.

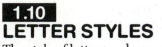

1.10 LETTER STYLES

The style of letters and numerals most commonly used in engineering applications is shown in Figure 1-17 and 1-18. The style is called *single-stroke Gothic* and is only one of many alphabet styles. It is used by engineers and drafters because of its legibility and ease of execution. The proper proportions of uppercase (capital) letters, lowercase (small) letters, and numerals are illustrated within square construction boxes. The squares are used as an aid in establishing the relative proportions of each letter in the same way that the construction box helps in sketching objects. The arrows denote a recommended direction for making the various strokes.

Uppercase letters are customarily used on engineering drawings and sketches and lowercase letters are used principally by the engineer for informal computations and for recording notes, as in Figure 1-19. When lowercase letters are used, the first letter of the first word of a sentence is capitalized. With this exception, and for capitalizing proper nouns, the uppercase and lowercase letter styles are not used together in a word. Either vertical or inclined letters can be used, but only one style should be used on a single drawing.

FIGURE 1-17 Engineering-style lettering: vertical uppercase and lowercase letters *(Section 1.10)*

FIGURE 1-18 Engineering-style lettering: inclined uppercase and lowercase letters *(Section 1.10)*

1.11 LETTERING PRACTICE

As contrasted with sketching, lettering requires finger and wrist motion with virtually no arm movement. Until you have developed a reasonable proficiency in this most important skill, you will need to form each letter and numeral in a slow, deliberate fashion, concentrating upon the development of the proper muscular control. To the beginner, lettering exercises may be somewhat tedious and boring, but professional competence is rarely achieved in any field without considerable practice and dedication. Lettering speed will improve as you learn the correct shape and proportion of each letter and develop a consistent stroke pattern in forming the letters. It is important to keep the form of each letter clearly fixed in your mind as titles and notes are lettered on a drawing or a sketch.

Try to develop the habit of forming each letter in the same way each time. Carefully study the directions for the strokes that are shown on each letter in Figures 1-17 and 1-18. In general the vertical strokes tend to be downward and the horizontal strokes tend to be in a left-to-right direction for right-handers. Left-handers should compensate for any minor differences and vary the direction of the strokes wherever necessary. For both right- and left-handers, developing a sequence of rapid, natural strokes is very important. Note that the strokes are directed in these illustrations so that the pencil is always *pulled* rather than pushed. As in sketching, the paper should be free to shift to a position that is most natural to the formation of acceptable letters.

Some letters are a full square in width, but other letters are slightly less. Only the letter *W* exceeds the width of the square. After trying each style of letter—the vertical and the inclined—you will very likely favor one style over the other. Concentrate your practice upon your favorite style. After having fully mastered one style, you will find that mastery of the other style is relatively easy.

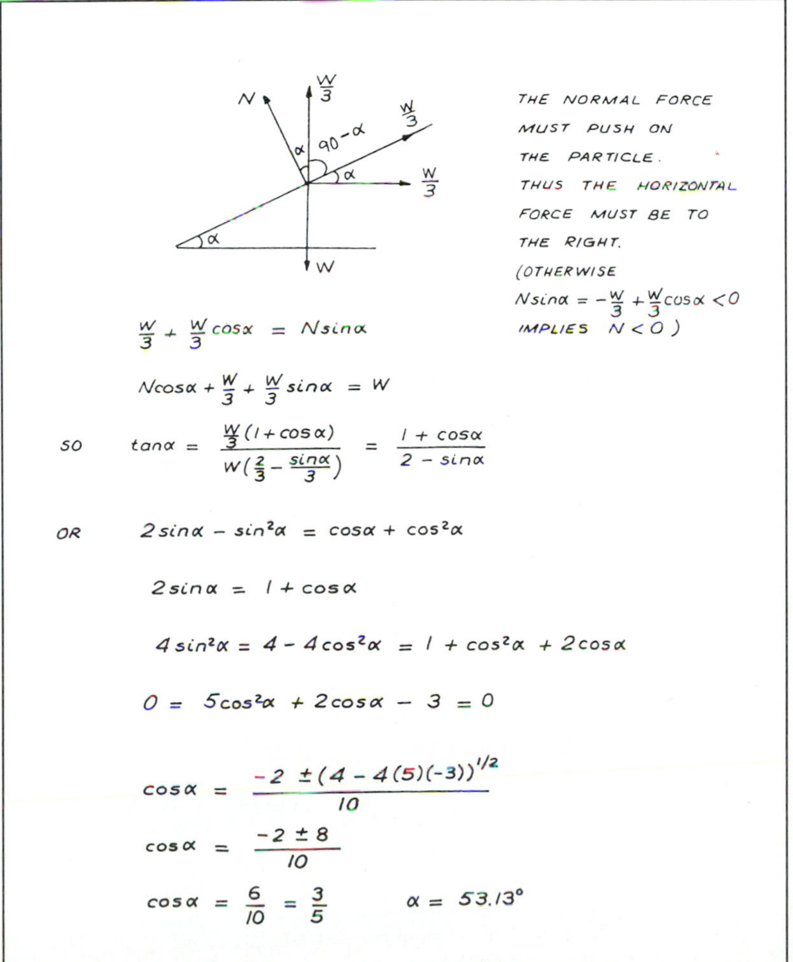

FIGURE 1-19 A sample page of engineering calculations in which both uppercase and lowercase letters are used *(Section 1.10)*

CHECKPOINT 11
LETTERING PRACTICE

Practice the letter forms in Figure 1-17 or Figure 1-18 by sketching them on ¼ inch cross-section paper. Work slowly and try to closely duplicate the given letter styles. Be patient, because the development of this skill is important to you in your professional life.

1.12 GUIDE LINES FOR LETTERING

The height for uppercase letters on most engineering documents is ⅛ inch, or about 3 mm, as shown in Figure 1-20(a). Lettering as high as 3⁄16 inch or even ¼ inch may be used in special cases (titles and subtitles, for example) for emphasis.

The standard height for lowercase letters is shown in Figure 1-20(b). The total height of letters such as *b*, *d*, and *f* is made the same as uppercase letters: ⅛ inch or 3 mm. These letters extend above the horizontal (body) guide line a distance of one third the uppercase letter height. Letters such as *a*, *c*, and *e* are made two thirds as high as the uppercase letters. Letters such as *g*, *j*, and *p* extend below the lower horizontal guide line a distance equal to one third the uppercase letter height.

Guide lines are used to ensure uniformity of lines of lettering and should be drawn very lightly using a 4H, 5H, or 6H grade pencil lead. Examples of horizontal and vertical guide lines for both the uppercase and the lowercase vertical style letters are given in Figure 1-20.

Before drawing guide lines, fasten the paper to the drawing board. When you use an 18 × 24 inch drafting board, for example, place the paper about 2 or 3 inches away from the left edge of the board, equally dividing the space above and below the paper. After aligning the upper edge of the T square blade or drafting machine scale with the top edge of the paper, fasten all four corners with short pieces of drafting tape. In practice, the position of the paper on an 18 × 24 inch drafting board is essentially governed by the size of the T square head. That is, if you place the paper too high on the drawing board, you will have difficulty using the upper edge of the T square blade. Similarly, it will be awkward for you to use the T square if you position the paper too low. The position of the paper on the board will also depend upon the size and types of other drafting equipment that will be used.

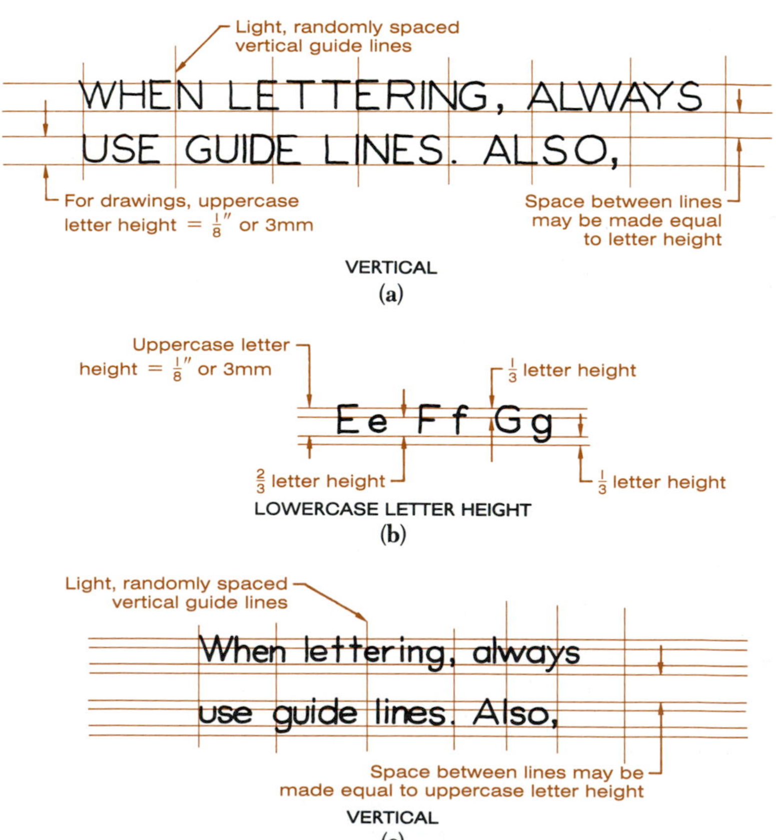

FIGURE 1-20 Guide lines, letter heights, and line spacing *(Section 1.12)*

Drawing Horizontal Guide Lines

If you are a right-handed drafter you should hold the head of the T square firmly against the *left* edge of the drawing board with the left hand. (Left-handers should hold the T square head against the right edge of the board.) Draw the lines on the paper by keeping the pencil lead in contact with the upper edge of the blade. Incline the pencil slightly in the direction of drawing, perhaps about 60° from the horizontal, and pull the pencil along the blade. After practice, some of your original awkwardness will be overcome, and in time you will be able to repeatedly produce crisp, uniform pencil lines with no slippage of the T square blade. Try to maintain a constant pressure on the pencil, and rotate it slightly to aid in retaining a sharp lead point. Figure 1-21 illustrates the method for drawing horizontal lines.

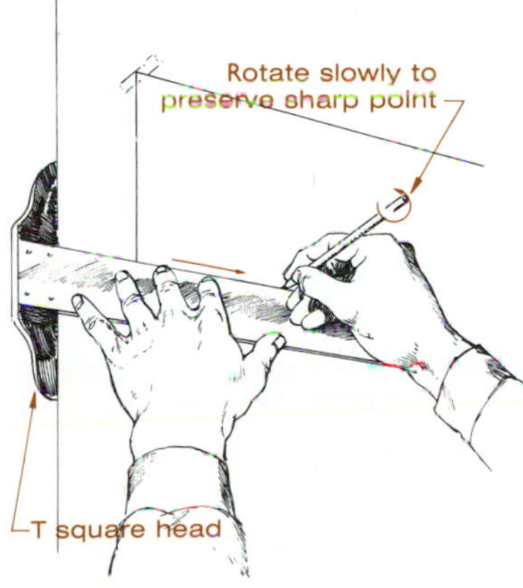

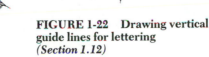

FIGURE 1-21 Drawing horizontal guide lines for lettering ***(Section 1.12)***

Drawing Vertical Guide Lines

Place one of the triangles on the T square blade with its vertical edge to the left. Bring the head of the T square firmly in contact with the left edge of the drawing board. Position your left hand as shown in Figure 1-22. Note that the palm of your hand should rest securely on the blade while your extended fingers rest on the triangle. Slide the triangle into the desired position with the fingertips of your left hand and draw the line by pulling the pencil upward along the left edge of the triangle. Tilt the pencil slightly forward and try to maintain a constant pressure.

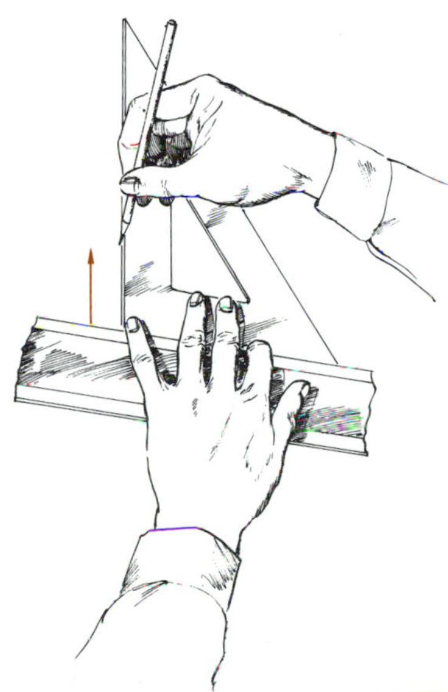

FIGURE 1-22 Drawing vertical guide lines for lettering ***(Section 1.12)***

Drawing Inclined Guide Lines

All inclined letters and numerals on a drawing should be made to the same slope. The standard angle for inclined letters is about 68° with the horizontal. It is helpful, at least at first, to establish the approximate slope by lightly drawing a small "slope triangle" at a convenient position on the paper. The ratio of the sides is 2:5, as shown in Figure 1-23. Draw randomly spaced inclined guide lines parallel to the slope triangle across the horizontal guide lines using the method shown in Figure 1-24. Begin by aligning one edge of the drafting triangle parallel to the hypotenuse of the slope triangle. Move the edge of the T square blade across the drawing until it contacts the drafting triangle. Hold it firmly in position and slide the triangle away from the given slope line to the various positions of the required inclined guide lines on the drawing. (A second triangle may be substituted for the T square.)

The use of guide lines is also recommended for making fractions, as shown in Figure 1-25. The height of fractions, for both the vertical and inclined lettering styles, is ¼ inch, or twice the height of a whole number. For clarity the numerals in the fractions should not touch the horizontal division bar. Each numeral should be made slightly less than ⅛ inch high.

The space between the lines of lettering is often a matter of convenience. Lines may be spaced one half of a letter height (1/16 inch), a full letter height, as shown in Figures 1-20(a) and 1-20(c), or as much as one and one-half times a letter height. Full letter height spacing is the most common. The space between the lines may be established by eye or marked off with a scale. All notes on a drawing or a sketch should be placed horizontally.

1.13 SPACING LETTERS AND WORDS

Because of the differences in the shape of the various letters, it is impossible to achieve a professional appearance by uniformly spacing each letter in a word. Figure 1-26(a) illustrates how putting equal spaces between each of the letters in the word *ALWAYS* produces an awkward result. Good letter spacing is accomplished by establishing the "white" space between the adjacent letters by eye so that each letter appears to be well balanced and equally spaced. Figure 1-26(b) shows how the appearance of the word *ALWAYS* is improved by making the letter spacing more compact. Some letters, such as *A*, *L*,

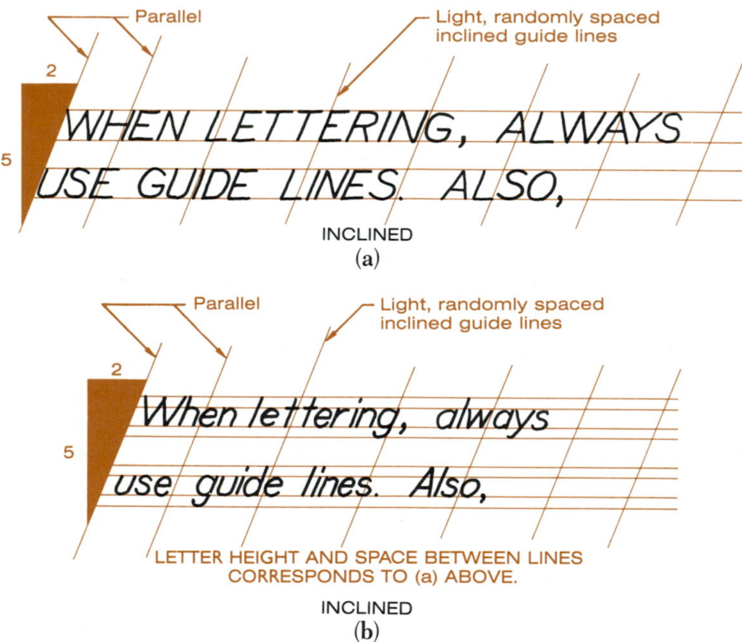

FIGURE 1-23 Use of a "slope triangle" for inclined lettering *(Section 1.12)*

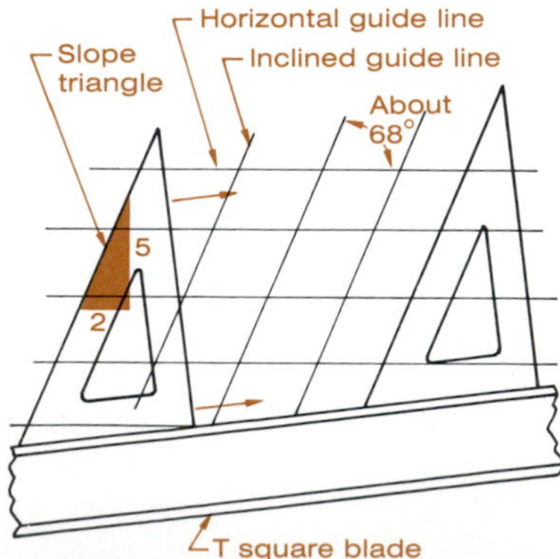

FIGURE 1-24 Drawing inclined guide lines for lettering *(Section 1.12)*

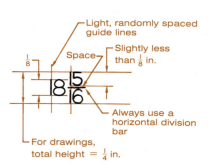

Light, randomly spaced guide lines

Space

Slightly less than 1/8 in.

1/8

18 15/16

Always use a horizontal division bar

For drawings, total height = 1/4 in.

FIGURE 1-25 Drawing guide lines for fractions *(Section 1.12)*

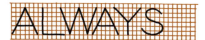

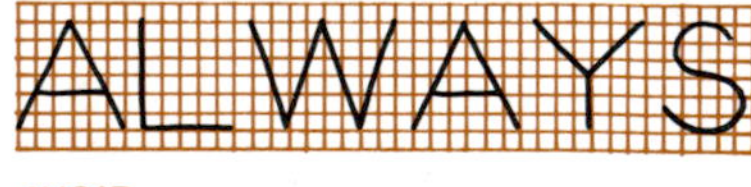

AVOID:
EQUAL SPACING BETWEEN LETTERS
(a)

ALWAYS

RECOMMENDED:
APPROXIMATELY EQUAL BACKGROUND AREAS BETWEEN LETTERS
(b)

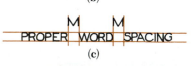

M M
PROPER WORD SPACING
(c)

FIGURE 1-26 Spacing letters and words *(Section 1.13)*

T, *W*, and *Y*, may have to be slightly interlocked for a more pleasing appearance when combined in a word with other letters.

Proper word spacing, or composition, also requires some careful attention. Words should be separated by a distance equivalent to the width of the letter *M*, as shown in Figure 1-26(c).

1.14 MECHANICAL LETTERING AIDS

Many types of lettering devices are available for mechanical lettering—including various instruments for spacing lines and for establishing the proper slant for letters—as well as a wide assortment of lettering templates and guides. There are also specially designed typewriters that can be used to letter information directly on drawings. One such device is shown in Figure 1-27. This type of equipment is used principally by drafters but not by engineers, whose lettering objectives are usually less precise. Figure 1-28 illustrates several different types of lettering aids.

SOURCE: DIAGRAM CORPORATION

FIGURE 1-27 A typewriter used for lettering *(Section 1.14)*

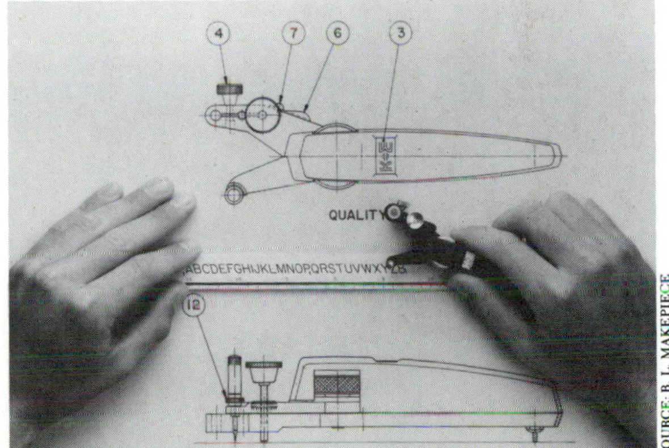

SOURCE: B. L. MAKEPIECE

FIGURE 1-28 Different types of lettering aids *(Section 1.14)*

1.15 COMPUTER-GENERATED LETTERING

Lettering presents no difficulties to the drafter or designer who uses computer-aided design and drafting equipment. Present models of automated drafting plotters and printers now on the market all have the capability of rapidly displaying lettered specifications on drawings. The Computervision Designer Interactive Graphics System produced by Computervision Corporation, for example, can display nine different fonts or assortments of letter styles. Figure 1-29 illustrates the various fonts and other characters available for this system. The lines of text are entered into the computer terminal in much the same manner as on a typewriter. The operator begins by selecting the particular lettering text and height desired and then commands the system to display the text strings or lines of lettering in a required location. Proper word spacing is automatically established, and multiple lines of lettering can be neatly justified or balanced on the drawing. This text can be created directly on a drawing or created separately as text and "merged" with a drawing.

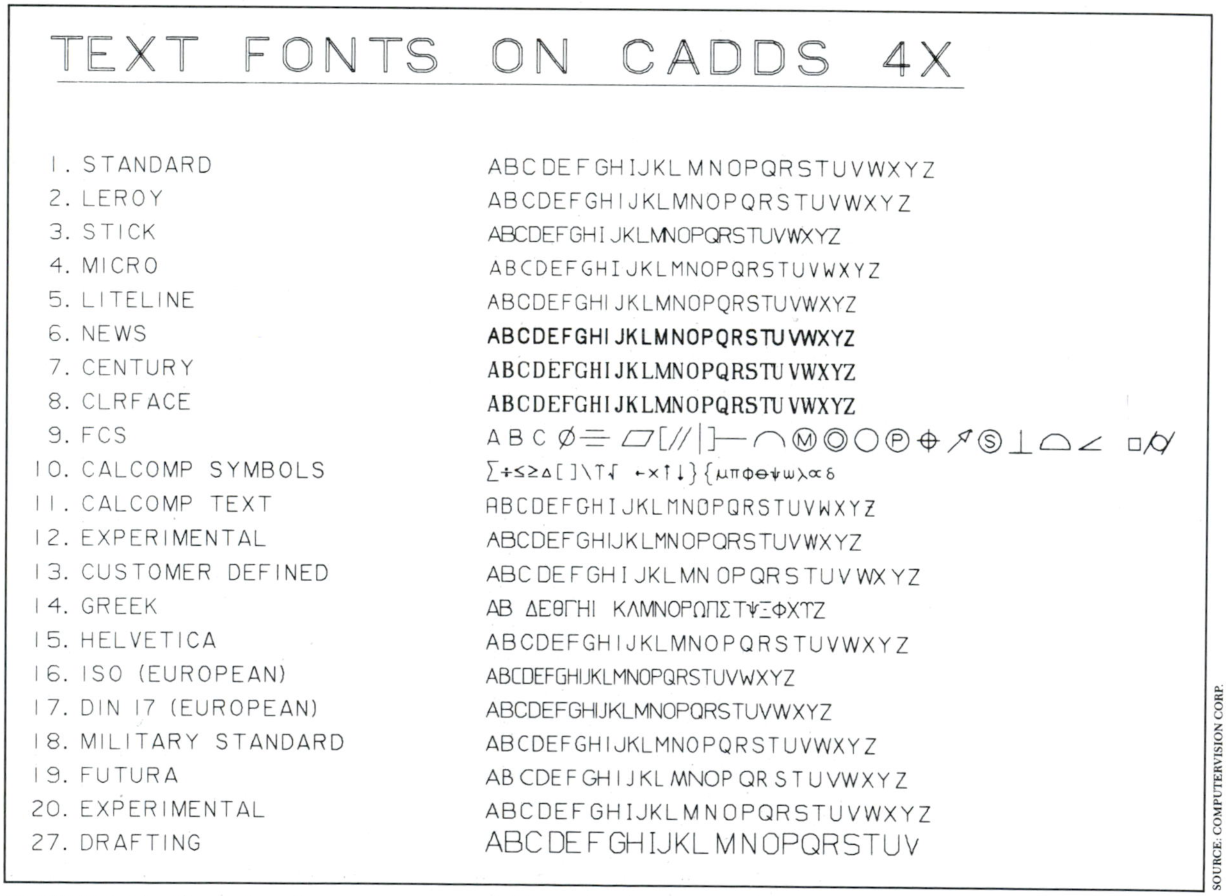

TEXT FONTS ON CADDS 4X

1. STANDARD	ABCDEFGHIJKLMNOPQRSTUVWXYZ
2. LEROY	ABCDEFGHIJKLMNOPQRSTUVWXYZ
3. STICK	ABCDEFGHIJKLMNOPQRSTUVWXYZ
4. MICRO	ABCDEFGHIJKLMNOPQRSTUVWXYZ
5. LITELINE	ABCDEFGHIJKLMNOPQRSTUVWXYZ
6. NEWS	ABCDEFGHIJKLMNOPQRSTUVWXYZ
7. CENTURY	ABCDEFGHIJKLMNOPQRSTUVWXYZ
8. CLRFACE	ABCDEFGHIJKLMNOPQRSTUVWXYZ
9. FCS	A B C ⌀ ⌯ ▱ [//|] ⌒ Ⓜ ◎ ○ Ⓟ ⌖ ↗ Ⓢ ⊥ ⌓ ∠ □ ⌭
10. CALCOMP SYMBOLS	∑+≤≥∆[]\T√ ←×↑↓{μπφθψωλα δ
11. CALCOMP TEXT	ABCDEFGHIJKLMNOPQRSTUVWXYZ
12. EXPERIMENTAL	ABCDEFGHIJKLMNOPQRSTUVWXYZ
13. CUSTOMER DEFINED	ABCDEFGHIJKLMNOPQRSTUVWXYZ
14. GREEK	AB ΔΕΘΓΗΙ ΚΛΜΝΟΡΩΠΣΤΨΞΦΧΥΖ
15. HELVETICA	ABCDEFGHIJKLMNOPQRSTUVWXYZ
16. ISO (EUROPEAN)	ABCDEFGHIJKLMNOPQRSTUVWXYZ
17. DIN 17 (EUROPEAN)	ABCDEFGHIJKLMNOPQRSTUVWXYZ
18. MILITARY STANDARD	ABCDEFGHIJKLMNOPQRSTUVWXYZ
19. FUTURA	ABCDEFGHIJKLMNOPQRSTUVWXYZ
20. EXPERIMENTAL	ABCDEFGHIJKLMNOPQRSTUVWXYZ
27. DRAFTING	ABCDEFGHIJKLMNOPQRSTUV

SOURCE: COMPUTERVISION CORP.

FIGURE 1-29 Examples of computer-generated lettering *(Section 1.15)*

PROBLEMS

Single-view Sketching and Engineering Lettering

Each of the problems in this chapter will fit on an 8½ × 11 inch sheet. Use layout A, shown on the inside of the front cover, and letter the required information in the title block. Begin all of your sketches and lettering sheets by carefully estimating the amount of space that you will need to neatly balance the work on your drawing sheet. Establish the proportions of each figure entirely by eye. Add center lines to your sketches wherever appropriate. Use either plain white or cross-section paper, as assigned by your instructor. For problems P1.1 to P1.47 prepare single-view sketches, two to each sheet.

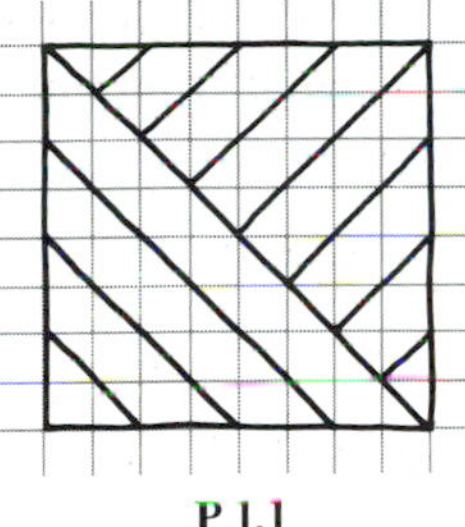
P 1.1

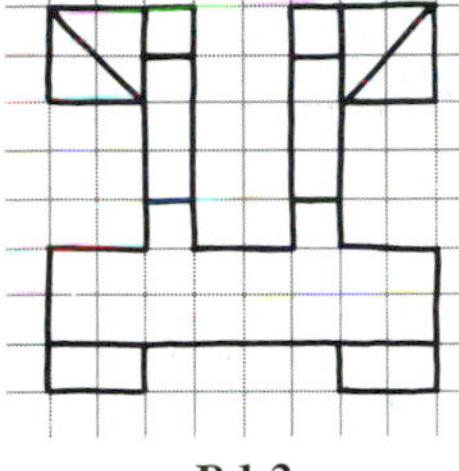
P 1.2

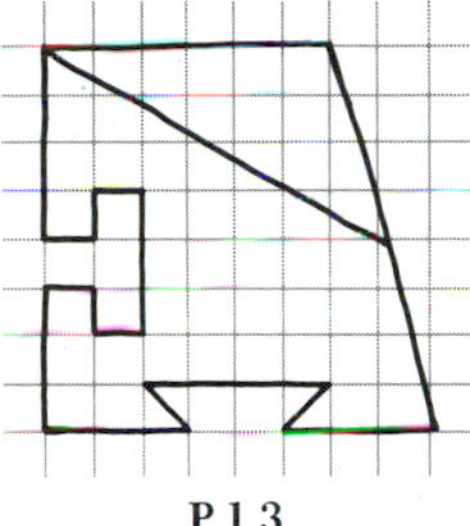
P 1.3

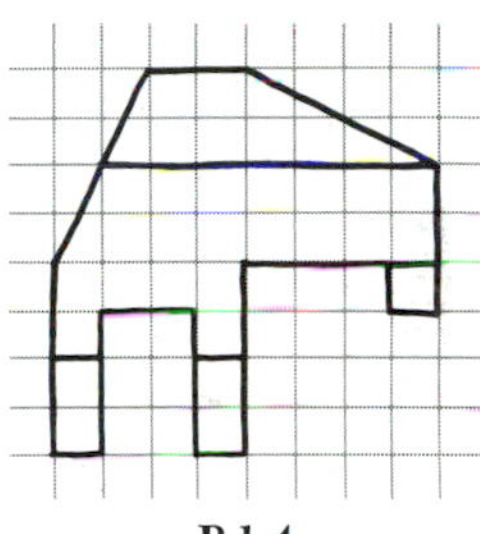
P 1.4

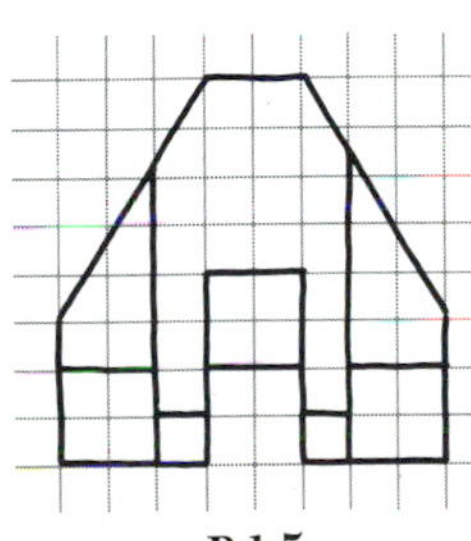
P 1.5

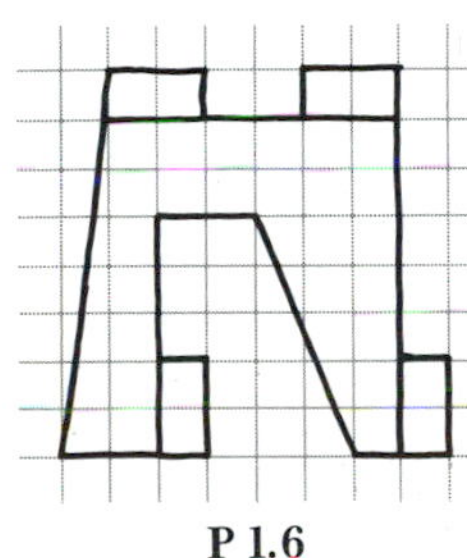
P 1.6

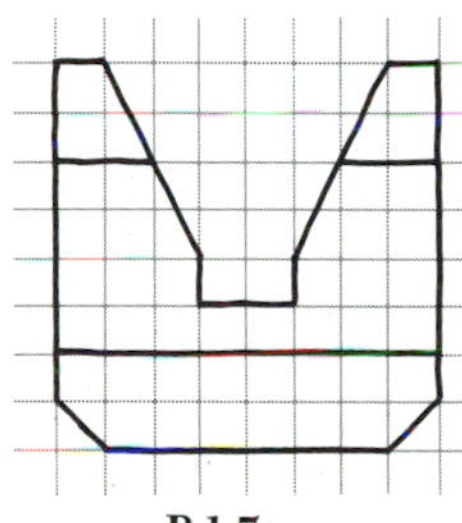
P 1.7

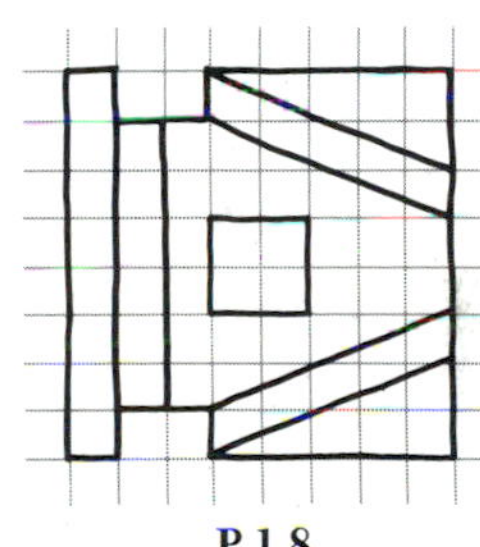
P 1.8

P1.1 to **P1.8:** Sketch a 4″ square (approximately) and then copy the given objects.
P1.9 to **P1.16:** Sketch a 2″ × 4″ rectangle (approximately) and then copy the given objects.

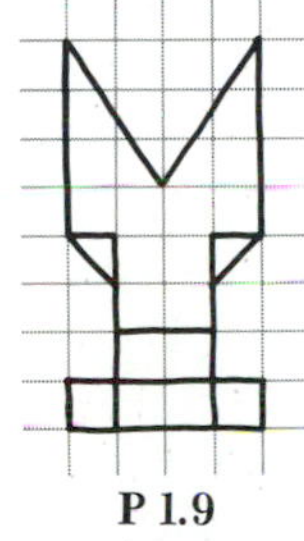
P 1.9

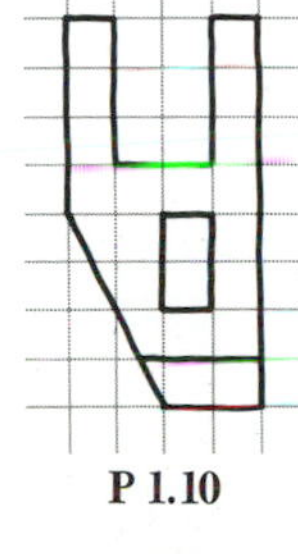
P 1.10

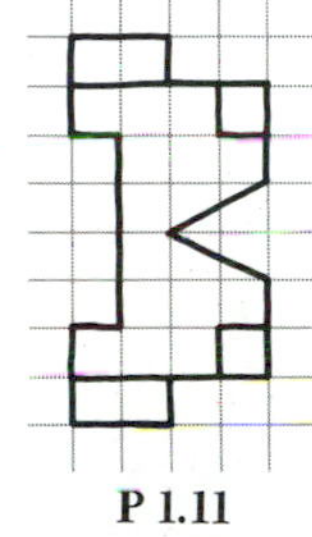
P 1.11

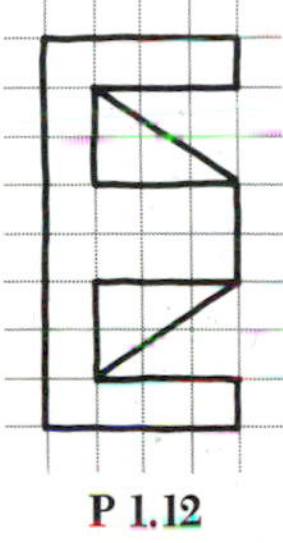
P 1.12

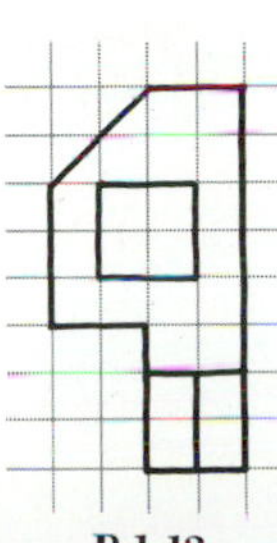
P 1.13

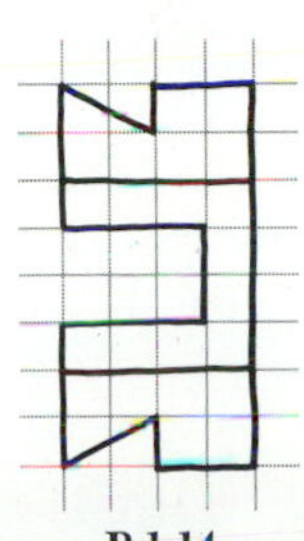
P 1.14

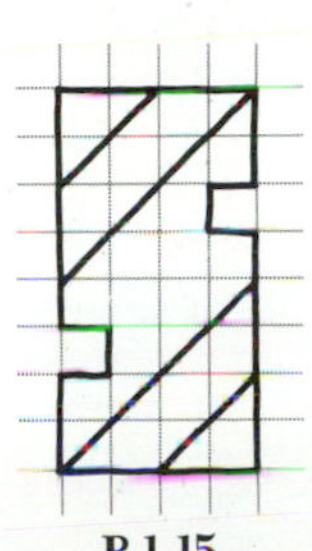
P 1.15

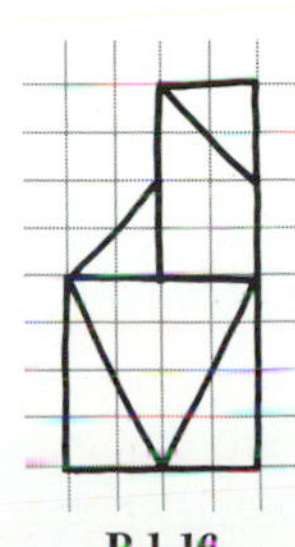
P 1.16

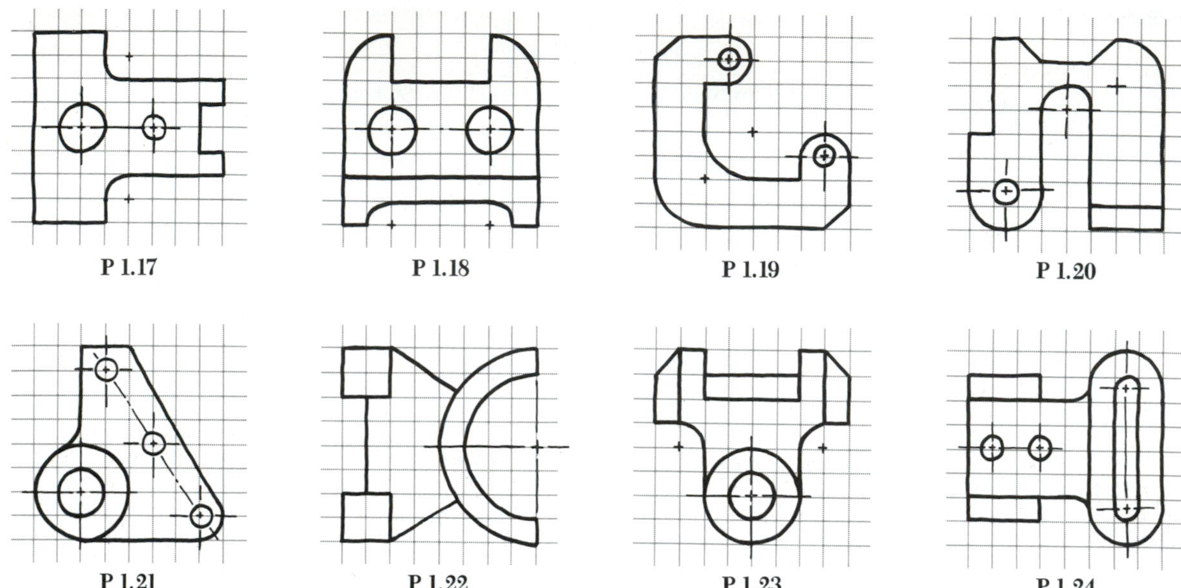

P1.17 to **P1.24:** Sketch a 4″ square (approximately) and then copy the given objects.

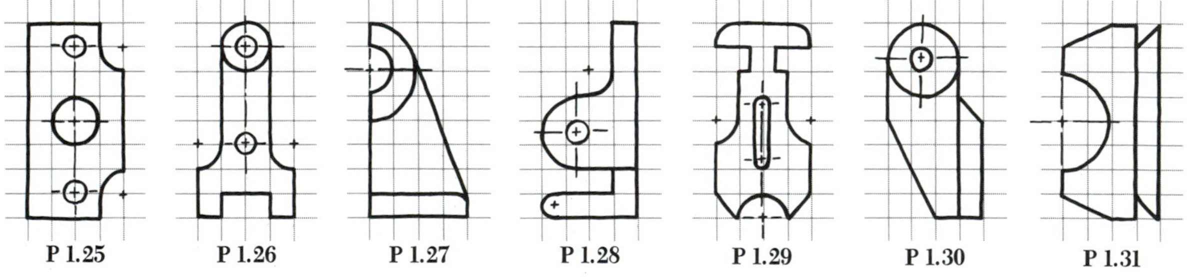

P1.25 to **P1.31:** Sketch a 2″ × 4″ rectangle (approximately) and then copy the given objects.

P1.32 to **P1.47:** The given figures represent only one-half of a symmetrical object. Sketch the entire object.

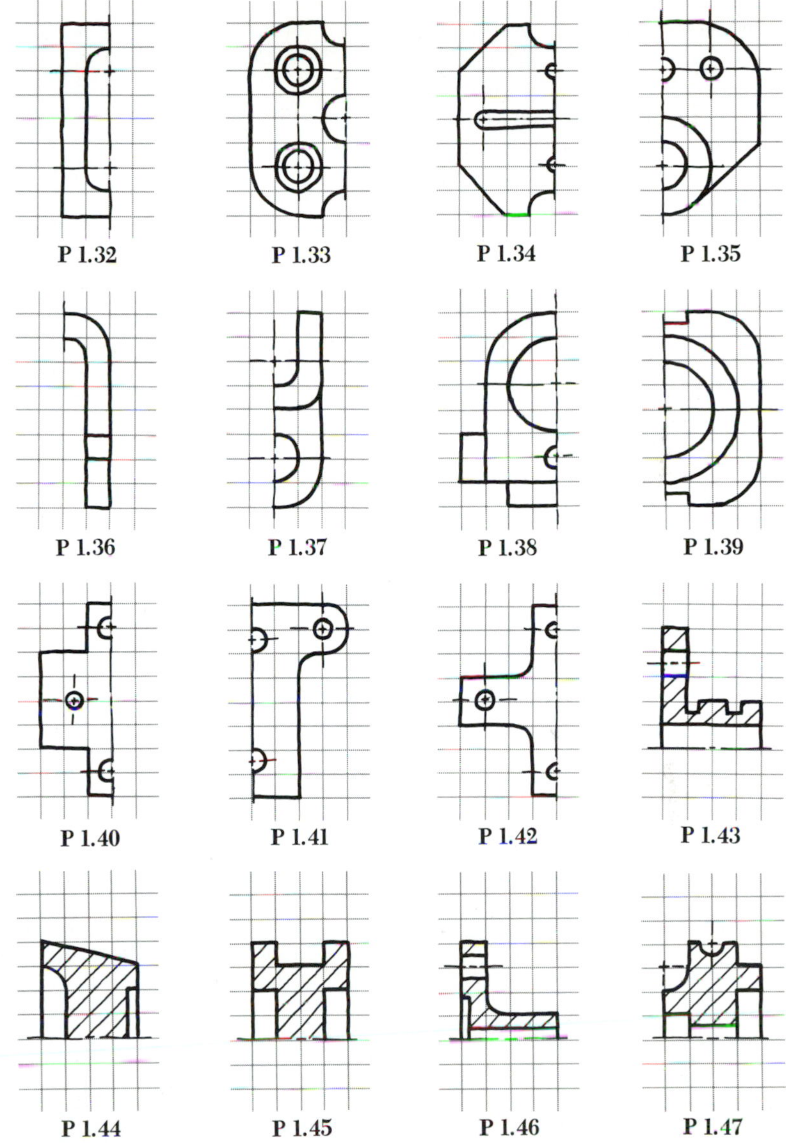

P1.48 to **P1.51:** Copy each of the given block diagrams to a size that will fit neatly on an 8½ × 11 inch sheet. Neatly letter the data given on each diagram on your sketch. Also, letter the following titles in a prominent position on your sketch:
P1.48 MODEL OF THE DESIGN PROCESS
P1.49 IMAGE PROCESSING SYSTEM
P1.50 TYPICAL INPUTS TO AND OUTPUTS FROM A COMPUTER
P1.51 A DNC SYSTEM PRESENTLY IN USE

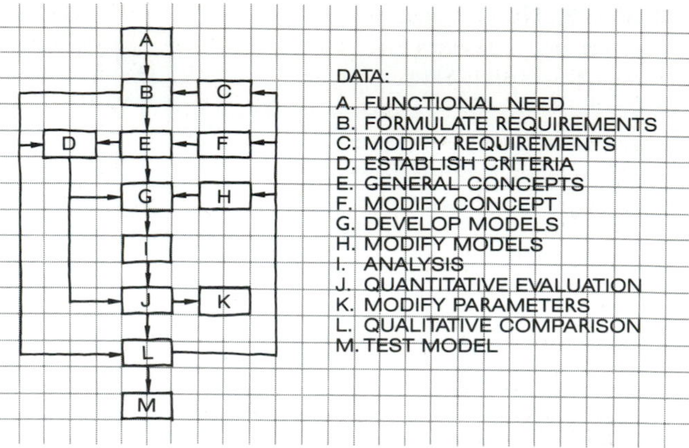

P 1.48

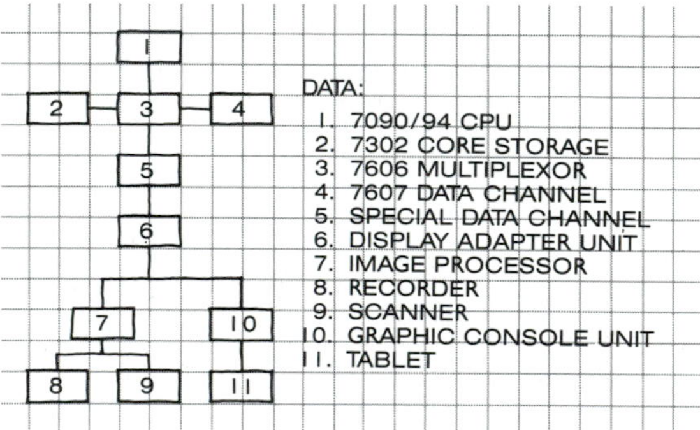

P 1.49

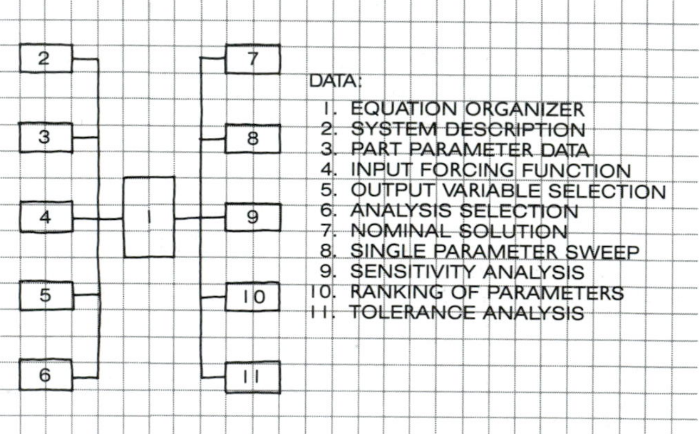

P 1.50

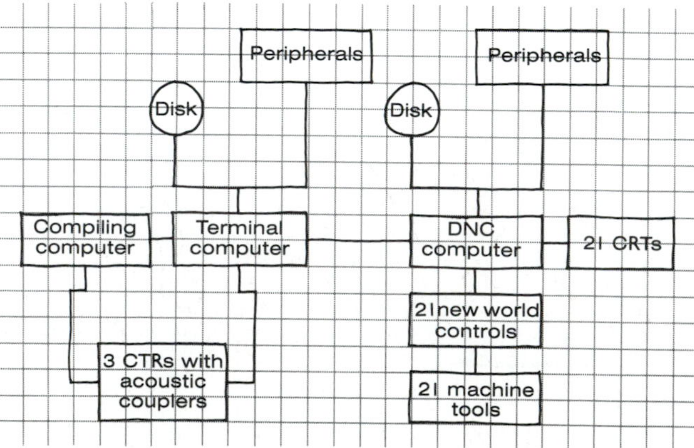

P 1.51

P1.52 to **P1.59:** Copy each of the given schematic layouts or diagrams to a size that will fit neatly on your sheet. Letter the data given on the diagrams on your sketch. Also, letter the following titles in a prominent position on your sketch:

P1.52 FLUID POWER CIRCUIT
P1.53 REACTOR FLOWCHART SYSTEM
P1.54 STEAM DISTRIBUTION SYSTEM
P1.55 WATER PURIFIER SYSTEM
P1.56 INDUSTRIAL POWER SUPPLY SYSTEM
P1.57 WHEATSTONE BRIDGE CIRCUIT
P1.58 DOMESTIC OIL BURNER SYSTEM
P1.59 SOLENOID OPERATED SAFETY LOCK

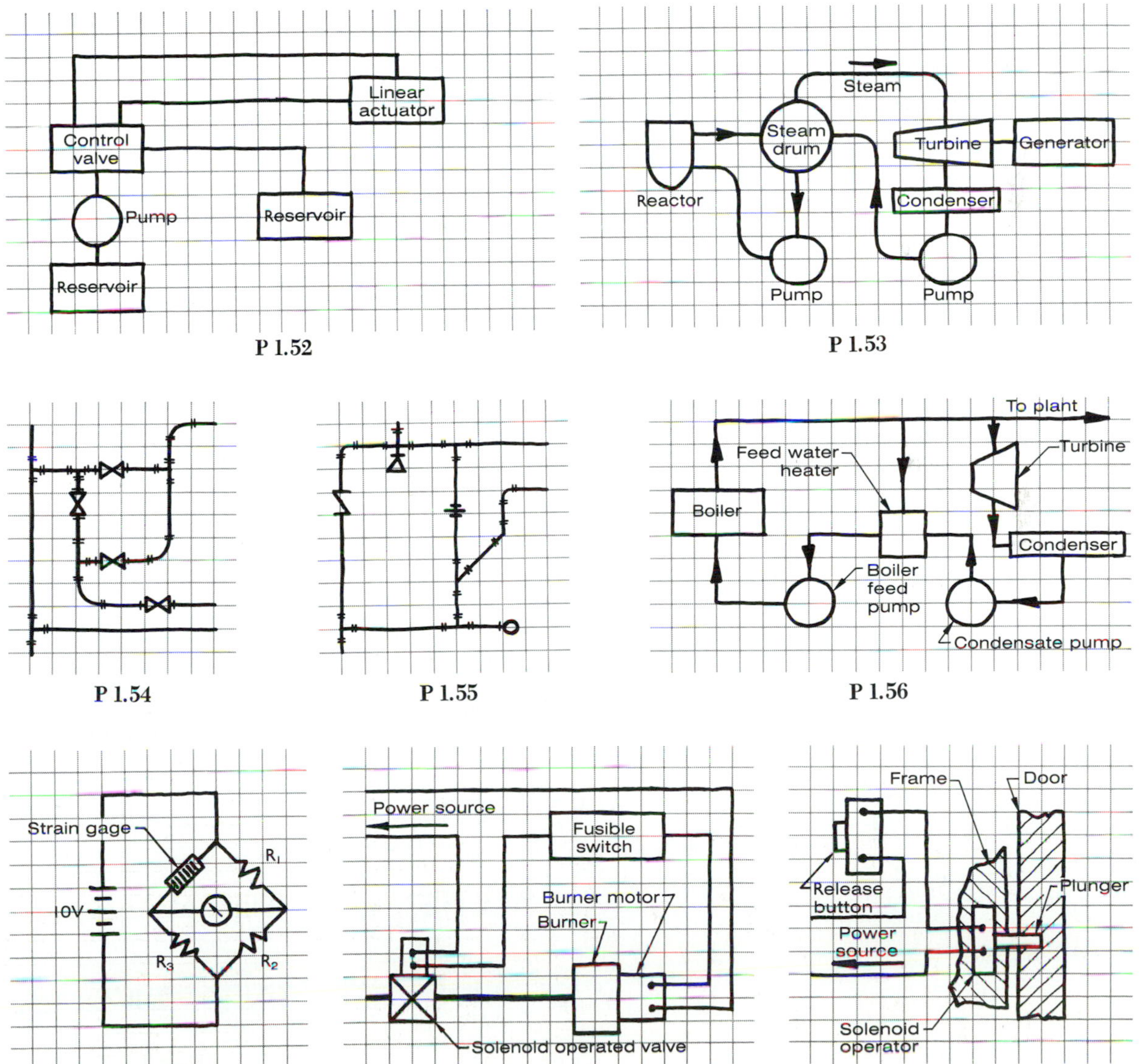

P 1.52

P 1.53

P 1.54

P 1.55

P 1.56

P 1.57

P 1.58

P 1.59

P1.60 to **P1.75:** Prepare a single-view sketch of the following objects, portraying the shape as accurately as you can recall or by referring to a catalog that illustrates the object. Sketch the objects to a reasonable size and neatly balance your sketch on the sheet.

P1.60 ENGINE VALVE

P1.61 AUTOMOBILE WATER PUMP GASKET

P1.62 C-CLAMP FRAME

P1.63 STOPPER FOR SINK DRAIN

P1.64 BASKETBALL BACKBOARD

P1.65 ROUNDHEAD MACHINE SCREW

P1.66 HEXAGONAL-HEAD MACHINE SCREW

P1.67 PADLOCK

P1.68 HEXAGONAL NUT

P1.69 D-SIZE FLASHLIGHT BATTERY

P1.70 BOAT DOCK CLEAT

P1.71 POOL TABLE TRIANGLE

P1.72 BIRD HOUSE

P1.73 DIVIDED PLASTIC PICNIC PLATE

P1.74 ONE-QUART MILK CARTON

P1.75 BUTT HINGE

Begin the following lettering exercises by taping each sheet to a working surface and then rule off light horizontal guide lines. Also rule off light vertical or inclined guide lines, depending upon the type of lettering style you select for practice. After you have drawn and spaced the guide lines according to the layout given for each problem, remove the tape so that you may freely move your sheet to a comfortable position for lettering. Practice lettering for about 10 minutes and then stop for a short period of time to allow your fingers to relax. When ready, begin again. Try to develop a smooth, repeated, and rapid sequence of letter strokes.

P1.76 to **P1.81:** Lay out each practice sheet as shown. Copy the given individual letters, figures, or words. Practice by lettering at least five or six individual letters and figures for each exercise.

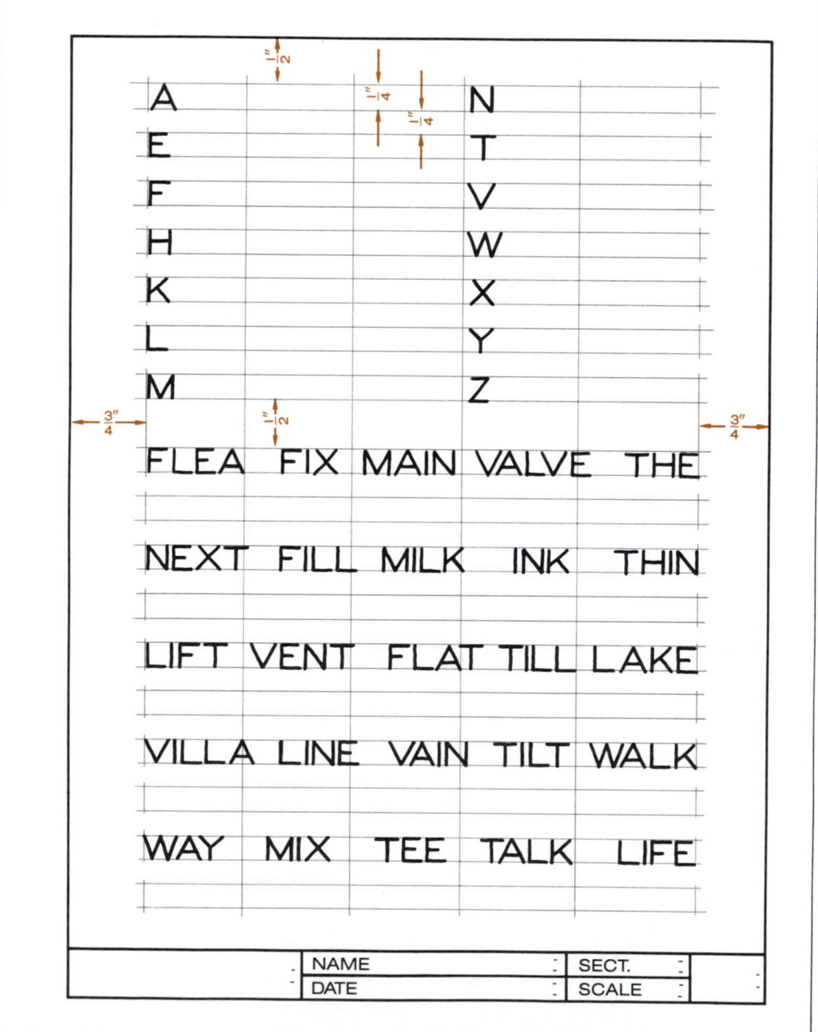

P 1.76

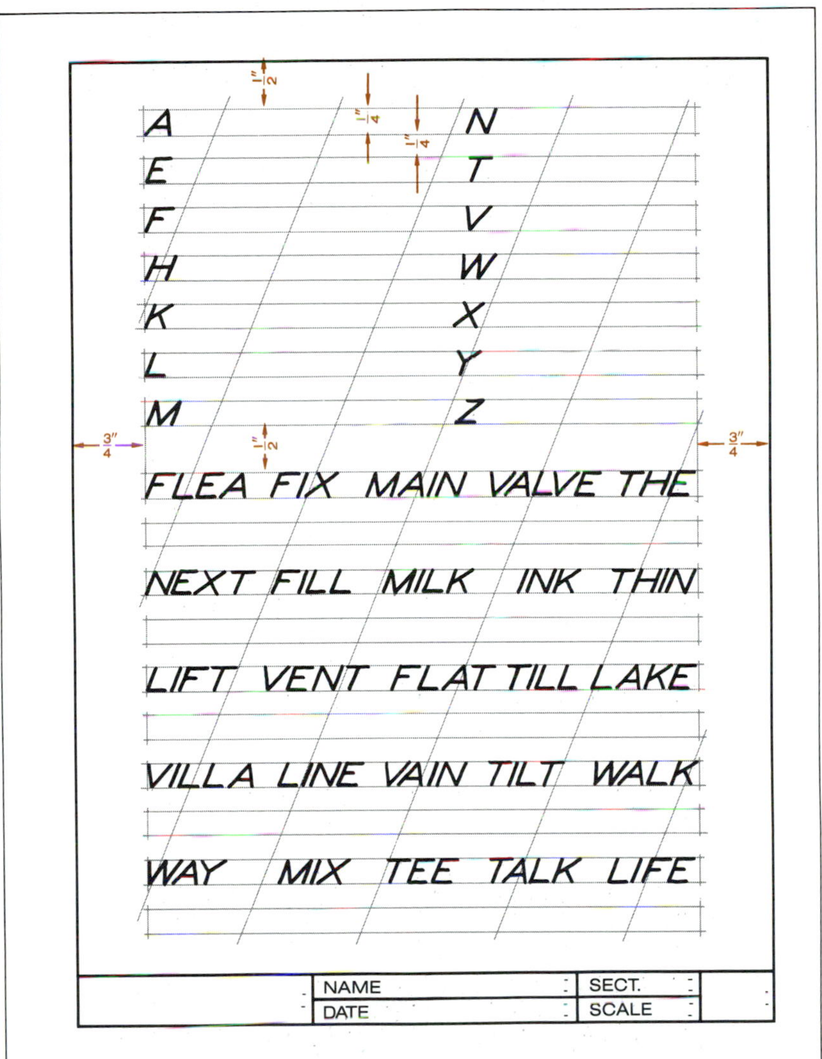
1"/2
1"/4
1"/4
A
E
F
H
K
L
M
N
T
V
W
X
Y
Z
3"/4
1"/2
3"/4
FLEA FIX MAIN VALVE THE
NEXT FILL MILK INK THIN
LIFT VENT FLAT TILL LAKE
VILLA LINE VAIN TILT WALK
WAY MIX TEE TALK LIFE
NAME
DATE
SECT.
SCALE

P 1.77

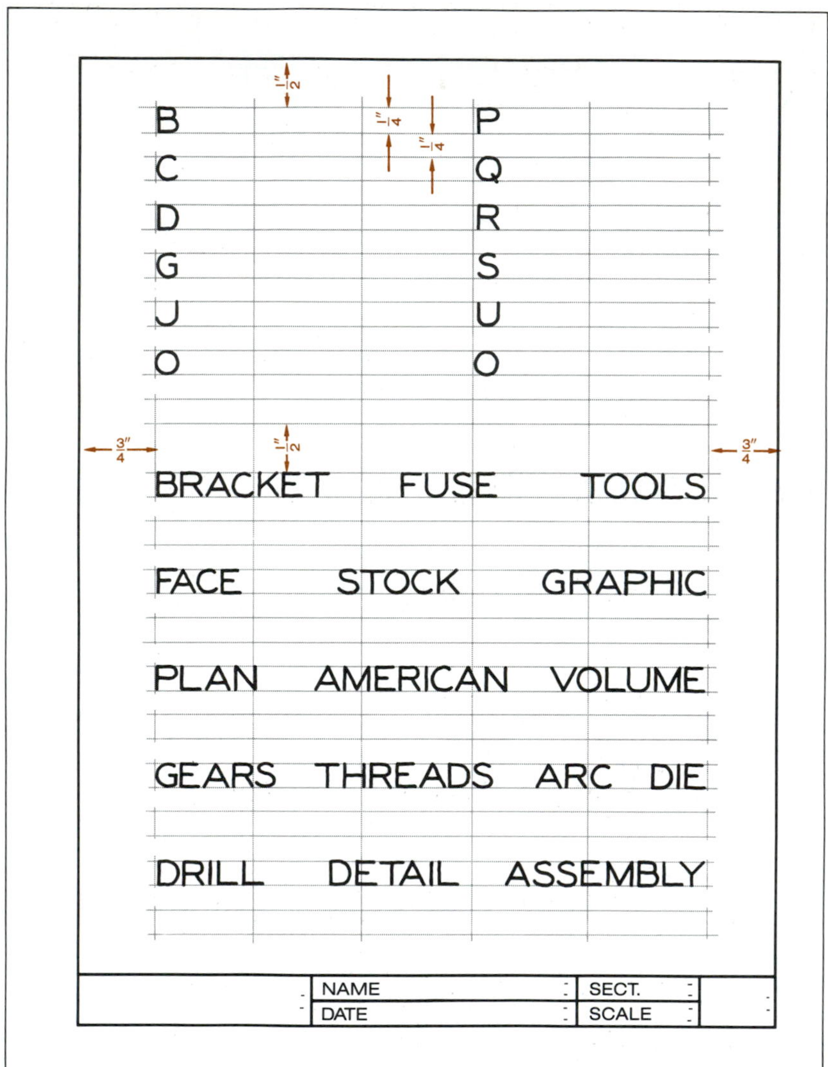

1"/2
1"/4
1"/4
B
C
D
G
J
O
P
Q
R
S
U
O
3"/4
1"/2
3"/4
BRACKET FUSE TOOLS
FACE STOCK GRAPHIC
PLAN AMERICAN VOLUME
GEARS THREADS ARC DIE
DRILL DETAIL ASSEMBLY
NAME
SECT.
DATE
SCALE

P 1.78

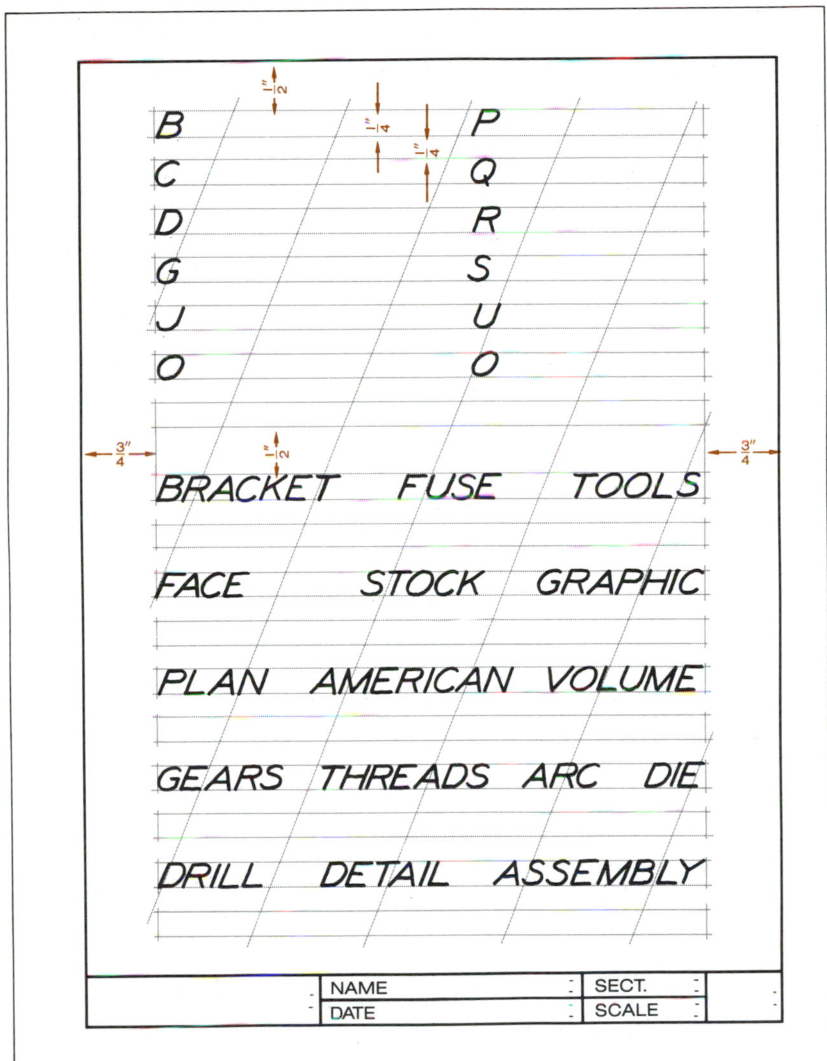
B
C
D
G
J
O
P
Q
R
S
U
O
BRACKET FUSE TOOLS
FACE STOCK GRAPHIC
PLAN AMERICAN VOLUME
GEARS THREADS ARC DIE
DRILL DETAIL ASSEMBLY
NAME
DATE
SECT.
SCALE

P 1.79

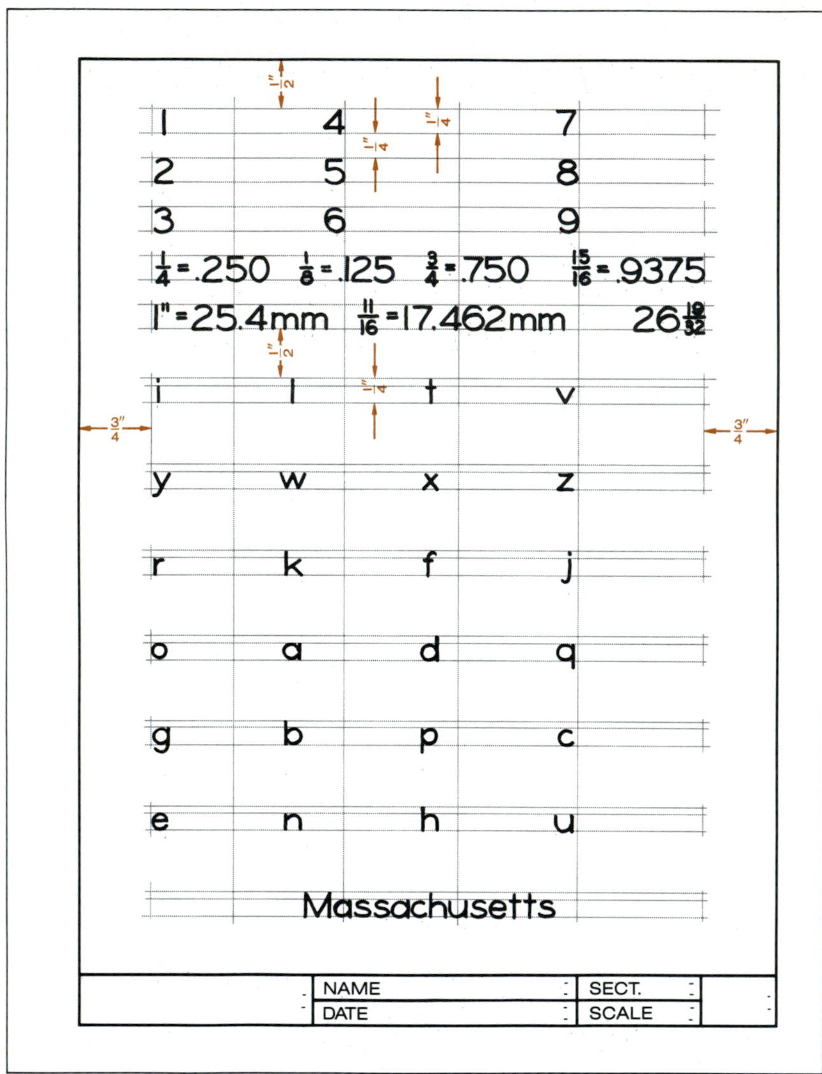
1
4
7
2
5
8
3
6
9
1/4 = .250
1/8 = .125
3/4 = .750
15/16 = .9375
1" = 25.4mm
11/16 = 17.462mm
26 19/32
1/2"
1/4"
3/4"
i
l
t
v
y
w
x
z
r
k
f
j
o
a
d
q
g
b
p
c
e
n
h
u
Massachusetts
NAME
SECT.
DATE
SCALE

P 1.80

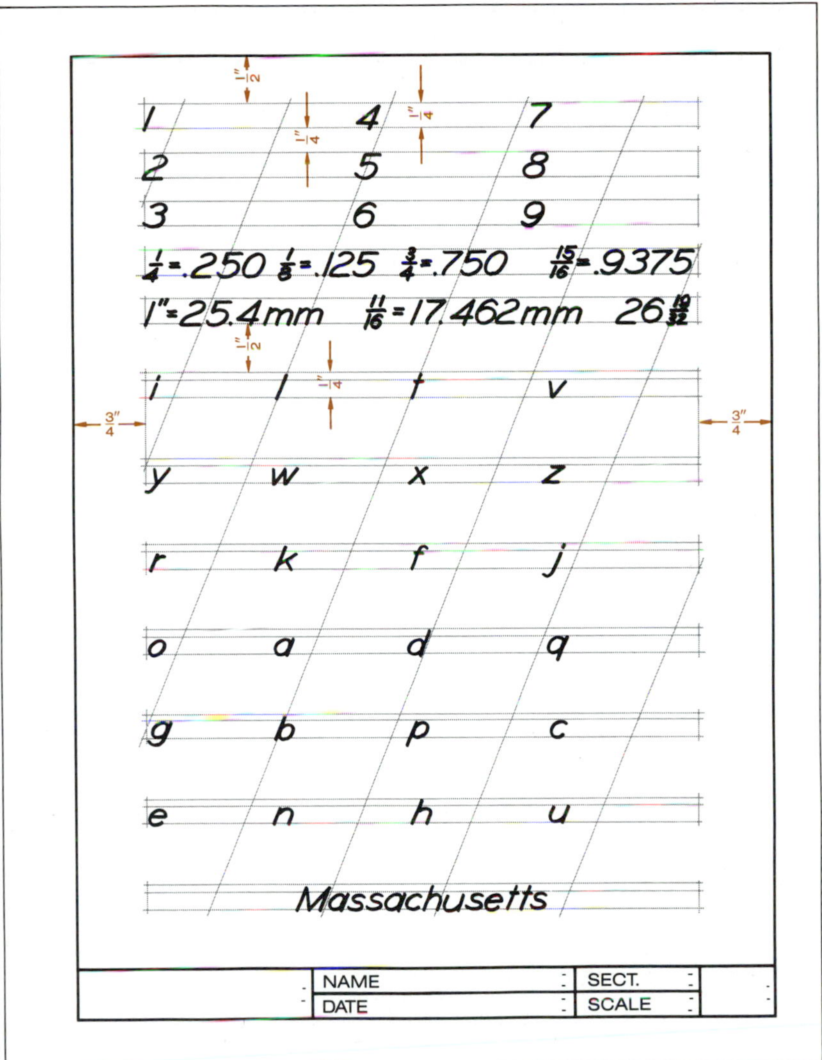
1/2"
1/4"
1/4"
1 4 7
2 5 8
3 6 9
1/4=.250 1/8=.125 3/4=.750 15/16=.9375
1"=25.4mm 11/16=17.462mm 26 10/32
1/2"
1/4"
3/4"
3/4"
i l t v
y w x z
r k f j
o a d q
g b p c
e n h u
Massachusetts
NAME
DATE
SECT.
SCALE

P 1.81

P1.82: The following specifications are typical of those used on engineering drawings. Copy these notes on as many problem sheets as necessary. Use two columns, laid out as shown in Figure 1-30. Use the style of engineering lettering you prefer.

1. DRILL AND COUNTERBORE ⅝″ DEEP FOR ½″ SOCKET HEAD CAP SCREWS, 2 HOLES.
2. 6.35 mm REAM THROUGH, 2 HOLES, PRESS FIT FOR ITEM #17.
3. ⅜ LAP THROUGH, 4 HOLES, SLIP FIT FOR ITEMS #6 AND #8.
4. MEAN CROWN AT 1.000 DIA. MUST BE WITHIN 0.0016 ± .0004. MAXIMUM RUNOUT AT 1.000 GAGE DIA. MUST NOT EXCEED 0.0010 F.I.M.
5. O.D. MUST BE STRAIGHT AND ROUND WITH 0.0003 F.I.M.
6. ALL RADII 1.5 mm UNLESS OTHERWISE SPECIFIED. FACE OF SPINDLE MUST BE SQUARE WITH AXIS WITHIN 0.005 mm F.I.M.
7. PACK BEARING S-5953 WITH 2 cm^3; BEARING S-686 WITH 3 cm^3 AND BEARING S-702 WITH 5 cm^3.

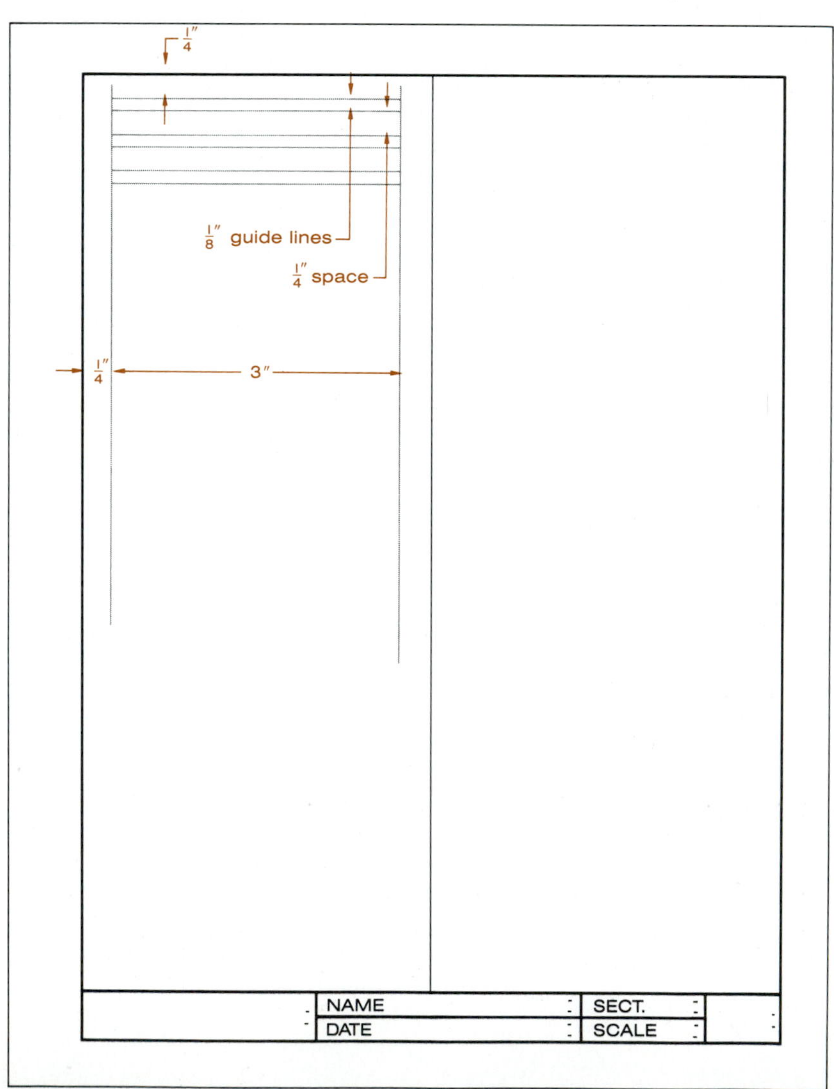

FIGURE 1-30 Layout for lettering practice sheets

8. 5⁄16–24 × 7⁄8 LG. BUTTON HEAD SOCKET CAP SCREWS, 5 REQUIRED.
9. MACHINE 3 BASE BOSSES TO 27⁄16″ ± 1⁄64″ OVERALL CASTING HEIGHT. BREAK ALL SHARP CORNERS TO 1⁄64″ MAX. RADIUS.
10. FOR CONTOURS AND SURFACES, WORK TO MODEL AS FURNISHED BY CADILLAC ENGINEERING DEPT. CONTOURS AND SURFACE DIMENSIONS SHOWN ARE INTENDED FOR REFERENCE ONLY.
11. ALL WALL SECTIONS 6 mm UNLESS OTHERWISE SPECIFIED. ALL FILLETS AND ROUNDS R3 mm UNLESS OTHERWISE SPECIFIED. ALL DRAFT ANGLES 5°.
12. .500 DIA. BOTTOM DRILL TO DEPTH AS SHOWN; 9⁄16–18-3B THREAD TO DEPTH SHOWN.
13. MODEL #B81 FEDERAL DIAL INDICATOR 0.0001 GRADUATIONS; SET TO ZERO WITH MASTER SET GAGE #D-G00090-S-H&R.
14. 1⁄2″ O.D. × 51⁄4″ LG HYD. STEEL TUBING: CUT TO SUIT AT ASSEMBLY. 3⁄4″ WIRE WOUND HOSE FURNISHED BY PIPING DEPT. ON FLOOR. REPLACE TWO NO. 12-5-S4 TUBE NUTS WITH ONE 12-5-C4 TUBE CAP ON SINGLE END MACHINE.
15. CARBURIZE, HARDEN AND GRIND TO ROCKWELL C 57-60. THE EXTENDED END OF THE KEYWAY MUST PASS AT LEAST 0.05 mm BEYOND THE INTERSECTION OF THE CENTER LINE AND THE 1.20 mm SLOT. THE CENTER LINE OF THE 1.260 mm SLOT AND THE CENTER LINE OF THE M 10 × 1.5 THREAD MUST BE PARALLEL WITHIN 0.13 mm F.I.M.
16. GROUND DIAMETERS MUST BE SQUARE WITH GROUND FACES WITHIN 0.003 IN 10 INCHES. HARDEN AND GRIND. ROCKWELL C 50–55. SHOT CLEAN.
17. ENCIRCLED MARKINGS AND REFERENCE LINES MUST NOT APPEAR ON THE GLASS PLATE. THE LINES MUST PROVIDE AN UNREVERSED IMAGE, BLACK, CLEAR CUT AND UNIFORM IN DENSITY AND WIDTH.
18. THE CHART MUST BE IMPERVIOUS TO OIL OR GREASE AND SHOULD NOT BE HARMED BY ROUTINE CLEANING WITH SOAP AND WATER. CHART SHOULD BE USED ON KODAK MODEL #3 COMPARATOR.
19. BLEND TO CONSTRUCTION SHOWN IN SECTION A-A FROM THIS LINE. POINT V APPEARS ON LINE X.
20. #WN-440 TO BE SECURELY WELDED PER MIL-W-6858. HOLE TO ACCOMMODATE WELD NUT AT 9 PLACES MARKED X.
21. THIS SURFACE IS TO BE GENERATED BY A SERIES OF HORIZONTAL LINES IN VIEW F-F PASSING THROUGH LINE T AND TANGENT TO BALL JOINT BOSS FROM LINE X. POINT C IS ON LINE U.
22. BRINELL HARDNESS 90–120. ROUGH TURN AFTER HEAT TREATING. DIMENSIONS APPLY TO 70°F AND BEFORE PLATING.
23. ±.015 ALLOWED ON ALL TWO-PLACE DECIMAL CASTING DIMENSIONS. IMMERSION OF TIN PLATE TO A MINIMUM THICKNESS OF .0001 AFTER FINISH GRIND.
24. BEFORE BALANCING REMOVE METAL PRIOR TO PRECISION-BORING THE PIN HOLES. THE TWO .3125 HOLES ARE FOR MANUFACTURING PURPOSES ONLY. THE FINISHED PISTON WEIGHT IS 1 LB 8.22 OZ, ± .026/.025 OZ.
25. THE TWO RING LANDS AND THE GROOVE BASES MUST BE CONCENTRIC WITH THE SKIRT AXIS WITHIN .003 F.I.M. HOLD .0007 TO .0017 CLEARANCE WITH CYLINDER.
26. 7.13 mm DRILL, 80° COUNTERSINK TO 13 mm DIA., 16 HOLES.

Pictorial Sketching

OBJECTIVES

After completing this chapter you should have gained the following abilities:

1. To develop the necessary level of competence to make a pictorial sketch quickly and clearly.

2. To decide upon which type of pictorial sketch, oblique or isometric, is best suited for your particular needs.

3. To apply the proper procedures for blocking in (constructing the framework) of various shapes of objects.

4. To maintain acceptable line quality.

5. To maintain the proper proportions when sketching an object.

6. To decide upon the position that shows the object to the best advantage in oblique or isometric sketches.

7. To understand the relationship of the width, height, and depth in the three principal planes of an object.

8. To decide upon an acceptable angle for a receding axis that results in the least distortion and in the most informative oblique sketch.

While single-view sketches are suited for graphically portraying flat objects or for laying out electrical circuits or piping systems, only two dimensions can be shown. Physical objects have three dimensions, not two. A single-view (two-dimensional) sketch of the front of a house, for example, can accurately describe the width and height of the house but cannot show its *depth*. The main advantage of a three-dimensional pictorial sketch is that all three dimensions (width, height, and depth) can be clearly displayed.

Three principal types of pictorial sketches are *oblique*, *axonometric*, and *perspective*. These forms are illustrated in Figure 2-1. Pictorial drawings prepared with instruments by technical illustrators or drafters are made with essentially the same sequence or steps that are used to produce a freehand sketch. Engineers use pictorial sketches because they can be made quickly and they represent an object quite realistically in an uncomplicated way. Pictorial sketches are relatively easy to make, while pictorial drawings require considerably more time to prepare.

Of the three principal types of pictorial sketches, the oblique and axonometric are most widely used by engineers. Perspective drawings, while appearing very natural, do not show true sizes of objects and are not ordinarily used as mechanical working drawings. Rather, perspective drawings are principally used by architects to portray interiors and exteriors of buildings. Perspective drawings are also used in some computer-graphics applications.

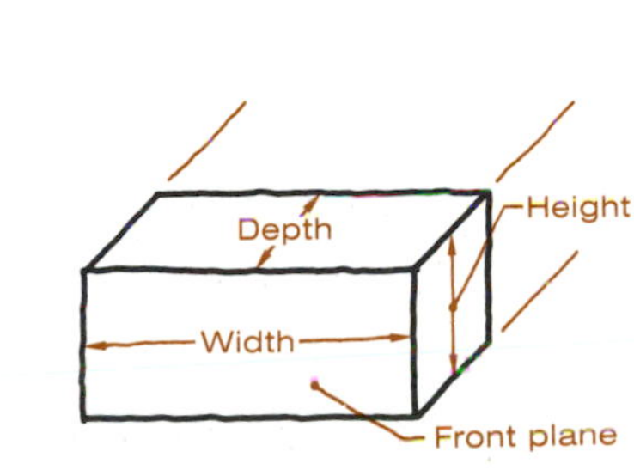

OBLIQUE: RECEDING PARALLEL LINES
(a)

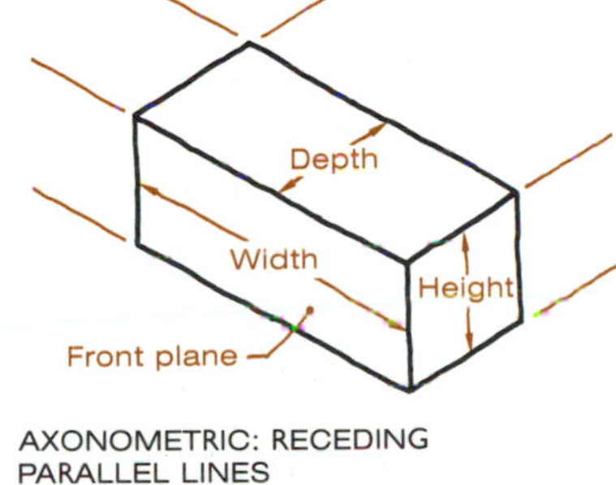

AXONOMETRIC: RECEDING PARALLEL LINES
(b)

FIGURE 2-1 Three kinds of pictorial sketches

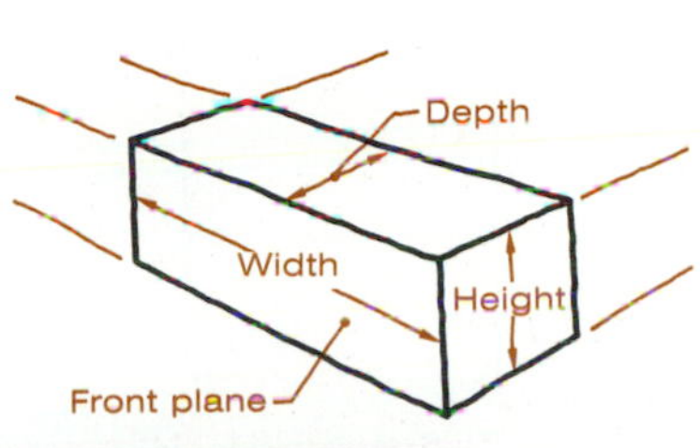

PERSPECTIVE: RECEDING CONVERGING LINES
(c)

2.1 OBLIQUE SKETCHING

Despite the limitations associated with its use, oblique sketching can greatly simplify the construction steps involved when making a three-dimensional picture of complicated and irregularly shaped objects. Practically all of the techniques discussed in Chapter 1 for preparing single-view sketches are also used for making oblique sketches. As in single-view sketching, the front plane of the object is shown true shape. The actual width and height of the object are shown in the front plane of an oblique sketch, just as in a single-view sketch. The major difference between a single-view sketch and an oblique sketch is that the third dimension of the object, the depth, is added to the picture. Oblique sketches are not recommended for objects having circular features or irregular contours on any but the front plane or in a plane parallel to it.

2.2 OBLIQUE SKETCHING AXES

The three axes of projection in oblique sketching, *vertical*, *horizontal*, and *oblique*, are shown in Figure 2-2(a). Because the horizontal and vertical axes are at right angles to each other, the front face of the object, shown in Figure 2-2(b), appears true shape. Any plane parallel to the front plane will also be true shape. In oblique sketching, the most irregular or complex surface is positioned on the paper so that it faces directly to the front. This position shows the object to the best possible advantage. The third axis recedes back at an oblique angle, which can be at any convenient angle between 0° and 90°. Generally an angle of 30° or 45° is selected for the receding lines. The receding edges of the object are drawn, as shown in Figure 2-2(b), as parallel lines.

Engineers often use commercially available oblique sketching paper, which is ruled with a network of equally spaced vertical and horizontal lines forming squares. In addition, diagonal lines are also ruled on the paper. These lines intersect the square grid lines at 30° or 45°. These intersections form the axes origins of the three faces (front, top, and side) of the object.

FIGURE 2-2 Three axes of projection for an oblique sketch *(Section 2.2)*

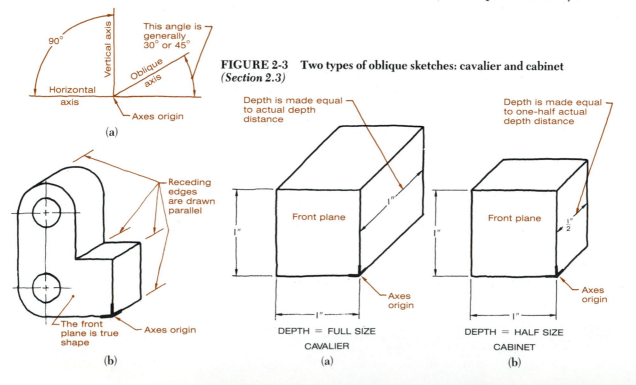

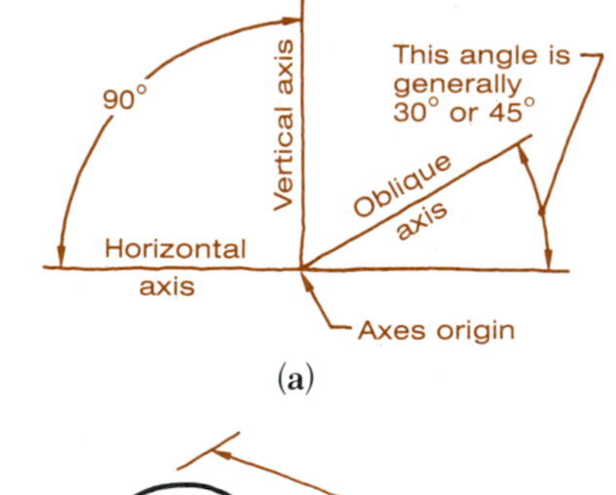

2.3 TYPES OF OBLIQUE SKETCHES

There are two main types of oblique pictorial sketches: *cavalier* and *cabinet*. A cavalier sketch, shown in Figure 2-3(a), is made with the depth equivalent to the actual depth dimension of the object. For most objects, when the depth is shown true distance, the picture of the object appears awkward and unnatural. When the depth scale is foreshortened to one-half the actual depth distance, as shown in Figure 2-3(b), the oblique sketch is known as a cabinet sketch. Reducing the depth dimension usually results in a less distorted picture that is more acceptable to the eye.

FIGURE 2-3 Two types of oblique sketches: cavalier and cabinet *(Section 2.3)*

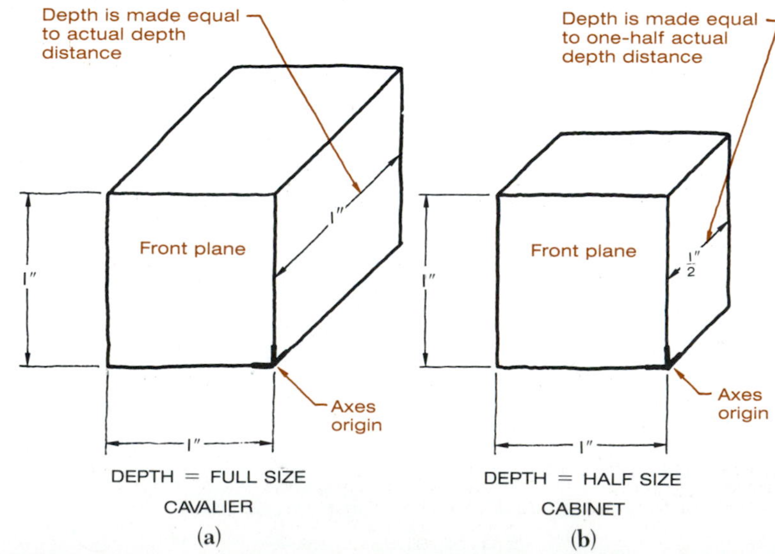

2.4 MAKING AN OBLIQUE SKETCH

Figure 2-4 illustrates the necessary steps for making an oblique sketch of the given object. Start by picturing in your mind how the shape of the main outline of the front face might appear "in the rough," before any of the cuts are made, and use the same construction methods that were previously used to make a single-view sketch. In Step 1 establish the correct proportions of the true-shape front plane of the object. Using the given object as a basis, the width may be divided into six equal divisions and the height into four equal divisions, or a ratio of 6:4. In this particular example each of the width and height divisions are equal. In Step 2 lay out the proportional spacings representing the various cuts on the front face. In Step 3 darken the lines representing the front edges of the object and establish the position of the axes origin. In this example the angle of the receding axes is 45°. Sketch all of the receding lines parallel to each other in the top and right-side planes. The receding lines represent the edges that form the third dimension of the object. In Step 4 estimate the desired depth, which, in our example, is made equal to one-half the actual depth distance. Darken the final lines. The pictorial is a cabinet oblique.

FIGURE 2-4 Making an oblique sketch ***(Section 2.4)***

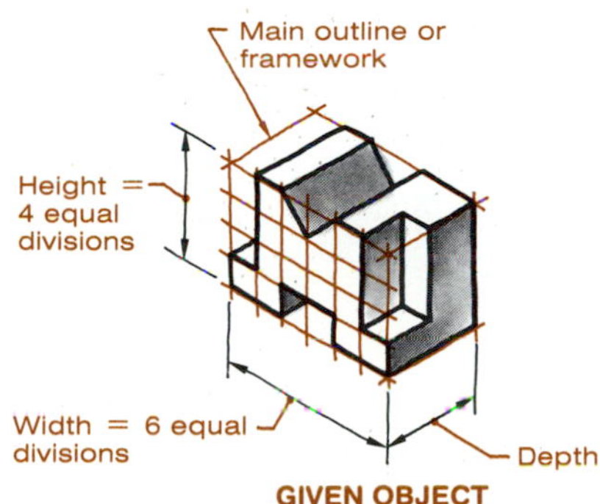

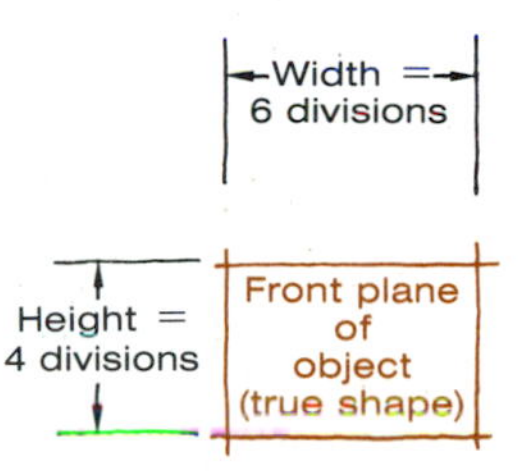

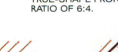

STEP 1
SKETCH THE OUTLINE OF THE TRUE-SHAPE FRONT PLANE TO A RATIO OF 6:4.

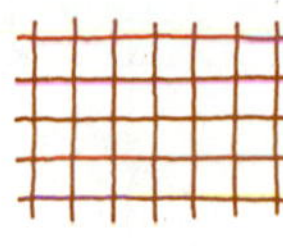

STEP 2
LAY OUT THE CUTS ON THE FRONT PLANE USING PROPORTIONAL DISTANCES.

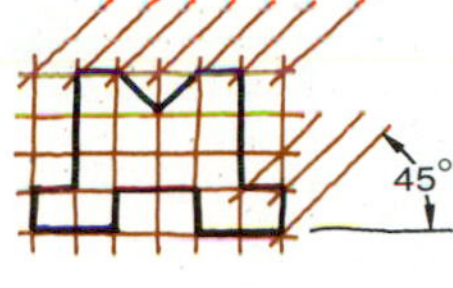

STEP 3
DARKEN THE EDGES ON THE FRONT PLANE AND ADD RECEDING PARALLEL LINES IN THE TOP AND RIGHT-SIDE PLANES.

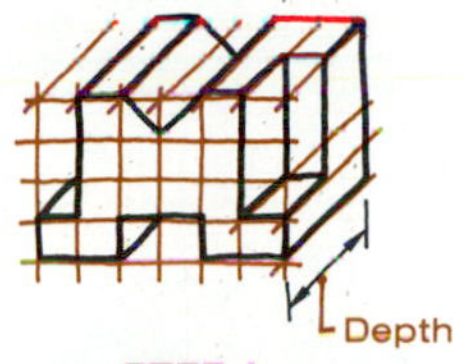

STEP 4
DARKEN FINAL OBJECT

2.5 HIDDEN LINES

It is customary to omit hidden lines on pictorial sketches except when representing features that cannot otherwise be shown. Figure 2-5 illustrates practices that apply to sketching hidden features. (A complete explanation of hidden lines is given in Chapter 5.)

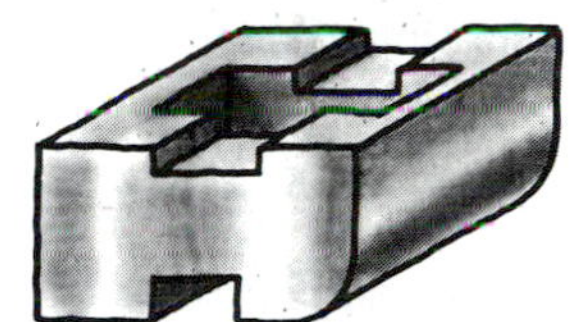

WHEN HOLES OR CUTS EXTEND ENTIRELY THROUGH THE OBJECT, OMIT HIDDEN LINES.
(a)

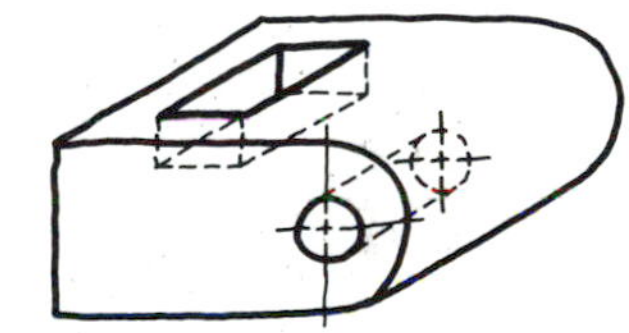

WHEN HOLES OR CUTS DO NOT EXTEND ENTIRELY THROUGH THE OBJECT, SHOW HIDDEN LINES.
(b)

FIGURE 2-5 Hidden lines in pictorial sketches ***(Section 2.5)***

CHECKPOINT 1
MAKING AN OBLIQUE SKETCH

Review Steps

1. Carefully examine the proportions of the figure to be sketched.
2. Lightly sketch a proportional outline representing the true shape of the front plane.
3. Gradually sketch the construction lines representing the cuts and features of the object.
4. Darken the lines that represent the shape of the front plane.
5. Establish the position of the axes origin.
6. Sketch all of the receding lines and select a suitable depth scale.
7. Sketch the features appearing on receding planes.
8. Darken the lines on the final figure.

Practice the steps in sketching the object shown in Figure 2-4 on a plain sheet of 8½ × 11 inch paper. Make the width about 4 inches and estimate the other distances in proportion. Use a 30° receding angle.

2.6 RECEDING LINE DIRECTIONS

One object may show more detail when the receding lines extend in a direction toward the upper left, but another object may be shown to a better advantage when sketched in a position with the receding lines extending toward the upper right. In certain cases, receding lines that extend downward to the left or to the right may also be sketched. Before you sketch any object, you should critically evaluate the overall shape and make sure that the sketch shows the object in the most revealing position. Figure 2-6 illustrates examples of objects drawn with receding lines in various directions. Note the various positions of the axes origins.

FIGURE 2-6 Receding line directions *(Section 2.6)*

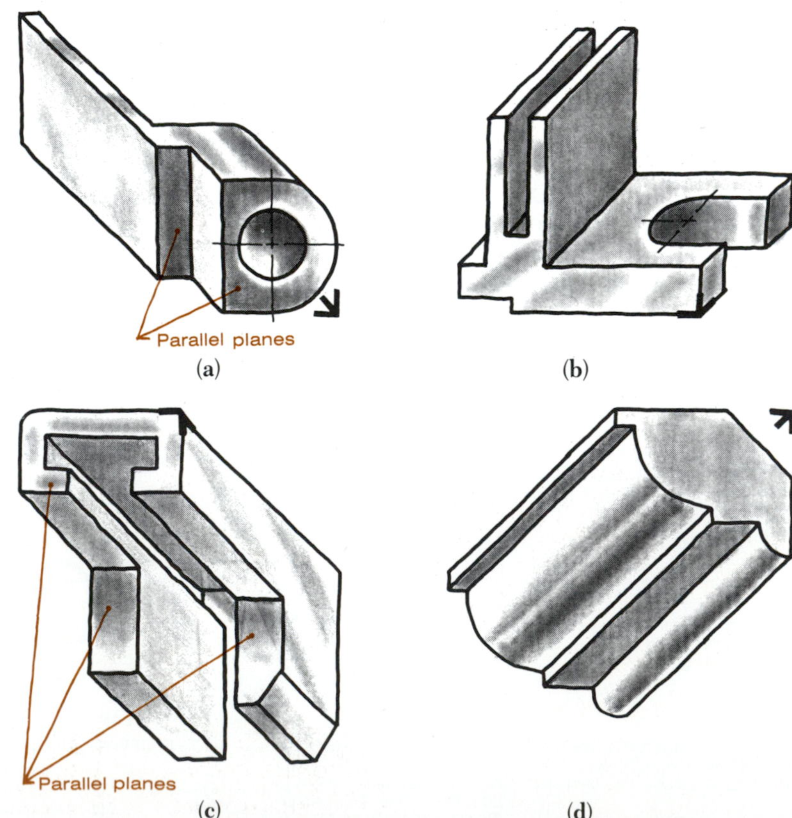

2.7 CIRCLES IN OBLIQUE SKETCHES

Figure 2-7 shows how to make an oblique sketch of a cylinder. When the circular end of the cylinder is positioned parallel to the front plane, that face will appear as a circle. Begin by visualizing how the cylinder would appear if it were enclosed in a framework. In Step 1 establish a convenient position for the axes origin and sketch an enclosing square construction box for the circle in the front and rear planes. Next sketch the receding cylinder axis (at 30° in this example) and then sketch the true shape of the circle in the front plane. In Step 2 sketch a diagonal construction line in each of the front and rear planes, and next sketch the lines representing the limiting elements of the cylinder parallel to each other and to the cylinder axis. Finally, sketch the visible half of the circle on the rear plane of the cylinder. Complete the figure by darkening the final lines and adding the center lines, as shown in Step 3.

CHECKPOINT 2
MAKING OBLIQUE SKETCHES OF CIRCULAR OBJECTS

Figure 2-8 shows two examples of oblique sketches of objects that are more complex. Note the positions of the centers of the various circles lying on the cylinder axis. Practice by copying these sketches on a plain sheet of 8½ × 11 inch paper. For both examples, make the largest circle about 3 inches in diameter and estimate the other distances in proportion.

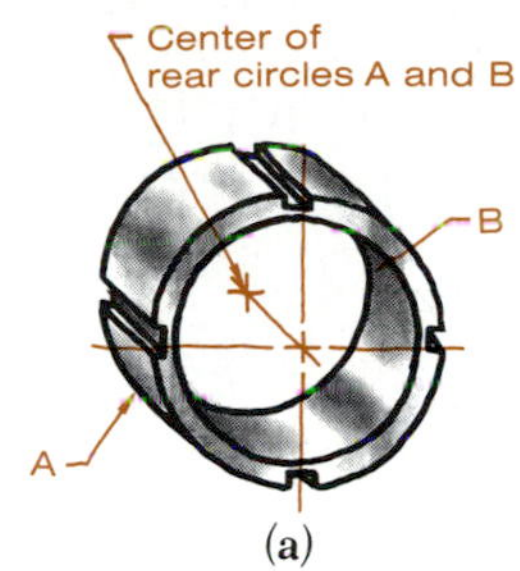

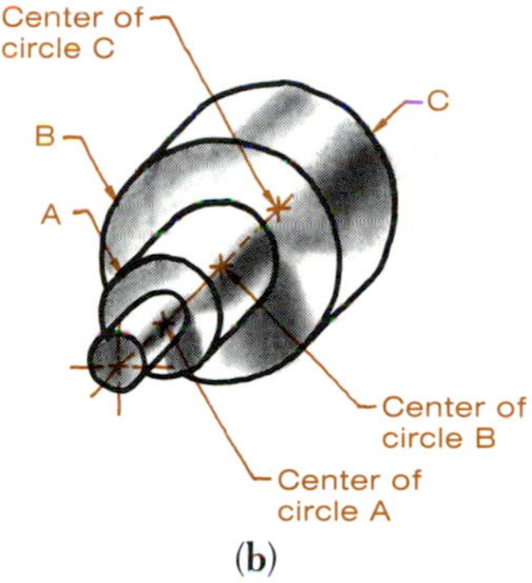

FIGURE 2-8 Checkpoint 2 problems

FIGURE 2-7 Making an oblique sketch of a cylinder *(Section 2.7)*

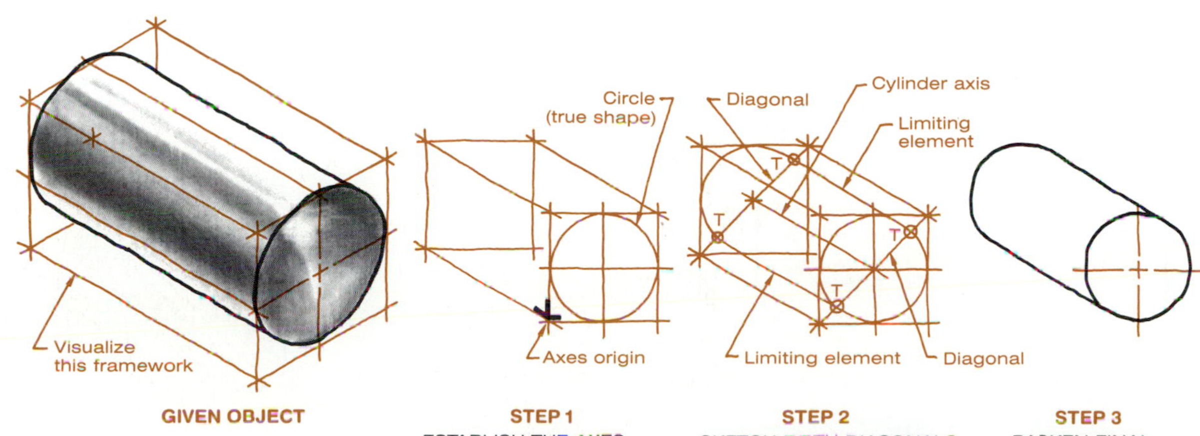

2.8 CIRCLES NOT PARALLEL TO THE FRONT, TOP, OR RIGHT-SIDE PLANES

As shown in Figure 2-9(a), parallelograms are constructed to aid in sketching the shape of the ellipses for the same purpose that squares are sketched as an aid in sketching circles. Ellipses should be shown in the proper position. The major axis of the ellipse (shown as xy) should coincide with the position of the longest diagonal of the parallelogram.

Figure 2-9(b) illustrates how circles should appear in the three planes of an oblique sketch. Only the circle that appears in the front face is true shape. The circles that lie in the top and the right-side planes appear elliptical. The center line of the ellipses, shown as conjugate diameters ab and cd, are parallel to the horizontal, vertical, and oblique axes. (Two diameters of an ellipse are *conjugate* if each diameter is parallel to the tangents drawn at the ends of the other.)

When arcs and ellipses are to be drawn on inclined planes, technical illustrators use either special drafting templates or construct the shapes with instruments using an "offset" or "coor-

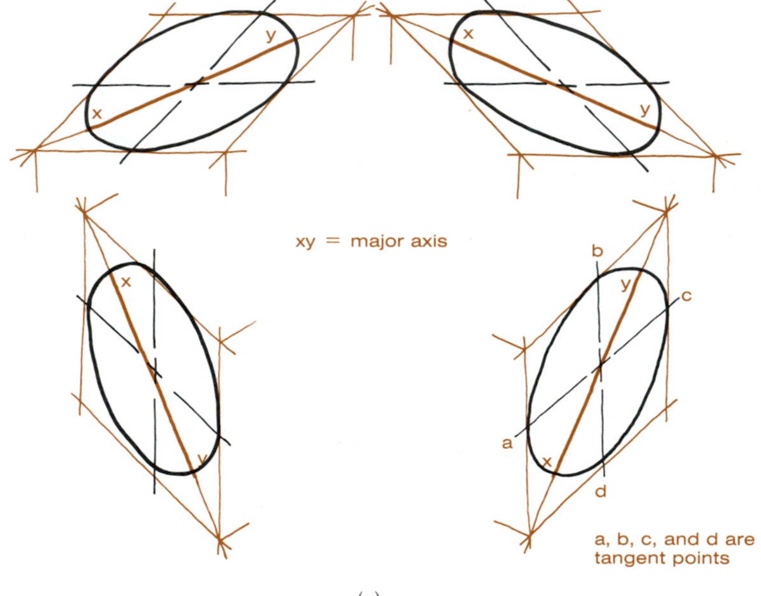

(a)

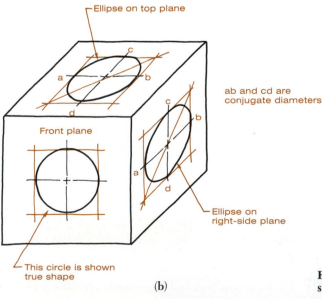

Ellipse on top plane
c
a
b
d
ab and cd are conjugate diameters
Front plane
c
b
a
d
Ellipse on right-side plane
This circle is shown true shape

(b)

FIGURE 2-9 Circles in oblique sketches ***(Section 2.8)***

dinate" method. Because an engineer's sketch has less exacting requirements, true circles and ellipses are approximated. Figure 2-10 shows the steps used to sketch the approximate shape of the ellipse that represents the circular hole on the inclined surface. When sketching an ellipse, take care to avoid excessively "fat" or "thin" elliptical shapes.

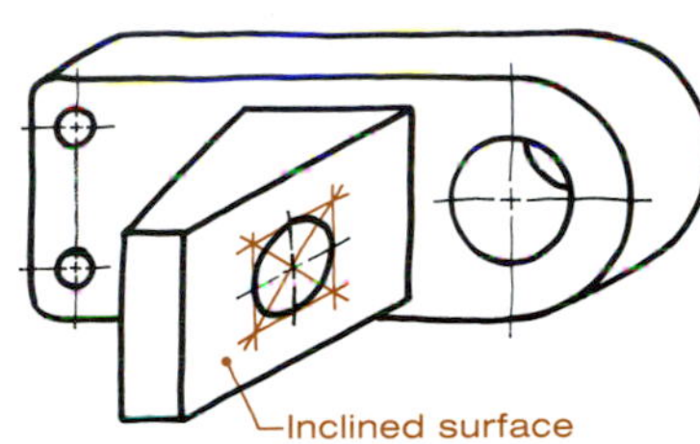

STEP 1
SKETCH THE TWO CENTER LINES.
STEP 2
SKETCH THE ENCLOSING PARALLELOGRAM.
STEP 3
TO AID IN POSITIONING THE ELLIPSE, SKETCH THE DIAGONALS.
STEP 4
SKETCH THE ELLIPSE.

FIGURE 2-10 Sketching a circle on an inclined surface *(Section 2.8)*

2.9 CHOICE OF POSITION

Objects, in most cases, appear most realistic when the more complex contours are shown in or parallel to the front plane. Figure 2-11 shows some objects drawn in different positions to illustrate this point. In every example the objects appear less distorted and more natural

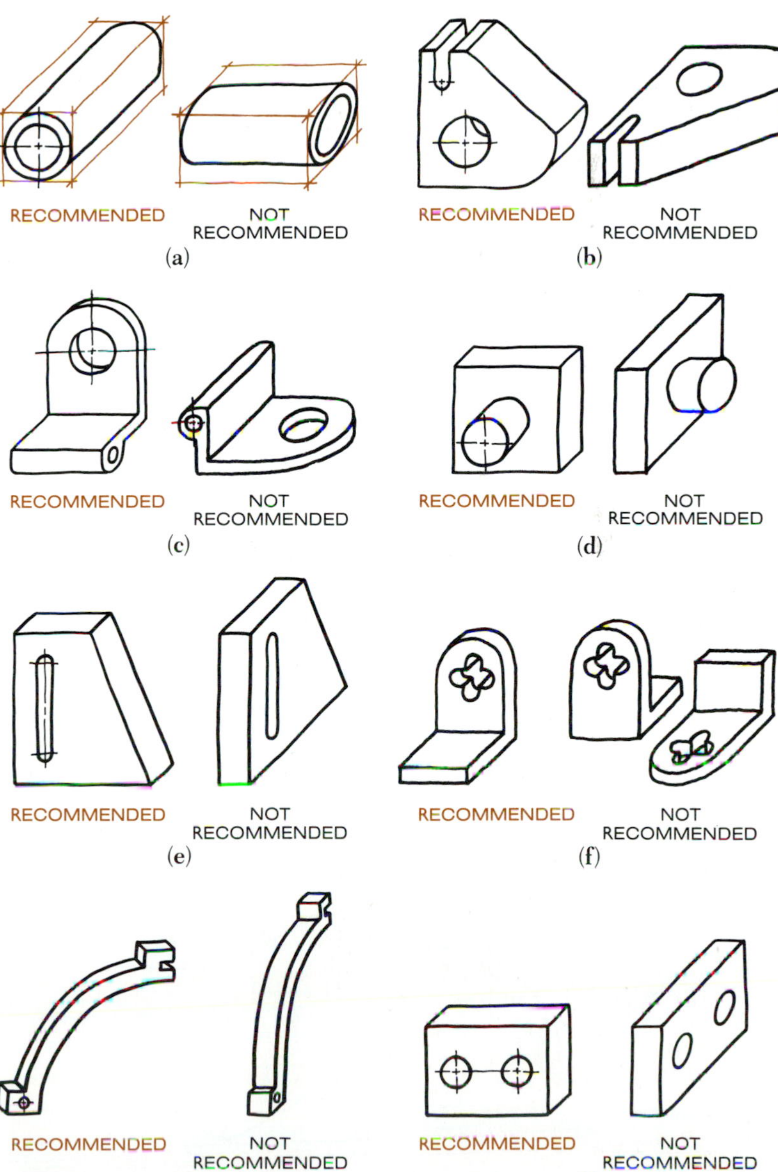

FIGURE 2-11 Choice of position for oblique sketches *(Section 2.9)*

when the object is placed in the recommended position. Also, objects in these positions are easier to sketch or to draw. By carefully selecting the position of the object to be sketched, you can usually avoid making sketches with objectional distortion. As a general rule, the *longest* dimension of an object should be placed in a position parallel to the front plane, as shown in the examples in Figure 2-12.

2.10 AXONOMETRIC POSITIONS

There are three types of axonometric positions: *isometric*, *dimetric*, and *trimetric*. In each case, the object is represented on the drawing by turning and tilting it in such a position that three surfaces—the front, top, and right or left side—are seen.

The type of axonometric position that is used almost exclusively for pictorial sketching by engineers is the *isometric* (equal measure) position. Visualizing a cube in an isometric position, as shown in Figure 2-13, may help you to better understand the meaning of axonometric drawing. In theory, suppose a cube is revolved 45° about a vertical axis and then tilted toward the viewer about 35° until the diagonal of the cube appears as a point. All edges are shown equal in length, and the receding edges are at a visual angle of 30°. Equal measurements are made along the three edges (axes). Unlike in the cabinet oblique position, the distance along the receding axis is *not* reduced in the isometric position. The same scale is used to lay off the various distances. Visual angles of 120° are formed wherever the three edges of the cube meet at the corner.

The major difference between isometric *projection* and isometric *drawing* is the measurements along the axes. The pictorial in an isometric projection will be smaller than the isometric drawing because all measurements are only about 82 percent of the actual distances.

Figure 2-14(a) shows a *dimetric* drawing of a simple object. Two of the axes make equal angles with the plane of projection. These angles are labeled A

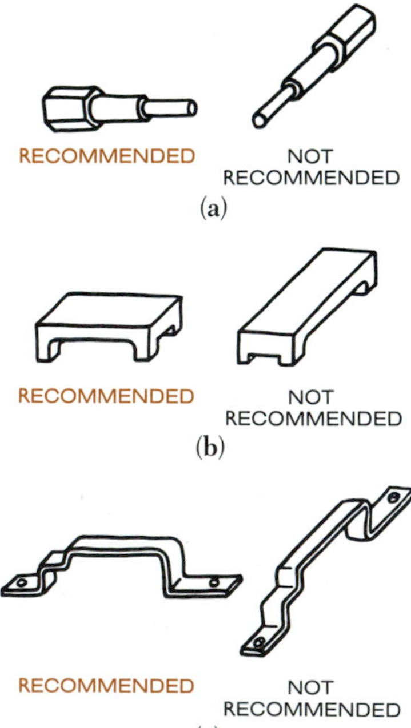

FIGURE 2-12 Choice of position for oblique sketches *(Section 2.9)*

FIGURE 2-13 Visualizing a cube in an isometric position *(Section 2.10)*

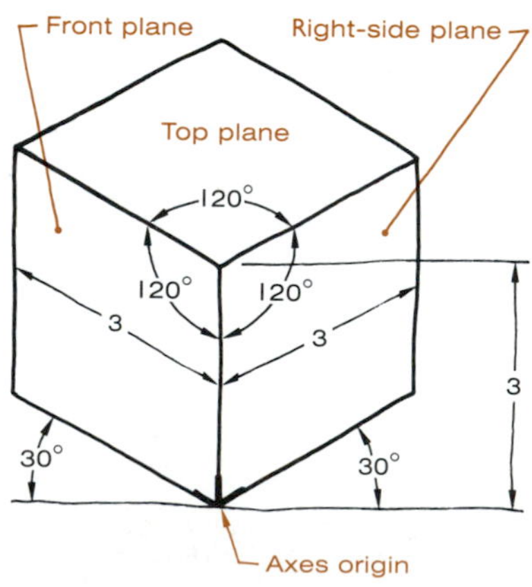

FIGURE 2-14 Two axonometric positions: dimetric and trimetric *(Section 2.10)*

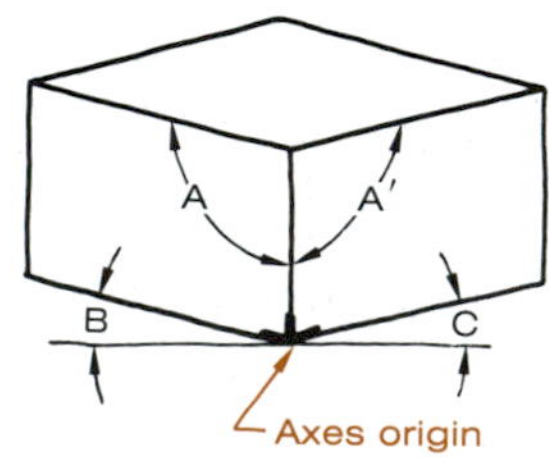

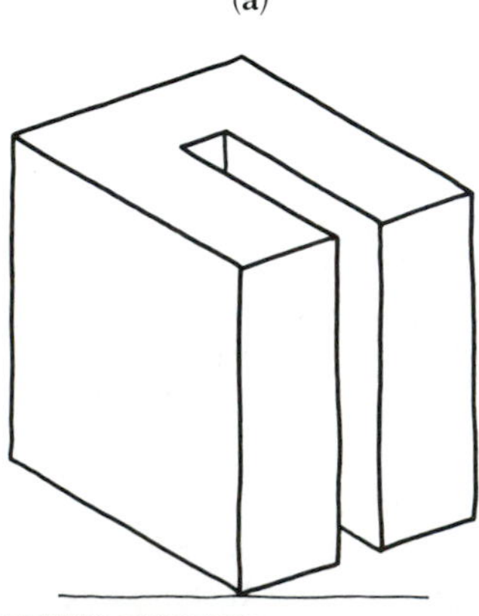

and A′. The third axis may make either a smaller or a larger angle. In practice, the object may be tilted at any convenient angle. An infinite number of combinations of angles is possible. The edges in dimetric drawings are frequently shortened to some convenient scale ratio. Dimetric drawings or sketches may be used for certain objects to specifically emphasize more detail in one plane.

A *trimetric* drawing of an object is shown in Figure 2-14(b). All three edges are unequally foreshortened and the angles between the planes are different. Trimetric sketches are usually too complex for general purpose drawing and sketching applications.

2.11 ISOMETRIC SKETCHING

The same scale is used on all three axes for isometric pictures. Many of the sketching techniques are the same as those used to produce single-view and oblique sketches, particularly with respect to blocking in the original object and estimating proportions.

Isometric sketches are usually made on plain paper, although commercially available isometric sketching paper may be used. The lightly printed grid on such paper is composed of a network of vertical lines running through the intersections formed by lines that recede 30° in right and left directions.

2.12 MAKING AN ISOMETRIC SKETCH

Figure 2-15 shows the various steps involved in making an isometric sketch of an object. Begin Step 1 by establishing the position of the axes origin, labeled point O. Sketch a short horizontal line and a vertical line. Next lay out two 30° axis lines at the intersection of the horizontal and vertical lines. Carefully estimate the relative proportions of the main outline (in this example the proportions are 3:3:4) and begin to sketch light construction lines blocking in the main framework of the object. Note that the width of the object can be conveniently divided into four equal divisions. Using the same distance for each of the three small divisions, set these divisions off

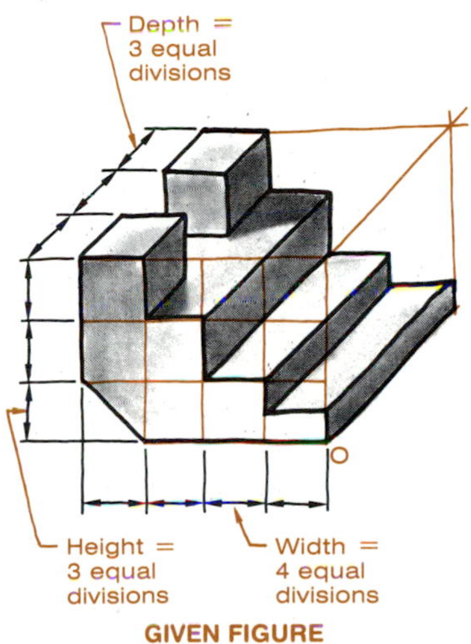

GIVEN FIGURE

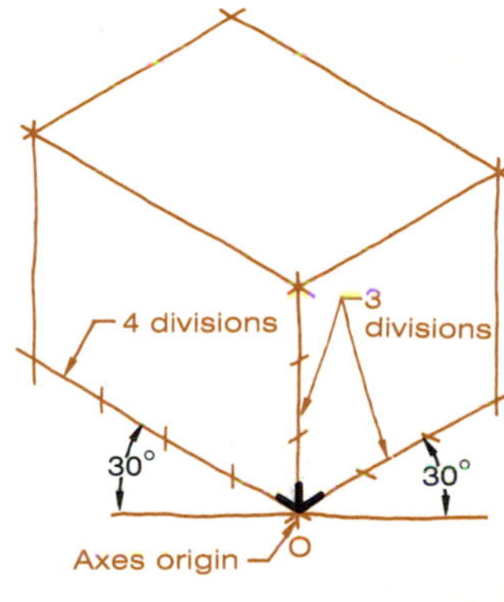

STEP 1
ESTABLISH AXES ORIGIN AND BLOCK IN THE OUTLINE.

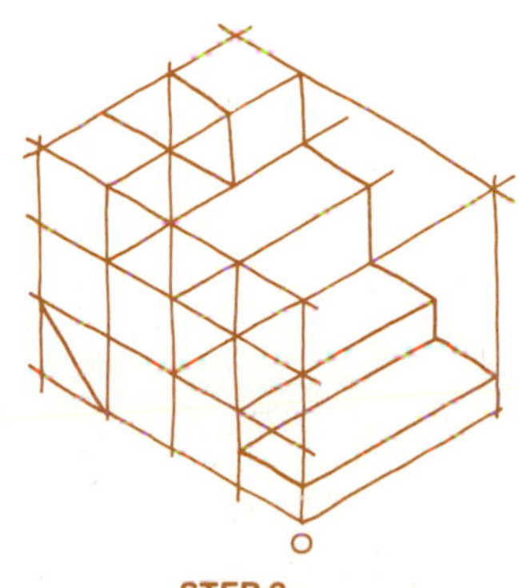

STEP 2
BLOCK IN DETAILS USING PROPORTIONAL DISTANCES.

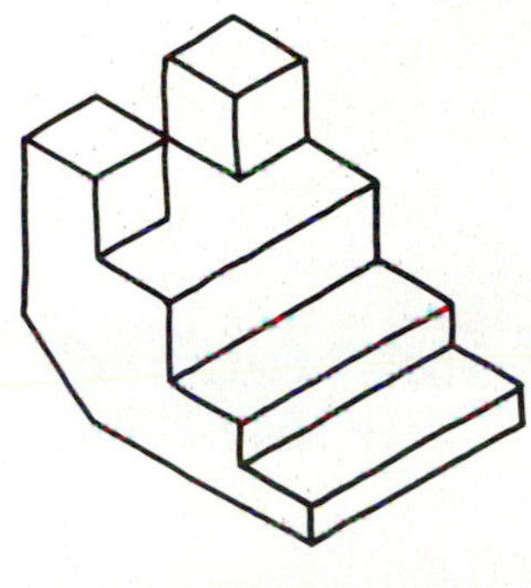

STEP 3
DARKEN FINAL OBJECT.

FIGURE 2-15 Making an isometric sketch *(Section 2.12)*

CHECKPOINT 3
MAKING AN ISOMETRIC SKETCH

Review Steps

1. Carefully examine the proportions of the figure to be sketched.
2. Establish the position of the axes origin and sketch the axes lines.
3. Lightly sketch a construction block enclosing the main outline of the object.
4. Gradually sketch the construction lines representing the other cuts and features of the object.
5. Darken the lines on the final figure.

Practice the steps in sketching the object shown in Figure 2-15 on a plain sheet of 8½ × 11 inch paper. Make the width about 4 inches and estimate the other distances in proportion. Critically examine your finished sketch by holding it away from you at arm's length. If you detect lines that are not satisfactory, erase them and replace them with acceptable lines.

along both the height and depth axes. Estimate and gradually sketch all of the lines that represent the locations of the proportionally smaller units and features of the object (Step 2). Finally, complete the sketch by darkening the visible lines (Step 3).

2.13 ISOMETRIC AND NONISOMETRIC LINES

An *isometric line* is any line that is parallel to an isometric axis. All measurements are made only on isometric lines. *Nonisometric lines* are those not parallel to the axes lines, shown by the diagonal lines OA, CB, BD, and CD in Figure 2-16. All other lines in Figure 2-16 are isometric lines.

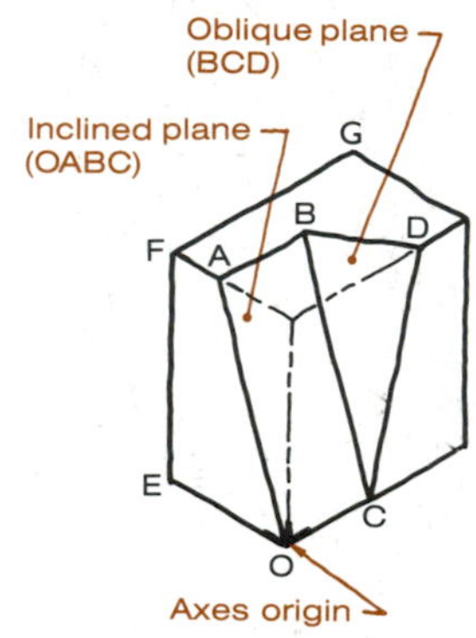
Oblique plane (BCD)
Inclined plane (OABC)
G
B
D
F
A
E
C
O
Axes origin
OA, CB, BD, and CD are nonisometric lines.

FIGURE 2-16 Isometric and nonisometric lines *(Section 2.13)*

2.14 INCLINED PLANES

Figure 2-17 shows the steps required to plot an inclined plane in an isometric sketch that contains a pair of inclined, nonisometric lines. In Step 1, after selecting a convenient position for the axes origin, sketch the main outline of the given object using construction lines. The end points of the nonisometric lines 1-4 and 2-3 (Step 2) are

FIGURE 2-17 Making an isometric sketch of an object with an inclined surface containing nonisometric lines *(Section 2.14)*

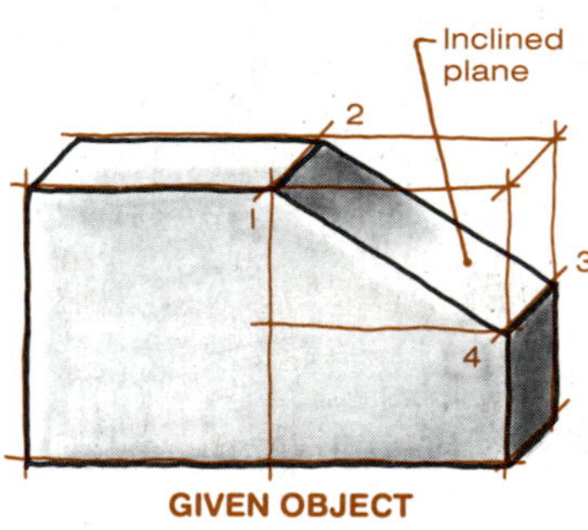
Inclined plane
2
1
3
4
GIVEN OBJECT

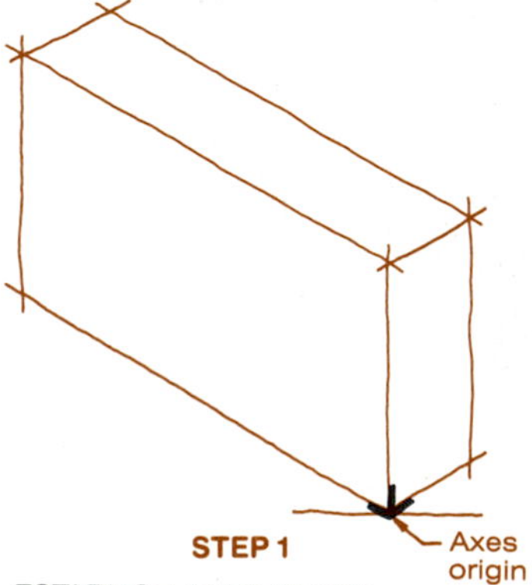
STEP 1
Axes origin
ESTABLISH AXES ORIGIN AND BLOCK IN MAIN OUTLINE.

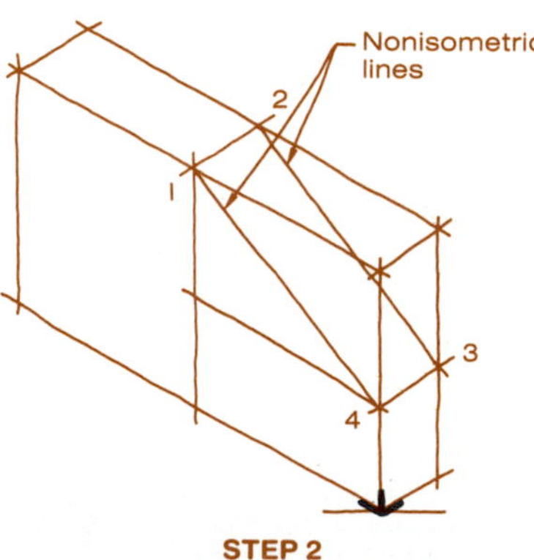
Nonisometric lines
2
1
3
4
STEP 2
PLOT THE NONISOMETRIC LINES USING PROPORTIONAL DISTANCES.

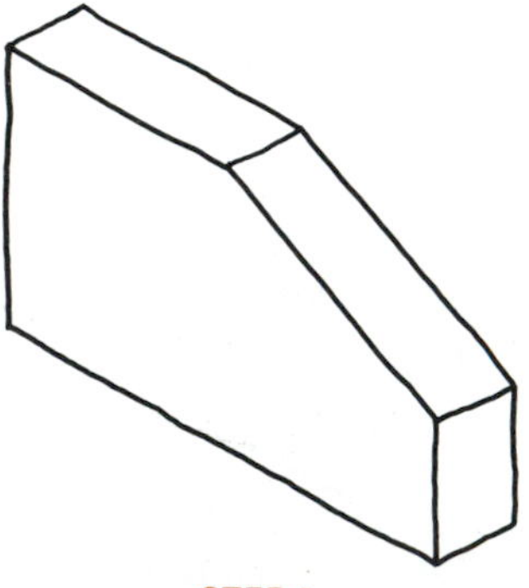
STEP 3
DARKEN THE LINES ON THE FINAL OBJECT.

plotted. Complete the isometric sketch by darkening the visible lines (Step 3). Note that points 1, 2, 3, and 4 form the boundary lines of the inclined surface. The reason an inclined plane must be plotted is that angles do not appear true size on an isometric drawing and thus cannot be measured.

> **CHECKPOINT 4**
> **SKETCHING AN OBJECT WITH AN INCLINED SURFACE**
>
> Practice the steps in sketching the object shown in Figure 2-17. Use the *top half* of a plain sheet of 8½ × 11 inch paper. Make the width about 4 inches and estimate the other distances in proportion.

2.15 OBLIQUE PLANES

Oblique planes are shown on isometric sketches by plotting the end points of the plane in a manner similar to that used in constructing inclined planes. In Figure 2-18 the two oblique surfaces shown on the view of the given object contain points 1, 4, 5, and 3 and points 6, 2, and 7. Begin the isometric sketch, as shown in Step 1, by establishing a convenient position for the axes origin and block in the main outline of the given object with construction lines. Add the construction lines for the slot. In Step 2 add the construction lines for the two oblique surfaces by plotting the end points 1, 2, and 3. Line 1-2 intersects the slot in the top plane at points 4 and 6. Line 2-3 intersects the slot on the right-side plane at points 5 and 7. Observe that lines 4-5 and 6-7 are parallel to each other and also to line 1-3. (A plane intersects parallel planes with parallel lines.) This observation is a check on your construction. Complete the sketch by darkening the visible lines, as shown in Step 3.

> **CHECKPOINT 5**
> **SKETCHING AN OBJECT WITH AN OBLIQUE SURFACE**
>
> Practice the steps in sketching the object shown in Figure 2-18. Use the space reserved on the lower half of the sheet that you previously used to sketch the problem for Checkpoint 4. Make the width about 4 inches and estimate the other distances in proportion.

2.16 CHOICE OF POSITION

In deciding how to position the object for an isometric sketch, keep in mind the important features of the object. Try to show

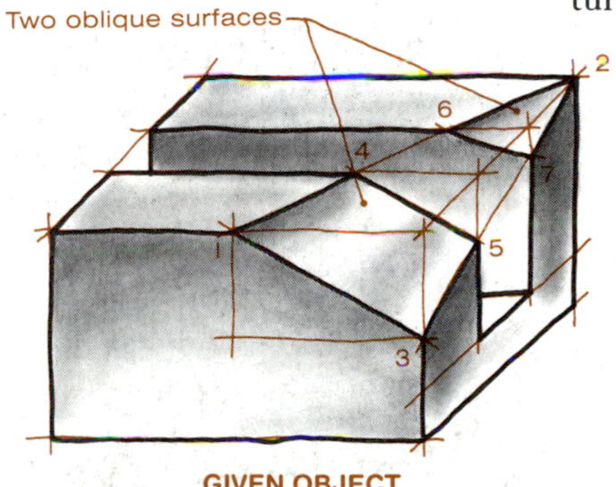
Two oblique surfaces
1
2
3
4
5
6
7
GIVEN OBJECT

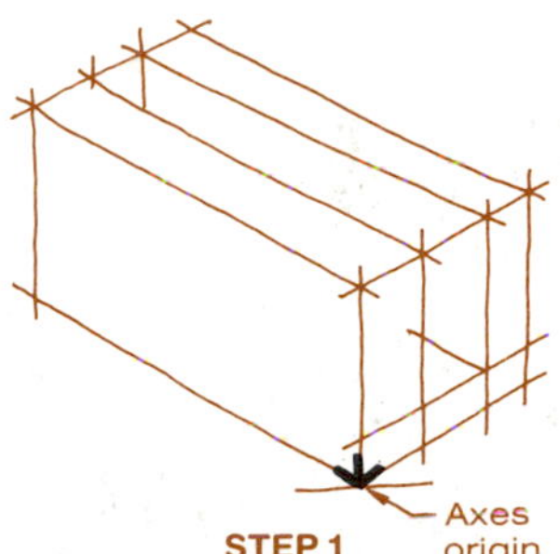
Axes origin
STEP 1
ESTABLISH AXES ORIGIN AND BLOCK IN MAIN OUTLINE AND SLOT DETAIL.

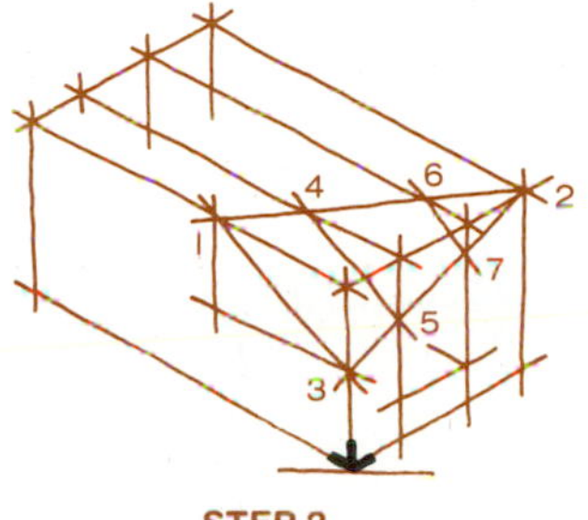
1
2
3
4
5
6
7
STEP 2
PLOT THE NONISOMETRIC LINES USING PROPORTIONAL DISTANCES.

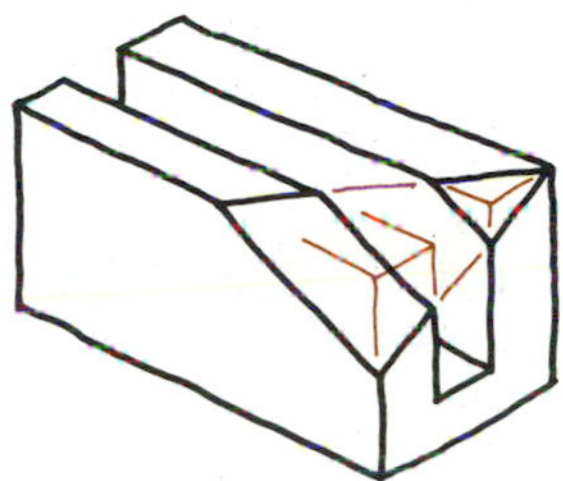
STEP 3
DARKEN THE LINES ON THE FINAL OBJECT.

FIGURE 2-18 Making an isometric sketch of an object with an oblique surface *(Section 2.15)*

the object in the most revealing position possible. With some objects, some features are hidden, regardless of which position you choose, because of the complexity of the configuration. Thus far, only objects with the isometric axes in one position have been shown. A number of other axes positions are possible, but the visual angle between the axes must remain 120°. Figure 2-19 shows a simple object that has been sketched in four common positions. At (a) one axis is vertical and at (b) one axis is horizontal. The best positions for viewing the object are shown in examples 1 and 4.

Sometimes the detail on the bottom of an object is more important to display than the detail on the top. For example, the best position to view the object in Figure 2-20 is shown at (b). The view of the object shown at (a) can be misleading and its use does not reflect good judgment. Practice sketching an isometric picture, approximately half size and viewed from below, of a $1 \times 6 \times 12$ inch shelf with two attached, simple triangular brackets of optional size and location.

2.17 ARCS AND CIRCLES IN ISOMETRIC SKETCHES

In isometric sketches *all* arcs and circles will appear as ellipses in each of the principal planes: front (or rear), top (or bottom), and sides. Figure 2-21(a) illustrates how an *approximate* ellipse may be conveniently sketched in a top plane using the following procedure: In Step 1 sketch two intersecting center lines to locate the center of the ellipse. In Step 2 sketch a parallelogram (or rhombus) with

FIGURE 2-19 A simple object sketched in four common positions *(Section 2.16)*

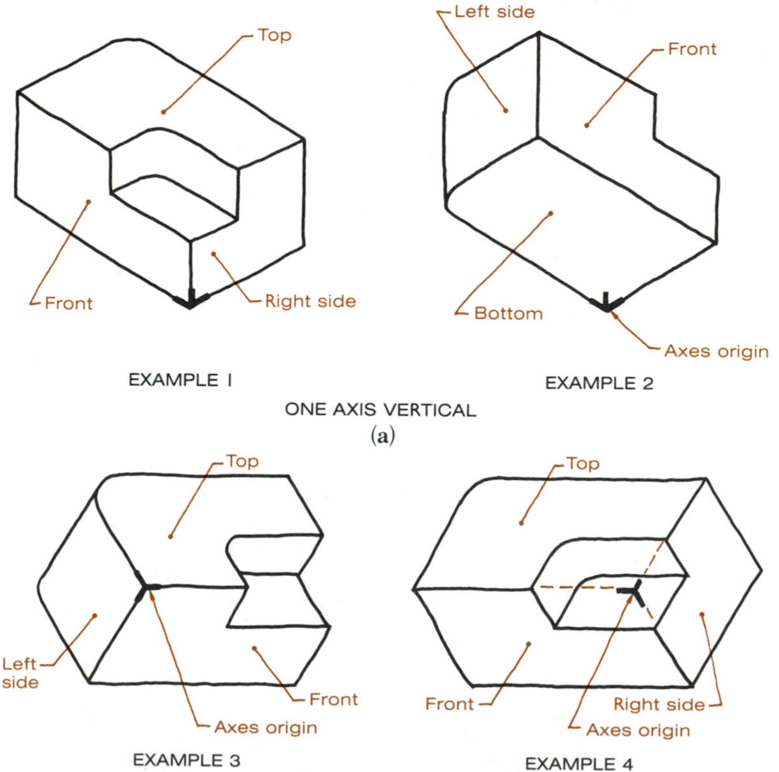

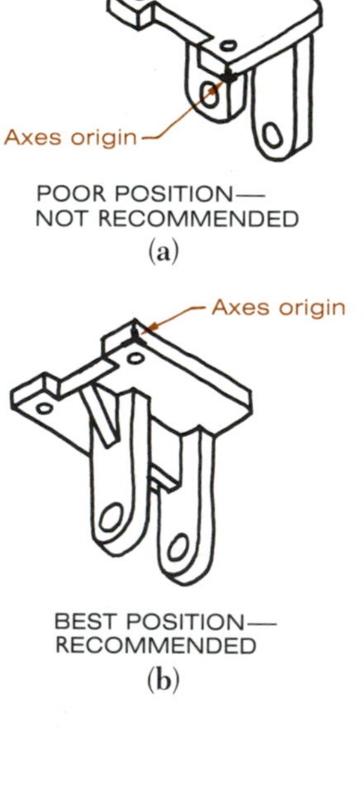

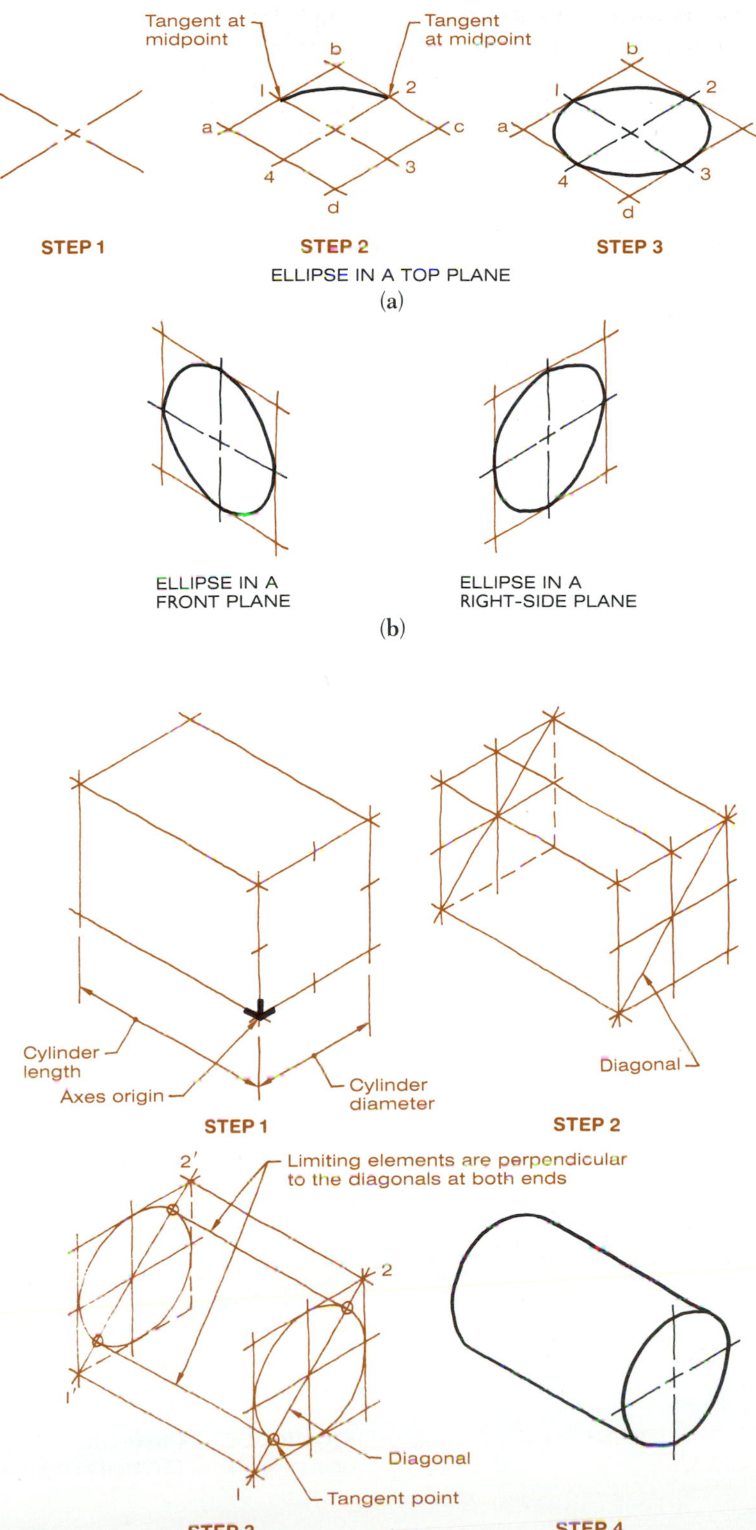

sides equal to the diameter of the desired circle. Points 1, 2, 3, and 4 are the midpoints of the sides. Next, sketch an arc that is tangent to ab at midpoint 1 and bc at midpoint 2. Finally, in Step 3 sketch another arc that is tangent to bc at midpoint 2 and cd at midpoint 3. Complete the remaining portion of the ellipse and darken the curve. With practice this method will soon produce surprisingly satisfactory results. Figure 2-21(b) shows how the ellipse appears in the front and the right-side planes. (This "four-center" method will be illustrated further in Chapter 4.)

CHECKPOINT 6
SKETCHING ISOMETRIC ELLIPSES IN THREE CONVENTIONAL POSITIONS

Review Steps

1. Sketch intersecting center lines.
2. Sketch the enclosing parallelogram.
3. Lightly sketch the arcs in each of the four quadrants.
4. Darken the final ellipse.

Practice this construction by sketching isometric ellipses of various sizes in four representative planes (front, top, right- and left-side). Use a plain sheet of 8½ × 11 inch paper.

2.18 RIGHT CIRCULAR CYLINDER

Figure 2-22 shows the method for sketching a right circular cylinder. Before beginning the

FIGURE 2-22 Sketching a cylinder in an isometric position *(Section 2.18)*

sketch, determine the length and diameter of the required cylinder. In Step 1, beginning at the axes origin, sketch a proportional construction framework that corresponds to the length-to-diameter ratio of the required cylinder. Mark off the midpoints on both ends and lightly sketch lines representing center lines and the longest diagonal (Step 2). In Step 3 sketch the arcs to form a complete ellipse on both ends of the cylinder. The shape of the partially hidden left end of a right circular cylinder is a mirror image of the nearest end but only one half of the ellipse will be visible on the hidden end. The parallel sides (or limiting elements) of the cylinder are perpendicular to the diagonals at the tangent points. Complete the isometric sketch by darkening the visible lines of the cylinder and adding the center lines (Step 4).

CHECKPOINT 7
MAKING AN ISOMETRIC SKETCH OF A CYLINDER

Review Steps

1. Determine the length and diameter of the desired cylinder.
2. Block in a construction framework, enclosing the entire figure.
3. Mark off the midpoints on the rhombuses.
4. Sketch center lines and the long diagonals.
5. Lightly sketch the ellipse curves.
6. Sketch the limiting elements perpendicular to the diagonals and parallel to the axis.
7. Darken the final figure.

Practice this construction by sketching cylinders of several different sizes on a plain sheet of 8½ × 11 inch paper. As a suggestion, use each of the three ellipse positions illustrated in Figure 2-21.

2.19 ISOMETRIC CIRCLE POSITIONS

If you conscientiously practice sketching ellipses, you will undoubtedly develop a mental picture of the correct appearance of circles lying in each of the isometric planes. After a great deal of practice the correct position for these circles will very likely become intuitive. The examples in Figure 2-23 show correct and incorrect positions for circles in isometric planes. The center lines and the outlines of the enclosing parallelograms should be aligned with the boundaries of the surfaces in which they are placed. Note how readily your eye discriminates between the correct and incorrect ellipse examples.

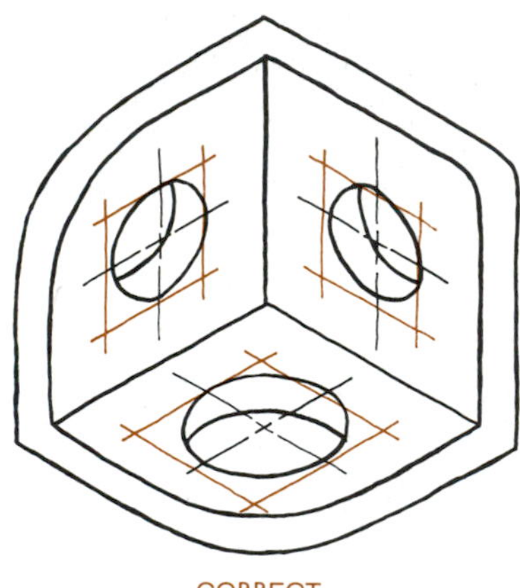

CORRECT
(a)

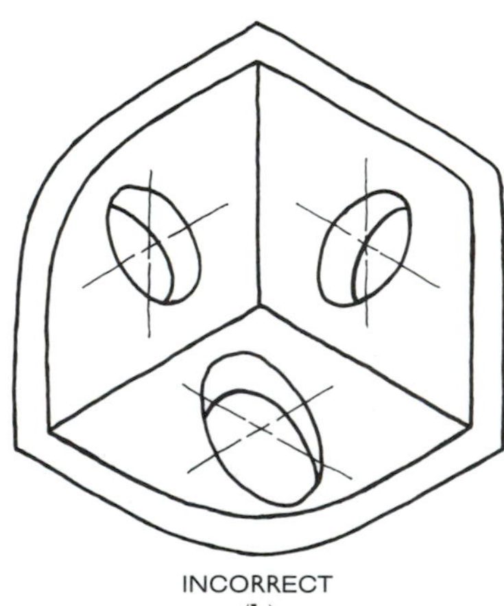

INCORRECT
(b)

FIGURE 2-23 Circles in isometric planes ***(Section 2.19)***

2.20 CENTERING A PICTORIAL DRAWING

Figure 2-24 shows how to center an isometric pictorial drawing in a given space. Begin the procedure by drawing the diagonals (Step 1). From point O, draw a vertical line OA equal to one-half the height of the object (Step 2). Draw line AB downward to the right, 30° to the horizontal. Make line AB equal to one-half the width of the object (Step 3). Draw line BO′ downward to the left, 30° to the horizontal. Make this line equal to one-half the depth (Step 4). Point 0′ is the axes origin of the isometric drawing. Finally, draw or sketch the isometric drawing (Step 5).

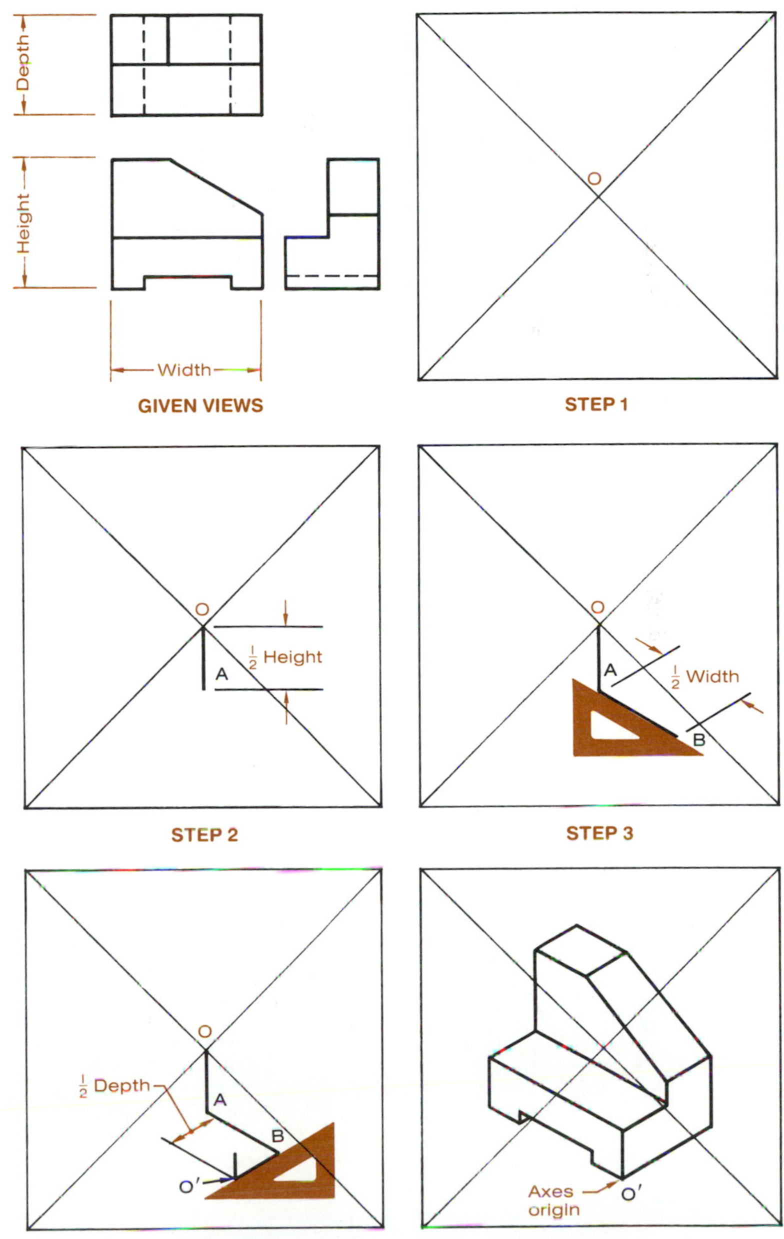

FIGURE 2-24 Centering a pictorial sketch in a given space ***(Section 2.20)***

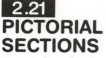

2.21 PICTORIAL SECTIONS

Even when displayed in the best possible position, some objects have interior cuts and features that are hidden from view. The sketch of the cylindrical object in Figure 2-25(a), for example, cannot possibly reveal either how far the small hole in the top or the larger hole on the right extend into the object, nor is there any way of knowing that there is a larger hole on the opposite end. Figure 2-25(b) shows a cutaway view of the part, exposing the interior construction as if an imaginary cutting plane was passed entirely through the object, in effect "sawing" the part in half lengthwise. For clarity the nearest half of the object should be removed. When the cutting plane passes fully through an object such that *half* of it is removed, the sectional view is called a *full section*. The cross-hatched lines, called *section lining*, denote the surfaces of the object that are "cut" by the cutting plane. The small 30°-60° drafting triangles shown on the figure indicate the suggested angle to be used when sketching the section lining.

Figure 2-26 is an example of the same cylindrical object shown as a pictorial *half section*, which is produced when the imaginary cutting plane removes the nearest quarter of the object. In this case, the "cut" surfaces are made more evident by sketching the section lining in opposite directions.

As a general rule, pictorial sections may be sketched most efficiently by first blocking in the entire object as usual and then gradually removing the cutaway portion. Begin by sketching light construction lines, erasing the "cut" portion, and adding newly exposed interior detail as a result of the cut. Add the section lining and complete the pictorial sectional view by sketching the visible lines. Pictorial sections are generally used only on symmetrical objects.

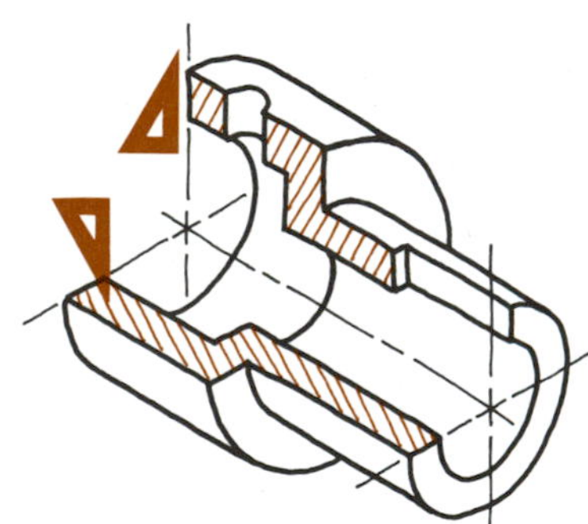

FIGURE 2-26 Pictorial half-section view of a cylindrical object *(Section 2.21)*

FIGURE 2-25 Pictorial full-section view of a cylindrical object *(Section 2.21)*

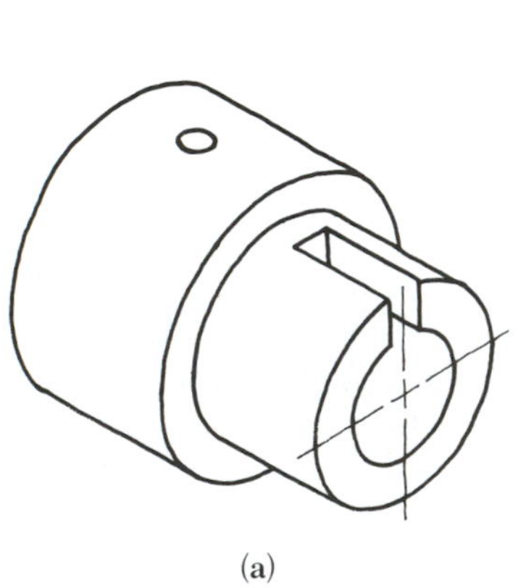

(a)

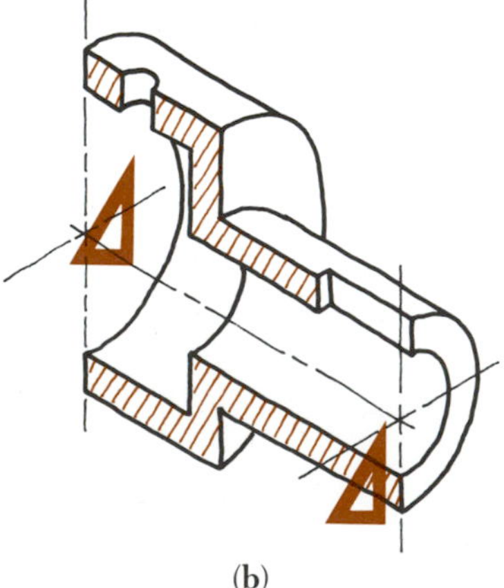

(b)

CHECKPOINT 8
SKETCHING PICTORIAL SECTIONS

Review Steps

1. Carefully study the object to determine the best direction for the cutting plane.
2. Sketch the construction framework enclosing the entire figure.
3. Gradually work in additional construction lines representing the various cuts, holes, and other details of the object.
4. Erase the "cut" portion.
5. Darken the lines representing the exposed interior detail and all other visible lines.
6. Sketch the section lining and necessary center lines.

Figure 2-27(a) shows a pictorial half section of an object. After closely studying the object, sketch a pictorial full section of this object on a plain sheet of 8½ × 11 inch paper. Figure 2-27(b) shows an object with a hexagonal hole to a depth equivalent to distance A. The depth of the small circular hole in the bottom is equal to ½A. Sketch a pictorial half section of this object.

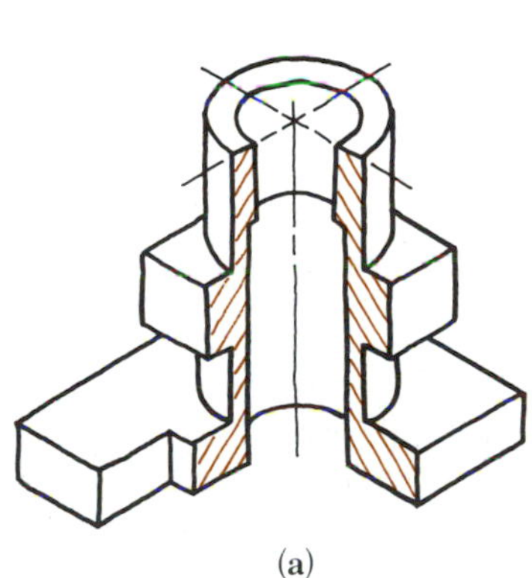

(a)

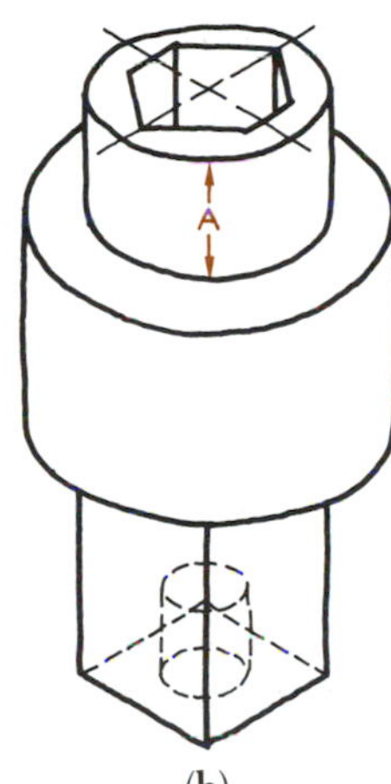

(b)

FIGURE 2-27 Checkpoint 8 problems

2.22 COMPUTER GENERATED PICTORIALS

Several digital computer graphics systems are presently available that can generate various forms of projected pictorial images. Three-dimensional drawings can be produced in various ways. In some systems the machine traces engineering detail drawings, automatically selects the axes, and tilts the object to a convenient position. An interesting development is three-dimensional modeling, which allows the user to display an accurate representation of the actual part being designed. The design data are stored by the computer, and once a three-dimensional model has been achieved, a number of useful operations are possible. For example, the graphic model can be rotated to be viewed from any direction. The model can also be displayed in cross-sections, combined with other parts in assemblies, and checked for clearances, interferences, and tolerances. Three-dimensional computer modeling also makes it possible to calculate such data as the centroid and the volume of parts and to create tool paths around the part for numerical-control machining operations.

Among the many advantages of computer-generated pictorials is that they can be readily stored and recalled. In addition, portions of the picture can be rapidly deleted, moved, and repositioned; the drawings may be readily generated from any view of the part; new parts can be easily added to fit with existing parts; and the object may be rotated and displayed over an infinite number of positions.

Miraculous as today's CAD (computer-aided design) systems may seem, they have certain shortcomings. The images sometimes consist of a network of lines that looks something like a wire mesh. Despite the fact that this type of structural representation mathematically describes the outline of an object, interaction between complex parts is sometimes difficult to discern.

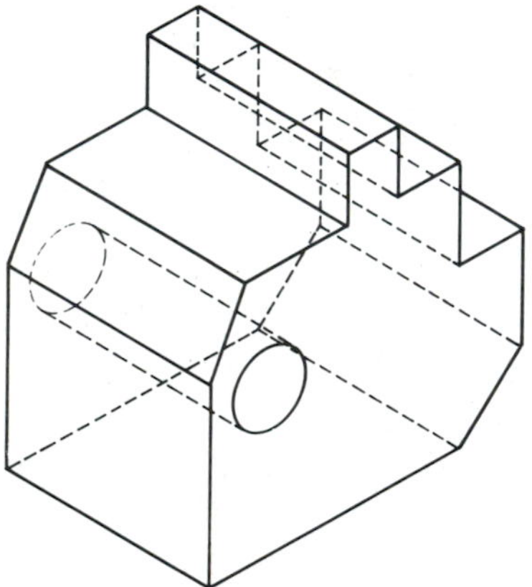

FIGURE 2-28 A computer-generated pictorial ***(Section 2.22)***

Figure 2-28 shows a computer drawing that was produced in a fraction of the time required for the same drawing to be done by hand.

PROBLEMS

Pictorial Sketching

You will generally make a pictorial sketch while working from the actual object, a mental image, or from engineering drawings. The engineering sketches that follow will help you develop understanding and competence in the techniques of pictorial sketching. Of course, it would not be common practice in the "real world" to prepare an oblique pictorial sketch working from an isometric drawing of the same object or to make an isometric sketch from an oblique sketch. These methods are simply a convenient way of assigning the following problems.

Prepare pictorial sketches of the following problems as directed. Use layout A, shown on the inside of the front cover. Letter the required information in the title block. Begin each problem by carefully estimating the amount of space that you will need to balance the figure neatly on the drawing sheet. If your sketches are made about twice as large as the figures in the book (unless otherwise directed) you will be able to fit several sketches on one sheet. Establish the proportions entirely by eye. Strive for clean, crisp final lines. Add center lines to your sketches wherever appropriate. Do not erase your lightly sketched construction lines.

P2.1 to **P2.12:** Make oblique sketches of the given objects. Assume a depth distance equal to about one-half the width of each object. (To prevent confusion, a typical width distance is denoted on the object in Problem 2.1.) Select a receding line direction and an angle that will reveal as much detail of the object as possible.

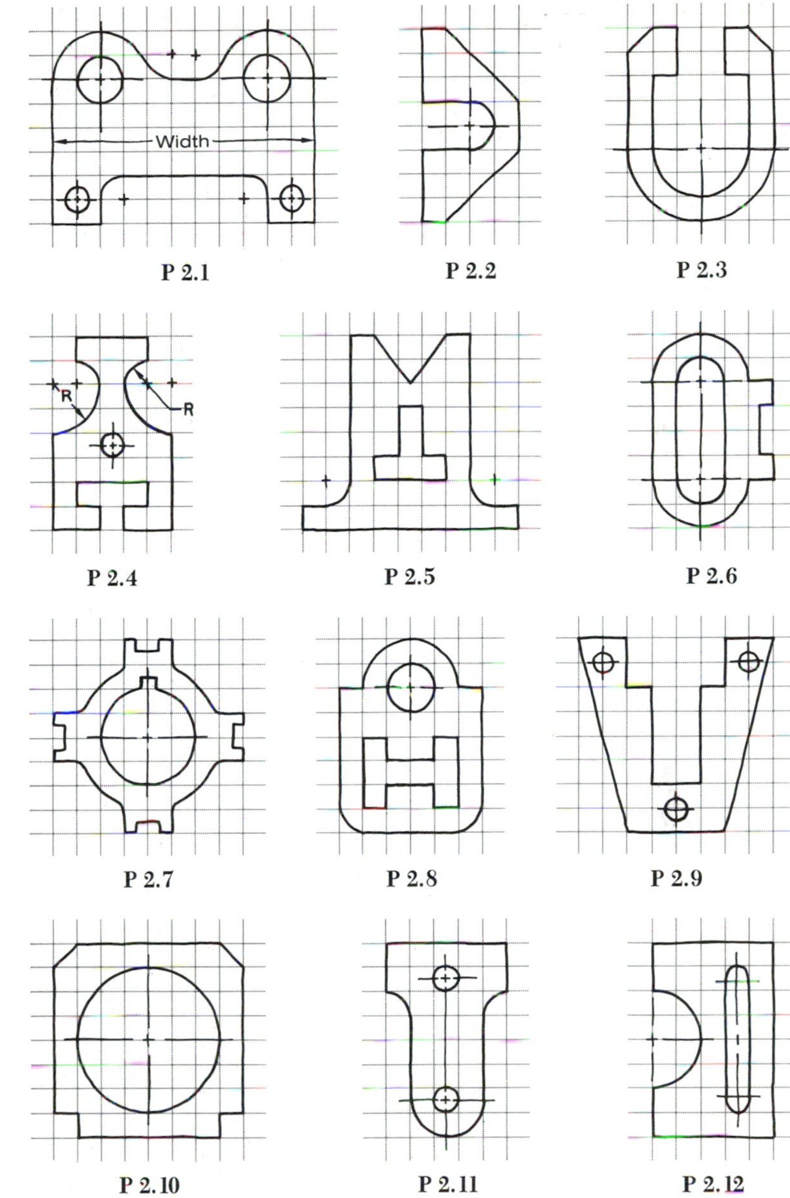

P 2.1 P 2.2 P 2.3 P 2.4 P 2.5 P 2.6 P 2.7 P 2.8 P 2.9 P 2.10 P 2.11 P 2.12

P2.13 to **P2.30:** All of the objects shown are drawn in isometric. Make an oblique sketch of each of these objects. Select a receding line direction and an angle that will reveal as much detail of the object as possible.

P 2.13 P 2.14 P 2.15 P 2.16

P 2.17 P 2.18 P 2.19 P 2.20

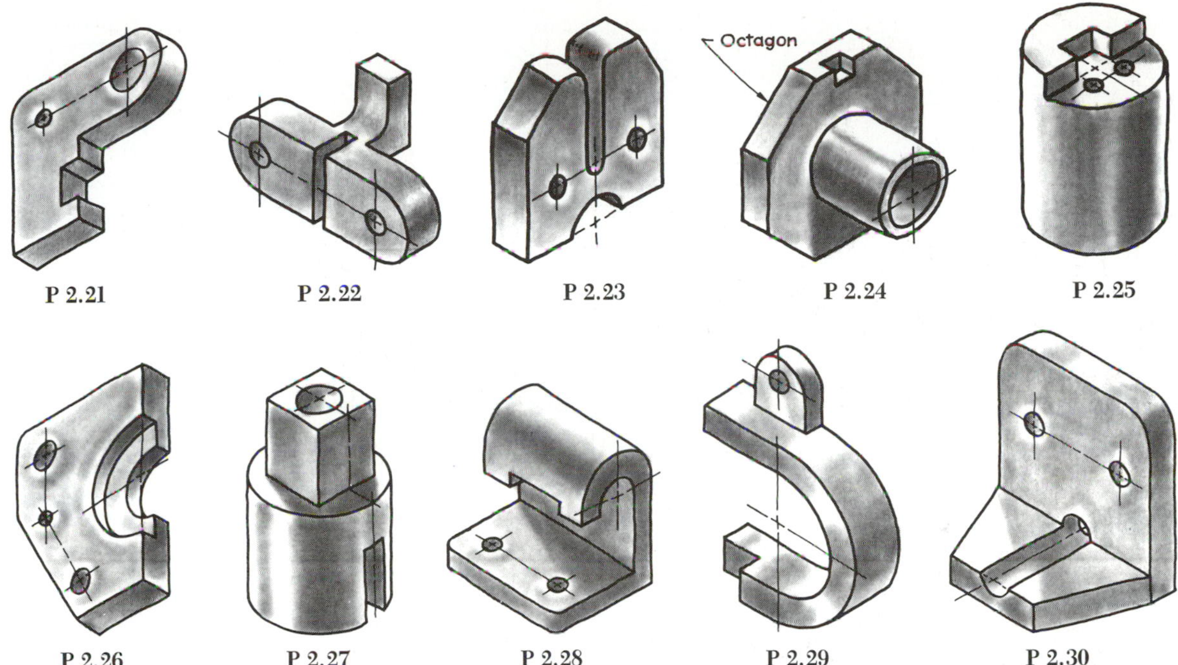

P 2.21 P 2.22 P 2.23 P 2.24 P 2.25

P 2.26 P 2.27 P 2.28 P 2.29 P 2.30

P2.31 to **P2.48:** All of the objects shown are drawn in oblique. Make isometric sketches of these objects. Some problems are shown with a surface labeled X. Draw these objects with surface X in the horizontal or "down" position. For all other problems, select a position that will reveal as much detail of the object as possible.

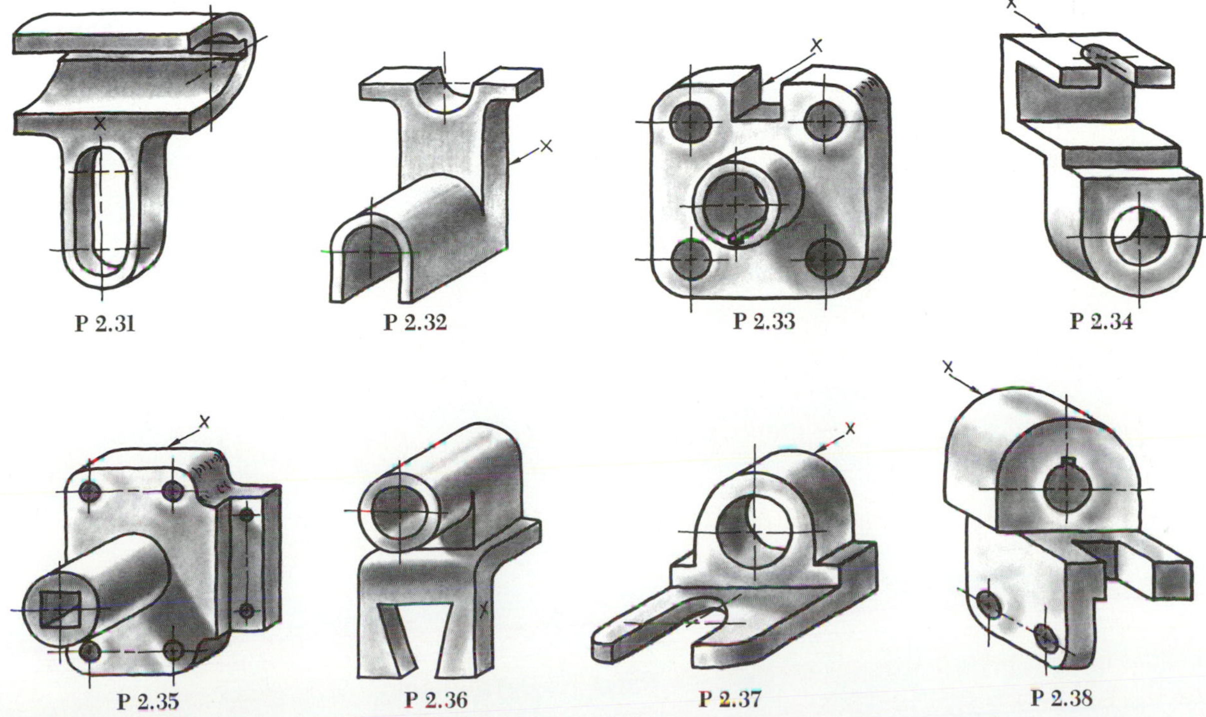

P 2.31 P 2.32 P 2.33 P 2.34

P 2.35 P 2.36 P 2.37 P 2.38

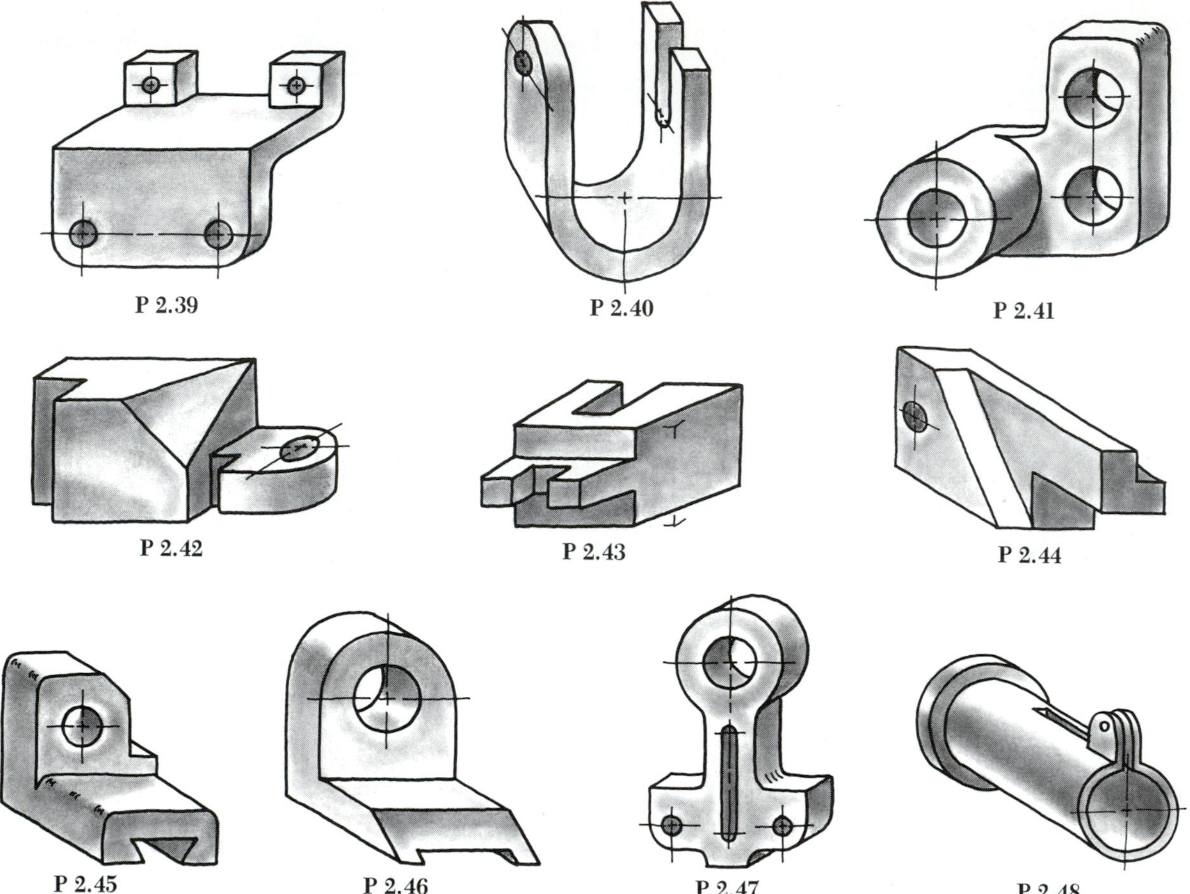

P 2.39 P 2.40 P 2.41

P 2.42 P 2.43 P 2.44

P 2.45 P 2.46 P 2.47 P 2.48

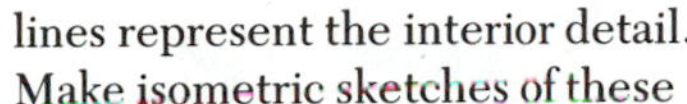

P2.49 to **P2.54:** The objects shown are drawn in oblique. The dashed lines represent the interior detail. Make isometric sketches of these objects, showing each of them as either a full section or a half section.

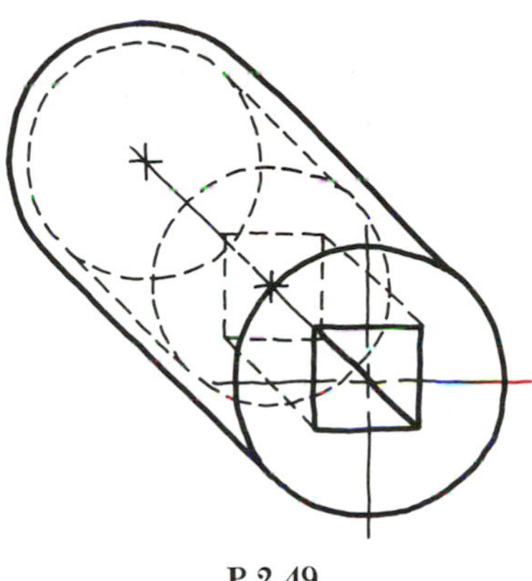

P 2.49

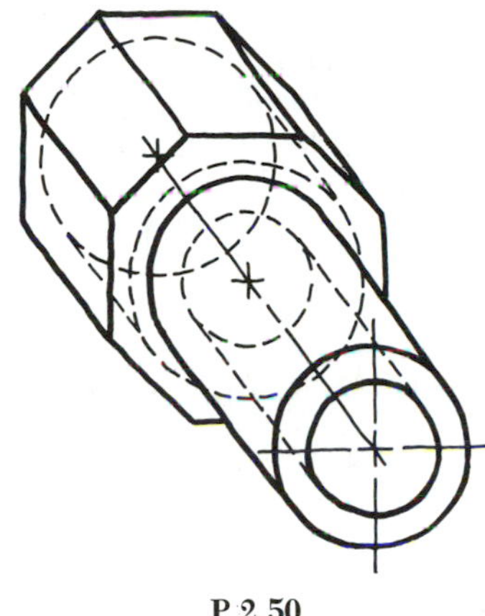

P 2.50

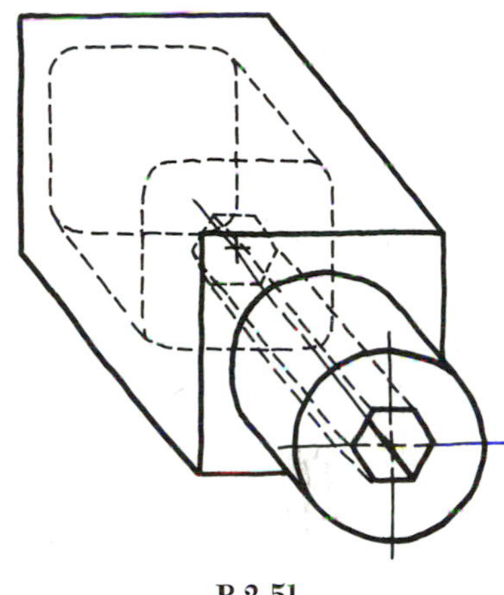

P 2.51

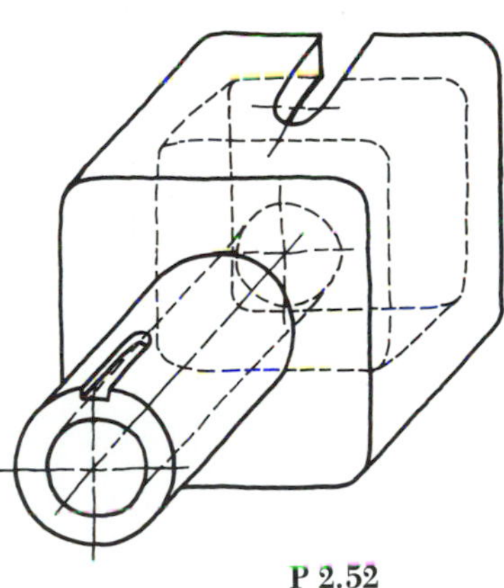

P 2.52

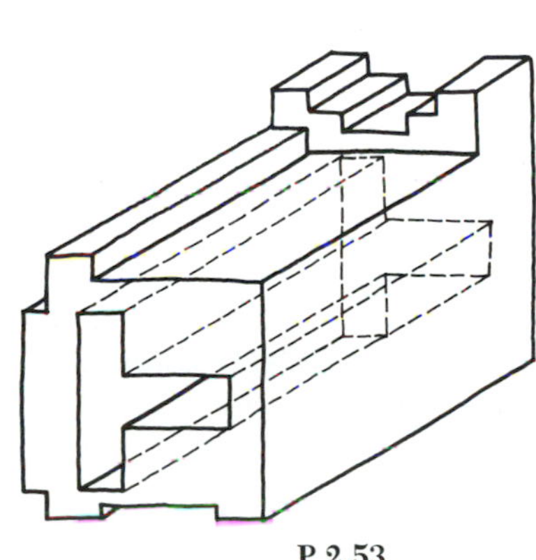

P 2.53

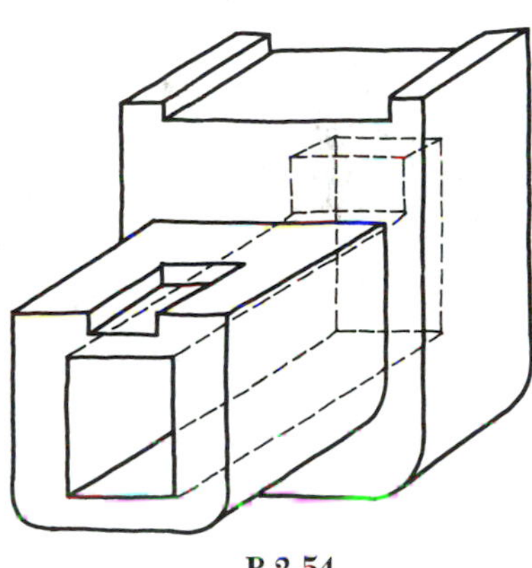

P 2.54

Use of Instruments

OBJECTIVES

After completing this chapter you should have gained the following abilities:

1. To select and properly use individual pieces of drafting equipment.

2. To mount a sheet of drawing paper on a drawing board and draw border lines and a title block.

3. To select the recommended grade of pencil lead for various line weights.

4. To use a parallel ruling straightedge or T square with triangles, or to use a drafting machine, to draw the following kinds of lines:

- **a.** horizontal
- **b.** vertical
- **c.** inclined at 15°, 30°, 45°, 60°, and 75° increments
- **d.** parallel
- **e.** perpendicular

5. To select and use a drawing scale in accordance with measurement requirements.

6. To sharpen, adjust, and properly use a drafting compass for drawing arcs and circles.

7. To use a protractor for measuring or laying out various angles.

8. To effectively use an irregular curve as an aid in drawing a smooth curve through plotted points.

9. To skillfully use dividers to transfer distances or to divide line segments into equal parts.

10. To use various templates.

11. To use an erasing shield to effectively make changes in crowded areas on drawings.

In spite of dramatic developments in computer-aided drafting and design, a definite need for manual drafting skills continues. The advent of computer graphics has not rendered drafting instruments obsolete. Manual drafting operations continue to be performed in companies not yet committed to automated drafting and design operations and in cases where low-volume, nonrepetitive graphical work is needed. Relatively inexpensive drafting instruments can be used to produce accurate drawings in a minimum amount of time and often with very little effort. Knowing the advantages and limitations of geometrical constructions and manual drafting operations will enable you to make the most effective use of the capabilities of computer-aided design equipment.

3.1 TYPICAL EQUIPMENT AND MATERIALS

The following list is a basic selection of equipment and materials necessary for making drawings. The sizes of the board, T square, and drafting triangles depend upon the range of drawing sheet sizes to be used. Your selection of a particular drafting scale to work in should conform with the measurement requirements of the work you are doing.

1. Drawing board
2. T square (parallel-ruling straightedge or drafting machine)
3. Drawing pencils
   a. mechanical lead holder and leads
   b. conventional wood pencils
4. Pencil lead pointer
5. 8- or 10-inch 30°–60° triangle
6. 6- or 8-inch 45° triangle
7. Scales
   a. civil engineer
   b. mechanical engineer
   c. architect
   d. metric
8. Protractor (or adjustable triangle)
9. 6-inch bow compass
10. 6-inch friction divider
11. Irregular curve
12. Templates
    a. circle
    b. ellipse
    c. other
13. Erasing shield
14. Eraser
15. Drafting tape
16. Drafting brush

Work Stations

Engineering departments in most companies establish work stations for drafters and designers. These special designer-desks are made with attached drafting boards. In most cases the desks contain space for personal equipment and for drawing storage. Some work stations include a flat table located behind the drafter for reference material. Two typical design-drafting work stations are shown in Figure 3-1.

SOURCE: HUSKY INJECTION MOLDING SYSTEMS

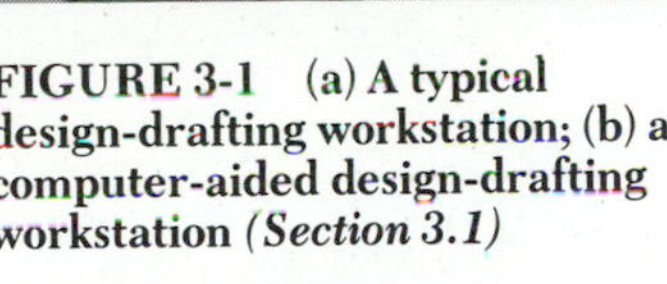

FIGURE 3-1 (a) A typical design-drafting workstation; (b) a computer-aided design-drafting workstation *(Section 3.1)*

T Squares

The material used in a T square may be wood, stainless steel, plastic, or a combination of wood and plastic. The T square accomplishes the same function as a straightedge. A T square, shown in Figure 3-2, is composed of two parts: the head and the blade. The upper edge of the blade is used for drawing horizontal lines and as a guide for supporting the drafting triangles. The lower edge must not be used because it may not be parallel to the top edge. For right-handed drafters, a T square is used by placing the head firmly against the left edge of the drawing board. T squares and drawing boards are rarely used in industrial engineering departments. The exception might be, perhaps, as a part of an engineer's personal equipment, occasionally used at his or her desk for limited graphical work.

SOURCE: KOH-I-NOOR RAPIDOGRAPH, INC.

FIGURE 3-2 A T Square *(Section 3.1)*

Drafting Machines

A drafting machine, shown in Figure 3-3(a) combines the functions of the T square, triangles, scale, and protractor. The device is conveniently mounted on any drafting surface and is positioned by one hand, leaving the other hand free for drawing. Considerable drafting time is saved with this versatile machine.

Parallel-Ruling Straightedges

Drafting tables may also be equipped with a parallel-ruling straightedge, which is used to draw horizontal lines and as a support for the drafting triangles for drawing vertical and inclined lines. The parallel-ruling straightedge, shown in Figure 3-3(b), operates by means of pulleys and is easily moved up and down the drafting surface.

(a)

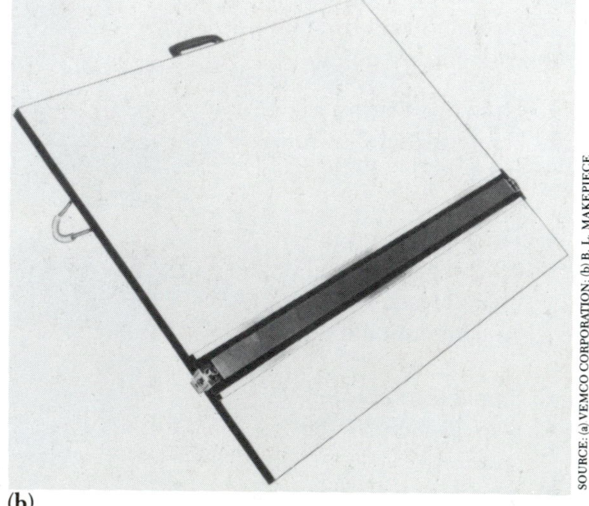

(b)

SOURCE: (a) VEMCO CORPORATION; (b) B. L. MAKEPIECE

FIGURE 3-3 (a) A Drafting Machine; (b) a parallel straightedge mounted on a drafting board. The straightedge has clear acrylic edges with a solid-color acrylic body. Spring-loaded end clips hold the straightedge firmly and evenly but permit easy use *(Section 3.1)*

3.2 PENCILS

Drawing pencils are available in 17 grades, or degrees of lead hardness, ranging from a 9H, which is very hard, to 6B, a very soft and black grade. The choice of the pencil grade is influenced by the kind of paper used and the existing humidity, which will vary the condition of the surface of the paper. The choice of pencil lead is also a matter of the individual drafter's preference and "touch." Most drafters use only two or three grades of pencil leads for drawing all lines. The harder grades, 5H or 6H, for example, are often used for drawing layout lines; and the medium grades, 2H or 3H, are selected for drawing final lines. The medium-soft grades of leads, H and F, are used for lettering, drawing arrowheads, and doing freehand work.

Three styles of drafting pencils are illustrated in Figure 3-4. The precision drafting pencil, shown at (a), has been widely adopted by drafters and engineers. A wide variety of lead grades are available in 0.3, 0.5, 0.7, and 0.9 mm diameters. The pencil features a nonslip metal grip and has a partially sliding sleeve that retracts gradually as a line is drawn. The metal sleeve prevents the thin lead from breaking and contacts the edge of the T square, triangles, drafting machine scales, templates, and other tools. The chief advantage of the precision drafting pencil is that the thin diameter leads do not require pointing.

The *mechanical lead holder*, shown at (b), has a knurled grip and a push-button action for feeding out the lead. Standard lead diameters may be used in lengths as short as about ½ inch. Also available are conventional *wood-cased pencils*, shown at (c).

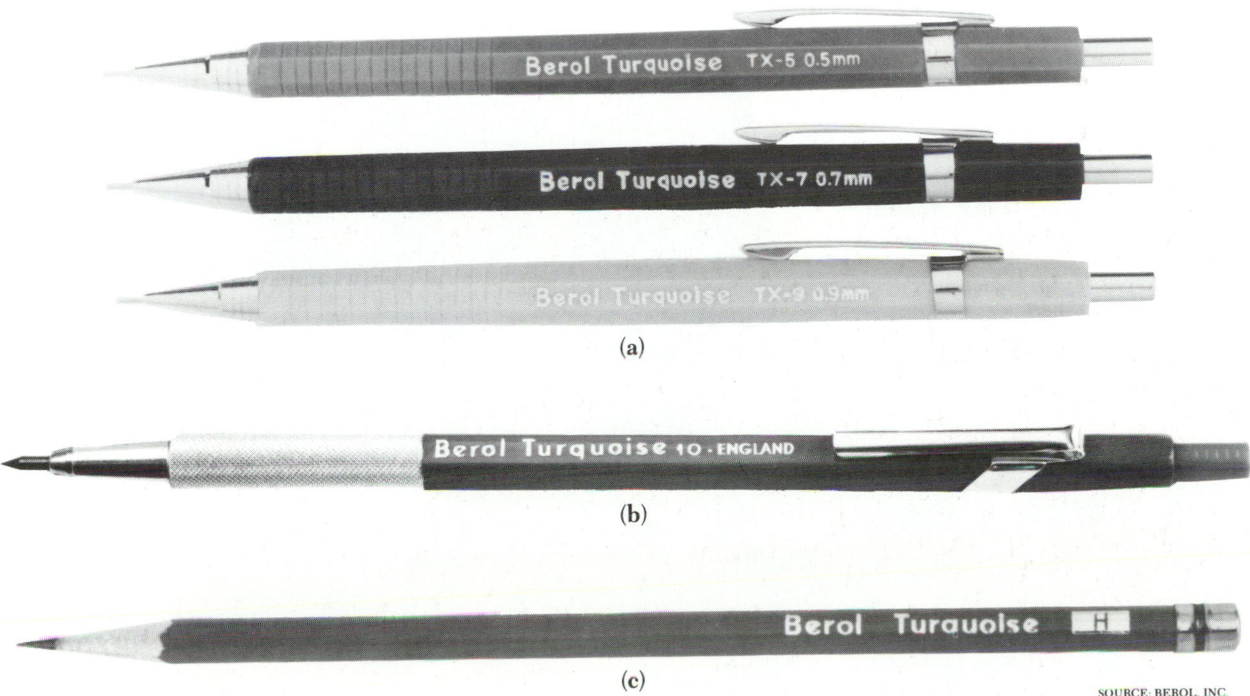

FIGURE 3-4 Three styles of drafting pencils *(Section 3.2)*

3.3 POINTING THE PENCIL

The standard office-type pencil sharpener produces a short, comparatively blunt point that is unsuitable for drafting purposes. Wood-cased drafting pencils must be sharpened in a special drafting pencil sharpener that leaves about ⅜ inch of lead exposed but unpointed, as shown in Figure 3-5(a). Following sharpening, the drafting leads are pointed to a long, tapering, conical shape. Figure 3-5(b) shows the desired conical shape of the point obtained by using a sandpaper pad or a file. The unmarked end should be sharpened to preserve the grade mark printed on the opposite end. Commercial mechanical lead pointers, similar to the one illustrated in Figure 3-6, may also be used for pointing leads in mechanical lead holders. It is important to maintain a sharp point, because dull pencil leads produce fuzzy, indefinite lines that not only detract from the appearance of a drawing but also yield a hazy, indistinct copy.

3.4 TRIANGLES

Two drafting triangles, the 45° and the 30°–60°, are illustrated in Figure 3-7. They are plastic and may be purchased in trans-

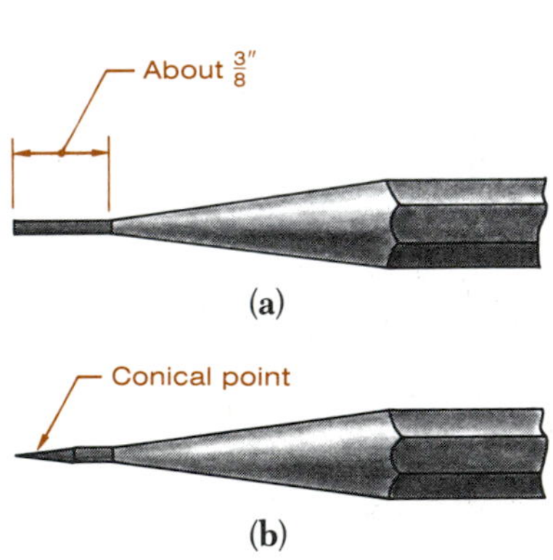

FIGURE 3-5 Pencil lead points ***(Section 3.3)***

FIGURE 3-6 A commercial lead pointer ***(Section 3.3)***

parent form or in several colors. They should be handled with care to ensure that the edges are not nicked. They should also be washed frequently with soap and water. The size of a 45° triangle is determined by the length of one of its equal legs. The size of a 30°–60° triangle is designated by the length of the longer leg, *not* the hypotenuse. Triangles are available in lengths of 4, 6, 8, 10, and 12 inches and longer.

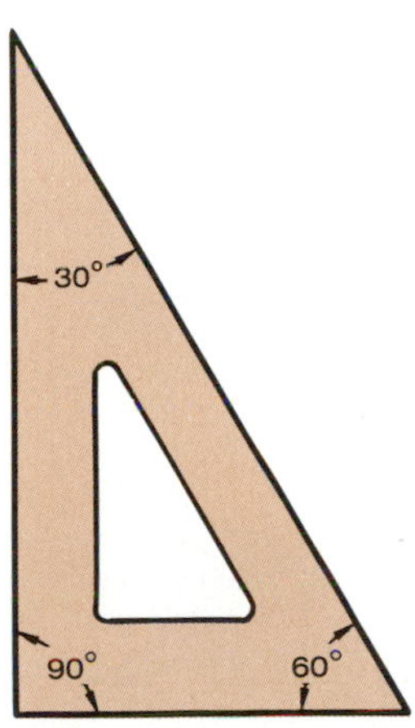

(a) 30°–60° TRIANGLE

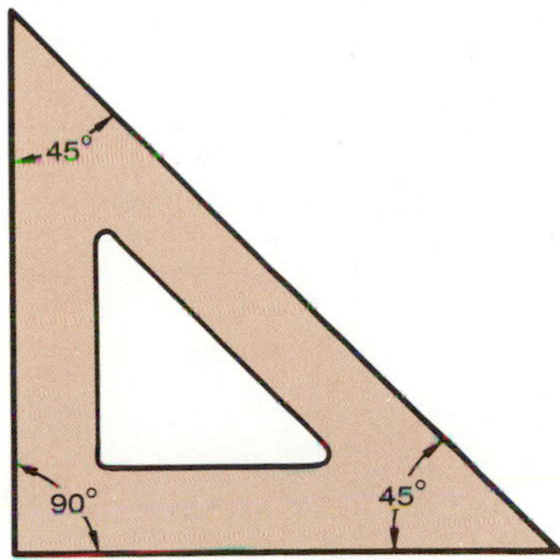

(b) 45° TRIANGLE

FIGURE 3-7 Drafting triangles ***(Section 3.4)***

3.5 DRAWING HORIZONTAL AND VERTICAL LINES

CHECKPOINT 1
DRAWING HORIZONTAL AND VERTICAL LINES

1. Fasten a plain sheet of 8½ × 11 inch paper to the drawing board.
2. Prepare an F, a 2H, and a 5H grade drawing pencil for drafting work.
3. Draw the border lines as follows: Begin by measuring a distance of ½ inch in from *each edge* of the paper. Using the T square and a 5H grade pencil, draw a light horizontal line parallel to the upper and lower edges of the sheet and then, with a triangle and a T square, draw two light vertical lines. Go over the lines using the 2H pencil to obtain dark border lines neatly terminating at each corner.
4. Referring to layout A, shown on the inside of the front cover, lay out the various divisions shown for the sheet title block using the given dimensions. Construct the title block with light, thin lines using the 5H pencil and go over the lines with a 2H pencil. For the lines of lettering, use a T square to draw pairs of very light, horizontal guide lines spaced about ⅛ inch apart. Use the F pencil to carefully letter the required information in the title block.
5. Working downward from the upper border line, measure and mark a series of 1-inch divisions on the left vertical border line. Using the T square and the 5H pencil, draw light, thin horizontal lines through these division marks extending across the entire sheet between the border lines.
6. Working across from the left to the right border line, set off a series of 1-inch division marks on the upper horizontal border line. Using the T square and triangle, draw a series of light, thin vertical lines through these division marks extending upward across the entire sheet between the border lines.
7. In a similar manner to steps 5 and 6, lay off a series of ½-inch divisions on the upper horizontal and left vertical border lines. Through these points, use your 2H pencil to draw *dark* lines extending across the sheet. Your final drawing should now consist of a checkerboard pattern of parallel horizontal and vertical lines, alternately light and dark, ½ inch apart.

3.6 DRAWING INCLINED LINES

When the T square is combined with the 45° triangle, the lines making an angle of 45° with the horizontal may be drawn. Similarly, angles of 30° and 60° may also be drawn to a vertical or horizontal line using the T square in combination with the 30°–60° triangle. Figure 3-8 shows how to obtain 15° spacings around a circumference by combining both of the drafting triangles and the T square. Inclined lines may be drawn at

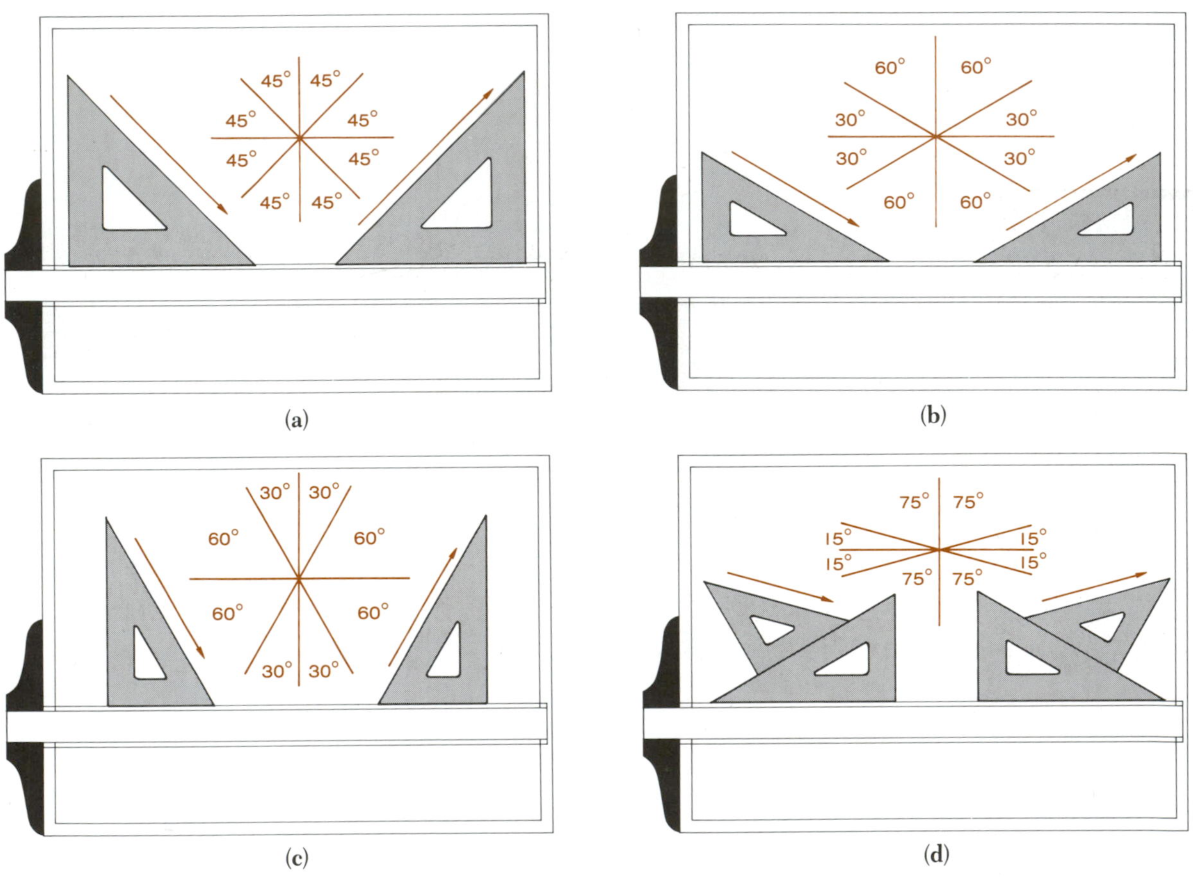

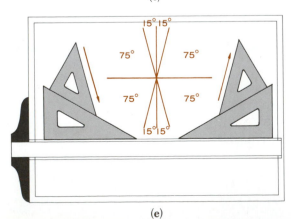

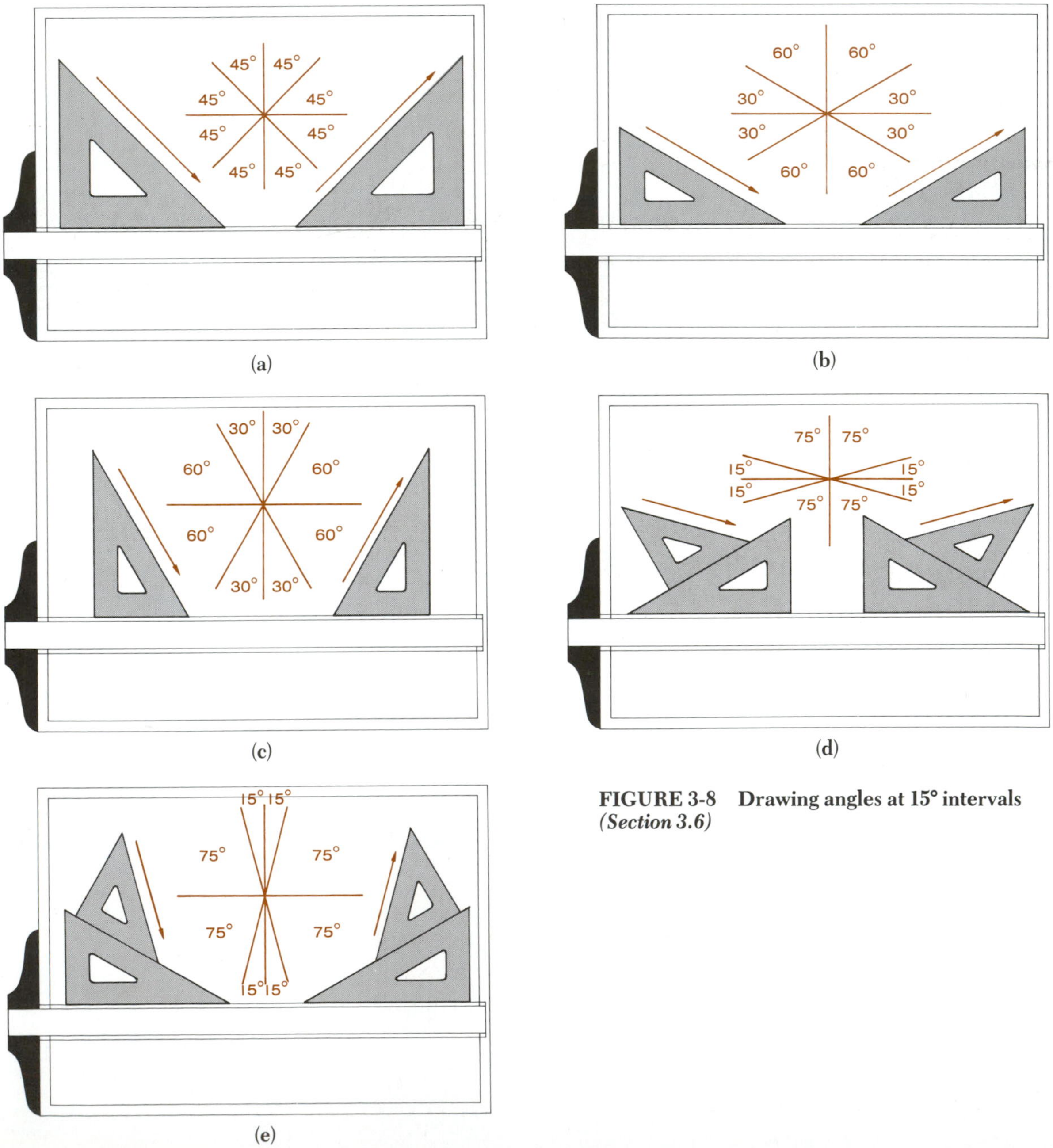

(e)

FIGURE 3-8 Drawing angles at 15° intervals ***(Section 3.6)***

random angles by aligning the edge of the triangle or the T square with any two given points. As a convenience, place the pencil at one of the points and slide the straightedge against the pencil point. Finally, pivot the straightedge about the pencil point until it aligns with both points and then draw the line.

CHECKPOINT 2
DRAWING INCLINED LINES

Reproduce the exercises shown in Figure 3-9 with a 5H grade pencil but do not copy the dimensions and notes. Use a plain sheet of 8½ × 11 inch paper. Point O is in the center of the space.

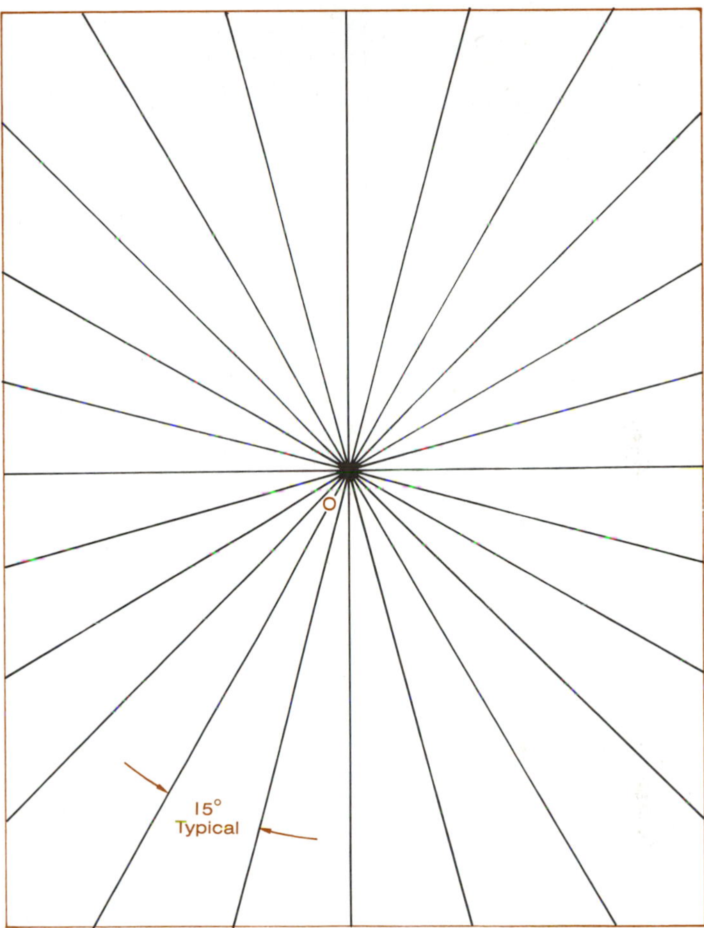

FIGURE 3-9 Checkpoint 2 problem

3.7 DRAWING PARALLEL LINES

A series of lines may be drawn parallel to a given line by using the T square and one of the triangles, as illustrated in Figure 3-10. Begin by placing one of the edges of the triangle in a position parallel to the given line on the drawing. Move the edge of the T square blade across the drawing surface until it contacts the triangle. Hold it firmly in position and slide the triangle away from the given line to the new position of the required parallel line. Many engineers substitute a second triangle for the T square straightedge, enabling them to perform acceptable graphical work while sitting at their desks.

3.8 DRAWING PERPENDICULAR LINES

The construction for drawing perpendicular lines is performed in essentially the same manner as for drawing parallel lines. Either the T square or a triangle may be used as a straightedge. Position the triangle, as shown in Figure 3-11, with one edge placed parallel to the given line on the drawing. Now move the T

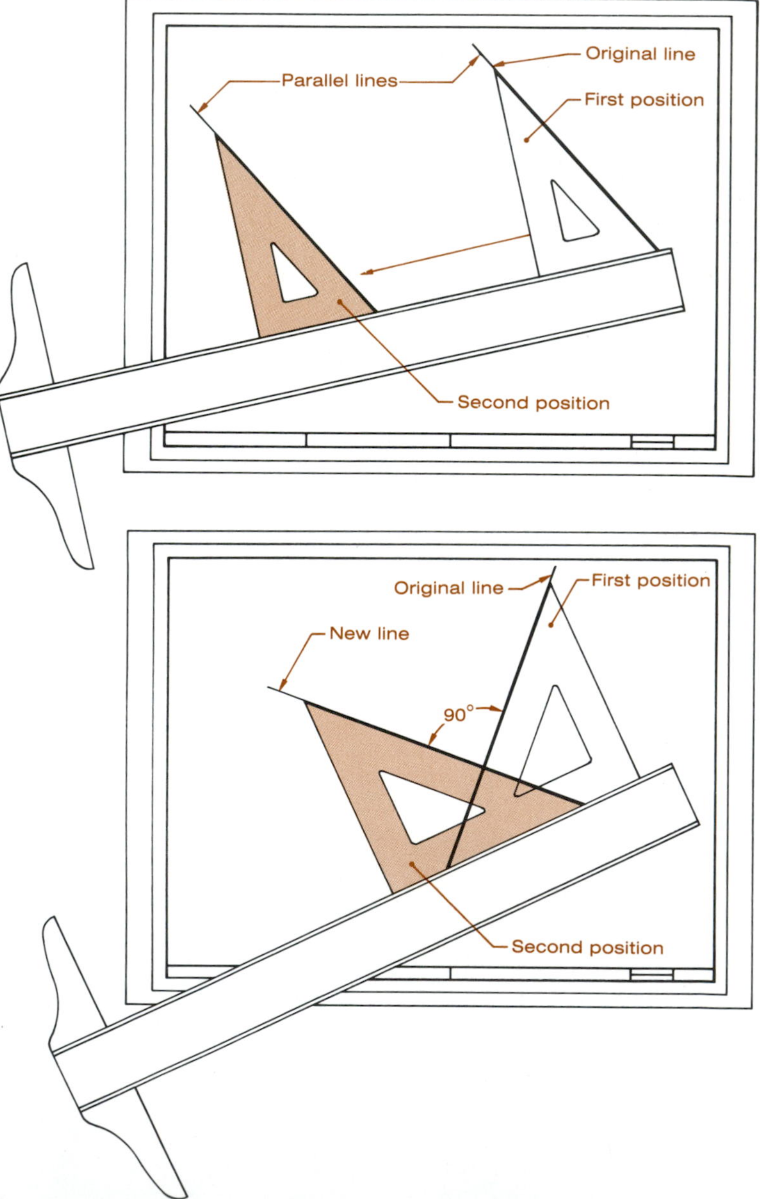

FIGURE 3-10 **Drawing parallel lines *(Section 3.7)***

FIGURE 3-11 **Drawing perpendicular lines *(Section 3.8)***

square until it contacts the triangle. Finally, reverse the triangle and move it along the straightedge until it corresponds to the position of the required perpendicular line.

CHECKPOINT 3
DRAWING PARALLEL AND PERPENDICULAR LINES

Reproduce the exercises shown in Figure 3-12 with a 5H grade pencil but do not copy the dimensions and notes. Use a plain sheet of 8½ × 11 inch paper, with point O in the center of the drawing space. When drawing lines, be careful to hold the T square and the triangle firmly to prevent slipping.

FIGURE 3-12 Checkpoint 3 problem

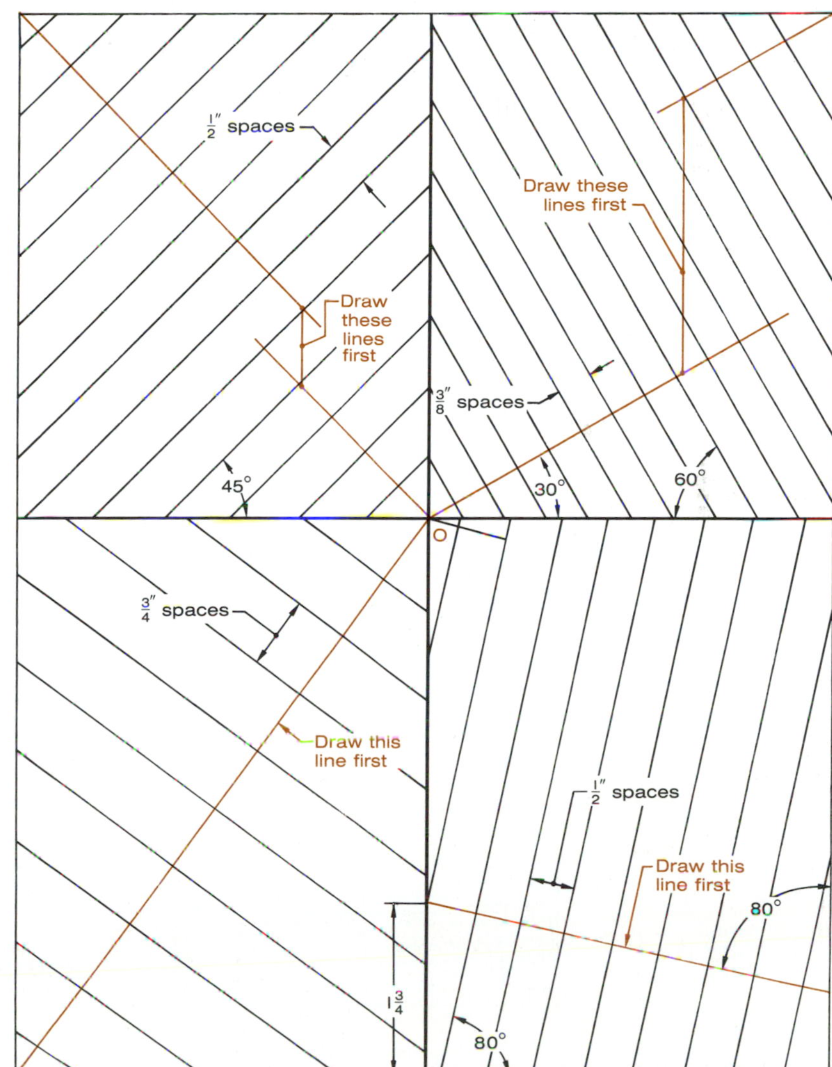

3.9 SCALES

Whenever possible, drawings of objects are prepared full, or actual, size. Objects too large to be shown full size on a drawing are drawn to a reduced size and objects too small to be clearly represented are drawn to an enlarged scale. Figure 3-13 illustrates a relatively large part that is drawn to a reduced size. The amount of reduction depends upon the relative size of the object and the size of the drawing sheet to be used. An enlarged drawing scale is generally necessary when portraying relatively small objects so that all of the various cuts, holes, and other features may be accurately represented. Figure 3-14 shows a drawing of a tiny part that has been drawn to an enlarged size.

The scale of a drawing is indicated in the title block when all of the objects drawn on that sheet are the same scale. If dif-

FIGURE 3-13 A part drawn at a reduced scale ***(Section 3.9)***

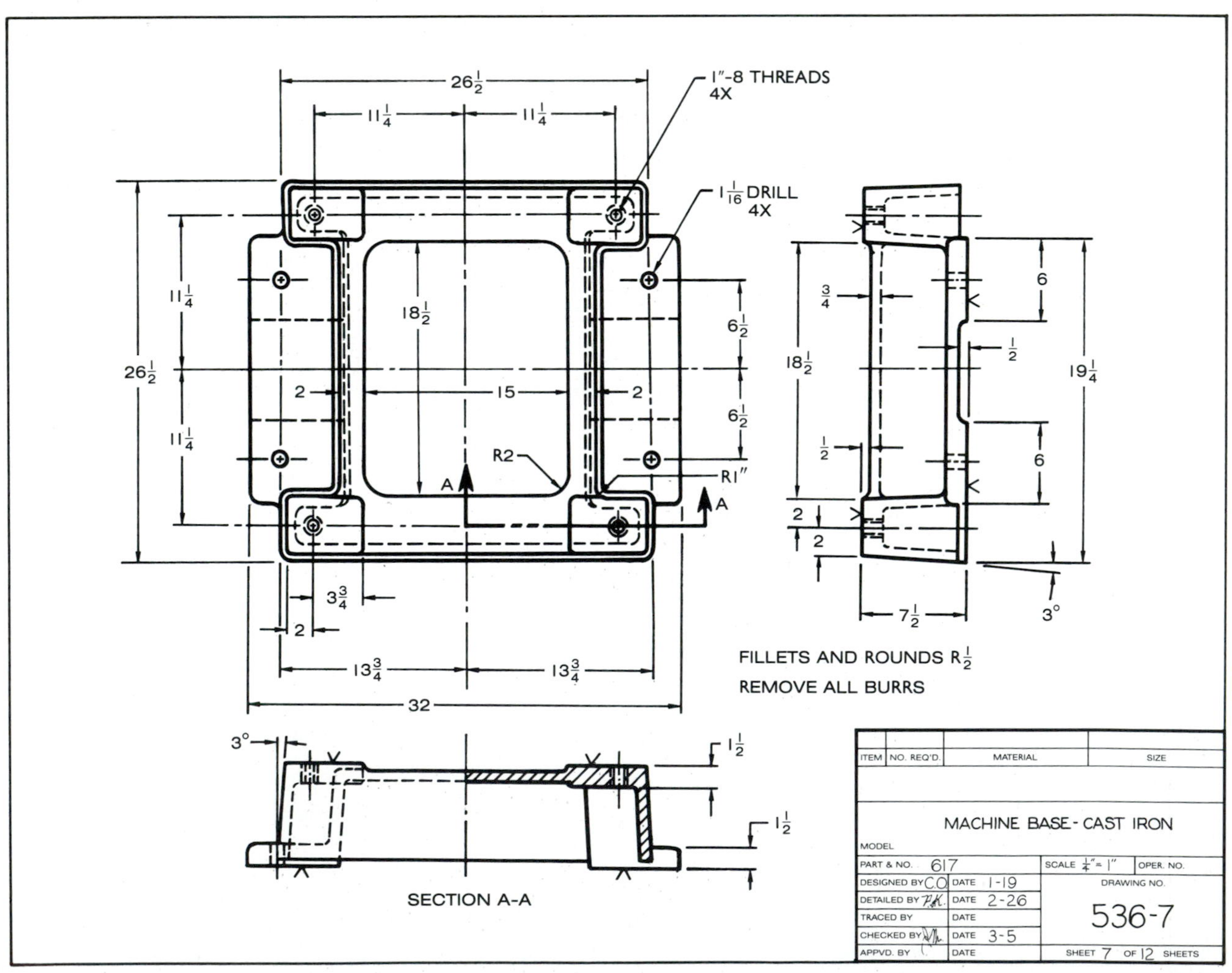

ferent scales are used to draw the various objects on a drawing sheet, the scale is given in a prominent position below or near each of the corresponding views.

In the U.S. Customary system, the equal sign is used in the title block to denote the scale as, 1 = 2 (half scale), or, as shown in Figure 3-13, ¼″ = 1″ (quarter scale). Unless otherwise specified, the unit of measurement is inches. If metric units are used, a colon is used instead of an equal sign (as shown in the title block in Figure 3-14) and the large letters *SI* or the word *METRIC* is placed near the title block.

Various types of scale shapes are shown in Figure 3-15. Most drafting scales are 12 inches long and are made of wood, plastic, or metal. A *fully divided scale* is one in which the basic units are

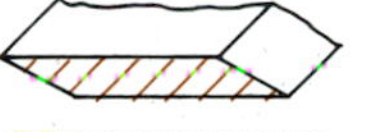
OPPOSITE BEVEL

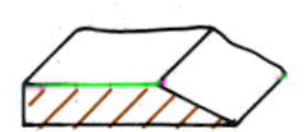
SINGLE BEVEL

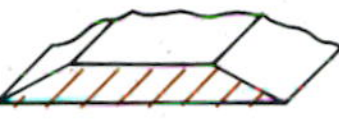
FLAT

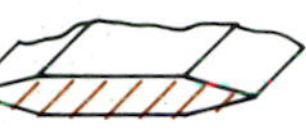
DOUBLE BEVEL

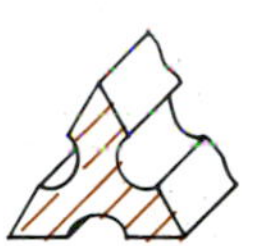
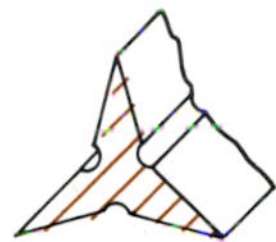
TRIANGULAR

FIGURE 3-15 Various scale shapes ***(Section 3.9)***

FIGURE 3-14 A part drawn at an enlarged scale ***(Section 3.9)***

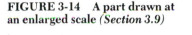

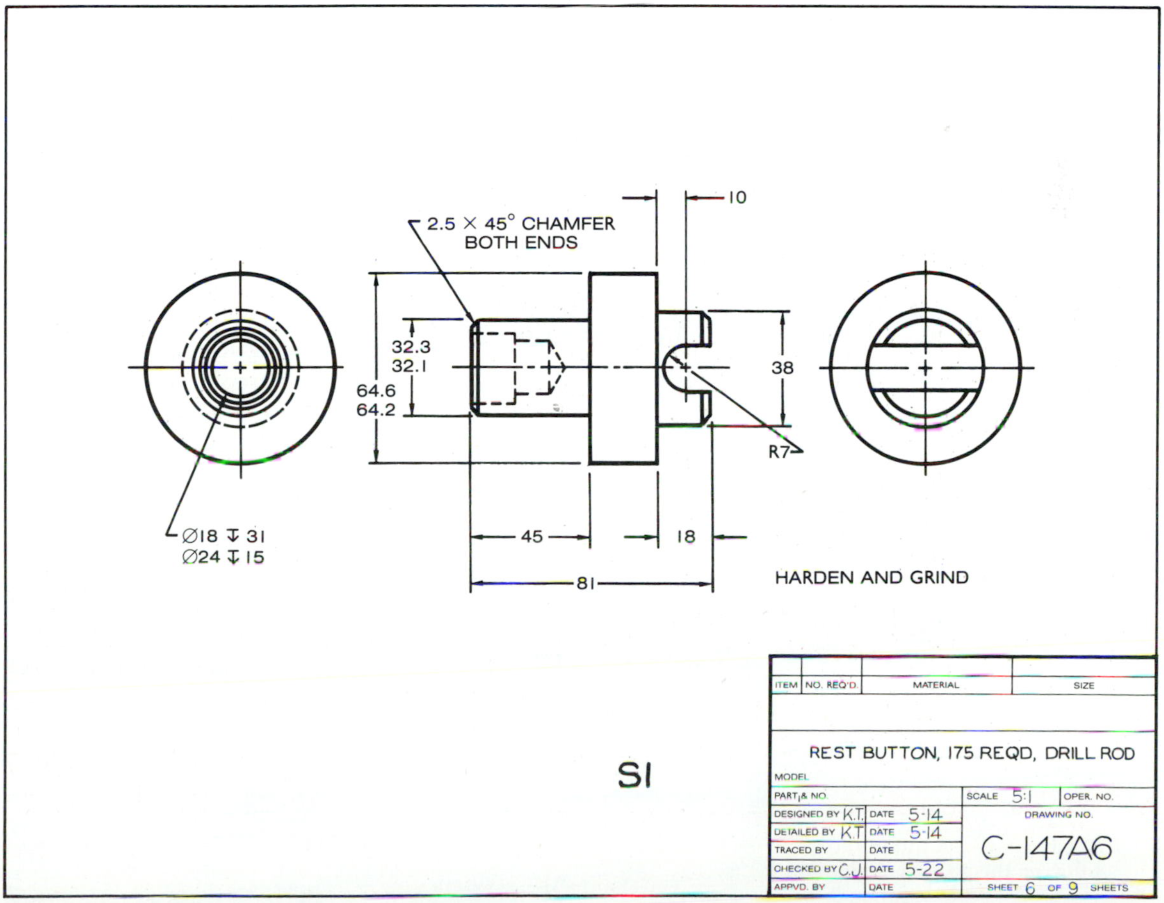

2.5 × 45° CHAMFER BOTH ENDS
10
32.3
32.1
64.6
64.2
38
R7
Ø18 ↧ 31
Ø24 ↧ 15
45
18
81
HARDEN AND GRIND

SI

ITEM	NO. REQ'D.	MATERIAL	SIZE

REST BUTTON, 175 REQD, DRILL ROD

MODEL
PART & NO. — SCALE 5:1 — OPER. NO.
DESIGNED BY K.T. DATE 5-14
DETAILED BY K.T. DATE 5-14
TRACED BY DATE
CHECKED BY C.J. DATE 5-22
APPVD. BY DATE
DRAWING NO. C-147A6
SHEET 6 OF 9 SHEETS

subdivided throughout the entire length of the scale edge, as shown in Figure 3-16. (An *open divided scale* is one in which the end unit is subdivided, as in Figure 3-18.)

Civil Engineer Scales

Civil engineer scales are graduated into decimal units. Each division is a multiple of 10 units. One inch is divided into 10, 20, 30, 40, 50, 60, and 80 parts. The numbers stamped on each scale face refer to the number of divisions in one inch, letting each major division represent 1 inch. The 10 scale is *full size*, with 10 divisions to the inch; the 20 scale is *half size*, with 20 divisions to the inch; the 40 scale is *quarter-size*, with 40 divisions to the inch, and so forth. Figure 3-16 shows how to lay off a distance of 1.55 inches on both a 10 scale and a 20 scale.

Letting each major division represent different units, the civil engineer scale may be conveniently used to set off map distances of 1 inch = 50 feet, 1 inch = 5 miles, and so on. The first figure given in the scale of 1 inch = 5000 feet, for example, is the true distance in inches on the drawing. The second figure, 5000, represents the *actual* distance represented on the earth. Figure 3-17 shows how the 30 civil engineer scale is used to measure a quantitive value of 650 units. In this example, each inch equals 300 units. Civil engineer scales are also employed for drawing stress diagrams or for other graphical constructions when representing decimal quantitative values such as force, velocity, or acceleration.

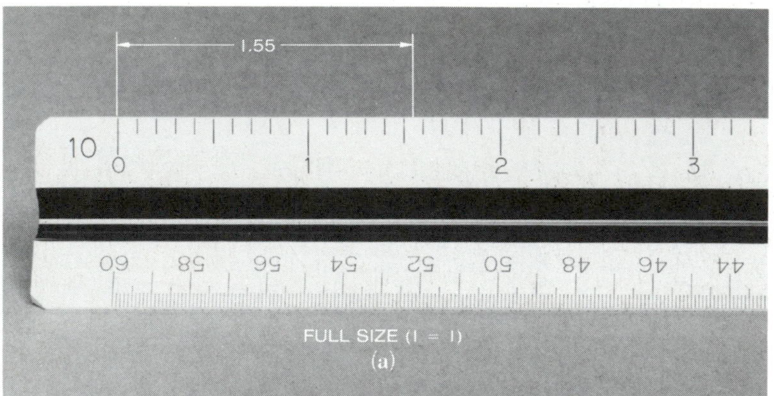

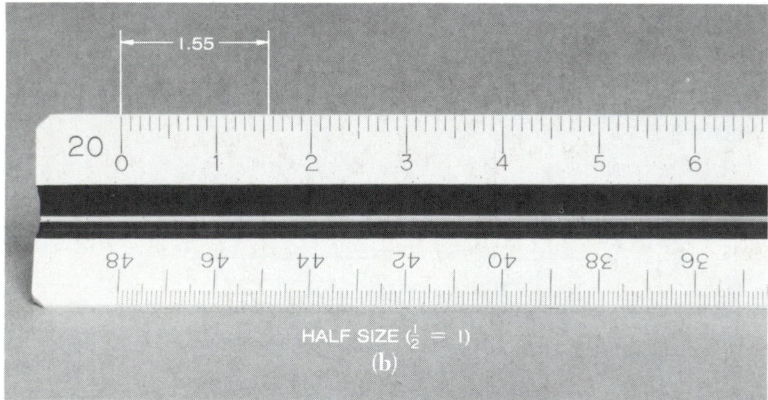

FIGURE 3-16 Decimal divisions on the civil engineer scale *(Section 3.9)*

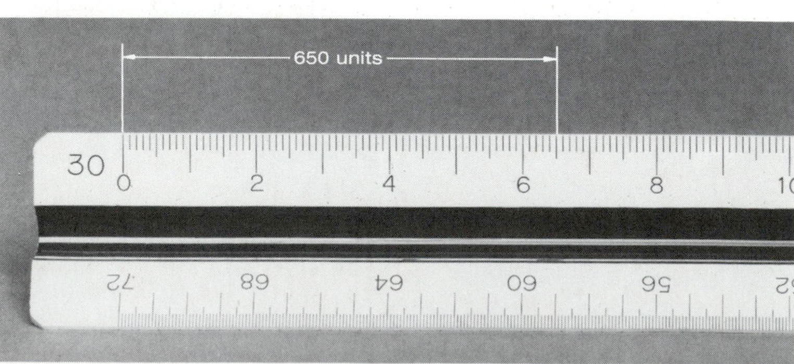

FIGURE 3-17 A quantitative value on the 30 civil engineer scale *(Section 3.9)*

Mechanical Engineer Scales

The chief application of the mechanical engineer scale is for drawing mechanical systems and machine parts. An open-divided scale is shown in Figure 3-18(a), and a fully divided scale is shown in Figure 3-18(b). The open-divided portion of the scale is graduated in inches. Fractions of inches are represented on the fully divided unit preceding the "0." The mechanical engineer scale is available in ratios of full-size (1 inch = 1 inch), half-size (½ inch = 1 inch), quarter-size (¼ inch = 1 inch), and eighth-size (⅛ inch = 1 inch). Some mechanical engineer scales are divided into decimal inches rather than fractional inches.

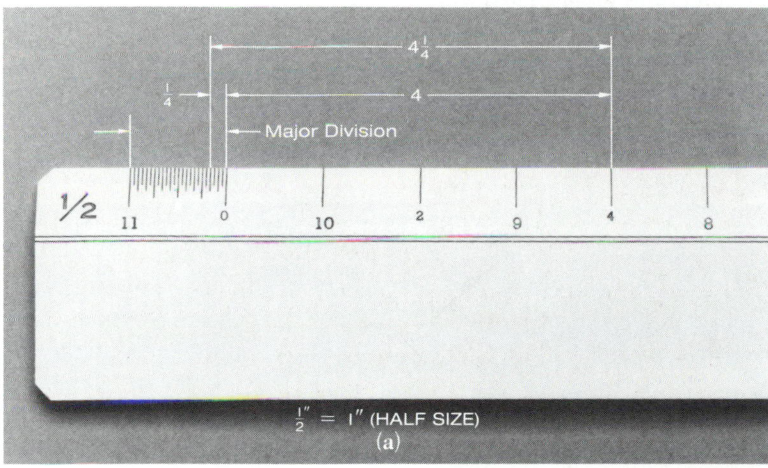

FIGURE 3-18 The mechanical engineer scale *(Section 3.9)*

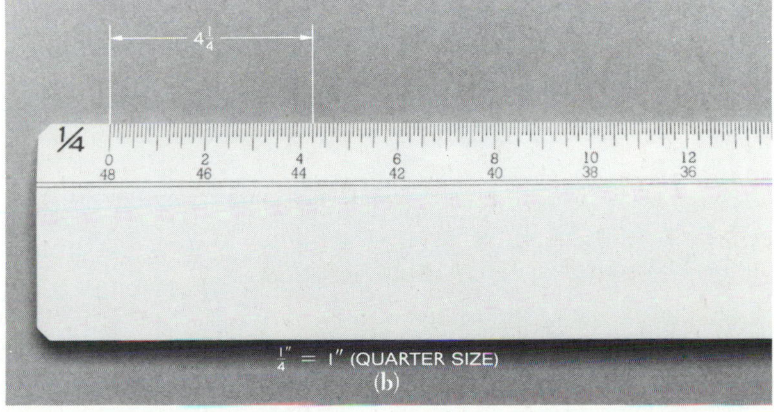

Architect Scales

The architect scale is used to lay out large objects that must be reduced to a smaller scale so that they will fit on a sheet of drawing paper. The full-size scale, stamped with the number 16, is divided into sixteenths of an inch. When this full-size scale is used to lay out half-size measurements, as in Figure 3-19(a), each dimension is mentally divided by two. Each of the major divisions on the other ten reducing scales on a triangular architect scale represent one foot, and the open divided area is divided into twelve parts to represent inches. The measurements are made in feet and inches. Ordinarily, triangular scales consist of: 12″ = 1′-0″ (#16), 3⁄32″ = 1′-0″, 1⁄8″ = 1′-0″, 1⁄4″ = 1′-0″, 3⁄8″ = 1′-0″, 1⁄2″ = 1′-0″, 3⁄4″ = 1′-0″, 1″ = 1′-0″, 1½″ = 1′-0″, and 3″ = 1′-0″. Figures 3-19(b), 3-19(c), and 3-19(d) illustrate, respectively,

FIGURE 3-19 The architect scale *(Section 3.9)*

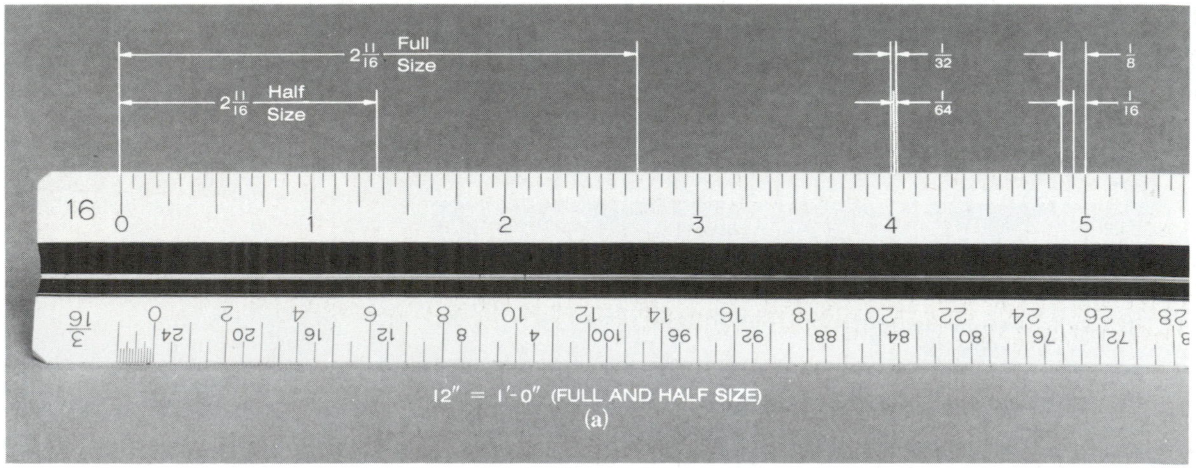

12″ = 1′-0″ (FULL AND HALF SIZE)
(a)

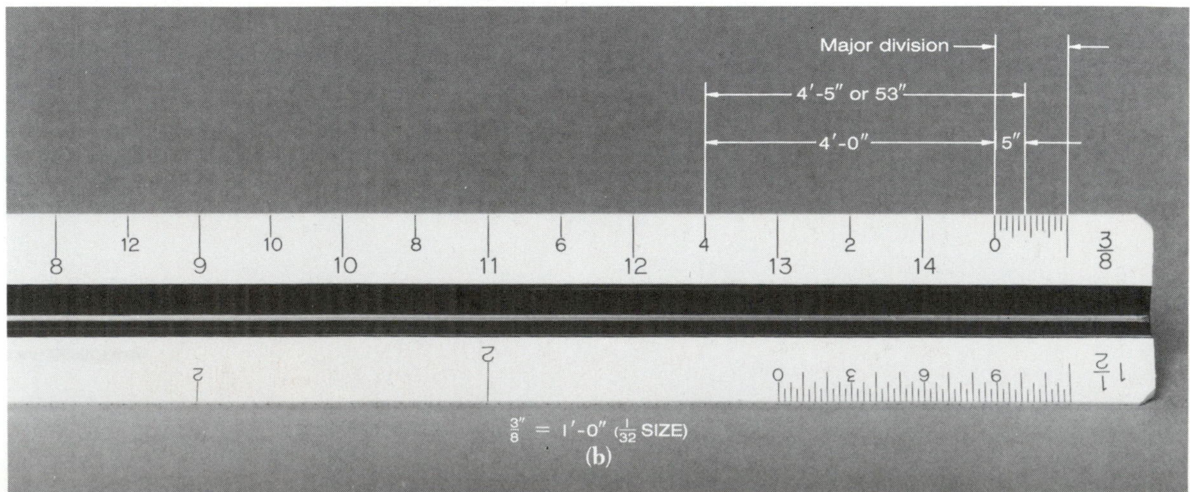

3/8″ = 1′-0″ (1/32 SIZE)
(b)

how the scales 3/8" = 1′-0", 1/2" = 1′-0", and 1" = 1′-0" are used. It is important to observe that, except for the 16 scale, the *major* divisions represent one foot, not one inch, and that the fully divided subdivisions represent inches and fractions of inches.

Metric Scales

Many American companies, particularly those engaged in international trade with other industrialized countries, are now using the millimeter as a common unit of length. The United States may soon join Great Britain in converting to the Systeme International d'Unites (International System of Units) metric system.

In the Systeme International (SI), the meter is the base unit of length. The millimeter (mm) equals one-thousandth of a meter (m), or 1 m = 1000 mm. One hundred centimeters (cm) equals one meter, or 1 m = 100 cm. One millimeter equals one-tenth of a centimeter, or 1 cm =

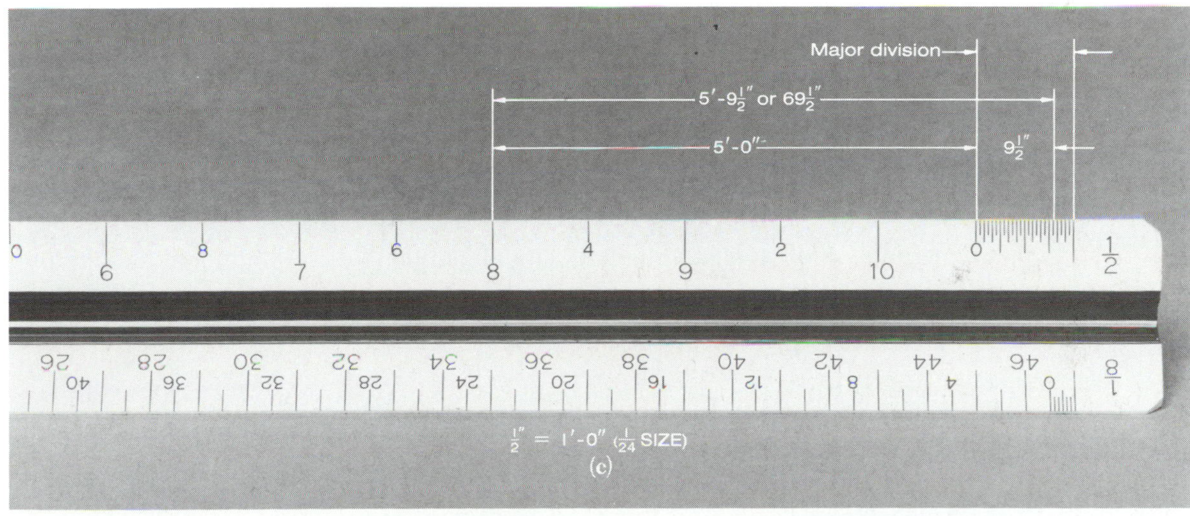

½" = 1′-0" (1/24 SIZE)
(c)

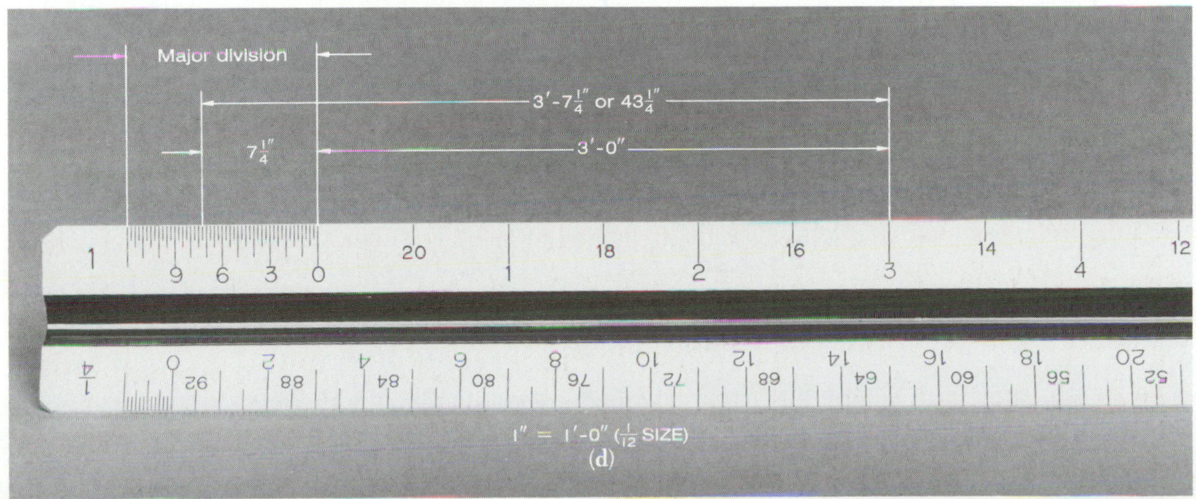

1" = 1′-0" (1/12 SIZE)
(d)

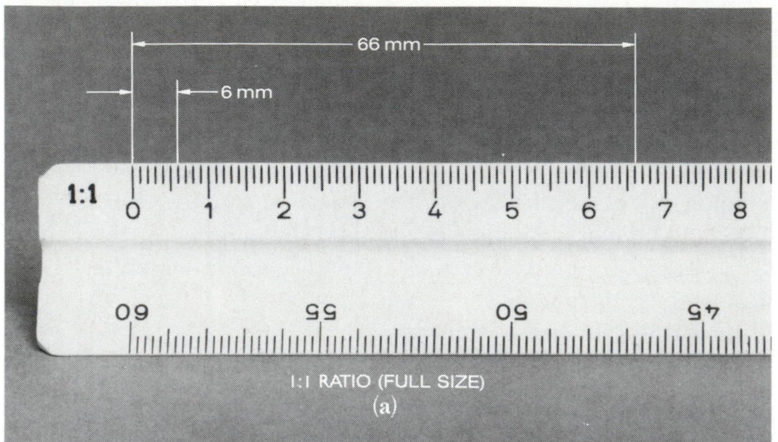

(a)

10 mm. Dimensions in centimeters, although occasionally found on drawings, are not standard units, and their use should be avoided. The millimeter is used almost exclusively for linear dimensions on product engineering drawings, even when their values lie outside the range of 0.1 mm to 1000 mm.

Figure 3-20 illustrates some standard metric scales. The

FIGURE 3-20 Metric reduction scales *(Section 3.9)*

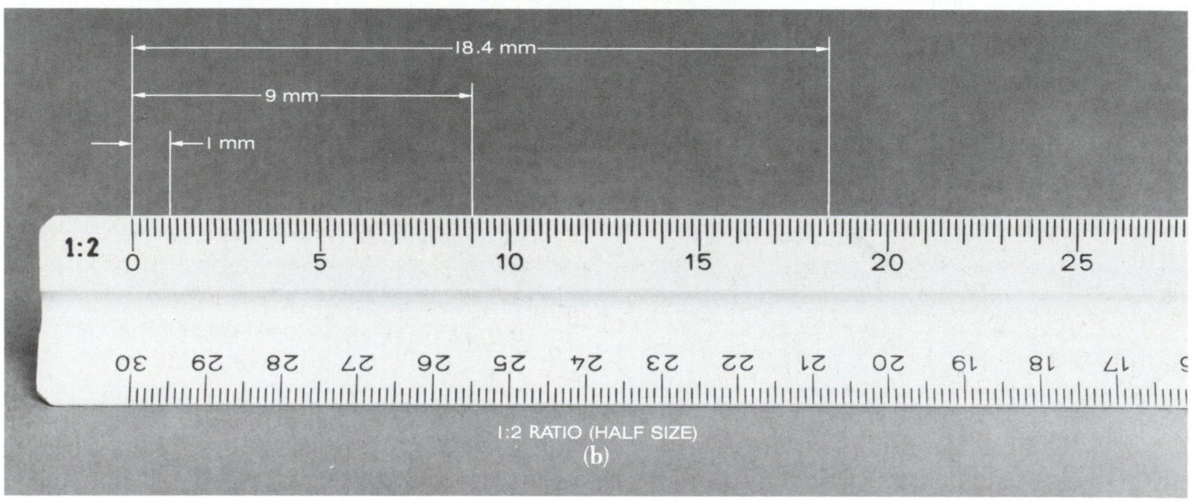

(b)

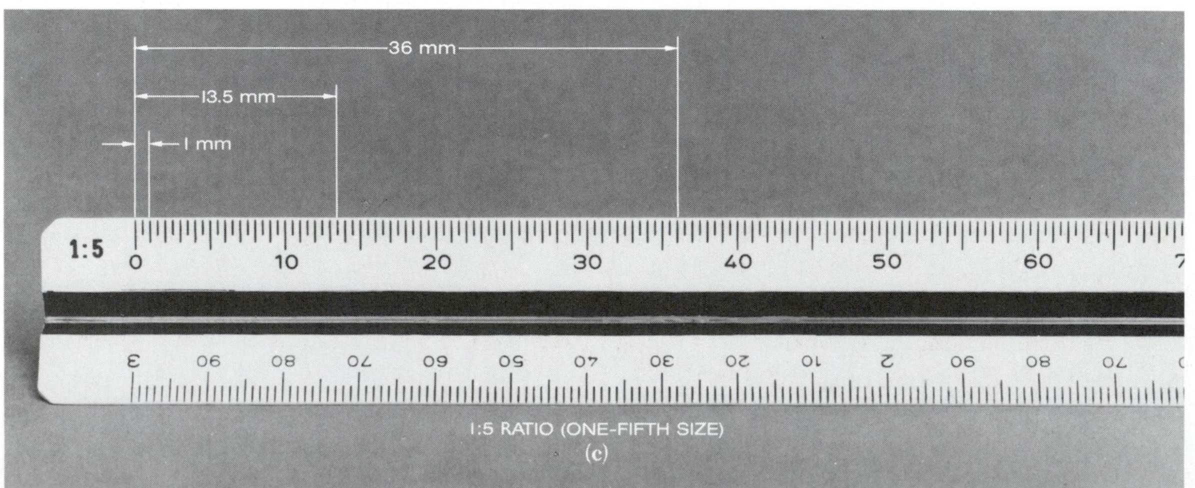

(c)

numerals on the 1:1 scale, shown in Figure 3-20(a), denote millimeters, and the distance between the smallest increments on the scale is one millimeter. The scale ratio is *full size*. The reduction scale ratio of 1:2, shown in Figure 3-20(b), is *half size* and indicates that a distance of 1 mm on the drawing equals 2 mm on the object. Similarly, a reduction scale ratio of 1:5, shown in Figure 3-20(c), indicates a drawing that is equivalent to *one-fifth* the size of the actual object. Larger scale reductions, ratios of 1:100 and 1:200, for example, are used on drawings of large structures or in civil engineering applications to lay out distances that are usually given in meters. The major divisions on the 1:100 instrument scale shown in Figure 3-21 are equal to centimeters. The scale factor of .01 indicates that each numbered unit length is .01 meters, or 1 centimeter. At a drawing scale of 1:1, or full size, the scale may be used to lay off a distance of 9.5 cm (or 9500 mm). At a drawing scale of 1:10000, the same distance would represent 95 m, because the drawing scale is 100 times the 1:100 scale ratio. In enlargement scale ratios of 2:1 or 5:1, for example, the first number in the ratio, 2 mm or 5 mm, represents the distance in millimeters on the drawing. The second number in the ratio equals 1 mm on the object.

Because the metric system is conveniently broken down into units of 10, the interchangeability of metric scales is much easier than for U.S. customary linear unit scales. A 1:1 scale ratio, for example, can be used to lay off a 1:100 reduction on a drawing, provided the proper conversions are made from one

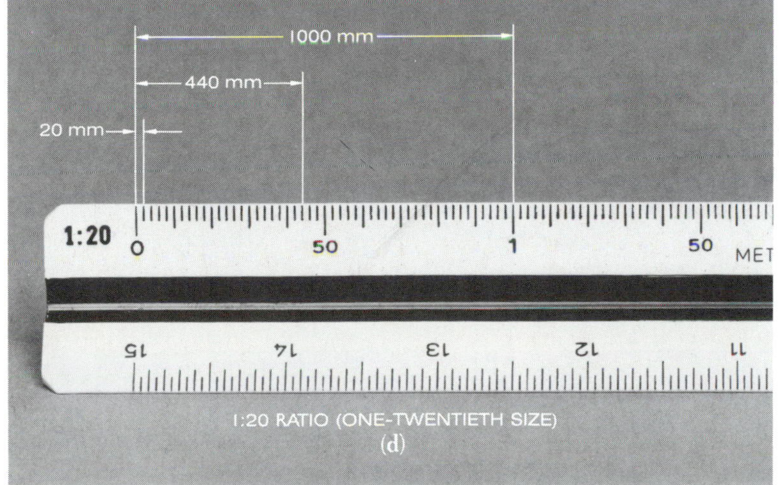

1:20 RATIO (ONE-TWENTIETH SIZE)
(d)

FIGURE 3-21 1:100 ratio metric scale ***(Section 3.9)***

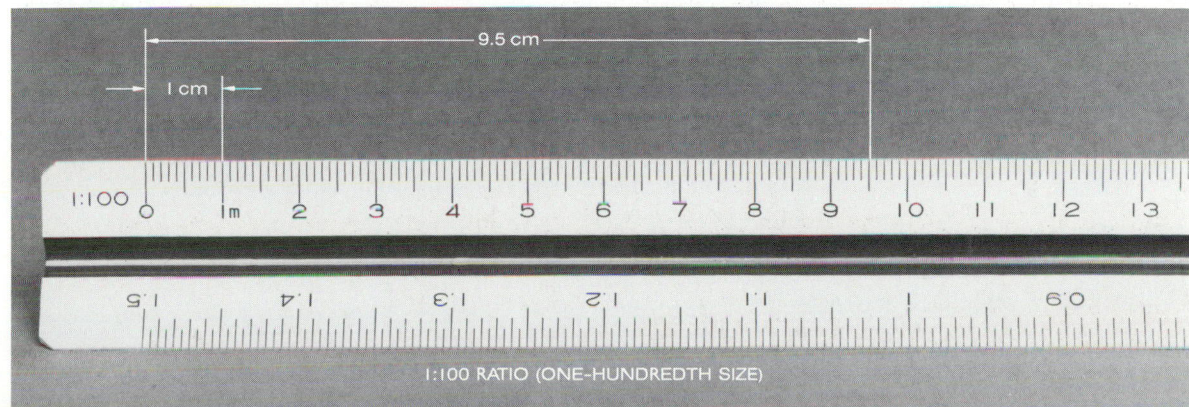

1:100 RATIO (ONE-HUNDREDTH SIZE)

ratio to another (multiplying or dividing by 10), as shown in Figure 3-22.

Figure 3-23 shows how a scale of 1:1 is used to set off both millimeter and centimeter values. Despite the relative ease in converting from one scale to another, most people purchase the desired standard metric scale in accordance with their particular needs. Metric drafting scales are available in various graduations of 1:1 to 1:100 and in lengths of 150 or 300 millimeters.

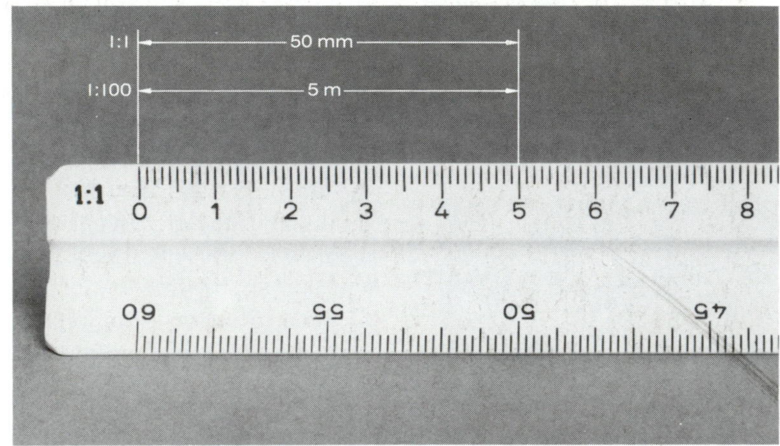

FIGURE 3-22 Use of a 1:1 scale for a 1:100 reduction ***(Section 3.9)***

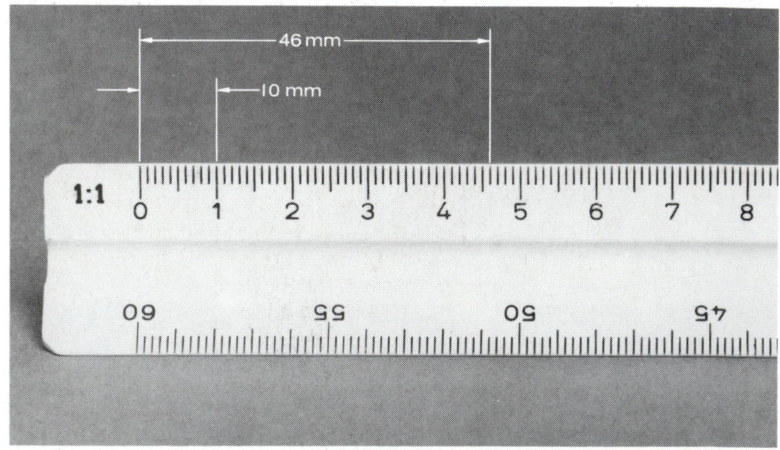

FIGURE 3-23 Use of a 1:1 scale for millimeter and centimeter values (1 cm = 10 mm) ***(Section 3.9)***

CHECKPOINT 4
USING VARIOUS DRAWING SCALES

Using Figure 3-24, measure and record the distances AB, AC, and AD on each of the given lines 1 through 9. Use the following scales:

- Line 1: Civil engineer #10 (1″ = 1″)
- Line 2: Civil engineer #20 (1″ = 2″)
- Line 3: Architect ¼″ = 1′-0″
- Line 4: Architect ⅜″ = 1′-0″
- Line 5: Architect 1″ = 1′-0″
- Line 6: Mechanical engineer 1″ = 1″
- Line 7: Mechanical engineer ¼″ = 1″
- Line 8: Metric 1 mm = 1 mm (1:1)
- Line 9: Metric 1 mm = 5 mm (1:5)

FIGURE 3-24 Checkpoint 4 problem

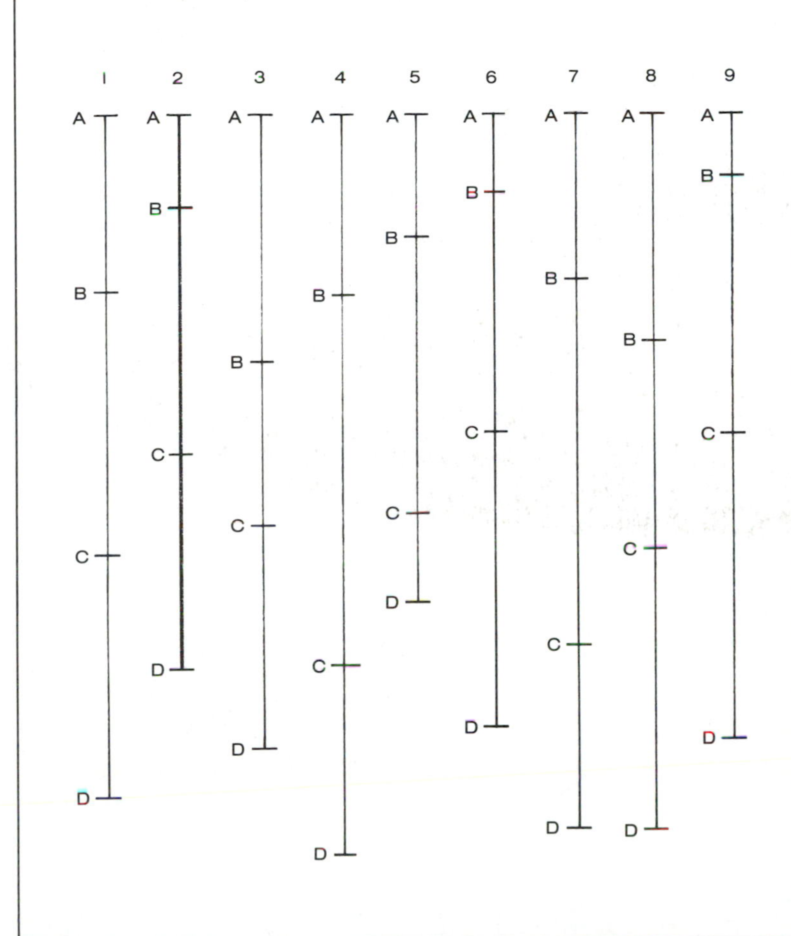

3.10 THE PROTRACTOR

A protractor is used for laying out or measuring angles. Most protractors are graduated in ½° or 1° increments and are numbered at 10° intervals through 180°. Figure 3-25(a) illustrates a semicircular 180° protractor. A circular protractor is shown in Figure 3-25(b). You must be careful to place the actual center of the protractor scale exactly over the vertex (intersection) of

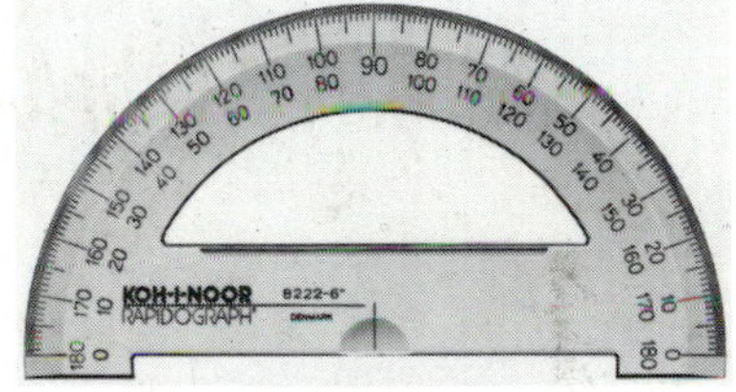

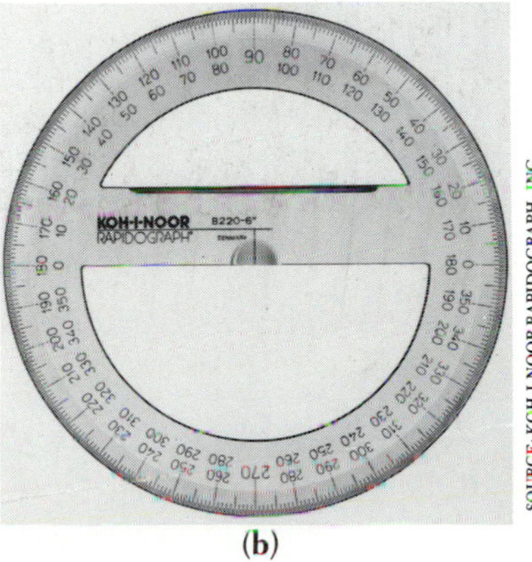

(b)

SOURCE: KOH-I-NOOR RAPIDOGRAPH, INC.

FIGURE 3-25 Protractors: semicircular and circular *(Section 3.10)*

the two sides of the angle you are measuring. You can also use the adjustable triangle, shown in Figure 3-26, to lay out angles.

SOURCE: KOH-I-NOOR RAPIDOGRAPH, INC.

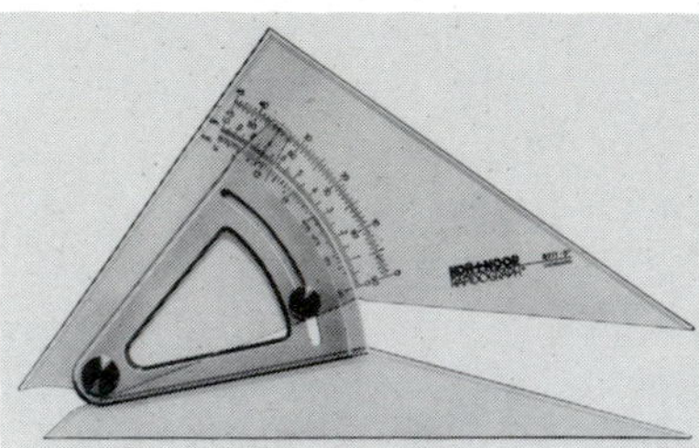

FIGURE 3-26 An adjustable triangle ***(Section 3.10)***

CHECKPOINT 5
LAYING OUT ANGLES

Reproduce the exercises shown in Figure 3-27 with a 5H grade pencil but do not copy the dimensions and notes. Use a plain sheet of 8½ × 11 inch paper. Draw the diagonal of the lower exercise. Using a protractor, work around center point O and divide the space into the required number of angles. Draw the angular lines as indicated. Draw the diagonal of the upper exercise and lay out the various angles with a protractor. Draw the angular lines as indicated.

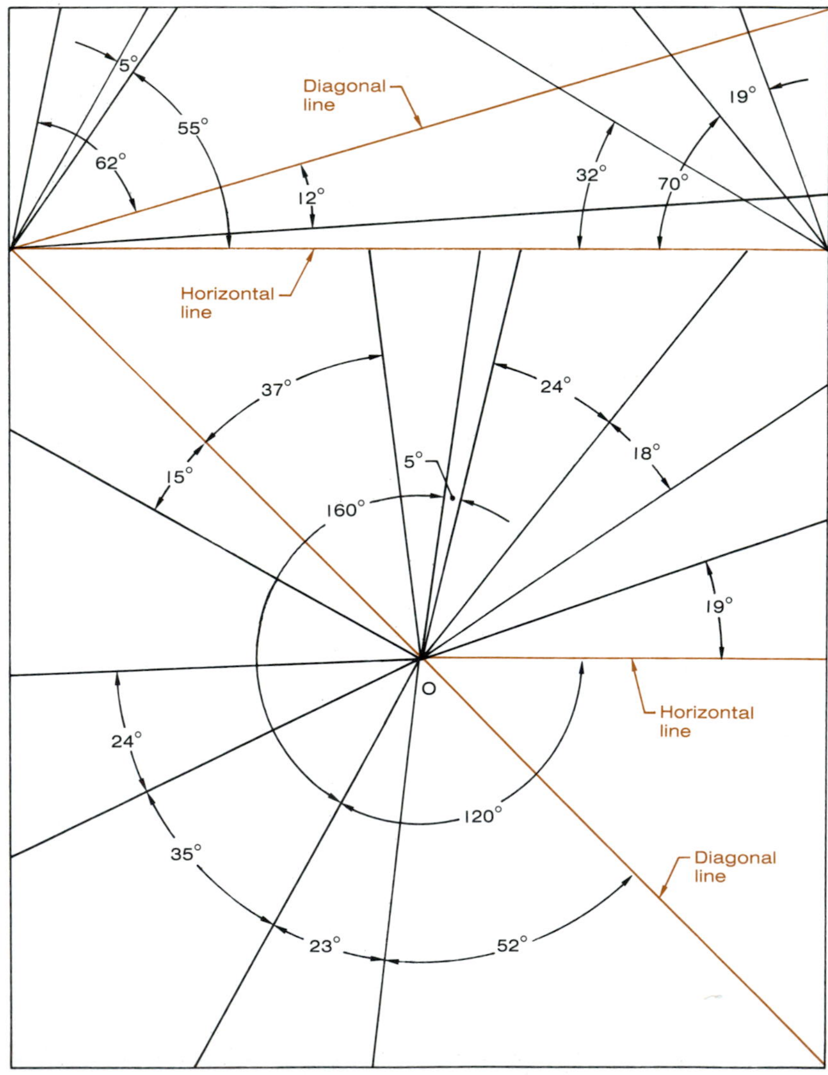

FIGURE 3-27 Checkpoint 5 problem

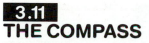

3.11 THE COMPASS

The compass is used for drawing arcs and circles. Several basic styles are available. Figure 3-28 shows the center-wheel compass, perhaps the most typical in current use, particularly for drawing circles and arcs up to 10 inches in diameter.

Before using a compass, you must sharpen the lead and adjust the steel shoulder point needle. Using a sandpaper pad or a fine file, sharpen the pencil point, as in Figure 3-29(a), with a long, flat bevel cut along the outside of the lead. Extend the lead from the compass leg so that at least 3/8 inch is exposed and adjust the steel point to about the same length, as illustrated in Figure 3-29(b). For good results use a compass lead that is a grade or two softer than the pencil you use to darken the final lines on the figure. A line drawn with a pencil results in a bolder line because it is generally easier to exert more pressure on a pencil than on a compass leg.

Before drawing an arc or a circle, lay off the desired radius distance with a scale on a thin line drawn on a scrap piece of paper, on the border line, or on some line in an inconspicuous place on the drawing. Place the compass needle at one point on the line and take a trial swing adjusting the compass until the setting corresponds to the position of the second point. When using the compass, hold it between the thumb and the forefinger. Begin by carefully placing the shoulder point needle at the precise intersection of the two center lines that delineate the position of the desired arc or circle. Bring the pencil point down on the paper and rotate the compass with a twist of the thumb and the finger. As the arc is being drawn, incline the compass slightly in the direction of travel. For darker lines, go over the line once or twice more, gradually increasing the pressure upon the compass leg with the pencil point.

FIGURE 3-28 A centerwheel compass *(Section 3.11)*

FIGURE 3-29 Sharpening and adjusting a compass *(Section 3.11)*

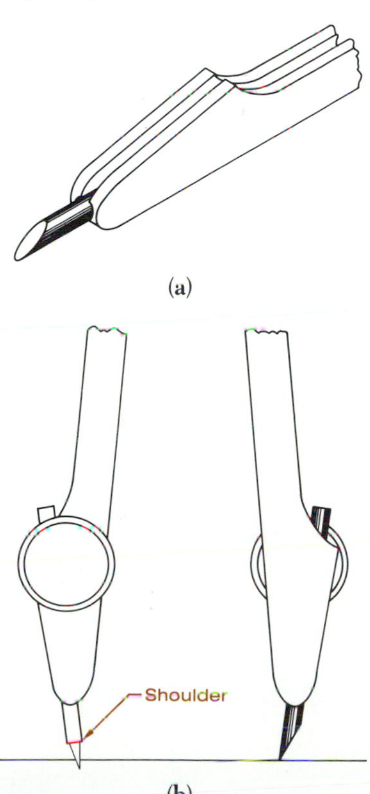

CHECKPOINT 6
USING DRAFTING INSTRUMENTS

Copy the objects shown in Figures 3-30(a) and 3-30(b) using a 5H grade pencil but do not copy the dimensions and notes. Use a plain sheet of 8½ × 11 inch paper. Go over all the construction lines with a 2H pencil to obtain bold, dense lines.

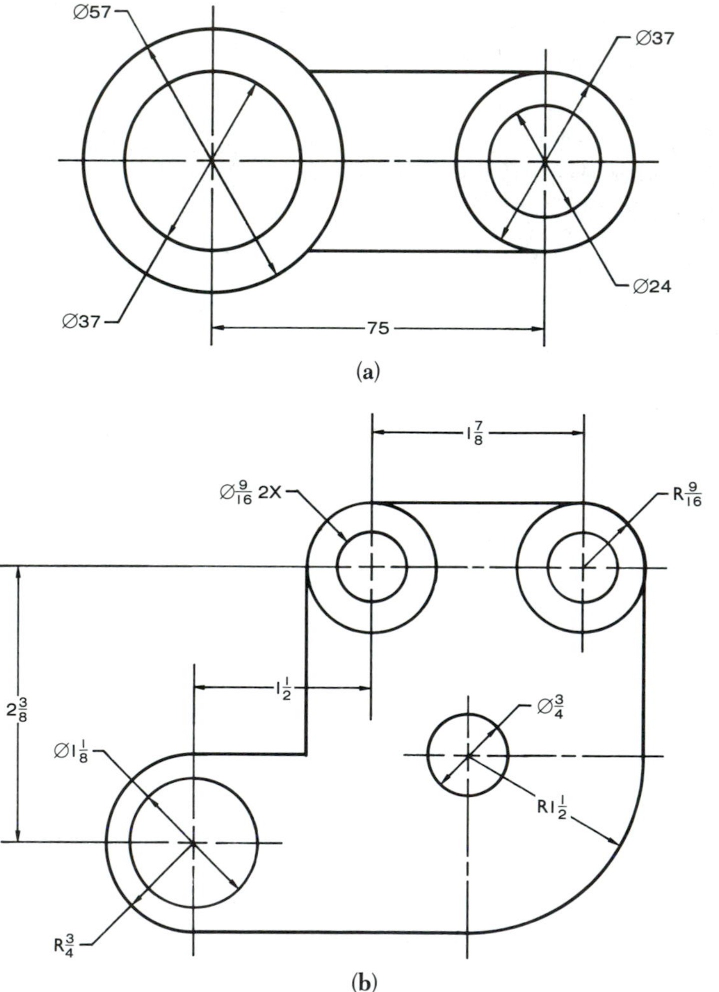

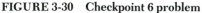

FIGURE 3-30 Checkpoint 6 problem

3.12 TEMPLATES

Many drafters and engineers find it easier and faster to use plastic templates, which are available for specialized drafting needs. Figure 3-31 illustrates a typical circle template, which has accurately cut openings that are used as guides for quickly drawing small circles, fillets, and rounds. In many cases templates are used for all common circle sizes and for all circles smaller than 1 inch in diameter. Templates are available individually or in sets that contain all common sizes for drawing arcs and circles, ellipses, boltheads and nuts, thread forms, hexagons, squares, triangles, architectural symbols, pipe and electronic symbols, and a wide variety of other repetitive features.

3.13 IRREGULAR CURVES

Many kinds and sizes of irregular curves exist. One typical style is shown in Figure 3-32. Most professional drafters find that two or three different sizes and styles are sufficient for practically all drafting applications. Irregular curves are used to draw curves that are not true arcs. The procedure for using an irregular curve, sometimes called "curve fitting" (Figure 3-33), begins with a series of plotted points on the drawing. Start by sketching a light, freehand trial line through the points in such a way that the general trend or direction of the desired curve line is established. Then draw a mechanical line over the freehand line in a series of steps by aligning the edge of the irregular curve with *at least four consecutive points.* Select a portion on the edge of the irregular curve that will match the three consecutive points and one or two points beyond the segment to be drawn. Always go back to a previously drawn point to assure a smooth curve. In this way, as you shift or overlap the irregular curve from one position to the next, successive tangent arcs will be smooth and continuous with no abrupt changes in curvature. The proper use of the irregular curve as it is fitted to the sketched line requires considerable skill, patience, and good visual judgment. You must take into account the fact that a line is characterized by *curvature* and also the *rate of change* of curvature.

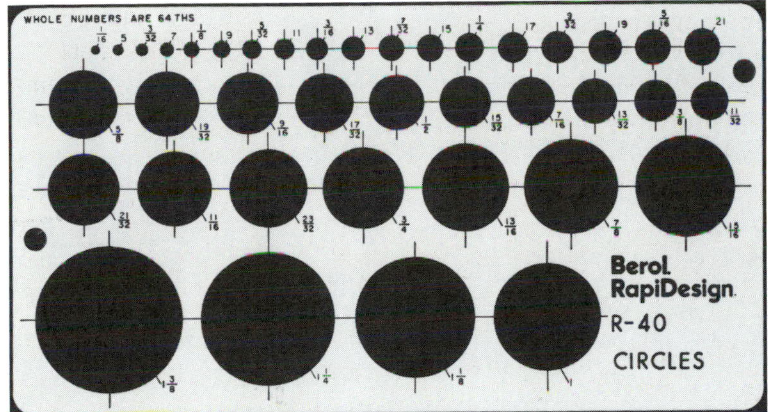

SOURCE: BEROL, INC.

FIGURE 3-31 A circle template *(Section 3.12)*

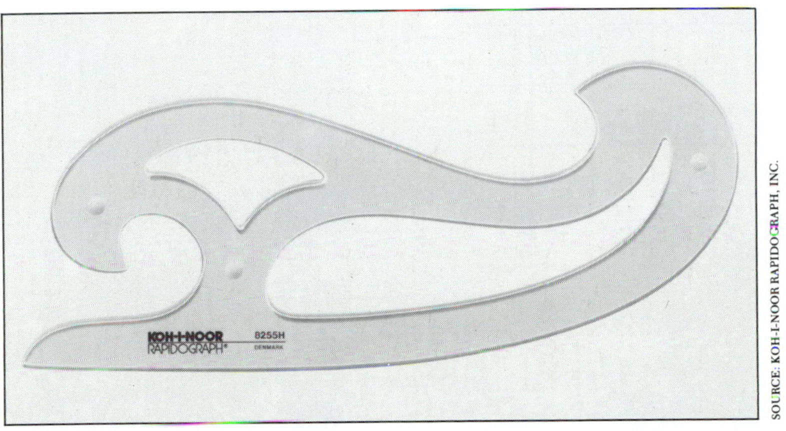

SOURCE: KOH-I-NOOR RAPIDOGRAPH, INC.

FIGURE 3-32 A typical irregular curve *(Section 3.13)*

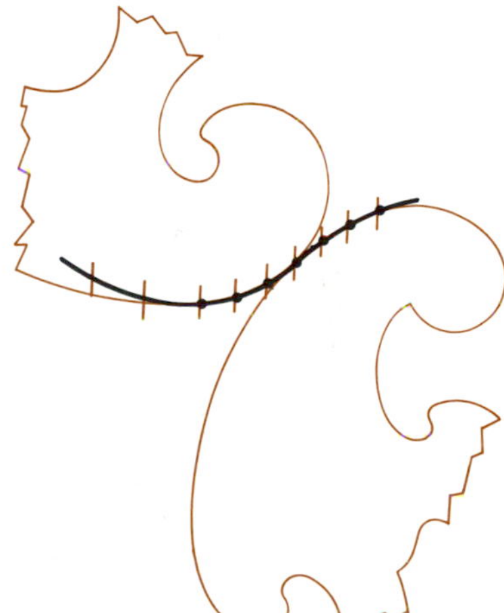

FIGURE 3-33 Using an irregular curve *(Section 3.13)*

CHECKPOINT 7
USING IRREGULAR CURVES

1. Use a plain sheet of 8½ × 11 inch paper.
2. Referring to Figure 3-34, lay out the graph full scale, plot the points, and draw a smooth curve. Copy all letters and figures. When drawing the final lines, use a 2H grade pencil for the curve and a 5H grade for all other lines.

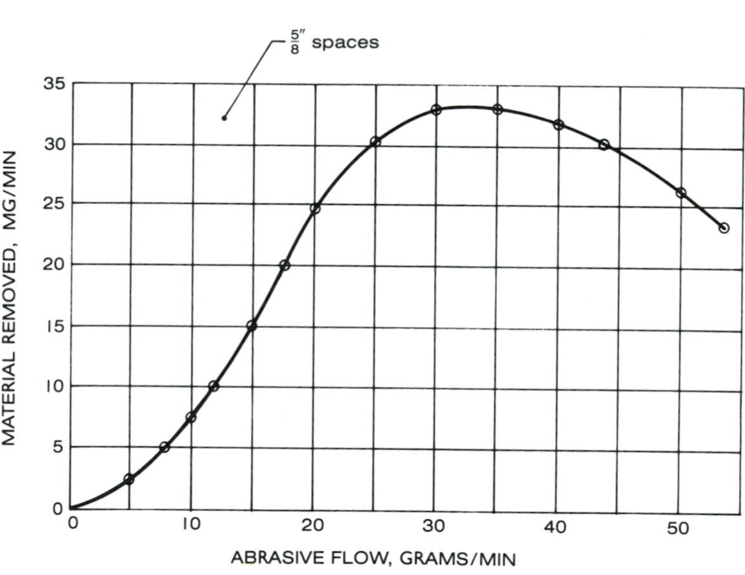

FIGURE 3-34 Checkpoint 7 problem

3.14 DIVIDERS

Two styles of dividers are illustrated in Figure 3-35. These instruments are used to transfer distances and to divide straight and curved lines into a desired number of equal spaces. Dividers are particularly useful for transferring dimensions from view to view during the process of constructing multiviews of an object. The dividers resemble a compass except that both legs contain a finely tapered steel needle point. The replaceable steel points must be maintained sharp and straight and must extend an equal distance from each divider leg.

As an example in using the dividers, assume a straight line XY is to be divided into five

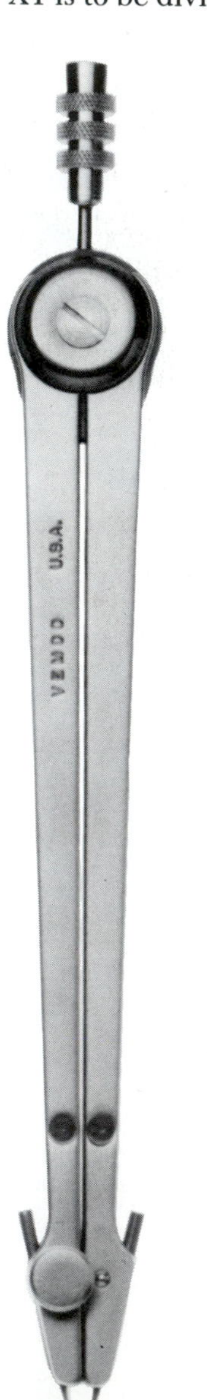

(a)

(b)

FIGURE 3-35 Dividers *(Section 3.14)*

SOURCE: (a) VEMCO CORPORATION; (b) KOH-I-NOOR RAPIDOGRAPH, INC.

equal parts. A trial and error procedure, shown in Figure 3-36, may be followed. Begin by setting the dividers until the distance between the divider points corresponds to about one-fifth of the total distance XY. Step off the spacings and make adjustments to the divider setting until the correct divisions are established. As you step off the distances, rotate the dividers alternately in an opposite direction on either side of the line. (See Section 4.3 for additional instructions on dividing a line.)

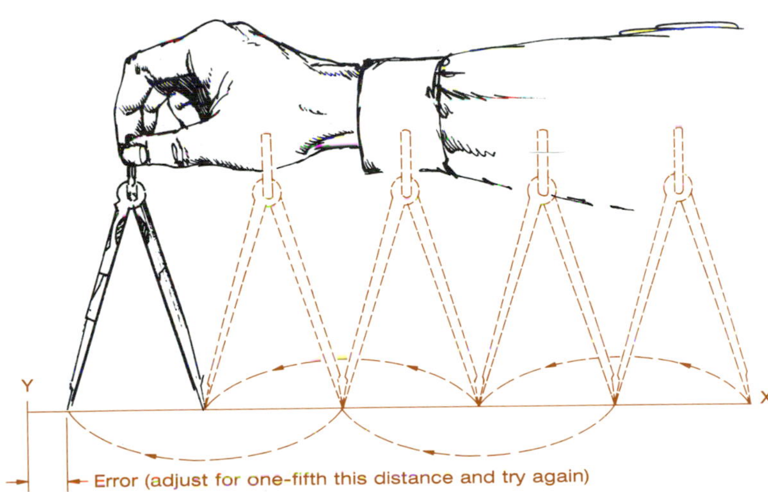

FIGURE 3-36 Using the dividers to divide a line into five equal parts ***(Section 3.14)***

3.15 ERASING SHIELD

The erasing shield is a very useful tool for selectively removing unwanted lines, numerals, and letters in crowded areas on the drawing while protecting adjacent lines. In addition, the use of an erasing shield prevents wrinkling the drawing. The thin metal shield is positioned as shown in Figure 3-37, with an opening aligned over the unwanted line, and is held firmly to prevent slipping.

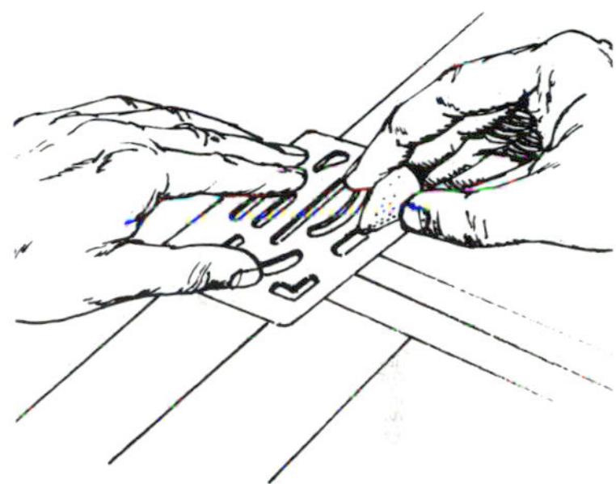

FIGURE 3-37 Using an erasing shield ***(Section 3.15)***

PROBLEMS

Use of Instruments

P3.1 Measure the distances AB, AC, AD, etc., on the given lines 1 through 16 using the following scales. Develop a neat format on an 8½ × 11 inch sheet for recording your measurements.

Line 1: Civil engineer scale #10
Line 2: Civil engineer scale #20
Line 3: Civil engineer scale #30
Line 4: Architect scale #16
Line 5: Architect scale ¼″ = 1′-0″
Line 6: Architect scale 1″ = 1′-0″
Line 7: Architect scale ⅜″ = 1′-0″
Line 8: Architect scale ½″ = 1′-0″
Line 9: Architect scale ⅛″ = 1′-0″
Line 10: Mechanical engineer scale 1″ = 1″
Line 11: Mechanical engineer scale ¼″ = 1″
Line 12: Mechanical engineer scale ½″ = 1″
Line 13: Metric scale 1:1
Line 14: Metric scale 1:2
Line 15: Metric scale 1:5
Line 16: Metric scale 1:20

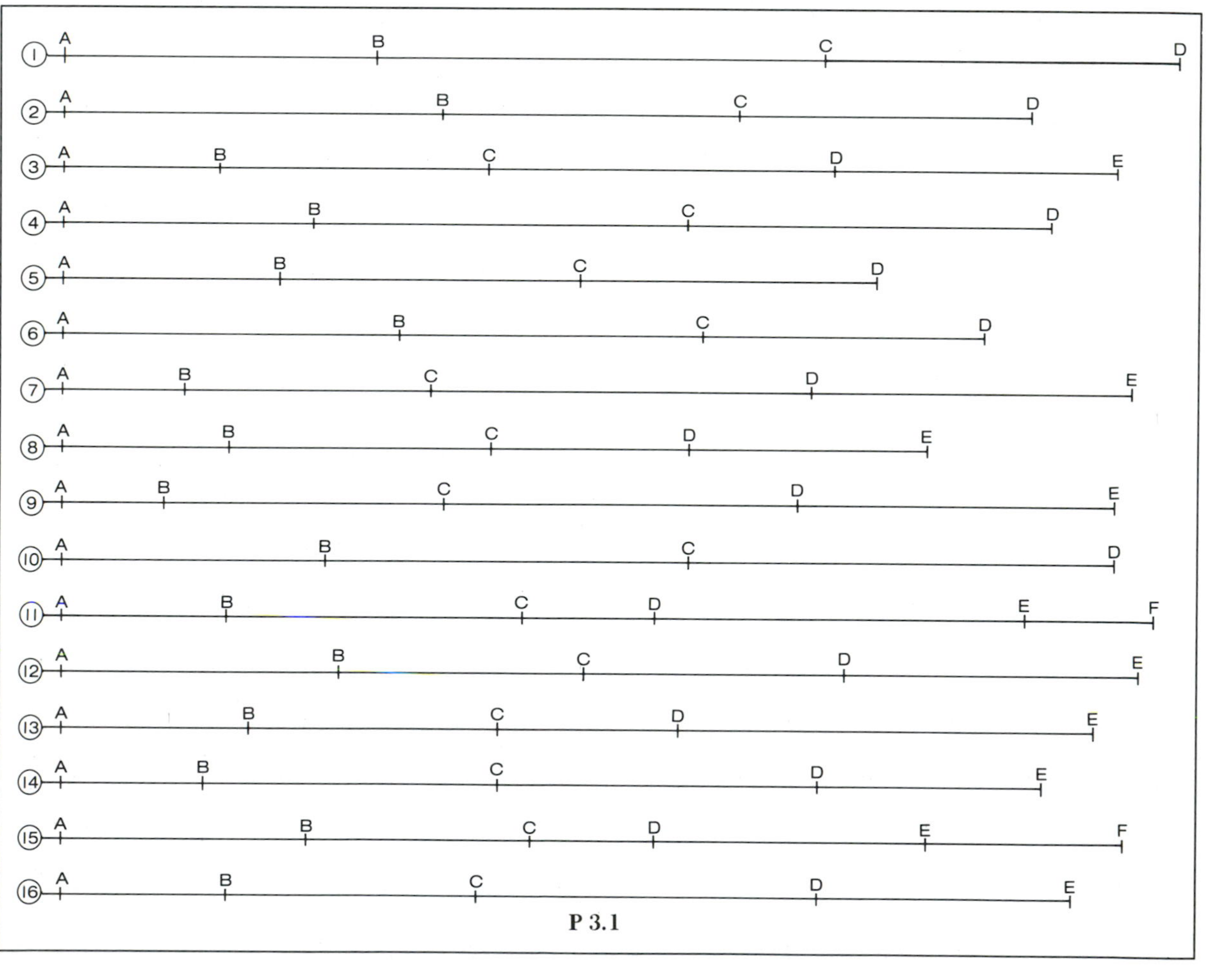

P 3.1

The following problems will fit on an 8½ × 11 inch sheet. Use layout A, shown on the inside of the front cover. Letter the required information in the title block. Begin each problem by carefully estimating the amount of space that is needed to balance the figure placed neatly in a vertical or a horizontal position on the drawing sheet. The required drawing scale is given for each problem. Unless otherwise specified, all dimensions given on the problems are in inches. Use a 5H or 6H grade pencil for light construction lines and center lines, a 2H or 3H grade pencil for drawing the final lines, and a H or F grade for lettering and figures. *Do not copy the figures given in the book, because they are not drawn to scale.* It is unnecessary to erase light construction lines. Do not copy the dimensions and notes.

Symbols for specifying geometric characteristics and other dimensional requirements on drawings are used on some of the following problems. These practices are in conformance with the symbols listed in ANSI Y14.2M (1982). Although a detailed discussion of symbology will not be given in this text until Chapter 10, Size Description, the symbols are used on selected problems in this chapter to promote familiarity with the recent changes in the standards. Below is a list of symbols used and the meaning of each:

Symbol (All linear values in mm)	***Meaning***
2 ×	Two times, two places, two required, or two holes
⌀40	40 diameter
R5	5 radius
□12	12 square
↧ 20	20 deep
⌴⌀14	14 diameter counterbore or spotface
⌵ ⌀10	10 diameter countersink

P3.2 Scale: Full (A symmetrical object)
P3.3 Scale: One-half (A symmetrical object)

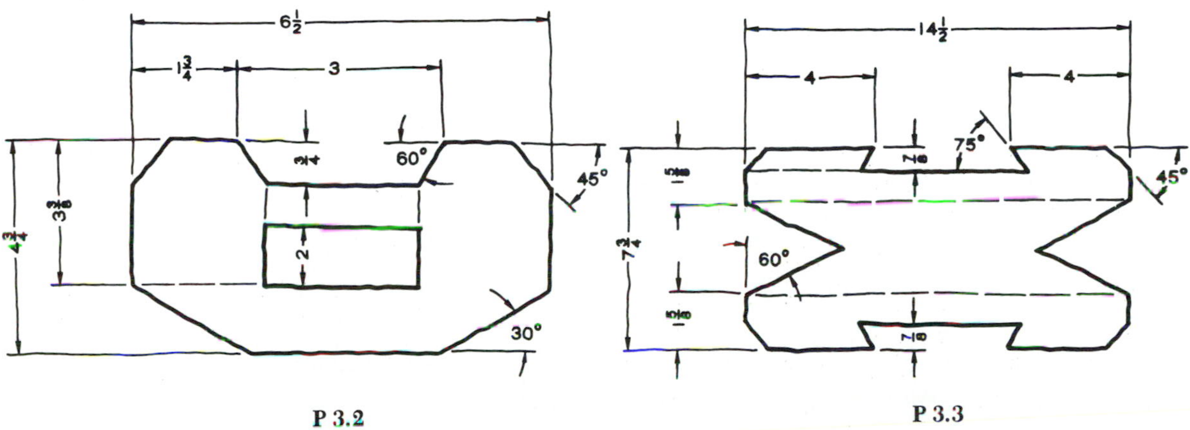

P 3.2

P 3.3

P3.4 Scale: ⅛″ = 1′-0″
P3.5 Scale: 1:1 (All diagonal lines are perpendicular to BC; all vertical lines are perpendicular to AC)

P3.6 Scale: Full. MNOP is a parallelogram. Line MP = 5.35″ and NP = 6.75″.
P3.7 Scale: ⅜″ = 1′-0″
P3.8 Scale: 1″ = 4′-0″

P3.9 Scale: ½″ = 1′-0″. All structural members are either perpendicular or at 45° to one another.

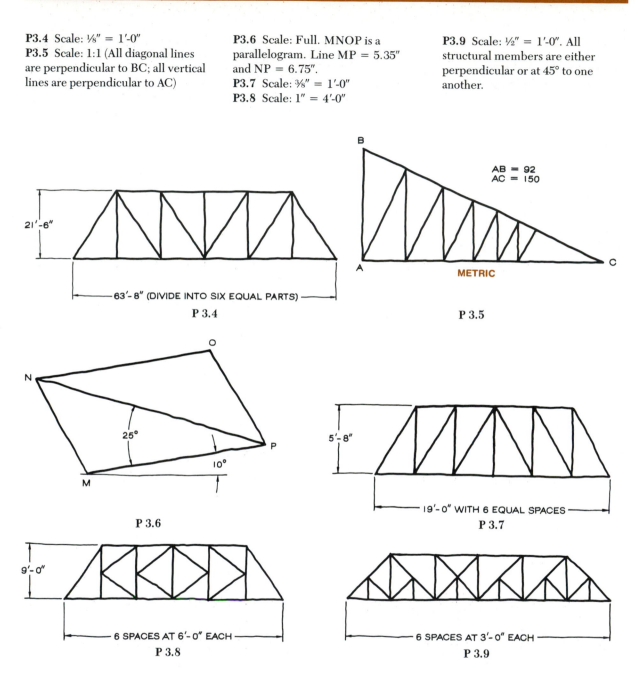

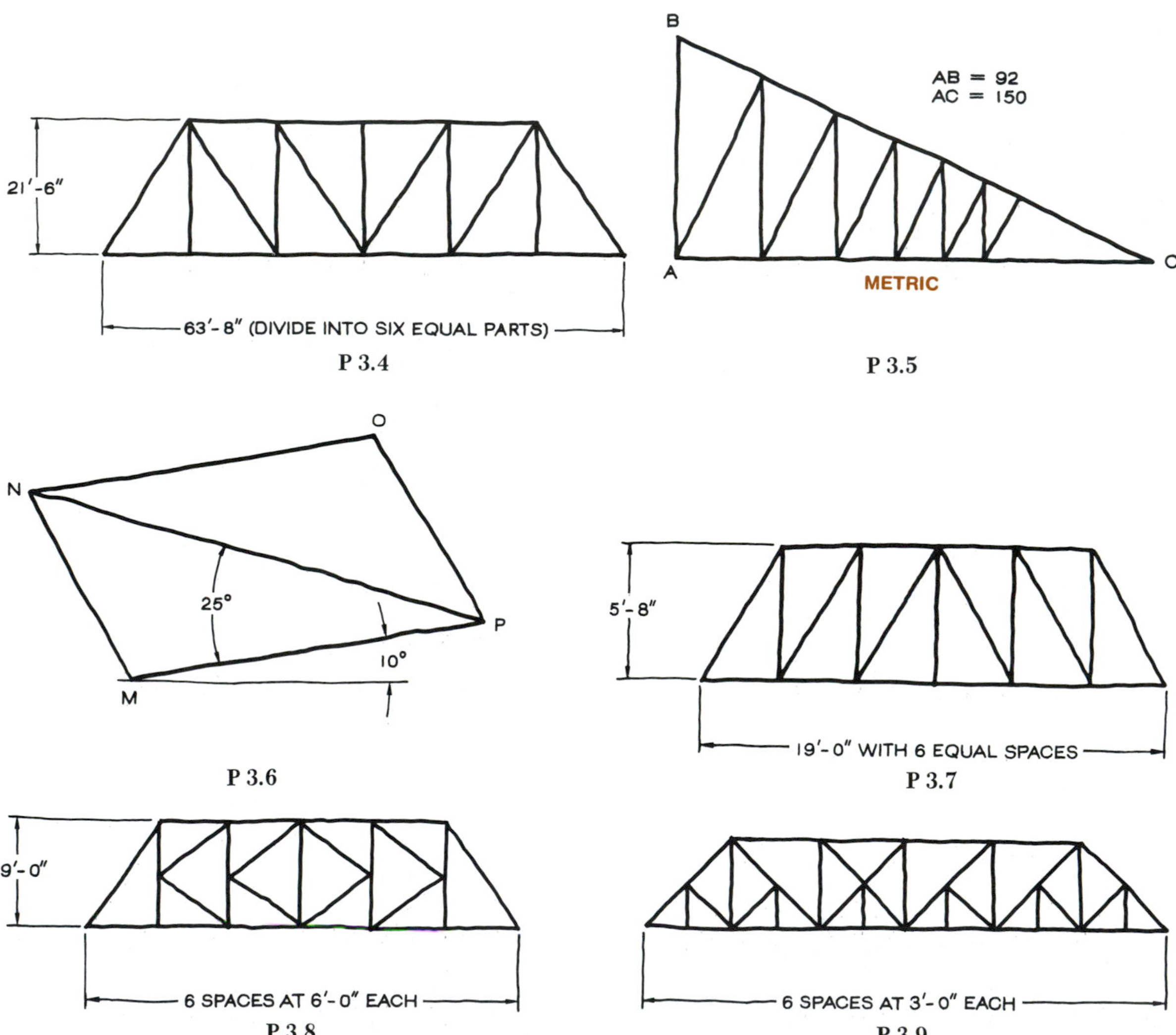

P 3.4

P 3.5

P 3.6

P 3.7

P 3.8

P 3.9

P3.10 Scale: ½″ = 1′-0″. Line ACD is a straight line. Angle ACB = 90°. AB = 6′-3″, BC = 4′-11½″, BD = 7′-4″, BE = 9′-6″, and DE = 6′-6″. Record the value on the drawing for angle X and line AE.

Working entirely from the following descriptions for P3.11 to P3.14, draw the figures full scale starting at a point A.

P3.11: 1″ downward to the left at 30° to the horizontal; 1½″ vertically downward; 3 7/16″ downward to the right at 15° to the horizontal; 1 11/16″ upward perpendicular to the preceding line; 19/32″ downward to the right at 60° to the preceding line; 1 44/64″ upward to the right at 30° to the vertical; a line to starting point A.

P3.12: Side AB extends 3¾″ upward to the left at 60° to the vertical and side BC 2″ extends upward to the right at 20° to the horizontal. Side AE extends upward to the right 4″ at 60° to the vertical. Side DE extends upward to the left 3½″ at 25° to the horizontal. Complete the polygon by connecting C and D.

P3.13 (Metric): Side AB is 95 long and extends upward to the right making an angle of 15° with the horizontal. Side BC is 108 long, extends from B, and makes an angle of 15° to the left of a vertical line through B. Side AE is 76 long and extends upward to the right from A making an angle of 60° with the horizontal. Side ED is 38 long and extends upward to the left from E making an angle of 30° with the horizontal. Connect points D and C to complete the polygon.

P3.14: Side AE extends 5″ upward to the right at 45° to the horizontal. Side ED extends upward to the left 2″, perpendicular to the preceding line. Side DC extends down 2½″ vertically. Side CB extends horizontally to the left 4½″. Complete the polygon by connecting B to A.

P3.15 Scale: Metric 1:1

P3.16 Scale: 2″ = 1″. (A cross-sectional view of an American National screw thread form.)

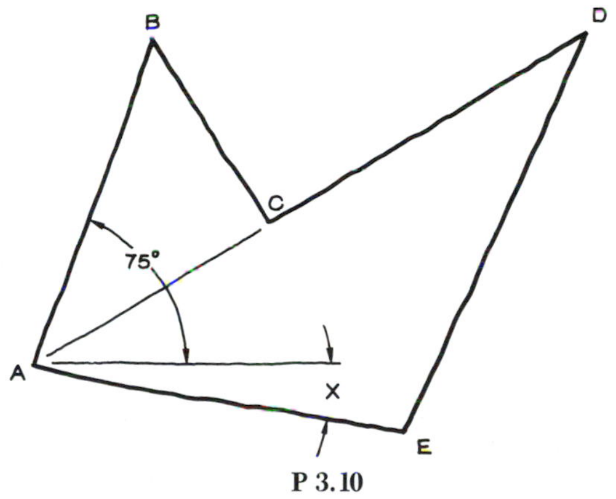

P 3.10

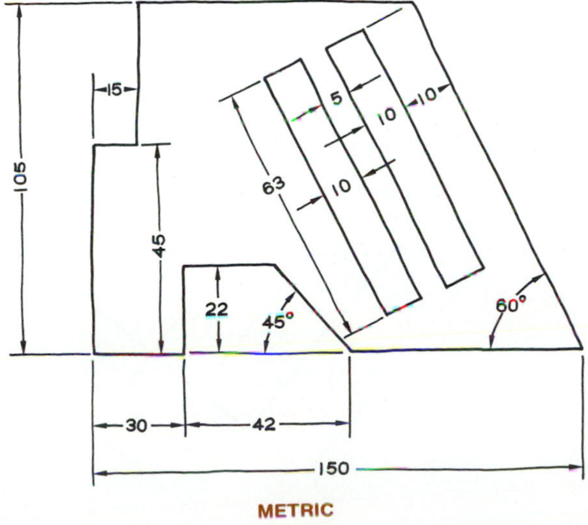

METRIC
P 3.15

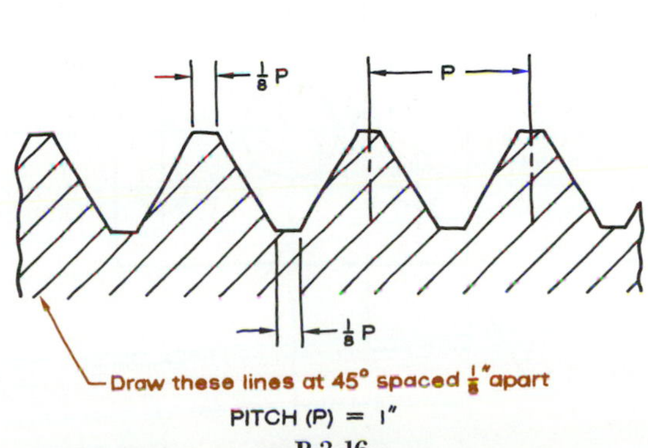

P 3.16

P3.17 Scale: Full (inches). (A cross-sectional view of a buttress screw thread form.)
P3.18 Scale: 1:1
P3.19 Scale: 10:1
P3.20 Scale: 1:2
P3.21 Scale: 10:1
P3.22 Scale: ¼″ = 1′-0″
P3.23 Scale: ¼″ = 1″
P3.24 Scale: ½″ = 1″

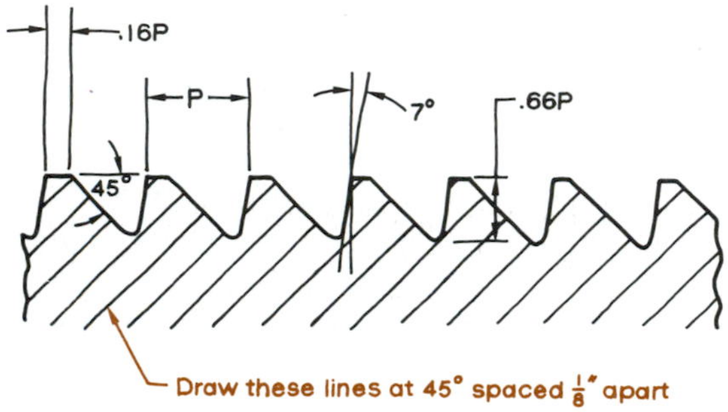

Draw these lines at 45° spaced $\frac{1}{8}$″ apart

PITCH (P) = 1.50″

P 3.17

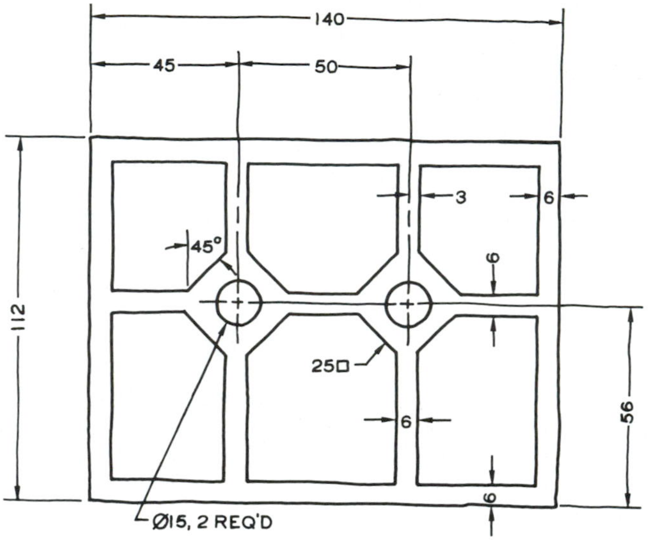

METRIC

P 3.18

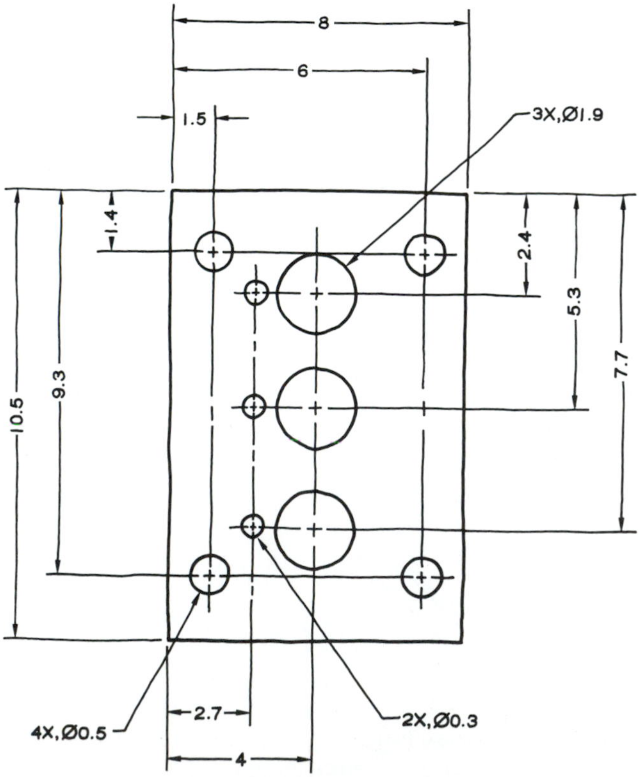

METRIC

P 3.19

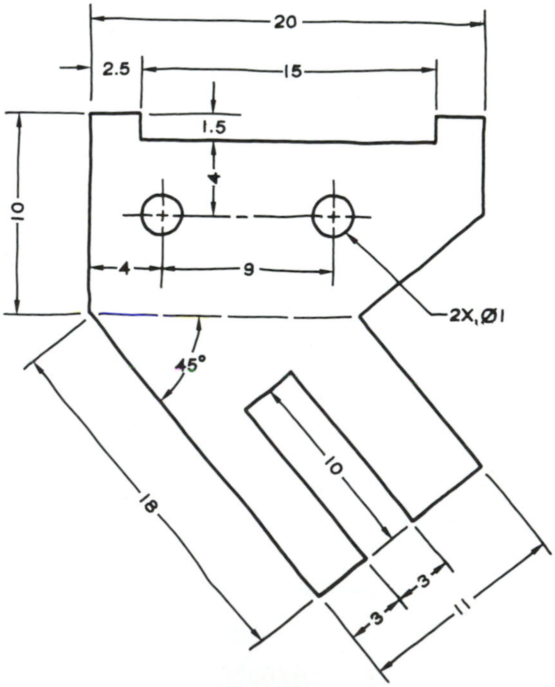

METRIC

P 3.20

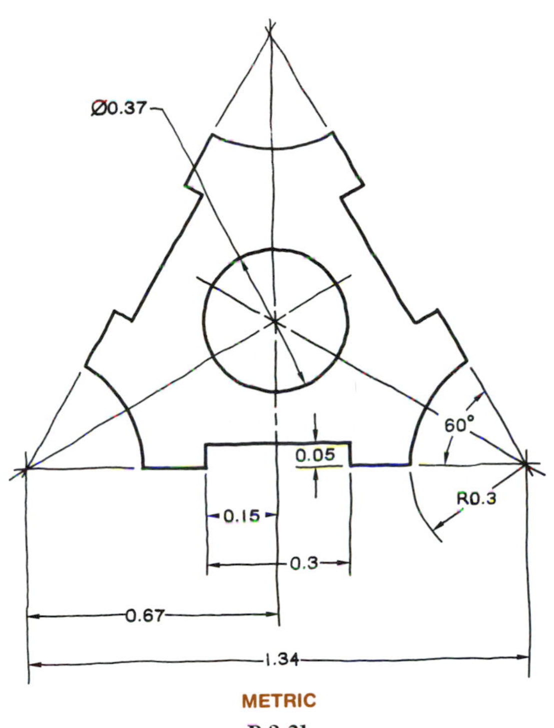

METRIC

P 3.21

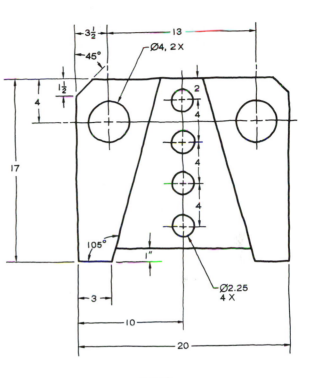

$3\frac{1}{2}$
13
Ø4, 2 X
45°
$1\frac{1}{2}$
4
2
4
4
4
17
105°
1″
Ø2.25
4 X
3
10
20

P 3.22

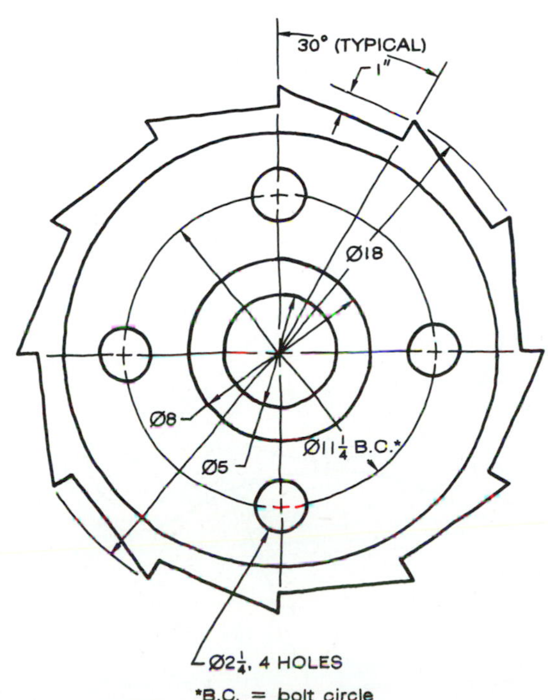

P 3.23

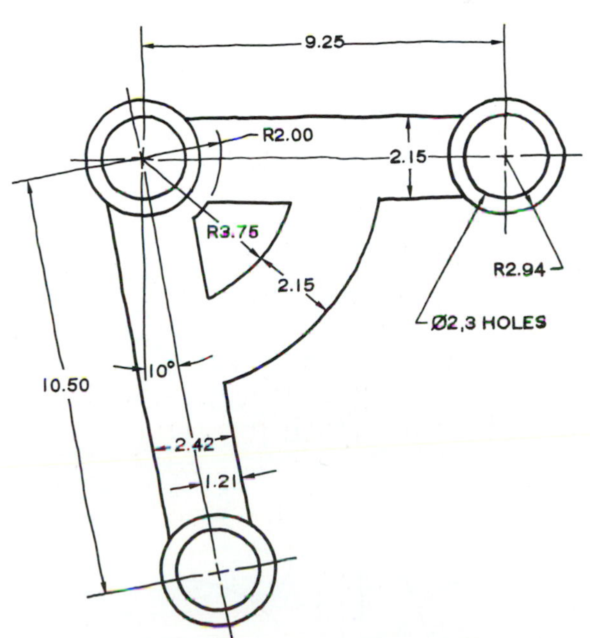

P 3.24

P3.25 Scale: Full (inches)
P3.26 Scale: Full (inches)
P3.27 Scale: ½″ = 1″
P3.28 Scale: Full (inches)
P3.29 Scale: 1:1
P3.30 Scale: 1:1
P3.31 Scale: Full (inches)
P3.32 Scale: Full (inches)

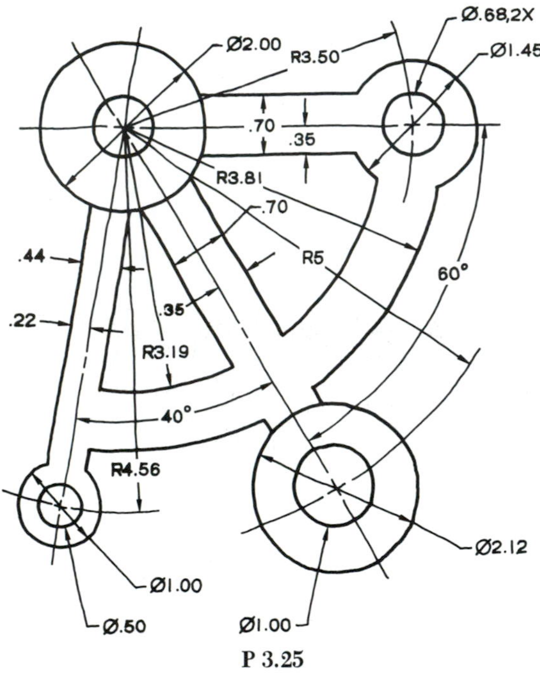

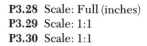

P 3.25

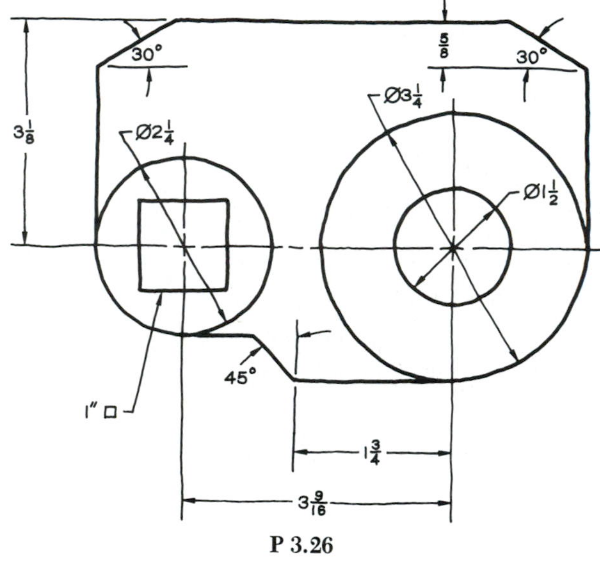

P 3.26

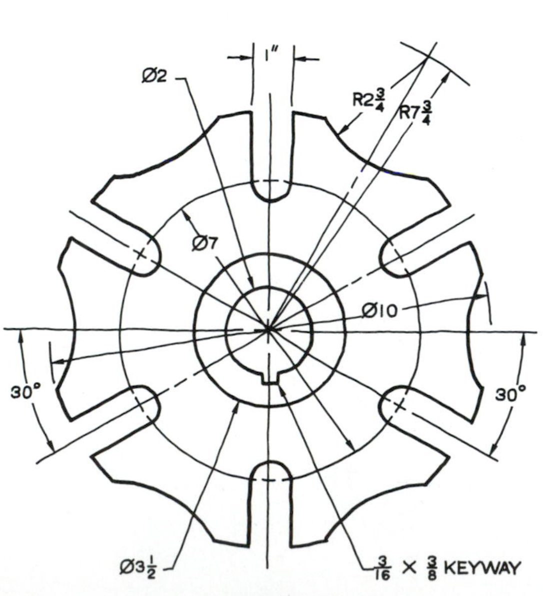

P 3.27

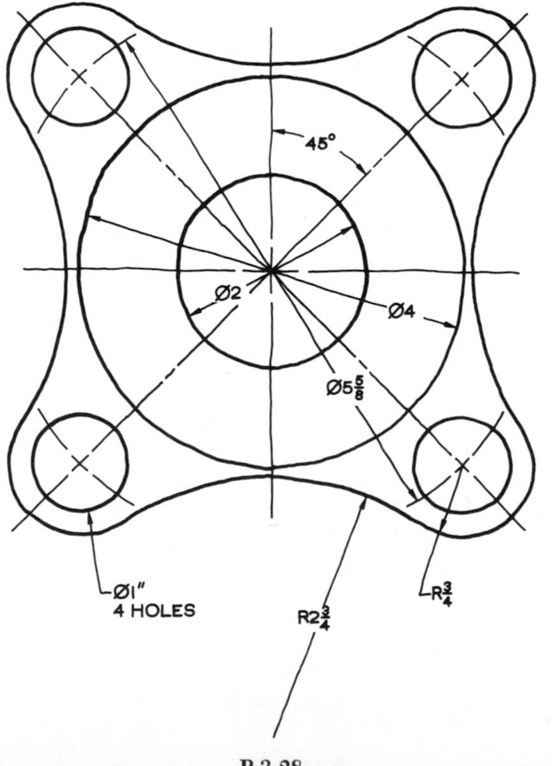

P 3.28

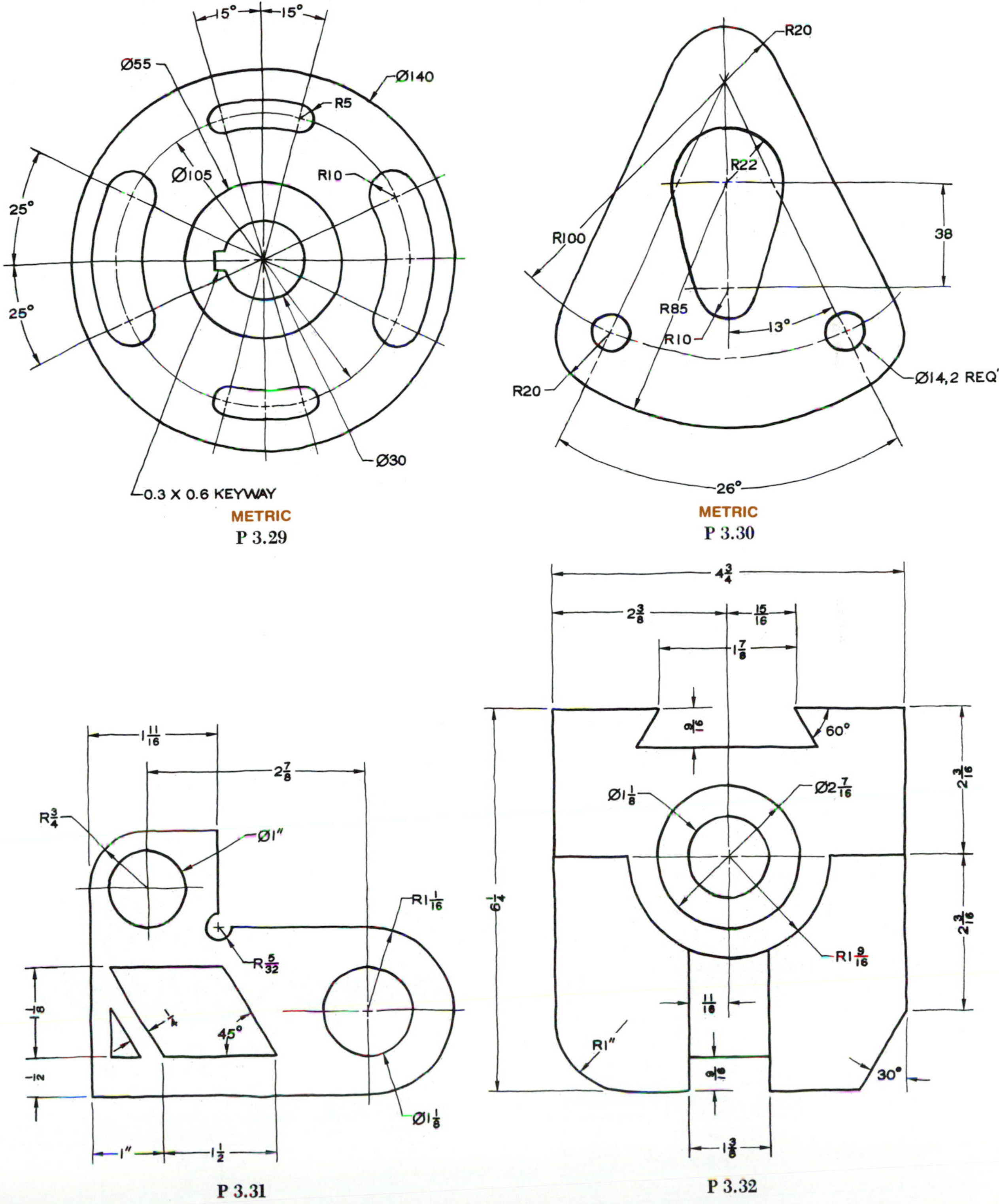
15°
15°
Ø55
Ø140
R5
Ø105
R10
25°
25°
Ø30
0.3 X 0.6 KEYWAY
METRIC
R20
R22
R100
38
R85
R10
13°
Ø14,2 REQ'D
R20
26°
METRIC
R$\frac{3}{4}$
Ø1"
1$\frac{11}{16}$
2$\frac{7}{8}$
R1$\frac{1}{16}$
R$\frac{5}{32}$
1$\frac{1}{8}$
$\frac{1}{4}$
45°
$\frac{1}{2}$
Ø1$\frac{1}{8}$
1"
1$\frac{1}{2}$
4$\frac{3}{4}$
2$\frac{3}{8}$
$\frac{15}{16}$
1$\frac{7}{8}$
$\frac{9}{16}$
60°
Ø1$\frac{1}{8}$
Ø2$\frac{7}{16}$
2$\frac{3}{16}$
6$\frac{1}{4}$
2$\frac{3}{16}$
R1$\frac{9}{16}$
$\frac{11}{16}$
R1"
$\frac{9}{16}$
30°
1$\frac{3}{8}$

P 3.29

P 3.30

P 3.31

P 3.32

P3.33 Scale: Full (inches)
P3.34 Scale: ¼″ = 1″
P3.35 Scale: Full (inches). Draw the *disk cam* with a reciprocating roller follower according to the dimensions given. The construction for graphically obtaining the cam curve has been simplified for this example, and a portion of the cam curve has been drawn. Begin by drawing the intersecting vertical and horizontal center lines. Next divide the figure into twelve equal angular divisions and draw the hub, the hole circle, and the keyway. Draw the roller follower assembly. Strike a 3.18 inch radius arc at opposite points A and A′, strike a 3.31 inch radius arc at opposite points B and B′, and so on. Continue in this way until all of the pairs of points, and point F, have been located. Working entirely around the figure, strike a ½-inch arc (equal to the roller radius) from each of these points. The intersections of these short roller radius arcs and the corresponding center lines establish twelve points on the required cam curve. Use an irregular curve to draw a smooth curve through these points.
P3.36 Scale: ⅛″ = 1″
P3.37 The bending stress concentration factor for a shaft with a shoulder fillet. Lay out the graph (full scale) using ½-inch squares and plot the points on the three curves. Using an irregular curve, draw a smooth curve through the plotted points. Complete the graph by copying all of the letters and numerals.
P3.38 The endurance limit versus tensile strength for wrought steel. Lay out the graph (full scale) using ⅜-inch squares and plot the points on the four curves. Using an irregular curve, draw a smooth curve through the plotted points. Complete the graph by copying all of the letters and numerals.

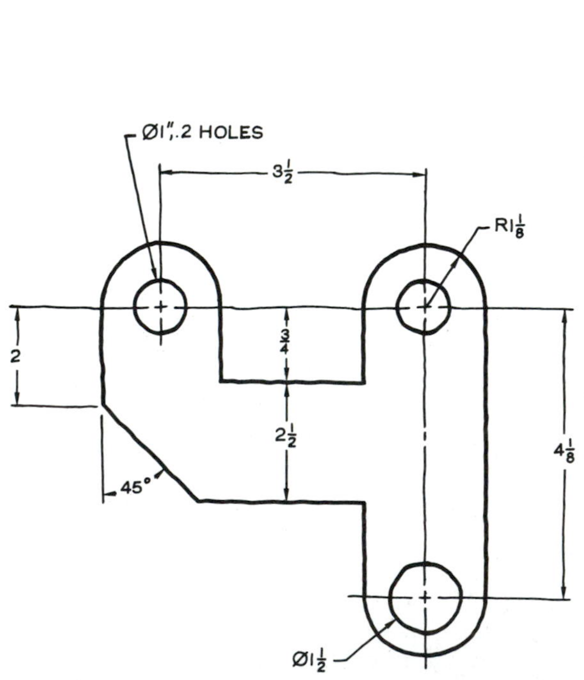

P 3.33

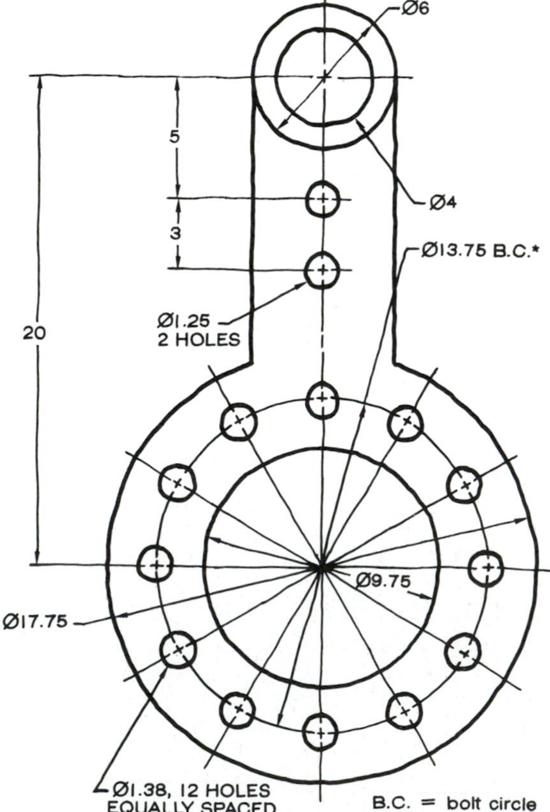

P 3.34

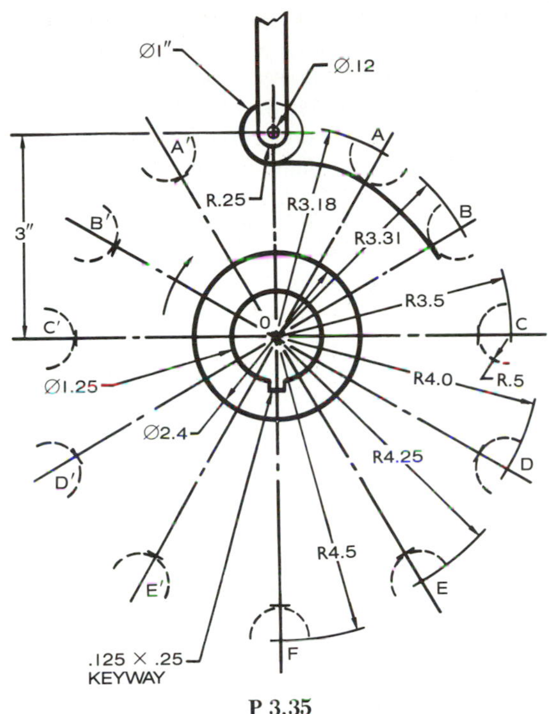
Ø1"
Ø.12
3"
A'
A
R.25
R3.18
B'
B
R3.31
0
R3.5
C'
C
Ø1.25
R4.0
R.5
Ø2.4
R4.25
D'
D
R4.5
E'
E
.125 × .25
KEYWAY
F

P 3.35

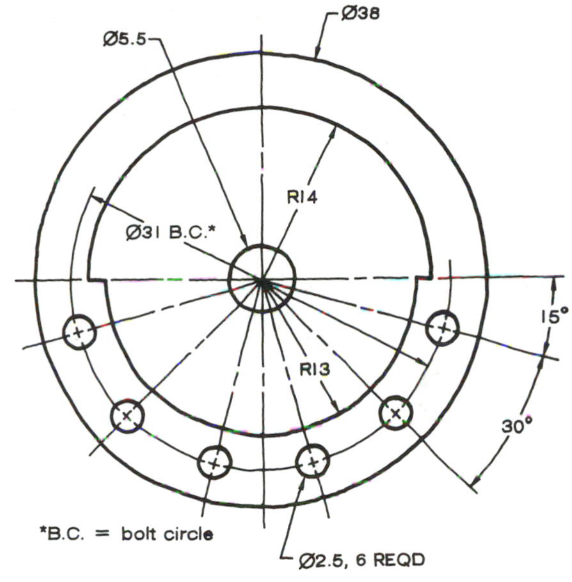
Ø38
Ø5.5
R14
Ø31 B.C.*
15°
R13
30°
*B.C. = bolt circle
Ø2.5, 6 REQD

P 3.36

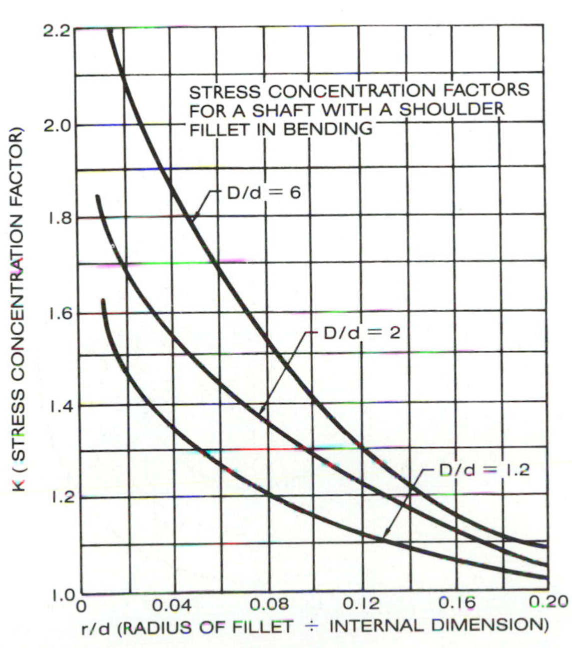
STRESS CONCENTRATION FACTORS
FOR A SHAFT WITH A SHOULDER
FILLET IN BENDING
D/d = 6
D/d = 2
D/d = 1.2
K (STRESS CONCENTRATION FACTOR)
2.2
2.0
1.8
1.6
1.4
1.2
1.0
0
0.04
0.08
0.12
0.16
0.20
r/d (RADIUS OF FILLET ÷ INTERNAL DIMENSION)

P 3.37

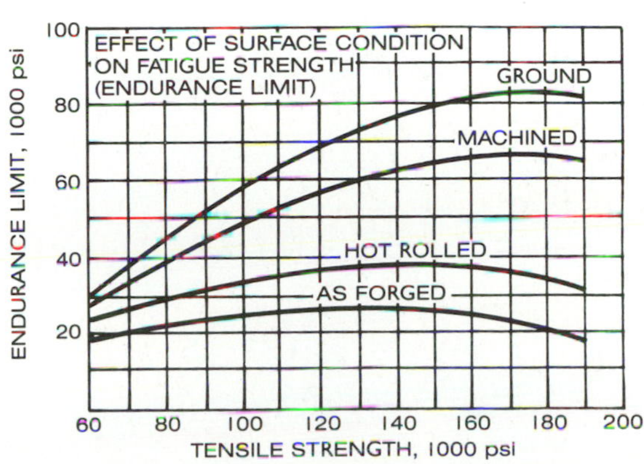
EFFECT OF SURFACE CONDITION
ON FATIGUE STRENGTH
(ENDURANCE LIMIT)
GROUND
MACHINED
HOT ROLLED
AS FORGED
ENDURANCE LIMIT, 1000 psi
100
80
60
40
20
60
80
100
120
140
160
180
200
TENSILE STRENGTH, 1000 psi

P 3.38

P3.39 Copy the given half of the figure at a scale of 2:1 and draw the identical opposite half. Use an irregular curve to draw a smooth curve through the plotted points.

P3.40 Scale: 1:2. A survey plan with distances in meters. Lay out and record the *compass direction* (to the nearest half degree) of the property line from position 1 to position 2 and show the correct plot dimensions of this line. Draw the property lines by using a repeated short line with two dashes and show the lot corners by small, solid circles. Complete the figure by lettering in all of the information on the survey plan.

P3.41. Scale: ⅛″ = 1′-0″. Follow the directions given in the preceding problem.

P3.42 Scale: ⅛″ = 1′-0″. An architectural floor plan. Using the given dimensions, lay out the floor plan. Estimate the sizes of the omitted dimensions. Add the lettering but omit the dimensions from your final drawing.

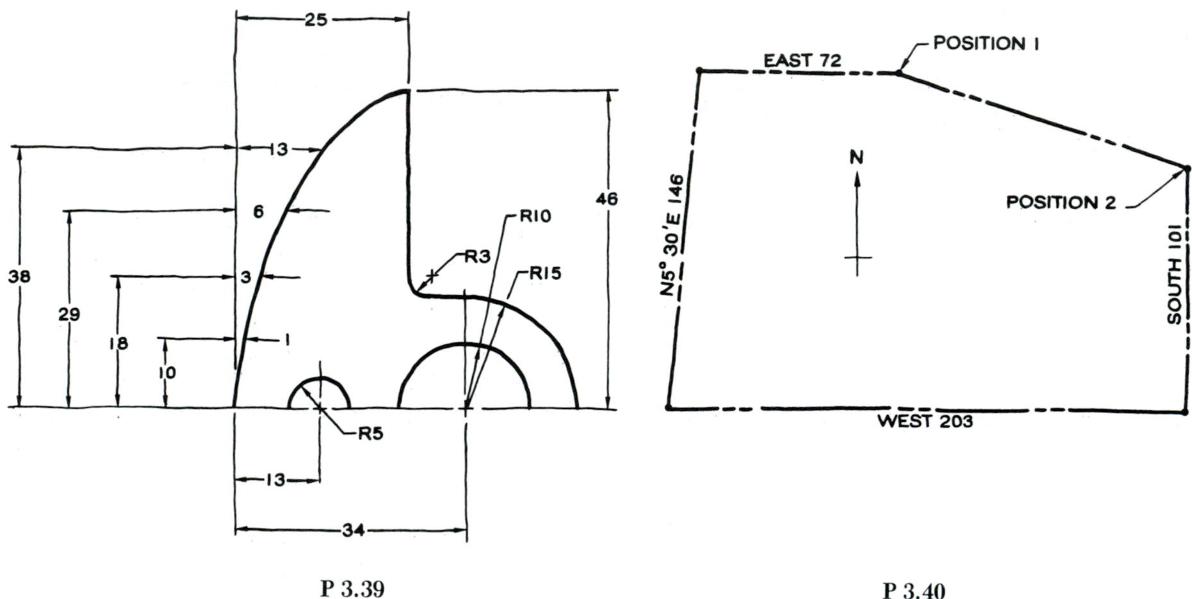

P 3.39

P 3.40

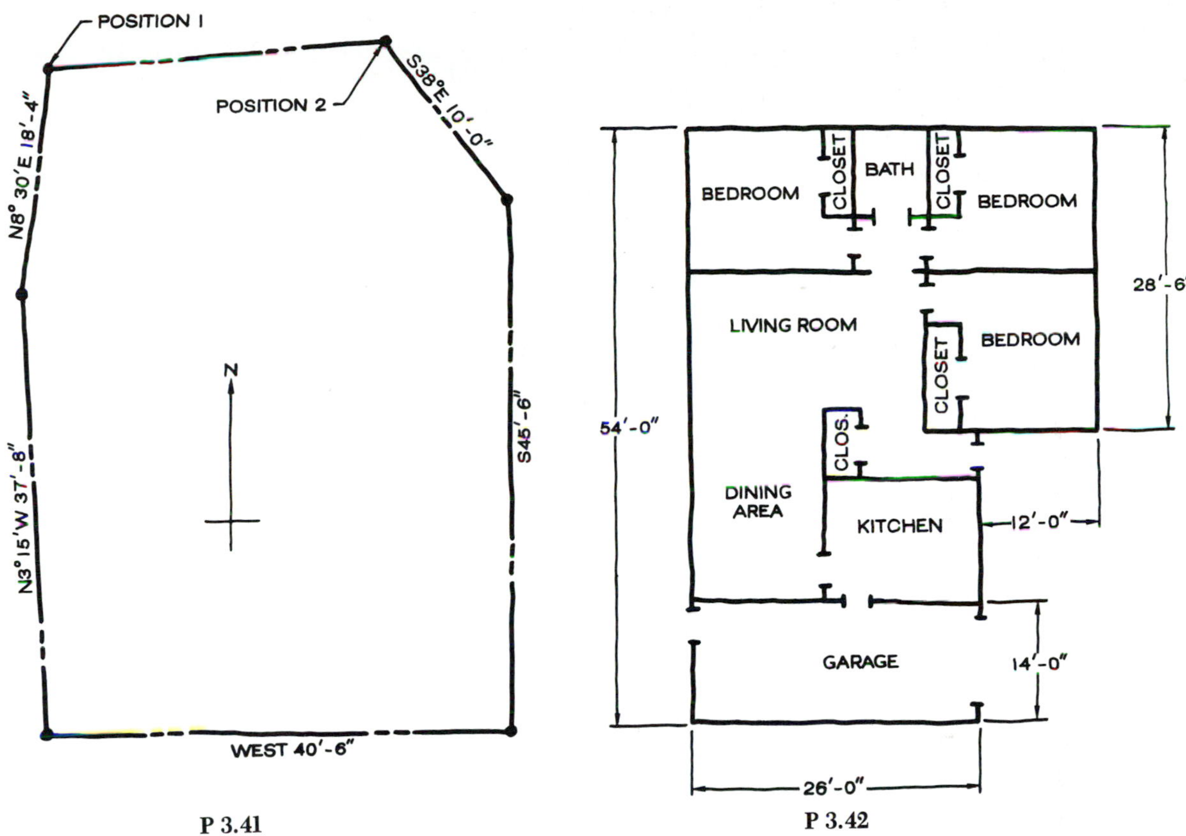
POSITION 1
POSITION 2
S38°E 10'-0"
N8° 30'E 18'-4"
N
N3° 15'W 37'-8"
S45'-6"
WEST 40'-6"
BEDROOM
CLOSET
BATH
CLOSET
BEDROOM
28'-6"
LIVING ROOM
BEDROOM
CLOSET
54'-0"
CLOS.
DINING AREA
KITCHEN
12'-0"
GARAGE
14'-0"
26'-0"

P 3.41

P 3.42

Engineering Geometry

OBJECTIVES

After completing this chapter you should have gained the following abilities:

1. To develop the necessary level of competency in performing accurate geometrical constructions.

2. To recognize the basic relationships of lines and circles, the combinations most commonly encountered in working out solutions to engineering design problems.

3. To understand the principles of plane geometry involved in the various constructions required for practical applications in graphical design layout.

4. To recognize the basic shapes of geometric figures: triangles, quadrilaterals, polygons, circles, and solids.

This chapter deals with engineering geometry, a subject about which all engineers must have adequate preparation. The constructions that follow are based upon the principles of plane geometry but have been modified to correspond with the use of drafting instruments for drawing geometric figures.

In spite of dramatic advances in computer-aided design (CAD), it is still necessary to work out designs on the drafting board in most cases. For the immediate future, the relatively high cost of terminal time will limit CAD to applications involving intricate geometry and/or highly repetitive drafting activities. Computer programs alone will not turn a poor design into a good one, and the competent, trained engineer will feel no threat from CAD. Most knowledgeable people insist that pencil drawings will continue to be prepared in the traditional way on the drawing board—particularly for conceptual work, preliminary layouts, and one-of-a-kind designs—for some years to come. While advances in pictorial graphical display coupled with the use of color offer enormous potential for the designer, it may be many years before such techniques are viable for normal, everyday drafting and conceptual design work. An engineer cannot attempt computerized design until he or she has become thoroughly familiar with basic methods of graphical construction. The graphical constructions that are discussed in this chapter are important, basic-skill tools of the drafter and designer.

4.1 GEOMETRIC FIGURES

Geometric figures include such basic shapes as triangles, quadrilaterals, polygons, circles, and solids.

Angles

An angle is formed by two intersecting lines. The unit of angular measurement is the degree (°), and a circle has 360 degrees. A degree is divided into 60 minutes (′), and a minute is divided into 60 seconds (″). An angle of 35 degrees, 37 minutes, and 12 seconds, for example, is expressed as 35°37′12″. An angle may be given in degrees and decimal parts of a degree as 42.5°. When the measurement is expressed in minutes alone, the number of minutes should be preceded by 0°, as in 0°49′. See Figure 4-1 for examples of different types of angles.

The procedure for constructing angles with the T square and triangles was explained in Chapter 3, as was the procedure for using the protractor for drawing odd angles or those other than multiples of 15°.

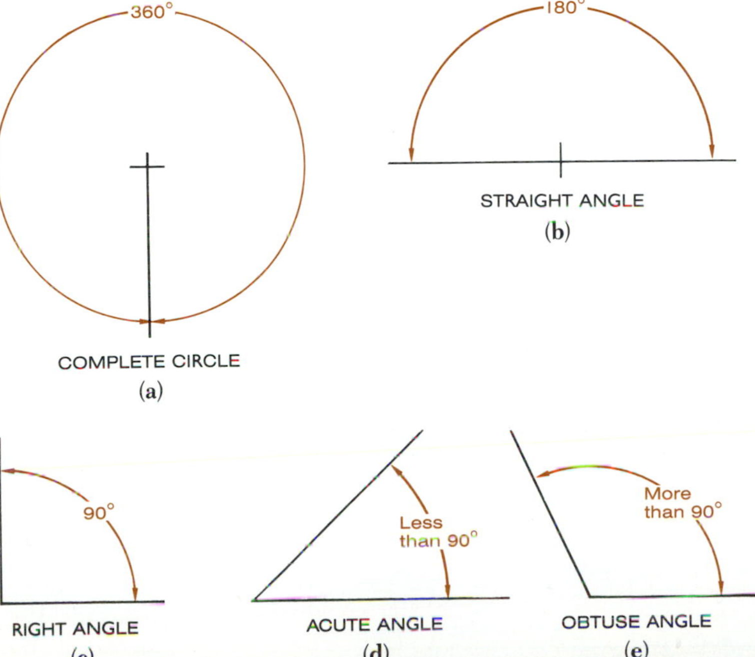

FIGURE 4-1 **Angles** ***(Section 4.1)***

Triangles

A triangle is a plane figure bounded by three straight sides. The sum of the interior angles is 180°. There are three kinds of triangles; *equilateral*, *isosceles*, and *scalene*. The *right triangle*, which may be isosceles or scalane, has one 90° angle. These triangles are shown in Figure 4-2.

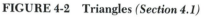

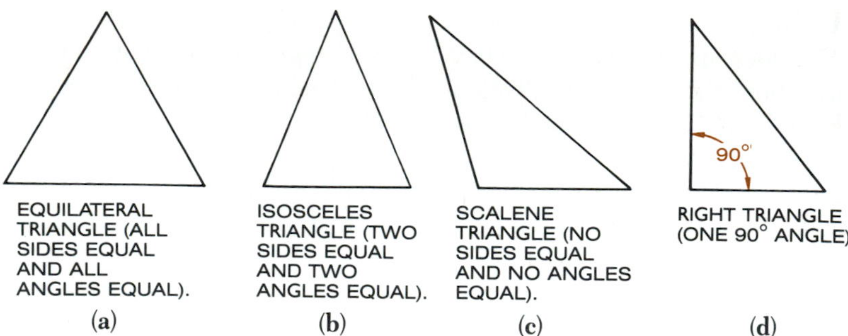

FIGURE 4-2 **Triangles *(Section 4.1)***

Quadrilaterals

A quadrilateral is a plane figure of any shape bounded by four straight sides (see Figure 4-3). If the opposite sides are parallel, the quadrilateral is also a *parallelogram*. If two opposite sides are parallel, the figure is a *trapezoid*.

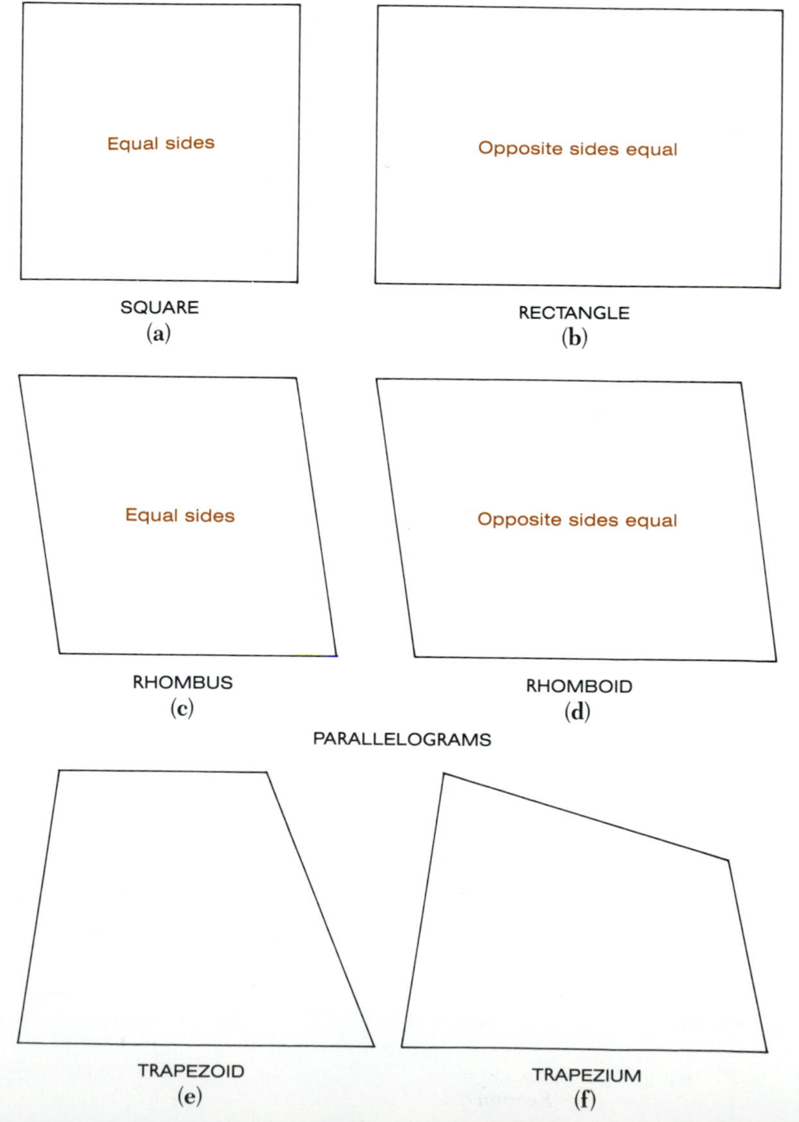

FIGURE 4-3 **Quadrilaterals *(Section 4.1)***

Polygons

A polygon is a plane figure of any shape bounded by straight lines. *Regular polygons*, shown in Figure 4-4, are plane figures bounded by any number of straight sides in which the angles and sides are all equal. A regular polygon can be inscribed in a circle or circumscribed *around* a circle.

Circles and Arcs

A circle is a plane figure bounded by a curved line every point of which is the same distance from a point called the center. The curved line is called the *circumference* of the circle. A line drawn through the center of the circle and terminating at the circumference is called the *diameter* (the largest chord). The diameter of a circle divides the circle into two equal parts called semicircles. A line drawn from the center of a circle to the circumference is called the *radius* and is equal to one-half the diameter. Other definitions are illustrated in Figure 4-5.

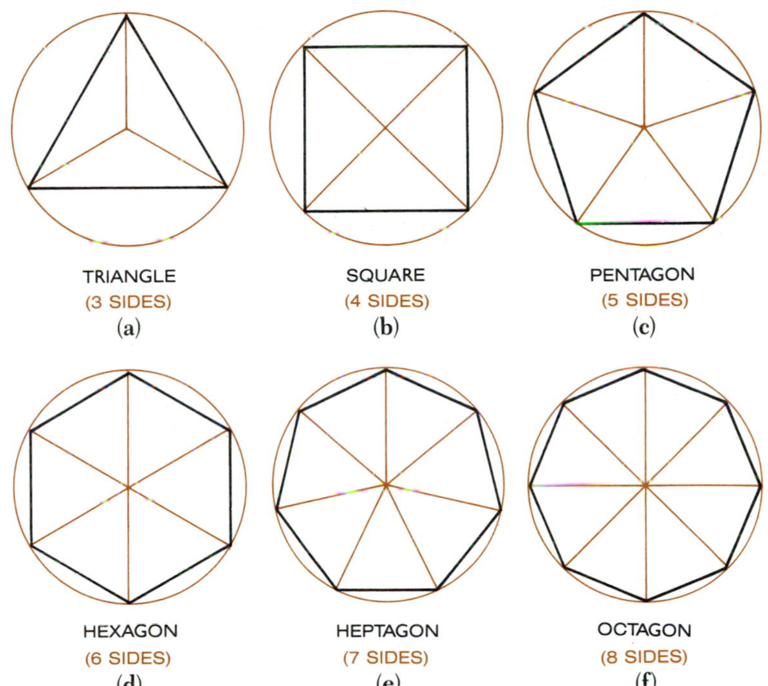

FIGURE 4-4 Regular polygons *(Section 4.1)*

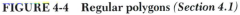

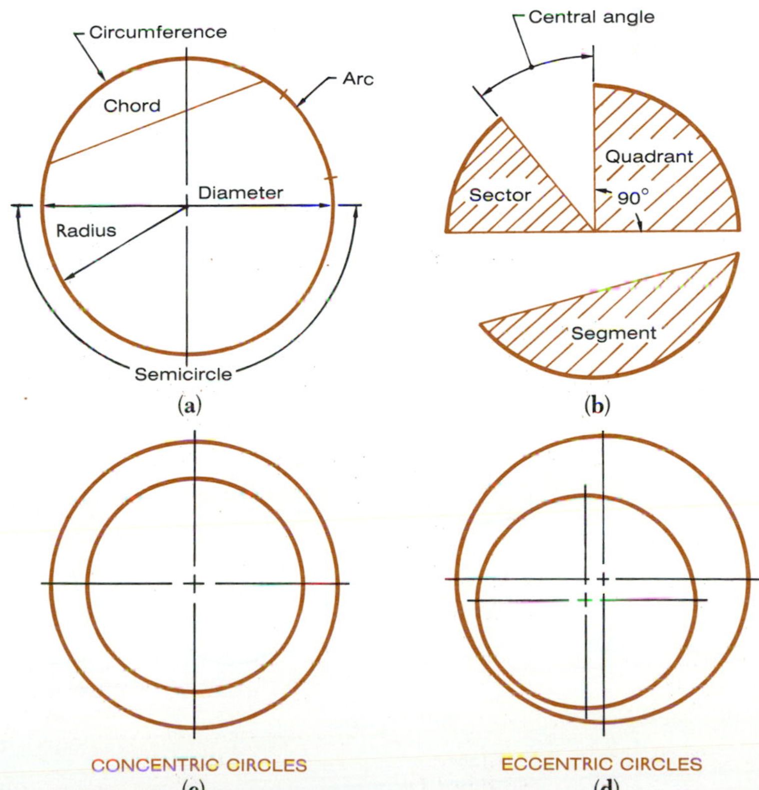

FIGURE 4-5 Circles and arcs *(Section 4.1)*

Solids

Solids, or polyhedrons, are three-dimensional forms whose faces or plane surfaces are polygons (see Figure 4-6). The most common examples of solids are prisms and pyramids. A *prism* has two parallel or congruent bases joined by lateral faces that are parallelograms. A triangular prism has a triangular base, a square prism has a square base, and so on. A *right prism* has faces and lateral edges perpendicular to the bases. An *oblique prism* has faces and lateral edges oblique to the bases. A *truncated prism* has one end cut off forming a base not parallel to the other.

A *pyramid* has a polygon for a base and triangular lateral faces intersecting at a common point called the vertex. The line from the vertex to the center of the base is called the axis. If the axis is perpendicular to the base, it is the altitude of the pyramid, and the pyramid is a *right pyramid*. A triangular pyramid has a triangular base, a rectangular pyramid has a rectangular base, and so on. If the axis of a pyramid is not perpendicular to the base, that pyramid is an *oblique pyramid*. A *truncated pyramid* is created by cutting off a portion near the vertex.

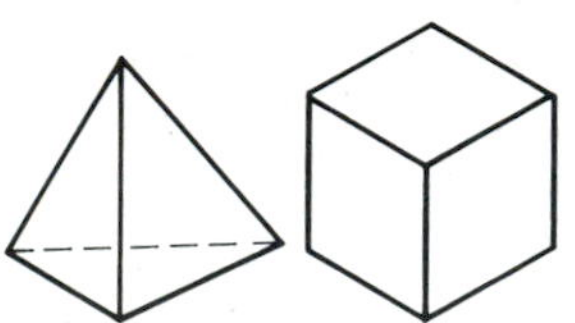

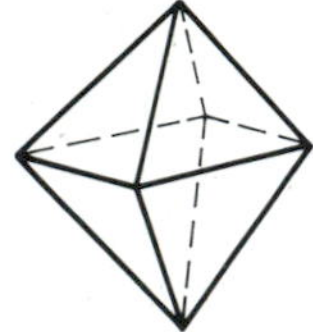

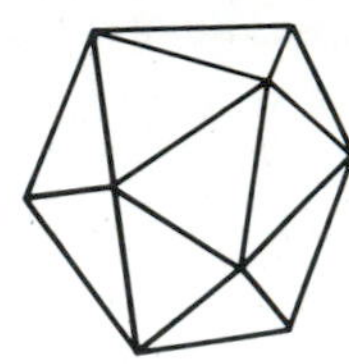

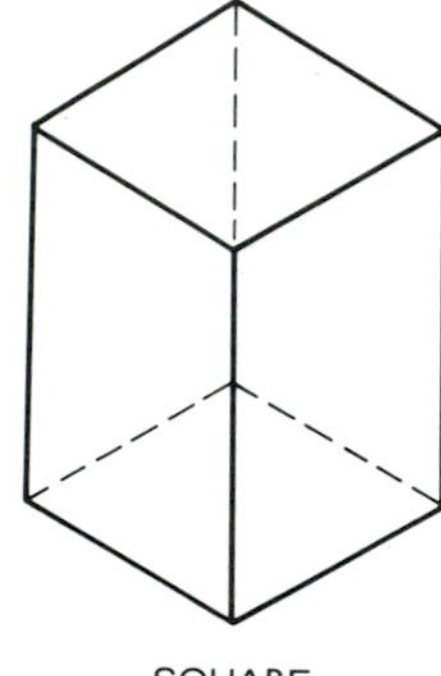

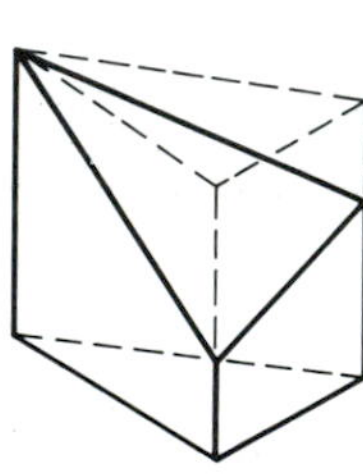

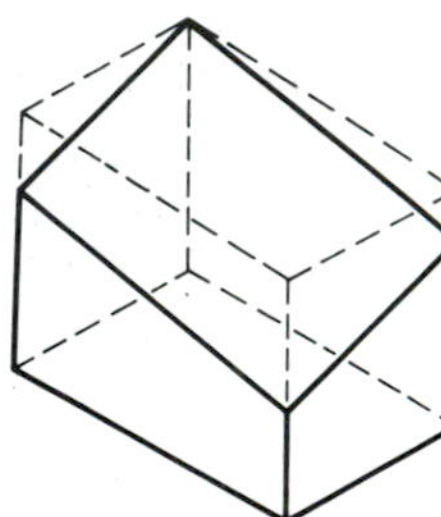

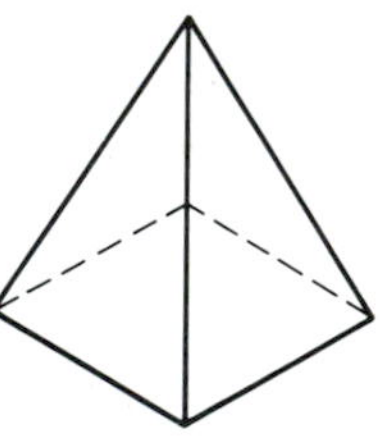

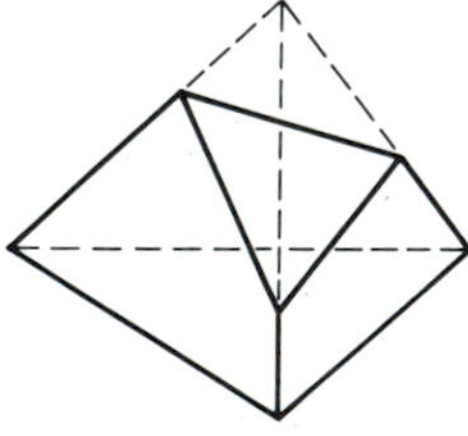

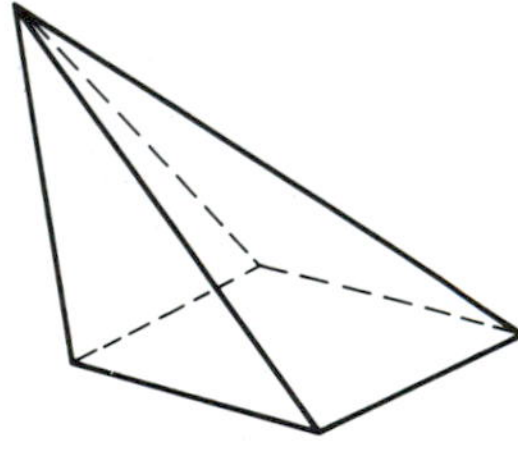

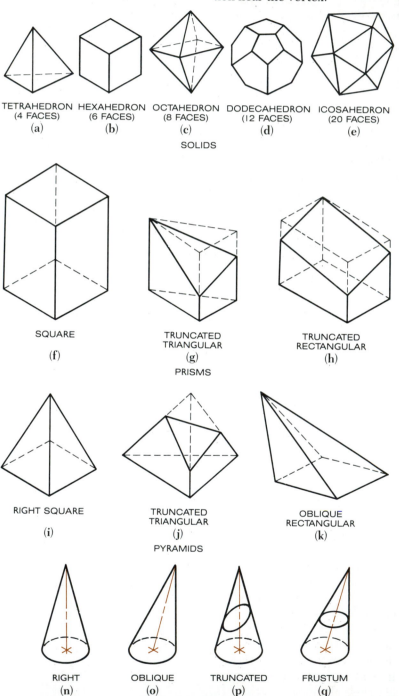

FIGURE 4-6 Solids
(Section 4.1)

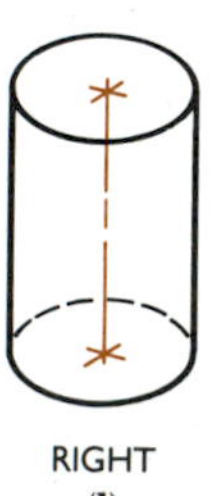

OBLIQUE
(m)
CYLINDERS

RIGHT
(n)

OBLIQUE
(o)

TRUNCATED
(p)

FRUSTUM
(q)
CONES

A *cylinder* is a figure that is formed by rotating a straight line, called a generatrix, in contact with the base circle while always keeping it parallel to the axis. The axis is the line joining the centers of the bases. Each position of the generatrix is called an *element* of the cylinder. A cylinder is a *right cylinder* when its elements are perpendicular to the bases and an *oblique cylinder* when they are not. A *right-circular* cylinder has a circular base. A *truncated* cylinder is the portion that lies between one of its bases and the cutting plane that cuts all of its elements. A cylinder does not necessarily have a closed base surface.

A *cone* is a figure formed by moving a straight line, called a generatrix, in contact with the base circle, and passing through a fixed point of the cone, the vertex. Each position of the generatrix is called an *element* of the cone. A cone is *right* if the axis and the altitude coincide; it is *oblique* if they do not coincide. A *truncated* cone is that portion lying between the base and the cutting plane that cuts all the elements. The *frustum* of a cone is that portion lying between the base and a cutting plane parallel to the base that cuts all the elements. A cone does not necessarily have a closed base surface.

A *sphere* is a solid with all points on its surface equally distant from the center. It is formed by revolving the plane of a circle about one of its diameters.

4.2 DIVIDING A LINE

In addition to the following methods of dividing a line into equal parts, you may also use the divider method, previously explained in Section 3.15.

Two Equal Parts

To *bisect* (divide into two equal parts) line AB, as shown in Figure 4-7(a), use the following procedure:

1. From A and B, draw equal arcs with a radius greater than ½AB.
2. Join intersections X and Y with a straight line to locate center point C on line AB. Line XY bisects and is perpendicular to AB.

Here is an alternate procedure, illustrated in Figure 4-7(b), for bisecting the same line:

1. Draw lines through points A and B with a standard 45° drafting triangle to intersect at point X.
2. Slide the triangle along the T square and draw a perpendicular line through point X locating the center point C on line AB. Line XC bisects and is perpendicular to AB.

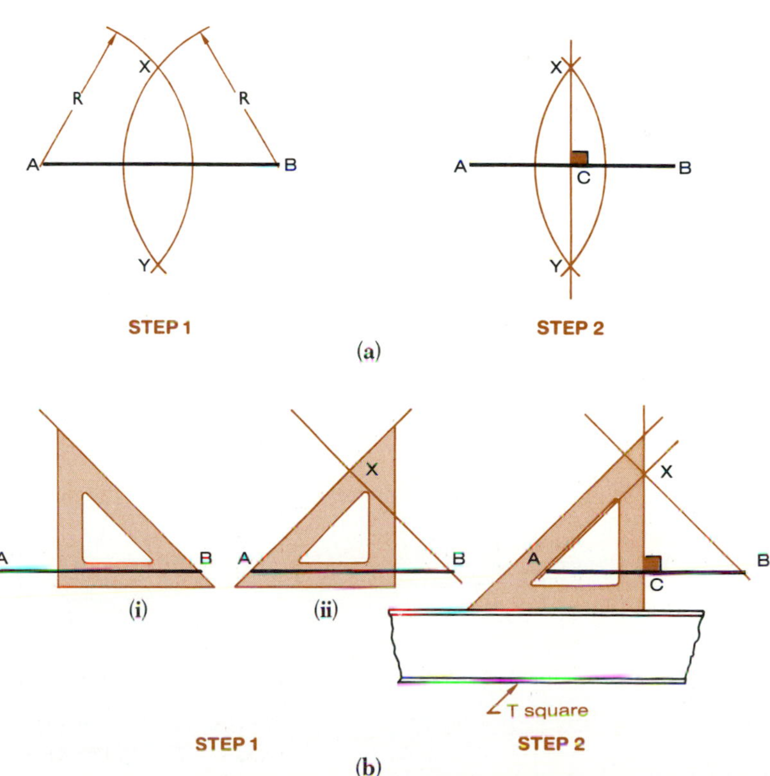

FIGURE 4-7 Dividing a line into two equal parts *(Section 4.2)*

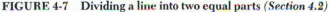

Any Number of Equal Parts

Referring to Figure 4-8(a), use the following procedure to divide line AB into equal parts:

1. Draw construction line AC at any convenient angle.
2. Using dividers or a scale, step off as many equal divisions as needed; in this example, five.
3. Connect the last division point on AC with a straight line to point B.
4. Draw parallel lines through other division points as shown to divide line AB into five equal divisions.

An alternate procedure is shown in Figure 4-8(b):

1. Working from point B, construct a perpendicular line downward from point B at any convenient length.
2. Set the zero mark of the scale at point A and incline the scale until the division mark corresponding to the number of divisions needed (in this case, six) lands on the perpendicular line you have just drawn.
3. Mark off six equal divisions along the scale edge and draw parallel lines through the divisions as shown to divide line AB into six equal parts.

Five equal divisions

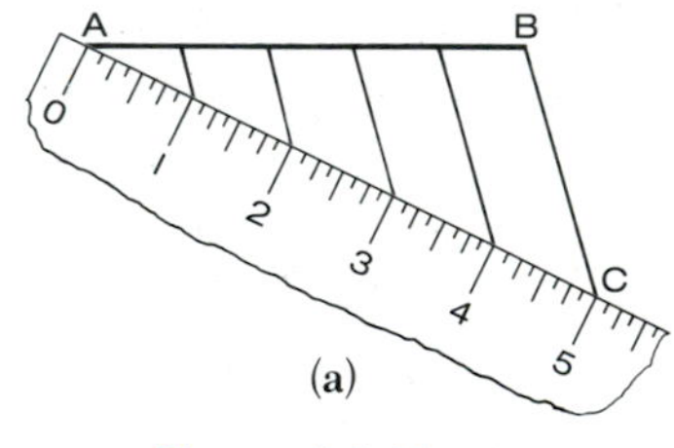

(a)

Six equal divisions

(b)

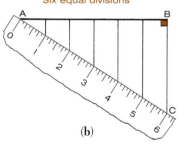

FIGURE 4-8 Dividing a line into any number of equal parts ***(Section 4.2)***

Proportional Parts

Figure 4-9 illustrates the following procedure for dividing line AB into five parts, proportional to 1, 3, 5, 3, and 1:

1. Draw a perpendicular construction line downward from point B at any convenient length.
2. Select a scale of any convenient size for a total of thirteen parts (1 + 3 + 5 + 3 + 1 = 13).
3. Set the zero mark of the scale at point A and the thirteenth unit on the vertical line.
4. Step off the points corresponding to divisions at 1, 3, 5, 3, and 1.
5. Draw parallel lines through the divisions as shown to divide line AB into proportional parts.

Proportional parts

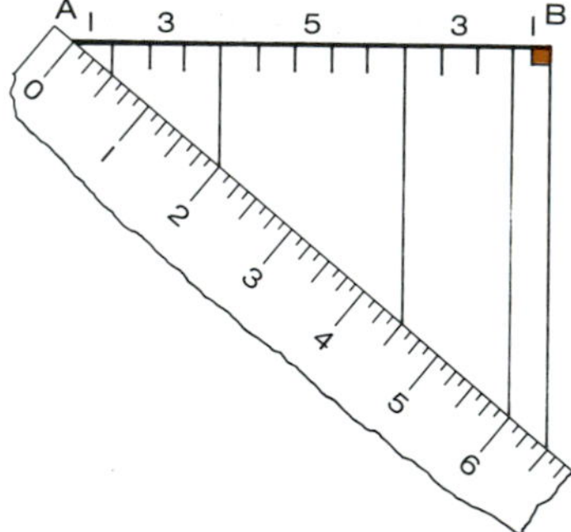

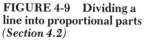

FIGURE 4-9 Dividing a line into proportional parts ***(Section 4.2)***

4.3 DIVIDING THE SPACE BETWEEN TWO LINES

The following procedure, illustrated in Figure 4-10, can be used to divide the space between parallel lines AB and CD into any number of equal divisions:

1. Select a scale whose total length exceeds the actual distance between the given lines.
2. Set the zero mark of the scale on point A and, in this example, the twenty-second unit on the other line (CD).
3. Step off the points corresponding to the twenty-two divisions.
4. Draw lines parallel to AB and CD through the divisions shown to divide the space between the two given lines into twenty-two spaces.

CHECKPOINT 1
DRAWING HORIZONTAL CONSTRUCTION LINES AND DIVIDING LINES INTO A REQUIRED NUMBER OF PARTS

Using a plain sheet of 8½ × 11 inch paper, work out a construction in accordance with the following information:

1. Draw a horizontal construction line across the sheet spaced about 1 inch down from the top.
2. Working from the left side of the sheet, mark off to the right a distance of 6¼ inches on this line. Using dividers, divide the line into nine equal parts.
3. Using the same line and one of the scale methods shown in Figure 4-8, divide the line into five equal parts.
4. Draw a second horizontal construction line across sheet approximately halfway down the sheet.
5. Make this line 6¹¹⁄₁₆ inches long. Using the method shown in Figure 4-9, divide the line into parts proportional to 2, 3, 4, 4, 3, and 2.

On a second sheet of 8½ × 11 inch paper, work out the following construction:

1. Draw four horizontal object lines across the sheet (labeled A, B, C, and D) as follows:
   Line A: 2 inches down from the top.
   Line B: 4⁷⁄₁₆ inches down from the top.
   Line C: 6½ inches down from the top.
   Line D: 8⁹⁄₁₆ inches down from the top.
2. Using the method shown in Figure 4-10, divide the space between the lines A and B into 14 equal parts, the space between lines B and C into 3 equal parts, and the space between lines C and D into 24 equal parts. Using the T square, draw construction lines entirely across the sheet through all of these divisions.

FIGURE 4-10 Dividing the space between two parallel lines
(Section 4.3)

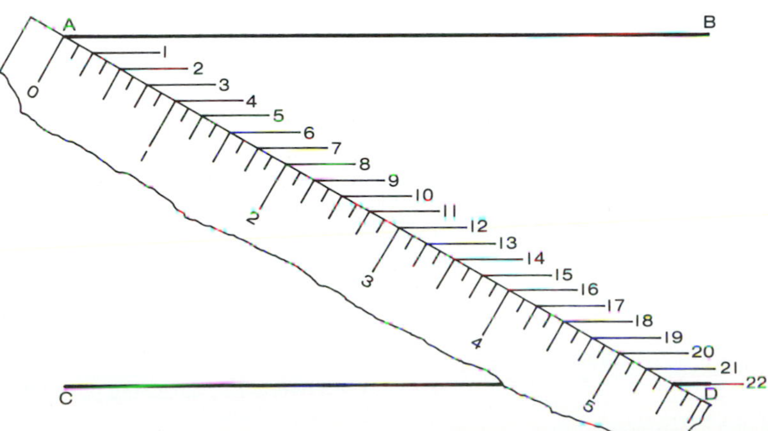

4.4 DRAWING PERPENDICULAR LINES

Use the following procedure, illustrated in Figure 4-11, to draw a line perpendicular to a given line (AB) and through a point (P) on that line:

1. With point P as center, swing an arc with any convenient radius to intersect line AB at points C and D.
2. With points C and D as centers, and with a radius slightly greater than ½CD, strike equal arcs to intersect at point O.
3. Draw line PO, the required perpendicular.

The following procedure, shown in Figure 4-12, is an alternate way of drawing a line perpendicular to line AB:

1. With point P as the center, strike an arc to intersect AB at C and D.
2. With points C and D as centers, and with a radius slightly greater than ½CD, swing equal arcs to intersect at point O.
3. Draw line PO, the required perpendicular.

Perpendicular lines may also be drawn using the T square and a triangle, as explained in Section 3.8.

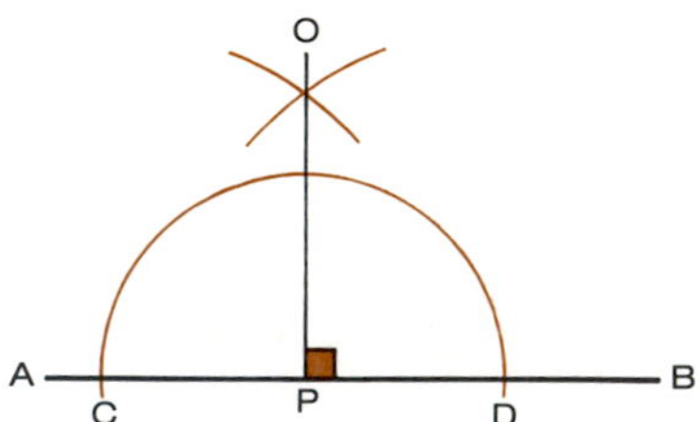

FIGURE 4-11 Drawing a perpendicular to a given line through a point on the line *(Section 4.4)*

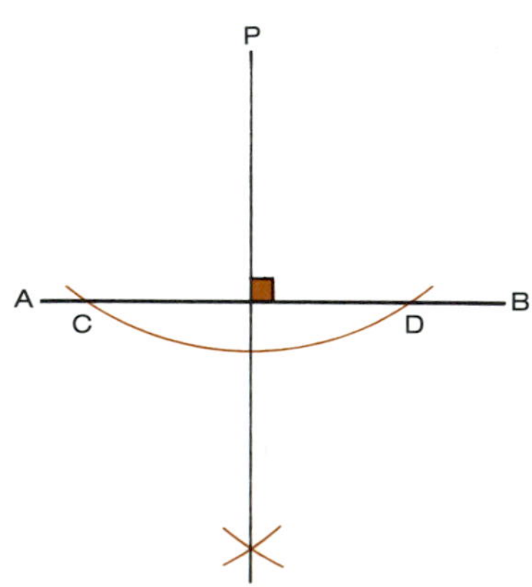

FIGURE 4-12 Drawing a perpendicular to a given line through a point off the line *(Section 4.4)*

4.5 DRAWING PARALLEL LINES

Figure 4-13(a) illustrates the following method for drawing parallel lines:

1. Draw a perpendicular construction line to AB at any position along the line.
2. Measure off a distance CD on the construction line equivalent to the desired distance the two parallel lines are to be set apart. Through point D, draw a line parallel to line AB.

Another method for drawing parallel lines, shown in Figure 4-13(b), is as follows:

1. Swing an arc with a radius R, equivalent to the desired distance the two parallel lines are to be set apart, from a center on line AB.
2. Use a straightedge to draw a parallel line tangent to the arc.

Parallel lines may also be drawn using the T square and a triangle, as explained in Section 3.7.

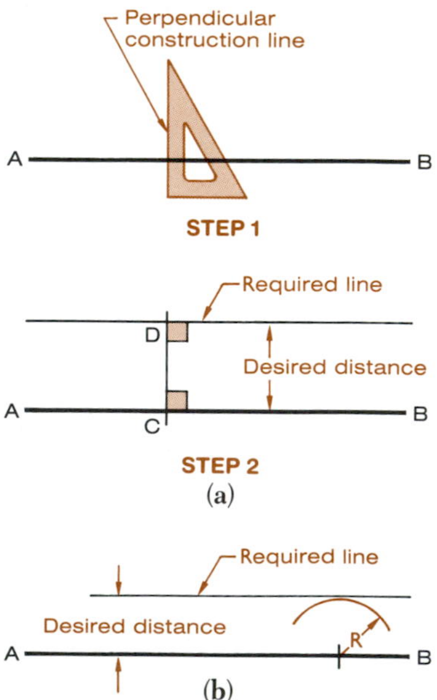

FIGURE 4-13 Drawing parallel lines *(Section 4.5)*

CHECKPOINT 2
DRAWING PERPENDICULAR AND PARALLEL LINES

Divide an 8½ × 11 inch sheet of ¼-inch-grid paper into four equal spaces, as shown in Figure 4-14, to use for the following exercises. (Scale: full size.)

1. Draw lines AB and CD. Locate points P_1, P_2, P_3, and P_4. Using the method shown in Figure 4-11, draw lines perpendicular to line AB through points P_1 and P_2 and perpendicular to line CD through points P_3 and P_4.
2. Draw line EF and locate points P_5 and P_6. Using the method shown in Figure 4-12, draw lines perpendicular to line EF through points P_5 and P_6.
3. Draw line GH. Using the scale method, shown in Figure 4-13(a), draw a parallel line LM 1 7/16 inches above line GH and a third parallel line XY 1¾ inches above line LM.
4. Draw line JK. Using the compass method, shown in Figure 4-13(b), draw a parallel line PQ 1 1/16 inches to the left of line JK and a second parallel line RS 15/32 inch to the right of line JK.

FIGURE 4-14 **Checkpoint 2 problem**

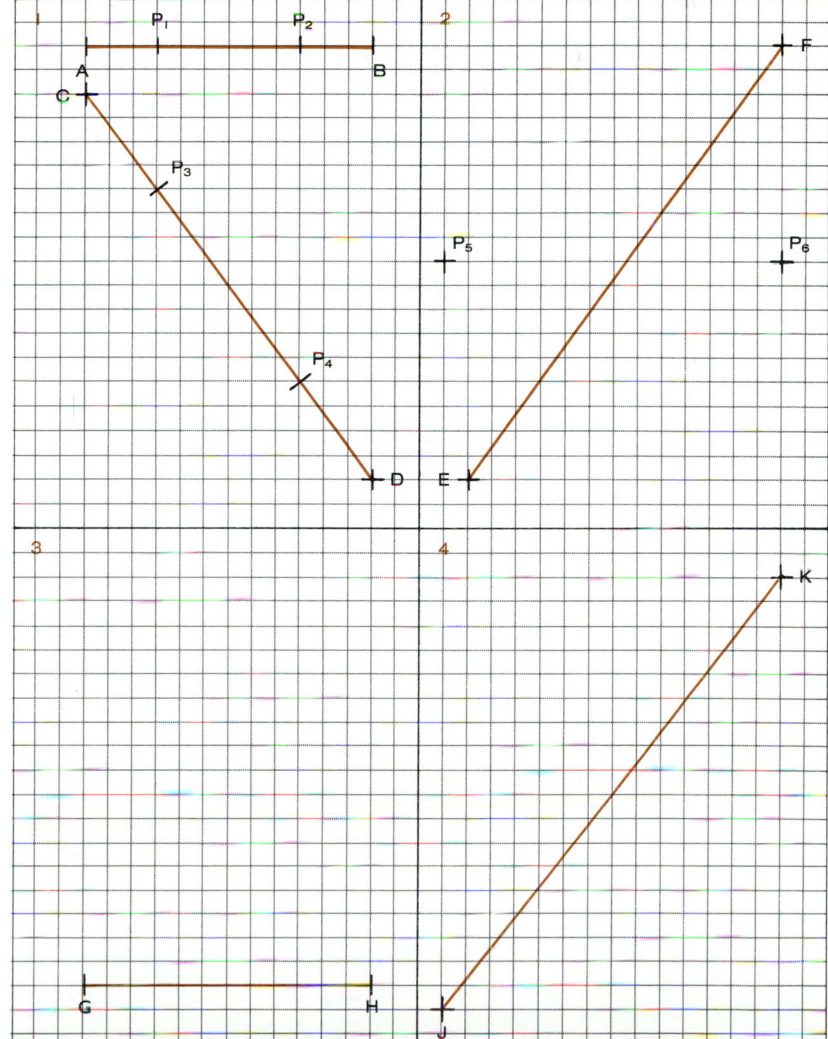

4.6 LAYING OUT AN ANGLE

Lay out given angles using your protractor or your drafting triangles. Follow the procedures outlined in Sections 3.6 and 3.10.

4.7 BISECTING AN ANGLE

Figure 4-15 shows how the following procedure can be used to bisect given angle AOB:

1. Swing an arc, using any convenient radius with point O as center, locating points X and Y.
2. Using the same radius, if greater than ½XY, swing two arcs with X and Y as centers, locating point C.
3. Draw line CO, the bisector.

With repeated bisection the angle can be further divided into four parts, eight parts, sixteen parts, and so on.

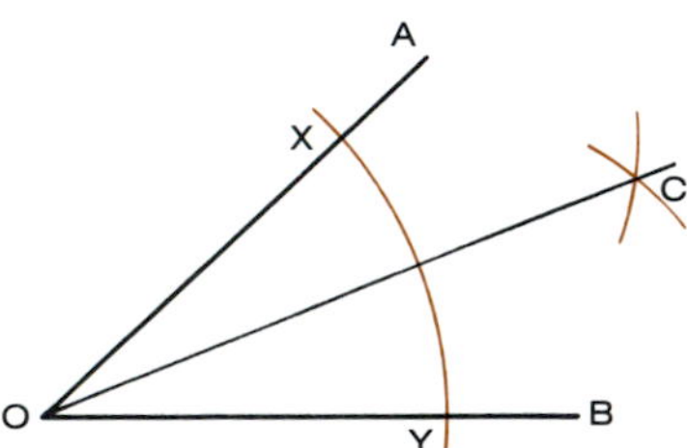

FIGURE 4-15 Bisecting an angle *(Section 4.7)*

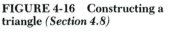

4.8 CONSTRUCTING A TRIANGLE

Figure 4-16 illustrates the procedure for drawing a triangle when all three sides are given:

1. Draw one side in the desired position; side AB, for example.
2. With A as a center, swing an arc with a radius equal to side AC.
3. With B as a center, swing an arc with a radius equal to side BC.
4. Draw sides AC and BC.

Figure 4-17 illustrates the procedure for drawing a right triangle when the hypotenuse and one side are given:

1. Draw a semicircle with a diameter equal to AB.
2. With B as a center, swing an arc with a radius equal to side BC.
3. Draw sides AC and BC.

Note that any angle inscribed in a semicircle will be a right angle.

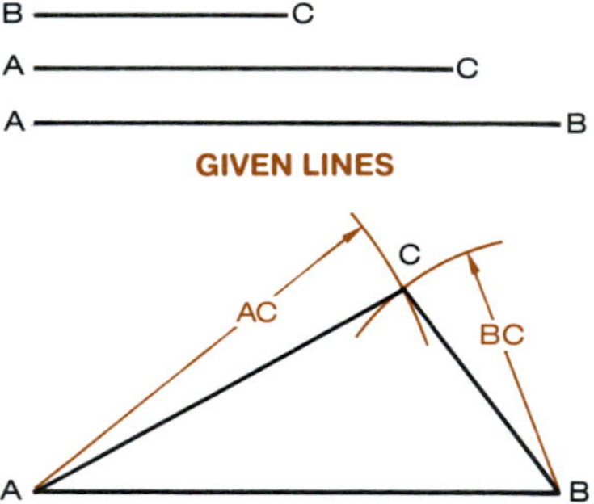

FIGURE 4-16 Constructing a triangle *(Section 4.8)*

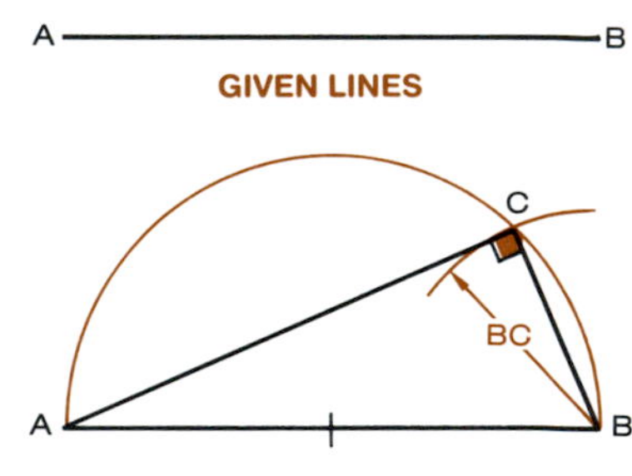

FIGURE 4-17 Constructing a right triangle *(Section 4.8)*

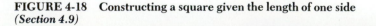

4.9 CONSTRUCTING A SQUARE

Given the length of one side, you can construct a square using the following procedure (shown in Figure 4-18):

1. Using the T square and the 45° triangle, draw perpendiculars to line AB through points A and B.
2. From point A, draw a line at 45° to locate point C.
3. Draw DC parallel to AB through point C.

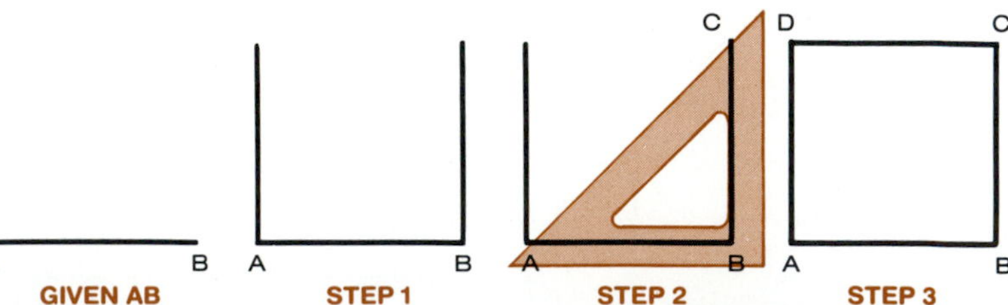

FIGURE 4-18 Constructing a square given the length of one side *(Section 4.9)*

Figure 4-19 shows how to draw a square when the distance across the flats is given:

1. Draw the circle with a diameter equal to the distance across the flats.
2. Using the T square and a 45° triangle, draw the four sides tangent to the circle as shown.

To draw a square when the distance across the corners is given, use the following procedure, shown in Figure 4-20(a):

1. Draw the circle with a diameter equal to the distance across the corners.
2. Using the T square and a 45° triangle, draw the four sides at right angles to each other at the intersection of the diameters and the circle.

Figure 4-20(b) illustrates the procedure for drawing the same square in a different position:

1. Draw the circle with a diameter equal to the distance across the corners.
2. Using the T square and a triangle, draw the diagonals.
3. Draw the four sides at the intersections of the diameters and the circle.

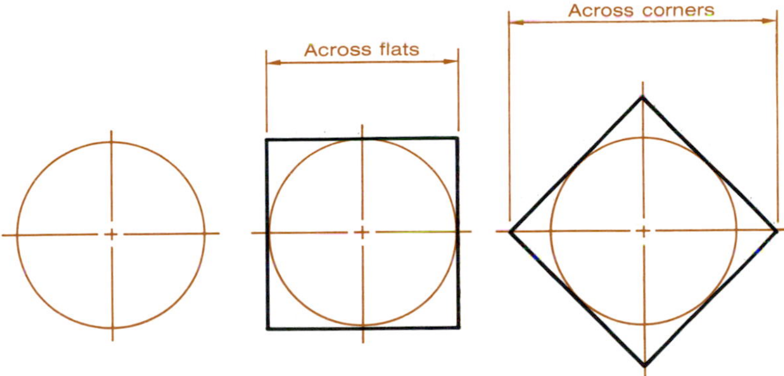

FIGURE 4-19 Constructing a square given the distance across the flats *(Section 4.9)*

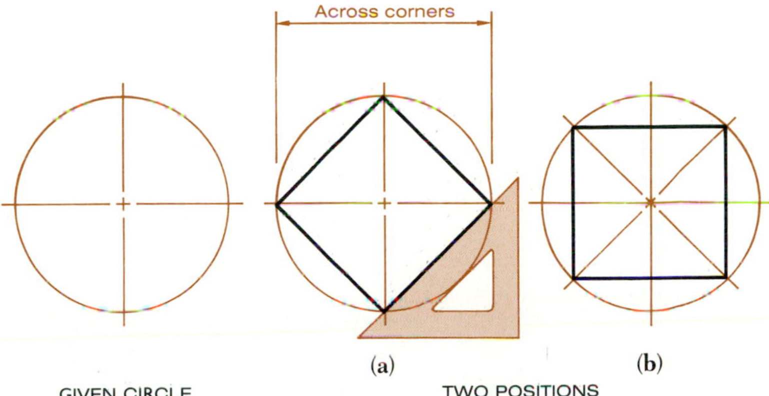

FIGURE 4-20 Constructing a square given the distance across the corners *(Section 4.9)*

CHECKPOINT 3
CONSTRUCTING GEOMETRIC FIGURES

Divide an 8½ × 11 inch sheet of quarter-inch graph paper into four equal spaces, as shown in Figure 4-21. Draw each of the following exercises at the scale indicated:

1. Construct triangle ABC with point A in the position shown. Triangle ABC is not shown true size. AB = 115 mm, AC = 82 mm, and CB = 120 mm. Scale 1:1 (metric).
2. (a) Construct triangle ABC with point A in the position shown. AB = 23′-6″, AC = 28′-3″, and angle BAC = 75°. (b) Using the method shown in Figure 4-15 bisect angle CBA. Scale ⅛″ = 1′-0″.
3. Inscribe a right triangle ABC in a semicircle with point A in the position shown. AB = 475 mm and BC = 400 mm. Scale 1:5 (metric).
4. Construct a 26½-inch square with point O in the position indicated. Scale 1″ = 1′-0″.

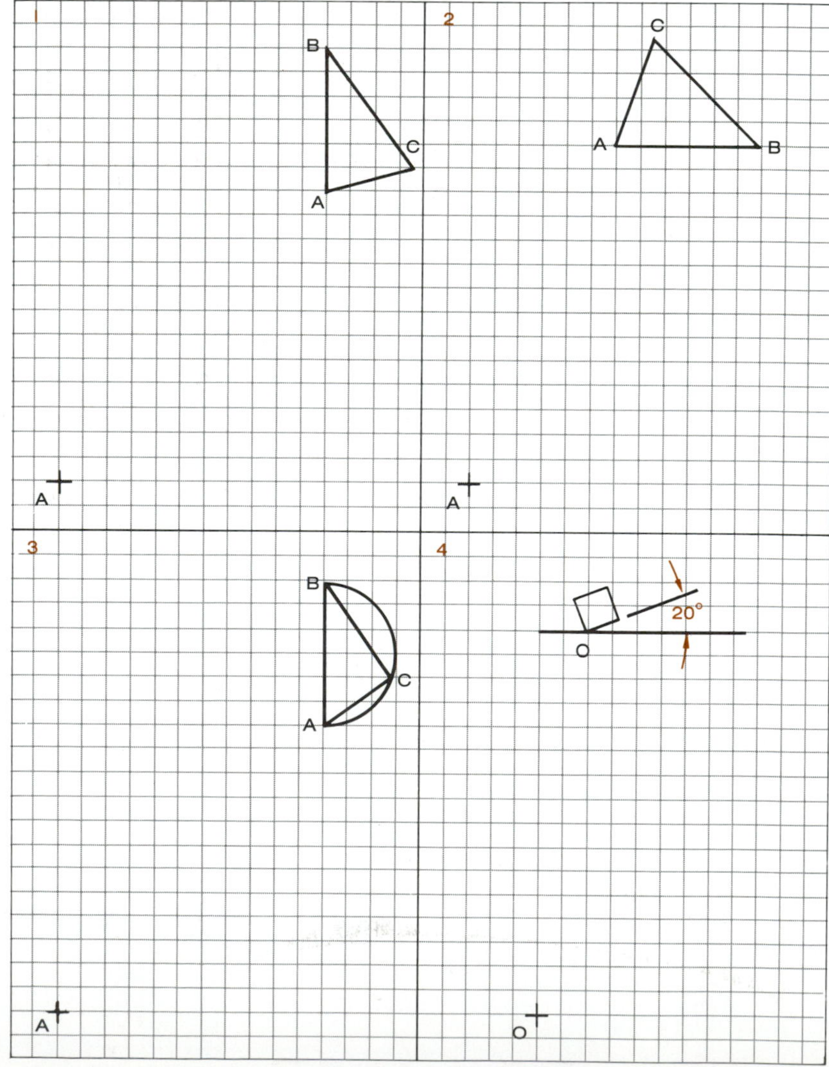

FIGURE 4-21 Checkpoint 3 problem

4.10 CONSTRUCTING A PENTAGON

Given the diameter of a circumscribed circle, construct a pentagon using the following method, shown in Figure 4-22(a):

1. Using the dividers, set off the circumference of the circle into five equal parts.
2. Join the points with straight lines.

Figure 4-22(b) shows an alternate method of constructing a pentagon when the diameter of the circumscribed circle is given:

STEP 1

1. Draw center lines AB and CD.
2. At point O, draw the circumscribing circle.
3. Bisect the radius OD to locate point E.
4. With point E as a center and EA as a radius, swing arc R_1 to locate point F.
5. With point A as a center and AF as a radius, swing arc R_2 to locate point G on the circle.

STEP 2

1. Draw line AG, which is one side of the pentagon.
2. Set off distances AG around the circumference of the circle.
3. Join these points with four straight lines.

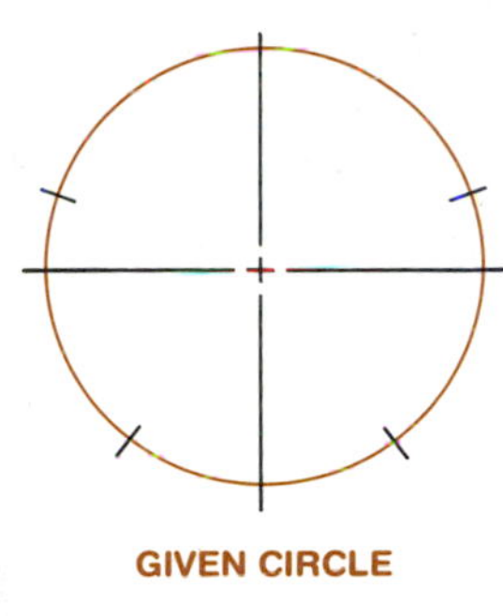

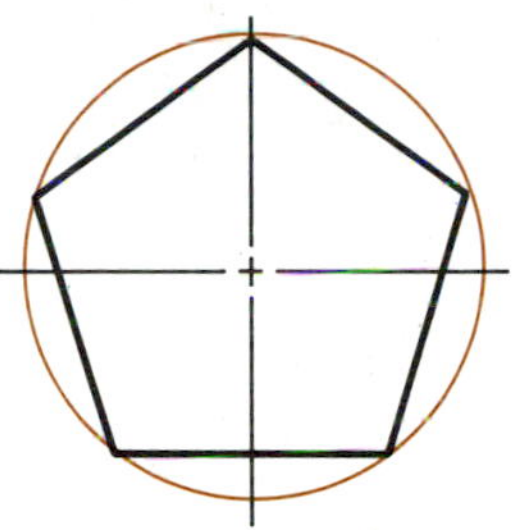

(a)

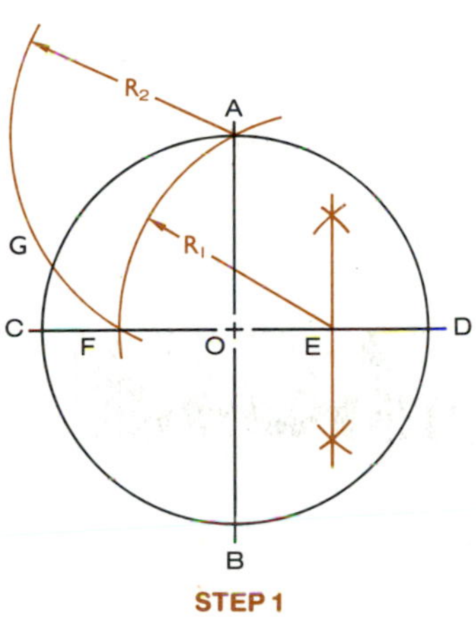

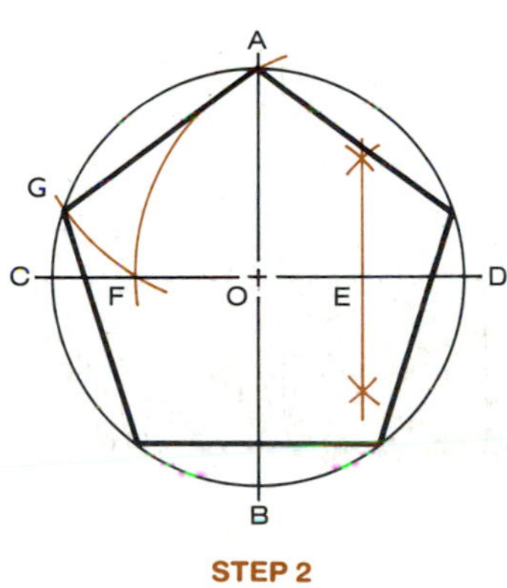

(b)

FIGURE 4-22 Constructing a pentagon ***(Section 4.10)***

4.11 CONSTRUCTING A HEXAGON

When given the diameter of the circumscribed circle, or the distance across the corners, you can construct a hexagon using the following procedure (see Figure 4-23):

STEP 1

1. Draw vertical and horizontal center lines AB and CD.
2. From point O, draw the circumscribing circle.
3. With points A and B as centers and a radius equal to AO, swing arcs R_1 and R_2 intersecting the circle at points EF and GH.

STEP 2

Using a triangle, join these points with six straight lines.

Given the distance across the flats, you can use the following method for constructing a hexagon, as shown in Figure 4-24(a):

1. Draw the vertical and horizontal center lines.
2. Draw the circle with the diameter equal to the distance across the flats.
3. Using the T square and a 30°–60° triangle, draw construction lines through the center of the circle at 60° intervals.
4. Draw the six sides perpendicular to the construction lines as shown.

The following procedure, illustrated in Figure 4-24(b), will enable you to draw a hexagon when given the distance across the corners:

1. Draw the vertical and horizontal center lines.
2. Draw the circle with the diameter equal to the distance across the corners.
3. Using the T square and the 30°–60° triangle, draw the diagonals at 30° or 60° angles with the horizontal.
4. Draw the six sides as shown.

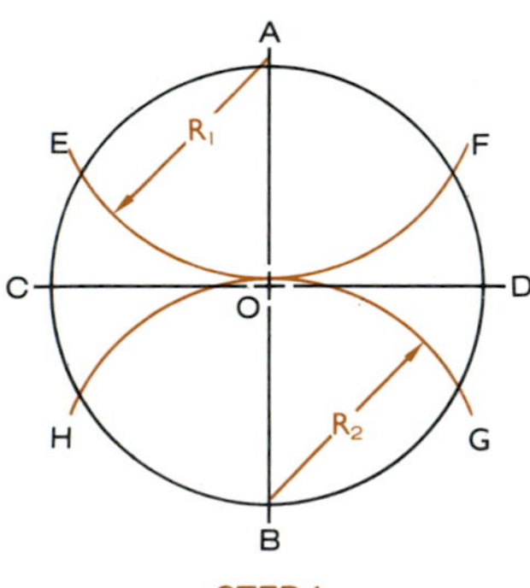

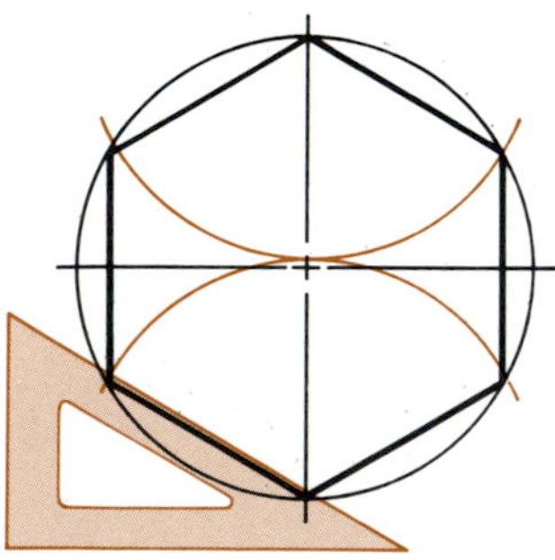

FIGURE 4-23 Constructing a hexagon given the diameter of the circumscribed circle or the distance across the corners *(Section 4.11)*

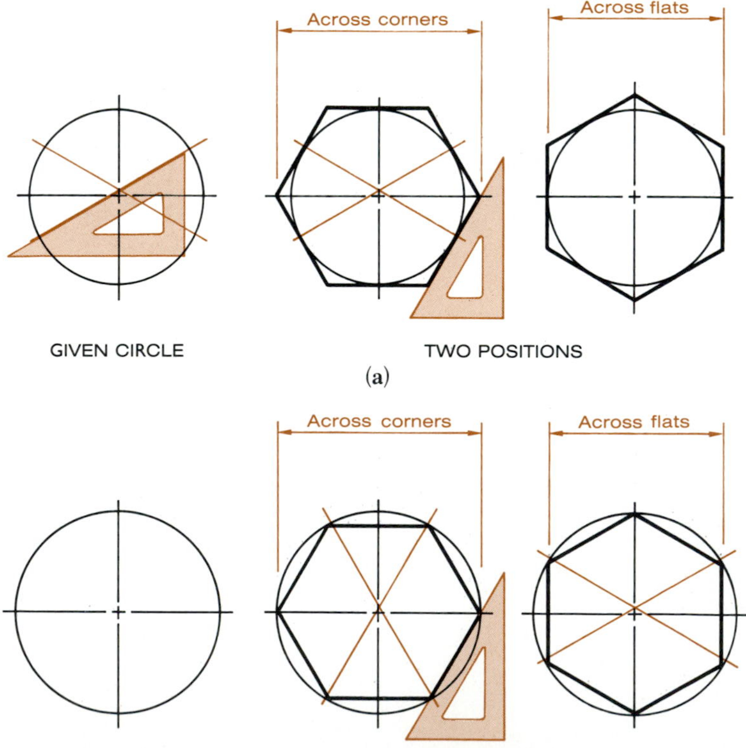

FIGURE 4-24 Constructing a hexagon *(Section 4.11)*

4.12 CONSTRUCTING AN OCTAGON

Figure 4-25(a) shows how to construct an octagon when the distance across the flats is given:

1. Draw the vertical and horizontal center lines.
2. Draw the circle with the diameter equal to the distance across the flats.
3. Using the T square and the 45° triangle, draw construction lines through the center of the circle at 45° intervals.
4. Draw the eight sides perpendicular to the construction lines and tangent to the circle as shown.

Given the distance across the corners, use the method shown in Figure 4-25(b) to construct an octagon:

1. Draw the vertical and horizontal center lines.
2. Draw the circle with the diameter equal to the distance across the corners.
3. Using the T square and the 45° triangle, draw the diagonals.
4. Draw the eight sides by joining the intersection of the diagonals of the circumference of the circle.

Figure 4-25(c) shows how to draw an octagon when given the circumscribed square or the distance across the flats:

1. Draw a diagonal of the square.
2. With the corners of the given square as centers and with one-half the diagonal as the radius, swing arcs as shown.
3. Using the T square and the 45° triangle, draw the eight sides.

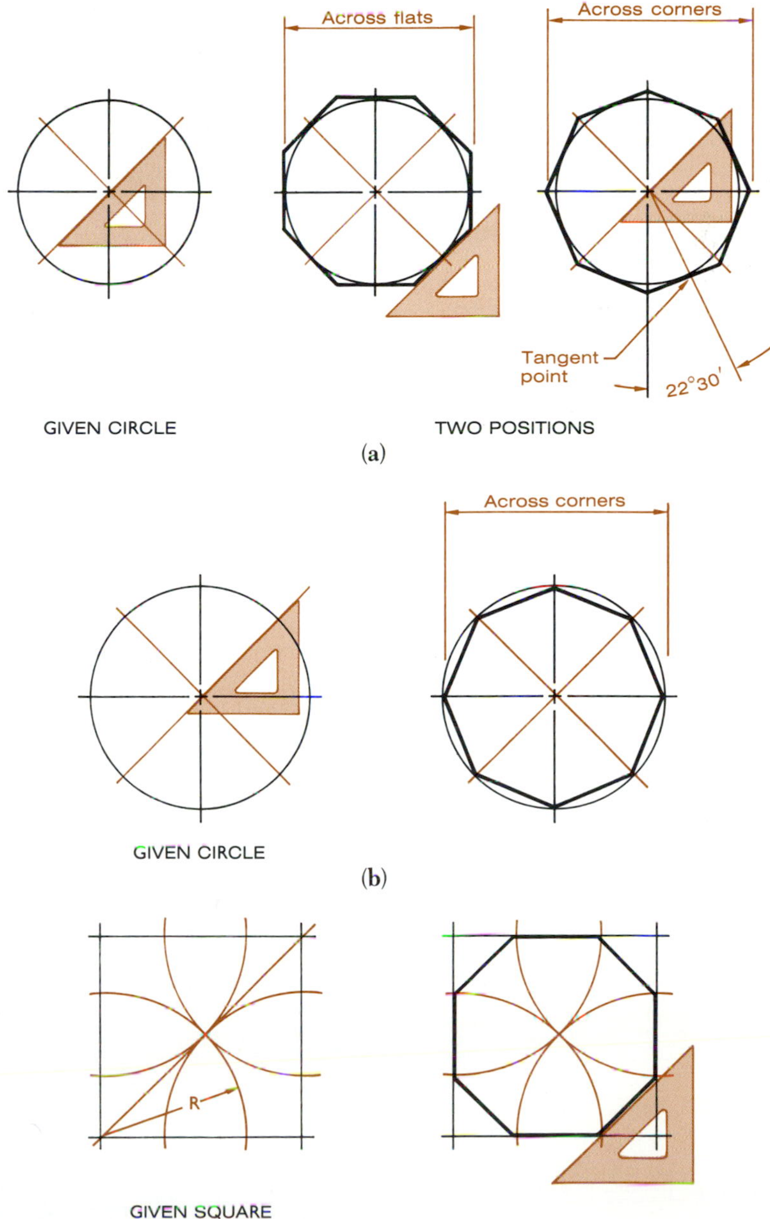

FIGURE 4-25 **Constructing an octagon** ***(Section 4.12)***

CHECKPOINT 4
CONSTRUCTING A PENTAGON, A HEXAGON, AND AN OCTAGON

Using plain, 8½ × 11 inch paper, lay out the sheet into six equal spaces, as shown in Figure 4-26. Prepare each drawing to the indicated scale. Neatly balance each drawing within the available space.

1. Using the method shown in Figure 4-22(a), inscribe a pentagon in a 4⅝-inch diameter circle. Make one side of the pentagon horizontal.
2. Using the method shown in Figure 4-22(b), inscribe a pentagon in a 2¼-inch diameter circle. Make one side of the pentagon vertical.
3. Using the method shown in Figure 4-23, construct a hexagon 2⅞ inches across the corners with one side 15° to the horizontal.
4. Using the method shown in Figure 4-24(a), construct a hexagon 50 mm across the flats with one pair of flats horizontal.
5. Using the method shown in Figure 4-25(a), construct an octagon 1¾ inches across the flats with one pair of flats horizontal.
6. Using the method shown in Figure 4-25(b), construct an octagon 450 mm across the corners with one pair of flats vertical.

FIGURE 4-26 Checkpoint 4 problem

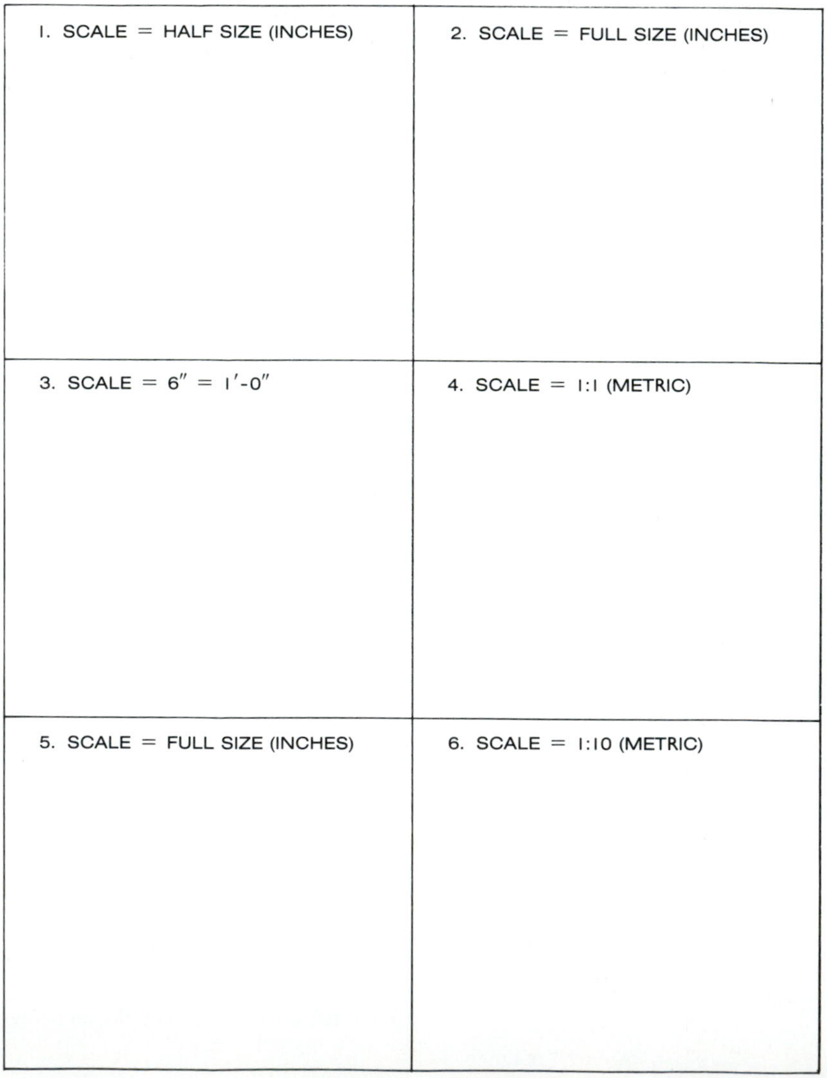

1. SCALE = HALF SIZE (INCHES)	2. SCALE = FULL SIZE (INCHES)
3. SCALE = 6″ = 1′-0″	4. SCALE = 1:1 (METRIC)
5. SCALE = FULL SIZE (INCHES)	6. SCALE = 1:10 (METRIC)

4.13 DRAWING A CIRCLE THROUGH THREE POINTS

The following procedure, shown in Figure 4-27, will enable you to draw a circle through given points A, B, and C:

1. Draw lines AB and BC.
2. Construct perpendicular bisectors OX and OY.
3. With point O as center and with a radius OA, draw the circle through the given points. (Note: Lines AB and BC are chords of the required circle.)

4.14 FINDING THE CENTER OF A CIRCLE

Given any circle's diameter, you can find it's center by using the following method (illustrated in Figure 4-28):

1. Draw any horizontal or vertical chord AB.
2. Draw perpendiculars from A and B intersecting the given circle at X and Y.
3. Draw lines AY and BX, whose intersections will be at the center of the circle, point O.

4.15 DRAWING A CIRCLE TANGENT TO A LINE AT A GIVEN POINT

Given line AB and point P on the line, you can use this method (shown in Figure 4-29) to construct a circle of any given radius tangent to point P:

1. Construct a perpendicular to the line AB at point P, and mark off the radius of the required circle (PO) on the perpendicular.
2. Draw the circle with a radius equal to PO.

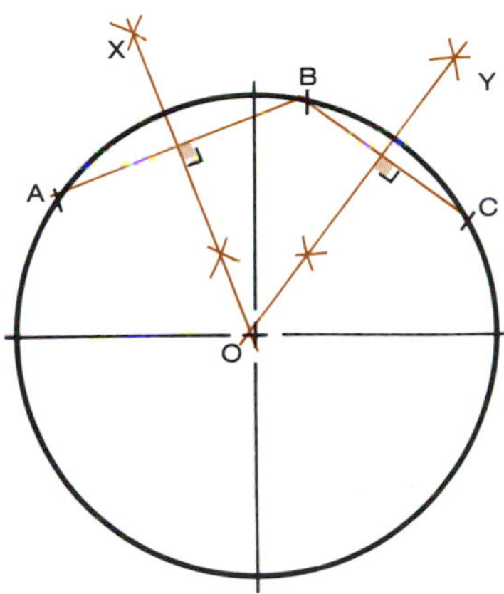

FIGURE 4-27 Drawing a circle through three points *(Section 4.13)*

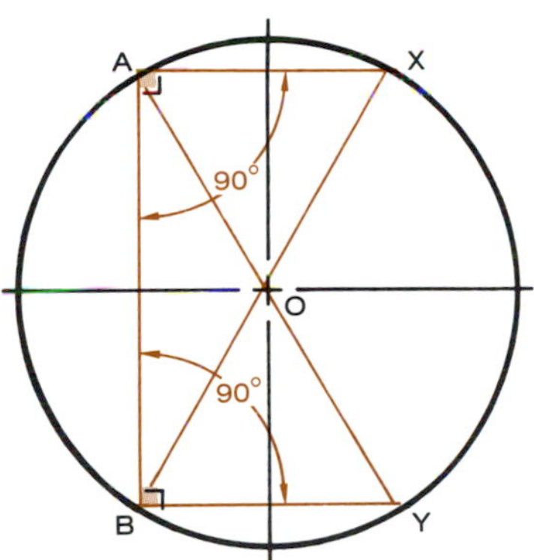

FIGURE 4-28 Finding the center of a circle *(Section 4.14)*

FIGURE 4-29 Drawing a circle tangent to a line at a given point *(Section 4.15)*

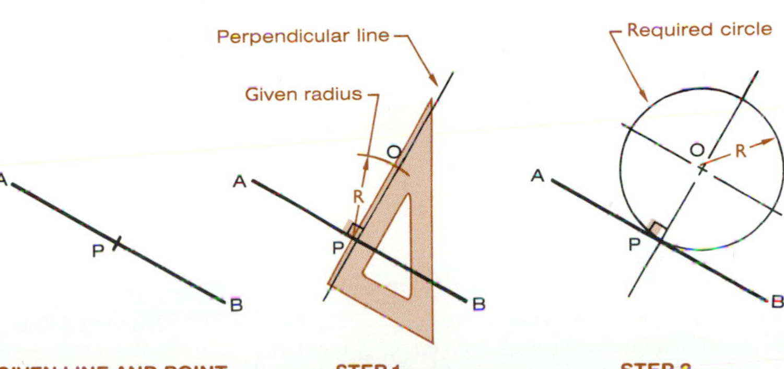

4.16 DRAWING A LINE TANGENT TO A CIRCLE THROUGH A POINT ON A CIRCLE

Figure 4-30 shows how to draw a line tangent to a given circle through point P on the circumference of the circle:

1. Set the triangle so that one side connects points P and O, then move the T square into contact with the triangle as shown and slide the triangle until the other side is perpendicular to line PO.
2. Draw the tangent line.

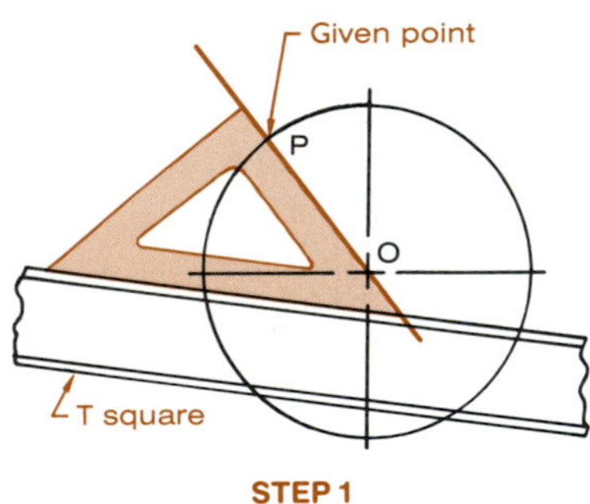

STEP 1

Required tangent line
P
O

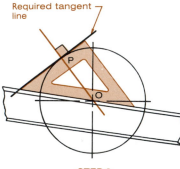

STEP 2

FIGURE 4-30 Drawing a line tangent to a circle through a given point on the circle *(Section 4.16)*

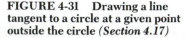

4.17 DRAWING A LINE TANGENT TO A CIRCLE THROUGH A POINT OUTSIDE A CIRCLE

Figure 4-31 illustrates this procedure for drawing a tangent to a given circle through point P outside the circle:

1. Set the triangle so that one side passes through point P and is tangent to the circle. Draw a light construction line.

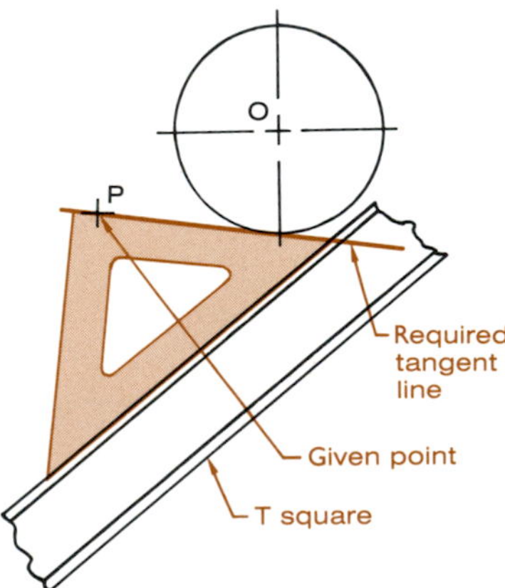

STEP 1

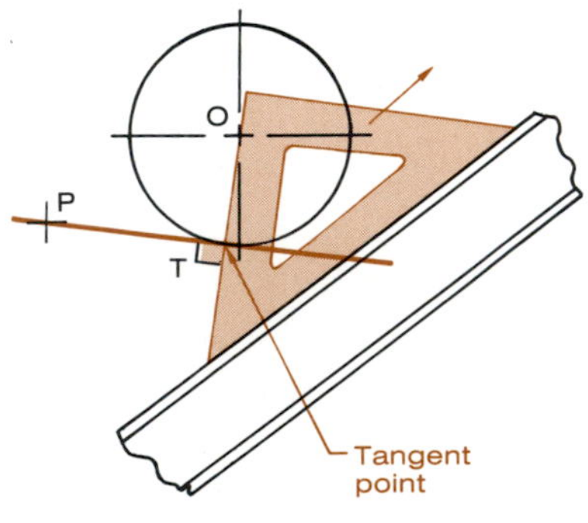

STEP 2

FIGURE 4-31 Drawing a line tangent to a circle at a given point outside the circle *(Section 4.17)*

2. Move the T square into contact with the triangle as shown and slide the triangle until the other side passes through point O.
3. Mark T, the point of tangency.

4.18 DRAWING TANGENT ARCS

There are several methods you can use to draw tangent arcs. The first, illustrated in Figure 4-32, is for use when you are given two lines intersecting at right angles at point O and radius R:

1. With point O as the center, swing an arc locating tangent points T_1 and T_2.
2. With points T_1 and T_2 as centers and given radius R, swing two arcs that intersect at point C. (Note: A line connecting points O and C is the bisector of angle O.) With point C as center and given radius R, swing the required tangent arc.

When given radius R and two lines intersecting at acute or obtuse angles at point O, you can use this method (see Figures 4-33 and 4-34) to construct a tangent arc:

1. Draw lines parallel to the given lines at a distance R to intersect at point C, then from C drop perpendiculars to the given lines respectively to locate points of tangency T_1 and T_2.

2. With C as center and with given radius R, draw the required tangent arc between the points of tangency.

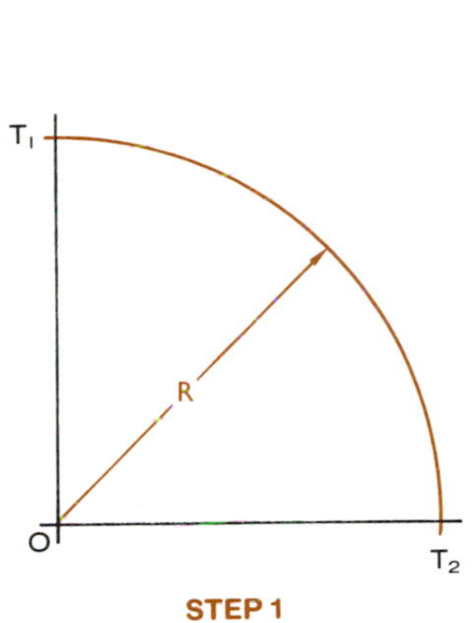

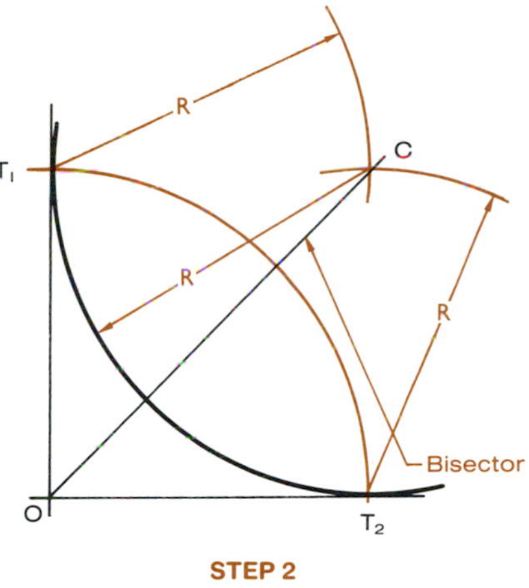

FIGURE 4-32 Drawing an arc tangent to two lines at right angles *(Section 4.18)*

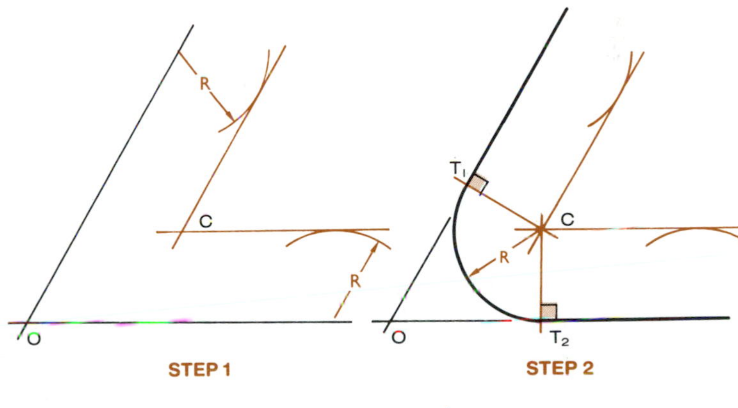

FIGURE 4-33 Drawing an arc tangent to two lines at acute angles *(Section 4.18)*

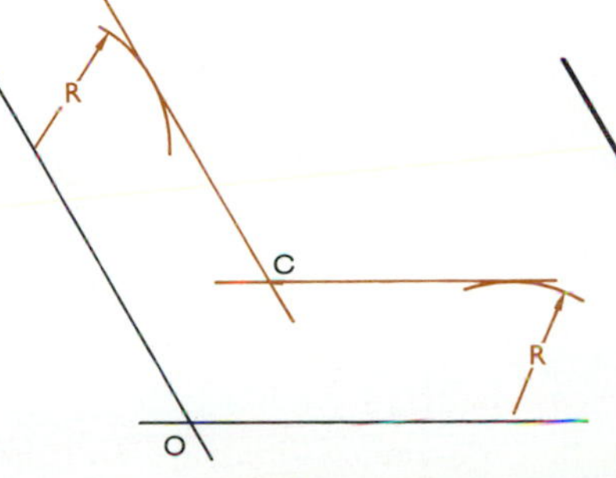

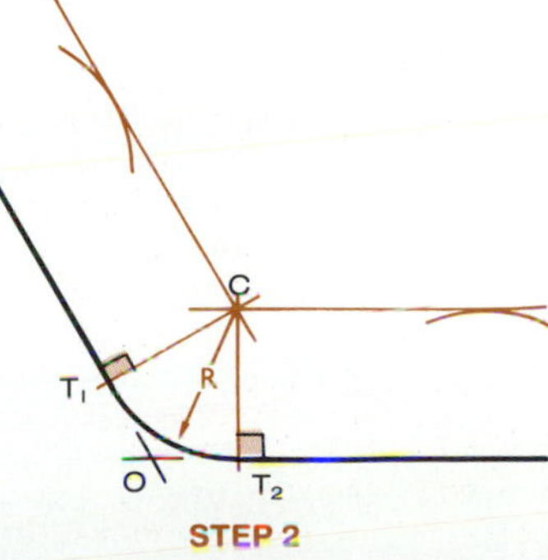

FIGURE 4-34 Drawing an arc tangent to two lines at obtuse angles *(Section 4.18)*

CHECKPOINT 5
CONSTRUCTING ARCS, CIRCLES, TANGENT LINES, AND POLYGONS

Use a plain sheet of 8½ × 11 inch paper for the following drawings. Make each drawing full size. Measure the given dimensions from the *edges* of the paper. Begin by dividing the sheet into four equal spaces, as shown in Figure 4-35. Do not erase your construction lines.

1. Lay out points A, B, and C in the positions shown in the upper left space of Figure 4-35. Draw a circle through these points using the method shown in Figure 4-27.
2. Locate point O at the center of the upper right space of your sheet by drawing diagonals. With O as center, draw a 3⅟16-inch diameter circle. Verify the location of point O as the center of the circle by using the construction shown in Figure 4-28.
3. Lay out line AB in the position shown in the lower left space of Figure 4-35. Locate point P midpoint on line AB. Draw a 2¹³⁄16-inch diameter circle tangent to line AB at point P using the method shown in Figure 4-29.
4. Locate point O at the center of the lower right space of your sheet by drawing diagonals. With O as center, draw a 2¾-inch diameter circle. Locate point A on the top half of the circumference of the circle 35° to the right of the vertical center line. Draw a line tangent to the circle through point A using the method shown in Figure 4-30. Locate point B 1⅞ inches below and ⅝ inch to the left of point O. Draw two lines, each tangent to the circle, through point B using the method shown in Figure 4-31.

Show both tangent points. On a second sheet of 8½ × 11 inch paper, lay out the problem shown in Figure 4-36 to scale of ⅛″ = 1′-0″. Begin at point A, located 1¾ inches in from the left edge of the sheet and 1¾ inches up from the lower edge of the sheet and lay out the figure using the given dimensions. At each of the intersection points on the figure, draw a tangent arc with the following radii using the methods shown in Figures 4-32, 4-33, or 4-34: A = 12′-6″; B = 4′-3″; C = 4′-0″; D = 4′-6″; E = 6′-6″; F = 8′-0″; G = 7′-9″; H = 10′-0″; I = 10′-0″; J = 6′-6″; K = 2′-0″. (All angles are 90° unless otherwise specified. Do not erase your construction lines and do not copy the dimensions.)

FIGURE 4-35 Checkpoint 5 problem

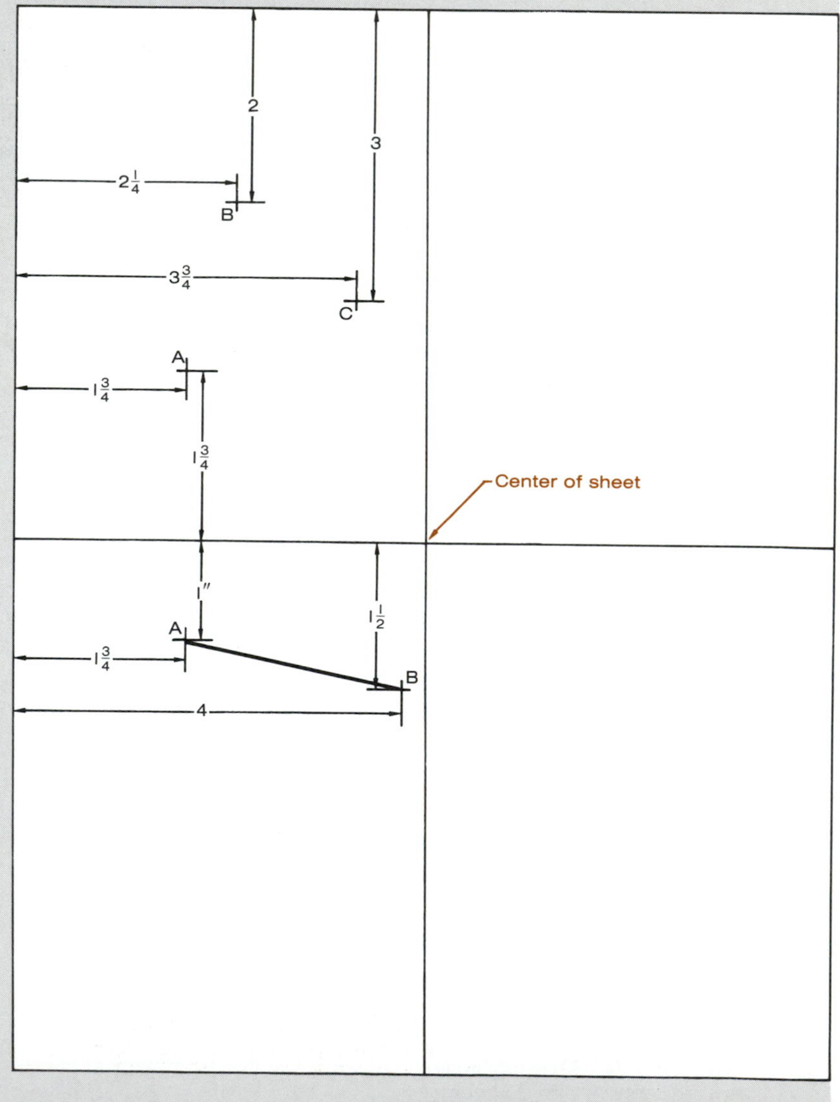

Given an arc of radius G with center at O and a straight line AB, you can construct a tangent arc of radius R using the following method (see Figure 4-37):

1. Construct a line parallel to AB at a distance R.
2. Using O as the center, swing a concentric arc (radius G plus radius R) locating point X on the parallel line.
3. Draw line XO connecting the centers of the given and the required arcs to locate tangent point T_1. Draw line XT perpendicular to AB to locate tangent point T_2. Using X as the center with radius R, swing the tangent arc.

Given the same information, but referring to Figure 4-38, you can construct a tangent arc as follows:

1. Construct a line parallel to AB at the required distance R.
2. Using O as the center, swing a concentric arc (radius G minus radius R) parallel to the given arc of radius G, locating point X on the parallel line.
3. Draw lines XO and XT (perpendicular to AB) to locate the tangent points T_1 and T_2 and then, using X as the center with radius R, swing the tangent arc.

FIGURE 4-36 Checkpoint 5 problem

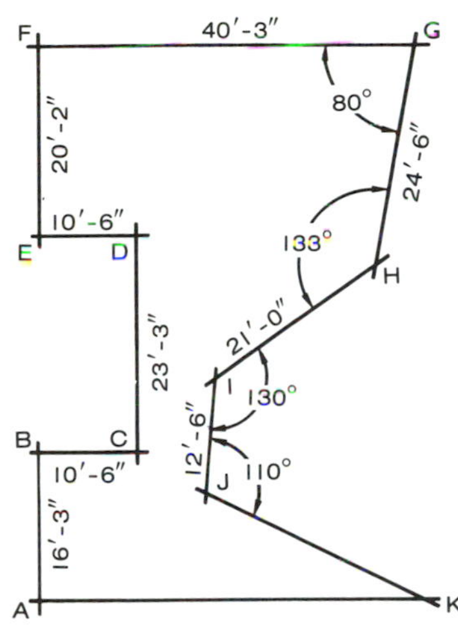

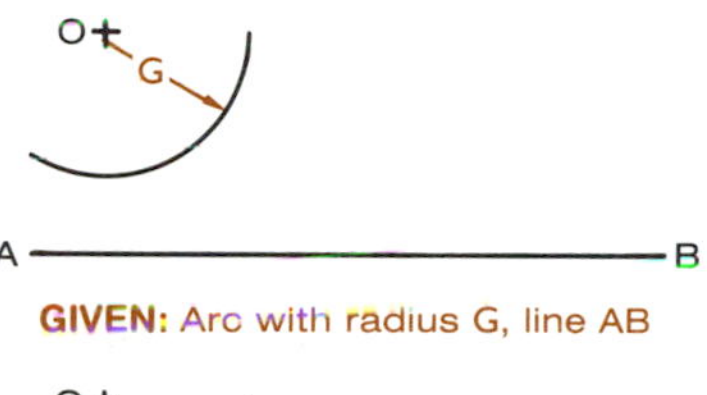

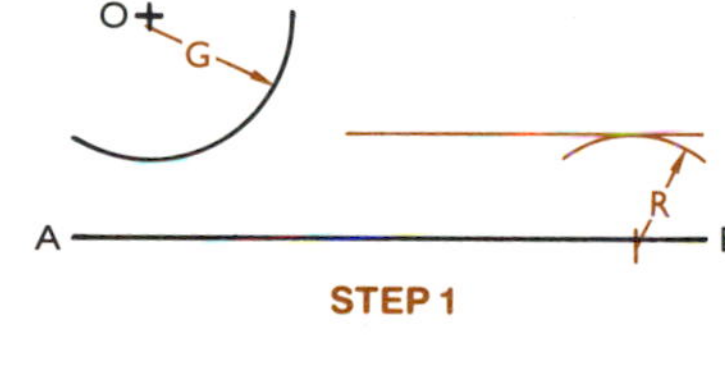

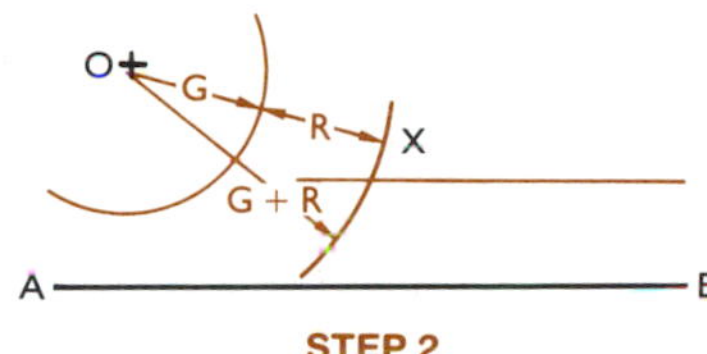

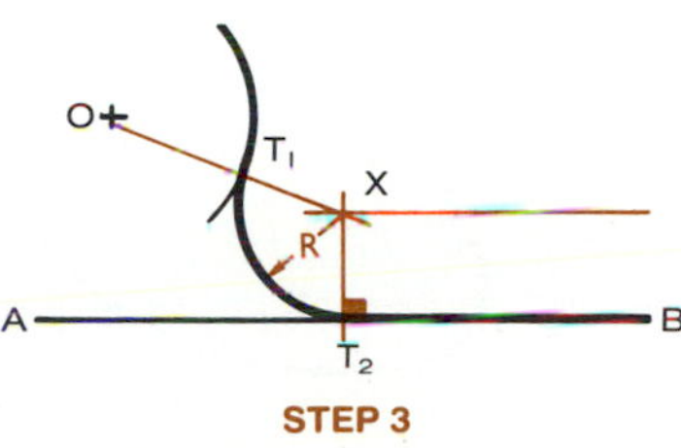

FIGURE 4-37 Drawing an arc of radius R tangent to a given arc and straight line *(Section 4.18)*

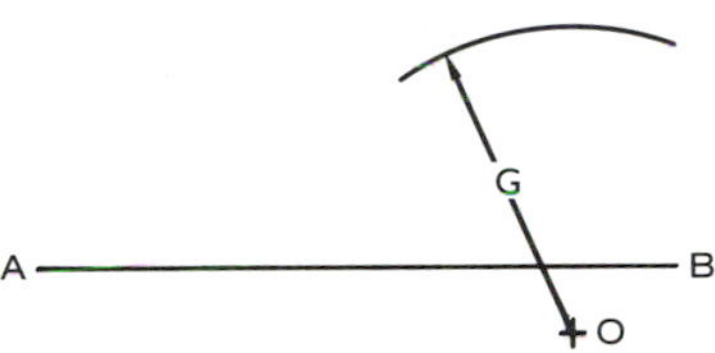

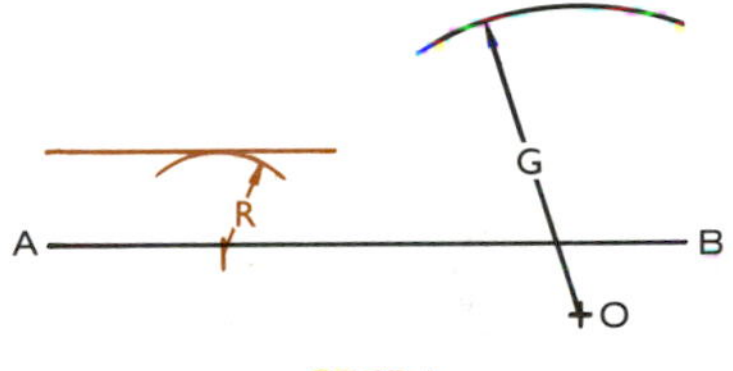

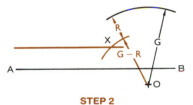

STEP 2

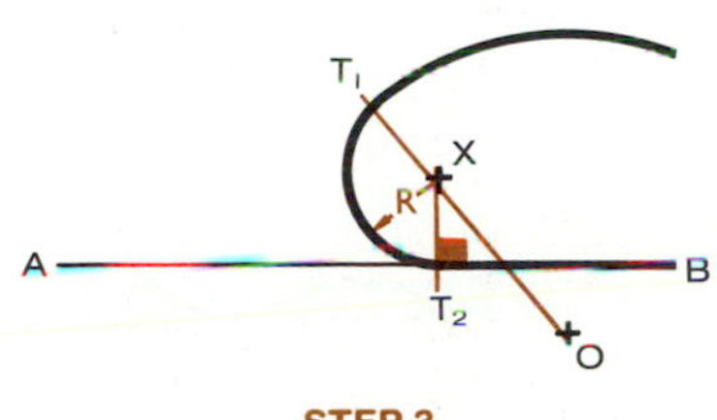

FIGURE 4-38 Alternate method for drawing an arc of radius R tangent to a given arc and straight line *(Section 4.18)*

To construct an arc of radius R tangent to two given arcs, use the following procedure (shown in Figure 4-39):

1. Using A and B as centers of the given arcs with radius G_1 and G_2, respectively, *add* the required radius R (radius G_1 plus R and radius G_2 plus R) and swing concentric intersecting arcs, locating point X.
2. Draw lines XA and XB to locate the tangent points T_1 and T_2. Then, using X as the center with radius R, swing the tangent arc.

You can construct an arc of radius R tangent to two given arcs and enclosing one of them by using the following method, as shown in Figure 4-40:

1. Using A and B as centers, swing concentric arcs to intersect at point X. (The radius drawn from point A is equal to radius G_1 *plus* R and the radius drawn from point B is equal to radius G_2 *minus* R.)
2. Draw lines XA and XB to locate the tangent points T_1 and T_2 and, using X as the center with radius R, swing the tangent arc.

You can construct an arc of radius R tangent to two arcs and enclosing both given arcs by using the following method (illustrated in Figure 4-41):

1. Using A and B as centers of the given arcs with radius G_1 and G_2, respectively, *subtract* radius R (radius R minus G_1 and radius R minus G_2). Swing intersecting arcs locating point X.
2. Draw lines XA and XB. Extend them to intersect each

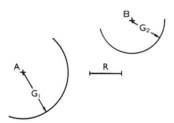

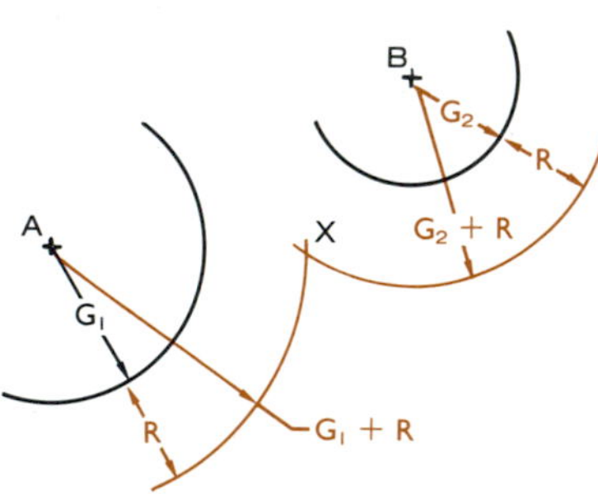

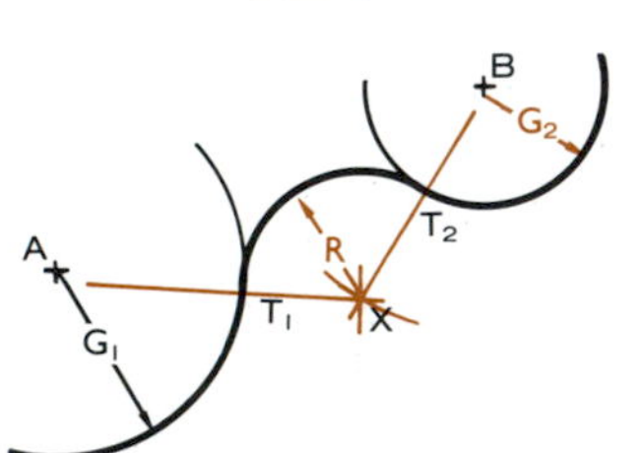

FIGURE 4-39 **Drawing an arc of radius R tangent to two given arcs *(Section 4.18)***

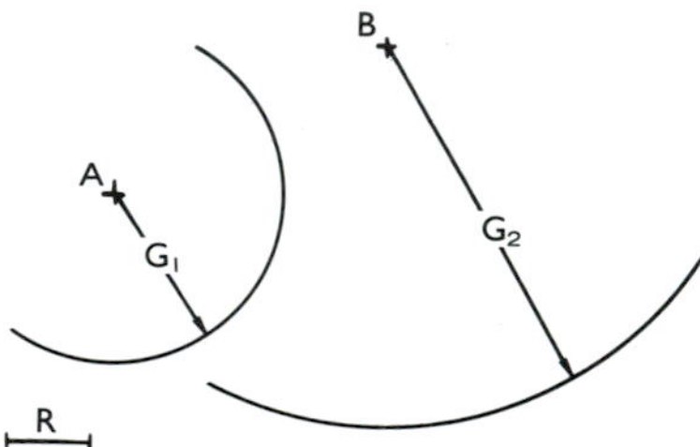

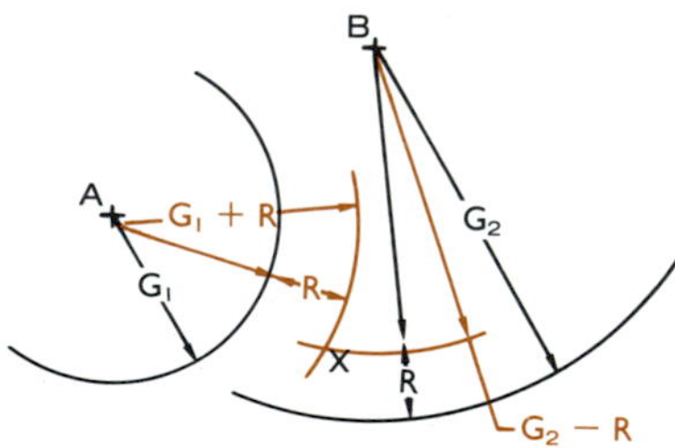

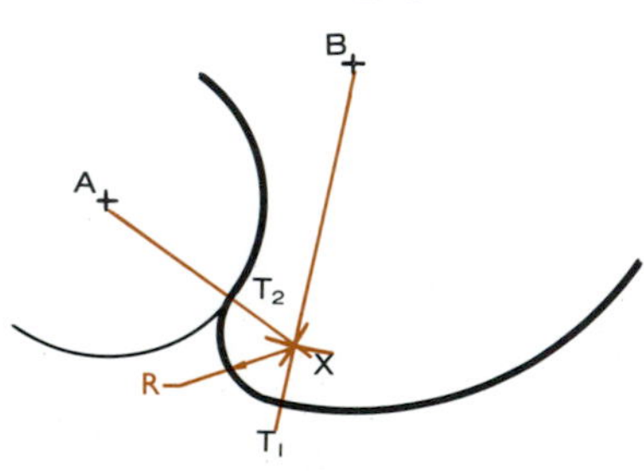

FIGURE 4-40 **Drawing an arc of radius R tangent to two given arcs and enclosing one of them *(Section 4.18)***

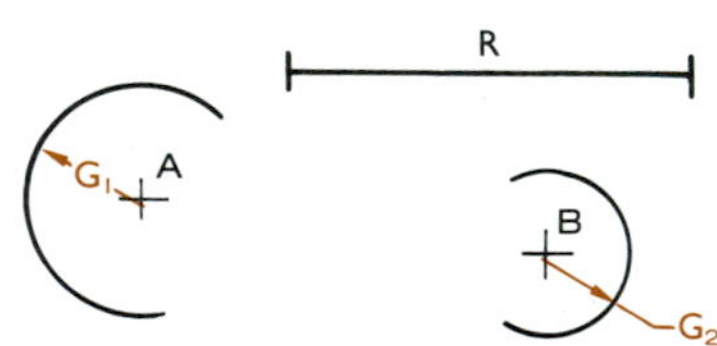

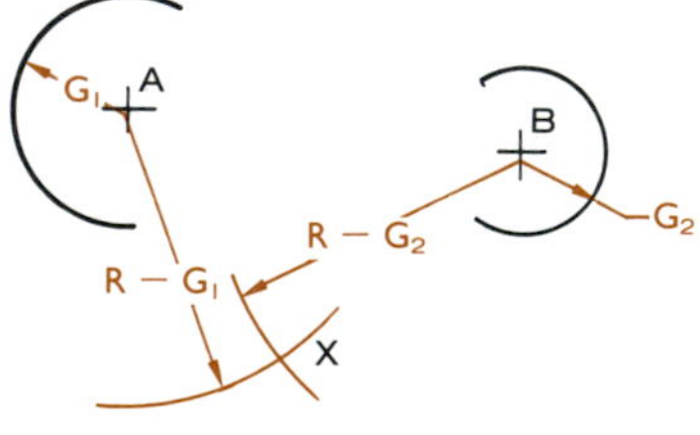

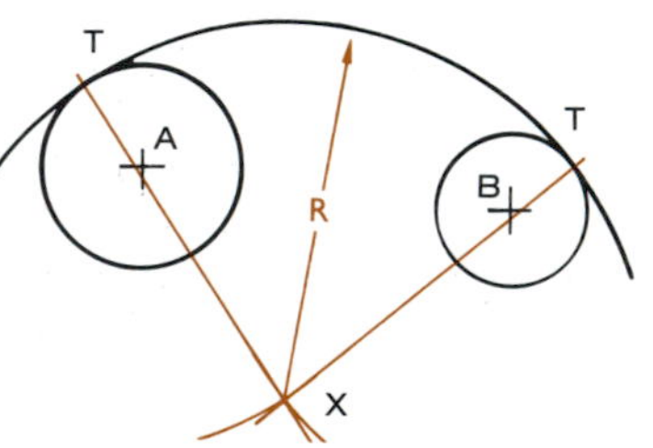

FIGURE 4-41 **Drawing an arc of radius R tangent to two given arcs and enclosing both of them *(Section 4.18)***

given arc to locate the tangent points. Using X as the center with radius R plus G_1 (or R plus G_2), swing the tangent arc.

Finally, you can construct an arc with radius R tangent to two arcs and enclosing one given arc, by using the method shown in Figure 4-42:

1. Using A and B as centers of the given arcs with radius G_1 and G_2, respectively, *subtract* radius R (radius G_1 minus R) and *add* radius R (radius G_2 plus R). Swing intersecting arcs locating point X.
2. Draw lines XA and XB to locate the tangent points. Extend line XA to intersect the given arc. Using X as center with radius G_1 minus R (or G_2 plus R), swing the tangent arc.

4.19 DRAWING A REVERSE, OR OGEE, CURVE

Use the following procedure to connect parallel lines AB and CD with tangent arcs of equal or unequal radii (see Figure 4-43):

1. Assume point P on line BC at any convenient location. If two equal arcs are required, assume point P at midpoint.
2. Construct a perpendicular at B on line AB and C on line CD to locate the tangent points and construct perpendicular bisectors of BP and PC.
3. Draw the tangent arcs using the intersections X and Y of the bisectors and the perpendiculars as the centers.

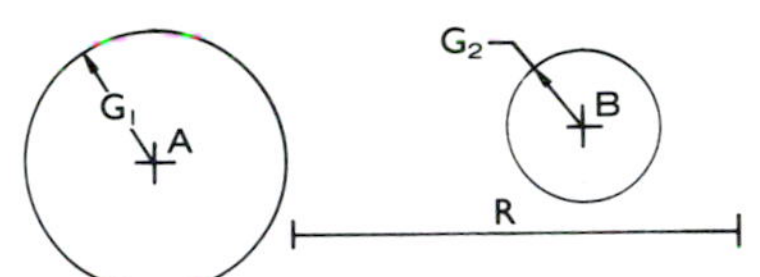

GIVEN: Arcs G_1 and G_2, radius R

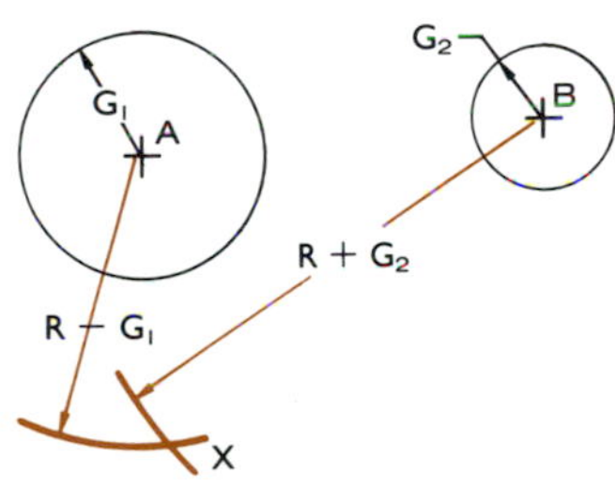

STEP 1

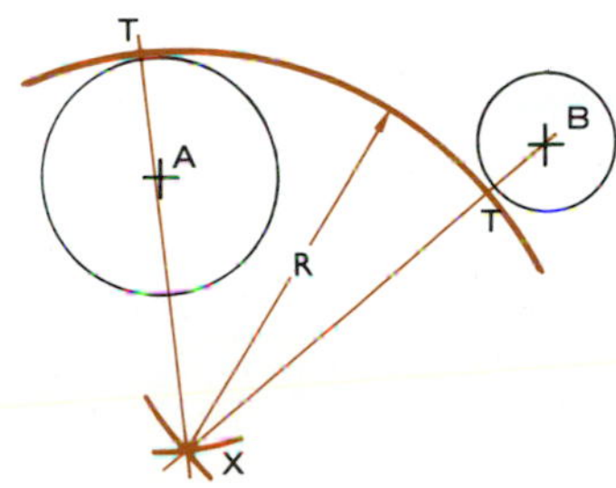

STEP 2

FIGURE 4-42 Drawing an arc of radius R tangent to two arcs and enclosing one given arc *(Section 4.18)*

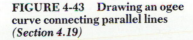

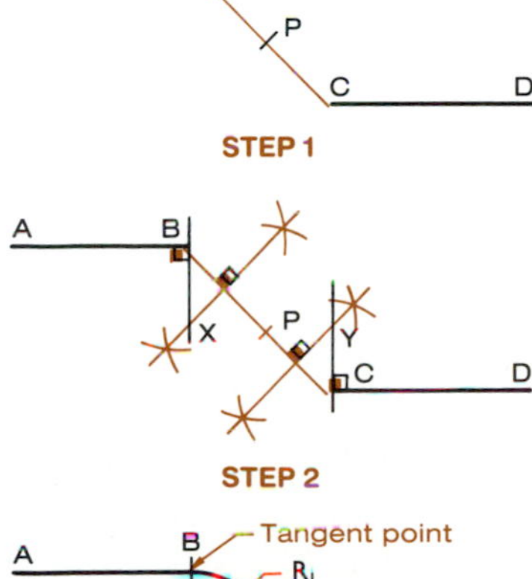

STEP 2

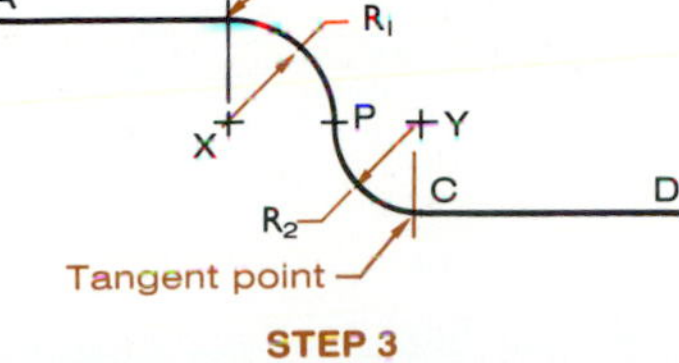

STEP 3

FIGURE 4-43 Drawing an ogee curve connecting parallel lines *(Section 4.19)*

Use the procedure shown in Figure 4-44 to connect two nonparallel lines AB and CD.

1. Construct a perpendicular at B. Locate point P so that PB equals the desired radius R_1 and swing the first of the two tangent arcs.
2. Construct a perpendicular at C and make CE equal to BP. Draw line PE and bisect it.
3. The intersection F of the bisector and the perpendicular is the center of the second tangent arc, with a radius R_2. (Note: The inflection point T of the curve is found by drawing line PF.)

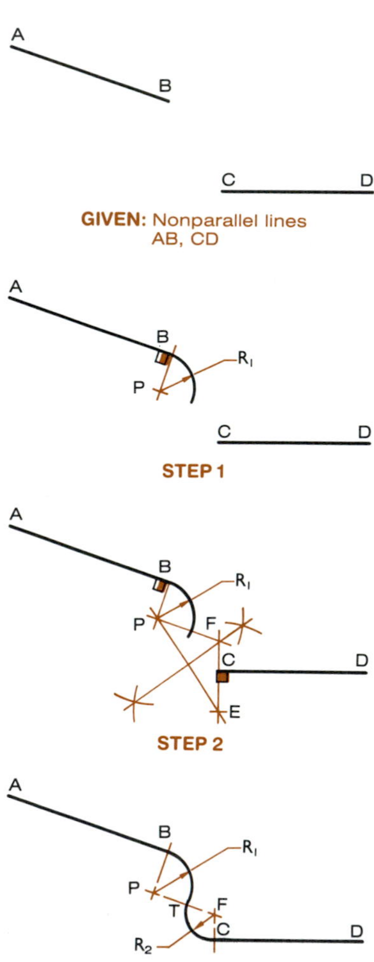

FIGURE 4-44 Drawing an ogee curve connecting nonparallel lines *(Section 4.19)*

CHECKPOINT 6
CONSTRUCTING ARCS

Lay out an 8½ × 11 inch sheet of ¼-inch-grid paper as shown in Figure 4-45. Prepare the following drawings full size. Show all construction with light lines and indicate all tangent points with a medium weight dashed line drawn across the point.

1. Construct an arc with radius of 1 11/16 inches tangent to an arc with radius G and line AB.
2. Construct an arc with radius of 13/16 inch tangent to an arc with a radius G and line AB.
3. Construct two arcs with radii of 2.12 inches tangent to arcs G_1 and G_2.
4. Construct an arc with a radius of .68 inch tangent to arcs G_1 and G_2.

Lay out a second sheet of ¼-inch-grid paper as shown in Figure 4-46. Prepare the following drawings full size.

1. Construct an arc with a radius of 3.25 inches tangent to and enclosing the two given circles with radii G_1 and G_2.
2. Construct an arc with a radius of 2 5/16 inches tangent to two circles (with radii G_1 and G_2) and enclosing circle G_2.
3. Construct an ogee curve connecting two parallel lines AB and CD with equal radii.
4. Construct an ogee curve connecting two nonparallel lines AC and CD with one radius equal to ½ inch.

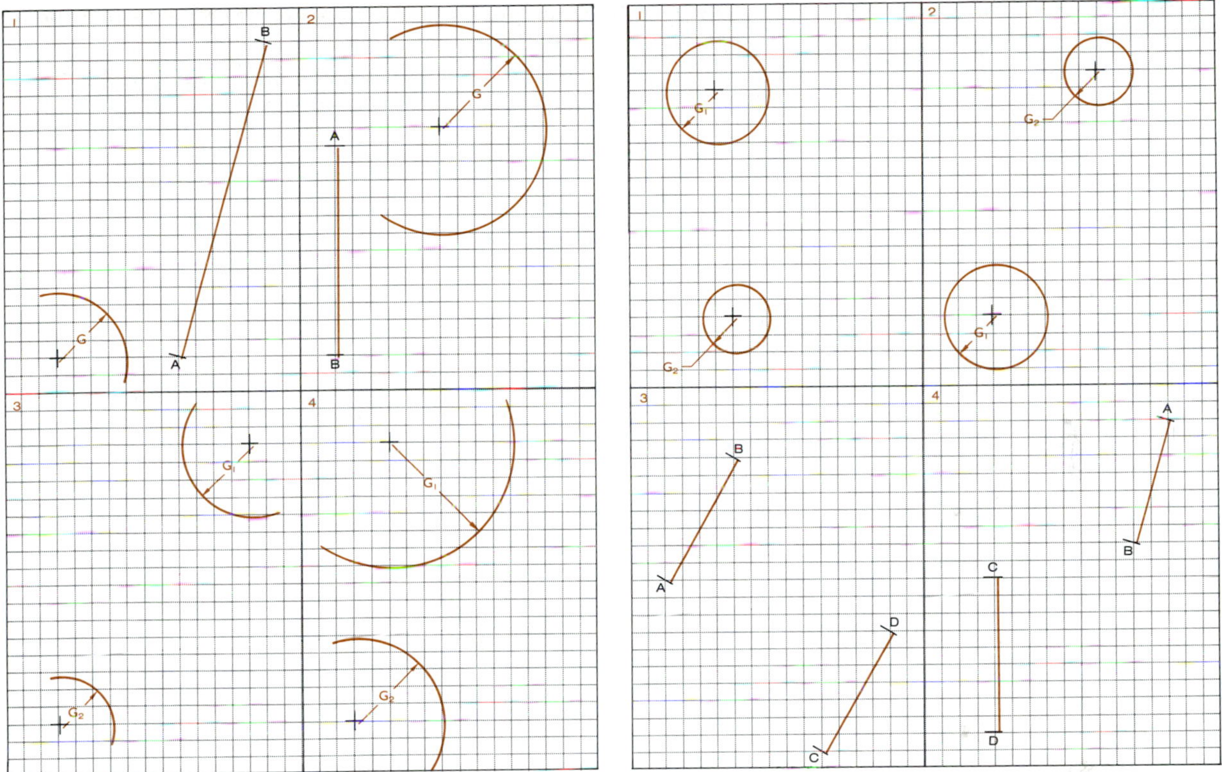

FIGURE 4-45 Checkpoint 7 problem

FIGURE 4-46 Checkpoint 7 problem

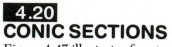

4.20 CONIC SECTIONS

Figure 4-47 illustrates four types of curves produced by cutting a right circular cone at various angles with a cutting plane: The *circle*, *ellipse*, *parabola*, and *hyperbola*. A circle is formed when the intersecting plane is perpendicular to the axis. An ellipse is formed when the intersecting plane makes a greater angle with the axis than do the elements. A parabola is formed when the intersecting plane makes the same angle with the axis as do the elements. A hyperbola is formed when the intersecting plane makes a smaller angle with the axis than do the elements or when it is parallel to the axis. Each of these plane figures may be defined as the locus of a point moving in a particular way.

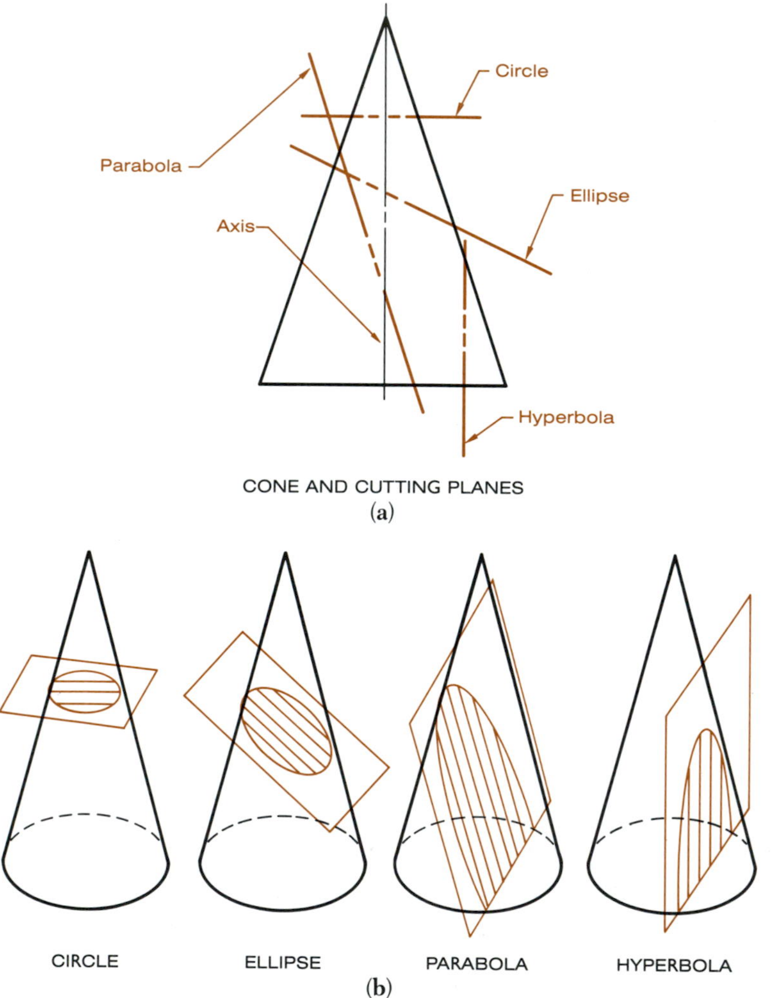

FIGURE 4-47 Conic sections ***(Section 4.20)***

4.21 CONSTRUCTING AN ELLIPSE

An ellipse is the locus of points, the sum of whose distances from two fixed points is a constant. Figure 4-48 illustrates the *foci method* for constructing an ellipse:

1. Using the given ends of the minor axis as centers (C and D), swing intersecting arcs with a radius equal to one-half the given major axis, locating the foci E and F.
2. Mark off a number of division points between E and O. (In this example, five points are sufficient to ensure an acceptable curve.) Using E and F as centers and radius A1 and B1, respectively, swing arcs to intersect at a corresponding point in each quadrant. Continue to swing pairs of arcs using the remaining division points, 2, 3, 4, and 5, in the same manner as before. Using an irregular curve, draw the ellipse through points A, 5′, 4′, and so on.

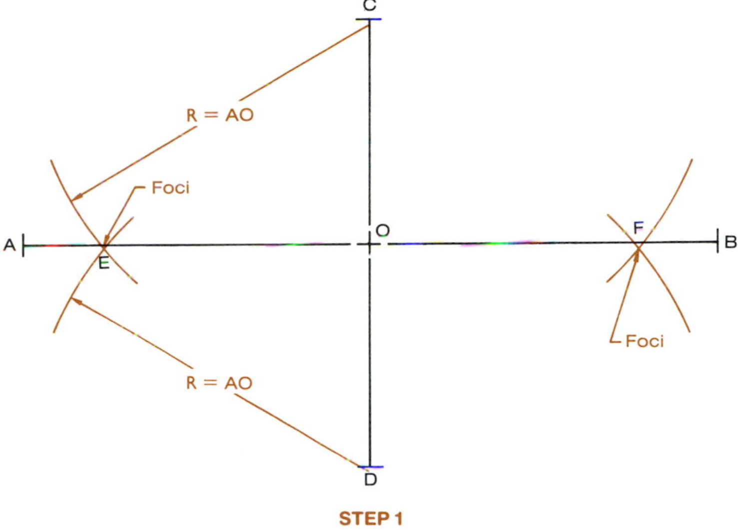

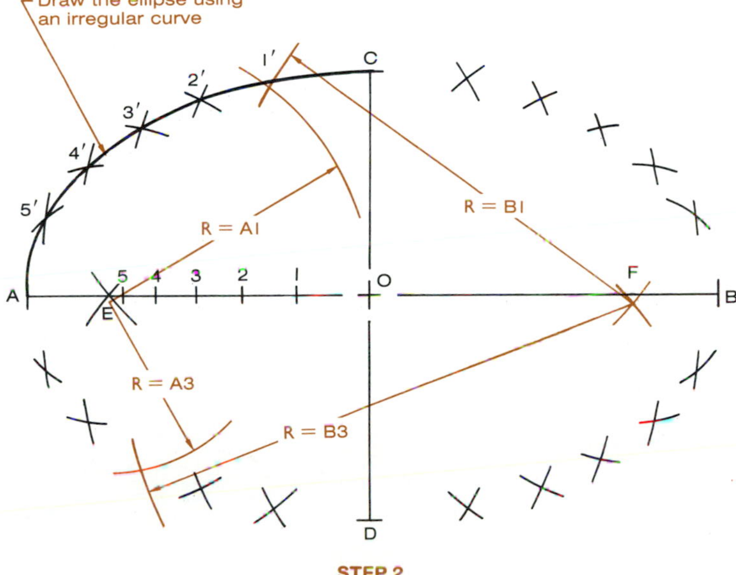

FIGURE 4-48 Ellipse construction: the foci method *(Section 4.21)*

The *concentric circle method* for constructing an ellipse is shown in Figure 4-49:

1. Draw two concentric circles with a center at O. The diameter of the large circle should equal the given major axis and the diameter of the small circle should equal the given minor axis. Divide the circles into any convenient number of sectors and draw diagonals. Points on the curve of the ellipse in each quadrant—point E, for example—may be plotted as shown by projecting downward from P_1 on the large circle to intersect with horizontal construction lines drawn from P_2 on the small circle.
2. Find additional points by repeating the procedure in each quadrant. Draw the ellipse using an irregular curve.

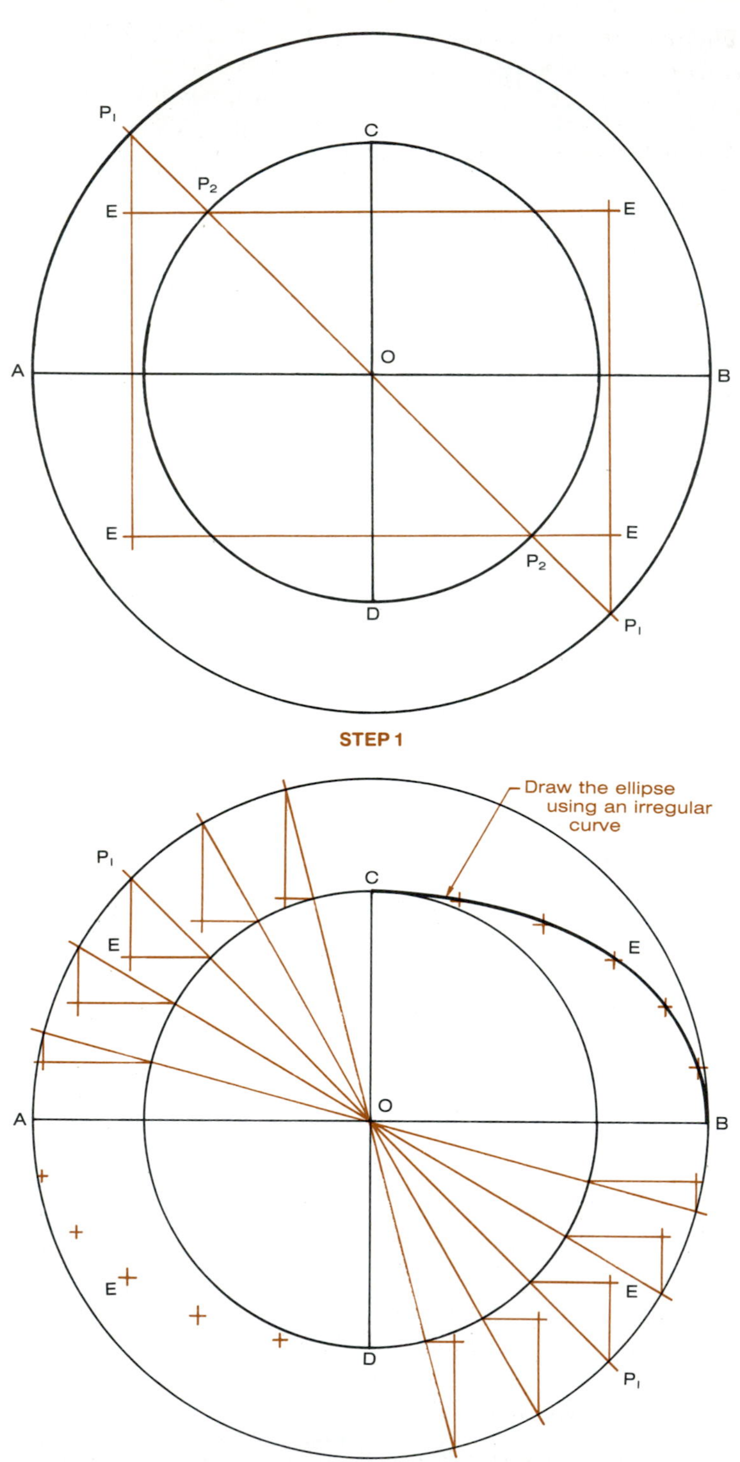

FIGURE 4-49 Ellipse construction: the concentric circle method *(Section 4.21)*

Figure 4-50 shows the *parallelogram method* for constructing an ellipse:

1. Given major axis AB and minor axis CD, draw a rectangle EFGH with the sides parallel and respectively equal to the axes.
2. Divide OA and AE into the same number of equal parts.
3. Draw lines C1, C2, C3, and C4 and lines D1, D2, D3, and D4. The intersection points of these lines fall on the ellipse. (Other points on the ellipse may be formed in each quadrant in a similar manner or they may be transferred from corresponding locations in the first quadrant.)
4. Draw the ellipse using an irregular curve.

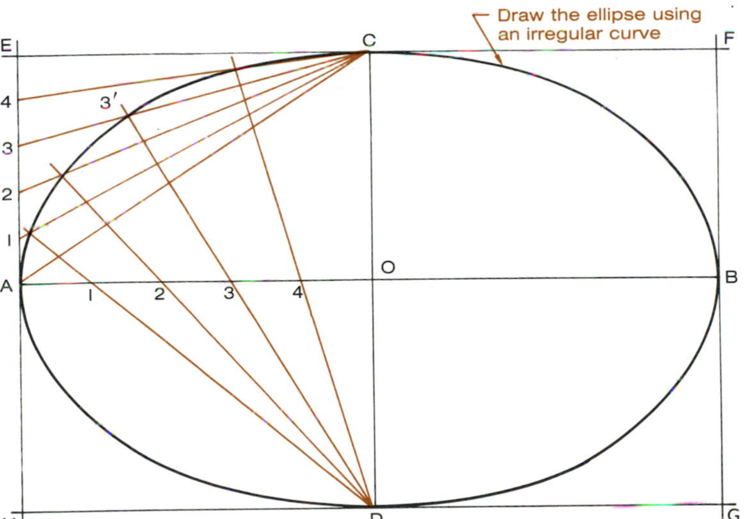

FIGURE 4-50 Ellipse construction: the parallelogram method ***(Section 4.21)***

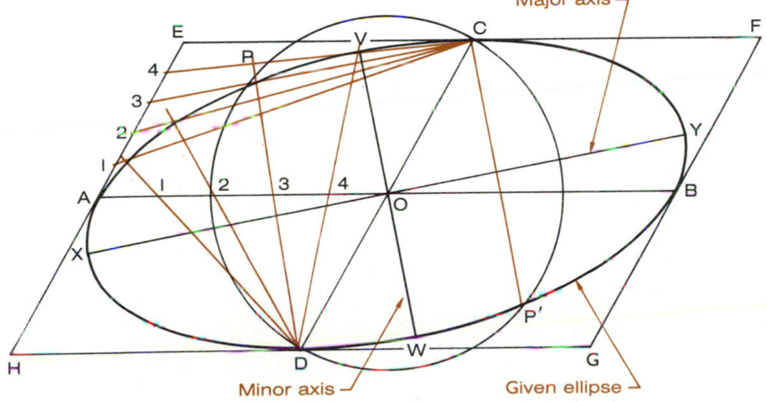

FIGURE 4-51 Ellipse construction: the parallelogram or conjugate diameter method ***(Section 4.21)***

The *parallelogram method*, or *conjugate diameter method*, shown in Figure 4-51, is used to find the major and minor axes of a given ellipse when only the conjugate diameters AB and CD are known:

1. Draw a circle with center at O and diameter CD intersecting the ellipse at P and P′.
2. Draw line VW through the center of the ellipse parallel to CP′. This line is the minor axis.
3. Draw the major axis, XY, perpendicular to VW.

The *trammel method* for drawing an ellipse is shown in Figure 4-52.

1. Draw the given major and minor axes.
2. Prepare a strip of paper or thin cardboard by marking off on one edge a distance AO equal to one-half the length of the major axis AB and a distance OC equal to one-half the length of the minor axis CD.
3. Place the trammel on the drawing across the axes so that point A is on the minor axis and point C is on the major axis. Point O will fall on the curve of the ellipse.
4. Find additional points in each quadrant by successively moving the trammel to various other positions, always keeping point A on the minor axis and point C on the major axis.
5. Draw the ellipse using an irregular curve.

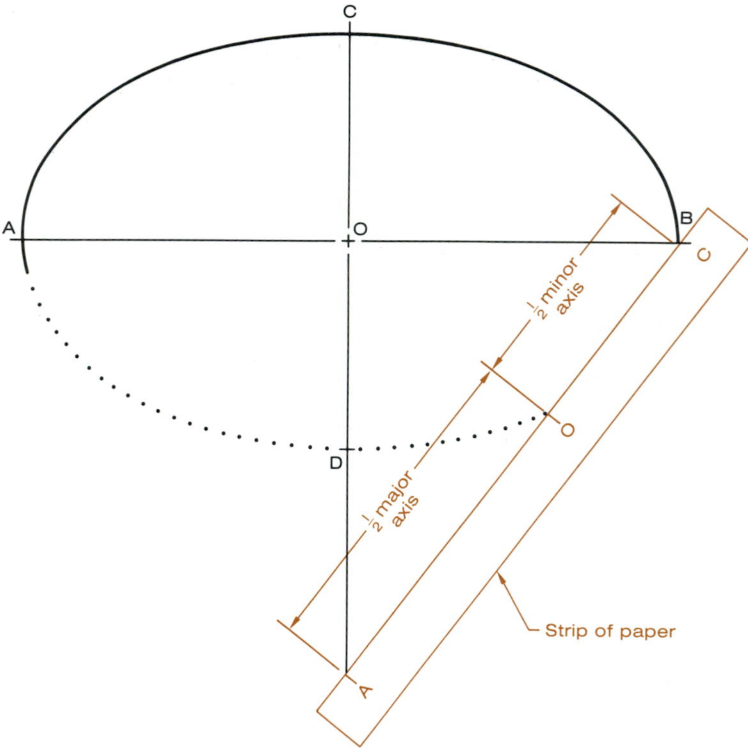

FIGURE 4-52 Ellipse construction: the trammel method *(Section 4.21)*

The figure formed by the following procedure, known as the *four-center method*, approximates a true ellipse. For many drafting purposes, particularly for small ellipses, this is a preferred method (see Figure 4-53).

1. Given major axis AB and minor axis CD, draw line AC.
2. Draw an arc with OC as a radius and center at O to intersect OA at E.
3. Draw an arc with EA as a radius and center at C to intersect AC at F.

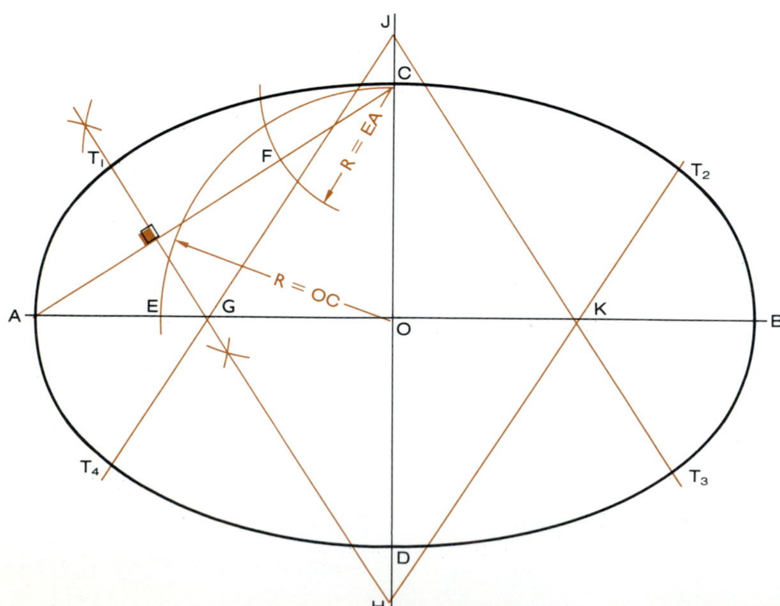

FIGURE 4-53 Ellipse construction: the four-center method *(Section 4.21)*

4. Draw the perpendicular bisector of line AF. Points G and K are the centers of two of the arcs forming the ellipse.
5. To locate the other two centers, make OJ equal to OH and OK equal to OG.
6. Locate the tangent points T_1, T_2, T_3, and T_4 by drawing lines through the centers of the tangent arcs.

4.22 ELLIPSE TEMPLATES

Many design engineers and drafters use templates to draw ellipses. A typical ellipse template is illustrated in Figure 4-54. The use of templates can reduce the effort required to draw ellipses, resulting in a considerable savings of time. These thin plastic sheets are made in sets in a wide variety of full or half ellipses in sizes ranging from ⅛ inch to 9 inches in sizes of about every ¼ inch. The precisely cut elliptical openings are available with line-of-sight angles from 5° to 80° in 5° intervals, which are useful for preparing pictorial drawings. To draw an ellipse in a line-of-sight position at 22°, for example, you must slightly shift the opening of the template nearest in size—in this case, the 20° template.

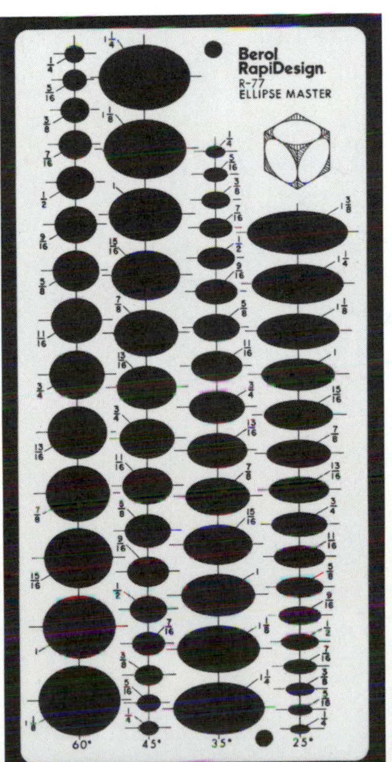

SOURCE: BEROL, INC.

FIGURE 4-54 Ellipse templates *(Section 4.22)*

CHECKPOINT 7
ELLIPSE CONSTRUCTION

On each of three plain sheets of 8½ × 11 inch paper, draw a horizontal line dividing the sheet into two equal halves. Prepare the following drawings full size, neatly balancing all constructions in the available space.

1. On the top half of sheet 1, use the trammel method to construct an ellipse with a major axis of 4 inches and a minor axis of 2.5 inches. Draw the major axis in a horizontal position. On the lower half of the sheet, use the foci method to construct an ellipse with a major axis of 130 mm and a minor axis of 75 mm.
2. On the top half of sheet 2, use the concentric-circle method to construct an ellipse with a major axis of 4.15 inches and a minor axis of 2.95 inches. Draw the major axis in a vertical position. On the lower half of the sheet, use the parallelogram method to construct an ellipse with a major axis of 6.05 inches and a minor axis of 3.25 inches. Draw the major axis in a horizontal position.
3. On the top half of sheet 3, use the parallelogram method to construct an ellipse and find the major and minor axes. The long conjugate diameter is 5.05 inches and should be placed horizontally. The short conjugate diameter is 4.25 inches. Draw the top half of the short conjugate diameter inclined to the right at an angle of 70° to the long conjugate diameter. On the lower half of the sheet, use an ellipse guide to draw at least ten different sizes of ellipses in different positions on the sheet. Mark the end points stamped on the ellipse guide for each ellipse that you draw and draw the major and minor diameter for each figure through these points.

4.23 CONSTRUCTING A PARABOLA

A parabola is the locus of a point whose distance from a fixed point, the focus, is always equal to its distance from a fixed line, the directrix. Given focus F and directrix AB, you can draw a parabola using the *locus method*, shown in Figure 4-55 (note: The axis of the curve passes through F and is perpendicular to AB at O):

1. Draw CD parallel to AB at any convenient distance away.
2. Using F as the center and radius OP, swing arcs to intersect line CD at points 1 and 1′. Because these points are equidistant from the focus and the directrix, they are points on the parabola.
3. Continue the construction and find other points on the curve by drawing additional lines parallel to AB and swinging arcs of varying radii with F as center. The vertex V is the midpoint of OF.
4. Draw the parabola using an irregular curve.

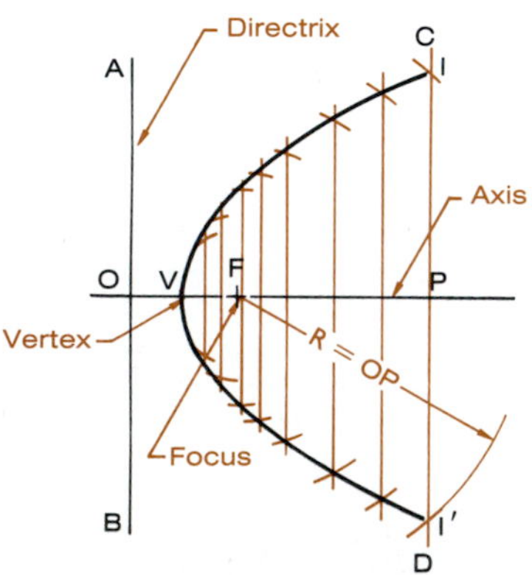

FIGURE 4-55 Parabola construction: the locus method *(Section 4.23)*

Given the rise AB and the span BC, use the *parallelogram method* (shown in Figure 4-56) to draw a parabola:

1. Draw the parallelogram ABCD.
2. Divide AO and AB into the same number of equal parts. Number the divisions, as shown.

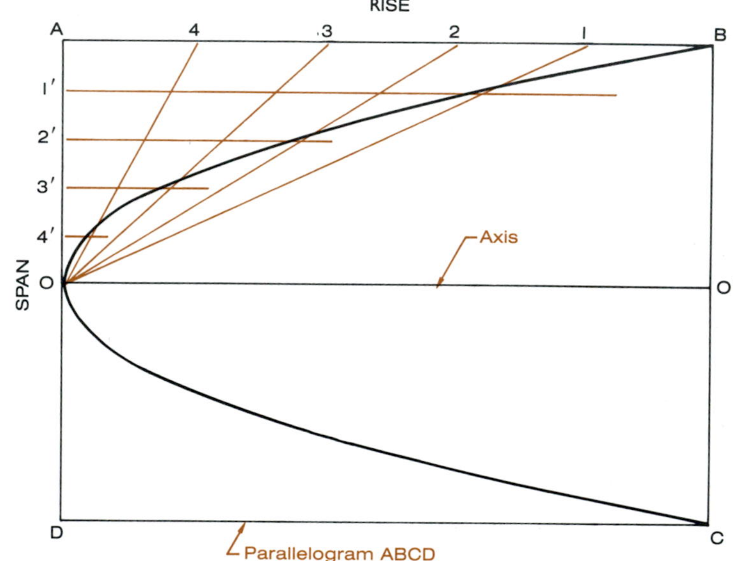

FIGURE 4-56 Parabola construction: the parallelogram method *(Section 4.23)*

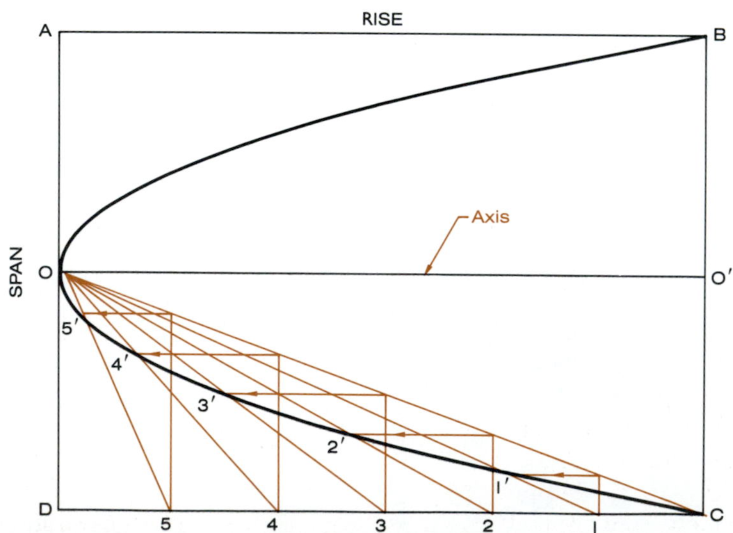

FIGURE 4-57 Parabola construction: alternate parallelogram method *(Section 4.23)*

3. Draw lines to O from points 1, 2, 3, and 4 on BA.
4. Draw lines parallel to axis 00′ from the points 1′, 2′, 3′, and 4′ on AO.
5. Locate the intersections of correspondingly numbered lines (points on one-half of the parabola). Repeat the process for the points on the other half.
6. Draw the parabola using an irregular curve.

An alternate parallelogram method for drawing a parabola is illustrated in Figure 4-57:

1. Draw parallelogram ABCD.
2. Draw diagonal OC
3. Divide CD into a number of equal parts. Number the divisions, as shown.
4. Draw lines to O from points 1, 2, 3, 4, and 5.
5. Project points, 1, 2, 3, 4, and 5 to intersect line OC and draw horizontal lines from the intersections on line OC to locate corresponding points 1′, 2′, 3′, 4′, and 5′ on the curve.
6. Draw the parabola using an irregular curve.

The *envelope method* for drawing a parabola, when given points B and C on the parabolic curve, is shown in Figure 4-58:

1. Assume point A and draw tangents AB and AC. If the tangents are equal, the axis of the parabola will bisect the angle between them.
2. Divide AB and AC into the same number of equal parts—in this example, six. Number the divisions as shown.
3. Draw lines to AB from points 1 through 5 on AC and to AC from points 1 through 5 on AB. These lines are tangent to, and thus envelop, the required curve.
4. Draw the parabola using an irregular curve.

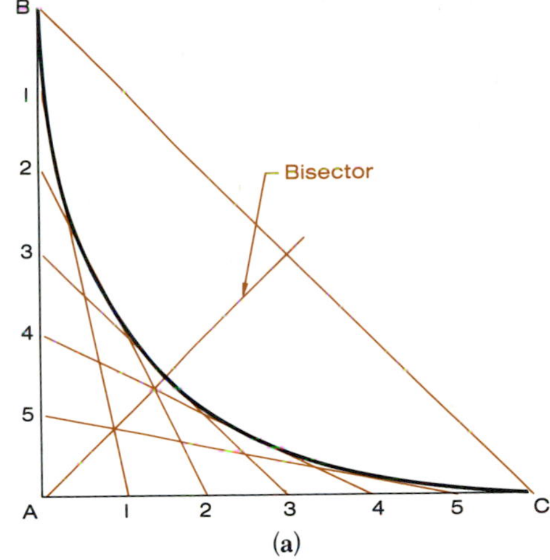

(a)

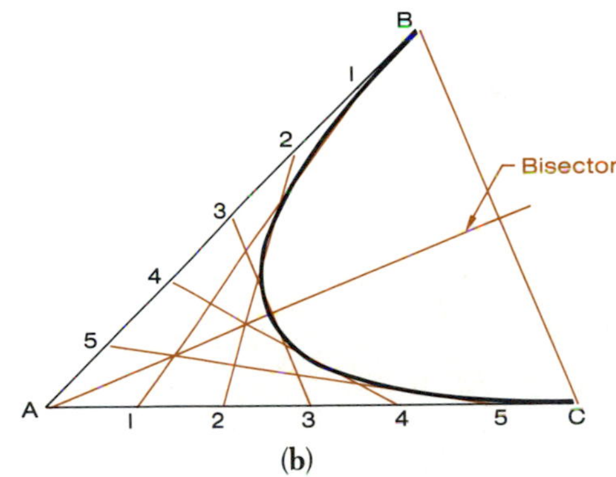

(b)

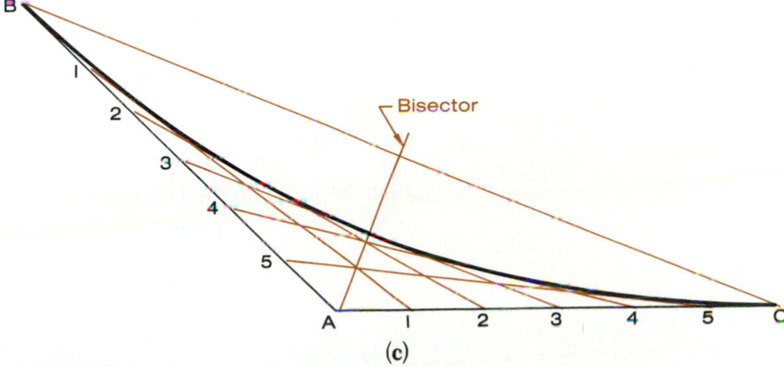

(c)

FIGURE 4-58 Parabola construction: the envelope method
(Section 4.23)

4.24 CONSTRUCTING A HYPERBOLA

The hyperbola is the locus of a point that moves so that the difference of its distances from two fixed points, called foci, is a constant and is equal to the transverse axis. Given foci F_1 and F_2, the transverse axis, and vertices A and B, construct a hyperbola using the *locus method* (Figure 4-59) as follows:

1. Select point 1 beyond F_2 on the transverse axis.
2. Using F_1 and F_2 as centers and radius A1, swing arcs.
3. Using F_1 and F_2 as centers and radius B1, swing intersecting arcs locating P_1 and P_1'. Points on the opposite half are located in the same manner using the same pairs of radii.
4. Repeat the construction and find all other points on the required hyperbola by selecting additional points on the transverse axis.
5. Draw the hyperbola using an irregular curve.
6. Locate asymptotes (lines that the curve approaches but never meets) by drawing the conjugate axis perpendicular to AB through the midpoint O. With O as center, draw a circle with diameter F_1F_2. Draw perpendiculars to A and B to intersect the circle at C, D, E, and G. The diagonals CE and GD are the asymptotes of the hyperbola.

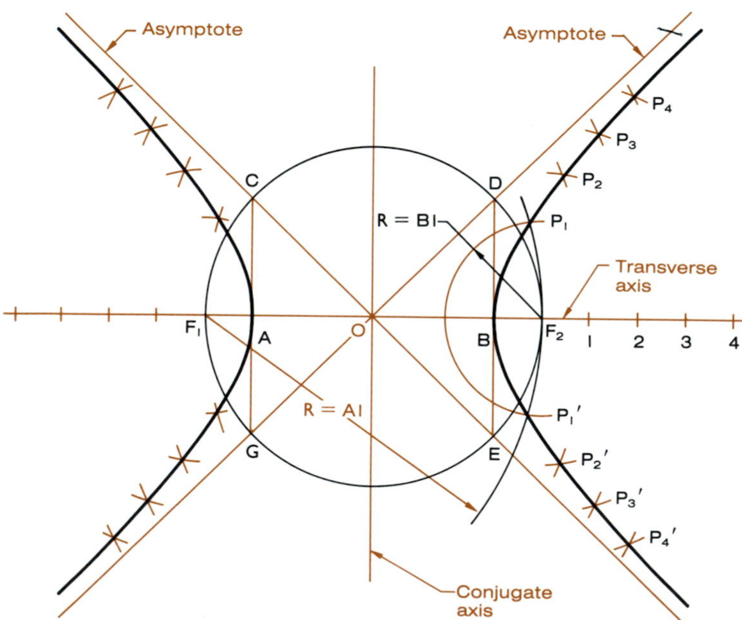

FIGURE 4-59 Hyperbola construction: the locus method *(Section 4.24)*

Given the transverse axis, vertices A and B, and one point P on the hyperbola, use the *rectangle method* (Figure 4-60) to construct the figure:

1. Draw the rectangle PEFG.
2. Divide PO and PG into the same number of equal parts—in this example, five. Number the divisions as shown.
3. Draw lines to B from points 1 through 5 on PG and draw lines to A from points 1 through 5 on PO.
4. Locate the intersections of correspondingly numbered lines. These are points on one-quarter of the required hyperbola. Repeat the process for points on the other three quarters.
5. Draw the hyperbola using an irregular curve.

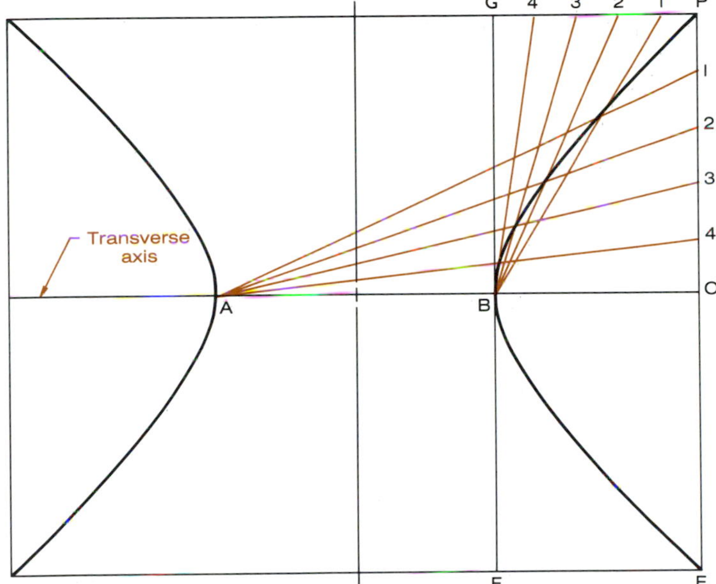

FIGURE 4-60 Hyperbola construction: the rectangle method *(Section 4.24)*

Given asymptotes OA and OB at right angles to each other and point P on the curve, construct an equilateral hyperbola as follows (see Figure 4-61):

1. Draw PC and PD.
2. On PC mark off any number divisions as shown.
3. Draw horizontal lines through these points parallel to OB.
4. Draw angular lines from O through points 1, 2, 3, 4, 5, and 6 on PC to intersect PD extended.
5. At these intersection points, draw lines perpendicular to PD extended. The intersections of these perpendicular lines and the corresponding horizontal lines give points 1′, 2′, 3′, 4′, 5′, and 6′ on the hyperbola.
6. Draw the hyperbola with an irregular curve.

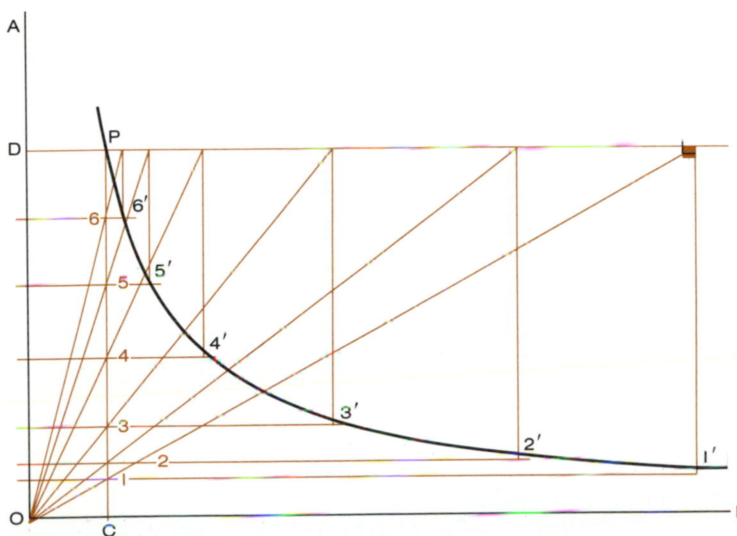

FIGURE 4-61 Hyperbola construction: the equilateral hyperbola *(Section 4.24)*

CHECKPOINT 8

PARABOLA AND HYPERBOLA CONSTRUCTIONS

On each of three plain sheets of 8½ × 11 inch paper, draw a horizontal line dividing the sheet into two equal halves. Prepare the following drawings full size, neatly balancing all constructions in the available space.

1. On the top half of sheet 1, use the locus method to construct a parabola with the axis in a horizontal position. Locate the focus 1⅛ inches to the right of the directrix. The vertex is midpoint between the directrix and the focus. On the lower half of the sheet, use the envelope method to construct a parabola in any convenient position with tangent lines AB and AC each 4¼ inches long and making an angle of 65°.
2. On the top half of sheet 2, use the parallelogram method to construct a parabola with a rise of 5⅛ inches and a span of 4⅛ inches. On the lower half of the sheet, use the parallelogram method to construct a parabola with a rise of 6 inches and a span of 3⅜ inches.
3. On the top half of sheet 3, use the locus method to construct a hyperbola with the transverse axis in a horizontal position. Locate the vertices A and B a distance of 1¾ inches apart with F_1 located ⅜ inch to the left of A and F_2 located ⅜ inch to the right of B. On the lower half of the sheet, use the rectangle method to construct a hyperbola with the transverse axis in a horizontal position. Construct a rectangle with a span of 5¾ inches and a rise of 3⅞ inches. Locate the vertices A and B 1¼ inches apart.

On a fourth sheet of 8½ × 11 inch paper, construct an equilateral hyperbola with asymptotes OA and OB at right angles to each other. Make OA, the horizontal axis, 5 inches long and OB 7½ inches long. Locate point P on the curve 2 inches to the right of OB and 1⅝ inches above OA.

PROBLEMS

Engineering Geometry

The following problems will provide you with practice in applying the principles of applied geometry discussed in this chapter. Draw each problem very accurately with instruments to the required scale, giving particular attention to achieving high quality linework. Use a 5H or 6H grade pencil lead for drawing light construction lines and for center lines. Use a 2H or 3H grade pencil lead for the final object lines. Indicate all points of tangency with a short dashed line drawn across the points. Do not erase construction lines.

For each problem, use layout A, shown on the inside of the front cover. Fill out the information shown in the title block and neatly balance each problem within the available space. Do not copy the given dimensions. The given sketches are not drawn to scale. All sizes are in inches unless otherwise specified. Before attempting to solve the problems in this chapter, refer to the table of symbols on page 85. For additional information about these symbols, see Chapter 10.

For problems 4.1 through 4.7, divide the sheet horizontally into two halves. Draw two problems, full size, on one sheet.

P4.1a Divide line OA into the proportions of 1, 3, 5, 3, and 1 and the line OB into six equal parts. Locate six points on the curve as shown and draw a smooth curve through these points. Make AO = 3¼″ and OB = 5⅜″.

P4.1b Construct triangle ABC in the position shown with AB = 4⅝″, BC = 3¹⁵⁄₁₆″, and AC = 4¹³⁄₁₆″. Draw arcs 1⅛″ R with centers on line AC, tangent to AB and BC. Bisect angle ABC and draw arcs of 1⅜″ R tangent to the bisector and the 1⅛″ R arcs.

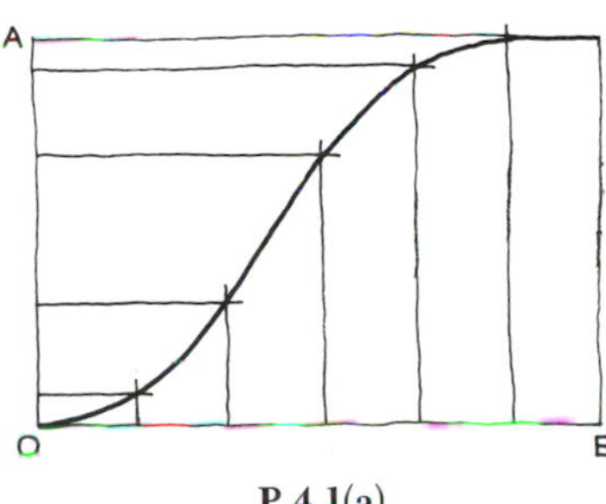

P 4.1(a)

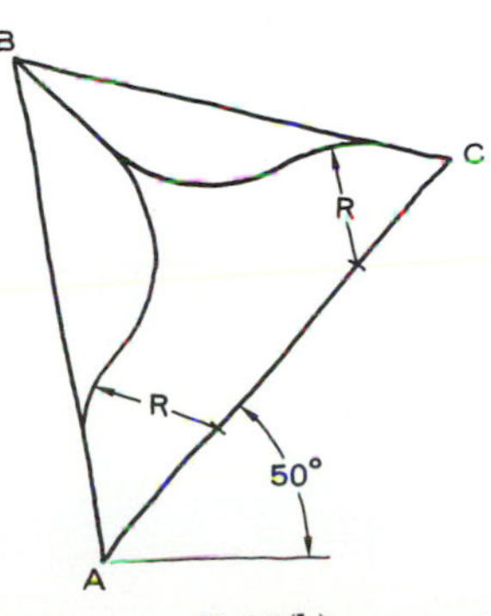

P 4.1(b)

P4.2a Draw the figure shown. Points A, B, C, and D are the centers of 8 mm diameter holes spaced as follows: AB = 66 mm, BC = 48 mm, CD = 70 mm, AD = 65 mm, and AC = 109 mm.

P4.2b Draw the figure shown.

P4.3a Draw the figure shown and then construct a 2½″ diameter circle externally tangent to the shaft circle and the table surface AB.

P4.3b Draw the figure shown.

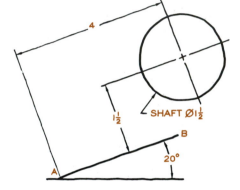

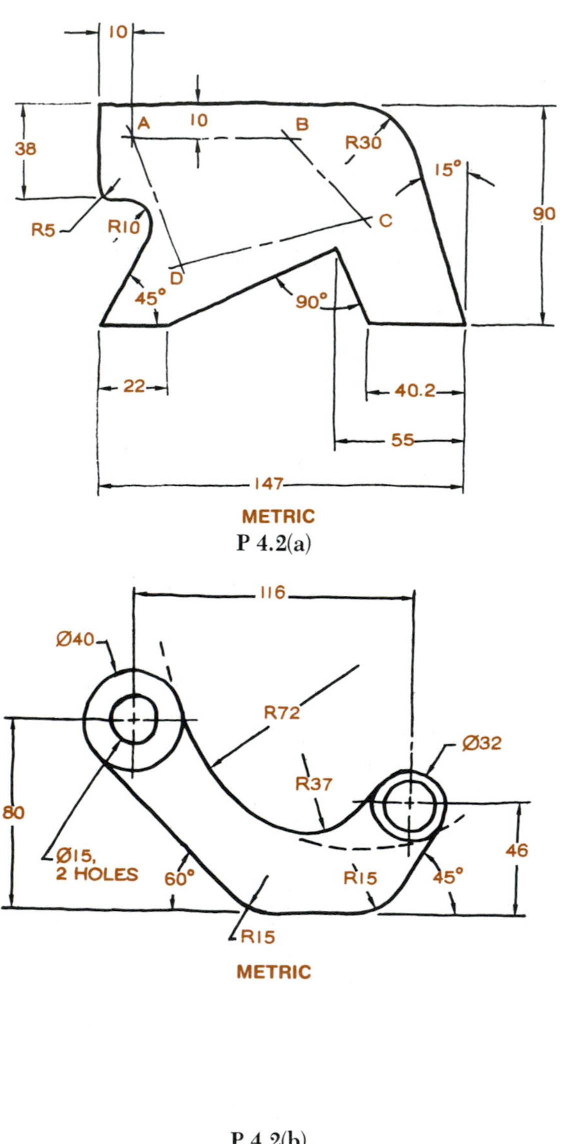

P 4.2(a)

P 4.2(b)

P 4.3(a)

P 4.3(b)

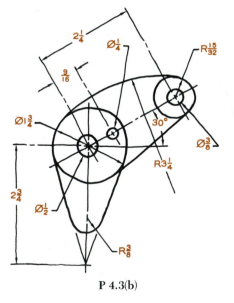

P4.4a Draw the figure shown.
P4.4b Draw the figure shown.
P4.5a Draw the figure shown.
P4.5b Draw the figure shown.

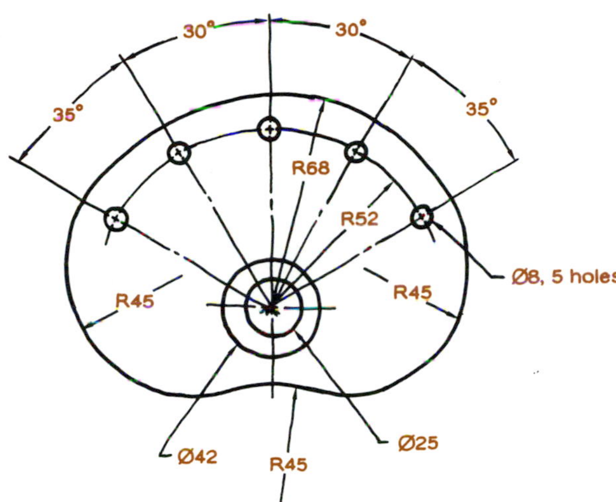

METRIC
P 4.4(a)

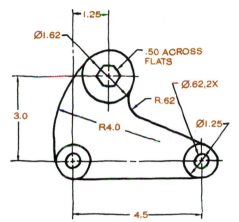

P 4.5(a)

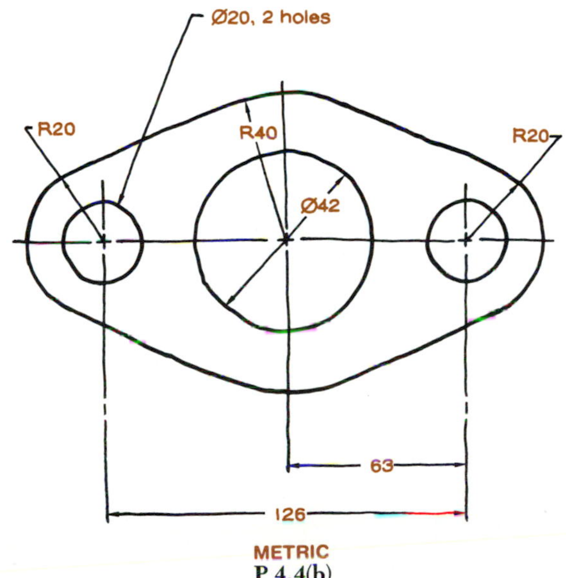

METRIC
P 4.4(b)

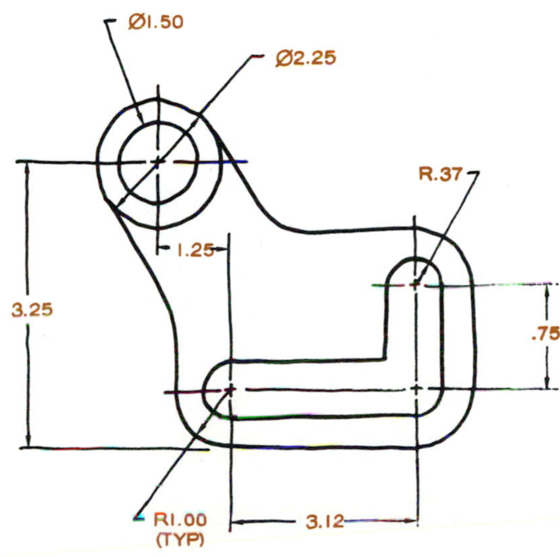

P 4.5(b)

P4.6a Draw the figure shown.
P4.6b Draw the figure shown.
P4.7a Draw the figure shown. The three spokes are equally spaced.
P4.7b Draw the figure shown.

For problems 4.8 through 4.21, use a full sheet for each problem. Unless directed differently, draw all problems full size.

P4.8 Draw the figure shown to a scale of 1″ = 100′.
P4.9 Draw the figure shown.
P4.10 Draw the figure shown.
P4.11 Draw the figure shown.

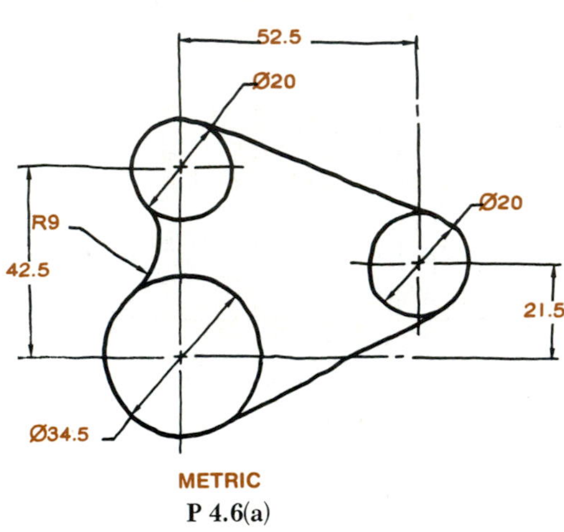

P 4.6(a)

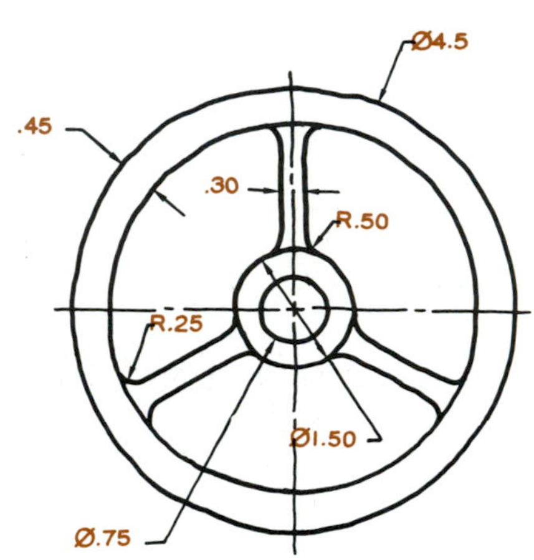

P 4.7(a)

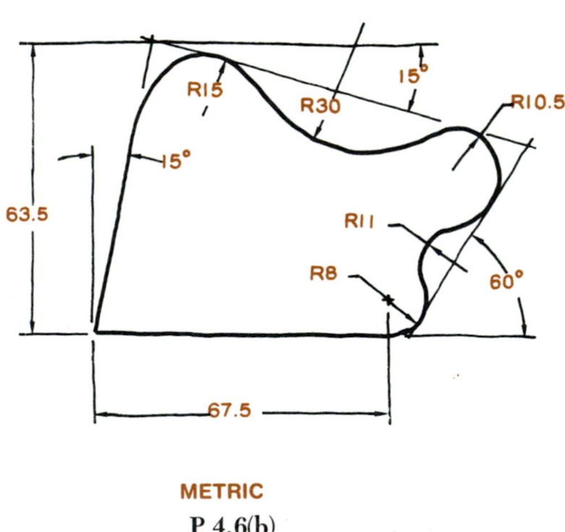

P 4.6(b)

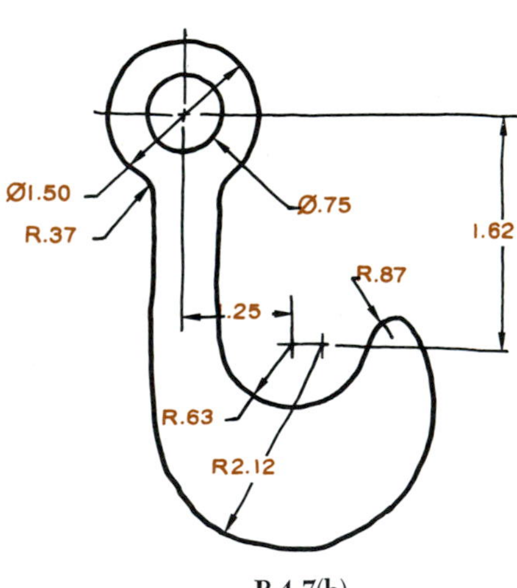

P 4.7(b)

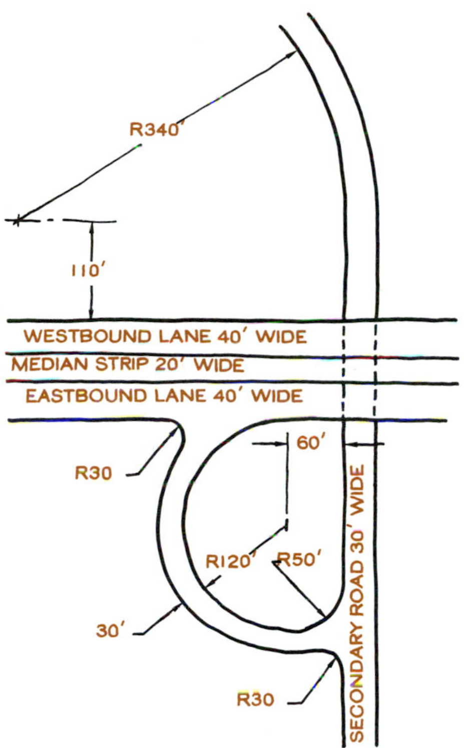

P 4.8

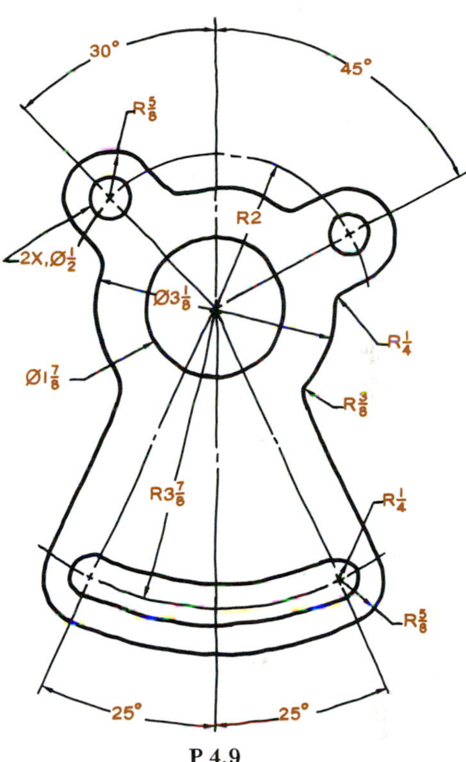

P 4.9

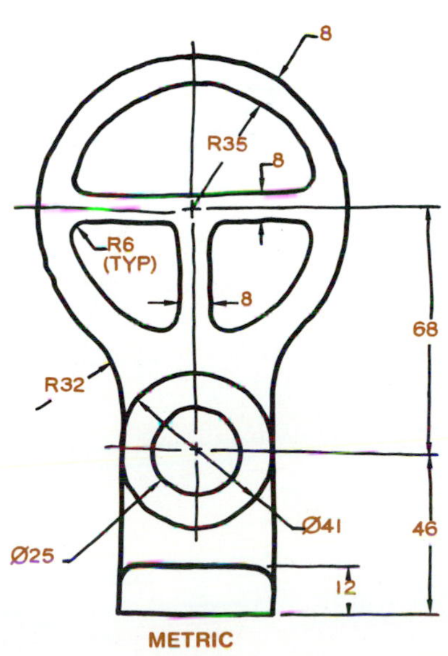

P 4.10

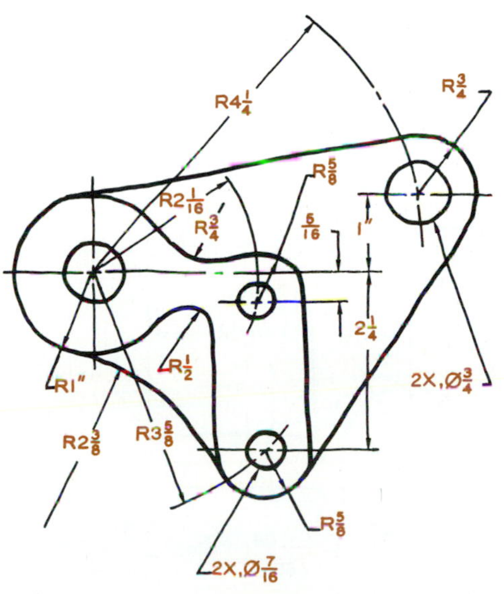

P 4.11

P4.12 Draw the figure shown.
P4.13 Draw the figure shown twice size.
P4.14 Draw the figure shown.
P4.15 Draw the entire symmetrical figure shown twice size.
P4.16 Draw an ellipse using the foci method shown in Figure 4-48. Make the major axis 4¼″ long and the minor axis 2¾″ long.
P4.17 Draw an ellipse using the foci method shown in Figure 4-48. Make the major axis 140 mm long and the minor axis 115 mm long.
P4.18 Draw an ellipse using the concentric circle method shown in Figure 4-49. Make the major axis 5½″ long and the minor axis 3¹¹⁄₁₆″ long.
P4.19 Draw an ellipse using the parallelogram method shown in Figure 4-50. Use the same axes lengths as in problem 4.18.
P4.20 Draw an ellipse using the trammel method shown in Figure 4-52. Make the major axis 6¼″ long and the minor axis 3⁹⁄₁₆″ long.
P4.21 Draw an ellipse using the trammel method shown in Figure 4-52. Make the major axis 125 mm long and the minor axis 92 mm long.

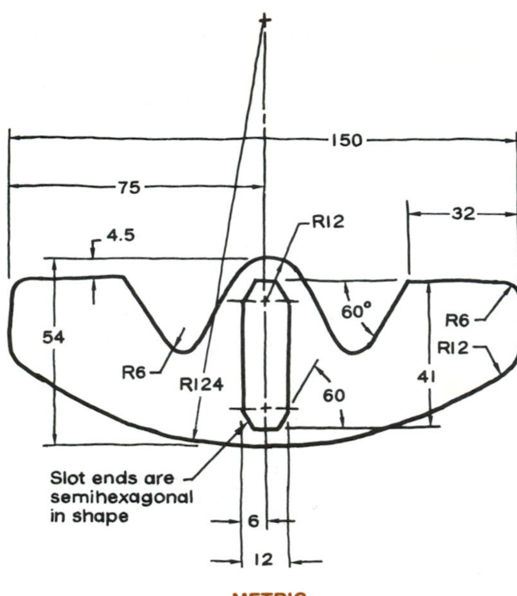

METRIC
P 4.14

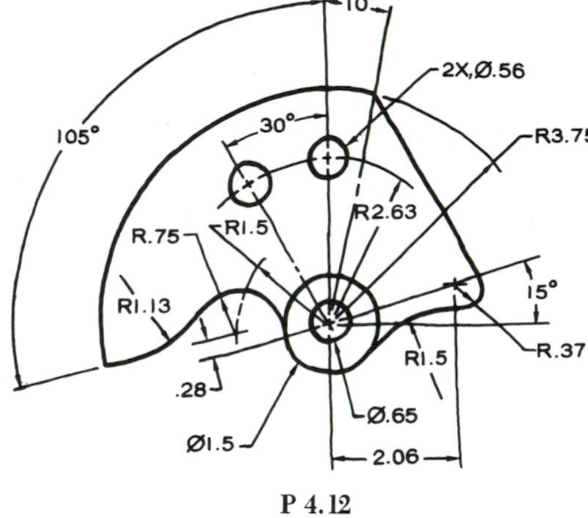

P 4.12

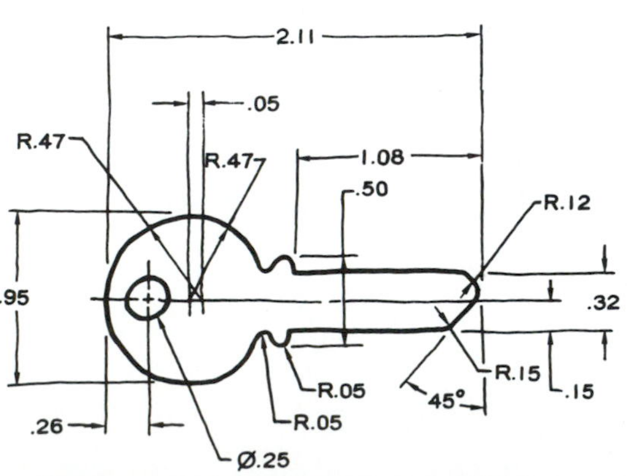

(Radii less than .15 may be drawn freehand)
P 4.13

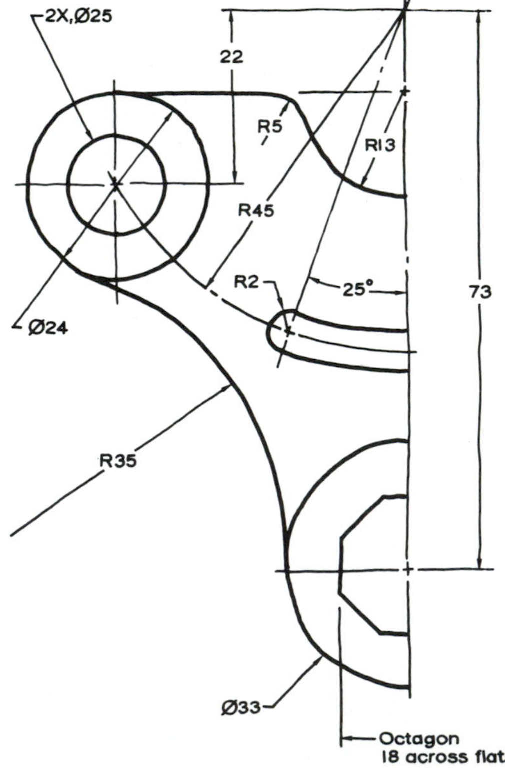

METRIC
P 4.15

Shape Description

OBJECTIVES

After completing this chapter you should have gained the following abilities:

1. To understand the principles of multiview projection and develop a level of competence to select and draw the necessary multiviews of an object.

2. To choose only the minimum and most descriptive combination of views to represent objects while maintaining the proper alignment of the views.

3. To continue to develop an acceptable level of competence in using the T square, drafting triangles, compass, dividers, scale, and protractor with speed and accuracy in working out graphical solutions on a drawing.

4. To apply the proper line symbol when using center lines, hidden lines, and phantom lines.

5. In the analysis of multiviews:

a. to recognize the various characteristics that distinguish normal, inclined, and oblique surfaces.

b. to identify adjacent surfaces to aid in reading or drawing the multiviews.

c. to use, where appropriate, surface identification labels.

d. to use the principles of configuration (similarity of surfaces).

e. to prepare, where appropriate, either a partial or a complete pictorial sketch to aid in analyzing the various surfaces of an object.

6. To properly display cylindrical shapes and intersecting cylinders.

7. To properly represent the fillets, rounds, and runouts on castings or forgings.

Orthographic projection, or multiview drawing, is a method used to represent the shape of objects on a drawing. Because they cannot show true distances and angles, pictorial sketches and drawings are limited, except for measurements in or parallel to one of the three axes directions. Also, in most cases pictorial sketches cannot completely reveal all features of an object (features on the back, for example) because the object must be portrayed in a single position. Most objects require a more satisfactory method of graphical representation to enable the viewer to visualize the complete shape of the object. The system of multiview drawing is based upon a simple, logical arrangement of two or more views of an object.

5.1 ORTHOGRAPHIC PROJECTION OR MULTIVIEW DRAWING

The basis of orthographic projection is illustrated in Figure 5-1. An imaginary transparent box has been placed over the object so that the walls of the box are parallel to those of the object. Each of the six viewing planes of the box is called a *projection* plane. These projection planes are called *frontal*, *horizontal*, and *profile* planes. You see the true shape of a side of an object when the line of sight is in a direction perpendicular to one side of the transparent box. Projecting this true shape image of the object into the corresponding viewing plane of the transparent box is the basis of orthographic projection.

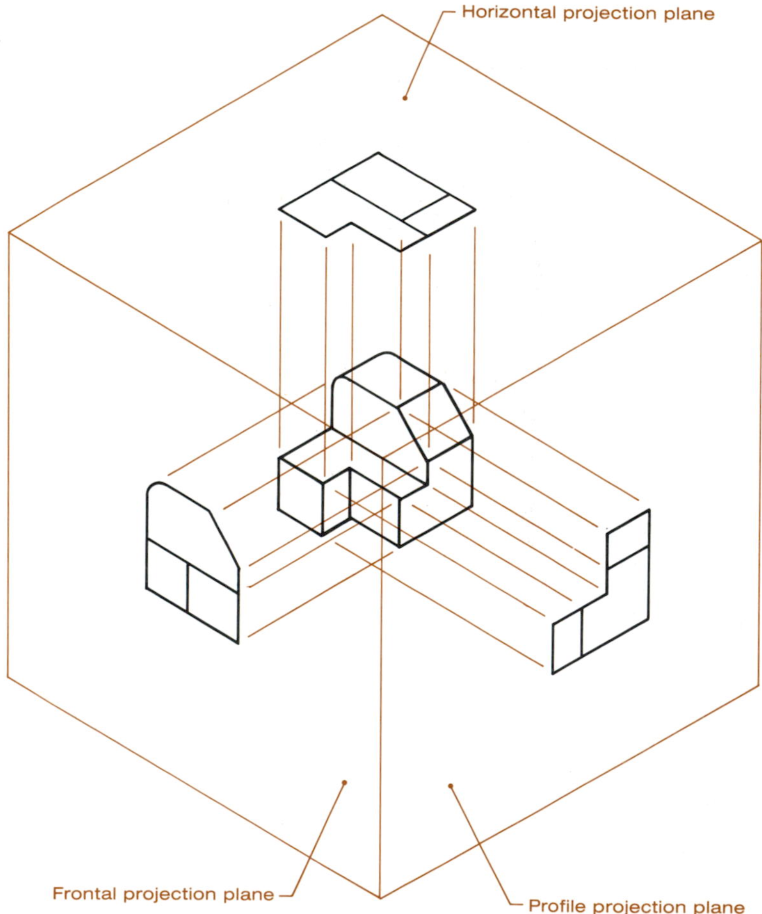

FIGURE 5-1 Orthographic projections. Three mutually perpendicular planes are used, on which the front, top, and right-side views are projected *(Section 5.1)*

The process of finding the front view of an object is illustrated in Figure 5-2. The front view is projected on a frontal projection plane using projection lines that are parallel to each other and perpendicular to the image representing the front of the object. In a similar manner, the top and the right-side views of the same object are projected in Figures 5-3(a) and 5-3(b). The top view is projected onto a horizontal projection

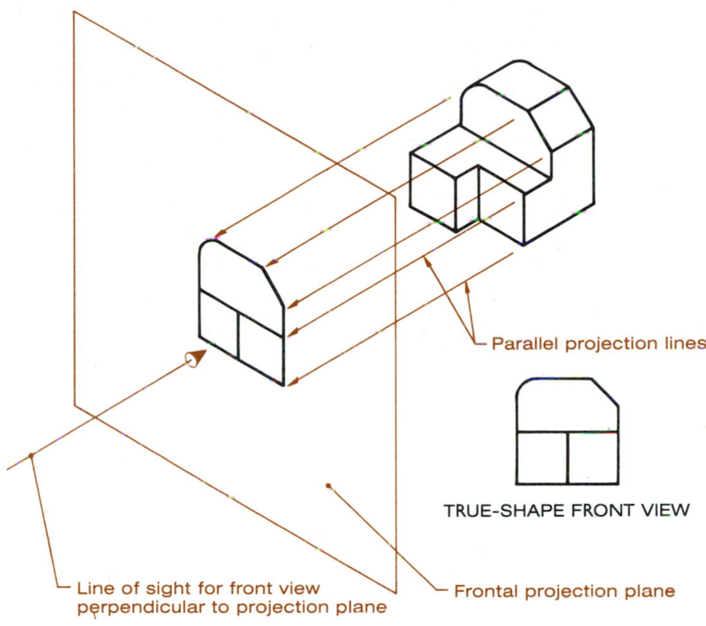

FIGURE 5-2 The frontal plane is a vertical plane on which the front view of an object is projected. The parallel projection lines are perpendicular to the frontal plane. The two-dimensional orthographic front view is shown true shape *(Section 5.1)*

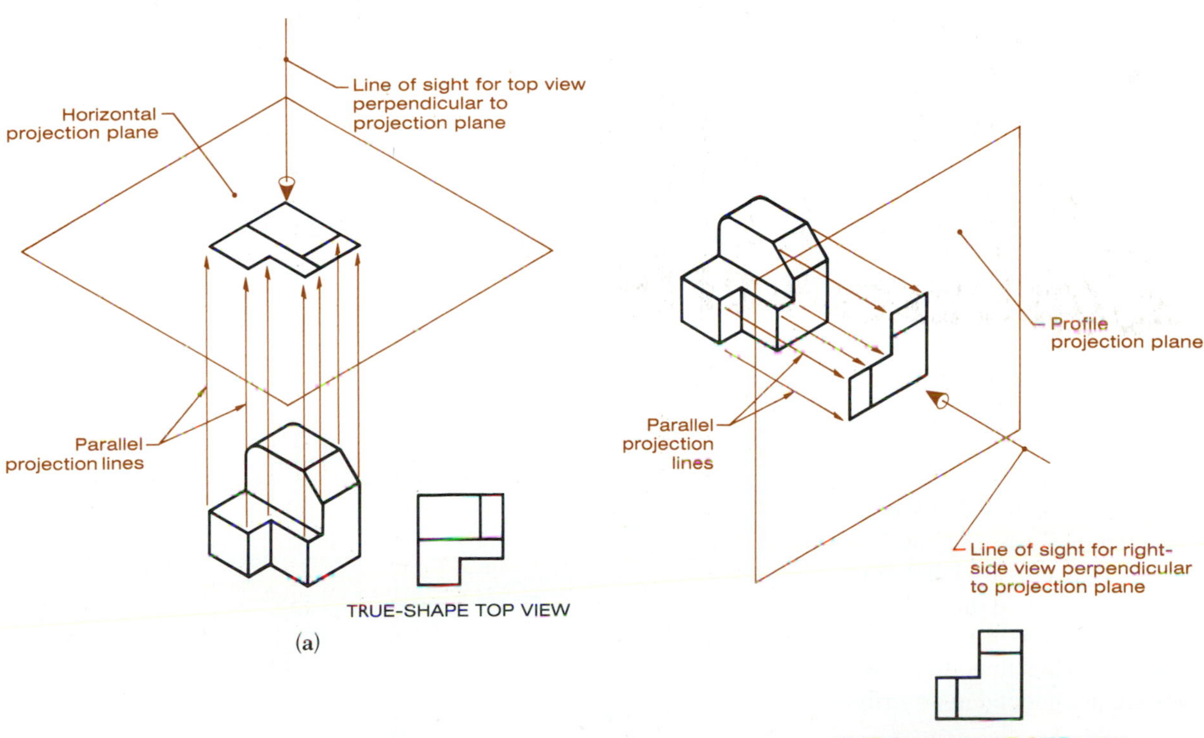

plane and the right-side view is projected onto a profile projection plane. Figure 5-4 shows how the projection planes of the transparent box are theoretically folded or opened outward. In summary, orthographic projection, or multiview drawing, is based upon the theory of images projected to viewing planes that are unfolded to a flat, common plane represented by the surface of the drawing paper.

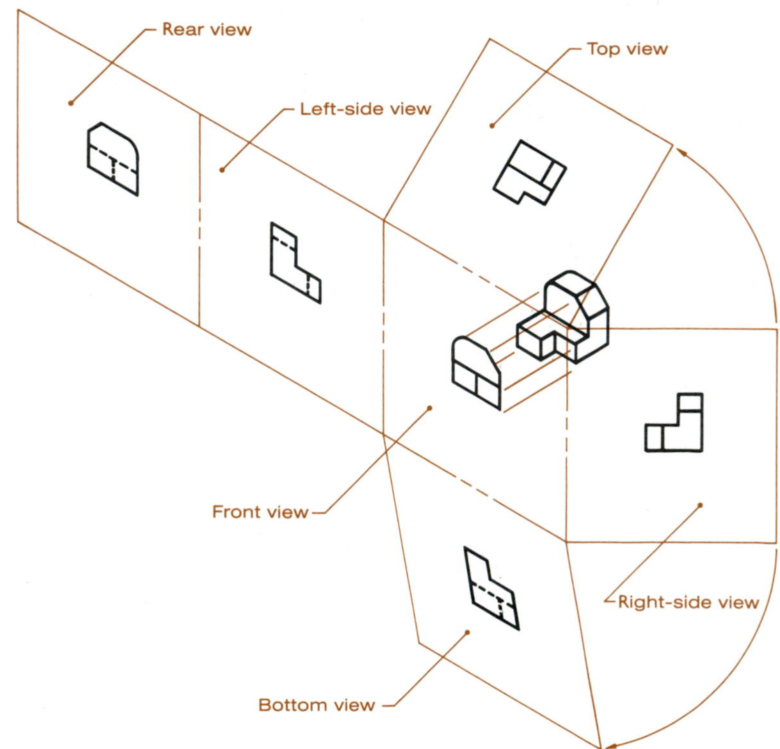

FIGURE 5-4 The projection planes of a transparent box opened into the plane of the drawing surface *(Section 5.1)*

5.2 THE SIX PRINCIPAL VIEWS

Any view that is projected onto one of the projection planes is called a *principal view.* The standard arrangement of the six principal views for a simple object is shown in Figure 5-5. It is unusual, however, for a part to be so complex in shape that all six views are required on an engineering drawing. Note that the left-side view is placed to the *left* of the front view and that the bottom view is aligned *below* the front view. The rear view may be placed in the position shown or aligned with the right-side view, the top view, or the bottom view. The top view is aligned directly above the front view and the right-side view is aligned with the front view. The projection planes are labeled H, F, and P at the fold lines.

In Figure 5-5 three of the views are so like the other three they are not needed in describ-

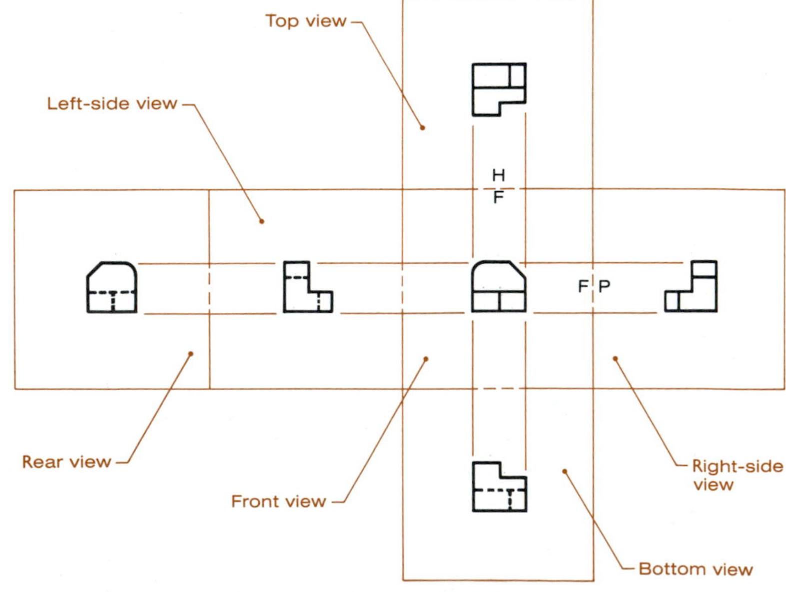

FIGURE 5-5 The six principal views of an object *(Section 5.2)*

ing the shape of the object. The remaining three views (top, front, and right-side), shown in Figure 5-6, are entirely adequate. The three dimensions that are used to show the three-dimensional form of an object are *height*, *width*, and *depth*. The distance representing the width of the object is the same in the front view and in the top view, and the distance representing the depth is the same in the top view and in the right-side view. The distance between the views may vary as needed. The outlines of the planes have been omitted but the lines representing the edges of the projection planes are shown and labeled accordingly.

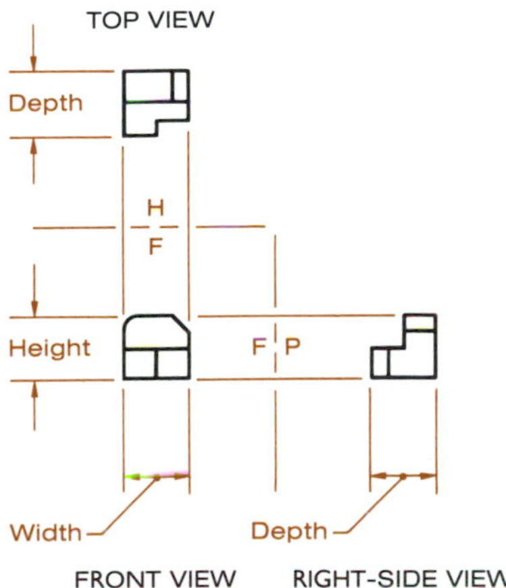

FIGURE 5-6 This standard arrangement of views shows the relationship of the height, width, and depth of an object *(Section 5.2)*

5.3 THE RELATIONSHIP BETWEEN THE VIEWS

Two terms, *adjacent views* and *related views*, are used throughout this text. Adjacent views are any two views placed in a position to align above and below (top and front views) or side by side (front and side views). Related views are views adjacent to the same views. For example, the combinations of top and left-side or right-side views or bottom and left-side or right-side views are related views.

5.4 LINE TECHNIQUES

The use of proper line techniques on a drawing cannot be overemphasized. The views must unambiguously describe in every detail the precise shape of an object. Each line has a definite meaning and should be used in a consistent way and should be clearly shown.

Visible Lines

Visible lines, shown in Figure 5-7, are used to represent edges or contours of an object that are visible to the observer. Visible lines should be dark and uniformly wide. The recommended width for visible lines is from .030 to .035 inch (about 0.8 to 0.9 mm).

Hidden Lines

Hidden lines are used to represent edges of an object that are not visible to the observer. Hidden lines, sometimes called "dotted lines," are drawn as a series of closely spaced dashes. Each dash is made about 1/8 inch long with the space between the dashes about 1/32 inch. The dashes are spaced entirely by eye. The line should be about medium weight or, about .020 to .025 inch (about 0.5 to 0.6 mm) wide. Hidden line conventions are shown in Figure 5-7 at the junctions in the encircled areas.

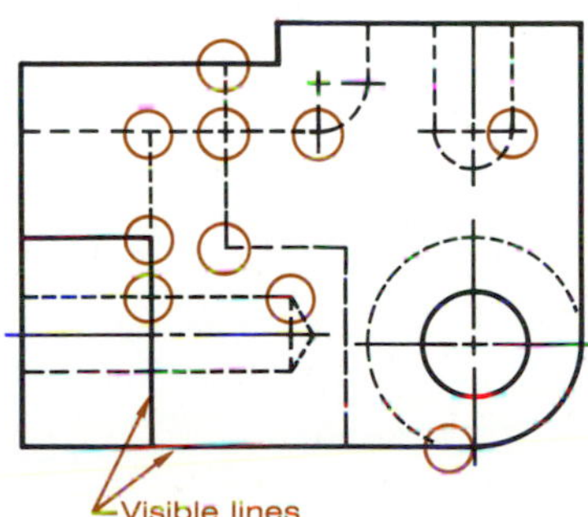

FIGURE 5-7 Line techniques: visible and hidden lines *(Section 5.4)*

Center Lines

As illustrated in previous chapters, center lines are drawn as a series of long and short dashes. The technique for showing center lines is given in Figure 5-8. Make the short dashes about 1/8 inch long with a space of about 1/16 inch on either side. The long portions of the lines may vary in length, depending upon the distance that the center line extends across the drawing. As in the case of the hidden line, the center line is spaced entirely by eye. The line weight should be thin (about one third the width of a visible line) but dark enough to contrast well with the other lines.

Precedence of Lines

Positions of visible lines, hidden lines, and center lines often coincide on a drawing. Examples of coinciding lines are shown in Figure 5-9. A visible line always takes precedence over any other type of line and a hidden line always takes precedence over a center line.

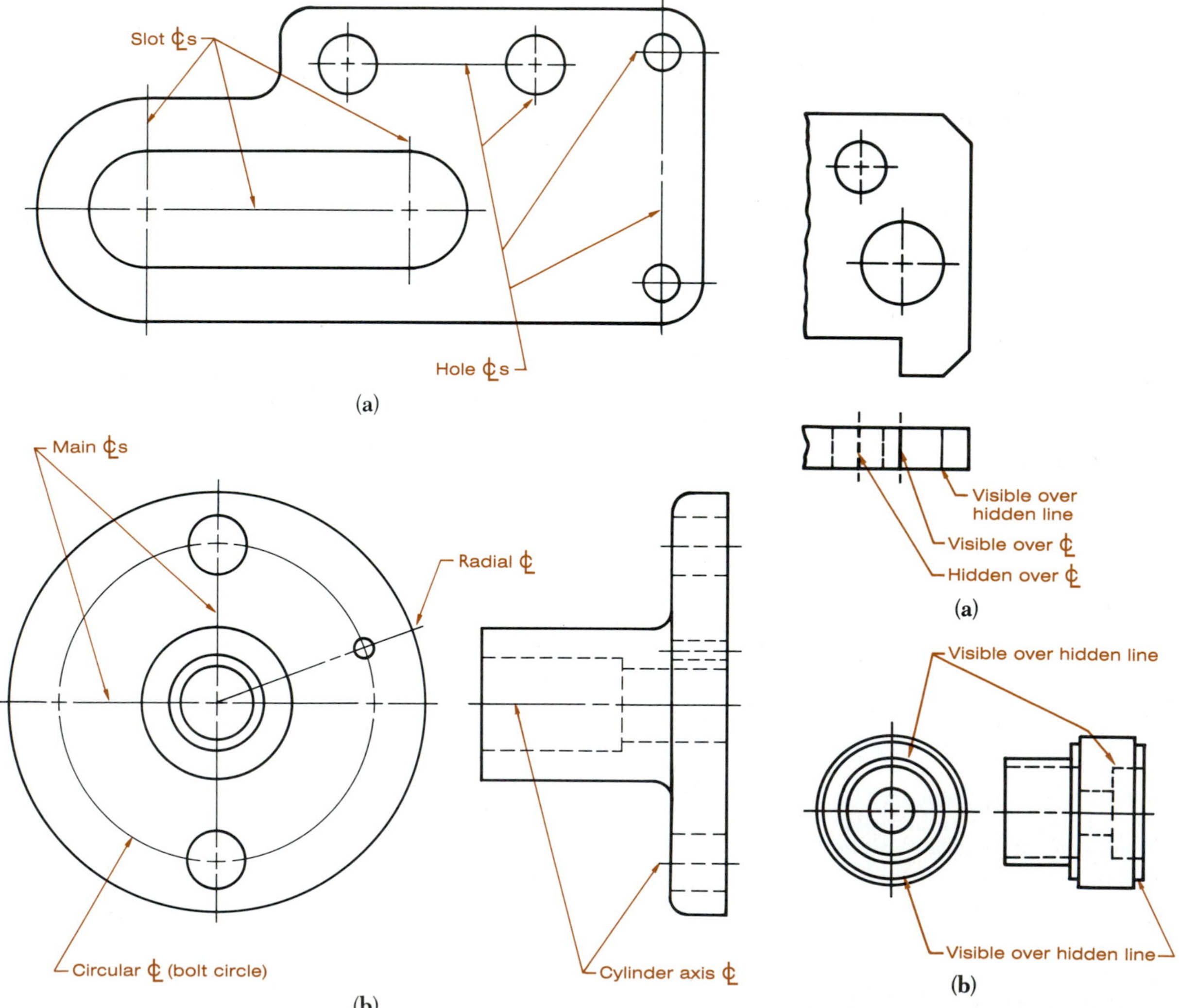

FIGURE 5-8 Center line techniques *(Section 5.4)*

FIGURE 5-9 Precedence of lines *(Section 5.4)*

CHECKPOINT 1
SKETCHING SIX PRINCIPAL VIEWS OF AN OBJECT

Sketch the six principal views of the objects shown pictorially in Figure 5-10 to a scale approximately full size. Each grid equals ⅛ inch, or about 3 mm. Label all views: front, top, and so on. Pay particular attention to the correct alignment of the views and show all hidden and center lines. Use a separate plain sheet of 8½ × 11 inch paper for each problem.

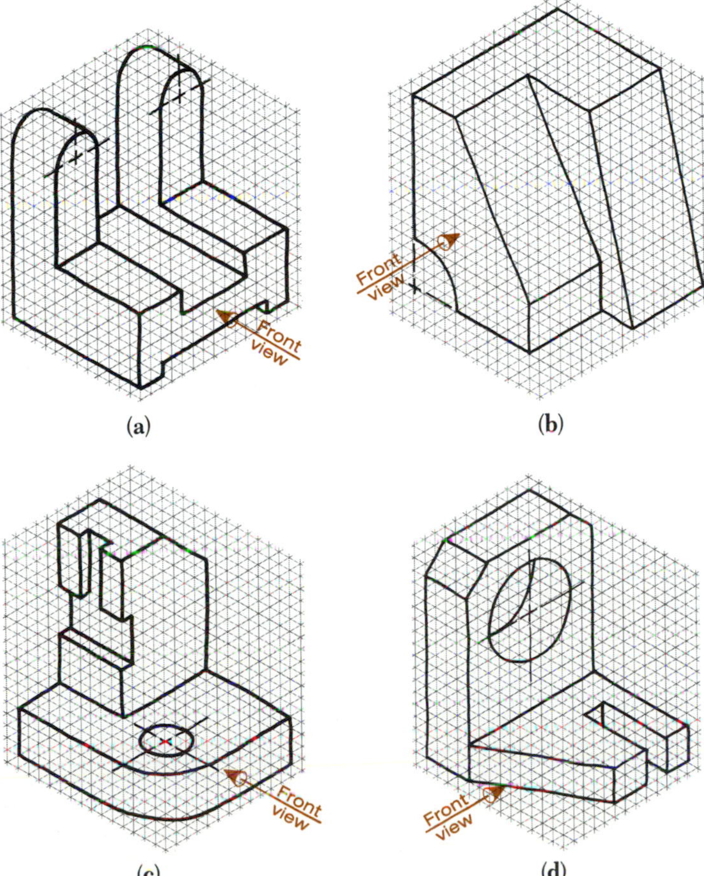

FIGURE 5-10 Checkpoint 1 problem

5.5 CHOICE OF VIEWS

Many objects, such as airplanes, houses, or telephones, for example, have fixed positions with a definite relationship to the earth that is universally accepted as the normal front, top, or side. Such objects are always placed in their natural position—otherwise they would appear unnatural. For most machine parts, however, there is no single, obvious position that characterizes the top, front, or any of the other principal views. Accordingly, the position selected to display a machine part on a drawing may be arbitrarily chosen unless its ultimate position in a machine or a structure is known. Generally, one face of an object shows the most detail, and this face should be chosen as the front view. Often this face will be the largest one. Most long, slender objects are placed horizontally in the front view.

Of the six principal views of an object, the top, front, and right-side views are the most preferred. If the left-side view is more descriptive (shows the contour of the part more clearly, for example, or has less hidden detail) it may be given in preference to the right-side view. The choice of views is always governed by the requirements that the object must be portrayed on the drawing in the best and most revealing position and that a view must be provided in each direction that is needed for a complete description of every detail.

5.6 ONE-VIEW DRAWINGS

While the objective in choosing views is to provide a complete and accurate description of the shape of an object, superfluous views should be avoided. Only the necessary views for a complete description of an object should be drawn. One-view drawings are often preferred for two classifications of part shapes: flat and cylindrical. For parts having a constant thickness, such as the flat part made from plate stock shown in Figure 5-11(a), a single view clearly shows the characteristic shape in two dimensions. The thickness is specified in the title note. Figure 5-11(b) shows an example of a cylindrical object that is completely described in one view. In this case the unnecessary side view would show only a number of concentric circles. The single view selected shows the axis of the cylindrical part as a center line, and the diameter symbol, ∅, is given along with each of the diameter dimensions. (The thread specification ¾-10 UNC-2 is explained in Sec. 9.10.)

5.7 TWO-VIEW DRAWINGS

Frequently two adjacent views of an object are sufficient to ensure clarity. (Adjacent views are placed side by side to align their common dimensions.) Two views are adequate for displaying simple objects resembling the symmetrical conical or pyramidal shapes shown in Figure 5-12. If a third view of these examples had been drawn, two of the views would be identical. The front view is usually selected as the central view for depicting the largest face of the object or the face which shows the most detail. The adjacent view that is selected to supple-

FIGURE 5-11 One-view drawings *(Section 5.6)*

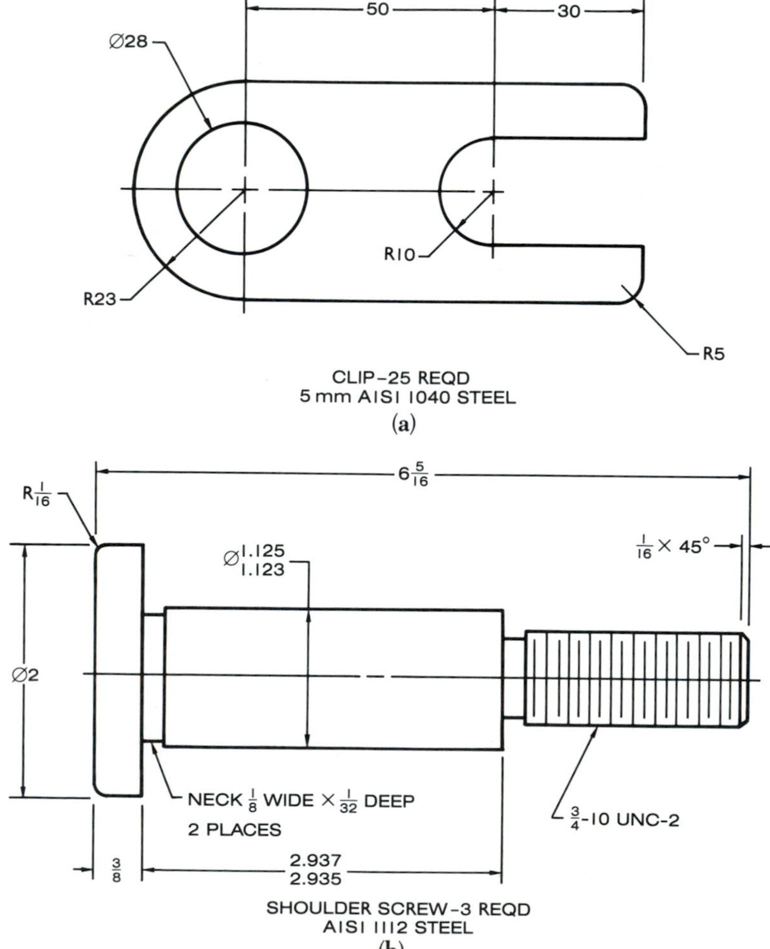

ment the front view may be either the right-side or the left-side view or the top view. Two-view drawings are selected when two of the views convey all of the necessary detail without ambiguity. Figure 5-13 shows some typical objects requiring only two views. Note how the views are arranged in proper alignment. Also, in Figures

FIGURE 5-12 Objects requiring only two views ***(Section 5.7)***

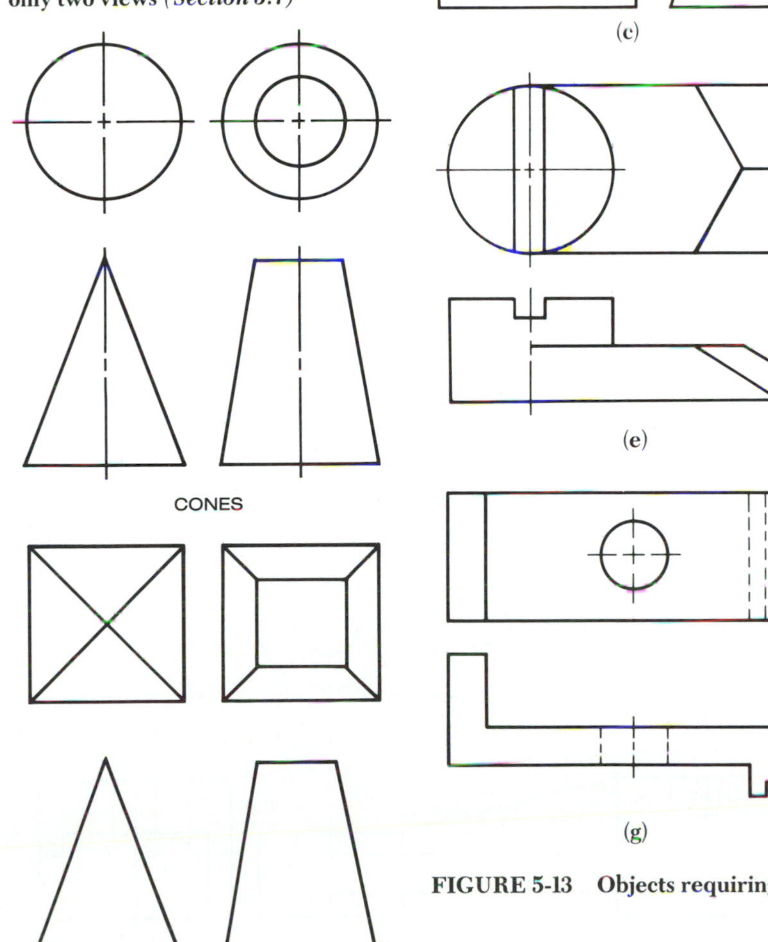

(a) (b)

(c) (d)

(e) (f)

(g) (h)

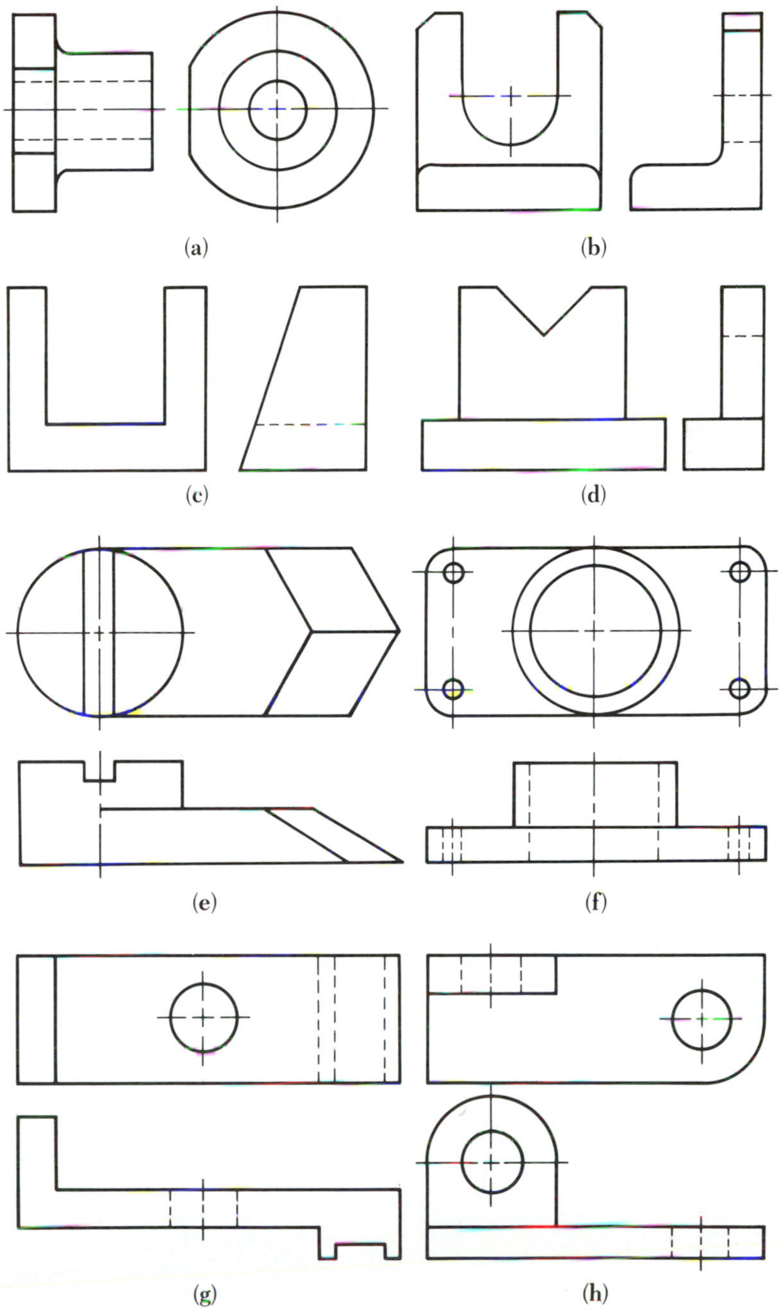

FIGURE 5-13 Objects requiring only two views ***(Section 5.7)***

5-14(a) and 5-14(b), two views are adequate to describe the object. In these examples the unnecessary views are confusing.

Considerable care must be observed in selecting the best

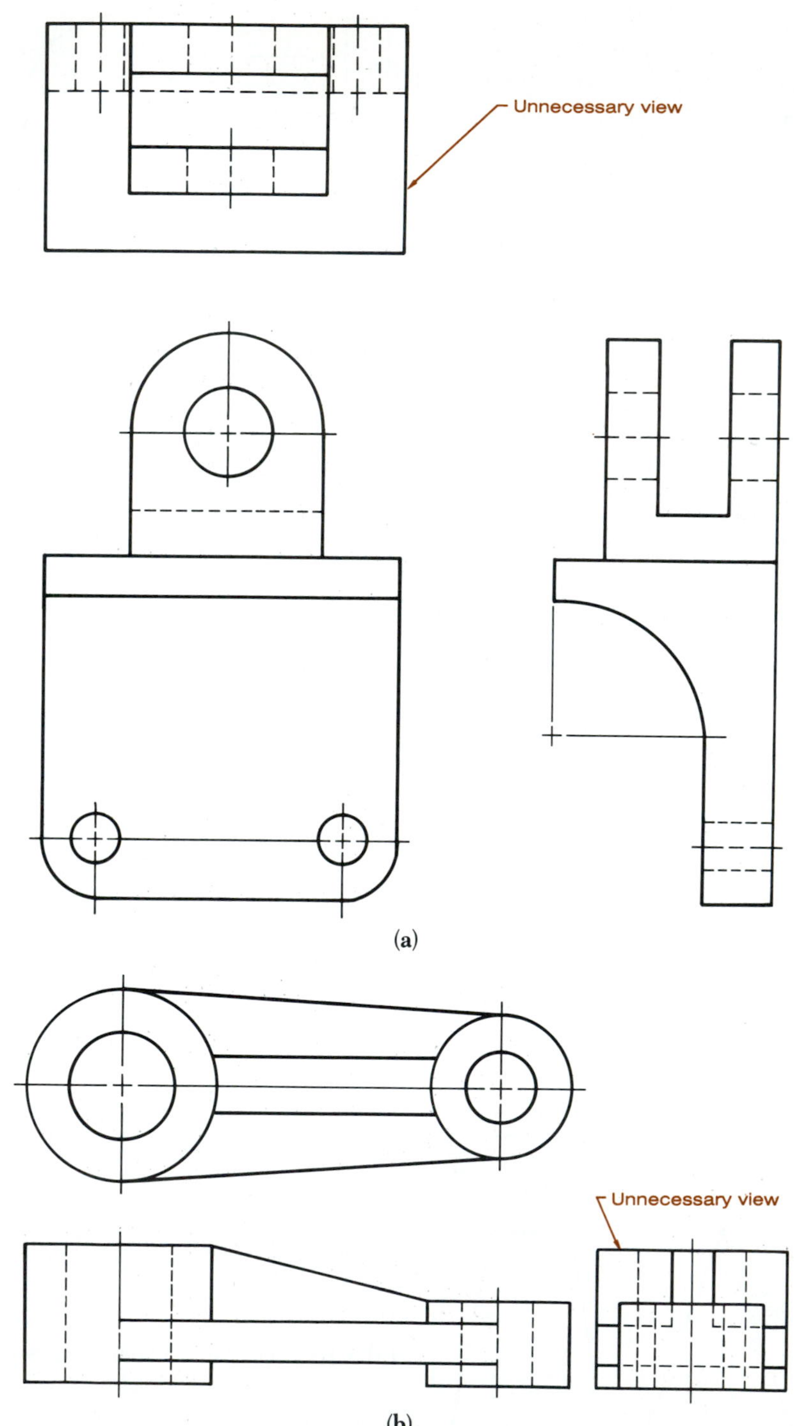

FIGURE 5-14 Avoiding unncessary views ***(Section 5.7)***

combination of two views. Figure 5-15 shows how an unwise selection of a pair of adjacent views may be improved by the selection of a less ambiguous combination of adjacent views.

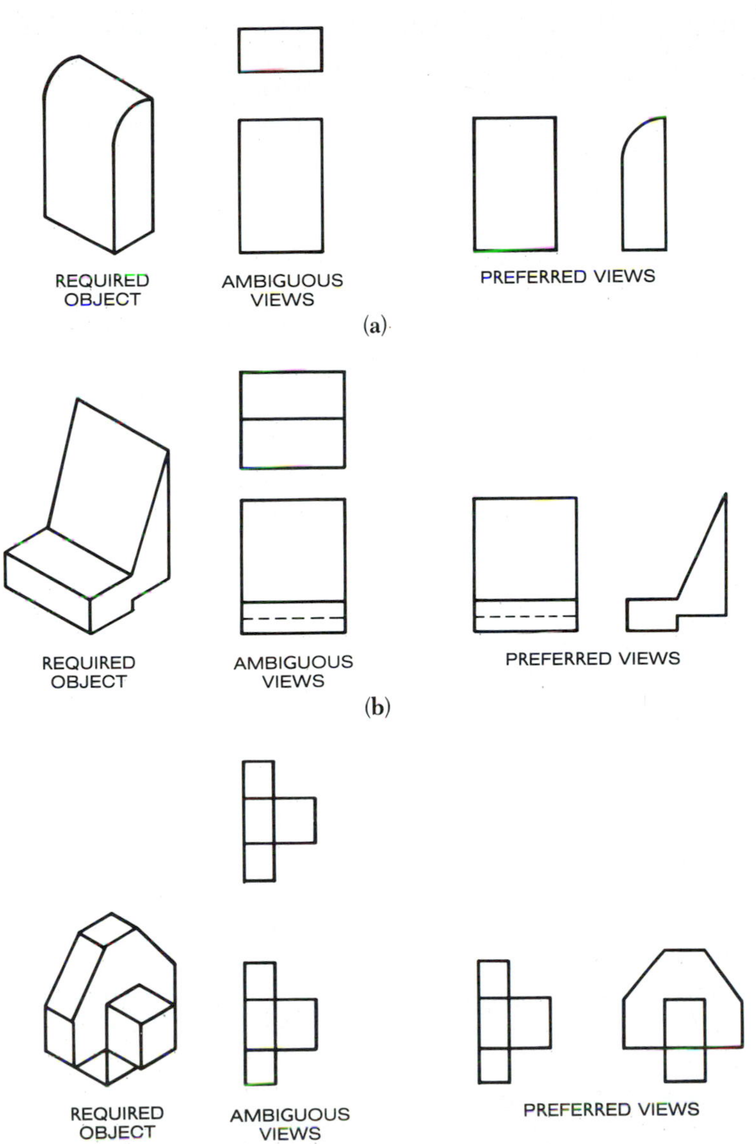

FIGURE 5-15 Avoiding ambiguous views ***(Section 5.7)***

Figure 5-16 illustrates the advantages of selecting views with the least hidden lines.

5.8 THREE-VIEW DRAWINGS

It is seldom necessary to draw more than three views for most parts. In general, the front, top, and right-side or left-side view are preferred for describing most objects. Figure 5-17 is an example of an object that requires three views. Carefully study this illustration and check the following points:

1. Think of the object as being stationary and you, the observer, as mentally changing position as you observe one view after another.
2. The most descriptive face is the front view, depicting what you see as if you were in a position directly in front of the object. Your line of sight is in a direction perpendicular to the front face of the object.
3. The top view is aligned directly above the front-face view and depicts what you see as if you were in a position directly above the object looking down. Your line of sight is in a direction perpendicular to the top face of the object.

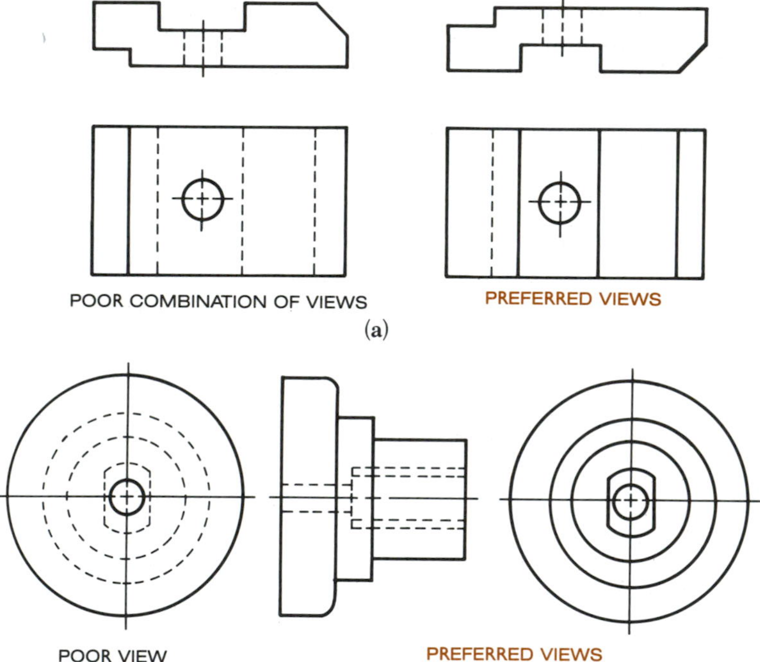

FIGURE 5-16 Selecting views with a minimum number of hidden lines ***(Section 5.7)***

4. The right-side view is aligned directly to the right of the front view and depicts what you see if you change to a position directly to the right of the object. Your line of sight is in a direction perpendicular to the right face of the object.
5. A constant relationship between the views exists. For example, the front face, or the face that is nearest you in the front view, appears in the top and right-side views as an edge. Note that the edge is in a position, in both the top and right-side views, that is nearest the front view.
6. Adjacent views are those in alignment in a side-by-side position, such as the front and top views or the front and right-side (or left-side) views. (Note: The top and right-side or left-side views are designated as *related* views.)
7. Draw the fewest number of views that will accurately describe the object. In this example each of the three views is needed for a complete description.

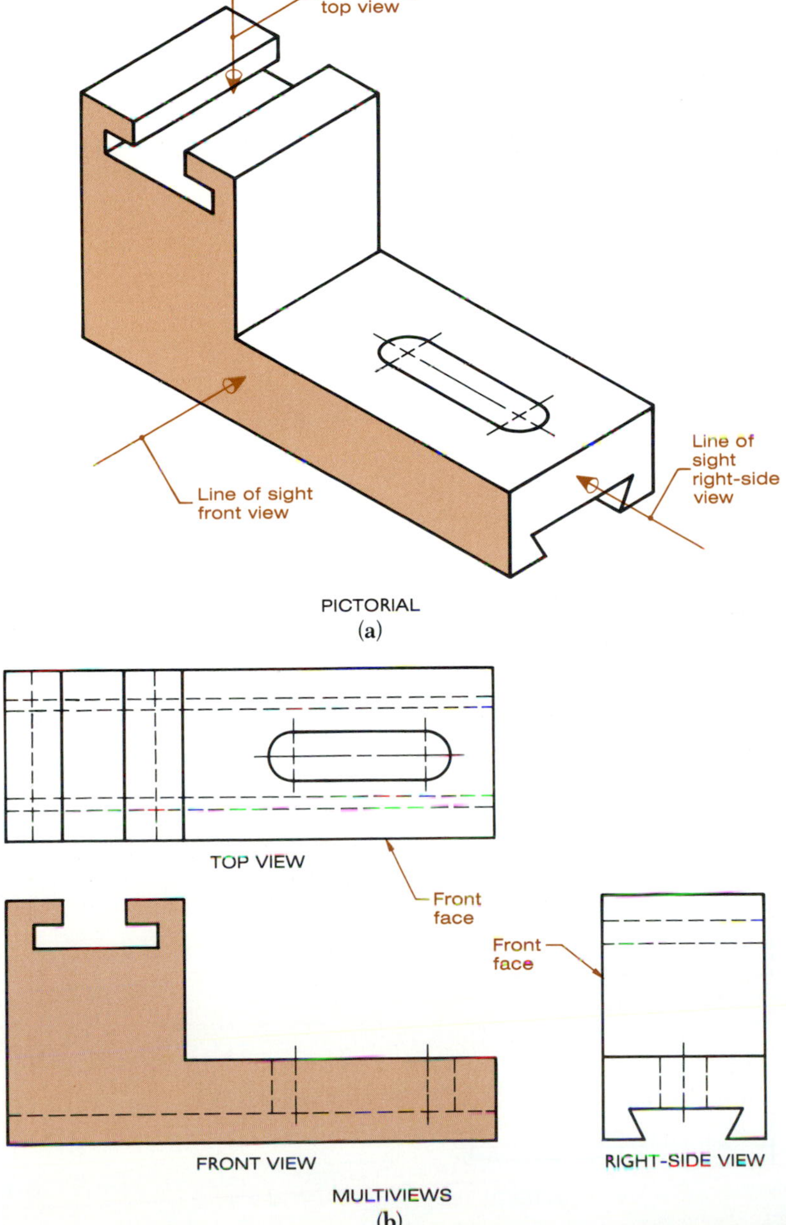

FIGURE 5-17 An object for which three views are required *(Section 5.8)*

5.9 ALTERNATE VIEW POSITION

Once a careful study of the necessary views of an object is made, the views may be rearranged to save space on the drawing. Figure 5-18 shows how the standard arrangement of views may be altered by placing the right-side view in alignment with the top view. This practice is generally reserved for broad, flat objects in which the height of a part is relatively small compared to the depth. Otherwise there is no significant gain in space.

Relatively simple examples of objects have been used thus far to explain the principles of multiview projection. In dealing with more complex shapes, you must learn to methodically analyze each point, line, and surface that forms an object. You can only obtain a clear mental picture of an object by paying close attention to the individual features and by using orderly reasoning. To completely visualize a three-dimensional shape, you must carefully inspect every line and surface in each of the views. Guesswork and trial and error will not lead to an acceptable solution.

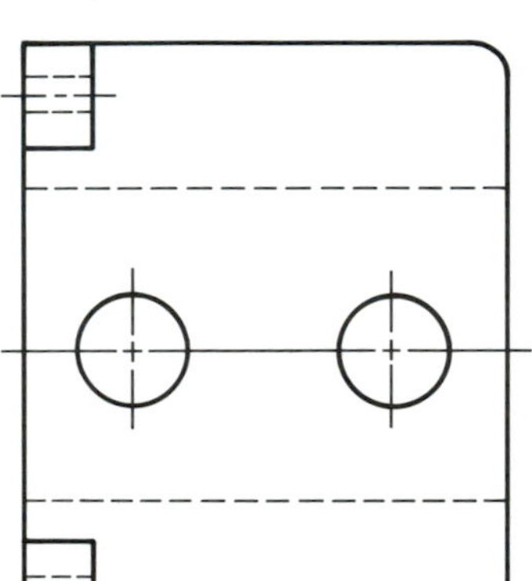

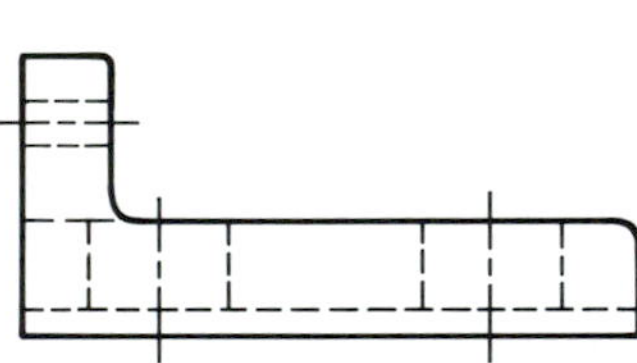

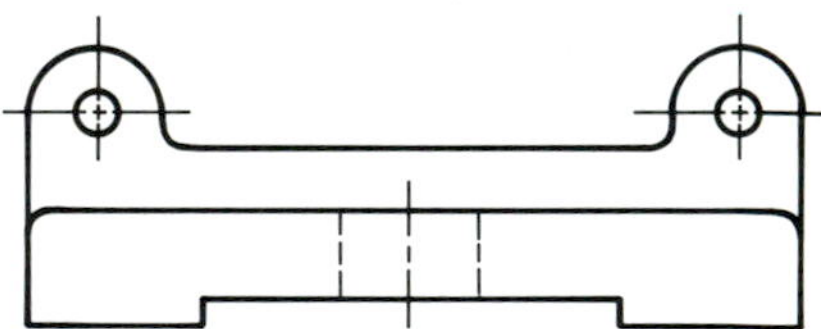

STANDARD ARRANGEMENT OF VIEWS
(a)

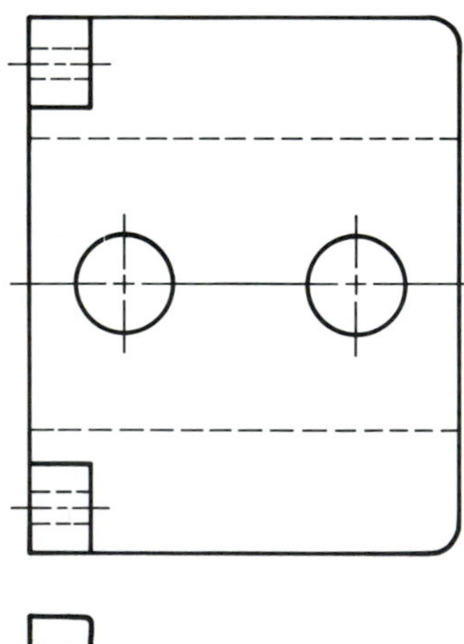

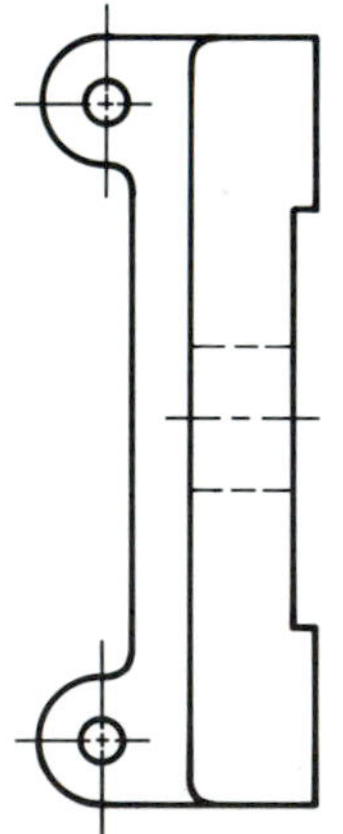

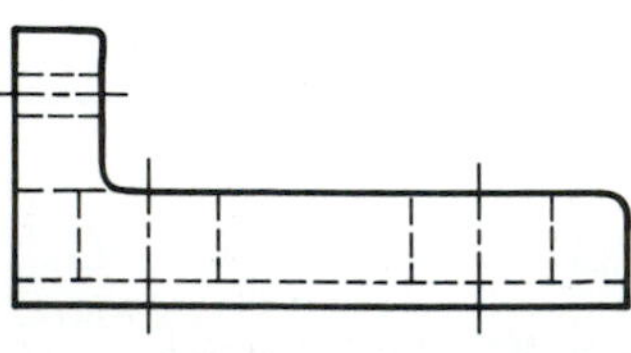

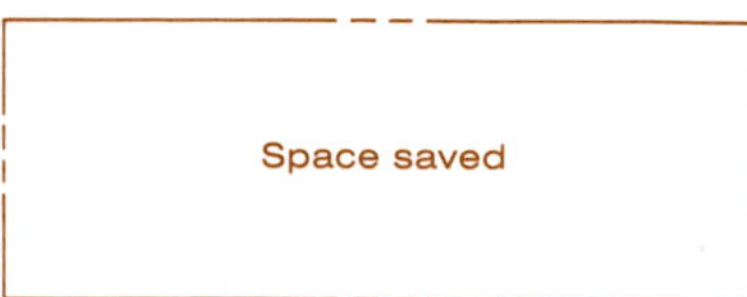

ALTERED ARRANGEMENT OF VIEWS
(b)

FIGURE 5-18 Alternate positions for views *(Section 5.9)*

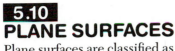

5.10 PLANE SURFACES

Plane surfaces are classified as *normal*, *inclined*, and *oblique*.

Normal Surfaces

A normal surface is a plane surface that is normal (perpendicular) to the line of sight and therefore parallel to a projection plane. The object shown in Figure 5-19 is composed entirely of normal surfaces. In the three principal views each normal surface appears as a true-shape plane in the view parallel to the plane of projection and as a vertical or a horizontal line representing the edge of the plane in each of the other views.

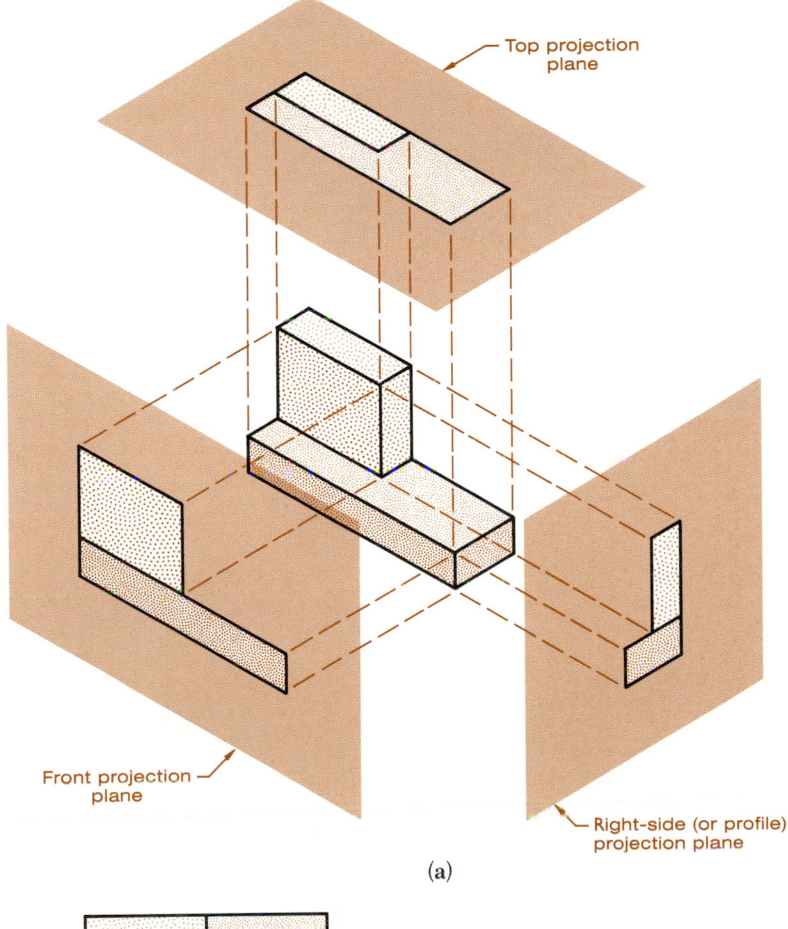

(a)

(b)

FIGURE 5-19 Normal surface. A normal surface appears as an edge in two views and true size in the third view *(Section 5.10)*

Inclined Surfaces

An inclined surface is a plane surface that is neither horizontal nor vertical. Figure 5-20 illustrates an object with an inclined surface. In the three principal views an inclined surface is perpendicular to one projection plane and will appear as a straight line representing the edge of the plane *only* in that view. In each of the other two views the inclined surface will *always* appear foreshortened but *similar in shape*. Note how the rectangular shape of the inclined surface in the top view is repeated in the right-side view as a figure of similar shape. In the three principal views, an inclined plane will appear as an area in two views and as an edge in the other view. An inclined plane never appears true shape in any of the principal views.

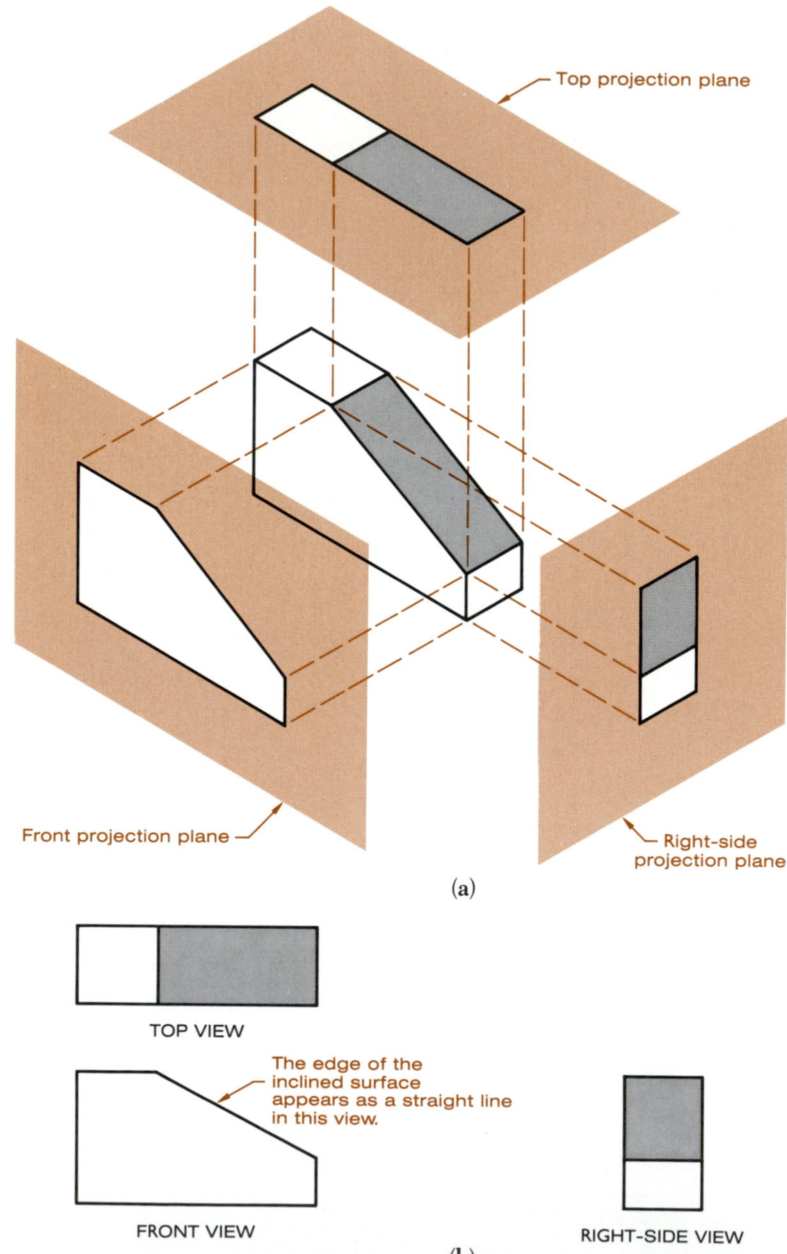

FIGURE 5-20 Inclined surface. An inclined surface appears as an edge in one view and as a figure of similar configuration in the other two views *(Section 5.10)*

Oblique Surfaces

An oblique surface is a plane that appears as a surface in all projection planes. Figure 5-21 illustrates an object with an oblique surface, labeled 1-2-3. An oblique surface is not perpendicular to any projection plane and appears in every view as a foreshortened surface of similar configuration. It does not appear as an edge in any of the principal views.

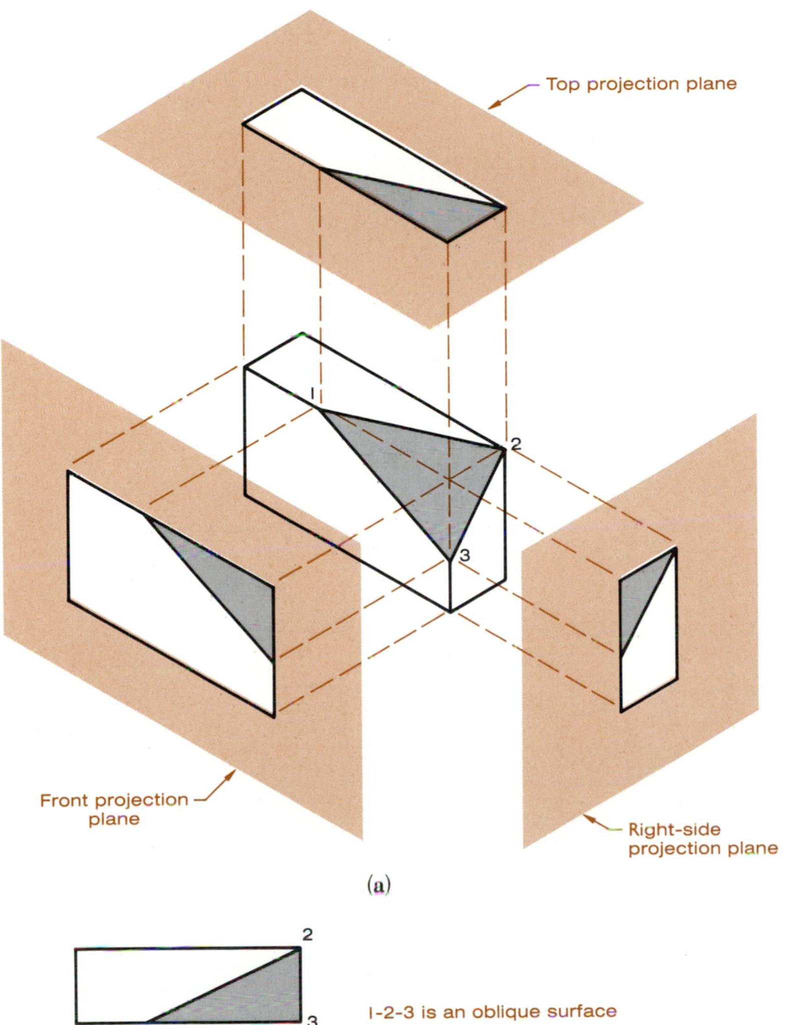

(a)

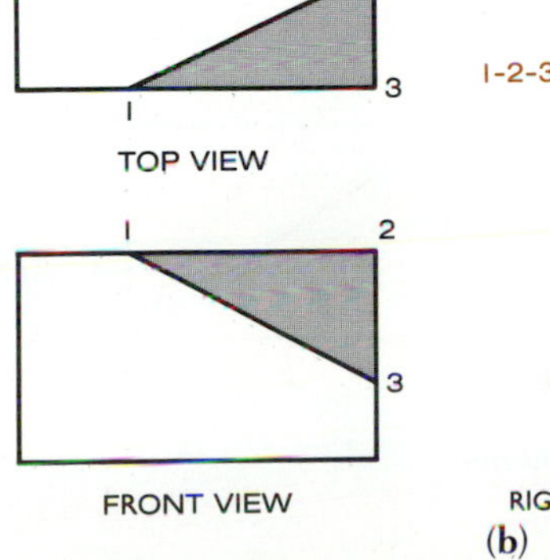

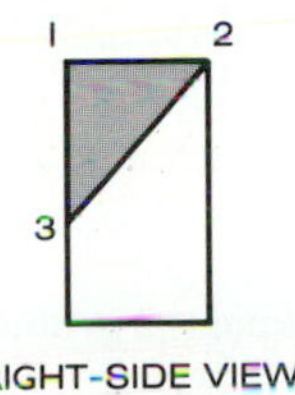

(b)

FIGURE 5-21 Oblique surface. An oblique surface never appears as an edge in a principal view. It always appears as a surface of similar configuration in all three views, but it never appears true size *(Section 5.10)*

CHECKPOINT 2
ANALYZING PLANE SURFACES

Use a plain sheet of 8½ × 11 inch paper and prepare an answer sheet as shown in Figure 5-22.

Carefully study the objects shown in Figure 5-23 and give the quantity and the identifying letter of each of the three kinds of surfaces (normal, inclined, and oblique). As an aid in analyzing the various surfaces, make a pictorial sketch of each object. The three given views are complete in each example, and all visible surfaces are identified with capital letters.

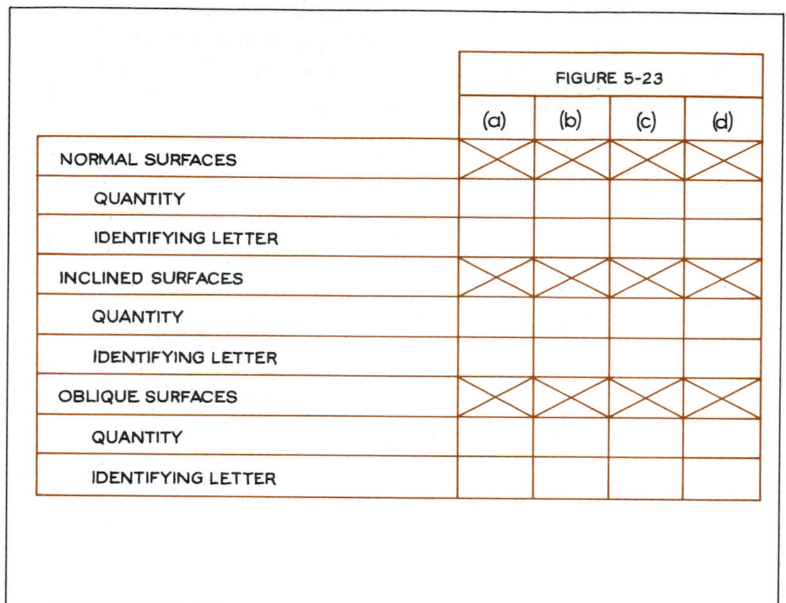

	FIGURE 5-23			
	(a)	(b)	(c)	(d)
NORMAL SURFACES				
QUANTITY				
IDENTIFYING LETTER				
INCLINED SURFACES				
QUANTITY				
IDENTIFYING LETTER				
OBLIQUE SURFACES				
QUANTITY				
IDENTIFYING LETTER				

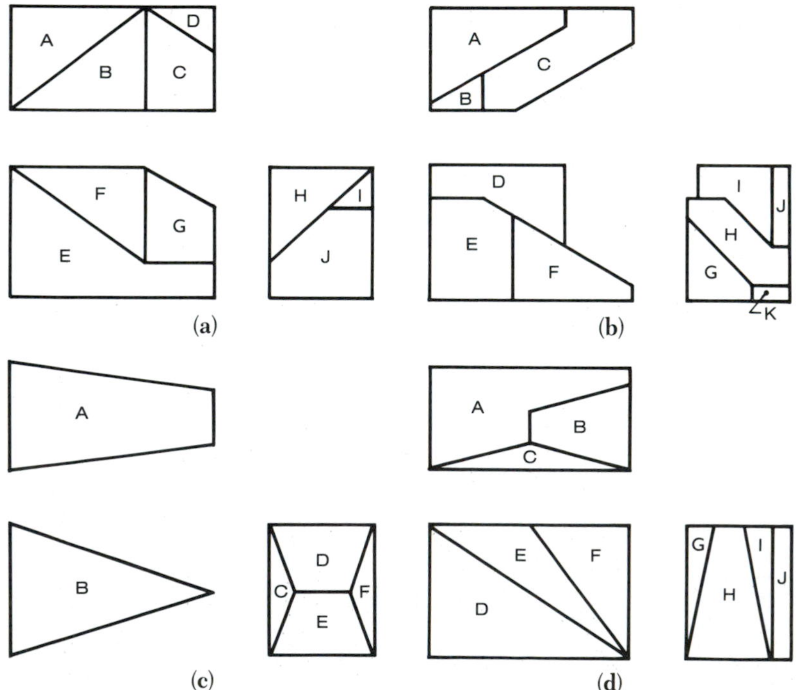

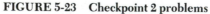

FIGURE 5-23 **Checkpoint 2 problems**

5.11 ADJACENT SURFACES

No two adjacent surfaces may lie in the same plane. Figure 5-24 shows four objects, each with an identical top view made up of visible adjacent surfaces labeled A, B, and C. Numerous possibilities for the three different surfaces exist. The surfaces may be normal, inclined, oblique, or cylindrical, or each surface may be higher or lower than the other two. Using the same top view, four possible shapes of objects are shown in Figure 5-24. Pictorial views have been given for clarity. When you attempt to determine the shape of any object, carefully examine *all* of the views. You can never be certain of the complete shape of an object by concentrating solely upon the features of a single view and neglecting the important details shown in the other given views.

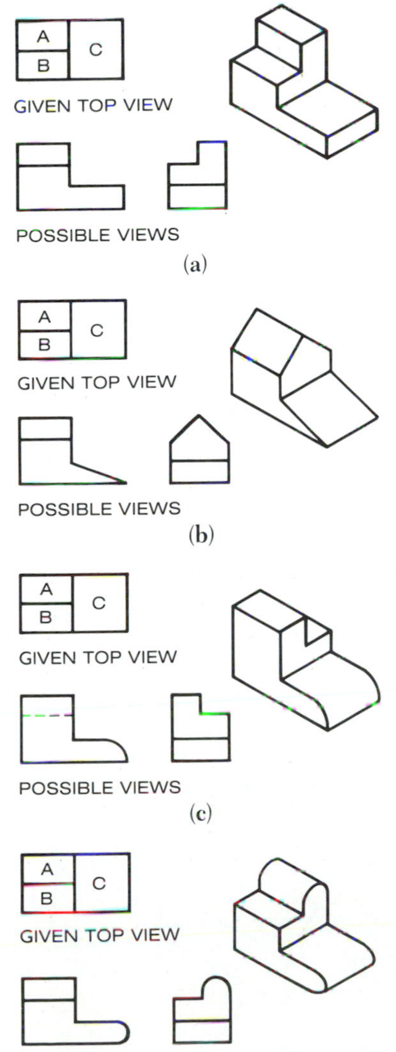

FIGURE 5-24 Surface identification *(Section 5.11)*

5.12 SURFACE IDENTIFICATION

The practice of numbering or lettering the corners of various surfaces on the views is recommended as an aid in developing a clear mental picture of an object. This procedure is particularly helpful for visualizing a third view when working from two given views. However, too many corner numbers or letters are often confusing. Try to confine the use of corner identification to only the most bewildering surfaces. As you correctly solve and project one surface in each of the required views, you may find it helpful to erase the identifying characters before solving another surface. Figure 5-25 illustrates the use of surface identification.

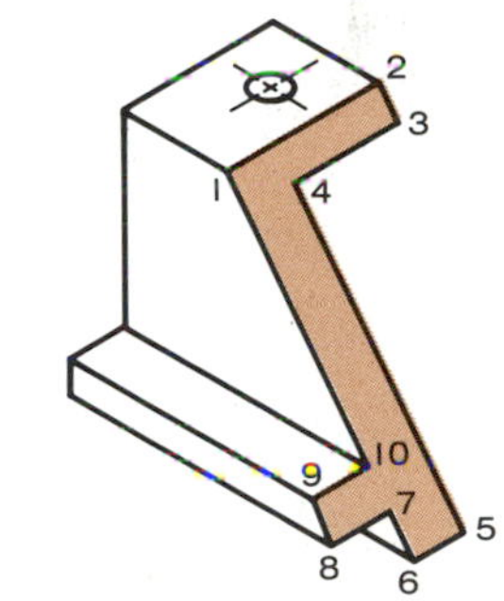

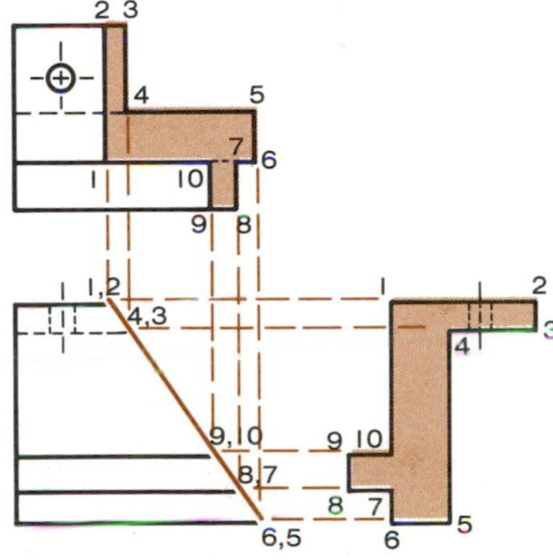

FIGURE 5-25 The use of surface identification *(Section 5.12)*

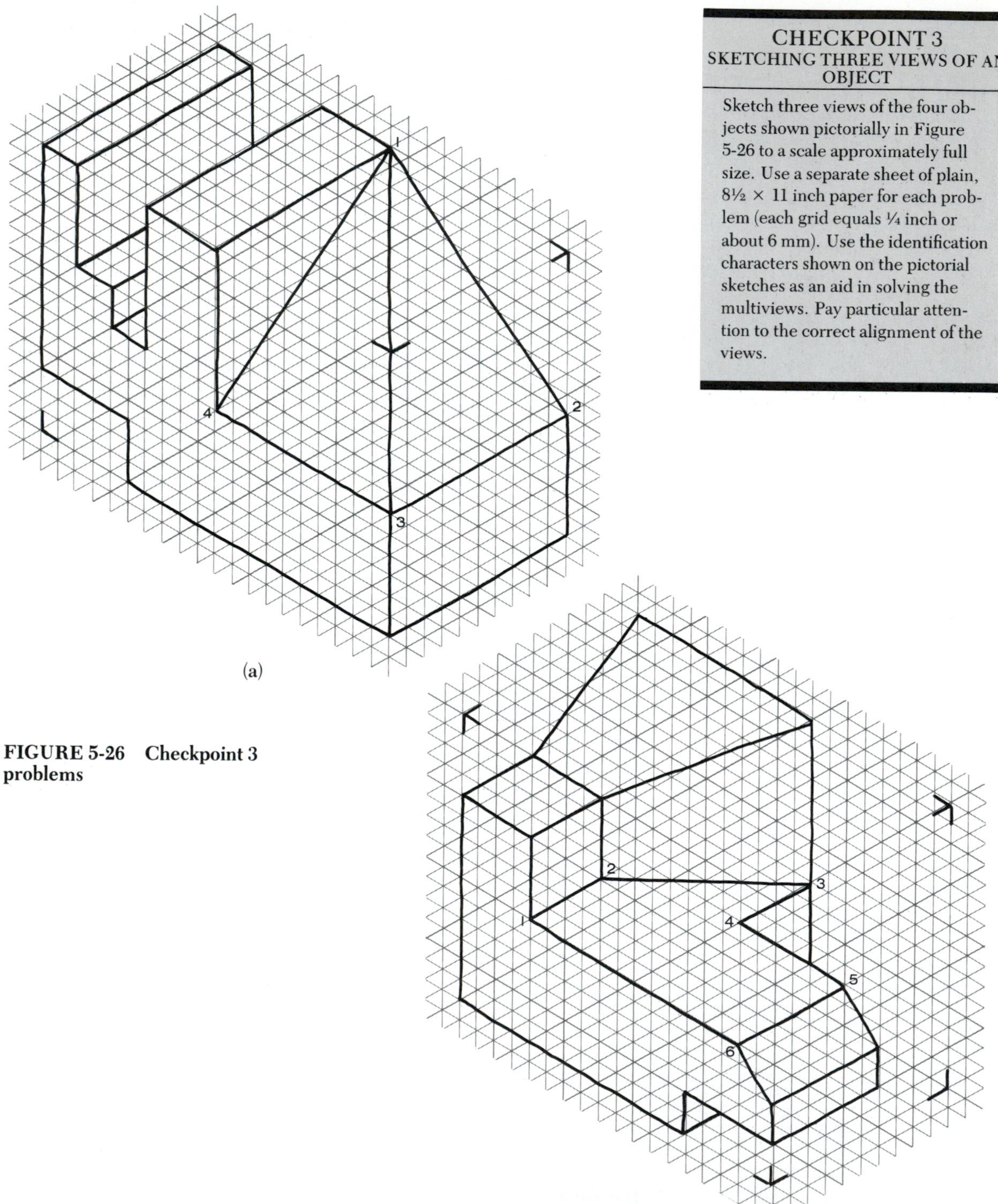

FIGURE 5-26 Checkpoint 3 problems

CHECKPOINT 3
SKETCHING THREE VIEWS OF AN OBJECT

Sketch three views of the four objects shown pictorially in Figure 5-26 to a scale approximately full size. Use a separate sheet of plain, 8½ × 11 inch paper for each problem (each grid equals ¼ inch or about 6 mm). Use the identification characters shown on the pictorial sketches as an aid in solving the multiviews. Pay particular attention to the correct alignment of the views.

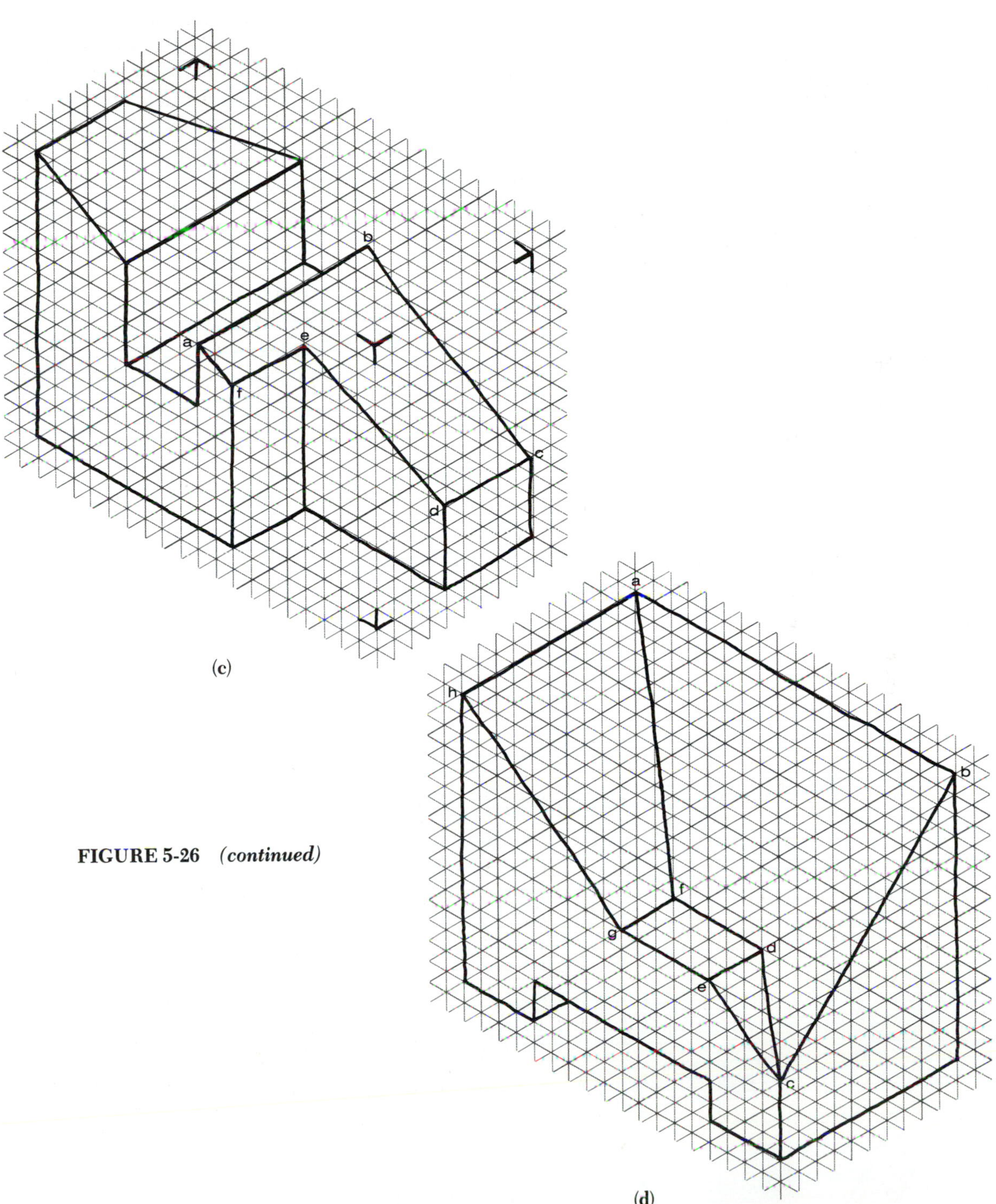

FIGURE 5-26 *(continued)*

5.13 SIMILARITY OF SURFACE CONFIGURATION

Any surface, regardless of its configuration, will appear in the three principal views either as an *edge* or as a *surface of similar configuration*. For example, the inclined surface 1-2-3-4-5-6 shown in Figure 5-27(a) appears as an L-shaped figure in every view except where it appears as an edge. Furthermore, parallel lines on the surface, such as 1-2, 5-6, and 3-4, will appear parallel in every view. The oblique surface 1-2-3, shown on the object in Figure 5-27(b), appears as a surface of similar configuration in all three views. Other examples of similarity of surface configuration are shown in Figures 5-27(c) through 5-27(f).

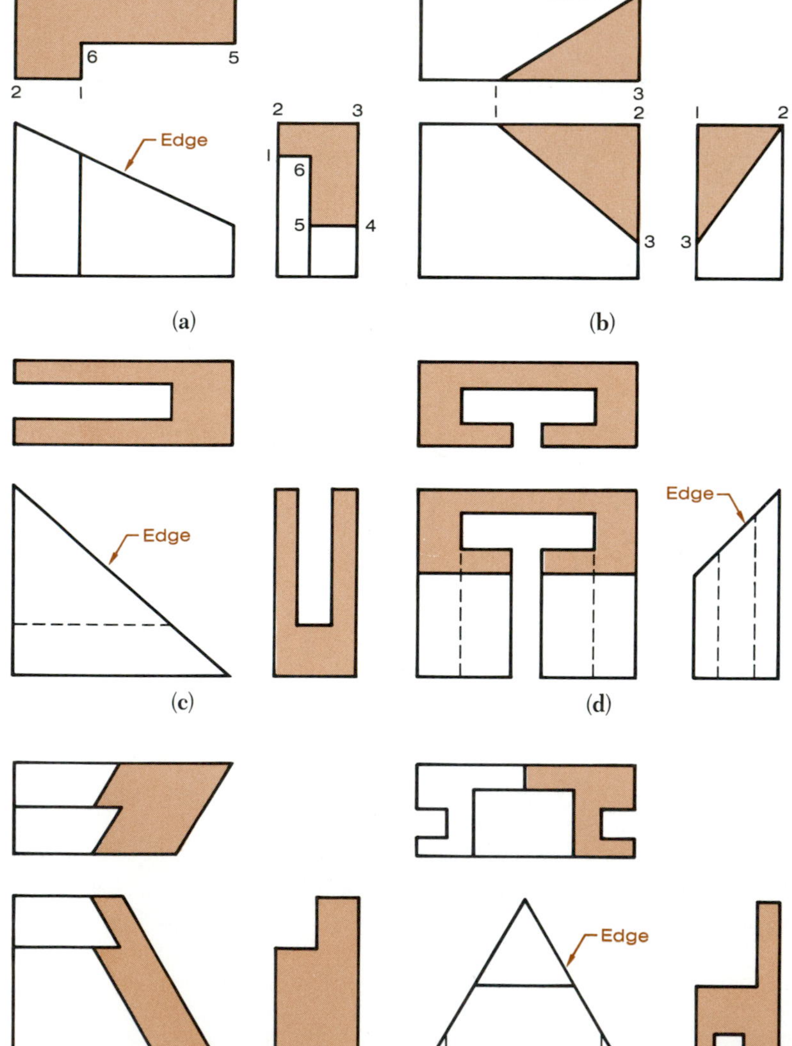

FIGURE 5-27 **Similarity of surface configuration** ***(Section 5.13)***

5.14 THE USE OF PICTORIAL SKETCHES

A pictorial sketch often serves as a valuable aid in the solution of a multiview problem. The pictorial sketch may be made in its entirety prior to attempting the solution of a missing view or it may be progressively developed along with the construction of the corresponding lines and surfaces on a missing view. For some problems just a partially completed sketch may be all that is needed for analysis. The shape of some objects may be so complex that there is little advantage in spending the time needed to develop the entire picture of the object.

5.15 CONSTRUCTING A THIRD PRINCIPAL VIEW

Given the front and right-side views of an object (see Figure 5-28), use the following procedure to construct the third principal view:

STEP 1
Analyze the given views and carefully survey the overall configuration. Think of the object as having been cut from a solid block of material. Try to visualize how the solid block might have appeared before the various machining cuts were made and lightly block in the framework with a pictorial construction box to a convenient size.

STEP 2
Project horizontal lines connecting the given views. These lines are called *projection lines*. Lightly block in a rectangular framework for the outline of the top view, establishing the correct depth. Maintain an approximately equal space between the views. Note that in the front view the right end of the object has been cut at an angle to the base. Add the necessary angular

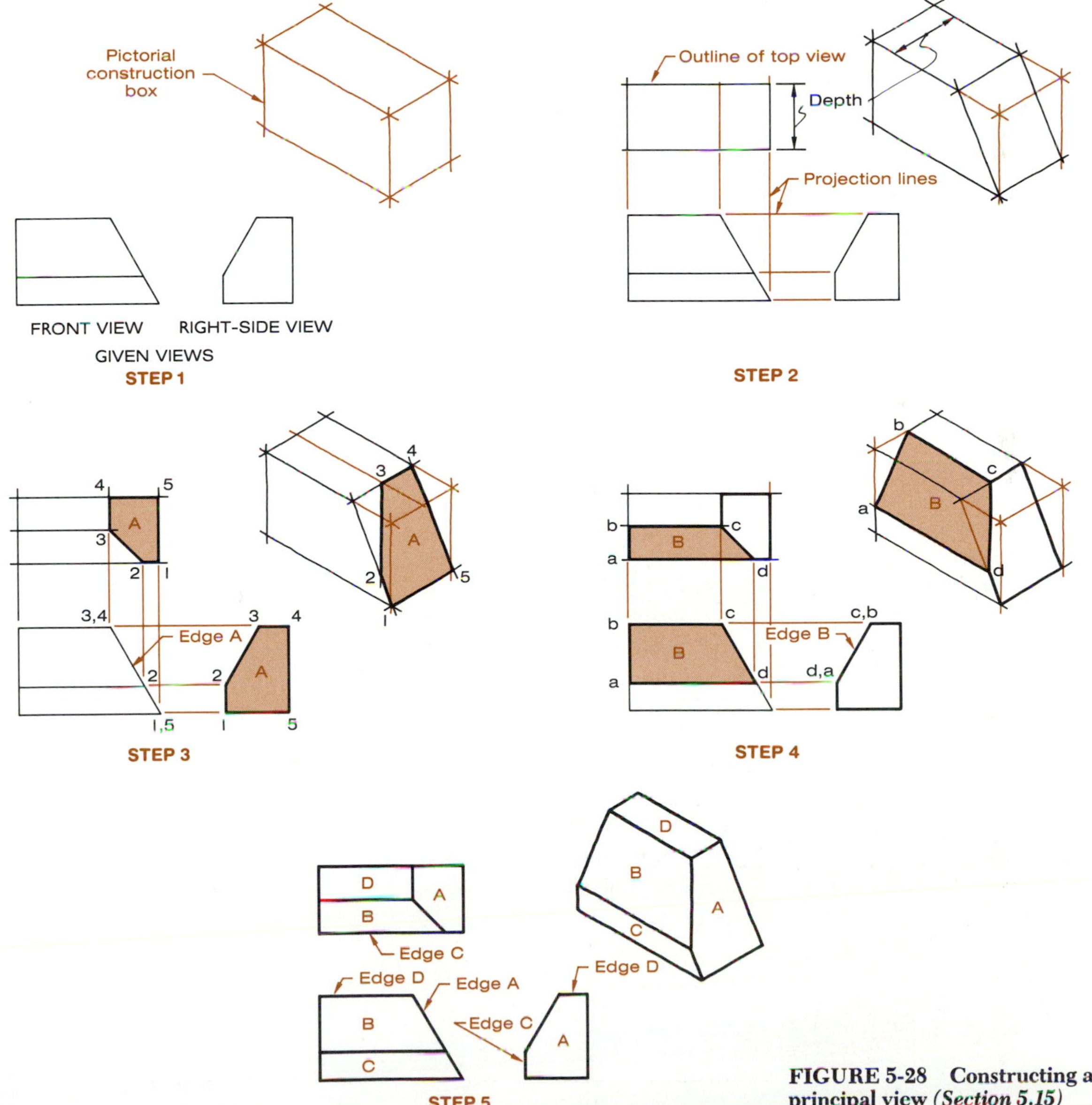

FIGURE 5-28 Constructing a principal view ***(Section 5.15)***

lines to the pictorial sketch and to the top view, as shown, to represent this cut.

STEP 3

An outstanding feature of the object appearing in the right-side view is surface A, which represents the angular cut. No other surface of similar configuration in the front view exists, therefore surface A must appear in that view as an edge. This line is labeled A. Starting with the right-side view, apply identification numbers to each point on surface A. Project these numbers to the inclined line in the front view. Apply the correspondingly numbered points to the correct locations on the pictorial sketch and darken the lines, enclosing surface A. Note that as an aid in plotting surface A, points 1 and 2 are on the *front* edge and points 4 and 5 are on the *back* edge. Lines 1-2 and 4-5 are also parallel. Plot the numbered points on surface A in the top view by alignment and measurement and draw the surface lines.

STEP 4

Before proceeding with the analysis of other surfaces, erase all of the identifying numbers previously used to plot surface A. Surface B in the front view is another odd-shaped surface. Since it does not appear as a surface of similar configuration in the right-side view, it must appear as an edge (labeled B in the right-side view). Starting with the front view, apply a set of identifying letters (or numbers) and plot surface B in both the pictorial sketch and in the top view. Note that line bc is parallel to ad and that points d and c are coincident with points previously numbered as 2 and 3.

STEP 5

The vertical surface C in the front plane and the horizontal surface D in the top plane are normal surfaces. Remember that a normal surface always appears true size in the view parallel to the projection plane and as a vertical or a horizontal line in each of the other views. Add the missing lines for surfaces C and D in the pictorial sketch and in the top view. As a final check on the accuracy of the solution, make sure that adjacent surfaces B and C in the front view and surfaces B, C, and D in the top view are separated by a line and lie in different planes.

Figure 5-29 illustrates the following procedure for drawing the third principal view when the top and front views are given:

STEP 1

Analyze the given views and carefully survey the overall configuration. An examination of the front view shows the object to be tentlike, or triangular in shape, when viewed from this position. The position from the top view shows a wedge-shaped object with the point of the wedge on the right. Surface A in the front view corresponds with the line labeled A in the top view. This line represents the edge of surface A. Starting with the front view, apply identification numbers (1-2-3) on the three corners of surface A and project these numbers to the inclined line labeled A in the top view.

STEP 2

Draw horizontal projection lines to the right and lightly block in a rectangular space for the right-side view. The depth dimension must be the same in the top and the right-side views. Maintain an approximately equal space between the views. Project and plot the positions of the numbered points 1-2-3 in the right-side view, establishing surface A as a figure of similar (triangular) configuration.

STEP 3

Before proceeding with the analysis of other surfaces, erase the identification numbers previously used to denote surface A. There are two visible surfaces, labeled B and C, in the top view. Surface B is trapezoidal in shape and surface C is triangular. No surfaces of similar configuration to B or C exist in the front view so they each appear as edges labeled B and C. Apply identifying numbers to surfaces B (4-5-6-7) and C (7-6-8). Project by alignment and plot the numbered points in the right-side view.

As a final check on the correctness of the views, note that the inclined triangular surfaces A and C should appear as figures of similar configuration in two of the views and as an edge in the other view. In this particular example surfaces A and C appear as visible planes in both the top and right-side views. Surface B, trapezoidal in shape, appears as a figure of similar configuration in the top and right-side views and as an edge in the front view. Due to its viewing position, in the right-side view, surface B (4-5-6-7) is a hidden plane.

FIGURE 5-29 Constructing a principal view *(Section 5.15)*

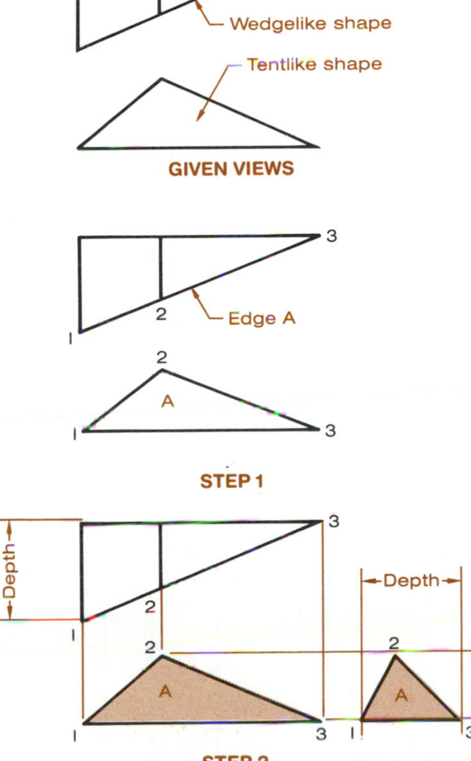

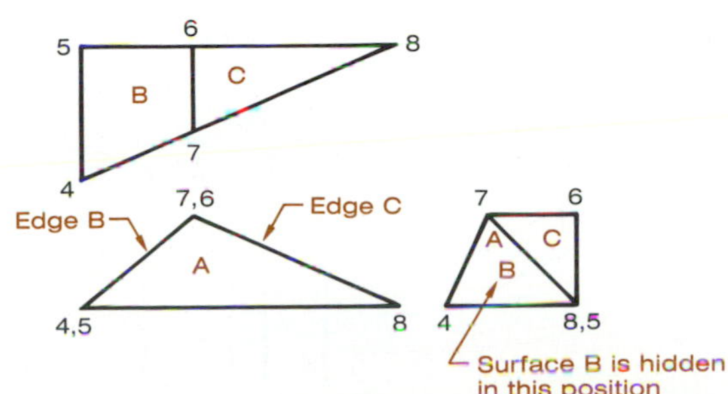

In most on-the-job applications a combination of the top and front views or the top and right-side views is available. However, to find the front view, you may sometimes need to use the following procedure (illustrated in Figure 5-30) when only the top and right-side views are given. Working with this unusual combination of views will provide you with practice in applying the principles and logic of visualization.

FIGURE 5-30 Constructing a principal view ***(Section 5.15)***

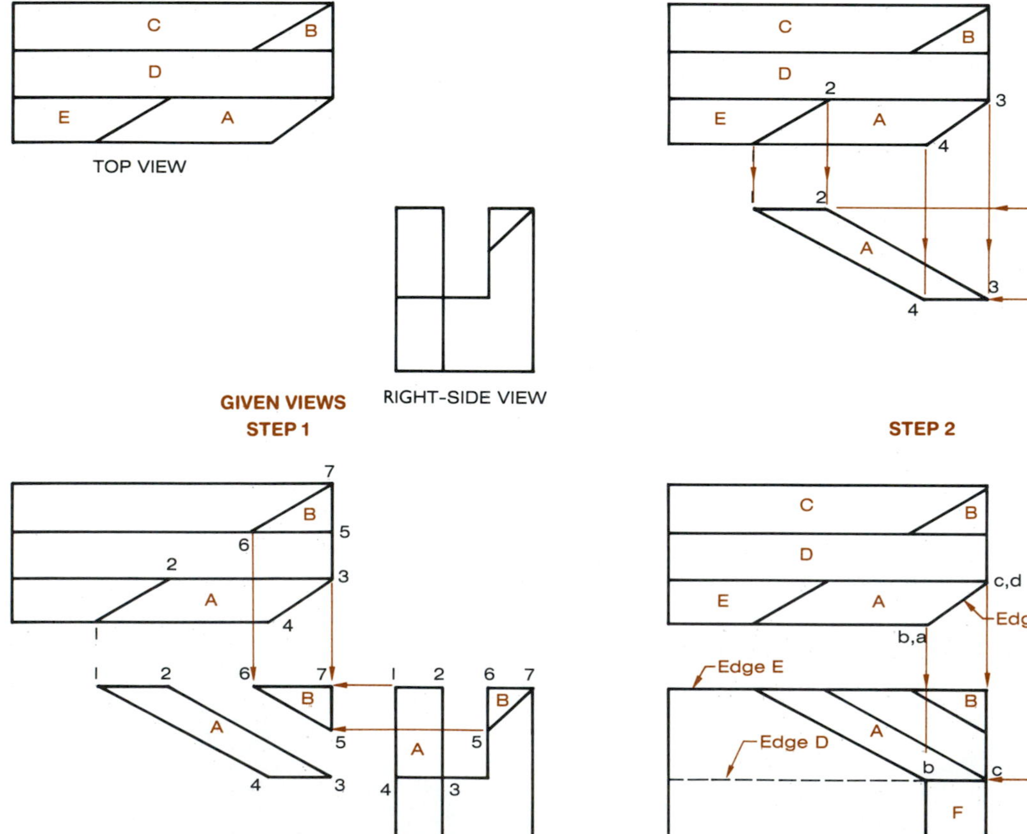

STEP 1

As in the previous examples, begin by analyzing the given views and carefully survey the overall configuration. Your initial examination, for example, will show a slot cut into the top of the object, extending from the left side to the right side. The right end of the top view shows that the object is cut away at an angle.

STEP 2

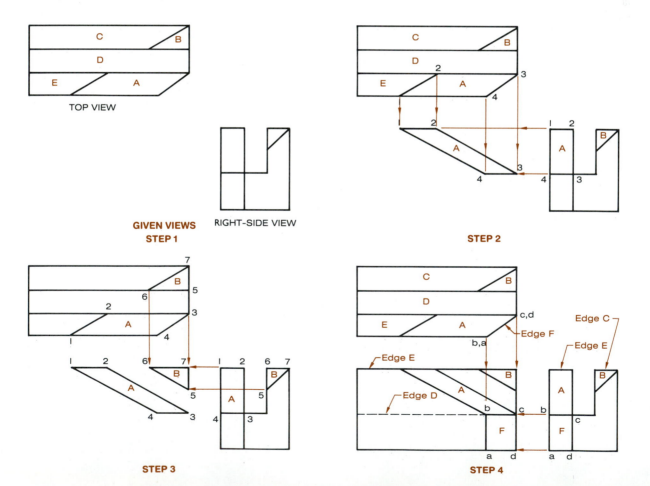

The five visible surfaces in the top view are labeled A, B, C, D, and E. Certain combinations of these surfaces are adjacent to each other (E and D, A and D, and C and B, for example). Accordingly, each of these surfaces must lie in different planes. Because of their odd shapes, surfaces A and B are appropriate surfaces to use for applying the principles of similarity of configuration (see Section 5.13). Locate and identify surfaces A and B in the right-side view. Beginning with surface A, apply identifying numbers to this surface in both of the given views. If surface A plots as a line in the front view, it is an inclined surface; if it plots as a figure of similar configuration in the front view, it is an oblique surface. Remember that an oblique surface will appear as a figure of similar configuration in *every* principal view. As the various points of surface A are plotted in the front view, it becomes apparent that the surface is oblique, not inclined.

STEP 3

Note the similarity of the triangular configuration of surface B in the top and right-side views. Number surface B in both the given views and plot it in the front view. Like surface A, surface B is oblique.

STEP 4

Identify surface F in the right-side view and place the corresponding identification letters on its edge-view position in the top view. Using the top and right-side views, plot the inclined surface F in the front view. Finally, observe that all of the remaining surfaces on the object are normal surfaces. You may now complete the front view by alignment and measurement. (The hidden edge of surface D is represented in the front view by a dashed line.)

CHECKPOINT 4
SKETCHING THREE VIEWS OF AN OBJECT

Divide a plain sheet of 8½ × 11 inch paper into two equal spaces and sketch three views of the objects shown pictorially in Figure 5-31 to a scale approximately full size. In the given drawing let each grid equal ⅛ inch, or about 3 mm. Pay particular attention to the correct alignment of the views.

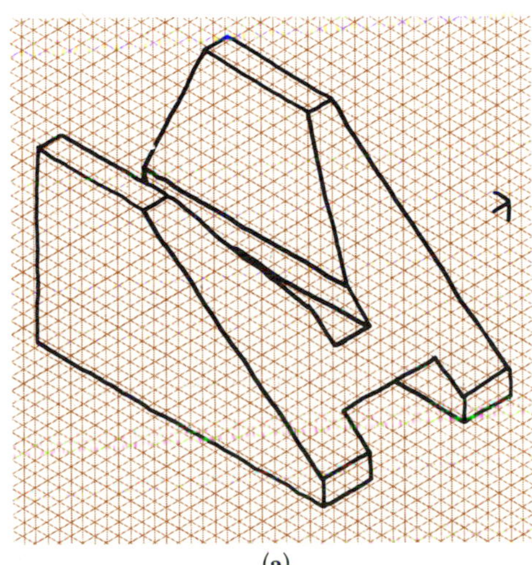

(a)

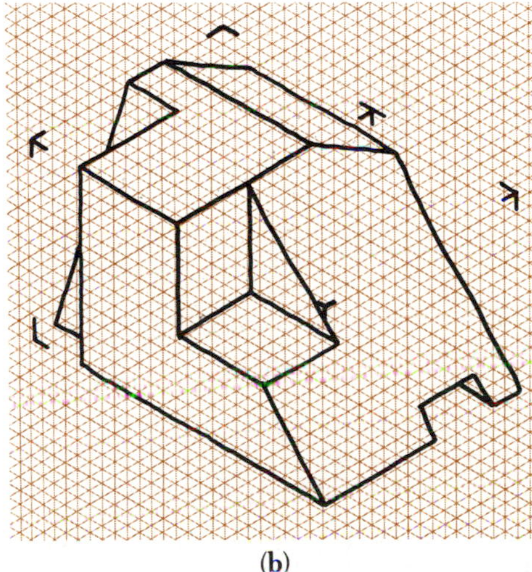

(b)

FIGURE 5-31 Checkpoint 4 problems

FIGURE 5-32 Machine parts having a cylindrical shape *(Section 5.16)*

FIGURE 5-33 Views of cylindrical parts cut in various ways *(Section 5.16)*

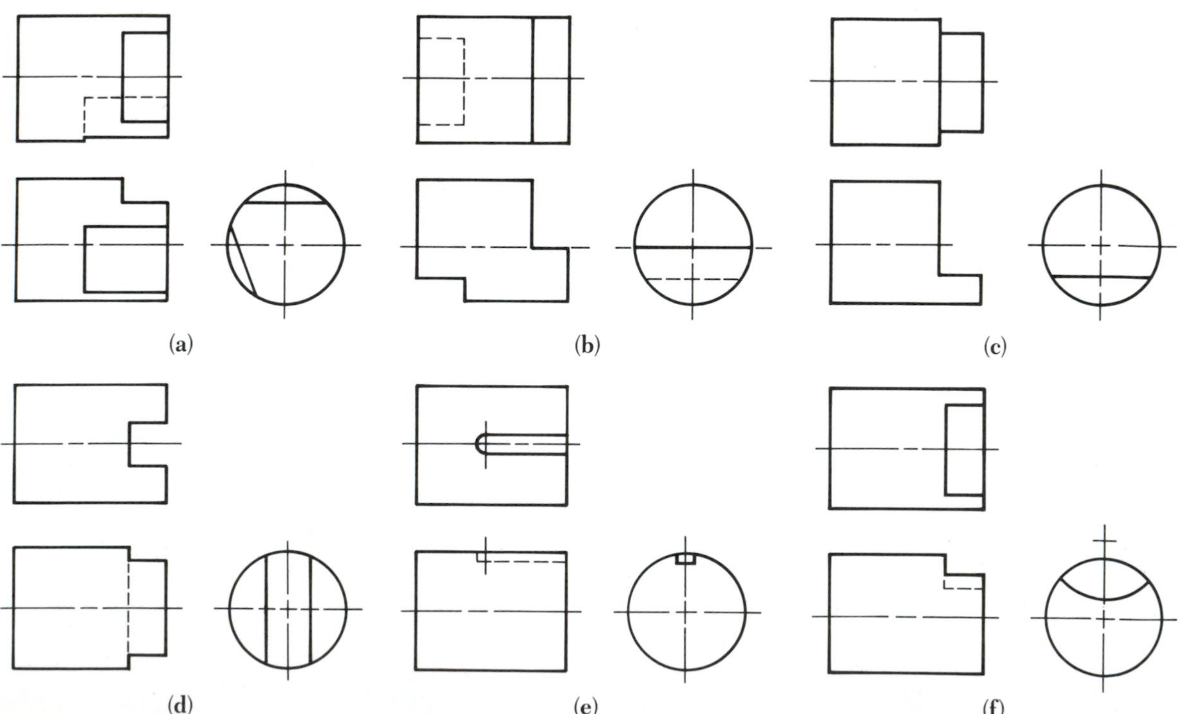

5.16 CYLINDRICAL SURFACES

Figure 5-32 illustrates several examples of machine parts with cylindrical shapes. Most uncomplicated, straightforward cylindrical shapes may be clearly represented in two views on a drawing. In such cases a circular view with one adjacent view is generally sufficient to describe the shape unambiguously. Views of some cylindrical objects, however, may become confusing when a cylindrical surface is cut by a machining operation. More complicated shapes may require three views.

Figure 5-33 illustrates views of objects that have been cut in various ways by the action of a cutting tool. Most of the surfaces thus formed are plane surfaces. Figure 5-33(i) shows a cylinder cut at an angle of 45° with the horizontal. In this special case the inclined surface appears as a circle in both the top and right-side views. If a cylinder is cut by an inclined plane at an angle other than 45°, as shown in Figure 5-33(j), the inclined surface appears as an ellipse in the top view. Regardless of the angle of the cut, the right-side view will appear as a circle because the major and minor axes in that view are equal. Since the major and minor axes ab and cd can be found by projecting them from the front view, the ellipse can be drawn by any one of the methods explained in Section 4.21 or by the aid of an ellipse guide (see Figure 4.54).

FIGURE 5-33 *(continued)*

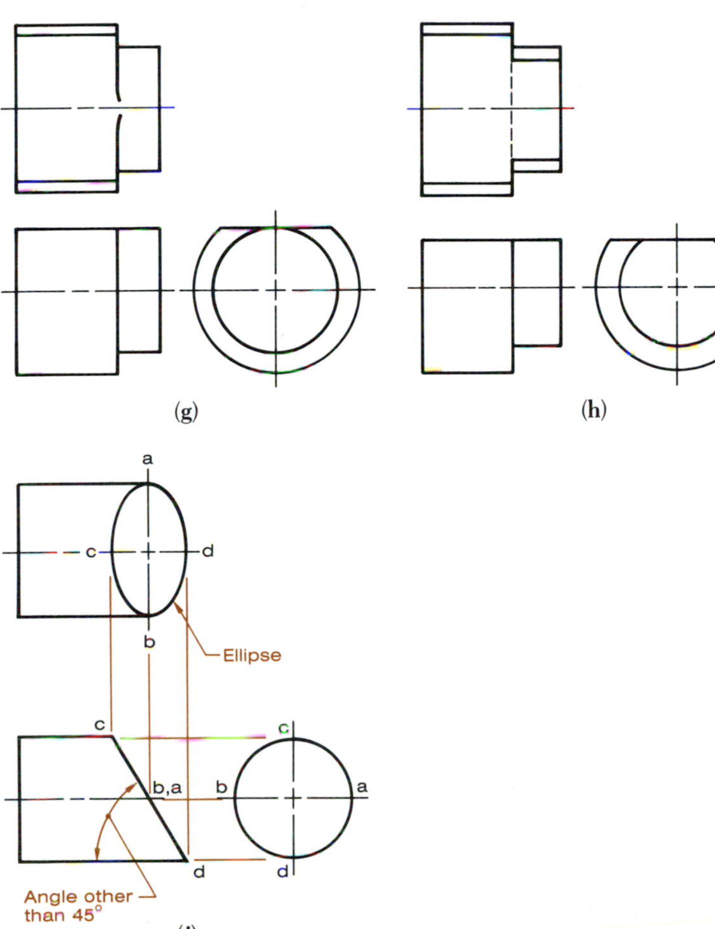

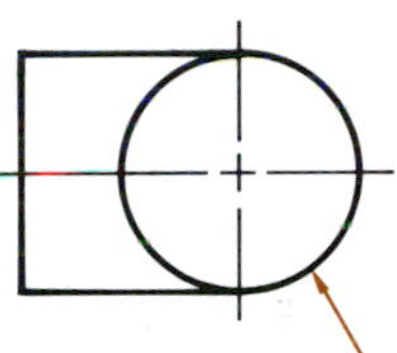

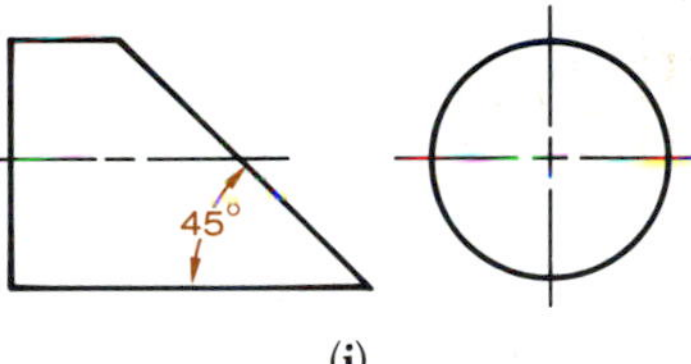

(g) (h) (i)

(j)

5.17 INTERSECTING CYLINDERS

Figure 5-34 illustrates some recommended methods for showing intersections of cylinders. In Figure 5-34(a) the curve of intersection is omitted when a small cylinder intersects a large one. When the intersection is large enough to be important, a true curve is plotted, as shown in Figure 5-34(b). In Step 1, points 1 through 5 are selected in the top view as shown. The points may be selected in the circle at random or, as in this example, uniformly spaced points may be taken. A sufficient number of points should be taken to ensure an accurate curve of intersection. These points are projected downward to the vertical cylinder in the front view as elements. In Step 2, to find point 4 on the line of intersection the measurement y is taken in the top view and layed off as an element on the vertical cylinder in the right-side view. Point 4 is then projected across to the front view to intersect with the corresponding element. The remaining other points are plotted in the front

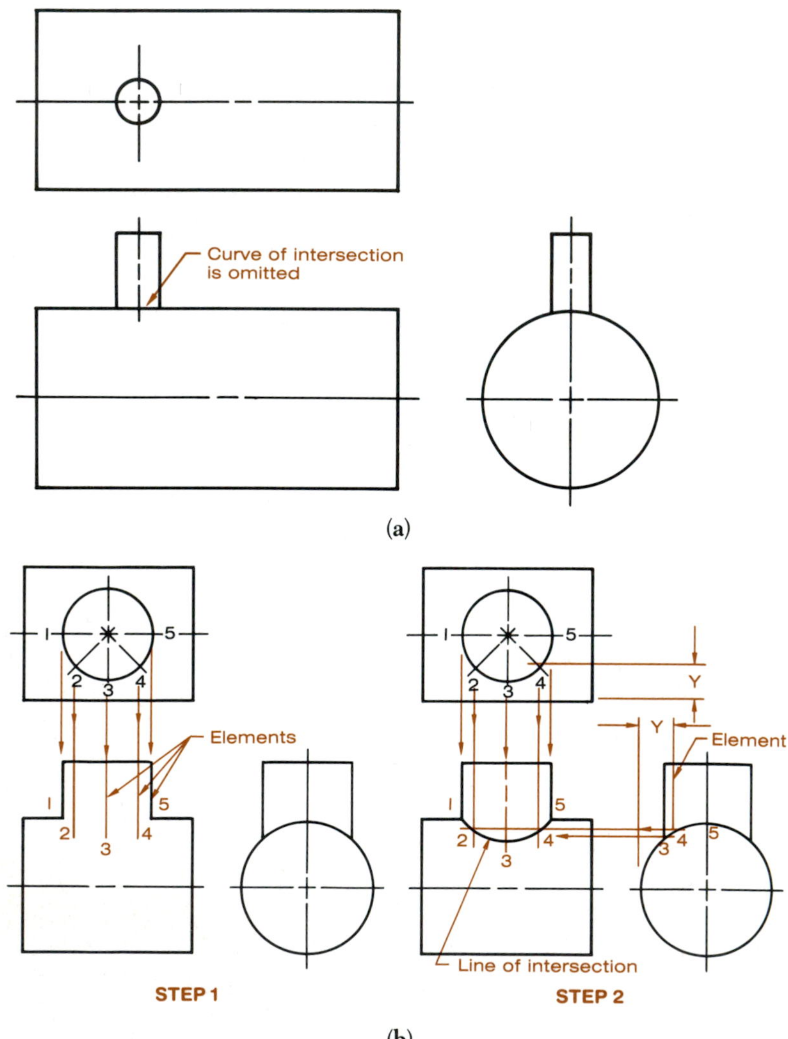

FIGURE 5-34 **Intersecting cylinders** ***(Section 5.17)***

view in the same way. The line of intersection is represented by a smooth curve drawn through these points. The lines of intersection of two cylinders of the same diameter appear as straight lines in the front view, as shown in Figure 5-34(c).

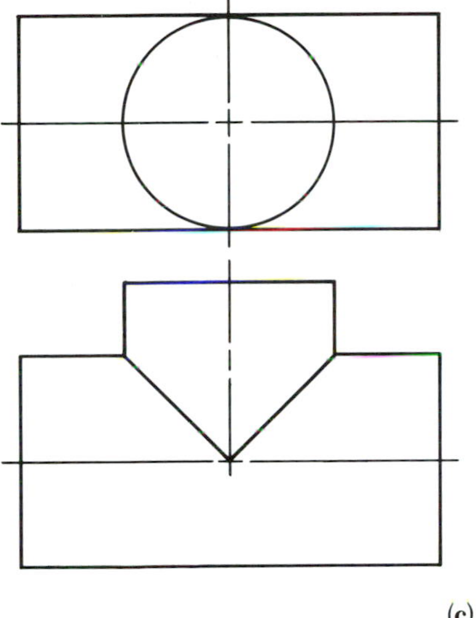

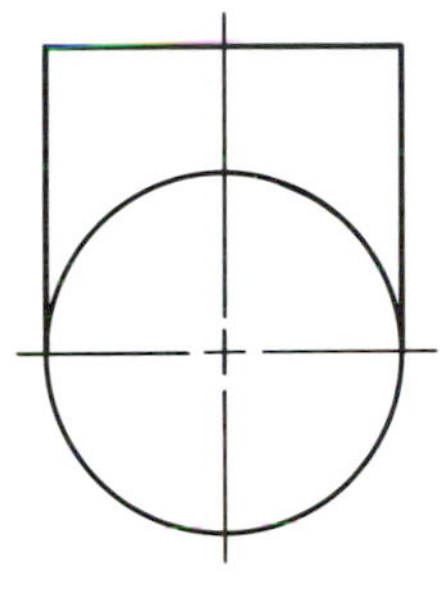

(c)

FIGURE 5-34 *(continued)*

5.18 INTERSECTIONS OF CYLINDERS AND PRISMS

In general, as shown in Figure 5-35(a), a relatively small or insignificant intersection formed by a prism and a cylinder may be disregarded. Draw the lines of intersection, however, when the size of the prism is larger, as shown in Figure 5-35(b).

5.19 FILLETS, ROUNDS, AND RUNOUTS

Parts produced by casting or forging have rounded corners. As shown in Figure 5-36, a rounded internal corner is called a *fillet* and a rounded external corner is called a *round.* Fillets

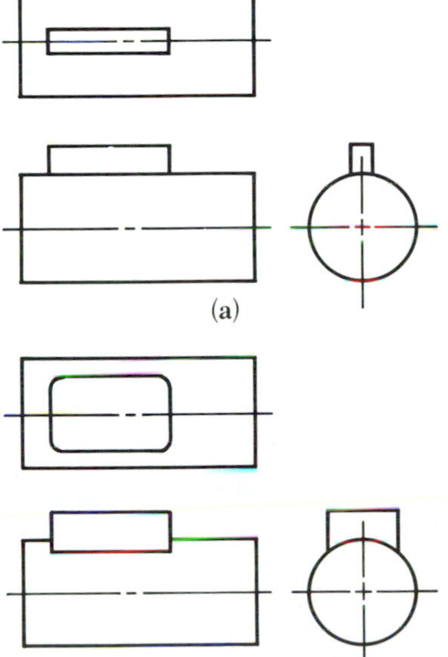

FIGURE 5-35 Intersecting cylinders and prisms *(Section 5.18)*

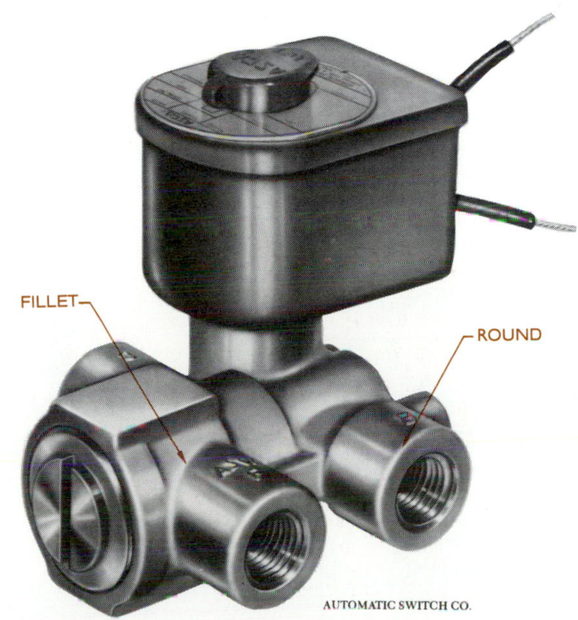

FIGURE 5-36 Fillets and rounds *(Section 5.19)*

and rounds are more easily produced on cast or forged parts than are sharp corners. While rounds improve the appearance of a part, fillets increase the strength and soundness of a part. When heat is used in the production of a casting or a forging, a small amount of shrinkage takes place as the metal cools. This action results in potential high stress concentration in the internal corner regions of the part, often leading to a weakness or to actual cracking. As shown in Figure 5-37(a), most designers make the fillet radius equal to the wall thickness for equal sections. For sections of unequal thicknesses, the fillet radius is usually made equivalent to the mean dimension of the two walls, as shown in Figure 5-37(b).

All parts made in a forge shop or in a foundry have relatively rough as-forged or as-cast surfaces. When a smooth surface is required, the surfaces are finished (machined). Before machining, a round corner is always present on forgings or castings at the intersections of two surfaces. If one or both of these surfaces is finished, as shown in Figure 5-38, the corners become sharp. Figure 5-39 illustrates a casting that has been finished in this manner. Note how the corners become sharp because of the machined surfaces. When a part is made entirely from stock, each of the surfaces is likely to be formed by machining and all of the corners are sharp. When a rounded corner is required on a part made in this way, special cutting tools must be used.

FIGURE 5-37 Recommended fillet radii ***(Section 5.19)***

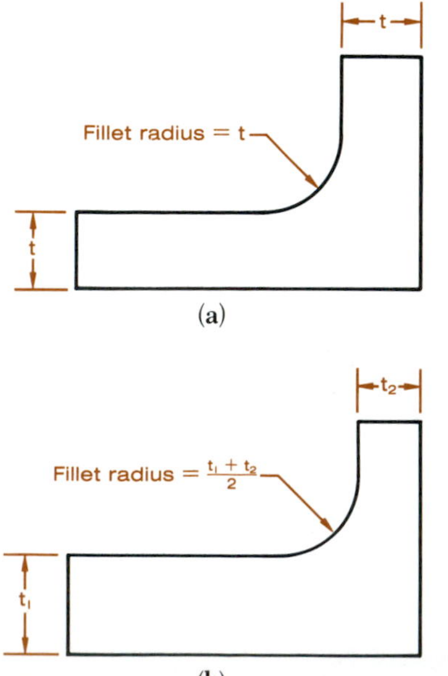

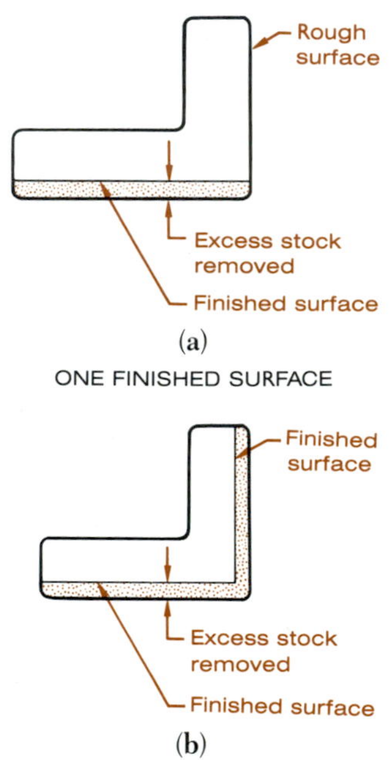

FIGURE 5-38 Unfinished and finished surfaces ***(Section 5.19)***

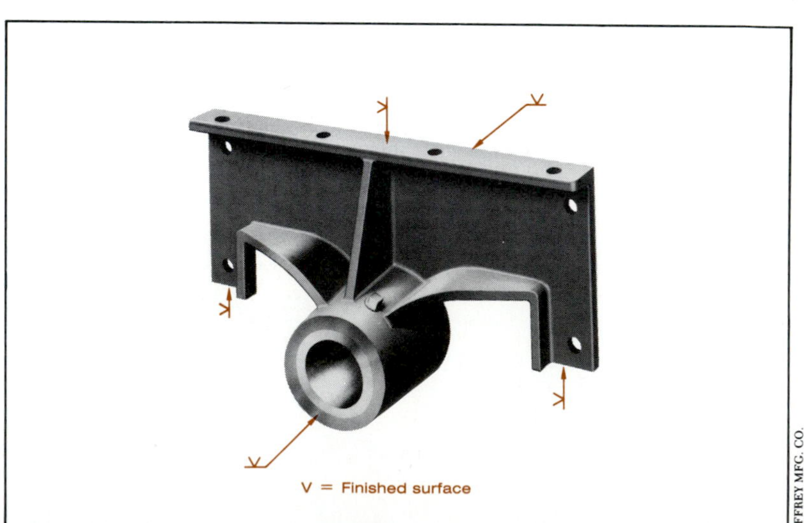

FIGURE 5-39 Fillets and rounds on a casting ***(Section 5.19)***

Fillets and rounds ⅛ inch or less may be drawn freehand. Larger radii should be drawn with a compass or with a template. Conventional representations of fillets and rounds are shown in Figure 5-40.

Runouts are short, curved lines that are formed at the intersection of two or more fillets or rounds. Typical methods of representing runouts on forged or cast parts are also shown in Figures 5-39(d) through 5-39(g). The runout radius is commonly made equal to the fillet radius.

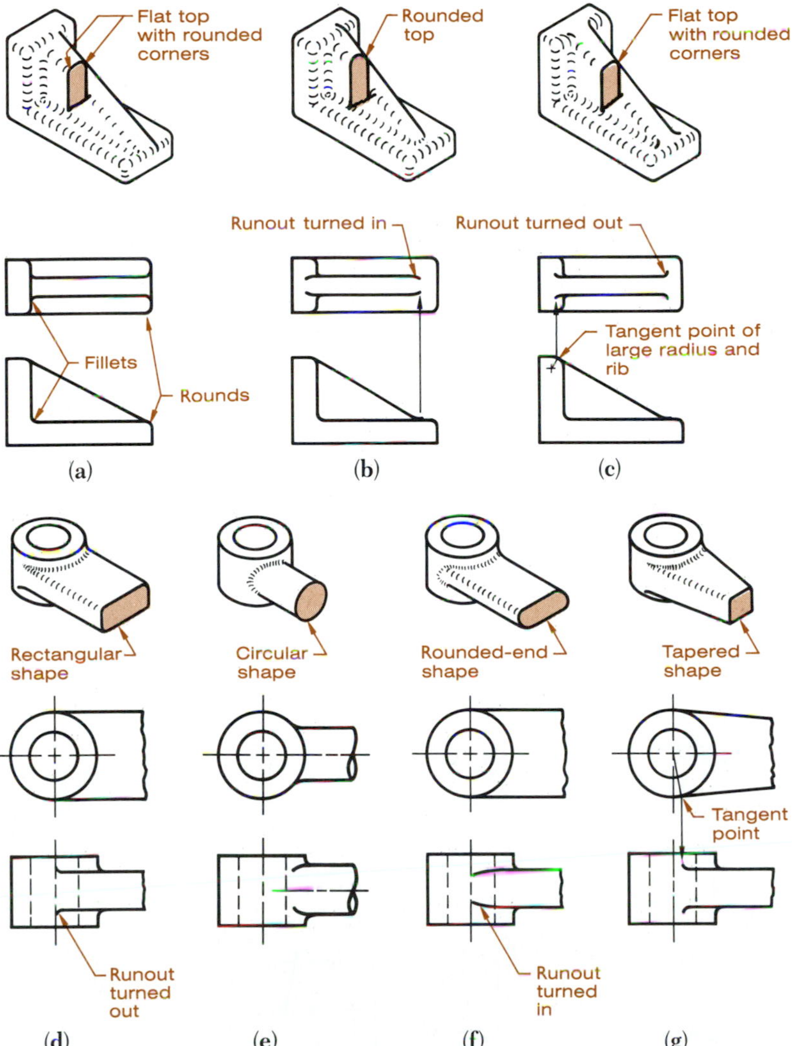

FIGURE 5-40 Conventional representation of fillets, rounds, and runouts ***(Section 5.19)***

5.20 PHANTOM LINES

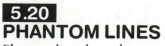

Phantom lines have the same weight as center lines, but are interrupted by pairs of long dashes. These long dashes may vary in length depending upon the size of the drawing. Phantom lines are used to indicate alternate positions of moving parts, as shown in Figure 5-41(a), or to represent repeated detail on a gear, spring, screw thread, or similar part, as shown in Figure 5-41(b). Phantom lines are also used to indicate features such as bosses, lugs, or excess stock that are later removed. See Figure 5-41(c) for an example of this use.

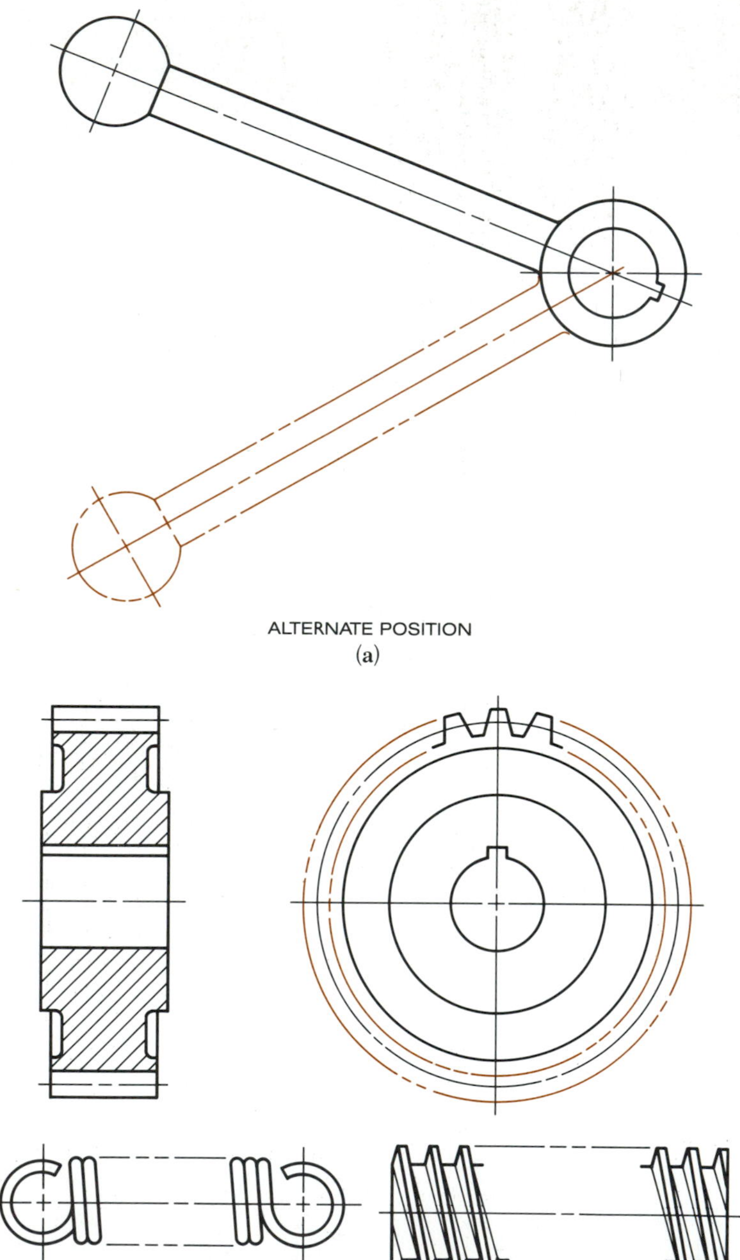

FIGURE 5-41 The use of phantom lines *(Section 5.20)*

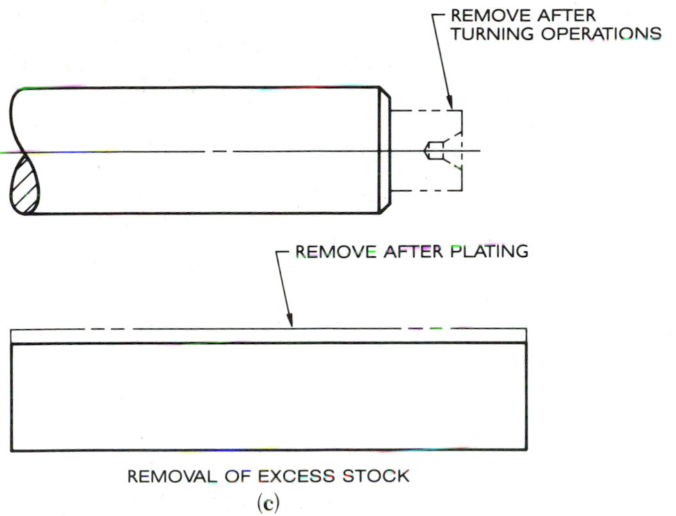

REMOVAL OF EXCESS STOCK

(c)

FIGURE 5-41 *(continued)*

PROBLEMS

Shape Description

The following problems will provide practice in applying the principles of multiview projection discussed in this chapter. Use layout A, shown on the inside front cover. Letter the required information in the sheet title block. Neatly balance each problem within the available space. Before attempting to solve the problems in this chapter refer to the table of symbols on page 85. For additional information about these symbols, see Section 10.26.

Problems **P5.1** through **P5.24** are missing-line sketching problems. For each of these problems, sketch the given views to a scale approximately full size, with two problems per sheet. Assume that each grid in the given drawings equals ¼ inch or 6 mm. Where necessary, add missing lines to the given views.

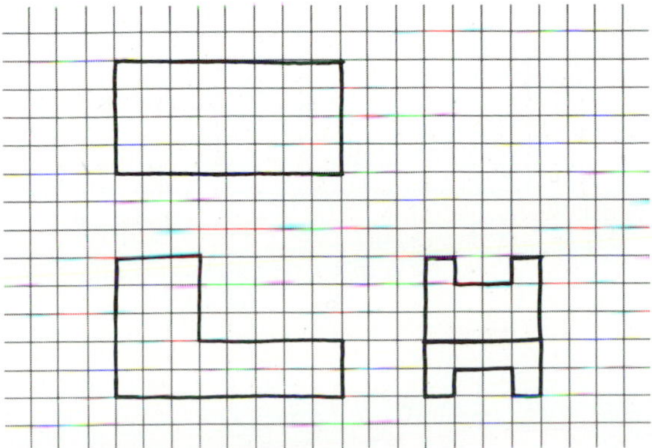

P 5.1

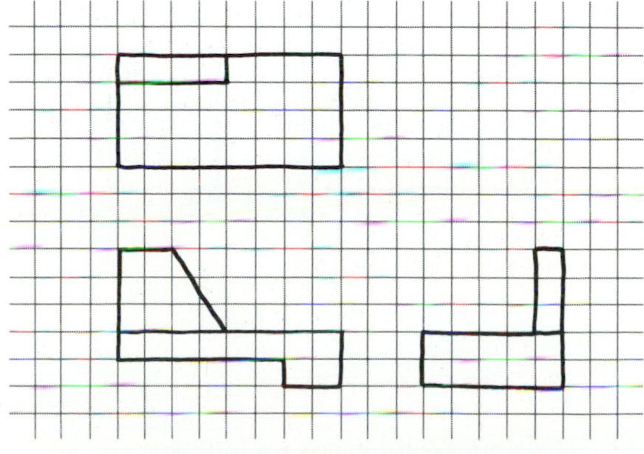

P 5.2

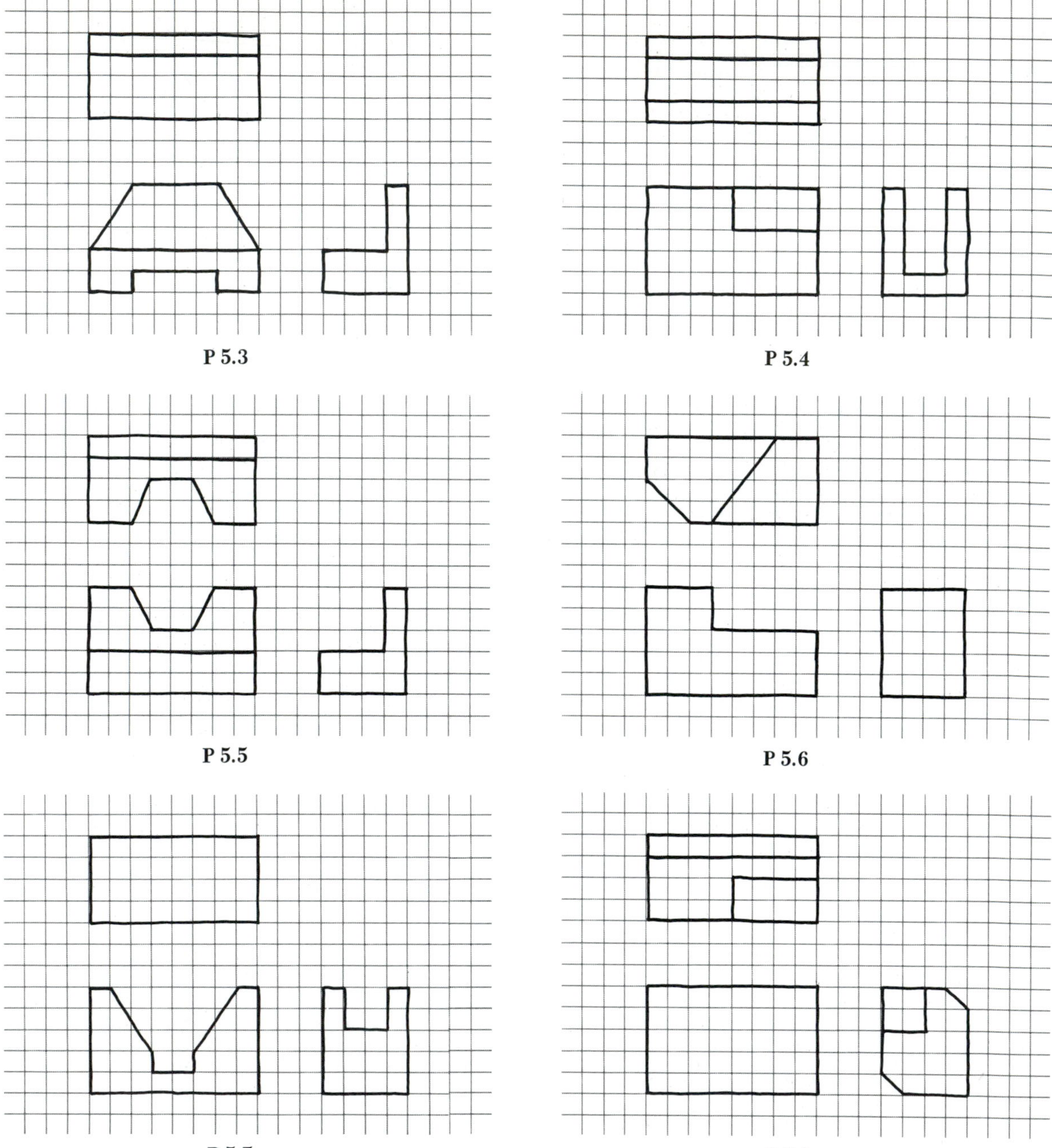
P 5.3
P 5.4
P 5.5
P 5.6
P 5.7
P 5.8

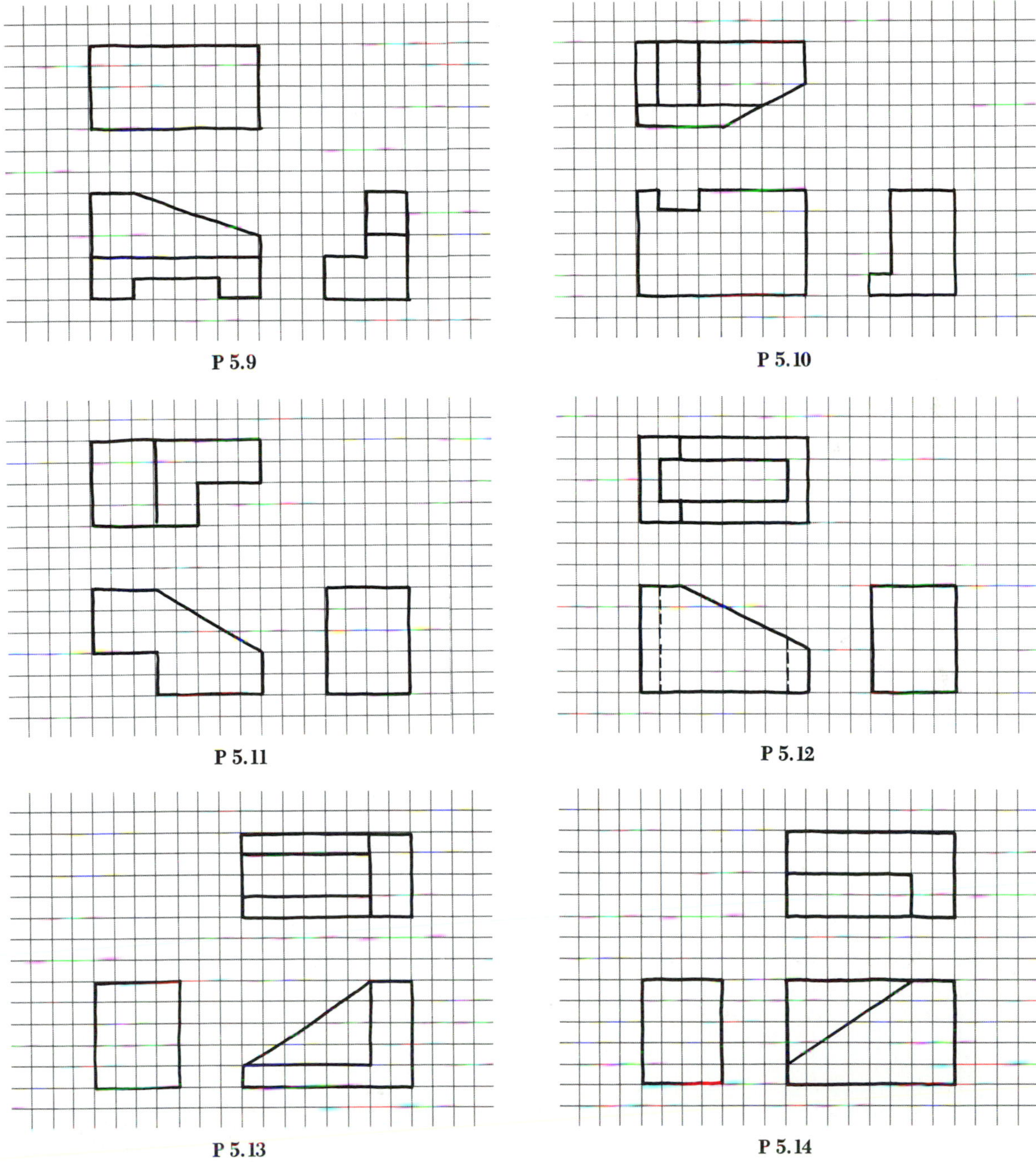
P 5.9
P 5.10
P 5.11
P 5.12
P 5.13
P 5.14

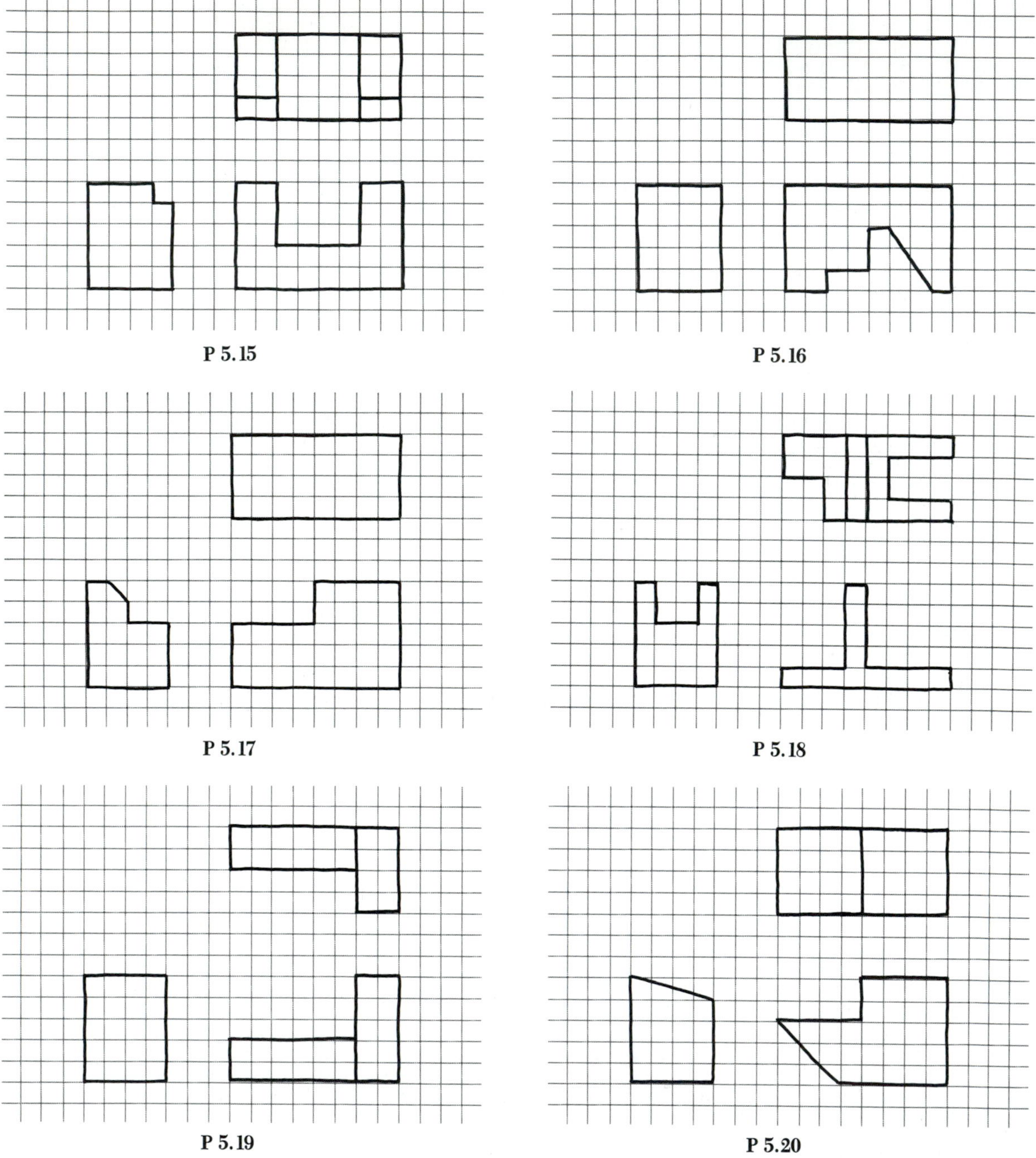

P 5.15

P 5.16

P 5.17

P 5.18

P 5.19

P 5.20

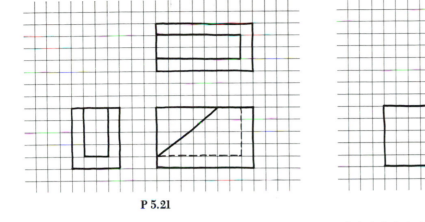

P 5.21

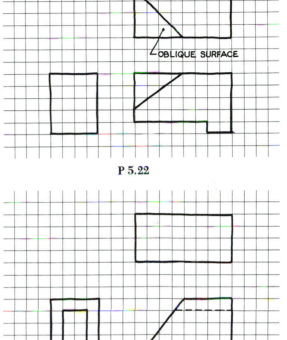

OBLIQUE SURFACE

P 5.22

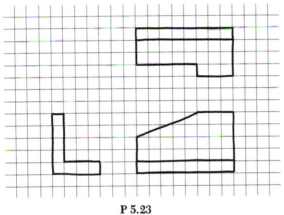

P 5.23

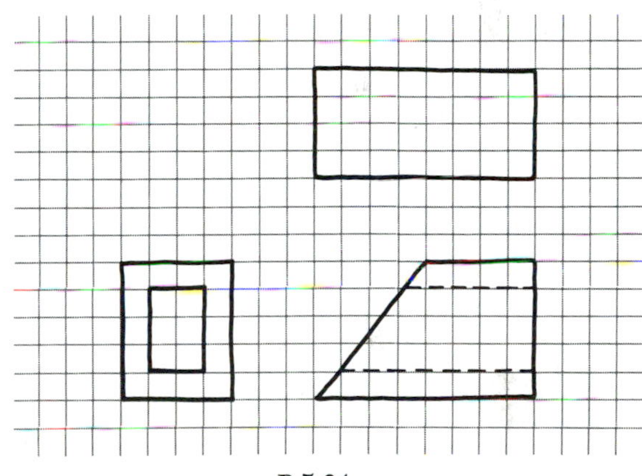

P 5.24

Problems **P5.25** through **P5.39** are missing-view drawing problems. Draw the given views with instruments, full size, with two problems per sheet. Assume that each grid equals ¼ inch or 6 mm. Draw the missing third view.

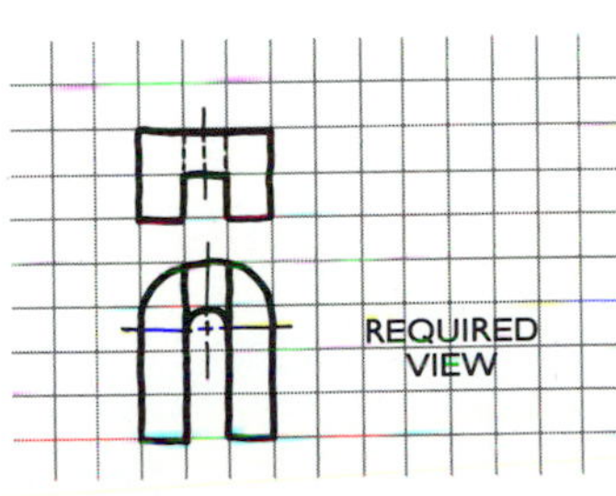

P 5.25

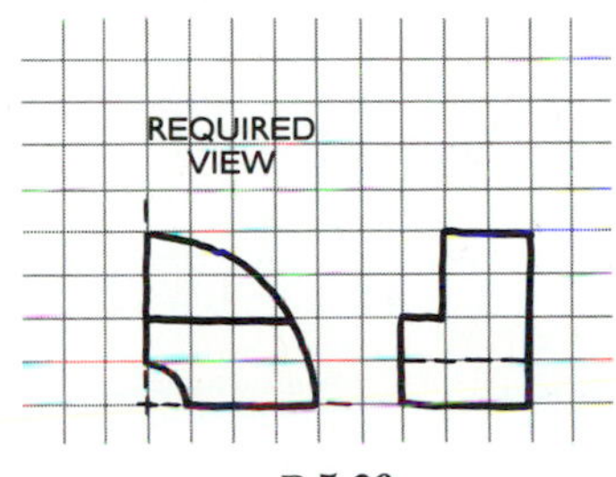

P 5.26

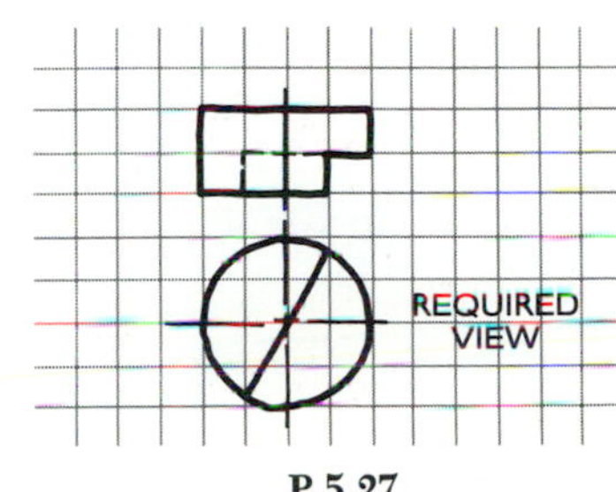

P 5.27

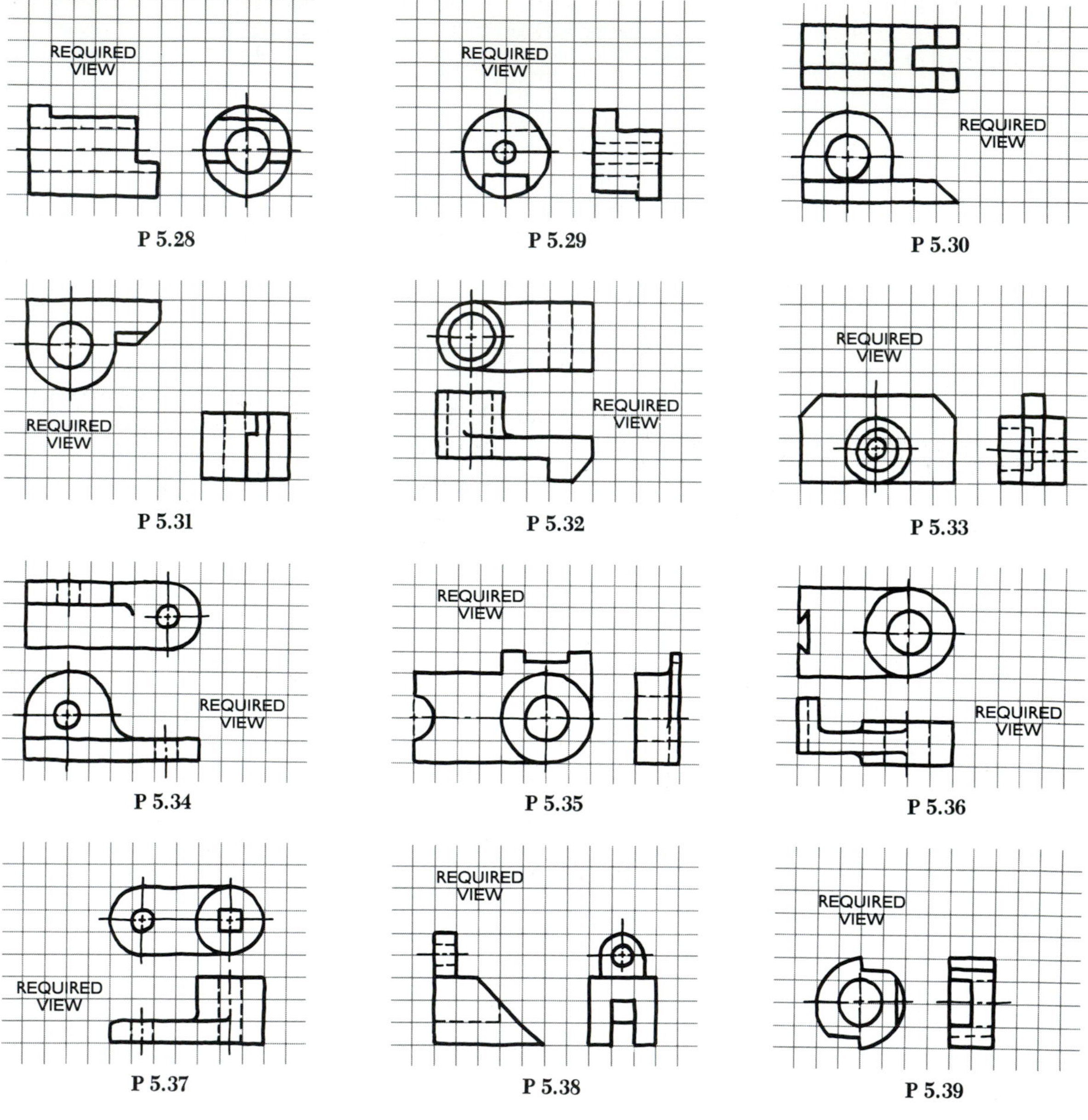
REQUIRED VIEW
P 5.28
REQUIRED VIEW
P 5.29
REQUIRED VIEW
P 5.30
REQUIRED VIEW
P 5.31
REQUIRED VIEW
P 5.32
REQUIRED VIEW
P 5.33
REQUIRED VIEW
P 5.34
REQUIRED VIEW
P 5.35
REQUIRED VIEW
P 5.36
REQUIRED VIEW
P 5.37
REQUIRED VIEW
P 5.38
REQUIRED VIEW
P 5.39

Problems **P5.40** to **P5.55** are multiview sketching problems. Sketch the necessary views of the objects shown pictorially to a scale approximately full size, with two problems per sheet. For convenience the overall length, height, and depth of each object has been made proportional. Assume that each grid equals ⅛ inch or 3 mm.

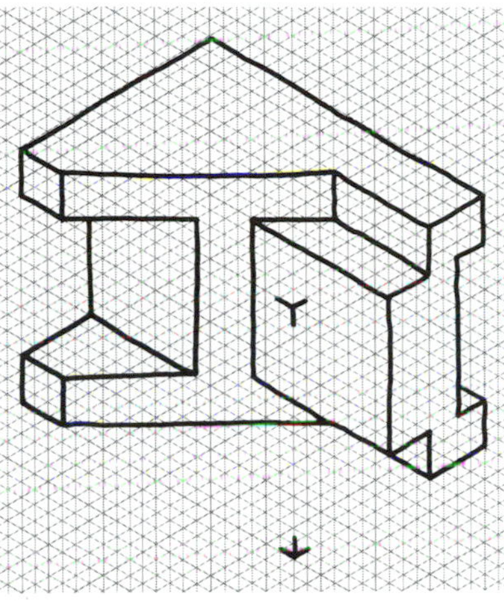

P 5.40

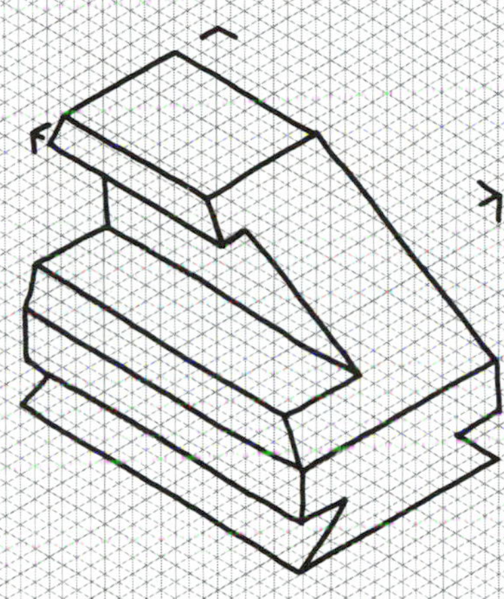
P 5.41

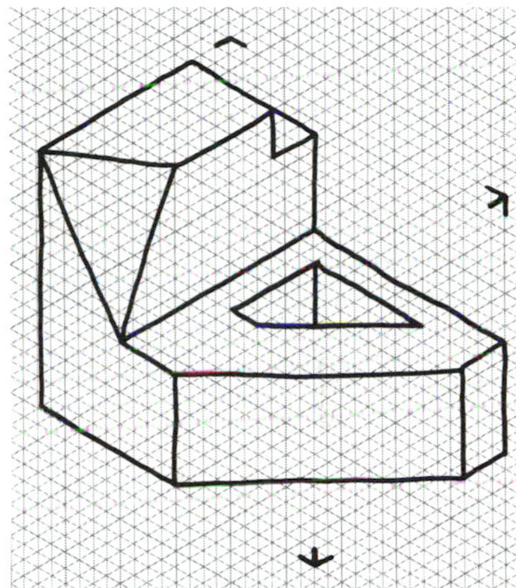
P 5.42

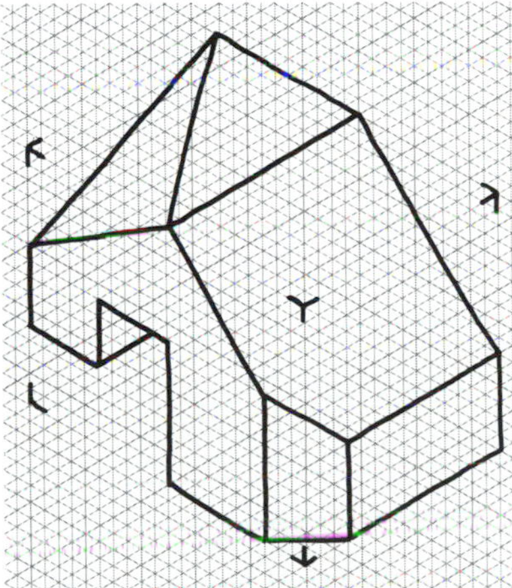
P 5.43

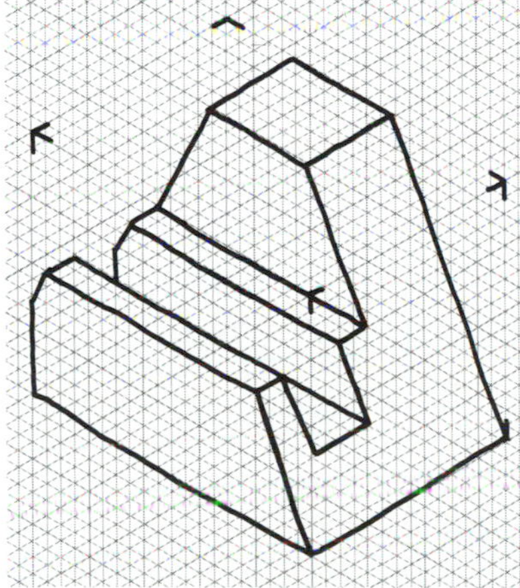
P 5.44

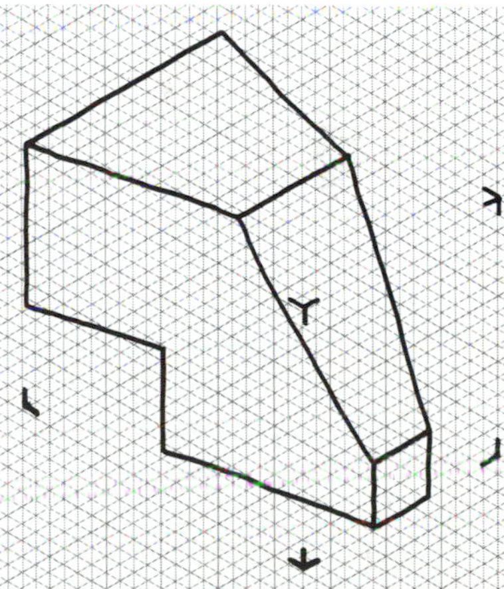
P 5.45

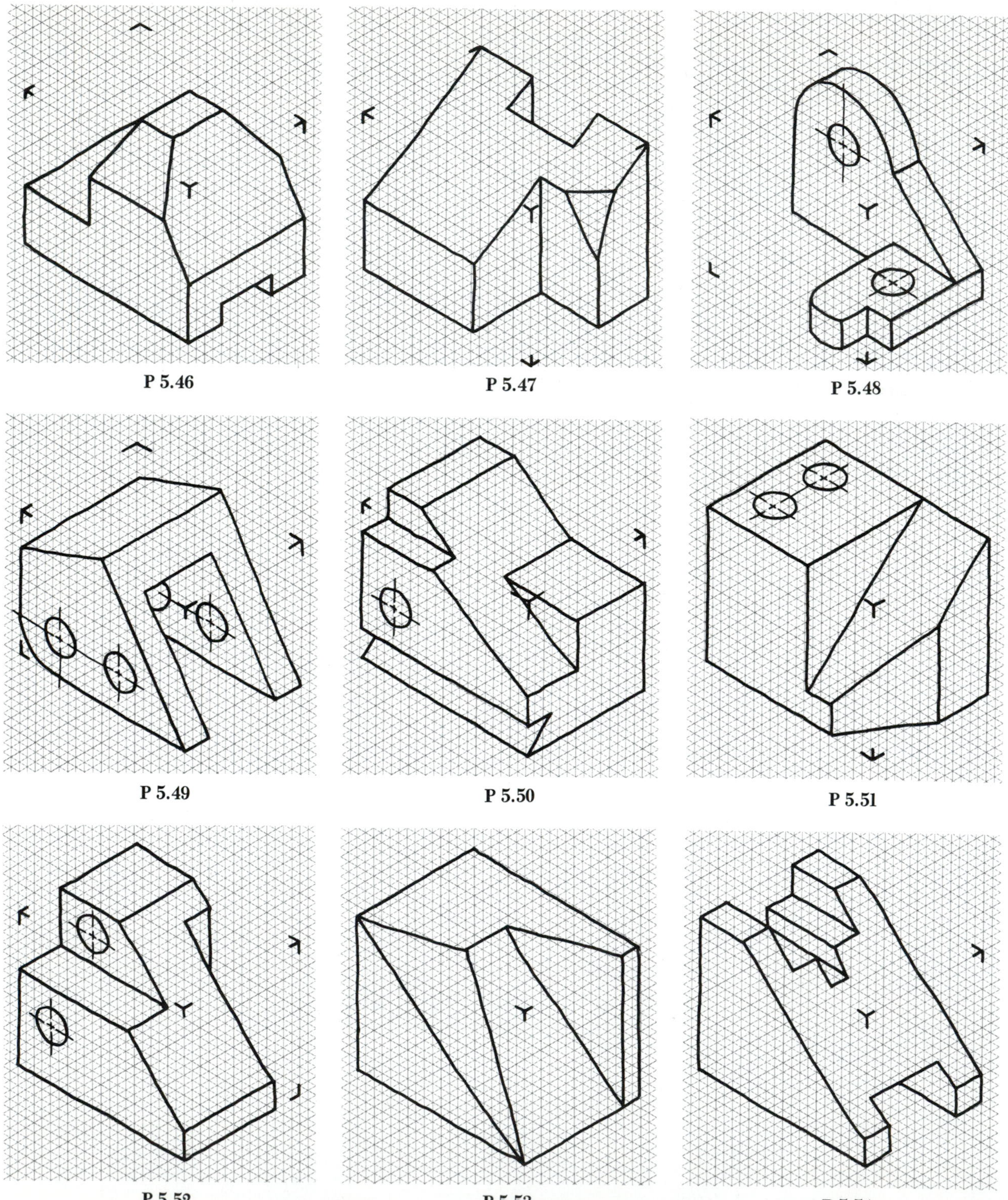

P 5.46

P 5.47

P 5.48

P 5.49

P 5.50

P 5.51

P 5.52

P 5.53

P 5.54

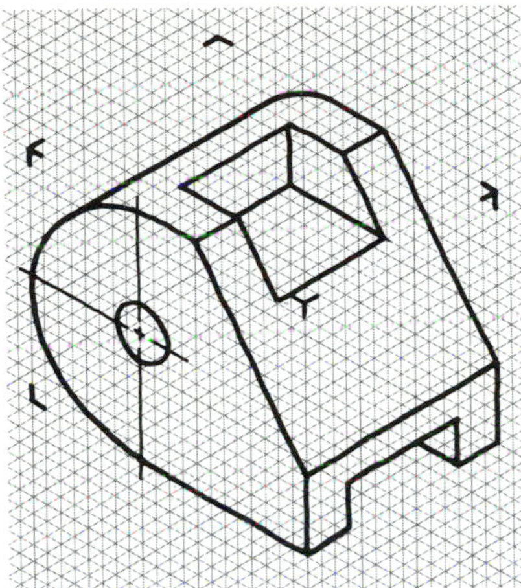

P 5.55

Problems **P5.56** through **P5.81** are multiview drawing problems. Use instruments to draw the necessary multiviews of each of the objects shown pictorially. Select a suitable scale so that two problems will fit neatly on one sheet. For unspecified sizes of fillets and rounds, select an appropriate size (see Section 5.19).

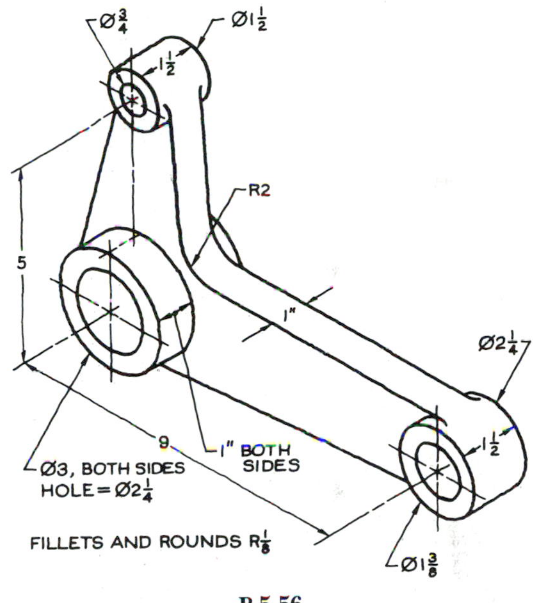

P 5.56

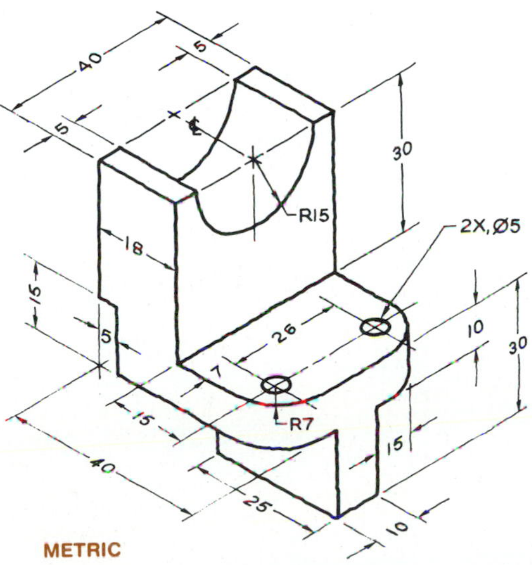

P 5.57

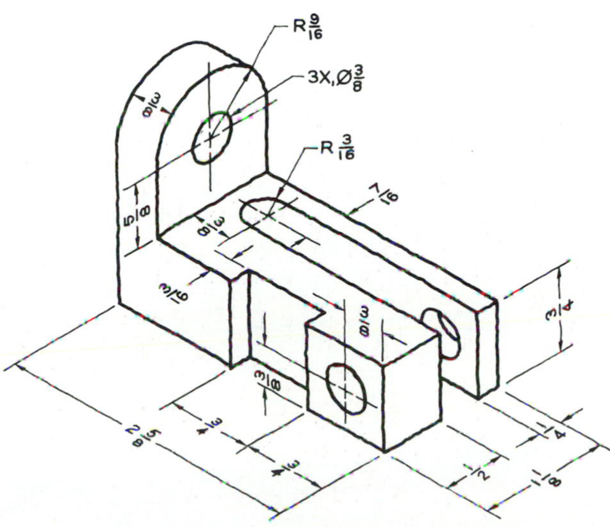

P 5.58

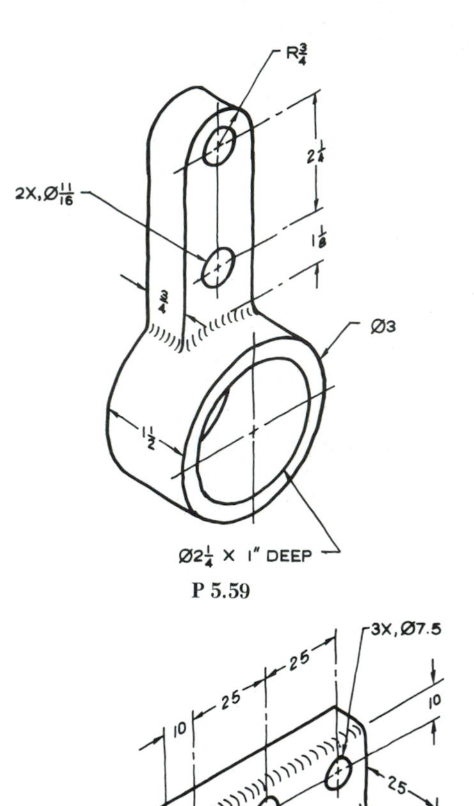

P 5.59

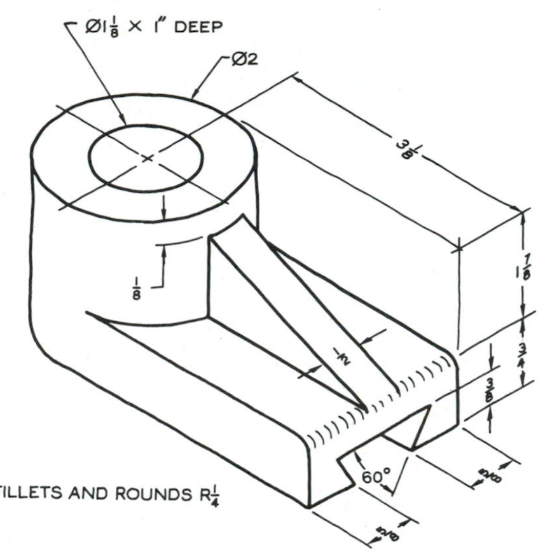

P 5.60

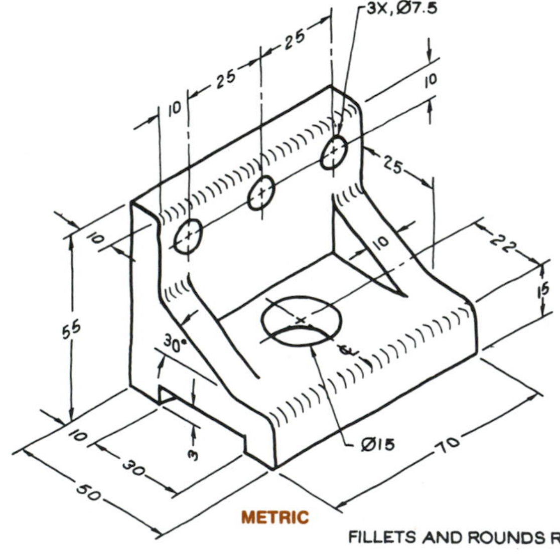

P 5.61

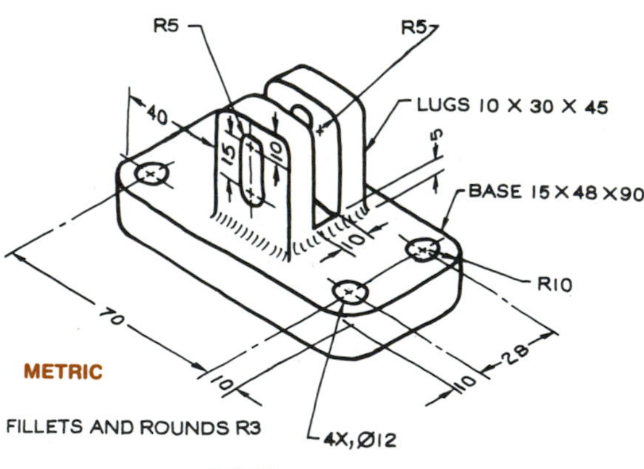

P 5.62

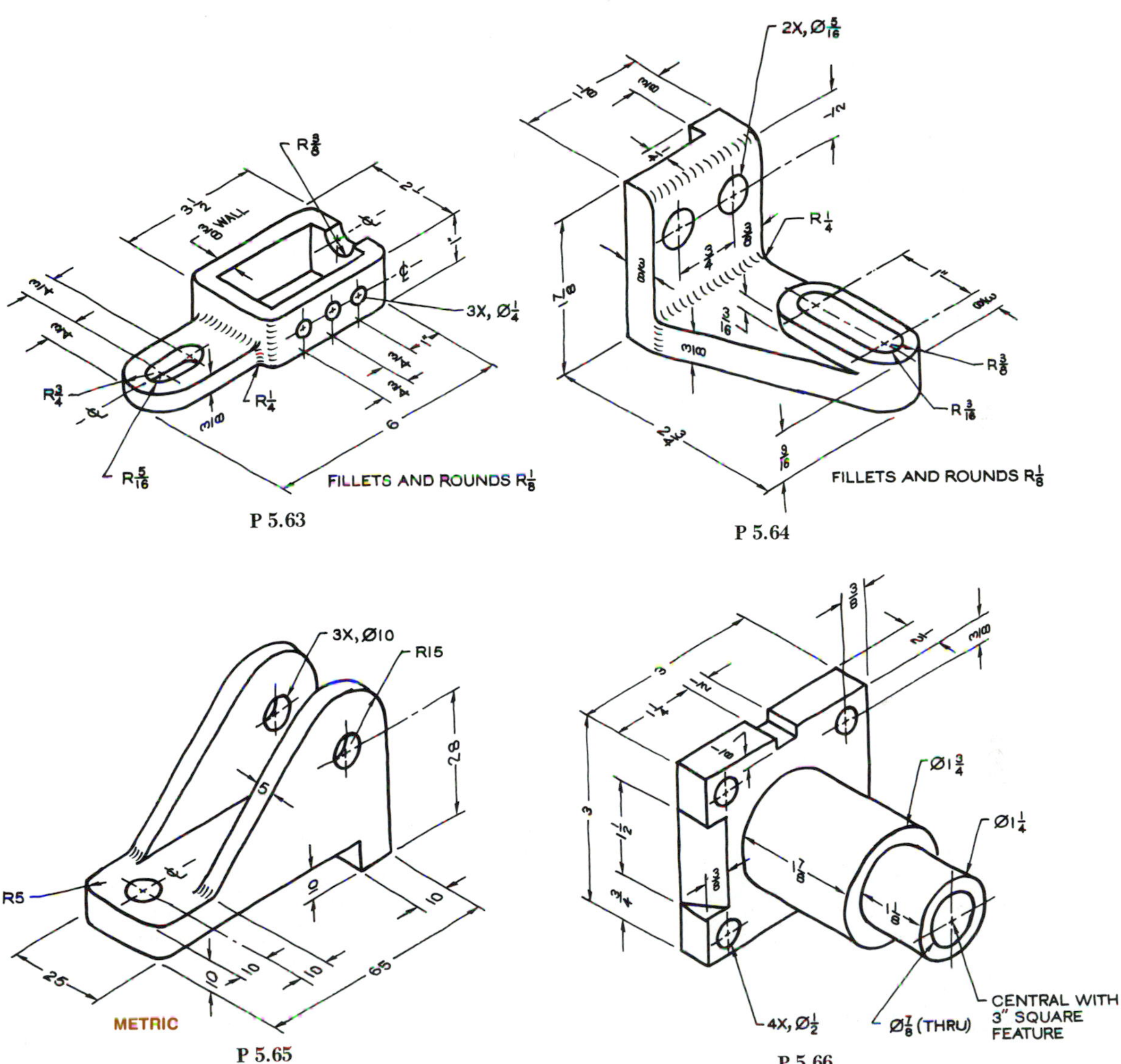

P 5.63

P 5.64

P 5.65

P 5.66

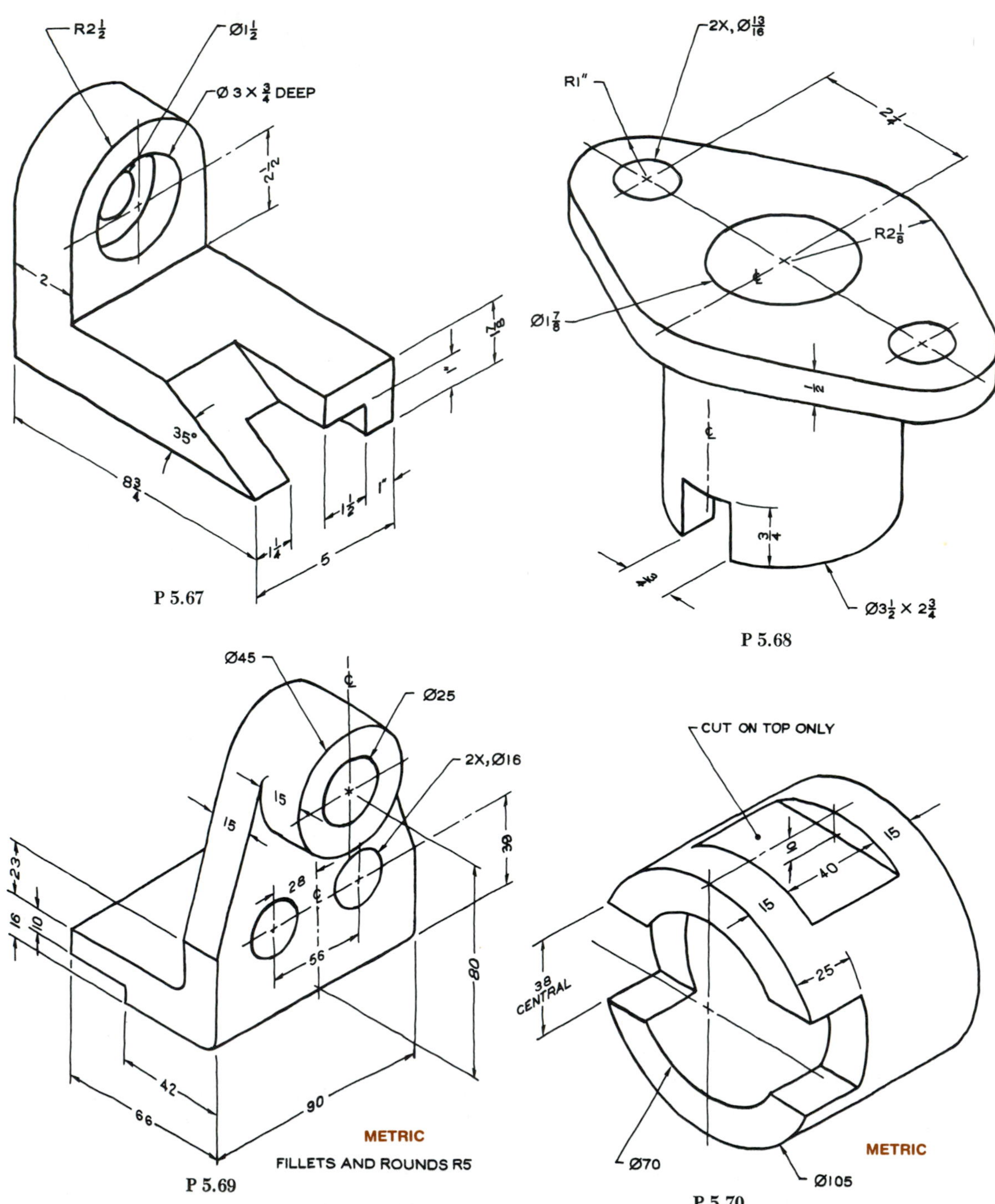

P 5.67

P 5.68

P 5.69

P 5.70

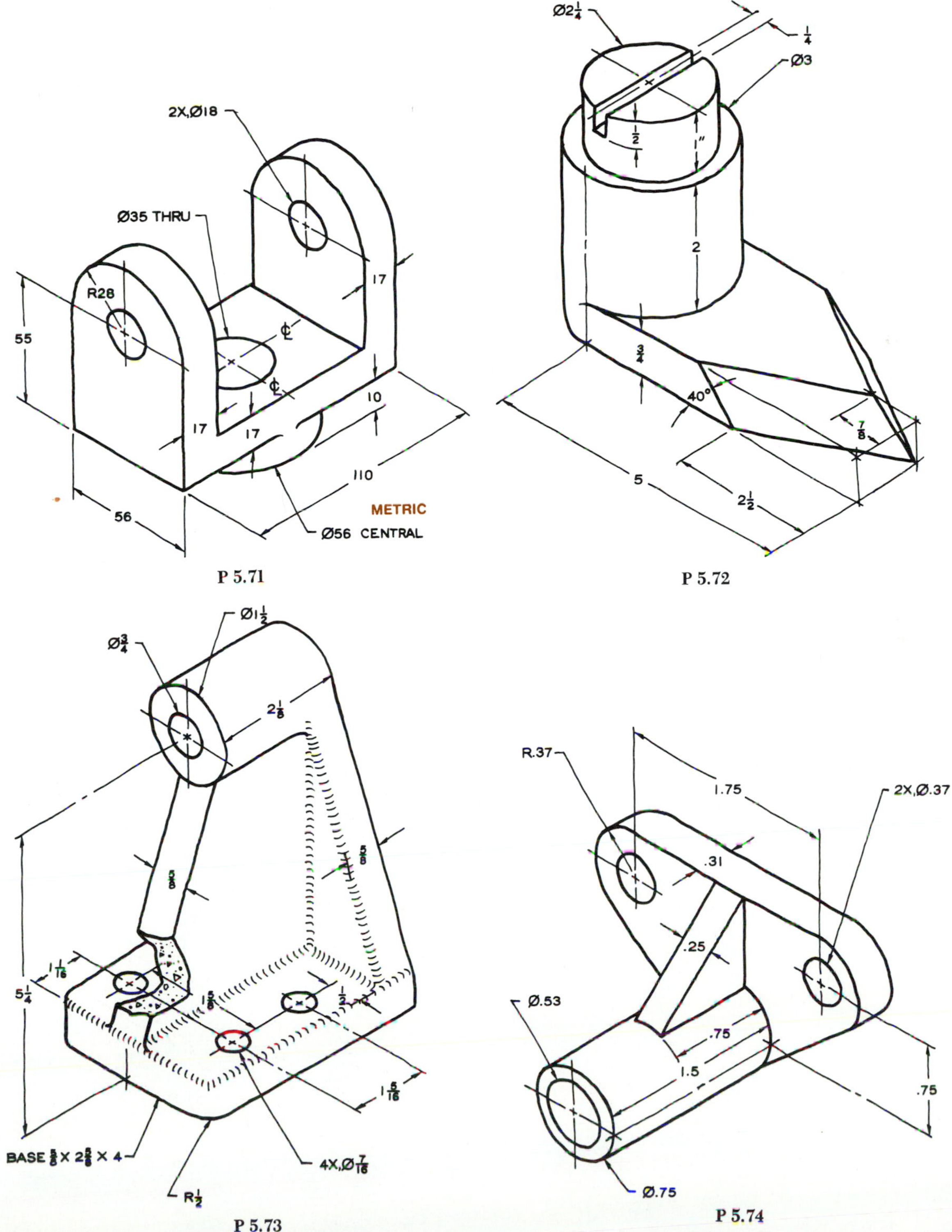

P 5.71

P 5.72

P 5.73

P 5.74

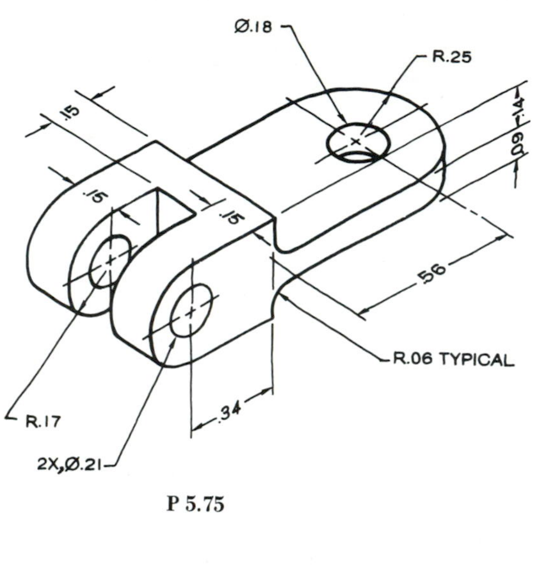

P 5.75

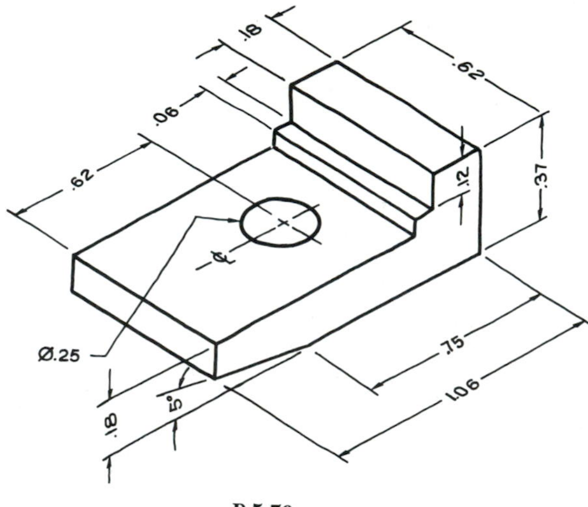

P 5.76

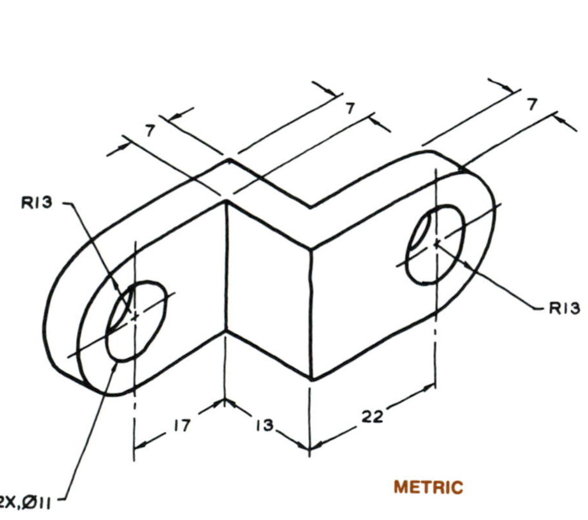

P 5.77

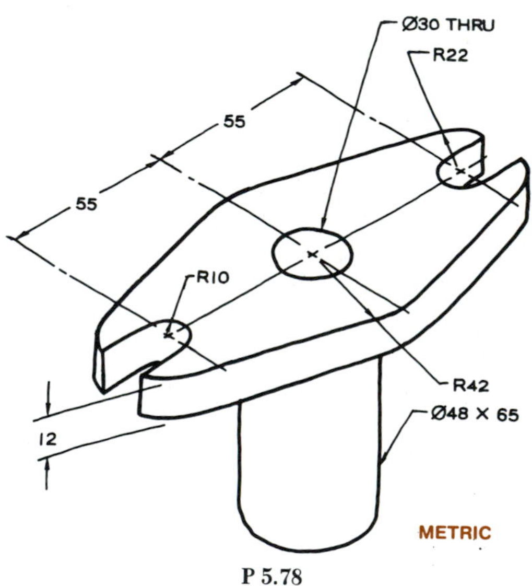

P 5.78

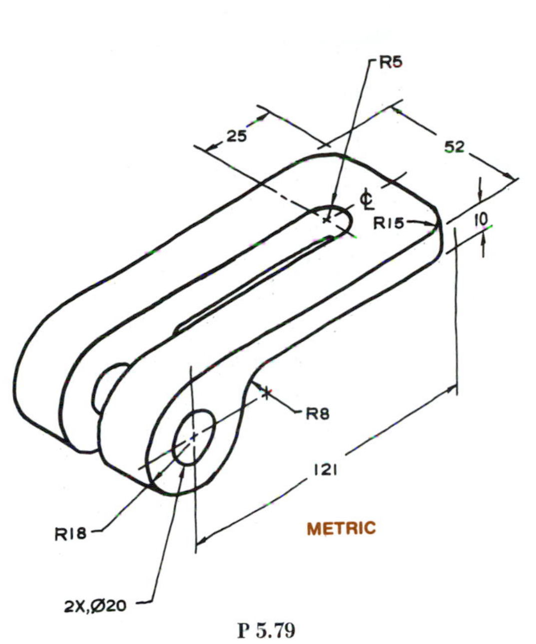

P 5.79

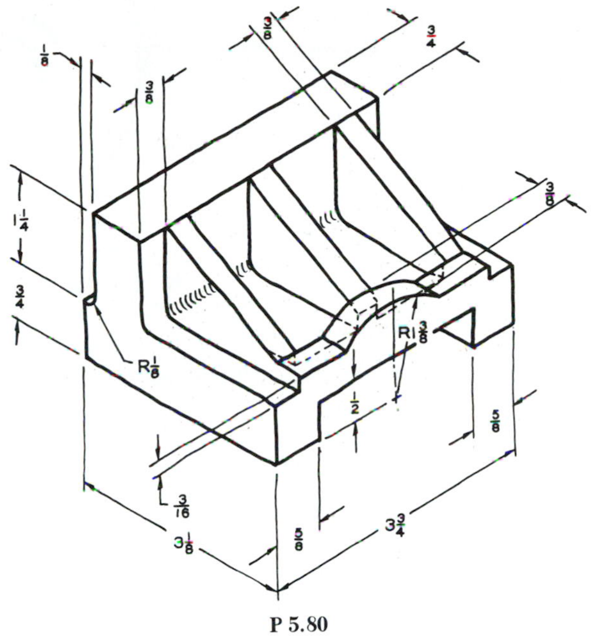

P 5.80

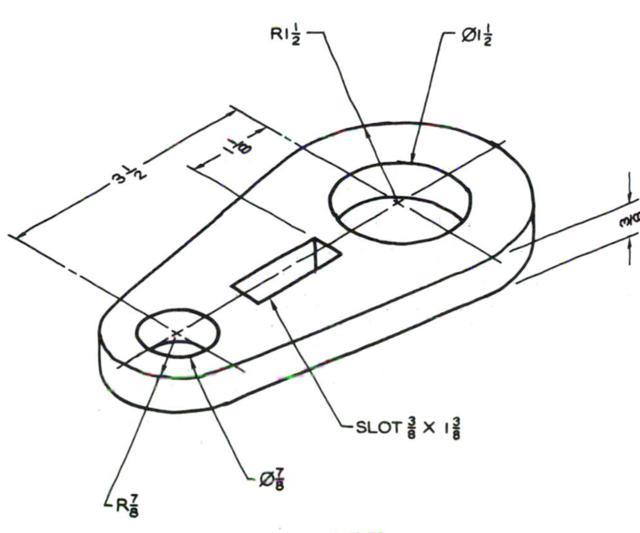

P 5.81

Auxiliary Views

OBJECTIVES

After completing this chapter you should have gained the following abilities:

1. To recognize the need for, and to draw properly, primary and secondary auxiliary views.

2. To use reference lines properly as an aid in constructing auxiliary views.

3. To use proper notation as an aid in clearly identifying all aspects of auxiliary views.

4. To draw, where appropriate, partial auxiliary views.

In Figure 5-5 the six principal views of a simple object were shown. Because the configuration of the object was so easily understood, only three of the views were needed to describe the shape (as in Figure 5-6). The other three views were not needed. As we saw in Section 5.1, the true shape of a surface appears only when the line of sight is in a direction perpendicular to the plane of projection. An object having an inclined surface (one that is not parallel to a plane of projection) does not project its true shape on any of the principal views. Such surfaces always appear shortened or distorted in the principal views. Since none of the principal views in orthographic projection has an inclined line of sight, the true shape of an inclined surface can only appear in a special view. Such a view, drawn in addition to the principal views, is called an *auxiliary view*, and can be drawn to show an object from any desired direction.

6.1 AUXILIARY VIEWS

Auxiliary views are used to show the true shape and relationship of features on an object that are not parallel to any of the principal planes of projection. In Figure 6-1 the inclined surface labeled A does not appear true shape in either the top or the right-side views. In Figure 6-2 surface A is shown true shape because, in this auxiliary view, the line of sight is perpendicular to the inclined surface. Auxiliary views are aligned with the views from which they are projected.

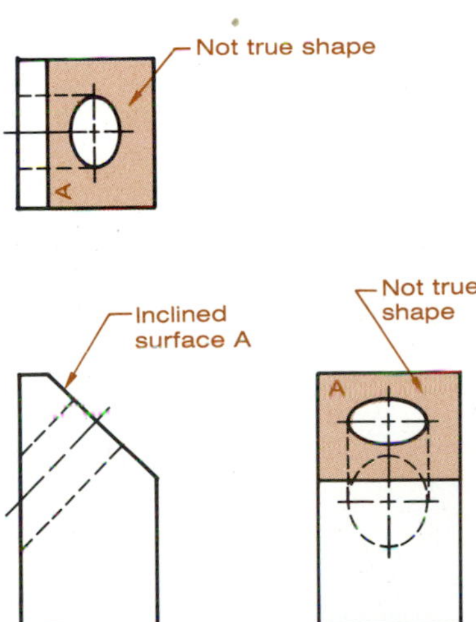

FIGURE 6-1 Principal views of an object having an inclined surface *(Section 6.1)*

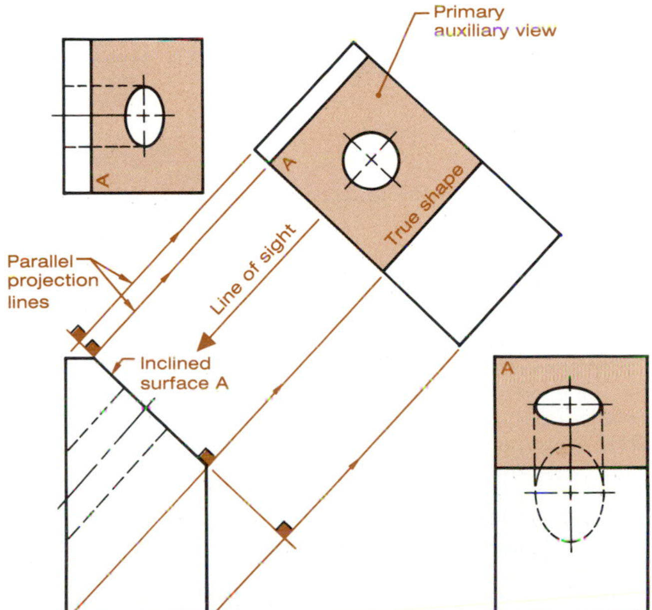

FIGURE 6-2 Use of a primary auxiliary view to obtain the true shape of an inclined surface *(Section 6.1)*

6.2 PRIMARY AUXILIARY VIEWS

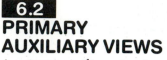

A *primary* auxiliary view is one that is adjacent to and aligned with a principal view. A primary auxiliary view is projected onto a plane that is perpendicular to one of the principal planes of projection and is inclined to the other two. For most constructions, two principal views of an object are required (top and front or front and side, for example) before a primary auxiliary view can be drawn.

Our discussion of orthographic views in Chapter 5 included the concept of projecting principal views by placing an imaginary transparent box over an object. This theory is based upon projecting an image onto viewing planes positioned parallel to the sides of the object. The line of sight for the image is perpendicular both to the sides of the object and to the principal projection planes. This same theory of projecting principal views applies to drawing auxiliary views. When a surface is inclined to two of the principal planes and is perpendicular to the third, it can be shown true shape in an auxiliary plane that is "folded" from one of the principal planes. In Figure 6-3 the true-shape view of the inclined surface is projected onto an auxiliary plane that is parallel to the edge of the inclined surface of the object. The fold line between the principal planes is labeled H-F, representing the horizontal and the frontal planes of projection. The fold line between the frontal and the auxiliary plane is labeled A-F (A is an abbreviation for the auxiliary plane). The fold line is always parallel to the edge view of the inclined surface.

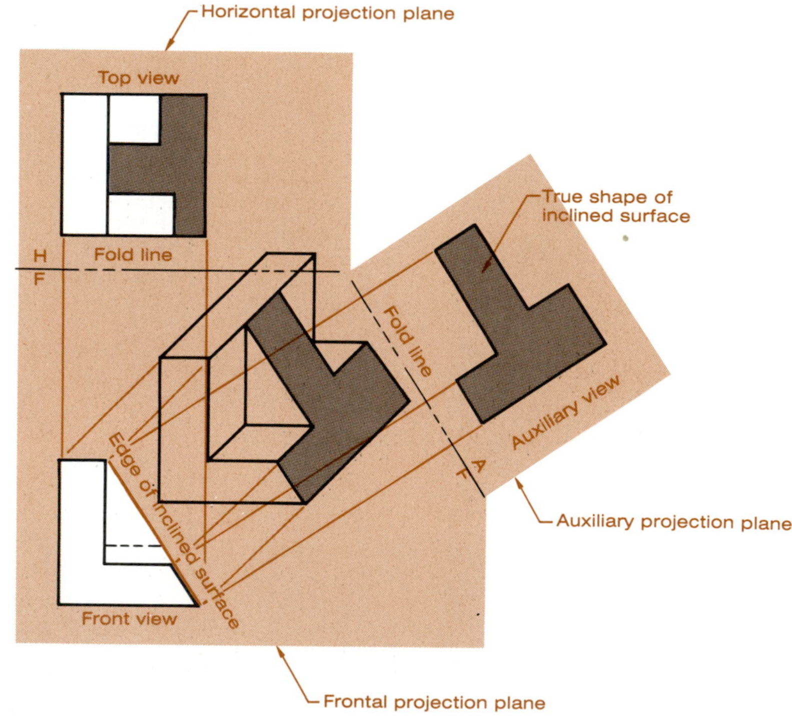

FIGURE 6-3 The basis of auxiliary views. When a surface is inclined to two of the principal planes and is perpendicular to the third, it can be shown true shape in an auxiliary projection plane that is "folded" from one of the principal planes. The view is called an auxiliary view *(Section 6.2)*

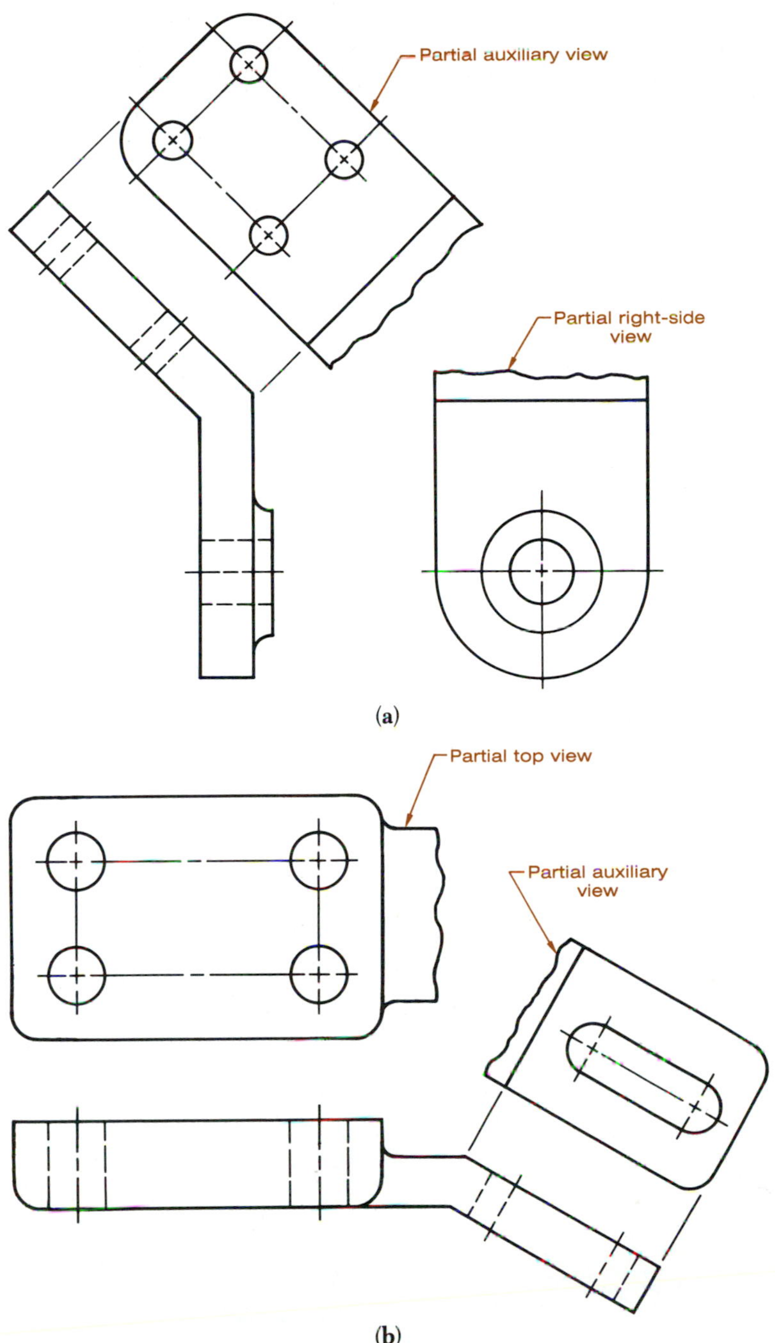

FIGURE 6-4 The use of partial principal and partial auxiliary views *(Section 6.3)*

6.3 PARTIAL AUXILIARY VIEWS

Most auxiliary views are only partial views that show the true shape of a particular surface. Generally only the inclined surface—not the entire object—is drawn in the auxiliary projection plane. In these views hidden lines are usually omitted for clarity. Similarly, foreshortened and distorted features not lying in the auxiliary plane may also be omitted. A freehand break line is used to indicate the imaginary break in the views. The examples shown in Figure 6-4 illustrate how partial auxiliary and partial principal views, when used together with other views of the object, clearly describe the size and shape of an object. An auxiliary view often eliminates the need for drawing one of the principal views. Note how the use of partial views simplifies the reading of a drawing but still conveys the necessary information.

6.4 TYPES OF PRIMARY AUXILIARY VIEWS

There are three types of primary auxiliary views: *front-adjacent*, *top-adjacent*, and *side-adjacent*. The views are classified into these types according to their position relative to the principal planes.

6.5 AUXILIARY VIEW CONSTRUCTION

Front-Adjacent Auxiliary Views

In the front-adjacent construction, the auxiliary plane is perpendicular to the frontal plane and inclined to the horizontal plane of projection. The auxiliary view is projected from the front view and the *depth* dimension is common to both the top and the auxiliary view. Figure 6-5 illustrates how to obtain the true size of the inclined surface 1-2-3-4 by using a front-adjacent auxiliary view. The inclined surface appears as an edge in the front view.

The steps of construction are as follows:

STEP 1

Draw a reference line, to be used as a base of measurement, between the given views. Make this a heavy line with sets of two short dashes and one long dash. Label the reference line H-F to represent the fold lines of the horizontal and frontal projection planes. Connect the given adjacent views with projection lines. Note that the projection lines are perpendicular to the reference line.

STEP 2

Draw a second reference line parallel to the edge of the inclined surface 1-2-3-4 at any desired distance from the front view between the adjacent view F and the desired auxiliary view. Label the line A-F. Extend projection lines from view F to view A. These lines are perpendicular to the reference line A-F.

STEP 3

With respect to view A, view H is *related*. (All views adjacent to the same view are related.) Transfer points 1-2-3-4 to view A, using distances X and Y from the related view H, to obtain the true size of the plane.

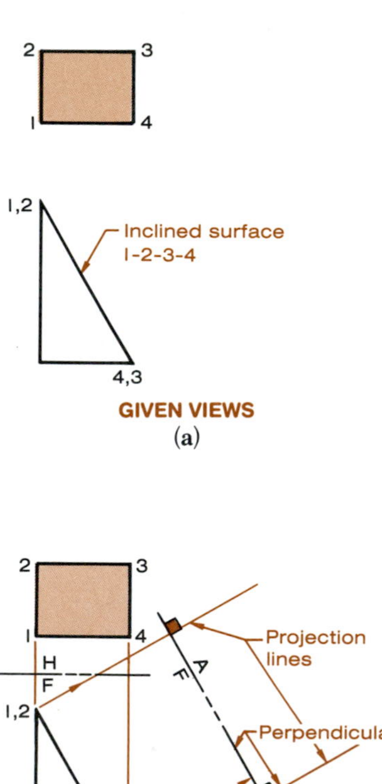

GIVEN VIEWS
(a)

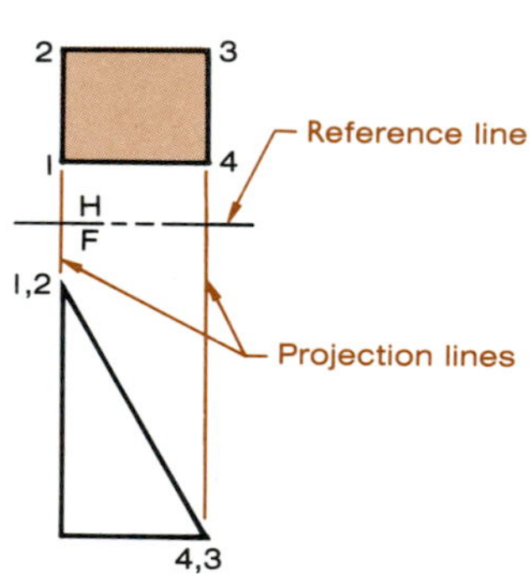

STEP 1
DRAW REFERENCE LINE H-F AND PROJECTION LINES.
(b)

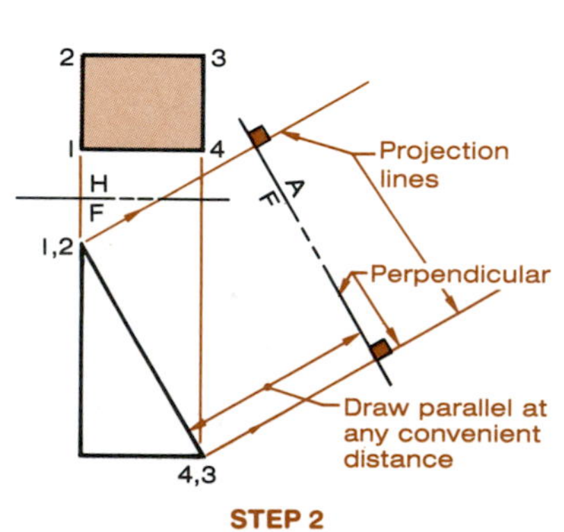

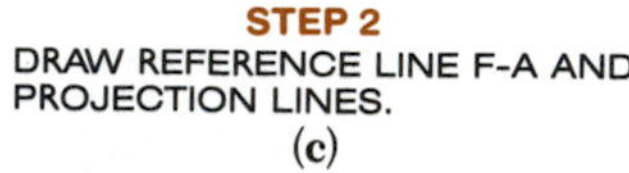

STEP 2
DRAW REFERENCE LINE F-A AND PROJECTION LINES.
(c)

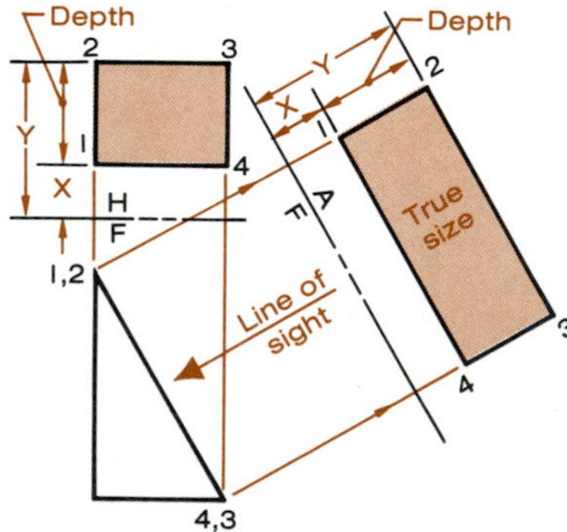

STEP 3
PLOT POINTS 1-2-3-4 IN VIEW A USING DISTANCES X AND Y. DRAW THE TRUE-SIZE SURFACE.
(d)

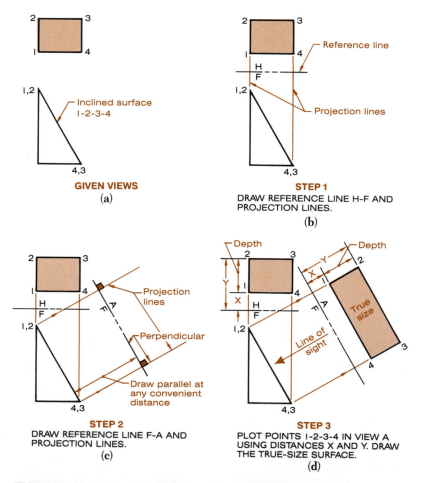

FIGURE 6-5 Constructing a front-adjacent auxiliary view *(Section 6.5)*

Top-Adjacent Auxiliary Views

In the top-adjacent construction, the auxiliary plane is perpendicular to the horizontal plane and inclined to the frontal and profile planes of projection. The auxiliary view is projected from the top view and the *height* dimension is common to both the front and the auxiliary view. As shown in Figure 6-6, the steps of construction for obtaining the true size of the inclined surface 1-2-3-4-5-6 by using the top-adjacent auxiliary view are exactly analogous to those used to construct a front-adjacent auxiliary view:

STEP 1
Draw the reference line H-F and connect the given adjacent views with projection lines.

STEP 2
Draw a second reference line H-A parallel to the edge view of inclined surface 1-2-3-4-5-6 at any desired distance from the top view. Draw projection lines aligning view A with adjacent view H.

FIGURE 6-6 **Constructing a top-adjacent auxiliary view *(Section 6.5)***

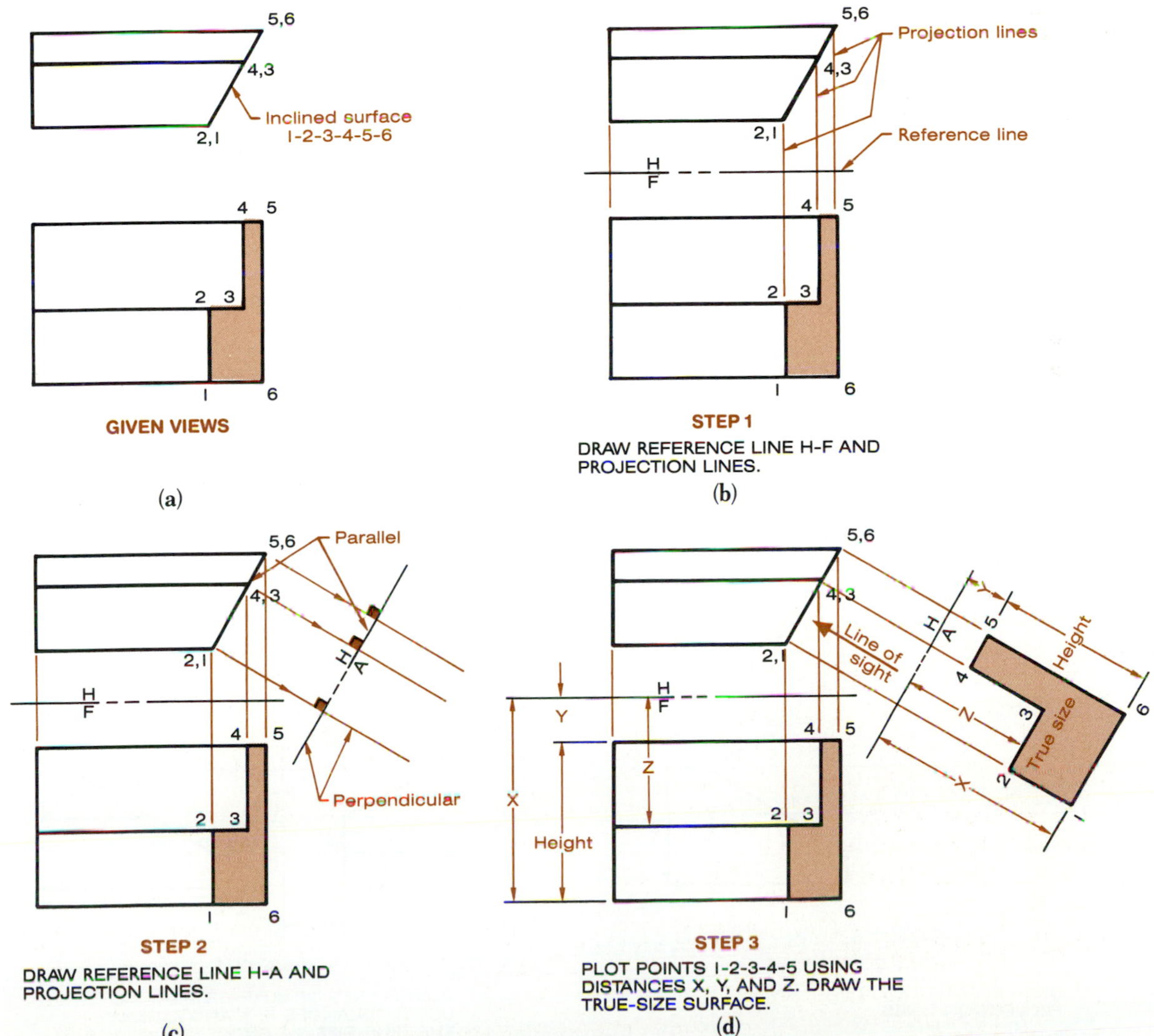

STEP 3

Transfer the points of the required surface to view A using distances X, Y, and Z transferred from the related view F. Connect these points to obtain the true size of surface 1-2-3-4-5-6.

Because all top-adjacent views show the true height or elevation of an object, they are called *elevation* views.

Side-Adjacent Auxiliary Views

In the side-adjacent construction, the auxiliary plane is perpendicular to the profile plane and inclined to the frontal and horizontal planes of projection. The auxiliary view is projected from a side view and the *width* dimension is common to both the front and the side-adjacent auxiliary view. As shown in Figure 6-7, the steps of construction for obtaining the true size of the inclined surface 1-2-3-4-5 by using a right-side-adjacent auxiliary view are exactly analogous to those used to construct the front-adjacent and top-adjacent auxiliary views. The inclined surface appears as an edge in the right-side view.

FIGURE 6-7 Constructing a side-adjacent auxiliary view *(Section 6.5)*

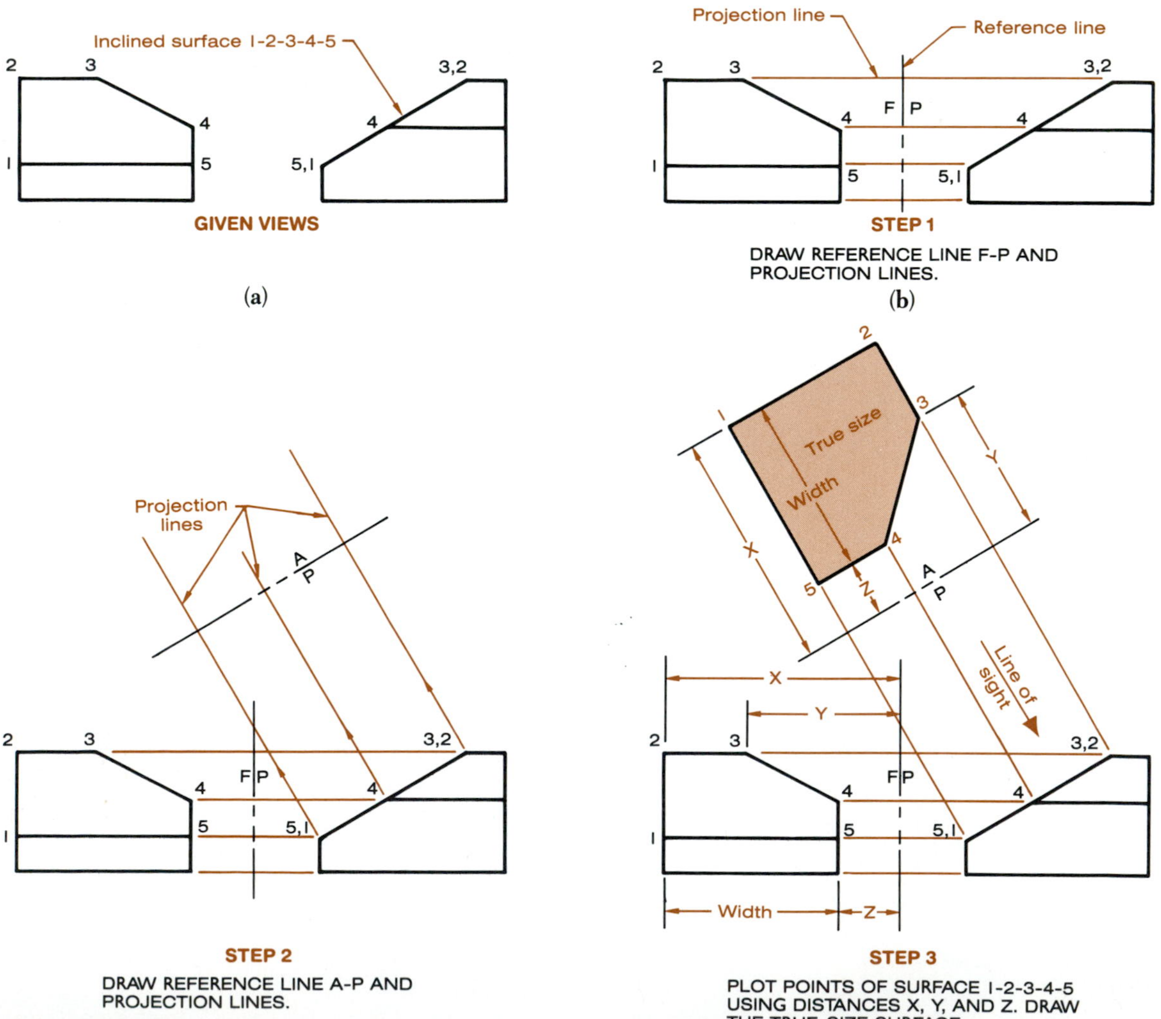

6.6 PLOTTING CURVES IN AUXILIARY VIEWS

Figure 6-8 illustrates a method for drawing a curve in an auxiliary view. The true-size projection of a curved surface is obtained by plotting a sufficient number of points to ensure a smooth curve. The perimeter of the circle in the top view is divided into any convenient number of equal parts and projected to the inclined plane representing the curved surface in the front view. The respective points on the curve in the auxiliary view are plotted on projection lines extending from the inclined line in the front view with distances transferred from the reference line H-F.

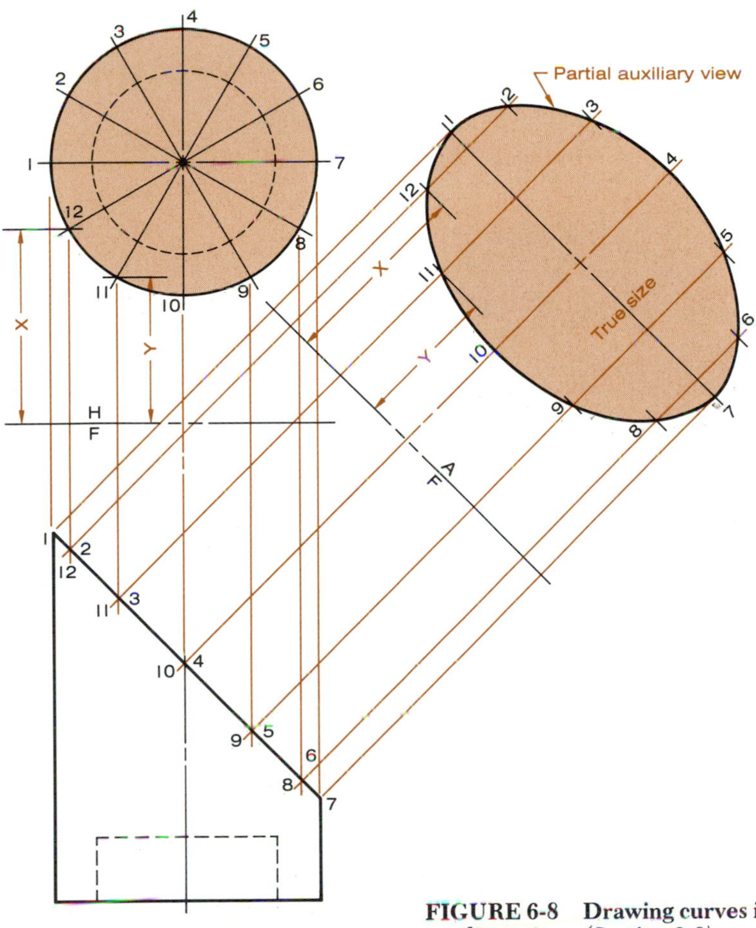

FIGURE 6-8 Drawing curves in auxiliary views *(Section 6.6)*

The shape of the inclined curved surface in the regular view shown on the object in Figure 6-9 is plotted in a similar manner. In this case it is necessary to first draw a partial (true size) auxiliary view of the object and then to use the front view, in combination with the auxiliary view, as a means of constructing the curve in the right-side view. The curve is plotted by projecting selected points from the true-size curve in the auxiliary view to the front view and from there to the right-side view.

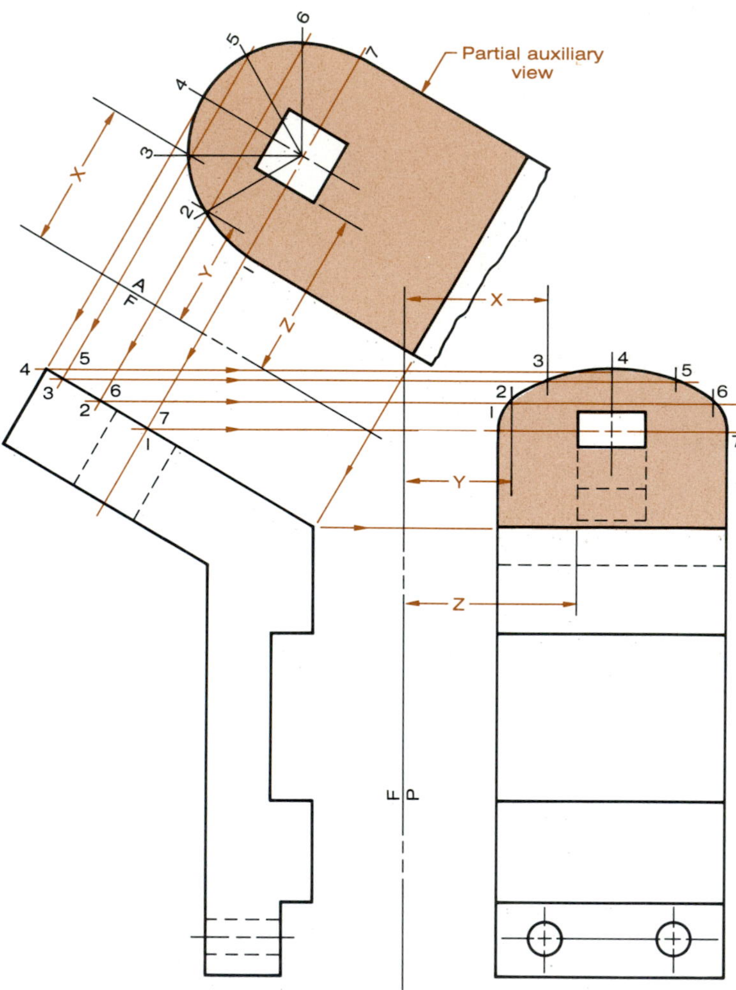

FIGURE 6-9 Drawing curves in auxiliary views ***(Section 6.6)***

CHECKPOINT 1
PRIMARY AUXILIARY VIEWS

For the examples shown in Figures 6-10 and 6-11, use instruments to draw the given views full size and then draw a primary auxiliary view showing the true size of the inclined surface. Use a plain sheet of 8½ × 11 inch paper for each problem. Use correct notation for the reference lines. Omit dimensions. (The given location dimensions are measured from the edge of the paper.)

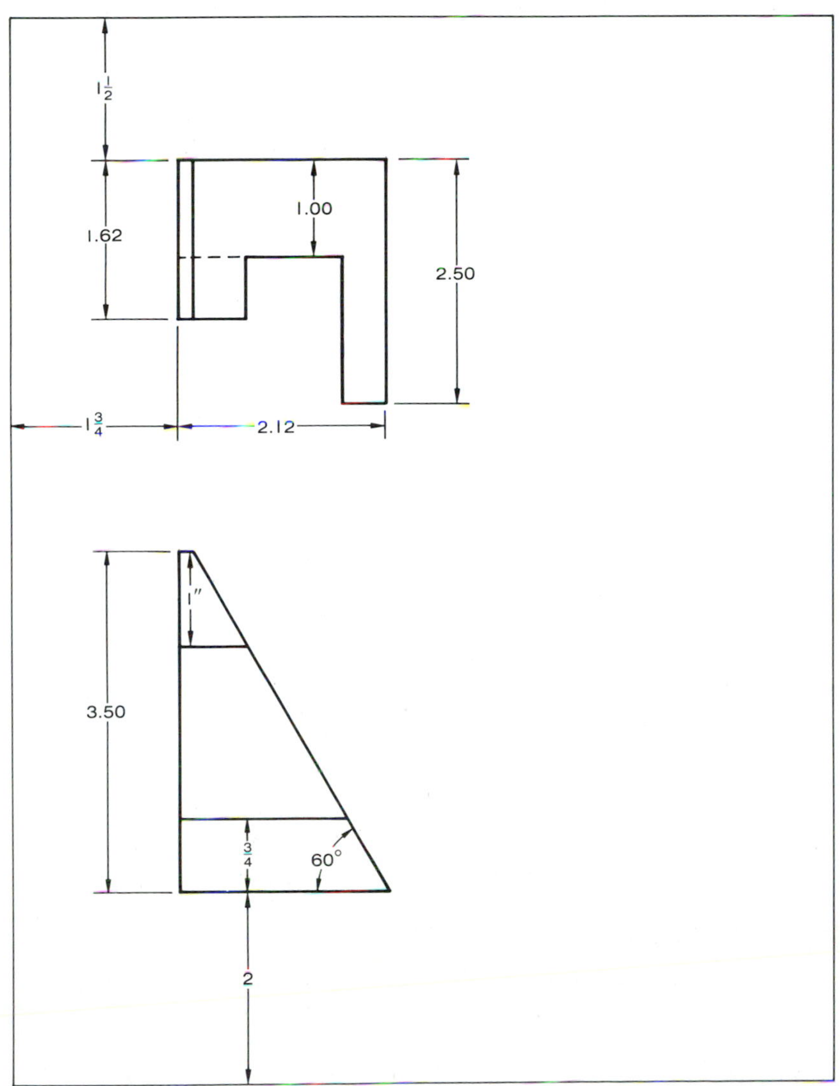

FIGURE 6-10 Checkpoint 1 problem

6.7 SECONDARY AUXILIARY VIEWS

For some objects it is impossible to use primary auxiliary views to show the true size of planes, the distance between points and lines, or the angles between lines and planes. In such cases you will need to use *secondary* auxiliary views. A secondary auxiliary view is one that is adjacent to and aligned with a primary auxiliary view or with another secondary auxiliary view.

FIGURE 6-11 Checkpoint 1 problem

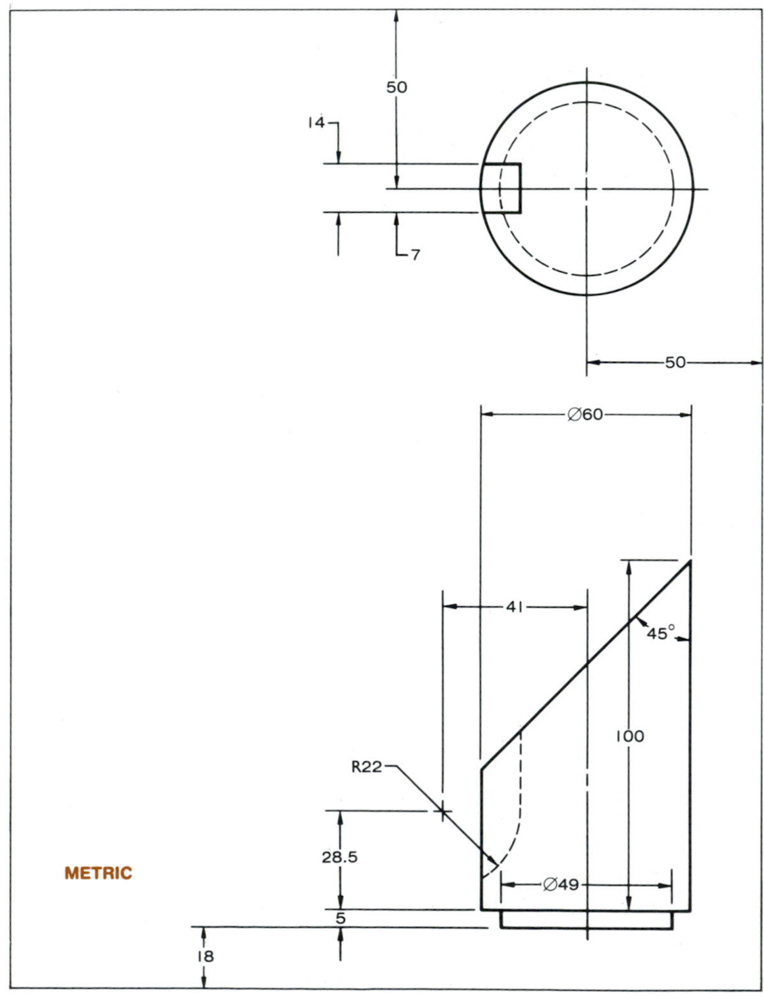

Figure 6-12 shows an example of an object in which a secondary auxiliary view (view B) is projected onto a plane that is inclined to the principal planes. View B shows the true size of the oblique surface 1-2-3. Note that, in the given views, this surface is inclined to each of the principal planes and does not appear as an edge in either of the principal views. The construction for obtaining the true shape of surface 1-2-3 is as follows:

1. Draw the reference line H-F between the given views.
2. Draw the reference line F-A perpendicular to the *true-length* line 1-3. Line 1-3 is true length in the front view because it lies in a position that is parallel to the reference line H-F in the adjacent (top) view. In other words line 1-3 appears true length in the front view because the observer is equidistant from the end points of the line. (Note: A more complete explanation for finding the true length of a line and for determining true length lines in a plane is given in Sections 12.5 and 12.14.)
3. Extend projection lines from surface 1-2-3 in view F to adjacent view A.
4. Transfer points 1, 2, and 3 to view A using distances X and Y, measured in the related view H.
5. To obtain the true size of surface 1-2-3, draw a secondary auxiliary view (view B) with a line of sight that is perpendicular to the edge view of the surface. Begin by drawing reference line B-A parallel to the edge view of surface 1-2-3.
6. Extend projection lines from view A to adjacent view B.
7. Transfer points 1, 2, and 3 to view B using distances K, L, and M respectively, measured in the related view F. Connect points 1-2-3 to obtain the desired true size of the plane. The plane is true size in view B because all points in the plane are equidistant from the observer.

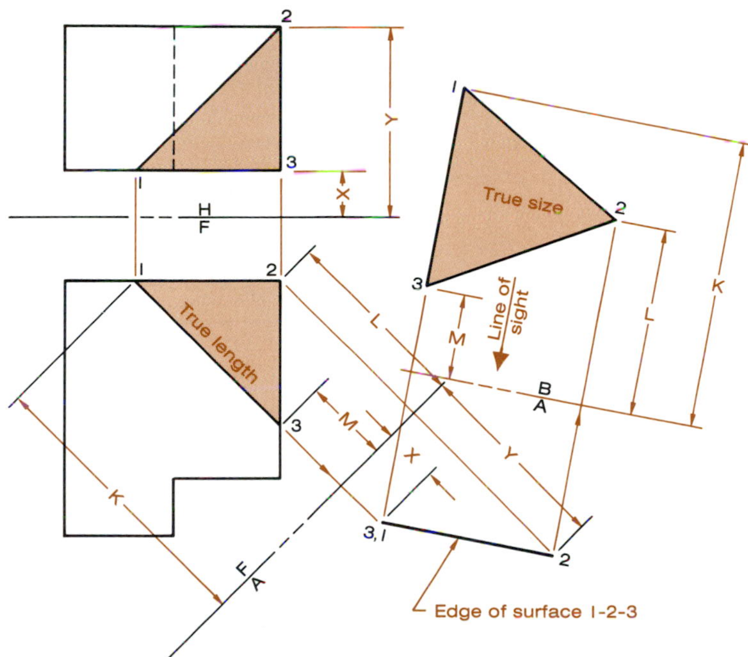

FIGURE 6-12 The use of a secondary auxiliary view to obtain the true size of an oblique surface *(Section 6.7)*

CHECKPOINT 2
SECONDARY AUXILIARY VIEWS

Use instruments to make full-size drawings of the views shown in Figure 6-13 and then construct a secondary (true size) auxiliary view of the oblique surface labeled M.

Use a plain sheet of 8½ × 11 inch paper. Use correct notation for the reference lines. (The given location dimensions are measured from the edge of the paper.)

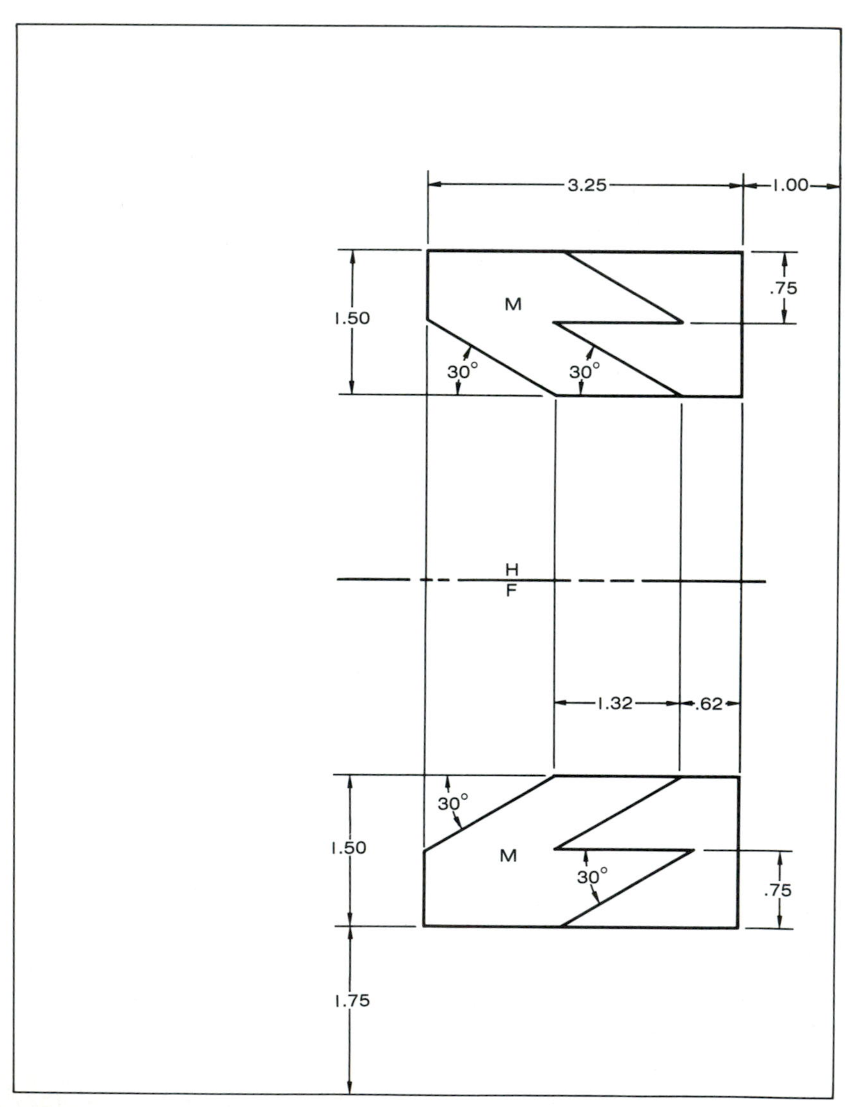

FIGURE 6-13 Checkpoint 2 problem

PROBLEMS

Auxiliary Views

In each of the following problems, lay out the given views full size. Each grid equals ¼ inch (about 6 mm). When positioning P6.18 to P6.23 and P6.30 to P6.32 on the sheet, use the small circles as locations for the end points of the various lines on the grid. The dimensions are given from the border lines of layout A, shown on the inside of the front cover. Fill out the required information in the sheet title block.

Primary Auxiliary Views

P6.1–P6.17 Draw a partial auxiliary view of the objects showing the true size of surface A.

P6.1–P6.12 Views F and P are given.

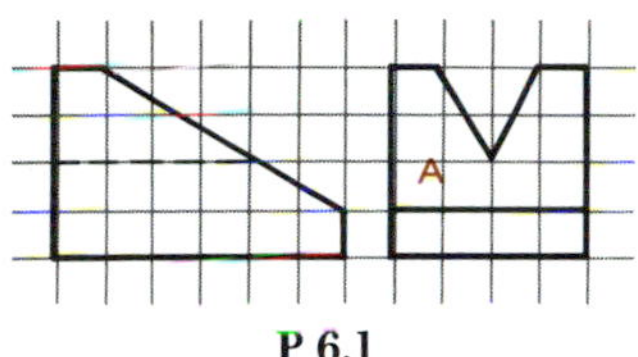

P 6.1

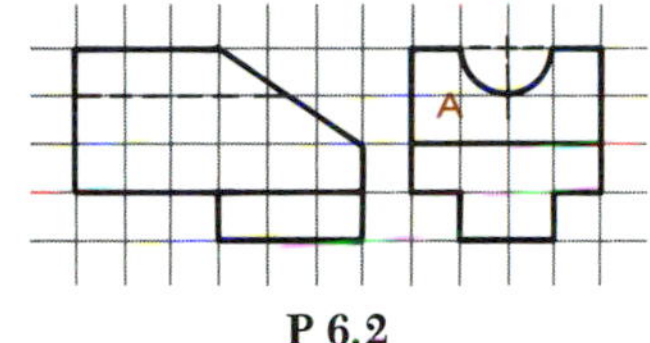

P 6.2

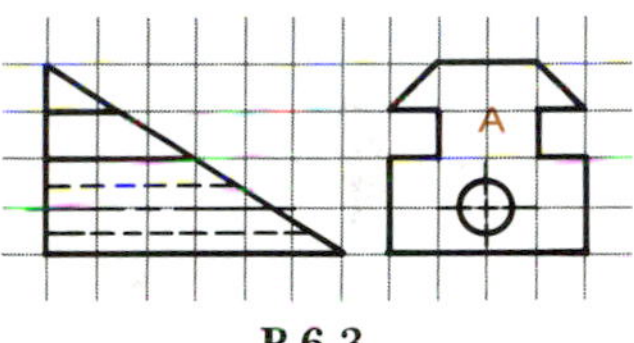

P 6.3

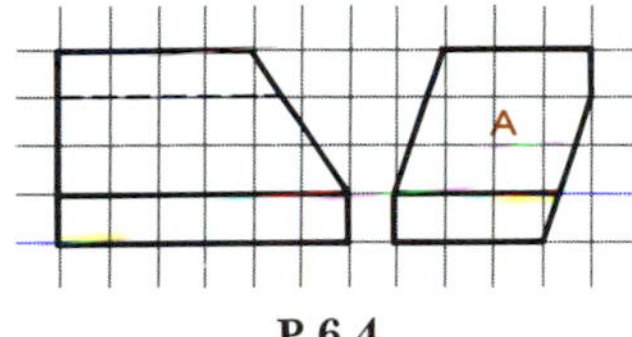

P 6.4

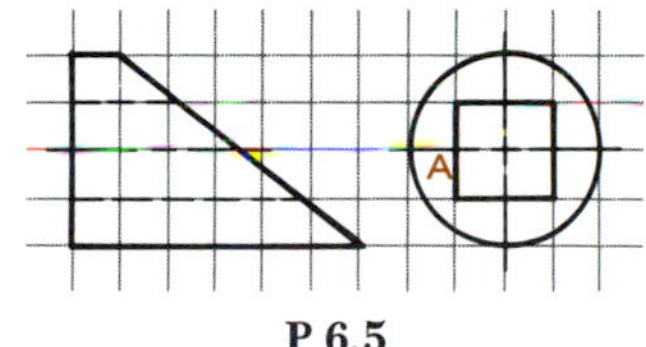

P 6.5

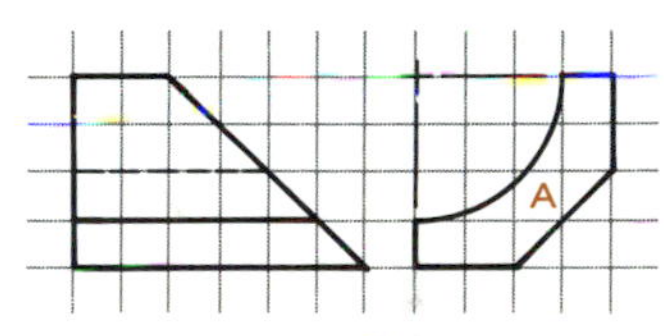

P 6.6

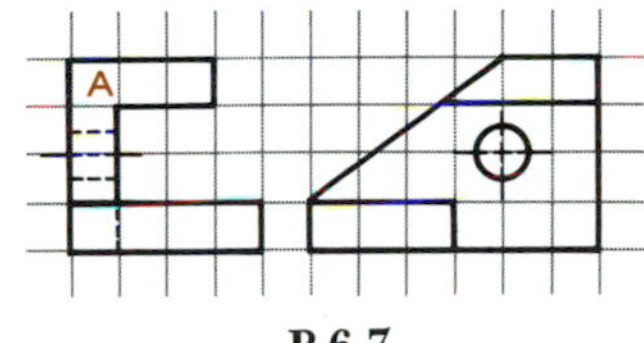

P 6.7

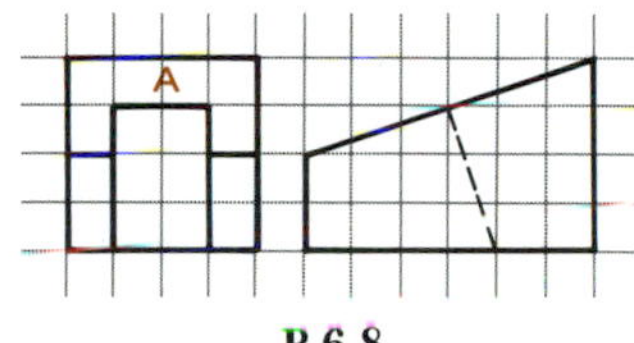

P 6.8

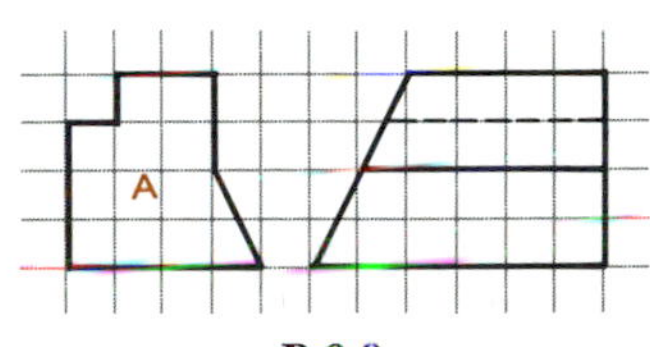

P 6.9

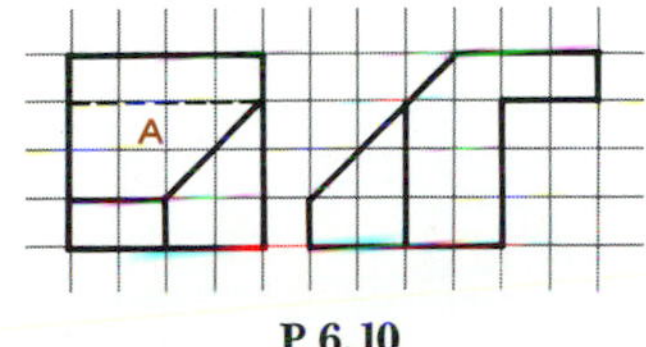

P 6.10

P 6.11

P 6.12

P6.13–P6.17 Views H and F are given.

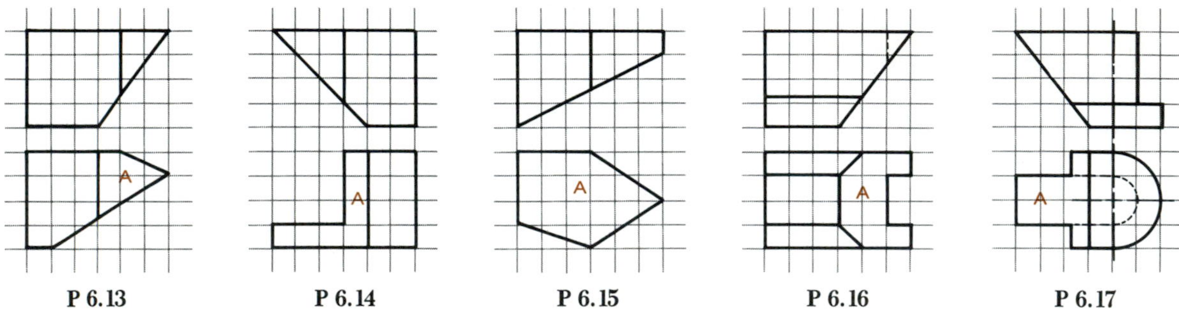

P 6.13 P 6.14 P 6.15 P 6.16 P 6.17

P6.18, P6.19 Draw complete auxiliary views of the solid object abcd in the position shown. Views H and F are given.

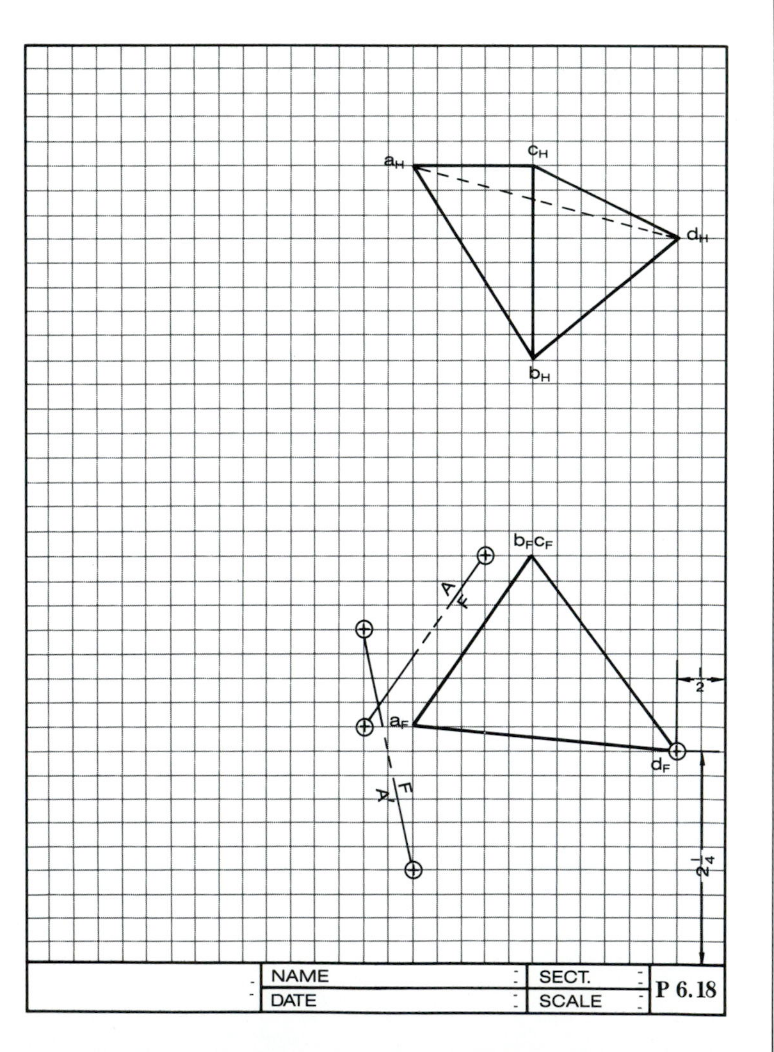

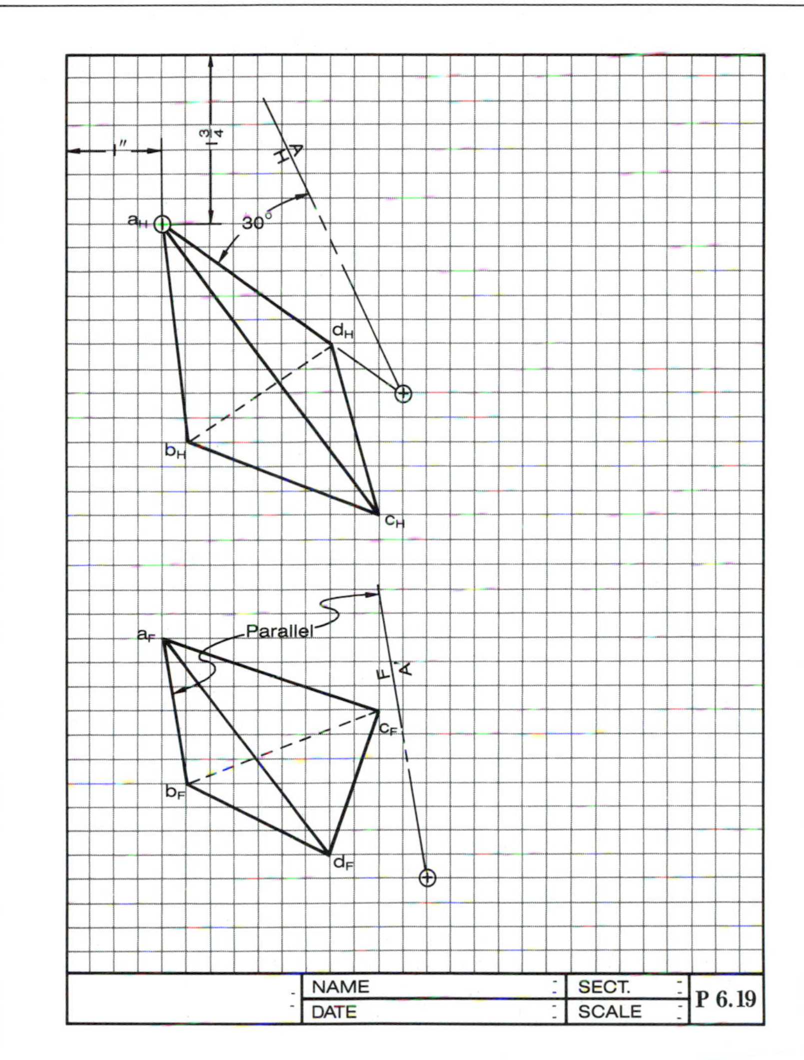
1"
3/4
H A
30°
a_H
d_H
b_H
c_H
Parallel
F A
a_F
c_F
b_F
d_F
NAME
DATE
SECT.
SCALE
P 6.19

P6.20–P6.23 Draw a partial auxiliary view of the parts showing the true size of the inclined surface.

P6.20A Views F and P are given.
P6.20B Views H and F are given.

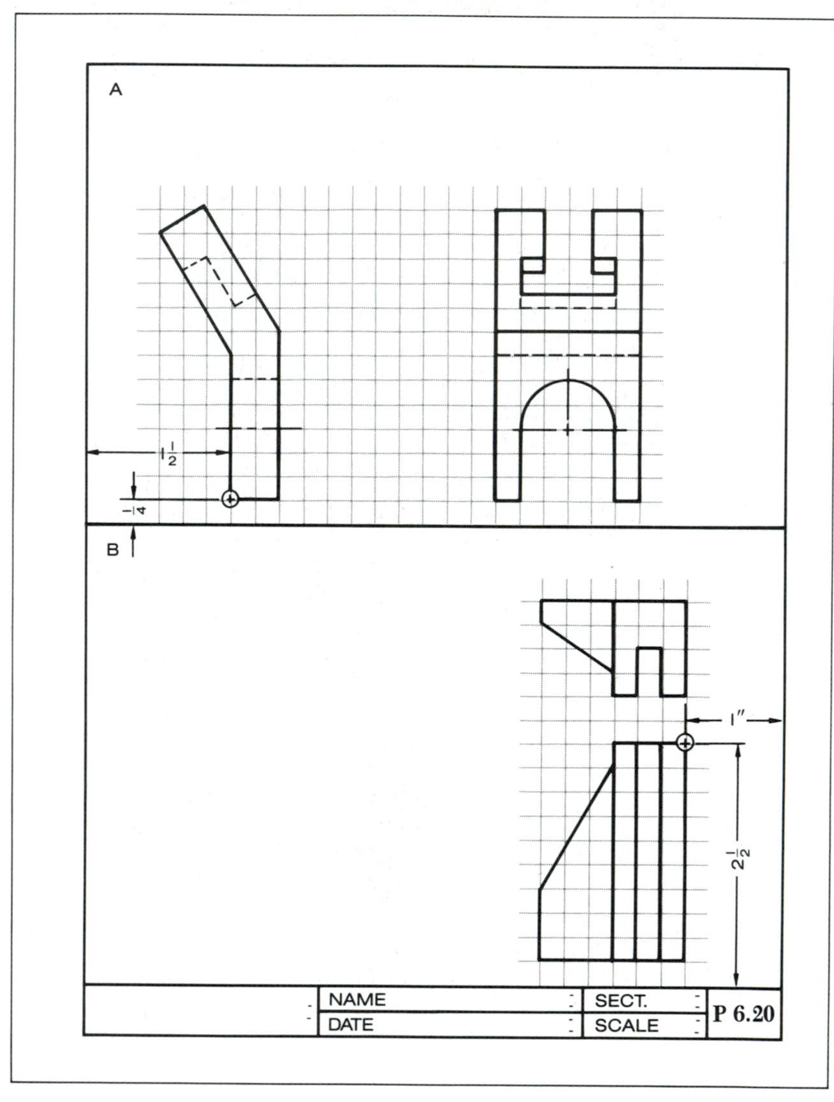

P6.21A Views H and F are given.
P6.21B Views F and P are given.

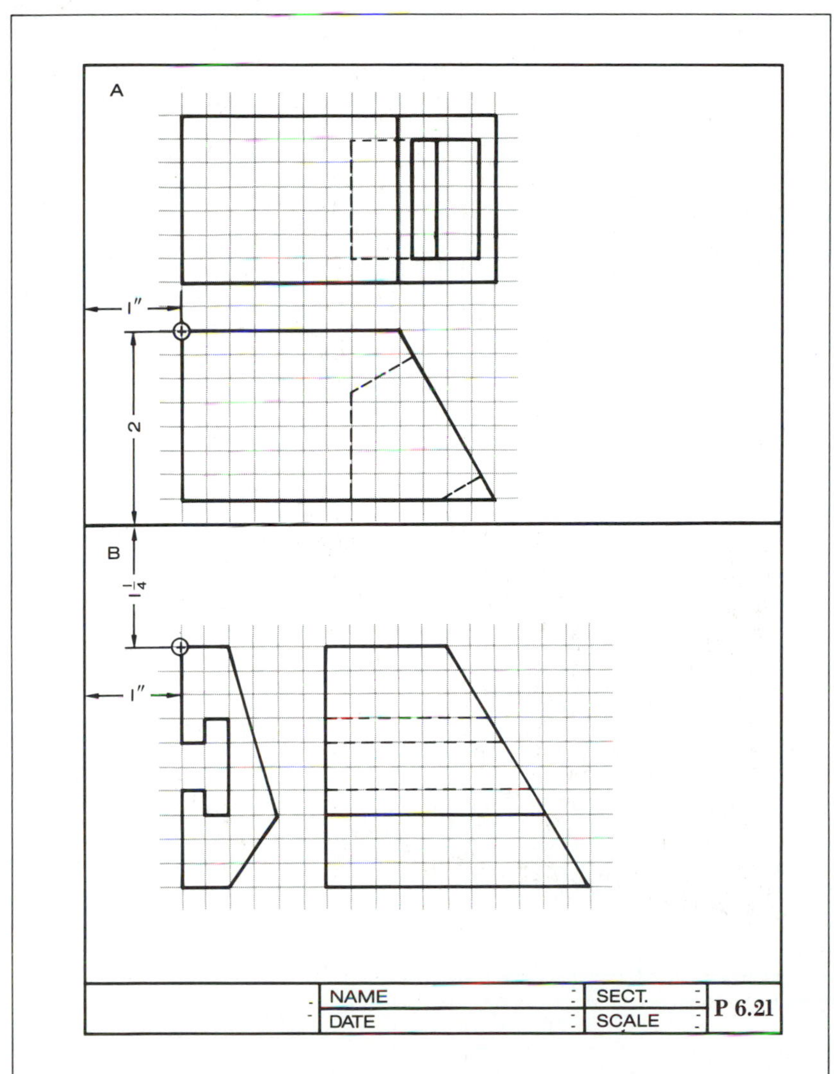

P6.22A Views H and F are given.
P6.22B Views H, F, and P are given.

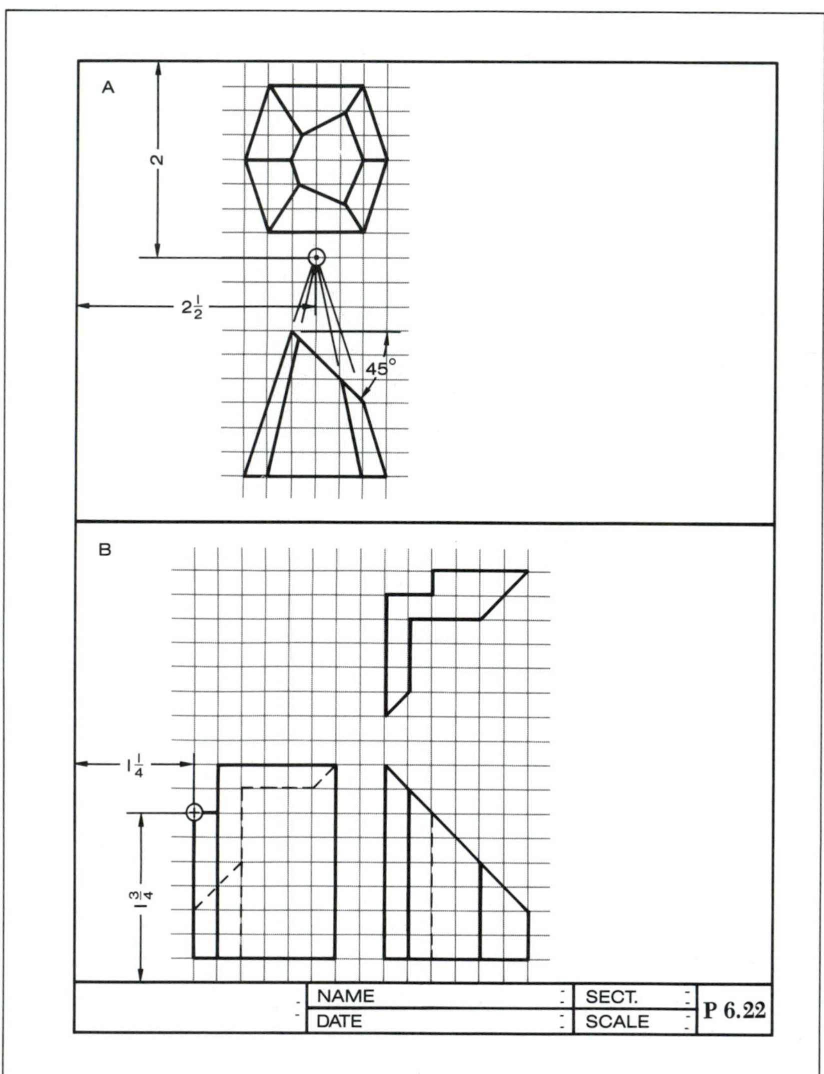

P6.23A Views H and F are given.
P6.23B Views F and P are given.

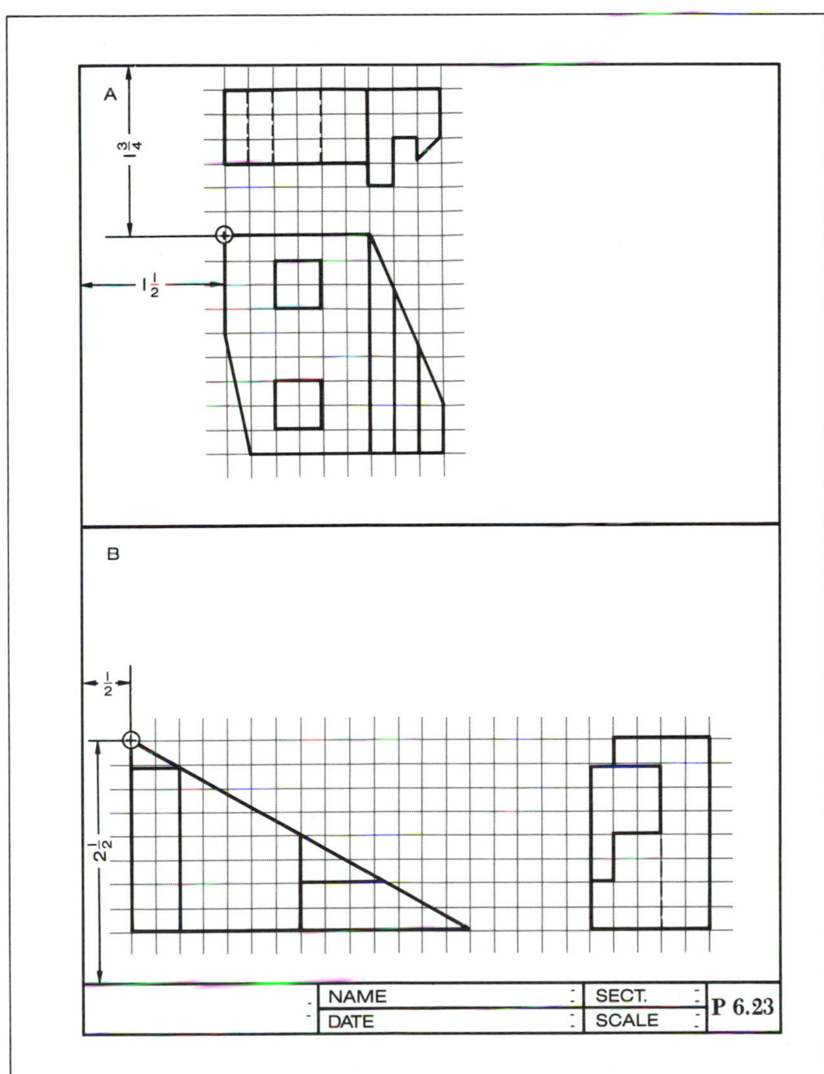

P6.24–P6.29 Draw the necessary principal views and partial auxiliary views to describe the parts shown. Adjust the scale of each problem to fit the space on layout A.

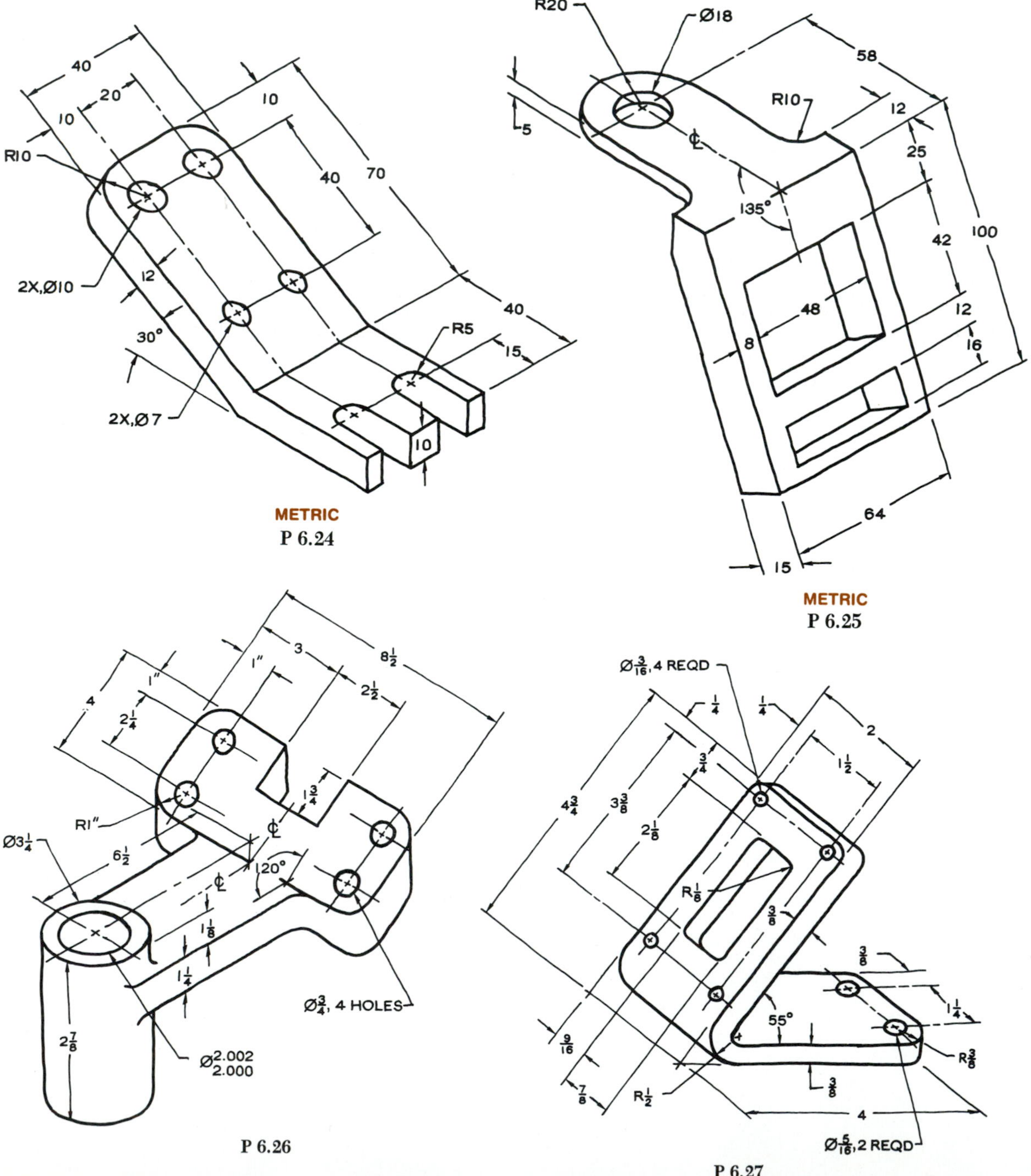

METRIC
P 6.24

METRIC
P 6.25

P 6.26

P 6.27

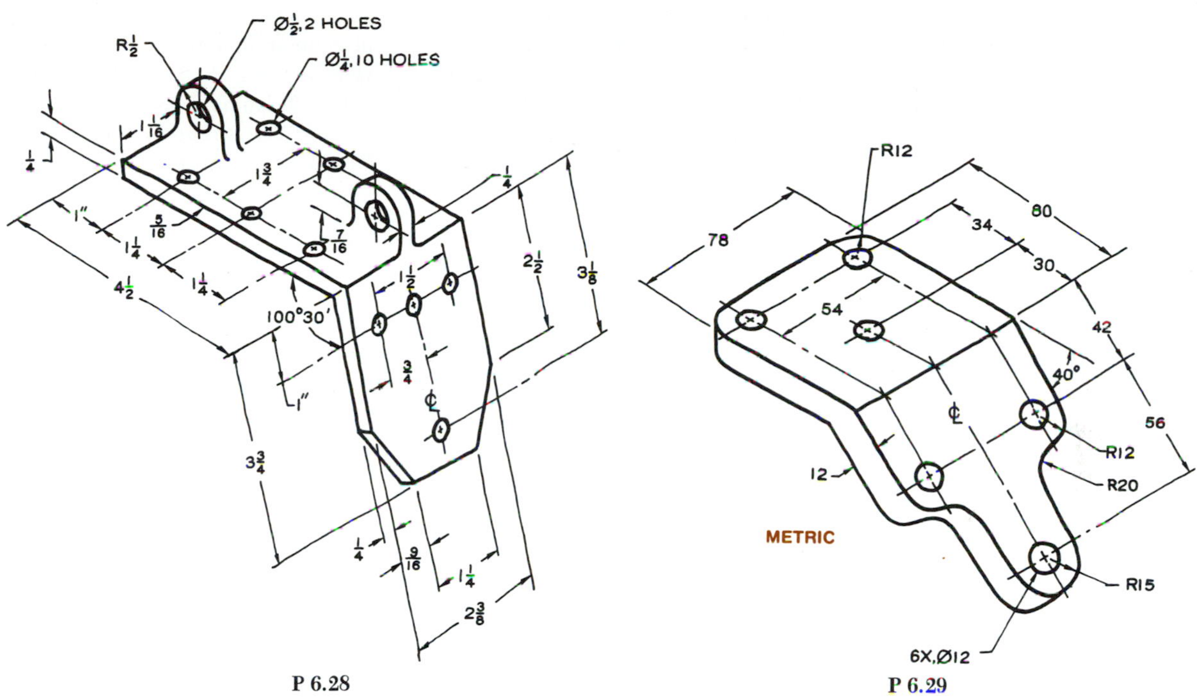
Ø1/2, 2 HOLES
R1/2
Ø1/4, 10 HOLES
1 1/16
1/4
1 3/4
1/4
1"
5/16
7/16
1 1/4
1 1/4
4 1/2
2 1/2
3 1/8
1 1/2
100°30'
3/4
1"
3 3/4
1/4
9/16
1 1/4
2 3/8
R12
78
80
34
30
54
42
40°
56
R12
R20
12
METRIC
R15
6X, Ø12

P 6.28

P 6.29

Secondary Auxiliary Views
P6.30–P6.33 Construct the indicated auxiliary views. In each problem the direction of the primary and the secondary auxiliary views has been indicated by giving one end-point position and the direction of the reference line between adjacent views. Locate one point on the views as the origin of measurements and place this point according to the conditions of the problems. Draw the given views full size on layout A using the placement dimensions to locate the view from the *border* of the sheet. Show hidden lines in all views.
P6.30A Views H and F are given.
P6.30B Views H and F are given.

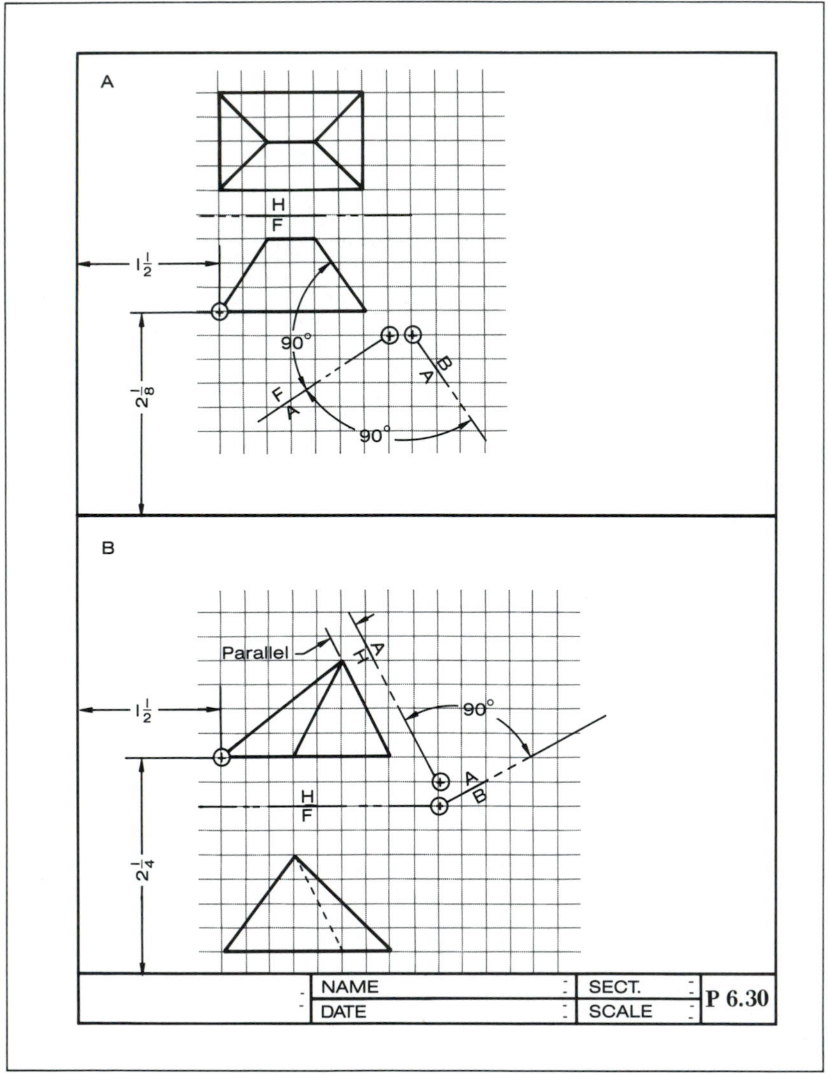

P6.31A Views H and F are given.
P6.31B Views B and C are given.

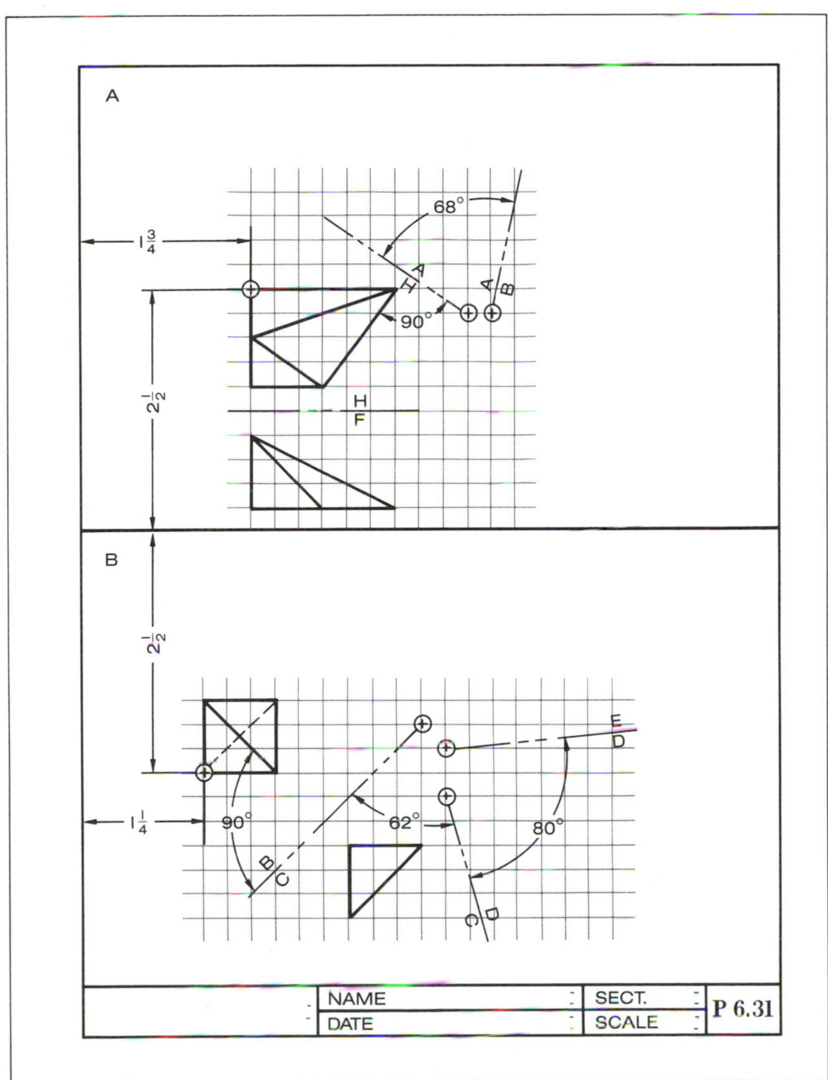

P6.32A Views H and F are given.
P6.32B Views H and F are given.

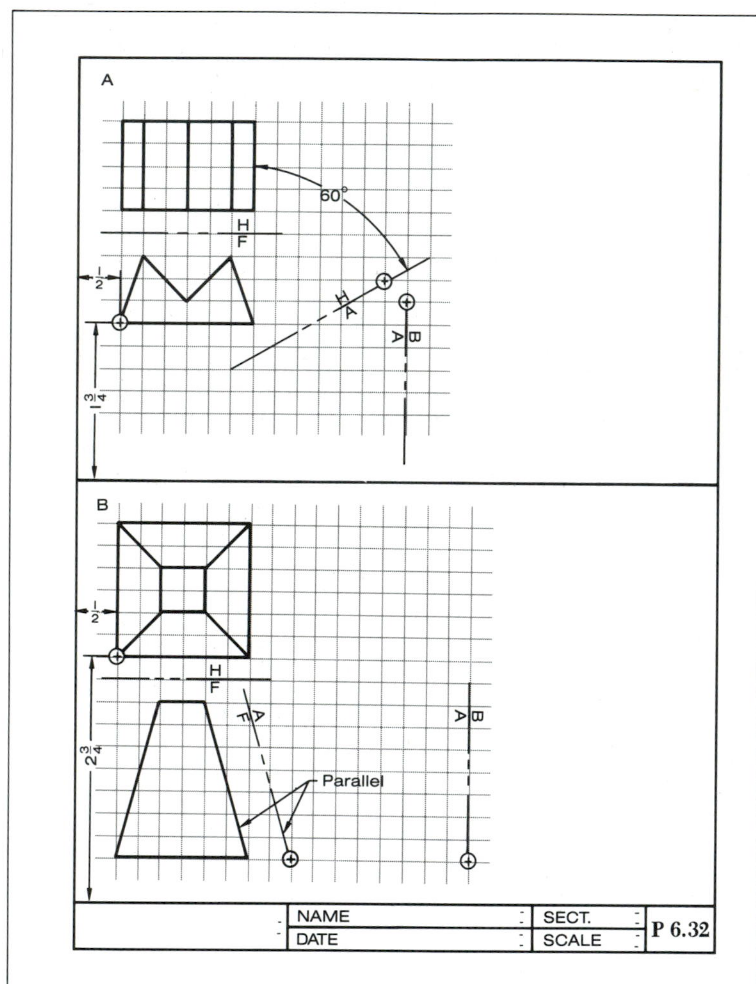

7 Sectional Views and Conventional Representation

OBJECTIVES

After completing this chapter you should have gained the following abilities:

1. To apply properly the conventional graphical techniques relating to sectional views with the use of cutting planes and section lining.

2. To recognize and use the following types of sections for representing parts:
   a. full section
   b. half section
   c. offset section
   d. aligned section
   e. revolved section
   f. removed section
   g. auxiliary section
   h. broken-out section

3. To apply the following types of conventional representation relating to sectioning practices:
   a. cutting plane passing through ribs, lugs, webs, and so on
   b. spokes in sectional views
   c. partial views
   d. rotated features
   e. breaks

Many times you will be called upon to draw parts that have an interior construction too complex to be clearly described by hidden lines in standard views. In such cases you will need to prepare *sectional views*, often simply called *sections*, in which imaginary cuts are made through objects, exposing their interior features. This chapter discusses the general procedures for preparing sections. Sectional views on assembly drawings are discussed in Chapter 13.

Often cases arise in the preparation of sectional views in which a literal use of standard drafting techniques can give a false impression of the object being drawn or can result in unnecessary drafting time. In these cases it can be helpful to use certain "short cut" techniques, including the use of breaks when drawing long, uniform objects and partial views when drawing symmetrical objects. These and other techniques discussed in this chapter are known as *conventional representation*.

7.1 SECTIONAL VIEWS

Sectional views, also called *sections*, are used to clarify the interior construction of an object that cannot be clearly described by hidden lines in the standard views. Figure 7-1(a) shows pictorially how a sectional view is obtained by passing an imaginary cutting plane through the object perpendicular to the line of sight. The portion of the object between the cutting plane and the observer is removed, and the exposed "cut" surfaces of the object are indicated by section lining or cross-hatching. The standard top and front views are shown in Figure 7-1(b). In the front view the hollow interior shapes are indicated by hidden lines that, on complicated machine parts, are difficult to interpret. The use of a section view is shown in Figure 7-1(c). Note that in this example the section view replaces the regular front view.

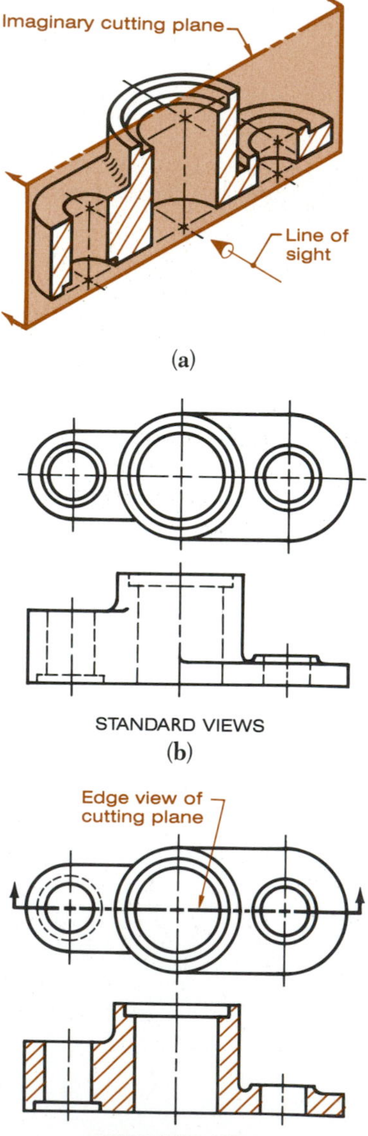

FIGURE 7-1 The basis of sectional views. An imaginary cutting plane is passed through an object perpendicular to the line of sight. The portion of the object between the cutting plane and the observer is assumed to be removed and the exposed "cut" portion is called a sectional view *(Section 7.1)*

Figure 7-2 is an industrial drawing of a coolant pump. Considerable interior detail exists on this complex assembly of parts. If hidden lines were used, as in a regular view, the interior shape would be difficult to read and interpret. Note how the sectional view provides clarity by showing the interior detail with visible lines.

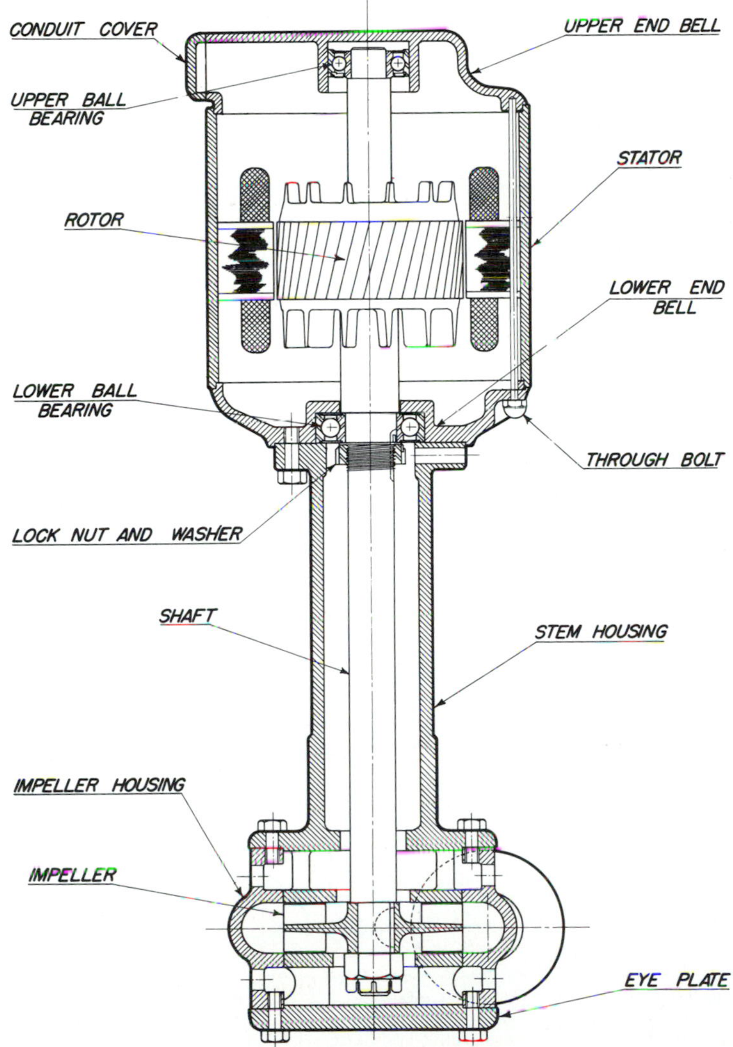

FIGURE 7-2 An industrial drawing of a coolant pump ***(Section 7.1)***

The Cutting Plane

The position of the imaginary cutting plane is shown by a heavy line, representing the *edge view* of the plane. Figure 7-3 illustrates alternate styles of cutting-plane lines. The spacing of the two dashes depends upon the size of the drawing. Arrowheads are used to indicate the direction of sight for the sectional view. Capital letters are placed at the bend on the cutting-plane line near the arrowheads to identify the cutting-plane line with the indicated section. The corresponding sectional views are identified by using double letters (SECTION A-A, SECTION B-B, and so on). The cutting-plane line takes precedence over all coinciding lines on the view except visible object lines.

Section Lining

The lines used to represent the material that has been "cut" by the imaginary cutting plane are called section lines. Section lining symbols for some commonly used engineering materials are shown approximately full size in Figure 7-4 for average size areas on drawings. Ordinarily, on detail drawings (views of single parts), the general-purpose (cast-iron) section lining is used for *all* materials. Symbolic section lining is particularly effec-

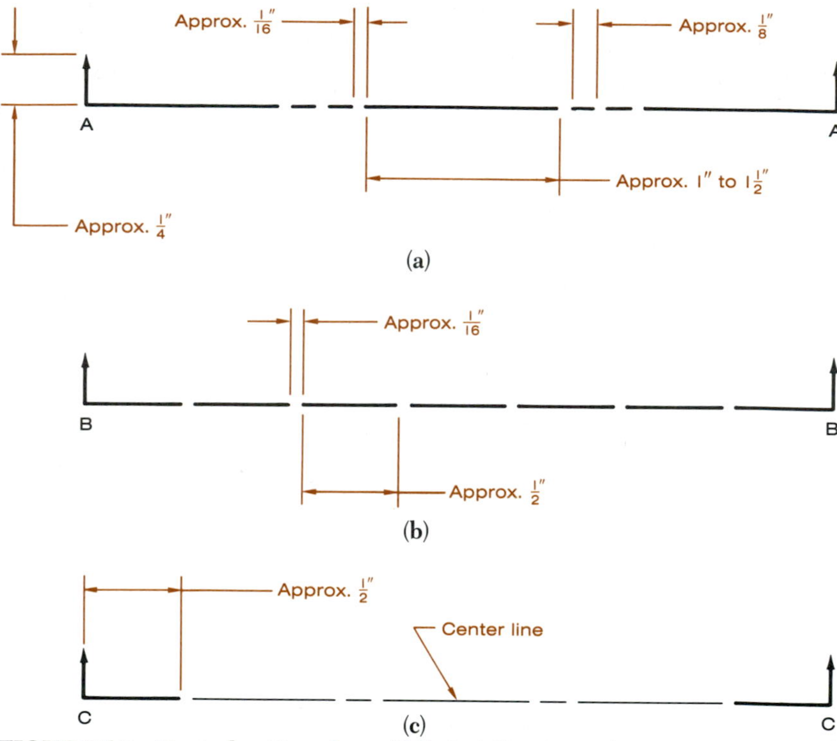

FIGURE 7-3 Typical cutting planes (full size) *(Section 7.1)*

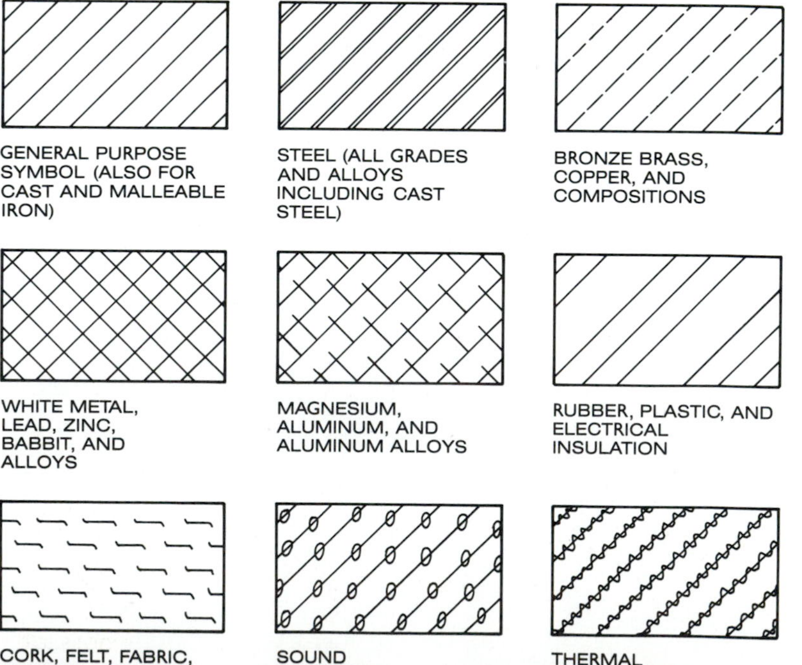

FIGURE 7-4 Symbols for section lining (full size) *(Section 7.1)*

tive when used on assembly drawings as a means of distinguishing parts of different materials. Examples of this practice are shown in Chapter 13.

An example of correct section lining technique is shown full size in Figure 7-5(a). Figure 7-5(b) shows several incorrect techniques. Use a 2H or 3H pencil lead for drawing section lining. The general-purpose section lining consists of thin, uniformly spaced, parallel lines usually slanted at 45°. The width of the spacing varies with the size of the drawing, but spacing lines 3/32 to 1/8 inch (about 2 to 3 mm) is common. The section lines are drawn closer together on views of small parts than on views of larger parts.

When drawing section lines, carefully plan the direction so the lines are not parallel or perpendicular to nearby object lines. Figure 7-6 shows examples of correct and incorrect directions for section lines. Using section lining on extremely thin parts is impractical. Instead, the areas are "blacked in," as shown in Figure 7-7.

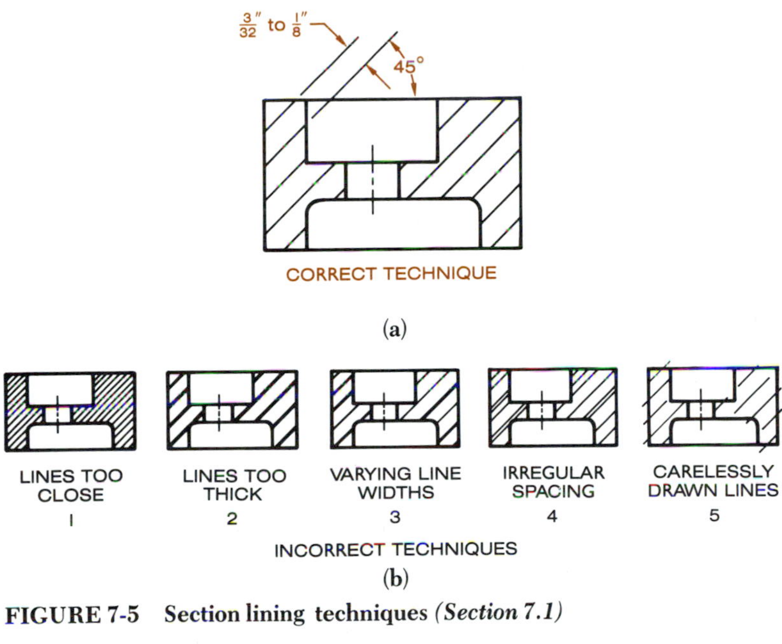

FIGURE 7-5 **Section lining techniques** ***(Section 7.1)***

FIGURE 7-6 **Preferred direction for section lines** ***(Section 7.1)***

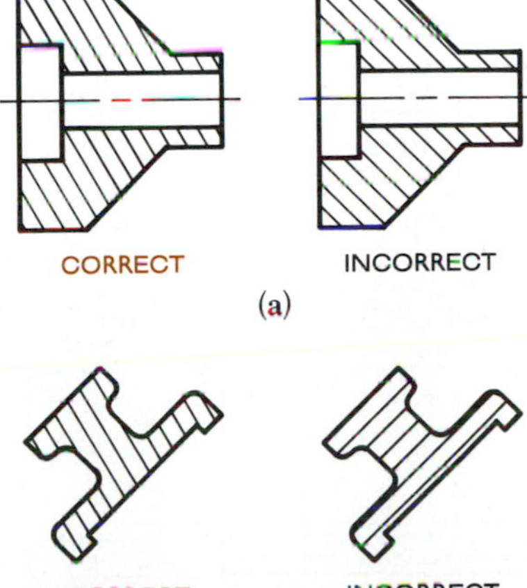

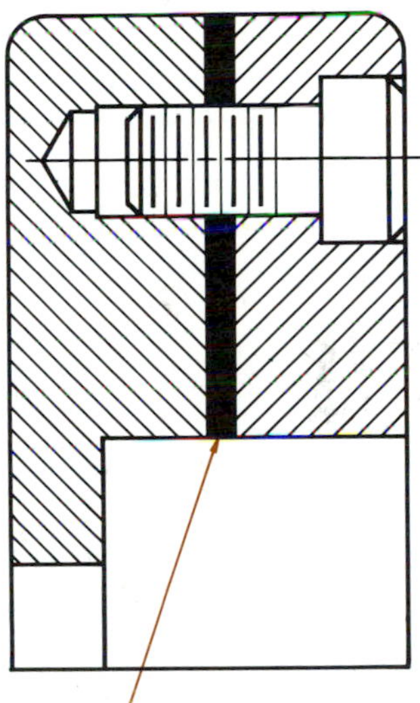

FIGURE 7-7 **Sectioning thin parts** ***(Section 7.1)***

Dimensions should not be placed in a crowded area in a sectional view. The section lining may be partially omitted for clarity when a dimension is placed in a larger area, as shown in Figure 7-8.

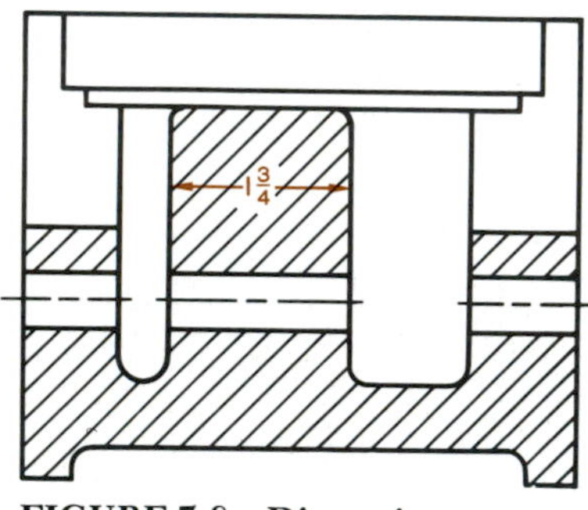

FIGURE 7-8 Dimensions on sectional views ***(Section 7.1)***

7.2 VISUALIZING A SECTIONAL VIEW

Visible Lines

Figure 7-9 is a pictorial drawing of a simple, cylindrical object cut by a full sectional cutting plane. Figure 7-9(a) shows two regular views of the same object. In Figure 7-9(b) the object is "cut" along the horizontal center line in the top view, and the object is shown with the front half theoretically removed. The view is incorrect because the *visible* lines beyond the cutting plane *must* be shown. Contrast this incorrect example with the correct example shown at 7-9(c), where the visible lines representing the back half of the object are shown.

FIGURE 7-9 Visualizing a sectional view ***(Section 7.2)***

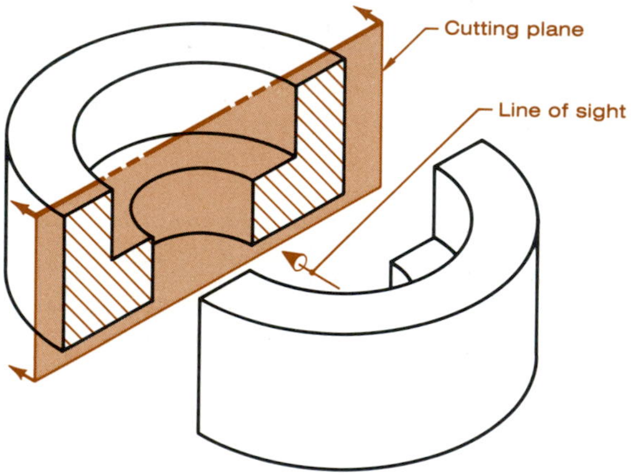

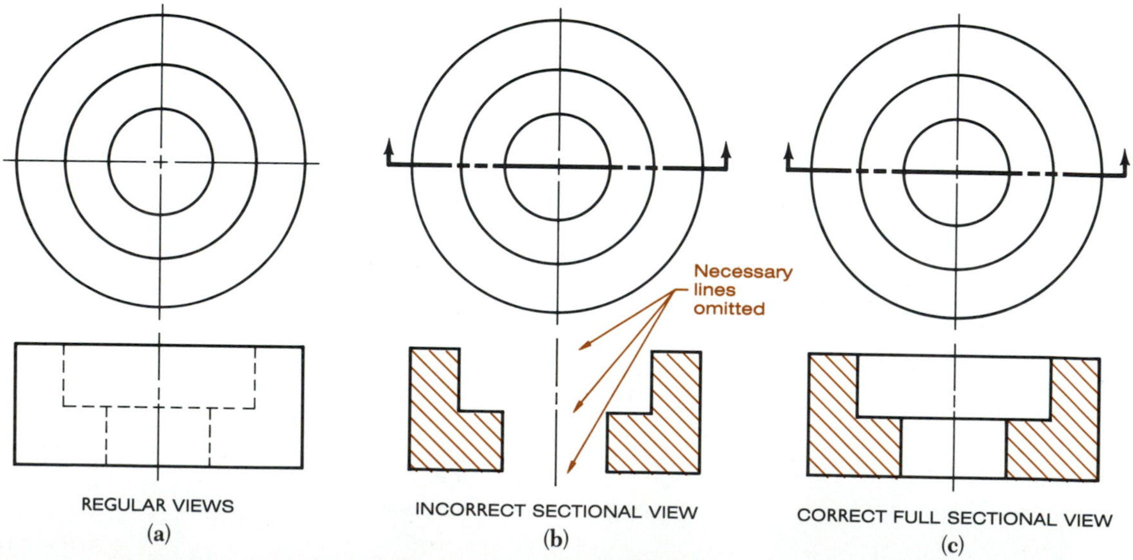

Hidden Lines

In general, *hidden* lines beyond the cutting plane are omitted in sectional views. The exception to this practice is where hidden lines are needed for clarity or where their use eliminates the need for drawing an additional view. Figure 7-10 shows sectional views of objects with necessary hidden lines in section.

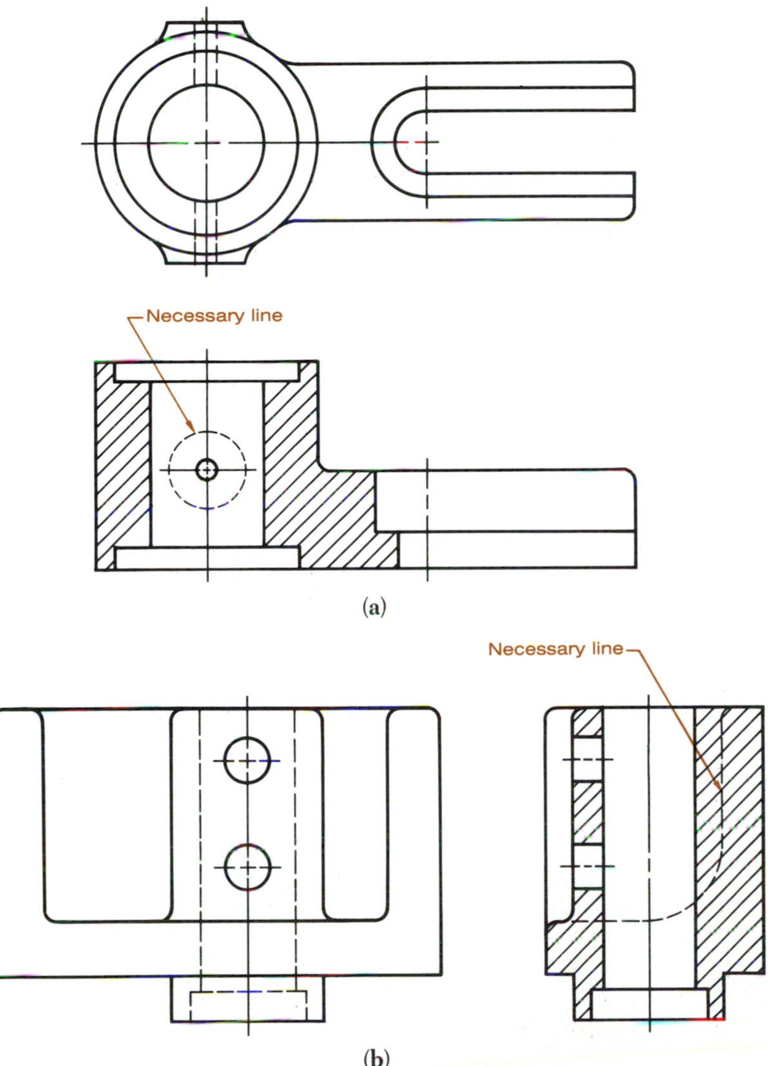

FIGURE 7-10 Hidden lines in sectional views *(Section 7.2)*

7.3 FULL SECTIONS

Where the cutting plane extends *fully* through the object, usually on the center line of symmetry, a *full section* is obtained, as shown in Figure 7-11. For clarity, the cutting plane is shown in this example, but it may be omitted when its location is obvious. Sectional views should always be "cut" along a plane to show the interior detail of the object with the greatest possible clarity. Three examples of full sections are shown in Fig-

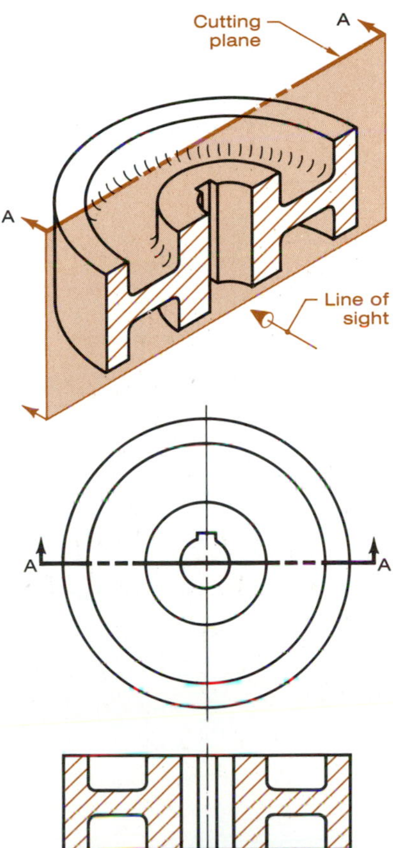

FIGURE 7-11 A full section *(Section 7.3)*

ure 7-12. In 7-12(a) the imaginary cutting plane (not shown) passes through the top view. In this case the front half of the top view is theoretically removed, exposing the "cut" surface on the rear half of the object. The sectional view replaces the regular front view. Similar examples of full sectional views are illustrated in 7-12(b) and 7-12(c).

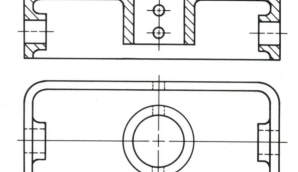

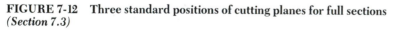

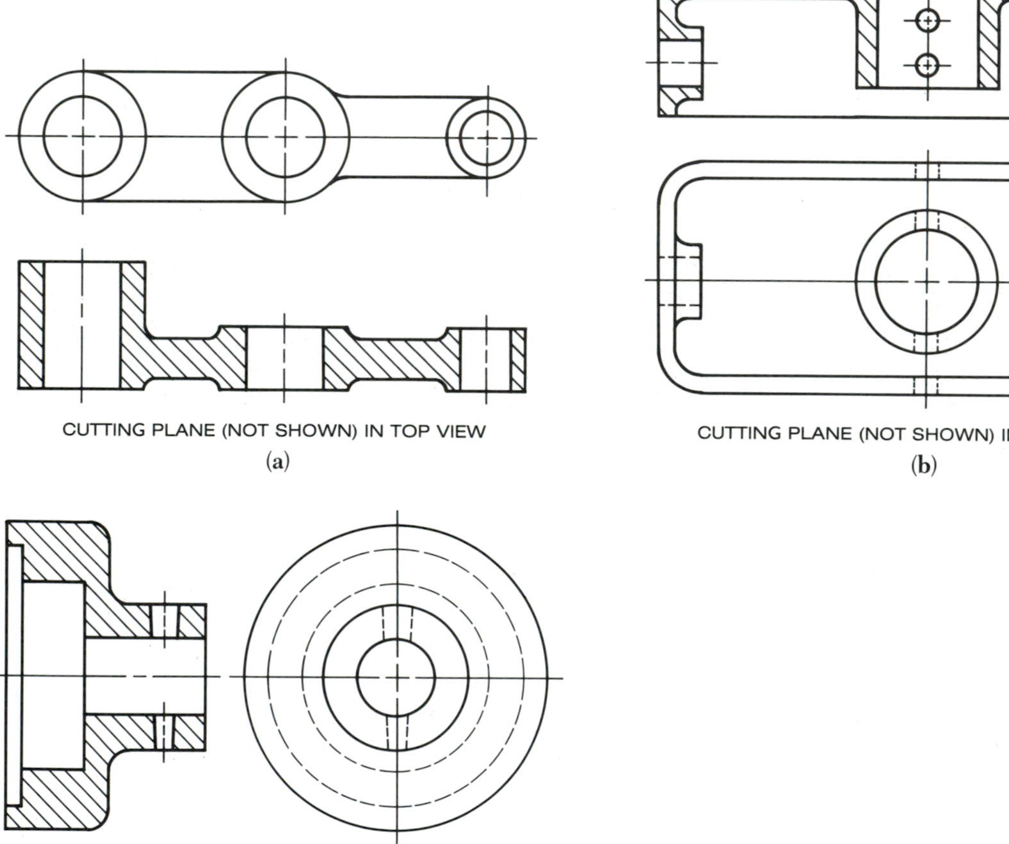

FIGURE 7-12 Three standard positions of cutting planes for full sections ***(Section 7.3)***

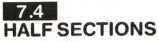

7.4 HALF SECTIONS

When an object is *symmetrical*, both the interior and exterior details can be shown in the same view by using a *half section*. The half section is obtained by passing two cutting planes at *right angles* to each other through the object so that the intersection line of the cutting planes is coincident with the plane of symmetry of the object. The manner of obtaining a half section is illustrated in Figure 7-13.

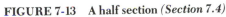

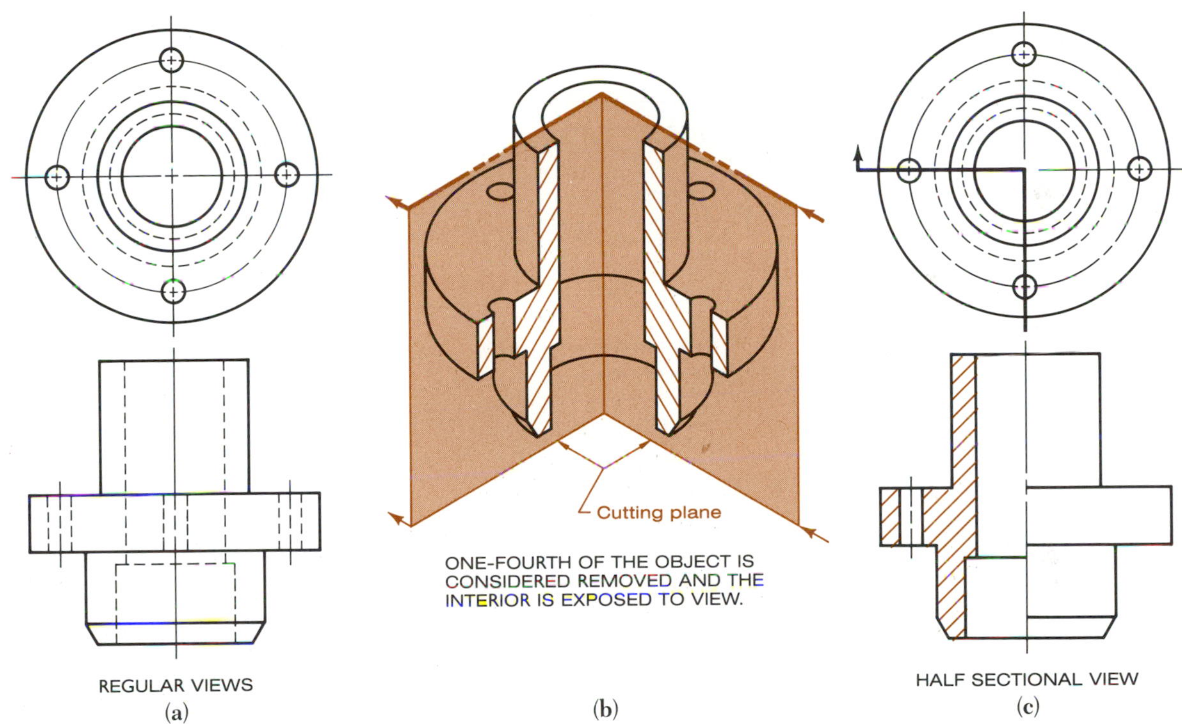

FIGURE 7-13 A half section ***(Section 7.4)***

The main advantages of a half section are that the interior detail of one-half of the object may be clearly represented and that the lines showing the exterior shape may be retained. The dashed (or dotted) lines representing the hidden edges are usually omitted on the exterior portion of half-sectional views unless they are needed for dimensioning purposes or for showing features that cannot otherwise be clearly represented. The cutting plane is omitted on half sections because its position in most cases is obvious. A center line is used in the sectional view to separate the unsectioned half from the

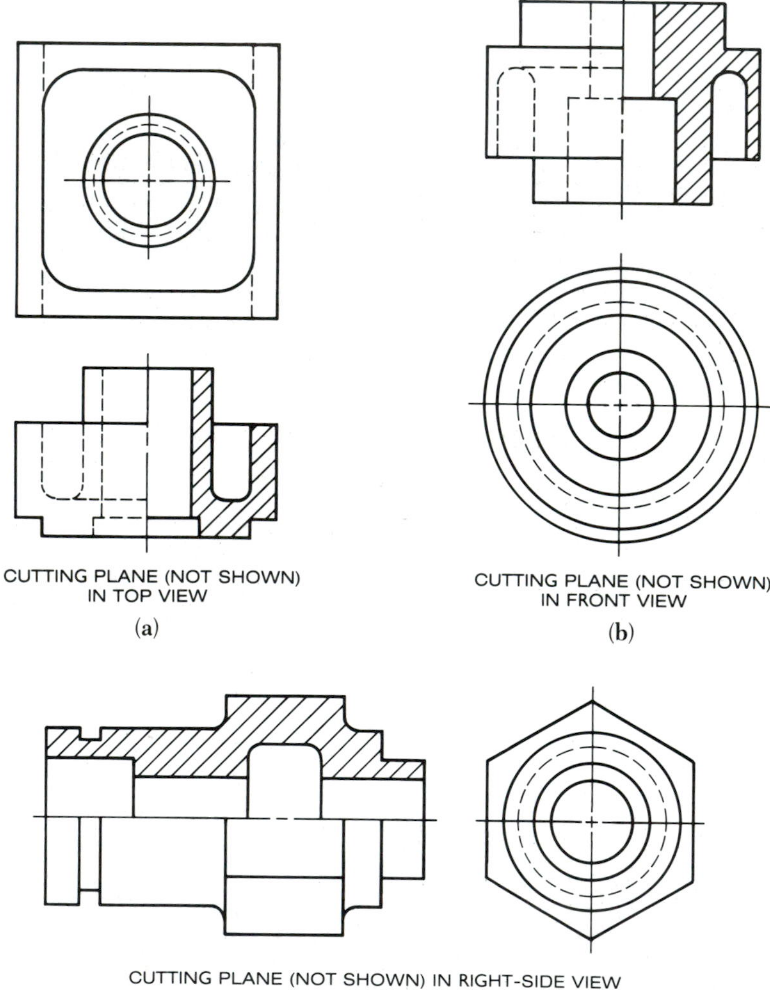

FIGURE 7-14 Three standard positions of cutting planes for half sections *(Section 7.4)*

sectioned half, as shown in Figure 7-13(c). Figure 7-14 illustrates other examples of half sectional views shown in the three standard positions.

CHECKPOINT 1
FULL AND HALF SECTIONS

Figure 7-15 consists of two principal views of four different machine parts. Sketch or draw the given views with sections as indicated, placing two problems on a plain sheet of 8½ × 11 inch paper. In the given drawings assume that each grid equals ¼ inch or about 6 mm. Omit the cutting-plane line.

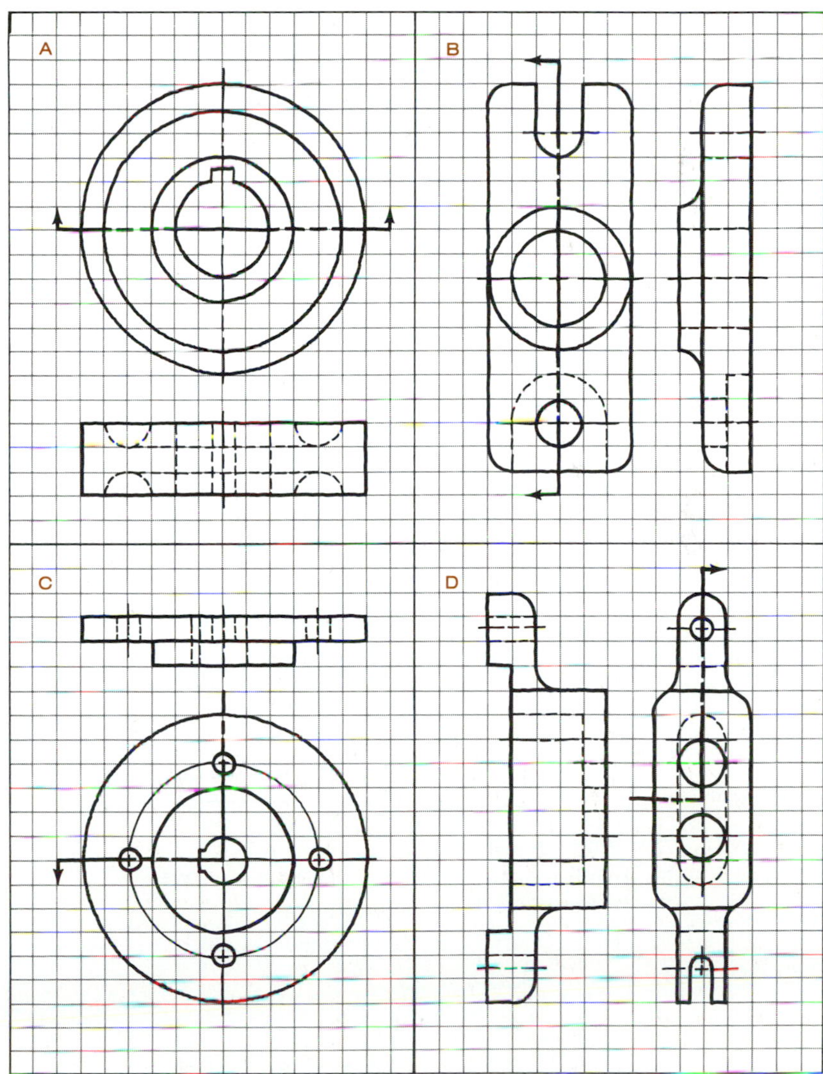

FIGURE 7-15 Checkpoint 1 problems

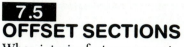

7.5 OFFSET SECTIONS

When interior features are not located on a straight line, the cutting plane may be *stepped* or *offset* (generally at right angles) to pass through these features, as shown in Figure 7-16. The section is drawn as if the offsets were in one continuous plane.

Note the similarity between *offset sections* and *full sections*. Because the "cut" is imaginary, no line is formed by offsetting the cutting plane in the sectional view, as shown in 7-16(b). Other

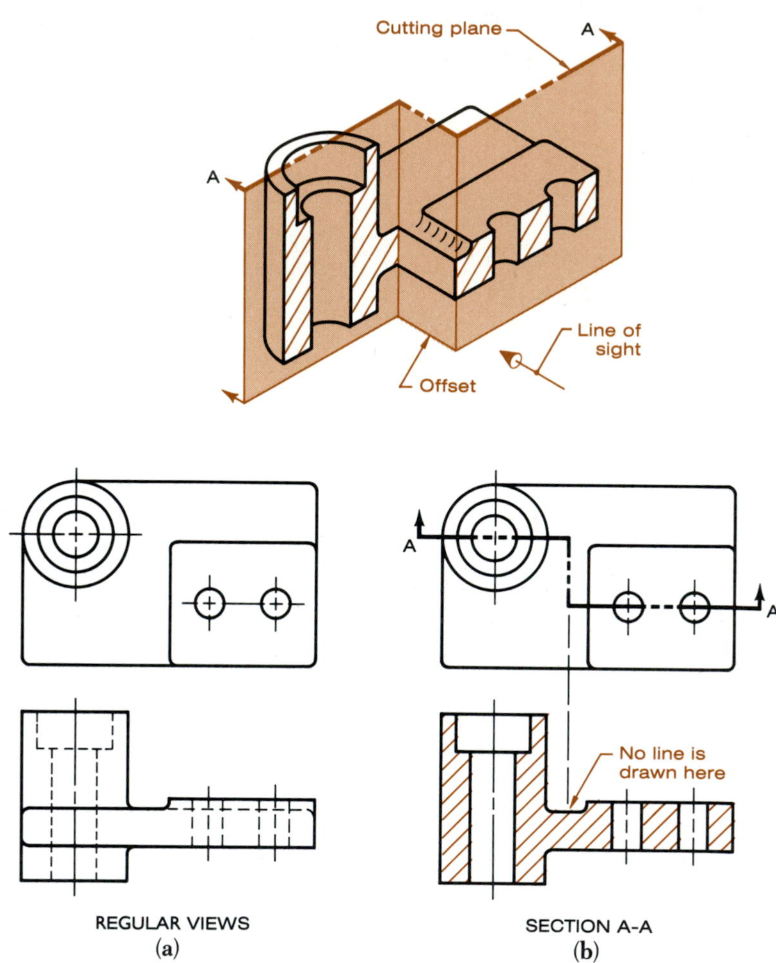

FIGURE 7-16 An offset section ***(Section 7.5)***

examples of offset sections are shown in Figure 7-17. The offset cutting plane should be drawn and labeled and the corresponding cross-sectional view identified.

CHECKPOINT 2
OFFSET SECTIONS

Figure 7-18 consists of two principal views of two different machine parts. Sketch or draw the given views with sections as indicated, placing each problem on a plain sheet of 8½ × 11 inch paper. In the examples each grid equals ¼ inch or about 6 mm. Show the cutting-plane line. (See page 226.)

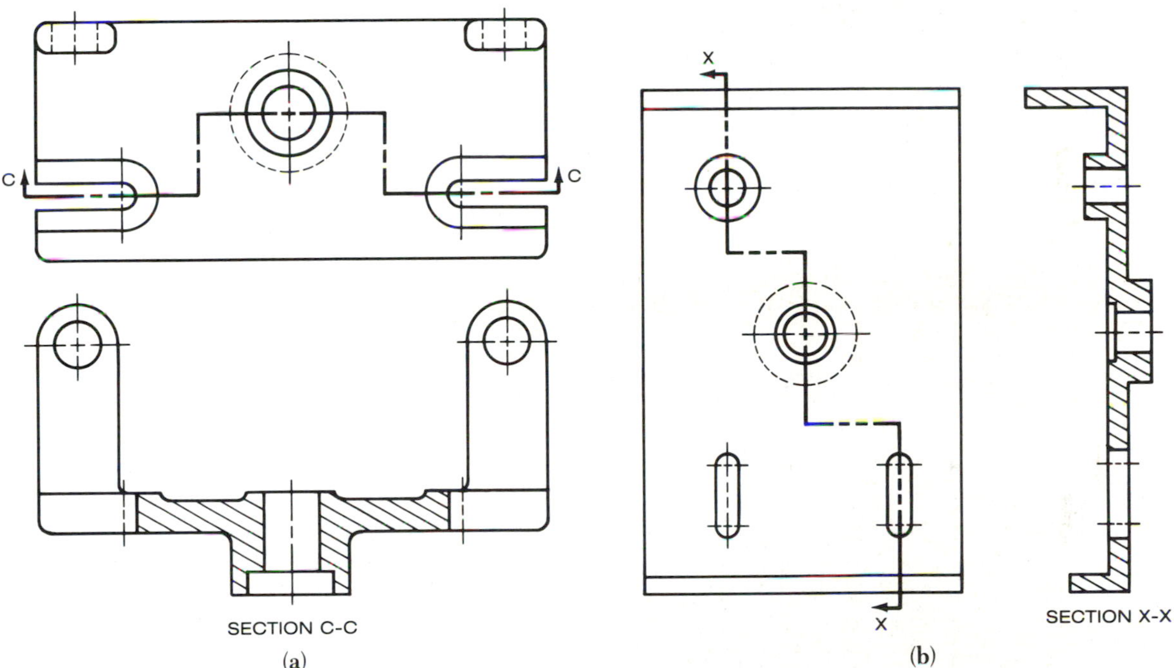

FIGURE 7-17 **Offset sections** ***(Section 7.5.)***

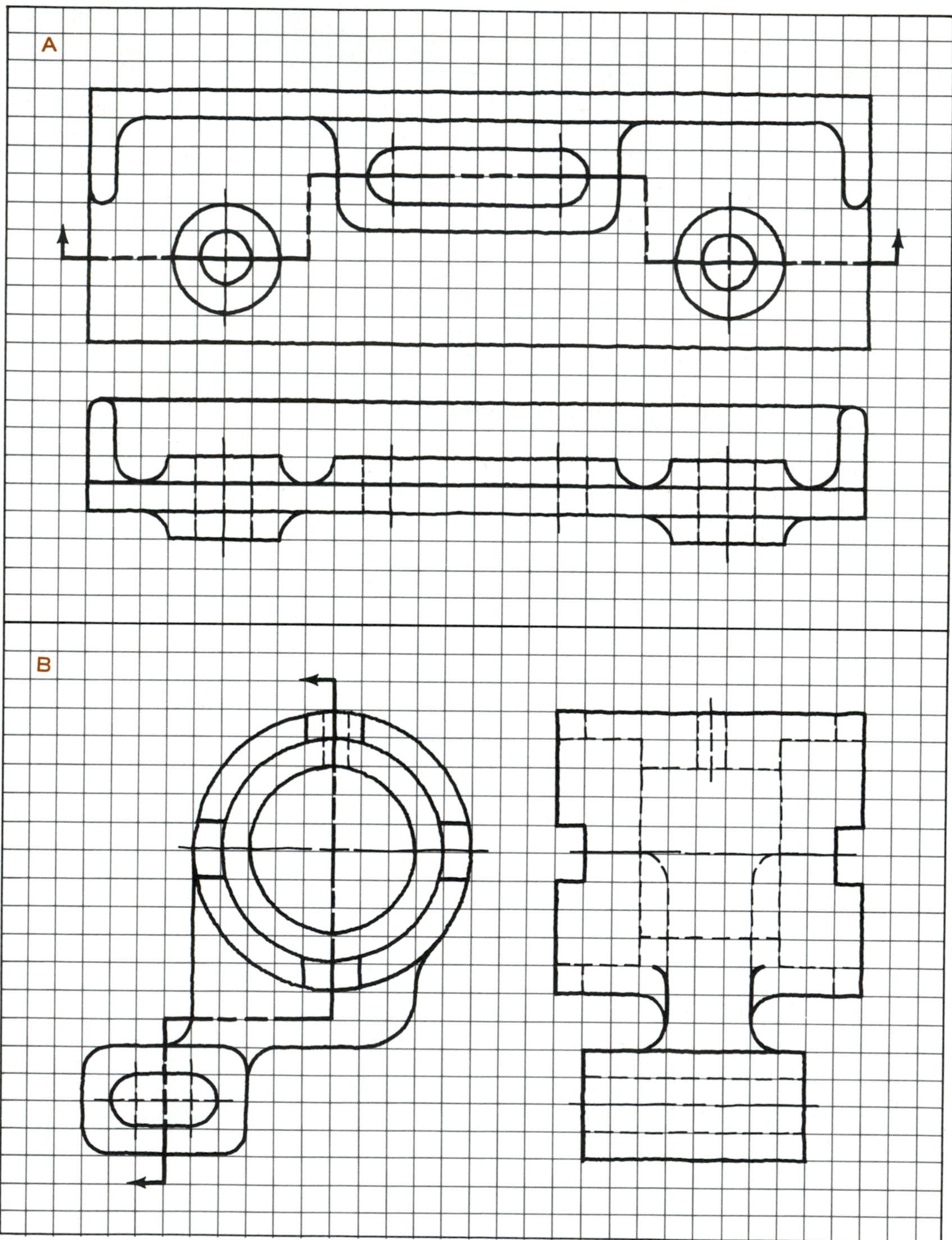

FIGURE 7-18 Checkpoint 2 problems

7.6 ALIGNED SECTIONS

The cutting plane may be bent or offset in *radial* directions to show interior features not lying in a straight line. Figure 7-19 shows how the sectional view appears when the cutting plane and features of the object are rotated into a plane perpendicular to the line of sight. Note how the cutting plane is selectively bent in an *aligned section* to pass through the radial features. The angle of revolution or angle through which the cutting plane is bent is always less than 90°. To prevent misunderstanding, a sectional view should never appear foreshortened, as shown in 7-19(c). Ordinarily it is not necessary to show the cutting plane location, but for aligned sections the cutting plane is shown and labeled. The note "SECTION A-A" is placed below the corresponding sectional view, as shown in 7-19(b). Figure 7-20 shows two examples of machine parts with aligned sections.

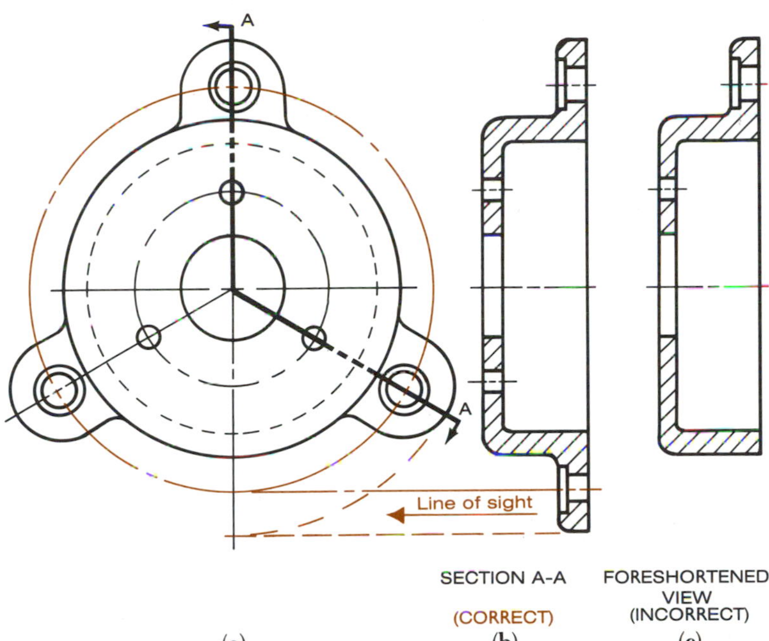

FIGURE 7-19 An aligned section *(Section 7.6)*

FIGURE 7-20 Aligned sections *(Section 7.6)*

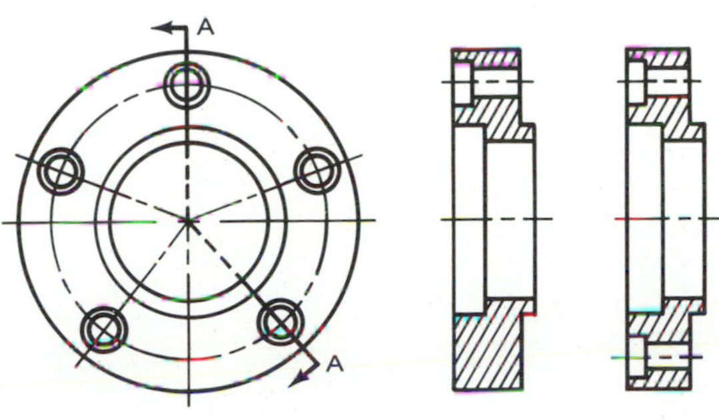

TRUE PROJECTION (AVOID)

SECTION A-A (PREFERRED)

(a)

A

A

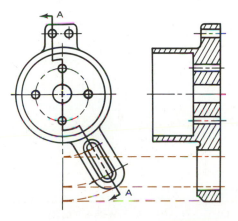

SECTION A-A

(b)

Features should not be bent or rotated into a plane perpendicular to the line of sight if the clarity of the object is affected. Carefully examine the shape of the object before drawing an aligned section. In Figure 7-21 a full sectional view is preferred over an aligned section. Note the loss of clarity in the view with the aligned features.

CHECKPOINT 3
ALIGNED SECTIONS

Figure 7-22 consists of two principal views of two different machine parts. Sketch or draw the given views with an aligned section, placing both problems on a plain sheet of 8½ × 11 inch paper. In the examples each grid equals ¼ inch or about 6 mm.

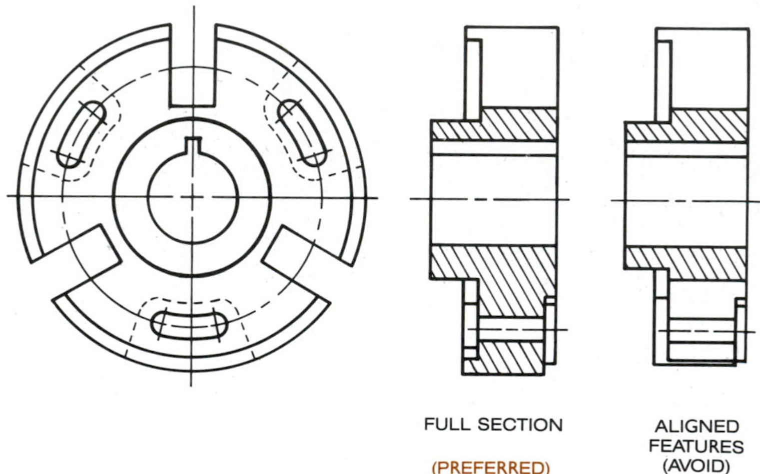

FIGURE 7-21 Selecting the proper sectional view ***(Section 7.6)***

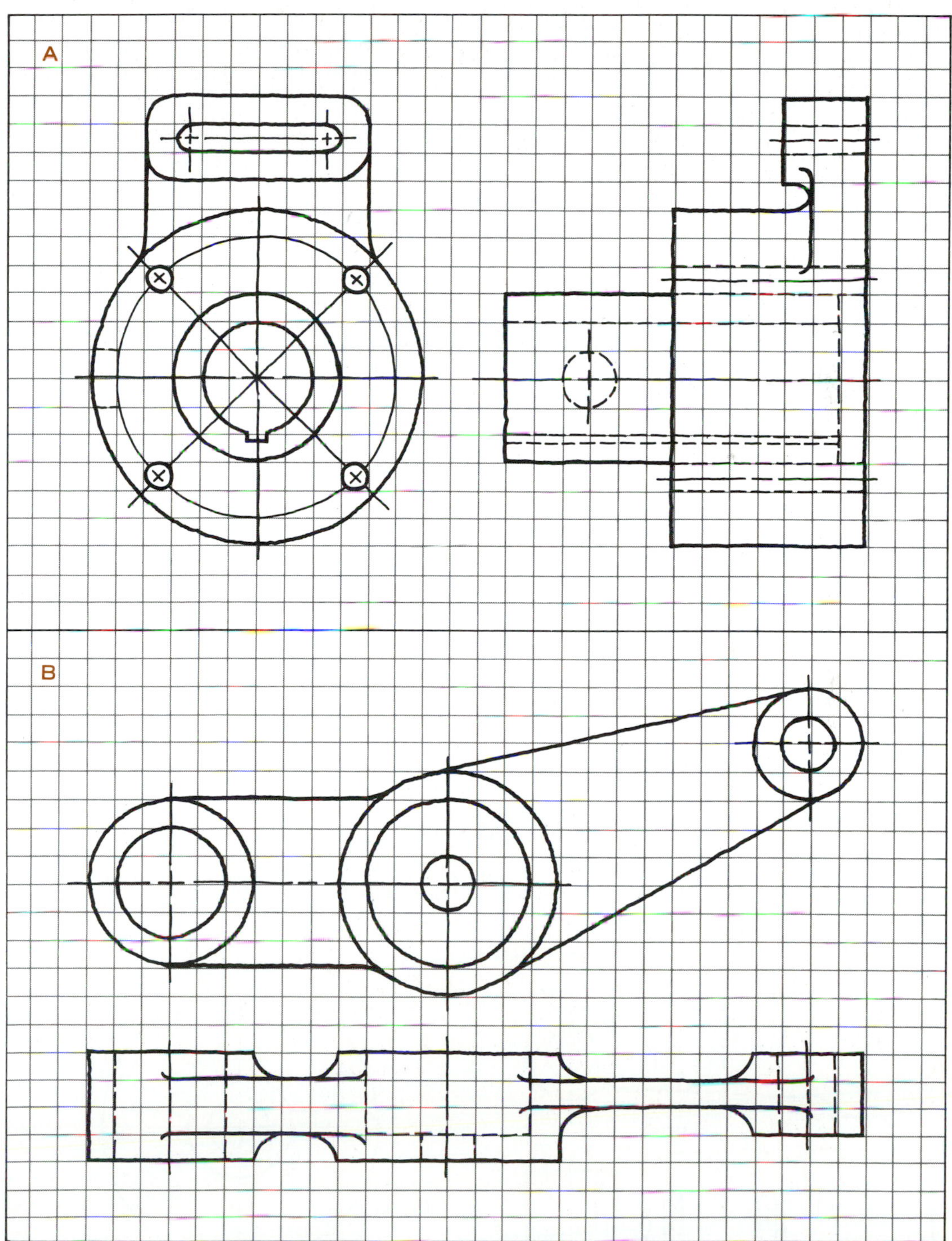

FIGURE 7-22 Checkpoint 3 problems

7.7 REVOLVED SECTIONS

If a cutting plane is passed perpendicularly to the axis of an elongated feature, such as a spoke, beam, or arm, and then revolved in place through 90° about a center line into the plane of the drawing, a *revolved section* is obtained. Figure 7-23 illustrates several examples of revolved sections. Object lines are omitted in the view covered by the superimposed true-shape revolved section. Revolved sections are used in the regular views when it is difficult to clearly show the true cross-sectional shape of an object. Figure 7-24(a) shows an alternate example of a revolved section using freehand break lines. Figure 7-24(b) illustrates a partial revolved section drawn to show the shape at the top of the ribs. A center line is drawn through revolved sectional views but no cutting plane is shown.

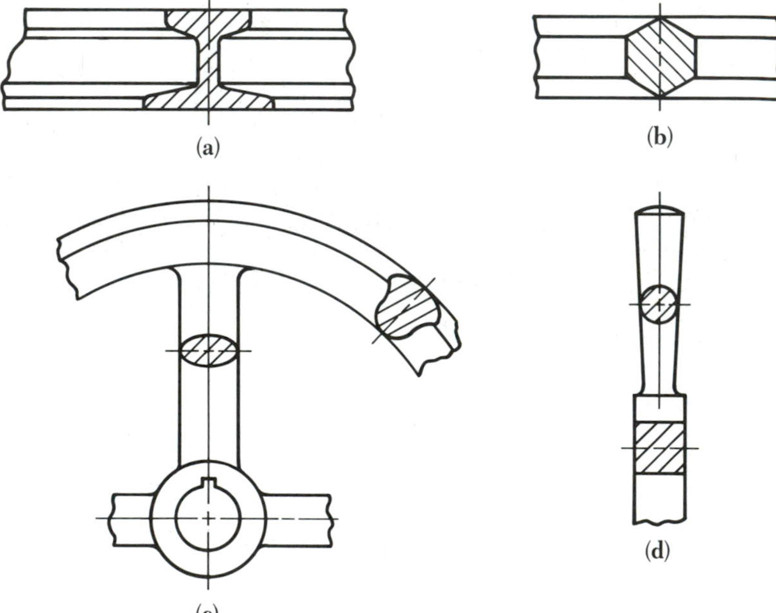

FIGURE 7-23 Revolved sections. The cross-sectional shape is superimposed on the view ***(Section 7.7)***

FIGURE 7-24 Examples of an alternate revolved section and a partial revolved section ***(Section 7.7)***

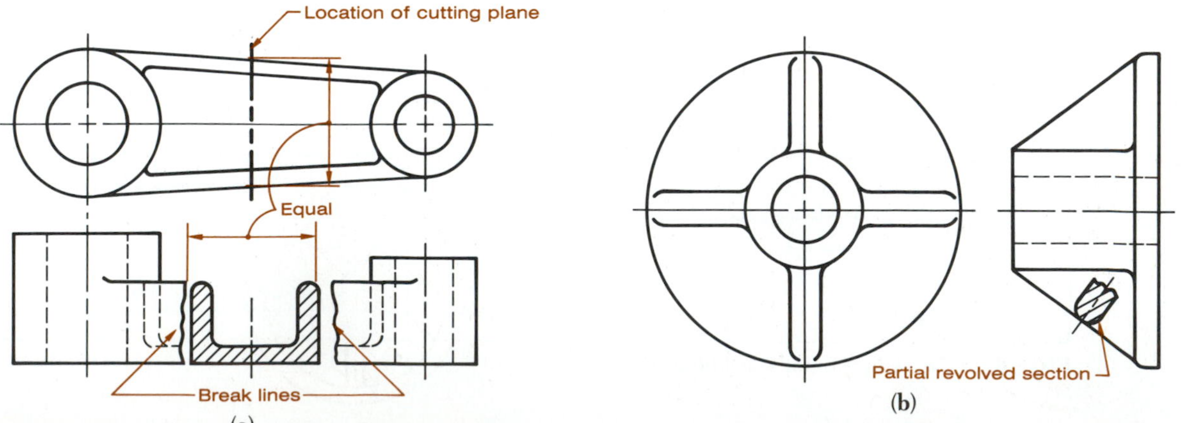

7.8 REMOVED SECTIONS

Removed sections, like revolved sections, are advantageous for showing the cross-sectional shape of complex parts. In some cases, due to the shape of the object, superimposing the revolved section on the regular views is impractical. In such cases a center line drawn on the regular view can be used as the axis of rotation to denote where the section is taken (see Figure 7-25). The corresponding removed section is projected along the axis extending from the desired "cut" position on the view. It is unnecessary to show a cutting plane or to label the sectional view. Generally the removed section shows only the cross-section shape, not any of the object behind the cutting plane.

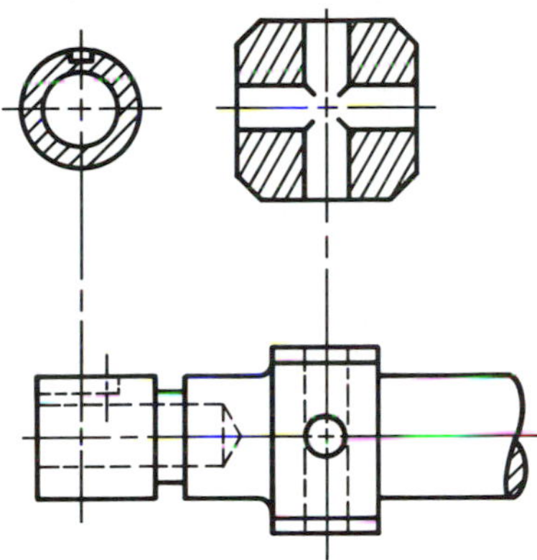

FIGURE 7-25 Removed sections taken along the axes of rotation *(Section 7.8)*

Where revolved sections on the given views would be crowded or where the method shown in Figure 7-25 is unsuitable, the sectional views can be *removed* to some convenient location on the drawing, as shown in Figure 7-26. In this method the cutting plane is drawn on the view and each removed section is labeled "SECTION A-A," "SECTION B-B," and so on, corresponding to the letters at the cutting-plane line.

FIGURE 7-26 Removed sections drawn in selected positions on the sheet *(Section 7.8)*

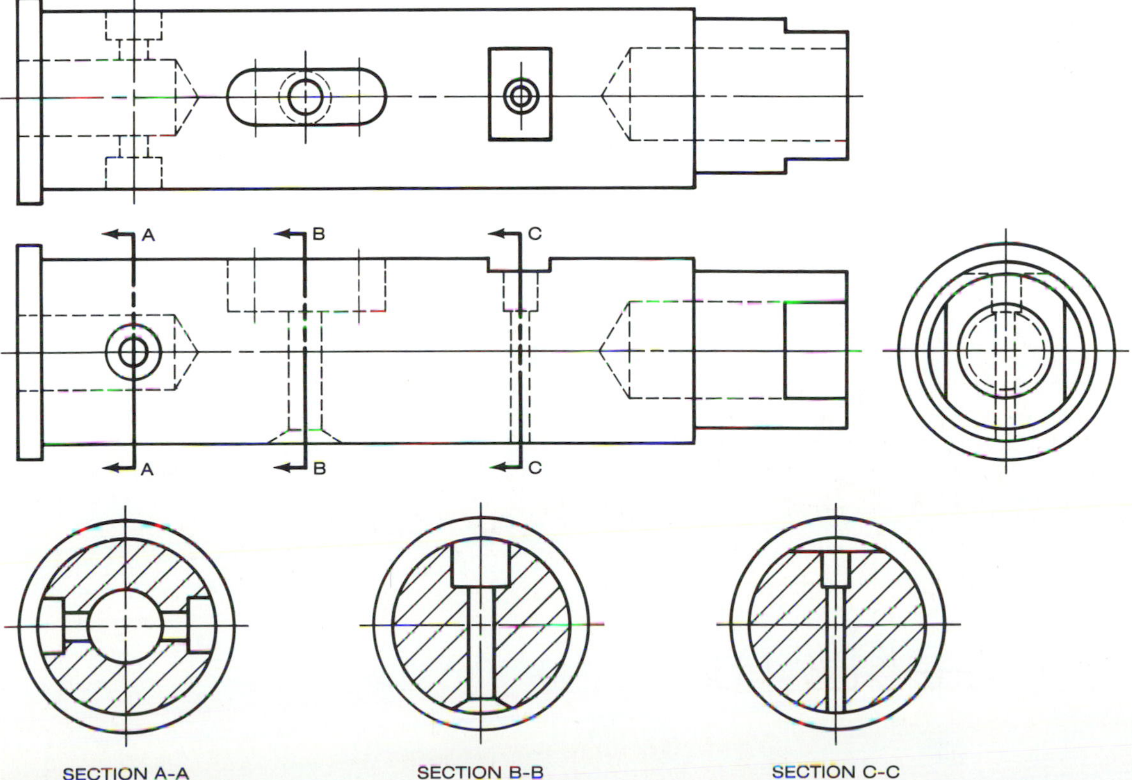

CHECKPOINT 4
REVOLVED AND REMOVED SECTIONS

Using a sheet of plain 8½ × 11 inch paper, sketch or draw the revolved sections indicated in Figure 7-27(a). Omit the cutting-plane line but show the center line of revolution. Assume that each grid equals ¼ inch or about 6 mm. On another sheet of paper, sketch or draw the views, shown in Figure 7-27(b), with the indicated section.

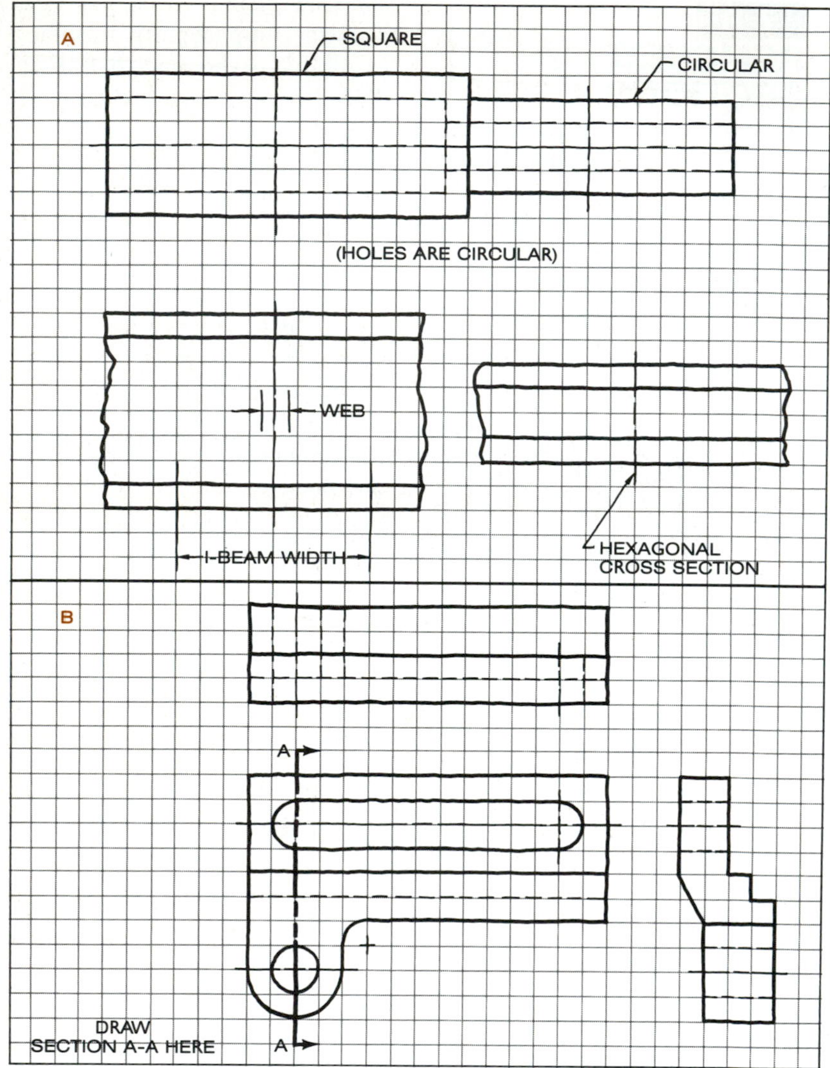

FIGURE 7-27 Checkpoint 4 problems

7.9 AUXILIARY SECTIONS

A sectional view appearing in other than a principal view is an *auxiliary section*. The auxiliary section is projected into a position on the drawing so that the line of sight for that view is perpendicular to the cutting plane. An example of an auxiliary section is shown in Figure 7-28.

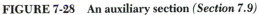

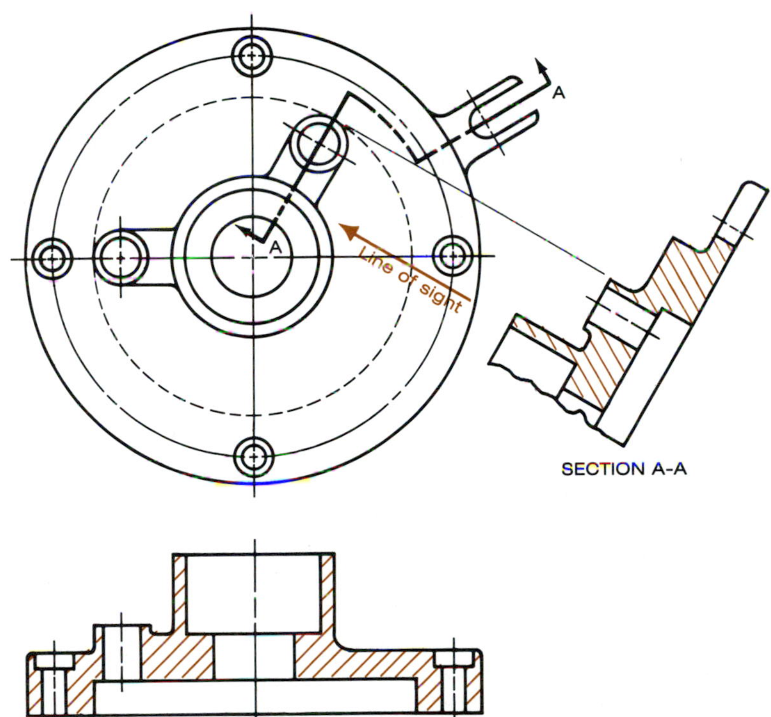

FIGURE 7-28 An auxiliary section *(Section 7.9)*

7.10 BROKEN-OUT SECTIONS

Sections may be used to show only a *portion* of the interior of the object without sacrificing important exterior features. As shown in the examples in Figure 7-29, a freehand break line limits the sectioned area. Such *broken-out sections* are time savers in that they eliminate the necessity for sectioning entire views of parts.

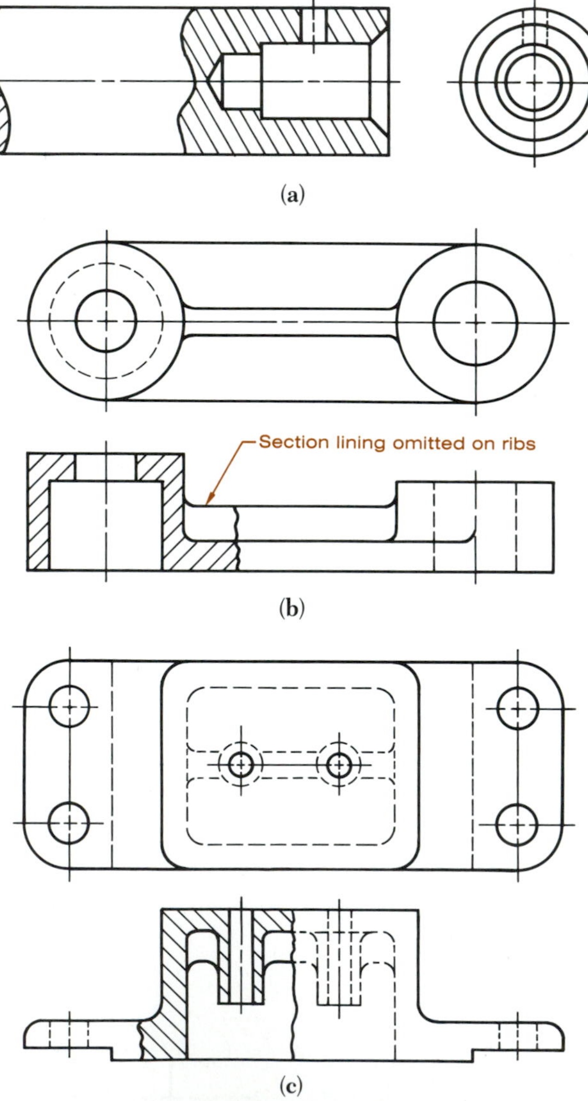

FIGURE 7-29 Broken-out sections *(Section 7.10)*

7.11 MEANING OF CONVENTIONAL REPRESENTATION

A number of widely recognized and accepted practices of abbreviated delineation are called *conventional representation*. Even though conventional representation violates certain rules of basic orthographic projection, it is widely accepted and is considered conventional drawing practice. Practices consisting of approximate or simplified representation result in drawings that are easier to read. Conventional representation generally reduces drafting time. Most companies develop engineering and drafting manuals for their own special needs that illustrate practices applying to conventional representation. The examples of conventional representation that follow are some of the most commonly used relating to sectional views.

7.12 CONVENTIONAL REPRESENTATION SECTIONS

Section Lining

Where the cutting plane passes through the center plane along the length of a rib, web, lug, or other relatively thin element, the section lining is omitted to avoid a false impression of thickness and solidarity. In the example shown in Figure 7-30(a) the cutting plane (not shown) passes through the center plane of the vertical rib. Note how the incorrect sectional view, showing the rib with section lines, gives the misleading impression that the area is solid. Compare the obvious improvement in the correct view.

In Figure 7-30(b) the cutting plane (not shown) passes through the projecting lugs. Section lining is used on the upper lug because the cutting plane passes through it crosswise, but it is omitted on the lower lug for the same reason that the vertical rib is not sectioned in 7-30(a).

When the cutting plane cuts across, or is perpendicular to, elements such as ribs, webs, lugs, bolts, and spokes, section lining is used as illustrated in Figure 7-31.

FIGURE 7-30 Ribs in sectional views ***(Section 7.12)***

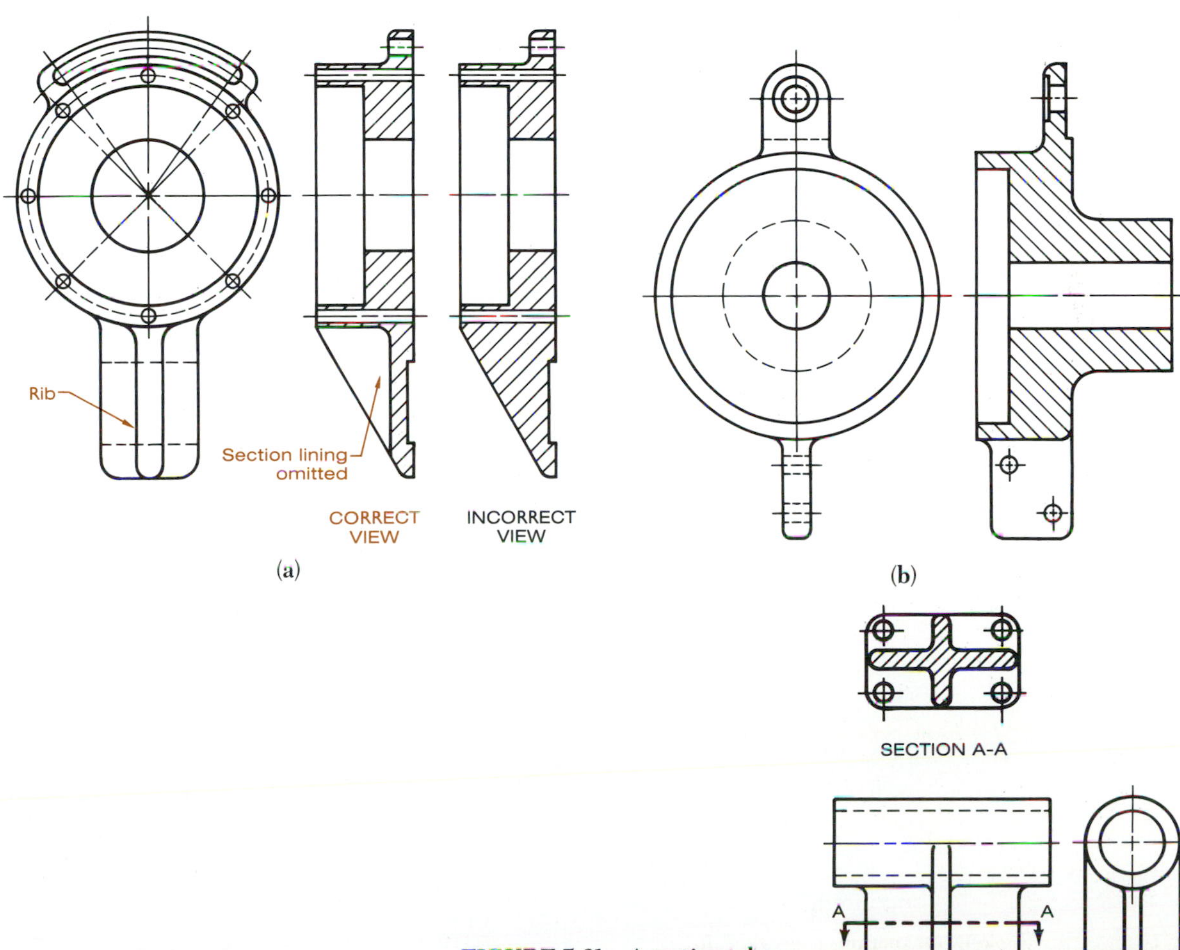

FIGURE 7-31 A section taken across ribs ***(Section 7.12)***

Spokes in Sectional Views

Where the true projection of an object results in foreshortening or in unnecessary drafting time, or in both, inclined elements such as lugs, ribs, spokes, and arms are rotated into a plane perpendicular to the line of sight of the sectional view. Alternatively, they may be entirely omitted. In Figure 7-32 the true projection is misleading. The actual spacing of the spokes is obvious in the circular front view. Section lining is omitted on the spokes.

Partial Views

To save space or drafting time and to simplify the drawing, partial views in connection with sectioning may be used for symmetrical objects. Figure 7-33 shows two examples of objects drawn with partial views. Note in both examples that the nearest half of the object with respect to the sectional view is removed.

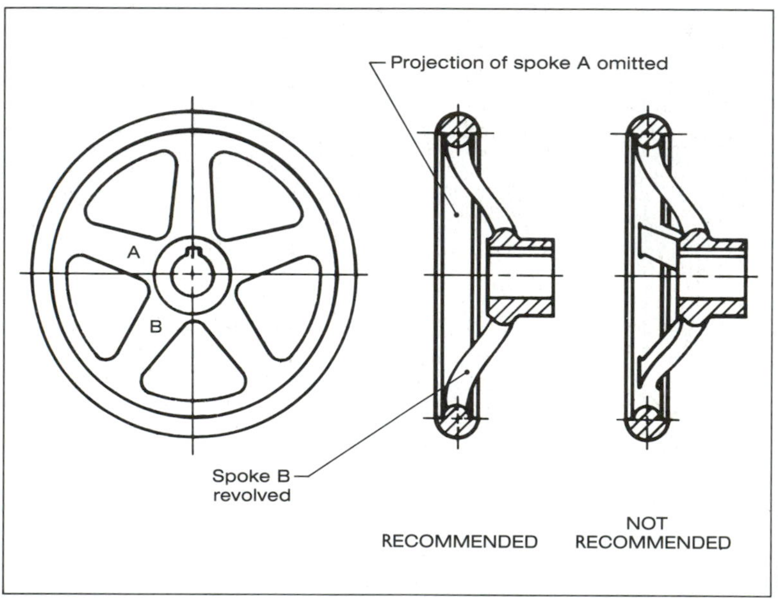

FIGURE 7-32 Spokes in section *(Section 7.12)*

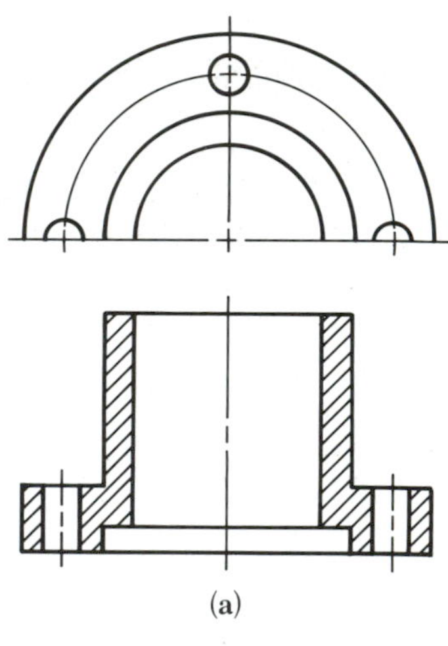

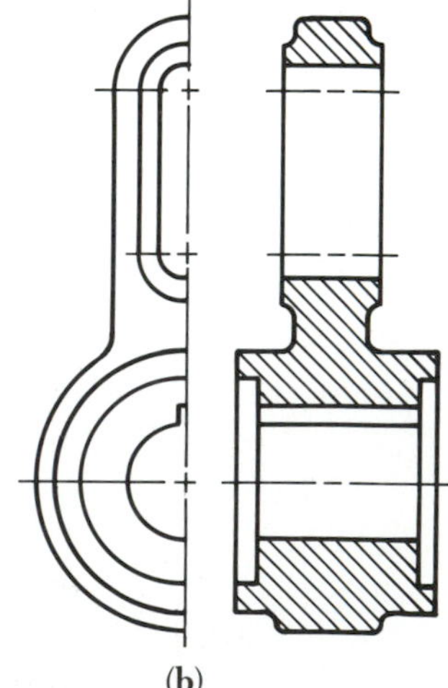

FIGURE 7-33 Partial views *(Section 7.12)*

Rotated Features

When an odd number of holes, slots, ribs, lugs, and other such features are spaced around a bolt circle or on a cylindrical flange, they are rotated to their true distance from the center axis. Figure 7-34 shows two examples of this convention.

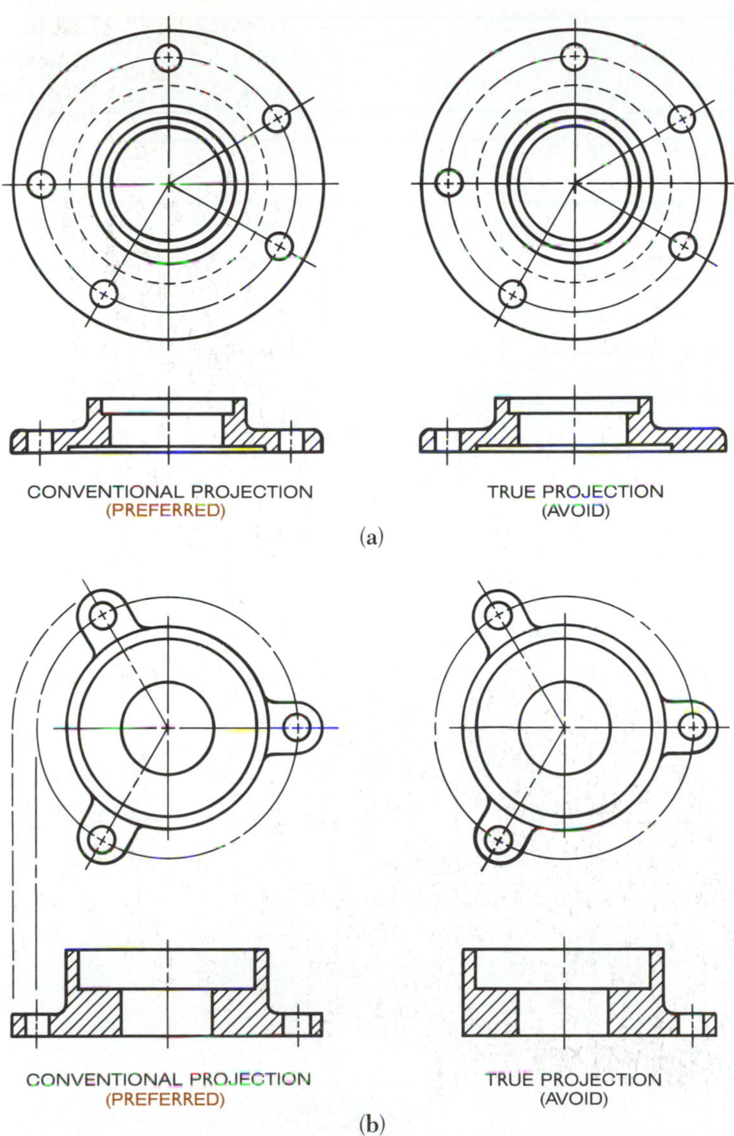

FIGURE 7-34 Rotated features ***(Section 7.12)***

Breaks

Views of long objects with a uniform cross section can be shortened on a drawing by using breaks, as shown in Figure 7-35. The dimension indicates the true length of the object. The long break line is drawn as a thin, straight line with freehand V intervals. The shorter break lines are drawn freehand. The S-shape break used on round stock may be drawn freehand, or for larger sizes an irregular curve may be used.

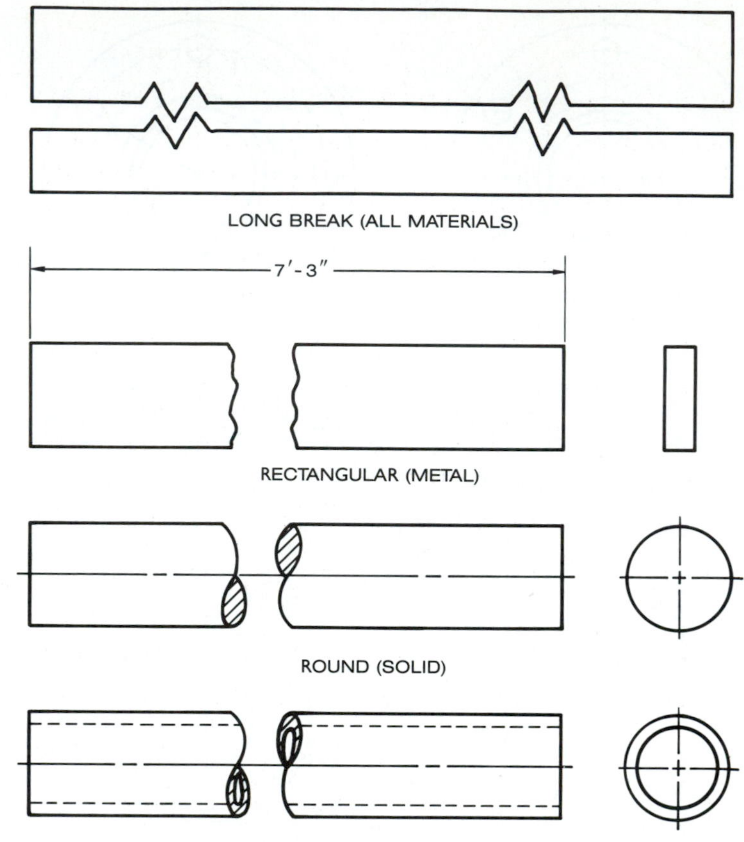

FIGURE 7-35 Conventional breaks ***(Section 7.12)***

PROBLEMS

Sectional views and conventional representation

The following problems will provide practice in applying the principles of sectional views discussed in this chapter. Draw each problem very accurately, with instruments, to the required scale, giving particular attention to achieving high quality linework. Use 5H or 6H grade pencil lead for drawing light construction lines, section lining, and center lines. Use 2H or 3H grade pencil lead for the final object lines.

For each problem, use layout A, shown on the inside of the front cover. Fill out the necessary information in the title block and neatly balance each problem within the available space. Assume that each grid equals ¼ inch or 6 mm. Draw the indicated sectional views and all other views that you consider necessary to completely describe the object.

P7.1 Draw the right-side view as a full section.

P7.2–P7.8 Draw the front view as a full section.

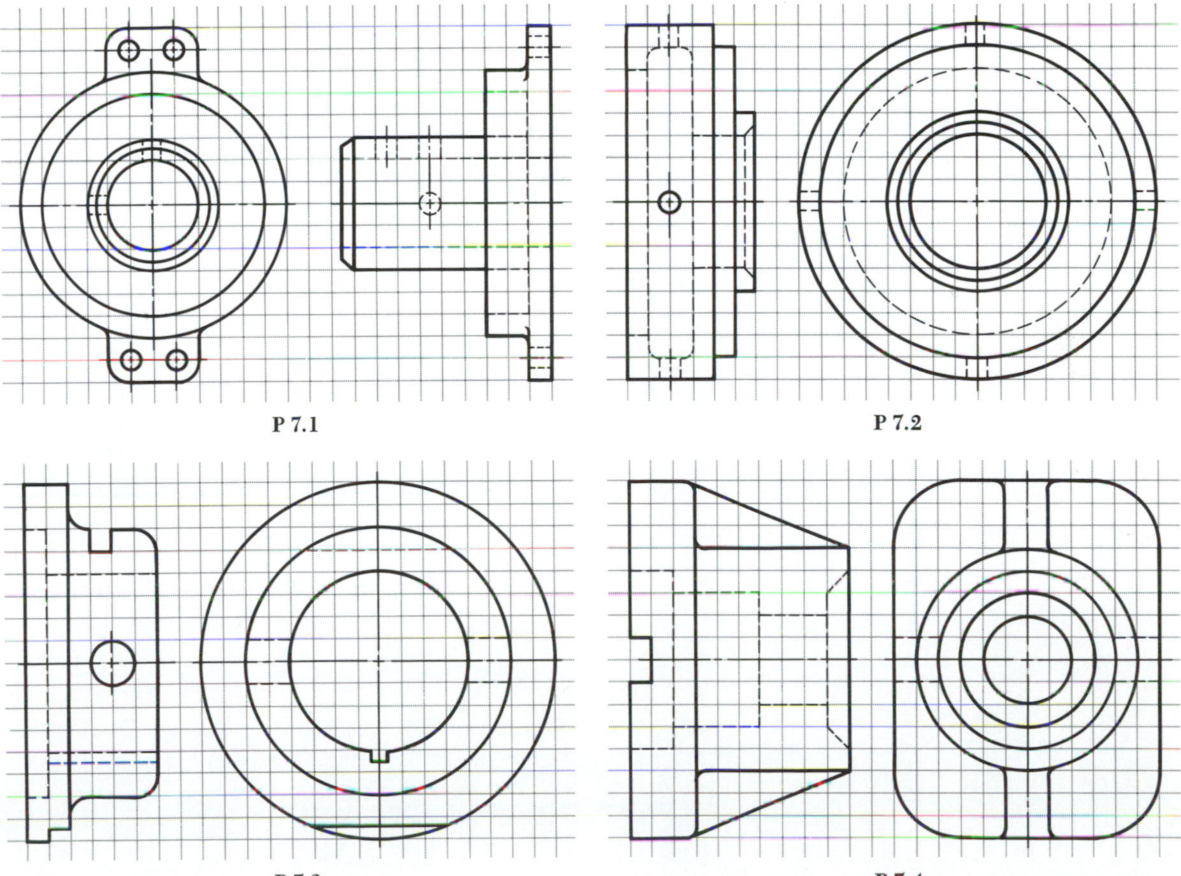

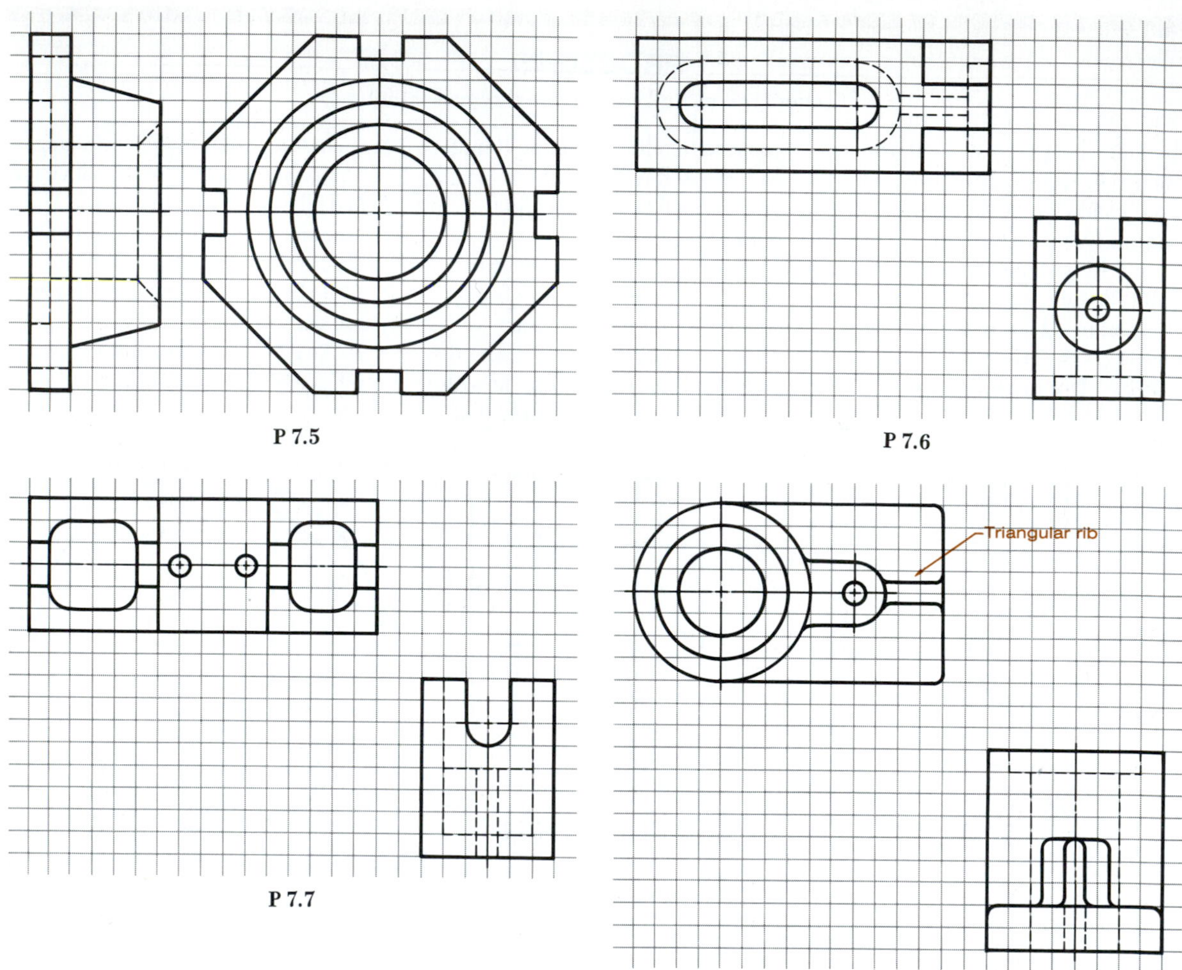

P 7.5

P 7.6

P 7.7

P 7.8

P7.9 Draw the right-side view as a full section.

P7.10 Draw the top view as a full section.

P7.11–P7.18 Draw the appropriate view as a half-section.

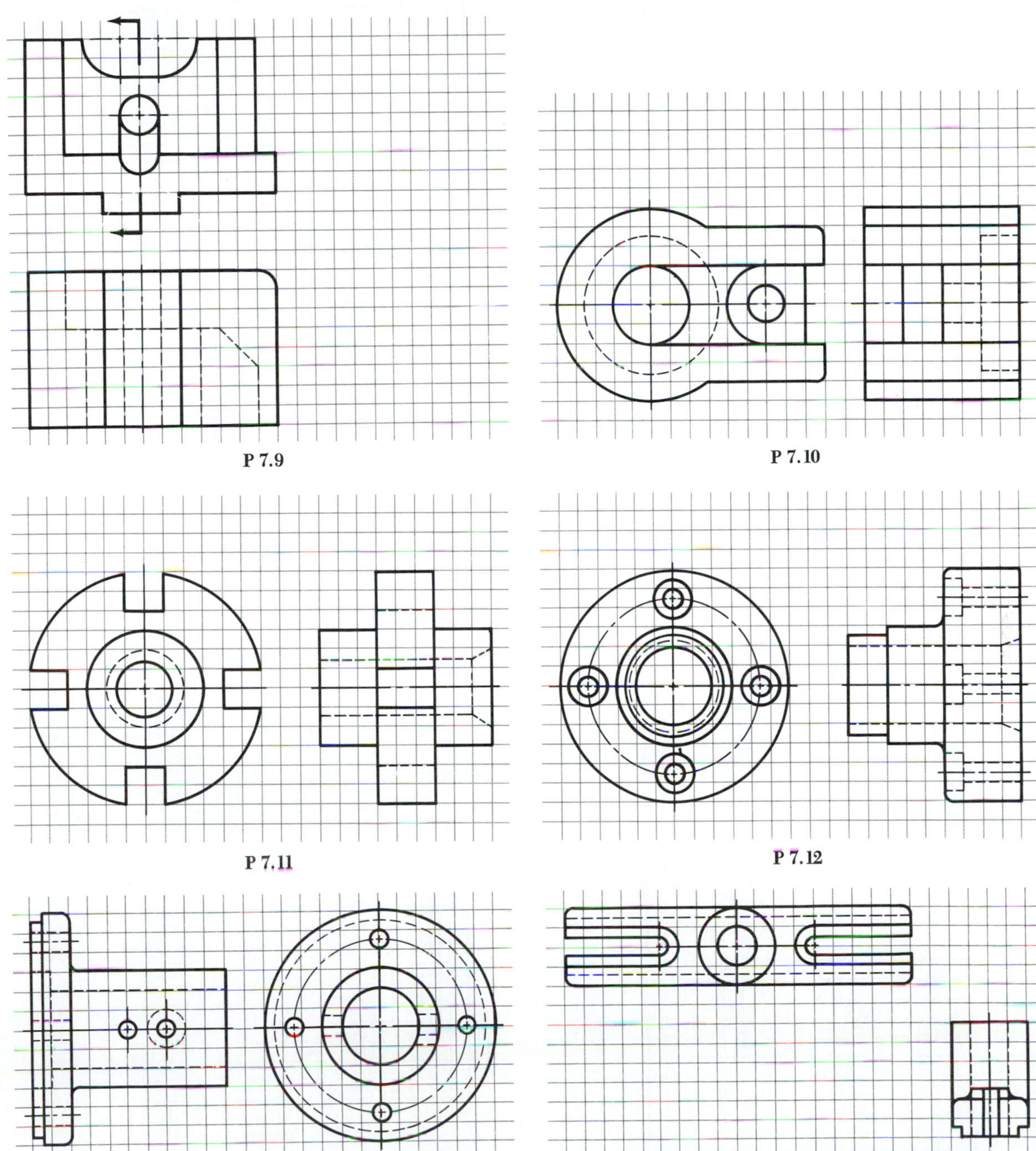

P 7.9

P 7.10

P 7.11

P 7.12

P 7.13

P 7.14

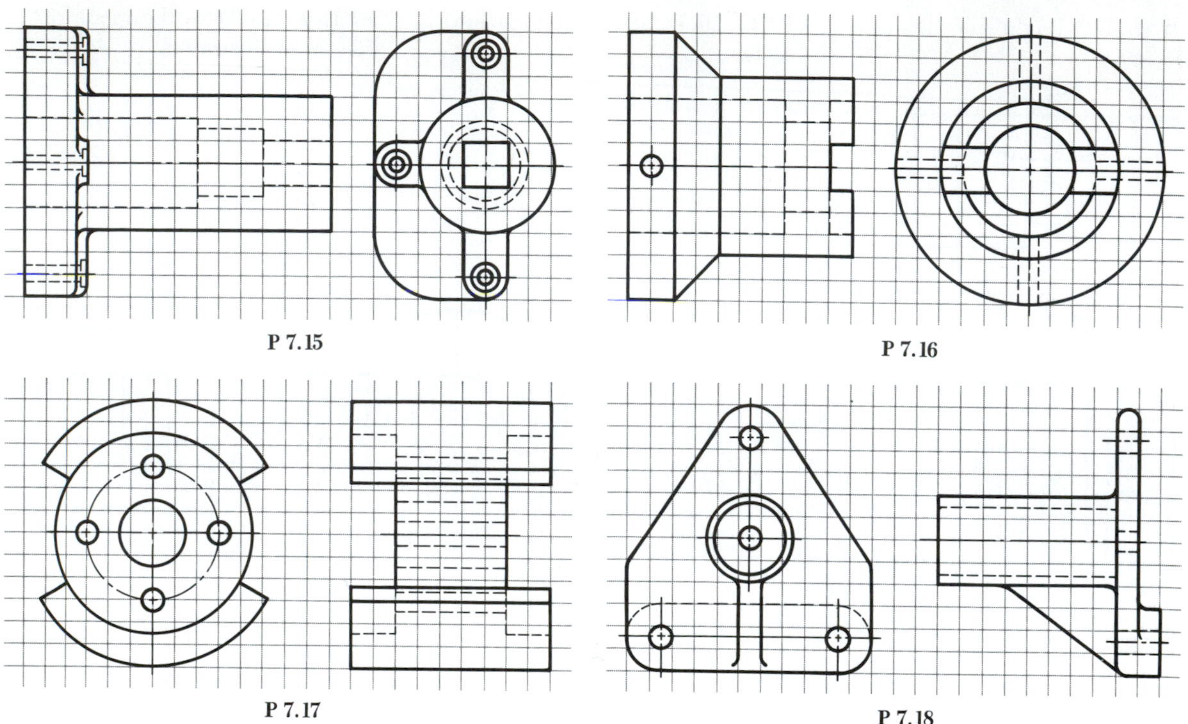

P 7.15

P 7.16

P 7.17

P 7.18

P7.19–P7.25 Draw the appropriate view as an offset section.

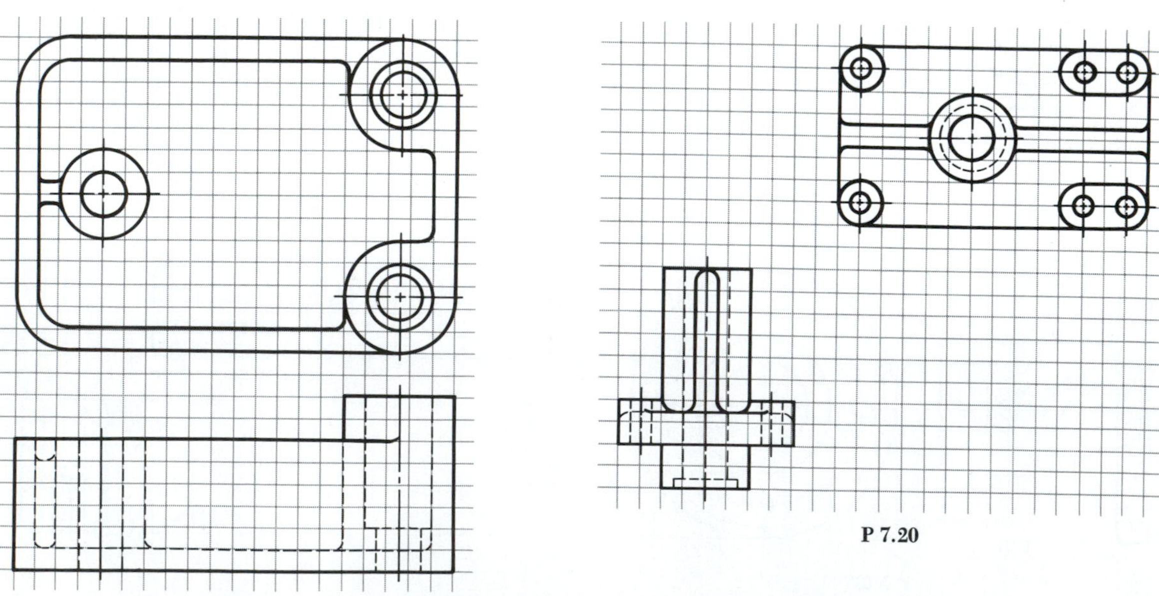

P 7.19

P 7.20

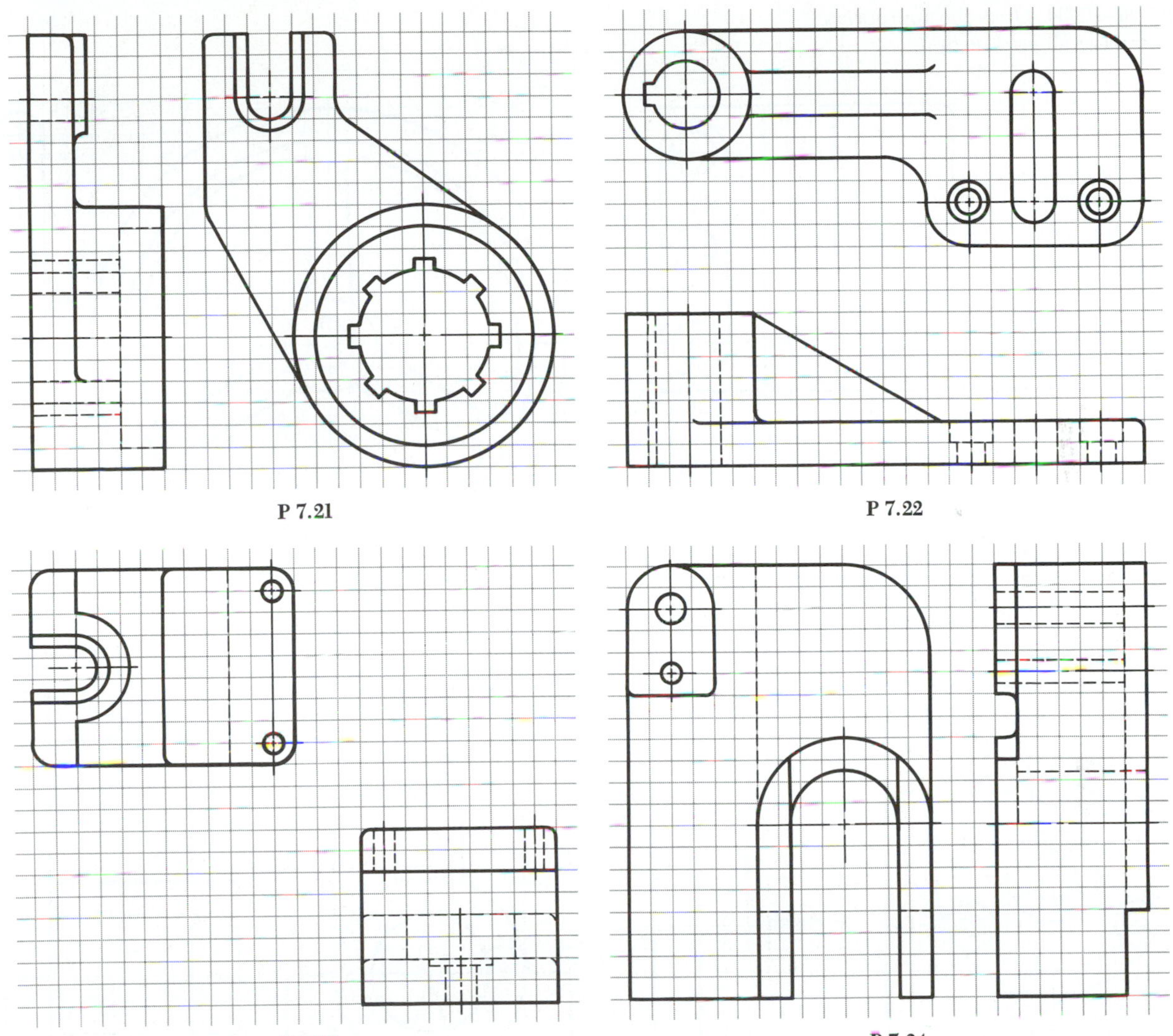

P 7.21

P 7.22

P 7.23

P 7.24

P7.26–P7.34 Draw the appropriate view as an aligned section.

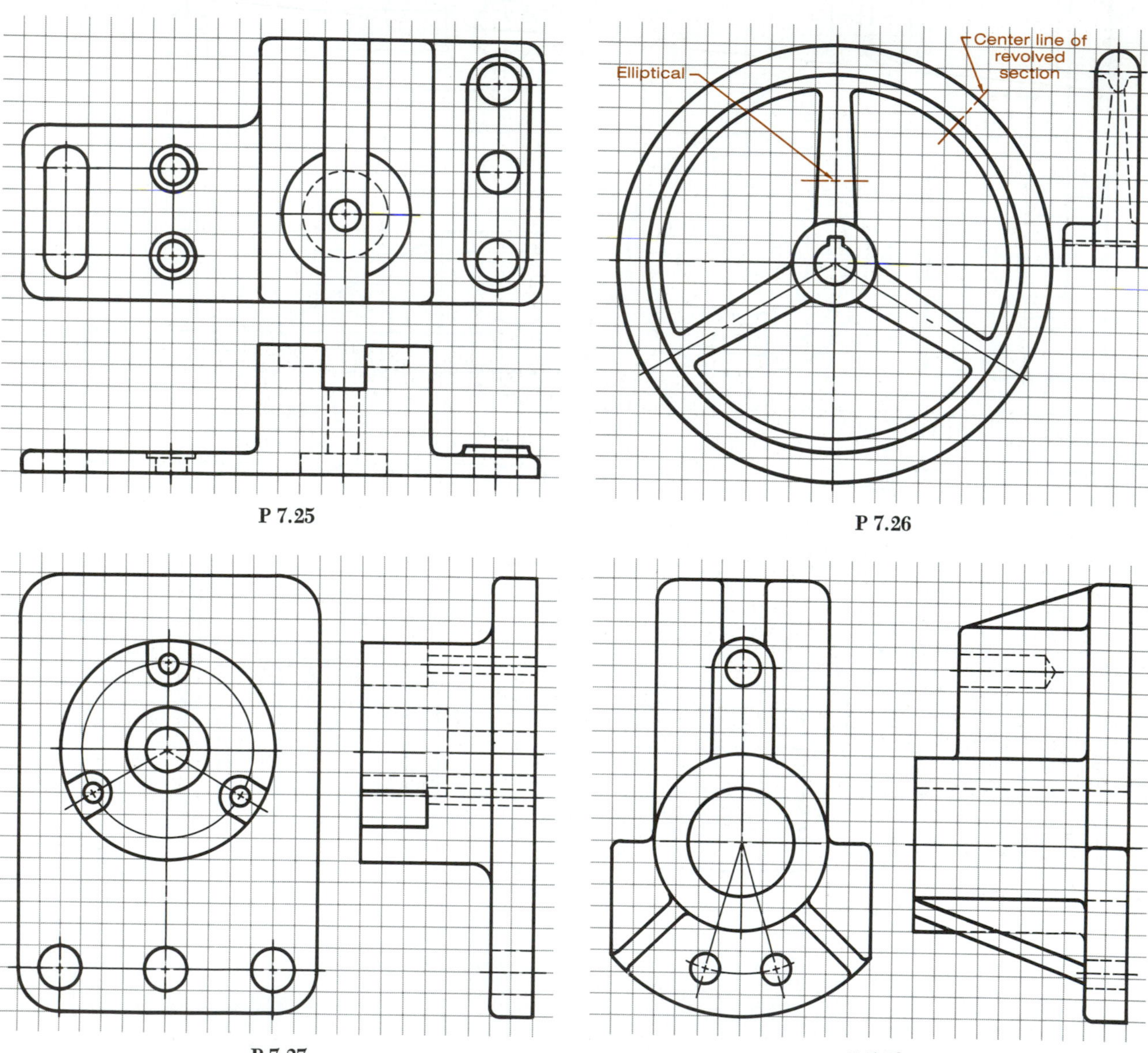

P 7.25

P 7.26

P 7.27

P 7.28

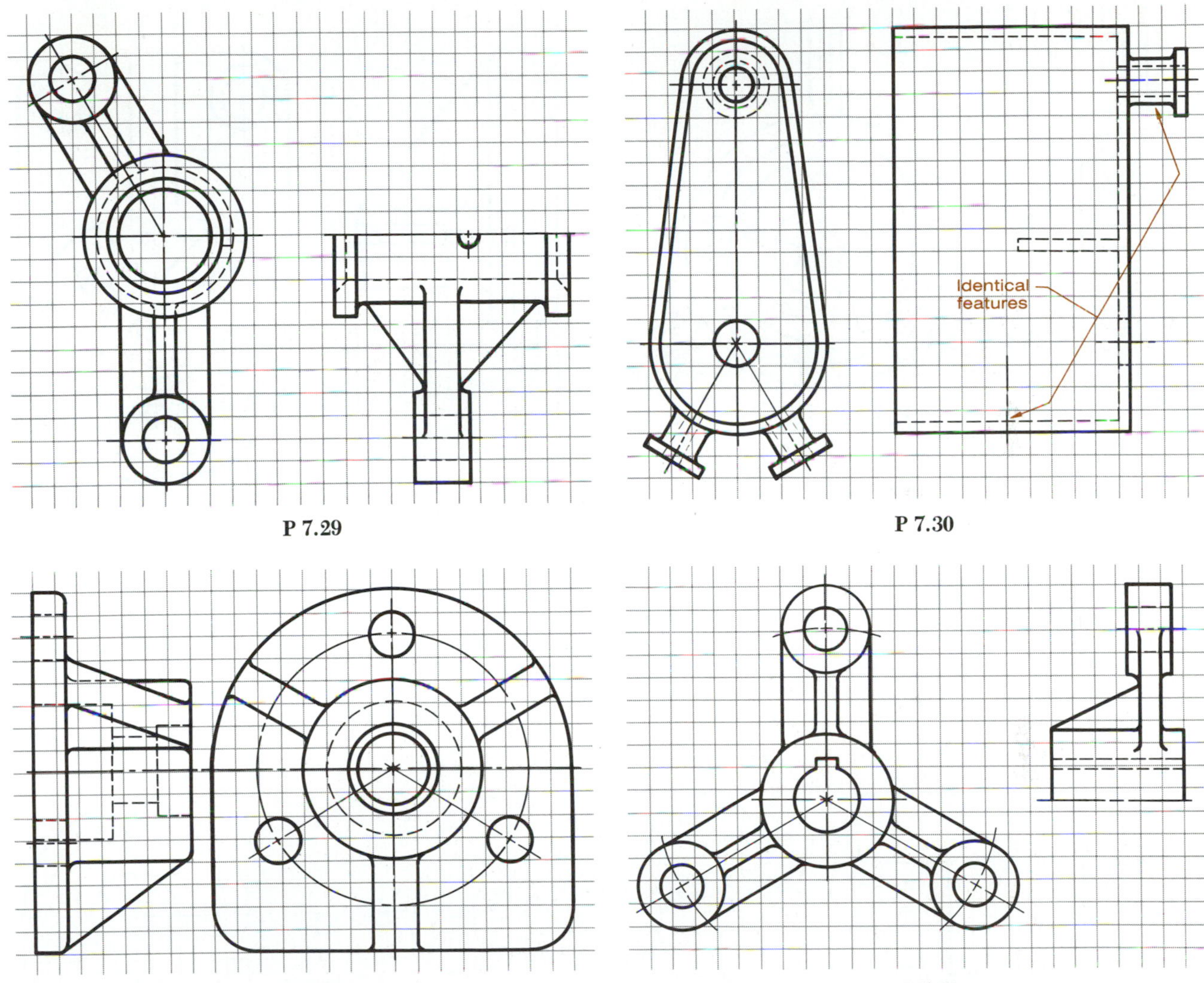

P 7.29

P 7.30

P 7.31

P 7.32

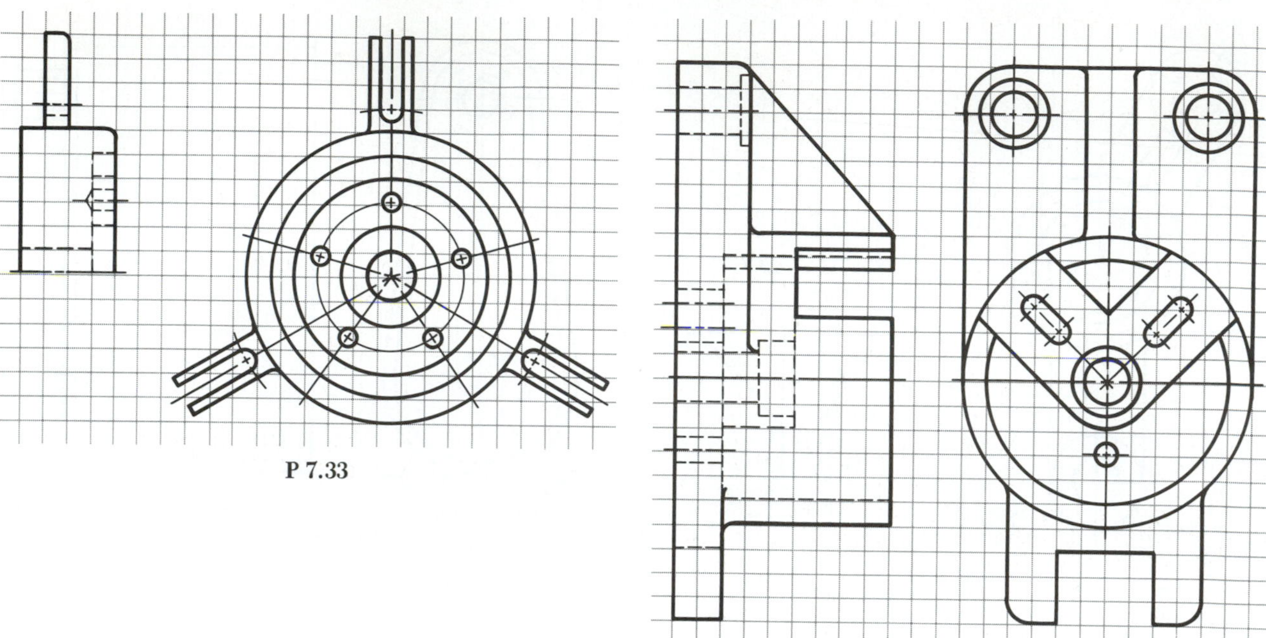

P 7.33

P 7.34

P7.35 Complete the broken-out section in the right-side view and draw the removed section A-A in the position indicated.

P7.36–P7.42 Complete the broken-out sections in the indicated view.

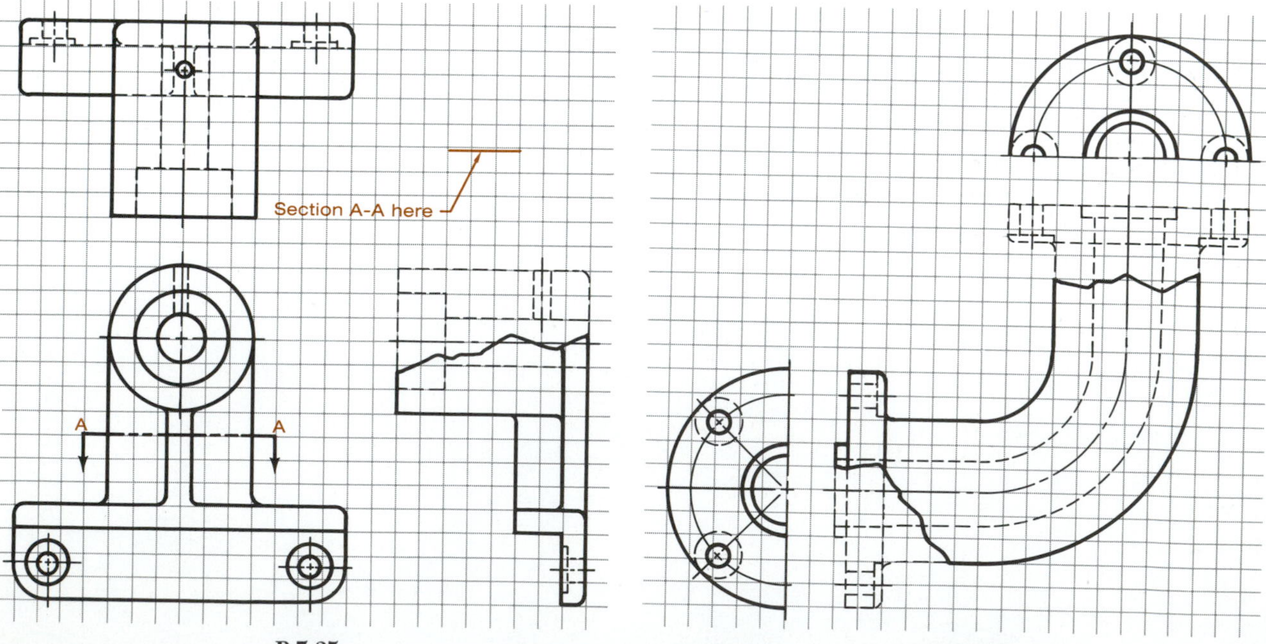

P 7.35

P 7.36

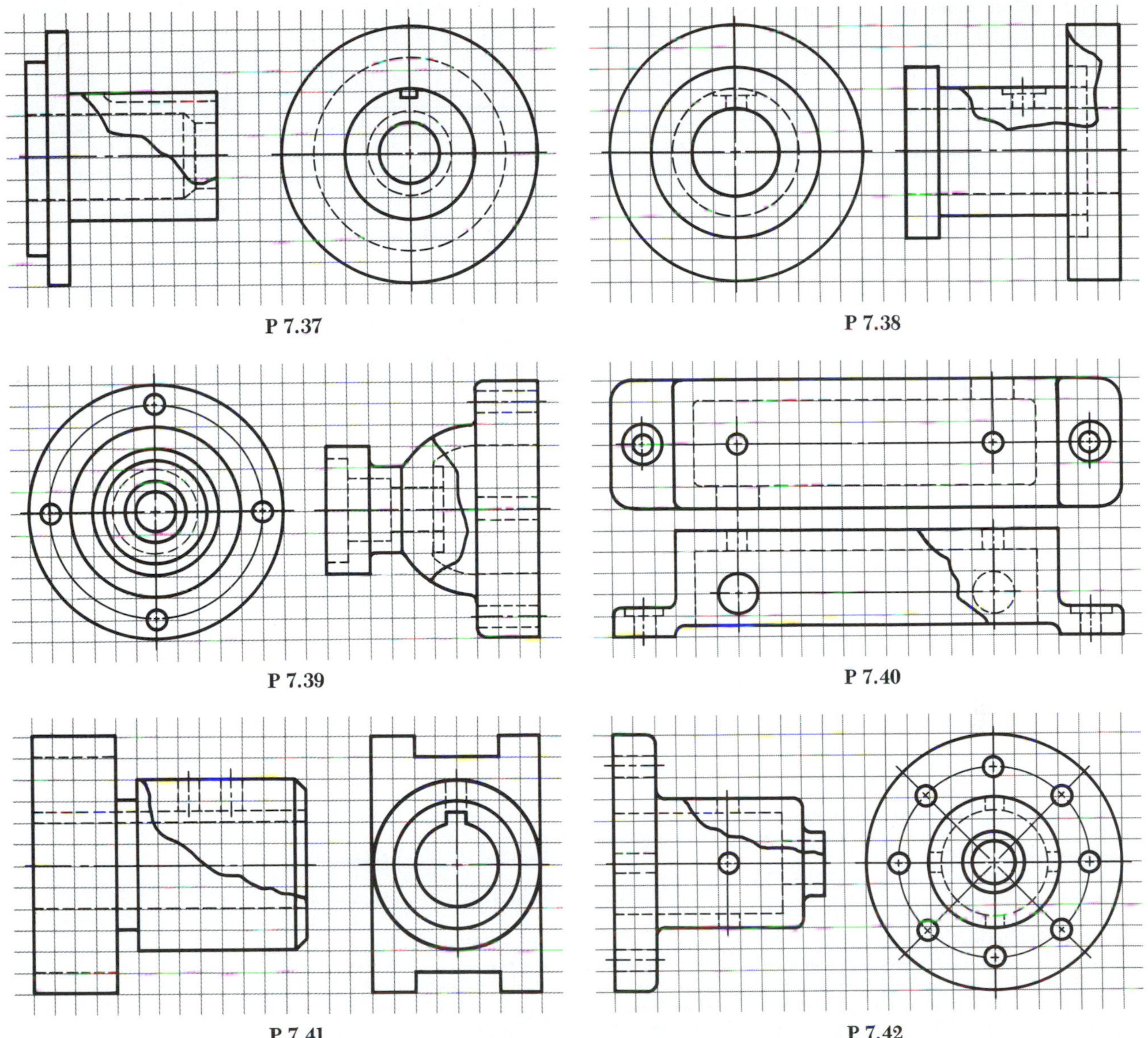

P 7.37

P 7.38

P 7.39

P 7.40

P 7.41

P 7.42

P7.43–P7.46 Complete the revolved sections as indicated.

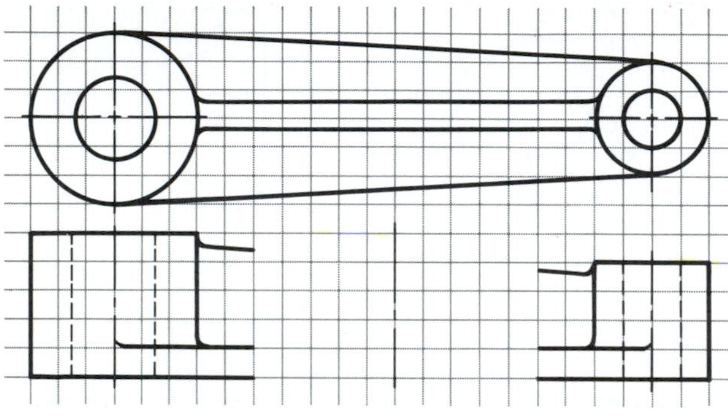

P 7.43

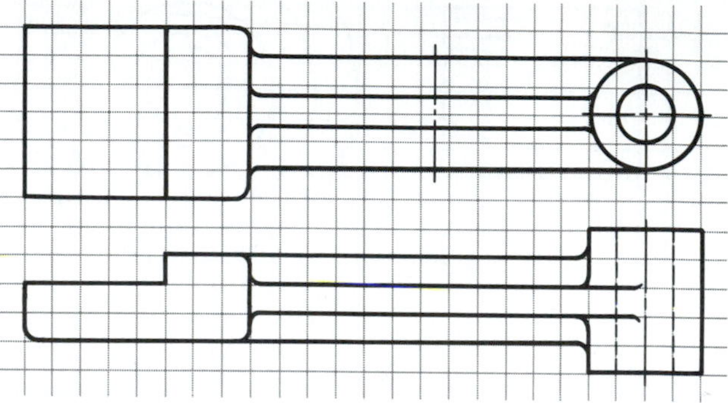

P 7.44

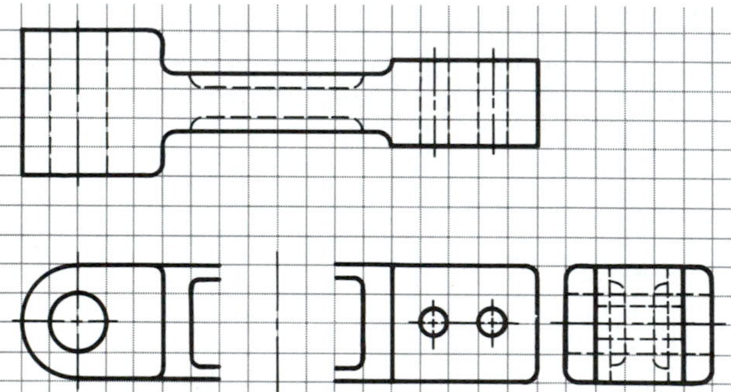

P 7.45

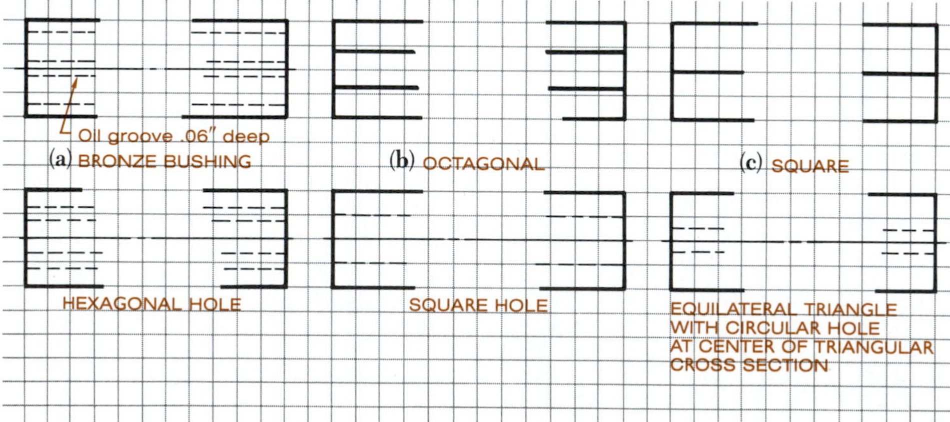

P 7.46

P7.47–P7.49 Draw the removed sections in the positions indicated.

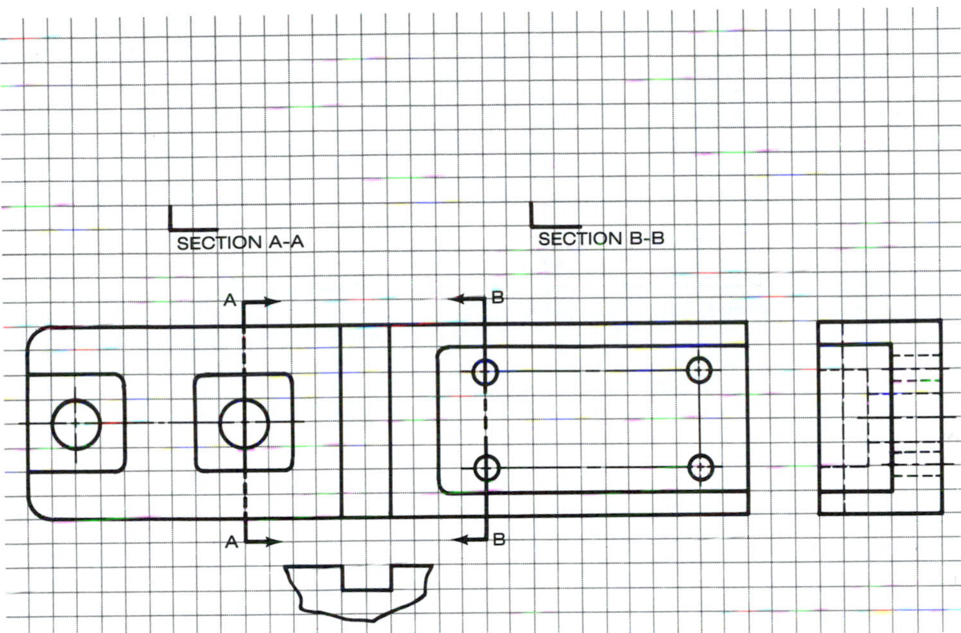

P 7.47

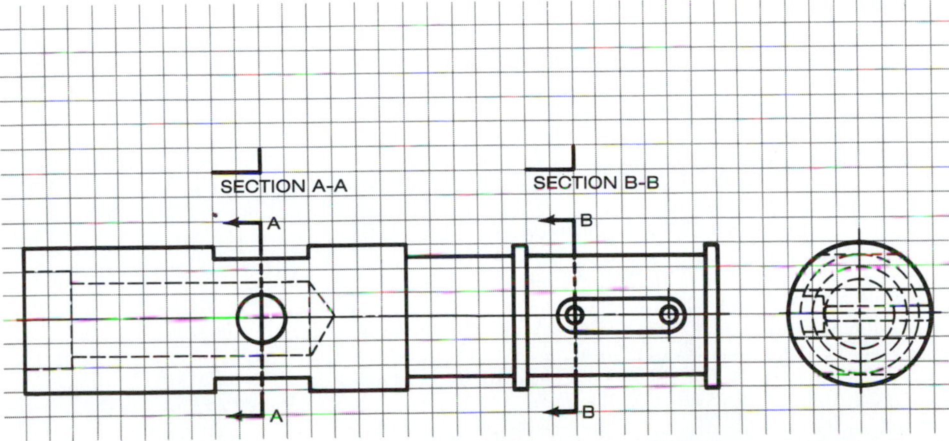

P 7.48

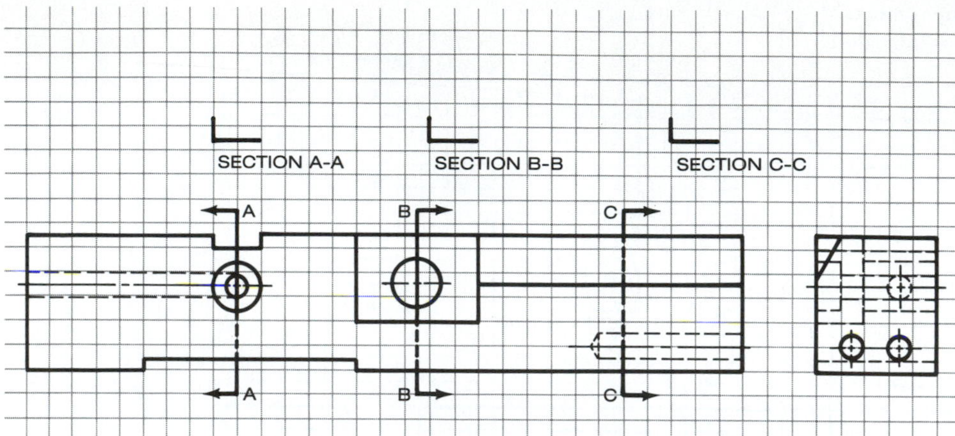

P 7.49

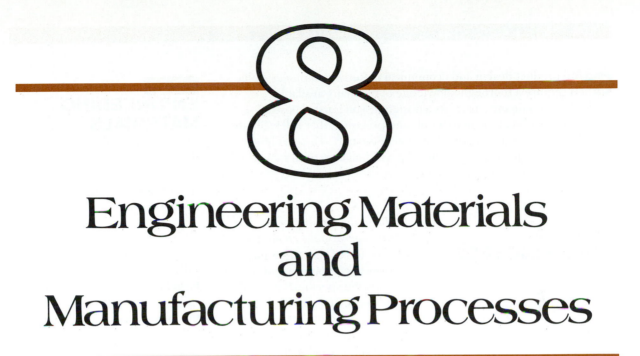

Engineering Materials and Manufacturing Processes

OBJECTIVES

After completing this chapter you should have gained the following abilities:

1. To understand the basic difference between a metal and an alloy.

2. To understand the various numbering systems commonly used in designating metals and alloys.

3. To know how to distinguish between ferrous and nonferrous metals.

4. To appreciate the influence that mechanical properties, a metal, or an alloy can have upon the design of a machine part.

5. To understand the basic differences between primary manufacturing processes and secondary manufacturing processes.

6. To recognize the advantages and limitations of machining operations that are performed on conventional and numerically controlled machine tools.

7. To be aware of the factors that must be considered before assigning the tolerances on a machine part.

8. To understand the nature of various surface finishing processes.

9. To know when and how to designate surface finish control and to understand the symbolism and related technology.

10. To know the advantages and disadvantages of various computer-aided manufacturing technologies.

Sound product design and economical material and process selection require a thorough knowledge of the many available methods of producing a part. As a designer, you must choose an acceptable combination of materials and processes that will facilitate the fulfillment of the intended function of the part. The part function must meet the necessary critical performance requirements after minimum manufacturing costs.

8.1 GENERAL GUIDELINES FOR DESIGN

The achievement of sound design is dependent upon your ability to apply certain basic rules:

1. *Keep the functional and physical characteristics of each part as simple and sturdy as possible.* Avoid unnecessarily complex shapes. Complicated part shapes usually increase production costs.
2. *Design for the most economical production method.* Impressive cost reductions can often be realized by limiting the ratio of the original stock weight to the weight of the final machined part. An economical production method usually results in significant savings in machining time and in subsequent scrap costs.
3. *Keep the number of machining operations to a minimum.* The fewer operations necessary to complete a part to drawing specifications, the lower the costs. Needless fancy or nonfunctional shapes requiring extra machining operations and material should be omitted from the design.
4. *Specify finish and accuracy no greater than actually needed.* The specification of an unreasonable degree of surface finish and tolerance always results in increased production costs.
5. *Select appropriate materials.* When you select materials, always keep in mind the intended production method as well as the functional design requirements. Pay special attention to selecting *free machining* materials. If you must specify a difficult-to-machine material for a given part, the part should not be cut from stock. Instead, consider ways of producing the part by some other process, where machining difficulty is not a factor. Parts processed by investment casting or powder metallurgy, for example, require little, and sometimes no, machining.

8.2 ENGINEERING MATERIALS

Most mechanical devices are mainly composed of metals and alloys because these materials provide an optimum balance of strength, toughness, dimensional stability, hardness, resistance to wear, appearance, and economy. When designing, you must select a material that will perform for a reasonable length of time and one that can be had at the lowest cost.

8.3 METALS AND ALLOYS

Metals belong to the class of chemical elements that includes iron, gold, aluminum, and so on. Pure metals are too soft and too weak to be used successfully for machine parts. An *alloy* (such as steel, cast iron, brass, or bronze) is a mixture of metals and other chemical elements. Alloys are substances with metallic properties. They are composed of two or more elements that are used to obtain certain desirable qualities not obtainable from pure metals. *Metallurgy* is the art and science of separating metals from their ores and preparing them for use. The metallurgist studies the highly complex behavior characteristics of metals and alloys. Most product designers work closely with metallurgists in selecting optimum metals and alloys for machine parts.

8.4 FERROUS METALS

Ferrous metals consist principally of iron. There are more engineering applications for ferrous metals than for all nonferrous metals combined.

Iron is a material that contains an excess of 2% carbon and that may be cast into any desired shape. It is used for a wide variety of cast machine parts (gears, cams, pulleys, levers, links) and for the great majority of machine frames and bases. Compared to steel, cast iron is a relatively inexpensive metal. In addition, iron responds to heat treatment in a way similar to steel in that it may be annealed, or hardened. Iron is a hard, strong metal with good wearing qualities.

There are three main types of iron: *gray iron*, *malleable iron*, and *ductile* (or nodular) *iron*. Gray cast iron is relatively brittle and lacks ductility and malleability. It has good machining characteristics and is used chiefly for parts that do not require toughness or high tensile strength. Typical product applications for cast iron include machine bases, pistons, engine blocks, and various machinery parts. Generally, gray cast iron is used for the lowest cost parts. Malleable iron castings have many of the tough characteristics of steel. Because they possess good ductility, strength, and machinability, malleable iron castings are used on plows, tractors, and earth-moving equipment. Ductile iron castings exhibit a wide variation in properties and uses. Basically, ductile cast iron is tough, shock resistant, high in tensile strength, and easily machined. Of all iron types, ductile iron results in the highest quality casting.

Steel is an alloy composed of iron, carbon (usually less than 1.7%), and a combination of other materials. The carbon content regulates the properties of a given type of steel and has a great effect on the final properties of the metal. Steels are classified into two broad types and these, in turn, into three types: *carbon steels*, consisting of low-, medium-, and high-carbon steels, and *alloy steels*, consisting of construction, tool, and special alloy steels.

8.5 NUMBERING SYSTEMS

AISI-SAE* System for Steel

Thousands of standard steel alloys are available, each of which has unique properties, making them the best choice for certain applications. In order for different manufacturers to produce the same alloys, it was necessary to standardize the alloy compositions. The *four-digit* system, shown in Table 8-1, is widely used. The first digit in the steel designation 1125, for example, signifies a resulphurized (free cutting) carbon steel. The second digit (1) gives the percentage content of the material represented by the first digit, or 1%. The last two digits represent the percentage of carbon in the alloy (0.25%). For the alloy steel 4440, for example, the first digit (4) indicates a molybdenum steel. The second digit indicates the approximate percentage of the principal alloying elements, (4% nickel). The last two digits indicate a carbon content of 0.40%. (Also shown in Table 8-1 are the general categories of uses for the various steel types.)

Many companies use a more complete specification, AISI (or SAE) C1040 steel, for example, to specify a material. The codes AISI or SAE may be used to indicate the origin of the numbering system. The prefix letter B, C, or E denotes the process that was used to make the steel. B signifies the Bessemer process, C is the open hearth process, and E represents high quality steel made in an electric furnace.

*AISI: American Iron and Steel Institute.
SAE: Society of Automotive Engineers.

TABLE 8-1 AISI-SAE System for Designating Steel Types and General Uses

Type	*Number*	*Uses*
Carbon steels		
Plane carbon	10XX	Fan blades, clutch fingers, chains, rivets, crane hooks, shafting, pressed-steel products
Resulphurized (free-cutting)	11XX	Nuts, bolts, studs, axles, gear blanks, machine parts
Manganese steels	13XX	Gears, cams, shafts
Nickel steels		Crankshafts, connecting rods, automobile parts, axles, structural shapes
0.50% nickel	20XX	
1.50% nickel	21XX	
3.50% nickel	23XX	
5.00% nickel	25XX	
Nickel-chromium steels		Ring gears, pinions, shafts, transmission gearing, forgings, piston pins, studs, screws
1.25% nickel, 0.65% chromium	31XX	
1.75% nickel, 1.00% chromium	32XX	
3.50% nickel, 1.57% chromium	33XX	
Corrosion- and heat-resisting steels	303XX	
Molybdenum steels		Axles, steering mechanism parts, aircraft forgings, transmission shafts, gears, cams, machine parts
Chromium	41XX	
Chromium-nickel (stainless steel)	43XX	Containers for food processing, dental and medical surgical instruments
Nickel	46XX and 48XX	
Chromium steels		Gears, propeller shafts, connecting rods, ball and roller bearing parts, thrust washers, springs
Low-chromium	50XX	
Medium-chromium	511XX	
High-chromium	521XX	
Chromium-vanadium steels		Shafts, transmission gearing, piston rods, dies, punches, rock drills
Tungsten steels	7XXX	
Triple alloy steels	8XXX	
Silicon-manganese steels	9XXX	Automobile springs

Unified System for Metals and Alloys

The unified numbering system (UNS) for metals and alloys is intended to provide a single, comprehensive designation system for all metals. The UNS system consists of 16 series of numbers; each series identifies certain metals and alloys. The UNS numbers are made up of a single letter followed by a five-digit number. Table 8-2 lists UNS categories and corresponding designations.

Whenever possible, the UNS uses numbers from other systems. For example, the aluminum alloy designated by 1100 by the Aluminum Association is assigned the UNS number A91100, carbon steel designated 1090 by AISI is assigned the UNS number G10900, and so on.

TABLE 8.2 Unified Numbering System

Type	*Number*
Aluminum and Aluminum Alloys	Axxxxx
Copper and Copper Alloys	Cxxxxx
Rare Earth and Rare Earthlike Metals and Alloys	Exxxxx
Cast Irons	Fxxxxx
AISI and SAE Carbon and Alloy Steels	Gxxxxx
AISI and SAE H-Steels	Hxxxxx
Cast Steels (except tool steels)	Jxxxxx
Miscellaneous Steels and Ferrous Alloys	Kxxxxx
Low-Melting Metals and Alloys	Lxxxxx
Miscellaneous Nonferrous Metals and Alloys	Mxxxxx
Nickel and Nickel Alloys	Nxxxxx
Precious Metals and Alloys	Pxxxxx
Reactive and Refractory Metals and Alloys	Rxxxxx
Heat- and Corrosion-Resistant Stainless Steels	Sxxxxx
Tool Steels, Wrought and Cast	Txxxxx
Zinc and Zinc Alloys	Zxxxxx

8.6 NONFERROUS METALS

Nonferrous metals do not contain iron except in small amounts as impurities and are seldom found in a pure state. They offer a wide variety of characteristics and mechanical properties. Nonferrous metals commonly specified for engineering applications include aluminum, magnesium, copper-base alloys like brass and bronze, zinc-base alloys, and nickel-base alloys.

Aluminum

Aluminum is chiefly used for products because of its lightweight characteristics and corrosion-resistant properties. Aluminum can be readily machined, forged, cast, spun, welded, and extruded. It has a wide range of applications in the aircraft industry and as a structural material for home and office buildings. For many industrial applications it is competitive with steel. Aluminum is usually alloyed with copper, manganese, or nickel.

The numbering system for identifying the types of aluminum is shown in Table 8-3. Four digits are used. The first digit indicates the principal alloying element; the second digit designates the degree of control exercised over impurities, with 0 indicating no control; and the last two digits indicate the aluminum purity. A designation of 1025, for example, signifies that the metal is aluminum with 99% or greater purity. The second digit (0) indicates there is no control over impurities, and the third and fourth digits (25) indicate that the purity is 99.25%. Table 8-4 illustrates temper designations used for aluminum. A designation "0" following the four digits (1100-0, for example) indicates the material is annealed. A temper designation consisting of a combination of a capital letter and a number, such as T3, may be used after the four-digit designation.

TABLE 8-3 Aluminum Numbering System

Composition	*Alloy number*
Aluminum (99% pure)	1XXX
Aluminum Alloys	
Copper	2XXX
Manganese	3XXX
Silicon	4XXX
Magnesium	5XXX
Magnesium and Silicon	6XXX
Zinc	7XXX
Other Elements	8XXX
Unused Series	9XXX

TABLE 8-4 Temper Designations for Aluminum

Designation	*Description*
F	As fabricated
O	Annealed (wrought alloys only)
H	Strain hardened (wrought alloys only)
	H1, + one or more digits, strain-hardened only
	H2, + one or more digits, strain-hardened, then partly annealed
	H3, + one or more digits, strain-hardened and stabilized
W	Solution heat-treated only, unstable temper
T	Thermally treated
	T2, annealed (casting alloys only)
	T3, solution heat-treated and cold-worked
	T4, solution heat-treated
	T5, artificially aged
	T6, solution heat-treated and artificially aged
	T7, solution heat-treated and stabilized
	T8, solution heat-treated, cold-worked, and artificially aged
	T9, solution heat-treated, artificially aged, then cold-worked
	T10, artificially aged and cold-worked

Magnesium

The most striking characteristic of magnesium is its lightness, the weight of a given volume being only 23% that of iron and 64% that of aluminum. The strength of pure magnesium is relatively low; for this reason, it is always used in the alloyed form. Common alloying elements are aluminum, zinc, and manganese. Magnesium alloys have excellent machinability characteristics and may be welded without difficulty.

Copper

Like aluminum, copper is seldom employed industrially in its pure state except for electrical conductors or similar uses. Instead, it has more strength and value when alloyed with other elements to produce brass and bronze, which are known as "copper-base alloys." Additions of lead, tin, aluminum, and manganese are made to improve the properties of copper.

Copper-Base Alloys

BRASS. Brass is essentially an alloy of copper and zinc. A small amount of lead is frequently added to improve the machinability of the alloy. Brass is generally used in engineering applications because of its strength, appearance, ductility, and resistance to corrosion.

BRONZE. Bronze is used in bearings because of its toughness, strength, and wear- and corrosion-resistance. It is also widely used for marine hardware. Bronze is an alloy of copper, tin, manganese, and several other elements, depending upon its ultimate product use.

Zinc-Base Alloys

Zinc-base alloys are widely used in the die-casting industry. These relatively strong, low cost materials can be readily cast with a good surface finish. Aluminum, copper, and magnesium are elements usually alloyed with zinc.

Nickel-Base Alloys

Materials in the large family of nickel-base alloys are used because of their properties of corrosion resistance, malleability, and ductility. They are sometimes used as a protective coating for other metals. The principal alloying elements are copper, aluminum, iron, silicon, and tungsten. Nickel-base alloys are commonly specified for parts in automotive, aircraft, and chemical-processing equipment and for many other product applications.

8.7 PROPERTIES OF METALS AND ALLOYS

As a designer, you must have an extensive knowledge of materials and their properties. Very early in the design process you must choose a material that satisfies the performance criteria commensurate with the part requirements. Since there are thousands of potential materials to be considered, you must make this choice very carefully. Metal parts may be formed by machining from stock, worked or "wrought" by forging, rolling, drawing, extruding, or powder metallurgy. You should know that the properties of a metal or alloy may differ greatly in the cast and wrought forms, and you must know how the properties of a given metal or alloy will affect your intended design.

Material choice is based upon the following guidelines: desired physical properties, end use of the product, shape of the part, availability of suitable materials and labor skills, raw material costs, basic method of forming or processing, ease of manufacturing (machinability), and quantity of required parts.

Strength is the ability to resist an applied stress without fracture (ultimate tensile strength) or permanent deformation (yield strength). Deformation is the disfigurement or changing of shape of a part by external forces. A metal may be very hard and have a high tensile strength and yet be unsuited for a use that requires it to withstand impact or a sudden load. Impact resistance is the ability to withstand shock loading. Shear strength is needed in the design of bolts, pins, and similar parts. The ultimate shear strength of a material is the maximum load it can withstand without rupture when subjected to a shearing action. The shear strength of a material is generally equivalent to about 50% of its tensile strength.

Toughness usually implies *impact* toughness, or the ability to resist impact without fracture. A tough metal is highly shock-resistant. There is no direct method by which the toughness of a metal can be measured. The property of toughness implies a metal that possesses high strength and malleability.

Ductility is the ability of a metal to stretch and deform plastically without fracturing. This property is often expressed in terms of elongation and reduction of area. Ductility is inversely related to strength: as strength increases, ductility decreases.

Malleability refers to the plasticity of a material when it is compressed, such as by rolling or hammering. Heat increases the malleability of thick metals.

Elasticity is the ability of a metal to return to its original shape after being stretched or squeezed.

Hardness is the ability of a metal to resist indentation. Hardness is directly proportional to strength and is related to the wear resistance of a material.

Brittleness is a characteristic limitation of hardened steels and cast irons. Materials that break with little or no plastic deformation are described as being brittle. Brittleness is the opposite of plasticity and ductility.

Machinability is a relative term that indicates the ease with which a metal may be machined. In some cases the choice of a metal is based on economical machining rather than on the strength requirements of the ultimate workpiece. Tables are available in various handbooks that list the *machinability index* of metals. Each metal is rated in terms of percentage points of 100 in comparison with the standard B1112, which is rated at 100%. Numbers in the table range from the very best machining material (those with the highest numbers) to the most difficult. These numbers are based upon an intensive study of the rate of metal removal, quality of the finished surface, and tool life for many materials.

8.8 PRIMARY MANUFACTURING PROCESSES

Primary manufacturing operations refer to methods of originally forming the geometry of a part. These methods may be performed hot or cold on a large or on a small scale. In some cases, because of the fine finish or dimensional accuracy possible in a primary manufacturing process, no subsequent machining operations are necessary. You must always plan the design of a part with the nature and limitations of the primary and secondary manufacturing processes in mind. The design decision on how to form some parts, because of their unique basic requirements, often seems to fall almost automatically to one or perhaps two primary manufacturing processing methods, but in other cases the selection of an appropriate production process that is both practical and economical can be a very difficult decision. Primary manufacturing processes include casting, forging, drawing, extruding, pressworking, powder metallurgy, machining from solid stock, and welding.

8.9 CASTING METHODS

Casting is the process of filling a mold or cavity by gravity or pressure with a liquid metal and allowing it to cool or solidify. There is no other metalworking process that gives the designer such unlimited freedom or as many options in terms of required product complexity, size range, choice of available materials, desirable properties, range of surface texture, close dimensional tolerances, and high-volume production.

Several different commercially important casting processes have been developed to fill specific manufacturing needs, each allowing potential advantages. The major casting methods employed for making machine parts include sand casting, plaster mold casting, shell molding, investment casting, permanent mold casting, and die casting.

Sand Casting

Sand casting is the most common of all casting methods. The process consists of making a cavity in a sand mold with a wood or metal pattern, removing the pattern, and filling the cavity with molten metal. Following a period of cooling, the metal solidifies, and the sand is broken away.

Sand molds are prepared in a boxlike container called a flask. Figure 8-1 shows a typical two-piece sand mold and related components. A basic requirement of the sand mold process is the pattern, the purpose of which is to form the cavity in the mold into which the metal is poured. The pattern is accurately constructed by a skilled pattern maker and is made slightly larger than the required final product to compensate for the shrinkage of the metal as it solidifies and cools in the mold. This condition is called *shrinkage allowance*. Patterns are also modified by the application of a slight taper, called *draft*, which is applied to the sides to facilitate the removal of the pattern from the mold. Excess stock, called *machining allowance*, is added to the pattern so that the castings will be oversize at surfaces requiring subsequent finishing operations.

FIGURE 8-1 A two-piece sand mold and related terminology ***(Section 8.9)***

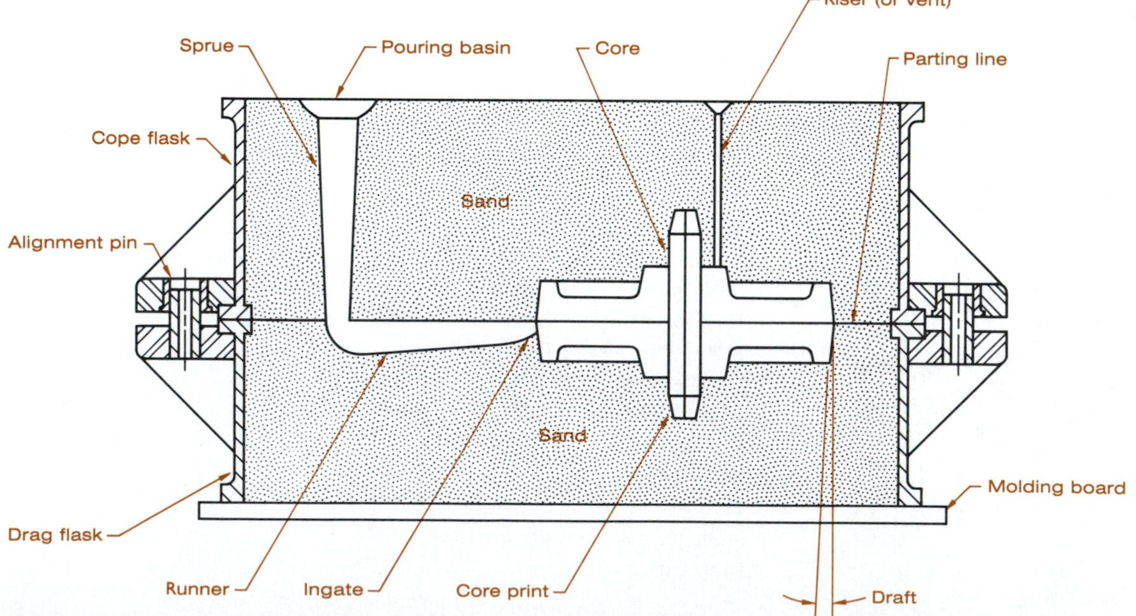

Cores are used wherever necessary to produce a hole or undercut in a casting. The core consists of a firmly oven-baked, rigid mixture of bonded synthetic sand. Cores are inserted into position after the pattern has been withdrawn from the mold. After the casting has cooled, the cores are broken away from their position in the casting.

As explained in Section 5.19, fillets are provided at all section junctions for strength. All external corners on the pattern are rounded because metal will not form a sharp corner in a mold.

All sand casting molds are expendable, or "sacrificial," and must be destroyed as the solidified casting is removed from the mold cavity. All ferrous and nonferrous metals may be sand cast. On the bench grinder shown in Figure 8-2, the grinding wheel covers, the motor casing, and the base are sand castings.

Plaster Mold Casting

The process of plaster mold casting requires that a new mold be prepared for each casting. The mold material is composed of a loose slurry mixture of gypsum plaster, a fibrous strengthener, and water. Patterns are made of highly polished wood, metal, phenolic resin, or molded rubber. One or more patterns are loosely mounted on a flat plate, which is placed in the flask. The plaster slurry mix is then poured into the flask directly over the pattern. The mold is vibrated slightly to allow the slurry mixture to compactly fill in and around the pattern. After the plaster has set, which is usually only a matter of minutes, the pattern is carefully removed from the mold. Both the cope and drag mold halves are made in this way. The molds are then baked and assembled, cores are set if necessary, and the molten metal is gravity fed into the mold cavity. After cooling and solidifying, the casting is removed by breaking the mold.

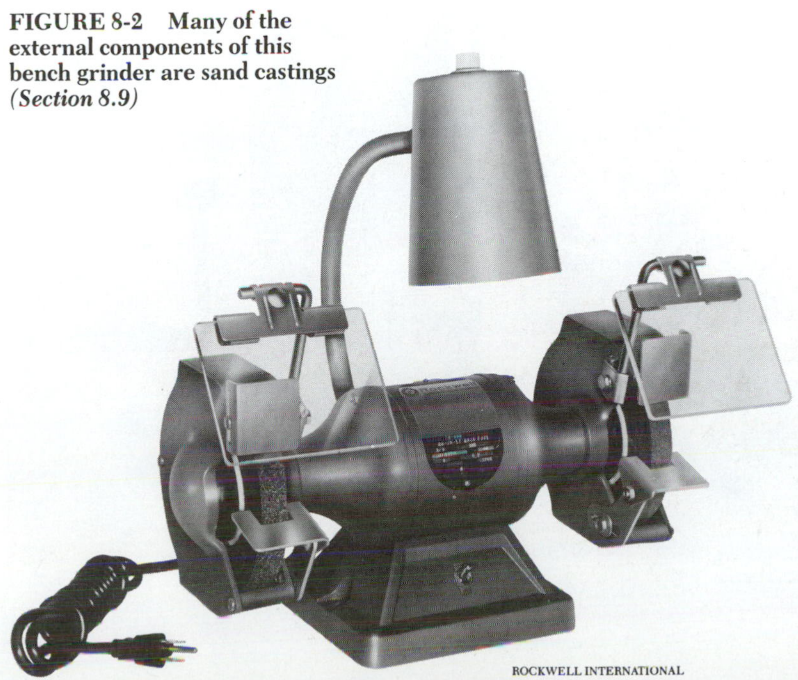

FIGURE 8-2 Many of the external components of this bench grinder are sand castings *(Section 8.9)*

ROCKWELL INTERNATIONAL

Shell Molding

Shell molding involves the use of metal match-plate patterns. Castings are formed by pouring molten metal into an opening in an oven-baked rigid shell that serves as the mold cavity. The mold cavity is formed by the space left between two matching shell halves that are clamped or glued together. As in sand casting, the mold is expendable. Shell molding is an important method for making cores.

Shell molded castings have as-cast surfaces that are considerably superior to sand castings. Shell molded products are used in practically every application in which sand castings are used except that large, heavy castings with thick walls and sections are not adapted to the shell mold casting process.

Investment Casting

The process of investment casting involves the use of expendable patterns surrounded by a shell of refractory material to create the mold. Castings are formed in cavities produced by melting out a wax pattern. In brief, the process consists of preparing a wax pattern, which is made by pouring the pattern material into a prepared metal mold. The pattern is then "invested" with successive coatings of a ceramic slurry mixture until the required thickness of the mold or shell is achieved. The flask, with its contents, is then placed in an inverted position in an oven, where the wax pattern is allowed to melt and drain out of the mold. Molten metal is then gravity poured into the cavity. Other pouring methods, including vacuum, centrifugal, and air pressure, are also used. Investment casting is a highly automated process. There are almost unlimited product applications for this high-production casting process. Figure 8-3 shows some parts that were made by investment casting.

Permanent Mold Casting

In the permanent mold casting process, molten metal is poured under gravity force into metal molds. Both ferrous and nonferrous metals may be cast in this manner. The process is especially adaptable to high-volume production of small, simple castings that have a reasonably uniform wall thickness, no undercuts, and limited coring requirements. Production runs of aluminum parts of up to 250,000 castings from the same mold are possible with this method. Molds are made in two or more sections and are securely clamped together before pouring the molten metal into the mold. Mold materials include graphite, cast iron, bronze, and ceramics. Unlike sand casting molds, which are made by ramming sand around a pattern, permanent mold cavities are machined to final size, giving the casting a superior surface finish with close dimensional tolerances. Most molds are constructed with an ejector pin system for casting removal. Whenever possible, molds of multiple cavities are used.

Die Casting

Of all the casting processes, die casting is the fastest. In die casting, molten metal is forced into cavities of alloy steel dies under high pressure. The molten metal is sometimes held for a short

HITCHINER MFG. CO., INC.

FIGURE 8-3 Typical investment castings *(Section 8.9)*

time under pressure in the dies as it solidifies. The die blocks are then opened and the casting is ejected by pins. The die is closed again and the same cycle is repeated. Die casting leads all other casting methods in the tremendous versatility of product applications. The chief restriction in this casting process is that it is limited to nonferrous metals. An illustration of a typical die casting is shown in Figure 8-4.

See Table 8-5 for a summary of casting principles.

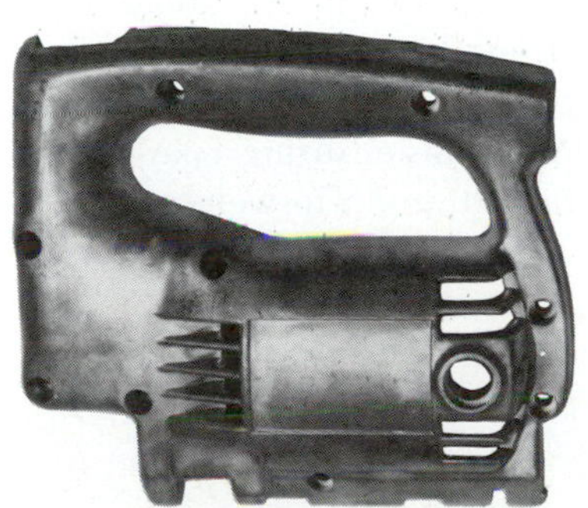

ARWOOD CORPORATION

FIGURE 8-4 Typical die castings *(Section 8.9)*

TABLE 8-5 Summary of Casting Principles

Process	*Advantages*
Sand casting	Inexpensive method of producing a mold; dimensional accuracy relatively good; almost no size or weight limit.
Plaster mold casting	Best possible as-cast surface is obtained; castings have high dimensional accuracy; low porosity; very intricate shapes possible.
Shell molding	Readily adapted to automation; better dimensional accuracy than sand castings; higher production rate than sand castings due to high-speed molding machines; low cleaning costs.
Investment casting	Castings are obtained with very fine detail, superior surface finish; more accurate parts may be produced by this method than by any other; almost any metal may be cast; parts require little or no machining; no flash is produced on castings; process lends itself to automation.
Permanent mold casting	More uniform grain structure than by casting; closer dimensional tolerances and superior surface finish.
Die casting	Very complex shapes may be produced; very thin walls are possible; good dimensional accuracy; very high production rates; little finishing, if any, is required.

Process	*Limitations*
Sand casting	Dimensional accuracy is worst of all casting processes; poor surface finish; slow production rate; maximum machining allowance is required; expendable molds.
Paster mold casting	Only nonferrous metals may be cast; molds are expendable; process is restricted to relatively small parts.
Shell molding	Size and weight of castings are limited; high cost of metal patterns.
Investment casting	Size and weight of castings are limited; initial tooling costs high.
Permanent mold casting	Process is best for small castings with relatively simple shapes; some shapes not possible due to difficulty in removal from mold; not all alloys are suitable; steel castings are not recommended; expensive for low quantity production; relatively low production rates.
Die casting	Process is limited for size and weight; equipment costs high; porosity may present special problems; die preparation costs high; limited to nonferrous metals with low melting points.

Process	*Principal casting materials*
Sand casting	Iron, low-melting-point steel alloys, copper-base alloys, aluminum, magnesium, and nickel alloys.
Plaster mold casting	Only nonferrous metals—aluminum, alloys, zinc, copper-base alloys.
Shell molding	Any castable metal.
Investment casting	High-melting-point steel alloys, aluminum, nickel, copper-base alloys, magnesium.
Permanent mold casting	Iron, aluminum, magnesium, copper-base alloys.
Die casting	Only nonferrous metals—aluminum, alloys, zinc, copper-base alloys.

Continued

TABLE 8-5 *Continued*

Process	*As-cast tolerances*	
Sand casting	1⁄16 to 11⁄64 in. depending upon metal being cast and size of casting.	
Plaster mold casting	±0.005 in./in.; ±0.010 in./in. across parting line.	
Shell molding	±0.002 to 0.005 in./in. across parting line.	
Investment casting	±0.003 to 0.005 in. for first inch; over 1 in. add ±0.003 in. for ferrous metals.	
Permanent mold casting	±0.005 in. for first inch; over 1 in. add ±0.002 in./in.; ±0.003 in./in. across parting line.	
Die casting	On average, ±0.003 in. for first inch; over 1 in. add ±0.0015 in.	
Process	***Section thickness***	
	Minimum	Maximum
Sand casting	Aluminum 3⁄16 in., magnesium 5⁄32 in., copper and iron 3⁄32 in.	Almost no limit.
Plaster mold casting	About 1⁄32 in.	Varies from 1⁄16 to 1⁄8 in.
Shell molding	Iron 1⁄8 in.; steel, aluminum, and magnesium 3⁄16 in.	3⁄8 in. is reasonable.
Investment casting	About 1⁄16 in.	1 to 3 in.
Permanent mold casting	Iron 3⁄16 in.; aluminum 3⁄32 to 1⁄8 in.; magnesium 5⁄32 in.; copper 3⁄32 to 5⁄16 in.	2 in.
Die casting	Varies with material from 0.015 to 0.080 in.	5⁄16 in. is preferable, but less than 1⁄2 in.
Process	***Minimum diameter of cored holes***	
Sand casting	3⁄16 to 1⁄4 in. but 1⁄2 in. is a more practical minimum.	
Plaster mold casting	1⁄2 in.	
Shell molding	1⁄8 to 1⁄4 in.	
Investment casting	0.020 to 0.050 in.	
Permanent mold casting	3⁄16 to 1⁄4 in.	
Die casting	Varies with material but 1⁄8 in. is a practical size.	
Process	***Bosses***	***Undercuts***
Sand casting	Yes.	Yes.
Plaster mold casting	Yes.	Yes.
Shell molding	Yes.	Yes.
Investment casting	Yes but sometimes with difficulty.	Yes.
Permanent mold casting	Yes.	Yes, with added cost and at a reduced production rate.
Die casting	Yes.	Yes, with added cost and at a reduced production rate.

8.10 FORGING

Forging is the process of plastically deforming metals and alloys to a specific shape by a compressive force exerted by a hammer, a press, rolls, or an upsetting machine of some kind. When metal is forged, both strength and ductility increase significantly along the lines of flow, resulting in a characteristic pattern of grain flow that is unbroken and that follows the contour of the part. The main objective of good forging design is to control the lines of metal grain flow so as to put the greatest strength and resistance to fracture where it is needed on a part, thus ensuring a part with a high strength-to-weight ratio.

Figure 8-5(a) illustrates the cross section of an etched crankshaft forging that shows the typical lines of grain flow. As shown in 8-5(b), the grain follows the direction of needed strength.

There are two main methods of forging: *hammering* and *pressing*. These differ only in their relative speed of pressure application. Hammering causes the metal to change shape by repeated blows of a freely falling ram. Pressing changes the shape of a metal bar or billet by the squeezing action of a slowly applied force.

Forging develops the quality of a metal to its greatest potential. When compared to sand castings, forgings are up to 1½ times stronger. The process is particularly appropriate for parts that are subject to vibrations and sudden shock loading.

In forging, a smooth, fine-textured surface is obtained, often requiring no secondary machining operations. Forgings can be made to close tolerances, which reduces or eliminates machining time. The process lends itself to the production of irregular shapes and complicated parts. Figure 8-6 illustrates a typical forged part. There is a wide range of forgeable metals.

Proper forging design requires that particular attention be given to the location of the parting line, draft angles, corner radii, and fillet radii; placement of webs and ribs; and allowances for shrinkage and finish.

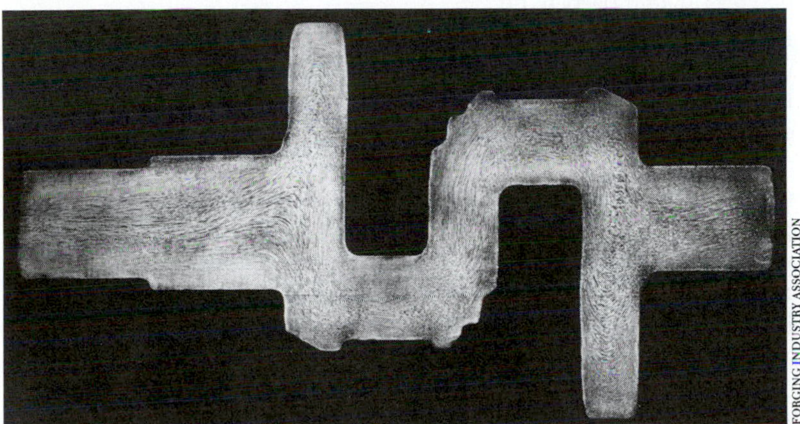

FORGING INDUSTRY ASSOCIATION

(a)

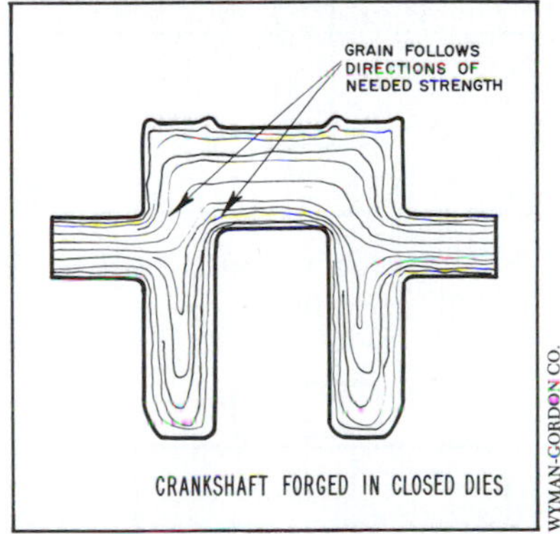

WYMAN-GORDON CO.

(b)

FIGURE 8-5 A cross section of a forged crankshaft showing typical lines of grain flow *(Section 8.10)*

FIGURE 8-6 A 4130 steel trunnion. An example of a typical forged part *(Section 8.10)*

SUNNEN PRODUCTS CO.

8.11 ASSEMBLY METHODS

The word *assembly*, as it is used here, refers to a mechanism or a structure that has been created by putting together individual components. The assembly methods included in this section are welds, adhesives, and screw threads.

Welds

Welding is one of the many ways to assemble metal parts. It consists of permanently fastening together two or more pieces into a single unit by the application of heat or pressure, or both. Three types of welding are most important: arc welding, gas welding, and resistance welding. In *arc welding* heat is generated by an electric arc struck between a consumable electrode and the workpiece. Intense heat is generated and melting and subsequent solidification of the metal occur very rapidly. The most common form of *gas welding* is oxyacetylene welding. Acetylene gas and oxygen are combined in the proper proportions to produce a concentrated high-temperature flame. The edges of the metal parts are joined without the need of pressure by fusion of the melted metal. A filler material in the form of a wire or rod is added to fill the gaps between the pieces being joined. *Resistance welding* is widely used in mass-production work, particularly in the joining of thin-gage metals up to about ⅛ inch (3 mm). Like arc welding, resistance welding employs electricity, but no arc is generated. Instead, heat is created from resistance losses as high current is sent across the joint between two firmly held parts to be joined. The process is widely used on automobile production lines or for building appliances, containers, and so on.

When designing, you must carefully consider a great many factors pertaining to welding as well as the influence of other factors that may not directly relate to welding operations. Simple, straightforward weldments (parts fabricated by welding) of mild steel are rarely troublesome to weld. Complex shapes, on the other hand, require compatible choices in equipment and metals to develop a suitable welding sequence.

Adhesives

Products may be permanently assembled by using adhesives. Various kinds of adhesive agents are currently available. They should be selected for a particular application only after a careful appraisal of all of the factors involved in their use. Joint strength is dependent upon surface bonding. Surface films on joints must be thoroughly removed. Increased strength is often obtained by roughening the surface to be joined.

Threaded Fasteners

Threaded fasteners, such as bolts and nuts, studs and nuts, and screws, are classified as semipermanent attachments. The application and representation of each of these fasteners is shown in Figure 8-7. Since these threaded fasteners are completely standardized, it is un-

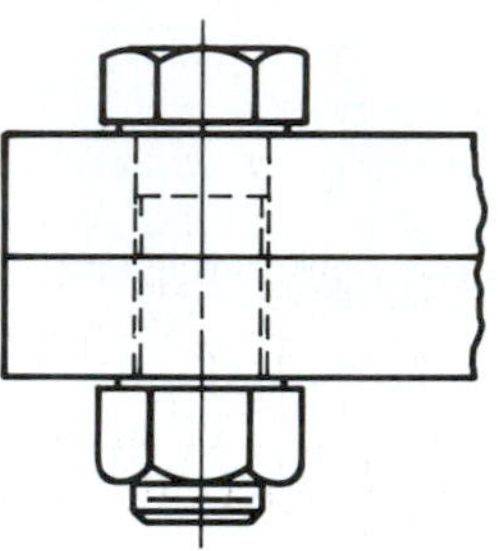

BOLT AND NUT
(a)

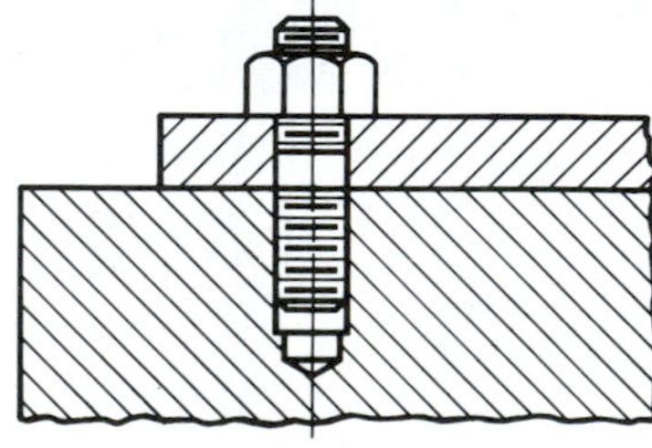

STUD AND NUT
(b)

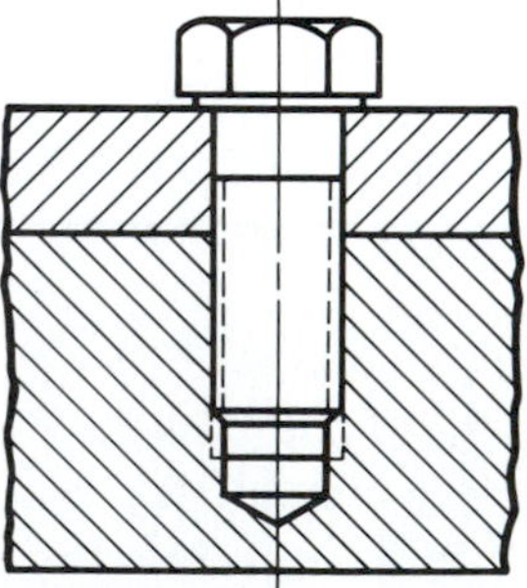

HEXAGON HEAD CAP SCREW
(c)

FIGURE 8-7 Applications and representations of threaded fasteners *(Section 8.11)*

necessary to prepare a detail drawing. Each of the parts is specified by giving the information as shown in the notes. Sizes of bolts, screws, studs, washers, and nuts are given in the appendix tables.

In the selection of a fastener, you must consider factors that include suitability for the intended functions, strength, relative costs, availability of "off-the-shelf" items, appearance, corrosion resistance, and ease of assembly and wrenching. Various methods of specifying and representing screw threads are shown in Chapter 9.

8.12 PRESSWORKING METHODS

Pressworking includes a wide variety of chipless processes by which workpieces are shaped from flat metal sheets. A general term, *stamping*, is used almost interchangeably with pressworking. Stamping is normally understood to include the general headings of cutting, bending or forming, squeezing, and drawing. It is common practice in many companies for product designers and die designers to confer in the design of pressworked parts to work out special considerations regarding bend radii, hole spacing, notching techniques on inside corners, stiffening techniques, and so forth. Except for the general discussion that follows, these processes are beyond the scope of this text.

Stampings are produced by the downward stroke of a ram in a machine called a *press*. A typical hydraulic press is shown in Figure 8-8. Parts produced by pressworking operations range in size from shoe eyelets to the ends of freight cars. Compared to other metalworking processes, pressworking techniques can offer an almost unlimited choice of metals and product shapes. Metal stampings can be rapidly produced in extremely large quantities. Products are economical, lightweight and strong, and have a superior strength-to-weight ratio. Figure 8-9 illustrates some typical sheet metal pressworked parts.

GREENERD PRESS AND MACHINE COMPANY, INC.

FIGURE 8-8 A 200-ton open-gap hydraulic press used to form a sheet metal panel for an electronic chassis *(Section 8.12)*

All pressworking operations require the use of a matching set of tools, called a punch and die, which are usually aligned by means of guide pins. When assembled, these tools are called a *die set*.

Cutting operations generally consist of *blanking* (cutting off, parting, shearing, lancing, notching, and nibbling); *piercing* or *punching* (slitting, perforating, and extruding); and *edge-improvement methods* (trimming and shaving). Regard-

A. M. BRUNING

FIGURE 8-9 Typical pressworked parts, including the various parts of a steel storage cabinet, the structure of a drafting stand, and the curved legs that form the base of a chair *(Section 8.12)*

less of the term used to describe the particular cutting operation, each method is associated with high-volume stamping of various outlines of flat workpieces.

Forming is a term that describes the manner by which flat metal blanks are used to make contoured shapes of parts. The general term *squeezing* includes cold-working processes such as peen forming, explosive forming, swaging, cold heading, thread rolling, and others. In each case the processes included in this forming category were developed specifically to overcome major drawbacks inherent in some of the other forming processes.

Drawing is a process for forming thin metal products such as cups, tubular shapes, and shell-like parts. Most drawing operations start with a flat blank of metal.

8.13 POWDER METALLURGY

Powder metallurgy is a production process in which parts are made by compressing metal powders in a mold on a press. Strength and other properties are added to the parts by subsequent sintering (heating) operations. The process may be used to produce porous parts and high-density part structures.

Product applications for powder metal parts are virtually unlimited in scope. Figure 8-10 illustrates some typical product applications. Parts range in size from a ball on a ballpoint pen to shapes as large as 10 inches (254 mm) in diameter by 8 inches (203 mm) in length. The tolerance on powder metal parts in a direction across the part can be held to ±.001 in./in. The concentricity for holes up to 1 inch in diameter is .003 inch, with .001 inch for each additional inch in diameter. Most ferrous parts may be heat treated. Powder metal parts are selected for in-service applications under conditions of light-to-medium loads. Because of the inability of metal powders to flow laterally into a die, parts cannot be produced with reentrant forms such as threads, peripheral grooves, undercuts, and so on. In this particular respect the limitations resemble those in pressworking processes. When considering this process, you should be alert to avoid sharp internal and external corners on parts. Corner radii permit uniform powder flow in the die.

8.14 SECONDARY MANUFACTURING PROCESSES

Secondary manufacturing processes take the products of some primary manufacturing processes (casting, forging, extruding, and so forth) and further transform them into a finished part. In modern industry almost any conceivable type of surface or

HOEGANAES CORP.

FIGURE 8-10 Typical powder metallurgy parts ***(Section 8.13)***

shape can be machined to very precise limits of finish and accuracy. Secondary manufacturing processes include the machining operations of sawing, milling, broaching, shaping, and others.

8.15 TYPES OF MACHINE TOOLS

Machine tools fall into two major categories: Conventional and numerically controlled (NC). Both machine types can be used to perform cutting operations on machine parts.

Conventional machine tools may be automatically controlled or they may be controlled by the operator, who manually loads and unloads the workpiece and activates the various levers and handwheels to guide the workpiece into contact with the cutting tool. The operator is also responsible for selecting the proper cutting speed and feedrate, applying coolant, and so on.

Numerically controlled machine tools are operated by means of magnetic tape or by punched holes on a tape containing the part program. After the part has been loaded into a vise or a specially designed fixture, the coded information stored in the moving tape controls the relative movement of the workpiece and the cutting tool. The principal duty of the operator is to monitor the machine and to operate the tape reader. NC machines are generally better designed and constructed and are often more accurate than conventional machine tools.

8.16 MACHINING OPERATIONS

Sawing

Most machine tools can perform cutting-off operations to a limited extent. When compared to dedicated or single-purpose power sawing machines, however, the operations are generally slow and often wasteful. In power sawing the cutting action is accomplished by passing a series of cutting teeth across the workpiece.

Three general types of power saws are in use: reciprocating saws (commonly known as power hacksaws), circular saws (or cold-sawing machines), and band saws. Reciprocating saws and circular saws are used principally for cutting-off operations. Band sawing machines are considered more versatile than any of the other sawing machines. Band saw machining is used for simple cutoff operations and also for cutting along straight and curved lines. Abrasive disk cutting is another method of cutting off stock.

Milling

With the possible exception of the lathe, the milling machine is considered to be the most versatile of all machine tools. Practically all shapes and sizes of both flat and curved surfaces on both the inside and the outside of workpieces can usually be machined by one of the milling methods. Milling operations are economical for producing parts in large quantities.

Machined surfaces are formed in milling by the cutting action of a rotating cutter, sometimes in a single pass of the work. The work is rigidly held and, in most cases, is fed against the rotating cutter. The metal removal rate of milling machines is usually considerably greater than that of lathe, shaping, or grinding operations.

Many different types of cutters are available for milling operations. Typical cutters are shown in Figure 8-11. The selection of the precise type of cutter is governed by the particular surface configuration required on the part to be machined, the surface quality, workpiece material, and available production

FIGURE 8-11 A representative sampling of milling cutters *(Section 8.16)*

ILLINOIS/ECLIPSE, DIV. OF ILLINOIS TOOL WORKS, INC.

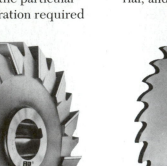

Staggered tooth cutter

Slitting saw cutter

Corner rounding form relieved cutter

Plain cutter

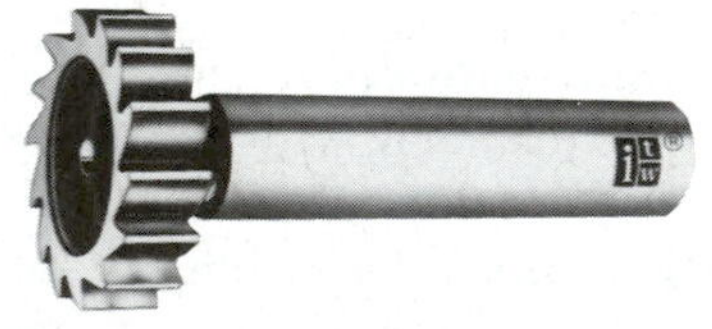

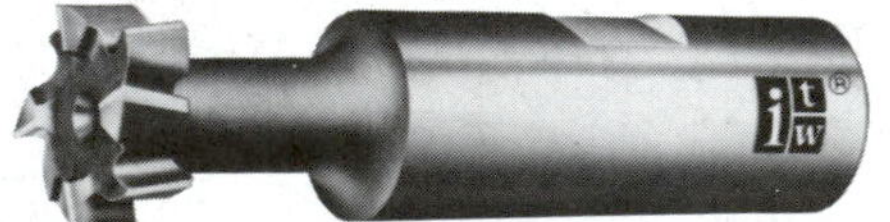

Solid T-slot cutter

Concave form relieved cutter

Convex form relieved cutter

Solid plain mill

equipment. Figure 8-12(a) illustrates a milling operation being performed on a forging die. A recent model of a CNC milling machine is shown in Figure 8-12(b).

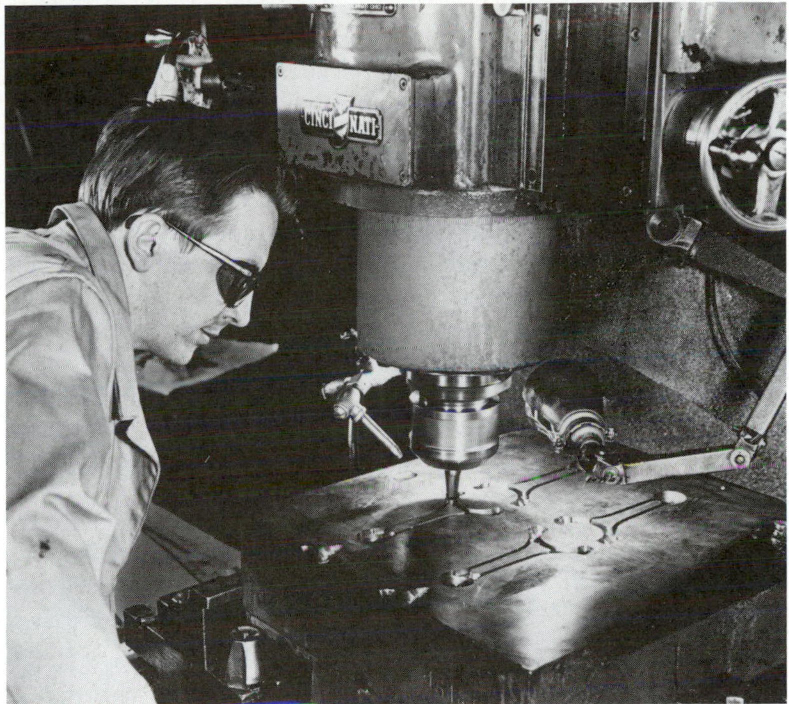

(a)

(b)

BRIDGEPORT MACHINES

FIGURE 8-12 **(a) A milling operation being performed on a forging die; (b) a modern CNC milling machine** ***(Section 8.16)***

Broaching

Broaching is an important method of stock removal and one that has many applications in manufacturing processing. The process of broaching employs a multitoothed tool that may be either passed across a fixed workpiece or held stationary with the work moved in a continuous stroke. The broach is a hardened steel bar with a series of cutting teeth that resemble those on a wood rasp. As each broaching tooth contacts the workpiece, it cuts a fixed thickness of metal. Both the cutting tool and the workpiece are held in rigid fixtures.

There are two types of broaching tools: *internal* and *external*. Internal broaches are used to enlarge and finish a wide variety of hole shapes previously formed by casting, forging, punching, drilling, boring, and so on. Figure 8-13 shows a variety of internal and external shapes that may be produced by broaching.

Broaching is a method that is usually employed for high-quantity production. One of the largest users of broaching is the automotive industry, which uses the process for finishing surfaces on engine parts. Other large users are rifle and gun manufacturers.

A surface cannot be broached if it has an obstruction that interferes with the path of the tool, such as a blind hole. Almost any irregular cross section can be broached as long as all surfaces of the section remain parallel to the direction of the broach travel. Parts of less than 1 ounce to workpieces weighing several hundred pounds may be broached. Most external surface configurations can be finished by other machining processes. The principal advantage of broaching is the speed of the operation.

FIGURE 8-13 A variety of external and internal shapes produced by broaching ***(Section 8.16)***

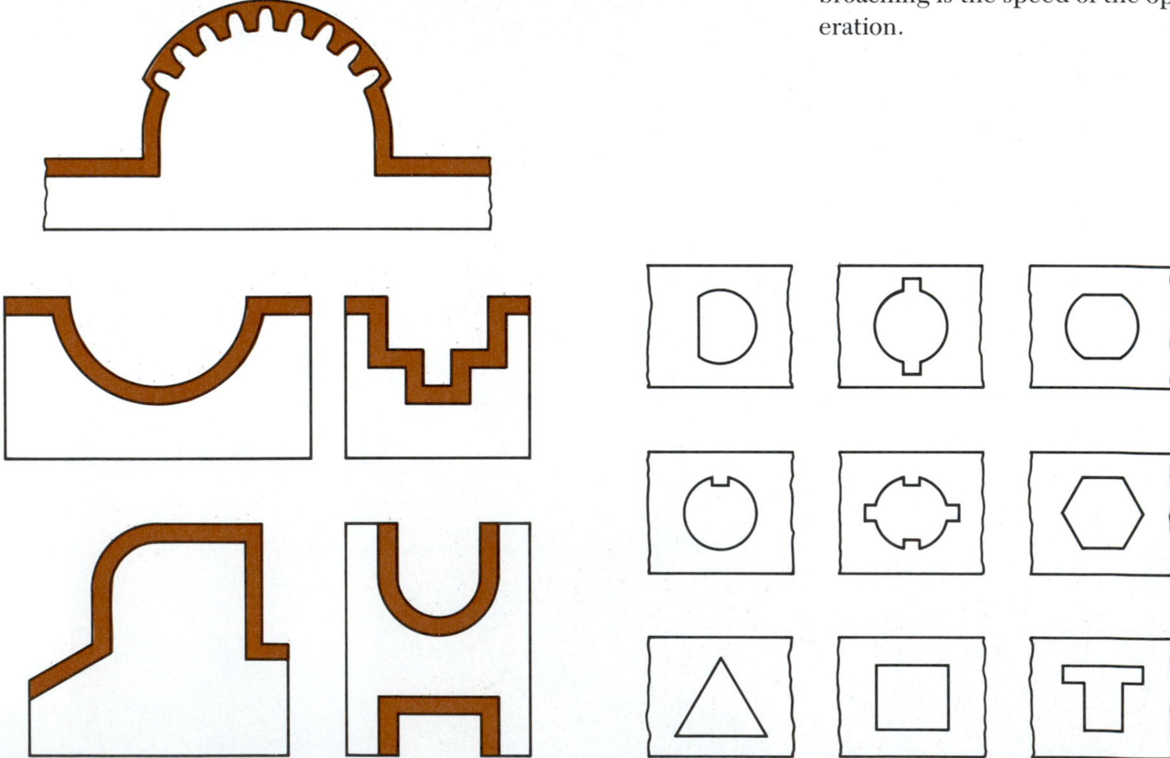

Shaping

In shaping, a reciprocating single-point tool bit held in a tool holder on a ram is used to produce a finished surface on a part. A shaping operation is illustrated in Figure 8-14. The tool moves back and forth in a straight line across a rigidly held workpiece. The machined surface is generated by the cutting action of the tool bit as it removes a metal chip on its forward stroke. As the ram returns, the tool bit feeds crosswise a preset incremental distance equal to the feed desired. This distance rarely exceeds $1/16$ inch. The length of the stroke is adjustable. Both horizontal and vertical types of shapers are available.

There are two main advantages to a shaper: They use relatively inexpensive tools and, for most types of work, they require only a comparatively short time for setup. Shaping is not ordinarily a high-volume machining process. Types of surfaces commonly produced by shaping are illustrated in Figure 8-15.

Turning and Boring

Both turning and boring are performed on a lathe. *Turning* is a machining process by which cylindrical, conical, or irregular-

ROCKFORD MACHINE TOOL CO.

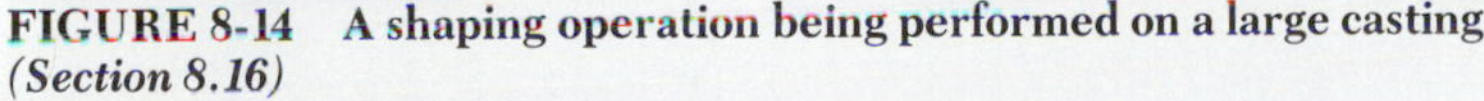
FIGURE 8-14 A shaping operation being performed on a large casting ***(Section 8.16)***

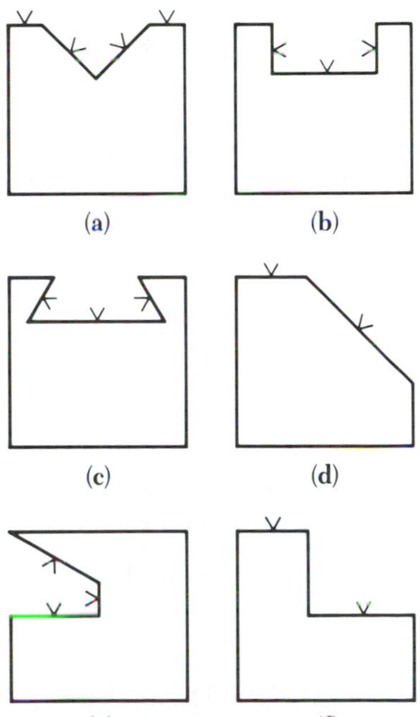

FIGURE 8-15 Types of surfaces commonly produced by shaping ***(Section 8.16)***

ly shaped external or internal surfaces are produced on a rotating workpiece. The cutting action is generated by one or more stationary single-pointed cutting tools, which are held at an angle to the axis of the workpiece rotation. A specialized production machine, a turret lathe, is illustrated in Figure 8-16. A turret lathe is designed to perform turning, boring, drilling, and many other operations rapidly and in sequence on one particular workpiece. Figure 8-17 illustrates several types of workpiece surfaces produced on a lathe.

Boring is the process of machining out large internal cylindrical holes on surfaces of parts. Such holes are produced at an end of the workpiece by feeding the tool in a direction that is perpendicular to the workpiece axis. Holes are generally bored when it is not practical to use a drill. Boring is often used to enlarge previously made holes formed by casting or by forging. Boring operations may also be performed on specially designed machines. One of them, known as a *jig borer*, resembles a drill press and is used for precision work. Other boring machines, such as a *vertical boring mill* and a *horizontal boring machine*, are adapted to large work.

BARDONS & OLIVER, INC.

FIGURE 8-16 A turret lathe. This versatile production machine has a six-sided turret on which a number of tools can be mounted *(Section 8.16)*

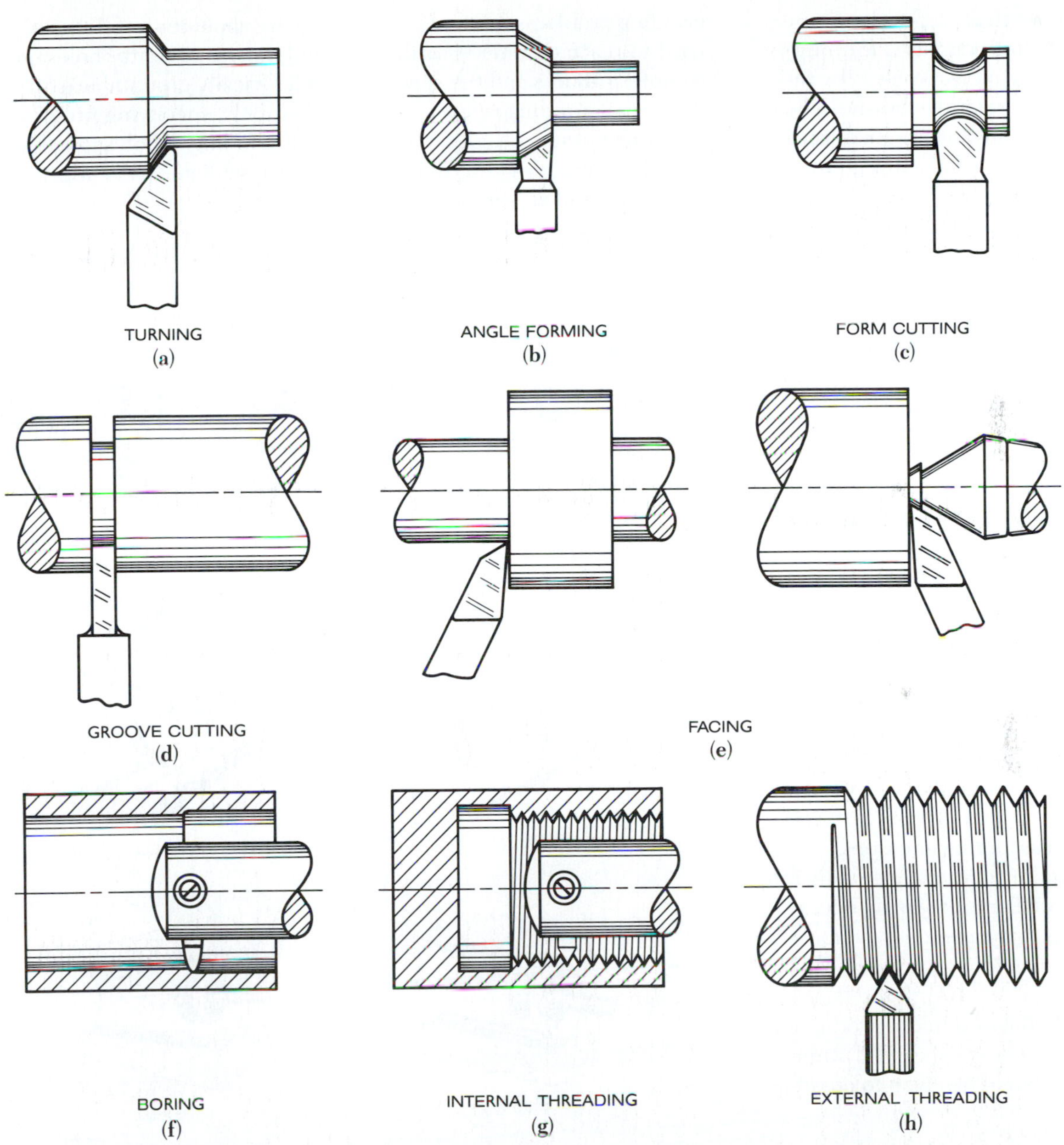

FIGURE 8-17 Types of surfaces commonly produced on lathes *(Section 8.16)*

Grinding

Grinding consists of forming surfaces on workpieces by the use of a rotating abrasive wheel composed of many small and hard-bonded abrasive grains. Grinding is the principal production method for machining materials that are too hard to be cut by other conventional tools. It is also the process employed to produce surfaces on parts to tolerance or finish requirements more exacting than can be obtained by other manufacturing processes.

Figure 8-18(a) shows three types of cylindrical surfaces that are finished by grinding. At 8-18(b) plane or formed surfaces are formed on flat workpieces, while at 8-18(c) several applications of internal grinding operations are shown. Grinding operations are performed on workpieces that have been previously rough shaped by some other primary forming process, such as forging or casting. Grinding operations usually follow one or more sizing operations that have removed most of the stock from the rough workpiece.

As the design of a production part progresses, careful attention should be given to possible grinding problems, such as hard-to-reach or unnecessarily complex surfaces and the specification of grinding reliefs. Other necessary considerations include techniques for grinding a diameter and an adjoining square shoulder or fillet radii and the use of center holes on cylindrically ground parts. Generally, specifying production grinding of holes less than ¼ inch in diameter is impractical.

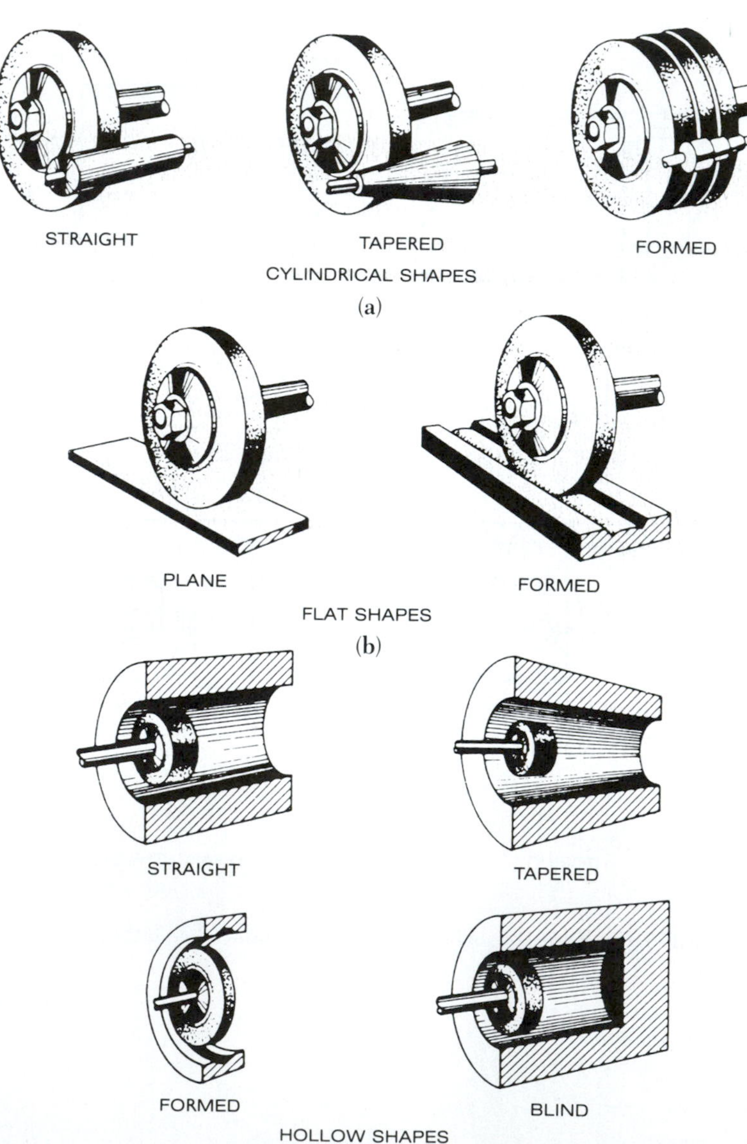

FIGURE 8-18 Types of surfaces finished by grinding operations ***(Section 8.16)***

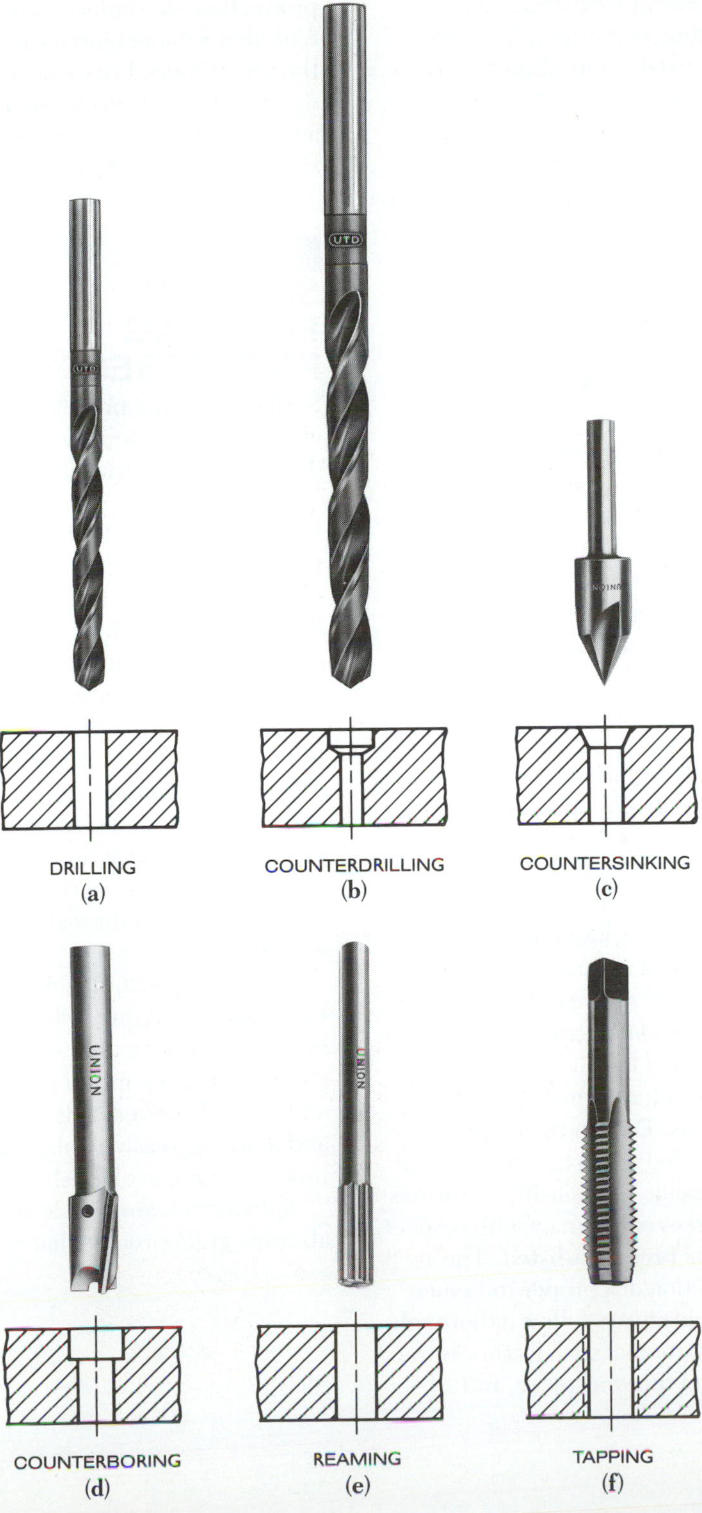

Hole Making Operations

Hole making operations are usually associated with a machine known as a *drill press*. The drill press is used to drill, counterdrill, countersink, counterbore, spotface, ream, and tap holes in workpieces. The tools used to perform these operations are shown in Figure 8-19. Drill sizes are indicated on the drill shank. Any one of the following systems may be used to identify drill sizes: numbers, letters, fractional inches, decimal inches, and millimeters. Drill sizes are given in Appendix Tables A-2, A-3, and A-4. Metal cutting requires that considerable pressure be applied to the cutting edges of the tool. A drill press provides the necessary feed pressure.

FIGURE 8-19 Examples of hole-making operations *(Section 8.16)*

(PHOTOS COURTESY UNION TWIST DRILL CO.)

8.17 TOLERANCES FOR VARIOUS PROCESSES

The correlation of mating dimensions is essential to the proper operation of a product. When all of the mating parts are brought together on the assembly line, each part should fit together within the desired degree of accuracy. As will be explained more completely in Chapter 11, precise maximum and minimum limit dimensions are used on drawings to control mating dimensions on parts that fit together.

The limit on the degree of accuracy to which parts are to be processed is based upon cost. Production costs, in turn, are largely dependent upon design simplicity. Tolerances on finish and on dimensions play an important part in manufacturing economy. The specification of needlessly close tolerances and an unreasonable degree of surface finish always result in excessive, and in some cases prohibitive, costs. Tolerances that are given on dimensions to within ten-thousandths of an inch are more costly to achieve than tolerances within thousandths of an inch. Parts with tolerances to within ± 1⁄64 inch are, of course, more economical to produce than parts with tolerances to within thousandths of an inch.

For precise dimensions on mating parts, limit dimensions are used. Limit dimensions are usually expressed, for example, as .376/.374 (inches) or 125.09/125.07 (millimeters). The limiting values in each of the dimensions are the maximum and minimum limits. The difference between the limits is called the *tolerance*. A common fraction should be used for dimensions on parts that have no precise size requirements. Unless otherwise specified on a drawing, all dimensions expressed as a common fraction or an untoleranced whole number should have a tolerance of 1⁄32 inch. A dimension of 1½ inches, for example, may be made as large as 1 33⁄64 inches or as small as 1 31⁄64 inches or any size value within this range. The amount of permissable variation in size is ± 1⁄64 inch. A common tolerance for two-place decimal dimensions is ± .01 inch.

The information given in Table 8-6 is based upon the degree of accuracy that can be expected as a result of a chip removal operation on a machine tool used for quantity production under normal shop conditions. Data in the table are intended to portray the general tolerance relationship of various degrees of accuracy with respect to the processes listed. The selection of a proper tolerance should always follow a thorough evaluation of such factors as the size of the workpiece, rate of production, workpiece material, tool, fixture, machine wear, and the use of special size-control devices. Operator skill can play a large part in the production of single or low-quantity parts.

8.18 SURFACE FINISHING PROCESSES

Surface finishing processes include grinding, as previously discussed, and barrel finishing, vibratory finishing, spindle finishing, abrasive belt finishing, polishing, and buffing.

Barrel finishing is a mass finishing process. Metal parts are batch-loaded into a metal barrel, which is rotated slowly. As the barrel turns, the contents—the parts, abrasive material, and compound—tumble and slide around in close contact. The rubbing action of the contents of the barrel removes unwanted sharp edges and burrs from the parts.

Vibratory finishing employs a tub-shaped container whose vibrating motion creates an abrasive action in the mix of parts and loose abrasives. It is a faster and more aggressive finishing process than barrel finishing.

Spindle finishing is a loose abrasive grain process that is

TABLE 8-6 Tolerances for Various Shop Processes (in Inches)

Drilling

DRILL SIZE (DIAMETER)	TOLERANCE +	TOLERANCE −
#60 to #30	0.002	0.000
#29 to #1	0.004	0.000
1/4 to 1/2	0.005	0.000
33/64 to 3/4	0.008	0.000
41/64 to 1	0.010	0.000
1 1/64 to 2	0.015	0.000

Lathe turning (rough cut)

WORKPIECE DIAMETER	TOLERANCE
1/4 to 1/2	0.005
1/2 to 1	0.007
1 to 2	0.010
2 and Over	0.015

Lathe turning (finish cut)

1/4 to 1/2	0.002
1/2 to 1	0.003
1 to 2	0.005
2 and Over	0.007

Milling

DESCRIPTION	TOLERANCE
Single Surface	0.002 to 0.003
Two or More Surfaces	0.002 to 0.005
In General	0.005

Grinding

TYPE	TOLERANCE
Cylindrical	0.0005
Surface	0.0005

Reaming

REAMER SIZE (DIAMETER)	TOLERANCE +	TOLERANCE −
Up to 1/2	0.0005	0.0000
1/2 to 1	0.001	0.0000
1 and Over	0.0015	0.0000

Broaching

WORKPIECE SIZES	TOLERANCE
Diameters	
Up to 1	0.001
1 to 2	0.002
2 to 4	0.003
Surfaces	
Up to 1 Apart	0.002
1 to 4 Apart	0.003
4 Apart and Over	0.004

Planning and shaping

Tolerance	±0.001 to 0.002

sometimes called "form grinding wheel finishing." In its simplest form, the work is positioned in a chuck and is then inserted into a tub. The chuck is slowly rotated to expose all surfaces of the workpieces to a high-velocity abrasive stream. During the finishing operation the tub also spins, covering the part with a wet abrasive wave at all times.

Abrasive belt finishing is an important, low-cost, relatively fast finishing process. It is especially adaptable for the precision finishing of flat, concave, and convex surfaces. Stock removal is accomplished by the abrasive grains on a moving belt as they continuously pass over the work area. In this way unwanted burrs, high spots, the coarse texture on cast or forged parts, parting lines, or machining marks are refined or totally removed from a workpiece.

Polishing, also called "flexible grinding," is generally preceded by grinding with a solid abrasive wheel and followed by buffing. In polishing, the surface is finished by the cutting action of millions of small abrasive grains adhering to an endless coated belt or flexible fabric wheel. The general practice is to polish the work with a succession of wheels that are set up with different grain sizes. Because most abrasive belts and polishing wheels are flexible, they will readily conform to the shape of a contoured workpiece when necessary.

Buffing generally follows polishing and is usually the final operation that is performed on a workpiece. In buffing, the rubbing action is more gentle than the vigorous and aggressive cutting action employed in polishing. Minute scratches left by polishing and other surface irregularities are reduced or entirely eliminated.

8.19 MICROFINISHING

Product designers constantly strive to design machinery that can run faster, last longer, and operate more precisely. Modern development of high-speed machines has resulted in higher loadings and increased speeds of moving parts. Bearings, seals, shafts, machine ways, and gears, for example, must be accurate—both dimensionally and geometrically. This section deals with three important microfinishing processes: honing, lapping, and superfinishing. Each of these processes was developed to generate a particular geometrical surface and to correct specific surface irregularities.

Honing is used to remove excess stock from internal and external surfaces. Honing is a mechanical means of stock removal that uses spring-loaded abrasive stones as the cutting tool. Abrasive stones are placed around the periphery of a holding device called a toolhead or a mandrel. As the abrasive stones are brought into contact with the workpiece, a driving shaft on the toolhead slowly reciprocates and rotates. Although the amount of stock removed by each cutting edge is small, the combined action of numerous cutting edges working simultaneously provides an efficient and accurate means of stock removal. Honing produces geometrically accurate cylindrical forms by correcting various inaccuracies left by previous operations, such as high spots, tool chatter marks, out-of-roundness, taper, or slight deviations in axial straightness. Honing cannot correct errors of location and alignment or improve concentricity with other diameters.

Lapping is performed on metal surfaces of precision parts by manual or by machine methods. It is specified principally when it is necessary to increase the accuracy of the workpiece, but it also corrects minor surface imperfections. Lapping is a gentle, final operation used to microfinish flat or cylindrical surfaces, but in special cases lapping may be used on spherical or contoured surfaces. Lapping is accomplished at a low speed. This process consists of rubbing the workpiece with an ever changing motion over an accurately finished surface of a lapping block or plate. The serrated lapping surface is charged with a fine-grain, loose abrasive compound.

Superfinishing produces the ultimate in the refinement of metal surfaces. It is an abrading process in which the cutting medium for cylindrical work is a loosely bonded abrasive stick or stone. An abrasive cup wheel is used for flat or spherical work. The process consists of removing fragmented or smear metal from the workpiece surface. The abrasive stone is applied over the surface in a multidirectional motion with a light but continuous pressure. Dimensional changes are principally restricted to removing minute surface projections or high spots. Superfinishing is not primarily a sizing operation. The major purpose of superfinishing is to produce a surface that is capable of sustaining an even distribution of a load by improving the geometrical accuracy of the workpiece.

8.20 SURFACE FINISH CONTROL

There are two principal reasons for surface finish control: to reduce friction and to control wear. A standard (Surface Texture Symbols, ANSI Y14.36-1978) designates controls for surface texture of solid materials. The standard establishes methods for specifying roughness, waviness, and lay and provides a set of symbols for use on drawings, specifications, and other engineering documents. Much of the following material has been abstracted from this standard.

Surface texture is defined as the overall condition of a surface. It consists of repetitive or random deviations from the normal surface that form the pattern of the surface. Surface texture is affected by roughness, waviness, lay, and flaws.

Roughness is a type of surface texture on a workpiece that is characterized by tiny nicks and dents resulting from the action of the cutting tool employed, such as a lathe or shaper tool, milling cutter, or grinding wheel. Fine irregularities may be generated by a casting process or by a shearing action. Roughness is sometimes referred to as a *primary texture*.

Roughness height is the average deviation from the mean plane of the surface and is measured in microinches (µin.) or in micrometers (µm). Figure 8-20 shows a full range of surface roughness heights obtained by

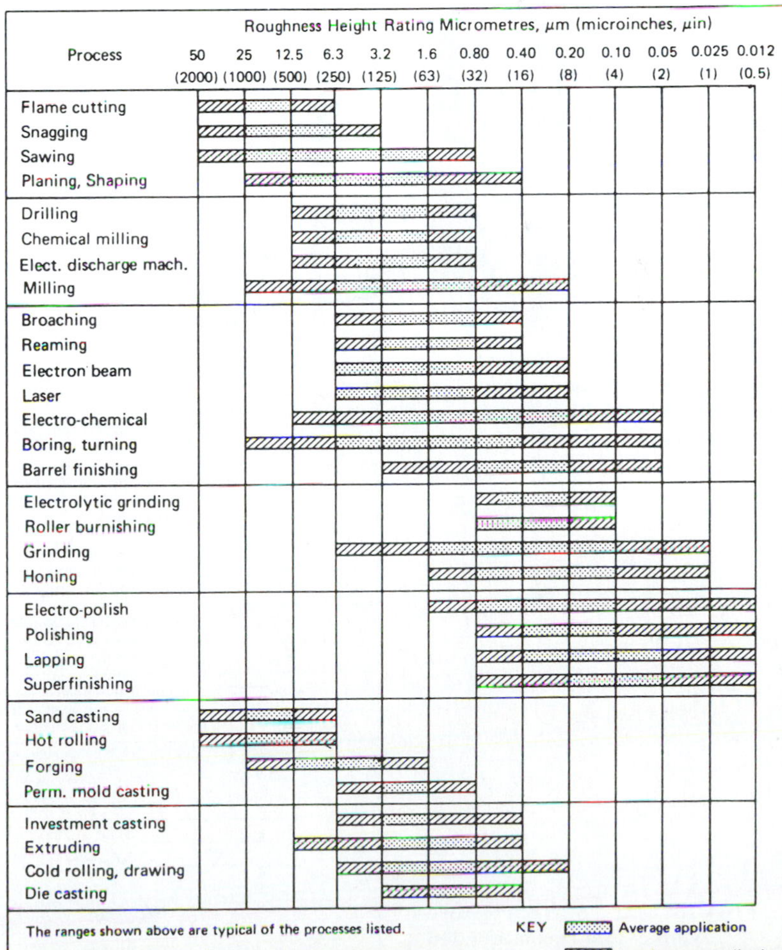

FIGURE 8-20 Typical surface finish values obtained by various production methods *(Section 8.20)*

various production methods. The darker band in each bar indicates the range of finish that can be normally expected in average production. Figure 8-21 illustrates some parts that have been superfinished.

Waviness is a condition of widely spaced surface variations, which may result from such effects as machine or workpiece deflections, vibrations, chatter, heat treatment, or warping strains. Waviness is sometimes referred to as a *secondary texture*.

Lay is the direction of the prominent surface pattern ordinarily determined by the production method used to form the surface.

Flaws are irregularities that occur in varying locations or at widely varying intervals in a surface. Flaws include such defects as cracks, blowholes, checks, ridges, and scratches. Unless otherwise specified, flaws are not included in the roughness-height measurements.

The *surface texture symbol* is used to designate the control of surface irregularities. Figure 8-22 illustrates the use of surface texture symbols and recommended drafting proportions. The symbol is always placed in an upright position with the point of the symbol touching a

BURGESS-NORTON MFG. CO.

FIGURE 8-21 Superfinished cylindrical parts. The average roughness height rating is about 4 microinches *(Section 8.20)*

	Symbol	Meaning
(a)	√	Basic Surface Texture Symbol. Surface may be produced by any method except when the bar or circle (Figure 1b or 1d) is specified.
(b)	√ with horizontal bar	Material Removal By Machining Is Required. The horizontal bar indicates that material removal by machining is required to produce the surface and that material must be provided for that purpose.
(c)	3.5 √ with horizontal bar	Material Removal Allowance. The number indicates the amount of stock to be removed by machining in millimeters (or inches). Tolerances may be added to the basic value shown or in a general note.
(d)	√ with circle	Material Removal Prohibited. The circle in the vee indicates that the surface must be produced by processes such as casting, forging, hot finishing, cold finishing, die casting, powder metallurgy or injection molding without subsequent removal of material.
(e)	√ with extended horizontal line	Surface Texture Symbol. To be used when any surface characteristics are specified above the horizontal line or to the right of the symbol. Surface may be produced by any method except when the bar or circle (Figure 1b and 1d) is specified.
(f)	3 X; 1.5 X; 3 X APPROX.; 00; 00; ⊥ 00; 60°; 60°; 0.00; 3 X; 1.5 X; LETTER HEIGHT = X	

ANSI Y14.36-1978

FIGURE 8-22 Surface texture symbols and recommended drafting proportions *(Section 8.20)*

line representing the surface, or an extension line of the surface. Figure 8-23 shows an example of a part with a variety of surface texture symbols. As shown, the symbol may be used following a diameter dimension.

Examples in Figure 8-24 illustrate designations of roughness, waviness, and lay by inserting the values in appropriate positions relative to the symbol. Symbols for designating the direction of lay are shown in

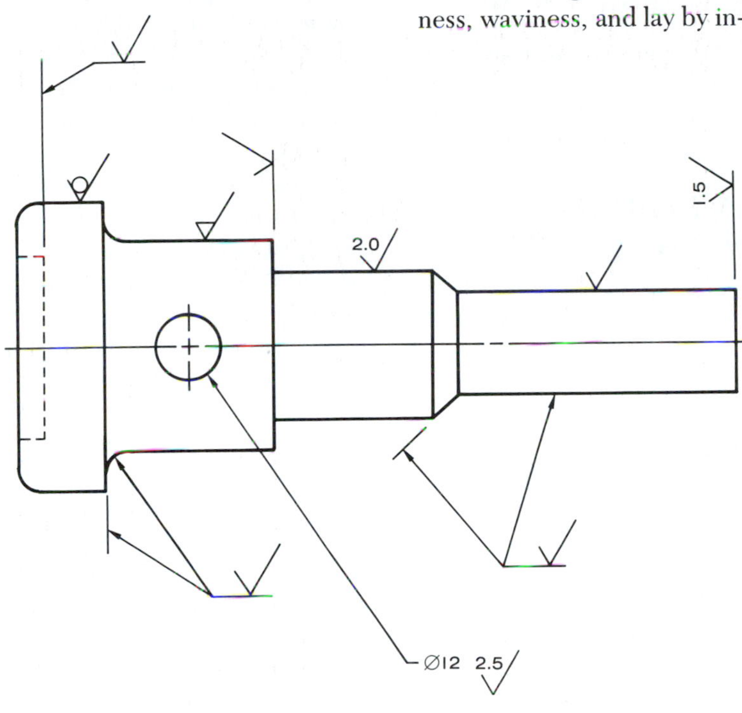

UNLESS OTHERWISE SPECIFIED, ALL SURFACES 3.0

FIGURE 8-23 Applications of surface texture symbols *(Section 8.20)*

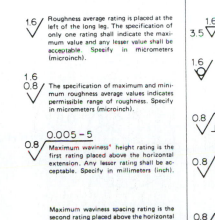

1.6	Roughness average rating is placed at the left of the long leg. The specification of only one rating shall indicate the maximum value and any lesser value shall be acceptable. Specify in micrometers (microinch).	1.6, 3.5	Material removal by machining is required to produce the surface. The basic amount of stock provided for material removal is specified at the left of the short leg of the symbol. Specify in millimeters (inch).
		1.6	Removal of material is prohibited.
1.6, 0.8	The specification of maximum and minimum roughness average values indicates permissible range of roughness. Specify in micrometers (microinch).	0.8 ⊥	Lay designation is indicated by the lay symbol placed at the right of the long leg.
0.8, 0.005 – 5	Maximum waviness[4] height rating is the first rating placed above the horizontal extension. Any lesser rating shall be acceptable. Specify in millimeters (inch).	0.8, 2.5	Roughness sampling length or cutoff rating is placed below the horizontal extension. When no value is shown, 0.80 mm (0.030 inch) applies. Specify in millimeters (inch).
	Maximum waviness spacing rating is the second rating placed above the horizontal extension and to the right of the waviness height rating. Any lesser rating shall be acceptable. Specify in millimeters (inch).	0.8 ⊥ 0.5	Where required maximum roughness spacing shall be placed at the right of the lay symbol. Any lesser rating shall be acceptable. Specify in millimeters (inch).

ANSI Y14.36-1978

FIGURE 8-24 Examples of designations of roughness, waviness, and lay *(Section 8.20)*

Figure 8-25(a). When appropriate, these symbols can be used with the surface texture symbols, as shown in 8-25(b).

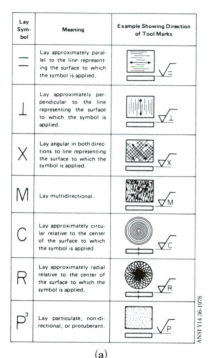

Lay Symbol	Meaning	Example Showing Direction of Tool Marks
=	Lay approximately parallel to the line representing the surface to which the symbol is applied.	=
⊥	Lay approximately perpendicular to the line representing the surface to which the symbol is applied.	⊥
X	Lay angular in both directions to line representing the surface to which the symbol is applied.	X
M	Lay multidirectional.	M
C	Lay approximately circular relative to the center of the surface to which the symbol is applied.	C
R	Lay approximately radial relative to the center of the surface to which the symbol is applied.	R
P[3]	Lay particulate, non-directional, or protuberant.	P

ANSI Y14.36-1978

(a)

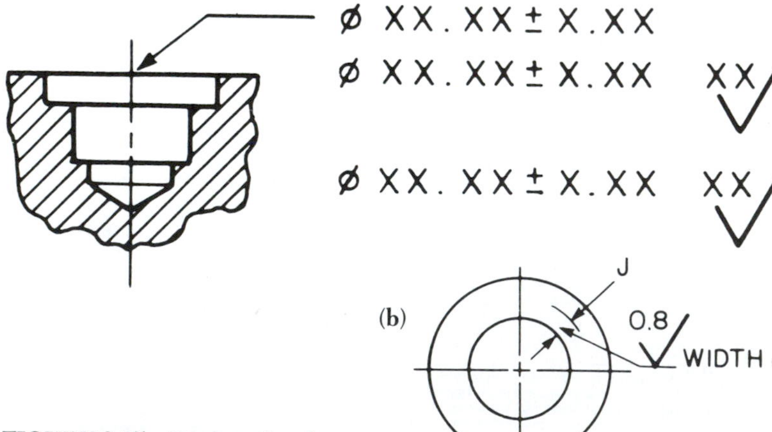

FIGURE 8-25 Designating the direction of lay *(Section 8.20)*

An example of special designations of surface roughness control is shown in Figure 8-26. In this case, the surface is initially produced by milling. A grinding operation follows, with the final surface texture produced by lapping.

8.21 NUMERICAL CONTROL (NC)

Numerical control is a technique for automatically controlling machine tools, equipment, or processes. It is generally agreed that NC provides one of the major technical innovations of our age and, in fact, some people have predicted that NC is the beginning of the second industrial revolution. It seems inevitable that NC will invade all repetitive manufacturing processes, large or small.

Principles

Information called the *part program* is supplied to the machine tool by means of punched or magnetic tape. The data are arranged in the form of blocks of information. Each coded numeric block consists of all the information necessary for processing one segment of the workpiece, including the tool path movements, spindle and worktable feed rates, and spindle speed. When machining a part, the tool bit cuts the workpiece in accordance with the information in one block and then pauses, or dwells, between blocks. Before the coded tape can be prepared, the programmer, often working with a planner or a process engineer, must select the appropriate NC machine tool, determine the workpiece material, calculate the necessary speeds and feeds, develop the sequence of coolant flow, determine the tool path

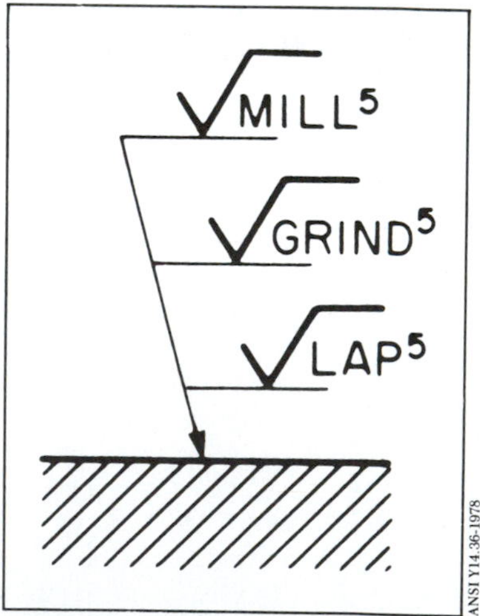

ANSI Y14.36-1978

FIGURE 8-26 Special designations of surface roughness control *(Section 8.20)*

and dwell cycle, and decide on the type of cutting tools needed for the various machining operations. The dimensions on the detail drawing are closely examined to determine suitable reference points from which to conduct the program. The machine control unit (MCU) receives and stores all coded data until a complete block of information has been accumulated. It then interprets the coded instruction and directs the machine tool through the required motions.

Advantages

The normal actions of the operator are largely eliminated in NC machining. The need for highly skilled and experienced operators is avoided. NC ensures identical part production, and since the accuracy is repeatable, inspection time is reduced. NC machine tools may function in as many as five or more different axis designations. Most NC applications, however, are accomplished by machine tools that function in three axes of motion: X, Y, and Z. As a result, contour or profile cutting in three dimensions is made possible by the remarkable freedom of the cutting tool movement. In many cases one NC machine tool can perform all of the machine operations formerly associated with several different conventional machine tools. A single machine tool used in one location eliminates the need for excessive material handling moves and repeated tool and workpiece setups. Finally, NC machining methods bring the benefits of mass production techniques to small-batch production of parts and even to the production of a single part.

Applications

The most widely used NC applications continue to be in the metal-cutting industries—in production operations such as turning, boring, counterboring, spotfacing, milling, cutter grinding, laser-beam cutting, and multifunction applications. The principal NC applications in the metal-forming industry are largely associated with conventional punching and shearing operations. Another important NC industrial application is in automatic welding.

8.22 COMPUTER-AIDED MANUFACTURING

An important advance in NC machine tools took place during the early 1970s in the form of a change toward the use of computers instead of the hardwired machine control units formerly employed. The development of the microprocessor is generally regarded as the greatest force in advancing the numerical control processing industry. The microprocessor serves as the control unit of a small computer. The development of this remarkable device allows microcomputers to be formed on a single printed-circuit board at a very low cost and operated as the main unit of the controller. The advent of microprocessors made computer numerical control (CNC) and direct numerical control (DNC) possible.

Computer Numerical Control (CNC)

The flexibility of an NC machine tool can be greatly increased by replacing the hardwired numerical machine controls with programmable mini- or microcomputer controls, commonly called CNC or computerized numerical control. In this system, tape or disk is the medium of transmitting the program from the programming computer to the dedicated computer of the machine tool. CNC is a self-contained NC system for a single machine tool that performs most or all of the basic NC functions. A main objective of the CNC system is to replace as much of the conventional NC hardware as possible with software and to simplify the remaining hardware.

CNC systems are very flexible and open-ended in that they may easily be expanded and modified as the state of the art continues to advance. CNC systems lend themselves to self-contained diagnostics and have editing capabilities for correcting errors. They can be used with cathode ray tube (CRT) readout options. A plotter may be utilized with the CNC for producing a line drawing of the part.

Direct Numerical Control (DNC)

Direct numerical control is fast reaching its full potential. The present DNC generation of NC machine systems is considered by some people to have reached the ultimate in data-processing capabilities. In manufacturing, disks are used for permanent storage of NC programs. System sensing devices automatically compensate for desired amounts of metal removal and preferred limits of accuracy. Special features are available that greatly expand the machine tool control functions that were not possible in previous technologies. In addition to the automatic production of parts, DNC can also be used for management and inventory control.

In DNC systems, a battery of NC or CNC machine tools are directly controlled by a central computer. Part programs are stored in the central computer. The maximum number of machine tools that can be controlled by a single DNC computer is governed primarily by the types of machines under control and the nature of the parts being produced. A range of five to ten NC machine tools is usually required before a system is considered economically justifiable. There are some DNC systems currently in use in which a single, general-purpose computer can control upwards of 200 NC machines. DNC systems equipped with a CRT permit a visual display that simplifies the "conversation" with the computer. When using a DNC system, rather than an individually controlled NC system, machine productivity is reportedly improved as much as 10 to 15% because of the comparative ease in optimizing the programs. Also, with DNC smaller lot sizes become practical or more profitable.

Chapter 15 covers other advanced design and manufacturing strategies, such as computer-aided design/computer-aided manufacturing systems (CAD/CAM).

Threaded Mechanical Fasteners

OBJECTIVES

After completing this chapter you should have gained the following abilities:

1. To select and apply threaded mechanical fasteners to meet design needs.
2. To understand the various thread forms and associated thread terms.
3. To choose correctly for a given application:
   a. right- or left-hand threads
   b. single-start or multiple threads
4. To understand the various methods used to produce external and internal threads.
5. To designate screw threads properly on a drawing, on a parts list, or on other engineering documents.
6. To establish a suitable distance for thread engagement.
7. To distinguish between a bolt and a screw and to select correctly the appropriate type of fastener for a given application.
8. To understand the recommended assembly applications for a stud.
9. To properly represent and designate American National Standard or Metric bolts, cap screws, machine screws, nuts, and studs.
10. To interpret and use correctly the data presented in appendix tables relating to threaded fasteners.

Most machines or structures contain threaded fasteners. The term "fasteners," as used in this chapter, broadly includes all types of removable threaded devices which may be used for a wide number of applications. Knowledge of fasteners and screw threads is so vitally important to the designer that omission of this subject in a text of this kind would be a serious injustice.

This chapter covers standardized practices that apply to screw thread nomenclature, definitions, and symbols for screw threads used on drawings. In addition various types of screw thread forms and recommended applications are discussed, along with procedures for representing and specifying them on a drawing.

Despite the countless engineering applications of screw threads, their uses can be conveniently grouped into three main categories: (1) to hold parts together (a nut and bolt, for example), (2) to adjust moveable parts relative to one another (such as the relative movement of parts on microscopes, binoculars, micrometers, and so on), and (3) to transmit power (a leadscrew on a milling machine or lathe, for example). Only the general types of standard industrial fasteners will be presented in this chapter; the variety that is actually available is considerably beyond the scope of this book. In fact, there are currently over 500,000 standard, off-the-shelf fasteners commercially available. Designers may select fasteners that offer virtually an infinite number of types, sizes, materials, and characteristics to match a wide range of design requirements. The appendix tables list the standard sizes and dimensions of some of the most commonly used threaded fasteners. These tables have been abstracted from various standardized documents published by the American Society of Mechanical Engineers. Most of the information contained in this chapter is based on published standards and specifications issued by technical organizations and societies, industrial groups, and individual companies.

9.1 SCREW THREADS

A *screw thread* is a ridge, usually of uniform section, in the form of a helix on the external or internal surface of a cylinder or in the form of a conical spiral on the external or internal surface of a cone or a frustum of a cone.

9.2 DEFINITION OF TERMS

Figure 9-1 illustrates the terms that apply to practically all screw thread forms. In the sections that follow, various screw thread forms and their uses and methods of representing and specifying them on the drawing will be given.

External thread: a thread on the external surface of a cylinder or cone, often referred to as a *male* thread.

Internal thread: a thread on the internal surface of a hollow cylinder or a cone, often referred to as a *female* thread.

Major diameter: for a straight thread, the diameter of the imaginary cylinder bounding the crest of an external thread. For a tapered thread the major diameter at a given position on the thread axis is that of the major cone at that position.

Minor diameter: for a straight thread, the smallest diameter of the thread.

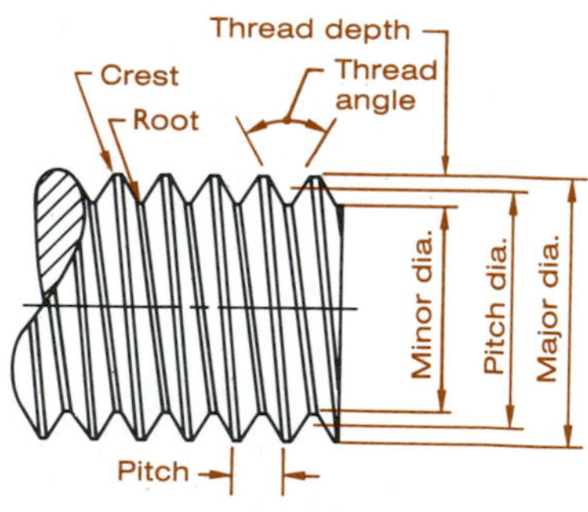

FIGURE 9-1 Screw thread terminology *(Section 9.2)*

Pitch diameter: the diameter of an imaginary cylinder passing through the threads where the width of the thread and the width of the space between the threads are the same.

Pitch: the distance from a point on one thread to the corresponding point on the next thread measured parallel to the axis. Pitch is equal to 1 divided by the number of threads per inch.

Lead: the distance a threaded part moves axially in one complete revolution with respect to a fixed mating part.

Crest: the outermost tip of an external thread as seen in a thread profile or the innermost tip of an internal thread.

Root: the bottom surface of the thread cut into a cylinder.

Angle of thread: the angle included between the sides of the thread measured in a plane through the axis of the screw.

Axis of thread: the longitudinal center line of the thread.

Depth of thread: the distance between the crest and the root of the thread measured in a direction normal to the axis.

Form of thread: the cross-sectional shape of the thread cut by a plane containing the axis.

Series of thread: standard numbers of threads per inch for various diameters.

9.3 THREAD FORM

Thread form refers to the profile of the thread for a length of one pitch in an axial plane. Figure 9-2 illustrates various thread forms, each of which was developed for a specific purpose.

The *Unified thread* is used for most of the threaded fasten-

FIGURE 9-2 Examples of screw thread forms ***(Section 9.3)***

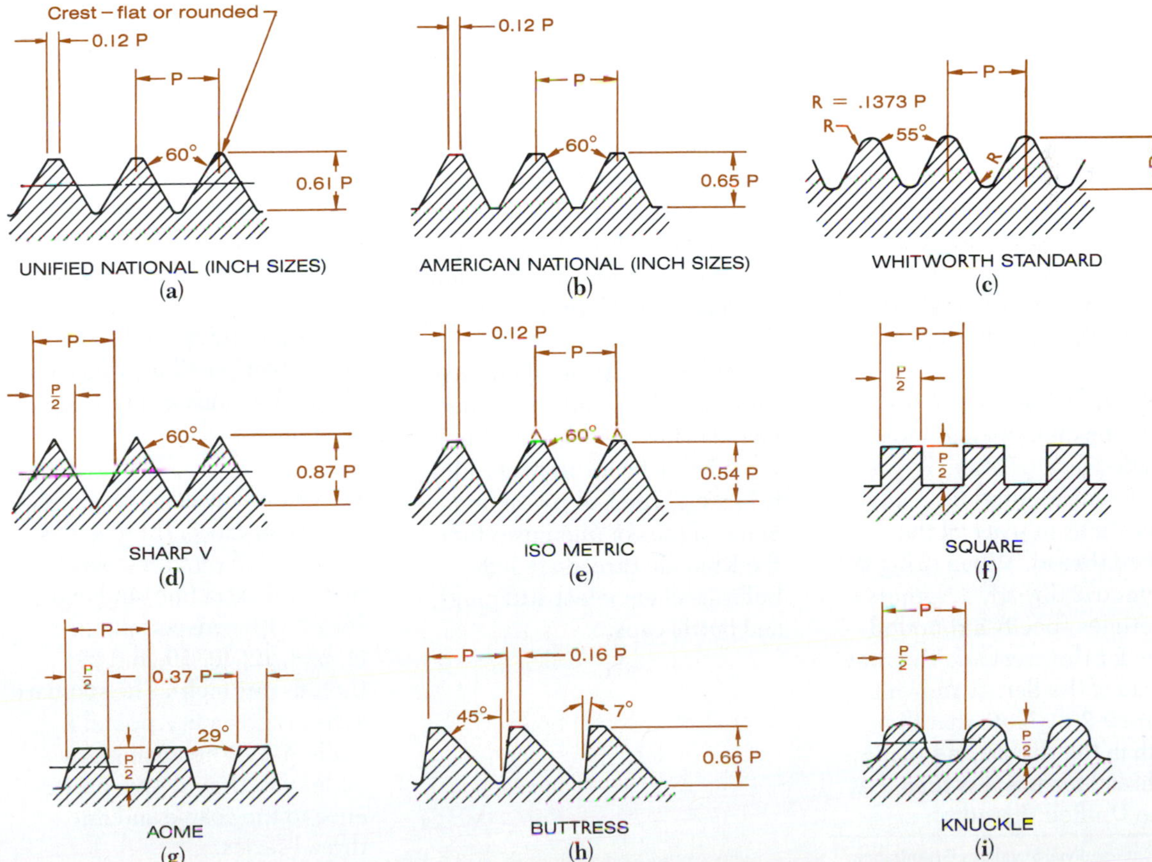

ers manufactured in the United States and is currently the United States standard. The Unified thread form was agreed upon as a standard by the United States, Canada, and Great Britain in 1948. The form adopted is a modification of the *American National thread* and the *Whitworth thread*. The crest of the external thread may be flat or rounded, but the root is always rounded. There is considerable recent evidence that many manufacturers are replacing the Unified thread with the Metric thread. Also, the American National thread and the Whitworth thread are gradually being replaced with Unified or Metric threads. Note that the angle of the Whitworth thread is 55°, not 60° as on the other V-threads.

Because of the full thread face, the 60° *Sharp-V thread* is used when friction and holding power are desired between working parts, as on setscrews, for example. Another use of the Sharp-V thread is for piping joints.

The *ISO* metric thread* is appearing on more and more threaded fasteners and is gradually being adopted for many applications in place of the Unified thread. When using the ISO metric thread, designers sometimes specify a rounded shape for the crest and the root instead of the flat, as shown in Figure 9-2(e). Note that the depth of the thread on the metric thread form is less than that of the Unified thread.

*ISO: International Standards Organization.

The *Square thread* and the *Acme thread* are used for power transmission. The nonstandard Square thread is rarely used, largely because of the difficulties in making the nearly square cut and the difficulties in engaging mating parts. To improve the thread engagement, the thread may be slightly modified by cutting each side with a 5° taper, making a 10° included angle.

The *Acme thread* is widely used for transmission of power and motion applications. A common application of an Acme thread is for feeding and operating screws on machine tools, vises, and so on.

The *Buttress thread* is designed to withstand high stresses in only one direction. Its major applications include breech locks on large artillery pieces, airplane propeller hubs, and screw jacks. The Buttress thread form is not standardized; consequently, you should expect to encounter various modifications in proportional sizes on existing threaded parts.

Although of general interest, the *Knuckle thread* has limited engineering applications. Such a thread may be formed by rolling thin metal stock or by casting. Some of the common uses for the Knuckle thread are light bulbs, sockets, electrical plugs, and bottle caps.

9.4 THREAD SERIES

A screw thread with a large pitch is characterized by deep grooves. If the screw were of a small diameter in proportion to the large pitch, the cross section would be reduced, resulting in a weak screw. The pitch of the thread must be properly proportioned to the screw diameter. To achieve standardization, various diameter-pitch combinations have been established for all forms of screw threads. These combinations are known as *thread series* and are distinguished from each other by the number of threads per inch, correlated to a series of specific diameters.

Unified Inch Screw Thread Series

There are two general thread series, classified as *standard* and *special*. Only the standard series will be discussed in this book. Use ANSI B1.1-1974 for details regarding the special series. Whenever possible, you should make selections from the standard series.

STANDARD SERIES. The standard series consists of three series with graded pitches (coarse, fine, and extra fine) and eight series with constant pitches (4, 6, 8, 12, 16, 20, and 28 threads per inch). The standard series is given in Appendix Table A-1. When using this table, you should give preference to the coarse and fine thread series.

COARSE THREAD SERIES (UNC). This series is used for the great majority of screws, bolts, and nuts. Coarse threads are commonly used for relatively low-strength materials, such as cast iron, aluminum, magnesium, brass, bronze, and plastic, because the coarse series threads provide more resistance to internal thread stripping than the fine or extra-fine series. Coarse thread series are utilized if rapid assembly and disassembly are required or if corrosion or damage from nicks because of handling or use is likely.

FINE THREAD SERIES (UNF). The fine thread series is commonly used for bolts, screws, and nuts in high-strength applications, particularly in the automotive and aircraft industries. Fine series threads have less tendency to loosen under vibration than coarse threads because of the smaller lead angle. (The lead angle is the angle made by the helix of the thread at the pitch line with a plane perpendicular to the axis.) This also allows for finer adjustment in cases such as a slotted nut and cotter pin assembly.

EXTRA-FINE THREAD SERIES (UNEF). This series is used particularly for equipment and threaded parts under high stresses that require fine adjustment, such as sheet metal parts, bearing retaining nuts, adjusting screws, thin walled tubing ferrules, and thin nuts. Extra-fine threads are often used when a maximum practical number of threads is required for a given length of thread engagement.

CONSTANT-PITCH THREAD SERIES. The various constant-pitch series (N or UN) offer a comprehensive range of diameter-pitch combinations for those purposes where the threads in the coarse, fine, or extra-fine series do not meet the particular requirements of the design. The primary sizes of the 8UN, 12UN, and the 16UN series shown in Appendix Table A-1 are the most commonly used. The 8-pitch series is often used for bolts on high-pressure pipe flanges, cylinder head studs, and similar fasteners. The 12-pitch thread series is used for large-diameter fasteners in boiler work, for thin nuts on shafts, and so on. The 16-pitch thread series is used if a fine thread is necessary, particularly on large-diameter fasteners such as retaining nuts and adjusting collars.

ISO Metric Thread Series

The diameter-pitch combinations for metric screw threads are given in Appendix Table A-2. This table lists both standard and nonstandard diameter-pitch combinations. In addition to a coarse- and fine-pitch series, a series of constant pitches is available. A comparison of thread sizes is shown in Figure 9-3.

9.5 THREAD CLASSES OF FIT

Fit is a term used to signify a range of tightness or looseness between mating threaded parts. The *classes of fit* are distinguished from each other by the amounts of tolerance and allowance. The required assembly of fits may be obtained by selecting the proper thread class for each component.

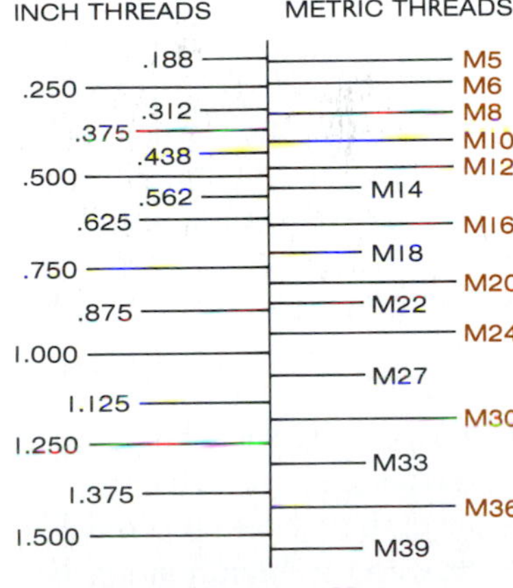

FIGURE 9-3 Comparisons of thread sizes—inch versus metric *(Section 9.4)*

Unified Inch Screw Threads

Classes of fit are denoted by the numerals 1, 2, or 3 followed by the letters A or B. (The higher the thread class number, the closer the fit.) The letter A signifies external threads on parts such as studs, screws, and bolts. Letter B represents internal threaded elements such as nuts. Examples given in Section 9.9 show how to specify the class of fit in a thread note.

CLASSES 1A AND 1B. These threads provide a loose fit and are used if parts must be assembled and disassembled quickly and easily. Liberal tolerance and allowance are required.

CLASSES 2A AND 2B. These medium fits are the most commonly used thread classes for general applications, including the high-volume commercial production of bolts, screws, nuts, and similar threaded fasteners.

CLASSES 3A AND 3B. These threads are used where a close fit and accuracy of lead and angle of thread are required. Parts with this high-quality class of fit are used in high-stress and vibration applications.

ISO Metric Screw Threads

A fit between mating metric threads is indicated in a manner that is quite different from that of a fit for Unified threads. A range of *tolerance grades* has been adopted to designate the degree of fit for external and internal threads. The numeral assigned to the tolerance grade reflects the size of the tolerance. Basically, grades 3, 4, 6, and 8 are used. Smaller grade numbers carry smaller tolerances. In other words, grade 4 tolerances are smaller than grade 6 tolerances, and grade 8 tolerances are larger than grade 6 tolerances. Grade 6 tolerances correspond to the Unified 2A and 2B classes of fit and are recommended for general purpose requirements with medium lengths of engagement. Tolerance grades below 6 are used for fine quality thread applications or short lengths of engagement.

In addition to the tolerance grade, a letter symbol designating the *tolerance position* is also required. This tolerance defines the maximum material limits of the pitch and major diameters for both external and internal threads. Lowercase letters are used for external threads and capital letters for internal threads. The symbols designating tolerance position are as follows:

External threads:
e = large allowance
g = small allowance
h = no allowance

Internal threads:
G = small allowance
H = no allowance

Used in combination, the tolerance grade number and the tolerance position letter are called the *tolerance class*. These classes are fine, medium, and coarse. In the tolerance class designation 4g6g, for an external thread, the first number and letter combination (4g) refers to the pitch diameter, and the second number and letter combination (6g) refers to the major diameter. In the designation 5H6H, for an internal thread, the first number and letter (5H) refer to the pitch diameter, but the second number and letter designate the *minor* diameter. The method of designating the tolerance class in a thread note will be explained in Section 9.10.

9.6 RIGHT-HAND AND LEFT-HAND THREADS

Almost all screw threads are right-hand. A screw with a right-hand thread advances into a threaded hole when turned clockwise; conversely, a screw with a left-hand thread advances when turned counterclockwise. Figure 9-4 shows the difference between right-hand and left-hand threads on a double-end

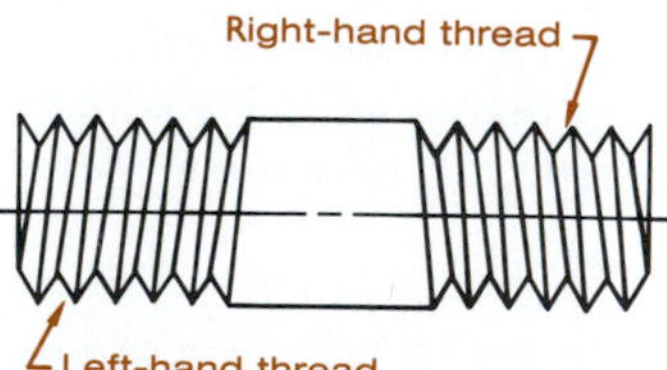

FIGURE 9-4 Right-hand and left-hand threads *(Section 9.6)*

stud. A right-hand thread on a stud or a bolt held horizontally slants upward to the left, and a left-hand thread slants upward to the right. Unless otherwise specified, a thread is right-hand. A left-hand thread is always labeled LH in the thread specification.

9.7 SINGLE-START AND MULTIPLE-START THREADS

A *single-start* thread is shown in Figure 9-5(a). Most screws in common use have single threads, and unless otherwise specified, the thread is assumed to be single. The single thread has one ridge in the form of a helix; for this thread, the lead is equal to the pitch (L = P). *Multiple-start* threads apply to double and triple threads. A double thread, shown in Figure 9-5(b), has two ridges. For this thread, the lead equals twice the pitch (L = 2P). A triple thread, shown in Figure 9-5(c), has three ridges. The lead equals three times the pitch (L = 3P).

Multiple-start threads are used for clamping screws, valves, fountain pens, and so on, when rapid movement is desired with a minimum number of rotations. Whenever double or triple threads are used, quick motion is obtained, but with less thrust.

9.8 HOW SCREW THREADS ARE PRODUCED

Threads may be produced by a number of processes and types of equipment. The process and/or equipment selected depend upon the workpiece material, the size, the required accuracy and finish of the threads, and the required production rate.

FIGURE 9-5 **Single and multiple threads** ***(Section 9.7)***

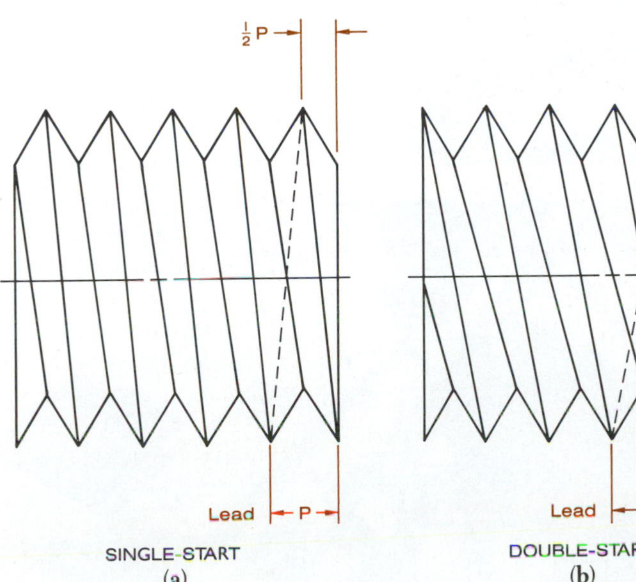

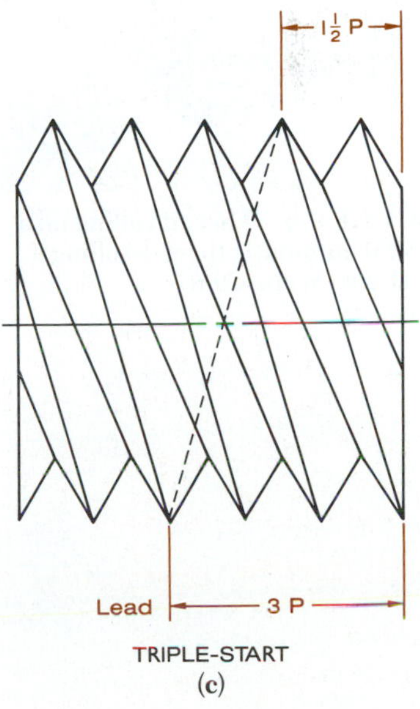

External Threads

Most of the threads produced today on commercial fasteners are *rolled* on special automatic thread-rolling machines that use flat or cylindrical dies. As shown in Figure 9-6, a thread is very rapidly cold-formed on a cylindrical workpiece blank between pairs of dies in one pass during the forward stroke of the machine. An engine lathe is commonly selected as a method to *cut* a thread, particularly if a relatively small quantity of parts is needed. Threads of various forms and sizes may be produced by taking a series of cuts on the workpiece blank with a single-point tool, as shown in Figure 9-7. Threads may also be cut on high-speed automatic threading machines by means of die heads and special inserted cutters called "chasers," which are held in tool bodies. This type of threading process is often re-

FIGURE 9-6 Thread rolling on a special automatic thread-rolling machine *(Section 9.8)*

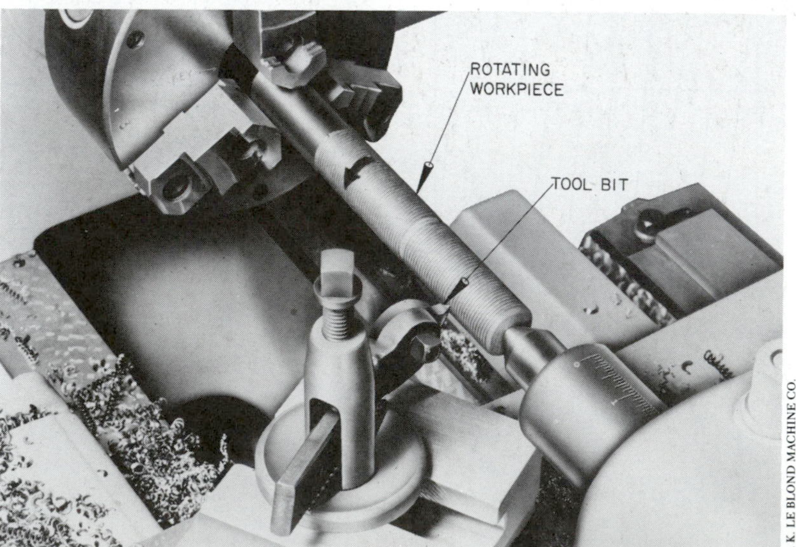

FIGURE 9-7 Thread cutting on a lathe *(Section 9.8)*

ferred to as "thread chasing." For small quantity work, threads may be manually cut by using an adjustable thread cutting die held in a die stock.

Internal Threads

Hardened steel cutting tools, called *taps*, are used to cut threads in holes. The actual cutting process is called *tapping* and can be performed on various machines or by hand methods. For tapping one or a limited quantity of small holes (1 inch in diameter or less), internal threads may be hand-tapped by using the tools shown in Figure 9-8. A tap wrench or a T-handle tap wrench is attached to the tap. The tap cuts threads on the hole wall as it is slowly wound into the hole. Figure 9-9 illustrates the following three styles of taps: *taper*, *plug*, and *bottoming*.

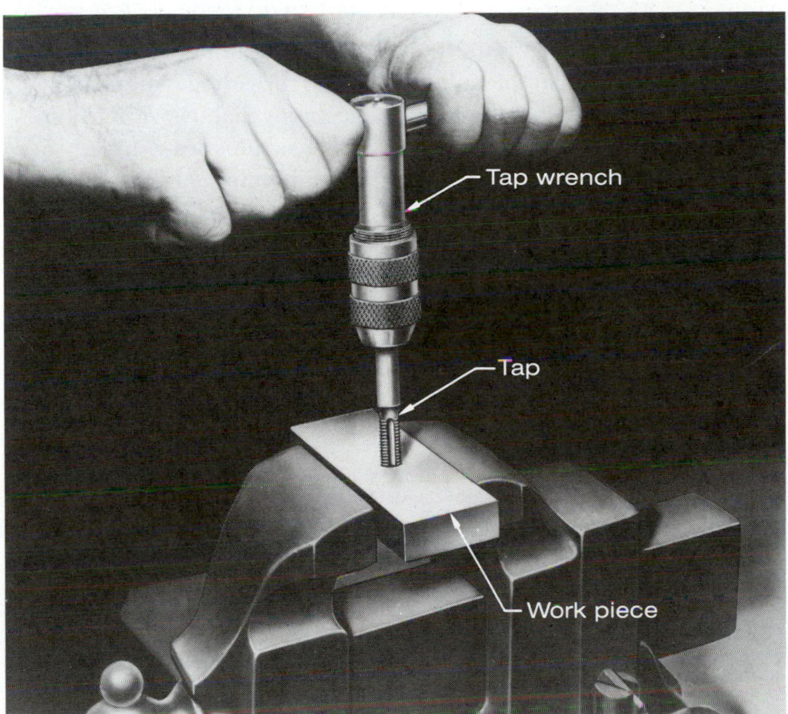

FIGURE 9-8 Cutting internal threads with a tap *(Section 9.8)*

FIGURE 9-9 Three styles of taps: taper, plug, and bottoming *(Section 9.8)*

Before internal threads can be produced, a hole slightly smaller in diameter than the major diameter of the thread must be drilled. This drilling operation is called *tap drilling*, and tap drill sizes for various thread sizes have been standardized. The drill size is dependent on the thread form and the nominal diameter of the thread. Tap drill sizes are listed in Appendix Tables A-1 and A-2.

Figure 9-10 illustrates the sequence of operations used to tap a hole. The first operation, shown in 9-10(a), consists of tap drilling. The bottom of the drilled hole is conical in shape. For drawing purposes, an angle of 30° is used, as shown. The drill depth does not include the cone point.

The taper and plug style taps are commonly used and are made with a slight taper at the cutting end—the taper tap having the longest taper. The use of the taper makes it easier to start the tap in the drilled hole but results in the formation of several imperfect threads. The second operation in tapping a hole is shown in Figure 9-10(b). The thread depth represents the length of *full* or *perfect* threads. Because the last of the threads cut by the end of the tap are imperfect, the drilled hole must extend into the material correspondingly deeper. Generally the drill depth is drawn (and specified) a distance equal to the thread depth plus about four or five times the pitch.

When it is necessary to extend a thread as deeply into a part as possible, as in Figure 9-10(c), a bottoming tap is used to finish the threads in the hole. Whenever possible, for manufacturing economy, threaded holes should extend completely through the material, as shown in 9-10(d).

Larger sizes of internal threads, from 1 inch in diameter and up, are often cut in a lathe. The hole to be threaded is first drilled or bored to the proper size. A boring bar with an inserted thread tool bit is then set up, and the cutting process proceeds in a similar manner to external threading. Automatic machines that use various types of special taps are generally used for high-volume hole threading. Holes can also be tapped by using a drill press with a special tapping attachment.

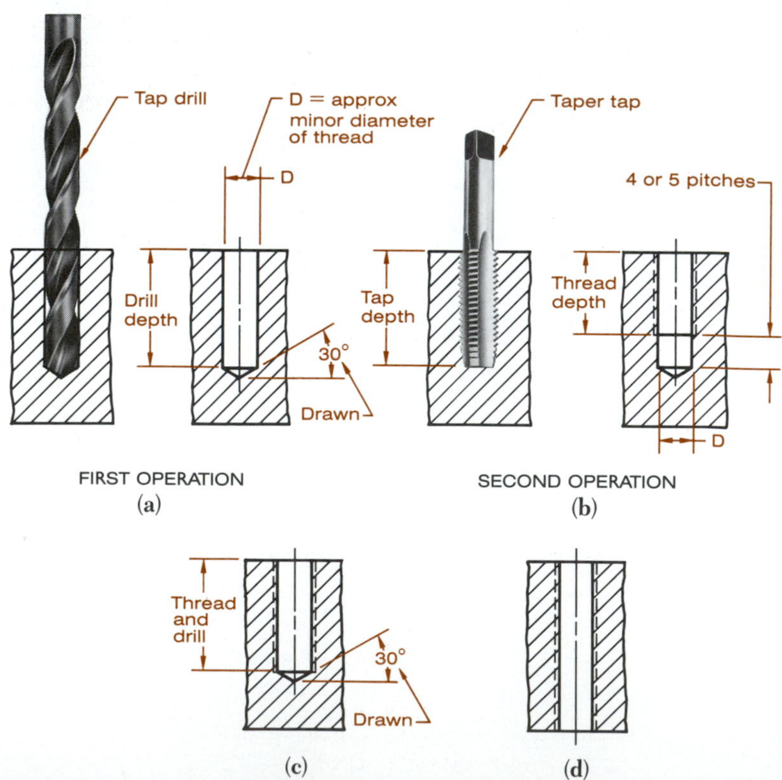

FIGURE 9-10 Drilling and tapping holes *(Section 9.8)*

9.9 REPRESENTATION OF SCREW THREADS

Screw threads are rarely shown on production drawings in *true representation*. Instead, they are shown on a drawing as a simple approximation or are represented with standard symbols. Unless the thread specification is given, it is not possible to identify the intended thread form or to produce the required thread form the way it appears on the drawing. With some minor exceptions, the same symbols are used to represent all forms of straight and tapered screw threads on a drawing.

Two systems of symbolic representation are in general use for representing screw threads on drawings. These are *simplified* and *schematic*. Only one method is used within any one drawing.

FIGURE 9-11 Representing external threads *(Section 9.9)*

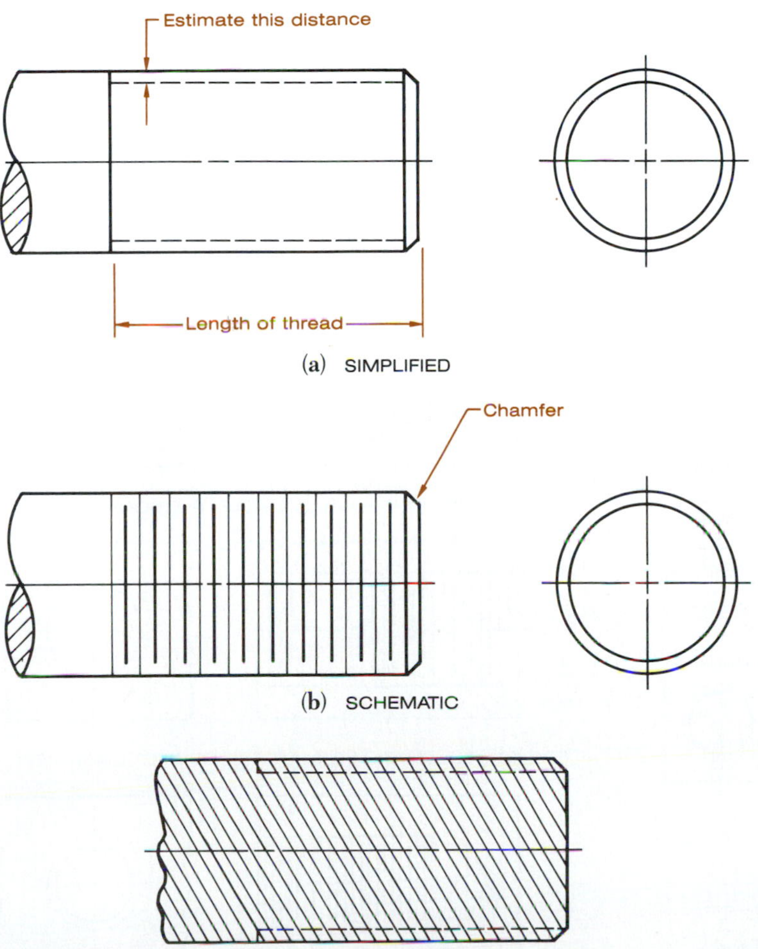

External Threads

The simplified thread symbol, shown in Figure 9-11(a), is the most commonly used representation. The minor-diameter dashed lines are estimated entirely by eye. For 1-inch-diameter screw threads or less, a satisfactory proportion can be obtained by spacing the depth of thread from 1/32 to 1/16 inch. The space may be increased for larger thread sizes.

The schematic representation, shown at 9-11(b), is entirely symbolic. All distances pertaining to the thread symbol are approximated by eye. Staggered lines, symbolic of the thread roots and crests, are indicated by alternate heavy and thin parallel lines. These lines may be spaced from 1/16 to 1/8 inch apart for 1-inch-diameter screw threads or smaller, with a wider space for larger thread sizes. No attempt should be made to draw the pitch to scale. The position of the root lines should correspond approximately with the root diameter. The preferred method for representing an external thread in section is shown in Figure 9-11(c). The schematic thread symbol is not used for sectional views of external threads. Except in special cases, standard parts such as bolts, screws, pins, rivets, and keys are left unsectioned.

A 45° chamfer is generally drawn on the end of a threaded element except on machine screws, which are not chamfered. For drawing purposes, the size of the chamfer circle should correspond to the minor or root diameter of the thread.

Internal Threads

Figure 9-12 shows various examples of internal threads drawn with the simplified and schematic representations. All drawing proportions previously given for external threads continue to apply. The schematic symbol is not used to represent hidden holes. Note that the visible circular view of a threaded hole is represented by a dashed-line circle that is equal in size to the major diameter and a visible concentric circle that is equal in size to the minor diameter.

CHECKPOINT 1
REPRESENTING EXTERNAL AND INTERNAL THREADS

Using a plain sheet of 8½ × 11 inch paper, draw a horizontal line dividing the sheet into two halves. In the upper half, in a manner similar to that shown in Figure 9-11, draw two neatly spaced partial cylinders labeled (a) and (b) according to the following directions: (a) Use the simplified thread representation for an external thread. Draw a cylinder with a ¾-inch major diameter, a 1⅝-inch thread length, and a 45° chamfer. (b) Use the schematic thread representation for an external thread. Draw a cylinder with a 1-inch major diameter, a 3-inch thread length, and a 45° chamfer.

On the lower half of your sheet, in a manner similar to that shown in Figure 9-12, draw three neatly spaced partial views to any convenient size labeled (a), (b), and (c) according to the following directions: (a) Use the correct thread representation for a hidden internal thread (a through hole) having an 11/16-inch major diameter. (b) Use the simplified thread representation for a sectional view of an internal thread (a through hole) having a ⅜-inch major diameter. (c) Use the schematic thread representation for a sectional view of an internal thread having a 1-inch major diameter with a thread depth of 1 inch. (Hint: The drill depth should extend a distance equal to 4 or 5P beyond the thread depth.)

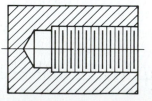

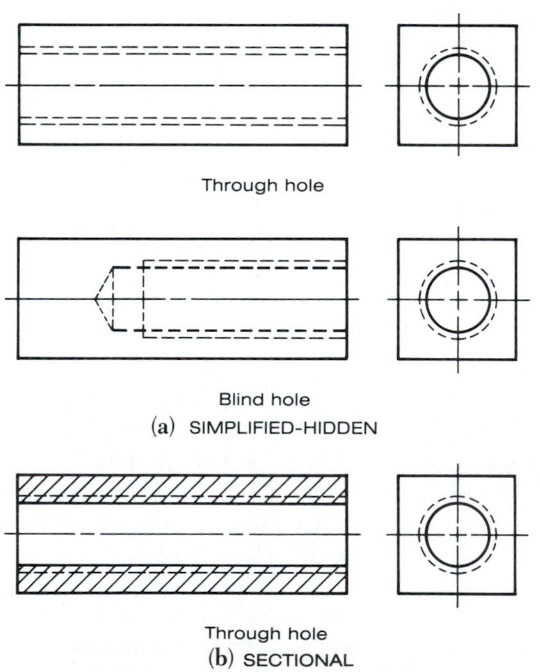

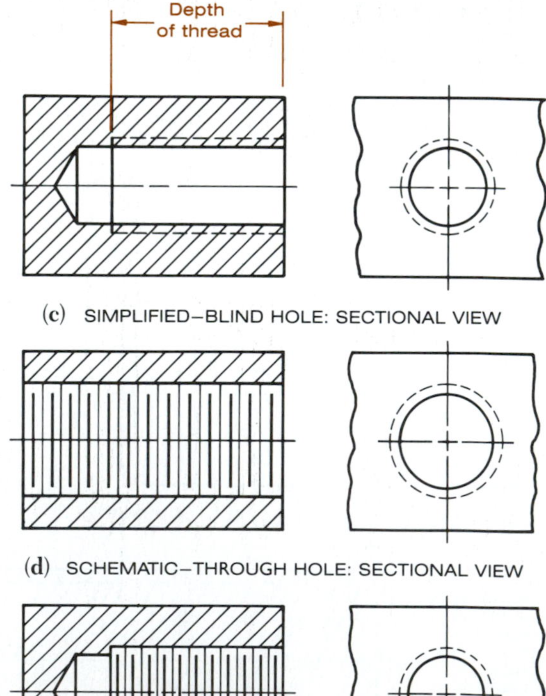

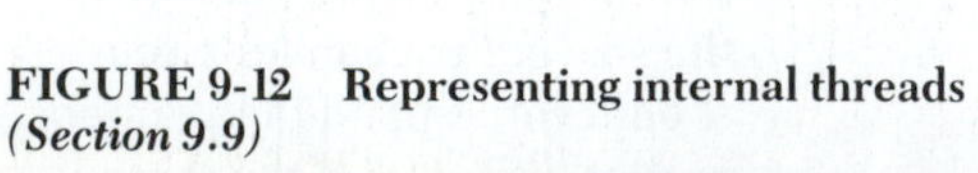

FIGURE 9-12 Representing internal threads
(Section 9.9)

9.10 DESIGNATION OF SCREW THREADS

As previously stated, it is impossible to identify the intended thread form or to produce the required thread form from the symbolic representation on the drawing. Threads are designated on a drawing by means of carefully worded *notes* or *callouts*. Generally the notes are connected to the view in which the thread is drawn by a leader. Thread notes for external threads should be given in the longitudinal view of the threaded shaft. Thread notes for internal threads should be given in the circular view of the hole.

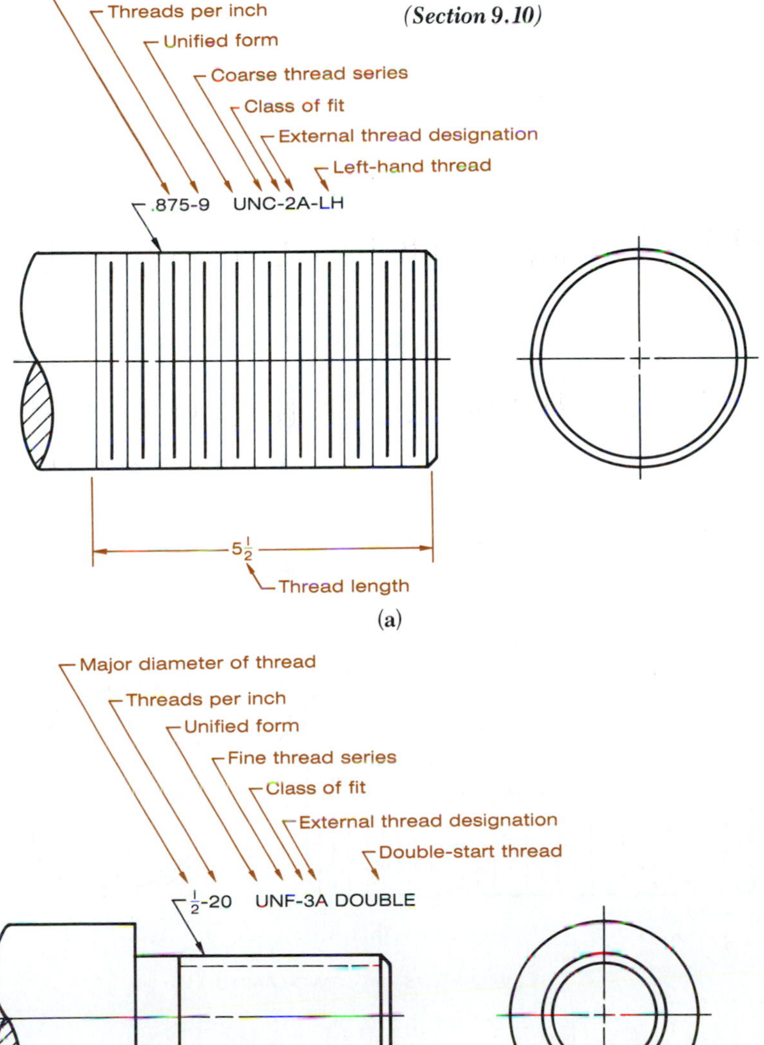

FIGURE 9-13 Designation of external threads—inch *(Section 9.10)*

External Threads—Inch

Figure 9-13 illustrates the elements of screw thread designations. At 9-13(a) the major diameter is expressed in decimal form. Unless otherwise specified, as in this example, the threads are right-hand. At 9-13(b) the major diameter is expressed in fractional form. The designation DOUBLE signifies a multiple thread. A more precise method of designating a multiple thread is to place the number of starts with the word "start" in parentheses following the pitch and lead values (2 START, for example). In both examples, the designation letter A for an external thread has been given. It is common practice to give the letter designation only in cases when the thread is not obviously external or internal. An example of a thread specification for a square thread is shown in 9-13(c). The nominal diameter is given first, followed by the number of

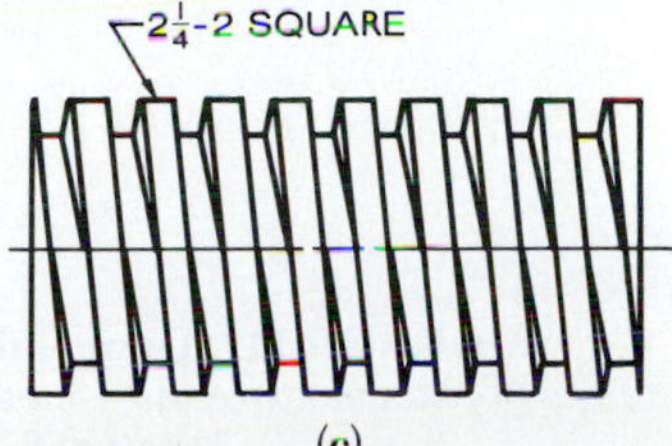

threads per inch and the type of thread. Standard pitch-diameter combinations for square and Acme threads are given in Appendix Table A-5. Specifications for general-purpose Acme threads are given in Appendix Table A-6.

External Threads—Metric

BASIC DESIGNATION. For many applications a basic designation of a metric thread, shown in Figure 9-14(a), may be used. This somewhat abbreviated designation consists of the letter M for the thread form profile, followed by the nominal diameter size, and the pitch is expressed in millimeters separated by the sign ×.

COMPLETE DESIGNATION. When it is necessary to show the tolerance class designation, a complete designation is given. The note shown in Figure 9-14(b) begins, as before, with the basic designation (M30 × 3.5), to which has been added the tolerance class designation 5g6g. As outlined in Section 9.5, three tolerance grades (4, 6, and 8) are generally used, with the smaller numbers denoting smaller tolerances. Letter symbols following the grade numbers are used to denote the tolerance positions.

Lowercase letters are used to designate external threads. The designation 5g, for example, signifies a thread having a medium tolerance with a small allowance for the pitch diameter. The 6g signifies a slightly wider tolerance with a small allowance for the major diameter.

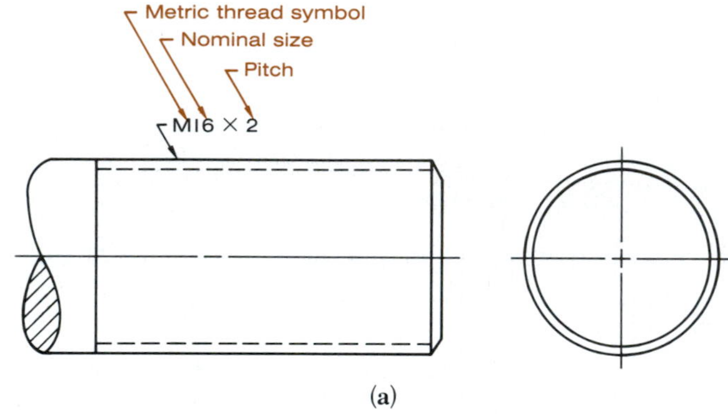

(a)

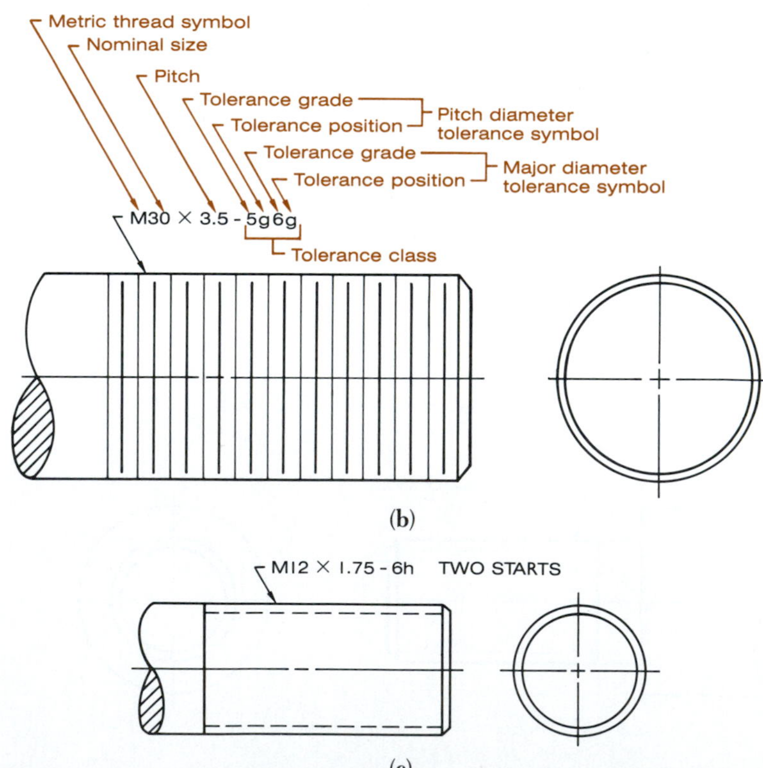

(b)

(c)

FIGURE 9-14 Designation of external threads—metric *(Section 9.10)*

If the pitch and major diameters have identical grades, the tolerance class symbol 6h is used, as shown in Figure 9-14(c). When a left-hand thread is specified, the tolerance class designation is followed by a space and the letters LH. A thread with a multiple start is designated by specifying TWO STARTS or THREE STARTS. The number of starts is the final item listed in the thread designation.

Because of the complex nature of the tolerance class designation for metric threads, the standard ANSI B1.13-1979 should be carefully studied before attempting to specify completely a metric thread on a drawing. Tables in this standard may be used to establish the length of engagement for various ranges of major diameters, recommended allowances, and tolerances for external and internal threads.

Internal Threads—Inch

A thread note for an internal unified screw thread that extends completely through the material is shown in Figure 9-15(a). The designation specifies, in sequence, the tap drill size, the nominal size, the number of threads per inch, the thread form, the series, and the class of fit. A thread note for a blind tapped hole is shown in Figure 9-15(b). The note follows the same sequence as 9-15(a) except that the depth of the tap drill and the thread are given. As previously stated for external threads, when it is not obvious that the thread is internal, the letter B is used following the number representing the class of fit. Left-hand threads and multiple-start threads should be indicated, if necessary, as shown in Figure 9-15(c). If the designation applies to more than one hole, the number of holes required should be indicated. Tap drill sizes are given in Appendix Table A-1.

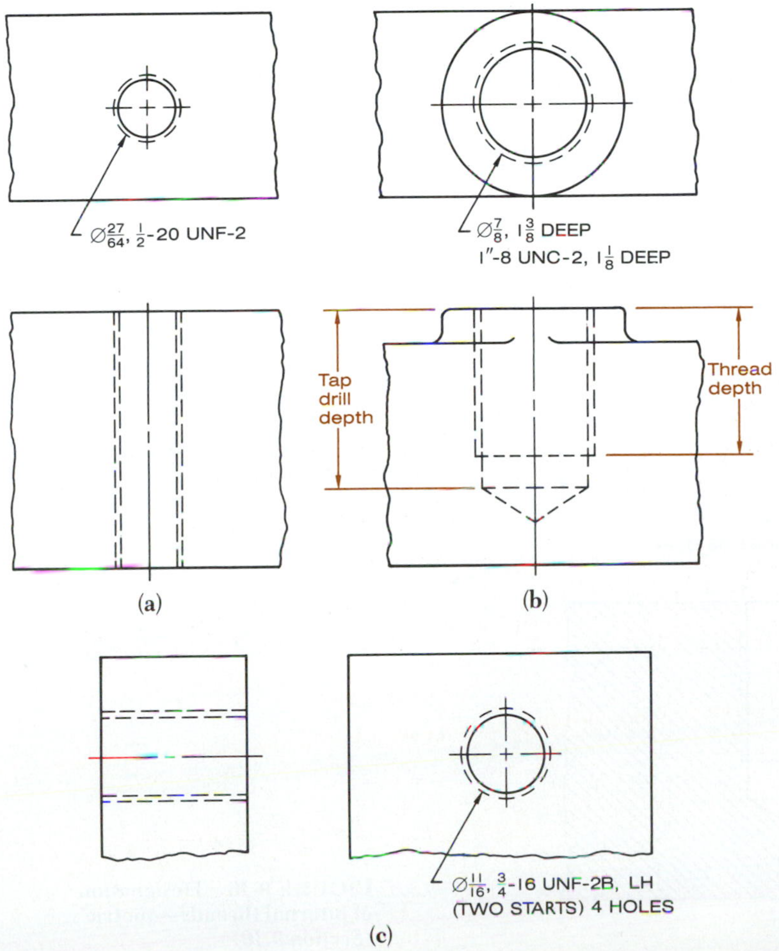

Ø$\frac{27}{64}$, $\frac{1}{2}$-20 UNF-2

(a)

Ø$\frac{7}{8}$, 1$\frac{3}{8}$ DEEP
1″-8 UNC-2, 1$\frac{1}{8}$ DEEP

Tap drill depth

Thread depth

(b)

Ø$\frac{11}{16}$, $\frac{3}{4}$-16 UNF-2B, LH
(TWO STARTS) 4 HOLES

(c)

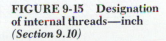

FIGURE 9-15 Designation of internal threads—inch *(Section 9.10)*

Internal Threads—Metric

The basic designation of an internal metric screw thread is shown in Figure 9-16(a). The sequence of information in this note is identical to that used to designate an external thread except the tap drill size is specified. Tap drill sizes are given in Appendix Table A-2. The complete designation for an internal metric screw thread is given in Figure 9-16(b). In this example, the depths of both the tap drill and the tap are given. The symbols in the tolerance class designation 5H6H mean the following:

5H = Pitch diameter tolerance symbol
5 = Tolerance grade (medium)
H = Tolerance position (internal thread, no allowance)
6H = Minor diameter tolerance symbol
6 = Tolerance grade (medium)
H = Tolerance position (internal thread, no allowance)

9.11 SPECIFYING THREAD LENGTH

The thread length dimension on the drawing should correspond to the distance along which all threads have full or perfect form. The imperfect or incom-

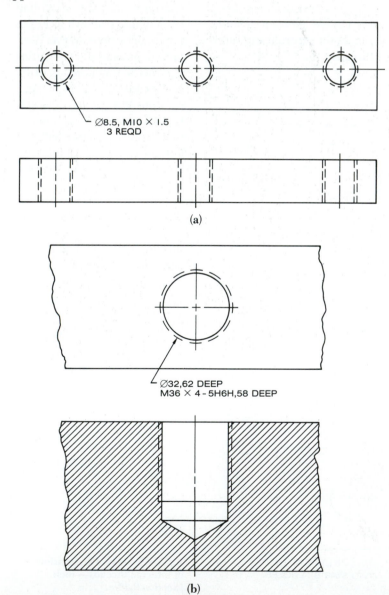

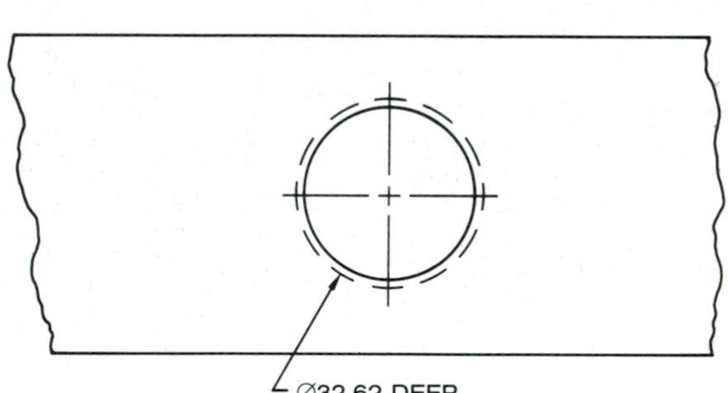

(a)

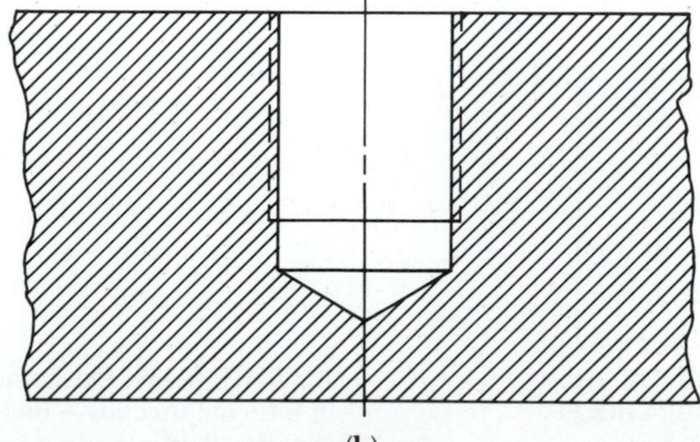

(b)

FIGURE 9-16 Designation of internal threads—metric ***(Section 9.10)***

plete threads should extend outside the length that is specified. When there is a reason to control or limit the number of incomplete threads on a given part, the overall thread length, including the "vanish threads" (meaning the runout or incomplete threads), should be represented and dimensioned as shown in Figure 9-17. The dimension for the full thread length should also be given.

CHECKPOINT 2
DESIGNATING SCREW THREADS

Using the data given for each of the following examples, neatly letter on a plain sheet of 8½ × 11 inch paper (using guidelines) the correct designation of the screw thread. Use the format shown in Figures 9-13, 9-14, or 9-16, and use the appendix tables to develop the correct values.

1. ***External Threads—Inch***

Thread Data	***Example 1***	***Example 2***
Major diameter	½	1
Thread form	U	U
Thread series	C	F
Class of fit	2	3
Left or right hand	RH	LH
Type of start	Single	Double

2. ***External Threads—Metric***

Thread Data	***Example 1***	***Example 2***	***Example 3***
Nominal size	20	10	5
Tolerance class	6g6g	6g	Not necessary

3. ***Internal Threads—Inch***

Thread Data	***Example 1***	***Example 2***
Type of threaded hole	Through	Blind
Major diameter	⅜	¾
Thread form	U	U
Thread series	C	F
Thread depth	—	1¼
Class of fit	2	2
Left or right hand	RH	RH
Type of start	Single	Single

4. ***Internal Threads—Metric***

Thread Data	***Example 1***	***Example 2***
Type of threaded hole	Through	Blind
Nominal diameter	36	16
Thread depth	—	54
Tolerance class	4H5H	—

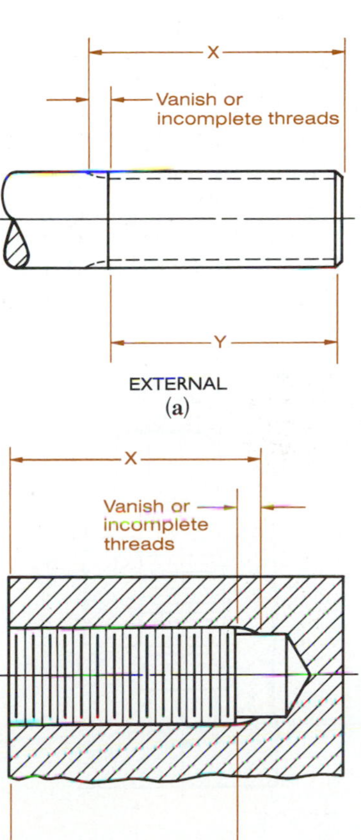

FIGURE 9-17 Specifying thread length *(Section 9.11)*

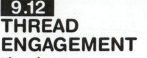

9.12 THREAD ENGAGEMENT

Thread engagement is measured by the distance the bolt or screw enters the material into which it is mated. When joining parts, vary the length of thread engagement according to the design requirements. Some of the factors that must be considered are in-service loads such as tension, torsion, shear, and bending or compression stresses, or a combination of these that may relate to the fasteners or the parts they attach. The type of fastener material and the material into which the fastener is threaded are also important.

Many designers and drafters use the empirical data shown in Table 9-1, as a guide for specifying the length of thread engagement for steel screws. The example shown in Figure 9-18 illustrates how the data in Table 9-1 can be used to assure adequate holding power for threads in mating parts. Assume a 1″-8 UNC steel hexagon-head cap screw is to be used to engage part A to part B in the position indicated by the center line in 9-18(a). Part B is cast iron.

TABLE 9-1 Recommended Thread Engagement

Steel screws	*Length of thread engagement (TE)*
Threaded into steel	Equal to major diameter
Threaded into cast iron, brass, or bronze	Equal to 1½ × major diameter
Threaded into aluminum, zinc, or plastics	Equal to 2 × major diameter

The tapered end of a plug tap (illustrated in Figure 9-9) will leave about four to five imperfect threads when screwed to the bottom of a drilled hole. Accordingly, a plug-tapped blind hole must be drilled at least four to five *pitches* deeper than the required thread depth.

As shown in Figure 9-18(a), Step 1, the thread engagement distance (1½ inches) has been set off as point X on the center line. The value for the thread engagement was obtained by multiplying 1½ times the major diameter, or 1½ × 1 inch = 1½ inches. Note that the thread engagement involves a distance that applies *only* to part B.

Step 2 illustrates two other distances that must be established. The depth of the thread contains only fully formed threads. It is customary to make this distance extend one or two pitches beyond the end of the engaging screw. Point Y denotes

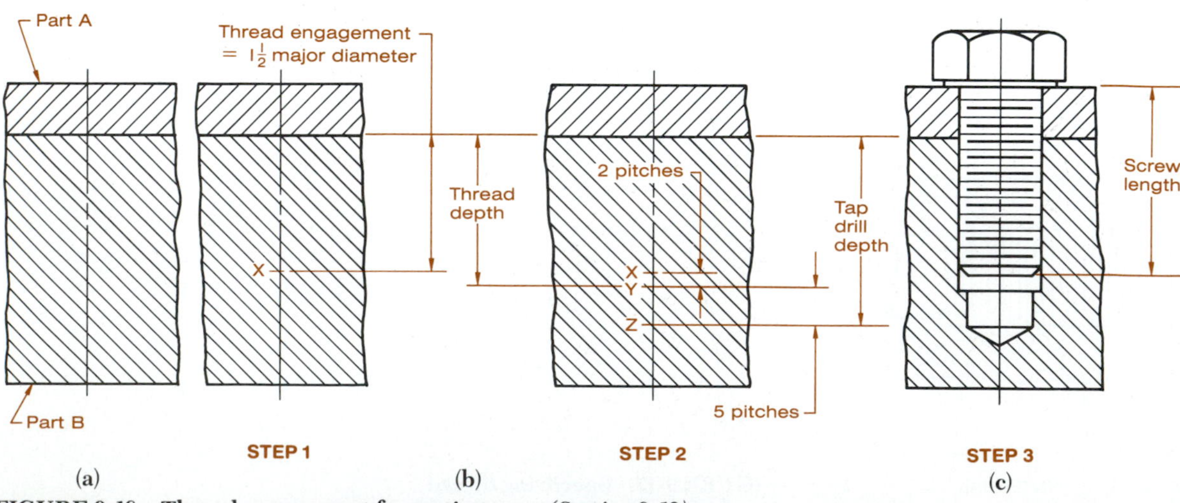

FIGURE 9-18 **Thread engagement for mating parts *(Section 9.12)***

a distance measured downward in part B that is the sum of the thread engagement and a distance equal to two pitches. The pitch of a 1″-8 thread equals .125 inches (2P = .250 inch) or 1½ + ¼ = 1¾ inches. Also shown in Step 2 is point Z, which establishes the depth of the tap drill. As previously mentioned, the tapered end of a plug tap (illustrated in Figure 9-9) will leave about five imperfect threads when screwed to the bottom of a drilled hole. Accordingly, a blind hole must be drilled at least five pitches deeper than the required thread depth if a plug tap is to be used. The distance representing the depth of the drilled hole (point Z) is equal to the sum of the thread engagement and a distance equal to five thread pitches (5P = .625 inch) or 1½ + ⅝ = 2⅛ inches. Note that point Z is also measured in part B.

Ordinarily it will be necessary to make a slight adjustment in distances X and Y, because screw lengths are generally selected in the nearest available commercial size (see Sections 9.20 and 9.24). It is helpful to keep all distances in even increments (to the nearest 1⁄16 inch, for example) as an aid to measurement in the shop. While the size of a clearance hole in part A should be specified, it is not usually drawn. Depending upon the design requirements, a clearance hole may be made as much as 1⁄32 to 1⁄16 inch larger in diameter than the screw. Many companies have adopted the practice of omitting the drawing of the threads in the bottom of the tapped hole so that the end of the cap screw may be clearly shown.

CHECKPOINT 3
CALCULATING VALUES WHEN JOINING TWO PARTS WITH A CAP SCREW

Using the following data, letter the given information and record the required values on a plain sheet of 8½ × 11 inch paper.

1. Assume a ¾ UNF steel hexagon-head cap screw is to be used to fasten a ½-inch-thick steel strap to a 2-inch-thick steel base. Supply the following information:
   Threads per inch
   Length of thread engagement in steel base
   Depth of thread in steel base
   Depth of tap drill
   Diameter of tap drill
   Length of cap screw
   Diameter of clearance hole (use 1⁄32-inch clearance)
2. Assume a ⅜ UNC steel round-head cap screw is to be used to fasten a ¼-inch-thick metal plate to an aluminum part 1 inch thick.
   Threads per inch
   Length of thread engagement
   Depth of thread in aluminum part
   Depth of tap drill
   Length of cap screw
   Diameter of tap drill
   Diameter of clearance hole (use 1⁄64-inch clearance)

9.13 SCREW FASTENERS

When making a preliminary function design layout (see Section 13.2) of a mechanism or a structure, it is common practice to represent screw fasteners by simply drawing a center line or by blocking in one typical fastener. The designer rarely draws fasteners in detail. In most cases the designer will identify the type of fastener and indicate the size of the thread diameter in a note. Later, when the design layout is turned over to the drafter so that a final assembly drawing and detail drawings of the parts to be manufactured may be prepared, more attention is devoted to the complete representation and specification of the required screw fasteners. Since most of the features of fasteners are standardized, they can be specified and ordered from a bill of materials without detail drawings.

The most commonly used types of threaded fasteners are bolts and nuts, studs, and screws. The following sections will deal with the various applications of these fasteners and with methods of representing and specifying them on engineering documents. Critical sizes of important features on fasteners for drawing purposes will be given in tables. There are certain drawing conventions that relate the proportion of the major diameter of a fastener to other features, such as the height of a bolt head, for example, which the engineer and the drafter need to know. The tables also provide this information. A wide selection of time-saving templates are available to aid in illustrating various types of fasteners.

9.14 BOLTS AND NUTS

A *bolt* is an externally threaded, headed fastener. The traditional distinction between a bolt and a screw is that a bolt is ordinarily used in a clearance hole in two or more aligned parts to be joined, but a screw is ordinarily used in a threaded hole without a nut. A typical installation of a bolt and nut is shown in Figure 9-19.

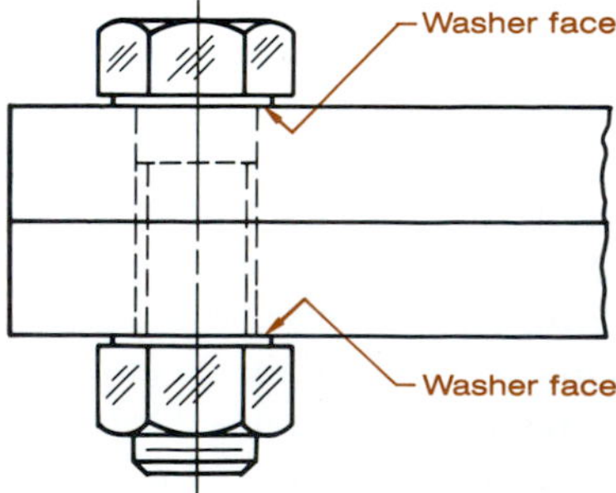

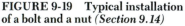

FIGURE 9-19 Typical installation of a bolt and a nut ***(Section 9.14)***

Bolt Types

The standard ANSI B18.2.1-1981 covers complete details of square and hex bolts and screws. American National Standard bolts are available in two head forms, square and hexagon, and in two series, regular and heavy. The principal difference between the two series is the thickness of the head. A further classification for a bolt is the finish. Two types of finish are available, unfinished and semifinished. Except for the threads, unfinished bolts and nuts are used as-forged. Semifinished bolts and nuts are machined on the bearing surface on the underside of the head. This 1/64-inch projecting surface is called a "washer face." Square-head bolts and nuts are made only in unfinished form. Dimensions for American National Standard hex bolts are given in Appendix Table A-7. See Appendix Table A-8 for American National Standard square bolts. Dimensions for metric hex bolts and cap screws are given in Appendix Table A-12.

A *nut* is an internally threaded fastener used on the threaded end of a bolt (and sometimes a screw) to hold an assembly of parts together.

Nut Types

The standard ANSI B18.2.2-1972 covers complete details of square and hex nuts. Forms, series, and finishes are similarly available for bolts. Dimensions for American National Standard square and hexagon nuts are given in Appendix Tables A-10 and A-9, respectively.

Metric Bolts and Nuts

There is a gradual trend toward increased production and use of metric fasteners. In the changeover that is currently in process, fewer thread sizes will be standardized and a greater uniformity of head styles will be adopted. Metric fasteners (bolts, screws, and studs) are classified and specified according to mechanical requirements. Steel fasteners have been divided into seven property classes that list values for the minimum tensile strength and yield strength for each class. Metric nuts classified as *style 1* and *style 2* correspond to the American National Standard regular and heavy series. Drawing dimensions of metric bolts are given in Appendix Table A-12 and drawing dimensions of metric hex nuts are given in Appendix Table A-13.

9.15 DRAWING OR SKETCHING A HEXAGON BOLT OR NUT

Before beginning the construction shown in Figure 9-20, consult the appendix tables that list the sizes of hexagon bolts and nuts. Reading across from the desired nominal diameter, 1/2 inch for example, record the values for the height of the bolt head (5⁄16 inch), the thickness of the nut (7⁄16 inch), and the width across the flats for the bolt and the nut (each 3⁄4 inch). The first step in drawing a hexagon bolt or nut is to draw lines in the front view that are spaced in accordance with the height of the head or the thickness of the nut. Next, lay out a hexagon (shown at the left in Figure 9-20) about an inscribed circle in the end view with a diameter equal to the distance across the flats. Project the necessary lines to the front view to form the corners of the flats. Using the 30°–60° triangle as shown, find the arc centers and draw or sketch the arcs to complete the front view.

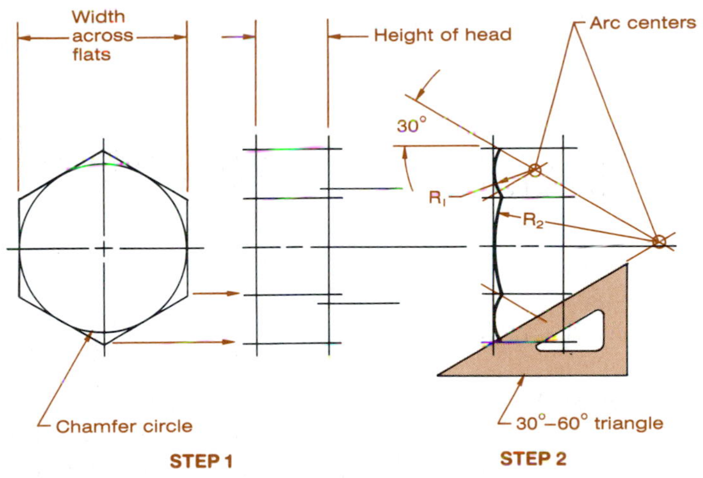

FIGURE 9-20 **Procedure for drawing hexagon bolt heads and hexagon nuts** ***(Section 9.15)***

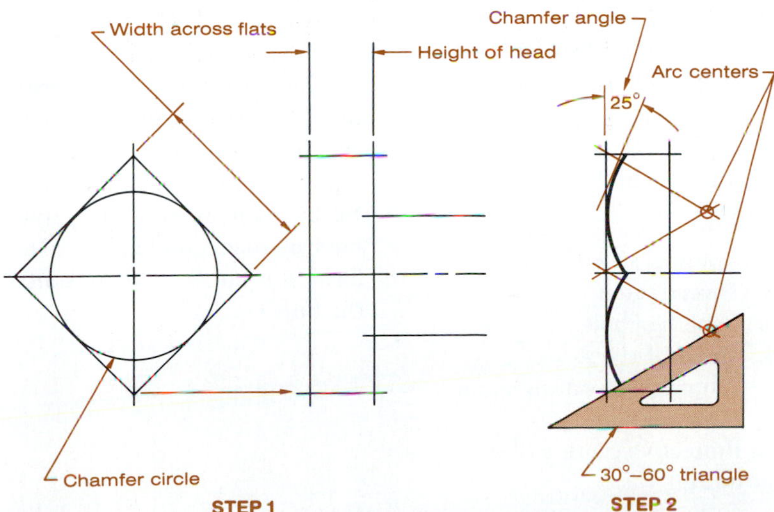

FIGURE 9-21 **Procedure for drawing square bolt heads and square nuts** ***(Section 9.15)***

Drawing or Sketching a Square Bolt or Nut

A square bolt head or square nut may be drawn or sketched as shown in Figure 9-21. Note that, except for laying out a square instead of a hexagon, the procedure is identical to the steps shown in Figure 9-20.

Preferred Position for Bolts and Nuts

Bolts and nuts are always represented in the "across corners," or widest, position. This conventional practice is followed in order to distinguish clearly between the hexagon and square form on drawings and to show actual required turning clearances.

Chamfer Angle

The top of the bolt head and the nut are chamfered at 30° for hexagons and 25° for square forms. All bolts greater than 1 inch in diameter are chamfered at 30°.

TABLE 9-2 Bolt and Cap Screw Lengths

	American National Standard							Metric		
	Hexagon-head bolt (inches)	*Square-head bolt (inches)*			*Hexagon-head socket cap and machine screw (inches)*			*Hexagon-head bolt (millimeters)*		
Range of lengths	¼	⅛	¼	½	⅛	¼	½	2	5	10
Increments	¾* to 8	½* to ¾	¾ to 5	Over 5	¼ to 1	1 to 3	3 to 6	8 to 20	20 to 50	50 to 150

*For bolt lengths less than this value, bolt is generally threaded over the entire length.

9.16 BOLT LENGTHS

The *bolt length* is the distance measured from the surface under the bolt head to the blunt end of the bolt. Bolt lengths are not standardized. The data shown in Table 9-2 may be used as a general guide. The given incremental values and the corresponding ranges of bolt lengths have been compiled from manufacturers' catalogs.

9.17 SPECIFYING BOLTS AND NUTS

American National Standard Bolts

Bolts are specified in parts lists and on drawings by giving the information as listed in the following sequence:

1. Nominal size of the bolt
2. Number of threads per inch
3. Series
4. Class of thread
5. Letter symbol (external = A)
6. Length of bolt
7. Finish
8. Type of head
9. Fastener name
10. Material (when required)

The following examples are typical:

⅜-24 UNF-2A × 2½
SEMIFIN HEX HD BOLT

½-13 UNC-3A × 1¼
SEMIFIN HEX HD BOLT

¾-10 UNF-2A × 3
HVY SEMIFIN HEX HD BOLT

¼-20 UNC-2A × 1¾
SQUARE HD BOLT

Metric Hexagon Bolts

Hexagon bolts are specified in a bill of material and in a note on a drawing by giving the following information in the sequence shown:

1. Nominal size of bolt
2. Thread pitch
3. Length of bolt
4. Material
5. Property class designation
6. Fastener name
7. Protective coating (if required)

The following examples are typical:

M12 × 1.75 × 100, CLASS 9.8
HEX BOLT

M6.3 × 1 × 25, CLASS 8.8
HEX BOLT, ZINC PLATED

American National Standard Nuts may be specified as follows:

⅝-11 UNC-2B SQUARE NUT

1-12 UNF-2B HEX NUT

Metric Nuts may be specified as follows:

M16 × 2, CLASS 9, NUT,
STYLE 1, M12 STEEL

The letter symbols representing external or internal threads should be used only when the type of thread is not obvious. Because square-head bolts and nuts are made only in unfinished form, it is unnecessary to specify the finish.

9.18 STUDS

As shown in Figure 9-22, a double-end stud is a rod threaded on both ends. Studs that are threaded over the entire length are called "continuous-thread studs." One end of the stud is tightly engaged into a tapped hole on one part, and the other end freely passes through a clearance hole in the second part. A nut used on the free end draws the parts together. For most applications the end that is screwed into the tapped hole has a coarse thread; the end that receives the nut has a fine thread. Studs are used if parts must occasionally be disassembled, as in the case of pumps, cylinder heads, and so forth.

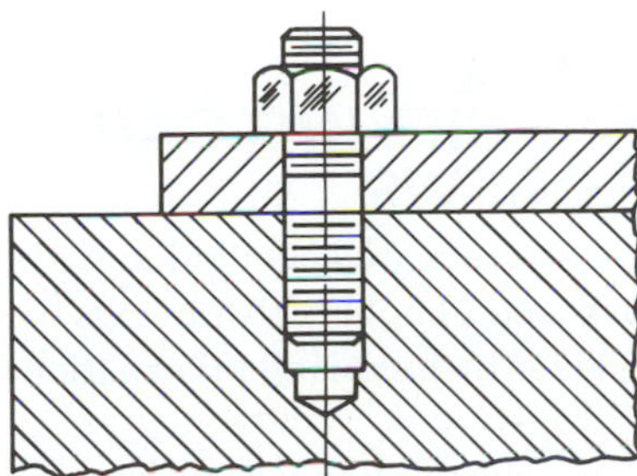

FIGURE 9-22 A stud *(Section 9.18)*

9.19 SPECIFYING STUDS

Studs are not standardized and therefore require explicit specification, including the length of the thread on each end. Studs are described by preparing a detail drawing or are specified as follows:

American National Standard:
⅜-24 UNF-2A × 2 Stud; thread length ½ and ¾.

Metric:
TYPE 2 DOUBLE-END STUD M12 × 1.75 × 75 STEEL; 25 THREAD LENGTH, BOTH ENDS
CLASS 9, CADMIUM PLATED

There are four basic types of studs in the metric system. Type 1 is unfinished and types 2, 3, and 4 are finished with slight differences in body tolerance. If, instead of a double-end stud, a continuous-thread stud is required, the specification should state "continuous-thread stud." The number 75 in the previous example refers to the overall stud length.

9.20 CAP SCREWS

Cap screws are finished headed fasteners that are usually used in a threaded hole without a nut. Figure 9-23 illustrates a typical installation of a hexagon-head cap screw. The two parts are held together by passing the screw through a clearance hole in one member (part A) and screwing it into the tapped hole in the other member (part B). The clearance hole is not shown on a drawing unless the details of installing the screw are not obvious. There are a number of

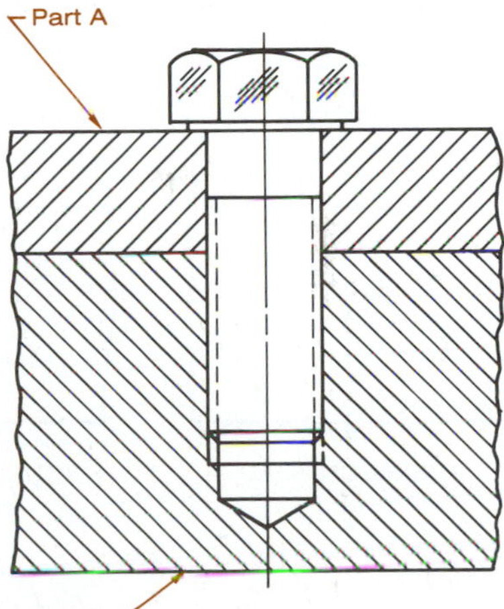

FIGURE 9-23 A typical installation of a cap screw *(Section 9.20)*

standardized head styles available for cap screws. In Figure 9-24 five commonly used types of American National Standard cap screws are illustrated. The actual sizes for American National Standard cap screws are given in Appendix Table A-14. Metric sizes for hex cap and hex socket cap screws are given in Appendix Table A-12.

The American National Standard hexagon-head cap screw is drawn in the same manner as shown for a hexagon bolt in Figures 9-20 and 9-21. Hexagon-head cap and machine screws have a washer face 1/64-inch thick with a diameter equal to the distance across the flats.

Hole making operations are explained in Section 8.13. Specifications for drilled, counterbored, spotfaced, and countersunk holes are given in Section 10.26. For the correct representation for thread engagement, see Section 9.12.

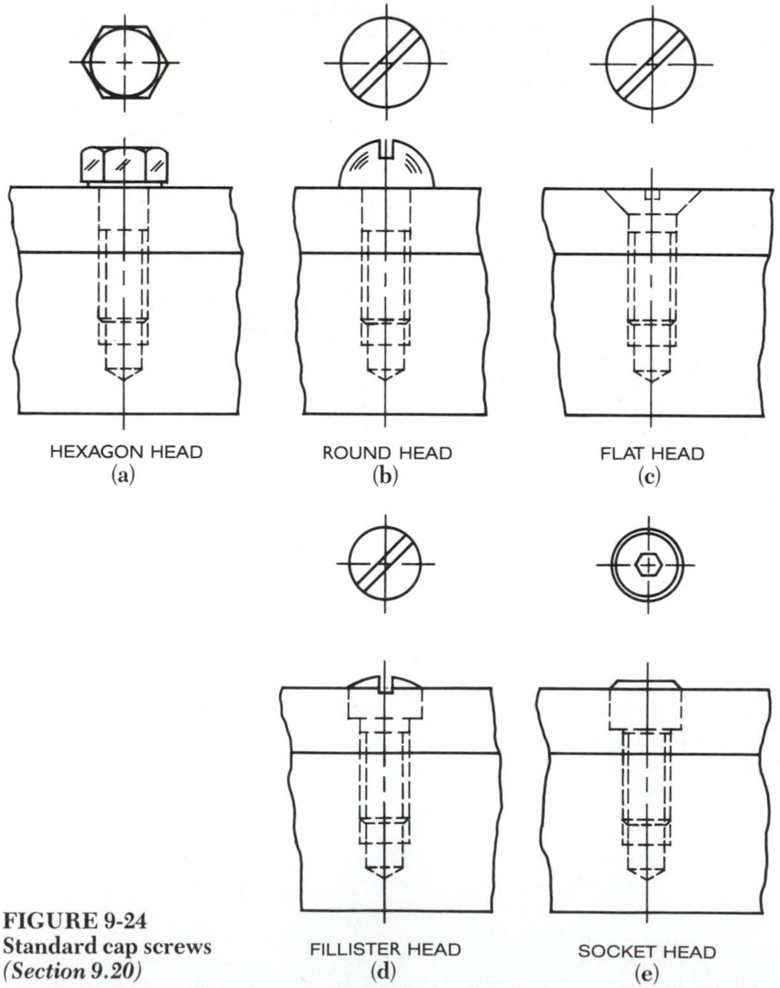

HEXAGON HEAD
(a)

ROUND HEAD
(b)

FLAT HEAD
(c)

FILLISTER HEAD
(d)

SOCKET HEAD
(e)

FIGURE 9-24
Standard cap screws
(Section 9.20)

9.21 CAP SCREW LENGTHS

As for bolts, cap screw lengths are unstandardized. The empirical data given in Table 9-2 may be used to obtain cap screw lengths.

9.22 CAP SCREW THREAD LENGTH

A good rule of thumb to use for hexagon-head and slotted-head cap screws with *coarse threads* is thread length = 2D + .50. For *fine threads*, thread length = 1½D + .50. All cap screws 1 inch or less in length are threaded within 2 to 3P of the head.

9.23 SPECIFYING CAP SCREWS

Cap screws are specified in parts lists and in notes on drawings by giving the following information:

American National Standard:
½-13 UNC-2A × 2½
SOC HD CAP SCR

Metric:
M8 × 1.25 × 30
HEX CAP SCR

M12 × 1.75 × 52
FIL HD CAP SCR, CLASS 8.8

TABLE 9-3 Machine Screw Thread Lengths

Nominal size (diameter)	*Lengths 3D and shorter*	*Lengths 3D and longer*
#5 (0.125) or smaller	Thread entire length up to within 1P of head.	Thread entire length up to within 2P of head.
#6 (0.138) or larger	Thread entire length up to within 1P of head.	Thread up to and including 2 inches to within 2P of head.

Thread length

9.24 MACHINE SCREWS

Machine screws resemble cap screws but are generally smaller, with a diameter normally ranging in size from #0 (.060) to ¾ inch. They have plain sheared ends (that is, not chamfered). The ranges of sizes and exact dimensions are given in Appendix Table A-15. Machine screws are generally used to fasten or clamp two or more parts together either by winding the threaded screw into a tapped hole in one of the parts or by using a nut. This type of fastener is used mainly for small work having thin sections. Sizes for American National Standard machine screws are given in Appendix Table A-15. Sizes for Metric machine screws are given in Appendix Tables A-16 and A-17.

9.25 MACHINE SCREW THREAD LENGTHS

The data extracted from ANSI B18.6.3-1972 (R1977) American National Standard Machine Screws and Nuts, given in Table 9-3, may be used to obtain machine screw thread lengths.

9.26 SPECIFYING MACHINE SCREWS

Machine screws are specified as follows:

American National Standard:
#10-32 UNF-2A × 1¼
FL HD MACH SCR

Metric:
M3 × 0.5 × 20
RD HD MACH SCR
ZINC PLATED

9.27 MACHINE SCREW NUTS

Sizes of American Standard machine screw nuts are given in Appendix Table A-11.

9.28 SPECIFYING MACHINE SCREW NUTS

Machine screw nuts are specified as follows:

American National Standard:
#6-32 UNF MACH SCR SQ NUT

Metric:
M8 × 1.25 CLASS 9
HEX NUT STYLE 1
WASHER FACE

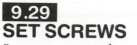

9.29 SET SCREWS

Set screws are used to prevent rotational or translational forces between two mating parts, such as a collar or a wheel and a shaft. In the assembly of a pulley and a shaft, for example, a set screw is screwed into a tapped hole in the hub of the pulley until its point bears firmly against the shaft. Figure 9-25 illustrates several types of standard set screws. Each head type is available in any one of the point styles shown. The dog and the cone point set screws are generally installed into a matching recess in the inner part. Sizes of various styles of American Standard set screws are given in Appendix Table A-18.

9.30 SPECIFYING SET SCREWS

Set screws are specified as follows:

American National Standard:
5/16-18 UNC-2A × 1½
SLOT HD FL PT SET SCREW

Metric:
M12 × 1.75 × 20 SOCK
FULL DOG PT SET SCR

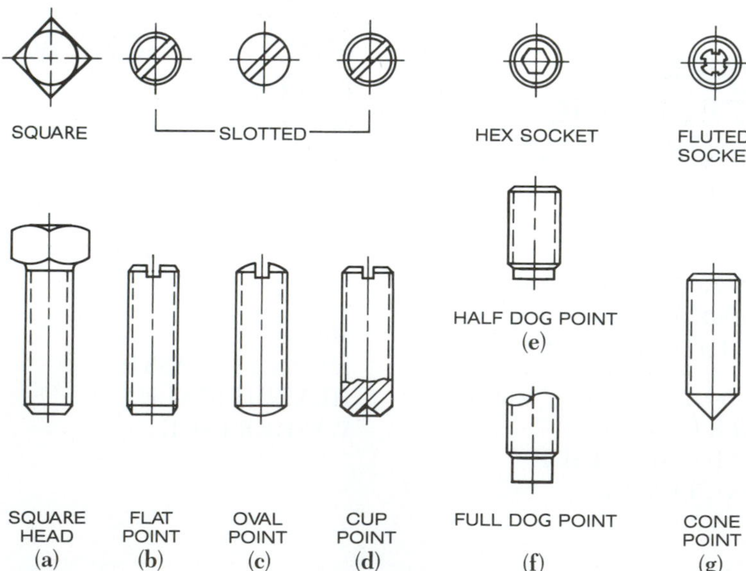

FIGURE 9-25 Standard set screws ***(Section 9.29)***

9.31 ADDITIONAL THREADED FASTENERS

On practically all assembled products there are opportunities to reduce costs associated with parts, machining operations, and assembly time. Choosing the optimum practical solution helps to reduce costs.

Thousands of fastener sizes, types, materials, and variations are available to the product designer. Before choosing the correct fastener for a particular application, you must be up-to-date with the offerings made available by the producers of mechanical fasteners. Take advantage of the design and application literature published by manufacturers, most of which is available on request.

Regular examination of advertising material and briefs in technical journals is also helpful. Many types of threaded-fastener standards provide helpful engineering data, dimensions, load ratings, and applications. Careful attention should be given to the many American National Standards topics pertaining to mechanical fasteners published by the American National Standards Institute. A catalog of American National Standards is available that will help you locate a particular topic. Many other types of standards are also available, including military standards, company standards, technical society standards, and other national and industry standards.

PROBLEMS

Threaded Mechanical Fasteners

For the problems in this chapter, use layout A, shown on the inside of the front cover. Letter the required information in the sheet title block.

No drawing is required for Problems 9.1 to 9.12. Using the appropriate appendix tables, record the missing values in the table at the right. Rule off suitable spaces for recording your answers. Letter the headings.

Unified Screw Threads:

	Nominal Diameter	*Series C = Coarse F = Fine*	*Threads Per Inch*	*Pitch*	*Tap Drill Diameter*
P9.1	1″	C	?	?	?
P9.2	½″	C	?	?	?
P9.3	¾″	F	?	?	?
P9.4	#10	F	?	?	?
P9.5	#2	F	?	?	?
P9.6	⅜″	C	?	?	?

Metric Screw Threads:

	Nominal Size	*Series C = Coarse F = Fine*	*Pitch*	*Tap Drill Diameter*
P9.7	M20	C	?	?
P9.8	M6.3	C	?	?
P9.9	M42	F	?	?
P9.10	M60	F	?	?
P9.11	M12	F	?	?
P9.12	M30	C	?	?

P9.13(a) On the top half of a sheet, neatly letter the definition of the following screw thread terms:

a. Major diameter
b. Minor diameter
c. Pitch diameter
d. Pitch
e. Lead
f. Crest
g. Root
h. Angle of thread
i. Axis of thread
j. Depth of thread

P9.13(b) On the lower half of the sheet make a freehand sketch of the partial view of a cylinder with a screw thread shown in Figure P9-1. Add dimensions (and notes, where applicable) to illustrate the given terms. Note: For sketching purposes, make the major diameter equal to about 2½ inches.

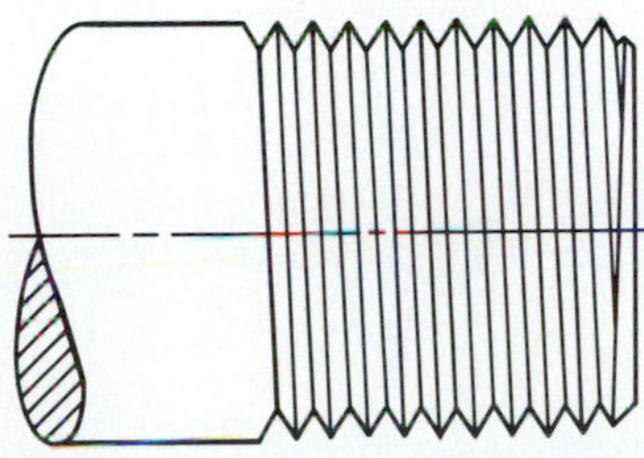

FIGURE P9-1

Use one sheet for Problems P9.14 to P9.17. Divide the sheet into four equal divisions. For each problem, sketch two views of a threaded cylinder, approximately full size, using the following specifications. (Hint: Refer to Figures 9-12 and 9-13.) Give the thread note and the dimensions for the thread length.

P9.14 Specifications: 1¼-12 UNF-2A, thread length = 3½ inches, left-hand thread. Use the schematic thread representation.

P9.15 Specifications: ¾-6 square, thread length = 6 inches. Use the simplified thread representation.

P9.16 Specifications: M20 × 2.5-5g5g, thread length = 40. Use the schematic thread representation.

P9.17 Specifications: M8 × 1.25, thread length = 125, double-start thread. Use the simplified thread representation.

Use one sheet for Problems P9.18 to P9.20. Divide the sheet into three equal divisions. Using the given sizes and the arrangement of views shown in Figure P9-2, sketch both views approximately full size. Show two views of the correct thread representation using the following specifications. (Hint: Refer to Figure 9-11.)

P9.18 (Upper third of sheet) A blind tapped hole is visible in the front view and hidden in the right-side view. Specification: ? DRILL, 1⅜ DEEP, ¾-16 UNF-2B, 1 DEEP

P9.19 (Middle third of sheet) A through tapped hole is visible in the front view. The right-side view is a sectional view. Use the schematic thread representation. Specification: ? DRILL, M20 × 2.5

P9.20 (Lower third of sheet) A through tapped hole is visible in the front view. The right-side view is a sectional view. Use the simplified thread representation. Specification: 1″-5 ACME, DOUBLE-START.

P9.21 Using the given scales, give the complete specifications for each of the threaded mechanical fasteners shown in Figure P9-3. On the *upper half* of a sheet, given the

FIGURE P9-2

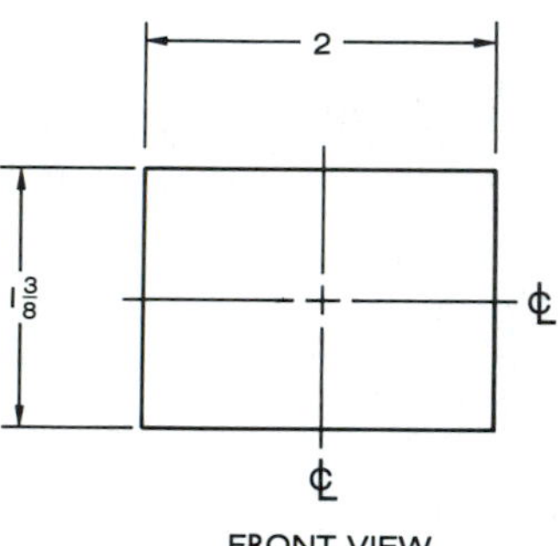

FRONT VIEW

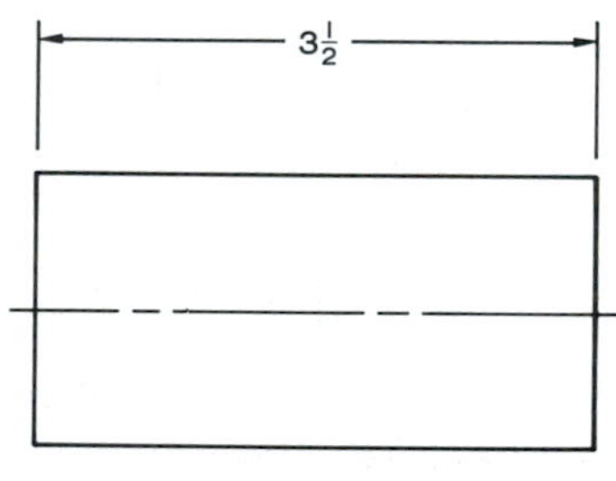

RIGHT-SIDE VIEW

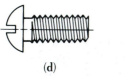

FIGURE P9-3

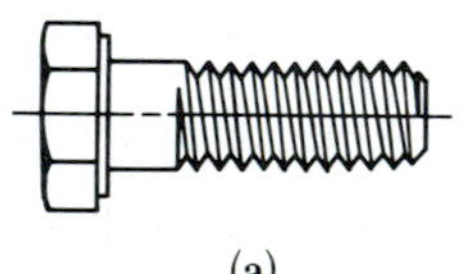
(a)

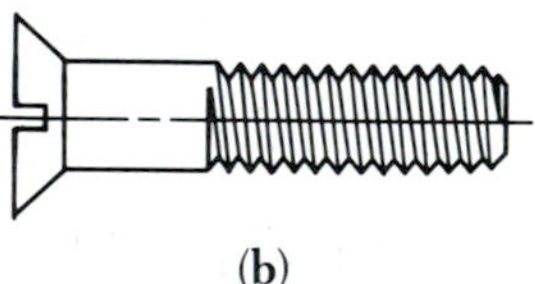
(b)

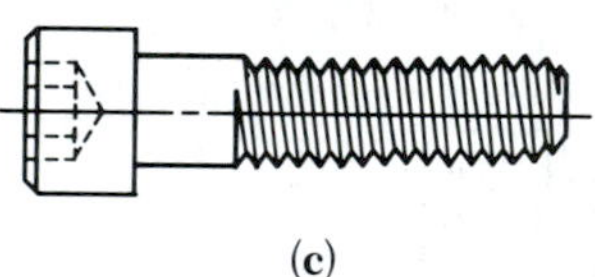
(c)

(d)

(e)

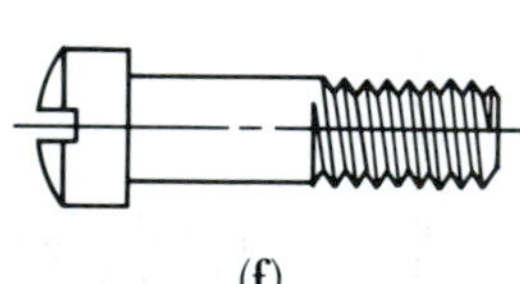
(f)

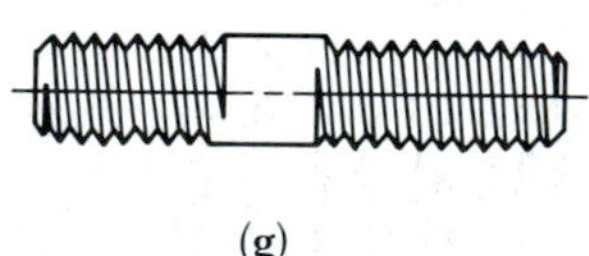
(g)

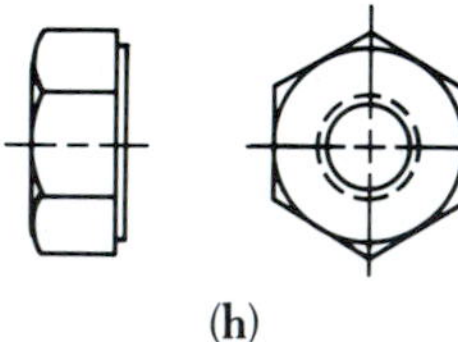
(h)

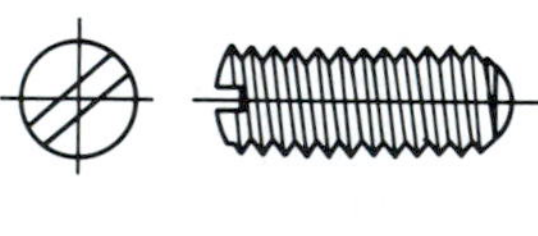
(i)

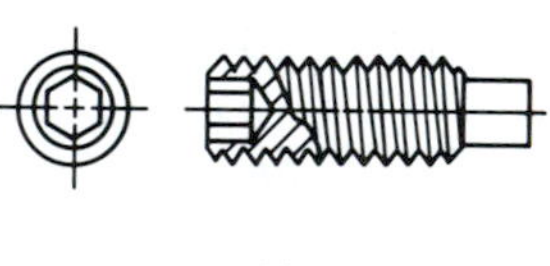
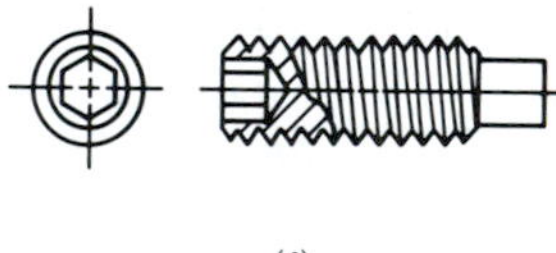

(j)

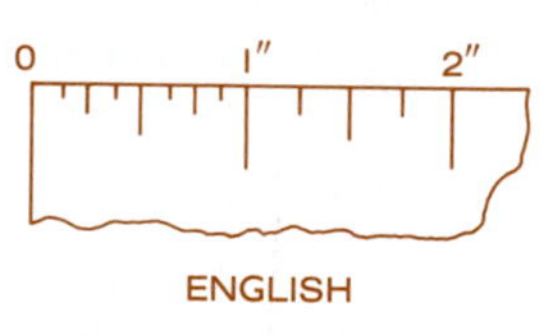

ENGLISH

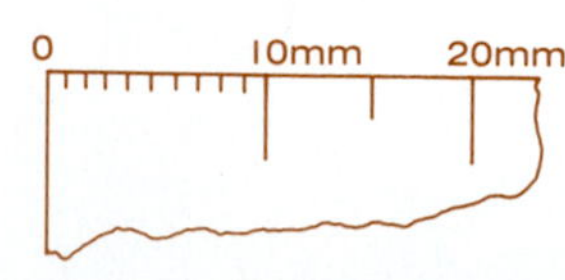

METRIC

specifications for unified inch threads. For each example, assume a coarse thread series with a #3 class of fit. On the *lower half* of the sheet, give the specifications for a metric thread.

No drawing is required for Problems P9.22 to P9.24. Figure P9-4 is given for information only. Rule off suitable spaces for recording your answers to the questions listed below Figure P9-4 that apply to the following problems.

P9.22 Assume a ¾ (coarse series) hexagon head cap screw is used to fasten part A to part B. Part B is cast iron.

P9.23 Assume a M8 × 1.25 roundhead machine screw is used to fasten part A to part B. Part B is steel.

P9.24 Assume a ¼ (fine series) pan-head machine screw is used to fasten part A to part B. Part B is plastic.

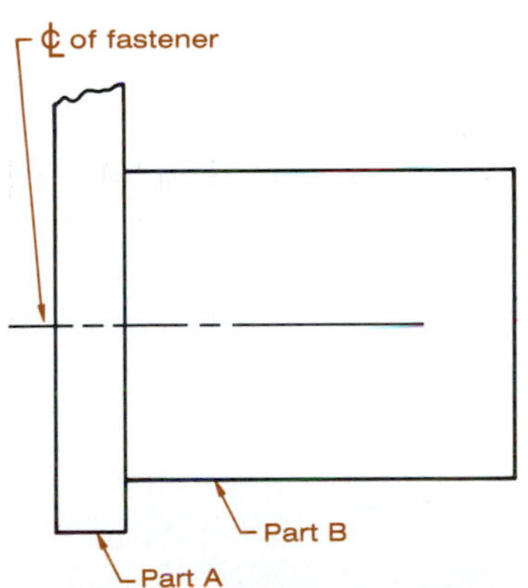
℄ of fastener
Part B
Part A
1. Size of drilled hole in part A?
2. Length of thread engagement?
3. Thread depth?
4. Tap drill depth?
5. Size of tap drill?

FIGURE P9-4

Use one sheet for Problems P9.25 to P9.28. Divide the sheet into four equal divisions. For each problem sketch two views approximately full size using the following specifications. (Hint: Refer to Figure 9-19 and to the appropriate appendix tables.)

P9.25 Specifications: ⅝-18 UNF-2A, semifinished hexagon head bolt, 3 inches long.

P9.26 Specifications: ½-13 UNC-2A, socket head cap screw, 1½ inches long.

P9.27 Specifications: ⅜-24 UNF-2A, squarehead bolt, 1¾ inches long.

P9.28 Specifications: M24 × 3 × 75, class 8.2, hexagon-head bolt.

Draw problems P9.29, P9.30, and P9.31 full size, with instruments. Make a two-view partial assembly drawing in section similar to Figure 9-17 using the following specifications. Draw the cap screw in the correct position.

P9.29 A ⅜ UNC socket-head cap screw is used to hold a 1-inch-thick plate to an aluminum object 2½ inches thick. Use the simplified thread representation shown in Figure 9-22. Neatly letter the correct notes for the cap screw and the required machining operations in the plate and in the aluminum object.

P9.30 A M24 × 3 roundhead cap screw is used to hold a ½-inch-thick plate to a steel object 2⅛ inches thick. Use the schematic thread representation. Neatly letter the correct notes for the cap screw and the required machining operations in the plate and in the steel object.

P9.31 A ⅝ UNF hexagon-head cap screw is used to hold a ¾-inch-thick plate to a cast iron object 1⅞ inches thick. Use the simplified thread representation. Neatly letter the correct notes for the cap screw and the required machining operations in the plate and in the cast iron object.

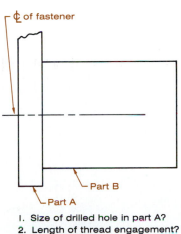

10
Limit Dimensioning

OBJECTIVES

After completing this chapter you should have gained the following abilities:

1. To select a drawing scale that will display the views of the part in the clearest possible manner.

2. To understand and use acceptable drafting techniques that apply to:

- **a.** the character of dimension lines, extension lines, center lines, and leaders
- **b.** arrowheads
- **c.** lettering and dimension figures
- **d.** dual dimensioning
- **e.** dimensioning in limited spaces
- **f.** overall dimensions
- **g.** dimensioning angles and arcs

3. To place dimensions in positions on the drawing where they will be easy to read.

4. To select dimensions that will enable parts to function as designed when assembled with mating parts.

5. To use, where appropriate, various units of measurement, including:

- **a.** fractional inches
- **b.** decimal inches
- **c.** SI values

6. To establish and identify properly finished surfaces to be used as bases of dimensioning.

7. To know standard dimensioning practices and techniques required to specify:

- **a.** various shapes of parts (rounded corners, regular and irregular curved outlines, rounded ends, cylinders, prisms, and so forth)
- **b.** machining operations (including chamfers, tapers, machining centers, keyseats and keyways, knurling, necks and undercuts, and thread reliefs)
- **c.** out-of or not-to-scale dimensions

8. To understand the principles of:

- **a.** geometric analysis in choosing dimensions
- **b.** size and location dimensioning
- **c.** contour dimensioning
- **d.** dimensioning sectional views
- **e.** tabular dimensioning

9. To know how to represent and specify machined holes.

10. To avoid the use of superfluous dimensions.

Thus far the material in this text has concentrated on the techniques of how to represent and produce the *shape* of an object. This chapter covers *size* description and explains the standardized techniques of applying dimensions and notes to the views so that drawings can be used to construct parts accurately and without question.

Figure 10-1 illustrates a typical *detail* drawing that clearly and precisely describes both the *shape* and the *size* of an object. A detail drawing provides a complete and explicit description of a single part, and one such drawing is prepared for every part that is to be manufactured. Detail drawings are often called *working* drawings. Every line, symbol, dimension, and note on the drawing should conform to the established, uniform practices that are used in

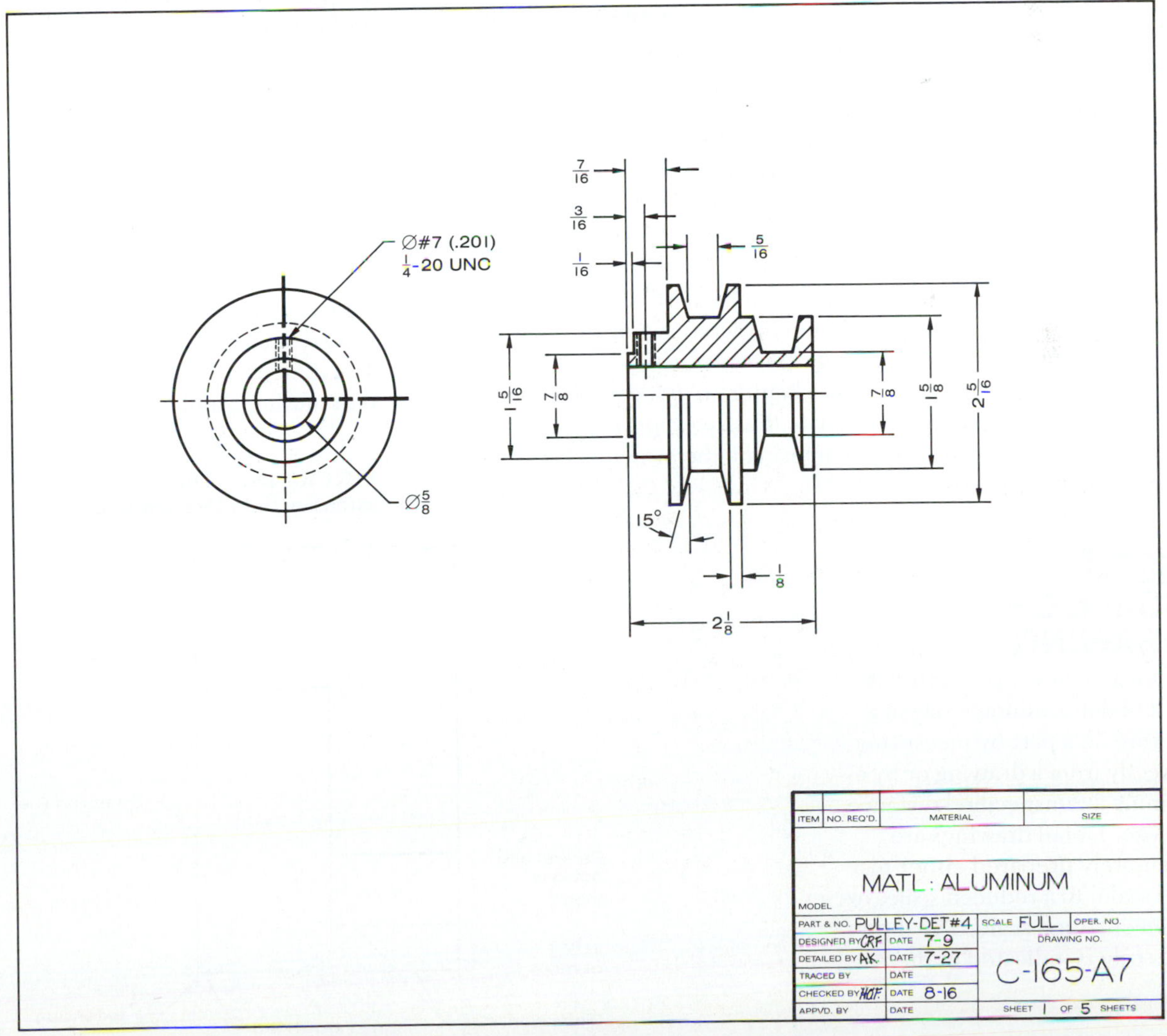

FIGURE 10-1 A typical detail drawing

portraying the required shape and stating the size of the part. The detail drawing shows the object in its completed condition. It also provides the manufacturer with all the information necessary to produce accurately the shape of the part together with all other dimensions that control the sizes of the various surfaces, cuts, and holes and the distances between them. A part that is accurately produced should function properly when assembled with other parts. In addition, the drawing supplies the specific requirements for the type of material, the kinds of surface finishes, the hardness, heat treatment, and the production quantity. Each detail drawing must be so complete that no further communication between the engineer and the manufacturer is necessary.

When specifying the size and shape of an object, it is helpful if drafters and engineers are well informed in as wide a variety of manufacturing processes and machine tool operations as possible. Some of the most commonly employed manufacturing processes and operations were covered in Chapter 8.

Many of the fundamental rules of size description that are discussed in this chapter are based upon a set of standards, developed by the American National Standards Institute, entitled *Dimensioning and Tolerancing*, ANSI Y14.5. These standards apply to recommended practices for engineering drawings and related documentation. Other helpful materials used as guidelines were various engineering drafting standards from the Ford Motor Company, General Motors Corporation, General Electric Company, and other major United States corporations.

Most of the illustrations in this chapter are not complete working drawings. They are intended only to illustrate a principle. Some of the details that are not essential to explain the principle have been intentionally omitted.

10.1 SCALE OF DRAWING

A practice never permitted is that of determining a size of a feature on a part by measuring directly from a drawing or by assuming an unspecified distance or size. Detail drawings are accurately prepared either to a full scale, to a reduced scale, or to an enlarged scale. The drawing scale is indicated in the sheet title block for reference. Whether drawn with instruments or sketched freehand, the dimension figures on the drawing always refer to the actual part sizes.

10.2 DIMENSIONING TECHNIQUES

Dimension Lines

Dimension lines, shown as 1¼ and 2½ in Figure 10-2, are drawn as a thin line. Arrowheads show the direction and extent of a dimension. Numerals placed near the middle of a dimension line indicate the size of the feature. Dimension lines are drawn parallel to the direction of measurement, and the shorter dimension lines are placed nearest the part outline. The dimension line nearest the part outline should be not closer

FIGURE 10-2 Dimension and extension lines *(Section 10.2)*

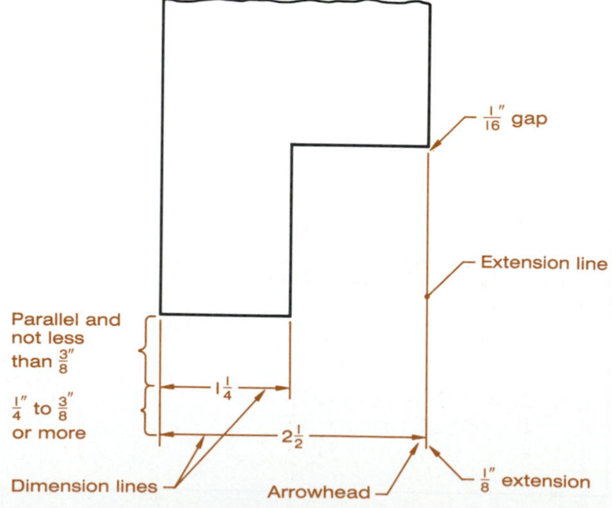

than ⅜ inch (10 mm) to the outline, and the space between one dimension line and other parallel dimension lines should be at least ¼ inch (6 mm), more if space is available.

Extension Lines

An extension line, drawn as a thin line, is used to indicate the point or line on the drawing to which the dimension applies. Extension lines, also shown in Figure 10-2, start with a short gap at the outline of the part or the feature being dimensioned and extend slightly beyond the outermost related dimension line. Except in special cases, extension lines are drawn at right angles to dimension lines.

Arrowheads

Figure 10-3 is an enlarged illustration showing the technique for making two styles of freehand arrowheads. Arrowheads are placed at the ends of each dimension line to indicate the endpoints of these lines. The size and style of arrowheads should be uniform on any one drawing. The length-to-width ratio is 3:1. The arrowhead length is generally made ⅛ inch, or about 3 or 4 mm. Arrowheads are made with the same grade of pencil leads as those used for lettering.

Center Lines

A center line, drawn as a thin line, is composed of alternate long and short dashes, as shown in Figure 10-4(a), and is used to represent axes of cylindrical parts and centers of symmetry. Center lines may be used as ex-

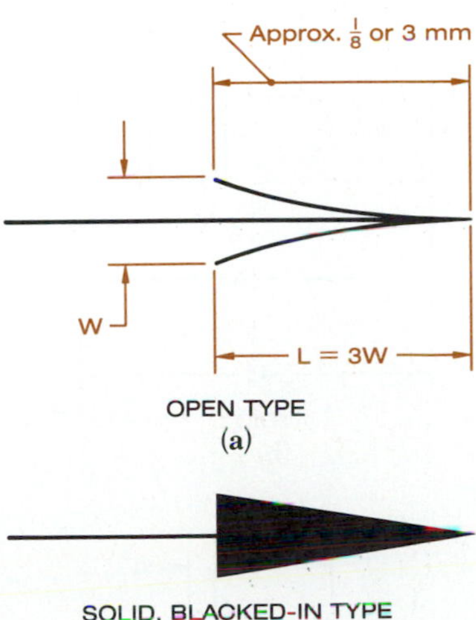

FIGURE 10-3 Two acceptable styles of arrowheads *(Section 10.2)*

FIGURE 10-4 Center lines used as extension lines in dimensioning *(Section 10.2)*

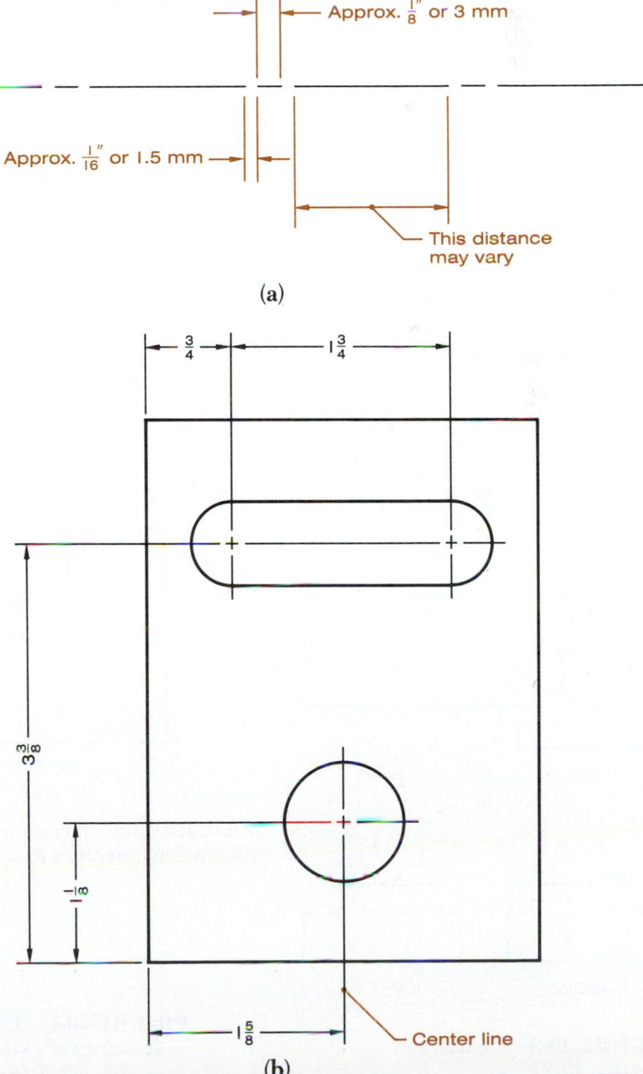

tension lines to locate holes and other cylindrical features, as shown in 10-4(b).

Leaders

A leader, shown in Figure 10-5(a), is drawn as a thin line and is used to direct a dimension, note, symbol, item number, or part number to the intended feature on the drawing. If the leader is terminated at a line, an arrowhead is used. A dot is used when the leader is terminated within the outline of the object. A leader consists of an inclined straight line with a short horizontal portion extending to the midheight of the first or last letter or digit of the note or dimension. Where a leader is directed to a circle or an arc, its direction should be radial so that, if extended, it would pass through the center, as shown in Figure 10-5(b). Two or more leaders to adjacent features on the drawing should be drawn parallel. For clarity, avoid the following:

1. Crossing leaders.
2. Disproportionately long leaders.
3. Leaders in a horizontal or vertical direction.
4. Leaders parallel to adjacent dimension lines, extension lines, or section lines.
5. Small angles between leaders and the lines where they terminate—see Figure 10-5(c).

10.3 ARRANGEMENT OF DIMENSION AND EXTENSION LINES

Dimension lines should be aligned or grouped as shown in Figure 10-6(a). Where extension lines unavoidably cross other extension lines (as shown in Figure 10-7) or other lines, such as dimension lines or object lines, they should be drawn continuously without gaps. Where a

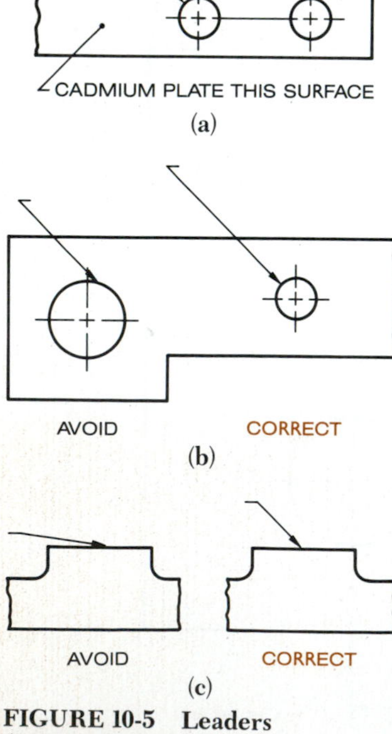

FIGURE 10-5 Leaders *(Section 10.2)*

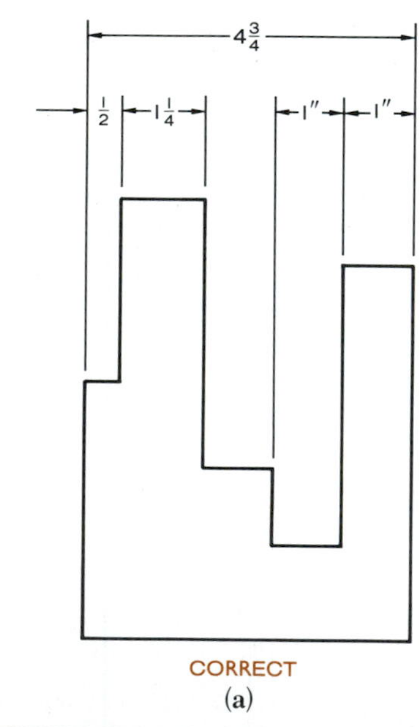

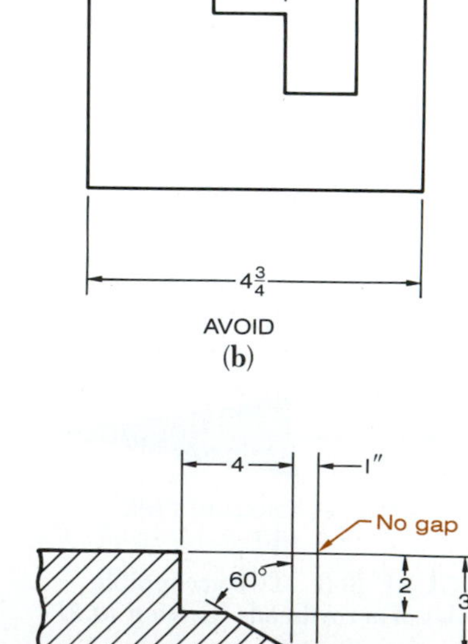

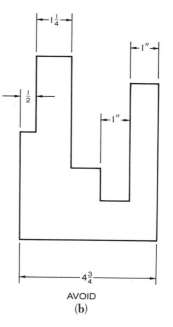

FIGURE 10-6 Grouping dimensions *(Section 10.3)*

FIGURE 10-7 Unavoidable crossings of extension lines *(Section 10.3)*

point is located by extension lines only, the extension line should pass through the point, as shown in Figure 10-8.

10.4 DIMENSION FIGURES

Figures and lettering on sketches and drawings must be legible. Clarity is of utmost importance on engineering documents. Requirements pertaining to lettering and figures are discussed in Sections 1.9 through 1.14. Either inclined or vertical figures are permissible, but only one style should be used throughout a drawing. As shown in Figure 10-9, the standard height for all figures and letters is ⅛ inch, or about 3 mm. The division line of a common fraction should be parallel to the direction in which the dimension reads and should be separated from the numerals by a minimum of 1/32 inch. Numerals in fractions should be slightly less than ⅛ inch high. The division line may be omitted for limit dimensions expressed in the form of a fraction, as shown in 10-9(c).

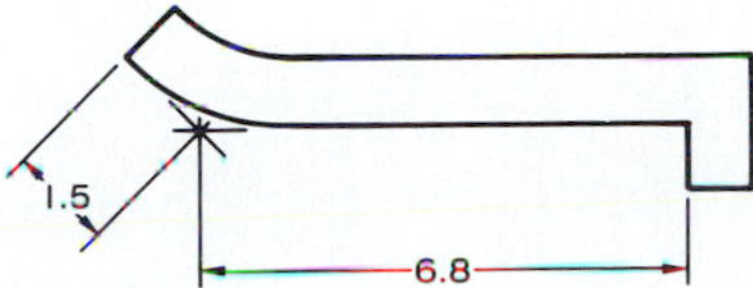

FIGURE 10-8 Point locations for extension lines *(Section 10.3)*

10.5 PLACEMENT OF DIMENSION FIGURES

Reading Directions

Two systems, aligned and unidirectional, may be used on drawings for placing dimension figures. These reading directions are illustrated in Figure 10-10. The unidirectional system is generally preferred because all of the dimension figures read from the bottom of the sheet. In particular, larger size drawings are easier to read when the unidirectional method is used. In both methods dimensions and notes shown with leaders should be lettered horizontally or aligned with the bottom of the drawing.

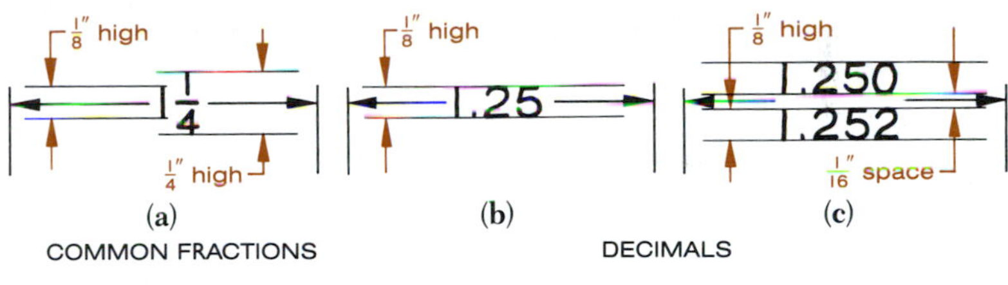

FIGURE 10-9 Standard height for figures *(Section 10.4)*

FIGURE 10-10 Two systems for placement of dimension figures: aligned and unidirectional *(Section 10.5)*

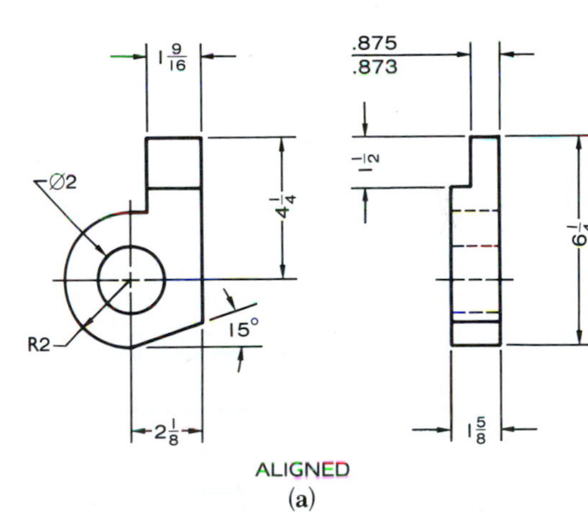

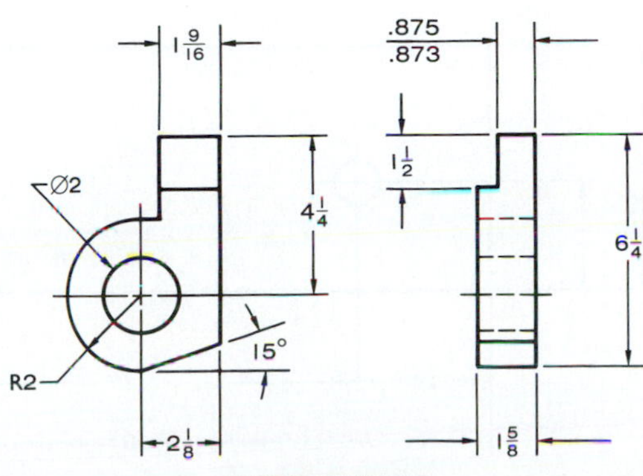

On or Off the Views

The preferred practice is to place dimensions off the views of a part whenever possible, as shown in Figure 10-11. If extension lines would be excessively long and result in the placement of a dimension too far from the feature it describes, a dimension may be placed within the outline of the view.

As shown on the sectional view of the complex part in Figure 10-12, placing dimensions directly on the view is often necessary. This practice should be executed judiciously with a *minimum* amount of crowding. When placing a dimension on a sectional view it is necessary to provide an opening for the dimension figure, as shown in Figure 10-12.

10.6 DIMENSIONING SYSTEMS

Fractional Inch

Most of the products and component parts manufactured today require drawings with more precise specifications than common fraction dimensions. A whole number of 3 inches or a common fraction of 1¾ inches,

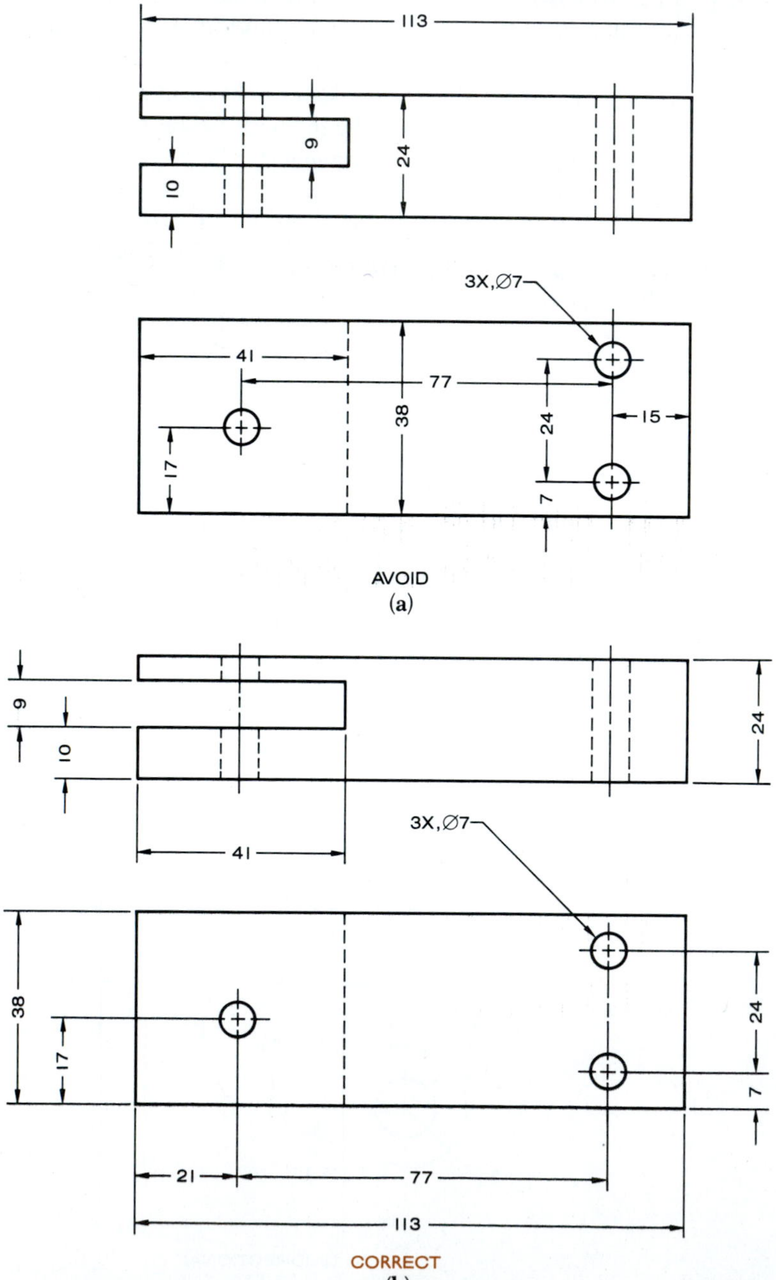

AVOID
(a)

CORRECT
(b)

FIGURE 10-11 **Dimensions on or off the views** ***(Section 10.5)***

for example, is considered accurate only within ± 1/64 inch. Common fractions are used solely for non-critical dimensions on a drawing and should always be reduced to their lowest denominator: for example, 2/8 equals 1/4, 10/16 equals 5/8, and so on. Errors are often more likely to occur when adding, subtracting, multiplying, or dividing common fractions than when performing similar calculations using decimal inches or millimeters.

Fractional and Decimal Inch Combinations

Common fractions or decimal dimensions may be given on the same drawing, but in many companies, decimal dimensioning is preferred. When the practice of combination dimensioning is employed, decimals are used for all dimensions except for the designations of most dimensions or nominal sizes of parts or features (such as bolts, screw threads, and keyseats) that use standardized fractional designations.

Decimal Inch

Today, due to increased precision manufacturing requirements, there is a tendency to use decimal inch dimensions entirely on drawings instead of common fractions. This trend is particularly noticeable in the aircraft, aerospace, and automotive industries. The two-place decimal inch system, where the smallest recommended division

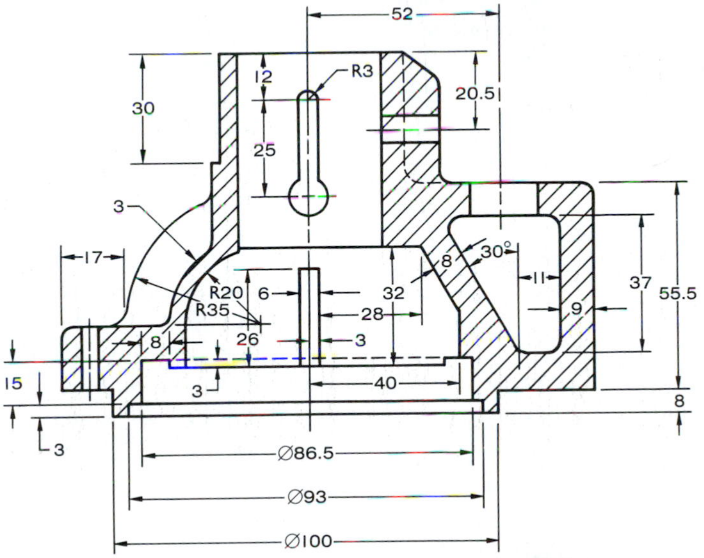

FIGURE 10-12 **Dimensions on sectional views** ***(Section 10.5)***

of the inch is 1/50 or .02, has been widely adopted. Figure 10-13 illustrates a drawing that is dimensioned with the two-place decimal inch system. Whenever possible, two-place figures should be an even digit (.02, .28, 8.66, for example) so that even when the dimension is halved, the result will remain a two-place decimal. Odd numbered two-place decimals (.37, 1.75, and so on) may be used for convenience in design for dimensioning points on a smooth curve or when halving dimensions, such as diameters to radii. When specifying decimal inch dimensions on a drawing for values of less than one, .44 for example, a zero is not used before the decimal point. Decimal points should be uniform, dense, and sufficiently large to be clearly visible.

Rounding Off a Decimal Inch

The rules of rounding off a decimal value to a lesser number of places than the total number available is explained in *Rules for Rounding Off Numerical Values*, ANSI Z25.1-1940(R1961). The following examples are based on this standard:

Rounding to three places

Example 1: Common fraction = 6 9/32
Decimal equivalent = 6.28125

Rule: When the figure beyond the last figure to be retained is *less* than 5, the last figure retained should not be changed. 6.28125 rounded to three places = 6.281.

Example 2: Common fraction = 1 31/32
Decimal equivalent = 1.96875

Rule: When the figure beyond the last figure to be retained is *more* than 5, the last figure retained should be increased by 1. 1.96875 rounded to three places = 1.969.

Rounding to two places

Example: Common fraction = 3 3/8
Decimal equivalent = 3.375

Rule: When the figure beyond the last figure to be retained is *exactly* 5, with only zeros following, the preceding number, if even, should be unchanged; if odd, it should be increased by 1. 3.375 rounded to two places = 3.38

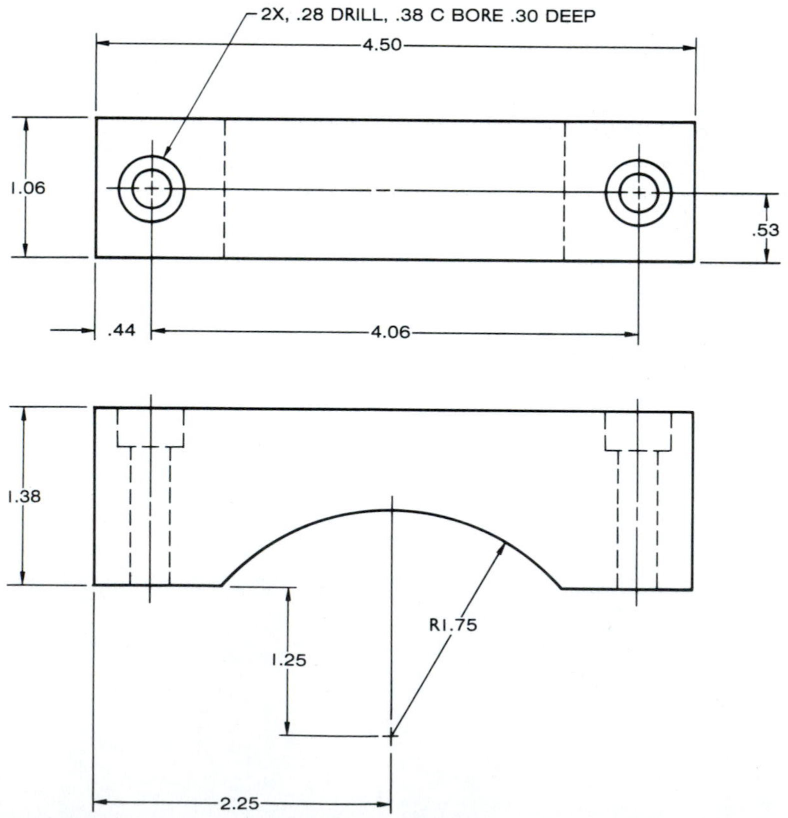

FIGURE 10-13 Example of a part dimensioned with two-place decimal-inch system *(Section 10.6)*

The Metric System

In general, numerical values should appear on engineering drawings in the manner prescribed by international drawing practices. The following paragraphs pertain to some of the more important practices that apply to the nearly universal metric system.

The standard unit of measurement on machine drawings is the millimeter (mm). Very large distances are dimensioned in meters. Millimeters are usually rounded off to the nearest whole unit without decimal remainders. A zero precedes a decimal point in an SI value of less than one unit, for example 0.76. Except for SI limit dimensions (explained in Section 11.11), zeros are not used following a significant digit to the right of the decimal point (12.4, for example).

Commas should not be used to denote thousands in an SI numerical expression, nor should a space be left between digits or in a position where a comma would otherwise be used to denote thousands. A space might be misinterpreted as a place where a decimal point is missing. The figure six thousand one hundred, for example, would appear on a drawing as 6100.

10.7 DUAL DIMENSIONING

Dual dimensioning is a procedure where dimensions on a drawing are expressed in *both* inch and metric units. This form of dimensioning, preferred by many companies, continues to be used on drawings prepared for the manufacture of interchangeable parts with the linear values given in both the U.S. Customary (Inch) and SI (Metric) units of measurement. Figure 10-14 gives two examples that show how dual dimensions may appear on engineering drawings, the *position* method and the *bracket* method. At 10-14(a) the millimeter dimension is located above the inch dimension but separated by a dimension line, or it may be placed to the left of the inch dimension but separated by a slash line. The millimeter dimension may also be enclosed in brackets, as at 10-14(b). In this system the location of the millimeter dimension is optional, but it should be uniform on any drawing. In other words, the millimeter dimension may be placed above, below, to the left, or to the right of the inch dimension. Only one of the methods shown in Figure 10-14 should be used to identify the inch and the millimeter dimensions on a single dual-dimensioned drawing. All drawings should illustrate or specify in a note how the inch and millimeter dimensions can be identified. The illustration for the position method would appear in this form:

$$\frac{\text{MILLIMETER}}{\text{INCH}}$$

or MILLIMETER/INCH.

The bracket method would appear as:

$$\frac{\text{[MILLIMETER]}}{\text{INCH}}$$

or [MILLIMETER]/INCH.

The noted specification would be in this form:

DIMENSIONS IN [ ] ARE IN MILLIMETERS.

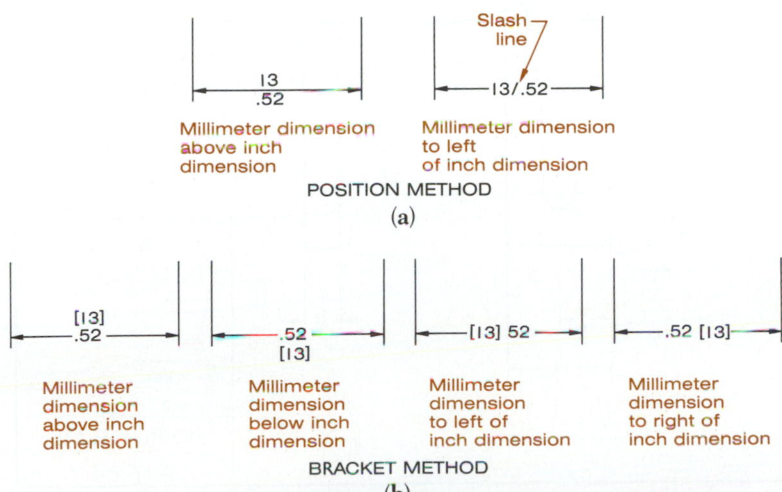

FIGURE 10-14 Dual dimensioning: inch and millimeter ***(Section 10.7)***

Computer-prepared drawings may require the use of parentheses rather than brackets because of character-set limitations.

Angles are stated in degrees (°), minutes (′), and seconds (″) or in degrees and decimals of a degree. No difference exists in the inch and metric system of angular measurement.

Designations such as nominal thread sizes or pipe sizes are not converted to the inch system. Instead the applicable metric values are used as the basis of designation.

10.8 DESIGNATION OF UNITS OF MEASUREMENT

On drawings where all dimensions are either in inches or in millimeters, the units are omitted from the dimension numerals. Unless otherwise specified, all units on a drawing are assumed to be in inches. The word METRIC (or the abbreviation SI) is given in bold letters on the drawing near the title block to signify that the units are metric.

The *scale* of the drawing is designated in the sheet title block. In the U.S. Customary system, the equal sign is used (1 = 1, 1 = 2, 1 = 5, 3″ = 1′-0″) or, where applicable, the words FULL(as shown in Figure 10-15) or HALF may be used.

In the metric system the colon is used (1:1, 1:2, 1:5). One exception to restricting the use of the colon to the metric system is on maps, where the colon is often used in ratios expressing various units of measurement, as 1″:10 miles.

To avoid the possibility of misinterpretation, the inch symbol is recommended for use on a drawing with the numeral 1 when it is used as a dimension (as in 1″) or as a size specified in a note (as in 1″ DRILL).

In special cases, only a limited number of metric values on an inch-dimensioned drawing may be needed. If so, the abbreviation mm should accompany each metric value. When some inch values are used on a millimeter-dimensioned drawing, the inch value should be followed by the abbreviation IN.

10.9 FINISHED SURFACES

Before applying dimensions to the views of a machine part, try to decide the manufacturing process that will be used to create the part. The choice of dimensions and the machining accuracy necessary is dependent upon the type of manufacturing processes that will later be used. The part shown in Figure 10-15 can be made efficiently by cutting it from a length of standard AISI 1020 cold-drawn steel. Such material is readily available in bars ½ inch thick by 1½ inches wide and in 10- to 20-feet lengths. The required machining operations are as follows:

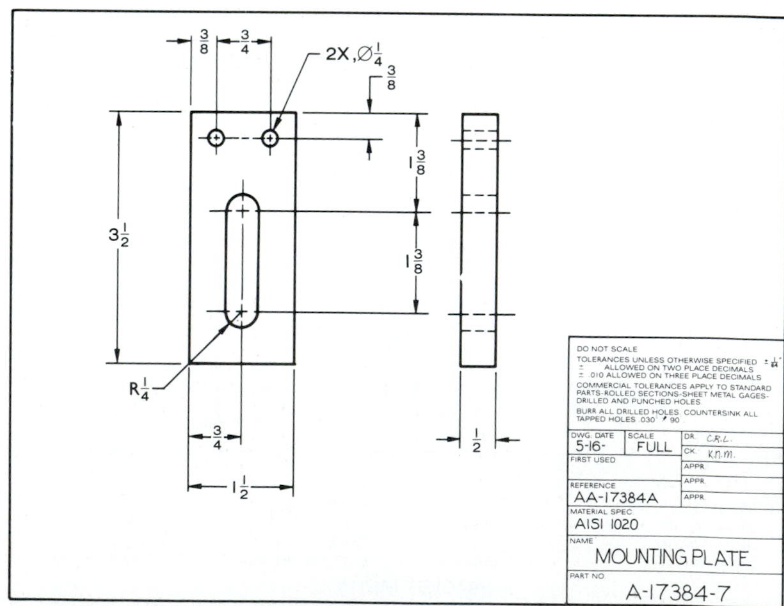

FIGURE 10-15 Detail drawing of a machine part *(Sections 10.8 and 10.9)*

1. Cut off the approximate length.
2. Mill ends square to desired length.
3. Lay out and mill slot.
4. Lay out and drill holes.
5. Deburr.

The only sizing operations necessary are to machine the ends. Machining the sides or the flat surfaces is unnecessary because the designer wisely specified available stock sizes. In addition, the quality of the surface finish (cold drawn) on the available stock is suitable for the final part requirements without the need for further machining. This part is manufactured by a process known as *machining from stock*.

A drawing of a part made by *sand casting* is shown in Figure 10-16. When obtained from the foundry, all surfaces of the castings are rough, and most surfaces are slightly tapered. Consequently, it is necessary to improve the surface finish, accuracy, squareness, and parallelism between one surface and another with machining operations. Like those on a sand casting, all surfaces on a *forging* are rough. The drawing of a forging, shown in Figure 10-17, portrays the completed machine part in its *finished* form. In some companies separate forging and machining drawings are prepared, one for the diemaker and one for the machinist. Unless the function of the part is known, it is generally impossible to distinguish, on a drawing, a sand casting from a forging by shape alone.

The methods for dimensioning drawings for sand castings and forgings are similar. Any surface that, when in use, contacts another surface must be machined smooth by removing a layer of metal. Both castings and forgings in the as-cast or as-forged condition (before machining) are purposely formed oversize by an amount ranging from 1/16 inch to 1/8 inch (about 1 mm to 3 mm) per surface. This oversize condition is called *stock allowance*. Stock allowance is provided for each surface that, because of its intended function (or perhaps solely for appearance), must be smooth and geometrically accurate. The patternmaker or the diemaker, working from the drawing that gives the finished size of the casting or forging, adds sufficient stock allowance for machining purposes.

If the part to be dimensioned is a sand casting or a forging, the drafter or engineer begins by applying *finish marks* to the

C
B
A

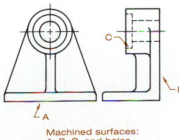

Machined surfaces: A, B, C, and holes

FIGURE 10-16 A sand casting with machined surfaces ***(Section 10.8)***

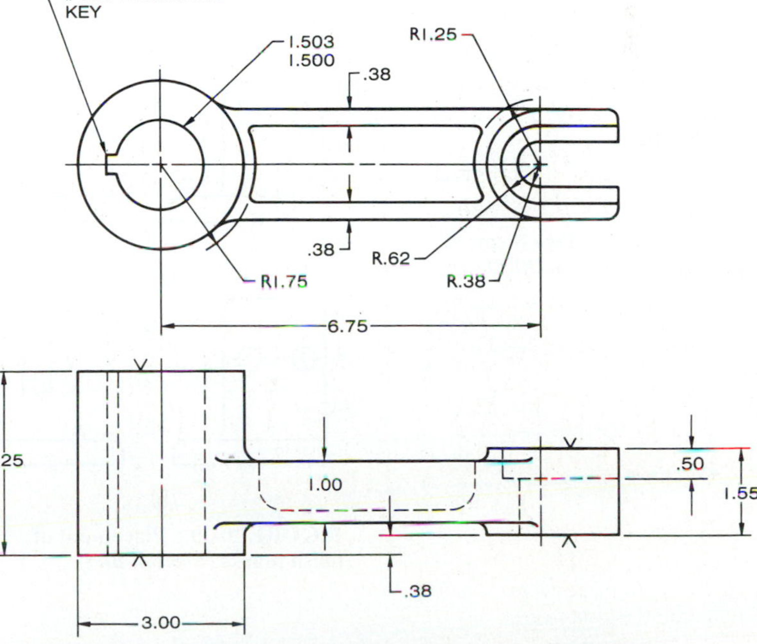

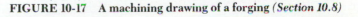

FIGURE 10-17 A machining drawing of a forging ***(Section 10.8)***

edge view of each surface that requires stock allowance. Two forms of finish marks are illustrated in Figure 10-18. This symbol on the drawing alerts the patternmaker (in casting) or the diemaker (in forging) to the location of the required excess stock on the rough casting or forging. The finish mark also identifies the required machined surfaces. Finish marks may be made freehand or mechanically. The V symbol is preferred, although some companies continue to use the older italic *f* symbol. As shown in Figure 10-19, finish marks are applied to each finished surface and are repeated in every view where the finished surface appears as an edge.

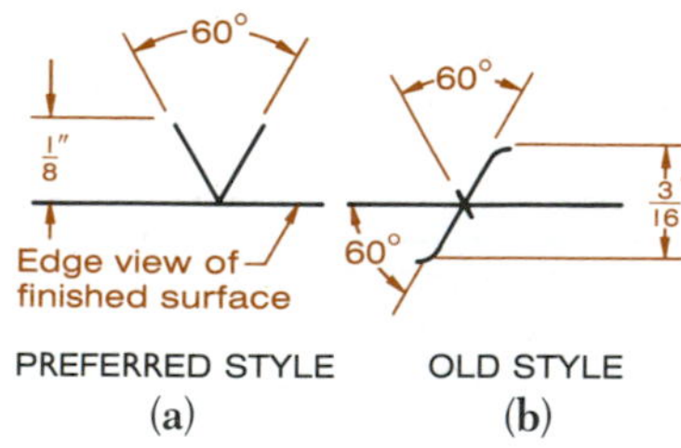

FIGURE 10-18 Two forms of finish marks ***(Section 10.8)***

Finish marks are not used on parts machined from stock, on weldments, or on preformed materials (including stampings, rolled shapes, and extrusions) because it is obvious that the surfaces of these parts are finished. Finish marks are also omitted on punched, drilled, countersunk, counterbored, or reamed holes that are designated by notes. If all surfaces on a part are to be finished, the finish marks are omitted and the general note FINISH ALL OVER, or the abbreviation FAO, is used.

Finish marks do not indicate *surface texture* that qualifies the smoothness or the quality of a machined surface. Special symbols that are used to denote surface texture are discussed in Section 8-6.

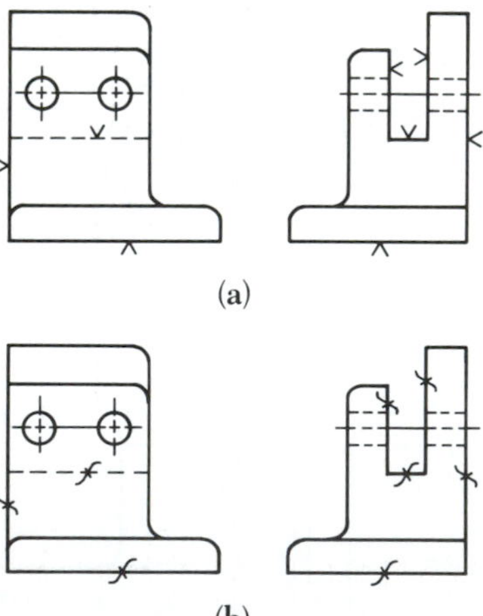

FIGURE 10-19 Placement of finish marks ***(Section 10.8)***

10.10 DEGREE OF ACCURACY

Regardless of the dimensioning method, some error or inaccuracy is inevitable. Unless otherwise stated on the drawing, all dimensions expressed as a common fraction may vary by plus or minus 1/64 inch. A part dimensioned 1½ inches, for example, may be made as large as 1 33/64 inches, as small as 1 31/64 inches, or any size within this range. The amount of permissible variation, ± 1/64 inch (a total of 1/32 inch), is called the tolerance. A customary tolerance for a two-place decimal dimension is ± .01 inch. Tolerancing for interchangeable manufacturing is covered in Chapters 11 and 12.

10.11 DIMENSIONING IN LIMITED SPACES

Dimensions must be placed in the clearest possible position on the drawing. Care should be taken to avoid crowded dimensions. Figure 10-20(a) gives several examples of dimensioning in limited spaces. Note how the practice of staggering numerals, shown in 10-20(b), improves the reading clarity of the drawing.

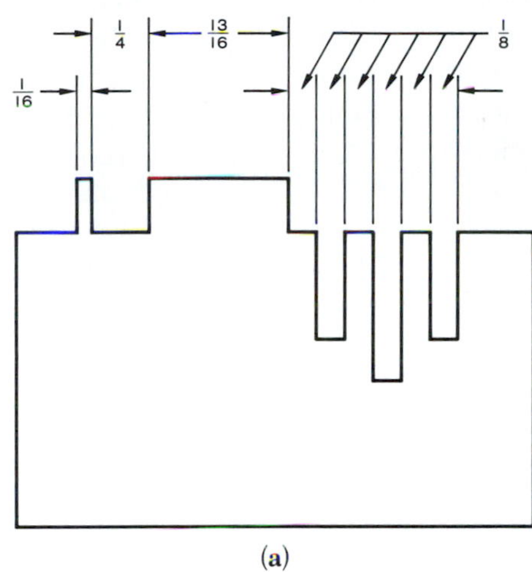

(a)

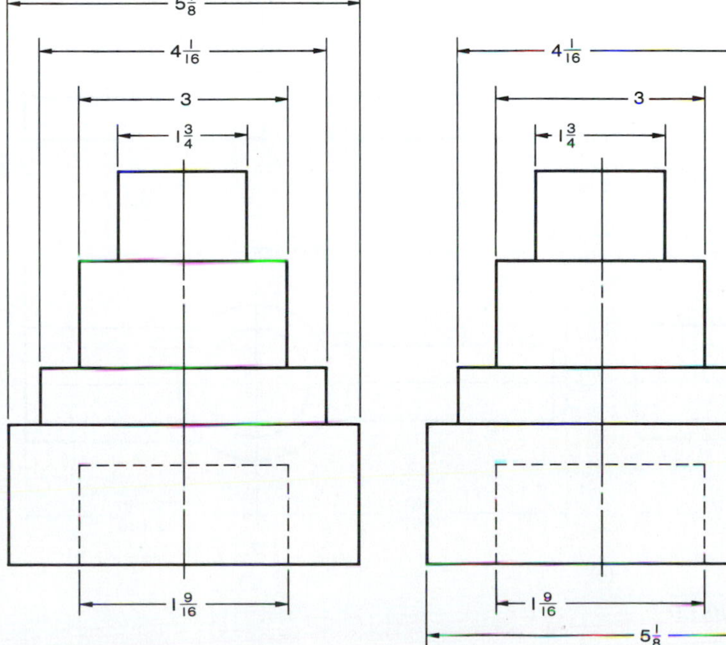

(b)

FIGURE 10-20 Dimensioning in limited spaces *(Section 10.11)*

10.12 OVERALL DIMENSIONS

Figure 10-21(a) shows a single view of a part with three encircled dimensions, 2½, ⅞, and 1⅝. The longest dimension, 2½ inches, is called an *overall* dimension and is generally given to denote the total distance along a feature of a part. Only one of the intermediate dimensions, ⅞ inch or 1⅝ inches, is needed. Depending upon the function of the part, the *least* important dimension in a consecutive series of dimensions should be omitted, as in Figure 10-21(b); identified as a *reference* dimension by placing it in parentheses, as in 10-21(c); or labeled REF, as in 10-21(d). Where the intermediate dimensions are more important to the function of the part than the overall dimension, as in 10-21(e), it is customary to qualify the overall dimension as a reference dimension.

10.13 DIMENSIONING REPETITIVE FEATURES

Equal spacing of features (holes, slots, and spokes, for example) in a series or a pattern may be specified by giving the required number of spaces and the letter X (meaning the number or quantity required) followed by the applicable dimension. Examples of dimensioning repetitive features are given in Figures 10-22(a) and 10-22(b). Where it is difficult to distin-

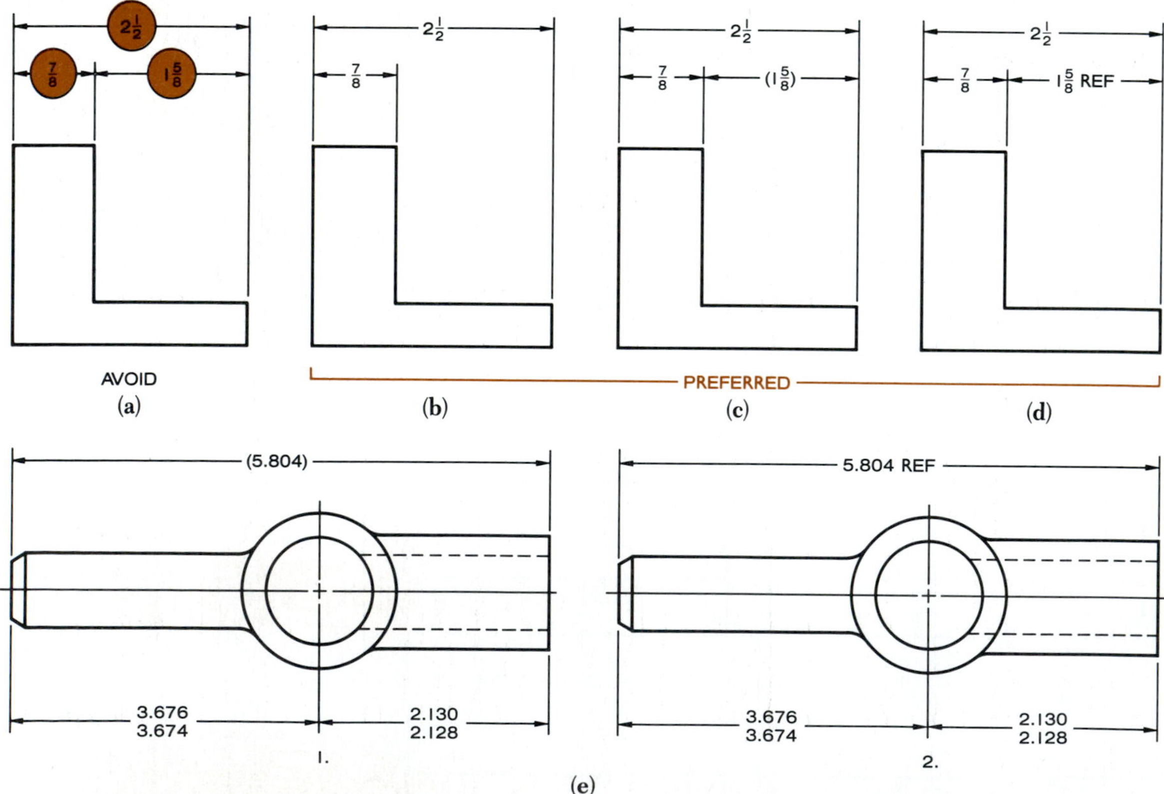

FIGURE 10-21 Overall dimensions *(Section 10.12)*

guish between the dimension and the number of spaces, as in 10-22(c), one space is dimensioned and placed inside parentheses, identifying it as a reference dimension.

10.14 DIMENSIONING ANGLES

The three methods of dimensioning angular features on parts are shown in Figure 10-23. At 10-23(a) coordinate dimensions of the two legs of a right triangle are given. At 10-23(b) one linear dimension and an angle in degrees are given. At 10-23(c) only the size of the angle in degrees is given. Depending upon the available space, the figure may be inserted between the extension and object line (15°) or placed outside the lines (5°10′). As stated in Section 9.8, the units for angular measurement are degrees (°), minutes (′), and seconds (″). An angle of 45 degrees and 30 minutes may be given as 45°30′ or 45.5°. It is customary to place angular dimension figures so they will read from the bottom of the drawing, as in the unidirectional system.

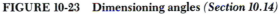

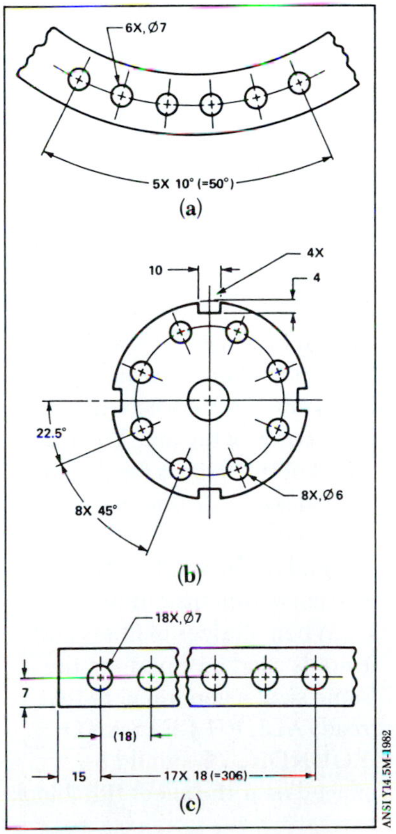

ANSI Y14.5M-1982

FIGURE 10-22 Dimensioning repetitive features *(Section 10.13)*

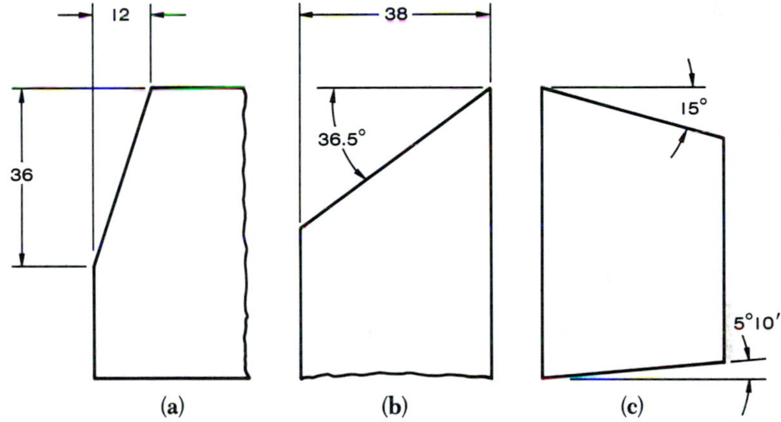

FIGURE 10-23 Dimensioning angles *(Section 10.14)*

10.15 DIMENSIONING ARCS

An arc is dimensioned by giving its radius. A radius dimension uses only one arrowhead, which must always touch the arc. Various methods for dimensioning arcs are shown in Figure 10-24. In both the customary inch system and the metric system, the letter R should precede the radius value, as shown in 10-24(a) and 10-24(b). Where space permits, a dimension line is drawn from the radius center to the arc, and the radius value is placed in a space between the arrowhead and the radius center. Where space is limited for the radius numeral, the radial dimension line may be extended through the radius center, with the radius value placed at a convenient position outside of the arc, as shown in 10-24(c), or a leader may be used, as shown in 10-24(d). When the center of the radius falls off the view or is in some inconvenient position, only a portion of the radius dimension line next to the arrowhead is drawn. A false radius center may be established and dimensioned with a zigzag line, shown in 10-24(e), to indicate that it is not a true center. The center of an arc may be dimensioned if necessary, as shown in 10-24(f). Note the use of the small intersecting dashed lines or center crosses placed at the center of the arc. The placement of the figures for the arc dimensions must be in accordance with the rules for the aligned or the unidirectional system of dimensioning.

When all sizes of fillets and rounds used on a part are the same size, a general note that reads ALL FILLETS AND ROUNDS __ R should be placed near the sheet title block.

FIGURE 10-24 Dimensioning arcs *(Section 10.15)*

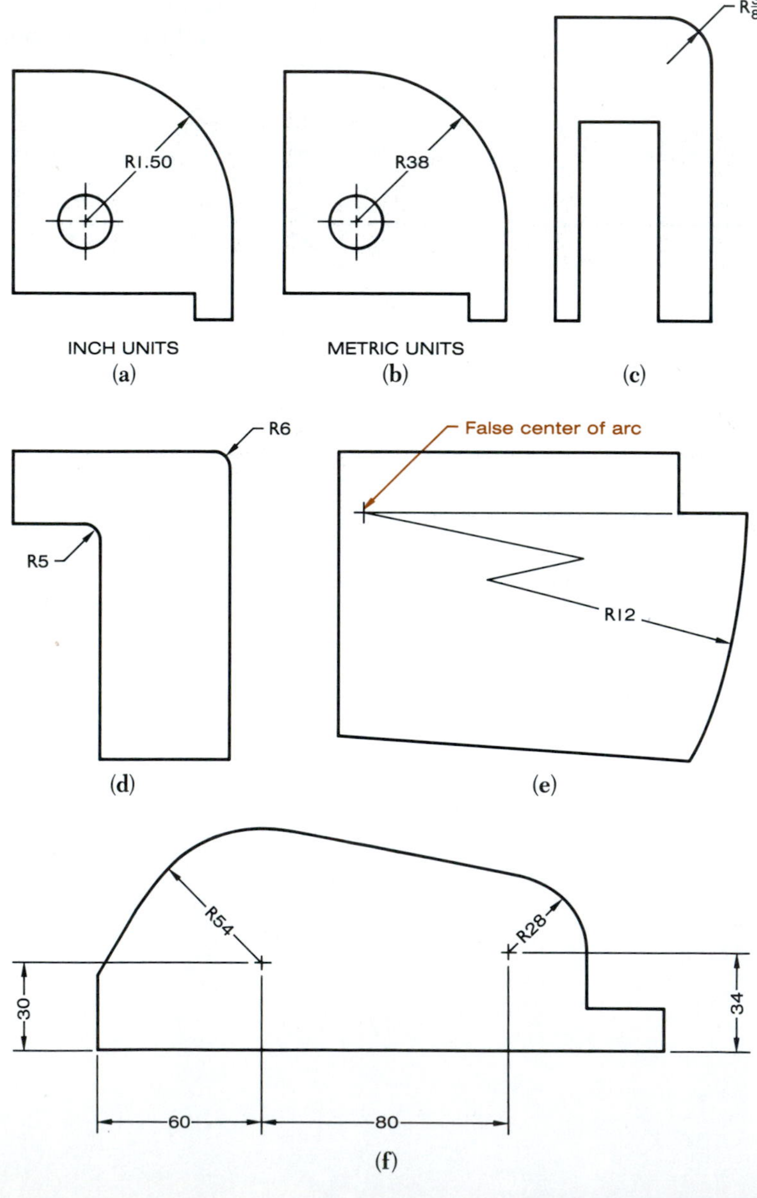

10.16 ROUNDED CORNERS

Where corners are rounded, radius dimensions are used, as shown in Figure 10-25. The arcs are understood to be tangent to the straight edges.

10.17 CURVED OUTLINES

When dimensioning parts having an outline consisting of two or more arcs, use the method shown in Figure 10-26(a). The radii of all arcs are given with the necessary coordinate dimensions locating the centers, or the arcs are located on the basis of their points of tangency. The coordinate method, shown at 10-26(b), is used to plot the positions of points along a noncircular or irregular curve. Each dimension line is extended to datum lines. Datums are points or edges of surfaces of an object that are assumed to be exact for purposes of dimensioning and from which the position of other features is established. Datums are explained more completely in Sections 12.1, 12.2, and 12.17. (Note: Neither of the objects shown in Figure 10-26 are completely dimensioned.)

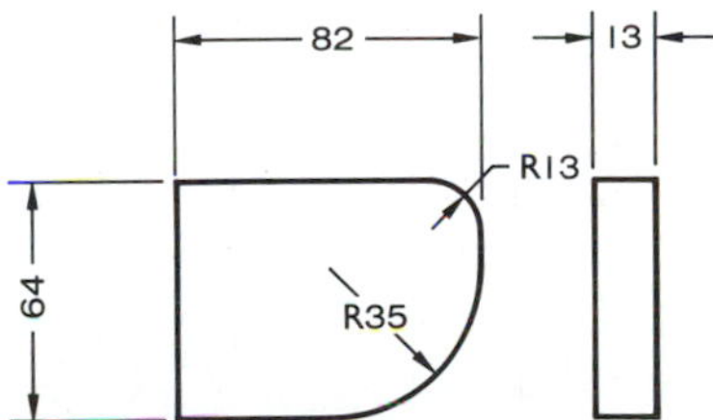

FIGURE 10-25 Dimensioning parts with rounded corners *(Section 10.16)*

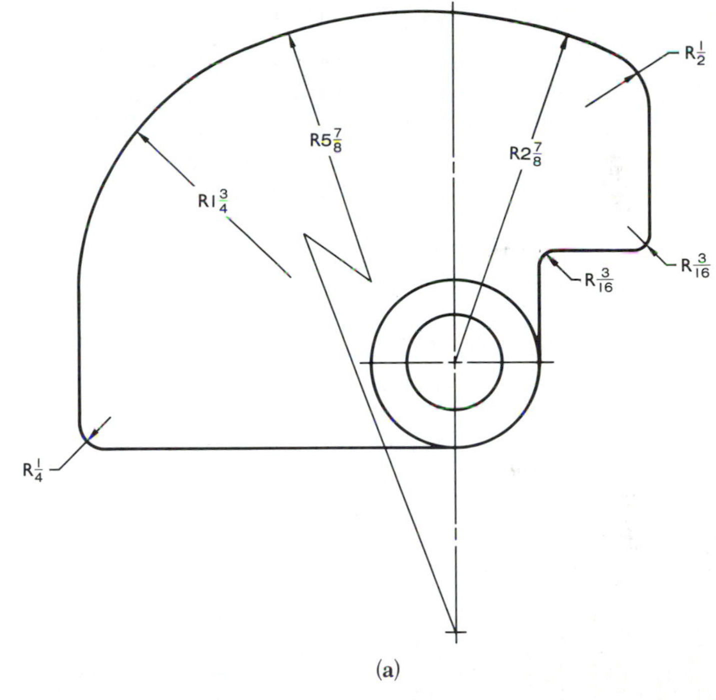

(a)

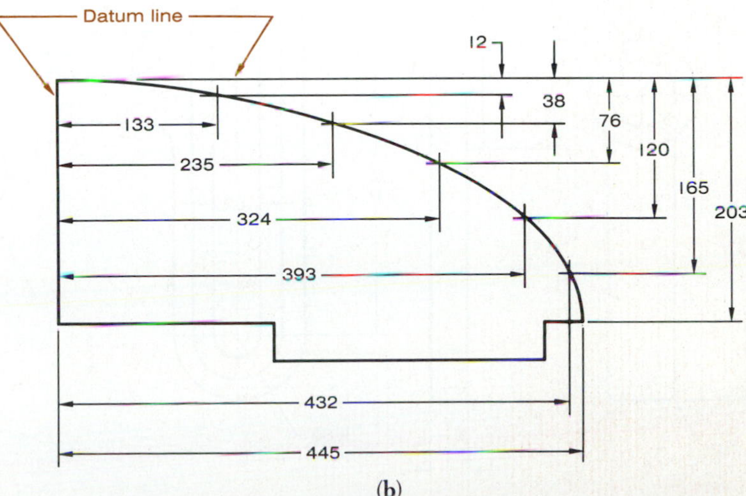

(b)

FIGURE 10-26 Dimensioning parts with curved outlines *(Section 10.17)*

10.18 PARTS WITH ROUNDED ENDS

The method selected to dimension parts with rounded ends or parts with cast (or cored) or milled slots depends upon the degree of accuracy required and the method of production selected. Some dimensions, particularly noncritical ones, are needed only by the patternmaker and diemaker. In other cases a dimension may be required by the machinist.

FIGURE 10-27 Low-precision dimensioning of parts with rounded ends ***(Section 10.18)***

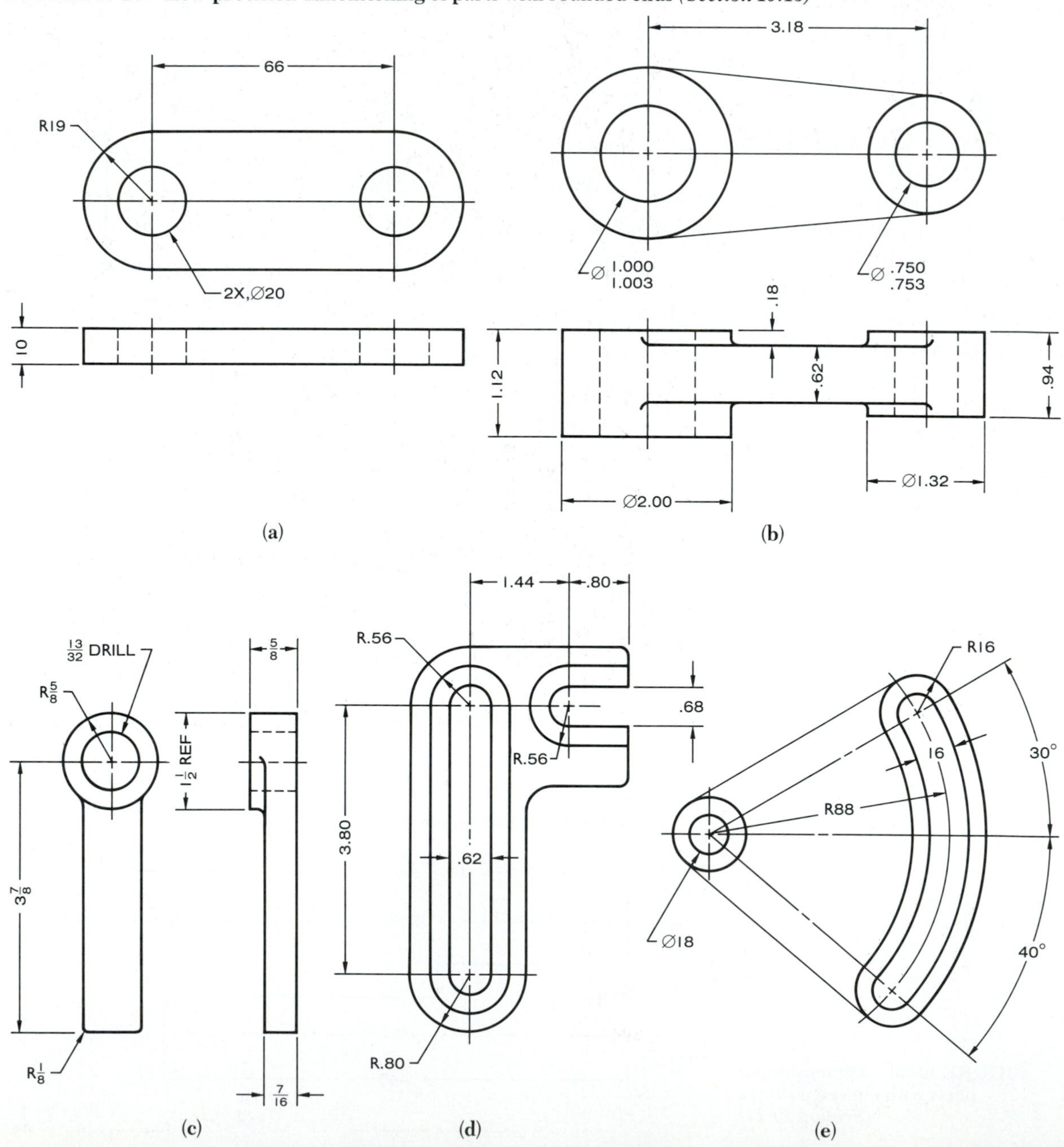

Low Precision

The examples shown in Figure 10-27 illustrate conventional dimensioning methods that are acceptable when manufacturing low precision parts. At 10-27(a) and 10-27(b) the center-to-center dimension and the end radius are given. At 10-27(c) the end radius and the hole share a common center and are located from the flat end. The single end radius dimensions given in these three examples apply to both ends. Overall dimensions are not required.

High Precision

Overall dimensions should be used for parts with rounded ends and for milled slots when close precision is required. Illustrations of customary dimensioning practices for various part shapes are given in Figure 10-28. Many companies have adopted the practice of using reference dimensions, which are shown below each example, to help reduce potential errors in the shop caused by calculating unspecified dimensions.

Partially Rounded Ends

As shown in Figure 10-29, the radii are dimensioned for parts with partially rounded ends.

FIGURE 10-28 High-precision dimensioning of parts with rounded ends *(Section 10.18)*

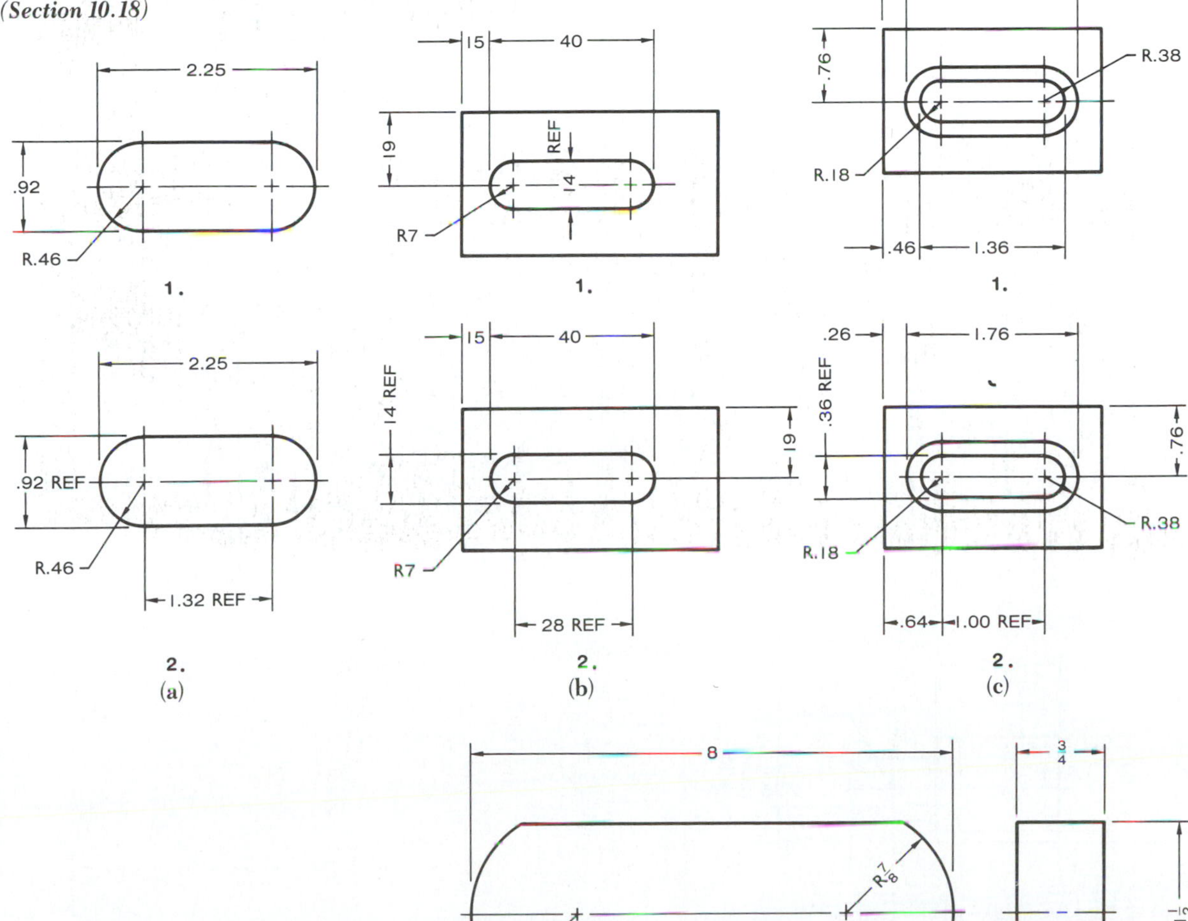

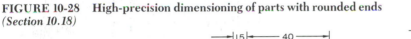

FIGURE 10-29 Dimensioning parts with partially rounded ends *(Section 10.18)*

10.19 DIMENSIONING CYLINDERS

Solid Cylinders

Dimensions that specify the length and diameter of a solid cylinder are usually given in the rectangular view, as shown in Figure 10-30(a). Cylindrical parts and round holes are measured in the shop with calipers or a micrometer. The diameter

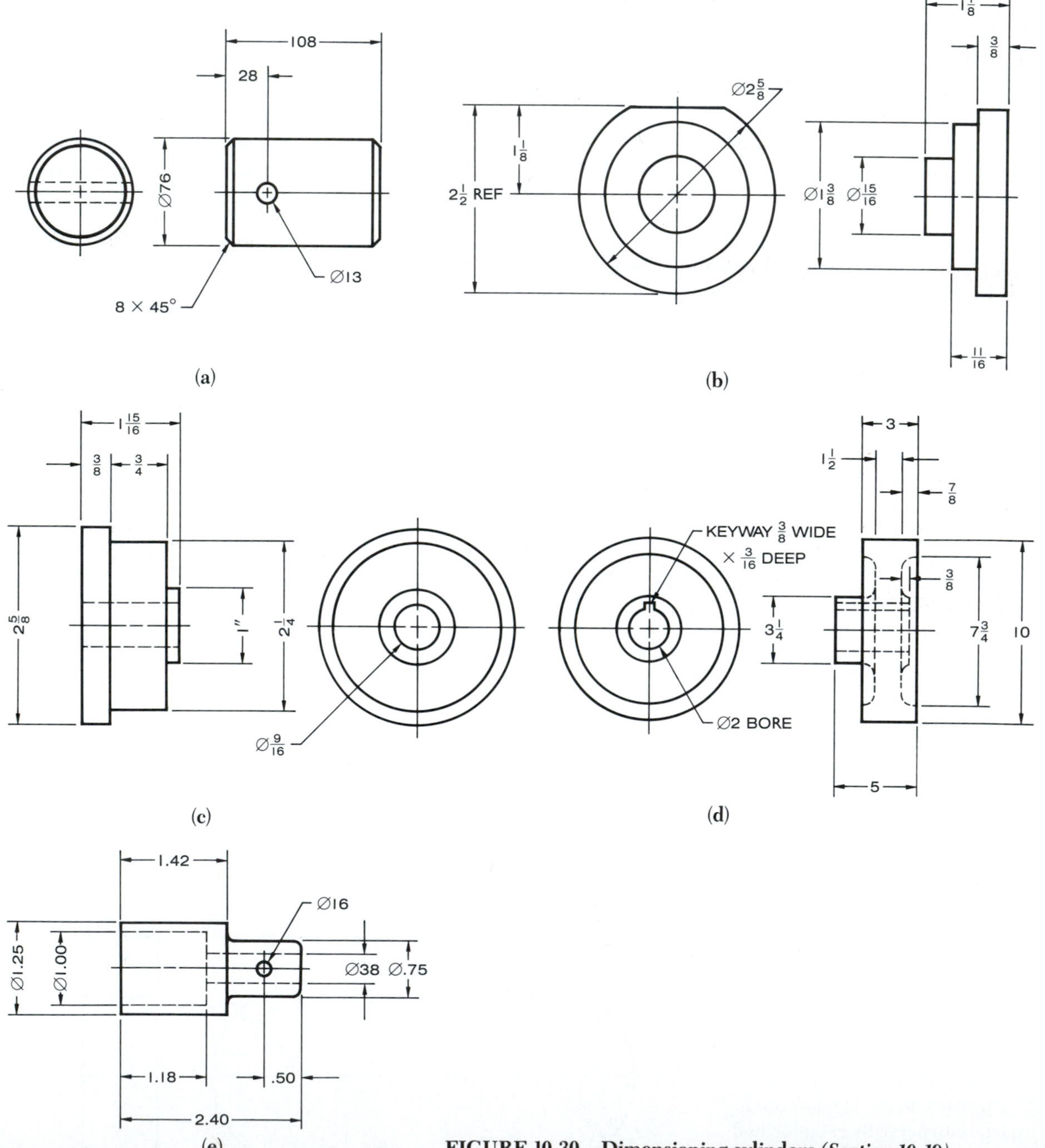

FIGURE 10-30 Dimensioning cylinders *(Section 10.19)*

should always be given, *not* the radius. Diagonal dimensions may be used on circular views in special cases. At 10-30(b), for example, the circular view is the preferred location for the 2⅝ diameter dimension because of a possible misunderstanding caused by the flat surface cut on the flanged end.

The radius is not given on cylindrical parts because the diameter dimension is required in the shop to measure the size. A part composed of several concentric cylinders is dimensioned as shown in Figures 10-30(c) through 10-30(e). In both the customary inch system and in the metric system, the diameter symbol (∅) or the abbreviation DIA is used to denote a diameter dimension. When there is no likelihood of misunderstanding, it is customary to omit the abbreviation or the symbol. On a cylindrical part, the diameter dimensions that are displayed in a single view should always show the ∅ symbol, as in Figure 10-30(e).

Hollow Cylinders

Holes formed by sizing tools such as drills, reamers, countersinks, and counterbores may be specified by a note similar to the one shown in Figure 10-31(a). (See also Section 10.26.) Larger holes that are produced by boring are specified as shown at 10-31(b).

FIGURE 10-31 Dimensioning hollow cylinders *(Section 10.19)*

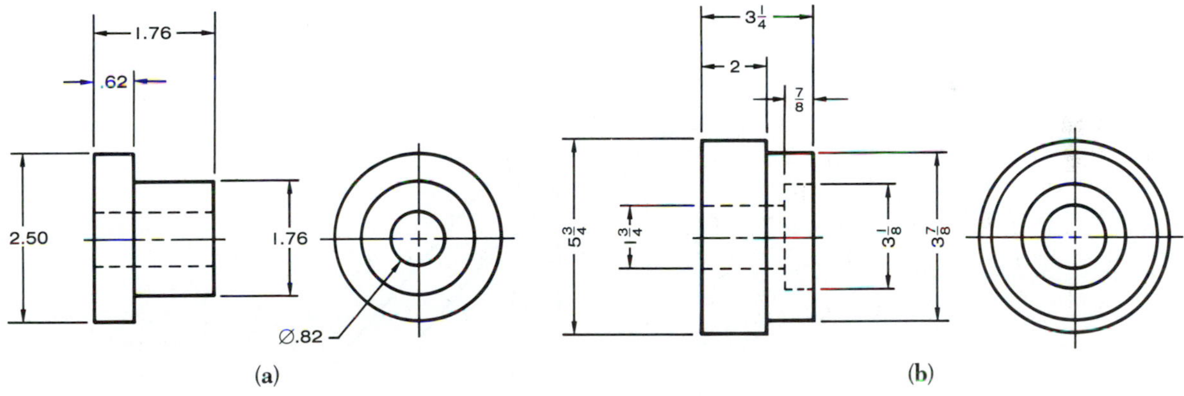

10.20 GEOMETRIC SHAPE ANALYSIS

A procedure that is frequently used by engineers to aid in selecting the necessary dimensions for the manufacture of a part begins with a careful analysis of the total *form* of the object. It is important to establish a clear, mental image of the object. For this procedure, carefully study the three views of the cast bracket shown in Figure 10-32(a). Next break down the object mentally into simple, geometric shapes such as prisms, pyramids, cylinders, or cones, as if each component were to be made separately. An example of the reduction of an object into simple geometric shapes, in this example, four, is illustrated in Figure 10-32(b). Fillets, rounds, and small holes are temporarily ignored in geometric shape analysis.

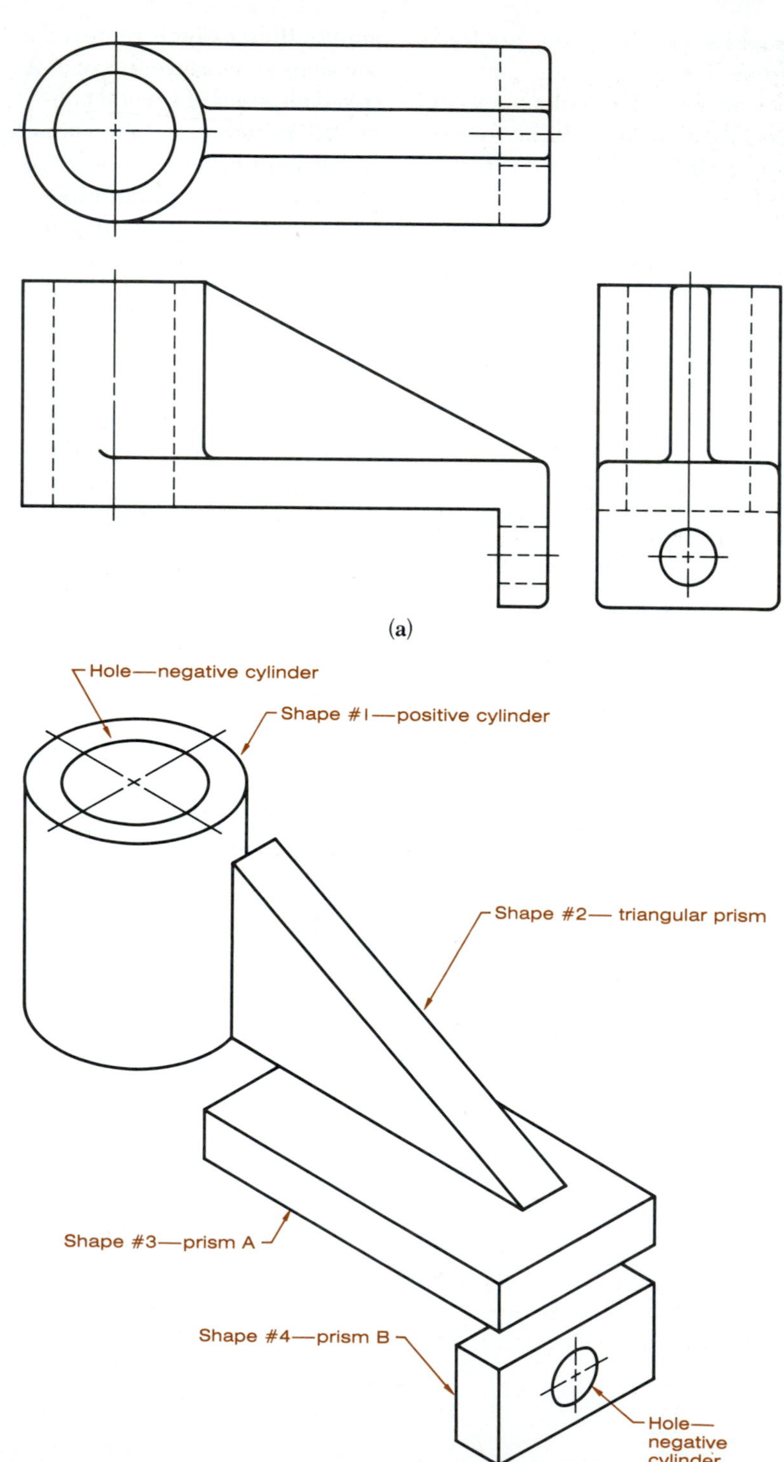

FIGURE 10-32 Geometrical shape analysis of a part ***(Section 10.20)***

10.21 SELECTION OF SIZE AND LOCATION DIMENSIONS

In the process of selecting dimensions for an object, both analyzing each component shape and then deciding the size dimensions needed to make that one shape is helpful. *Size dimensions* describe the size of the object with regard to its form and other features, such as holes and slots. This dimensional analysis is shown in Figure 10-33(a). For example, the size dimensions for the outside diameter, the hole, and the length are required to produce the cylinder shown as shape 1. The size dimensions for shape 2, the triangular prism, and shapes 3 and 4 (prisms A and B, respectively) are also shown. Other dimensions, called *location dimensions*, are also needed. Location dimensions give the *position* of holes, slots, and so on with respect to important finished surfaces, center lines,

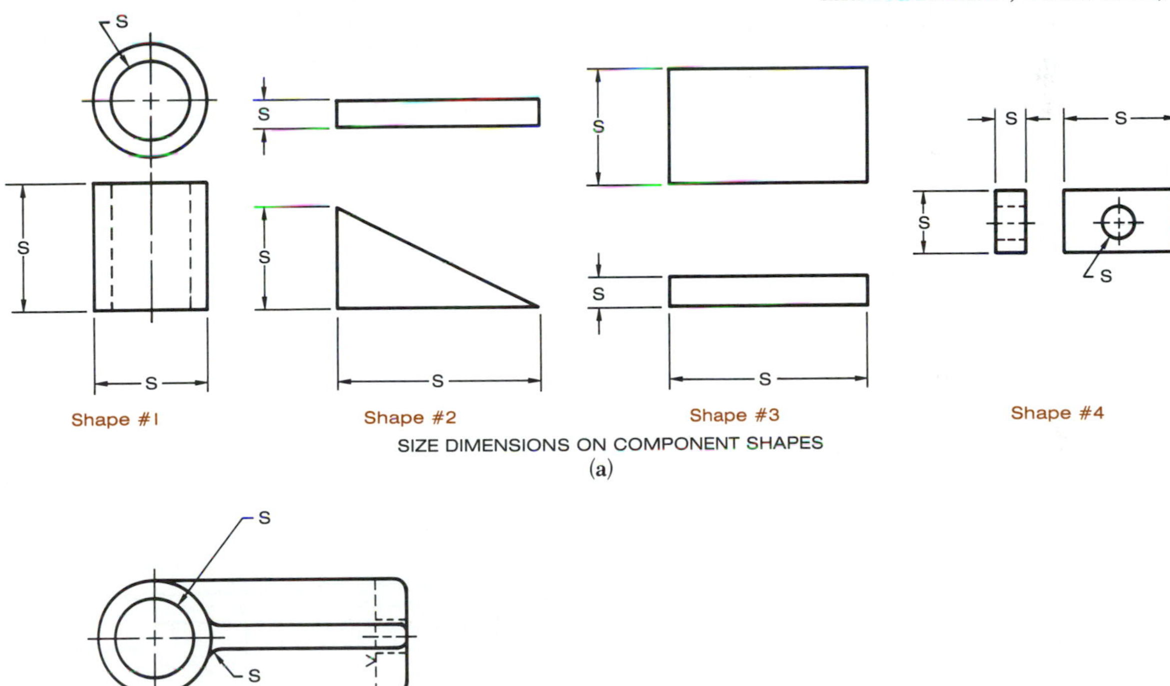

(a)

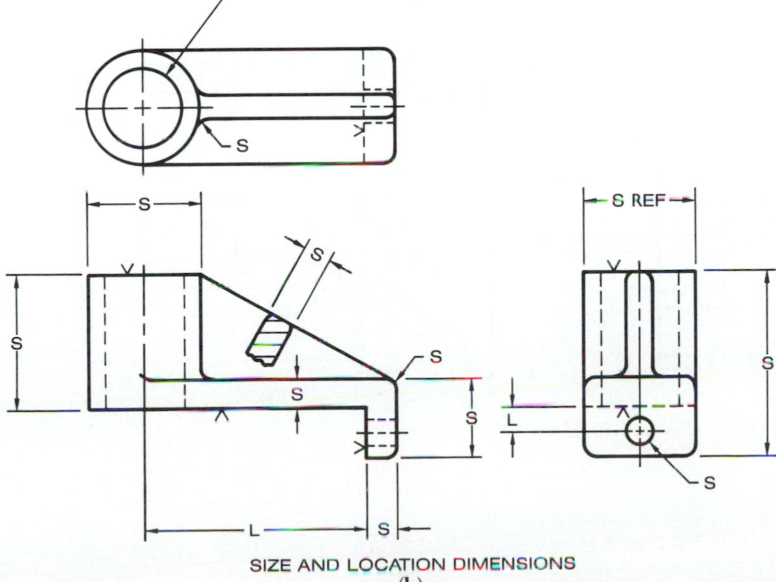

(b)

FIGURE 10-33 Size dimensions on component shapes *(Section 10.21)*

or unfinished surfaces. Note that some of the dimensions previously used in the geometrical analysis shown in Figure 10-33(a) have been rearranged, modified, or entirely omitted in 10-33(b) to accommodate the production requirements. In general the total number of dimensions needed for the final detail drawing is usually less than the number needed for the separate parts. Four examples of size (S) and location (L) dimensions applied to various geometric shapes are shown in Figure 10-34. An object may be dimensioned by indicating the size of each geometric shape and giving the relationship relative to a center line or to a finished surface.

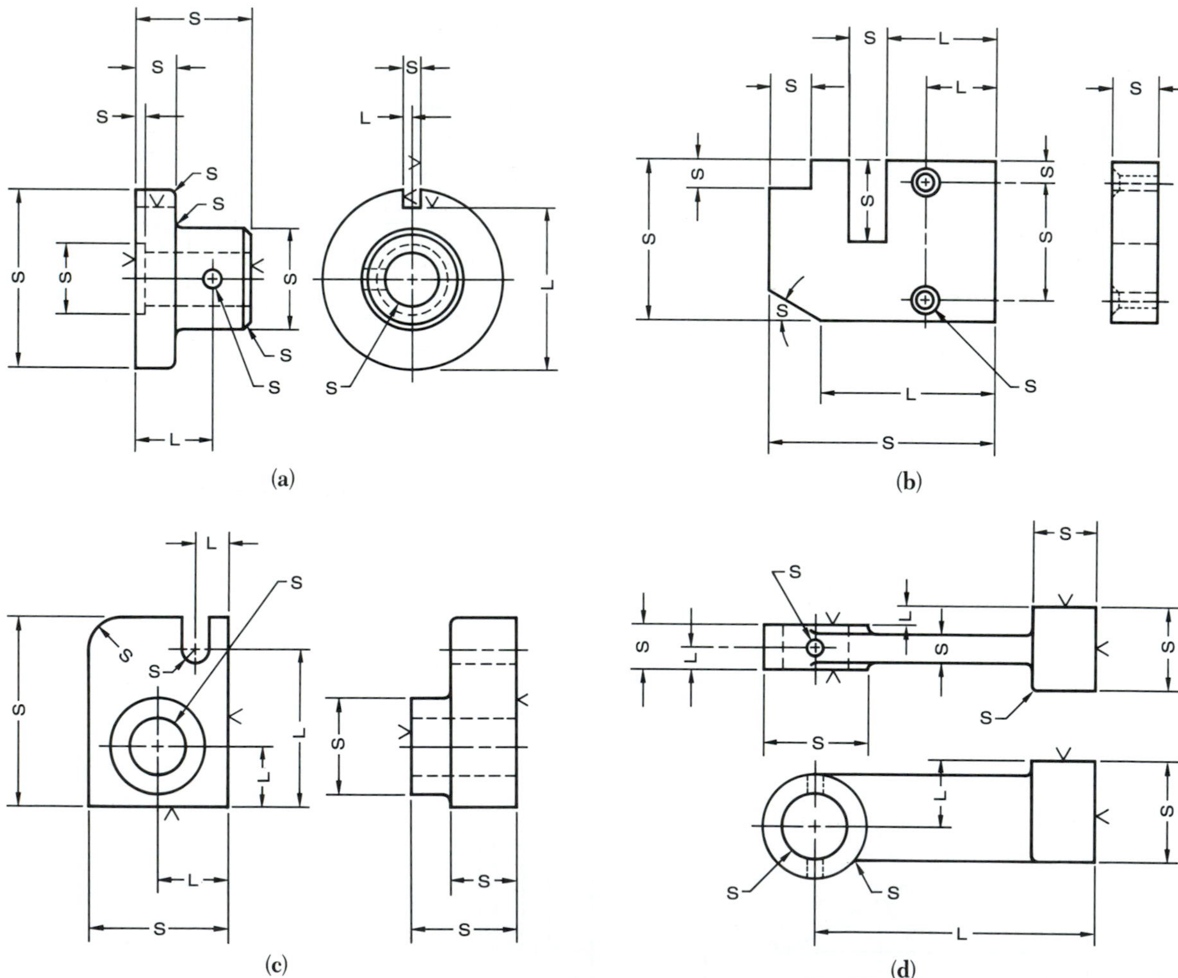

FIGURE 10-34 Size and location dimensions ***(Section 10.21)***

10.22 CONTOUR DIMENSIONING

Views are always selected and drawn with the intent of showing the shape of a part in the clearest possible manner. Dimensions are used to specify the size and location of the shape of the object. They should be selected carefully and placed in a position on the drawing where they will be most easily understood. A dimension should be drawn to a view that shows the contour of the feature to which it applies. The most descriptive view of an object is always selected for specifying the size and location of features, as shown in Figure 10-35(a). Many of the dimensions are incorrectly placed on the views shown in 10-35(b).

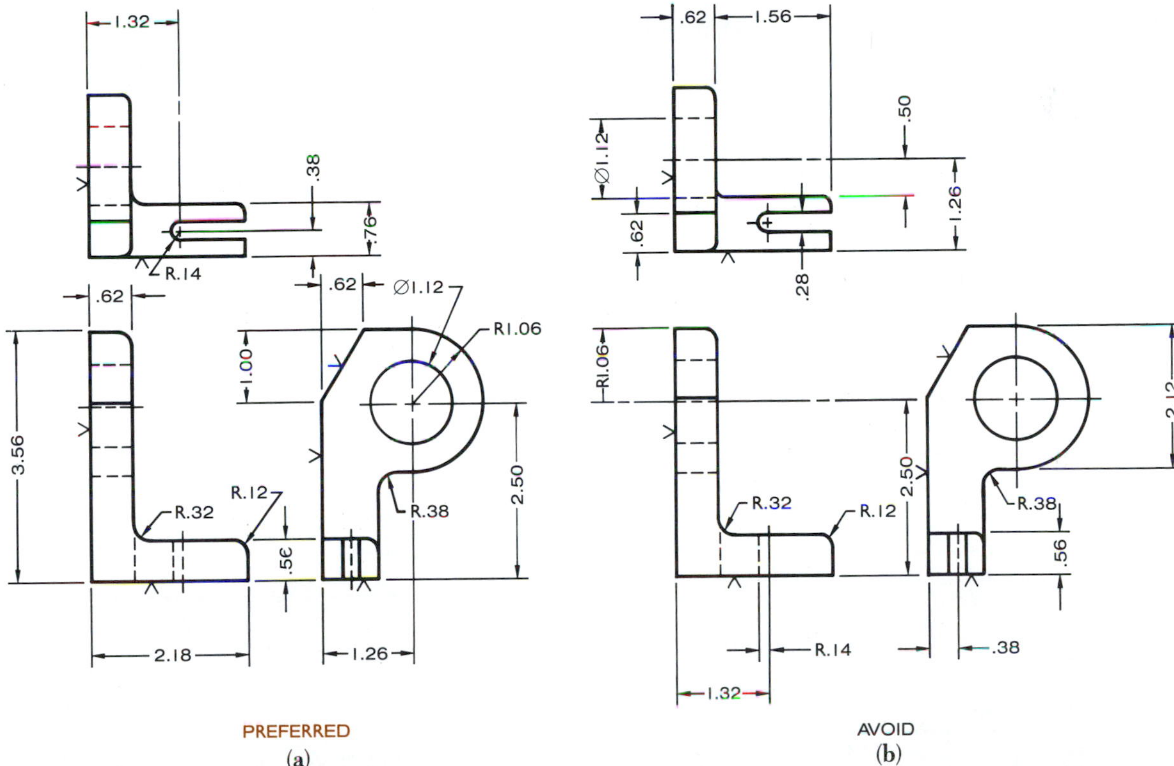

FIGURE 10-35 Contour dimensions *(Section 10.22)*

10.23 SYMMETRICAL CURVED OUTLINES

Parts with symmetrical curved outlines are dimensioned by using coordinate or offset dimensions that plot points along one-half of the curve, as shown in Figure 10-36(a). Because of the size of some symmetrical parts, one-half of the outline is often drawn to conserve space on a drawing. As shown in 10-36(b), the outline of the part is extended slightly beyond the center line and terminates with a break line. In both examples the symmetry is indicated by using the center line symbol (℄) or by an appropriate note, such as SYM ABOUT ℄.

10.24 IRREGULAR CURVED OUTLINES

As shown in Figure 10-37(a), each of a series of points along the curve of an irregular outline is located with respect to the two *datum planes*, which are at right angles to each other. Datum planes are edges of an object that are used as a reference base of measurements in the shop. The position of datum planes must be accessible dur-

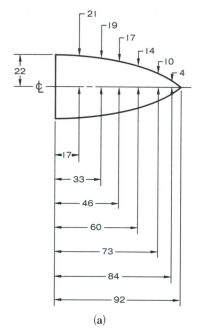

(a)

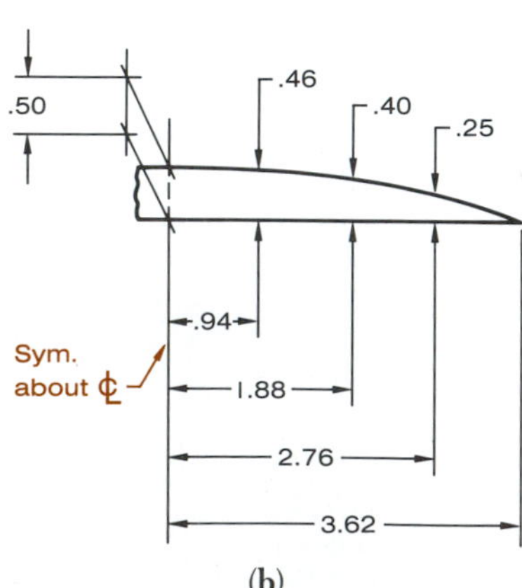

(b)

FIGURE 10-36 Dimensioning symmetrical curved outlines ***(Section 10.23)***

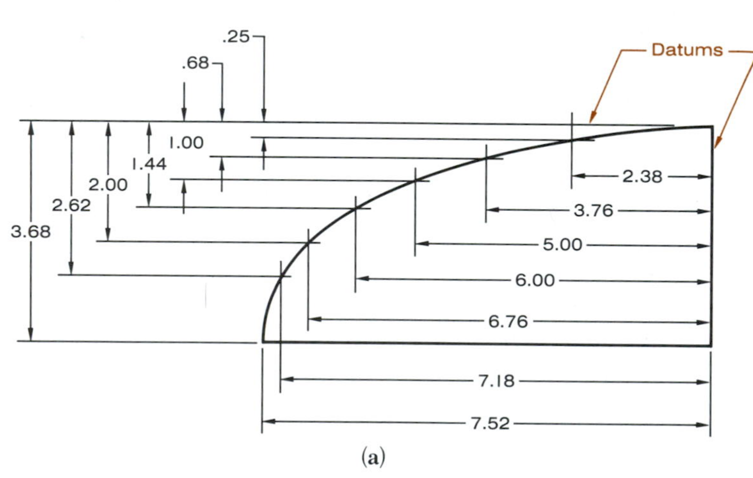

(a)

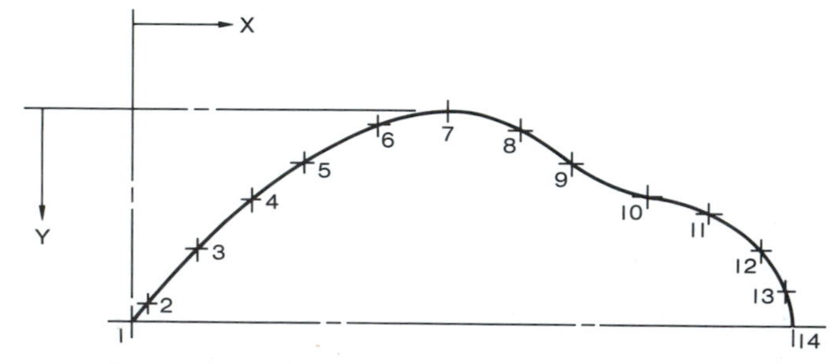

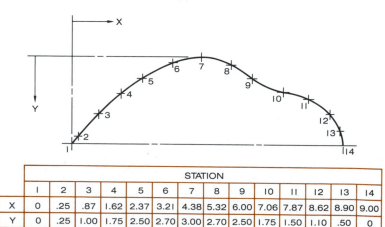

	STATION													
	1	2	3	4	5	6	7	8	9	10	11	12	13	14
X	0	.25	.87	1.62	2.37	3.21	4.38	5.32	6.00	7.06	7.87	8.62	8.90	9.00
Y	0	.25	1.00	1.75	2.50	2.70	3.00	2.70	2.50	1.75	1.50	1.10	.50	0

(b)

FIGURE 10-37 Dimensioning irregular curved outlines ***(Section 10.24)***

ing production. Rectangular coordinates or offset dimensions are used to plot the points on a curve. The spacing of the points should be carefully selected so that when the curve is plotted on the part, an accurate representation of the intended shape results. If possible the numerical values selected for the various coordinate dimensions should be in even units or those easily measured (.50, *not* .48).

Where many coordinates are required to describe a contour, the X- and Y-coordinates are tabulated as shown in Figure 10-37(b). This method is particularly effective for dimensioning parts with quick-changing curved outlines.

10.25 LOCATION DIMENSIONING OF HOLES

Holes are located on parts in various ways depending upon the accuracy required and the production methods used to establish the location of the holes. One rule that must be carefully followed is that, regardless of the method used to dimension the holes, *location dimensions for holes must always be given on the view on which they appear as circles.*

The method of laying out the precise location of the holes shown in Figure 10-38(a) would require the use of a highly accurate machine known as a *jig borer*. A jig borer has two tables that can be moved independently and at right angles to each other. The holes are accurately produced in one step without the necessity for prior layout and marking operations in the X and Y locations given on the drawing. By way of contrast, the holes located on the part shown in Figure 10-38(b) require less precision. Ordinarily the positions for the centers of the holes would be laid out in the shop by manual methods, center-punched, and then drilled to size in a drill press.

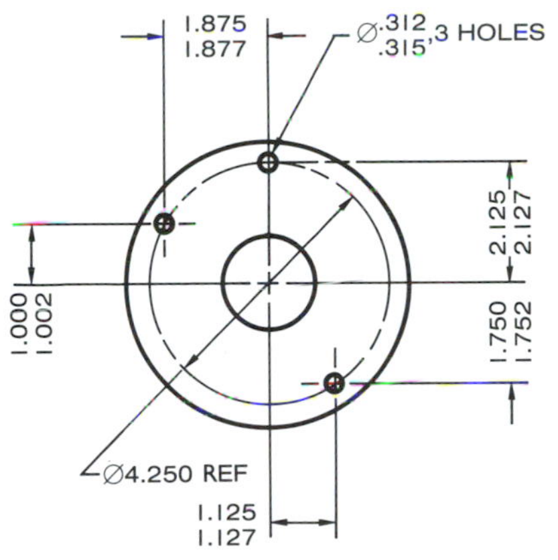

PRECISION LAYOUT OF HOLES
(a)

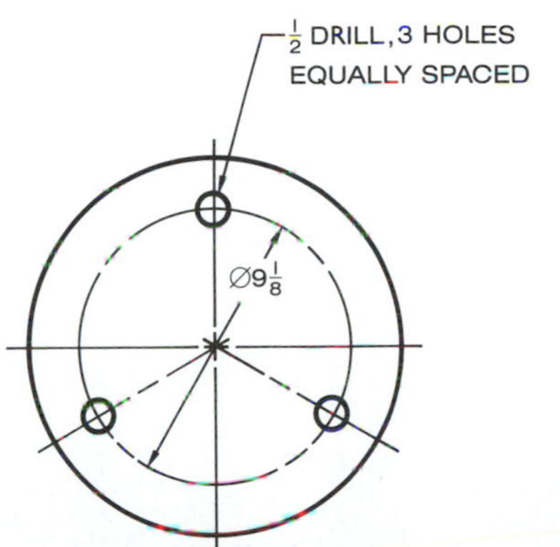

NONPRECISION LAYOUT OF HOLES
(b)

FIGURE 10-38 Location dimensions of holes ***(Section 10.25)***

The examples shown in Figure 10-39 are precision methods commonly used to dimension the location of holes. These dimensioning procedures are well suited to numerical control and automated part-tooling techniques. At 10-39(a) all hole location dimensions are taken from mutually perpendicular datum planes. At 10-39(b) a datum hole is used as a reference point to locate the holes. Dimensions from datum planes or datum holes are often used if a part with more than one critical dimension must mate with another part. Another method of using datum planes in dimensioning the location of holes is shown at 10-39(c). The first and the longest dimensions have two arrowheads. Each of the intermediate dimensions is given with respect to the datum plane, is placed along a single line, and has only one arrowhead.

Another precision-location method, shown in Figure 10-39(d), is to use coordinate dimensions, but without the dimension lines that originate from the datum planes indicated as zero coordinates. Dimensions are shown at the center lines and the extension lines. A variation of the method shown at 10-39(d) is the tabular method at 10-39(e). Here dimensions from mutually perpendicular datum planes are listed in a table placed on the drawing near the corresponding view. This method is often used on drawings that require the location of a large number of holes of different diameters. Tables are prepared in any suitable manner that adequately describes the location of the holes.

One or more pairs of holes may be dimensioned at an angle, shown in Figure 10-39(f), if the distance between the holes is of particular importance to the function of the parts.

The examples shown in Figures 10-39(g) and 10-39(h) show how repetitive dimensions are used on parts where a series of holes (or other features) are spaced equally. A dimension is given to locate the first and last hole and a note is given to specify that the holes are equally spaced. Holes on a circular part may be accurately located by using the rectangular coordinate method shown in 10-39(i) and previously shown in 10-38(a).

FIGURE 10-39 ***(opposite page)***
Precision layout of holes
(Section 10.25)

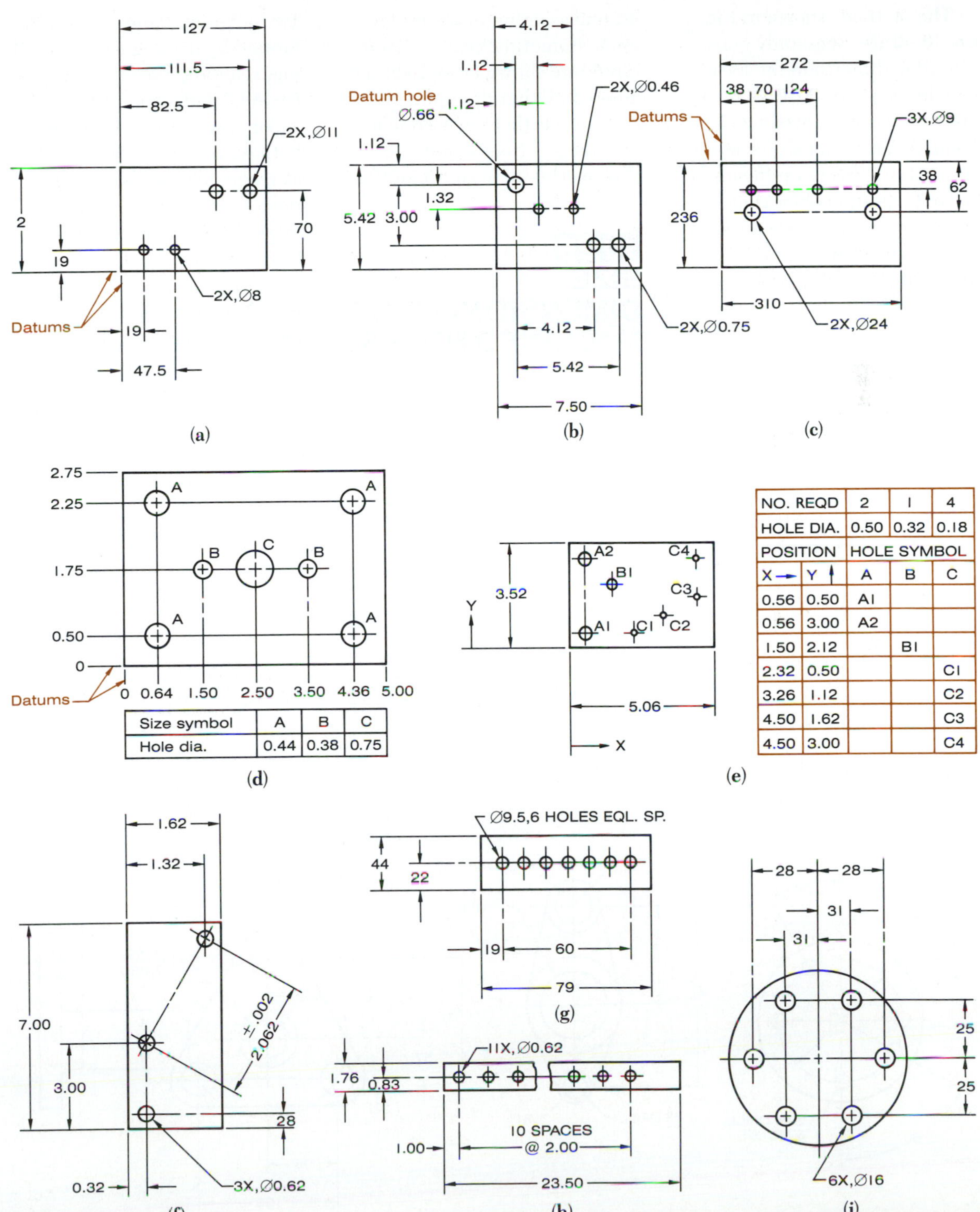

Size symbol	A	B	C
Hole dia.	0.44	0.38	0.75

NO. REQD	2	1	4
HOLE DIA.	0.50	0.32	0.18

POSITION		HOLE SYMBOL		
X →	Y ↑	A	B	C
0.56	0.50	A1		
0.56	3.00	A2		
1.50	2.12		B1	
2.32	0.50			C1
3.26	1.12			C2
4.50	1.62			C3
4.50	3.00			C4

The methods shown in Figure 10-40 are commonly employed to dimension the location of holes on parts. At 10-40(a) the holes are located by using consecutive dimensions. For many applications this is a sufficiently accurate method of locating holes. Polar coordinate dimensions are used at 10-40(b). Methods for locating holes around a center on a common circle (sometimes called a *bolt circle*) are shown at 10-40(c) and 10-40(d). Other examples of dimensions that are used to locate the positions of holes are shown at 10-40(e) and 10-40(f).

10.26 SIZE DIMENSIONING OF MACHINED HOLES

Small cylindrical holes within the range of sizes up to 1 and 1½ inches in diameter may be formed by drilling, counterdrilling, reaming, boring, counterboring, spotfacing, or countersinking. As outlined in Chapter 8, these operations are classified as *secondary machining operations* because they are performed *after* a primary manufacturing process has been used to develop the basic shape of the part. Primary manufacturing processes include various casting methods, forging, press-

FIGURE 10-40 **Nonprecision layout of holes** ***(Section 10.25)***

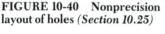

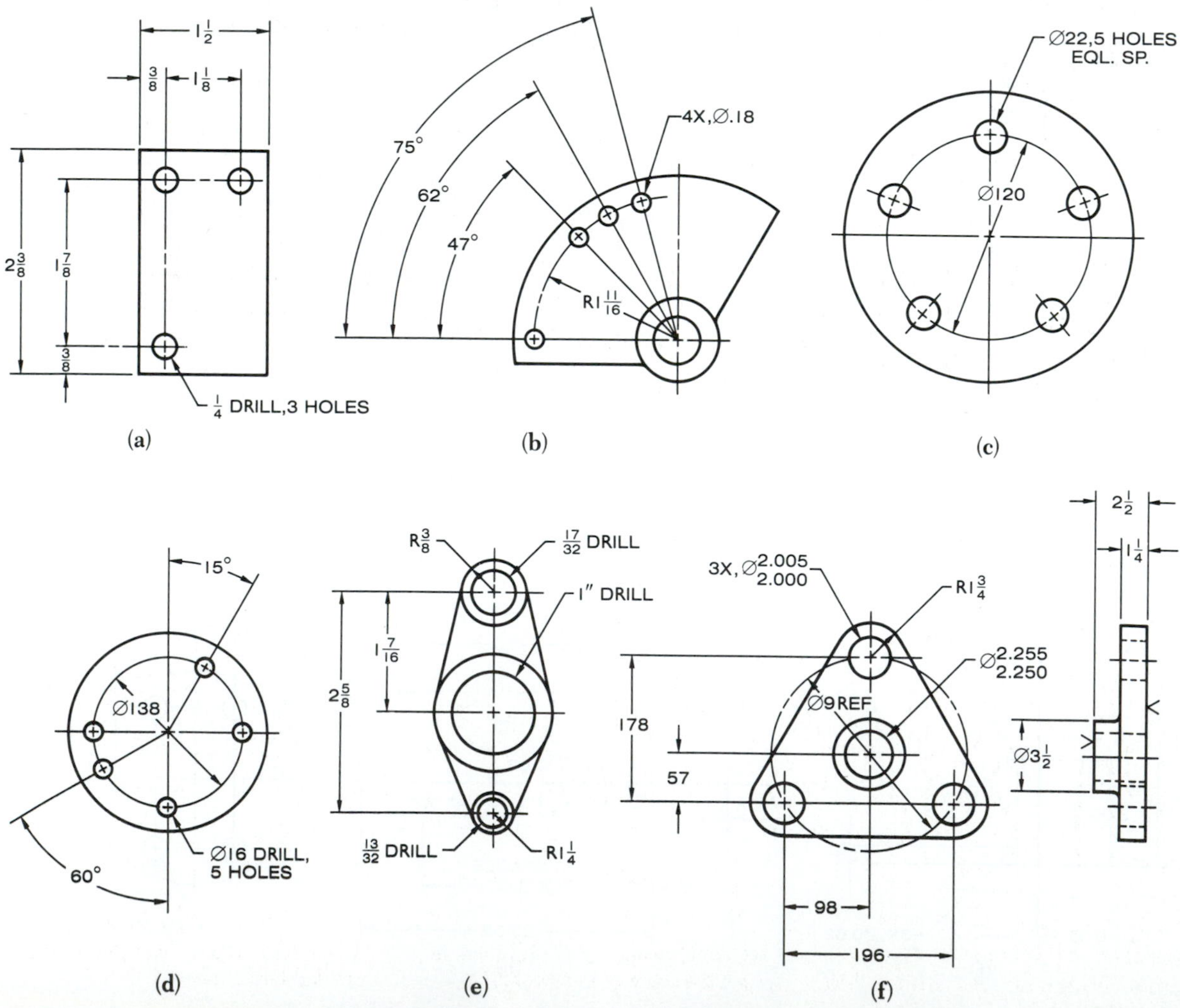

working, powder metallurgy, fabrication by welding, and so on.

Format for Specification

Instructions for secondary operations on holes are given to the shop on the drawing in the form of a note, by a dimension, or as a combination of both. The notes clearly explain the procedure and sequence to be followed in producing a hole. A custom of long standing has been to use a term that describes the machining method (such as drill, ream, or punch), as shown in Figure 10-41(a). Recently the practice recommended by ANSI standards and adopted by many companies is to specify holes without reference to the manufacturing method used to produce the hole, as shown at 10-41(b) and 10-41(c). As indicated in Section 10-20, the diameter symbol (∅) is used to precede the diametrical value. Alternatively, the abbreviation DIA may be used following the size value.

Proper composition of notes is important. For clarity, the format should be consistent and brief. Notes are always lettered horizontally on the drawing. Round hole sizes are always specified by giving the diameter, never the radius. A leader is used with the arrowhead terminating at the edge of the hole. The leader should point to the circular view of a hole. When specifying two or more concentric circles, the arrowhead should touch the outer circle. Although not recommended by ANSI, a comma or a dash is often used to separate each phrase in a note. Figure 10-41(c) illustrates a chamfer dimension in which the letter X is used to indicate the word "by" between coordinate dimensions.

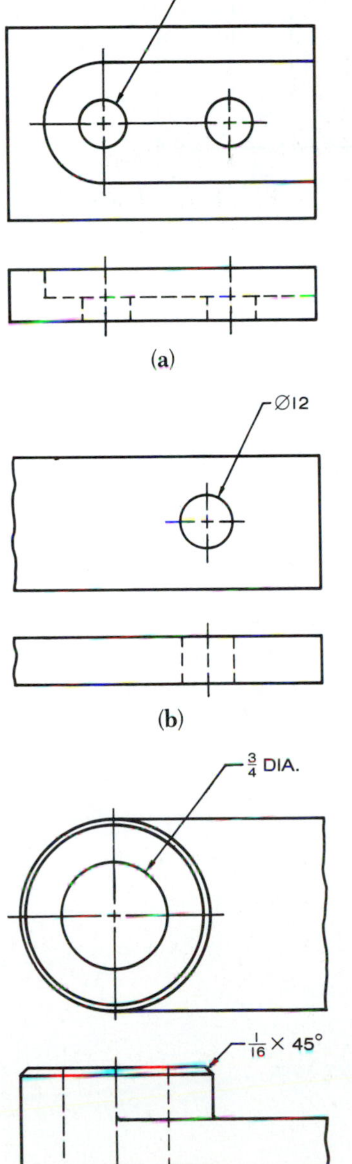

FIGURE 10-41 Specifications for repetitive holes *(Section 10.26)*

Repetitive Holes

Only one note is necessary for two or more identical holes. It has been customary for many years to indicate the quantity of identical holes by using the wording in the note shown in Figure 10-41(a). The standards for denoting repetitive holes have been recently changed in favor of a more abbreviated format in which repetitive holes are specified by use of the letter X in conjunction with a numeral to indicate the number or quantity of holes required. This practice is illustrated in Figure 10-42.

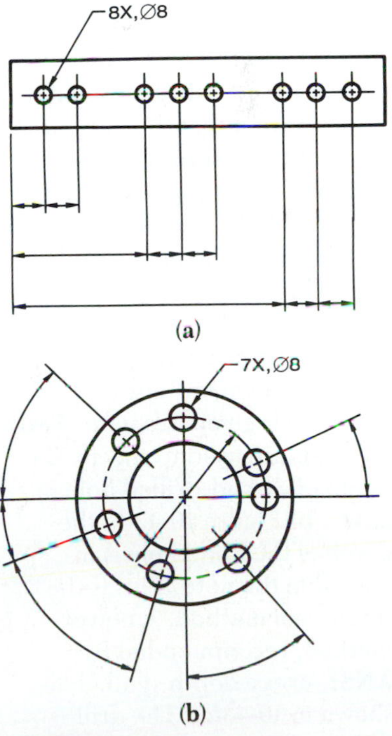

FIGURE 10-42 Specifications for drilled holes *(Section 10.26)*

Drilled Holes

A complete listing of standard inch and metric drill sizes is given in Appendix Tables A-3 and A-4. A drilling operation is illustrated in Figure 8-18(a). In the inch system, drill sizes are designated in common fractions or decimals, or a number or letter specification is given. Common fractions are often used to designate drill sizes on drawings (1/4 DRILL, for example). It is also customary to indicate the equivalent decimal size in parentheses when a numbered or letter-size drill is stated. The specification would read #10 (.1935) DRILL or "H" (.266) DRILL. Ordinarily a hole is drilled prior to counterdrilling, reaming, tapping, counterboring, spotfacing, and countersinking operations.

For through holes, the abbreviation THRU may be used if it is not evident that the hole extends entirely through the object. Such a note would read 3/8 DRILL THRU, for example.

Blind holes do not extend entirely through the object. The depth of the blind hole may be given as a dimension on the view, as shown in Figure 10-43(a), or given in a note. Two methods are used to specify the depth of a blind-drilled hole in notes. In Figure 10-43(b) the depth of the blind hole is included in the note. This is the traditional method. A newer method, recommended by ANSI, uses a depth symbol, as shown in 10-43(c). The drill depth is measured to the *full* diameter of the hole and does not include the conical impression formed by the drill point. The conical hole portion is drawn with a 30°–60° triangle, as shown in 10-43(d).

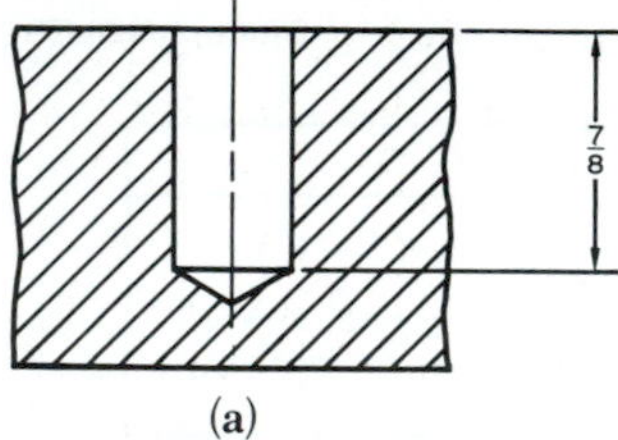

(a)

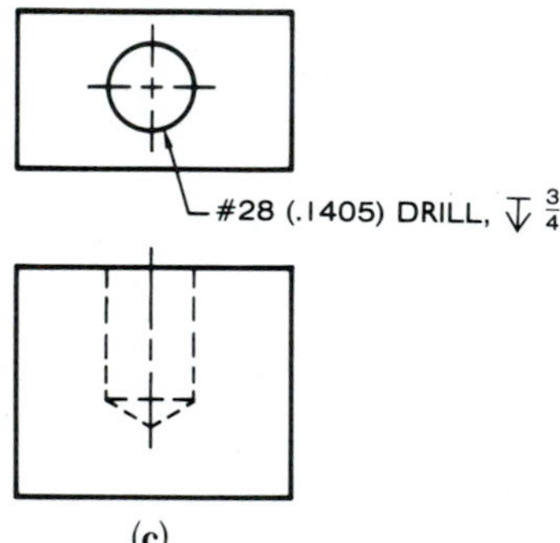

(c)

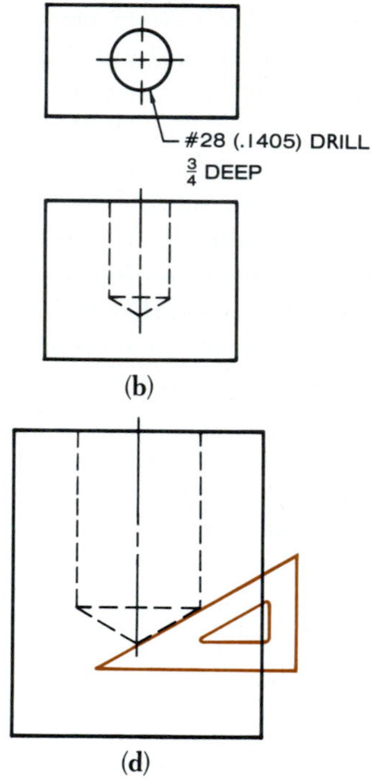

(b)

(d)

FIGURE 10-43 Blind holes *(Section 10.26)*

Counterdrilled Holes

For counterdrilled holes, the diameter, depth, and included angle of the counterdrill (actually 118° but drawn at 120°) are

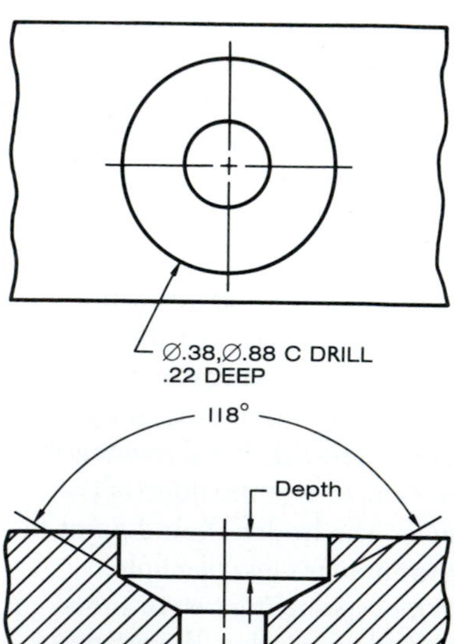

FIGURE 10-44 Specifications for counterdrilled holes *(Section 10.26)*

specified as shown in Figure 10-44. The depth dimension, as shown, is measured to the full diameter of the counterdrill. The word DEEP may be used in the note or the newer depth symbol may be used. A counter-drilling operation was shown in Figure 8-18(b).

Reamed Holes

A *reamer* is a relatively precise tool for sizing a hole. Reaming is used to improve the accuracy, finish, and roundness by enlarging a previously drilled hole. As was shown in Figure 8-18(e), a reamer has a blunt end. Consequently it cannot be used to produce a hole. The reamer diameter may be specified in a note like the one shown in Figure 10-45(a). The choice of the drill size and depth (for a blind-reamed hole) is left to the shop. At 10-45(b) a reaming operation is not specifically stated. An examination of the permissible difference in size or variation between the two limits, .628 minus .625, or .003, indicates that this hole size cannot ordinarily be obtained by a production method other than reaming. The size of reamed holes may also be specified on a drawing by using dimensions as shown at 10-45(c) and 10-45(d). Tapered holes may be finished to a precise size by using a tapered reamer. The sizes of tapers are well standardized to fit standard tools. Various methods of specifying taper sizes are shown at 10-45(e), 10-45(f), and 10-45(g).

FIGURE 10-45 Specifications for reamed holes ***(Section 10.26)***

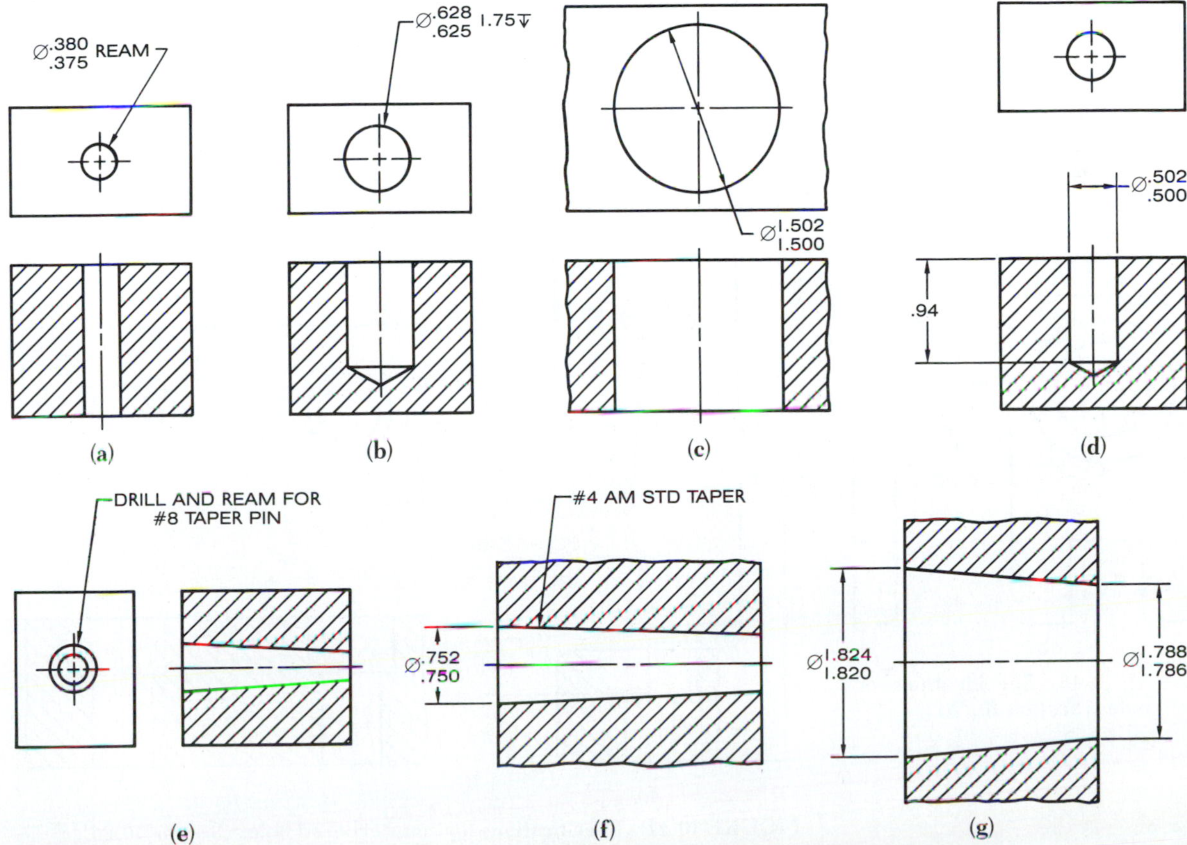

Bored Holes

Holes may be produced on a lathe or on a special boring machine by using a tool called a *boring bar.* Boring, like reaming, enlarges a previously formed hole. This process may be used to improve the finish, accuracy, and roundness of a drilled hole or a hole formed in a part by casting or by forging. The size of a bored hole may be specified in a note as shown in Figure 10-46(a) or by a dimension, as shown in 10-46(b).

Counterbored Holes

The standard procedure for producing a counterbored hole in a part is to begin by drilling a hole. The next step is to enlarge the size of the drilled hole with a tool called a *counterbore.* A counterboring operation was illustrated in Figure 8-18(d). Counterboring is performed on workpieces principally for the purpose of installing the head of a socket-head or fillister-head screw. Counterbored holes may be specified in any of the ways illustrated in Figure 10-47.

Spotfaced Holes

Spotfacing is similar to counterboring except that the enlarged hole is comparatively shallow, usually only about 1/16 inch, or 1 mm, deep. Spotfacing produces a flat, circular seat often

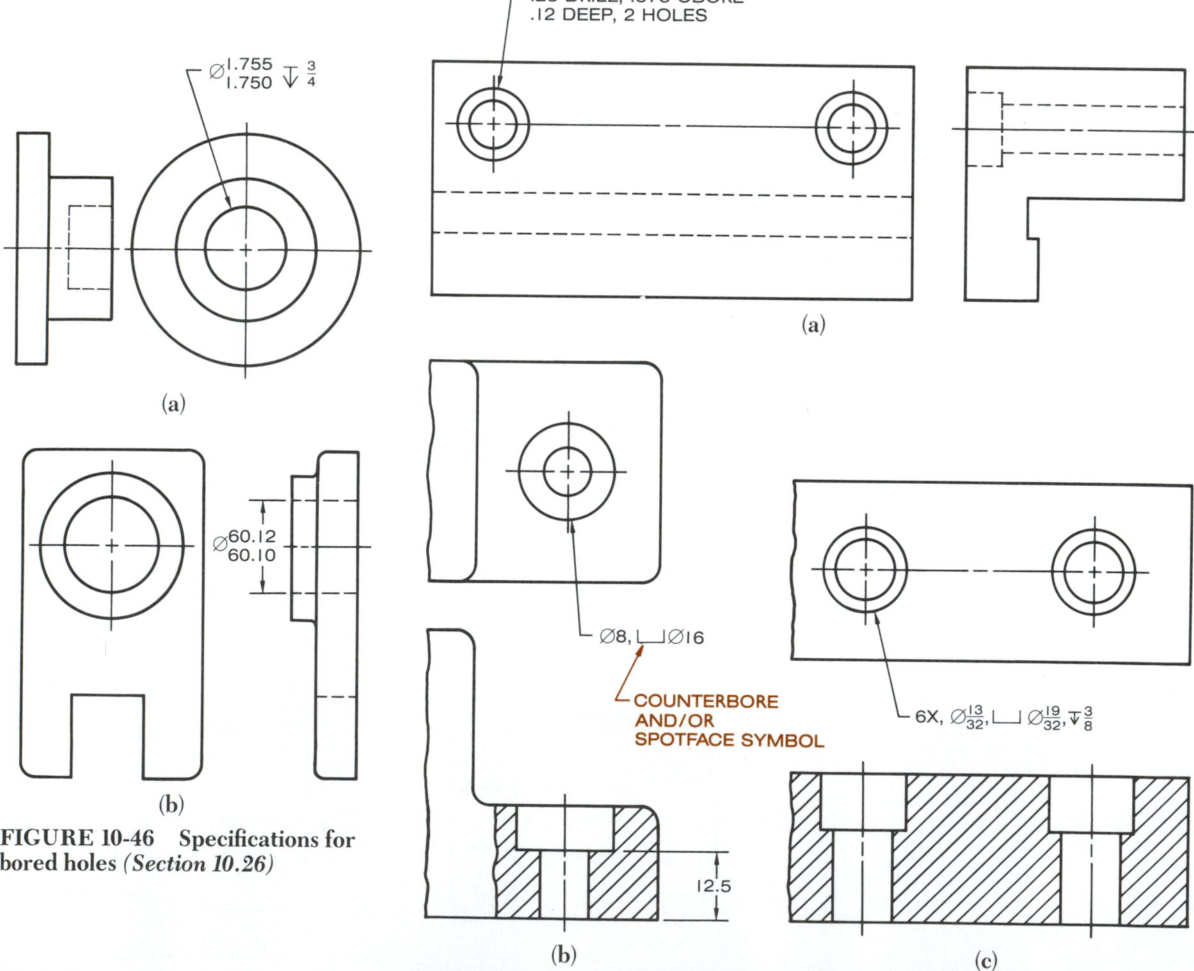

FIGURE 10-46 Specifications for bored holes *(Section 10.26)*

FIGURE 10-47 Specifications for counterbored holes *(Section 10.26)*

used as a bearing surface for the underside of a screw head or bolt head when these are installed on a rough or tapered surface of a cast or forged part. Various methods commonly employed when specifying a spotface are shown in Figure 10-48. If the depth is not specified, as at 10-48(c), the spotface is cut to the minimum depth necessary to clean the surface to the specified diameter. The drawing symbol used for a spotface, shown at 10-48(b), is identical to that used for a counterbore.

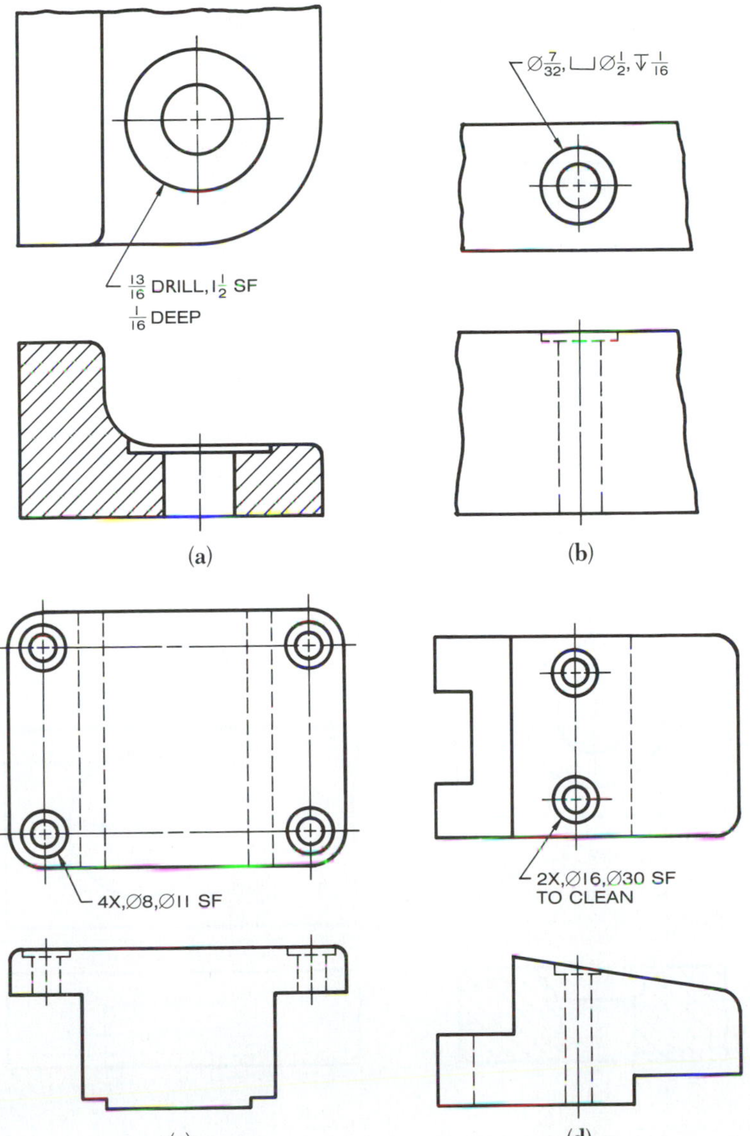

FIGURE 10-48 Specifications for spotfaced holes ***(Section 10.26)***

Countersunk Holes

Countersinking is the process of forming a conical seat at the end of a hole on a machine part for the installation of a flathead screw. A countersinking operation was illustrated in Figure 8-18(c). Methods of specifying countersunk holes are shown in Figure 10-49. Note the symbolic means of indicating the countersink at 10-49(d).

Tapped Holes

The specifications for tapped (threaded) holes are given in Section 9.10.

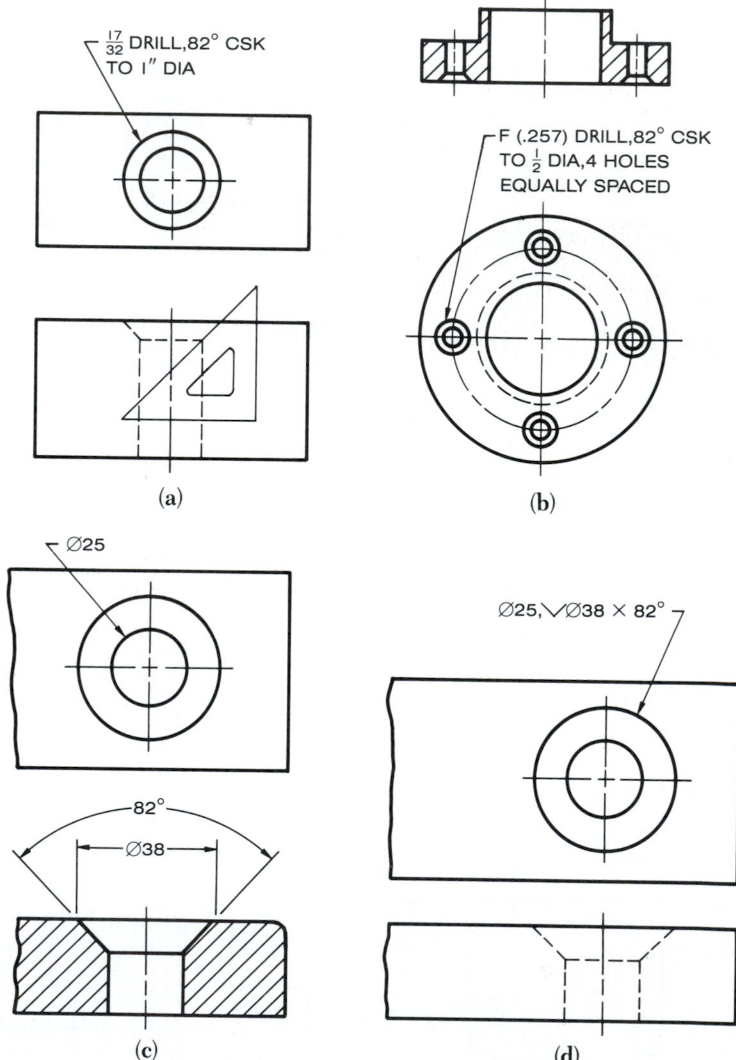

FIGURE 10-49 Specifications for countersunk holes ***(Section 10.26)***

10.27 DIMENSIONING CHAMFERS

Chamfers are beveled or sloping cuts along the intersection of two edges on a part. Methods for dimensioning chamfers are shown in Figure 10-50. At 10-50(a) a linear dimension along the length of the part is given and the angle is also specified. The methods shown at 10-50(b) are used only with 45° chamfers. For 45° angles the linear value applies to the distance measured along either of the right angular edges. Methods of dimensioning a chamfer on the end of a round hole are shown at 10-50(c) and 10-50(d).

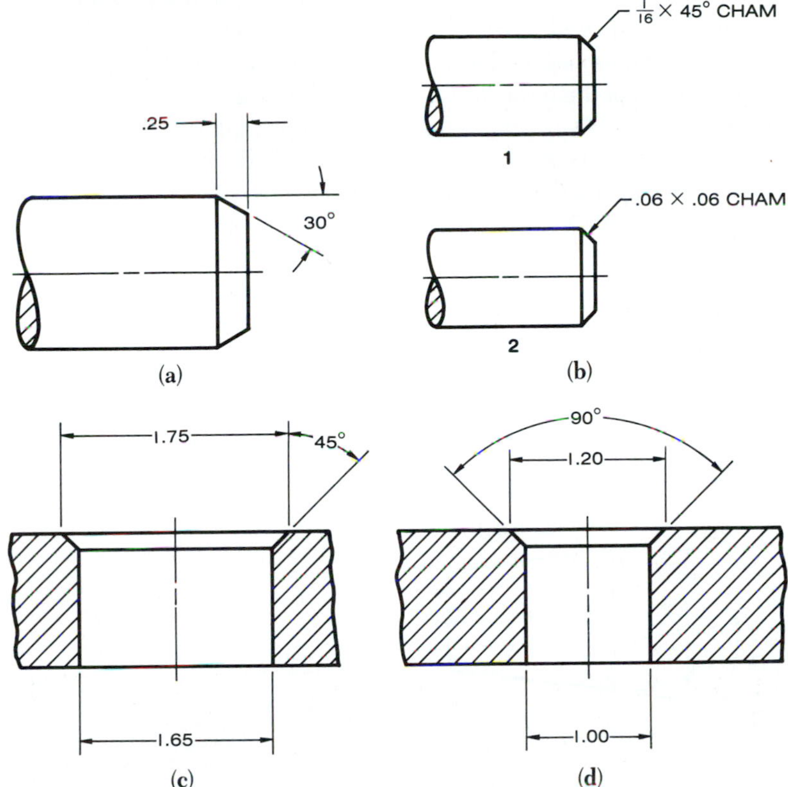

FIGURE 10-50 Dimensioning chamfers ***(Section 10.27)***

10.28 DIMENSIONING TAPERS

A taper can be a conical surface on a cylinder or a sloping surface on a flat plane. Standard conical tapers are specified by giving the large diameter and a note to denote the number of the taper, as shown in Figure 10-51(a). At 10-51(b) the basic taper (0.2:1) and the basic diameter (24) are specified. Note the use of the standard symbol denoting a conical-taper feature. The basic diameter controls the size of the tapered feature as well as its longitudinal position (10.1/9.9 mm) in relation to some other surface or, in this example, the left side of the taper. As shown at 10-51(c), the taper must fall within the zone (0.2 mm wide) created by the basic taper and the locating dimension of the basic diameter.

The profile of a conical feature may be specified as shown in Figure 10-51(d). A complete explanation of a feature-control frame and geometric characteristic symbols is given in Section 12.1. As shown in 10-51(e), the conical surface must lie between two coaxial boundaries 0.02 mm apart having an included angle of 15°. Additionally the surface must be within the specified limits of size between 29.8 and 30.2 mm diameters.

A flat taper may be specified by giving a toleranced height at one end and a toleranced slope angle, shown in Figure 10-51(f). The slope is expressed as the difference between the heights at each end. Note the use of the symbol denoting a flat-taper feature.

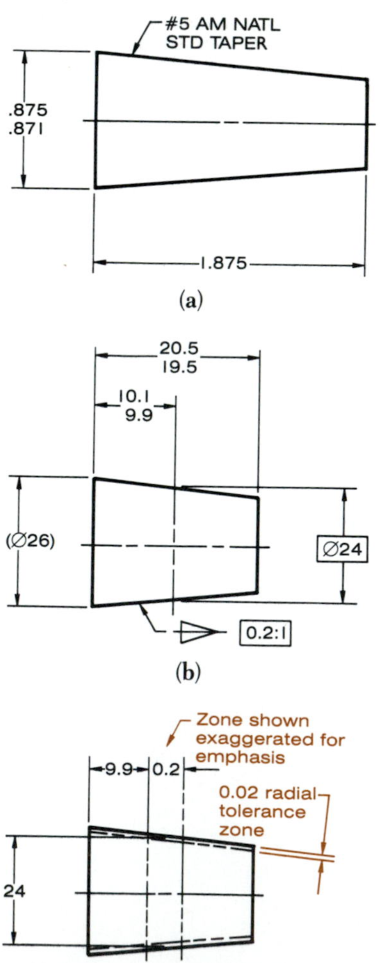

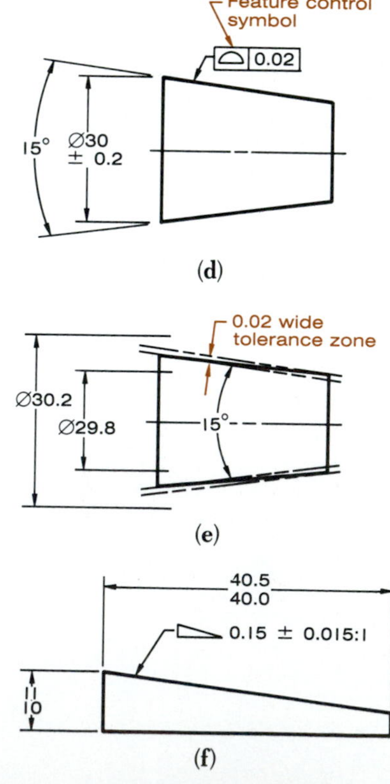

FIGURE 10-51
Dimensioning tapers
(Section 10.28)

10.29 DIMENSIONING MACHINING CENTERS

The combined drill and countersink or center drill, shown in Figure 8-18, is used to produce a hole called a *machining center* on the ends of cylindrical workpieces for turning and grinding operations. The size of the center drill to be used varies with the diameter of the workpiece. Shaft center sizes are listed in Appendix Table A-20. Where machining centers are to remain on the finished part, they are represented and specified as shown in Figure 10-52(a). If a machining center hole on a finished part is objectionable, the designer should allow extra stock on the workpiece, which can be later removed. The method for representing and specifying stock removal is shown in 10-52(b).

FIGURE 10-52 Dimensioning machining centers *(Section 10.29)*

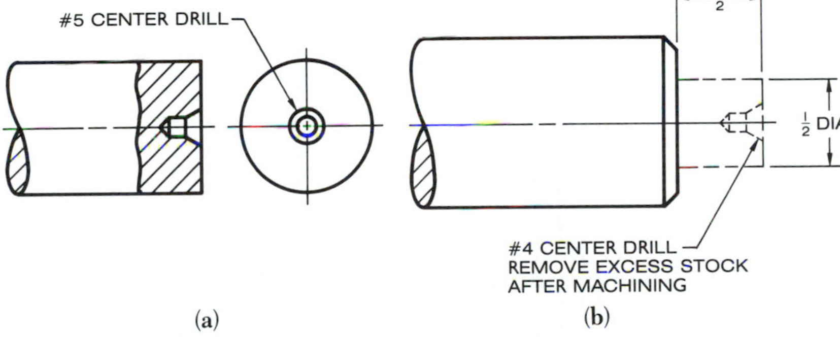

10.30 DIMENSIONING SQUARE SHAPES

The symbol used to indicate that a single dimension applies to a square shape is shown in Figure 10-53. The symbol for a square precedes the value of the square dimension.

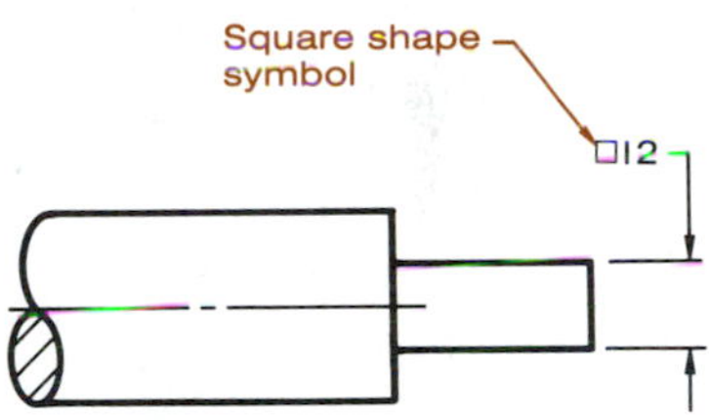

FIGURE 10-53 Dimensioning square shapes *(Section 10.30)*

10.31 DIMENSIONING KEYSEATS AND KEYWAYS

A *keyseat* refers to a slot or recess that is cut into a shaft to hold a key. A *keyway* is a slot cut in a hub. Keys are used to prevent slipping between mating parts. Various types of keys are listed in Appendix Tables A-21, A-31, and A-34. The method of dimensioning a keyseat and a keyway for a Woodruff key is shown in Figure 10-54(a) and for a square key in 10-54(b) and 10-54(c).

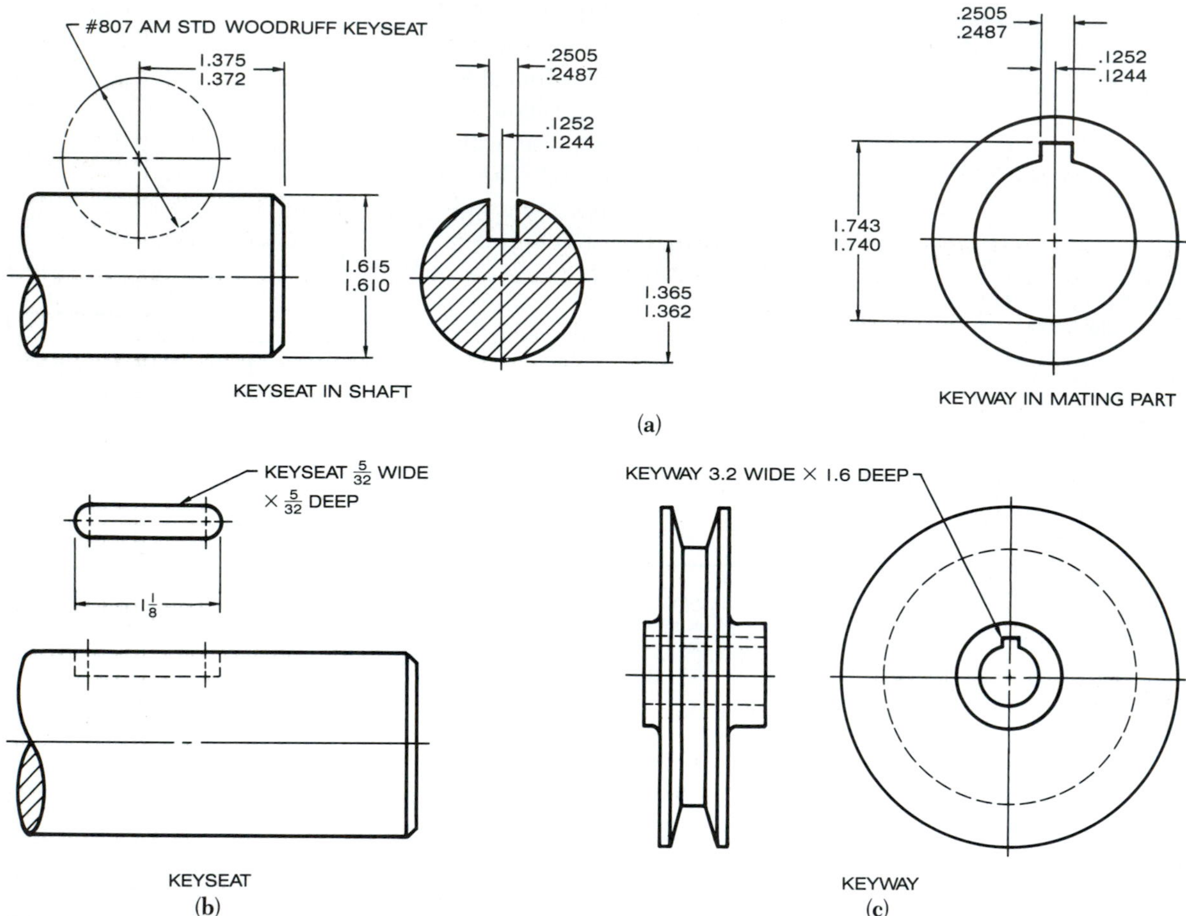

FIGURE 10-54 Dimensioning keyseats and keyways ***(Section 10.31)***

10.32 DIMENSIONING KNURLING

Knurling is the process of form-rolling depressions of various designs on cylindrical and sometimes flat surfaces. Knurling is used to provide a gripping surface on a handle or a knob or to provide a slightly raised and roughened surface on a shaft when a tight fit is required between the shaft and a mating part. Examples of representation and specifications for knurling a diamond pattern on a gripping surface are shown in Figure 10-55(a) and for a straight knurl pattern on a gripping surface at 10-55(b). The specification for a part knurled for the purpose of making a tight fit is shown at 10-55(c).

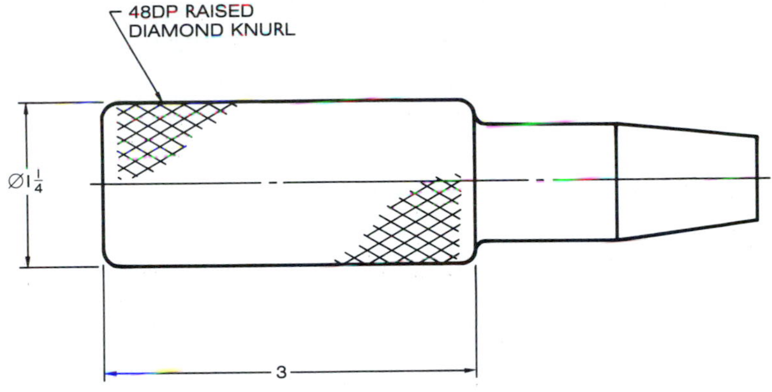

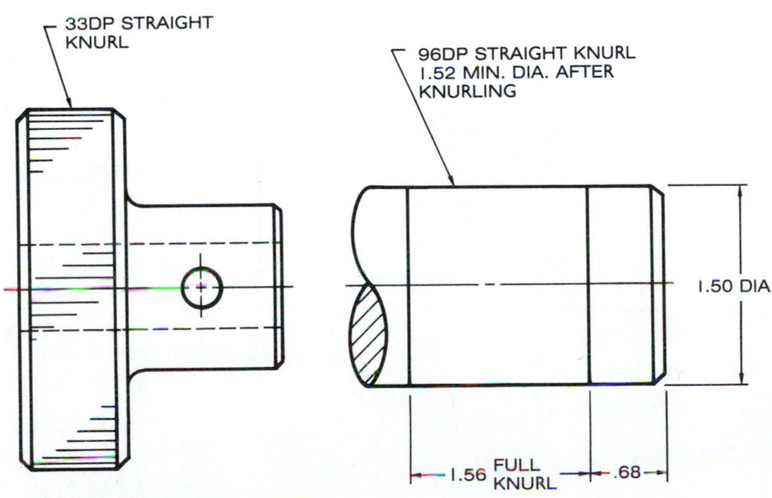

FIGURE 10-55 Dimensioning knurling *(Section 10.32)*

10.33 DIMENSIONING NECKS AND UNDERCUTS

A *neck* or *undercut* is a small square or curved groove cut at the shoulder on a cylindrical part or in the corner formed by two intersecting flat surfaces. These recesses are cut in the corners to eliminate the small fillet that is left in an otherwise sharp corner by a cutting tool bit or grinding wheel. Alert designers apply necks and undercuts to mating parts as a practical necessity so that each part will fit flush against the other. Figure 10-56 shows how to represent and specify these features on a drawing.

10.34 DIMENSIONING THREAD RELIEFS

Thread reliefs are used to eliminate nonexistent or imperfect threads left on the end of a threaded portion near a shoulder. This special form of neck is cut into the shoulder when it is necessary to screw a threaded part into the mating piece entirely along its threaded length.

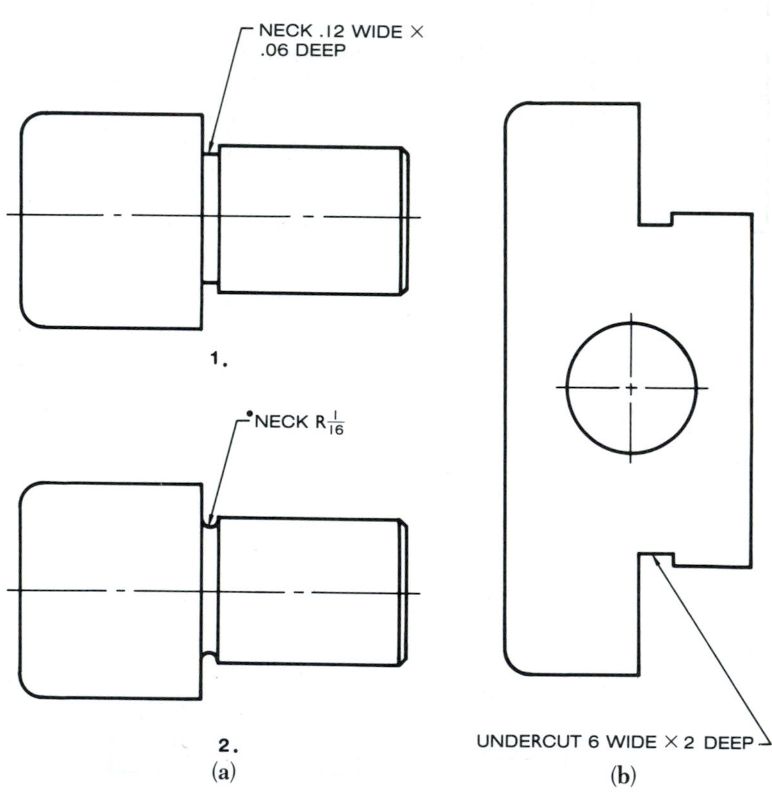

FIGURE 10-56 Dimensioning necks and undercuts ***(Section 10.33)***

A thread relief on an *external* thread is represented and specified in Figure 10-57(a). A thread relief on an *internal* thread is shown at 10-57(b). The width of the neck or the relief on an internal thread is made about three times the pitch, and the depth is equivalent to the major diameter plus .02 inches, or 0.5 mm. See Chapter 9 for a more complete discussion of screw threads.

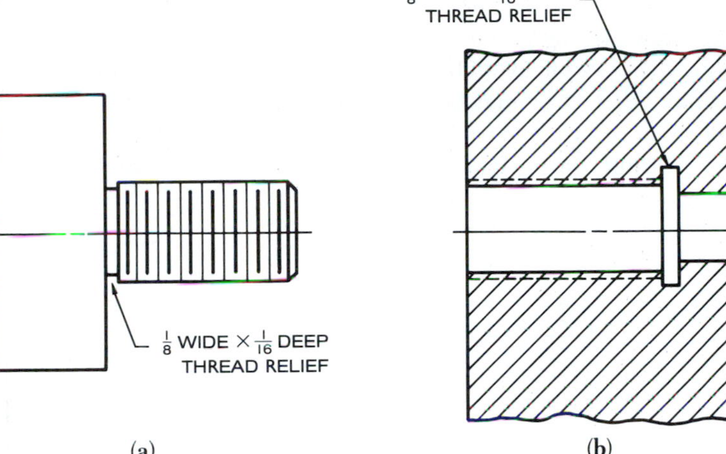

10.35 DIMENSIONING A HALF SECTION

Half sectional views are dimensioned as shown in Figure 10-58. For regular views the practice of using hidden lines for dimensioning is usually avoided. In this special case extension lines are drawn to the hidden lines on the unsectioned half for dimensioning purposes.

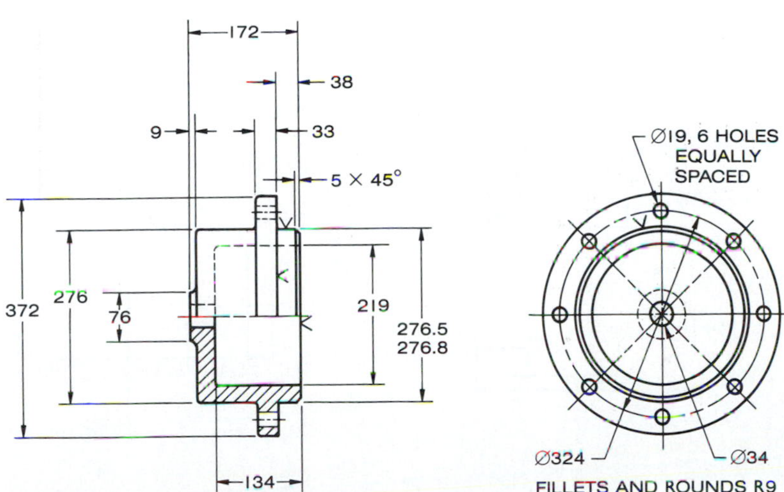

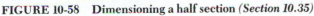

FIGURE 10-58 Dimensioning a half section ***(Section 10.35)***

10.36 SUPERFLUOUS DIMENSIONS

After considerable practice and time spent in design-related activities, you will become well versed in most of the dimensioning practices outlined in this chapter. An important fact to understand, however, is that some variance in applying these principles of dimensioning is permissible, depending upon the specific manufacturing needs.

Some dimensioning practices have become so well established that they have been adopted as firm rules. In these cases the practices are so well standardized that little if any variation is permitted. One such rule is: *duplicate or superfluous dimensions must never be given either on the same view or on another view.* Drawings should be scrupulously checked to avoid the possibility that a part may be made in more than one way due to excess dimensions. Figure 10-59 gives some examples of superfluous dimensions to avoid.

10.37 OUT-OF-SCALE DIMENSIONS

Design changes occurring some time after a drawing has been made may affect one or more dimensions on a part. Rather than make an entirely new drawing for a minor change, the standard practice in many companies is to change only the dimension value without actually changing the lines on the view affected by the dimensional

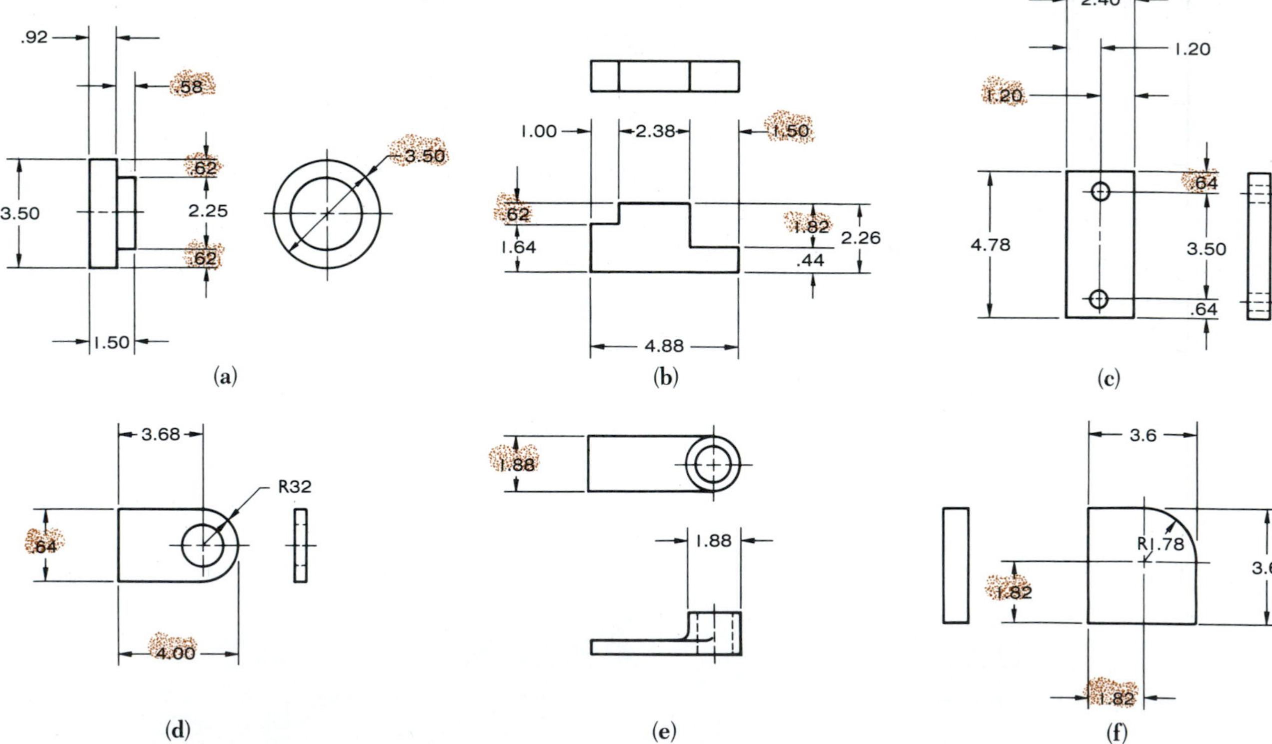

FIGURE 10-59 Superfluous dimensions *(Section 10.36)*

change. Two methods are commonly used to indicate that a dimension is out of scale. A short line may be drawn below the dimension, as shown in Figure 10-60(a), or the abbreviation NTS (NOT TO SCALE) may be added to the dimension, as shown in 10-60(b).

10.38 TABULAR DRAWINGS

When several similar parts that differ in only a few dimensions are required, only one drawing with a table of sizes is made, as shown in Figure 10-61. On the view, a letter is substituted for each different dimension. The table lists the various letters and the corresponding sizes for each part number. This method saves drafting time by avoiding the needless drawing of nearly identical parts. The size of the table will depend upon the available space and the conditions of the part. Such information as material, number of parts required, and so on may be added where appropriate.

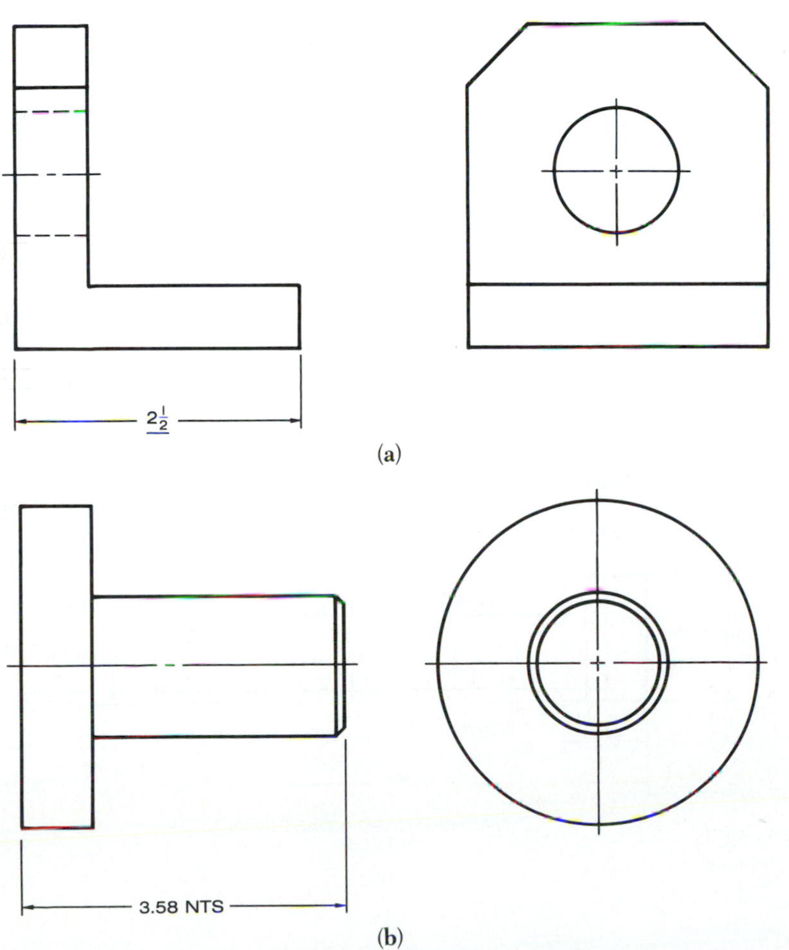

FIGURE 10-60 Out-of-scale dimensions *(Section 10.37)*

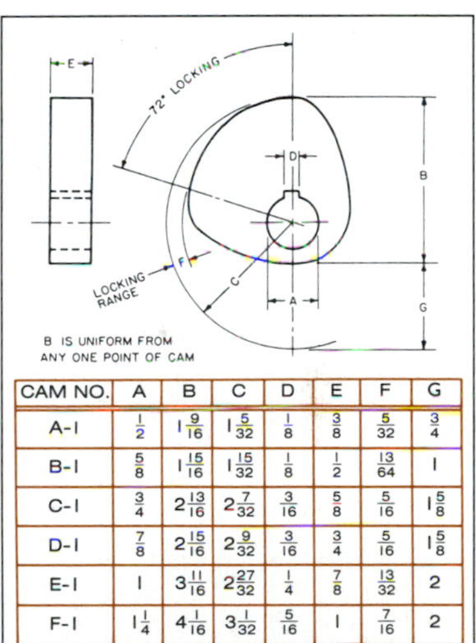

CAM NO.	A	B	C	D	E	F	G
A-1	1/2	1 9/16	1 5/32	1/8	3/8	5/32	3/4
B-1	5/8	1 15/16	1 15/32	1/8	1/2	13/64	1
C-1	3/4	2 13/16	2 7/32	3/16	5/8	5/16	1 5/8
D-1	7/8	2 15/16	2 9/32	3/16	3/4	5/16	1 5/8
E-1	1	3 11/16	2 27/32	1/4	7/8	13/32	2
F-1	1 1/4	4 1/16	3 1/32	5/16	1	7/16	2

FIGURE 10-61 Tabular dimensioning *(Section 10.38)*

10.39 DIMENSIONING FOR CAD AND CAM

Industry acceptance of computer-aided design (CAD) and computer-aided manufacturing (CAM) systems for use in component design and fabrication is rapidly accelerating. Collectively, these highly sophisticated systems can be used to describe a desired part as a geometric model, interactively insert manufacturing data, and deliver this information to a designated machine tool for execution of the finished part. Although computer-aided systems continue to require dimensions and tolerances for part definition, in many cases dimensioning is accomplished by means of algorithms that emulate manual dimensioning practices. In view of the changing state of the art, it is important that you understand where certain practices can be employed for expressing dimensional requirements most effectively.

The choice of the dimensioning system to be used and the tolerancing requirements are generally a responsibility of the engineer or the designer.

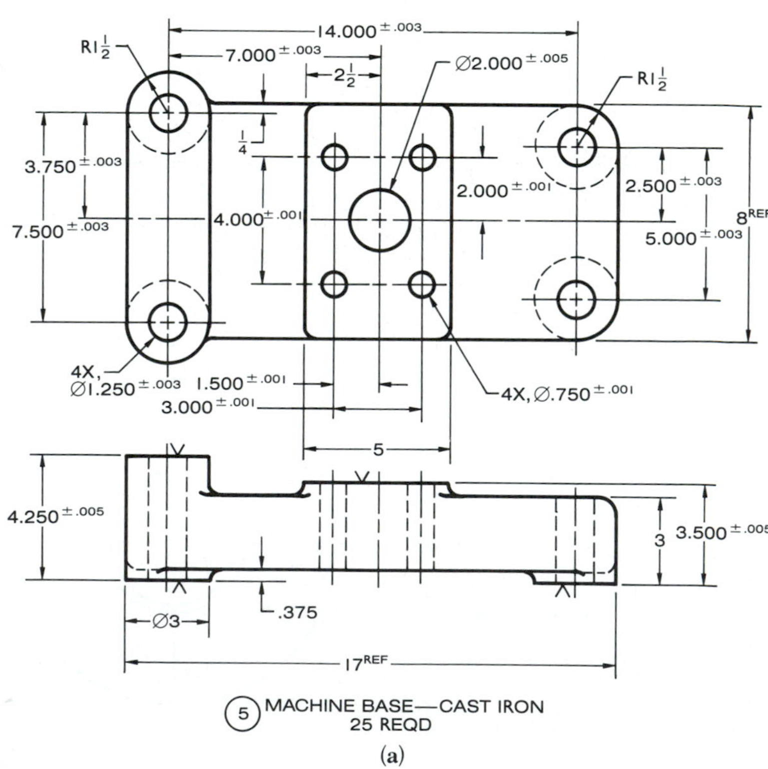

FIGURE 10-62 Comparison of dimensioning practices for a part made by conventional manufacturing to one made by numerical-control machining methods *(Section 10.39)*

Dimensioning systems vary with different industrial firms, and design engineering personnel should become familiar with the special practices that apply to their own numerical control (N/C) machine operations.

The rectangular coordinate or "datum dimensioning" method best conveys information for programming N/C machine tools and should be used exclusively. The use of datum planes or datum points, previously discussed in Sections 10.17 and 10.24, are used to indicate the location or relationship of the various features on a part. All related measurements of a part originate from these positions. If the positions of the datum planes are not obvious, they should be clearly identified on the drawing.

When a part is to be processed on a numerically controlled machine, it may be necessary to modify some of the traditional dimensioning and tolerancing techniques ordinarily used on a drawing. For example, the method of dimensioning the part shown in Figure 10-62(a) is preferred over the method shown in 10-62(b).

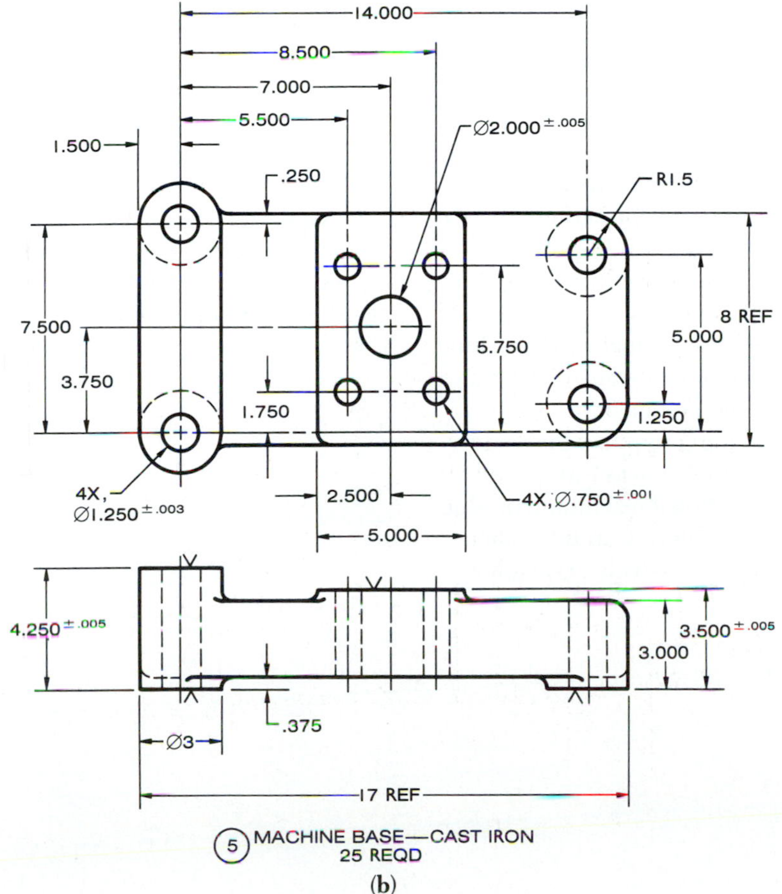

(b)

Figure 10-63 shows examples of typical differences in dimensioning practices for the same machine part. Each method is compatible with a type of N/C machine that may be used. At 10-63(a) dimensions are given that suit the requirements for producing a part on a Cincinnati MILACRON CINTIMATIC N/C machine. At 10-63(b) the part is dimensioned for manufacture on a Pratt & Whitney TAPE-O-MATIC N/C machine. At 10-63(c) the method of dimensioning is compatible with the requirements imposed by a Moog HYDRA-POINT N/C machine.

The operator of an automated graphics system must follow the same dimensioning rules as if the drawings were being prepared by manual drafting methods. For example, one of the two conventional systems of reading dimension figures on a drawing, unidirectional or aligned, should be selected. The N/C machine control unit requires that information input be in decimal form. Fractional dimensions would require the programmer to convert these values to decimals when writing the numerical-control program. Stating tolerances in N/C machining requires the same close attention as is required for conventional machining production methods. Careful evaluation of excessively close tolerances, which *always* increases manufacturing costs, is essential. It is unnecessary to specify tolerances for location dimensions, because the inherent accuracy of the machine tool will compensate for the untoleranced location dimension. Cumulative or chain dimensions that lead to an accumulation of tolerances should be avoided. Angular features on parts should be given in the form of coordinate dimensions rather than in degrees. All machining dimensions should be stated in the form best suited for the method of manufacture to be used.

Numerical control methods make use of conventional cutting tools such as drills, reamers, taps, and counterbores to remove material. Standard sizes of these tools should be specified whenever possible.

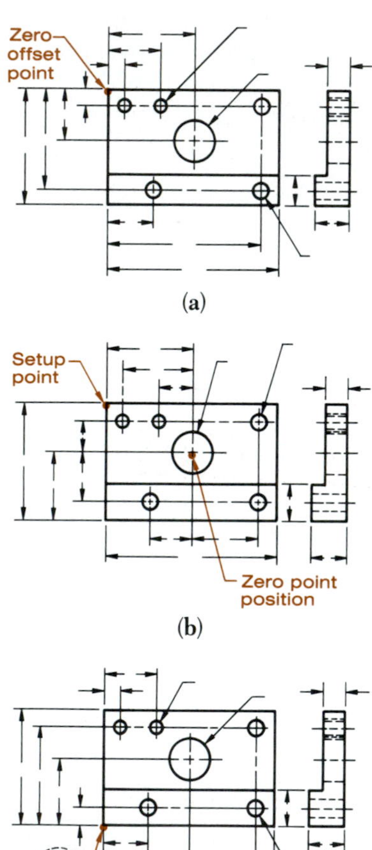

FIGURE 10-63 Dimensioning requirements based upon the type of numerical-control machining method to be used *(Section 10.39)*

PROBLEMS

Size Description

The following problems will provide practice in making detail sketches or drawings and in applying the principles of dimensioning discussed in this chapter. For each problem, use layout A, shown on the inside of the front cover. Place one problem on each sheet. Fill out the information in the sheet title block.

The problems may be solved freehand or with instruments, as directed by your instructor. Begin by studying the given problems to determine an appropriate selection and arrangement of views. If the solution is to be freehand, try to maintain good proportions in your sketches. For an instrument drawing, select a suitable scale so that the views will fit neatly on the drawing sheet. Be sure to allow sufficient space around the views for the appropriate dimensions and notes.

P10.1–P10.24 Two or three views of each problem are given. The dimensions for the views are obtained by using the dividers to transfer the drawing sizes to the scales shown on the page. The detail drawings may be prepared with either SI (Metric) or U.S. Customary linear units, as assigned by your instructor. (1″ = 25.4 mm; 1 mm = .0397″.)

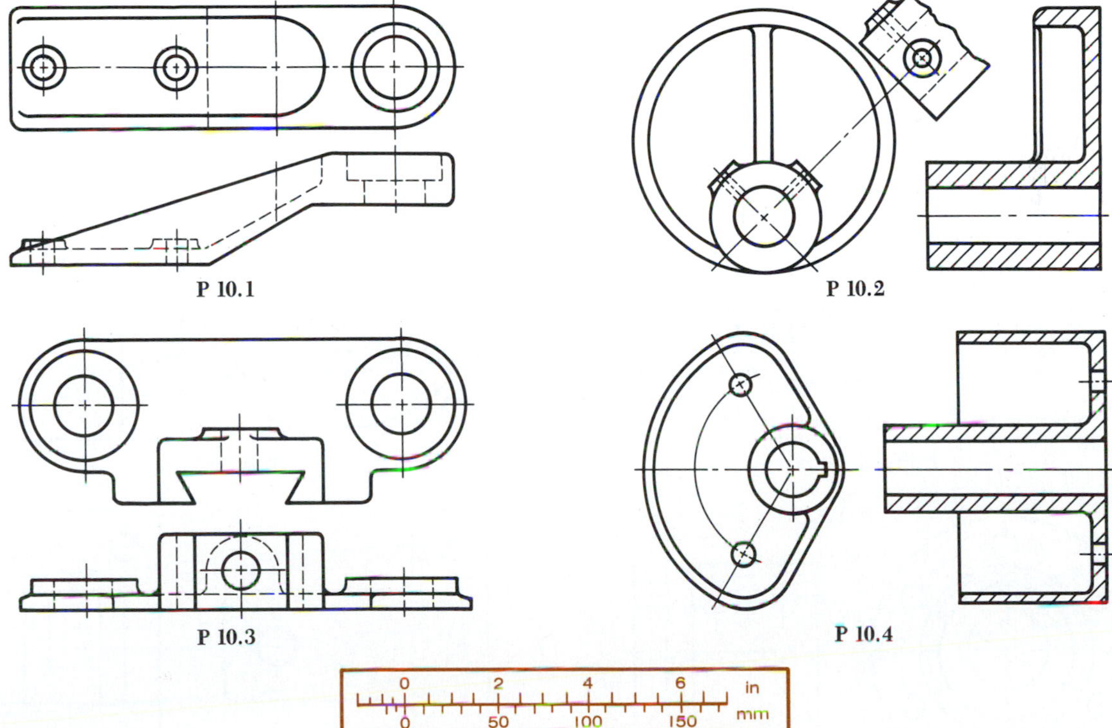

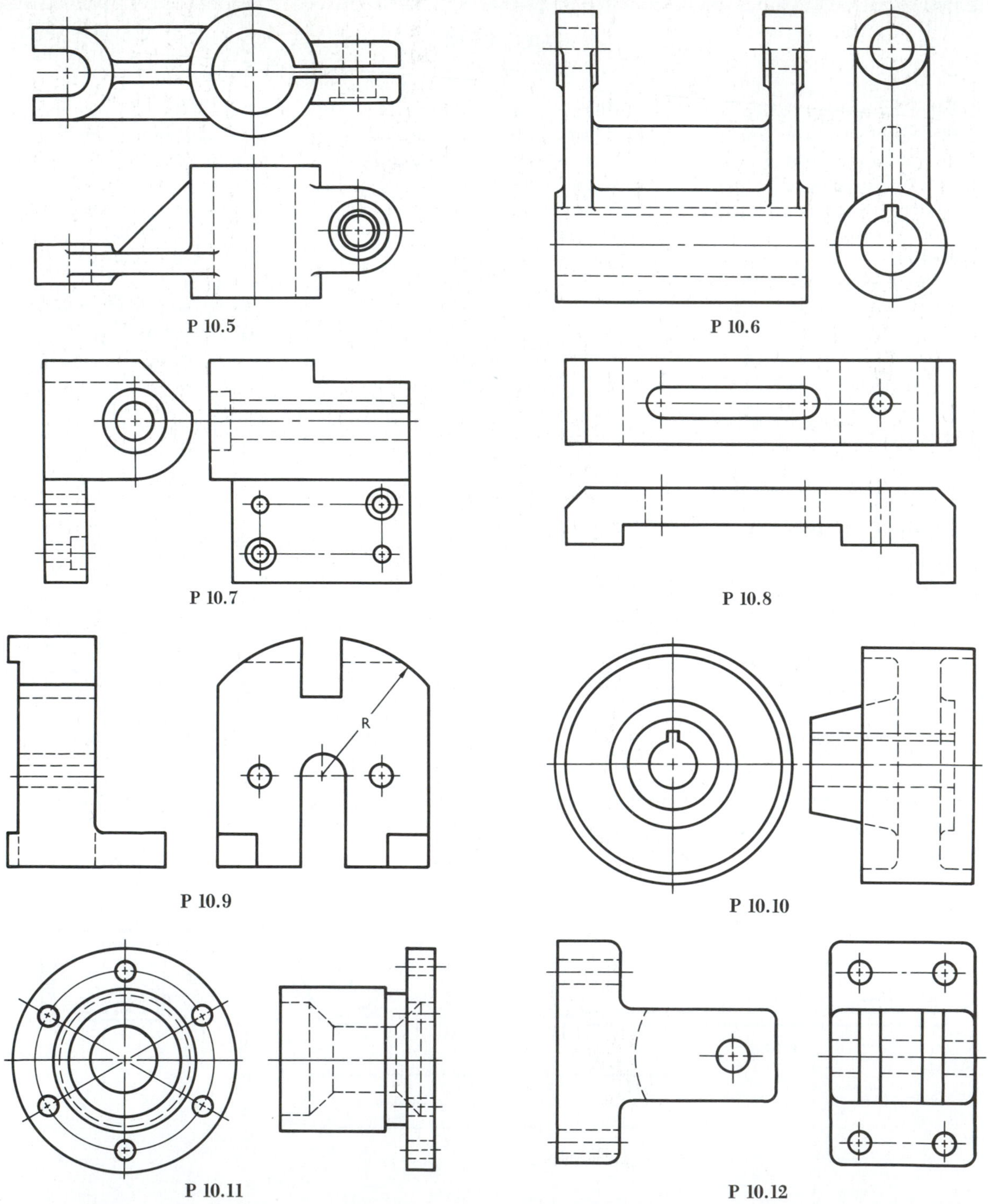

P 10.5

P 10.6

P 10.7

P 10.8

P 10.9

P 10.10

P 10.11

P 10.12

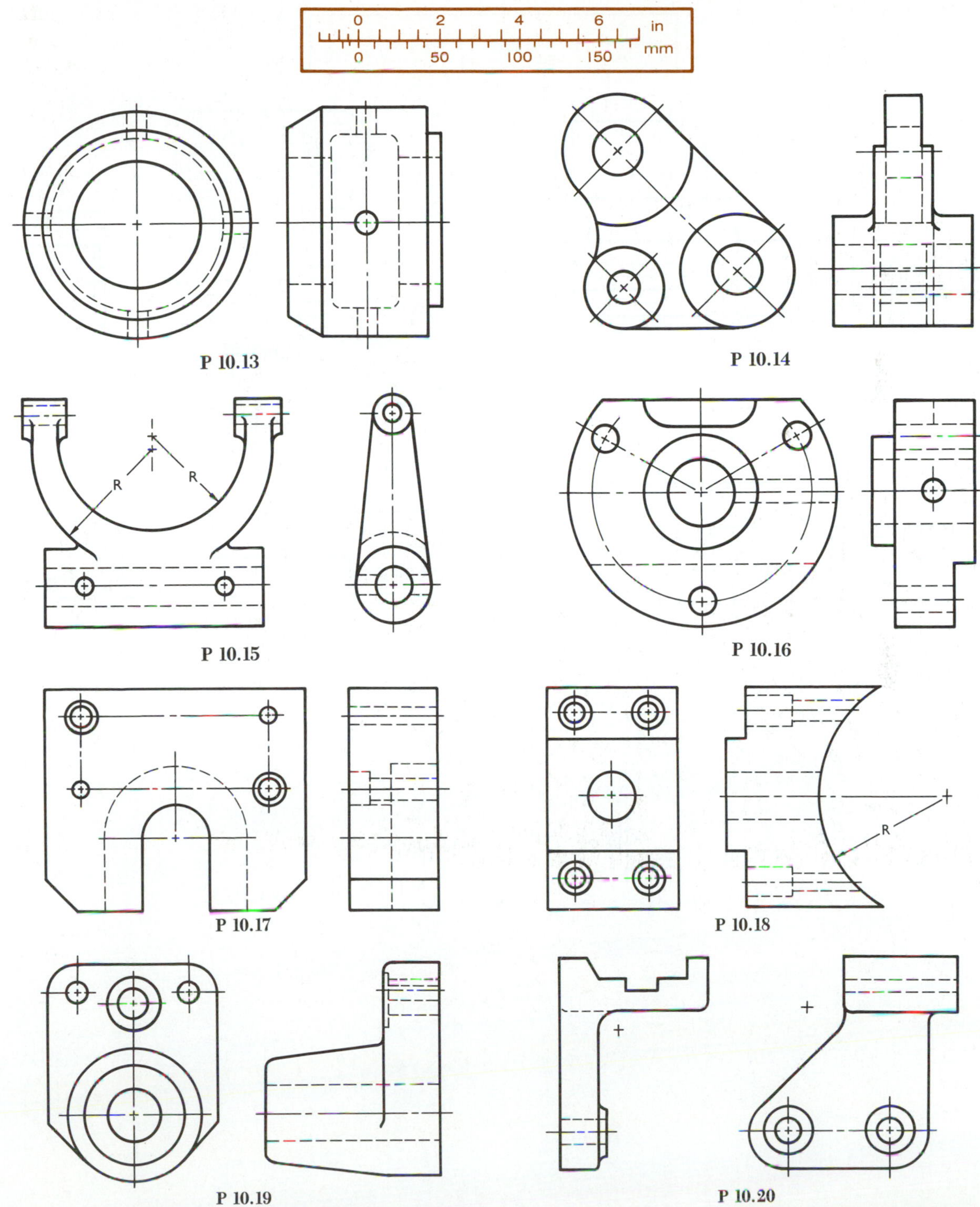

P 10.13

P 10.14

P 10.15

P 10.16

P 10.17

P 10.18

P 10.19

P 10.20

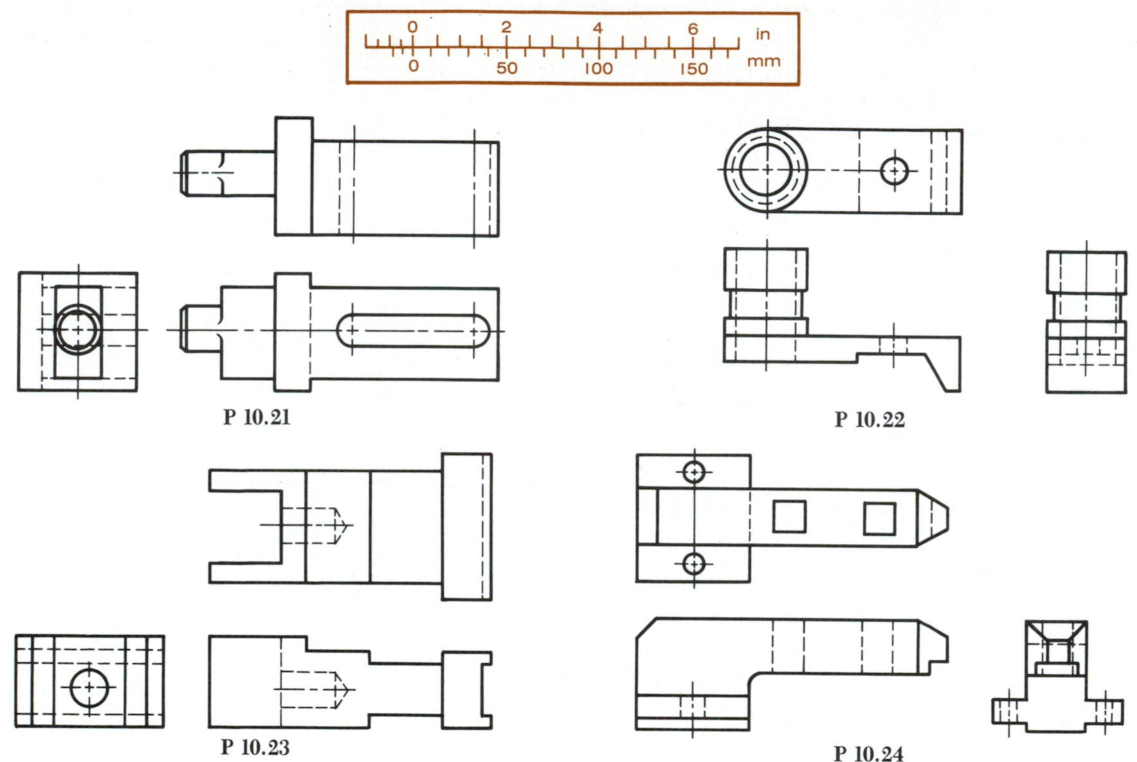

P 10.21

P 10.22

P 10.23

P 10.24

P10.25–P10.80 The dimensions on the drawing problems are not intended to represent conventional dimensioning practices because of space limitations, but they are adequate for the views to be correctly drawn. Do not copy the given dimensions. In many cases the given dimensions would not be suitable for use on a two- or a three-view detail drawing. Only the critical sizes of the parts have been given. Sizes of some of the less important features of the parts have been omitted, leaving the selection of these sizes to your discretion. For drawing purposes, features of parts that appear to be located on center with corresponding edges may be assumed to be central. It is important to give dimensions that locate all of the features of the part, thus eliminating the possibility of misunderstanding in the shop if your detail drawings were actually to be used to manufacture the parts.

The abbreviation TYP means "typical" and is used when more than one feature of the same size and kind are used on a problem. Note that fillets and rounds are not shown on some problems. The

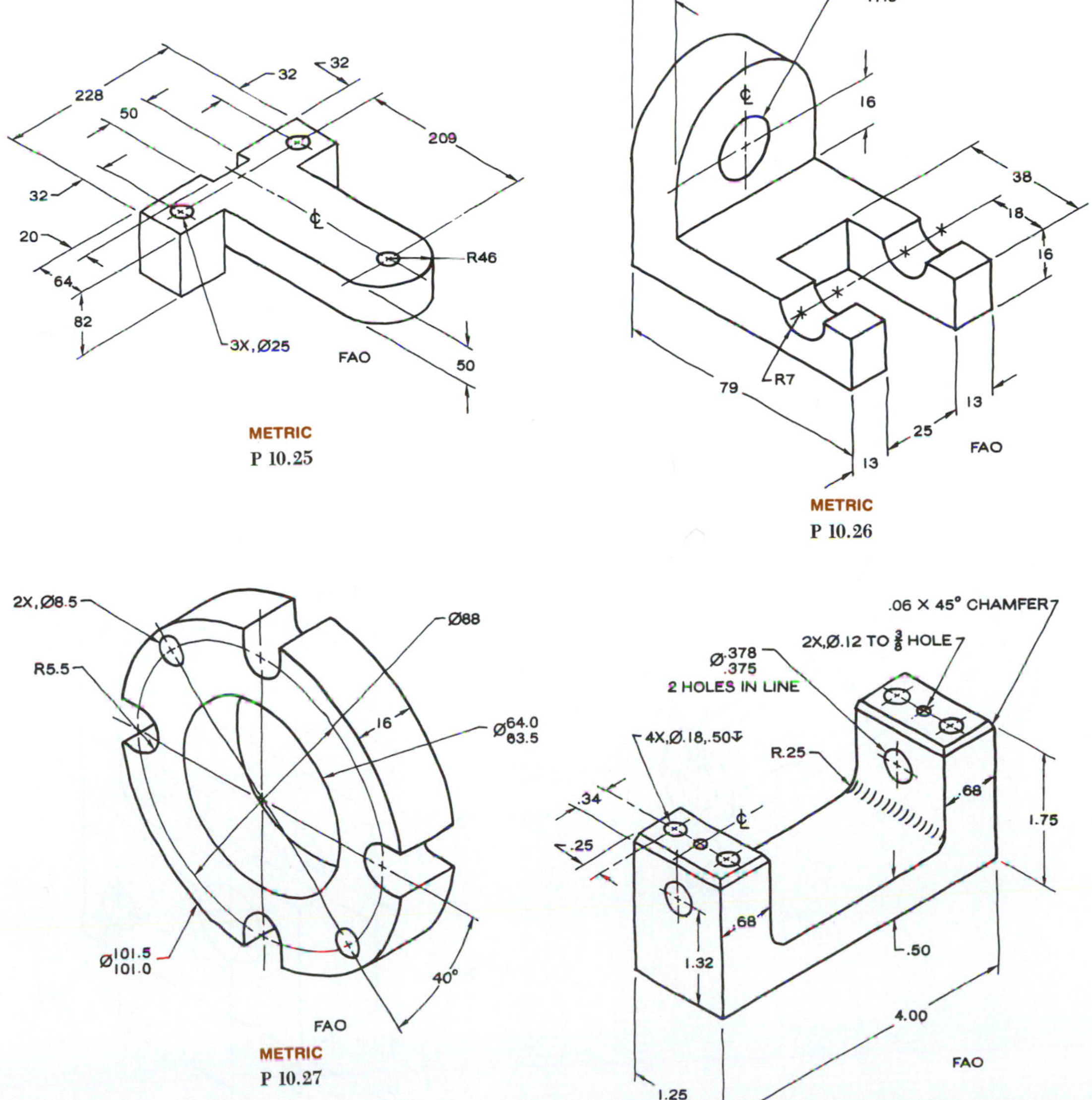

P 10.25

P 10.26

P 10.27

P 10.28

placement and (for some problems) the selection of the various sizes of these features have been left to your discretion. Avoid copying or measuring the problems, because the problems are not shown at an accurate scale. Use appropriate appendix tables where necessary.

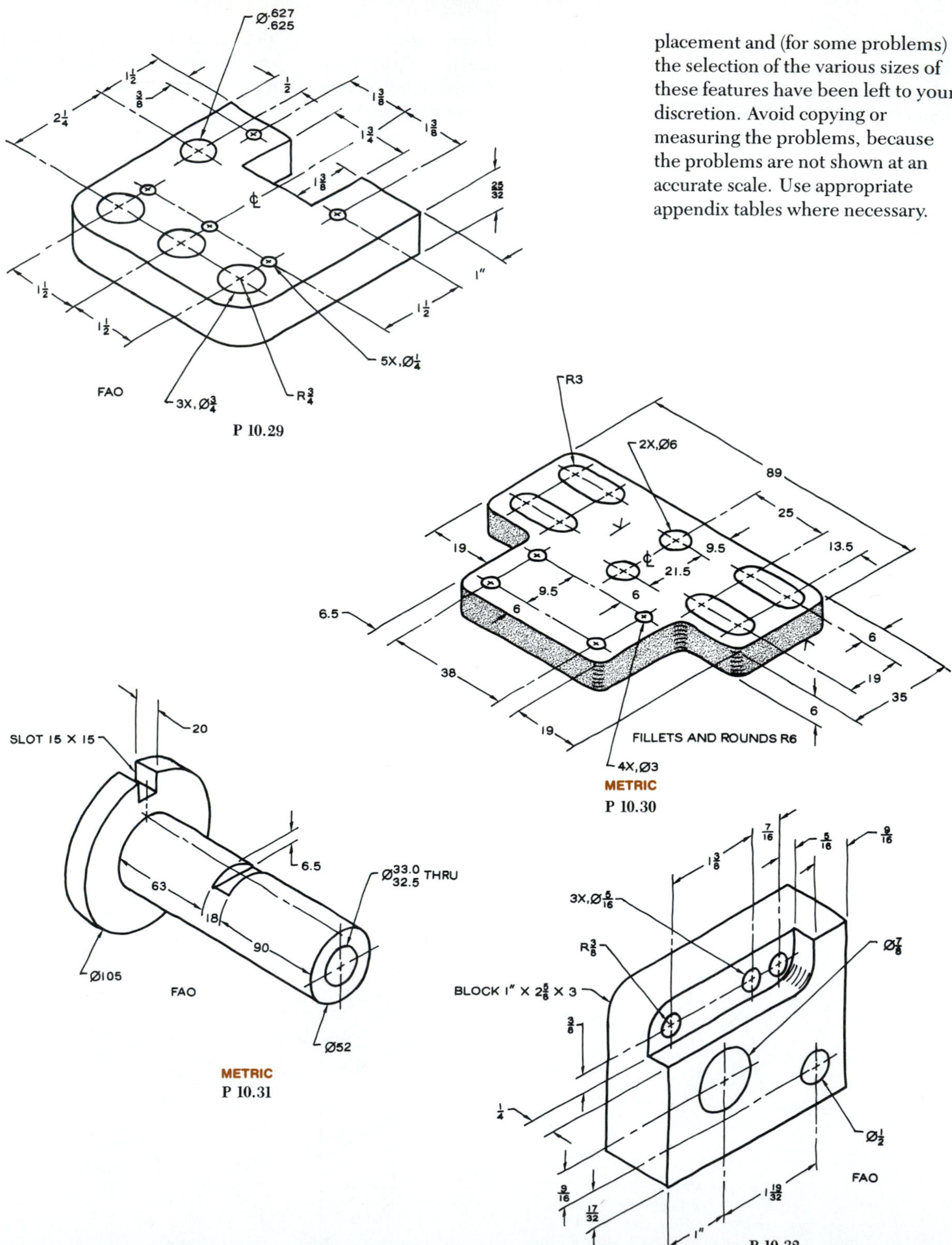

P 10.29

P 10.30

P 10.31

P 10.32

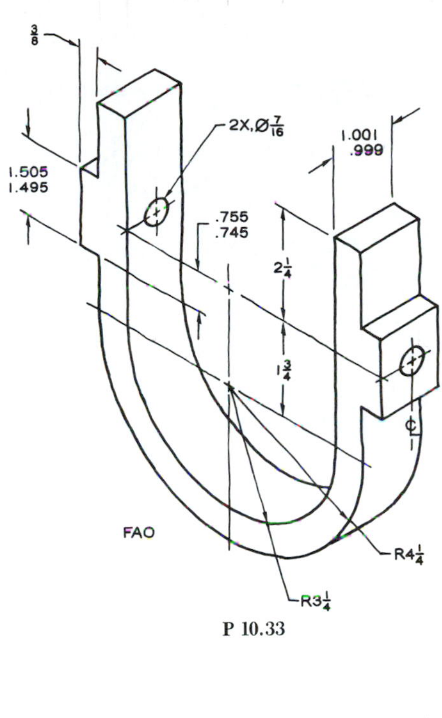

P 10.33

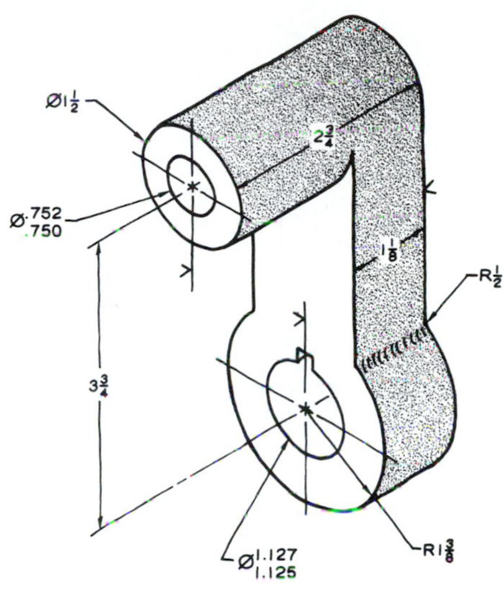

P 10.34

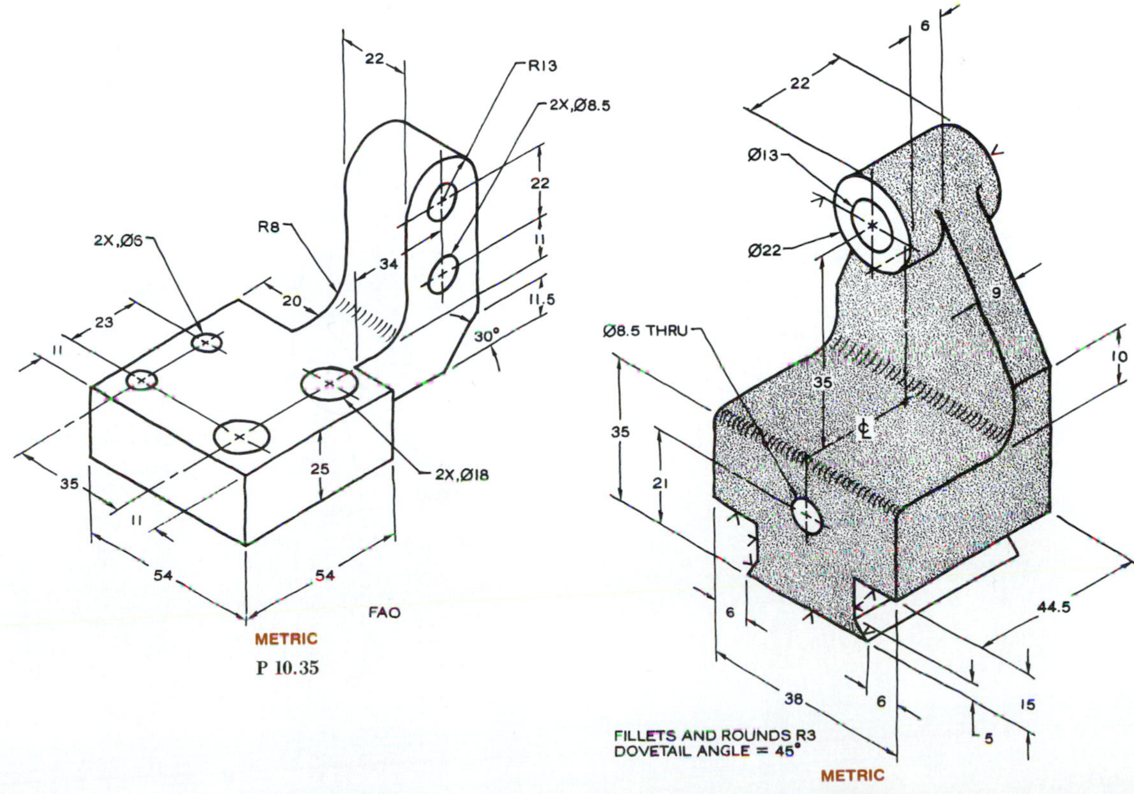

P 10.35

P 10.36

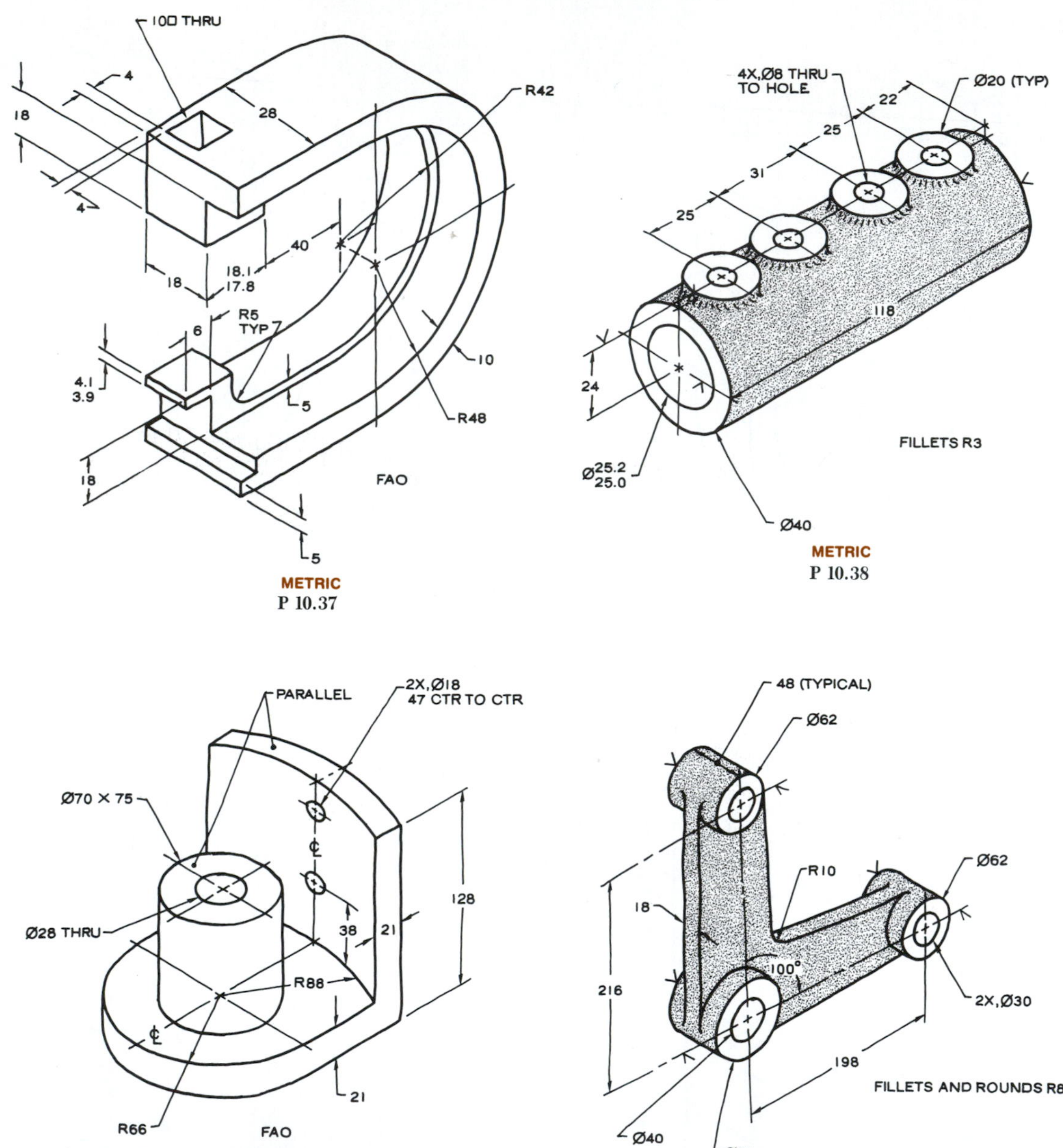

METRIC
P 10.37

METRIC
P 10.38

METRIC
P 10.39

METRIC
P 10.40

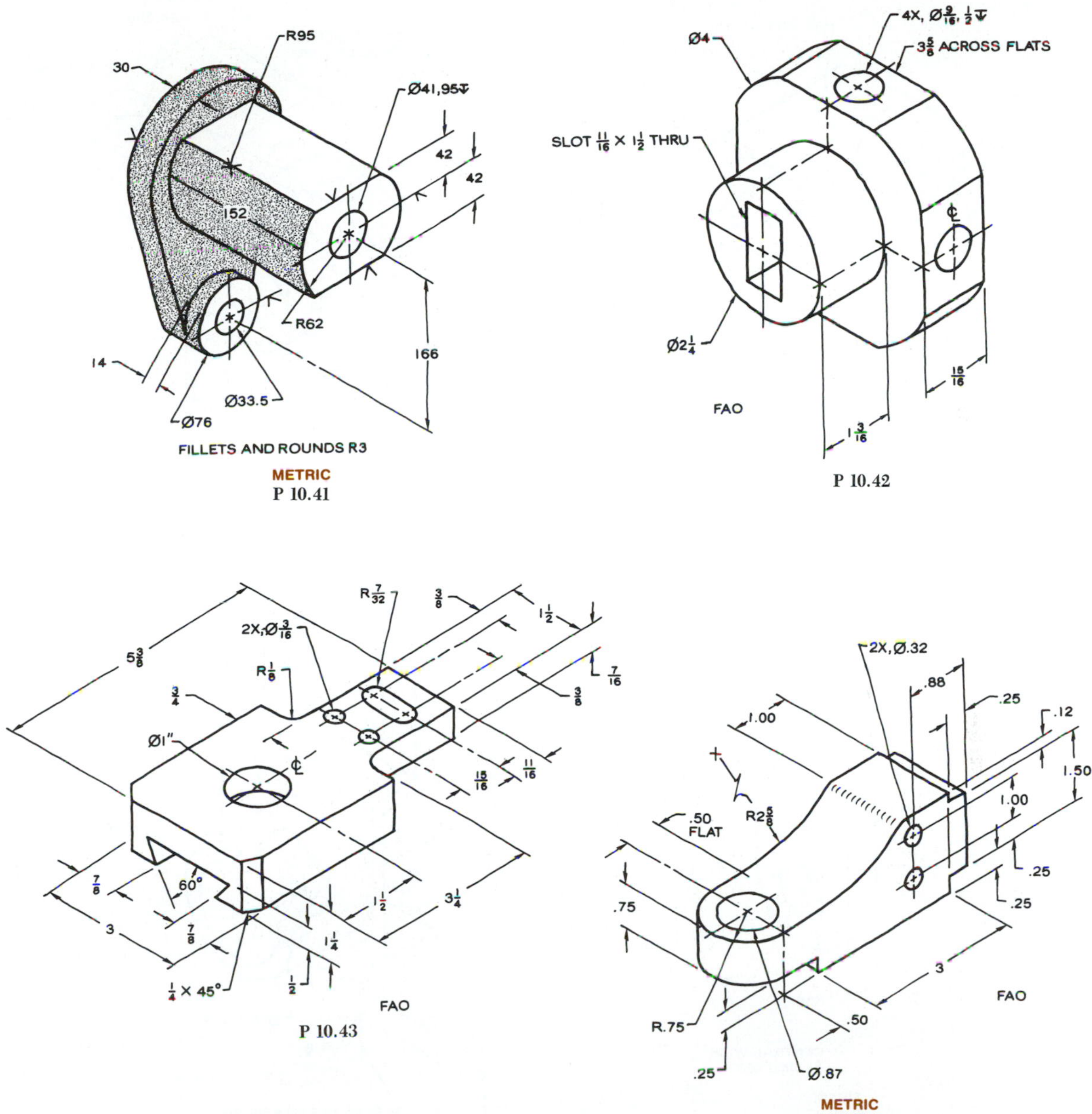

METRIC
P 10.41

P 10.42

P 10.43

METRIC
P 10.44

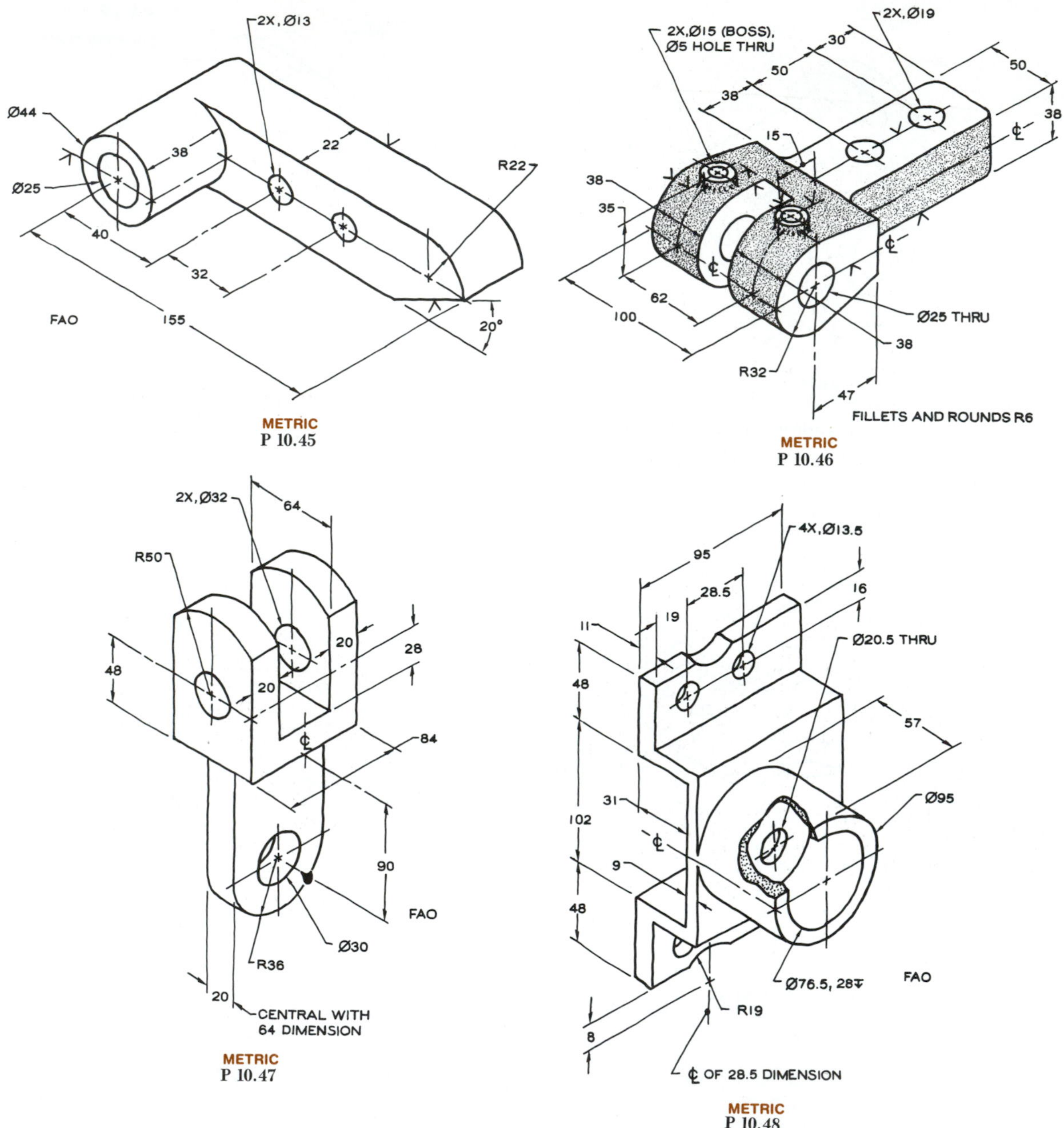

2X, Ø13
Ø44
38
22
R22
Ø25
40
32
FAO
155
20°
METRIC
P 10.45
2X,Ø15 (BOSS), Ø5 HOLE THRU
2X,Ø19
30
50
50
38
38
15
38
35
62
100
Ø25 THRU
38
R32
47
FILLETS AND ROUNDS R6
METRIC
P 10.46
2X,Ø32
64
R50
20
28
48
20
84
90
FAO
Ø30
R36
20
CENTRAL WITH 64 DIMENSION
METRIC
P 10.47
4X,Ø13.5
95
28.5
16
11
19
Ø20.5 THRU
48
57
31
Ø95
102
9
48
Ø76.5, 28↧
FAO
R19
8
℄ OF 28.5 DIMENSION
METRIC
P 10.48

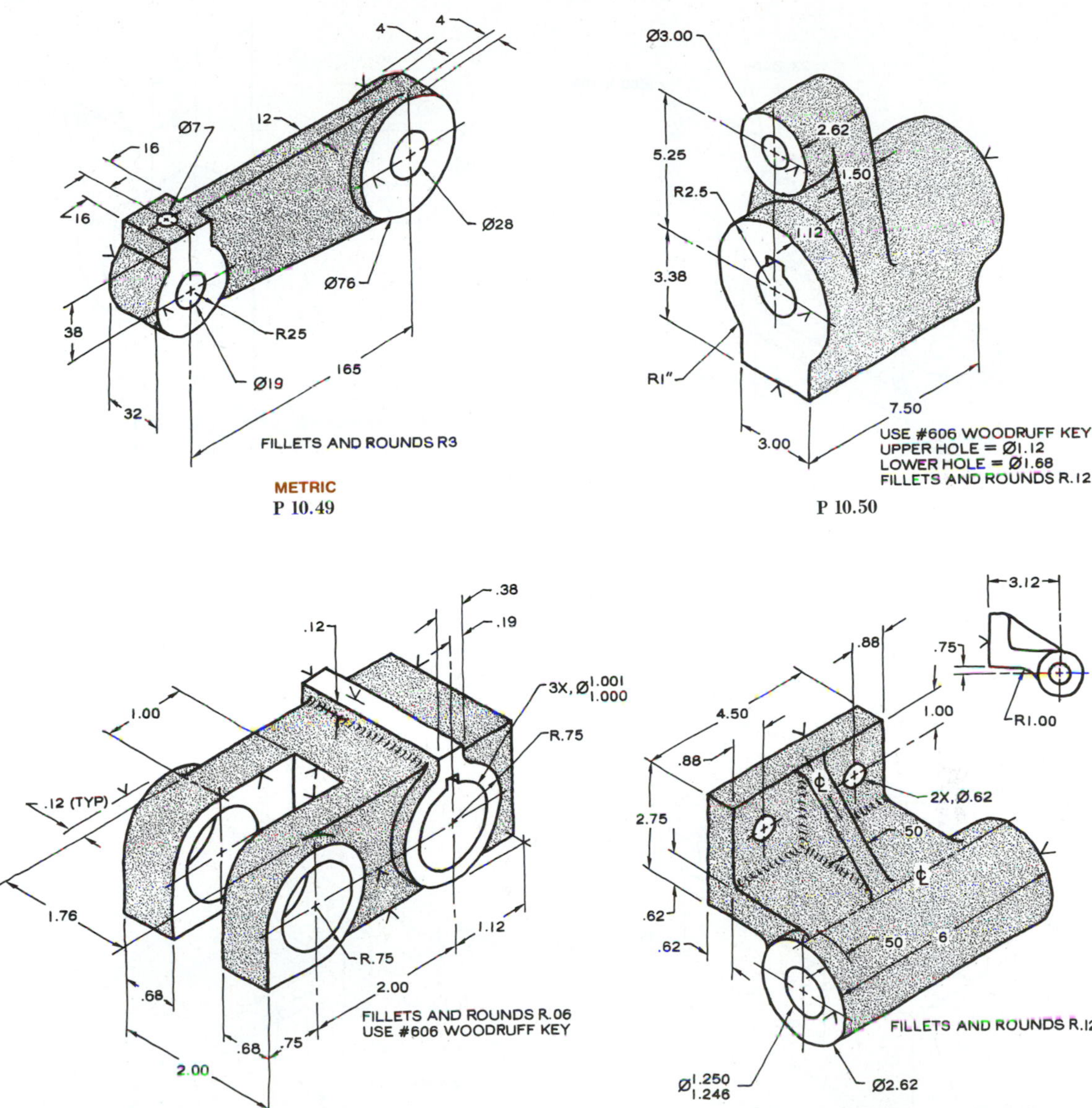
4
4
Ø7
12
16
16
38
32
165
R25
Ø19
Ø28
Ø76
FILLETS AND ROUNDS R3
Ø3.00
2.62
5.25
1.50
R2.5
1.12
3.38
R1"
7.50
3.00
USE #606 WOODRUFF KEY
UPPER HOLE = Ø1.12
LOWER HOLE = Ø1.68
FILLETS AND ROUNDS R.12
.38
.19
.12
3X, Ø1.001 1.000
1.00
R.75
.12 (TYP)
1.76
1.12
R.75
.68
2.00
.68
.75
2.00
FILLETS AND ROUNDS R.06
USE #606 WOODRUFF KEY
3.12
.88
.75
4.50
1.00
R1.00
.88
2X, Ø.62
2.75
.50
.62
.62
.50
6
FILLETS AND ROUNDS R.12
Ø1.250 1.246
Ø2.62

METRIC
P 10.49

P 10.50

P 10.51

P 10.52

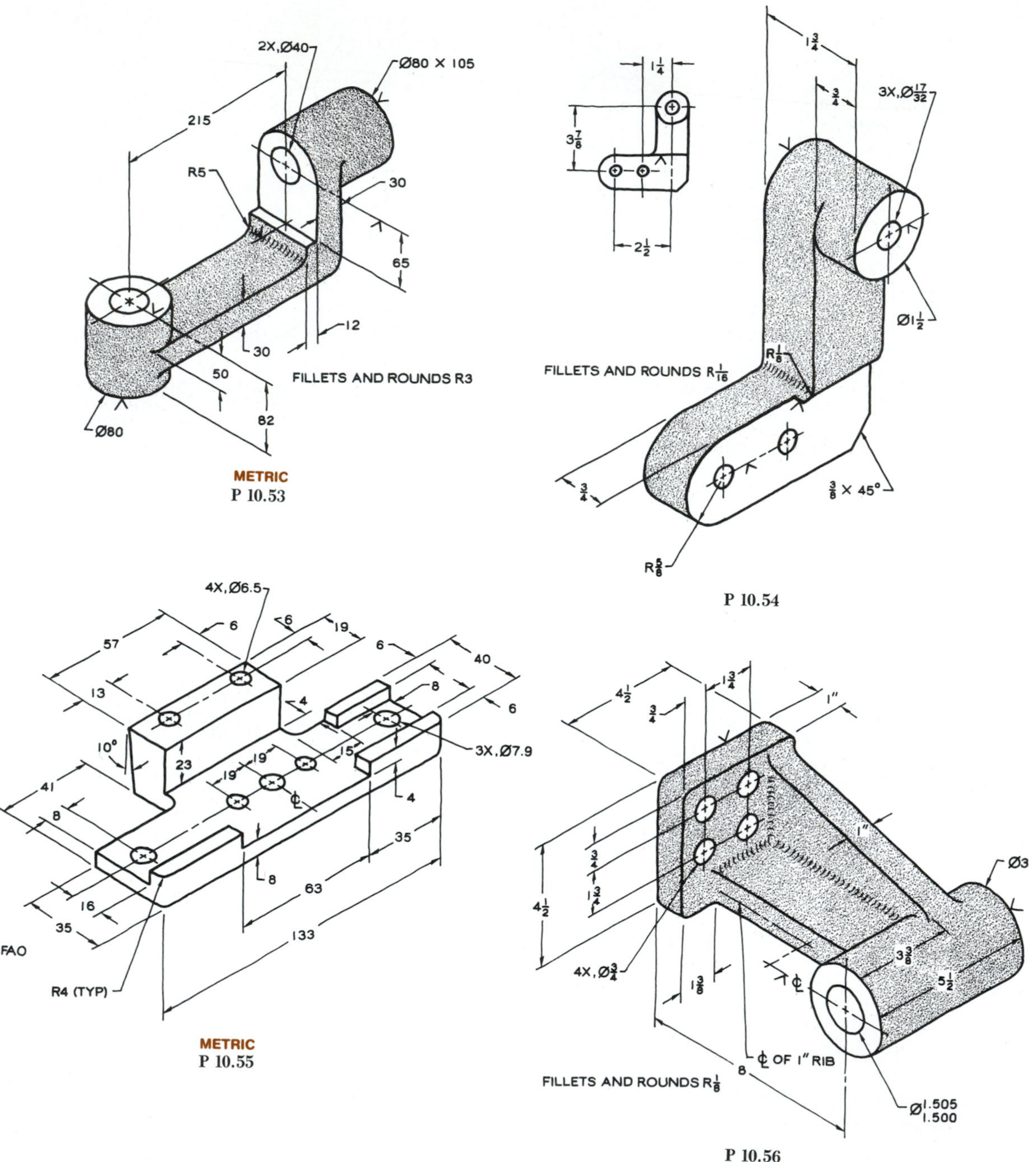

METRIC
P 10.53

P 10.54

METRIC
P 10.55

P 10.56

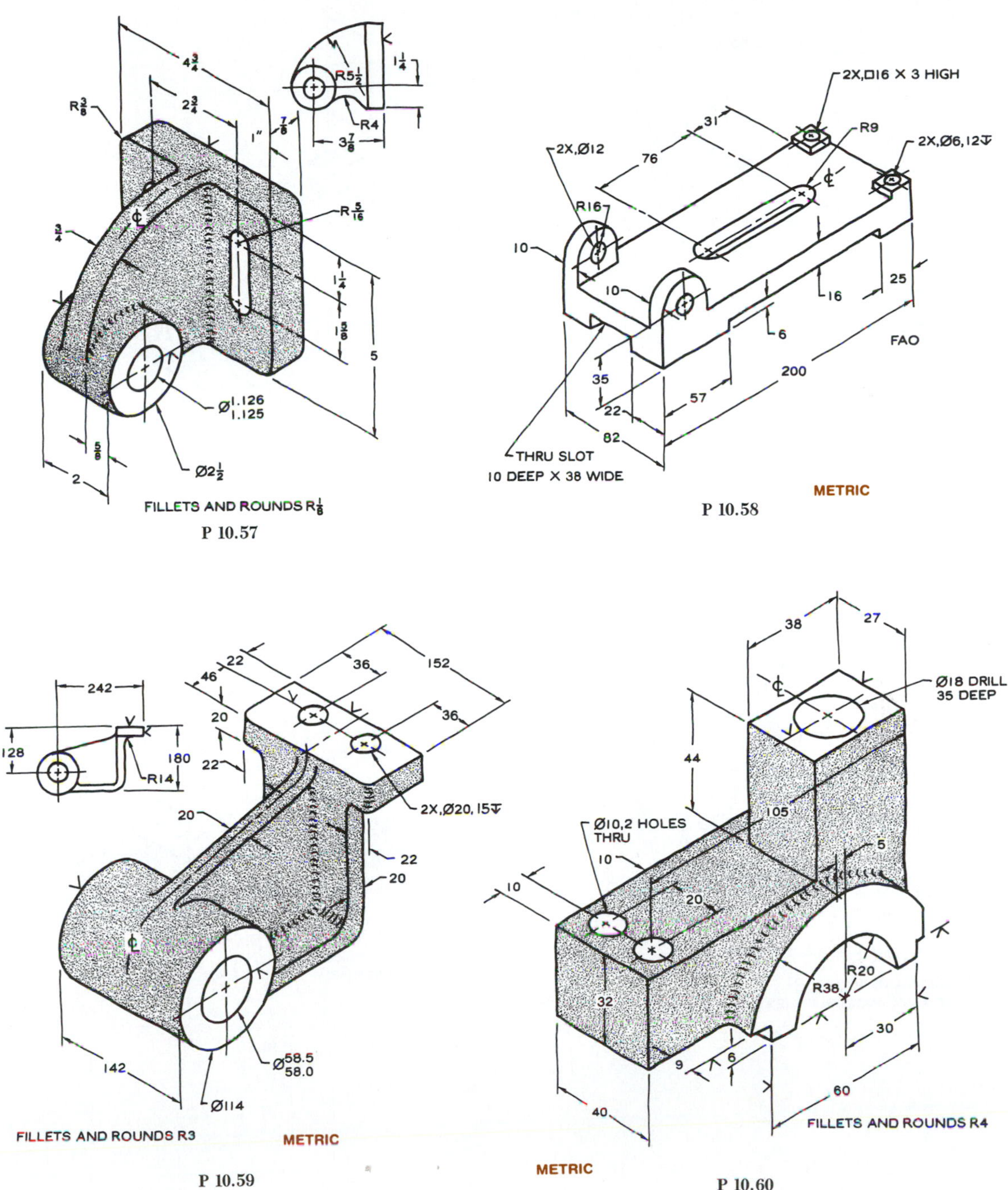

P 10.57

P 10.58

P 10.59

P 10.60

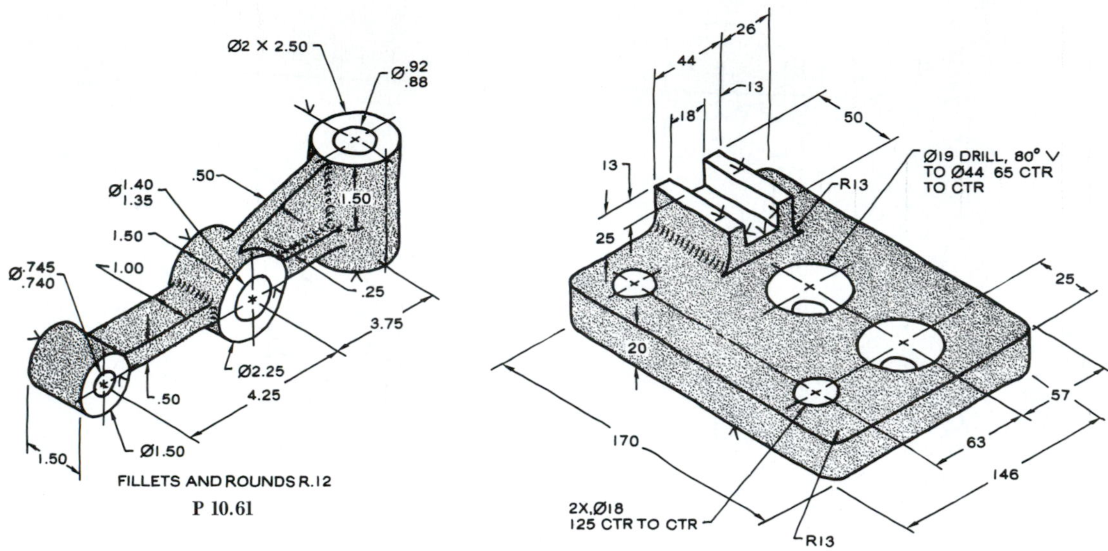

P 10.61

P 10.62

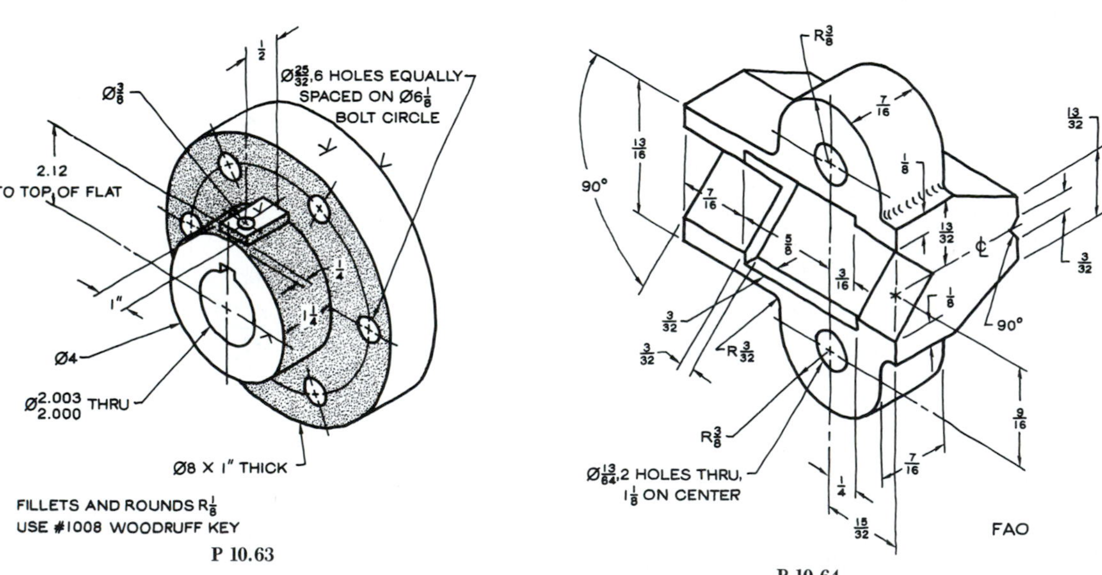

P 10.63

P 10.64

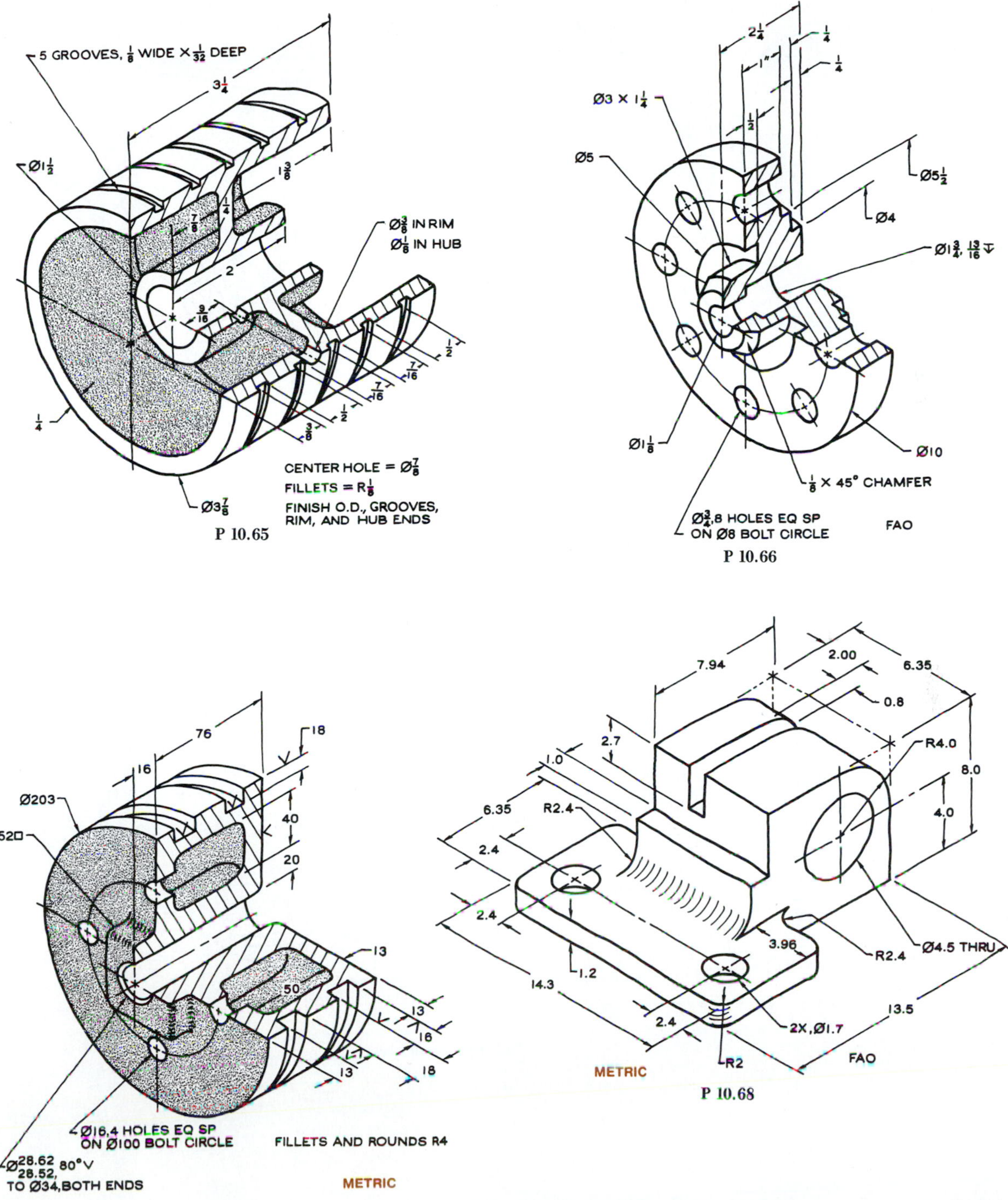
5 GROOVES, $\frac{1}{8}$ WIDE × $\frac{1}{32}$ DEEP
CENTER HOLE = Ø$\frac{7}{8}$
FILLETS = R$\frac{1}{8}$
FINISH O.D., GROOVES, RIM, AND HUB ENDS
Ø$\frac{3}{8}$ IN RIM
Ø$\frac{1}{8}$ IN HUB
Ø3 × 1$\frac{1}{4}$
Ø$\frac{3}{4}$,8 HOLES EQ SP ON Ø8 BOLT CIRCLE
$\frac{1}{8}$ × 45° CHAMFER
FAO
Ø16,4 HOLES EQ SP ON Ø100 BOLT CIRCLE
FILLETS AND ROUNDS R4
Ø28.62 28.52, 80°V TO Ø34,BOTH ENDS
METRIC
Ø4.5 THRU
2X,Ø1.7
METRIC
FAO

P 10.65

P 10.66

P 10.67

P 10.68

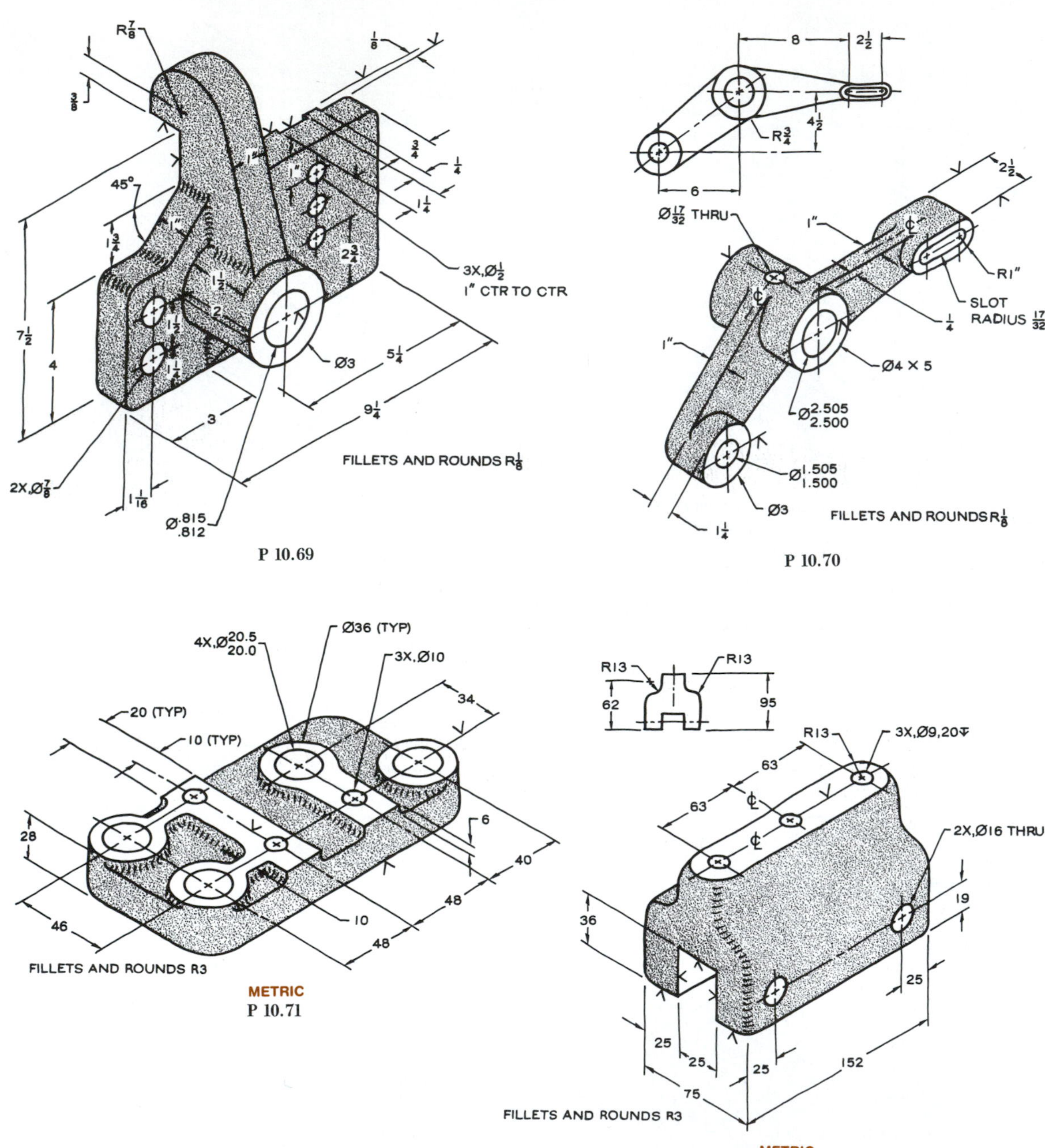

P 10.69

P 10.70

METRIC
P 10.71

METRIC
P 10.72

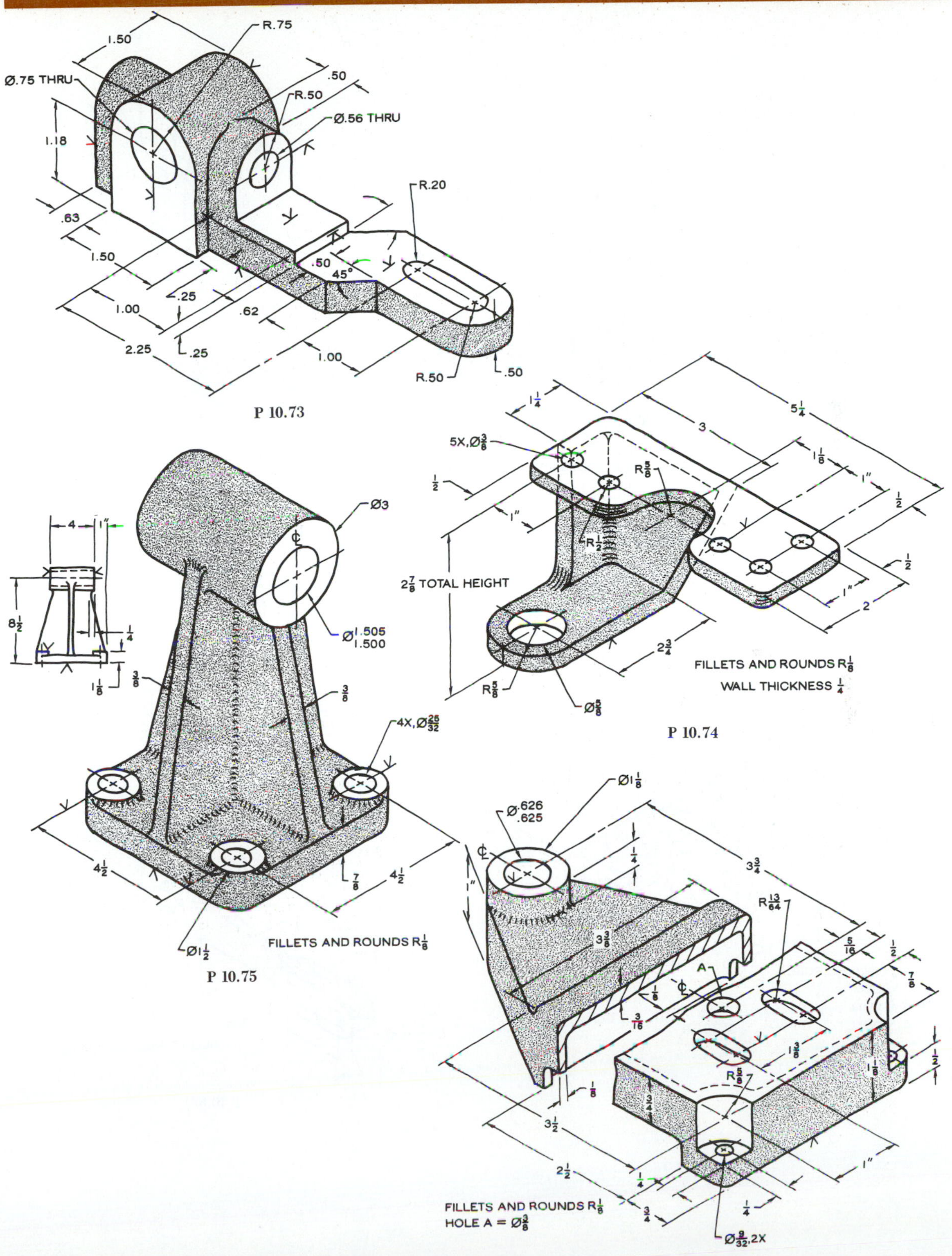

P 10.73

P 10.74

P 10.75

P 10.76

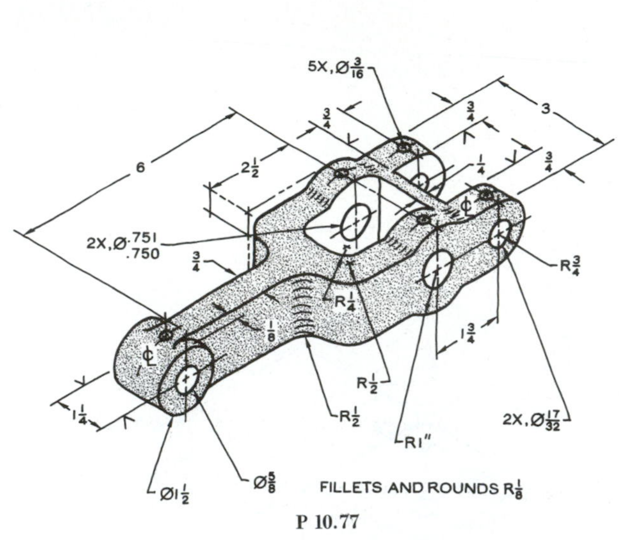

P 10.77

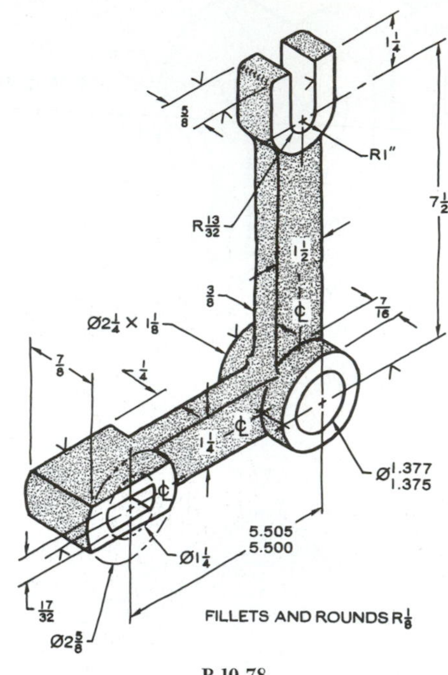

P 10.78

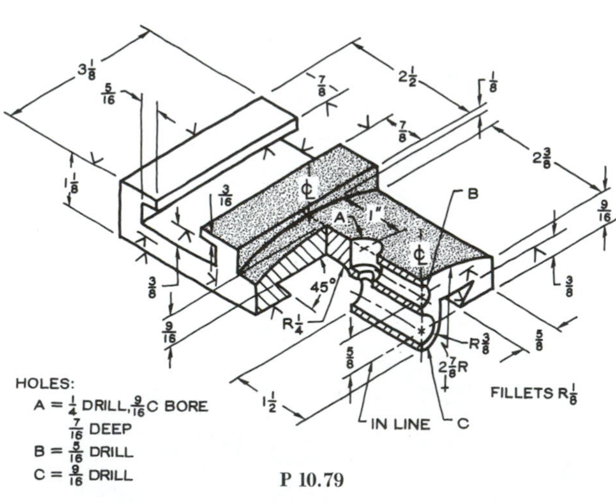

HOLES:
A = 1/4 DRILL, 9/16 C BORE 7/16 DEEP
B = 5/16 DRILL
C = 9/16 DRILL

P 10.79

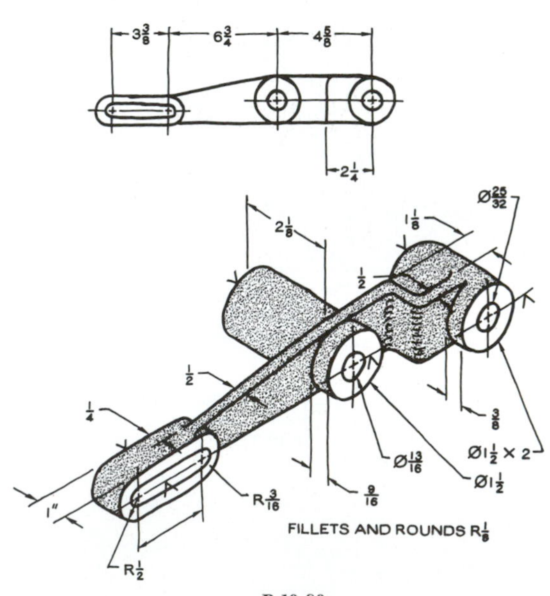

P 10.80

11 Size Description

OBJECTIVES

After completing this chapter you should have gained the following abilities:

1. To understand the English system of standard fits.
   - **a.** To know the definition of:
     - Size
     - Nominal size
     - Basic size
     - Limits of size
     - Tolerance
     - Allowance
   - **b.** To know how to use:
     - Limit-dimension tolerancing
     - Plus-and-minus tolerancing
     - Single limits
   - **c.** To know the three types of fits for mating parts.
   - **d.** To know how to compute allowance and tolerance using the tables for the basic hole size and the basic shaft size.
2. To understand the SI Metric system of standard fits.
   - **a.** To know the definition of:
     - Deviation
     - Upper and lower deviation
     - Fundamental deviation
     - Tolerance zone
     - International tolerance grade
   - **b.** To know how to compute allowance and tolerance using the tables for the hole basis system and the shaft basis system.
3. To understand the concepts of:
   - **a.** Tolerance accumulation
   - **b.** Selective assembly

No two parts can be made exactly alike. In the manufacture of every object there is a certain amount of variation or inaccuracy. There are a number of reasons why variations or inevitable errors occur on a part, including the type of machining operation that is used, the skill of the operators, inherent size variations in raw and in process materials, and certain environmental factors such as temperature and humidity. The objective in interchangeable manufacture is to produce all parts as nearly identical as necessary for proper function. Parts may be made to very close tolerances, even to millionths of an inch, but manufacturing to unnecessarily close tolerances is always costly and should be systematically avoided. Tolerances should be as large as possible without interfering with the function of the part.

11.1 SYSTEMS OF LIMITS AND FITS

There are two systems in use for specifying limits and fits between mating parts; the *English* system and the *SI Metric* system. A complete coverage of the English system is presented in ANSI Y15.5M-1982, *Dimensioning and Tolerancing*, and in USAS B4.1-1967(1974), *Preferred Limits and Fits for Cylindrical Parts*. The SI Metric system is presented in ANSI B4.2-1978, *Preferred Metric Limits and Fits*.

11.2 TOLERANCING

Tolerance is defined as the maximum permissible variation in the size, location, or form of a feature on a part. In tolerancing, each dimension is allowed a certain degree of variation within a precisely specified range.

Tolerances of size apply to the principles of size control of mating parts. These parts may be cylindrically shaped, as in the assembly of a shaft and a bushing, or they may be flat parts, such as an assembled key in a keyway.

11.3 ENGLISH SYSTEM—DEFINITION OF BASIC TERMS

A *dimension* is used on a drawing to indicate the size, location, and/or geometric characteristics of the various features of a part so that the part may be efficiently manufactured. Dimensions are used to define the size of a diameter, length, angle, or a center distance between features on a part. Chapter 10, "Size Description," deals with the principles of dimensioning.

Size is a designation of magnitude. When a value is assigned to a dimension it is referred to as the "size" of a dimension.

Nominal size is the approximate size designation that is used for purposes of general identification. For a .500-inch basic size hole, for example, the nominal size is ½ inch. In this case the value of the nominal size is identical to the basic size.

Basic size is the exact theoretical size from which the limits of variation on a part are computed.

A *tolerance limit* is the maximum or minimum size within which a feature on a part is permitted to deviate from the design size. Limits and directly applied tolerance values are specified as follows:

LIMIT-DIMENSION TOLERANCING. As shown in Figure 11-1(a), the high limit (maximum value) is placed above the low limit (minimum value). Although recommended by ANSI Y14.5M-1982, this convention has not been universally adopted by U.S. industrial firms for use on engineering drawings. Many companies prefer to place the minimum value above the maximum value when specifying limits for a hole size. When expressed as a dimension in a single line, as at 11-1(b), the low limit precedes the high limit and a dash is used to separate the two values. Angular tolerances are shown at 11-1(c).

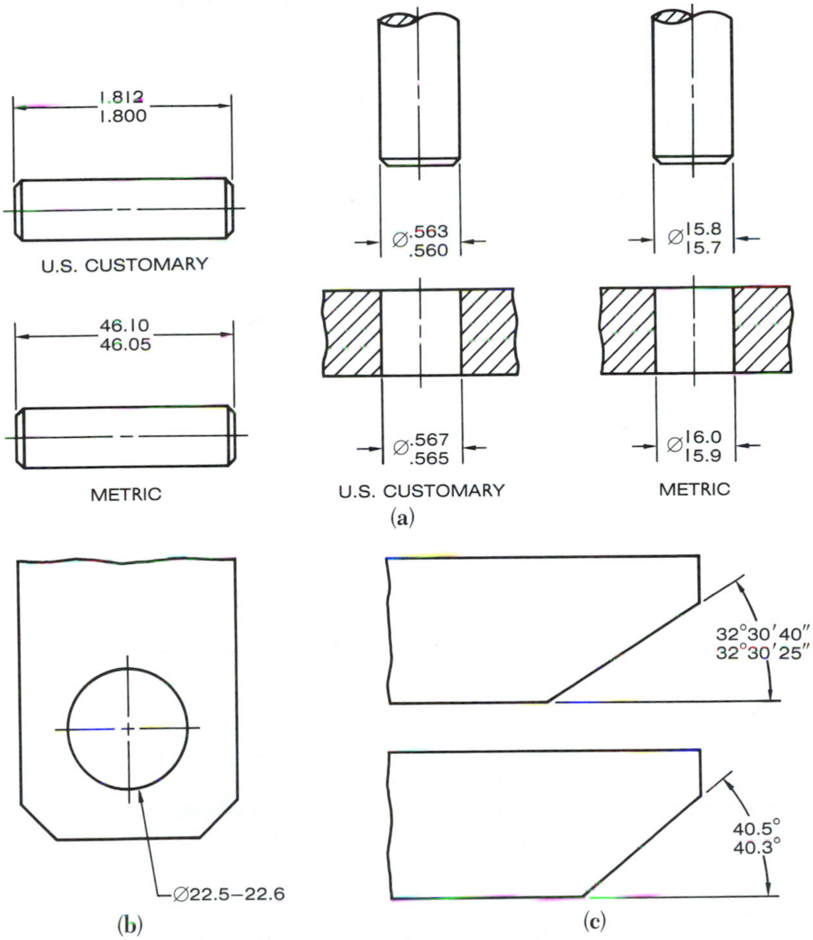

1.812
1.800
U.S. CUSTOMARY
46.10
46.05
METRIC
∅.563
.560
∅.567
.565
U.S. CUSTOMARY
∅15.8
15.7
∅16.0
15.9
METRIC
(a)
∅22.5–22.6
(b)
32°30′40″
32°30′25″
40.5°
40.3°
(c)

FIGURE 11-1 Limit dimensioning tolerancing ***(Section 11.3)***

PLUS-AND-MINUS TOLERANCING. The dimension is given first, followed by a plus-and-minus expression of tolerance. Two systems, *unilateral* and *bilateral*, are used as shown in Figure 11-2. Unilateral tolerances allow variation in only one direction from the specified basic size dimension. Bilateral tolerancing allows variation in either direction from the specified basic size dimension.

SINGLE LIMITS. Features such as depths of holes, lengths of threads, corner radii, chamfers, and so forth may be dimensioned as shown in Figure 11-3 by using a single limit. The abbreviations MAX or MIN are placed after the dimension. Single limits are used only on detail drawings when the intent will not be misunderstood.

LIMITS OF SIZE. The limits of size prescribe the extent within which the variations in sizes of a dimension on a part are allowed.

SELECTIVE ASSEMBLY. Inspection operations of closely fitting, interchangeable parts generally require the employment of manual or computer-controlled methods. In selective assembly all parts are accurately measured and sorted into several size

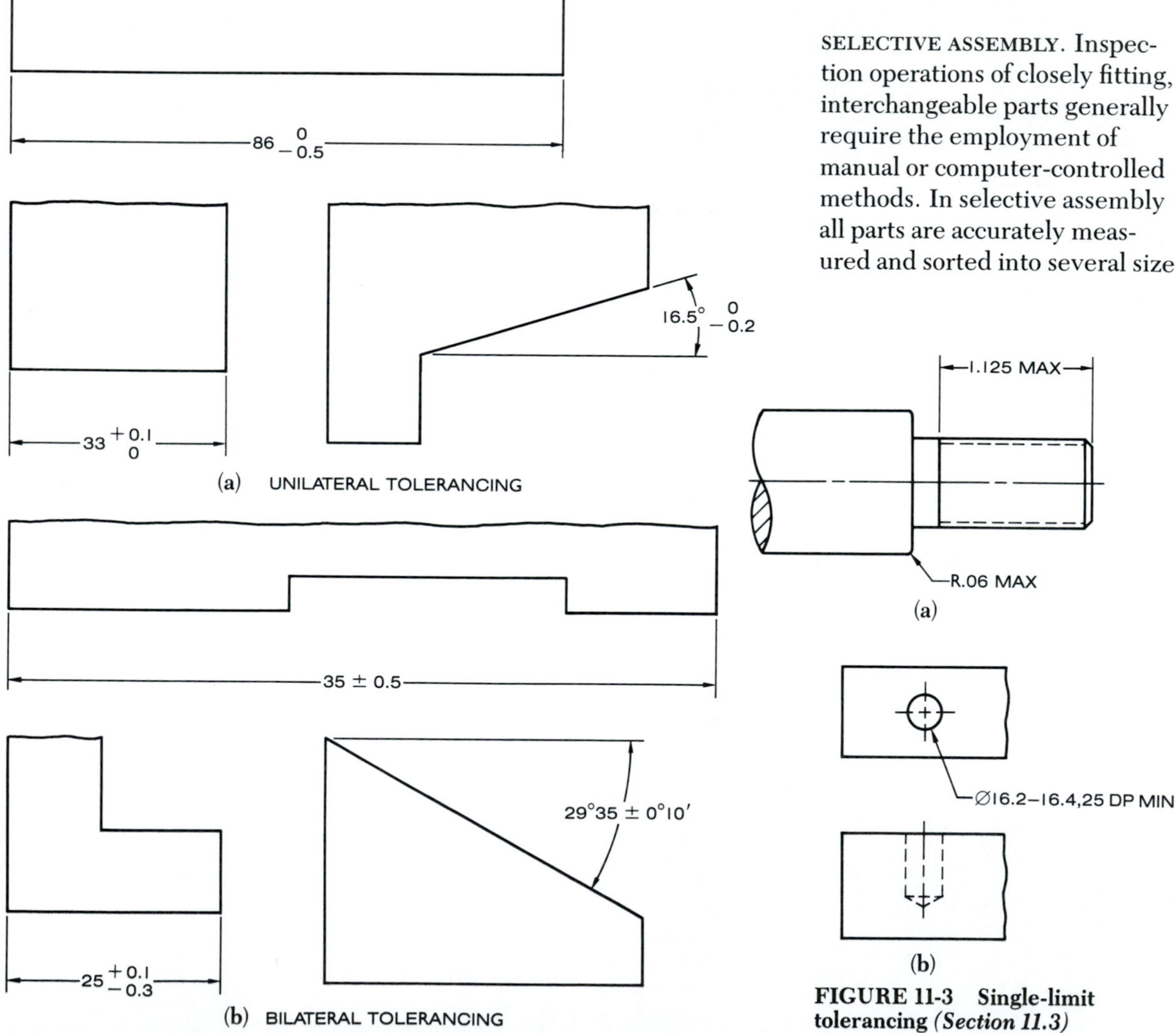

FIGURE 11-2 Plus and minus tolerancing: unilateral and bilateral *(Section 11.3)*

FIGURE 11-3 Single-limit tolerancing *(Section 11.3)*

grades. Later, as the parts are assembled, small shafts may be satisfactorily mated with small holes, medium shafts with medium holes, and so on. A proper fit between mating parts may thus be ensured to an acceptable degree of interchangeability within each size grade.

SPECIFYING TOLERANCES. Tolerances for noncritical features on a part may be specified on a detail drawing by a general note, placed in or near the sheet title block, such as UNLESS OTHERWISE SPECIFIED, ALL DIMENSIONS ±.015. Tolerances may also be specified on processing sheets, which often accompany the working drawing. In most companies the standard tolerance for fractional dimensions on a drawing in U.S. Customary units is ±1⁄64 inch. Dimensions for critical features on a part are specified by using tolerance limits, which are explained in the following section.

11.4 FIT

Fit is a general term that signifies the relationship between two mating parts with respect to the amount of clearance or interference that is present when they are assembled. There are three general types of fits: clearance, transition, and interference. A *clearance* (or running) fit is obtained when the prescribed limits of size result in looseness or a space between the mating parts. A *transition* fit is obtained when the prescribed limits of size result in either a clearance or an interference between the mating parts. An *interference* (or force) fit is obtained when the prescribed limits of size result in an interference between the mating parts.

11.5 ALLOWANCE

The relationship between toleranced mating parts is the *allowance*, which is defined as the prescribed difference in size between the maximum material limits of mating parts, or the *tightest* fit between any two parts. For a clearance fit, the allowance is classified as *positive*. For an interference fit, the allowance is classified as *negative*. In Figure 11-4, for example, the allowance is +.002 (1.000 − .998). In short, the allowance always equals the minimum (smallest) hole limit minus the maximum (largest) shaft limit.

11.6 DESIGN CONSIDERATIONS

There are many factors that you must take into consideration in the selection of fits for a particular assembly of parts. Some of these factors are length of engagement, bearing load, the speed of the moving parts, lubrication, temperature, humidity, surface texture, and material. A critical analysis of mating parts must be performed to insure that the specifications permit optimum manufacturing ease and economy.

The selection of the proper fit requires considerable practical experience. When designing, you must always specify practically attainable tolerances and an allowance that will result in mating parts that satisfy the critical functional requirements.

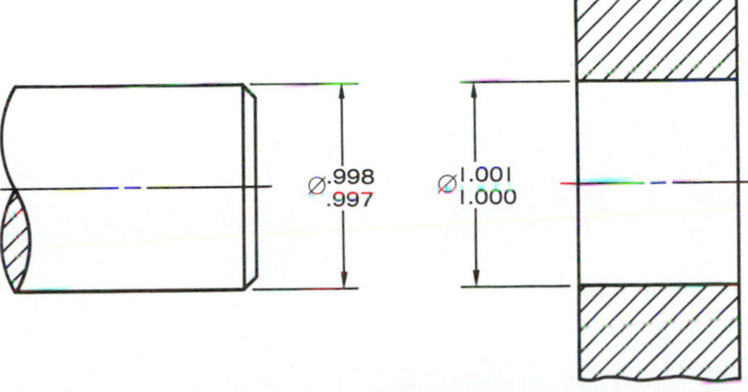

Minimum hole limit (1.000) minus maximum shaft limit (.998) or 1.000 − .998 = +.002 (allowance)

FIGURE 11-4 Calculating the allowance *(Section 11.5)*

11.7 THE BASIC HOLE SYSTEM

The basic hole system is an arrangement of fits in which the minimum hole diameter is taken as the basic size. Depending upon the conditions of the desired fit, you begin by selecting a suitable allowance, and then you assign tolerances and compute limits for the hole and the shaft diameters. The basic hole system is widely used because holes may be conveniently bored to a desired size or produced by standard-size cutting tools such as drills, reamers, broaches, or punches. The mating shafts can easily be machined to any size desired to obtain the required fit.

Figure 11-5 illustrates two examples of the basic hole system. At 11-5(a), a 1½-inch nominal size shaft is to run freely (clearance fit) in a reamed hole. The assigned values are as follows:

Allowance	+.002
Hole tolerance	.0025
Shaft tolerance	.0015

To find the hole limits:

Basic size (minimum hole limit)	1.5000
Assigned hole tolerance	.0025
Maximum hole limit	1.5025

To apply the allowance (clearance fit):

Assigned allowance	+.002
Basic size (minimum hole limit)	1.5000
Maximum shaft limit	1.4980

To find the shaft limits:

Maximum shaft limit (from above)	1.4980
Assigned shaft tolerance	.0015
Minimum shaft limit	1.4965

At 11-5(b), a 3⅞-inch nominal size shaft is to fit tightly (interference fit) in a bored hole. The assigned values are as follows:

Allowance	−.0019
Hole tolerance	.0006
Shaft tolerance	.0006

To find the hole limits:

Basic size (minimum hole limit)	3.8750
Assigned hole tolerance	.0006
Maximum hole limit	3.8756

To apply the allowance (interference fit):

Assigned allowance	−.0019
Basic size (minimum hole limit)	3.8750
Maximum shaft limit	3.8769

To find the shaft limits:

Maximum shaft limit (from above)	3.8769
Assigned shaft tolerance	.0006
Minimum shaft limit	3.8763

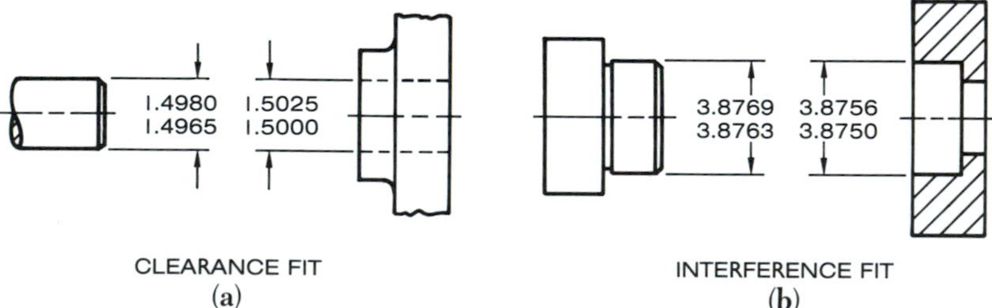

FIGURE 11-5 Basic hole system *(Section 11.7)*

CHECKPOINT 1
COMPUTING TOLERANCES USING THE BASIC HOLE SYSTEM

Using the basic hole system, compute the limits for a hole and a mating shaft using the following data:

	Nominal Size	*Tolerances*		*Allowance*
		Hole	*Shaft*	
1.	½	.0004	.0003	−.0008
2.	2¾	.0012	.0007	−.0029
3.	5⁄16	.0002	.0003	+.0001
4.	1⅞	.001	.001	+.002

11.8 THE BASIC SHAFT SYSTEM

Some companies find it advantageous to use standard-diameter cold-finished shafting stock, particularly when a number of parts requiring different types of fits must be mounted on the same shaft. In this case it is more practical to vary the sizes of holes on the parts into which the shaft will fit. Of the two systems, the basic shaft system is the least used. In this system the maximum shaft diameter is always taken as the basic size. The designer selects a suitable allowance, and the limits of the desired tolerances are applied to the hole and the shaft diameters.

Figure 11-6 shows two examples of the basic shaft system. At 11-6(a) a ½-inch nominal size shaft is to run freely (clearance fit) in a hole. The assigned values are as follows:

Allowance	+.0015
Hole tolerance	.0002
Shaft tolerance	.001

To find the shaft limits:

Basic size (maximum shaft limit)	.5000
Assigned shaft tolerance	.001
Minimum shaft limit	.4990

To apply the allowance (clearance fit):

Assigned allowance	+.0015
Basic size (maximum shaft limit)	.5000
Minimum hole limit	.5015

To find the hole limits:

Minimum hole limit (from above)	.5015
Assigned hole tolerance	.0002
Maximum hole limit	.5017

At 11-6(b) a 4¾-inch nominal size shaft is to fit tightly (interference fit) in a hole. The assigned values are as follows:

Allowance	−.0015
Hole tolerance	.0005
Shaft tolerance	.0005

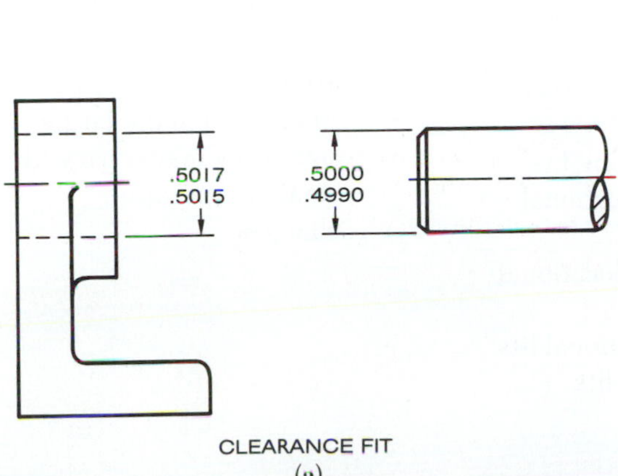

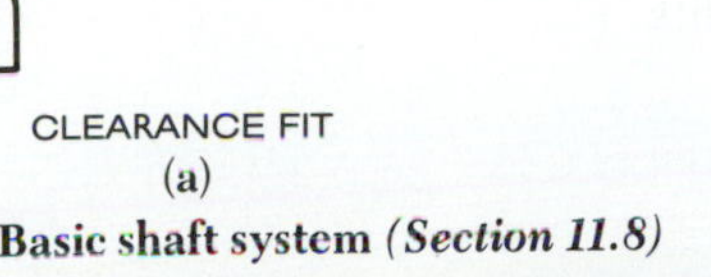
CLEARANCE FIT
(a)

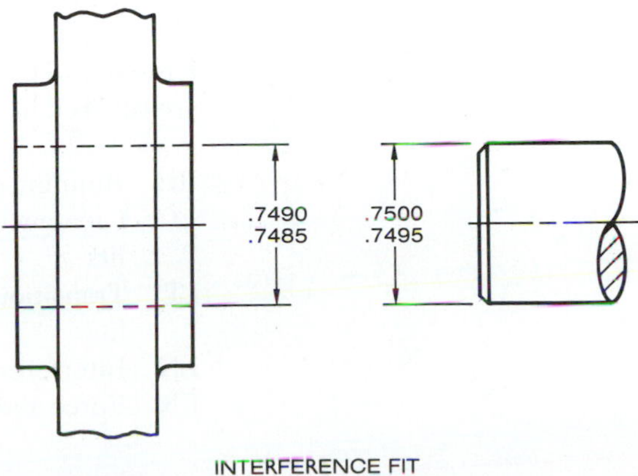

INTERFERENCE FIT
(b)

FIGURE 11-6 Basic shaft system ***(Section 11.8)***

CHECKPOINT 2
COMPUTING TOLERANCES USING THE BASIC SHAFT SYSTEM

Using the basic shaft system, compute the limits for a hole and a mating shaft using the following data:

	Nominal Size	*Tolerances*		*Allowance*
		Hole	*Shaft*	
1.	1¼	.003	.001	+.002
2.	⅝	.0015	.0005	+.0025
3.	2½	.0008	.0003	−.0015
4.	⅞	.0015	.0015	−.002

To find the shaft limits:

Basic size (maximum shaft limit)	.7500
Assigned shaft tolerance	.0005
Minimum shaft limit	.7495

To apply the allowance (interference fit):

Assigned allowance	−.0015
Basic size (maximum shaft limit)	.7500
Minimum hole limit	.7485

To find the hole limits:

Minimum hole limit	.7485
Assigned hole tolerance	.0005
Maximum hole limit	.7490

11.9 STANDARD FITS—ENGLISH SYSTEM

The English system establishes a series of standard fits of mating cylindrical parts that are based upon the basic hole system. Tables (see Appendix Tables A-35 through A-43) have been developed that give a series of standard types and classes of fits that will cover most applications. These tables can also be used to specify the fit between any set of parallel surfaces, such as a key and a slot, for example. The hole tolerance is unilateral and positive and the minimum hole limit is always equal to the basic size. The types of fits that are covered in this standard are:

- RC Running and sliding fits
- LC Clearance and locational fits
- LT Transitional and locational fits
- LN Interference locational fits
- FN Force and shrink fits

Letter symbols are used in conjunction with numbers representing the class of fit. There are several classes of fit included in each of the types given here. "LC2," for example, represents a class 2, locational clearance fit.

Running and Sliding Fits

Running and sliding fits (see Appendix Table A-35) are intended to provide a similar running performance, with suitable lubrication allowance, throughout the ranges of sizes. The clearances for the first two classes, used chiefly as slide fits, increase more slowly with diameter than the other classes, so that accurate location is maintained even at the expense of free relative motion.

RC1 *Close sliding fits* are intended for the accurate location of parts that must assemble without perceptible play.

RC2 *Sliding fits* are intended for accurate location but with greater maximum clearance than class RC1. Parts made to this fit move and turn easily but are not intended to run freely, and in the larger sizes they may seize with small temperature changes.

RC3 *Precision running fits* are about the closest fits that can be expected to run freely. They are intended for precision work at slow speeds and light journal pressures, but they are not suitable where appreciable temperature differences are likely to be encountered.

RC4 *Close running fits* are intended chiefly for running fits on accurate machinery, with moderate surface speeds and journal pressures, where accurate location and minimum play are desired.

RC5 RC6 *Medium running fits* are intended for higher running speeds, heavy journal pressures, or both.

RC7 *Free running fits* are intended for use where accuracy is not essential, where large temperature variations are likely to be encountered, or under both of these conditions.

RC8 RC9 *Loose running fits* are intended for use where wide commercial tolerances may be necessary, together with an allowance, on the external member.

Locational Fits

Locational fits (see Appendix Table A-37) are fits intended to determine only the location of the mating parts. They may provide rigid or accurate location, as with interference fits, or some freedom of location, as with clearance fits. Accordingly, they are divided into three groups: clearance fits, transition fits, and interference fits.

LC *Locational clearance fits* are intended for parts that are normally stationary but that can be freely assembled or disassembled. They run from snug fits for parts requiring accuracy of location, through the medium-clearance fits for parts such as ball, race, and housing, to the looser fastener fits where freedom of assembly is of prime importance.

LT *Locational transition fits* are a compromise between clearance and interference fits. They are used where accuracy of location is important but where a small amount of either clearance or interference is permissible.

LN *Locational interference fits* are used where accuracy of location is of prime importance and for parts requiring rigidity and alignment with no special requirements for bore pressure. Such fits are not intended for parts designed to transmit frictional loads from one part to another by virtue of the tightness of fit, as these conditions are covered by force fits.

Force Fits

Force or *shrink* fits (see Appendix Table A-39) constitute a special type of interference fit, normally characterized by maintenance of constant bore pressures throughout the range of sizes. The diameter, and the difference between its minimum and maximum value, is small to maintain the resulting pressures within reasonable limits.

FN1 *Light drive fits* are those requiring light assembly pressures. They produce more or less permanent assemblies, and they are suitable for thin sections or long fits, or in cast iron external members.

FN2 *Medium drive fits* are suitable for ordinary steel parts or for shrink fits on light sections. They are about the tightest fits that can be used with high-grade cast iron external members.

FN3 *Heavy drive fits* are suitable for heavier steel parts or for shrink fits in medium sections.

FN4 FN5 *Force fits* are suitable for parts that can be highly stressed or for shrink fits where the heavy pressing forces required are impractical.

11.10 METHOD OF COMPUTING THE ENGLISH SYSTEM OF FITS AND LIMITS

Assume an RC4 designation of fit is required between a mating shaft and a hole with a nominal size of 1⅝ inches. As explained in the previous section, an RC4 fit is a close running fit. To calculate the limits, allowances, and clearances shown in Figure 11-7, refer to Appendix Table A-35 for the values. The first column gives the range of the nominal hole or shaft sizes in inches. The corresponding basic size, 1.6250 (the decimal equivalent of 1⅝ inches), falls between the range of 1.19 and 1.97. All limits in the tables are given in thousandths, thus the value 1.6, for example, should be read as .0016 inch. Always move the decimal point three places to the left. The values given in the column "Limits of Clearance" for RC and LC types of fits are, in order, the minimum and maximum clearance. The *upper* figure represents the allowance. In the column "Limits of Interference" for the LN types of fits, the minimum and maximum interference are given in that order. The *lower* figure represents the allowance.

The upper and lower limits of the hole and shaft are calculated as follows:

Class of fit: RC4 (use Appendix Table A-35)

Basic Size: 1.6250 (the size falls between 1.19 and 1.97)

Table data:

Limits of clearance	*Hole*	*Shaft*
1.0	+1.6	−1.0
3.6	0	−2.0

Hole limits:

Minimum = 1.6250

Maximum 1.6250 + .0016 = 1.6266

Shaft limits:

Maximum 1.6250 − .0010 = 1.624

Minimum 1.6250 − .0020 = 1.623

In this example the values 1.0 and 3.6 in the column headed "Limits of Clearance" indicate that an RC4 fit will have an allowance of +.001 inch and a loosest fit of +.0036 inch. By inspection, the hole tolerance is .0016 inch and the shaft tolerance is .001 inch. Remember that *the tolerance is the difference between the limits.* Figure 11-7 illustrates how the size limits derived from the table for the hole and the shaft are applied to the views of the respective parts on a detail drawing. Note that the symbol for the particular class of fit, in this case RC4, does not appear on the drawing. Size limits in U.S. Customary units may be converted to millimeters by multiplying inches by 25.4, (1 inch = 25.4 mm).

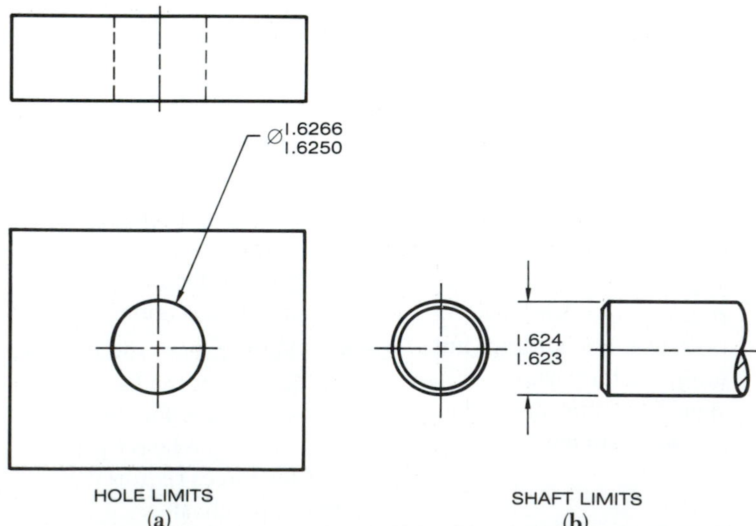

FIGURE 11-7 Limits for an RC4 fit (1⅝ inch nominal size) ***(Section 11.10)***

CHECKPOINT 3
COMPUTING LIMITS USING THE ENGLISH SYSTEM OF FITS AND LIMITS

Using the given set of values and the appropriate appendix tables, compute the limits for the mating parts.

	Nominal size	*Type of Fit*
1.	7/16	RC8
2.	10½	FN5
3.	2″	LC3
4.	1¾	LN2

11.11 SI METRIC SYSTEM—DEFINITION OF BASIC TERMS

Basic size is the primary size from which the tolerances are determined. The basic size is the same for both mating members of a fit. (See Figure 11-8 for illustrations of this and other terms used in the SI Metric system.)

Deviation is the difference between the maximum limit of size on mating parts, such as a hole and a shaft or a slot and a key, and the corresponding basic size.

Upper deviation is the difference between the maximum limit of size and the corresponding basic size.

Lower deviation is the difference between the minimum limit of size and the corresponding basic size.

Fundamental deviation is the deviation that is closest to the basic size. For the hole size, fundamental deviation is designated by the uppercase letter *H* in the tolerance symbol 40H7. For the shaft size, the fundamental deviation is represented by the lowercase letter *g* in the tolerance symbol 40g7.

Tolerance is the difference between the maximum and the minimum limits on a part.

Tolerance zone is the zone that represents the tolerance and its position in relation to the basic size.

International tolerance grade (IT) is a group of tolerances that vary depending upon the basic size but that provide the same relative level of accuracy within a given grade. The IT grade establishes the magnitude of the tolerance zone or the amount of part size variation allowed for internal and external dimensions. Tolerances are expressed in grade numbers such as 6(IT6) or 11(IT11).

Hole basis system is a system of fits where the minimum hole size is the basic diameter.

Shaft basis system is a system of fits where the maximum shaft size is the basic diameter. (Note: The terms "hole basis" and "shaft basis" correspond to the terms "basic hole" and "basic shaft" used in the English system of standard fits.)

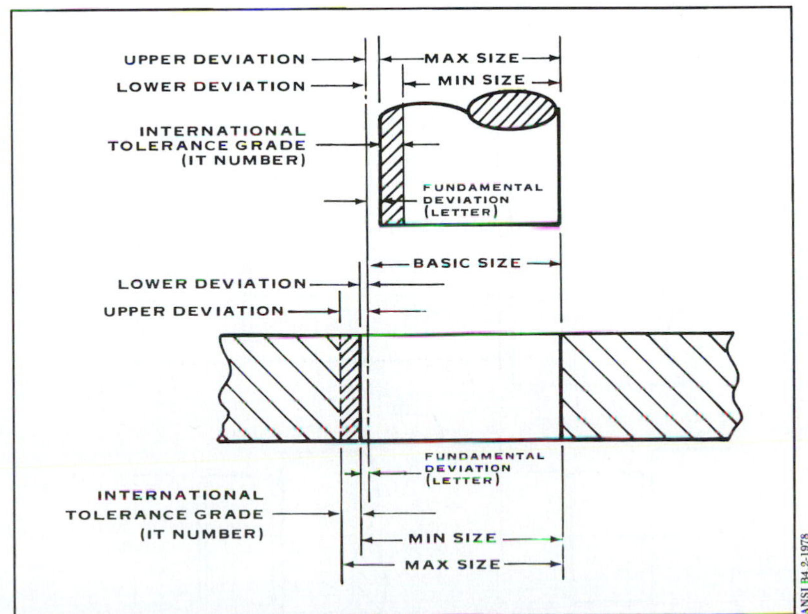

FIGURE 11-8 Preferred metric limits and fits ***(Section 11.11)***

11.12 FIT

As in the English system of standard fits, in the SI Metric system there are three general types of fits: clearance, transition, and interference. *Clearance* fit is the relationship between assembled parts when clearance occurs under all tolerance conditions. *Transition* fit is the relationship between assembled parts when either a clearance or an interference fit can result, depending upon the tolerance condition of the mating parts. *Interference* fit is the relationship between assembled parts when interference occurs under all tolerance conditions.

11.13 TOLERANCE SYMBOLS

Individual Part Limits

Figure 11-9(a) illustrates how a tolerance symbol is used to designate the diameter of a hole on a part: The diameter symbol (∅) is given first, followed by the basic size (25 mm), the fundamental deviation (uppercase letters are used for hole tolerances—H in this case), and, lastly, the IT grade (9). Taken together, the H and the 9 in the tolerance symbol represent the tolerance zone. At 11-9(b) the tolerance symbol for the shaft dimension is given in the following order: the diameter symbol (∅), the basic size (20 mm), the fundamental deviation (lowercase letters are used for the shaft tolerance—d in this case). In combination, d9 represents the tolerance zone. The figure 9 is the IT grade.

Alternate methods for specifying a tolerance on a drawing, shown at 11-9(c), may also be used. As before, the diameter symbol, the basic size, and the tolerance symbol are given. In addition, the upper and lower deviations (found in Appendix Tables A-40, A-41, A-42, and A-43) are given. Values in parentheses are given for reference only.

Mating Part Limits

The first component in the tolerance symbol for a fit, 40H8/f7 for example, specifies the basic size common to both components (40 mm). The next specifications (H8/f7) represent the symbols corresponding to each component. The internal part symbol (H8) should always precede the external part symbol (f7). Methods of computing the SI Metric system of fits and limits are covered in Section 11.17.

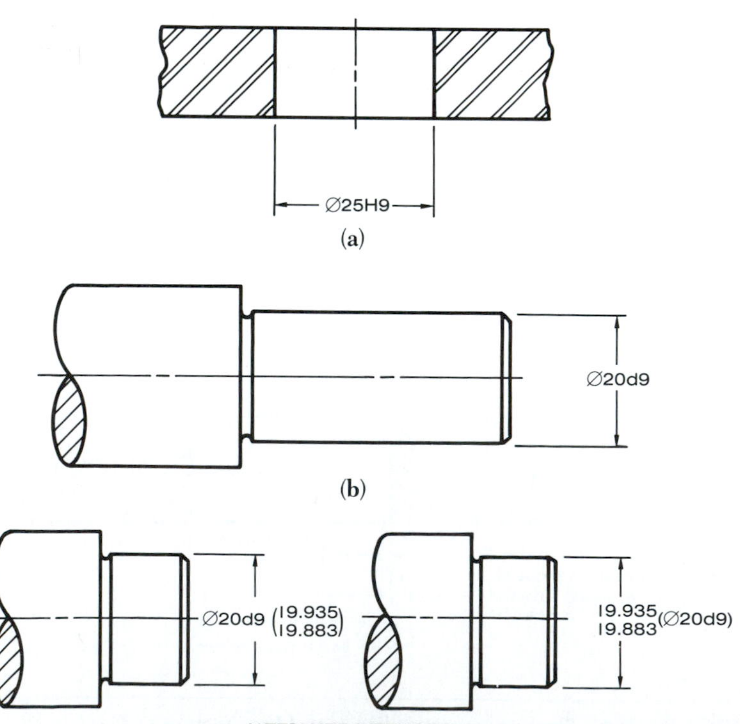

FIGURE 11-9 Tolerance symbols for shafts and holes *(Section 11.13)*

11.14 PREFERRED BASIC SIZES

Tolerances are computed using Table 11-1, which lists a series of preferred basic sizes. Where possible, you should select the basic size of mating parts from the "First Choice" column, since these values are based upon standard sizes of round, square, and hexagonal metal products. The values in the column entitled "First Choice" increase in increments by about 25% and values in the column entitled "Second Choice" increase by approximately 12%.

TABLE11-1 Preferred Basic Sizes

Basic size (mm)		*Basic size (mm)*		*Basic size (mm)*	
First Choice	*Second Choice*	*First Choice*	*Second Choice*	*First Choice*	*Second Choice*
1		10		100	
	1.1		11		110
1.2		12		120	
	1.4		14		140
1.6		16		160	
	1.8		18		180
2		20		200	
	2.2		22		220
2.5		25		250	
	2.8		28		280
3		30		300	
	3.5		35		350
4		40		400	
	4.5		45		450
5		50		500	
	5.5		55		550
6		60		600	
	7		70		700
8		80		800	
	9		90		900
				1000	

TABLE 11-2 Clearance, Transition, and Interference Fits

	ISO Symbol		Description	
	Hole Basis	*Shaft* Basis*		
Clearance fits	H11/c11	C11/h11	*Loose running* fit for wide commercial tolerances or allowances on external members.	More Clearance
	H9/d9	D9/h9	*Free running* fit not for use where accuracy is essential, but good for large temperature variations, high running speeds, or heavy journal pressures.	
	H8/f7	F8/h7	*Close running* fit for running on accurate machines and for accurate location at moderate speeds and journal pressures.	
	H7/g6	G7/h6	*Sliding* fit not intended to run freely, but to move and turn freely and locate accurately.	
	H7/h6	H7/h6	*Locational clearance* fit provides snug fit for locating stationary parts but can be freely assembled and disassembled.	
Transition fits	H7/k6	K7/h6	*Locational transition* fit for accurate location, a compromise between clearance and interference.	
	H7/n6	N7/h6	*Locational transition* fit for more accurate location where greater interference is permissible.	
Interference fits	H7/p6	P7/h6	*Locational interference* fit for parts requiring rigidity and alignment with prime accuracy of location but without special bore pressure requirements.	
	H7/s6	S7/h6	*Medium drive* fit for ordinary steel parts or shrink fits on light sections; the tightest fit usable with cast iron.	
	H7/u6	U7/h6	*Force* fit suitable for parts that can be highly stressed or for shrink fits where the heavy pressing forces required are impractical.	More Interference

*The transition and interference shaft basis fits shown do not convert to exactly the same hole basis fit conditions for basic sizes in range from 0 through 3 mm. Interference fit P7/h6 converts to a transition fit H7/p6 in the above size range.

11.15 PREFERRED FITS

The nature of clearance, transition, and interference fits are described in Table 11-2. Tolerance symbols for both the hole basis and shaft basis systems of fits are also given. Normally the hole basis system is preferred. However, when a standard diameter shaft is used to mate with several holes of different fits, the shaft basis system is recommended.

11.16 BASIS OF PREFERRED FITS

A series of preferred fits for the *hole basis system* is shown graphically in Figure 11-10. For all computations in this system, the *smallest* size of the hole is used as a basic size. The sizes in the table range from H11/c11 to a maximum interference fit of H7/u6.

Figure 11-11 graphically illustrates a series of preferred fits for the *shaft basis system*. In this system, the *largest* size of the shaft is used as the basic size. The sizes in this table range from a clearance fit of C11/h11 to an interference fit of U7/h6.

Note that the fits described in Figures 11-10 and 11-11 correspond to the fits given in Table 11-2.

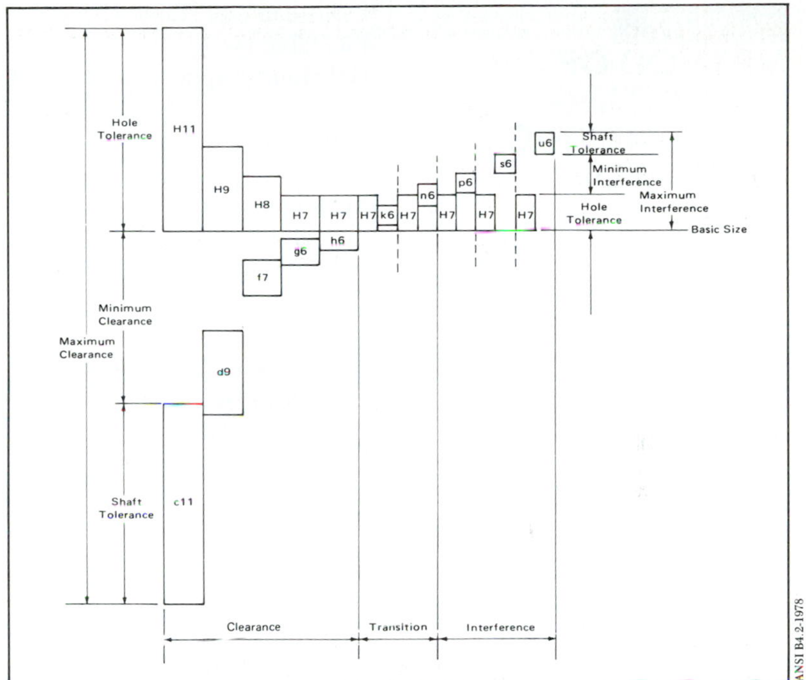

FIGURE 11-10 Preferred hole basis fits *(Section 11.16)*

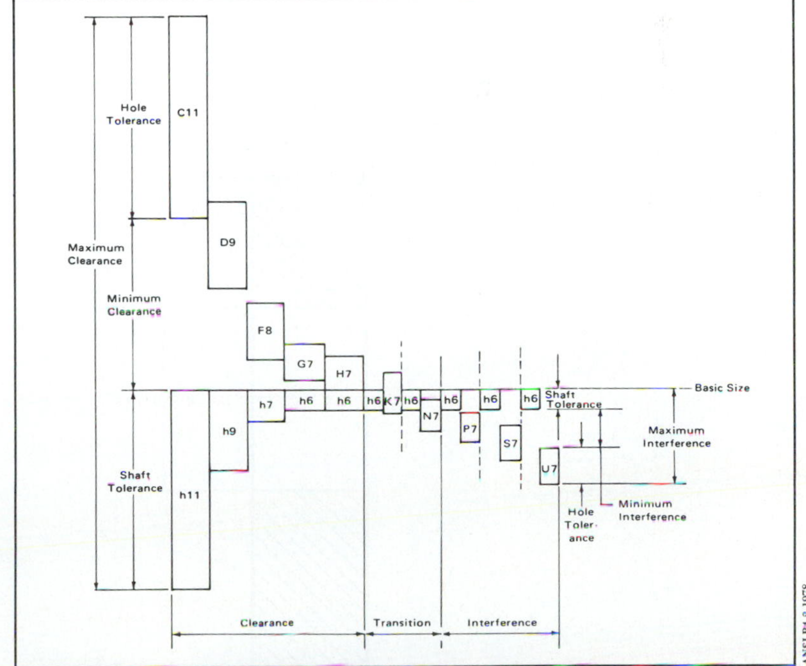

FIGURE 11-11 Preferred shaft basis fits *(Section 11.16)*

11.17 METHOD OF COMPUTING THE SI METRIC SYSTEM OF FITS AND LIMITS

Figure 11-12 illustrates two examples of the hole basis system. At 11-12(a) a shaft is to be assembled in a hole on a part with a loose running fit:

Basic diameter (Table 11-1)	=	20 mm
Fit (Table 11-2)	=	H11/c11
Hole limits (Appendix Table A-40)	=	20.130
		20.000
Shaft limits (Appendix Table A-40)	=	19.890
		19.760

At 11-12(b) a 50 mm diameter shaft is to be assembled in a hole on a part with a medium drive fit:

Basic diameter (Table 11-1)	=	50 mm
Fit (Table 11-2)	=	H7/s6
Hole limits (Appendix Table A-41)	=	50.025
		50.000
Shaft limits (Appendix Table A-41)	=	50.059
		50.043

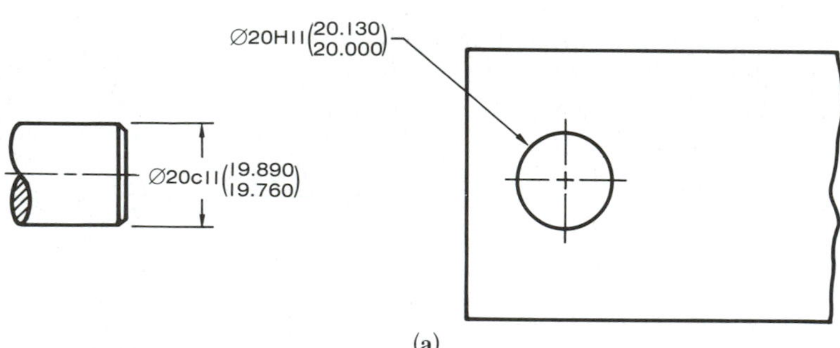

(a)

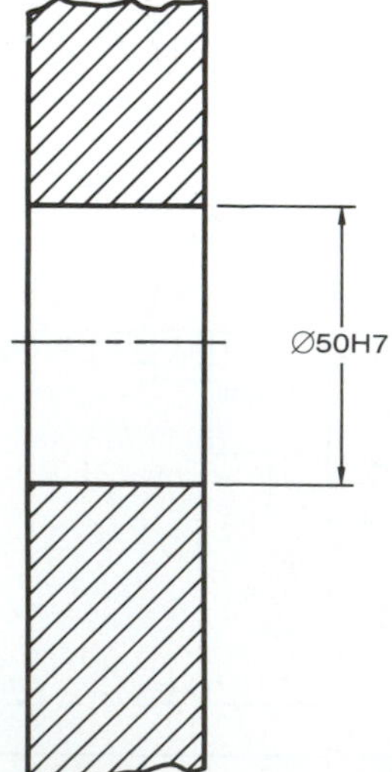

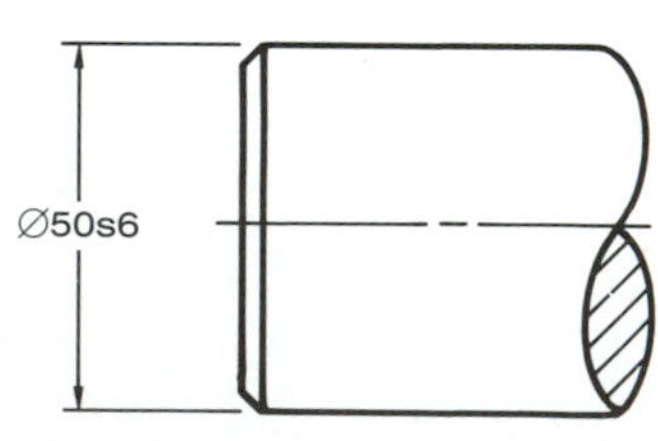

(b)

FIGURE 11-12 Example of the hole basis system ***(Section 11.17)***

Figure 11-13 illustrates two examples of the shaft basis system. At 11-13(a) a 76 mm diameter shaft is to be assembled in a hole on a part with a sliding fit:

Basic diameter (Table 11-1)	=	80 mm
Fit (Table 11-2)	=	G7/h6
Hole limits (Appendix Table A-42)	=	80.040
		80.010
Shaft limits (Appendix Table A-42)	=	80.000
		79.981

At 11-13(b) a 49 mm diameter shaft is to be assembled in a hole on a part with a force fit:

Basic diameter (Table 11-1)	=	50 mm
Fit (Table 11-2)	=	U7/h6
Hole limits (Appendix Table A-43)	=	49.939
		49.914
Shaft limits (Appendix Table A-43)	=	50.000
		49.984

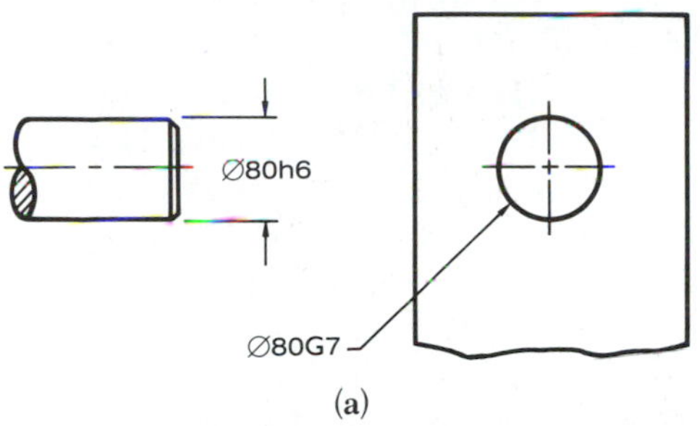

(a)

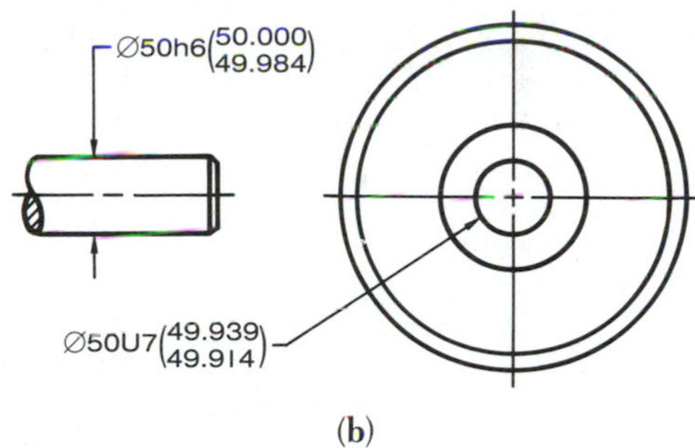

(b)

FIGURE 11-13 Example of the shaft basis system ***(Section 11.17)***

Two examples of tolerance symbols are given in Figure 11-13. Both methods are acceptable.

> **CHECKPOINT 4**
> **FINDING LIMITS USING THE SI METRIC SYSTEM OF FITS AND LIMITS**
>
> Using the given set of values, find the limits for a mating hole and shaft. It will be necessary to use Tables 11-1 and 11-2 and the appropriate appendix tables.
>
> *Hole basis system*
>
> **1.** Close running fit, basic diameter = 3.1 mm
> **2.** Medium drive fit, basic diameter = 103 mm
>
> *Shaft basis system*
>
> **3.** Locational clearance fit, basic diameter = 78.5 mm
> **4.** Free running fit, basic diameter = 21 mm

11.18 ACCUMULATION OF TOLERANCES

In Section 10.25 various methods of dimensioning the location of holes are discussed. The use of datum planes and datum holes are illustrated in Figure 11-39. Dimensioning from datum planes is recommended to prevent the accumulation of tolerances.

Chain Dimensioning

The maximum variation between two features on a part is equal to the sum of the tolerances on the intermediate distances. This condition results in the *greatest* tolerance accumulation. In Figure 11-14(a) the tolerance accumulation between the hole axis at X and surface Y is ±0.15, or 0.05 + 0.05 + 0.05 = ±0.15.

Baseline Dimensioning

The maximum variation between two features on a part is equal to the sum of the tolerances on the two dimensions from the origin to the features. This improved method of dimensioning results in a reduction of the tolerance accumulation. In Figure 11-14(b) the tolerance accumulation between the hole axis at X and surface Y is ±0.1, or 0.05 + 0.05 = ±0.10.

Direct Dimensioning

The maximum variation between two features on a part is controlled by the tolerance on the dimension between the respective features. This condition results in the *least* tolerance accumulation. In Figure 11-14(c) the tolerance between surfaces X and Y is ±0.05.

The dimension origin symbol, used to indicate that a toleranced dimension between two features on a part originates from one of these features, is shown in Figure 11-15.

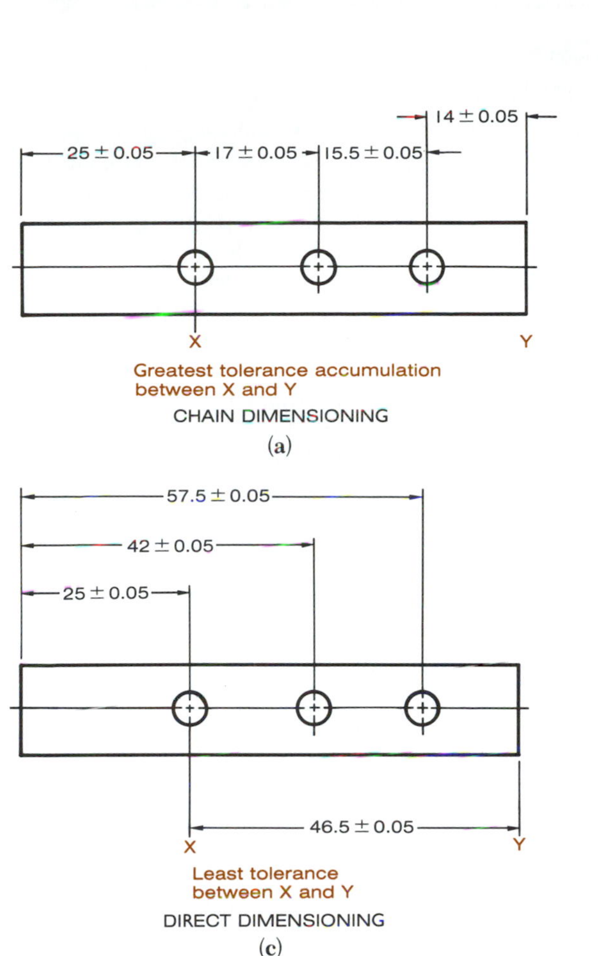

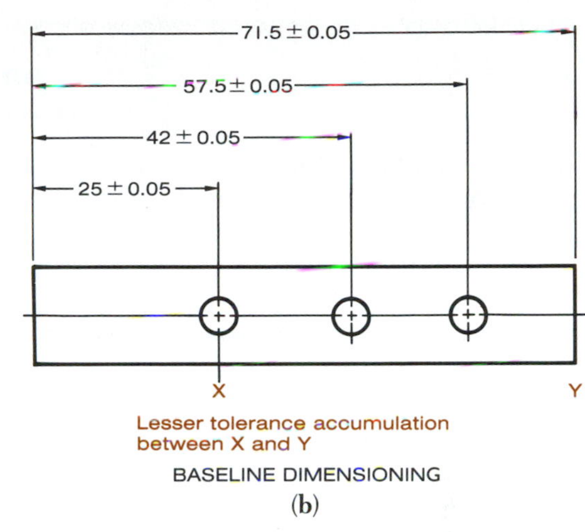

FIGURE 11-14 **Accumulation of tolerances** ***(Section 11.18)***

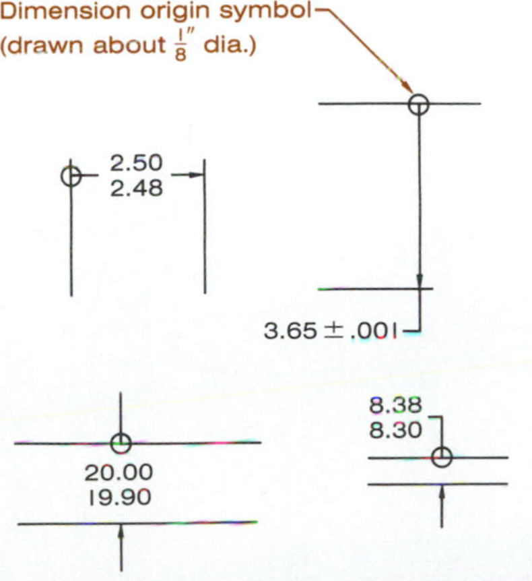

FIGURE 11-15 **Dimension origin symbol** ***(Section 11.18)***

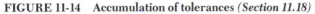

PROBLEMS

Limit Dimensioning

Using the basic hole system, compute the limits for a hole and a mating shaft using the following data:

	Nominal size	Tolerance		Allowance
		Hole	Shaft	
P11.1	⅝	.0005	.001	+.001
P11.2	1⅞	.0007	.0005	−.0015
P11.3	2³⁄₁₆	.0005	.0005	+.0003
P11.4	3¼	.0005	.001	+.0015
P11.5	1⅛	.0008	.0008	−.0013
P11.6	1½	.0008	.0025	−.0021
P11.7	¾	.001	.001	+.0013
P11.8	2⁹⁄₁₆	.0007	.0005	−.0006
P11.9	2″	.0015	.0015	+.0015
P11.10	⅞	.003	.003	+.002

Using the basic shaft system, compute the limits for a hole and a mating shaft using the following data:

	Nominal size	Tolerance		Allowance
		Hole	Shaft	
P11.11	1″	.0004	.0005	+.0006
P11.12	3⁷⁄₁₆	.0002	.001	−.0015
P11.13	1¹¹⁄₁₆	.0005	.0008	+.0005
P11.14	2½	.0008	.0007	−.001
P11.15	5⅝	.0005	.001	+.001

Using the set of values that follow and the appropriate appendix tables, compute the limits for the mating parts:

	Nominal size	Basic size	Type of fit
P11.16	⁹⁄₁₆	—	RC8
P11.17	3⅞	—	FN5
P11.18	—	5.763	LT1
P11.19	—	1.895	LN2
P11.20	2½	—	LC2
P11.21	1³⁄₁₆	—	RC5
P11.22	¹⁵⁄₁₆	—	LC5
P11.23	—	2.395	LN3
P11.24	—	8.555	FN2
P11.25	—	1.015	RC1

Using the following set of values, find the upper and lower limits for a mating hole and shaft (use Tables 11-1 and 11-2 and the appropriate appendix tables):

Hole basis system:

P11.26 Free running fit, basic diameter = 8.5 mm

P11.27 Locational interference fit, basic diameter = 36 mm

P11.28 Locational interference fit, basic diameter = 105 mm

P11.29 Loose running fit, basic diameter = 2.5 mm

P11.30 Medium drive fit, basic diameter = 52.8 mm

Shaft basis system:

P11.31 Medium drive fit, basic diameter = 25 mm

P11.32 Close running fit, basic diameter = 5 mm

P11.33 Sliding fit, basic diameter = 85 mm

P11.34 Force fit, basic diameter = 85 mm

P11.35 Free running fit, basic diameter = 11.5 mm

12 Geometric Tolerancing

OBJECTIVES

After completing this chapter, you should have gained the following ability:

To understand the techniques for specifying tolerance of position and tolerance of form and the corresponding methods of expressing each concept on an engineering drawing.

Geometric tolerancing, which first came into use in the late 1950s, is a precise technique for specifying the maximum amount of variation that can be permitted in position or form from the true geometry of a part. Geometric tolerancing consists of a series of well established techniques that are used for controlling such geometrical characteristics on parts as straightness, flatness, cylindricity, angularity, and so forth. The principal application of this system is in meeting functional and interchangeability requirements of manufactured parts. Geometric tolerancing has been perfected to the point where it is now widely adopted by companies both in the United States and internationally. The system for specifying tolerances of position and form of geometrical features on parts has been standardized by ANSI Y14.5M Standards, ISO R1101 (International Organization Standardization), and the Military Standards (MIL-Std) of the U.S. Department of Defense.

It is not necessary to use geometric tolerancing for every feature on a part drawing. Geometric tolerancing should only be used when there is a doubt regarding the adequacy of the manufacturing process and equipment to produce a given feature on a part within the desired limits of accuracy. This precise system of engineering tolerancing is most often used to control sizes on parts within closer limits than might ordinarily be required from a manufacturing process or for parts of such size or shape where bending or other distortion is likely to occur.

This modern system of engineering tolerancing may be conveniently divided into two main areas: tolerance of position and tolerance of form. *Tolerance of position* involves the specification of dimensions that control the desired function of a part with regard to various features such as holes, slots, or other cuts. *Tolerance of form* deals with controlling the form or shape of the geometrical characteristics of a part. Tolerances of form refer to specifications that apply to flatness, straightness, roundness, parallelism, and so on.

12.1 TOLERANCES OF POSITION

A *positional* tolerance defines a zone within which the center point, the axis, or the center plane of a feature of size is permitted to vary from the true or theoretically exact position. Basic dimensions are used to establish the true position from specified datum features and between interrelated features.

To understand the need for positional tolerancing in the manufacture of interchangeable parts, study the following example: The part shown in Figure 12-1(a) has coordinate dimensions that result in a square tolerance zone around the hole center position, as shown in 12-1(b). Each coordinate dimension has a tolerance of .002 inch. In this example the center of the ½-inch diameter hole can be located *anywhere* within the .002-inch square and the maximum variation will occur along the diagonal of the square. As shown at 12-1(c), an incorrect tolerance of .0028 inch is obtained ($1.4 \times .002 = .0028$). To control this difficulty, the preferred method of positional tolerancing discussed in the following sections has been developed.

12.2 METHOD OF EXPRESSING POSITIONAL TOLERANCES

Feature Control Frame

A geometric tolerance for an individual feature is specified as shown in Figure 12-2. It is shown actual size. The rectangular frame is divided into compartments containing a geometric characteristic symbol followed by a tolerance value. Where applicable, the tolerance is preceded by the diameter symbol.

Geometric Characteristic Symbols

The symbols denoting geometric characteristics, shown in Figure 12-3, are used to express positional and form tolerances. Tolerances of form are discussed in later sections in this chapter.

Material Condition Symbol

The material condition symbol is shown in the frame in Figure 12-4 following the tolerance value. One of three symbols may be used in positional tolerancing: M = at maximum material condition (MMC); S = regardless of feature size (RFS); and L = at least material condition (LMC). The implications of MMC, RFS, and LMC will be covered in more detail in Sections 12.4, 12.5, and 12.6.

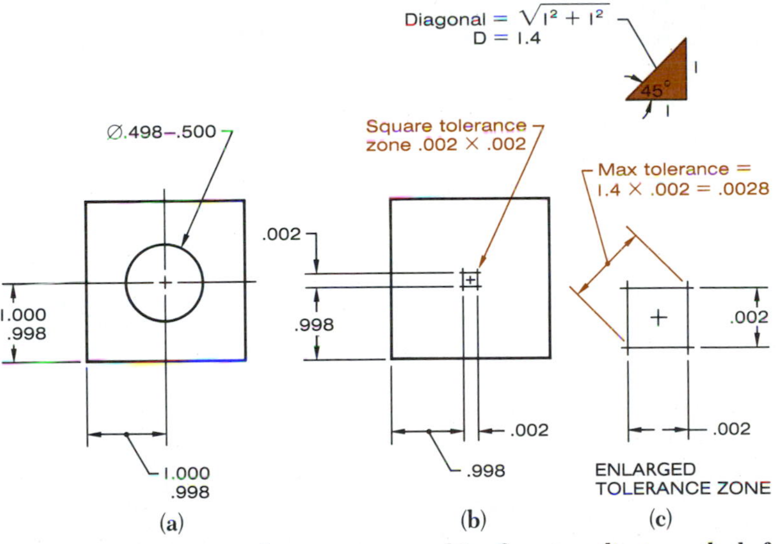

FIGURE 12-1 Square tolerance zone resulting from coordinate method of locating a hole position *(Section 12.1)*

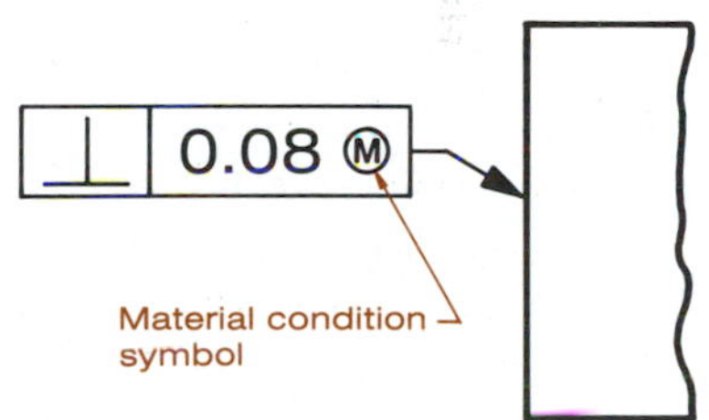

FIGURE 12-4 Material condition symbol applied to a feature *(Section 12.2)*

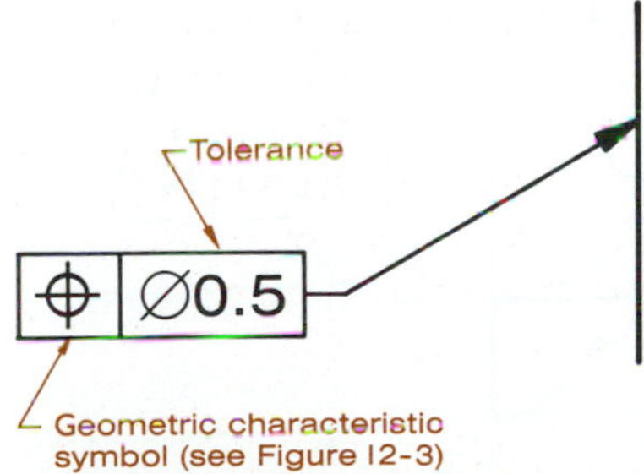

FIGURE 12-2 Feature control frame *(Section 12.2)*

CHARACTERISTIC	SYMBOL
STRAIGHTNESS	—
FLATNESS	▱
CIRCULARITY (ROUNDNESS)	○
CYLINDRICITY	⌭
PROFILE OF A LINE	⌒
PROFILE OF A SURFACE	⌓
ANGULARITY	∠
PERPENDICULARITY	⊥
PARALLELISM	//
POSITION	⌖
CONCENTRICITY	◎
CIRCULAR RUNOUT	↗*
TOTAL RUNOUT	⌰*

ANSI Y14.5M-1982

FIGURE 12-3 Geometric characteristic symbols *(Section 12.2)*

Datum Reference Letters

Letters, except I, O, and Q, are used as datum identifying letters. A different letter should be used to identify each datum feature. Where a datum is established by two datum features (an axis established by two datum diameters, for example), both datum letters, separated by a dash, are entered in a single compartment, as shown in Figure 12-5(a). When more than one datum is required, as at 12-5(b) and 12-5(c), the datum letters are entered in separate compartments in the desired order of preference from left to right. Datum reference letters need not be in alphabetical order in the feature control frame. The method of specifying datums is explained in Section 12.19.

Primary

0.08 | A–B

(a)

Secondary

0.05 | A | B

(b)

Tertiary

0.01 | A | B | C

(c)

FIGURE 12-5 Control frame ***(Section 12.2)***

Feature Control Placement

There are four acceptable methods of placing feature control symbols on the views of a detail drawing:

1. Locate the feature control frame below or attach it to a note pertaining to the feature, as shown in Figure 12-6(a).
2. Run a leader from the frame to the feature, as shown in 12-6(b).
3. Attach a side or an end of the frame to an extension of a dimension line, as shown in 12-6(c).
4. Attach a side or an end of the frame to an extension line from the feature (provided it is a plane surface), as shown in 12-6(d).

Datum Feature Symbol

The datum feature symbol consists of an identifying letter preceded and followed by a dash. The letter is placed in a frame as shown in Figures 12-6(a), 12-6(b), and 12-6(c).

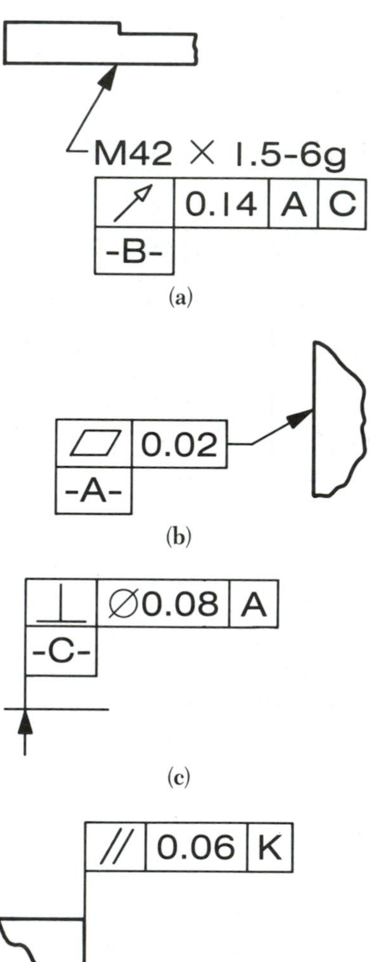

FIGURE 12-6 Datum reference letters ***(Section 12.2)***

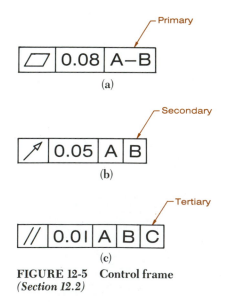

12.3 APPLICATION OF A POSITIONAL TOLERANCE

In Figure 12-1 the use of coordinate dimensions created a maximum error of .0008 inch (.0028 − .002), which resulted in a greater variation than the designer originally intended. Figure 12-7(a) illustrates how the method of positional tolerancing removes this inaccuracy. Note the use of datum letters A and B, which establish reference surfaces. The datum identification letters (preceded and followed by a dash) are placed in a rectangular frame at the end of the appropriate extension line. A dimension enclosed in a frame or a box is the symbol in positional tolerancing that denotes the basic, or the theoretically exact, size. The note for the hole is the same as that in Figure 12-1, but the feature control frame is placed below the note.

The positional tolerance symbol is given in the first compartment, followed by the diameter symbol and the required tolerance. Next, the material condition symbol is given. This particular symbol indicates "at maximum material condition" (see Section 12.4). Finally, the two reference datum letter symbols are given. In this example, instead of a square tolerance zone (as before), a *circular* tolerance zone is created, with a diameter of .002 inch, whose center is located by the two basic dimensions. A circular tolerance zone, illustrated at (b), assumes a .002-inch tolerance in all directions without detrimental effect upon the location of the feature. The axis of the hole at all points must lie within the specified circular tolerance zone. The center location of the hole is controlled by the basic dimensions.

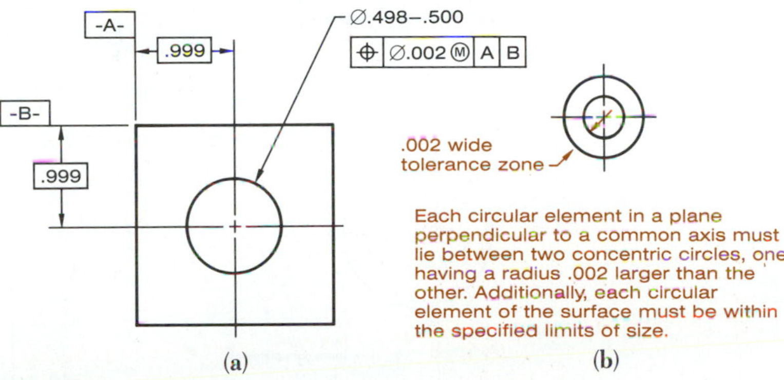
-A-
.999
-B-
.999
Ø.498–.500
Ø.002 Ⓜ A B
.002 wide
tolerance zone
Each circular element in a plane perpendicular to a common axis must lie between two concentric circles, one having a radius .002 larger than the other. Additionally, each circular element of the surface must be within the specified limits of size.
(a)
(b)

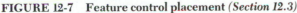

FIGURE 12-7 Feature control placement ***(Section 12.3)***

12.4 THE MMC PRINCIPLE

The positional tolerance and *maximum material condition* (MMC) of mating features must be considered in relation to each other. MMC by itself means a feature contains the maximum amount of material permitted by the toleranced size dimension for that feature. Thus for holes, slots, and other internal features, maximum material is the condition where these features are at their *minimum* allowable size. For shafts as well as for bosses (cylindrical projections usually associated with forgings or castings), lugs, tabs, and other external features, maximum material is the condition where these are at *maximum* allowable sizes.

A positional tolerance applied at MMC may be explained in either of the following ways:

(a) In terms of the *surface* of a hole: While maintaining the specified limits of a hole, no element of the hole surface shall be inside a theoretical boundary located at true position, as shown in Figure 12-8.

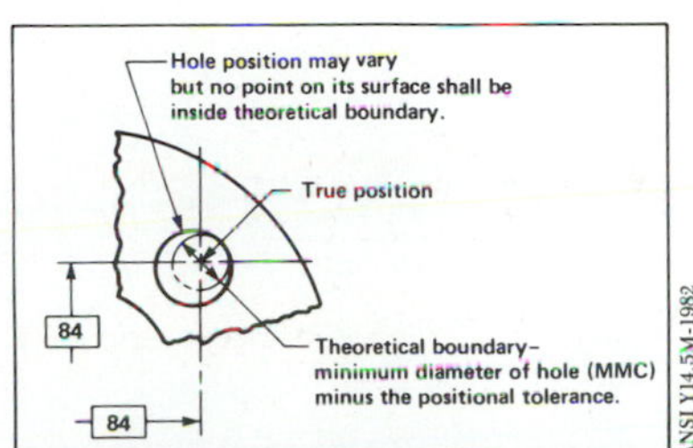
Hole position may vary but no point on its surface shall be inside theoretical boundary.
True position
Theoretical boundary–minimum diameter of hole (MMC) minus the positional tolerance.
84
84
ANSI Y14.5M-1982

FIGURE 12-8 Positional tolerancing—boundary for surface of hole at MMC ***(Section 12.4)***

(b) In terms of the *axis* of a hole: Where a hole is at MMC (minimum diameter), its axis must fall within a cylindrical tolerance zone whose axis is located at true position. The diameter of this zone is equal to the positional tolerance as shown in Figures 12-9(a) and 12-9(b). This tolerance zone also defines the limits of variation in the attitude of the axis of the hole in relation to the datum surface as shown at 12-9(c).

It is only when a feature is at MMC that the specified positional tolerance applies. Where the actual size of the feature is larger than MMC, additional positive tolerance results, as shown in Figure 12-10. This increase in positional tolerance is equal to the difference between the specified maximum material limit of size (MMC) and the actual size of the feature. Where the actual size is larger than MMC, the specified positional tolerance for a feature may be exceeded and still satisfy function and interchangeability requirements.

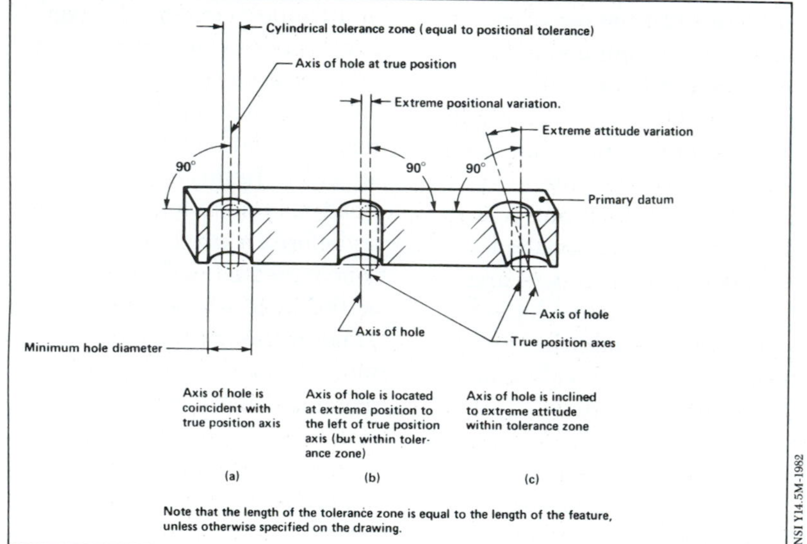

FIGURE 12-9 Hole axes in relation to positional tolerance zones ***(Section 12.4)***

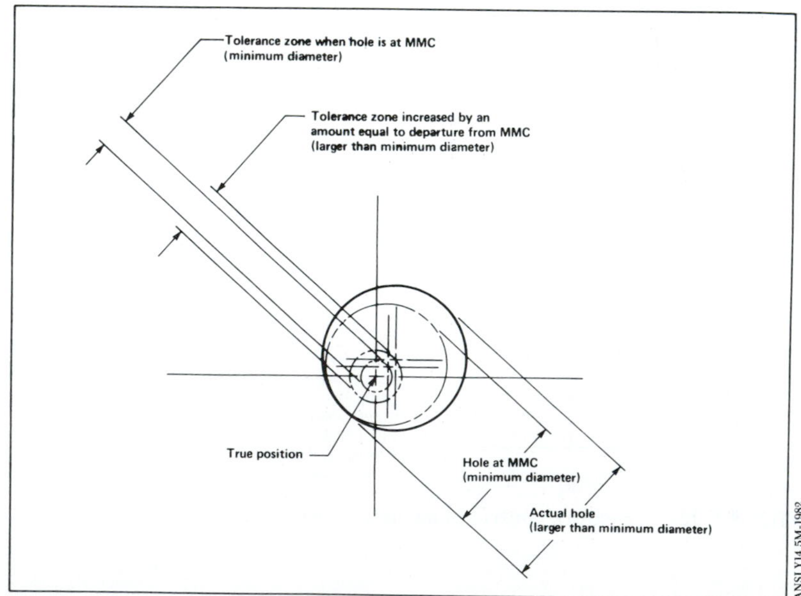

FIGURE 12-10 Increase in positional tolerance where hole is not at MMC ***(Section 12.4)***

12.5 RFS AS RELATED TO POSITIONAL TOLERANCING

In certain cases the design or function of a part may require the positional tolerance or datum reference, or both, to be maintained regardless of actual feature sizes. The term *regardless of feature size* (RFS), where applied to the positional tolerance of circular features, requires the axis of each feature to be located within the specified positional tolerance regardless of the size of the feature. This requirement imposes a closer control of the features involved and introduces complexities in verification.

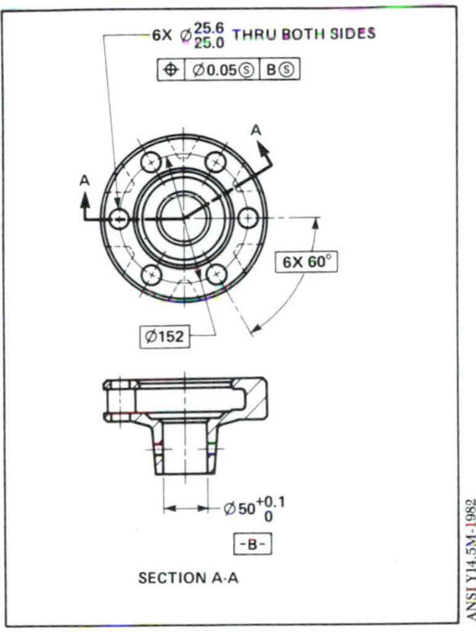

FIGURE 12-11 RFS applied to a feature and a datum ***(Section 12.5)***

In Figure 12-11 the equally spaced holes may vary in size from 25.6 mm to 25.0 mm in diameter. Each hole must be located within the specified positional tolerance regardless of the size of that hole. A hole that is at least material condition (LMC = 25.6 mm diameter) is as accurately located as a hole that is at maximum material condition (MMC = 25.0 mm diameter). The RFS positional control is more restrictive than the MMC principle.

The functional requirements of some designs may require that RFS be applied to both the hole pattern and the datum feature. That is, it may be necessary to require the axis of an actual datum feature (such as datum hole diameter B in Figure 12-11) to be the datum axis for the holes in the pattern regardless of the size of the datum feature. The RFS application does not permit any shift between the axis of the datum feature and the pattern of features, as a group, where the datum feature departs from MMC.

12.6 LMC AS RELATED TO POSITIONAL TOLERANCING

Where positional tolerancing at *least material condition* (LMC) is specified, the feature on the part contains the minimum amount of material permitted by its toleranced size dimension. Specification of LMC further requires perfect form at LMC. Where the feature departs from its LMC size, an increase in positional tolerance equal to the amount of such departure is allowed. This concept is shown in Figure 12-12. Specifying LMC is limited to positional tolerancing applications where MMC does not provide the desired control and RFS is too restrictive (see Figures 12-13 through 12-15). LMC is used to maintain a desired relationship between the surface of a feature and its true position at tolerance extremes. Considerations that are affected by the design of the part are usually involved in specifying LMC.

Figure 12-13 illustrates a boss (a cylindrical projection) and hole combination located by basic dimensions. The wall thickness is minimum where the boss and hole are at their LMC sizes and both features are displaced in opposite extremes. Since positional tolerances are specified on an LMC basis, as each feature departs from LMC the wall thickness increases. This permits a corresponding increase in the positional tolerance, thus maintaining the desired material thickness between the surfaces.

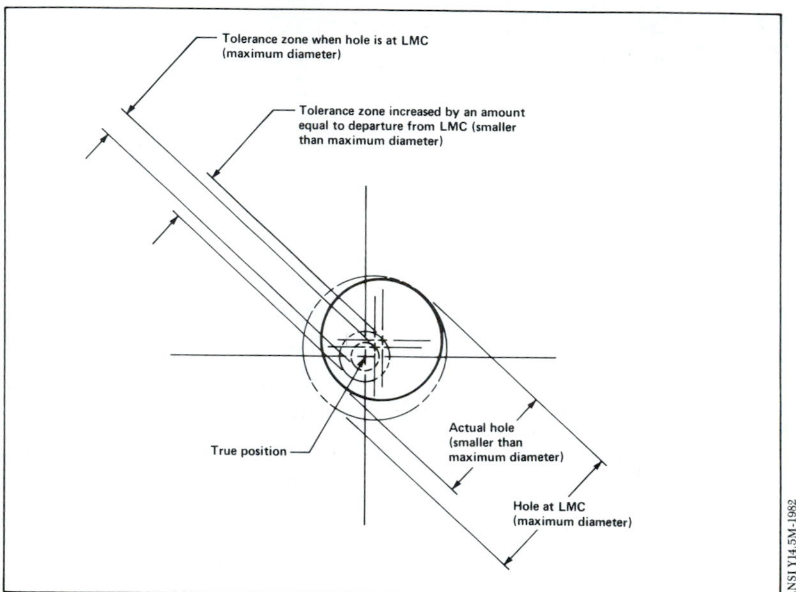

FIGURE 12-12 Increase in positional tolerance where hole is not at LMC ***(Section 12.6)***

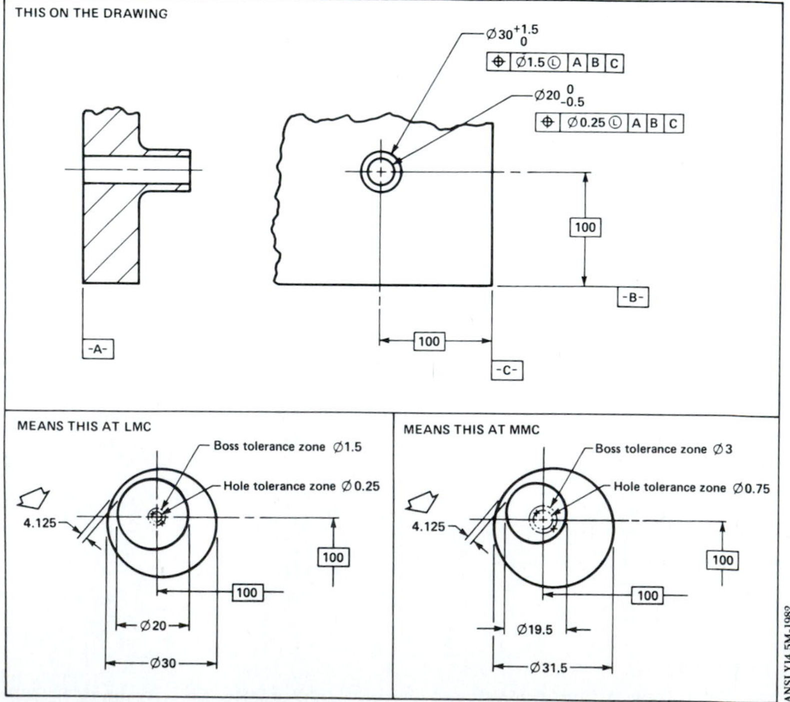

FIGURE 12-13 LMC applied to a boss and a hole ***(Section 12.6)***

In Figure 12-14 a radial pattern of slots is located relative to an end face and a center hole. LMC is specified to maintain the desired relationship between the side surfaces of the slots and the true position, where rotational alignment with the mating part may be critical.

LMC may also be applied to single features such as the hole shown in Figure 12-15. In this example the position of the hole relative to the inside web is critical. RFS can be specified. However, LMC is applied, which permits an increase in the positional tolerance in satisfying design considerations.

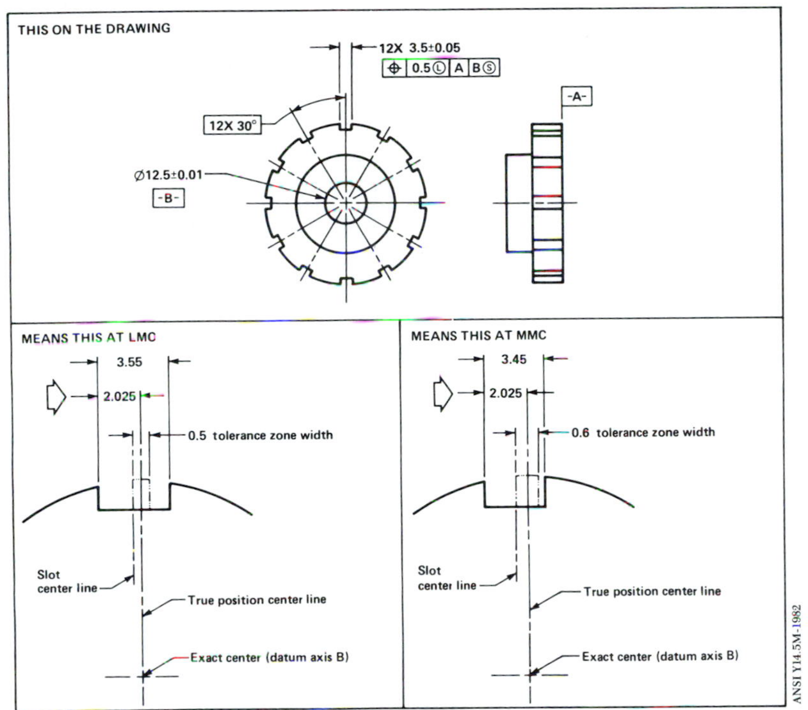

FIGURE 12-14 LMC applied to a pattern of slots *(Section 12.6)*

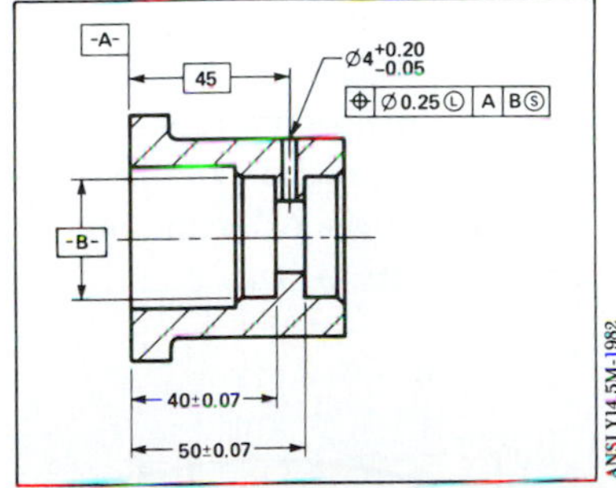

FIGURE 12-15 LMC applied to a single feature *(Section 12.6)*

12.7 CONCENTRICITY

Concentricity is a condition in which the axes of all cross-sectional elements of a surface of revolution share a common axis with the datum feature. A concentric tolerance, shown in Figure 12-16, specifies a cylindrical tolerance zone whose axis coincides with a datum axis and within which all cross-sectional

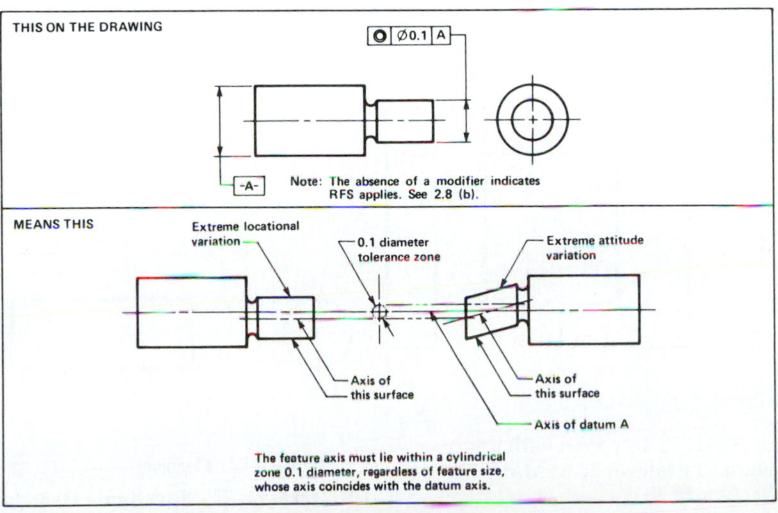

FIGURE 12-16 Concentricity tolerancing for coaxiality *(Section 12.7)*

axes of the feature being controlled must lie. Figure 12-17 illustrates another common example of a part shape, a hollow cylinder, wherein the diameters are to be concentric. In this example the center hole is designated as the datum feature.

12.8 SYMMETRY

Symmetry is a condition in which one or more features of a part are symmetrically arranged about the center plane of the part. A symmetrical tolerance specifies the width of a tolerance zone and indicates that the true position of open end slots, notches, tabs, and elongated holes, for example, are symmetrical. In Figure 12-18 the symmetrical relationship of the slot is controlled by specifying a positional tolerance at MMC. The control of symmetry for noncircular features serves the same purpose on parts as concentricity serves for circular features.

12.9 TOLERANCES OF FORM

Tolerance of form refers to the control of the geometrical characteristics of such features on a part as straightness, flatness, roundness, and so on. A form tolerance specifies a zone within which the line elements, the axes, or the center plane of a feature must be contained. Form tolerances are applicable to single features or elements of single features and, therefore, are not related to datums.

12.10 STRAIGHTNESS TOLERANCE

Straightness is a term that is used to describe a surface in which all of the elements are straight lines. A straightness tolerance specifies a tolerance zone within which the elements or axes must lie. As with positional tolerancing, a feature control frame is used. As shown in Figure 12-19, the straightness tolerance (0.15 mm) must be less than the size tolerance (0.20 mm). That is, no part of the cylindrical surface may lie outside the limits of size.

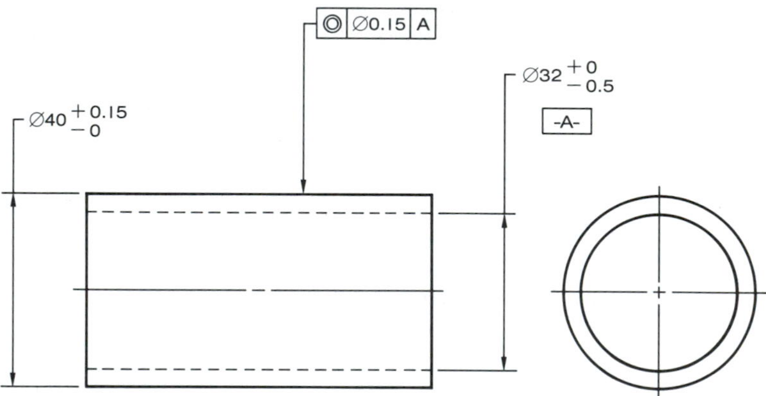

FIGURE 12-17 A concentricity tolerance ***(Section 12.7)***

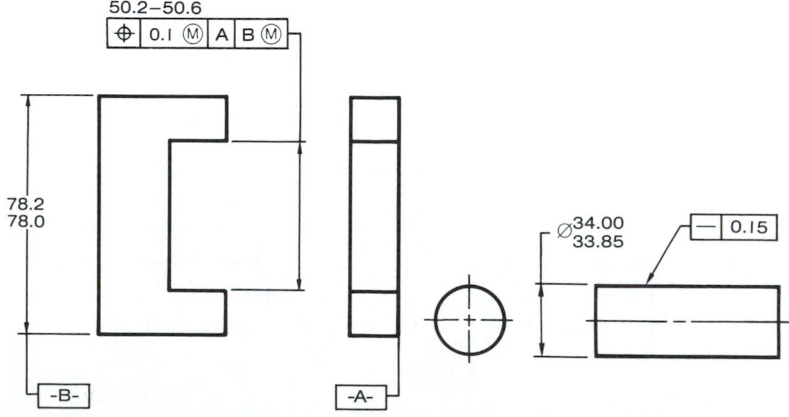

FIGURE 12-18 Part with a symmetry tolerance at MMC ***(Section 12.8)***

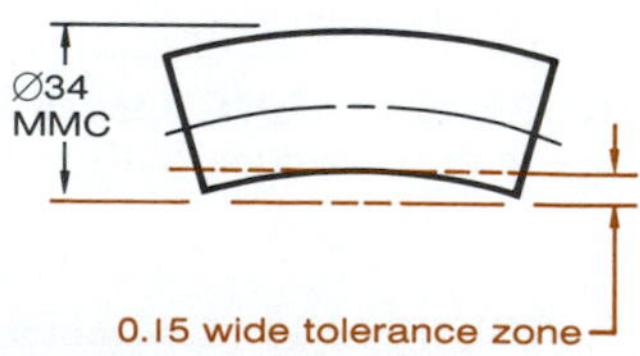

FIGURE 12-19 Specifying straightness ***(Section 12.10)***

12.11 FLATNESS TOLERANCE

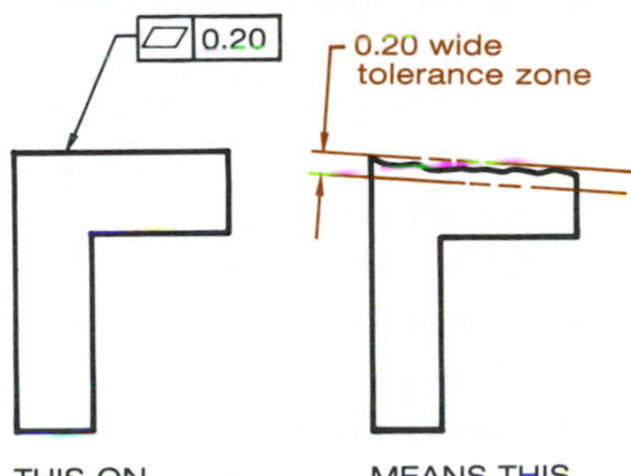

FIGURE 12-20 Specifying flatness *(Section 12.11)*

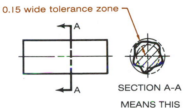

Flatness is a condition of a surface when all of the surface elements lie in one plane. A flatness tolerance specifies a tolerance zone defined by two parallel planes within which the surface must lie. The feature control frame must be directed to the surface or to an extension line of the surface, as shown in Figure 12-20. The flatness tolerance (0.20 mm) must be less than the size tolerance.

12.12 CIRCULARITY OR ROUNDNESS TOLERANCE

Circularity and *roundness* refer to the condition of a surface of revolution where, for a cylinder or a cone, all points of the surface intersected by a plane perpendicular to a common axis are equidistant from the axis. For a sphere, all points of the surface intersected by a plane passing through a common center are equidistant from that center. A circularity tolerance specifies a tolerance zone bounded by two concentric circles within which each circular element of that surface must lie. Figure 12-21 illustrates the method of specifying circularity for a cylinder and a cone. Figure 12-22 shows how to specify circularity for a sphere. The circular tolerance must be less than the size tolerance except with parts subject to a condition known as "free state variation," which refers to distortion of a part after removal of forces applied during manufacturing.

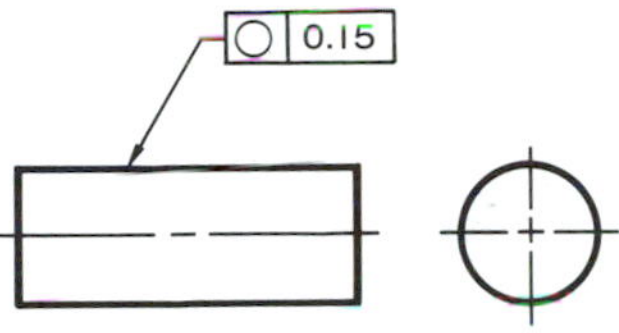

(a)

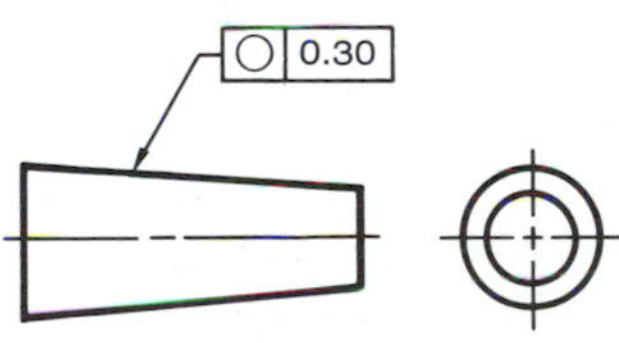
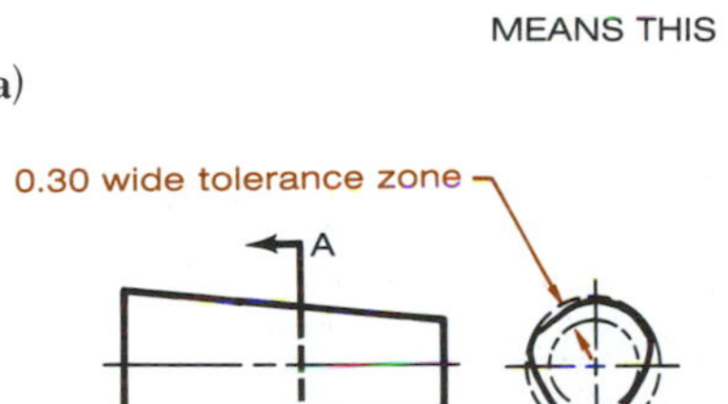

(b)

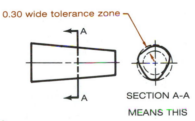

FIGURE 12-21 Specifying circularity for a cylinder and a cone *(Section 12.12)*

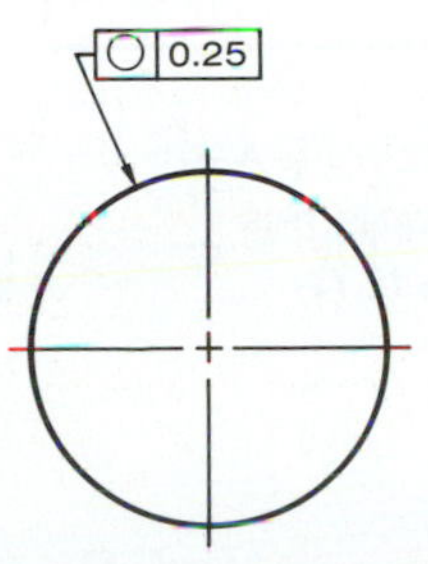

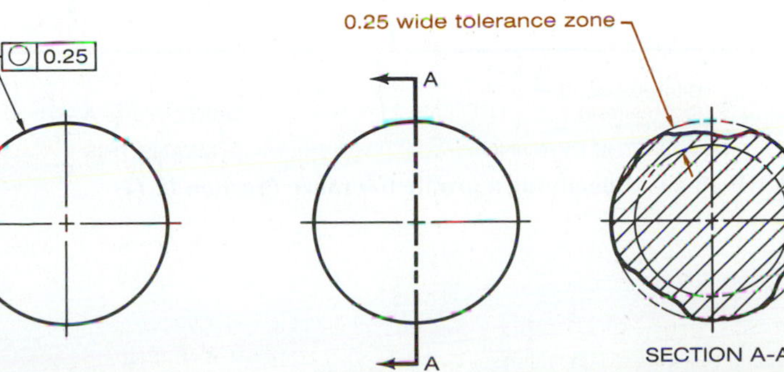

FIGURE 12-22 Specifying circularity for a sphere *(Section 12.12)*

12.13 CYLINDRICITY TOLERANCE

Cylindricity is a condition of a surface of revolution in which all points of the surface are equidistant from a common axis. As shown in Figure 12-23, a cylindricity tolerance specifies a tolerance zone bounded by two concentric cylinders within which the surface must lie. Unlike that of circularity, the cylindricity tolerance applies simultaneously to both circular and longitudinal elements of the entire surface. The cylindricity tolerance must be less than the size tolerance.

12.14 PROFILE TOLERANCE

The *profile* tolerance specifies a uniform boundary along the true profile within which the elements of the surface must lie. It is used to control form or a combination of form, size, and orientation. The contour may consist of an irregular curve, tangent arcs, or a combination of curves and straight lines. Points on the curve of an irregular outline are located by coordinate dimensions, as shown in Figure 10-37(a). A profile tolerance for a basic contour can be specified as shown in Figure 12-24. See ANSI Y14.5M-1982 for other applications of profile tolerancing.

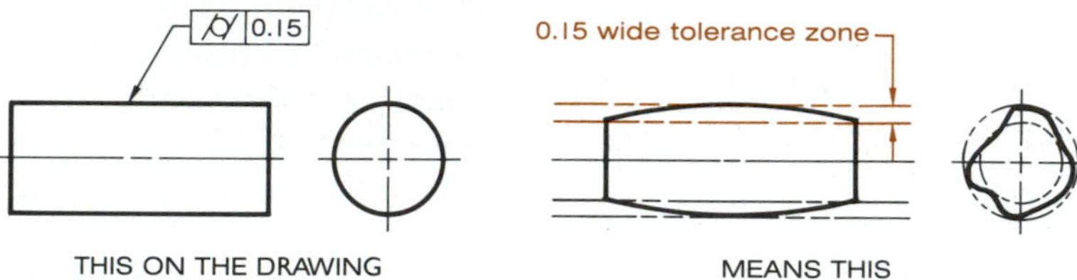

FIGURE 12-23 Specifying cylindricity *(Section 12.13)*

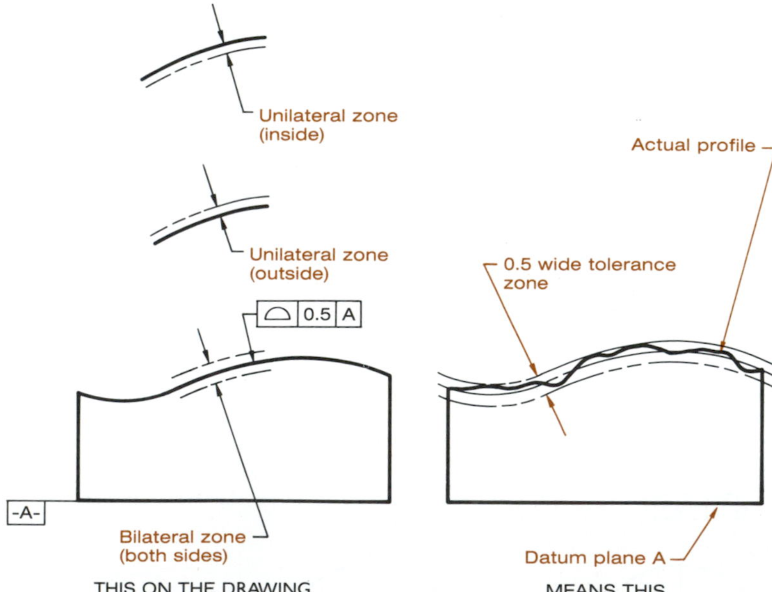

FIGURE 12-24 Specifying a profile tolerance *(Section 12.14)*

12.15 ANGULARITY TOLERANCE

Angularity is the condition of a surface or an axis at a specified angle (other than 90°) from a datum plane or axis. As shown in Figure 12-25, the angularity tolerance specifies a tolerance zone defined by two parallel planes at the specified basic angle from a datum plane or axis within which the surface of the feature must lie.

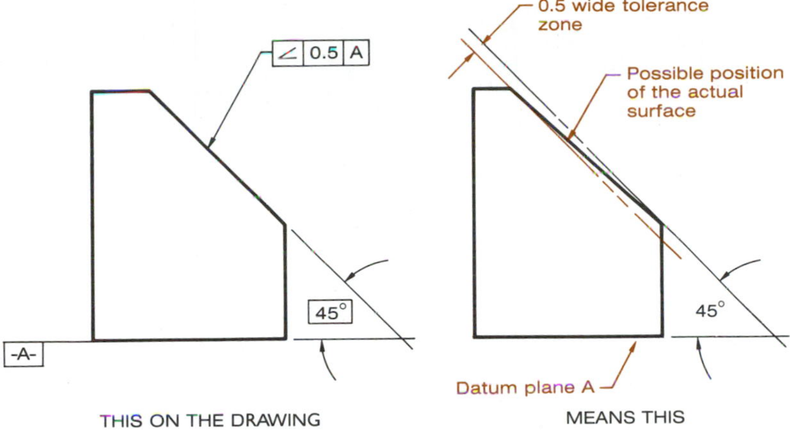

FIGURE 12-25 Specifying angularity *(Section 12.15)*

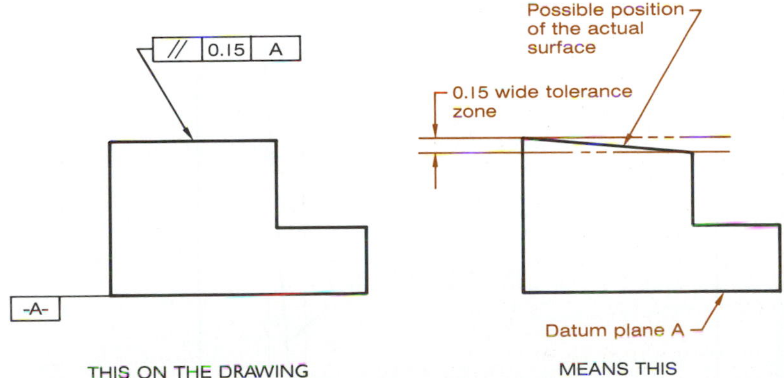

FIGURE 12-26 Specifying parallelism *(Section 12.16)*

12.16 PARALLELISM TOLERANCE

Parallelism is the condition of a surface or a line when all points are equidistant from a datum plane or an axis. A parallelism tolerance specifies one of the following:

(a) A tolerance between two planes or lines parallel to a datum plane or an axis within which the line elements of the surface or axis of the feature must lie (see Figure 12-26).

(b) A cylindrical tolerance zone whose axis is parallel to a datum axis within which the axis of the feature must lie (see Figure 12-27).

12.17 PERPENDICULARITY TOLERANCE

Perpendicularity is the condition of a surface, median plane, or axis at right angles to a datum plane or axis. In Figure 12-28 a perpendicularity tolerance specifies a tolerance zone that is defined by two parallel planes perpendicular to a specified datum plane. Figure 12-29 illustrates how perpendicularity can apply to the axis of a feature such as the center line of a cylinder.

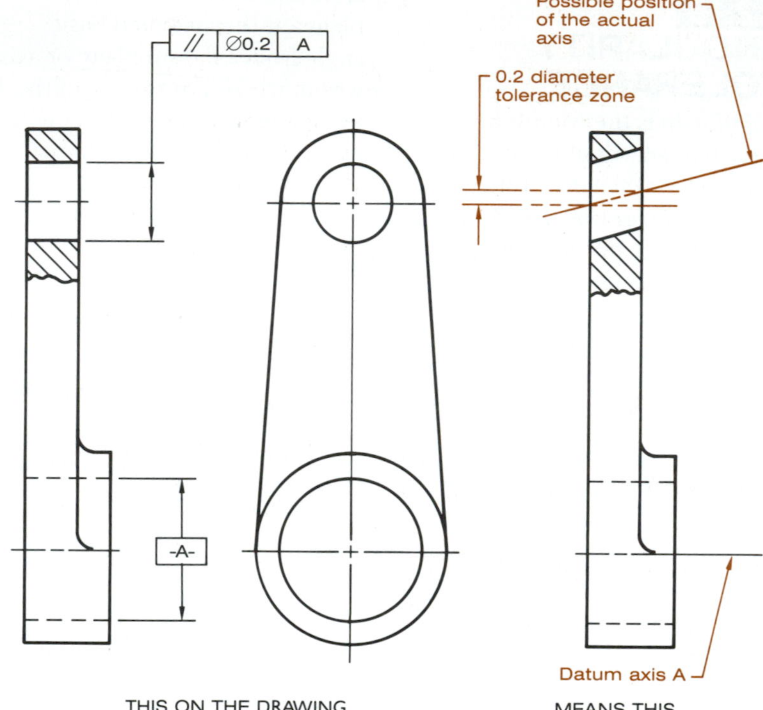

FIGURE 12-27 **Specifying parallelism for an axis** ***(Section 12.16)***

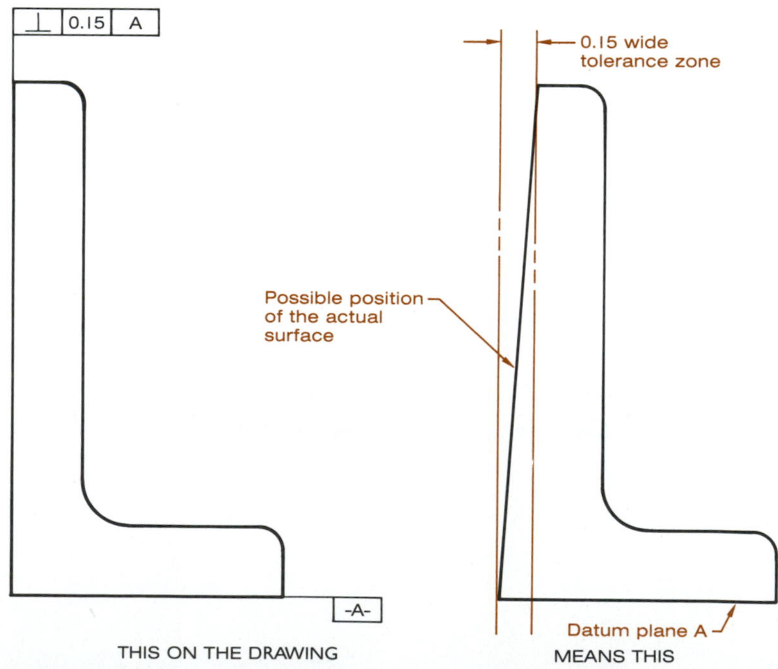

FIGURE 12-28 **Specifying perpendicularity** ***(Section 12.17)***

12.18 RUNOUT TOLERANCES

Runout is a composite tolerance used to control the relationship of one or more features of a part to a datum axis. The types of features controlled by runout tolerances are shown in Figure 12-30. These features include those surfaces constructed around a datum axis and those constructed at right angles to a datum axis. There are two types of runout control: circular runout and total runout. *Circular runout* applies to methods of controlling the relationship of the circular elements of a surface, while *total runout* provides composite control of all surface elements. The type of control used is dependent upon the particular design requirements and manufacturing considerations. ANSI Y14.5M-1982 should be used as a reference when applying the principles of runout control.

12.19 SPECIFYING DATUMS

A properly identified single datum is usually sufficient for orienting one feature of a part to another feature or to a datum. To properly position a feature on a part, three mutually perpendicular datum planes, jointly called a datum reference frame, are used. Datums must be specified in an order of preference. The surface may be flat or cylindrical, but other shapes may be used when necessary. Whenever possible, normally flat surfaces should be selected as datum surfaces although flat surfaces, when magnified, are seen to have irregularities. Datum planes may consist of physical surfaces of the part or they may correspond to planes on locating surfaces of machines, fixtures, or inspection equipment that the part contacts during manufacturing or inspection.

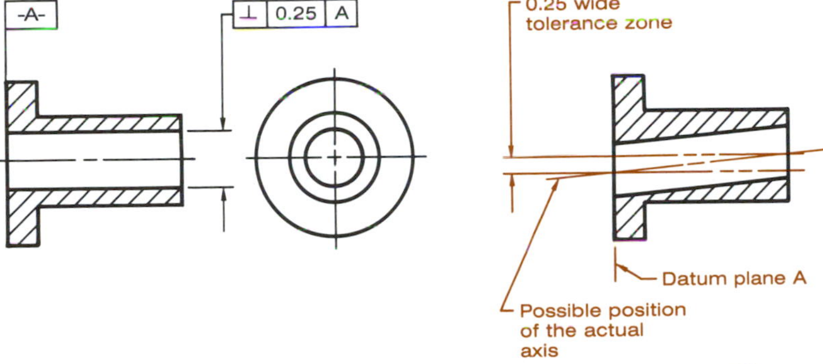

FIGURE 12-29 Specifying perpendicularity for an axis *(Section 12.17)*

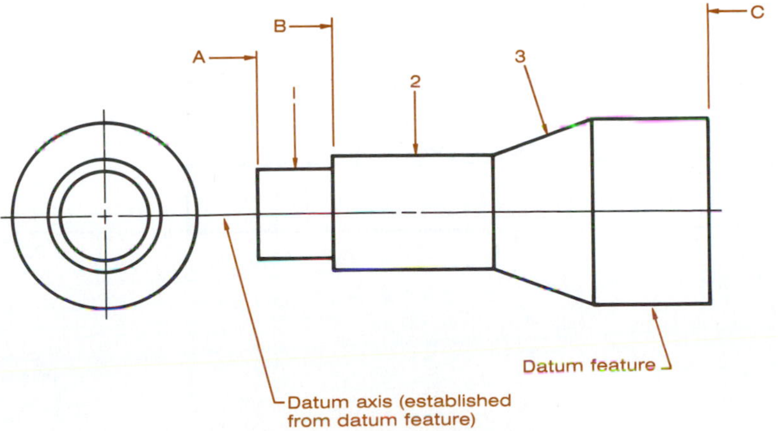

FIGURE 12-30 Features applicable to runout tolerancing *(Section 12.18)*

12.20 THE THREE DATUM PLANE SYSTEM

Flat Datum Features

Figure 12-31 illustrates a part where the datum features are plane surfaces. The desired order of precedence of the reference letters J, K, and L is indicated by the order in which they are listed in the feature control frame. Surfaces J, K, and L are the primary, secondary, and tertiary datum features, respectively.

The relationship of these features is important to the design and function of the part. As shown in Figure 12-32, the primary datum feature of the

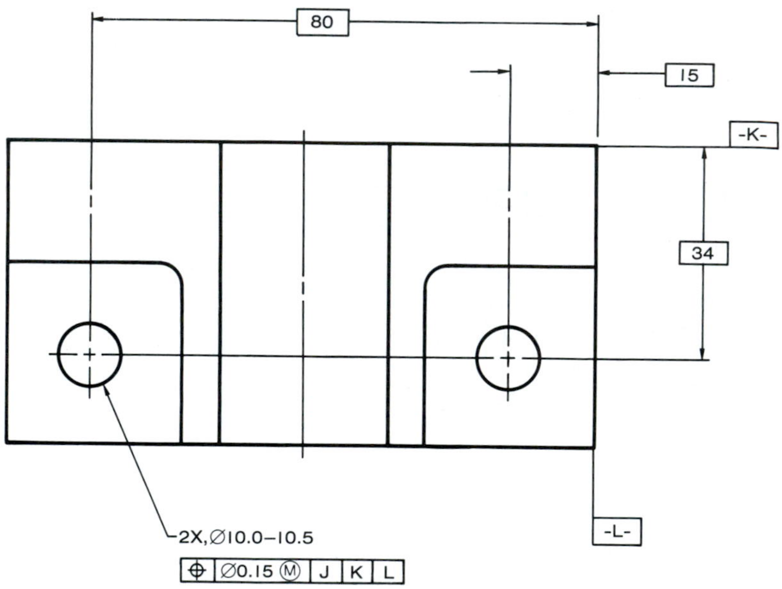

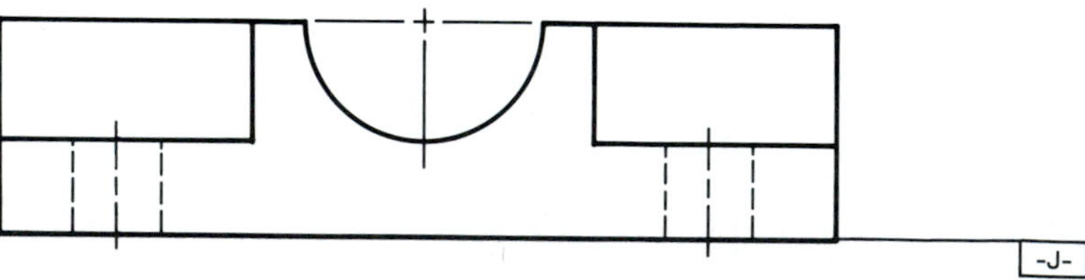

FIGURE 12-31 A part whose datum surfaces are plane surfaces ***(Section 12.20)***

part is a horizontal flat surface that is in contact with the first datum plane, such as a machine table on the processing equipment or a surface of an inspection gage. Theoretically, a minimum of three high spots on this surface should contact the primary datum feature. The second datum feature of the part in this position will contact a secondary datum plane at a theoretical minimum of at least two points. The relationship is completed by bringing at least one point of the tertiary datum feature into contact with the third datum plane.

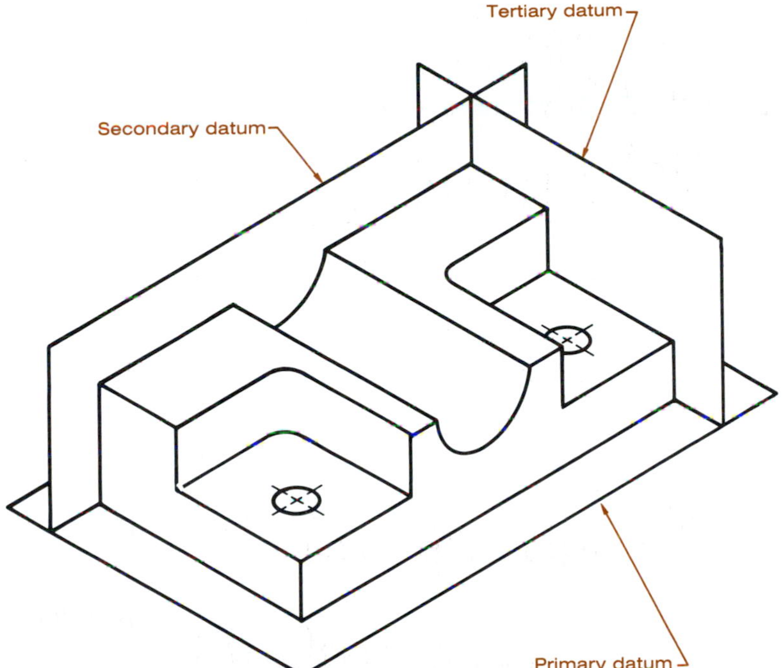

FIGURE 12-32 Three mutually perpendicular datum planes ***(Section 12.20)***

Cylindrical Datum Features

Figure 12-33 shows a method for establishing cylindrical datums. The axis of a cylindrical part is established by two datum planes (X and Y) that intersect at right angles at the center of the cylinder. These two theoretical planes are represented on a drawing by center lines in the circular view. The datum axis is the origin for related dimensions. The primary datum feature K, a flat surface, relates the part of the first datum feature. The secondary datum feature M is cylindrical.

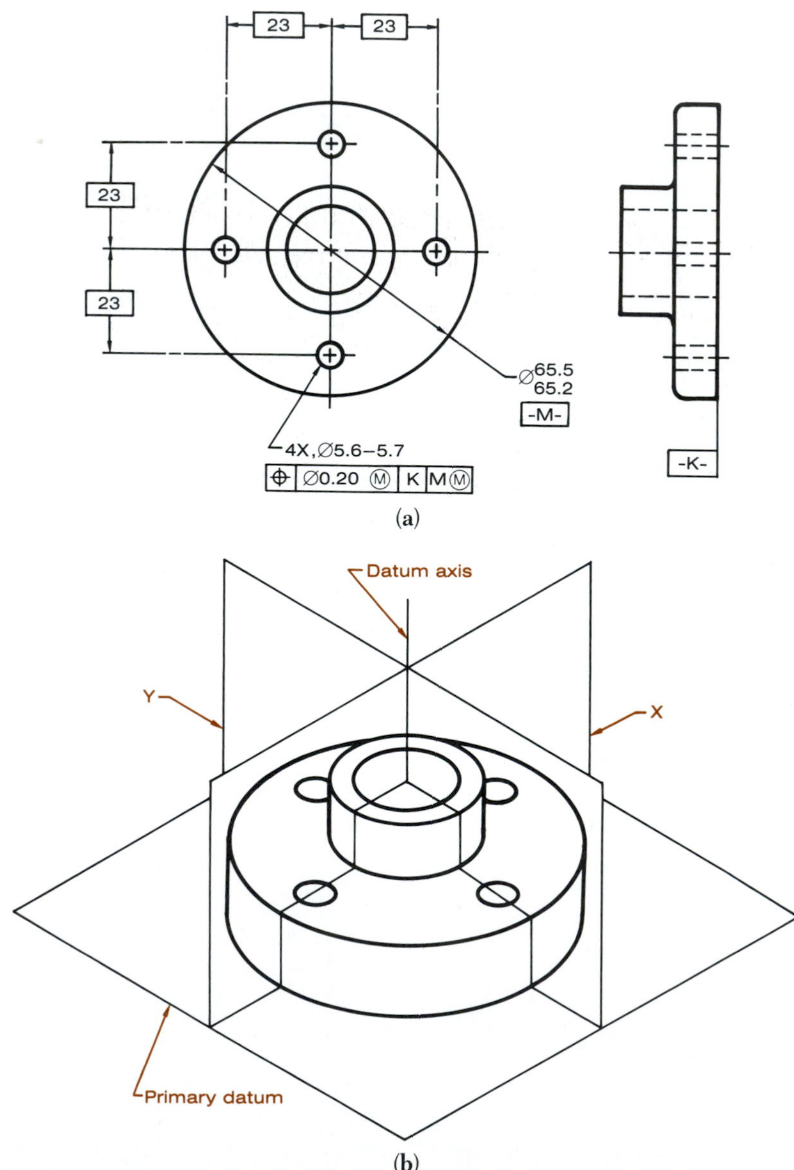

FIGURE 12-33 Three mutually perpendicular datum planes ***(Section 12.20)***

PROBLEMS

Geometric Tolerancing

Using instruments, draw the figures described in the table below on an 8½ × 11 inch sheet. Use layout A, shown on the inside of the front cover. Letter the required information in the title block. Carefully estimate the amount of space that is needed to balance the figures neatly on the drawing sheet. Using the given data listed in the table, draw a feature control frame for each problem using the correct geometric tolerance symbols placed in the appropriate compartments. For datum references, where applicable, use letter A for a primary datum, letter B for a secondary datum, and letter C for a tertiary datum. Use A, B, or C, as desired, for the datum feature symbol (where applicable). Do not draw a leader from the feature control frame.

		Diameter Symbol		*Tolerance value (millimeters)*	*Material condition*	*Datum Reference*			*Datum feature symbol*	
	Characteristic	*Yes*	*No*			*Primary*	*Secondary*	*Tertiary*	*Yes*	*No*
P12.1	Position	X		0.1	MMC	X			X	
P12.2	Perpendicularity		X	0.05	MMC		X		X	
P12.3	Cylindricity	X		0.02	RFS			X		X
P12.4	Angularity		X	0.15	RFS	X				X
P12.5	Concentricity	X		0.1	LMC		X		X	
P12.6	Profile (Line)		X	0.08	LMC			X		X
P12.7	Parallelism		X	0.03	MMC	X			X	
P12.8	Flatness		X	0.2	RFS		X		X	
P12.9	Straightness		X	0.05	LMC			X	X	
P12.10	Position	X		0.02	MMC	X				X

Production Drawings

OBJECTIVES

After completing this chapter you should have gained an understanding of the following:

1. Detail drawings and detail titles.
2. Functional design layout.
3. Production design assembly drawings.
4. Working assembly drawings.
5. Methods of identifying different parts on assembly drawings.
6. Procedures for specifying and identifying individual parts on bills of material.
7. Sectioning techniques associated with assembly drawings.
8. Standard sizes of drawing sheets, titles, and revision blocks.
9. Recommended checking procedures used to verify the correctness of production drawings.

Except for preparing the functional design layout and checking the various production drawings for correctness before releasing them to the shop, engineers are not generally directly involved in the performance of the activities discussed in this chapter. However, engineers do supervise these activities and must be familiar with them. Engineers are usually the quality assurance link between the original product design and the preparation of engineering drawings. Experienced engineers can often exert a definite educational influence upon the detailers, drafters, and designers through their direct and daily contact with them. A successful relationship between the engineer and drafting personnel is a vital factor in the efficient management of the day-to-day operations of an engineering section in an industrial firm.

Ordinarily, drafting personnel are responsible for preparing detail drawings, final design assembly drawings, and working assembly drawings. Working from the specifications provided by the design engineer, the drafter also develops the correct listing of each required item in the bill of material. Likewise, the drafter is accountable for using the proper detail drawing standards and sectioning conventions that apply to assembly drawings. The drafter is also responsible for entering the required information in the sheet title block and on the revision blocks.

13.1 DETAIL DRAWINGS

A typical *detail drawing* (sometimes called a working drawing) is illustrated in Figure 13-1. A detail drawing shows how a single part looks when completed. It explicitly describes the size and the shape of the part and gives complete directions for everything the shop needs, including the material, heat treatment (if necessary), tolerances, finished surfaces, and the quantity of required pieces. The views are selected and projected as described in Chapter 5, sectional views are prepared in accordance with the techniques given in Chapter 7, and the sizes are specified as outlined in Chapter 10.

It is standard practice in most companies to letter the detail title below the views of the part. The detail or part number is followed by the part name, and on the same line, or on a line below, the number or quantity of required parts is given, followed by the material designation. In addition, the required heat treatment and type of surface finish may also be indicated.

The detail drawing serves as an accurate and permanent record of the required part. Detail drawings are not prepared for such standard or purchased parts as screws, keys, bearings, and so on unless modifications are required for these parts.

FIGURE 13-1 A typical detail or working drawing (*Section 13.1*)

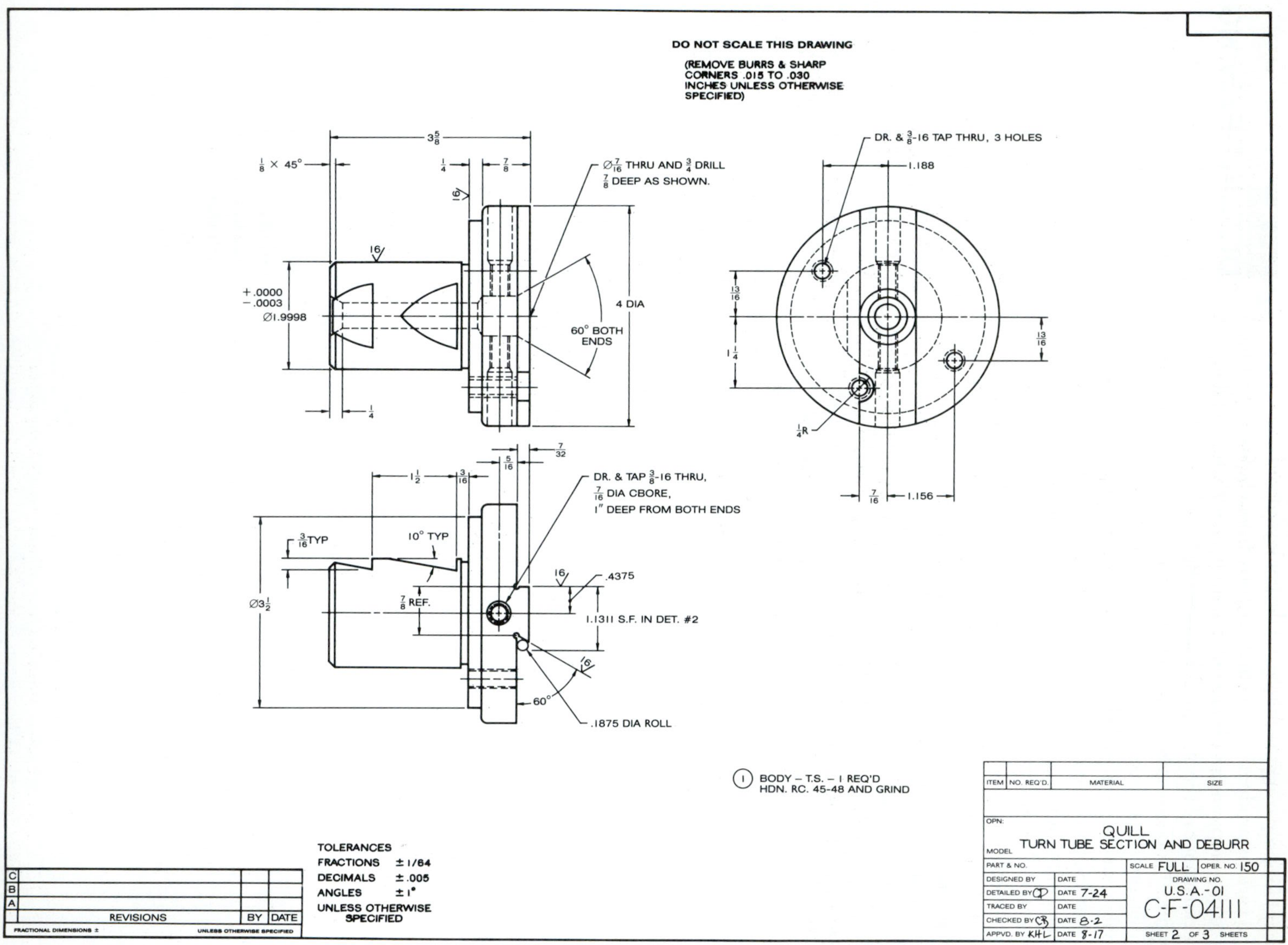

Each part may be detailed (drawn) on separate sheets or several different parts may be grouped on one or more large sheets. The practice varies with individual companies. Whenever practicable, details should be drawn full size.

Figure 13-2 is a detail drawing of a single part dimensioned by datum reference frames that are used to establish tolerance of position and form.

13.2 FUNCTIONAL DESIGN LAYOUT

The first phase of product development consists of a written problem statement or directive (ordinarily a management responsibility) that outlines the product specifications in as precise terms as possible. Once one or more designers have reduced the proposed product development ideas into a few tentatively acceptable ideas, usually in the form of freehand sketches, the product concepts are then ready for further evaluation by an accurately prepared *functional design layout. Layout* is a general term for a planning drawing that shows the relationship of the various parts of a system, a mechanism, or a structure. In the early stages of design, design engineers can critically analyze each component to assure that the product will function acceptably and will meet the necessary critical performance requirements. The design layout provides the design engineer with a graphic model that can be used to study the various component parts and their relative positions. There is considerable freedom during the preparation of this preliminary assembly drawing, and care should be taken to prevent the design from becoming needlessly complex.

Functional design layouts are prepared full size whenever possible. Only the critical features of the assembled parts are drawn, which may erroneously give the drawing a somewhat incomplete appearance. For example, center lines are used to represent the positions of nuts, bolts, screws, gears, and so on. In most cases detail drawings of the actual features of these parts are unnecessary and may be omitted. The preliminary design layout gives the designer a unique opportunity to be highly creative and analytical. Graphical drafting methods that employ basic geometric constructions and elements of descriptive geometry are used to establish critical functional relationships.

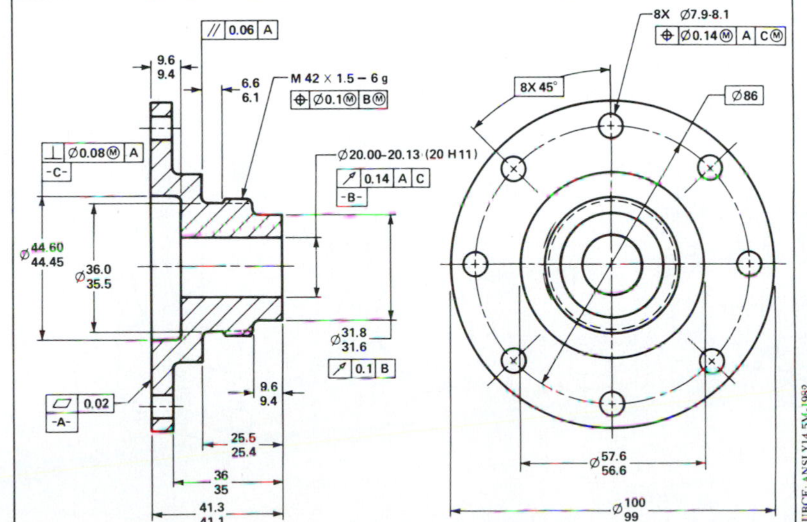

FIGURE 13-2 A detail drawing that has been dimensioned using datum reference frames *(Section 13.1)*

An example of a functional design layout is illustrated in Figure 13-3. As shown, the designer may use the layout to communicate with the engineering drafter (or the model maker) regarding certain desired critical sizes, thread specifications, materials, or additional details relating the specific parts, such as keys, bushings, gears, and so forth.

In some cases a product may be "proved out" by constructing an experimental mechanical model or a prototype that can be thoroughly tested and evaluated. The functional design layout may be readily changed as design alterations become necessary.

13.3 PRODUCTION DESIGN ASSEMBLY DRAWING

Production design follows functional design. A drawing such as the one illustrated in Figure 13-4 is called a *production design assembly drawing* and shows all the component parts assembled in proper working relationships. Basically, a production design assembly drawing involves adapting each component part to the simplest method of manufacture. Whereas both the layout drawing and the experimental model serve the designer in evaluating the functional requirements of a given product design, the alert designer will, from the beginning, mentally "process" the various individual functional parts in an effort to simplify manufacture and assembly requirements.

The process of accomplishing the very most for the least is often referred to as "optimizing." Optimizing consists of making design decisions from as many points of view as possible. This ongoing process is vital in achieving successful results in production design. Very often the final production design assembly drawing is the result of a gradual evolution of very sketchy original concepts displayed on the design layout. All avenues of product development, particularly those that apply to functional, production, and assembly considerations, must be considered prior to selecting a final design and before releasing the drawings for production.

FIGURE 13-4 *(opposite)* **A production design assembly drawing** *(Section 13.3)*

FIGURE 13-3 A functional design layout *(Section 13.2)*

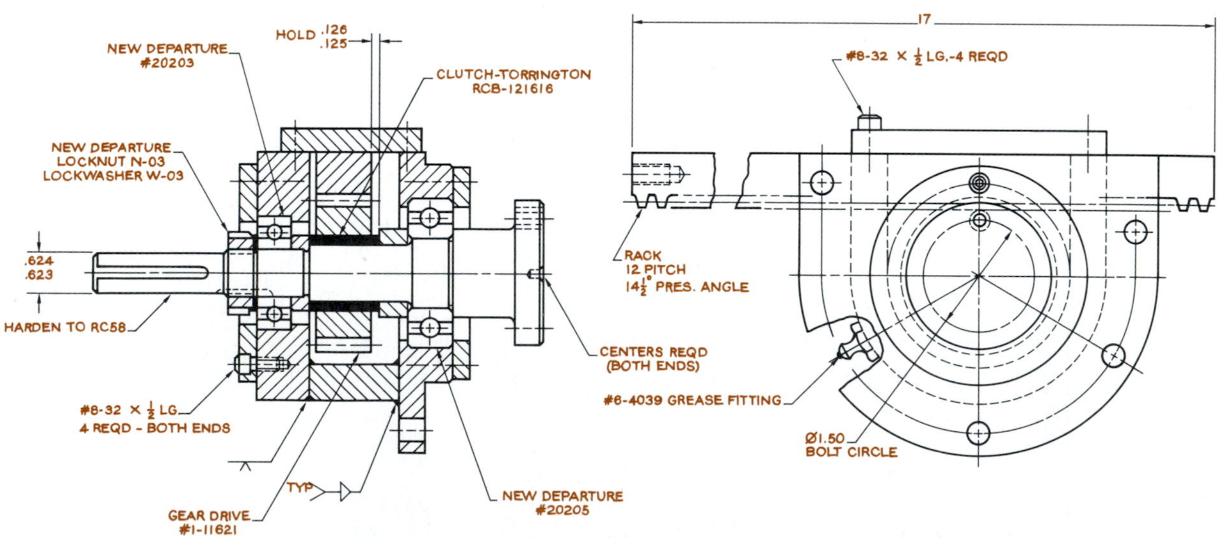

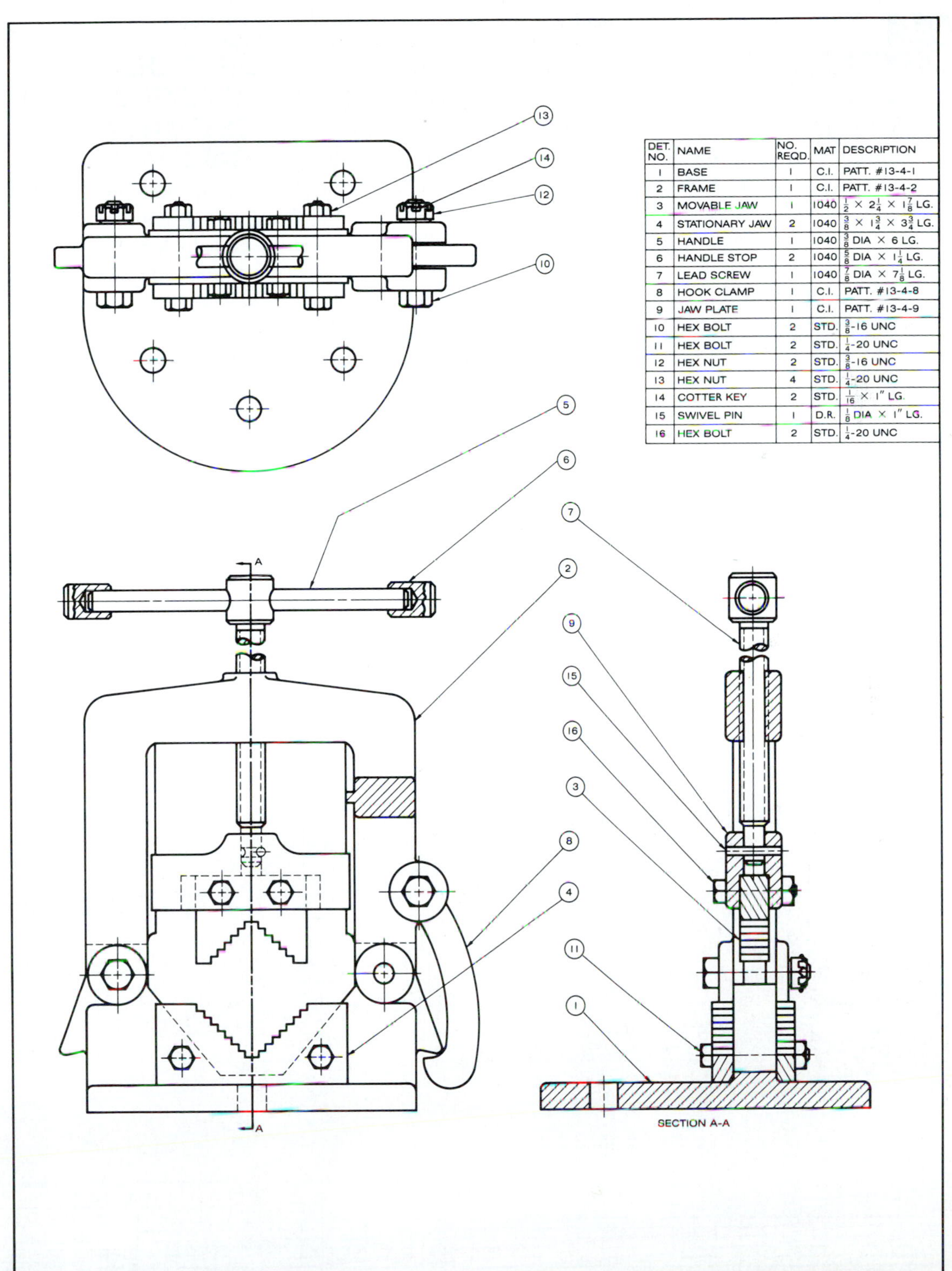

DET. NO.	NAME	NO. REQD.	MAT	DESCRIPTION
1	BASE	1	C.I.	PATT. #13-4-1
2	FRAME	1	C.I.	PATT. #13-4-2
3	MOVABLE JAW	1	1040	$\frac{1}{2} \times 2\frac{1}{4} \times 1\frac{7}{8}$ LG.
4	STATIONARY JAW	2	1040	$\frac{3}{8} \times 1\frac{3}{4} \times 3\frac{3}{4}$ LG.
5	HANDLE	1	1040	$\frac{3}{8}$ DIA × 6 LG.
6	HANDLE STOP	2	1040	$\frac{5}{8}$ DIA × $1\frac{1}{4}$ LG.
7	LEAD SCREW	1	1040	$\frac{7}{8}$ DIA × $7\frac{1}{8}$ LG.
8	HOOK CLAMP	1	C.I.	PATT. #13-4-8
9	JAW PLATE	1	C.I.	PATT. #13-4-9
10	HEX BOLT	2	STD.	$\frac{3}{8}$-16 UNC
11	HEX BOLT	2	STD.	$\frac{1}{4}$-20 UNC
12	HEX NUT	2	STD.	$\frac{3}{8}$-16 UNC
13	HEX NUT	4	STD.	$\frac{1}{4}$-20 UNC
14	COTTER KEY	2	STD.	$\frac{1}{16}$ × 1″ LG.
15	SWIVEL PIN	1	D.R.	$\frac{1}{8}$ DIA × 1″ LG.
16	HEX BOLT	2	STD.	$\frac{1}{4}$-20 UNC

13.4 WORKING ASSEMBLY DRAWINGS

Some companies use *working assembly drawings*, as shown in Figure 13-5. Generally this type of drawing is used when the product consists of only a limited number of parts and in cases where the unit will be produced singly or in small quantities. Dimensions and notes that apply to each of the individual parts are placed directly on the assembly views, thus eliminating the need to prepare separate detail drawings. This technique saves drafting time and allows all information to be grouped on a single sheet. Ordinarily, however, a working assembly drawing would not be prepared for mass produced products consisting of many and complex parts.

13.5 EXPLODED PICTORIAL ASSEMBLY DRAWINGS

Assembly drawings like that shown in Figure 13-6 are used principally for catalog or display purposes and also for instructional manuals. In some cases exploded drawings are very helpful to workers in repair shops for reordering or assembling component parts in a machine or a structure.

FIGURE 13-5 A working assembly drawing *(Section 13.4)*

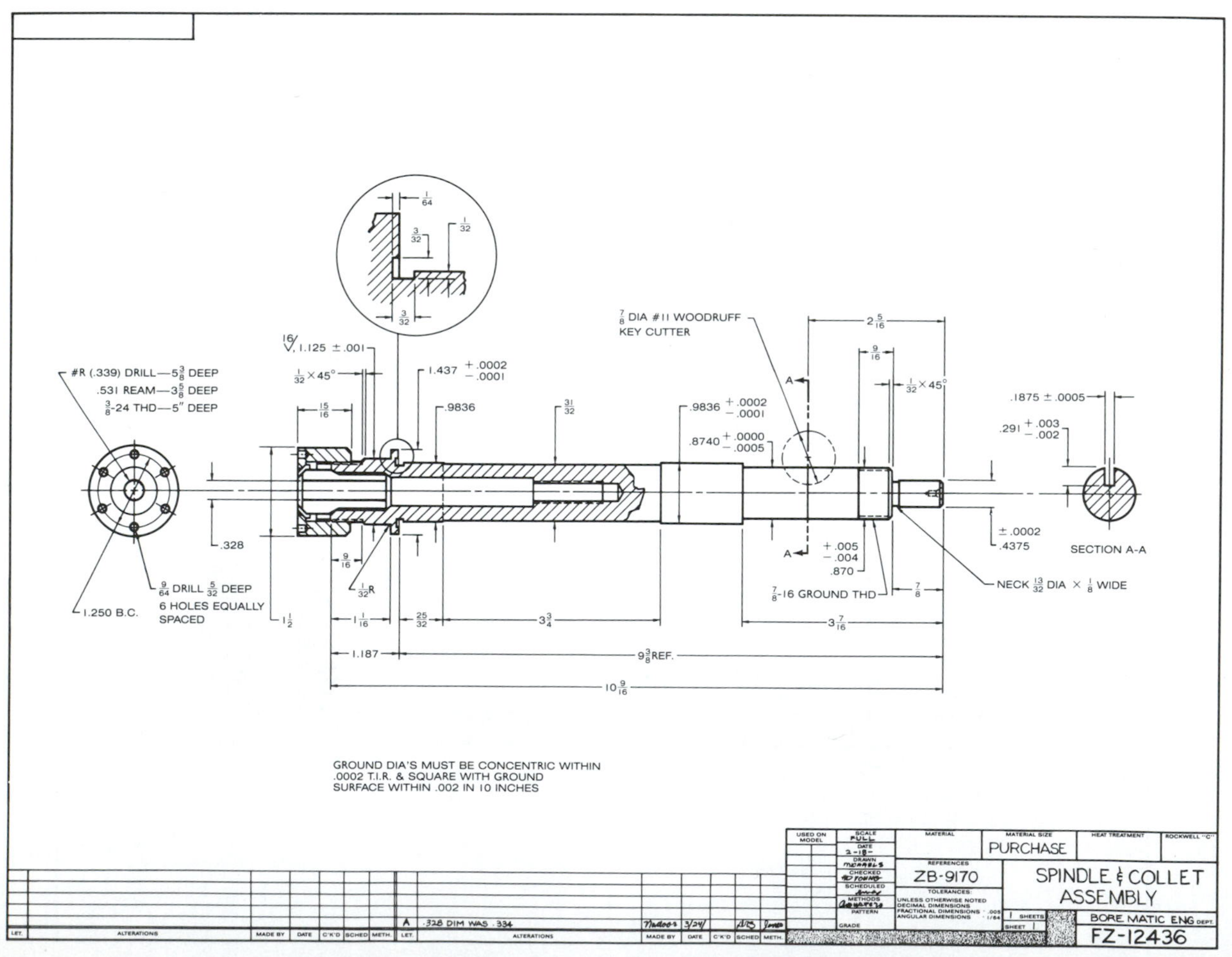

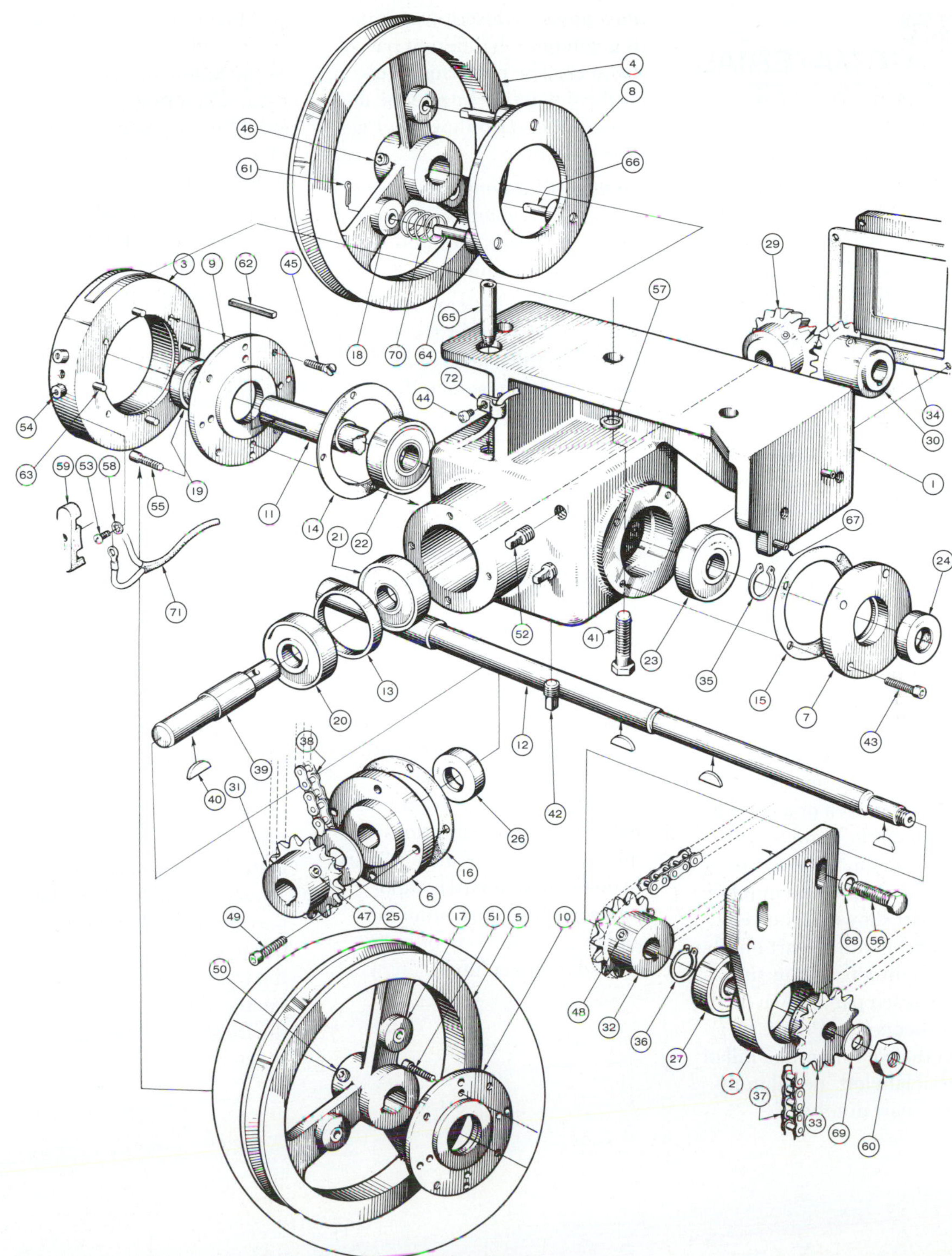

FIGURE 13-6 An exploded assembly drawing *(Section 13.5)*

13.6 BILL OF MATERIAL OR PARTS LIST

A *bill of material*, shown on the assembly drawing in Figure 13-4, consists of an itemized list of all of the parts required to assemble one complete unit. The list may be placed on the assembly sheet in the lower right-hand corner above the title block, in the upper right-hand corner of the sheet, or on a separate sheet. Bills of material are generally used by the purchasing department to order the necessary items for the design. While the contents of bills of material may vary slightly from company to company, they generally contain the following information:

Detail or part number: Each part is designated by a number on the assembly drawing. As shown in Figure 13-4, detail numbers are frequently enclosed in a ⅜ inch or a ½ inch diameter circle, sometimes called a balloon. A leader is drawn connecting the number to the part shown in the view. The corresponding part number is placed directly below the views of each part shown on the detail sheets (see Figure 13-1) and in the column in the bill of material labeled "detail number" or "part number."

Part name: A descriptive name is given to each different part of the unit. The part name is given in the detail title beside the detail number (also shown in Figure 13-1).

Number or quantity of parts required: This column lists the quantity of parts required for one complete unit.

Material: The general practice is to use standard abbreviations to designate the materials for required parts to be manufactured. These abbreviations include CI for cast iron, CRS for cold-rolled steel, and so on. In most cases it is also customary to omit the material specification for purchased parts such as nuts, bolts, keys, bearings, and so forth.

Description: The description column usually gives the required raw material or stock size rather than the finished size for processed parts. For castings, a pattern number is usually assigned and given in this column. Thread specifications and other descriptive data necessary to describe standard components (such as stock numbers and trade names) are often included in this column.

Most bills of material are organized by grouping the parts in the following order: (1) castings, (2) forgings, (3) parts machined from stock, and (4) commercially available or purchased parts. Castings and forgings are generally assigned the lowest part numbers, followed by the machined parts. The purchased parts are given the highest part numbers and are grouped at the end of the list. Bills of material placed above the title block should read *up*, or from the bottom to the top. Parts lists positioned at the top of the sheet should read *down*, from top to bottom. In this way parts may be conveniently added to the list at a later time if modifications are made to the design.

Most companies use standard size drawing sheets printed with a margin or border line that extends around the edges, a title block, and one or two lines to represent the spacings and columns of a bill of a material. Preprinted title blocks and bills of material in the form of gummed sheets (called decalcomanias) may be applied in the desired position on the reverse side of the tracing paper. These labor saving "decals" are visible from the front side of the drawing sheet.

13.7 SECTIONING TECHNIQUES

Sectioning is used extensively on assembly drawings to show the relationship of the various parts. Standardized symbolic section lining (shown in Figure 7-4) is used to distinguish between adjacent parts and to denote different materials. Figure 13-7(a) illustrates how section lining should be drawn for two parts whose positions are adjacent. Even though a part may be separated into different areas, as in this example, the section lining symbol remains the same. Any suitable angle may be selected to make one part distinguishable from another. As shown at 13-7(b), section lining is not used on shafts, bolts, screws, nuts, rivets, washers, or keys when the axis lies in the cutting plane except where internal construction must be shown. Where the cutting plane is perpendicular or cuts across these items, section lining is used on the sectional view in the usual manner.

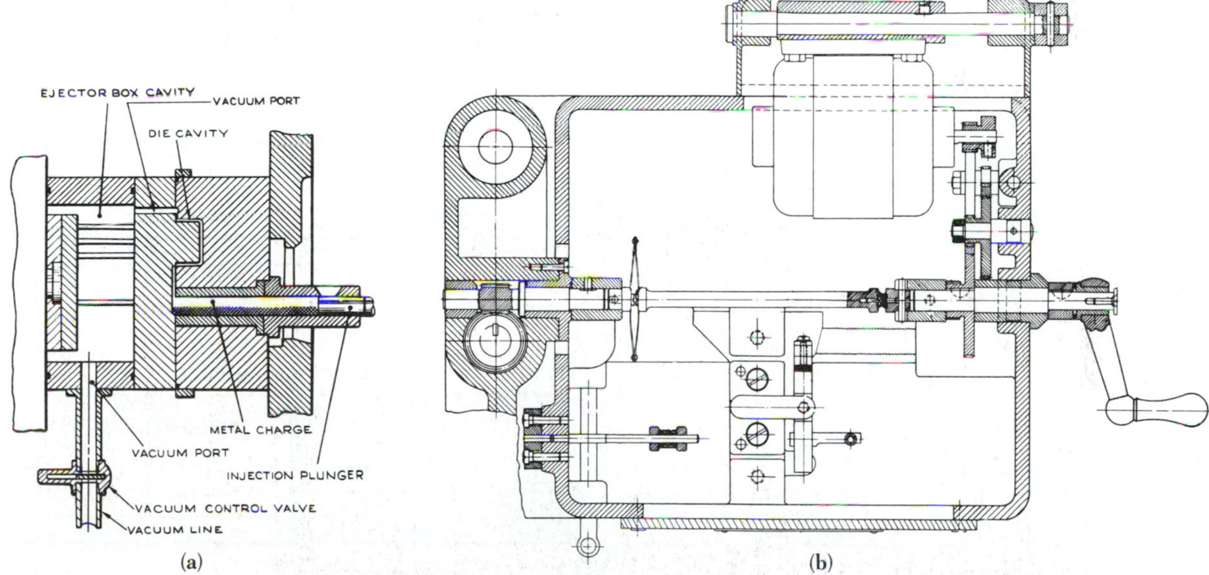

FIGURE 13-7 Section lining of adjacent parts on assembly drawings ***(Section 13.7)***

13.8 TITLE BLOCKS

Many companies develop their own standard format for title blocks. Figure 13-8 illustrates several types of commercial title blocks. The title block is located in the lower right-hand corner of the engineering drawing sheet. The following information is common to practically all title blocks:

1. Name and address of the company.
2. Drawing title.
3. Drawing number, sometimes in combination with identifying letters. These are placed in a large block in the extreme lower right-hand corner of the title block.
4. Name of part (for a detail drawing) or structure (for an assembly drawing).
5. Scale of the drawing. In the U.S. Customary system the equal sign is used: 1 = 2 (half scale). It is understood that the unit of measurement is in

FIGURE 13-8 Typical industrial drawing sheet title blocks *(Section 13.8)*

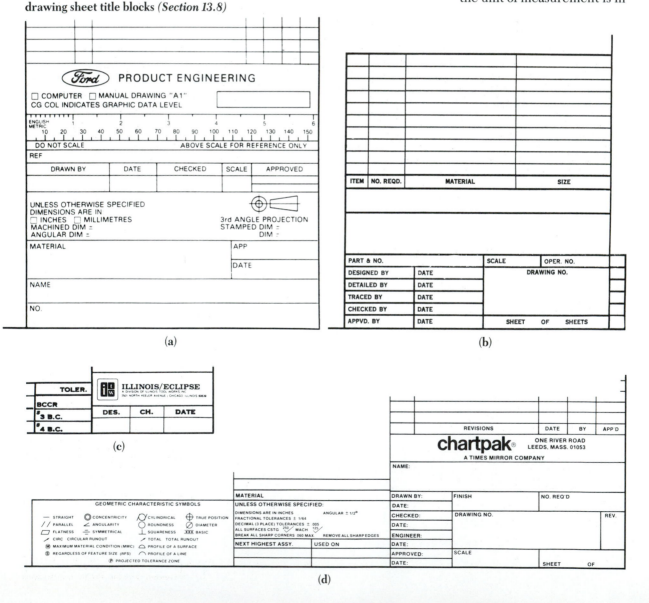

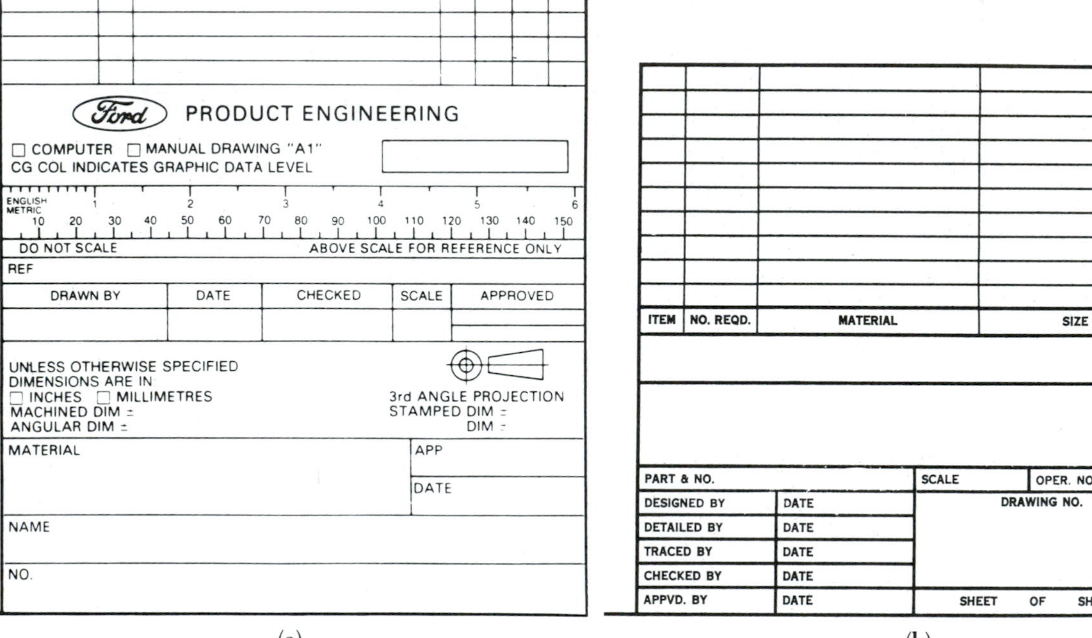

(a)

(b)

TOLER. | ILLINOIS/ECLIPSE A DIVISION OF ILLINOIS TOOL WORKS INC
BCCR | DES. | CH. | DATE
#3 B.C.
#4 B.C.

(c)

GEOMETRIC CHARACTERISTIC SYMBOLS

STRAIGHT, CONCENTRICITY, CYLINDRICAL, TRUE POSITION, PARALLEL, ANGULARITY, ROUNDNESS, DIAMETER, FLATNESS, SYMMETRICAL, SQUARENESS, XXX BASIC, CIRC CIRCULAR RUNOUT, TOTAL TOTAL RUNOUT, MAXIMUM MATERIAL CONDITION (MMC), PROFILE OF A SURFACE, REGARDLESS OF FEATURE SIZE (RFS), PROFILE OF A LINE, PROJECTED TOLERANCE ZONE

MATERIAL
UNLESS OTHERWISE SPECIFIED:
DIMENSIONS ARE IN INCHES — ANGULAR ± 1/2°
FRACTIONAL TOLERANCES ± 1/64
DECIMAL (3 PLACE) TOLERANCES ± .005
ALL SURFACES CSTG 250 MACH 125
BREAK ALL SHARP CORNERS .060 MAX — REMOVE ALL SHARP EDGES
NEXT HIGHEST ASSY. | USED ON

REVISIONS | DATE | BY | APP'D

chartpak® ONE RIVER ROAD LEEDS, MASS. 01053
A TIMES MIRROR COMPANY
NAME:
DRAWN BY: | FINISH | NO. REQ'D
DATE:
CHECKED: | DRAWING NO. | REV.
DATE:
ENGINEER:
DATE:
APPROVED: | SCALE
DATE: | SHEET OF

(d)

inches. In the metric system the colon is used: 1:2. The word METRIC is given on a drawing to signify that the units of measurement are in millimeters.

6. Record information relative to the preparation of the drawing, including signatures or initials of drafters, tracers, and checkers, dates of approving functions, completion or issue dates, and contract number if applicable.
7. Drawing size letter designation (see Section 13.10).
8. Sheet number for multiple-sheet drawings (sheet 1 of 6, for example). Some title blocks also include information describing the material, tolerances, surface finish, and actual or estimated weight.

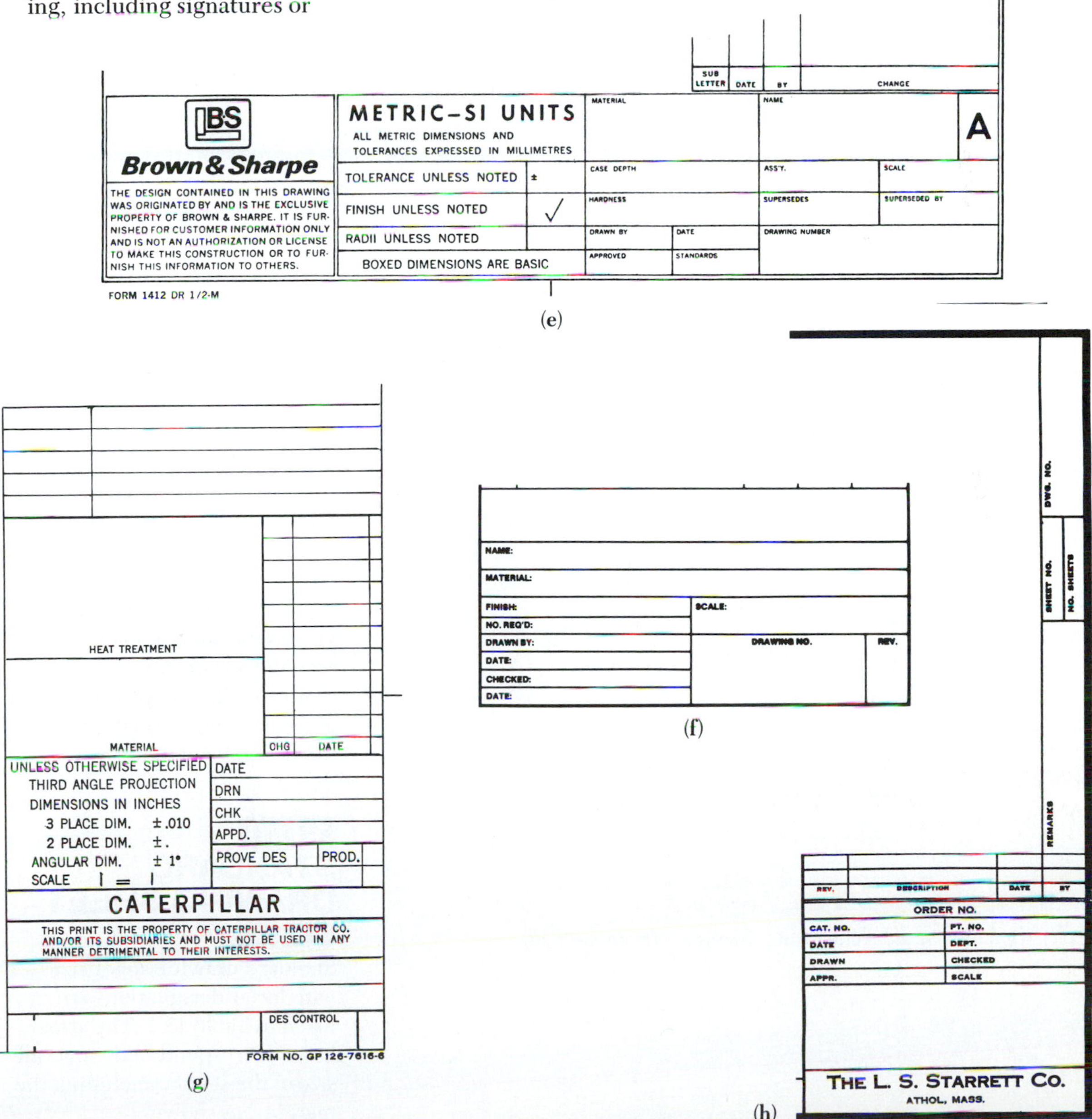

(e)

(f)

(g)

(h)

TABLE 13-1 Standard Drawing Sheet Sizes

Flat sizes (inches)

Size designation	*Width (vertical)*	*Length (horizontal)*	*Margin*	
			Horizontal	*Vertical*
A (Horiz)	8.5	11.0	0.38	0.25
A (Vert)	11.0	8.5	0.25	0.38
B	11.0	17.0	0.38	0.62
C	17.0	22.0	0.75	0.50
D	22.0	34.0	0.50	1.00
E	34.0	44.0	1.00	0.50
F	28.0	40.0	0.50	0.50

Roll sizes (inches)

Size designation	*Width (vertical)*	*Length (horizontal)*		*Margin*	
		Min	*Max*	*Horizontal*	*Vertical*
G	11.0	22.5	90.0	0.38	0.50
H	28.0	44.0	143.0	0.50	0.50
J	34.0	55.0	176.0	0.50	0.50
K	40.0	55.0	143.0	0.50	0.50

ANSI Y14.1-1980

13.9 REVISION BLOCK

When it is necessary to make changes on production drawings, a clear record of these changes must be registered on the drawing. A revision block is used for this purpose. The revision block provides space for a revision symbol, a description of the change, the date of the change, and the approval signature or initials. An encircled number or a letter should be used for each individual change and should be placed near the corresponding alteration on the drawing. Figure 13-9 illustrates recommended sizes of revision blocks for various sheet sizes and gives an example of a typical entry. The width of the block may be adjusted to provide for other columns as necessary. The revision block is located near the upper-right corner of the drawing sheet. Space should be reserved to extend the revision block downward as required. Where additional space for the revision block is needed, a supplemental revision block should be located to the left of and adjacent to the original revision block. In most companies revisions may not be made on engineering documents by the drafter without approval of an authorized engineering management representative.

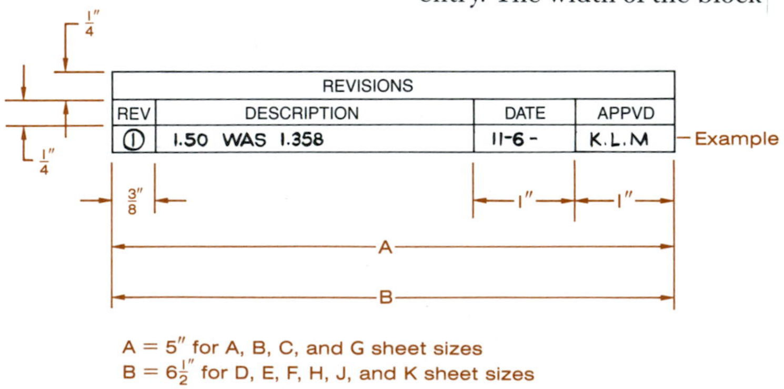

REVISIONS			
REV	DESCRIPTION	DATE	APPVD
①	1.50 WAS 1.358	11-6 -	K.L.M

— Example

$\frac{1}{4}''$ $\frac{1}{4}''$ $\frac{3}{8}''$ 1″ 1″ A B

A = 5″ for A, B, C, and G sheet sizes
B = $6\frac{1}{2}''$ for D, E, F, H, J, and K sheet sizes

FIGURE 13-9 Standard drawing sheet sizes *(Section 13.10)*

13.10 STANDARD DRAWING SHEET SIZES

Standard drawing sheet sizes with letter designations are listed in Table 13-1. The sizes listed correspond to the overall size of the sheets, including the margins, as shown.

13.11 CHECKING DRAWINGS

In most firms all engineering drawings are carefully examined as a matter of organizational procedure before releasing them to the shop. Highly qualified personnel, called checkers, review the drawings to verify their accuracy and completeness. Checking constitutes the final assurance that a newly designed machine or structure is in correct form.

Checking is considered an art, and not all people are suited to the intense detail work that is required to be a competent checker. A checker must be thoroughly familiar with all aspects of the product and must have a broad knowledge of the company drafting and manufacturing practices. Whereas the checker is not held responsible for the design, he or she is responsible for calling oversights to the designer's attention. These oversights may include possible unsafe conditions because of unguarded moving parts, lack of provisions for lubrication, or the possibility of fluid leakage because of insufficient seals.

Checkers are always on the alert toward reducing production costs. Generally, the greatest contributors to high manufacturing costs are unreasonably demanding specifications for surface finish and tolerances. A checker may be justified in proposing minor adjustments in the location of a surface or a reduction in surface finish and tolerance specifications if such modifications will lead to a decrease in the cost of production. Designs should be as simple as possible, with machined surfaces that are functional and readily accessible. Sometimes a checker can make important contributions in reducing production costs by recommending that a part be produced more economically by casting or that it be fabricated by welding instead of being entirely machined to shape from stock.

Standard practice in most companies is for checkers to use prints of the newly prepared drawings to avoid defacing the originals. Prints also provide a record from which the checker can make a final comparison after the changes have been made. Initially the checker will use a colored pencil to place a check mark near each feature, dimension, note, or line as it is verified on the print. Customarily corrections are indicated by ruling out the error and inserting the desired change or omission. The check print is then returned to the drafter, who makes the indicated changes on the original documents. As a final step following the changes, the checker will examine the original drawings for conformance with the checking changes on the check print. Although the procedures used in checking ordinarily follow a systematic order, the exact order will likely vary with the checker. A suggested checking sequence for production design assembly drawings is as follows:

1. The entire drawing should be examined for conformance to high quality drafting standards. All necessary views and sections must be shown. Notes and dimensions should be correct and properly located on the drawing. The drawing should be to scale and all lines should be sufficiently bold to reproduce satisfactorily.
2. The drawing should be inspected to see that all parts will function correctly and that proper clearances are maintained for adjacent parts.
3. Manufactured parts should be evaluated for proper relative proportions, strength, rigidity, and appearance. Each part must be designed for economical use and suitability of material.
4. The mechanism or structure should be examined for convenience of assembly and disassembly and for maintenance or repair work. Sufficient space should be provided for the use of a wrench on nuts and bolt heads.
5. The bill of material should be checked for accuracy and spelling. Purchased parts

should be studied to see if they are standard and readily obtainable from reliable suppliers. Proper stock sizes must be listed.

6. The title block and revision notes should be examined for completeness and correctness.

Here is a suggested checking sequence for detail drawings:

1. The entire drawing should be examined for conformance to high quality drafting standards. All necessary views and sections must be shown. Notes and dimensions should be correct and properly located on the drawing. The drawing should be to scale and all lines should be sufficiently bold to reproduce satisfactorily.
2. Each part shape should be clearly illustrated and the production methods required to manufacture the part should be evident from the views. The checker should evaluate the sequence of operations needed to make each part both in terms of potential manufacturing difficulties and in terms of the most economical production method.
3. Proper specifications should be given on castings and forgings for fillets and rounds and also for draft angles where appropriate. All finished surfaces on every view should be clearly indicated. If consistent with company practices, the checker should ascertain that the amount of finish material is specified and that the location of the parting line is shown.
4. Tolerances should be no more precise than actually needed. They should conform to the capabilities of the equipment in the plant where the parts will be made. Tolerances should be checked for function and accuracy. Corresponding dimensions on mating parts should also be checked.
5. The position of all holes, cuts, and surfaces should be checked to ensure that the necessary machining operations can be readily performed without having the shop calculate omitted distances. Only standard sizes of drills, counterbores, countersinks, taps, and reamers should be specified.
6. Proper finish specification must be given.
7. All notes must be given in proper form with the information listed in the correct sequence.
8. Some parts must be made right- or left-handed. The correct notation must be given.
9. The correctness of the detail title must be checked. The part name, number of pieces required, material, stock size, pattern, forging or die numbers, and so on, should all be given.
10. The appropriateness of special painting, plating, or heat treatment specifications should be checked.
11. The title block and revision notes should be examined for completeness and correctness.

PROBLEMS

Production Drawings

The following problems are not intended to be complete in every respect. Sizes of some problems must be approximated using your own judgment. Most problems will provide you with a realistic opportunity to thoroughly analyze the functional requirements of each part and to select and properly specify necessary design elements. These elements may include chamfers, necks and undercuts, thread reliefs, fillets and rounds, hole spacings, threaded fasteners, springs, keys, and pins. Appropriate tolerances and suitable clearances should be given for the proper function of mating parts. Where appropriate, surface finish symbols should be applied to the detail drawings. Do not copy the given dimensions because they are not always representative of approved dimensioning practices. It will be necessary to use data in the appendix tables for most of the problems.

When making the detail drawings, give all of the appropriate dimensions, symbols and notes so the part could be properly manufactured without question. Where directed, the detail drawings may be prepared in accordance with the principles outlined in Chapter 11, dealing with tolerances of form or tolerances of position. Product design assembly drawings should be prepared with a bill of material.

Use either layout A or layout B, both shown on the inside of the front cover, for the problems in this chapter. The size of the sheet will depend upon the drawing scale you select. All sizes given in the bills of material are *actual* sizes and do not provide for stock allowance. The following abbreviations are used:

DR *Drill Rod* (Round steel bars, ground and polished)

STD *Standard* (Used for standard, purchased parts)

CI *Cast Iron*

AL *Aluminum*

STL *Steel* (The kinds and uses of steel are many. A more precise specification may be given if desired. Numbering systems and general categories of steel uses are given in Section 8.4 and Table 8-1, respectively.)

P13.1 to **P13.6** Make detail drawings of every part except for commercially available components. Determine the dimensions by transferring them from the drawings to one of the given scales by means of the dividers.

P13.1 Adjustable jack.

P13.2 Punch and die set.

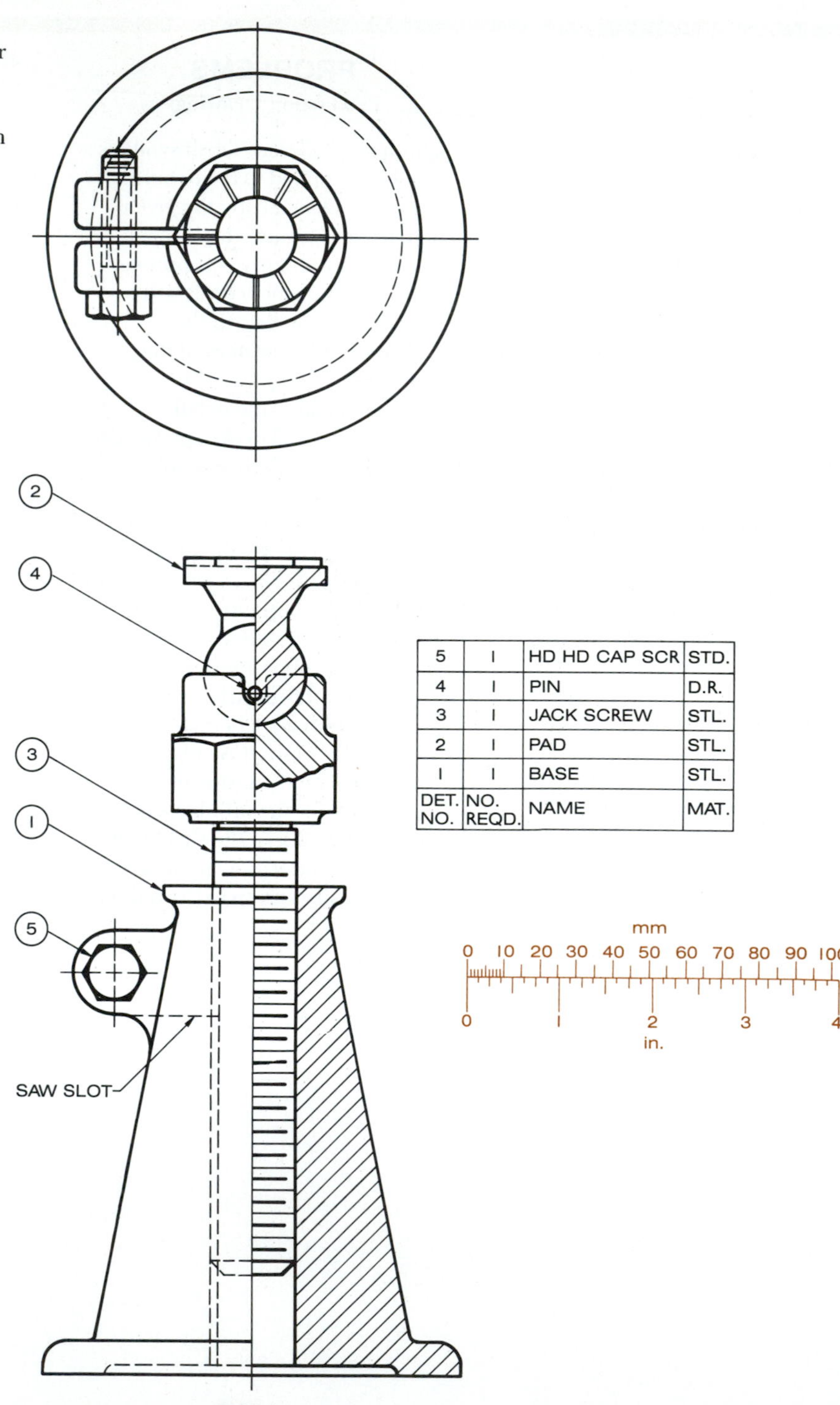

DET. NO.	NO. REQD.	NAME	MAT.
5	1	HD HD CAP SCR	STD.
4	1	PIN	D.R.
3	1	JACK SCREW	STL.
2	1	PAD	STL.
1	1	BASE	STL.

P 13.1

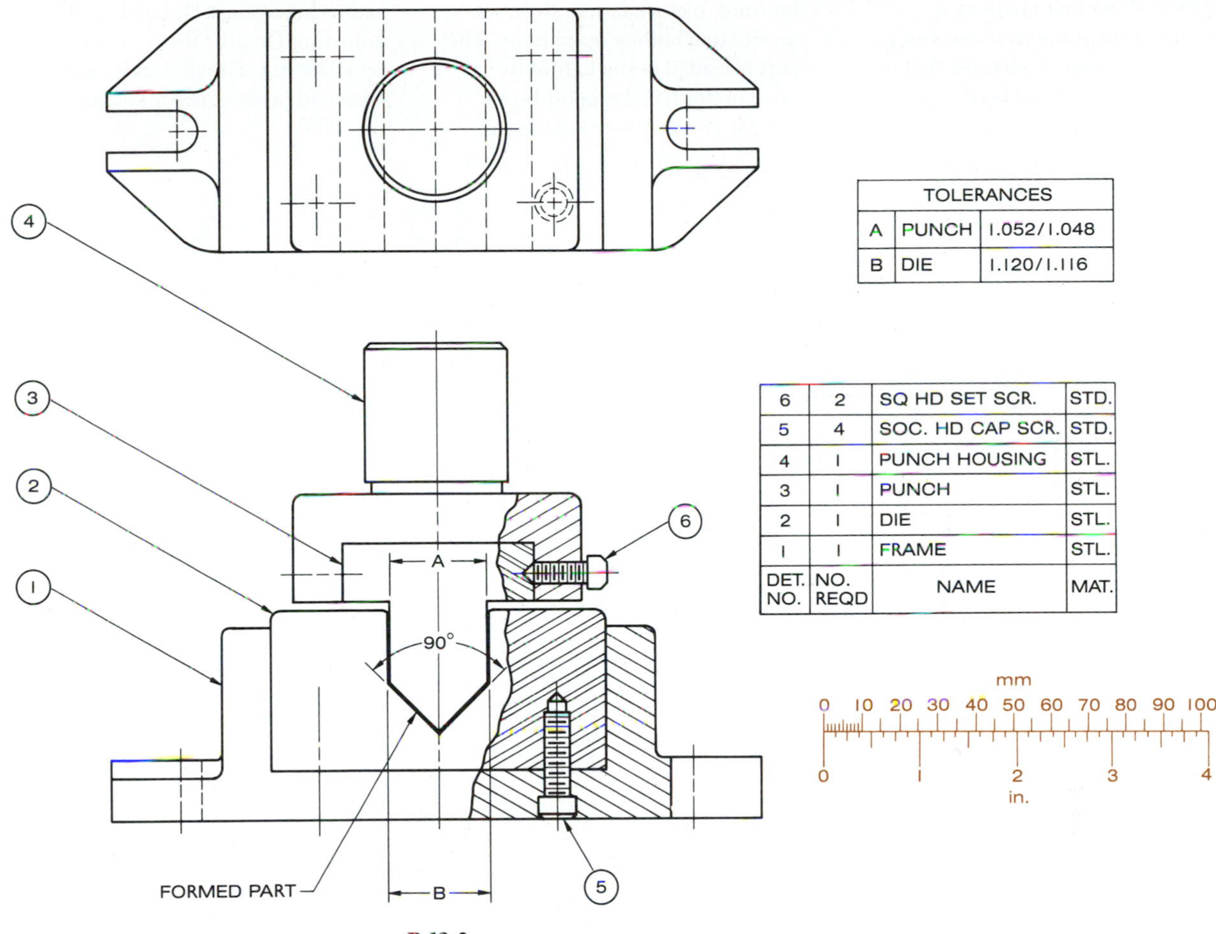

TOLERANCES		
A	PUNCH	1.052/1.048
B	DIE	1.120/1.116

DET. NO.	NO. REQD	NAME	MAT.
6	2	SQ HD SET SCR.	STD.
5	4	SOC. HD CAP SCR.	STD.
4	1	PUNCH HOUSING	STL.
3	1	PUNCH	STL.
2	1	DIE	STL.
1	1	FRAME	STL.

P 13.2

P13.3 Work holding device.
P13.4 This problem shows the top and front views of a fixture that is used to position and hold a shaft housing, pictured separately, for production milling operations. The part is loaded in the fixture by nesting the 1.12 diameter boss against Detail 6, nesting the .62 radius lug against Detail 2, and adjusting Detail 3 firmly against the opposite lug. Detail 3 slides in the base with a clearance allowance of .0005″.

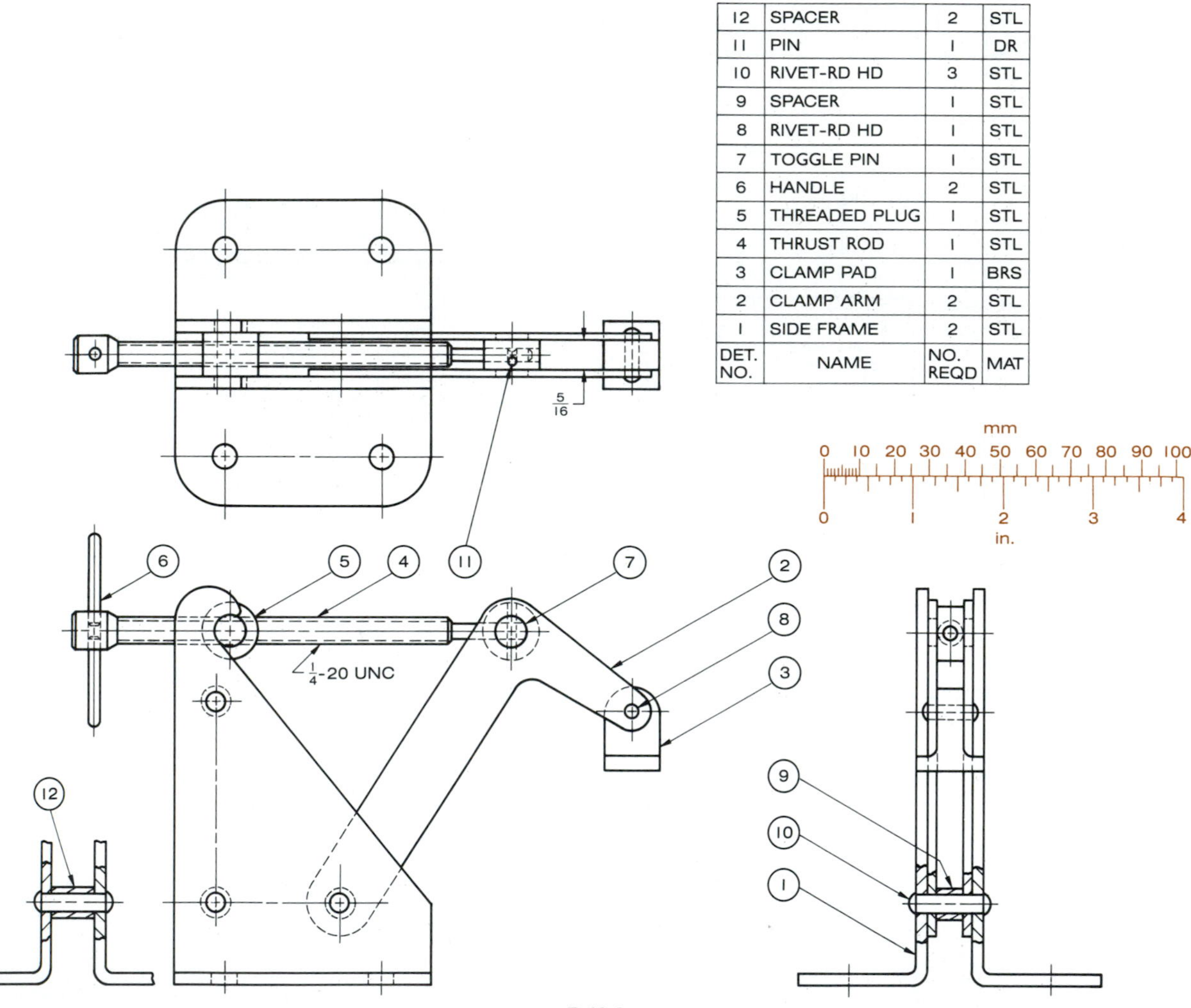

12	SPACER	2	STL
11	PIN	1	DR
10	RIVET-RD HD	3	STL
9	SPACER	1	STL
8	RIVET-RD HD	1	STL
7	TOGGLE PIN	1	STL
6	HANDLE	2	STL
5	THREADED PLUG	1	STL
4	THRUST ROD	1	STL
3	CLAMP PAD	1	BRS
2	CLAMP ARM	2	STL
1	SIDE FRAME	2	STL
DET. NO.	NAME	NO. REQD	MAT

P 13.3

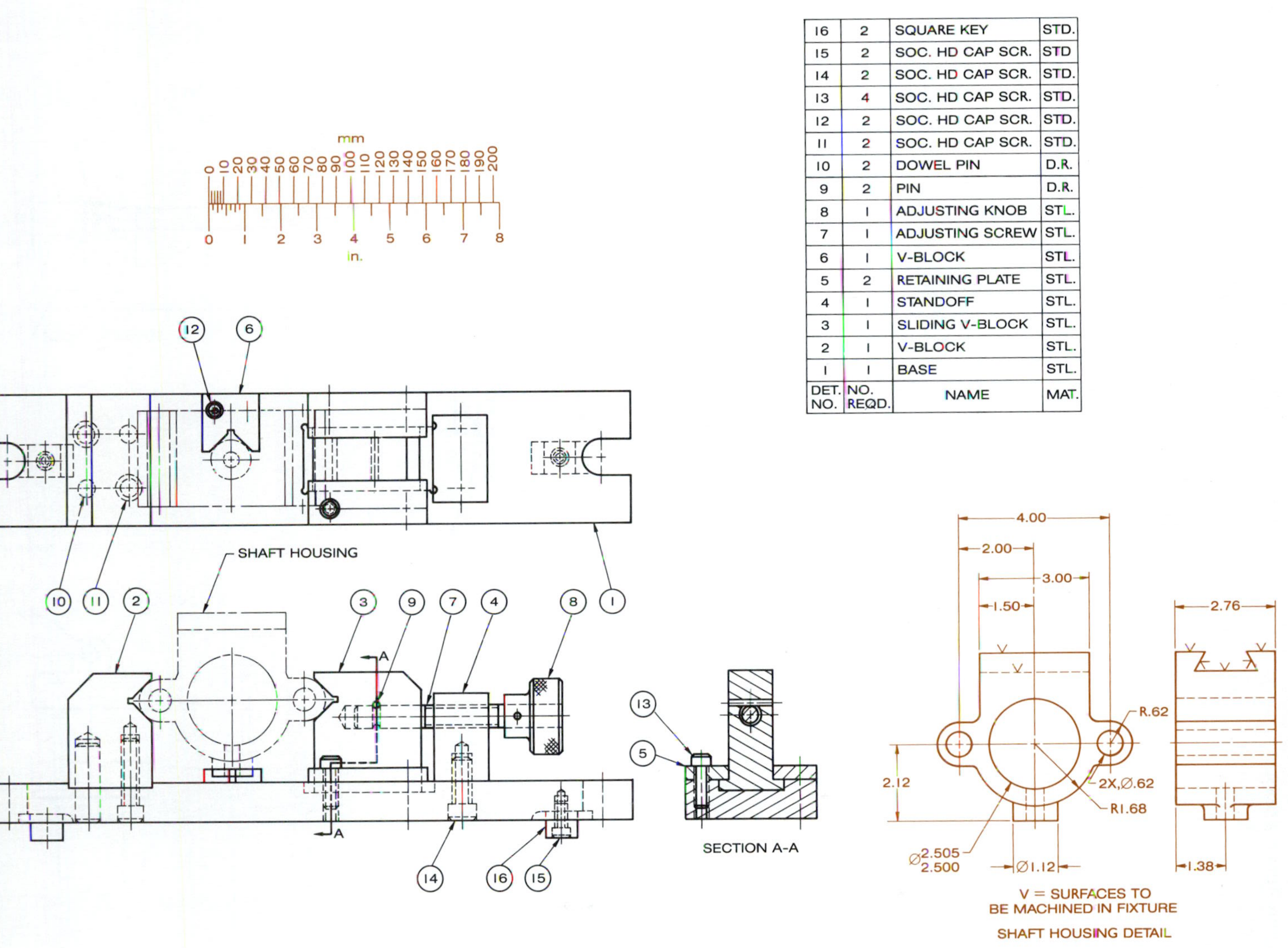

DET. NO.	NO. REQD.	NAME	MAT.
16	2	SQUARE KEY	STD.
15	2	SOC. HD CAP SCR.	STD
14	2	SOC. HD CAP SCR.	STD.
13	4	SOC. HD CAP SCR.	STD.
12	2	SOC. HD CAP SCR.	STD.
11	2	SOC. HD CAP SCR.	STD.
10	2	DOWEL PIN	D.R.
9	2	PIN	D.R.
8	1	ADJUSTING KNOB	STL.
7	1	ADJUSTING SCREW	STL.
6	1	V-BLOCK	STL.
5	2	RETAINING PLATE	STL.
4	1	STANDOFF	STL.
3	1	SLIDING V-BLOCK	STL.
2	1	V-BLOCK	STL.
1	1	BASE	STL.

P13.5 Picture-framing device.
P13.6 Vise.

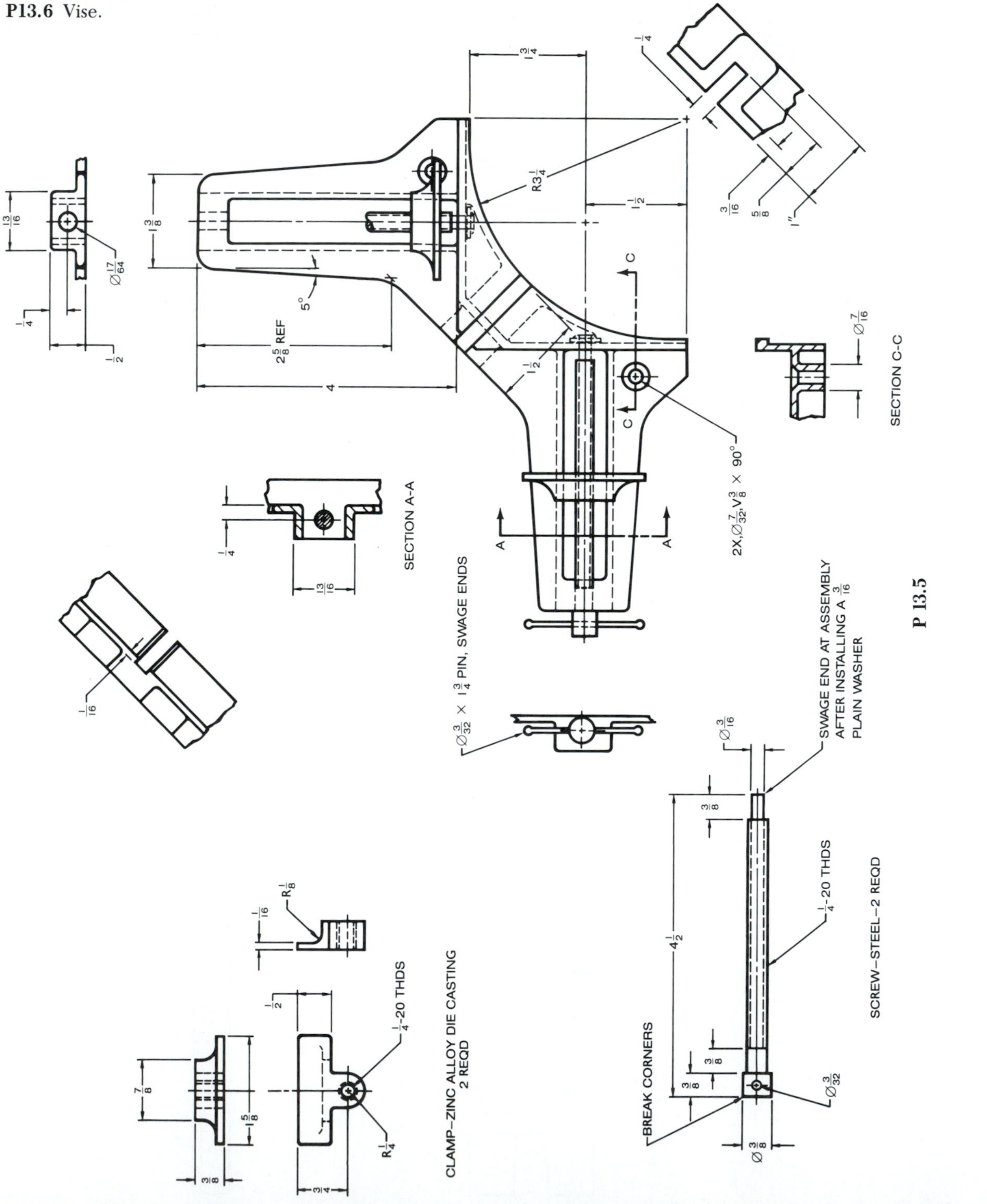

P 13.5

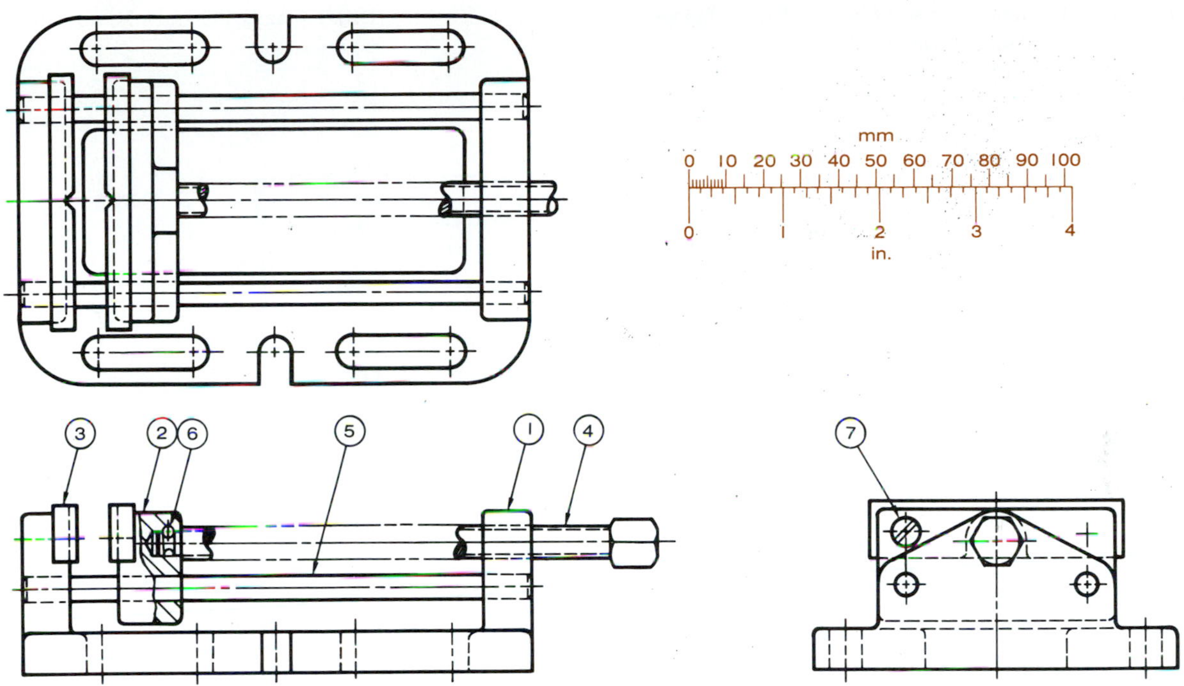

7	4	FL HD MACH SCR	STD
6	I	PIN	DR
5	2	RODS	DR
4	I	DRIVE SCREW	STL
3	2	JAW PLATE	BZ
2	I	MOVABLE JAW	CI
I	I	BASE	CI
DET. NO.	NO. REQD.	NAME	MAT.

P 13.6

P13.7 to **P13.12** Make an assembly drawing with a bill of material using the given details. Approximate all omitted sizes.

P13.7 Tap wrench.

P13.8 (pages 443 and 444) Polishing wheel arbor.

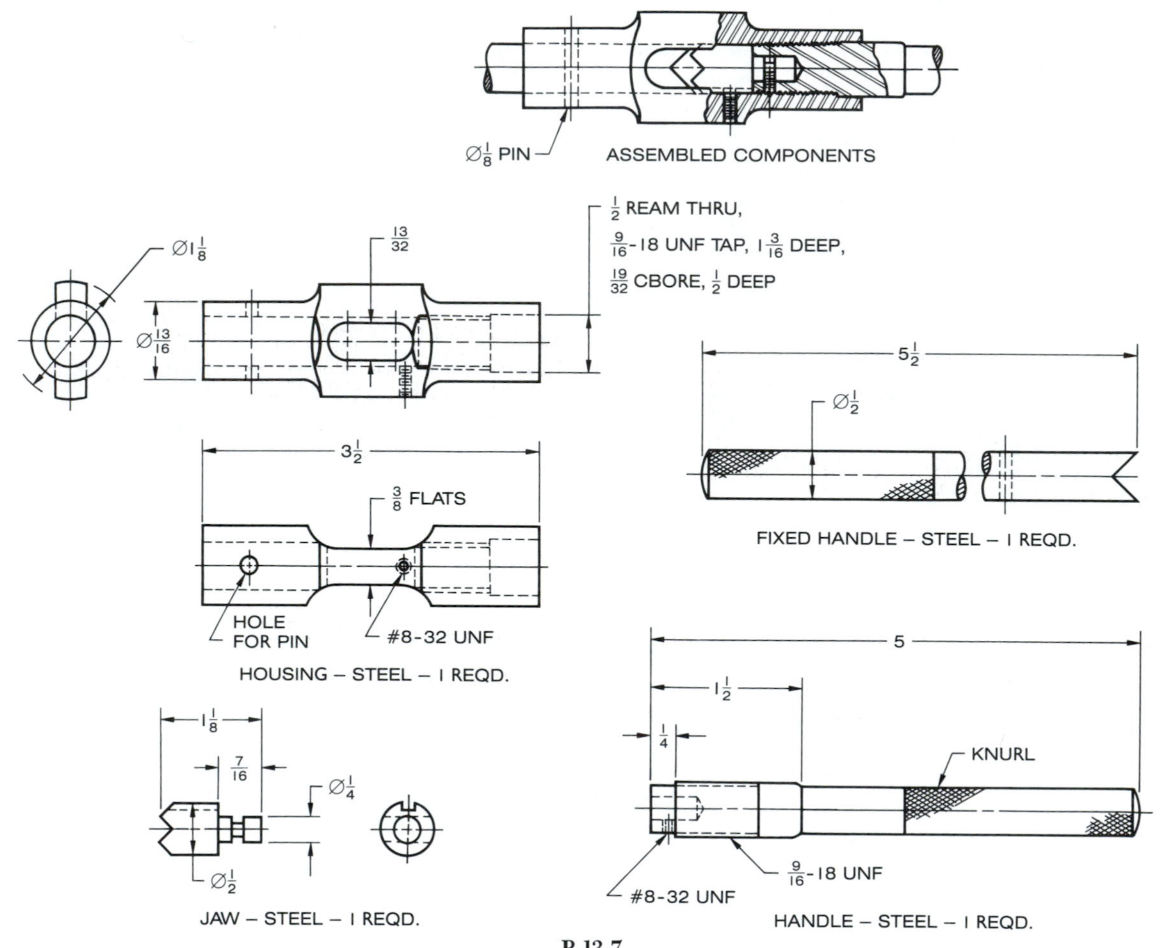

P 13.7

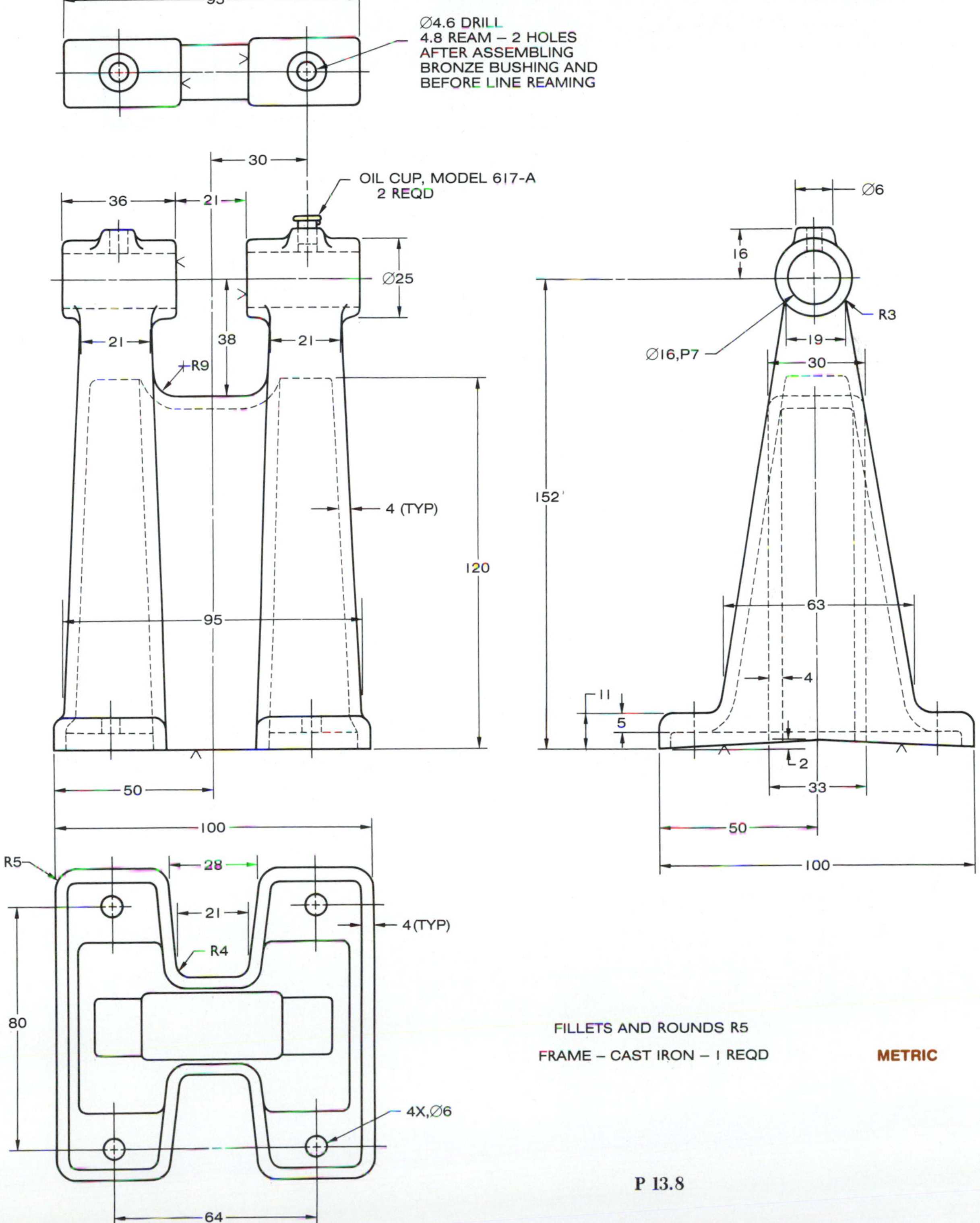
95
∅4.6 DRILL
4.8 REAM – 2 HOLES
AFTER ASSEMBLING
BRONZE BUSHING AND
BEFORE LINE REAMING
30
OIL CUP, MODEL 617-A
2 REQD
36
21
∅25
38
21
21
R9
4 (TYP)
120
95
50
100
∅6
16
R3
∅16,P7
19
30
152
63
4
11
5
2
33
50
100
R5
28
21
4(TYP)
R4
80
4X,∅6
64
FILLETS AND ROUNDS R5
FRAME – CAST IRON – I REQD
METRIC

P 13.8

P13.8 (continued)

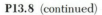

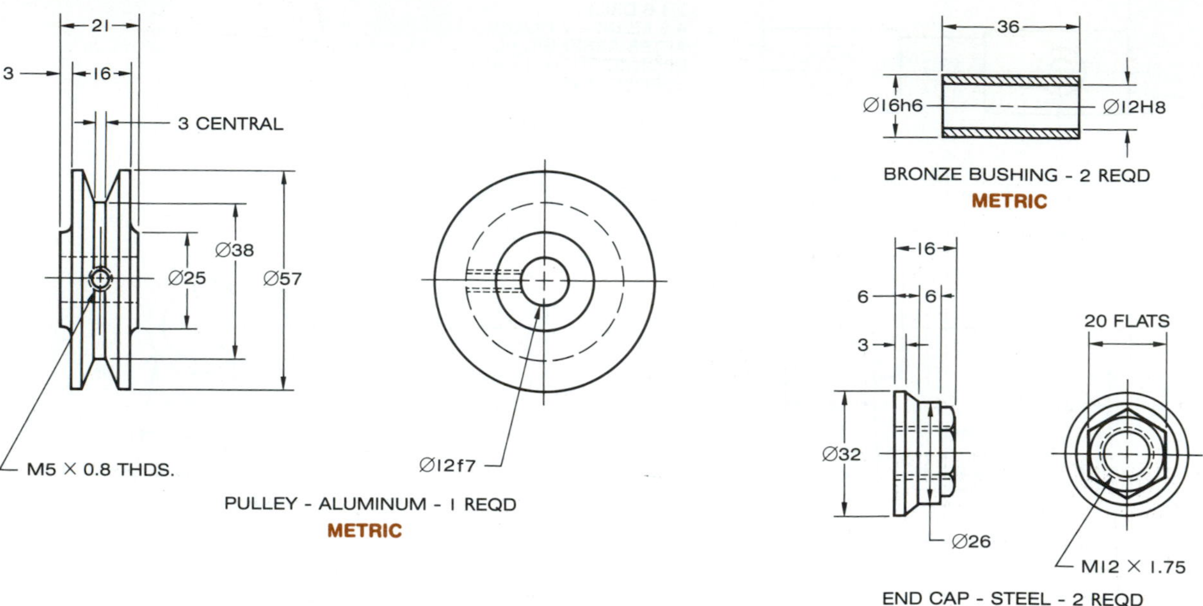

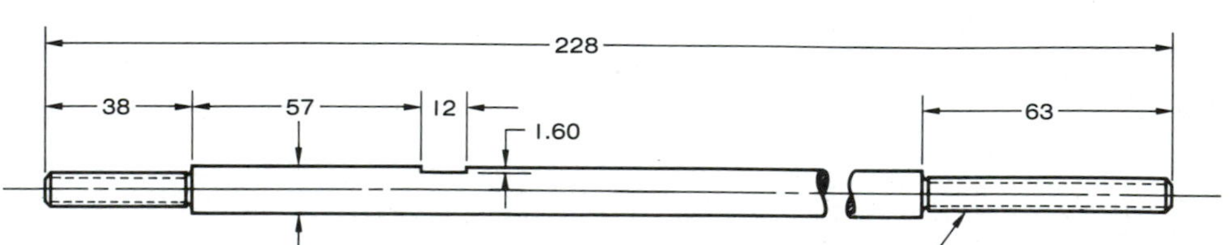

P13.9 C-hanger.

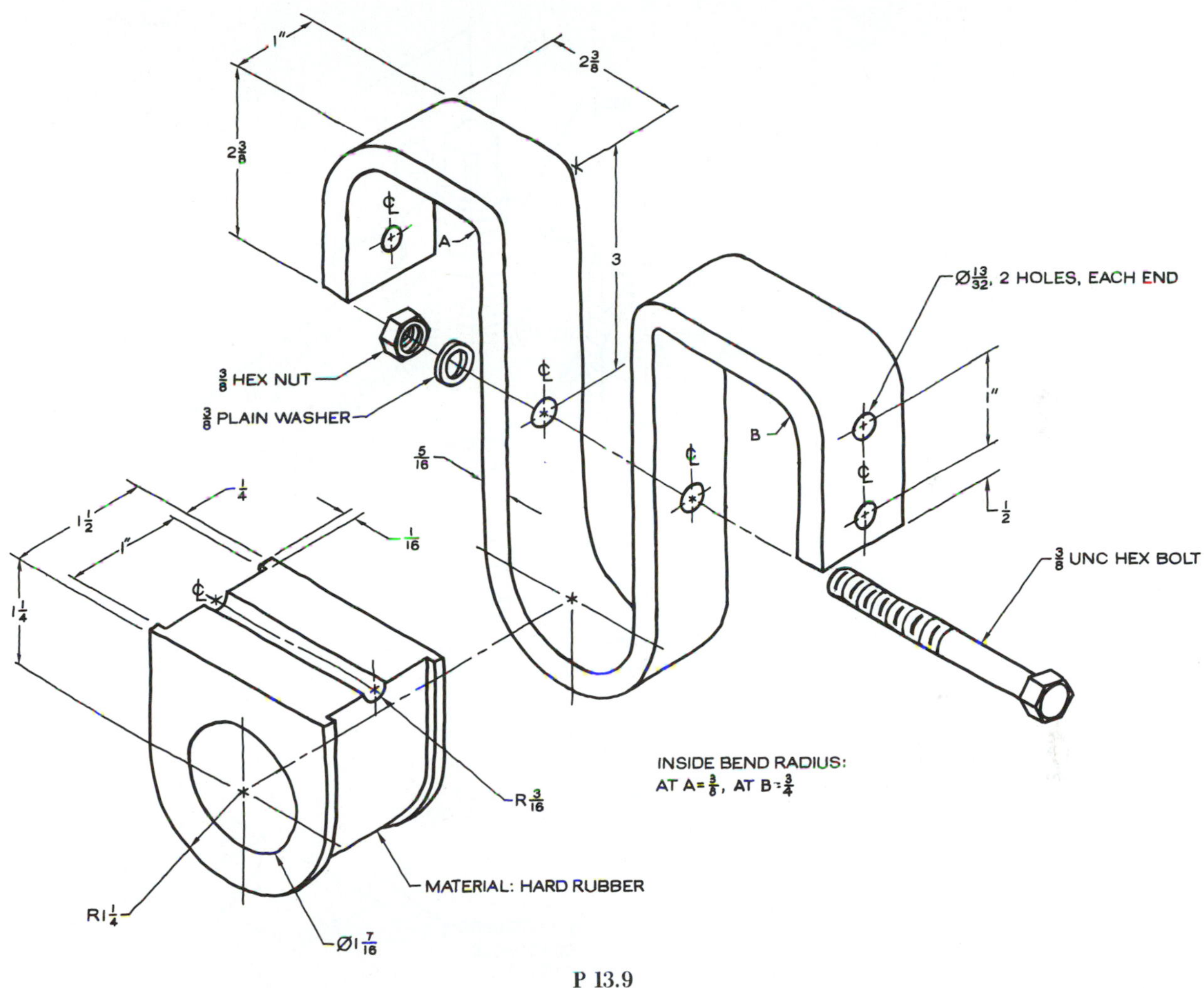
1"
2 3/8
2 3/8
A
3
Ø 13/32, 2 HOLES, EACH END
3/8 HEX NUT
3/8 PLAIN WASHER
5/16
B
1/2
3/8 UNC HEX BOLT
1 1/2
1/4
1"
1/16
1 1/4
R 3/16
INSIDE BEND RADIUS:
AT A = 3/8, AT B = 3/4
MATERIAL: HARD RUBBER
R1 1/4
Ø1 7/16

P 13.9

P13.10 Wheel puller.

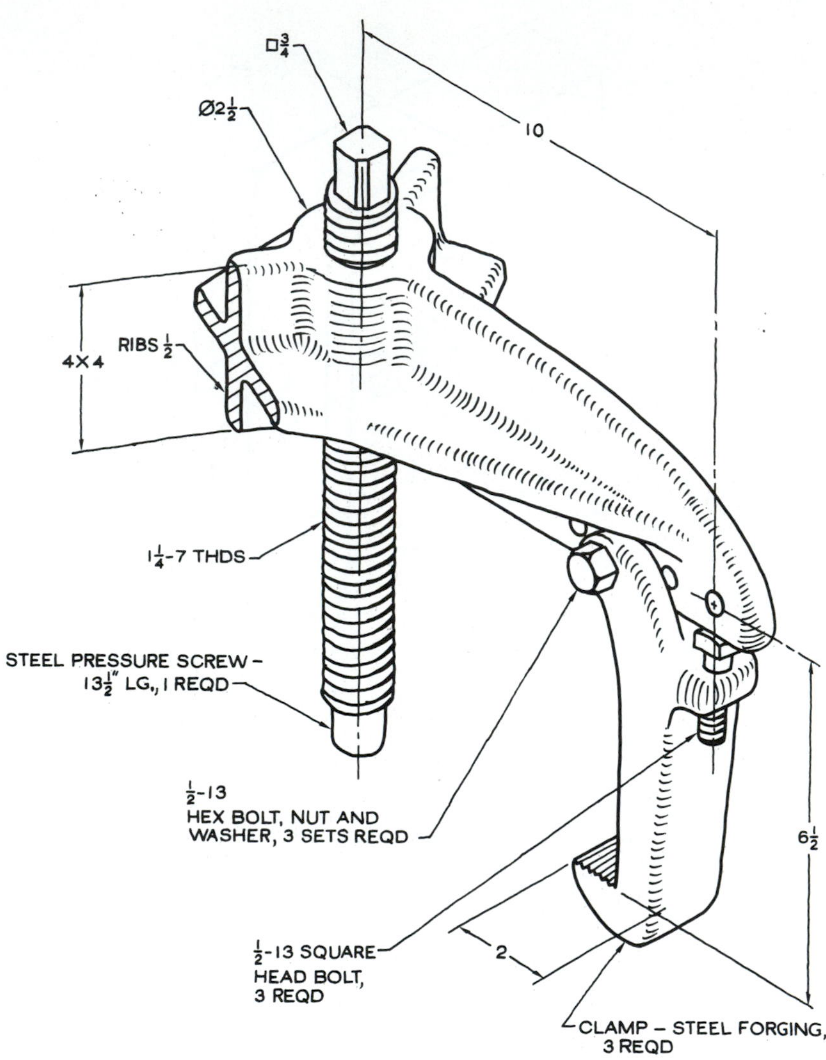

P 13.10

P13.11 Caulking gun.

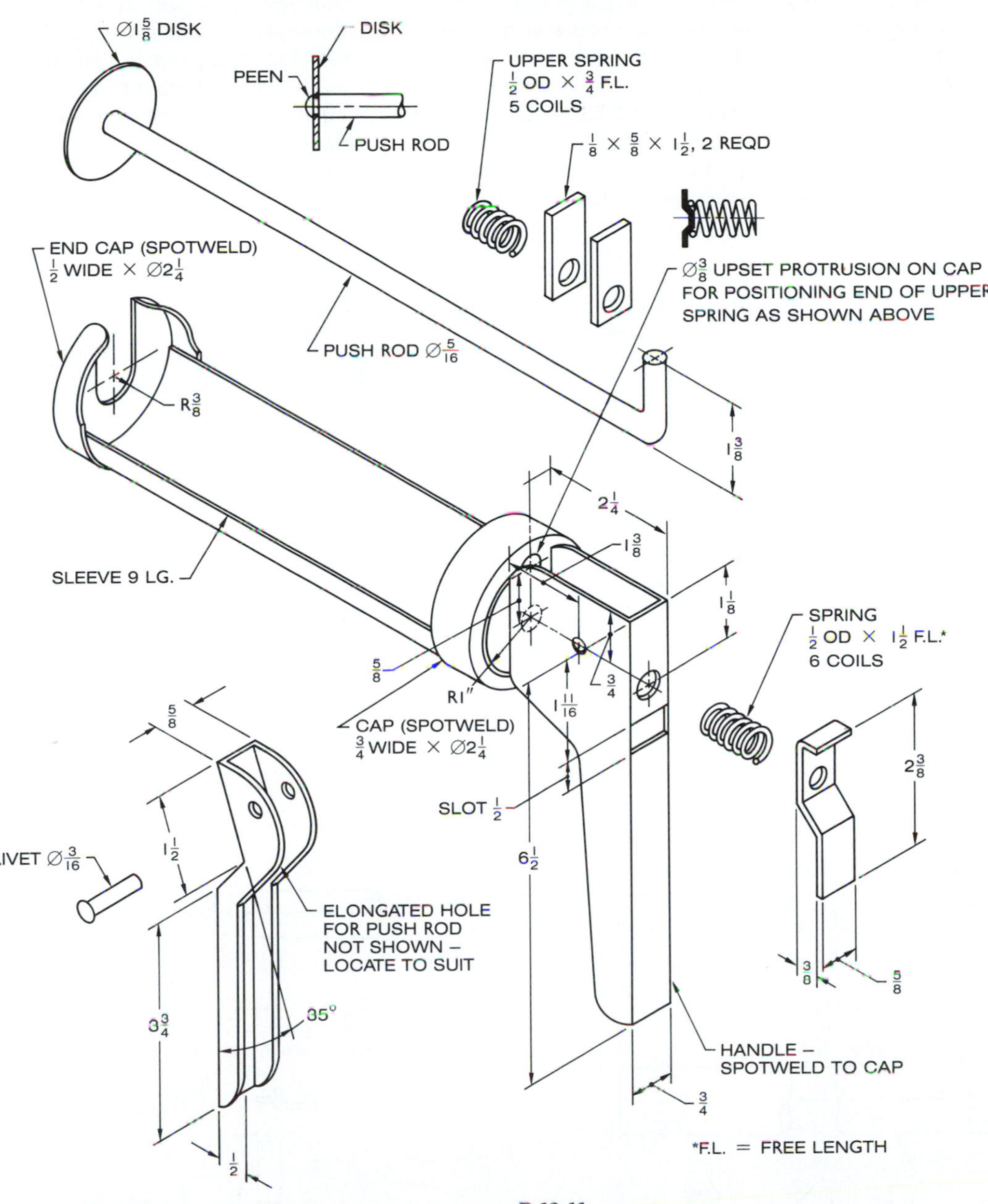

P 13.11

P13.12 Faucet wrench. This all-purpose wrench is used in confined areas in plumbing or automotive applications. It fits nut-flats in sizes ranging from ½ to 1⅛″. The serrations permit gripping around a smooth pipe. The right-angle feature is used in cramped locations for adjusting sink lock nuts or water-supply coupling nuts. The snapback spring action maintains a constant grip on fittings. The swivel jaw turns 180° for flexibility.

P13.13 to **P13.23** Make a complete set of working drawings of the problems in this group. The complete set should consist of detail drawings of the individual parts (with the exception of standard purchased parts) and an assembly drawing complete with a bill of material. Approximate all omitted sizes. Do not attempt to obtain omitted sizes by measuring the drawings because they are not drawn to scale.

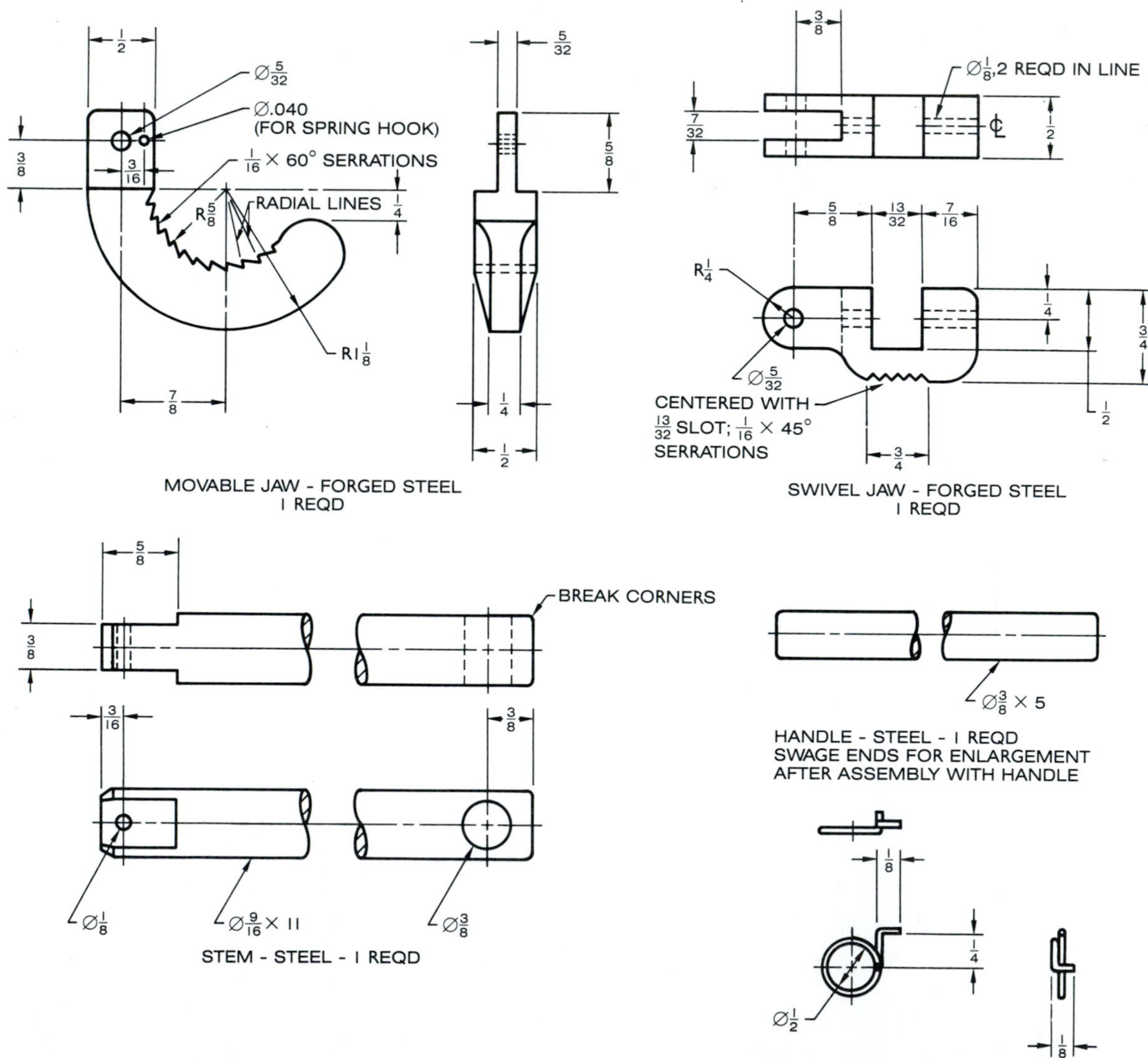

P 13.12

P13.13 (pages 449 to 453) Forming punch and die set.

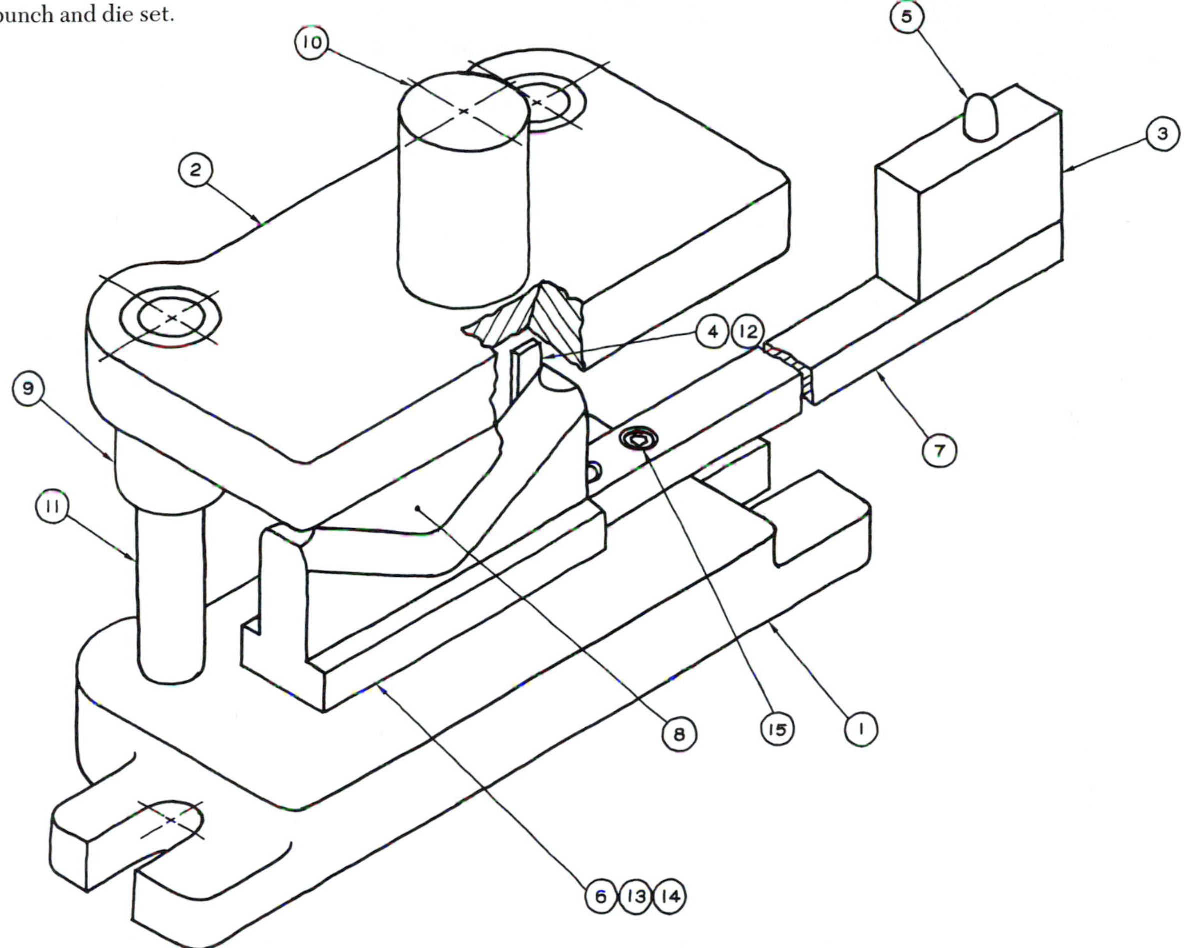

15	4	SOC. HD CAP SCR.	STD.
14	4	DOWEL PIN	D.R.
13	8	SOC. HD CAP SCR.	STD.
12	2	SOC. HD CAP SCR.	STD.
11	2	ALIGNMENT POST	D.R.
10	1	PLUG	STL.
9	2	GUIDE BUSHING	STL.
8	1	PUNCH	STL.
7	1	STANDOFF	STL.
6	1	DIE	STL.
5	1	PLUG	D.R.
4	1	STOP	STL.
3	1	BLOCK	STL.
2	1	PLATE	CI
1	1	BASE	CI
DET. NO.	NO. REQD	NAME	MAT.

P13.13 (continued)

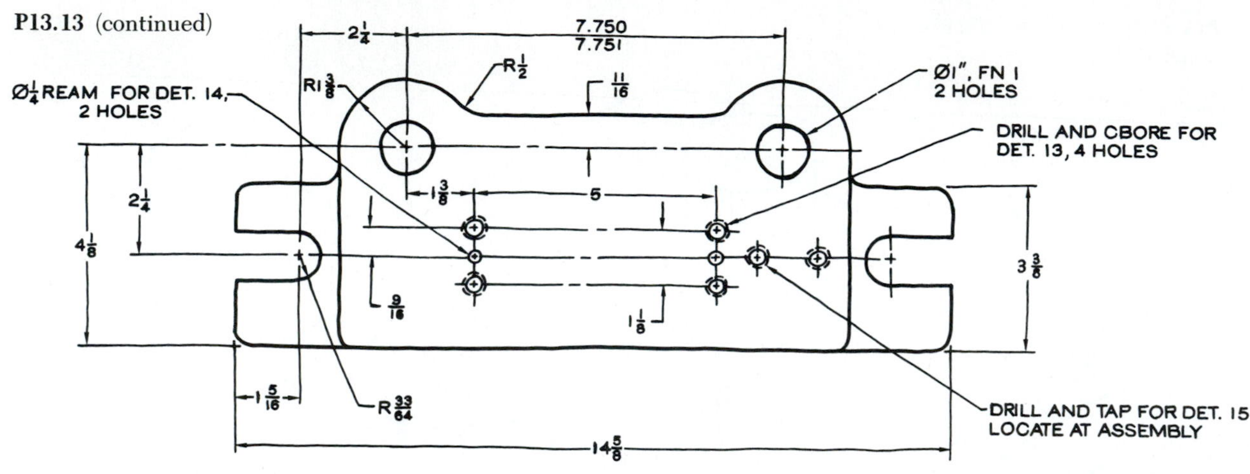

FILLETS AND ROUNDS R⅛

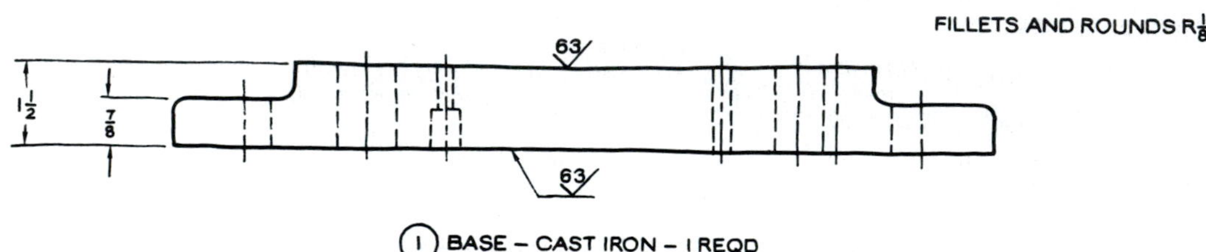

(1) BASE – CAST IRON – 1 REQD

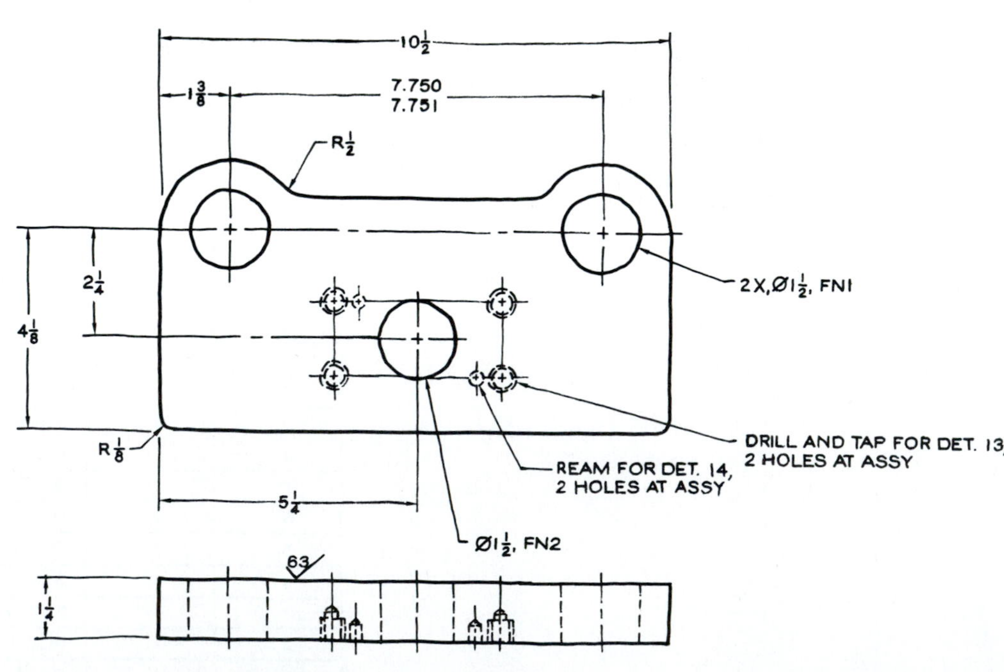

(2) PLATE – STEEL – 1 REQD

P13.13 (continued)

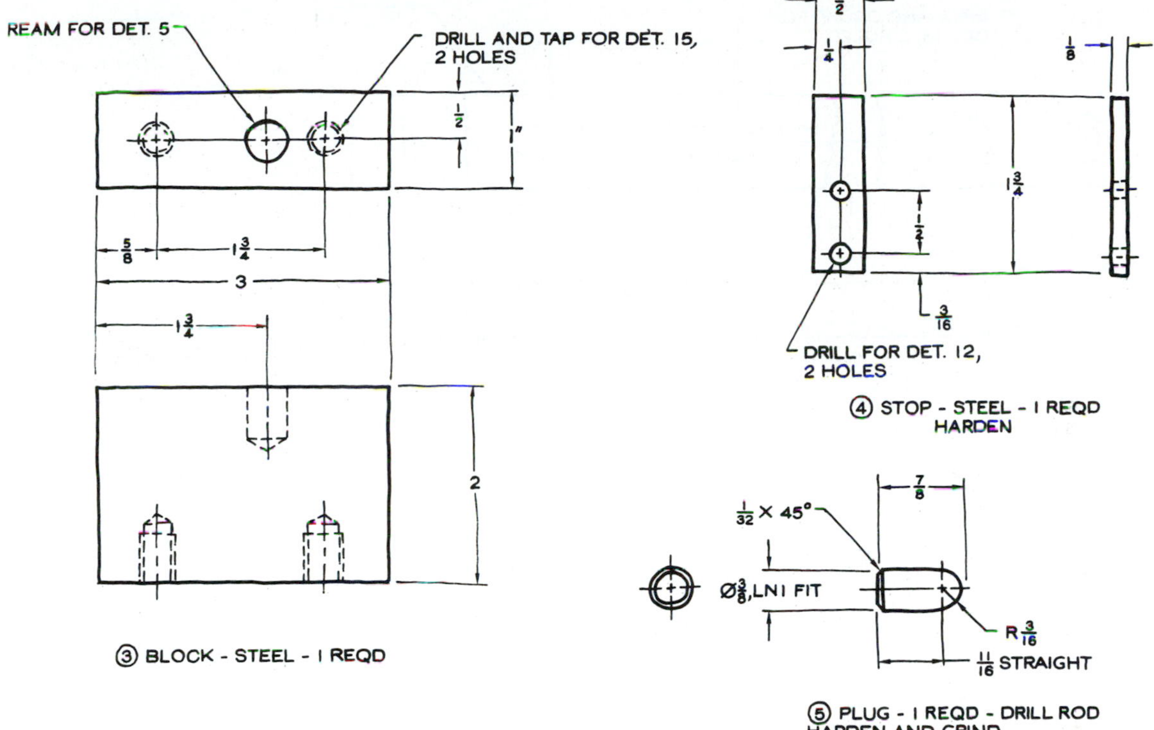
REAM FOR DET. 5
DRILL AND TAP FOR DET. 15, 2 HOLES
DRILL FOR DET. 12, 2 HOLES
③ BLOCK - STEEL - 1 REQD
④ STOP - STEEL - 1 REQD HARDEN
1/32 × 45°
Ø3/8, LN1 FIT
R 3/16
11/16 STRAIGHT
⑤ PLUG - 1 REQD - DRILL ROD HARDEN AND GRIND

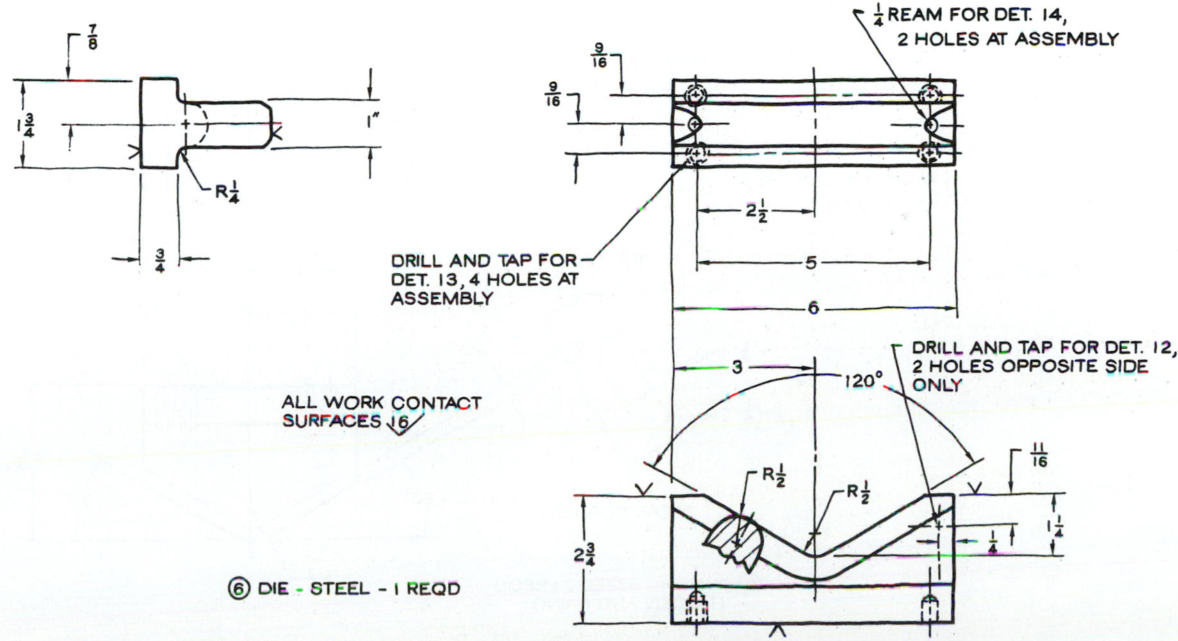
1/4 REAM FOR DET. 14, 2 HOLES AT ASSEMBLY
DRILL AND TAP FOR DET. 13, 4 HOLES AT ASSEMBLY
DRILL AND TAP FOR DET. 12, 2 HOLES OPPOSITE SIDE ONLY
ALL WORK CONTACT SURFACES 16
⑥ DIE - STEEL - 1 REQD

P13.13 (continued)

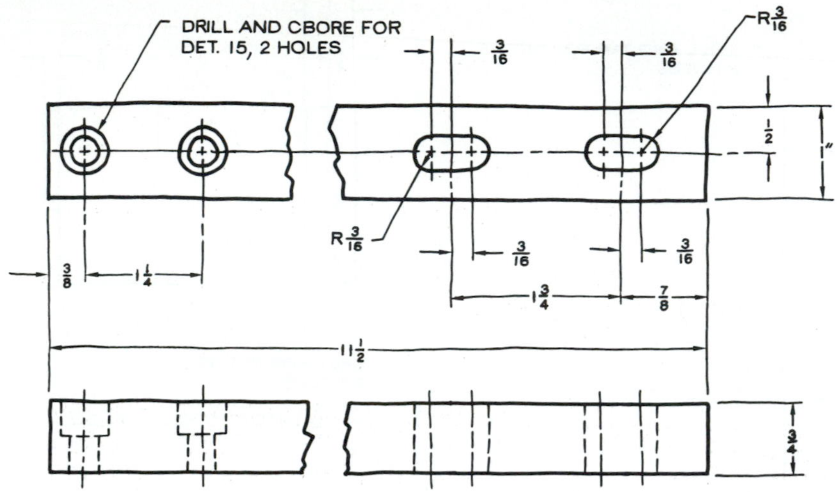

⑦ STANDOFF - STEEL - 1 REQD

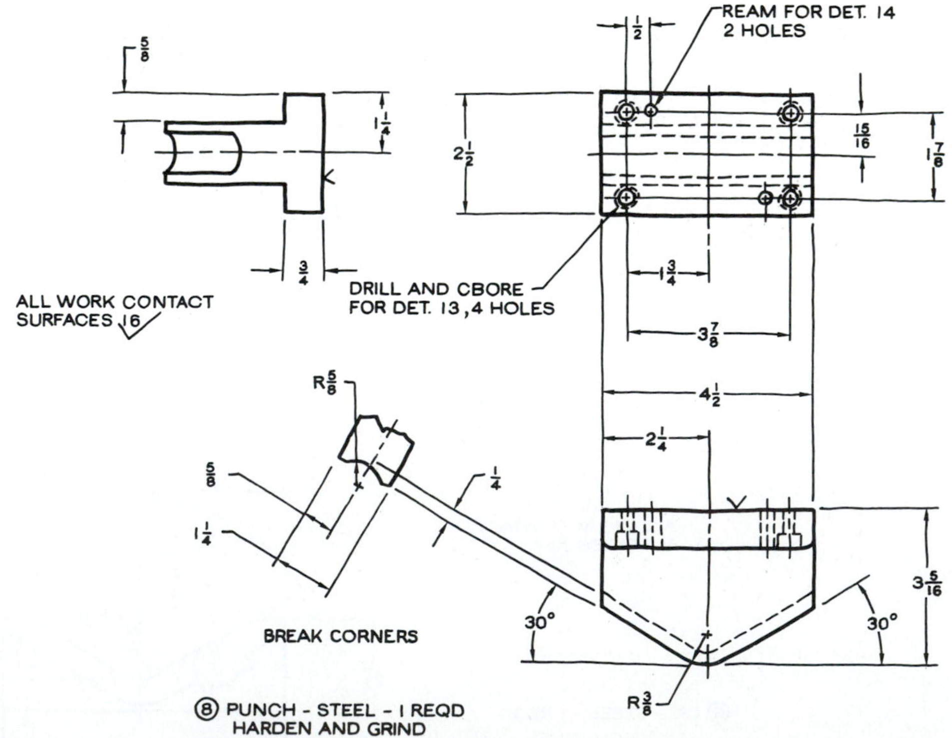

⑧ PUNCH - STEEL - 1 REQD
HARDEN AND GRIND

P13.13 (continued)

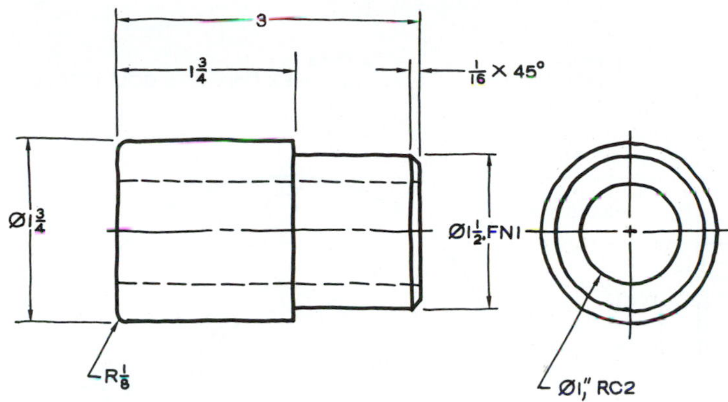

⑨ GUIDE BUSHING - STEEL - 2 REQD
HARDEN

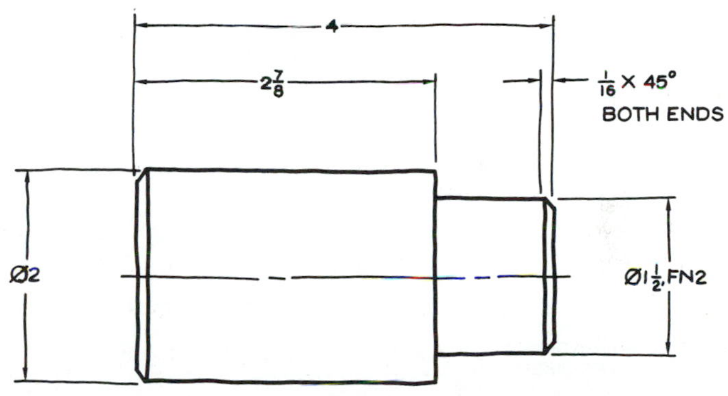

⑩ PLUG - STEEL - 1 REQD

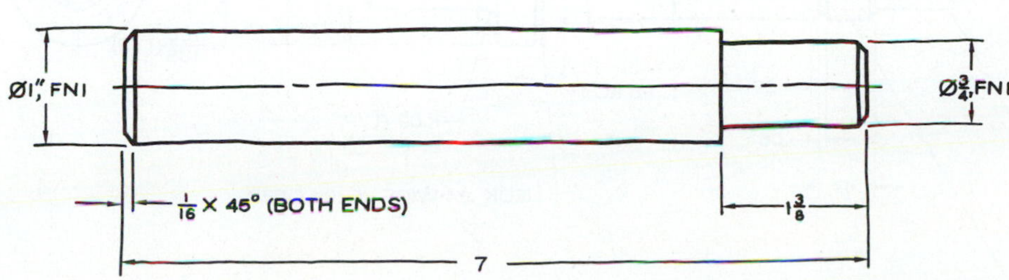

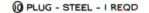

⑪ ALIGNMENT POST - DRILL ROD - 2 REQD

P13.14 (pages 454 and 455)
Housing assembly.

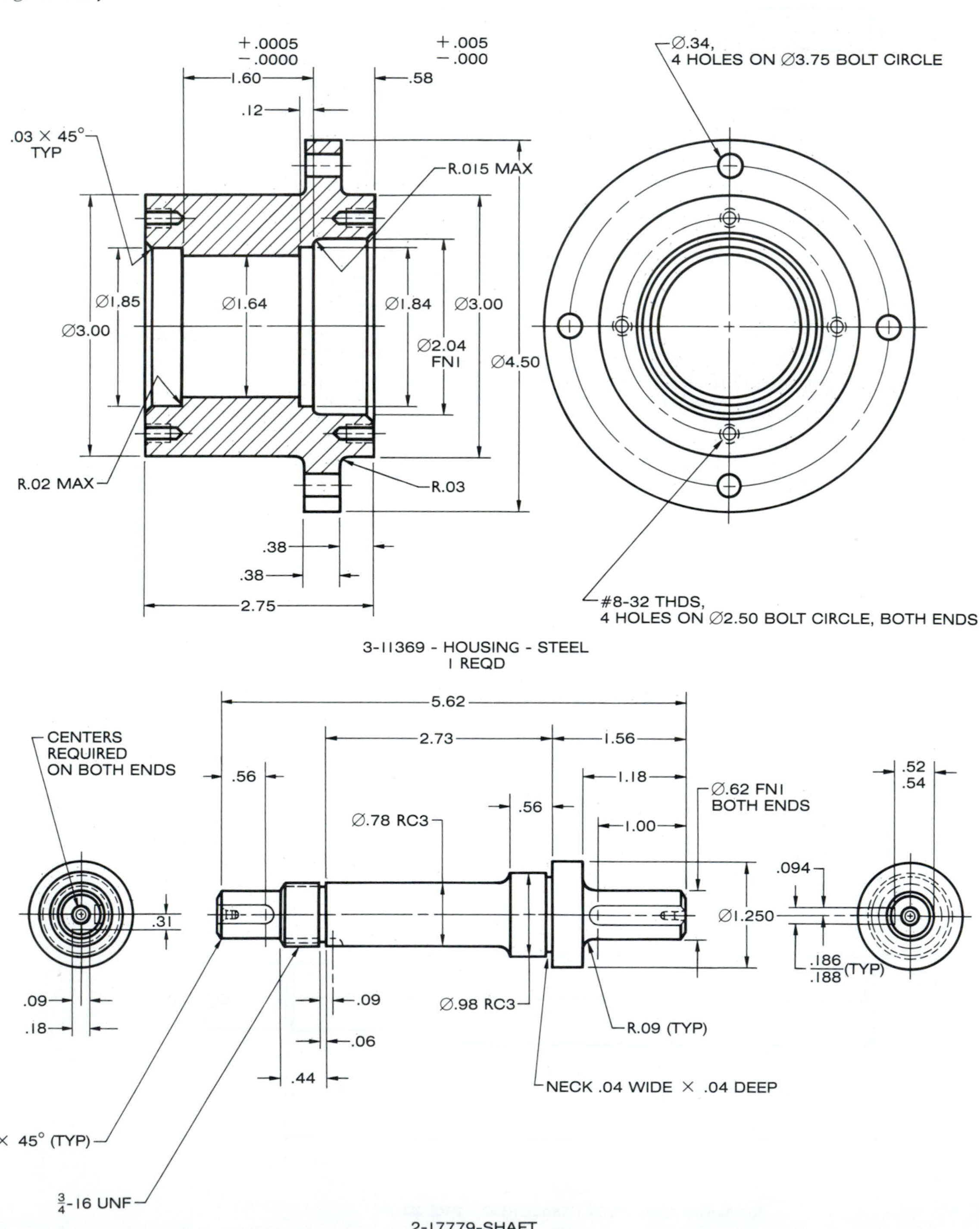
+.0005
−.0000
1.60
+.005
−.000
.58
.12
.03 × 45°
TYP
R.015 MAX
Ø.34,
4 HOLES ON Ø3.75 BOLT CIRCLE
Ø1.85
Ø1.64
Ø1.84
Ø3.00
Ø3.00
Ø2.04
FN1
Ø4.50
R.02 MAX
R.03
.38
.38
2.75
#8-32 THDS,
4 HOLES ON Ø2.50 BOLT CIRCLE, BOTH ENDS
3-11369 - HOUSING - STEEL
1 REQD
5.62
2.73
1.56
CENTERS
REQUIRED
ON BOTH ENDS
.56
1.18
Ø.62 FN1
BOTH ENDS
.52
.54
.56
Ø.78 RC3
1.00
.094
.31
Ø1.250
.186
.188 (TYP)
.09
.09
Ø.98 RC3
.18
.06
R.09 (TYP)
.44
NECK .04 WIDE × .04 DEEP
.03 × 45° (TYP)
3/4-16 UNF
2-17779-SHAFT
STEEL – 1 REQD

P13.14 (continued)

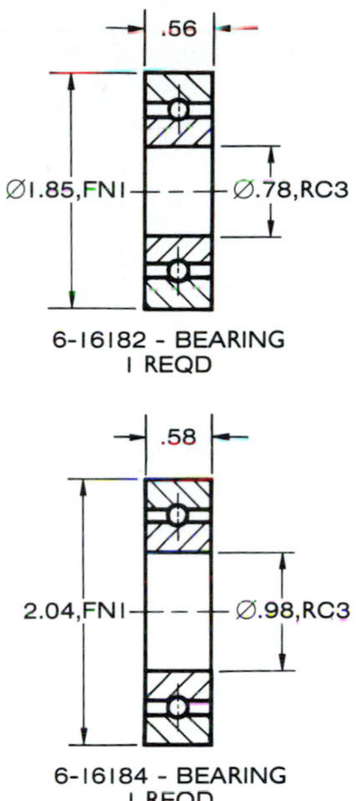
.56
Ø1.85,FN1
Ø.78,RC3
6-16182 - BEARING
1 REQD
.58
2.04,FN1
Ø.98,RC3
6-16184 - BEARING
1 REQD

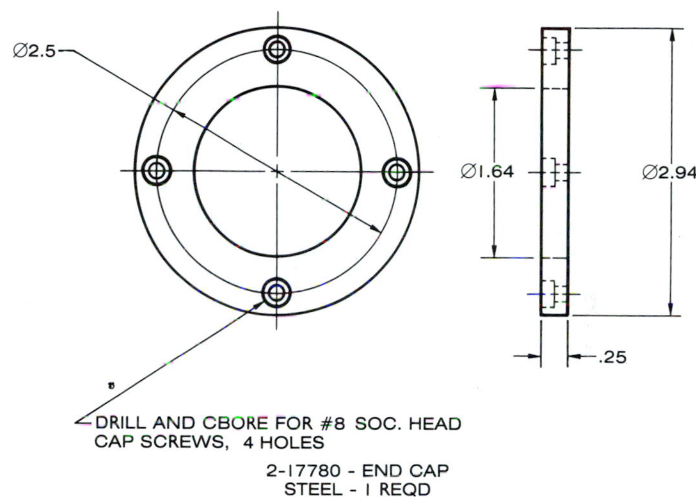
Ø2.5
Ø1.64
Ø2.94
.25
DRILL AND CBORE FOR #8 SOC. HEAD CAP SCREWS, 4 HOLES
2-17780 - END CAP
STEEL - 1 REQD

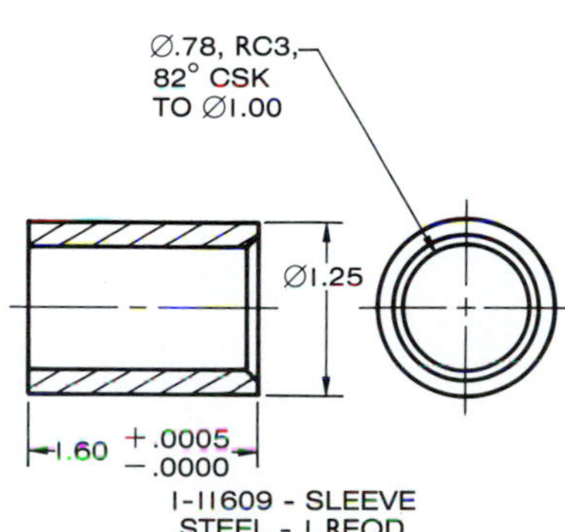
Ø.78, RC3, 82° CSK TO Ø1.00
Ø1.25
1.60 +.0005 −.0000
1-11609 - SLEEVE
STEEL - 1 REQD

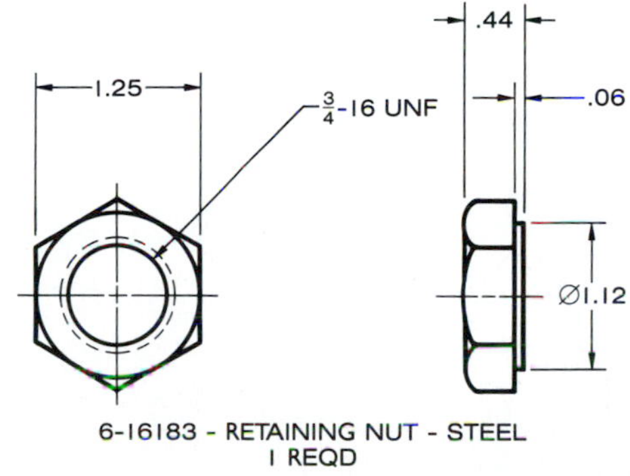
.44
1.25
3/4-16 UNF
.06
Ø1.12
6-16183 - RETAINING NUT - STEEL
1 REQD

P13.15 Work-holding hinge clamp.

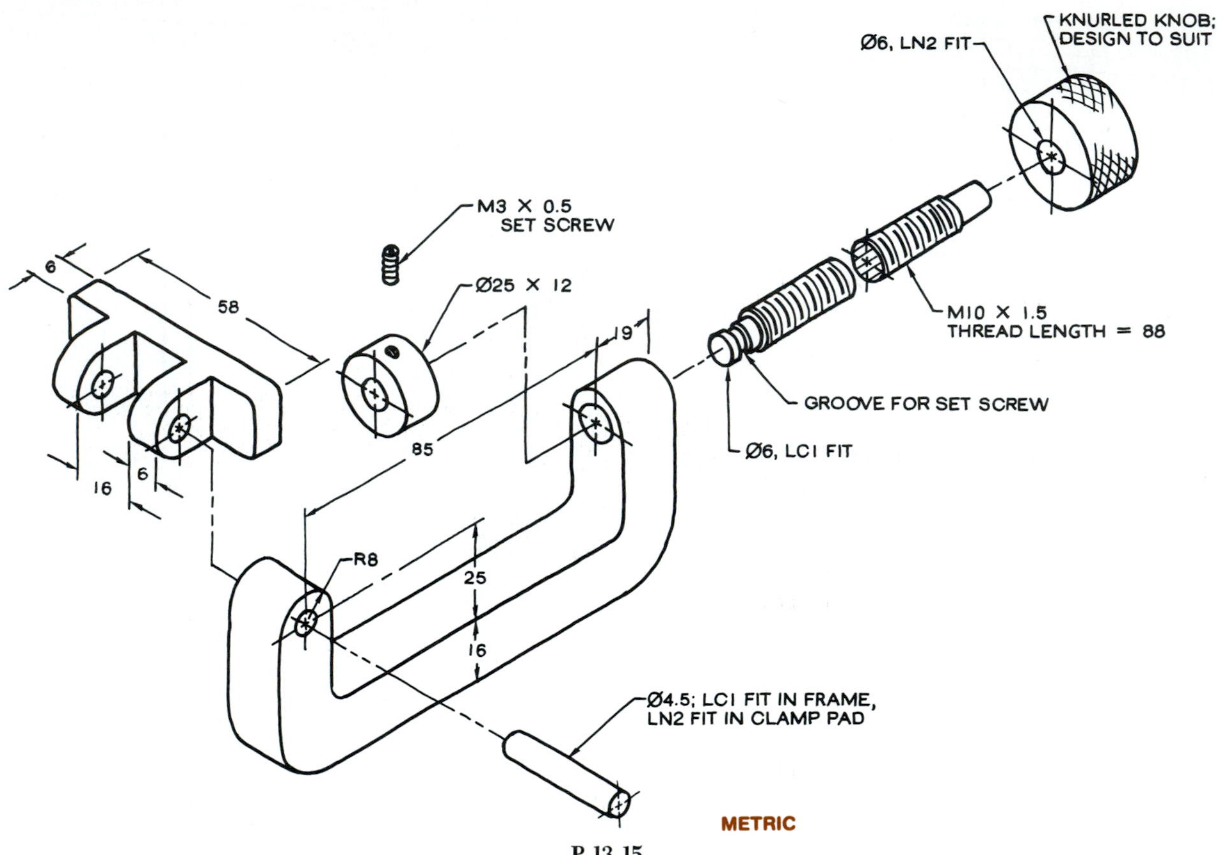

P 13.15

P13.16 Vise

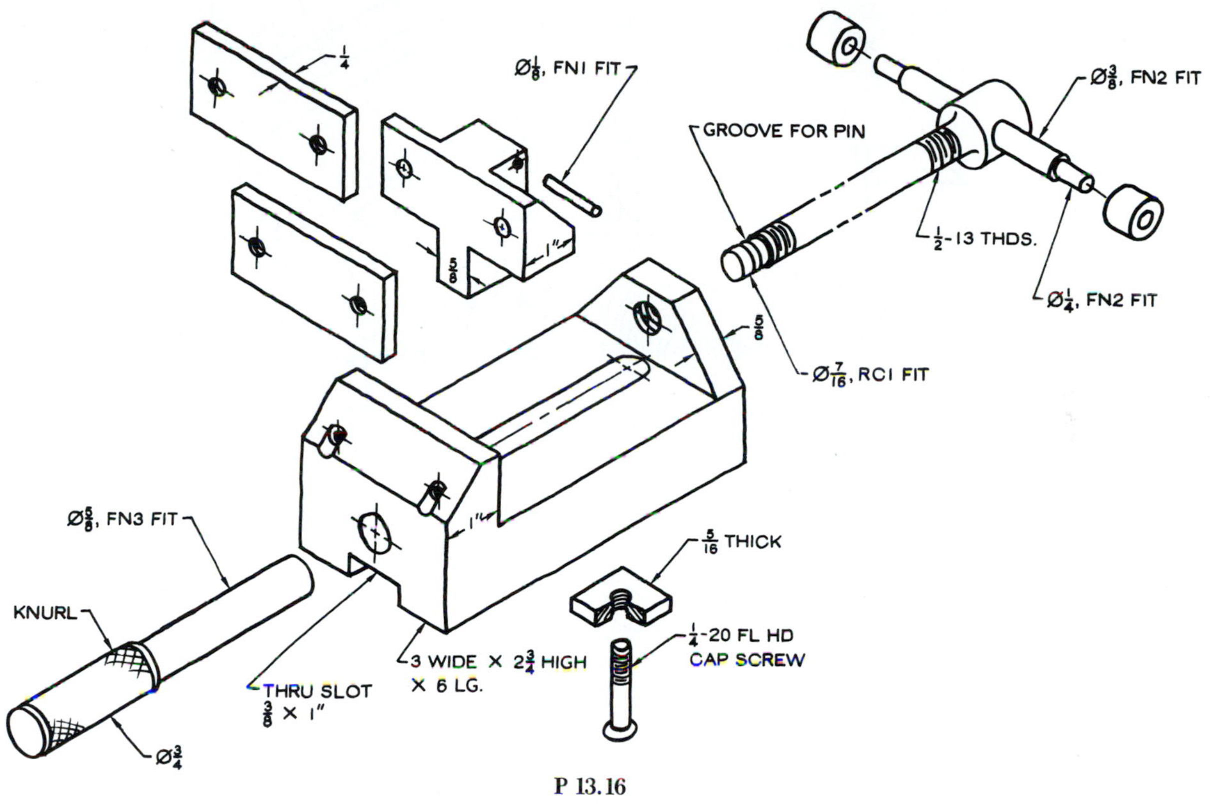

P 13.16

P13.17 Vise

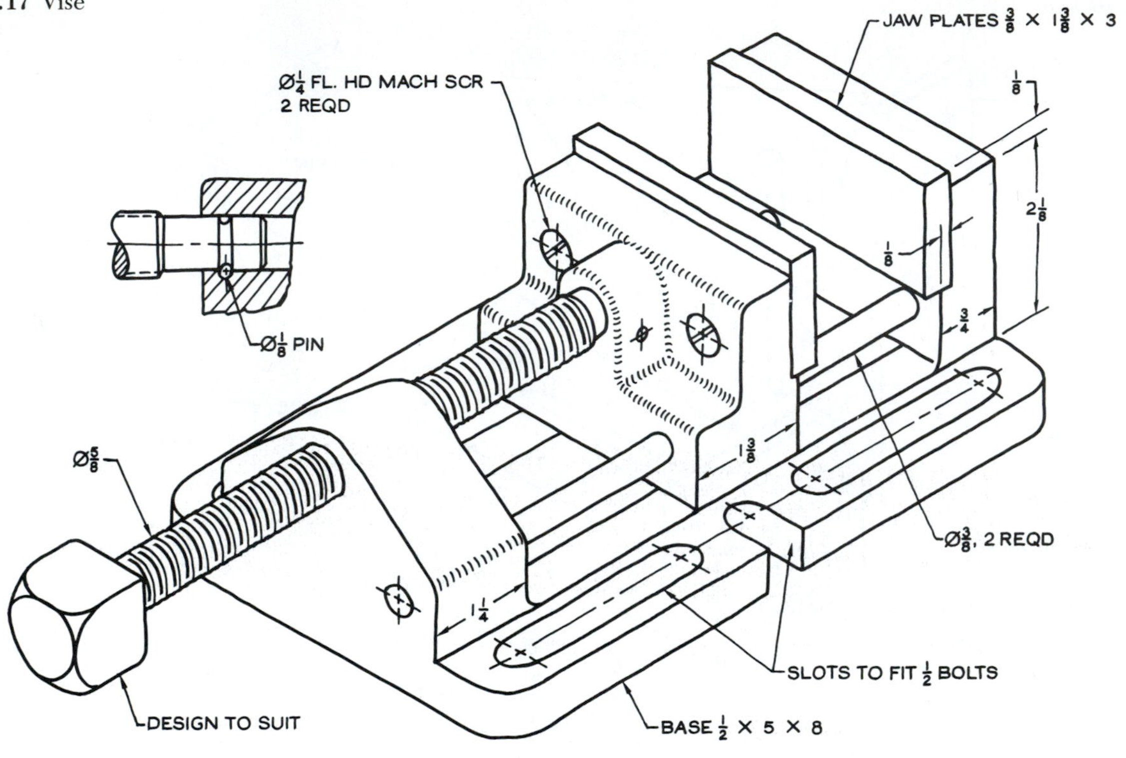

FILLETS AND ROUNDS R$\frac{1}{8}$

P 13.17

P13.18 Horizontal fixture clamp.

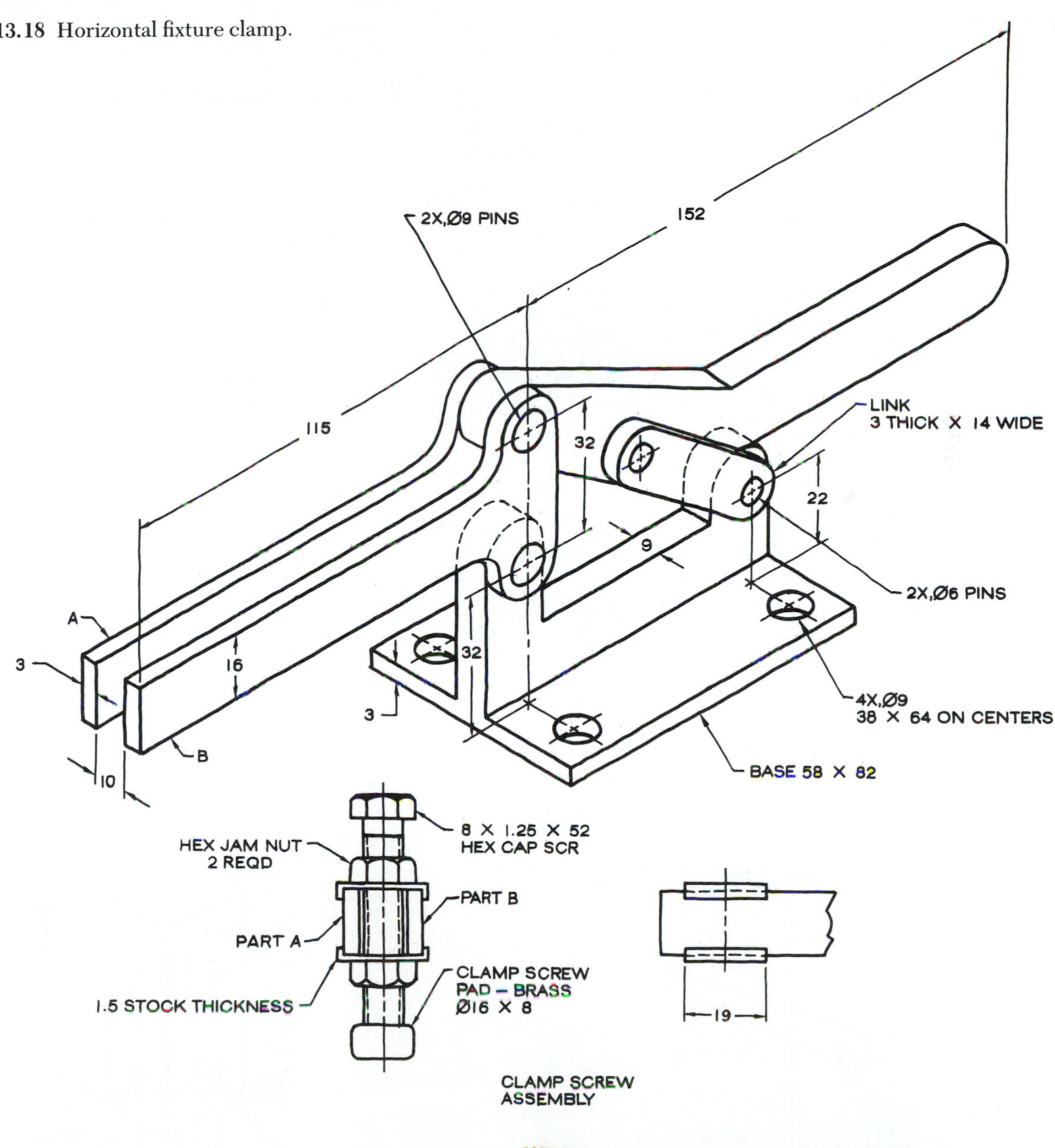

P 13.18

P13.19 Hobby vise.

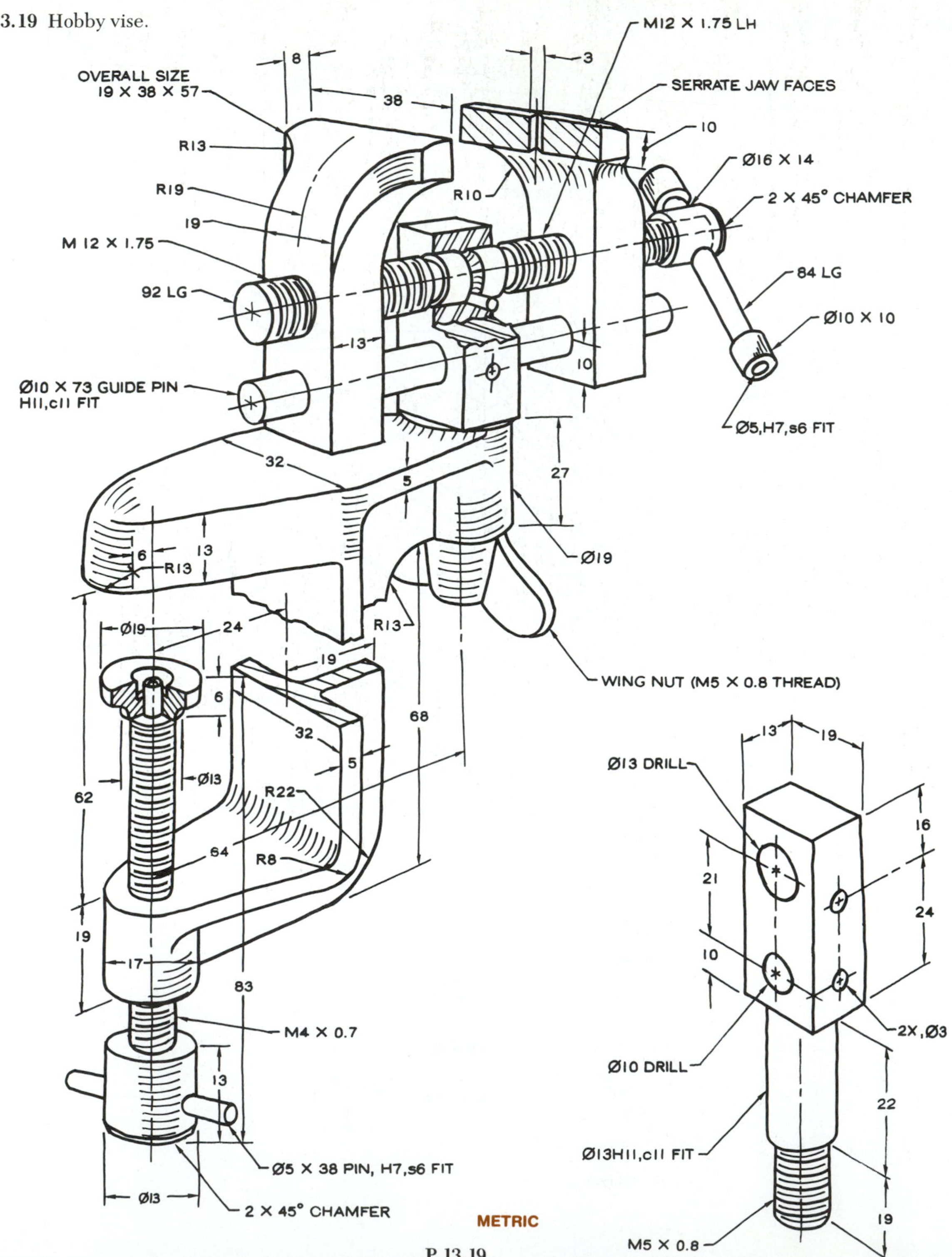

P 13.19

P13.20 Toolmaker's vise.

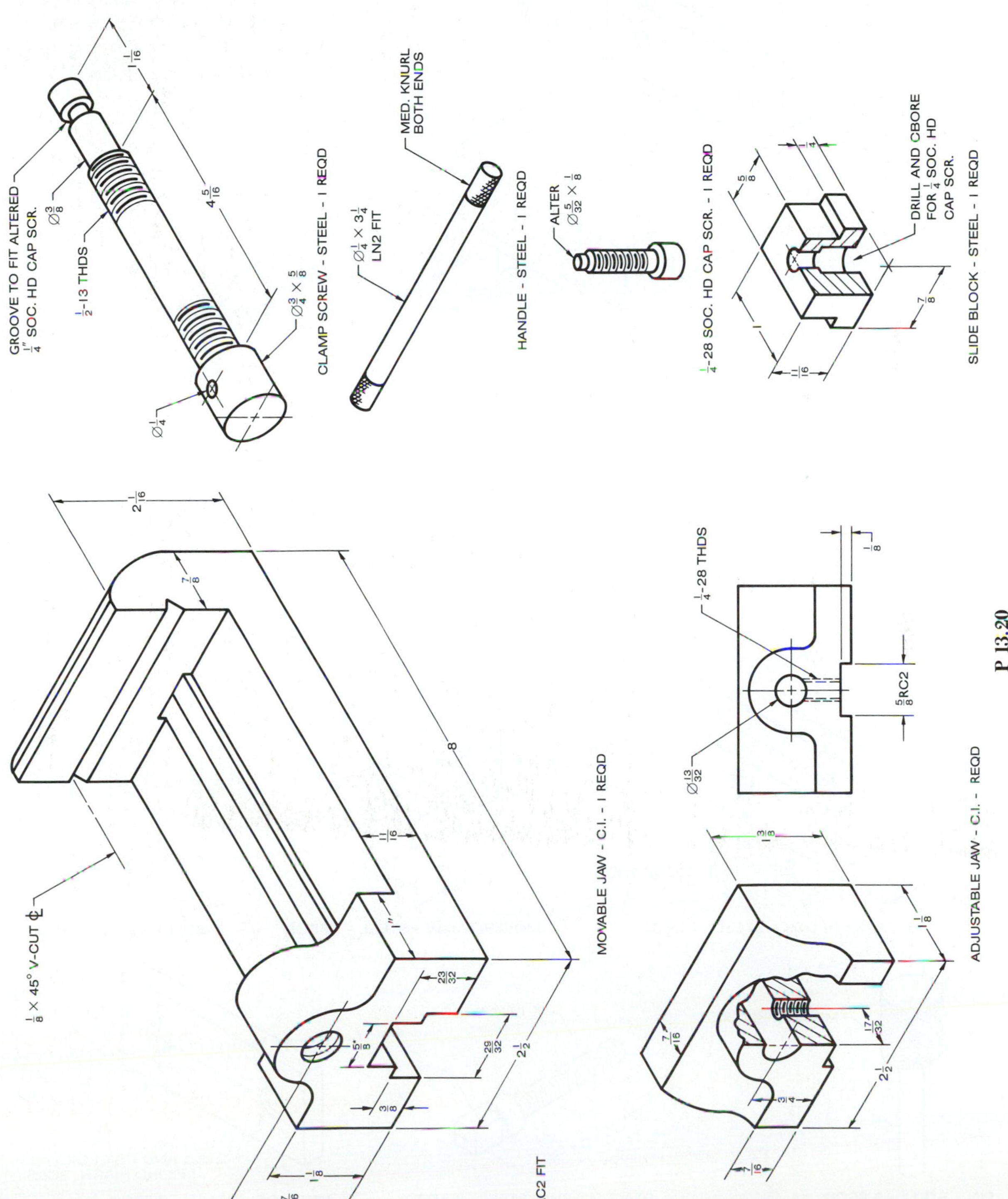
GROOVE TO FIT ALTERED 1/4" SOC. HD CAP SCR.
1/2-13 THDS
CLAMP SCREW - STEEL - 1 REQD
MED. KNURL BOTH ENDS
LN2 FIT
HANDLE - STEEL - 1 REQD
ALTER
1/4-28 SOC. HD CAP SCR. - 1 REQD
DRILL AND CBORE FOR 1/4 SOC. HD CAP SCR.
SLIDE BLOCK - STEEL - 1 REQD
1/8 × 45° V-CUT ℄
MOVABLE JAW - C.I. - 1 REQD
*RC2 FIT
1/4-28 THDS
5/8 RC2
ADJUSTABLE JAW - C.I. - REQD
P 13.20

P13.21 Toolmaker's vise.

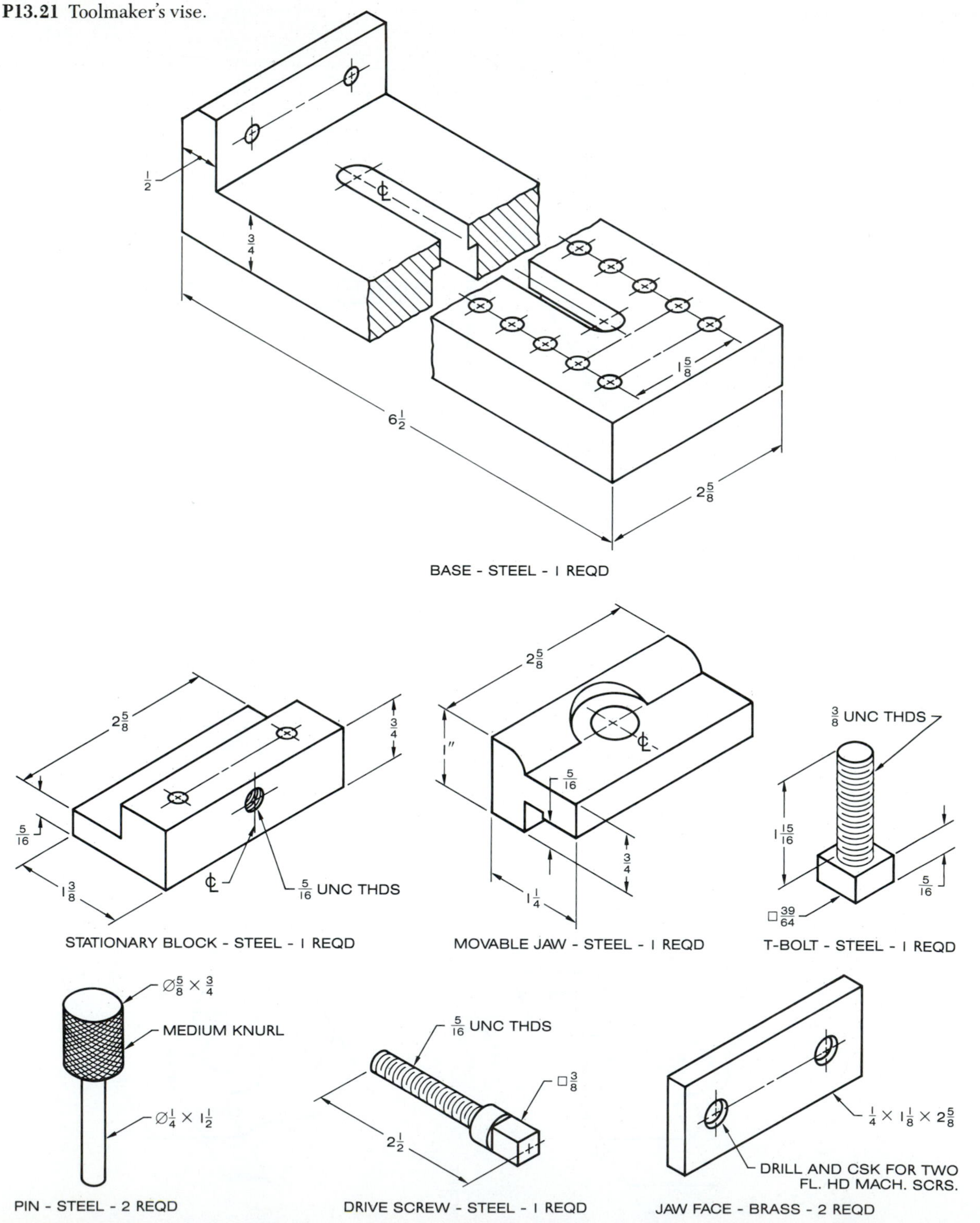

P13.22 Doweling jig.

DET. NO.	NAME	NO. REQD.	MAT.	DESCRIPTION
9	WASHER	2	STD.	$\frac{5}{32}$ I.D.
8	HEX HD MACH. SCR.	2	STD.	# 6-32 × $\frac{3}{8}$
7	SET SCR.	I	STD.	$\frac{1}{4}$-28 × $\frac{5}{8}$ SOC. HD
6	HANDLE	I	D.R.	$\frac{1}{4}$ DIA. × $2\frac{1}{2}$
5	CLAMPING SCR.	I	STL.	$\frac{5}{8}$ DIA. × $3\frac{3}{4}$
4	ALIGNMENT RODS	2	D.R.	$\frac{3}{8}$ DIA. × $3\frac{1}{2}$
3	DRILL GUIDE BLOCK	I	STL.	$\frac{5}{8}$ × $1\frac{1}{2}$ × $4\frac{3}{4}$
2	JAW	I	AL.	$\frac{5}{8}$ × $2\frac{3}{4}$ × $4\frac{3}{4}$
I	JAW	I	AL.	$\frac{5}{8}$ × $2\frac{3}{4}$ × $4\frac{3}{4}$

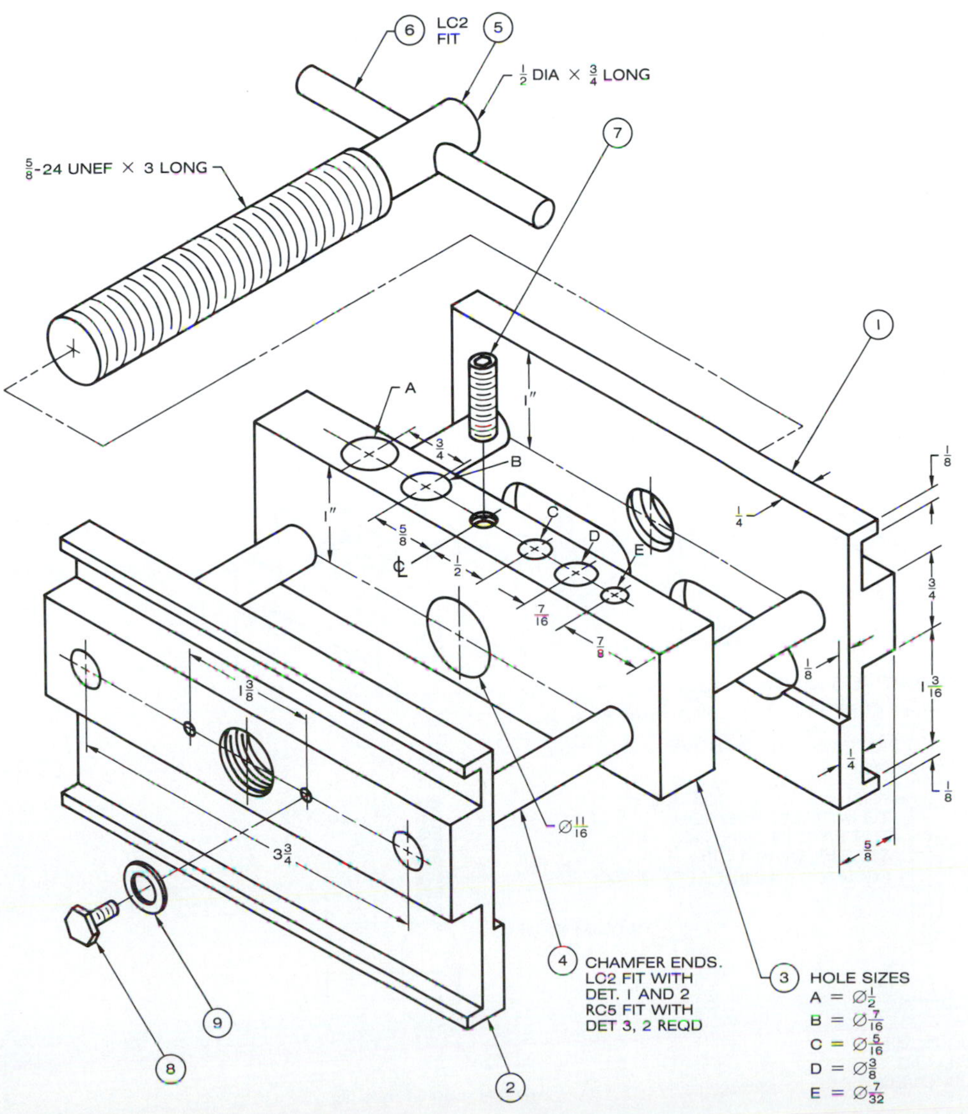

P13.23 Box-type drill jig.

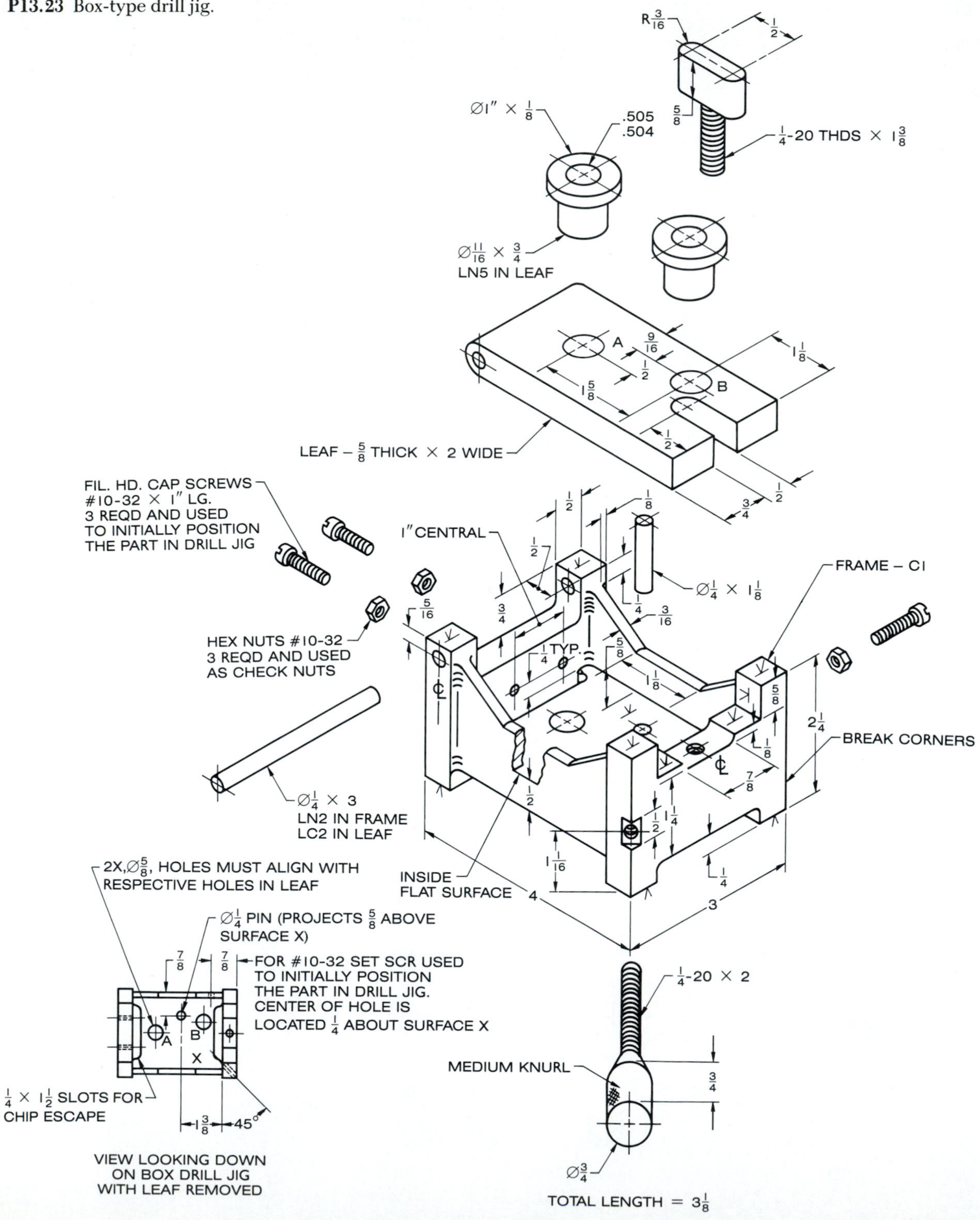

R 3/16
1/2
5/8
1/4-20 THDS × 1 3/8
Ø1″ × 1/8
.505
.504
Ø 11/16 × 3/4
LN5 IN LEAF
A
9/16
1/2
1 1/8
1 5/8
B
1/2
LEAF – 5/8 THICK × 2 WIDE
1/2
3/4
FIL. HD. CAP SCREWS #10-32 × 1″ LG. 3 REQD AND USED TO INITIALLY POSITION THE PART IN DRILL JIG
1″ CENTRAL
Ø 1/4 × 1 1/8
FRAME – C1
HEX NUTS #10-32 3 REQD AND USED AS CHECK NUTS
1/4 TYP.
3/16
5/16
BREAK CORNERS
2 1/4
Ø 1/4 × 3
LN2 IN FRAME
LC2 IN LEAF
2X, Ø 5/8, HOLES MUST ALIGN WITH RESPECTIVE HOLES IN LEAF
INSIDE FLAT SURFACE
1 1/16
1 1/4
4
3
Ø 1/4 PIN (PROJECTS 5/8 ABOVE SURFACE X)
FOR #10-32 SET SCR USED TO INITIALLY POSITION THE PART IN DRILL JIG. CENTER OF HOLE IS LOCATED 1/4 ABOUT SURFACE X
7/8
1/4 × 1 1/2 SLOTS FOR CHIP ESCAPE
1 3/8
45°
VIEW LOOKING DOWN ON BOX DRILL JIG WITH LEAF REMOVED
1/4-20 × 2
MEDIUM KNURL
Ø 3/4
TOTAL LENGTH = 3 1/8

Principles of Basic Descriptive Geometry

OBJECTIVES

After completing this chapter you should have gained the following abilities:

1. To select views that will describe:
- **a.** the position of a point
- **b.** the direction and position of a line

2. To identify various kinds of principal lines:
- **a.** horizontal
- **b.** frontal
- **c.** profile

3. To identify other kinds of lines:
- **a.** oblique
- **b.** parallel
- **c.** perpendicular
- **d.** intersection
- **e.** skew

4. To find the true length of a line.

5. To find the point view of a line.

6. To find the bearing and azimuth of a line.

7. To find the grade of a line.

8. To find the slope of a line.

9. To identify various kinds of principal planes:
- **a.** frontal
- **b.** horizontal
- **c.** profile

10. To identify other kinds of planes:
- **a.** inclined
- **b.** oblique

11. To use the proper method to:
- **a.** locate a point in a given plane
- **b.** find the edge view of a plane
- **c.** find the true size of a plane
- **d.** find the strike of a plane
- **e.** find the slope of a plane
- **f.** find the dip of a plane

12. To correctly use graphical methods in solving for:
- **a.** the shortest line from a point to a plane
- **b.** the angle between a line and a plane
- **c.** the intersection point of a line and a plane
- **d.** the shortest distance between two parallel planes
- **e.** the dihedral angle between two planes
- **f.** the shortest distance between two skew lines
- **g.** the line of intersection between two planes

Descriptive geometry is the theory of engineering graphics. The fundamentals of descriptive geometry employ the principles of orthographic projection, but in some cases it is impossible to obtain solutions with just the three principal views. An example would be an object with an inclined or an oblique surface where the surface would appear distorted in the principal views. An auxiliary view with the viewer's line of sight taken perpendicular to the surface would be required to show the true size of the surface. (Auxiliary views are discussed in Chapter 6.) True views of lines, angles, and planes are obtained by projecting three-dimensional figures onto a two-dimensional plane of paper.

Descriptive geometry is used as a problem-solving tool in many engineering branches, including mechanical, aeronautical, civil, electrical, chemical, and architectural engineering. As you will see, many descriptive geometry techniques used to solve problems are considerably simpler and more direct than pure mathematical methods.

The solutions of many engineering problems are greatly facilitated by computer-aided procedures. Many designers are now using graphical computers rather than tedious drawing methods to determine quickly and accurately various angles, lengths, and shapes. The material presented in this chapter consists of a brief description of the basic principles of descriptive geometry as they are applied to the solution of common engineering problems.

14.1 PROJECTION OF A POINT

A point is shown in the views by a small dot or by a cross. Figure 14-1 shows the projection of point a on the three principal planes of projection. The front view is labeled a_F, the top view is labeled a_H, and the right-side view is labeled a_P.

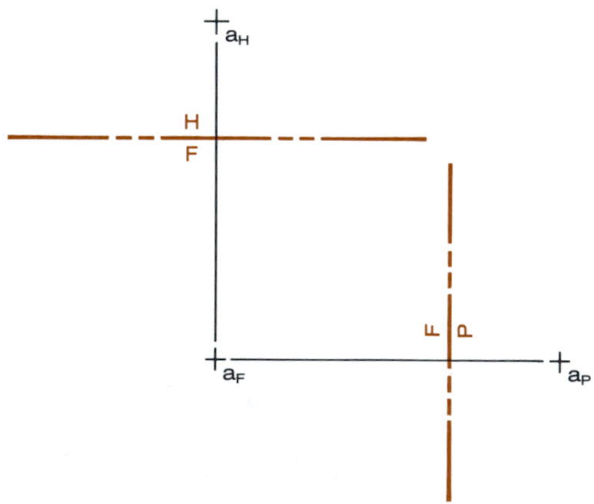

FIGURE 14-1 Projection of a point *(Section 14.1)*

14.2 POSITION OF A POINT

No one of the three principal views can completely describe the position of a point with respect to another point. In Figure 14-2 the coordinate position of point x with respect to the reference point y may be described as follows:

Front View: Point x_F is 60 mm below point y_F.

Top View: Point x_H is 110 mm to the left and 32 mm behind point y_H.

Any two adjacent views describe the position of one point with respect to another point. Note that in the top and front adjacent views, or in the front and right-side adjacent views, the coordinate position of point x with respect to point y may be located. Also, it is important to understand that the terms "above" and "below" broadly describe a position that only means "higher than" or "lower than" the reference point.

14.3 PROJECTION OF A LINE

The term "line" is understood to designate a *straight* line unless otherwise specified. Theoretically a straight line is indefinite in length. The end points in a straight line are labeled as shown in Figure 14-3.

14.4 DIRECTION AND POSITION OF A LINE

Any two points on a straight line will establish the *direction* and *position* of a line. A line may be constructed as shown in Figure 14-4. In this example the location of point a is given. The position of point b is 1.50 to the *right* of a, 1.90 *above* a, and 1.25 in *front* of a. The line is constructed by drawing a straight line connecting points a and b in each view.

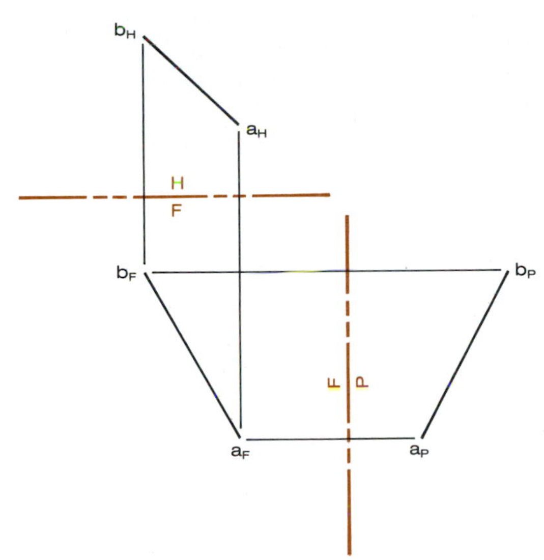

FIGURE 14-3 Projection of a line *(Section 14.3)*

FIGURE 14-2 Position of a point *(Section 14.2)*

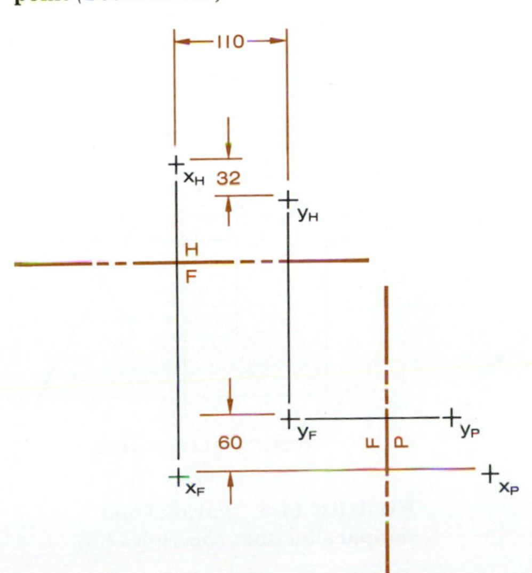

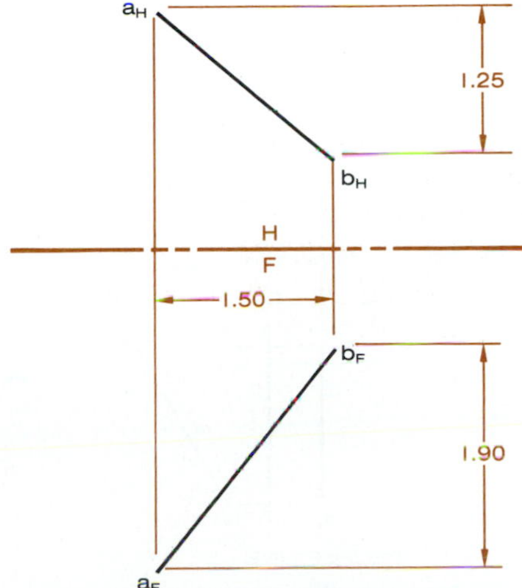

FIGURE 14-4 Direction and position of a line *(Section 14.4)*

14.5 KINDS OF LINES

If a line is parallel to a principal plane, it is called a *principal* line. There are three principal lines: horizontal, frontal, and profile. As shown in Figure 14-5, each of these lines appears true length in one of the principal views.

A *horizontal* line, shown at 14-5(a), appears true length in the top view because in the adjacent front view it appears parallel to the reference line.

A *frontal* line, shown at 14-5(b), appears true length in the front view because in the adjacent top view it appears parallel to the reference line.

A *profile* line, shown at 14-5(c), appears true length in the side view because in the adjacent front view it appears parallel to the reference line.

An *oblique* line, shown at 14-5(d), appears inclined in all principal views. Oblique lines do not appear parallel or perpendicular to any of the reference lines nor are they true length in any principal view.

Parallel lines are noncoinciding lines that have a common direction. They appear parallel in all views where they appear as lines, as in Figure 14-6(a). Lines may *appear* parallel in one or two views, as at 14-6(b), but they may not actually be parallel. An additional view should be drawn for proof of parallelism. Parallel lines appear parallel in all views except where they appear as points.

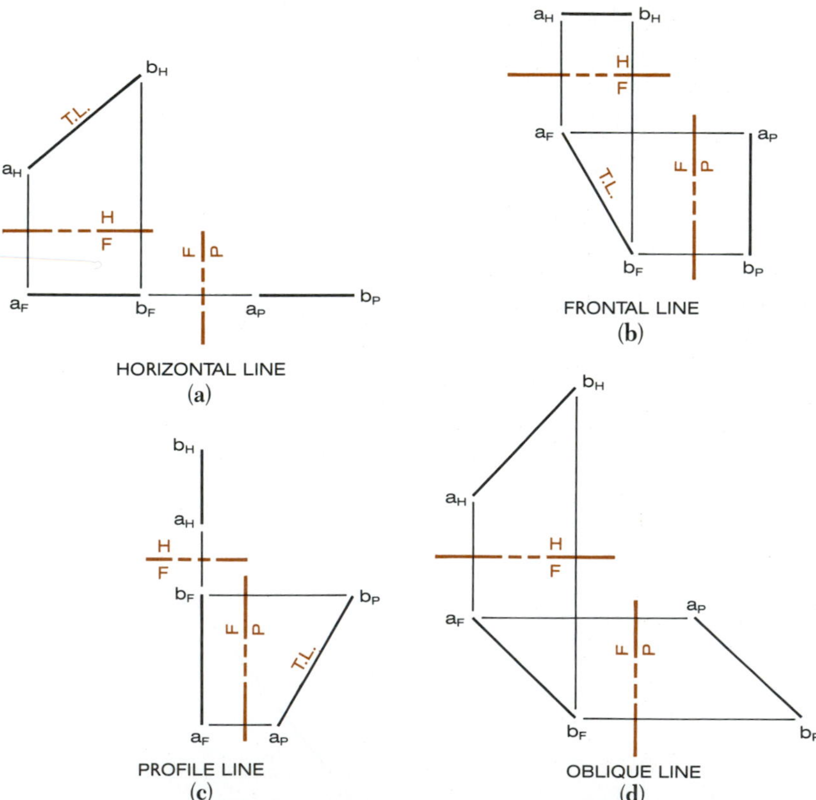

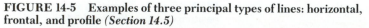

FIGURE 14-5 Examples of three principal types of lines: horizontal, frontal, and profile ***(Section 14.5)***

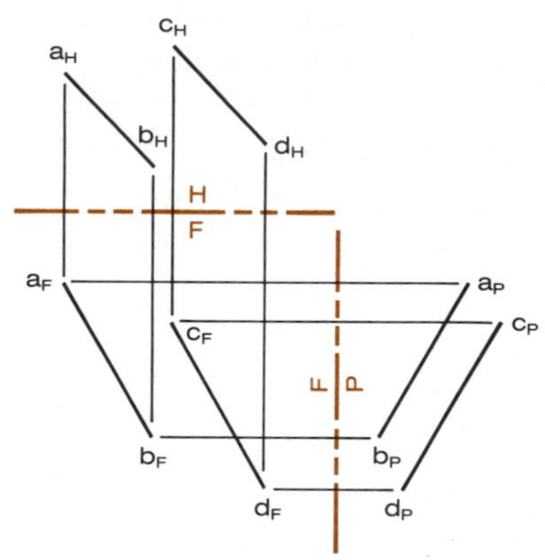

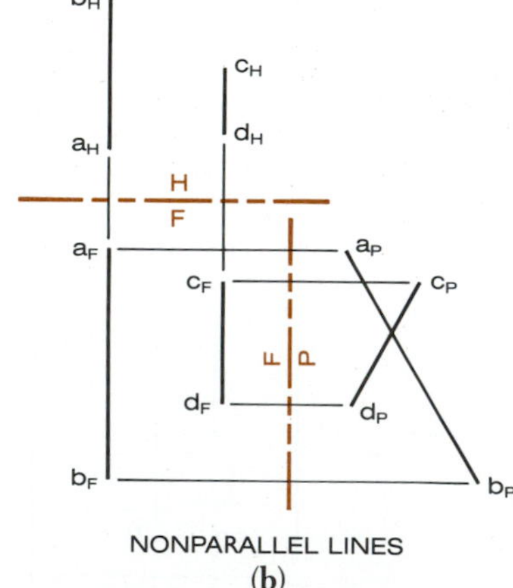

FIGURE 14-6 Parallel and nonparallel lines ***(Section 14.5)***

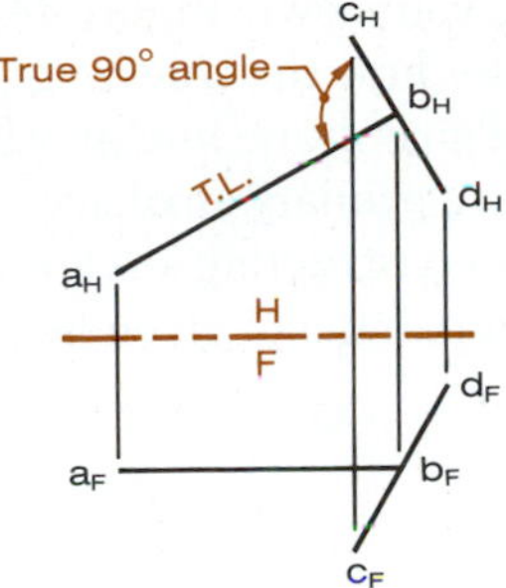

INTERSECTING PERPENDICULAR LINES
(a)

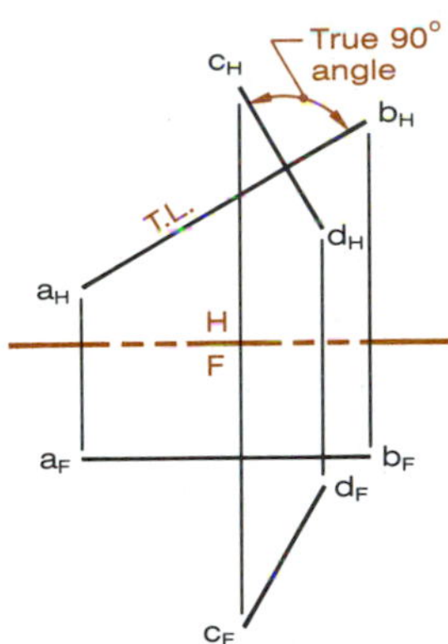

NONINTERSECTING PERPENDICULAR LINES
(b)

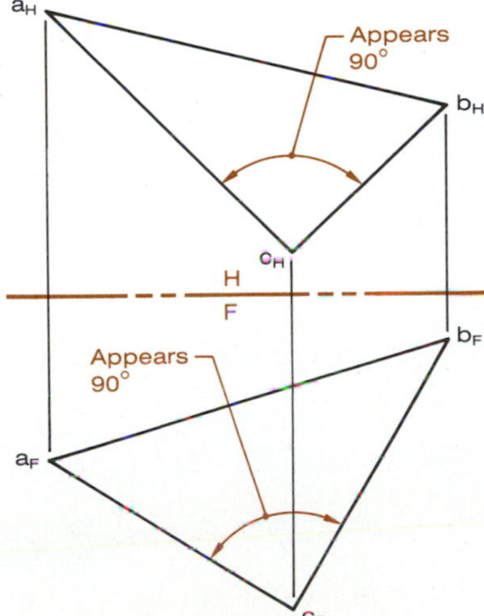

NONPERPENDICULAR LINES
(c)

FIGURE 14-7 Perpendicular and nonperpendicular lines ***(Section 14.5)***

Perpendicular lines may be intersecting, as shown in Figure 14-7(a), or nonintersecting, as in 14-7(b). Lines that are perpendicular will project at a true 90° angle when one or both of the lines are true length. Although the lines appear perpendicular in the plane abc, shown at 14-7(c), the lines are not perpendicular because neither line is true length.

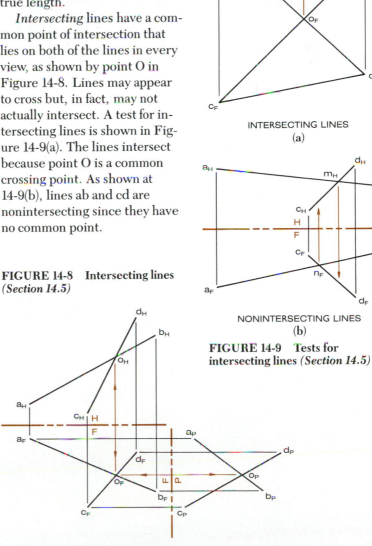

Intersecting lines have a common point of intersection that lies on both of the lines in every view, as shown by point O in Figure 14-8. Lines may appear to cross but, in fact, may not actually intersect. A test for intersecting lines is shown in Figure 14-9(a). The lines intersect because point O is a common crossing point. As shown at 14-9(b), lines ab and cd are nonintersecting since they have no common point.

FIGURE 14-8 Intersecting lines ***(Section 14.5)***

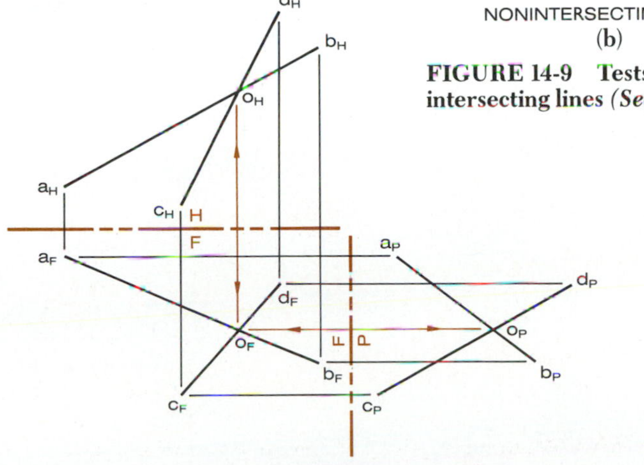

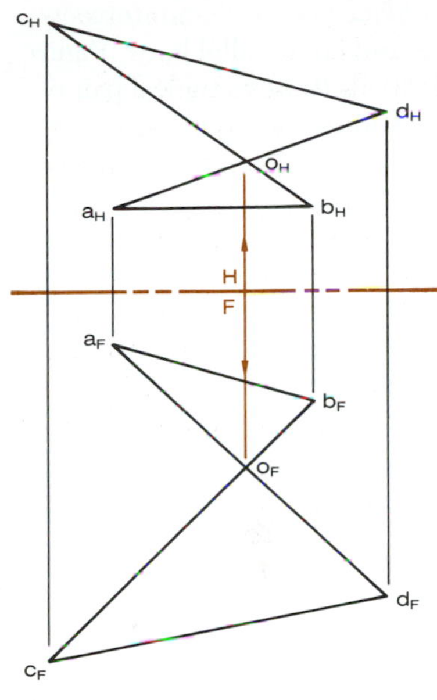

INTERSECTING LINES
(a)

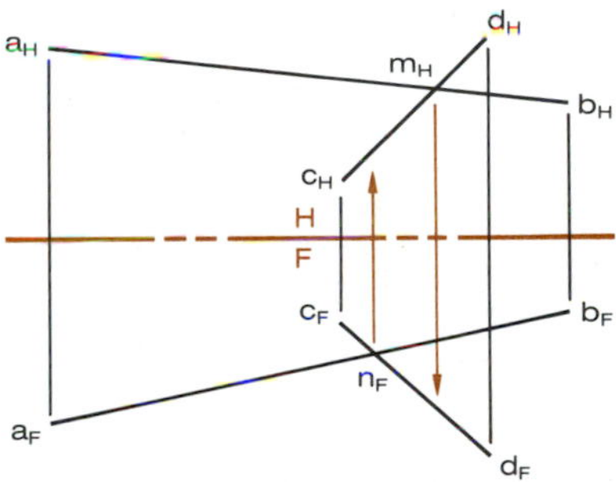

NONINTERSECTING LINES
(b)

FIGURE 14-9 Tests for intersecting lines ***(Section 14.5)***

Skew lines are nonintersecting and nonparallel lines. Figure 14-10 illustrates a typical pair of skew lines.

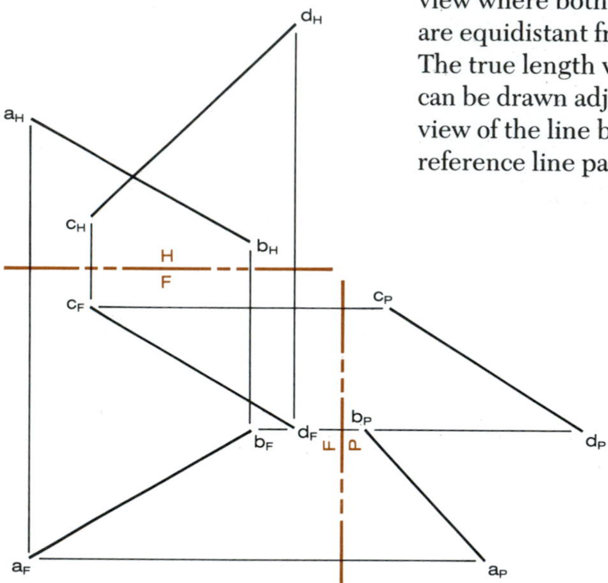

FIGURE 14-10 Skew lines *(Section 14.5)*

14.6 TRUE LENGTH OF A LINE

The *true length* of a line is the actual straight-line distance between its end points. The true length of a line appears in any view where both ends of the line are equidistant from the viewer. The true length view of a line can be drawn adjacent to any view of the line by choosing a reference line parallel to the line in the given views. Figure 14-11 illustrates how the true length (T.L.) of an oblique line may be found in a primary auxiliary view by constructing a reference line P-A parallel to given line ab

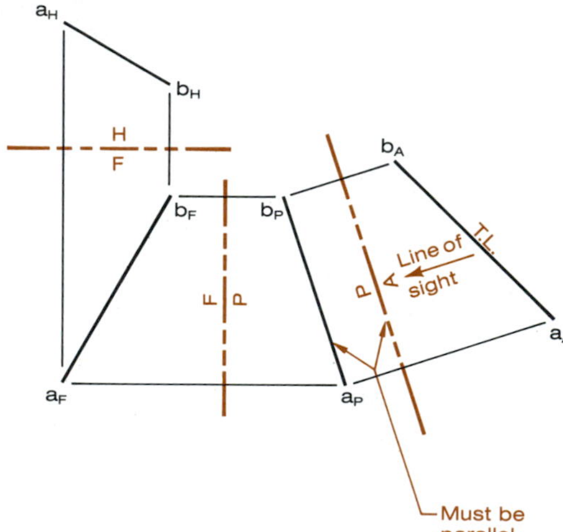

FIGURE 14-11 Finding the true length of an oblique line *(Section 14.6)*

and plotting the end points of the line. In Figure 14-12 the lines are true length in views where the observer is looking in a direction perpendicular to the line. Note that in each example, when the line appears parallel to a reference line, its adjacent view shows the line true length.

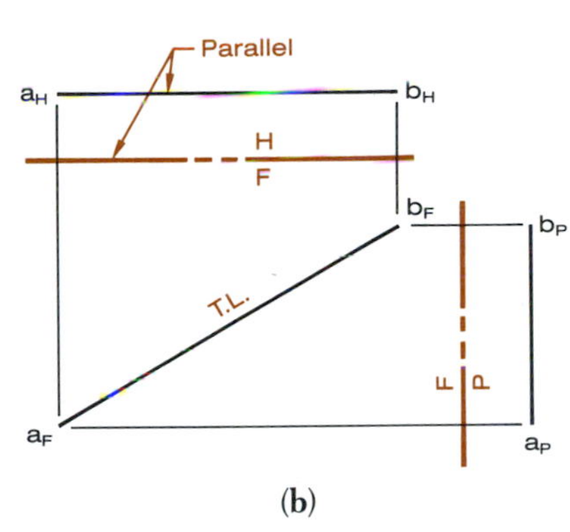

(a)

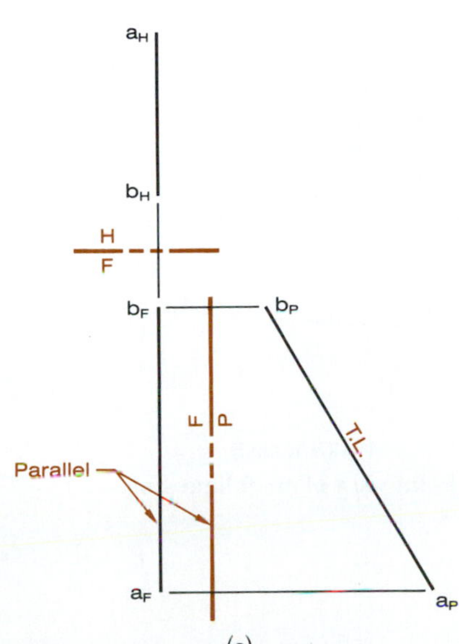

(b)

(c)

FIGURE 14-12 True length lines *(Section 14.6)*

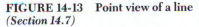

14.7 POINT VIEW OF A LINE

The point view of a line is obtained in any view where the line of sight is parallel to a true length line. As shown in Figure 14-13, the given lines appear true length in the principal views. The point view of each line is shown in the respective adjacent views.

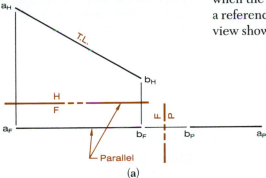

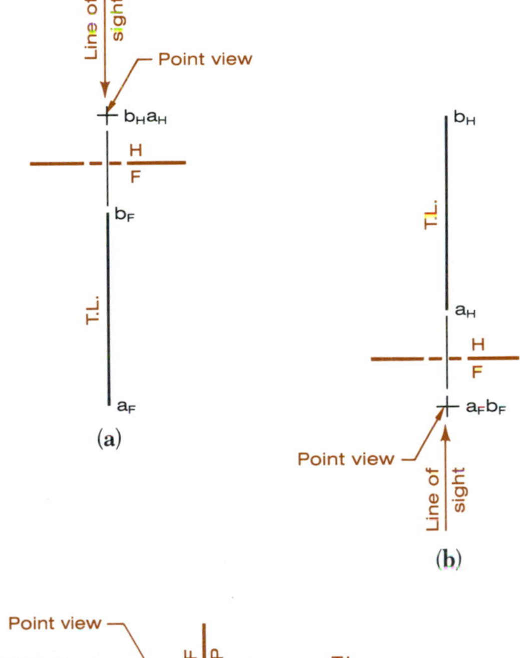

(a)

(b)

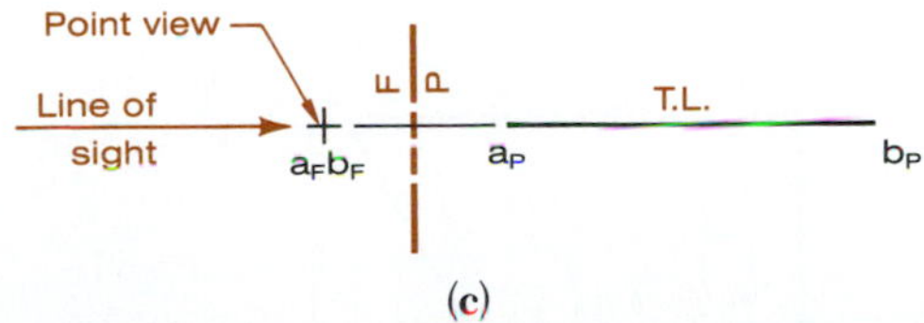

(c)

FIGURE 14-13 Point view of a line *(Section 14.7)*

In Figure 14-14 the point view of the principal lines (horizontal, frontal, and profile, respectively), is obtained in an auxiliary view where the line of sight is parallel to a true length line.

In Figure 14-15 view A is constructed to obtain the true length of the given oblique line, and the point view of the line is shown in view B. Since the line was viewed from the end labeled h, the point view is labeled h,g. The far end of the line (point g) lies directly behind the near end, point h.

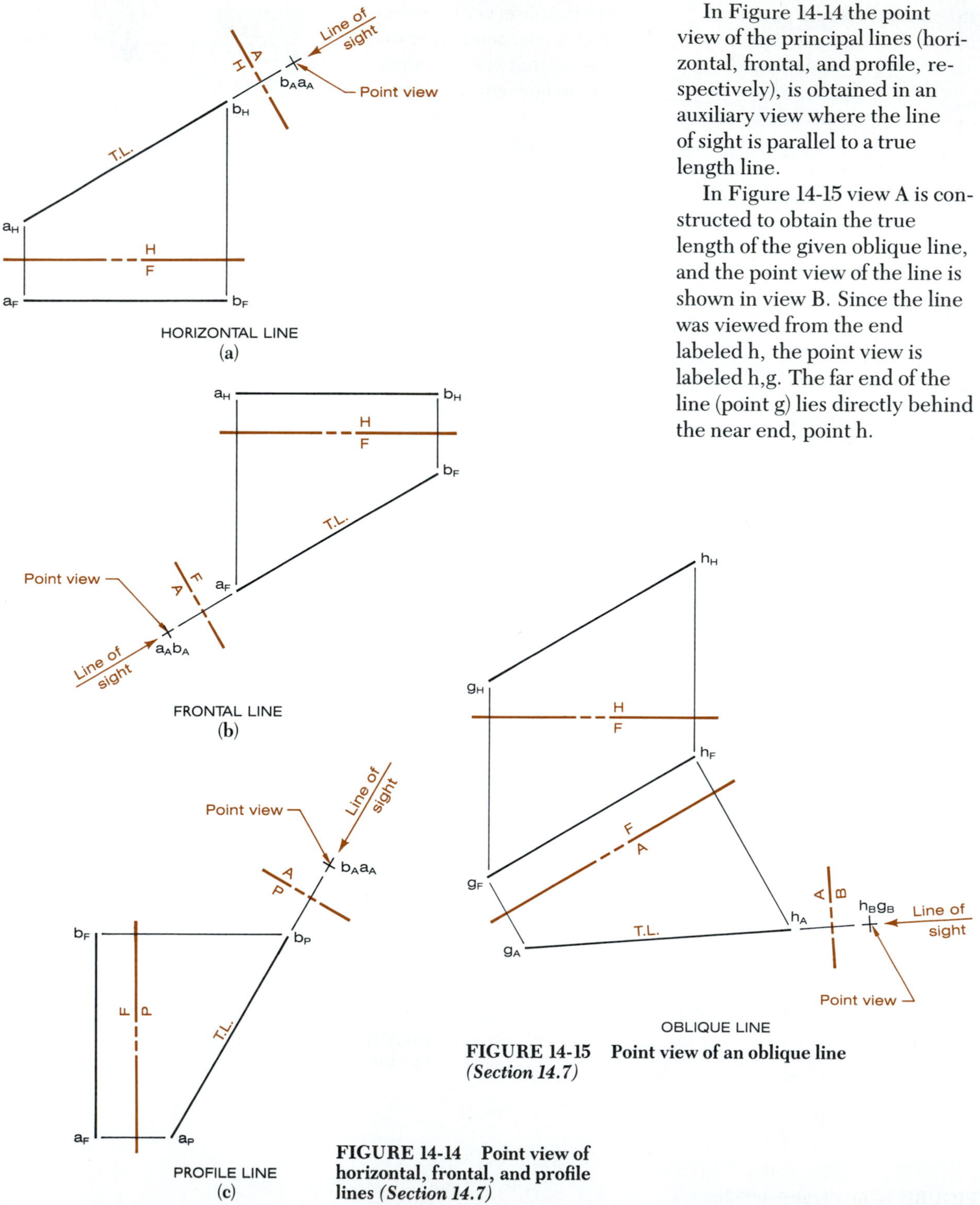

FIGURE 14-15 **Point view of an oblique line** *(Section 14.7)*

FIGURE 14-14 **Point view of horizontal, frontal, and profile lines** *(Section 14.7)*

14.8 PROJECTION OF A POINT ON A LINE

If a point is known to lie on a line, the point will appear on all views of the line. One coordinate distance is sufficient to fix its position. In Figure 14-16(a), point x lies on line $a_F b_F$. As shown at 14-16(b), x_H is located by intersection on line ab by projecting the point from the front view to the top view.

14.9 BEARING AND AZIMUTH OF A LINE

The *bearing* of a line is the angle by which it deviates east or west from a north-south line. North is assumed to be toward the top of the drawing or map. Bearing lines are map directions and can only be measured in the top view. The bearing of a line is entirely independent of the angle between the line and the horizontal plane. The bearing angle never exceeds 90°. In Figure 14-17(a) the horizontal line ab has a bearing of N60°E. The oblique line ac has a bearing of N35°W.

The *azimuth* of a line is the angle a line makes with the north-south line, measured *clockwise* from the north. At 14-17(b), line ef has an azimuth of 75° and line eg has an azimuth of 225°.

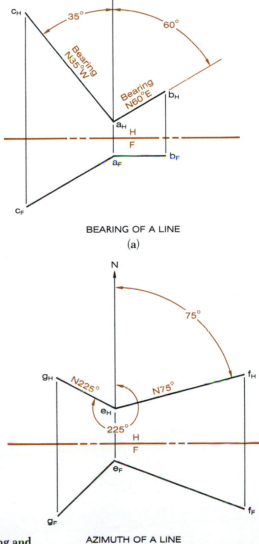

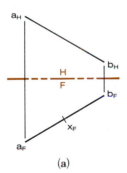

(a)

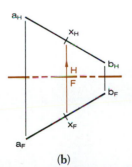

(b)

FIGURE 14-16 Projection of a point on a line ***(Section 14.8)***

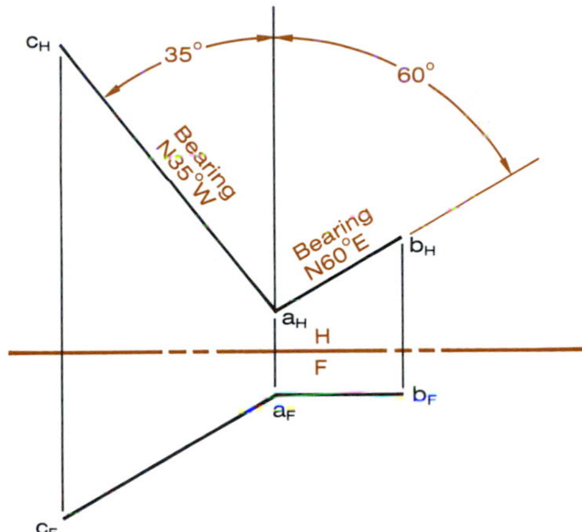

BEARING OF A LINE
(a)

AZIMUTH OF A LINE
(b)

FIGURE 14-17 Bearing and azimuth of a line ***(Section 14.9)***

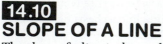

14.10 SLOPE OF A LINE

The slope of a line is the angle (usually measured in degrees) that a line makes with a horizontal plane. The slope angle of a line can be seen true size only when measured in an elevation view (front, side, top-adjacent) in which the given line appears true length. In elevation views the horizontal plane always appears as an edge. Figure 14-18(a) illustrates how the true size of the slope angle of the frontal line ab is determined in the front view. At 14-18(b) the slope angle of the profile line cd may be measured in the side view. In both examples the given line appears true length in an elevation view. At 14-18(c) the oblique line ef does not appear true length in the front view. Accordingly, top-adjacent view A is drawn as shown to obtain the true length of the line. Because view A is an elevation view, the true magnitude of the slope angle may be properly measured. A second true-length elevation view that shows the slope angle may be obtained by constructing top-adjacent view A′.

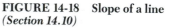

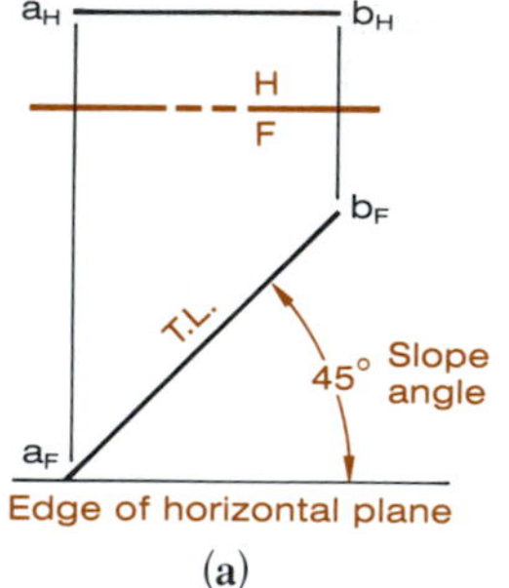

(a)

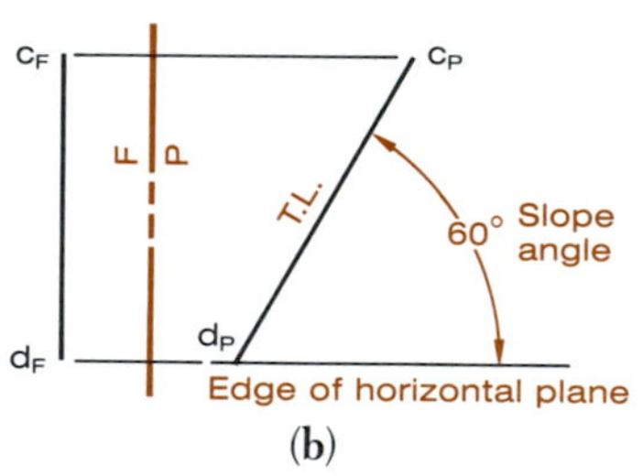

(b)

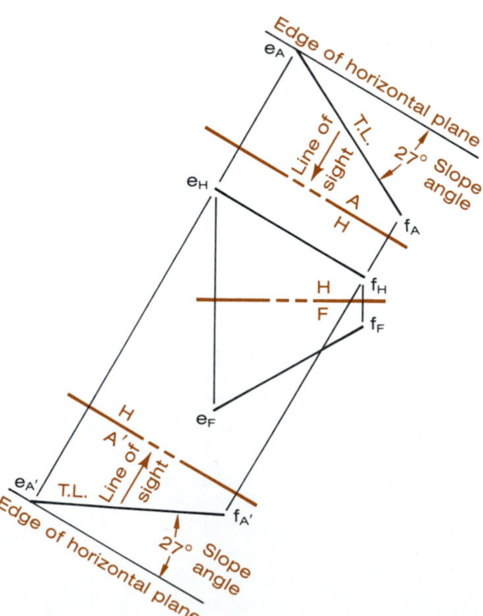

(c)

FIGURE 14-18 **Slope of a line** ***(Section 14.10)***

14.11 GRADE OF A LINE

The grade is a slope expressed in percentage. The percent of grade of a line is equal to the tangent of the slope angle multiplied by 100, or

$$\text{Percent of grade} = \frac{\text{Vertical rise}}{\text{Horizontal run}} \times 100$$

In Figure 14-19(a) the grade is 30%; at 14-19(b) the grade is 50%. Note that, in both examples, the grade is measured in a true length elevation view.

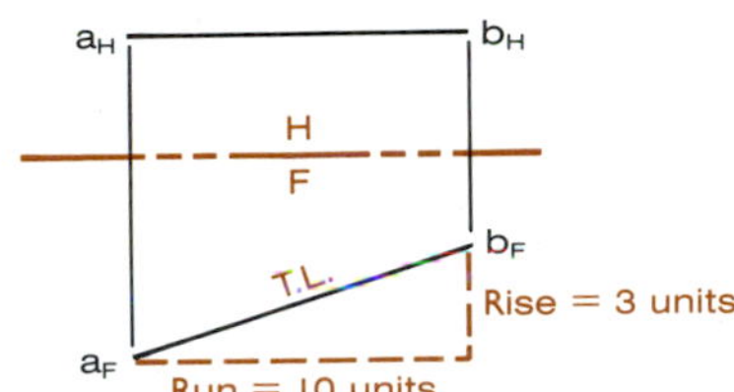

Percent of grade $= \frac{\text{rise}}{\text{run}} = \frac{3}{10} \times 100 = 30\%$

(a)

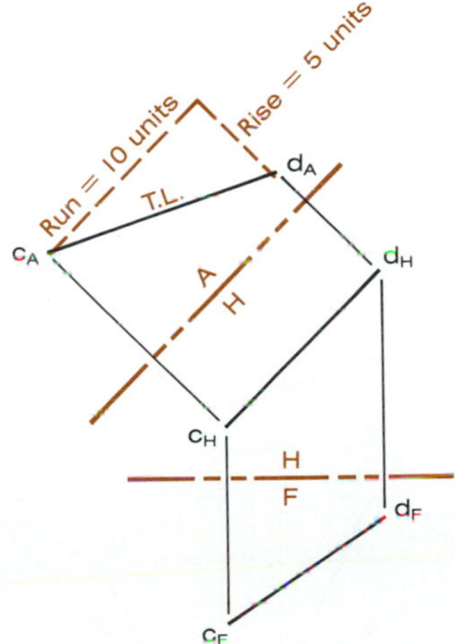

Percent of grade $= \frac{\text{rise}}{\text{run}} \times 100 = \frac{5}{10} \times 100 = 50\%$

(b)

FIGURE 14-19 Grade of a line
(Section 14.11)

14.12 PLANE SURFACES

A plane is considered to be flat, without thickness, and unlimited in extent. In many problem solving applications it is helpful to establish the bounds or limits of a plane, but in other cases the plane may be extended indefinitely. A plane may be represented or determined, as shown in Figure 14-20, by intersecting lines, shown at 14-20(a), two parallel lines, shown at 14-20(b), a line and a point, shown at 14-20(c), or by three points not in a straight line, shown at 14-20(d).

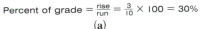

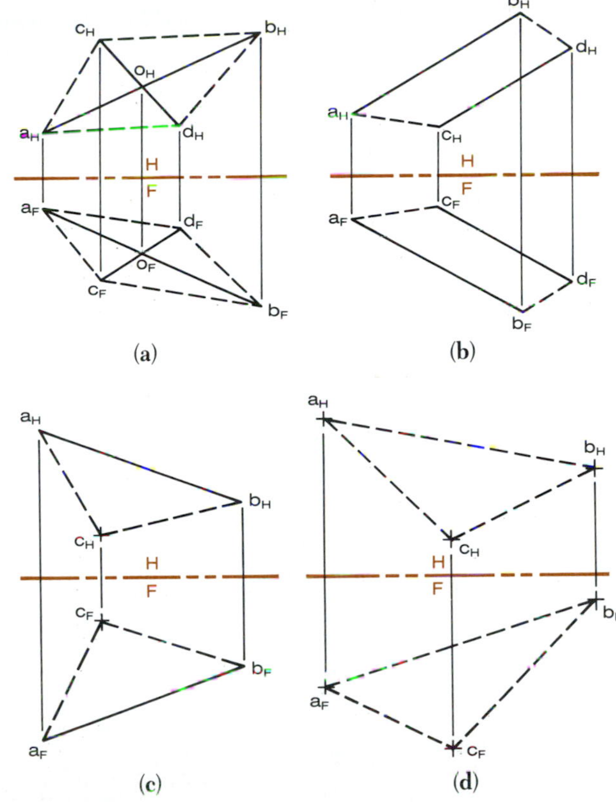

FIGURE 14-20 Plane surfaces
(Section 14.12)

14.13 TYPES OF PLANES

Principal planes (horizontal, frontal, and profile) are planes that are parallel to the principal projection planes, as shown in Figure 14-21. A *horizontal* plane, shown at 14-21(a), appears true size in the top view and as an edge in the other two views. A *frontal* plane, shown at 14-21(b), appears true size in the front view and as an edge in the other two views. A *profile* plane, shown at 14-21(c), appears true size in the side view and as an edge in the other two views.

Two other types of planes are illustrated in Figure 14-22. These are *inclined*, shown at 14-22(a), and *oblique*, shown at 14-22(b). An inclined plane appears distorted (that is, not true size) in two of the views and as an edge in the other view. The oblique plane is a plane whose shape appears distorted in all three views.

14.14 LOCATING POINTS IN A PLANE

To locate a point in a given plane, the point must lie on a line in that plane. In Figure 14-23(a) point x_H lies in plane abc but not on one of the given lines of the plane. Although an

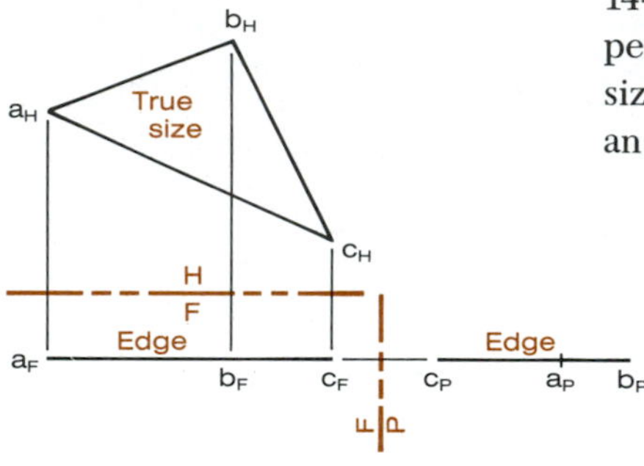

HORIZONTAL PLANE
(a)

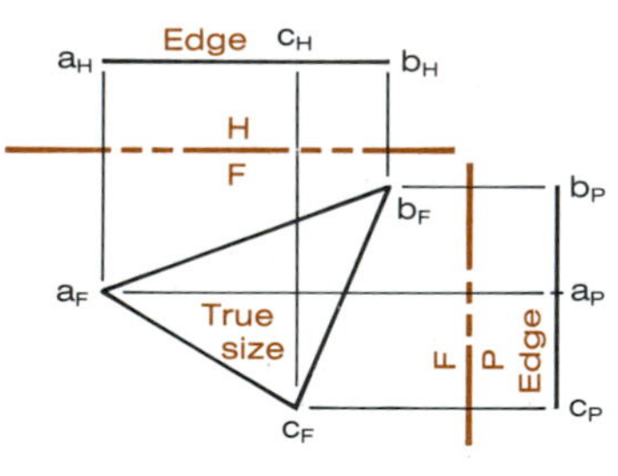

FRONTAL PLANE
(b)

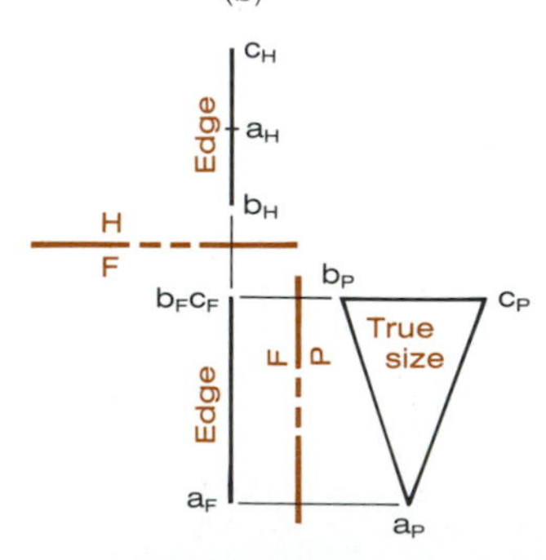

PROFILE PLANE
(c)

FIGURE 14-21 Examples of principal planes ***(Section 14.13)***

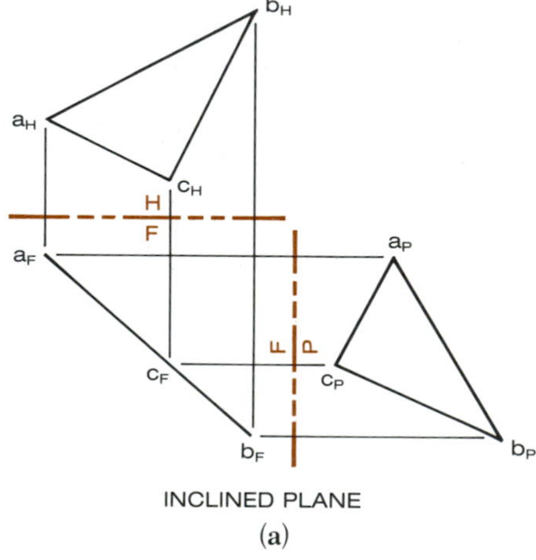

INCLINED PLANE
(a)

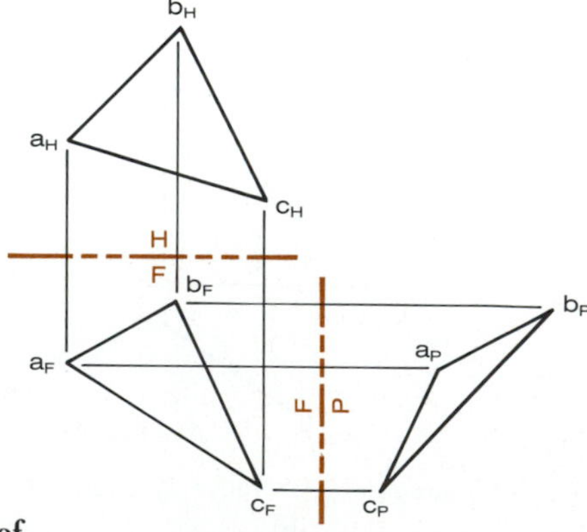

OBLIQUE PLANE
(b)

FIGURE 14-22 Examples of inclined and oblique planes ***(Section 14.13)***

infinite number of lines that contain the given point may be drawn in the plane, line $a_H d_H$ was selected, as shown at 14-23(b). After projecting line ad to the front view, point x_F is located by intersecting a projection line with the line ad. Figure 14-24 illustrates how to locate a point in a plane that lies outside the bounds of the plane.

FIGURE 14-23 Locating a point in a plane that lies within the bounds of the plane *(Section 14.14)*

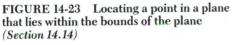

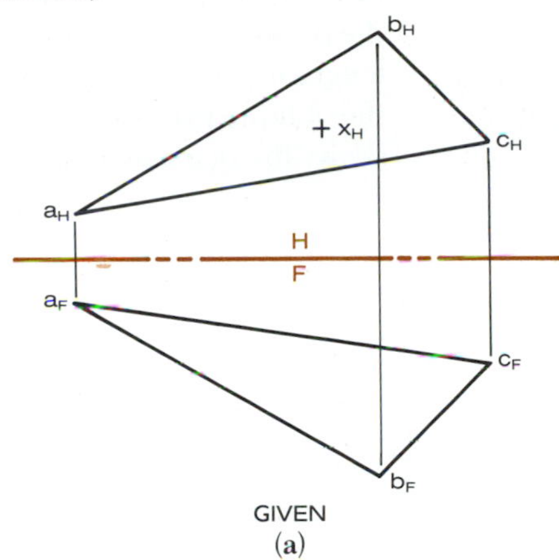

(a)

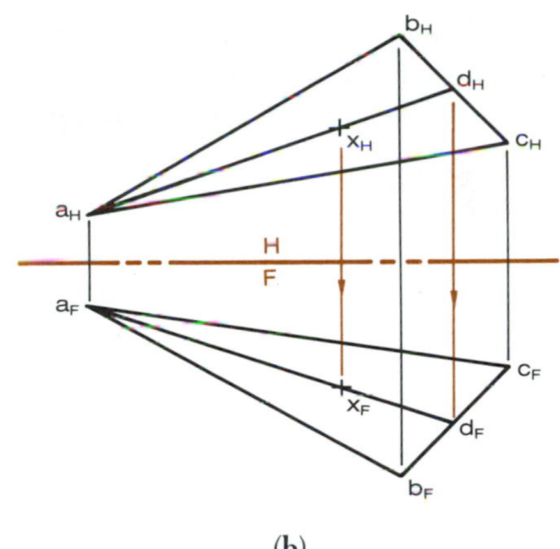

(b)

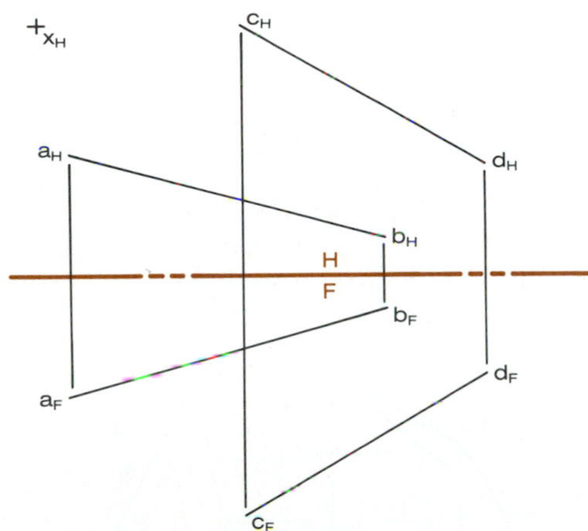

(a)

FIGURE 14-24 Locating a point in a plane that lies outside the bounds of the plane *(Section 14.14)*

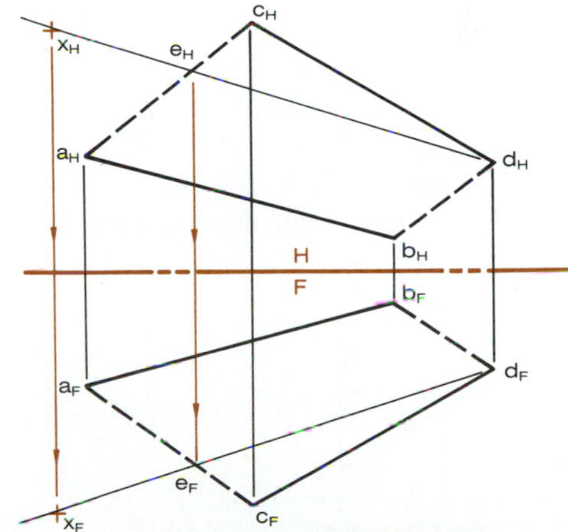

STEP 1 Establish the plane by drawing lines ac and bd in views H and F.

STEP 2 Draw line $e_H d_H$ in the plane and extend the line to contain point x_H.

STEP 3 To locate point x_F, draw line $e_F d_F$, extend the line, and project point x_H to intersect the line.

(b)

14.15 TRUE-LENGTH LINES IN PLANES

To establish a true length line in a plane, in any view, the line must lie in the plane in a position that is parallel to the reference line in the adjacent view. Constructions for obtaining frontal, horizontal, and profile true length lines are illustrated in Figure 14-25. As you will see, an infinite number of true length lines may lie in a plane. All true length lines in any one view will be parallel.

14.16 EDGE VIEW OF A PLANE

As previously illustrated in Figure 14-21, principal planes always appear as an edge in *two* of the three principal views. In Figure 14-22 you saw that an inclined plane will appear as an edge in *one* of the principal views. The edge view of a plane is obtained when the line of sight is parallel to a true length line in the plane, such as horizontal line bc in Figure 14-26. An oblique plane does not appear as an edge in any of the principal views. In Figure 14-27 the profile line bd appears true length in the right-side view of the oblique plane abc. View A shows the plane as an edge.

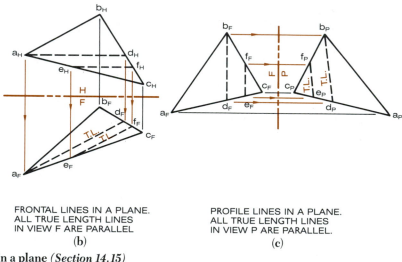

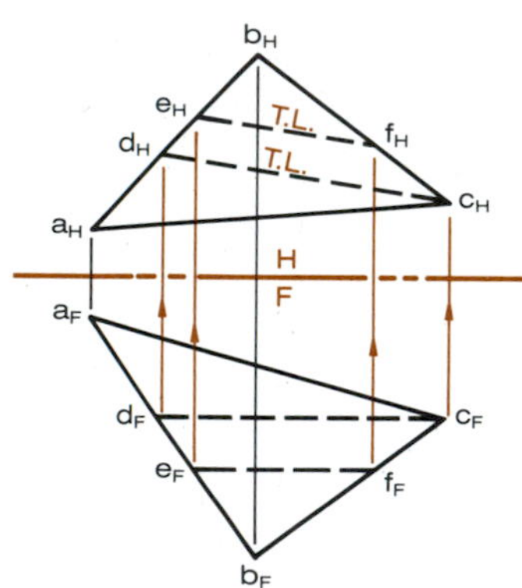

HORIZONTAL LINES IN A PLANE.
ALL TRUE LENGTH LINES
IN VIEW H ARE PARALLEL.
(a)

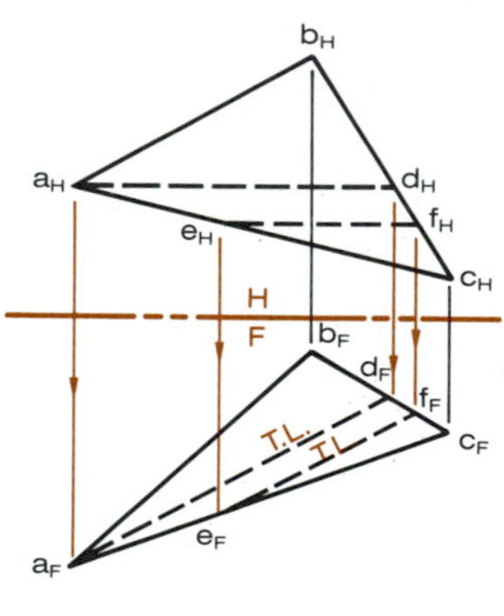

FRONTAL LINES IN A PLANE.
ALL TRUE LENGTH LINES
IN VIEW F ARE PARALLEL
(b)

PROFILE LINES IN A PLANE.
ALL TRUE LENGTH LINES
IN VIEW P ARE PARALLEL.
(c)

FIGURE 14-25 **True length lines in a plane** ***(Section 14.15)***

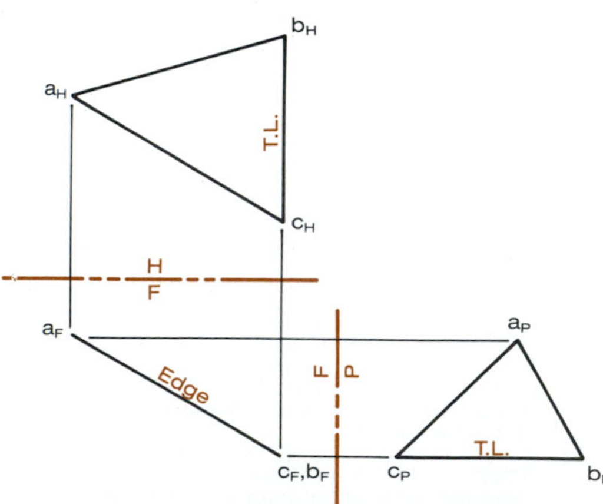

FIGURE 14-26 **Edge view of a plane** ***(Section 14.16)***

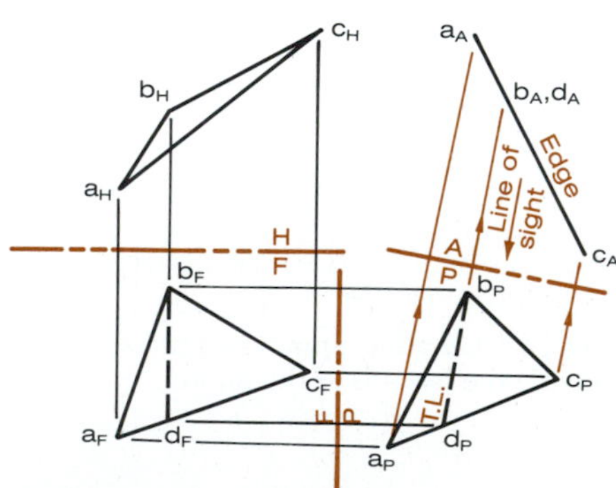

FIGURE 14-27 **Edge view of a plane** ***(Section 14.16)***

14.17 TRUE SIZE OF A PLANE

The *true size*, or normal, view of a plane is obtained when the line of sight is perpendicular to the plane. As was shown for principal planes in Figure 14-21, a plane will appear true size when it is adjacent to an edge view. Figure 14-28 illustrates the top and front views of oblique plane abcd. To show this plane true size, line ae must first be constructed true length (Step 1). For this example, either view A or B may be used to show the plane as an edge. In Step 2, view A is selected in a position that shows the point view of the true length line ae. Plane abcd appears as an edge in this view. In order to obtain a true size view of the plane (Step 3), the reference line A-B must be drawn parallel to the edge of the plane. In view B the line of sight is perpendicular to the edge view of the plane. Points a, b, c, and d are located using the procedures explained in Section 6.5 for constructing secondary auxiliary views.

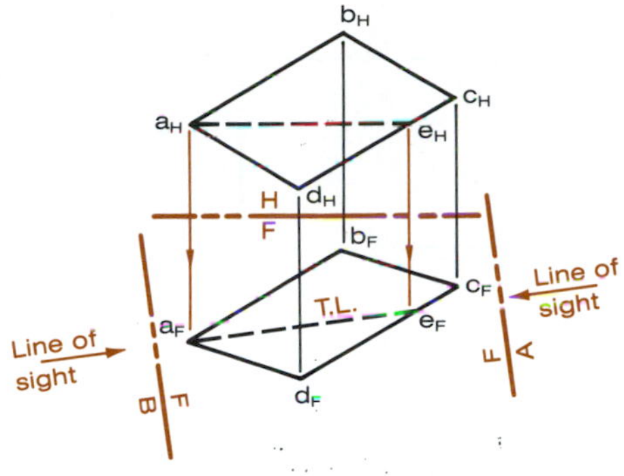

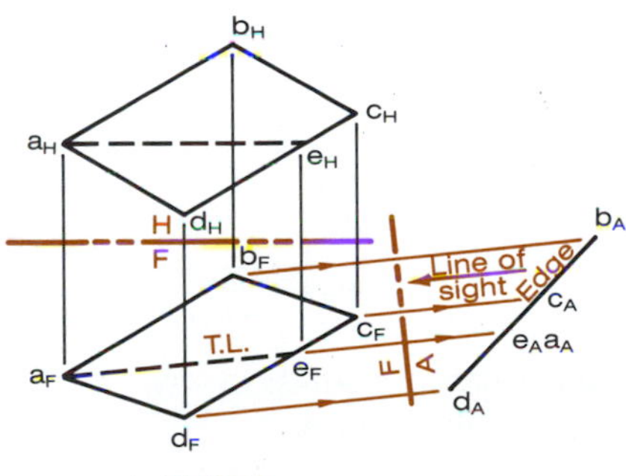

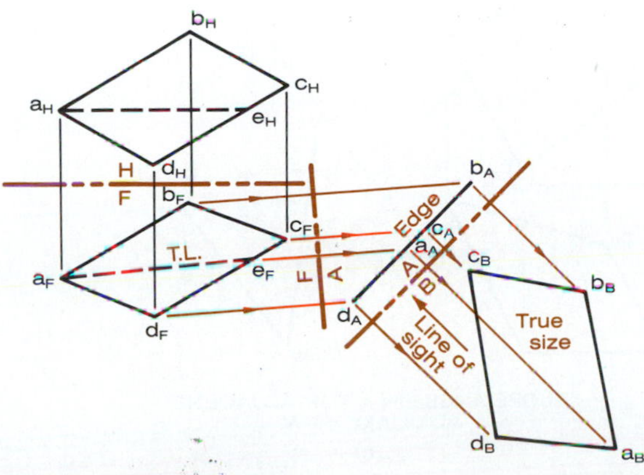

FIGURE 14-28 **True size of a plane** *(Section 14.17)*

14.18 STRIKE OF A PLANE

The *strike* of a plane is the bearing of a horizontal line in the plane. All horizontal lines in a plane are parallel and true length in the top view and have the same strike. The term "strike" is used by mining engineers and geologists to denote the direction of the various strata or layers under the earth's surface. Two views of plane abc are given in Figure 14-29. The horizontal strike line ad appears true length in the top view and has a strike of N75°E, as shown.

14.19 SLOPE OF A PLANE

The slope of a plane is the angle (measured in degrees or in percent of grade) that the given plane makes with a horizontal plane. The true slope may only be measured in an elevation view where the plane appears as an edge. Figures 14-30(a) and 14-30(b) illustrate how the slope angle of an inclined plane may be determined in the principal views. In both examples, the given plane appears as an edge in an elevation view. At 14-30(c) the plane does not appear as an edge in a principal view. Therefore top-adjacent view A is drawn to show the plane as an edge in an elevation view. The true slope angle of the plane may be measured in this view.

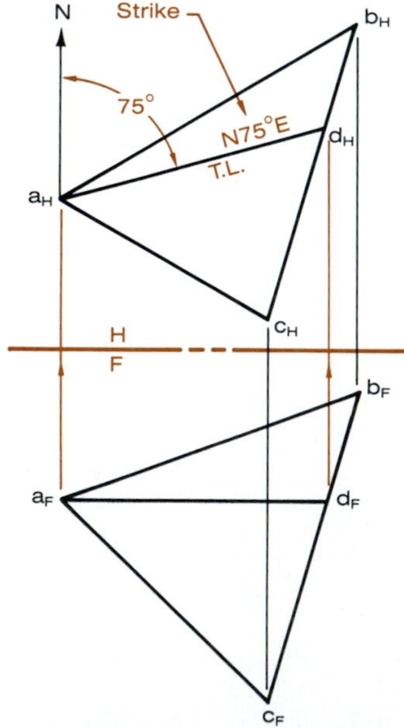

FIGURE 14-29 Strike of a plane *(Section 14.18)*

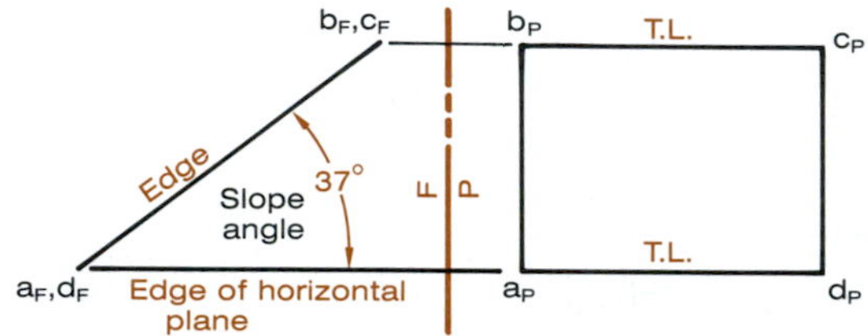

SLOPE ANGLE IN FRONT VIEW
(a)

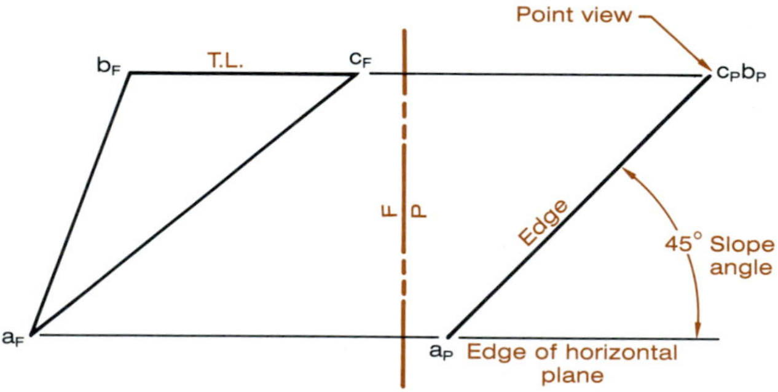

SLOPE ANGLE IN PROFILE (RIGHT-SIDE) VIEW
(b)

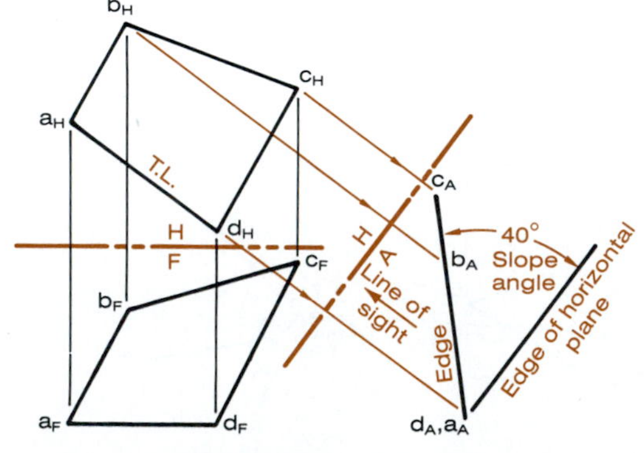

SLOPE ANGLE IN A TOP-ADJACENT AUXILIARY VIEW
(c)

FIGURE 14-30 Slope of a plane *(Section 14.19)*

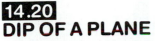

14.20 DIP OF A PLANE

The *dip* of a plane is a term used by geologists to denote the slope angle of a plane. In Figure 14-31 a top-adjacent auxiliary view (view A) of vein abc is projected with the line of sight parallel to the true length strike line, dc, to obtain the dip angle of the vein. The size of the dip angle, 50°, is measured, as shown in elevation view A, where the vein appears as an edge. The dip is also indicated in the top view by an arrow perpendicular to the strike line and pointing toward the downward side of the vein.

14.21 BASIC CONSTRUCTIONS

The constructions that follow, in combination with the preceding material dealing with basic geometric principles, will serve as a foundation for analyzing and solving many modern engineering problems. In engineering design, manually performed graphical methods are often sufficiently accurate to produce the desired results. There are countless practical examples of on-the-board and computer-aided applications that require an understanding of points, lines, and planes.

14.22 FINDING THE SHORTEST DISTANCE FROM A POINT TO A PLANE

The shortest line from a point to a plane is a line that is perpendicular to the plane. Figure 14-32 is an example showing a method of constructing the shortest line from the given point O to a plane. Any edge view of plane abc will show the desired line, but since line a_Hc_H is true length in the given plane, the most direct method is to draw an edge view in view A projected directly from the top view. Point x_A is located by drawing the desired shortest line o_Ax_A perpendicular to the edge of the plane. Because line o_Ax_A is true length in view A, the line will appear parallel to reference line H-A in the top view. Point x_F is located in the customary way by projection and measurement.

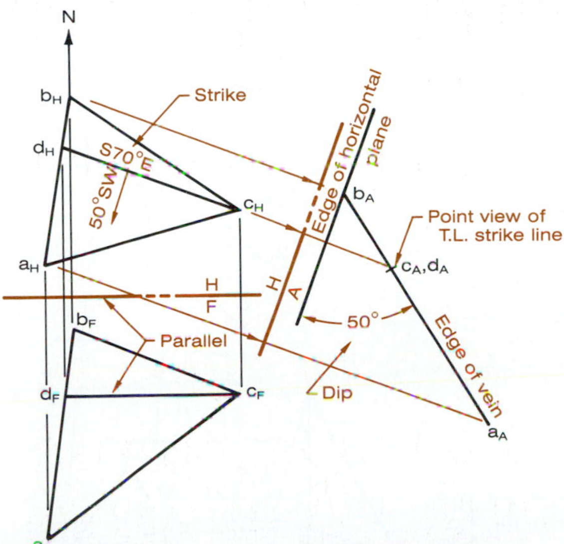

FIGURE 14-31 Dip of a plane *(Section 14.20)*

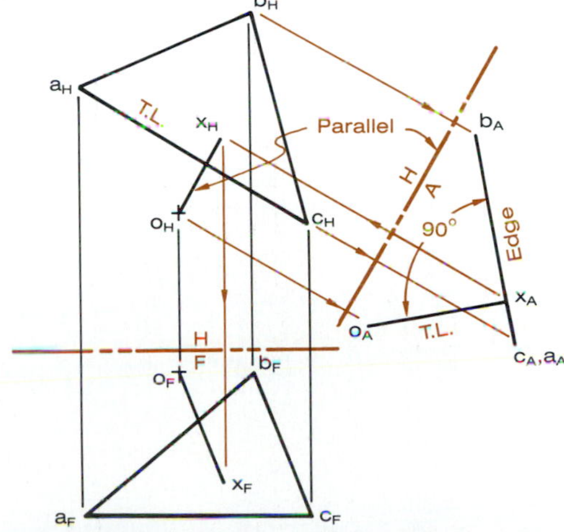

FIGURE 14-32 Construction for finding the shortest line from a point to a plane *(Section 14.22)*

14.23 FINDING THE ANGLE BETWEEN A LINE AND A PLANE

The angle between a line and a plane is measured in a view that shows the given plane as an edge and the line in true length. A construction sequence of auxiliary views must be carefully observed. The plane must first be shown as an *edge*, second as *true size*, and third as an *edge* again where the given line appears in true length. In Figure 14-33 two views of plane abc and line ox are given. In this case, to find the angle that the line makes with the plane, it is necessary to begin by constructing an edge view of the plane as in view A. (In some instances the given plane may appear as an edge in one of the given views, but unless the given line appears true length, the true angle between the given line ox and the plane *cannot* be measured in this view.) View B is next drawn to obtain the true size of the plane. Finally, the edge view of the plane is again drawn but in a position in view C where line o_Cx_C is true length. The required true size of the angle between the line and the plane can now be measured as shown.

14.24 FINDING THE INTERSECTION OF A LINE AND A PLANE

The location of the point of intersection (also called a piercing point) of a line and a plane is found in the view where the plane appears as an edge. In Figure 14-34(a) the inclined plane abc appears as an edge in the front view. The intersecting point of the given line, xy, and the edge appears in the front view and is labeled p_F. Point p_H is located by projecting p_F to in-

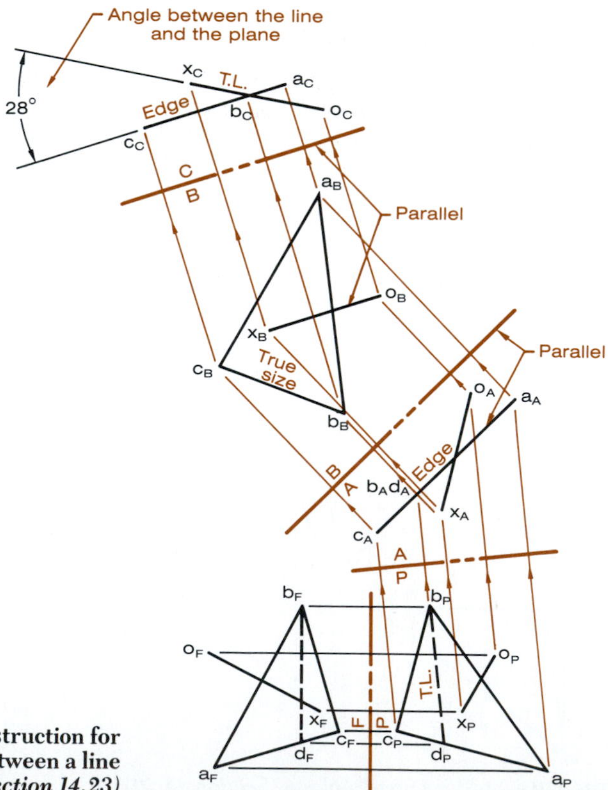

FIGURE 14-33 Construction for finding the angle between a line and a plane *(Section 14.23)*

tersect with line xy in the top view. The visibility of the line xpy in the top view may be easily determined by inspection. As seen in the front view, line segment xp lies *above* the plane and is, therefore, visible in the top view. Line segment py is hidden because it lies *below* the plane.

In Figure 14-34(b) the given plane is oblique. In this case it is necessary to construct an edge view A. The point p_A where line x_Ay_A intersects the plane $a_Ab_Ac_A$ is projected to the top and front views to locate p_H and p_F, respectively. The visibility of that part of the line y_Hp_H can be determined by observing in view A that y_Ap_A lies *above* the plane. (*Up* is measured in an elevation view in a direction that is perpendicular to the reference line.) Note the "up" arrow in views F and A. Line y_Hp_H is, therefore, visible. As an additional check, the visibility of the line y_Hp_H can be determined by labeling the apparent crossing points 1 and 2 of the line and the plane. Point 1 on line y_Fp_F is *above* point 2 on line a_Fb_F, proving that line y_Hp_H is visible. Similarly, apparent crossing points 3 and 4 can be used to determine the visibility of the line. Line x_Fp_F is visible because point 3 on line x_Hp_H is in *front* of point 4 on line a_Hc_H.

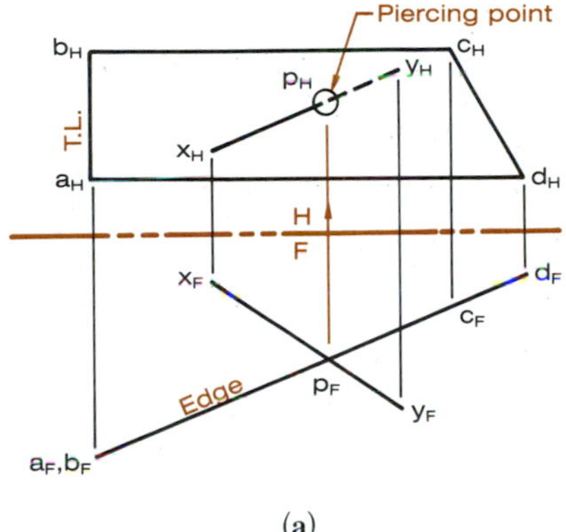

(a)

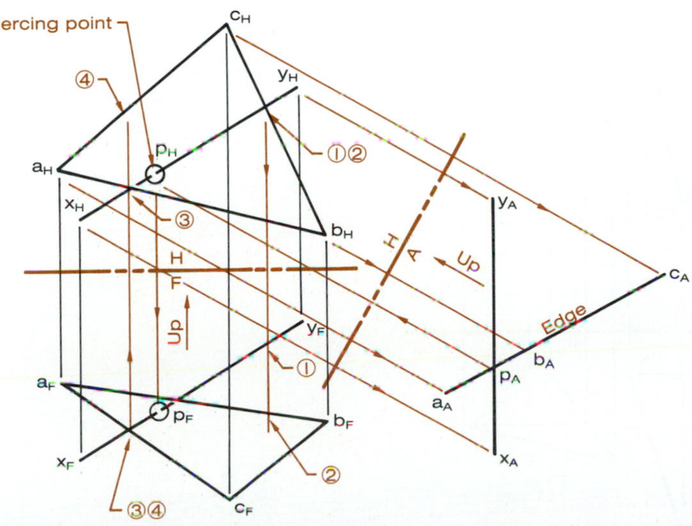

(b)

FIGURE 14-34 Construction for finding the intersection of a line and a plane *(Section 14.24)*

14.25 FINDING THE SHORTEST DISTANCE BETWEEN TWO PARALLEL PLANES

The shortest distance between two parallel planes is measured in the view that shows each of the planes as an edge. Two parallel planes are given in Figure 14-35. The distance between the planes can be measured in edge view A, where the two planes appear as parallel lines.

14.26 FINDING THE DIHEDRAL ANGLE BETWEEN TWO PLANES

The angle formed by two intersecting planes is known as the *dihedral* angle. The dihedral angle must be measured in a view that is perpendicular to each of the planes. Two intersecting planes are shown in Figure 14-36. The true dihedral angle can be seen only in a view

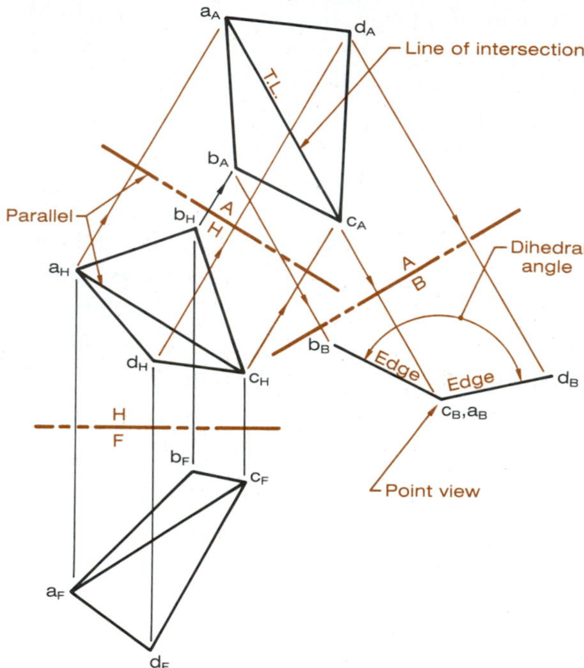

FIGURE 14-36 Construction for finding the dihedral angle *(Section 14.26)*

FIGURE 14-35 Construction for finding the distance between two parallel planes *(Section 14.25)*

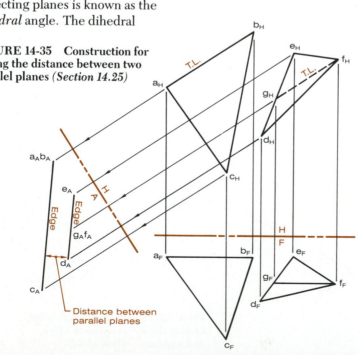

where the *line of intersection* of the two planes appears as a *point*. In this view both planes appear as *edges*. To find the dihedral angle, begin by finding the true length of the line of intersection in view A. In view B the line of intersection appears as a point and the required dihedral angle appears as shown.

14.27 FINDING THE SHORTEST DISTANCE BETWEEN TWO SKEW LINES

The shortest distance between two skew lines (lines that are nonintersecting and nonparallel) can be measured in a view where one of the lines appears as a point. The line connecting the two given lines must be perpendicular to both. In Figure 14-37 two skew lines are given. To find the shortest distance between them, begin by finding the true length of line ab, as shown in view A. Line cd should also be shown in this view. The required shortest distance must be perpendicular to ab in view A but its exact location is unknown. The point view of line ab is constructed in view B and, once again, line cd is also shown. The shortest distance between the two skew lines is shown by the true length line ef, which is drawn perpendicular to ab. Although line cd is not true length in view B, line ef is true length. As explained in Section 14.5, lines are perpendicular when one or both of the lines are true length. The shortest line ef is true length in view B and hence will appear perpendicular to line cd. To project line ef back to the given views, it is important to recall that a line appears true length in any view where both ends of the line are equidistant from the viewer. Line ef in view A is drawn parallel to the reference line A-B. Line ef is located in the top and front views by drawing projection lines from e and f that respectively intersect with the given skew lines.

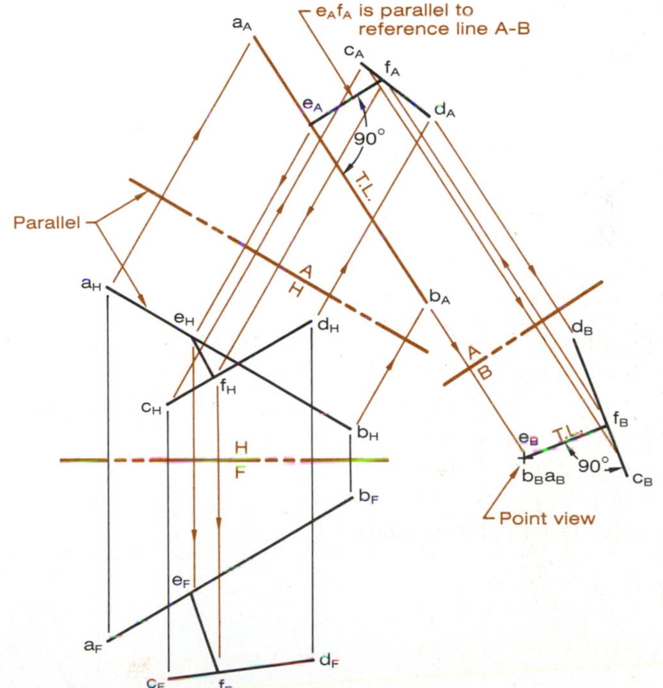

FIGURE 14-37 Construction for finding the shortest distance between two skew lines *(Section 14.27)*

14.28 FINDING THE LINE OF INTERSECTION BETWEEN TWO PLANES

The line of intersection between two planes is a straight line. This line can be found by locating two points that are common to both intersecting planes.

Cutting Plane Method

The cutting plane method is used to find the line of intersection xy between the given planes, as shown in Figure 14-38. The cutting planes selected must intersect *both* planes. For each cutting plane, one point on the required line of intersection is obtained. The first step consists of passing a cutting plane labeled CP1, containing line ad in the front view. Points 1 and 2 are projected to the top view to locate point x on the line of intersection. Point x, in turn, is projected to the front view. To obtain a second point on the line of intersection, CP2, which contains line bc in the front view, is used to locate points 3 and 4. These points are projected to the top view. Point y, the second point on the required line of intersection, is located on line 3-4 in the top view. Point y is projected to the front view. A check on the accuracy of the line of intersection may be made by taking one or two additional cutting planes in either the front or the top view. The position of the cutting planes may be randomly selected or, as in this example, the position of CP1 and CP2 may coincide with existing lines in the planes. Draw the line of intersection xy in the given views. In every case when the line of intersection falls within the boundaries of two nonparallel planes, the line of intersection will appear as a visible line. The visibility of the planes is established by analyzing the points where the lines of each plane cross, as outlined in Section 14.24.

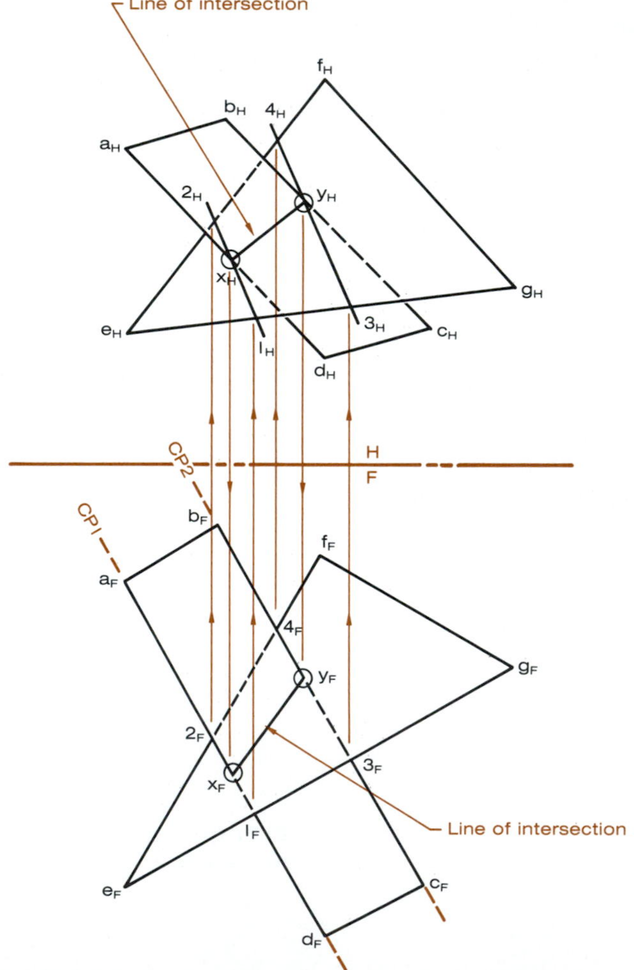

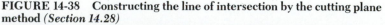

FIGURE 14-38 Constructing the line of intersection by the cutting plane method *(Section 14.28)*

Edge View Method

Another method for finding the line of intersection between two planes is the edge view method. In a view where one of the planes appears as an edge, the edge view plane may be considered a cutting plane. Figure 14-39 illustrates how the required line of intersection is found for the same two planes used in Figure 14-38. The first step consists of finding the edge view of one of the planes, as shown in view A. Points x and y, which are the extremities of the line of intersection, are projected back to the given views. These points are connected with a straight, visible line. The visibility of the planes is determined, as before, by analyzing the points where the lines of each plane cross.

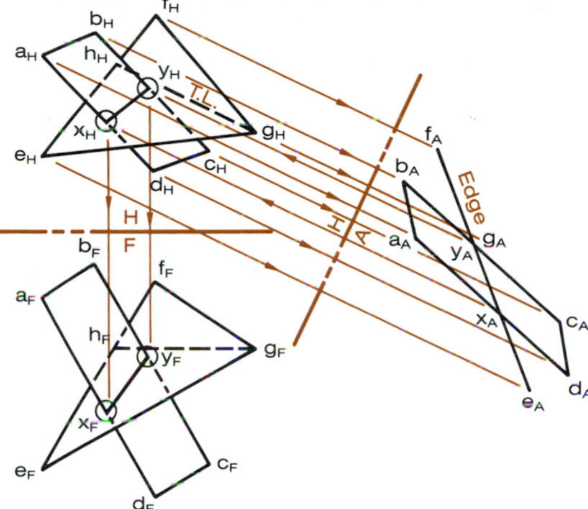

FIGURE 14-39 Constructing the line of intersection by the edge view method ***(Section 14.28)***

PROBLEMS

Basic Descriptive Geometry

For each of the following problems, lay out the given views full size. Assume that each grid equals ¼ inch (about 6 mm). When positioning the problems on the sheet, use the small circles as locations for the endpoints of the various lines on the grid. The full-scale dimensions are given from the border lines of layout A, shown on the inside front cover. Fill out the required information in the sheet title block. Use the proper notation for all reference lines, points, lines, slope angles, grades, edge views, or true size views.

Problems P14.1 through P14.9 deal with the position of a point on a line and the bearing, azimuth, true length, slope, and grade of lines. Problems P14.10 through P14.16 deal with plane surfaces. Problems P14.17 through P14.27 are applications of basic constructions.

P14.1(a) Given mast ab and guy wire ac in views F and P, find the bearing of ac, the true length, and the angle the guy wire makes with the mast. Draw a check view verifying the true length of ac. (Scale: 1″ = 10′)

P14.1(b) Given line ab and point x in views H and F, find the bearing, true length, and grade of line ab. Construct a view showing the point view of line ab and indicate the distance line ab is from point x. (Scale: 1″ = 50′)

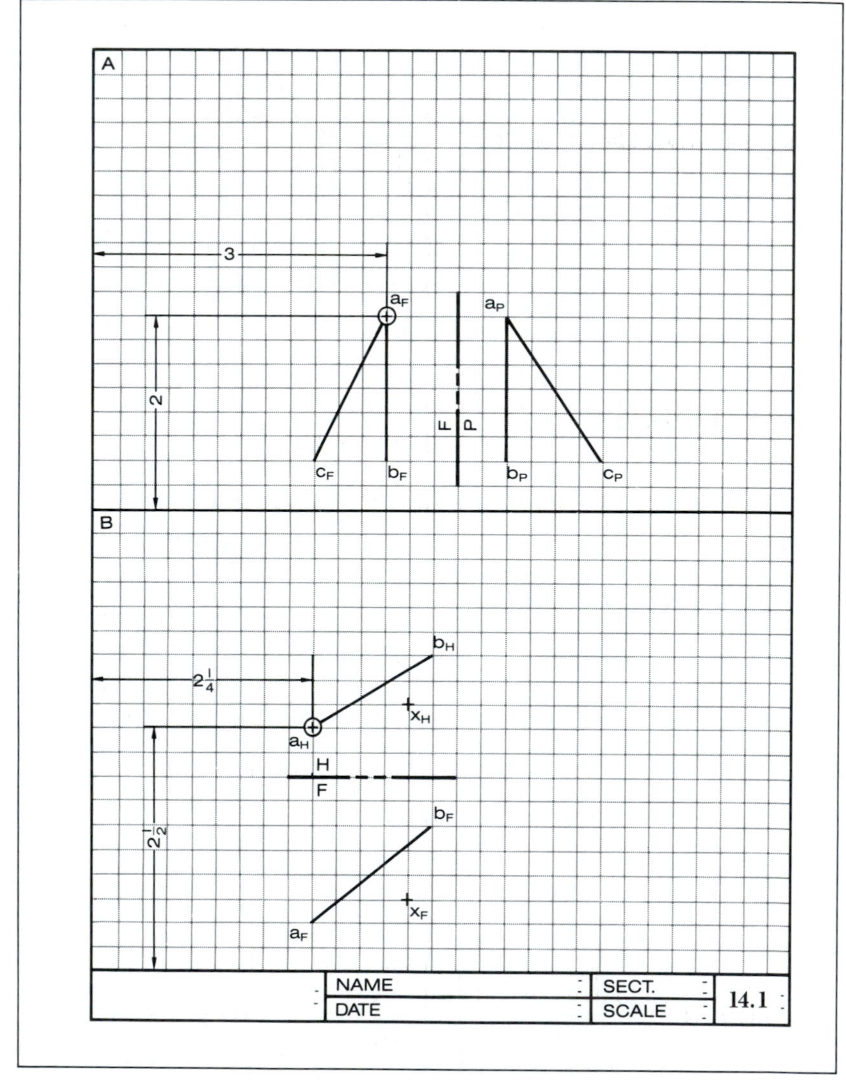

P14.2 Given views F and P of a structure represented by three members, determine and record the true lengths in millimeters of members oa, ob, and oc, the bearing of oc, and the angle member oc makes with a horizontal plane. (Full scale)

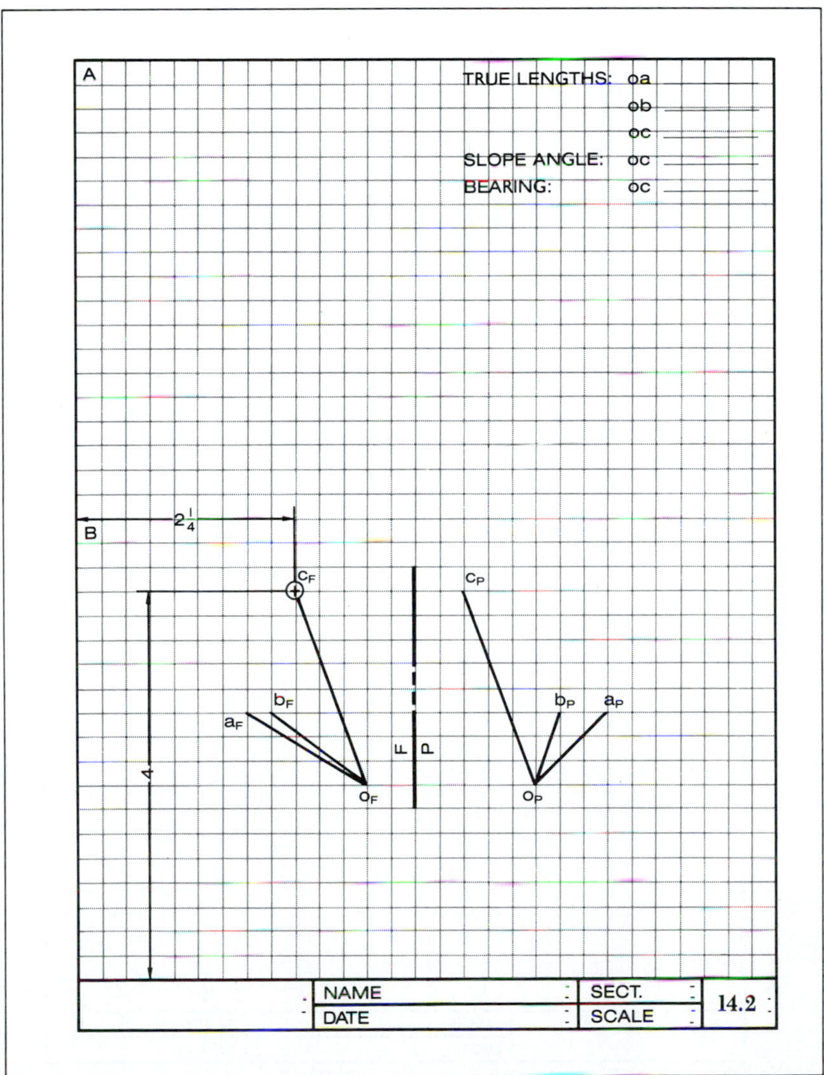

P14.3(a) Given line ab and point x on the line in views H and F, and given that line xy has a bearing of S45°E, is 25 mm long, and slopes down from x to y at 30°, show line xy in views H and F. Find the point view of line ab. (Full scale)

P14.3(b) Given pipe ab in views H and A, find the bearing, true length, and slope angle of the pipe. Locate point x at midpoint on line ab. A second pipe (xy), with a bearing of S45°E, slopes down on a forward direction from x to y at a grade of 50% and ends at the same elevation as point b. Find the true length of pipe xy. (Scale: 1″ = 10″)

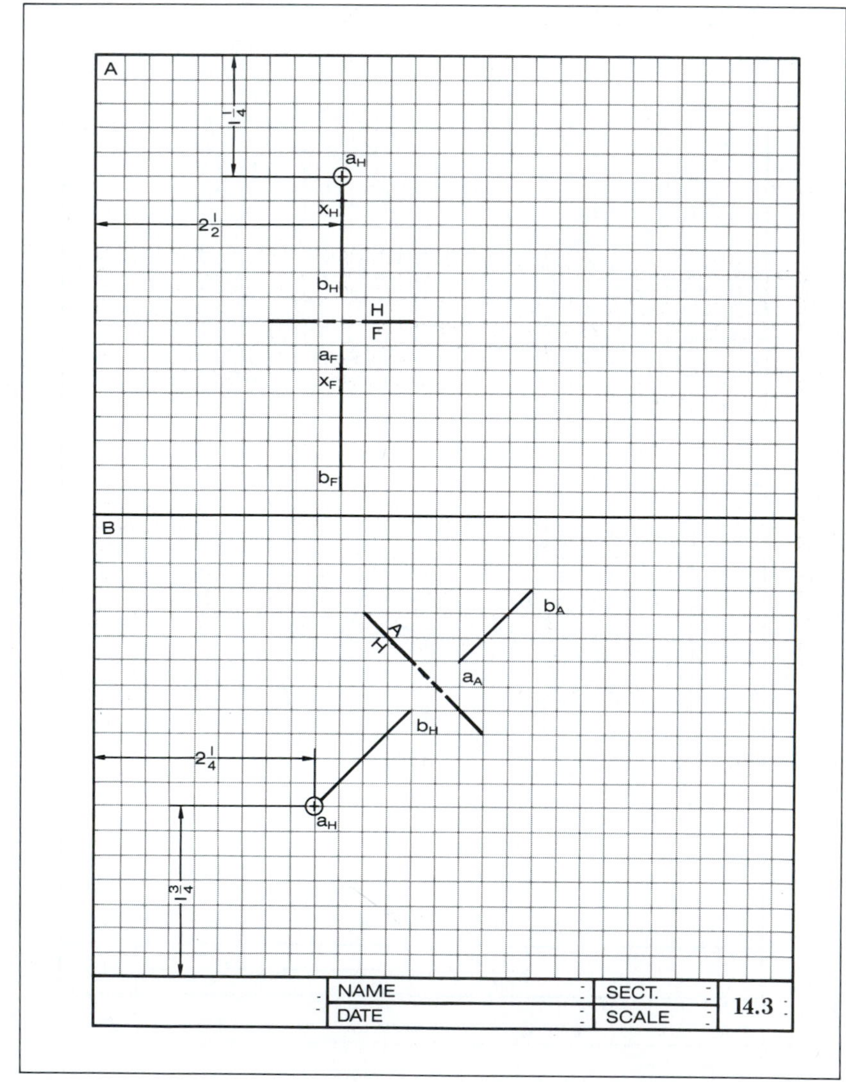

P14.4(a) Given line ab in views F and P and point x on line ab in view P, and given that line xy has a 15° azimuth, is 1.38″ long, and slopes down at 10° and that point y is rearmost of point x, show line xy in views H, F, and P. Label the azimuth of ab and xy. (Full scale)

P14.4(b) Given line ab in views F and P and point x on line ab in view P, find and label the bearing, true length, and slope angle of line ab. Also, show the point view of line ab. Line xy has a due south bearing. Point y is frontmost of point x. Line xy slopes up 30° and is .88″ long. Show line xy in views H, F, and P. (Full scale)

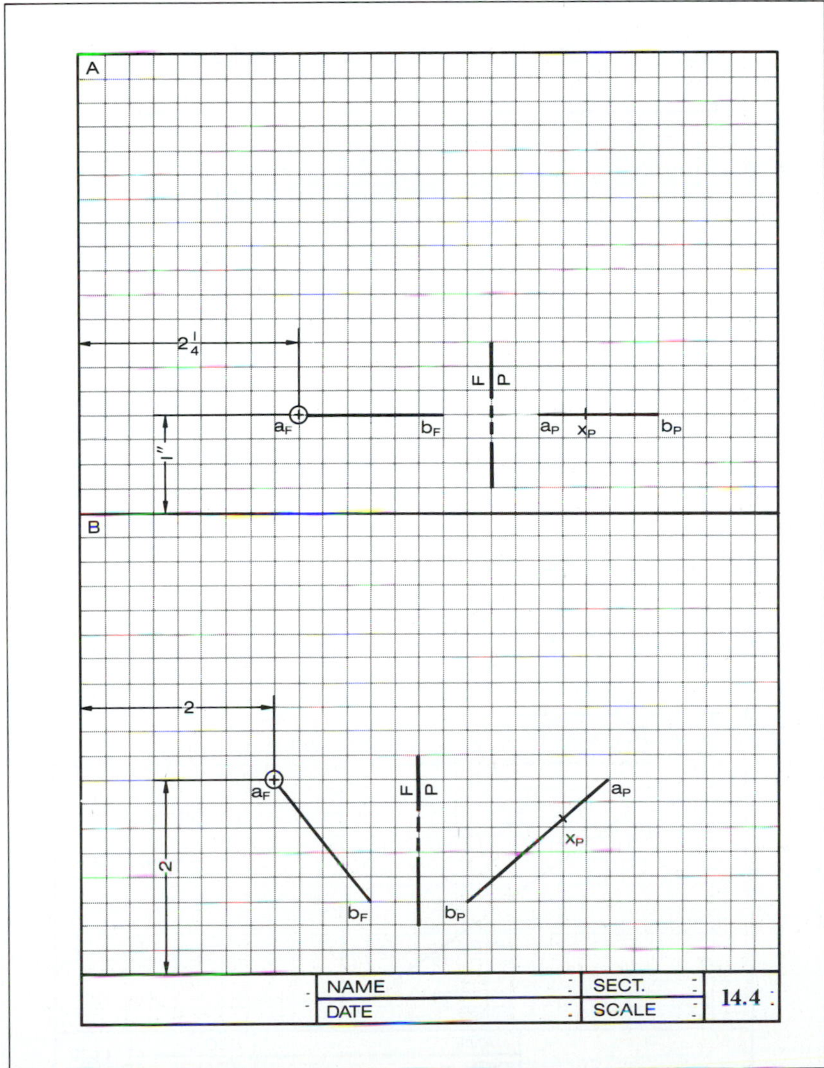

P14.5(a) Given line ab and point x on the line in views H and F, and given that line xy is 28′ long with a bearing of S15°W and slopes down from x to y and that point y is frontmost of point x, complete views H and F. Label the bearing of both lines. (Scale: 1″ = 20′)
P14.5(b) Given line ab in views H and F, point x on line ab in view H, and point y in views H and P, find and label the azimuth, true length, and grade of lines ab and xy. Give dimensions showing the relationship of point y to point a. Show line ab in view P and all points in views F and P. (Scale: 1″ = 100′)

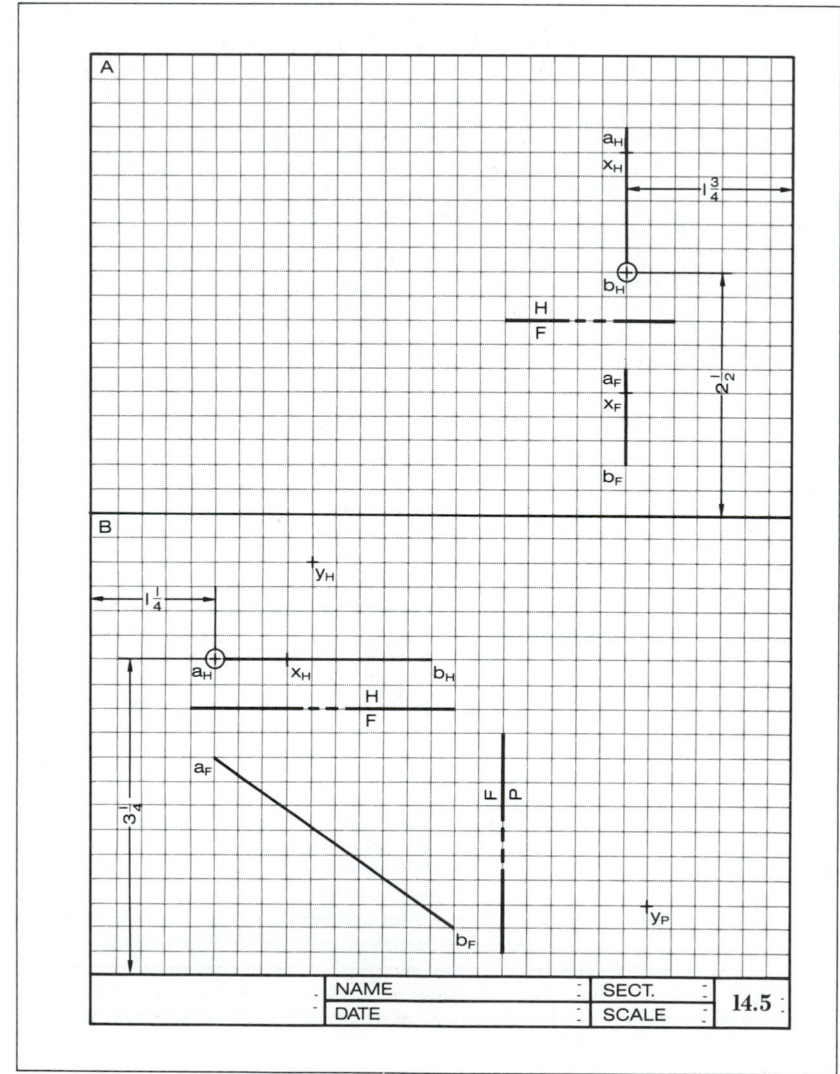

P14.6(a) Given point a in views F and P, and given that line ca has an azimuth of 315°, is 10′-6″ long, and slopes down from a to c at 25°, and that point c, with respect to point a, is rearmost, show and label the azimuth, true length, slope angle and point view of line ca. Show line ca in views H, F, and P. Give a dimension showing how much lower the position of point c is than point a. (Scale: ⅛″ = 1′-0″)

P14.6(b) Given line ab and views F and A, find and label the bearing, true length, and grade of line ab. Point x is on line ab, 1′-9″ from point a. Line xy has a S18°E bearing, is 3′-0″ long, and has an upward grade of 65% from x to y. Point y is frontmost of point x. Show lines ab and xy in views H and F. (Scale: ⅜″ = 1′-0″)

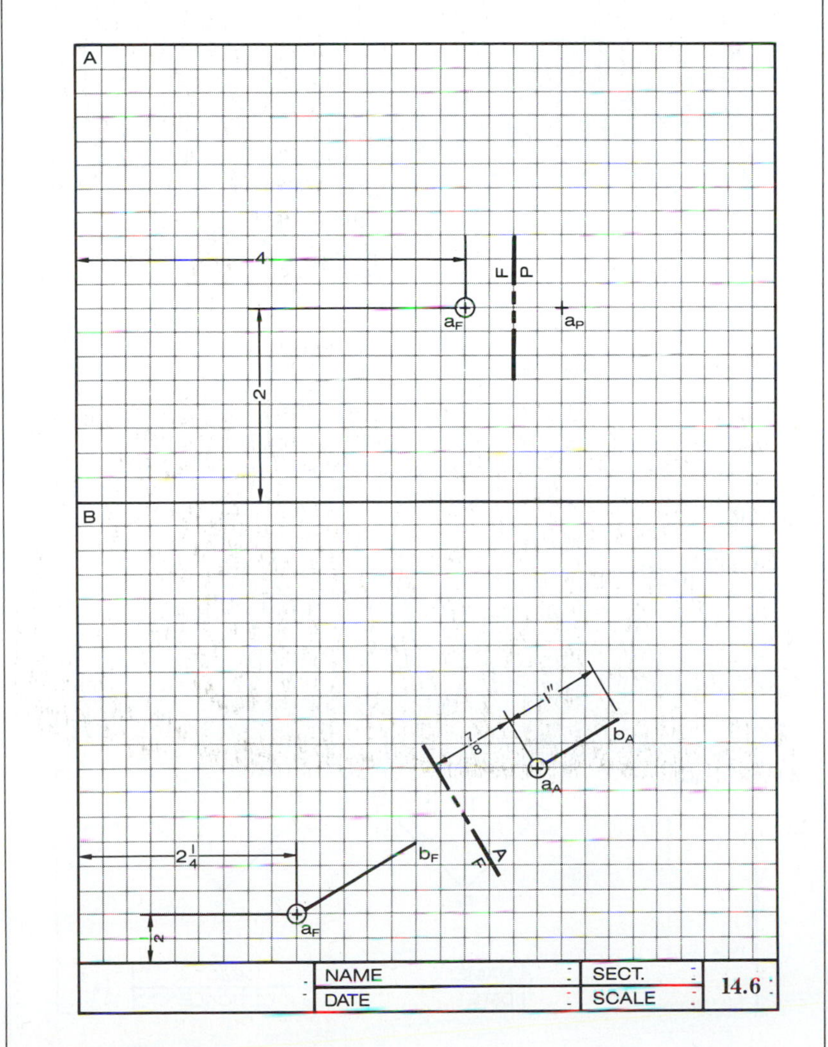

P14.7(a) Given line ab in view F and point a in view H, find the true length of line cd and show the line in views H and F. Line ab has a bearing of N65°E. Point c, on line ab, is 5″ below b. Line cd has a bearing of due north with a slope angle of 30°. Point d is rearmost of point c and at an elevation of 1′-9″. (Scale: 1″ = 1′-0″)

P14.7(b) Given struts oa, ob, and oc in view F and struts oa and ob in view P, find the bearing, true length, and slope angle of oa. Find the bearing and slope angle of oc and the bearing of ob. (Scale: 1/8″ = 1′-0″)

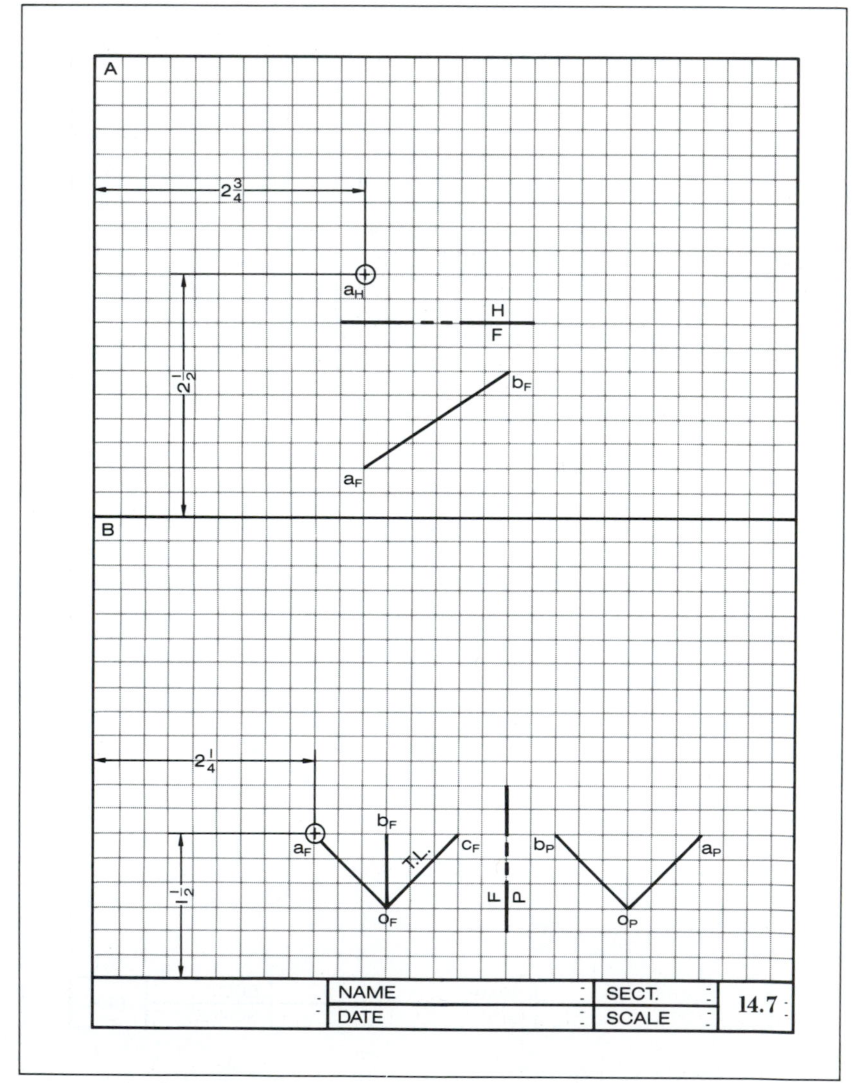

P14.8(a) Given line cd and point a in views F and P, and given that line ab has a bearing of S62°E and is 7′-6″ long with a 22°30′ slope angle from point a down to b, find the bearing and true length of line cd. Show line ab in views H, F, and P. (Scale: ¼″ = 1′-0″)

P14.8(b) Given point a in views H and F, and given that line ab has a N72°E bearing and is 14″ long with a 40% grade extending down from a to b, locate point x as follows: 9″ to the right, 7″ in back, and 11″ below point a. Show lines ab and ax in views H and F. Find the bearing and true length of line xb. (Scale: 1″ = 1′-0″)

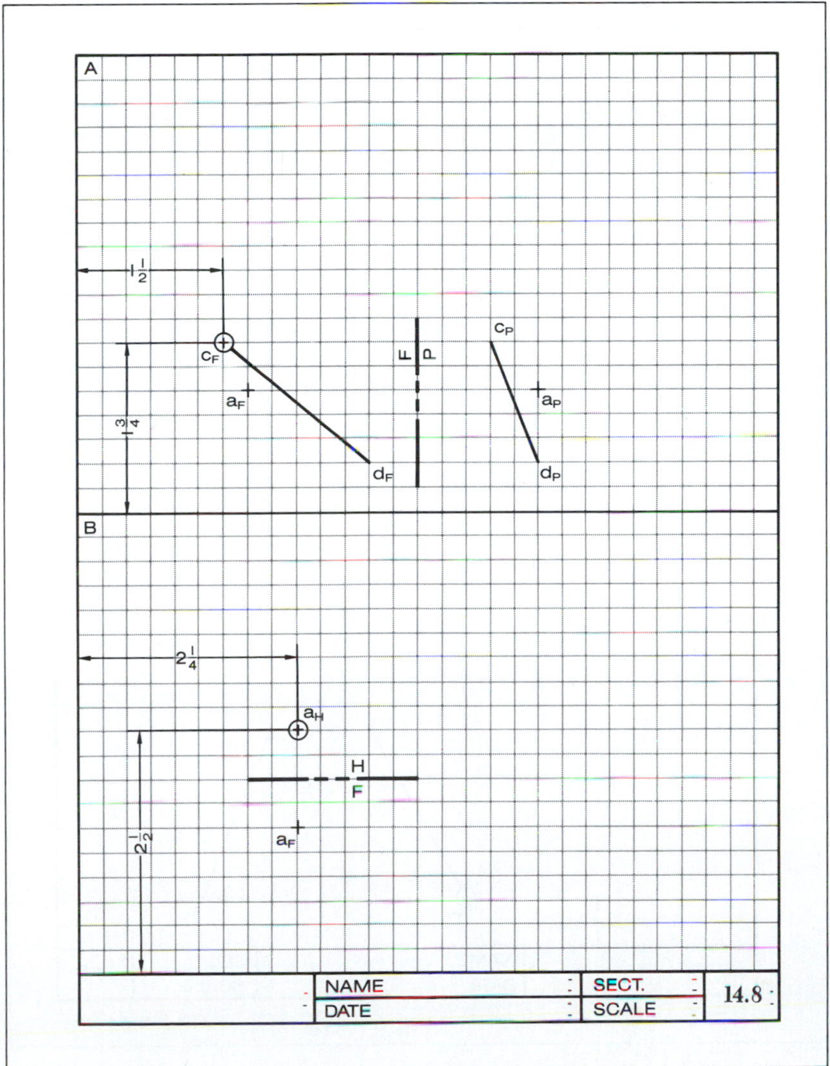

P14.9(a) Tubing is bent to the configuration shown by the center lines in views H and F. Find the angle the tubes bc and cd are bent with the horizontal. Indicate which of the tube sections appear true length in the given views. Show the tubes in view P.

P14.9(b) Given line ab in views F and P, draw line ab in view H and locate point x on the midpoint of this line in views H, F, and P. A second line, bc, with an identical bearing to line ab, is 50 mm long and slopes down from b to c at an angle of 25°. With respect to point b, point c is rearmost. Show line bc in all views. Construct a point view of line bc. Find the true length of line cx. (Full scale)

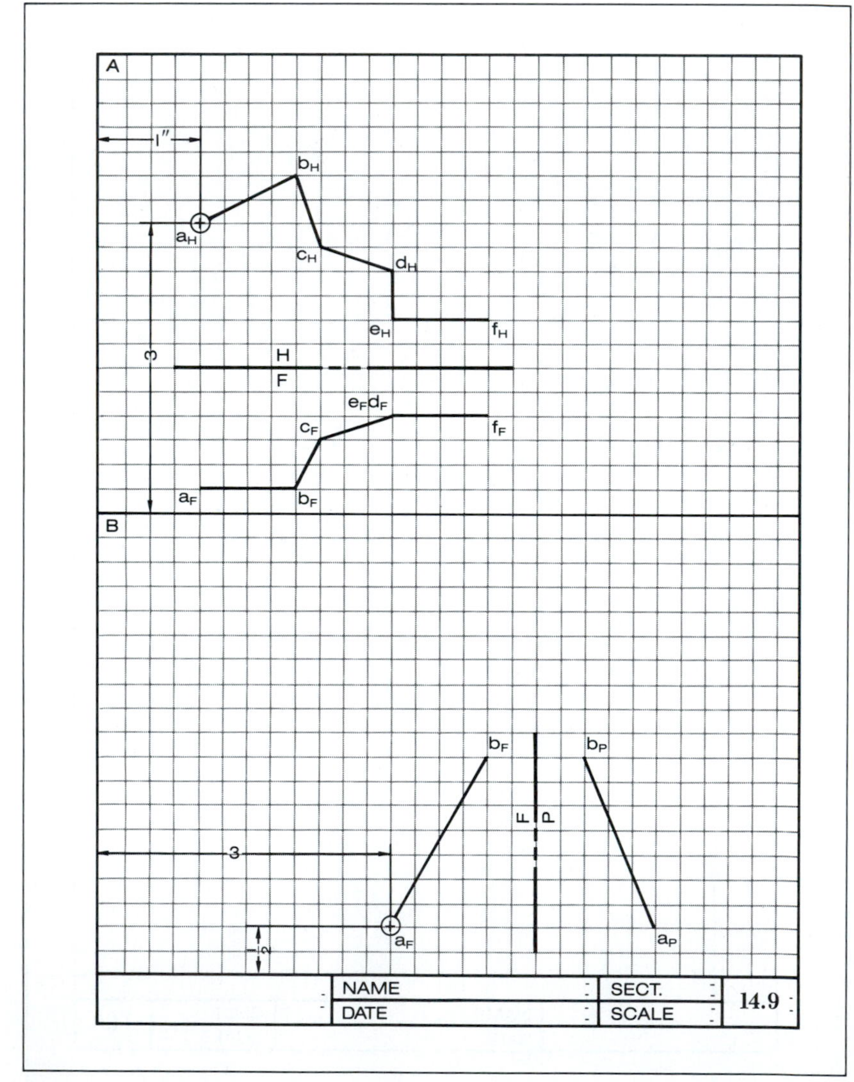

P14.10 Find and record the value in inches of a true length line in each view of the following:

a. Given views H and F of plane abc. (Full scale)

b. Given views F and P of plane abcd. (Full scale)

c. Given views H, F, and P of plane abc. (Full scale)

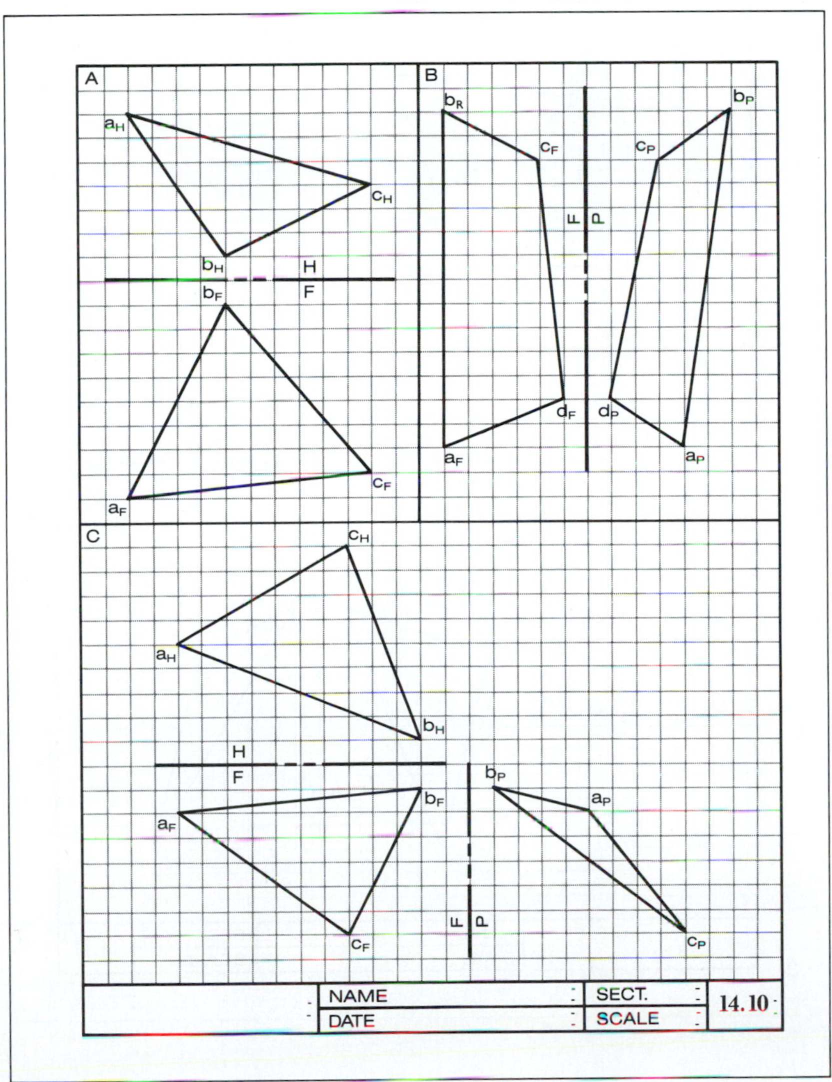

P14.11(a) Given view H of plane abc and view F of line ab, and given that the plane slopes down to c at an angle of 35°, show the plane in view F and label the bearing of lines ab, bc, and ca.

P14.11(b) Given views F and P of plane abc, show the edge view of the plane in a view adjacent to view F.

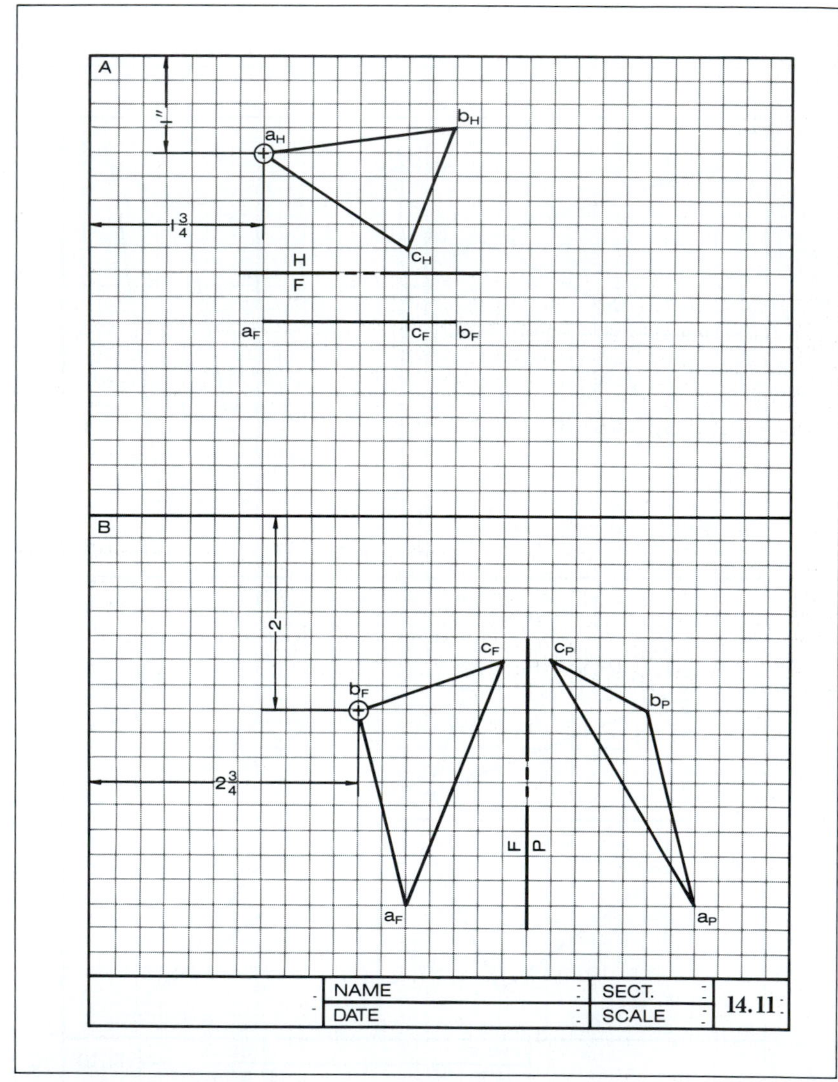

P14.12(a) Given views H and F of plane abcd, construct a true size view of the plane and measure and record the angles cba and adc. Label the slope of the plane.
P14.12(b) Given views H and F of plane abcd and view F of point x, which lies in the plane, construct a true size view and dimension in millimeters the position of point x with respect to point a. Label the slope of the plane. (Full scale)

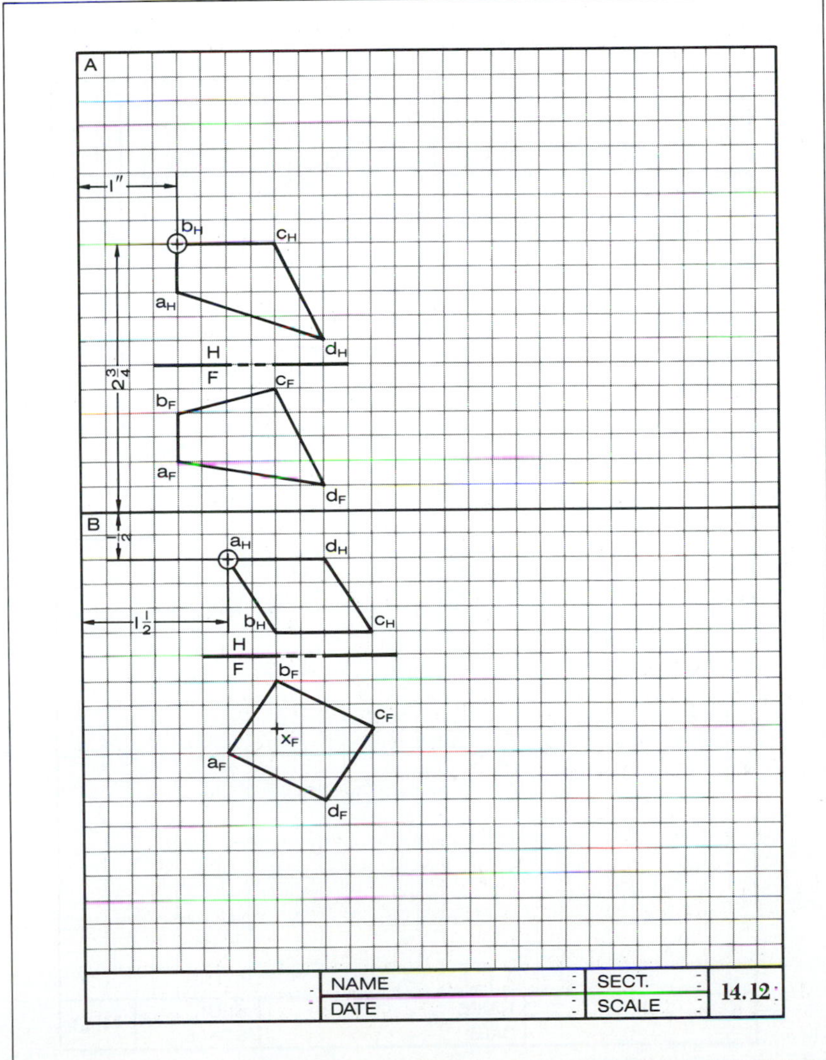

P14.13(a) Given views H and F of plane abc, locate points x and y, which lie in the plane. Point x is ¾″ in front of and 1″ to the left of point b. Point y is ½″ in front of and ½″ below point a. (Full scale)

P14.13(b) Given views H and F of points abcd, locate point x 1½″ to the right and ½″ behind a; point y ½″ to the left and ½″ higher than b; and point z ¾″ to the left and ¾″ higher than d. All points are in the plane. (Full scale)

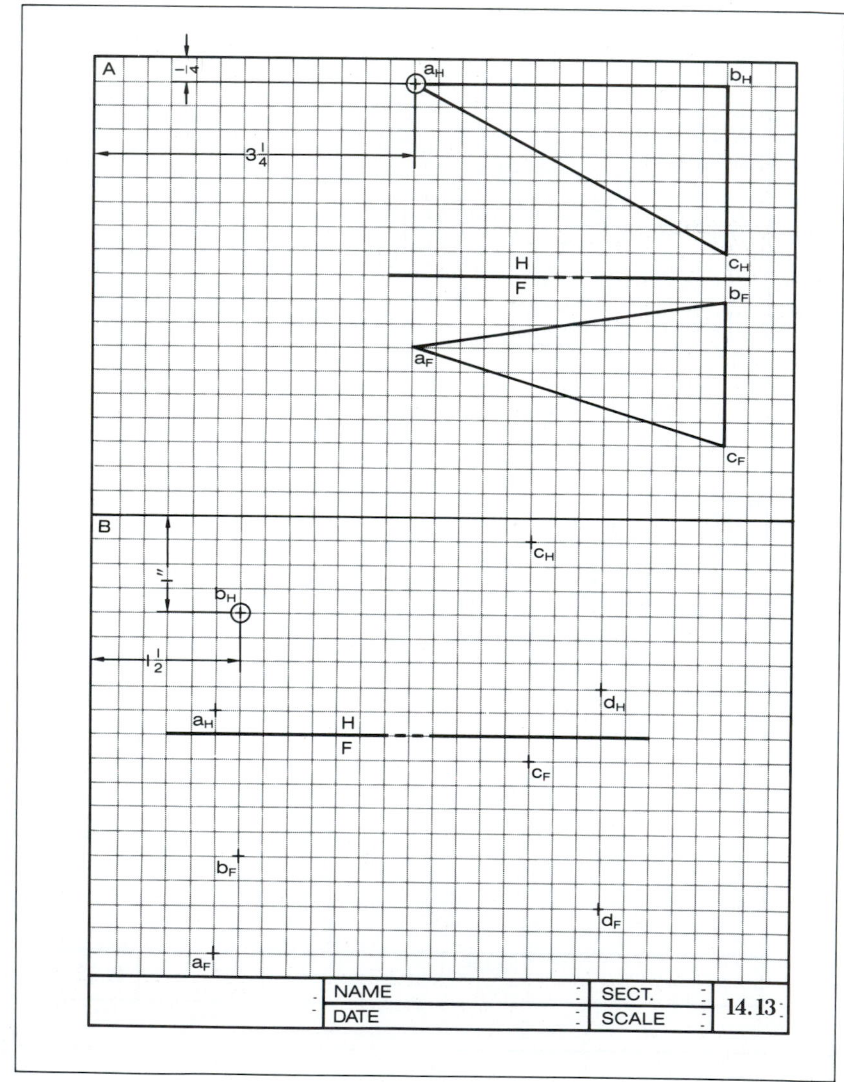

P14.14(a) Given views H and F of line ab and point x, show xy in the given views and show the shortest distance between line ab and line xy. Line xy is 180′ long with a bearing of N30°W. It has a grade of 65% sloping up from x to y. With respect to point x, point y is rearmost. (Scale: 1″ = 100′)

P14.14(b) Given views H and F of lines ab, cd, mn, and op, show graphical proof for the following questions:

a. Are lines ab and cd parallel?
b. Are lines mn and op parallel?
c. Do lines cd and op intersect?

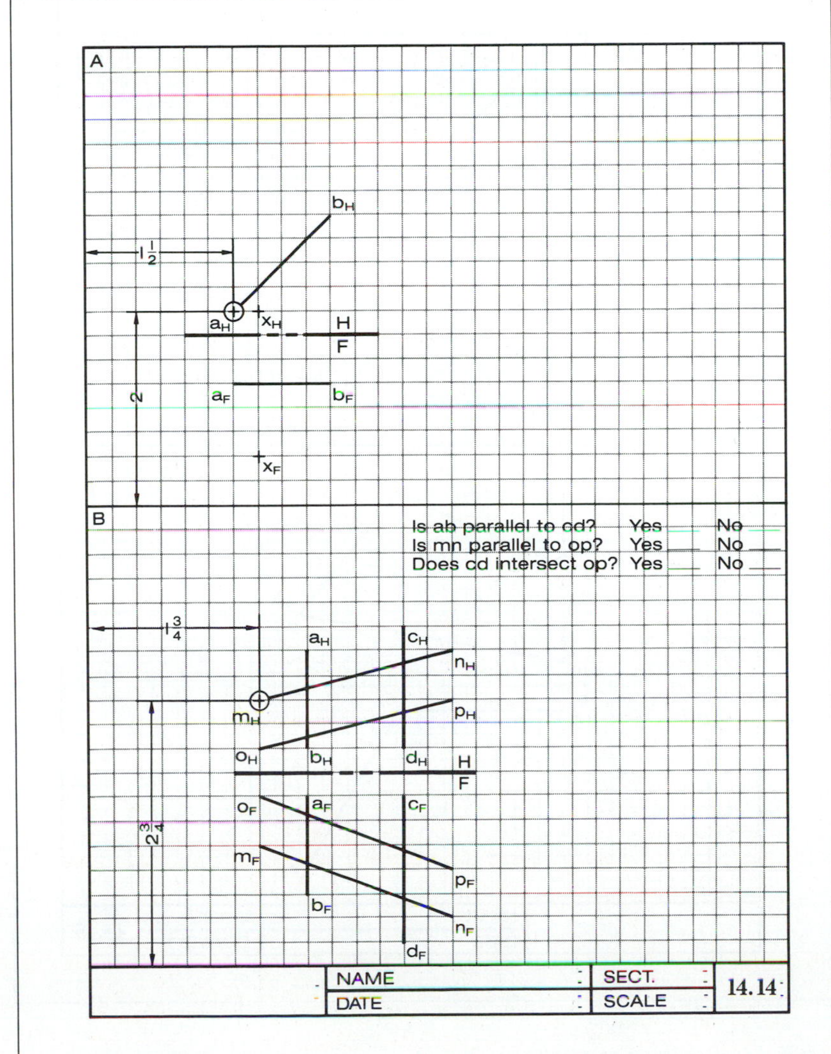

P14.15(a) Given views H and F of plane abcd, locate letters N and L in view F. (They lie in the plane.)

P14.15(b) Given a complete view H of plane abcd and a partial view of the plane in view F, complete the plane in view F. Locate point x in view H and point y in view F. Each point lies in the plane.

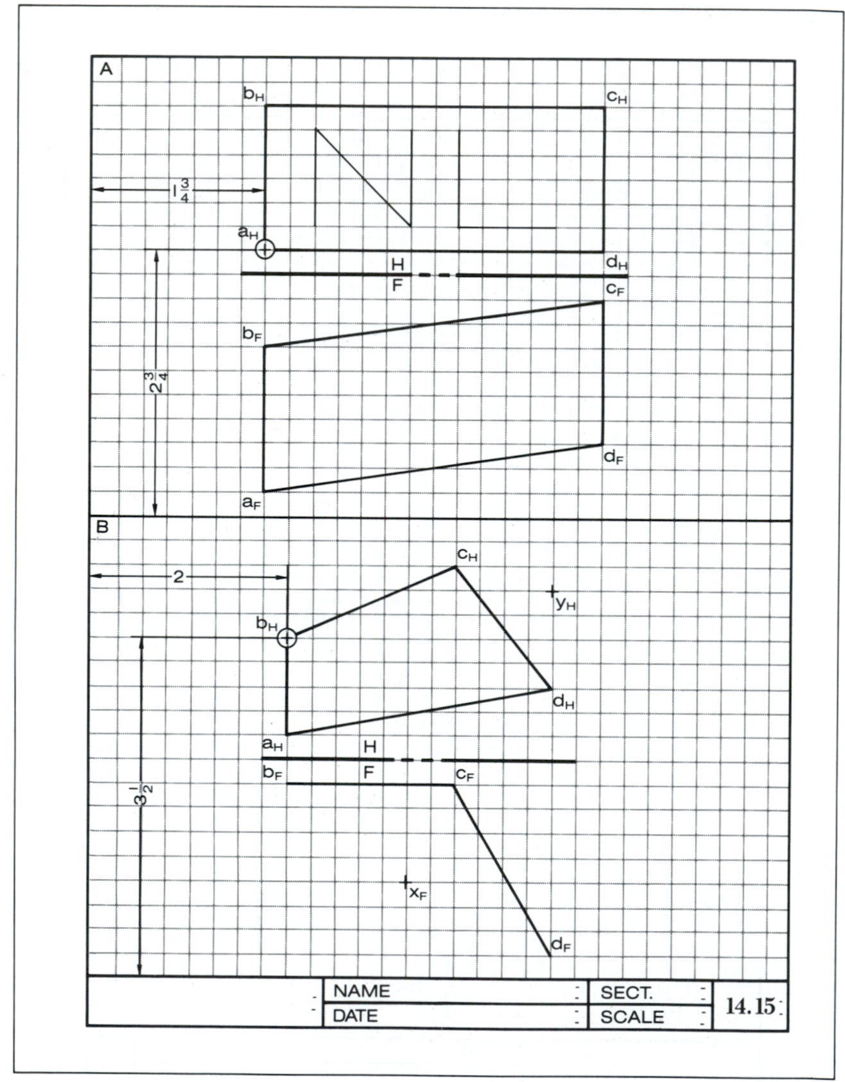

P14.16(a) Given views H and F of plane abc, and given that the letter V lies in the plane, locate letter V in view F.

P14.16(b) Given views F and P of plane abcd, locate point x, which lies in the plane. Point x is 30 mm behind and 10 mm below point b. (Full scale)

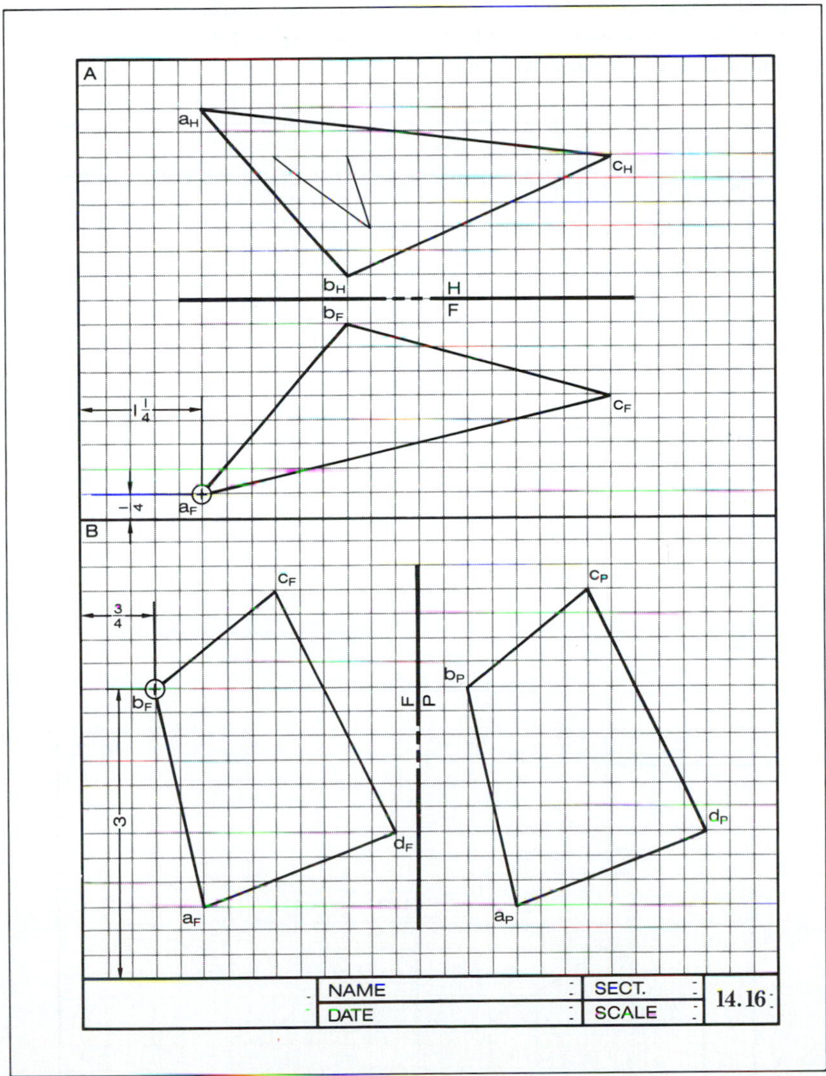

P14.17(a) A vein of uranium ore is formed by plane abc, which slopes 28° downward from b to c. Point c is due north from b and is 110′ deeper than point b. Given views H and F of line ab and point x, measure and record the bearing of line ac, the shortest distance (xy) from point x to the plane, and the magnitude of angle bac. Also give dimensions showing how far point x is from point c. (Scale: 1″ = 100′)

P14.17(b) Given views H and F of line ab and point x, and given that line ab represents a portion of a cable, tie into ab from point x with a second cable (xy) at an angle of 60° Show the point of the tie-in (y) in the given views and measure and record the length of xy and ab. (Scale: ¼″ = 1′-0″)

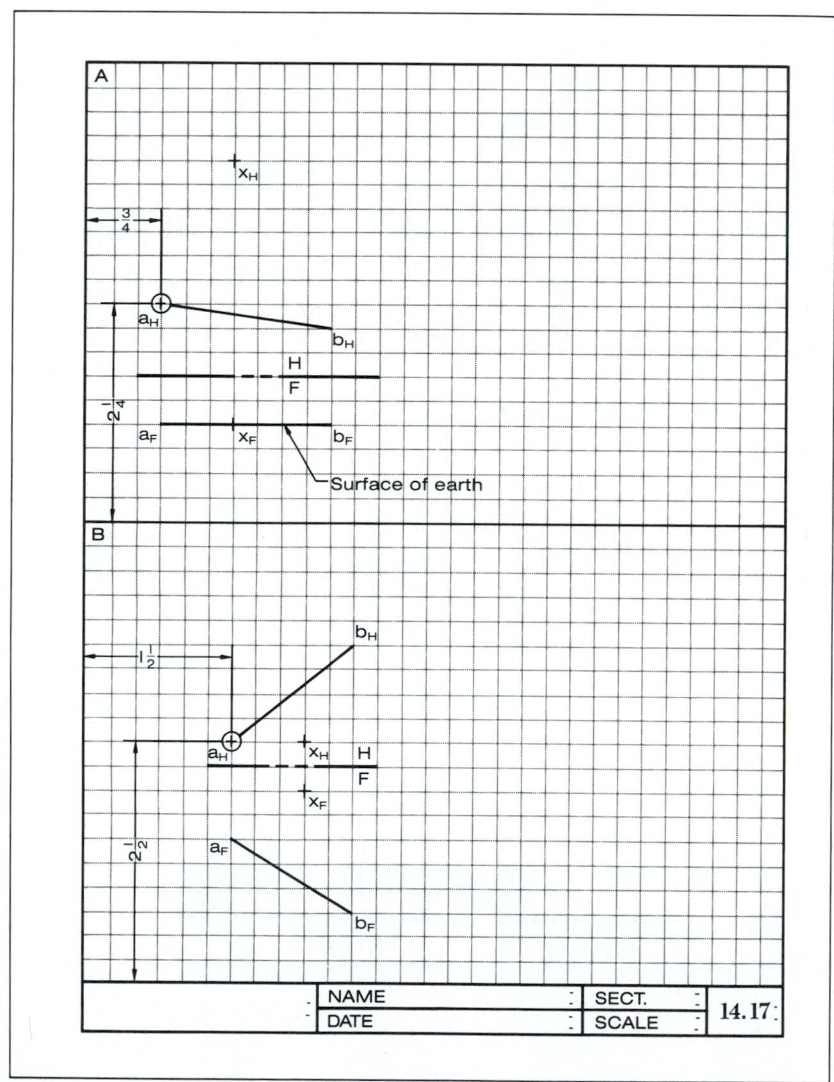

P14.18(a) Given views H and F of plane abc and point x, and given that points abc lie in a vein of copper ore, measure and record both the shortest distance from point x to the vein of ore and the shortest vertical distance. In addition, record the strike and dip of the plane and the bearing of line ab. (Scale: 1″ = 50′)

P14.18(b) Given views H and F of line ab and point x, and given that line ab represents the center line of a metal brace and that a second brace (xy) makes an angle of 60° with the brace ab, measure and record the length of the brace xy and show where the 60° angle is measured. Show xy in views H and F. (Scale: 3″ = 1′-0″)

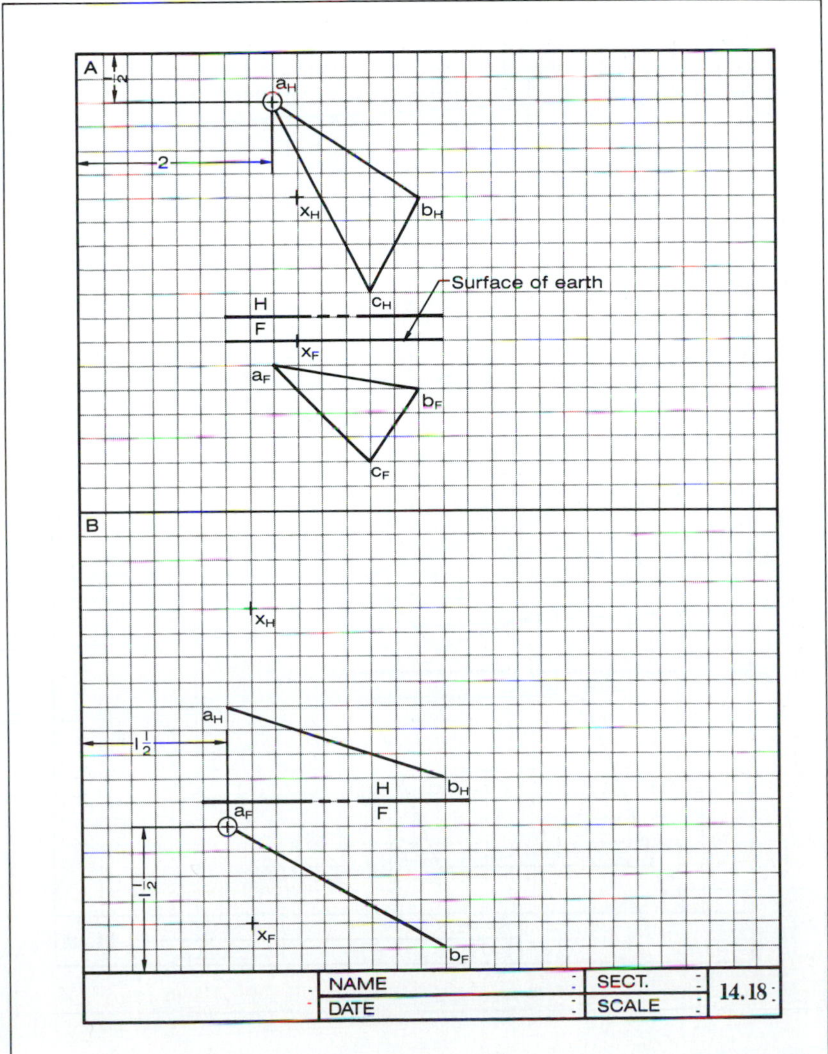

P14.19(a) Given views F and P of an opaque plane abc and line xy, pass a wire, represented by line xy, through the plane at point p. Using an auxiliary view, find the point of intersection of the wire and the plane and show the line xy in the correct visibility in the given views.

P14.19(b) Given views H and F of an opaque plane abcd and points x and y representing end points of a line that intersects the plane, find the point of intersection (p) of the line and the plane and show line xy in correct visibility in the given views. Use an auxiliary view.

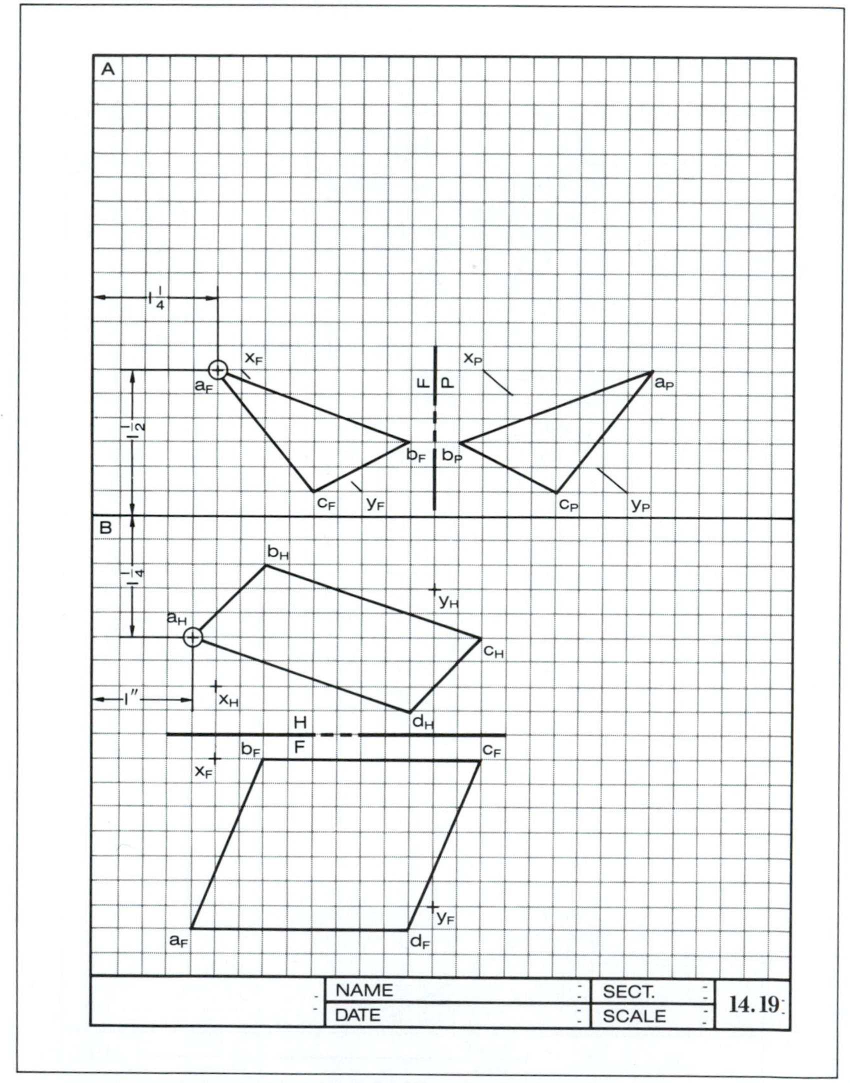

P14.20(a) Given views H and F of plane abc and view H of point o, and given that plane abc is one face of a locating block, drill a hole vertically through the block with a center at point o. Find the angle the center line of the hole makes with the plane abc and show point o in view F.

P14.20(b) Given views H and F of line ab, which represents a bracket that will be attached to the wall (shown as an edge in views H and F), find the angle the bracket makes with the wall.

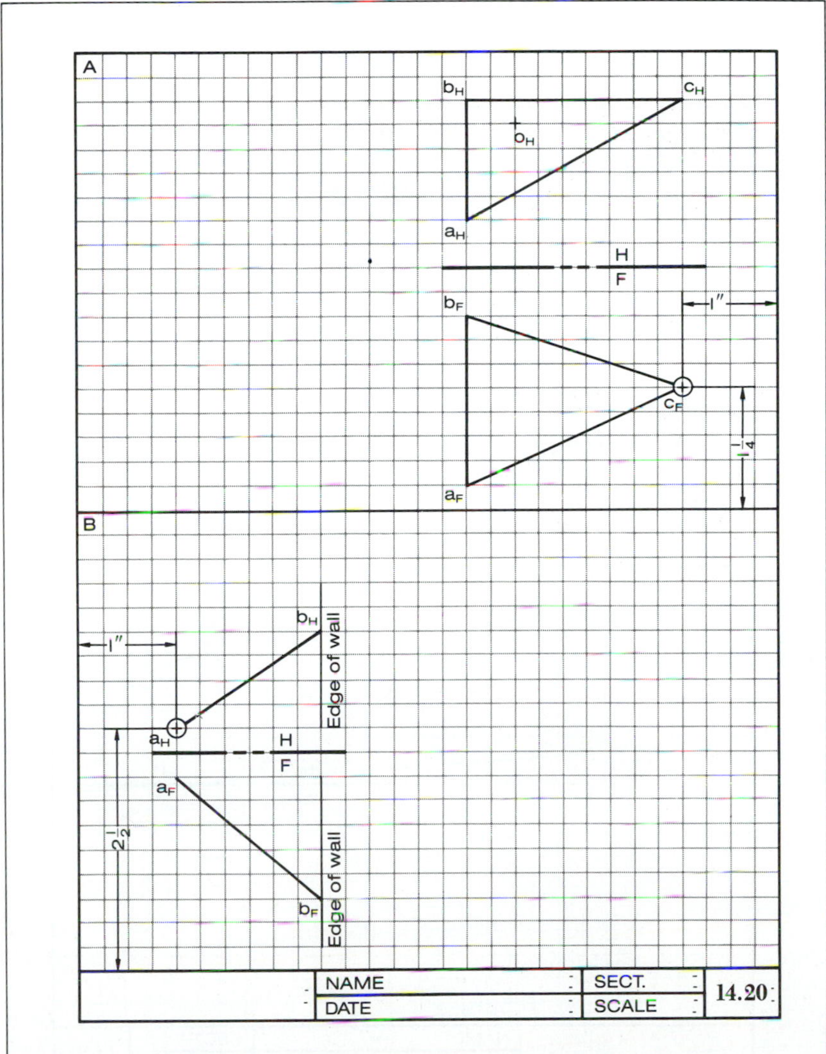

P14.21(a) Given view F of lines ab and cd and view H of point c (the lines represent the center lines of perpendicular pipes) and given that line cd has a bearing of N30°E, complete view H.
P14.21(b) Given views H and F of points abc, with plane abc representing three points that lie in the top surface of a vein of coal, give the bearing of ac, ab, and bc and the strike and the dip of the plane.

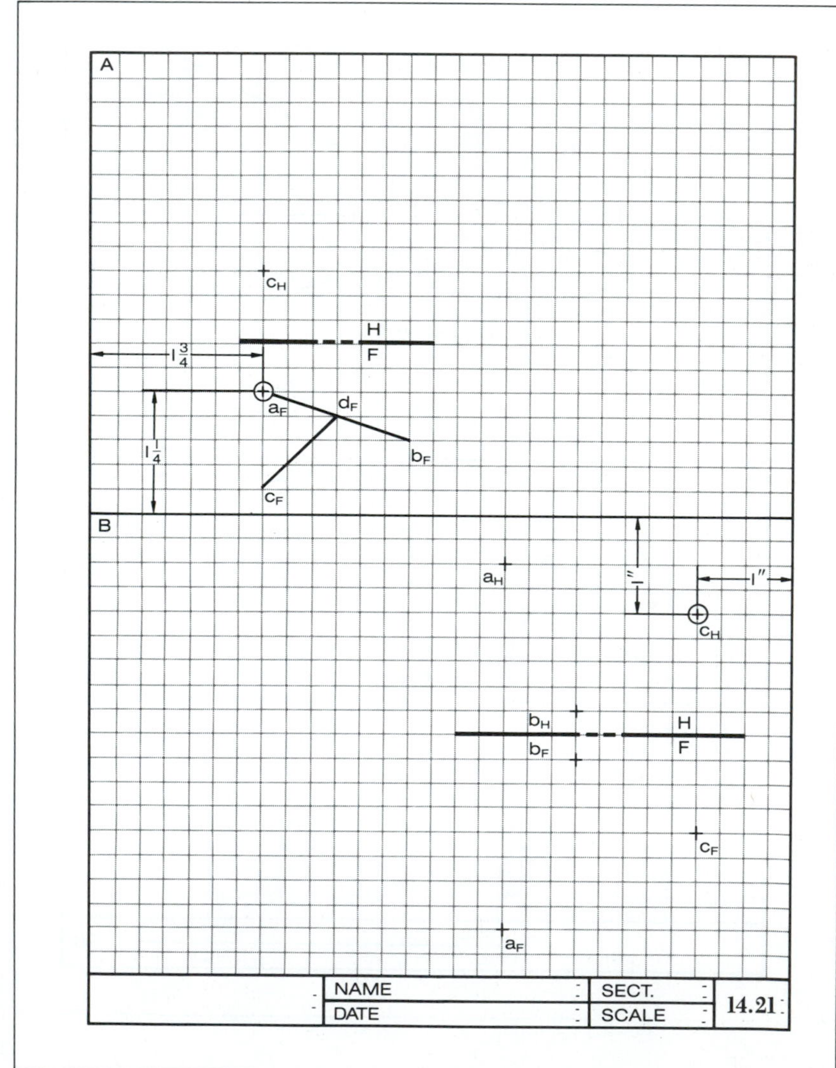

P14.22(a) Given views H and F of line ab in the framework of three welded members, bc is equal to ½ ab, is perpendicular to ab, and bears S65°E. Member cd is equal to bc, is perpendicular to bc, and slopes up at an angle of 45°. Point d is rearmost of point c. Complete the given views and find the distance between points a and d. (Scale: ¼″ = 1′-0″)

P14.22(b) In the given views H and F of line ab, line ax is perpendicular to ab, is 180′ long, and has a N45°E bearing. Line by is also perpendicular to ab, is 210′ long, and bears due east. Find the slope angle, bearing, and true length of xy. Complete the given views. (Scale: 1″ = 10′)

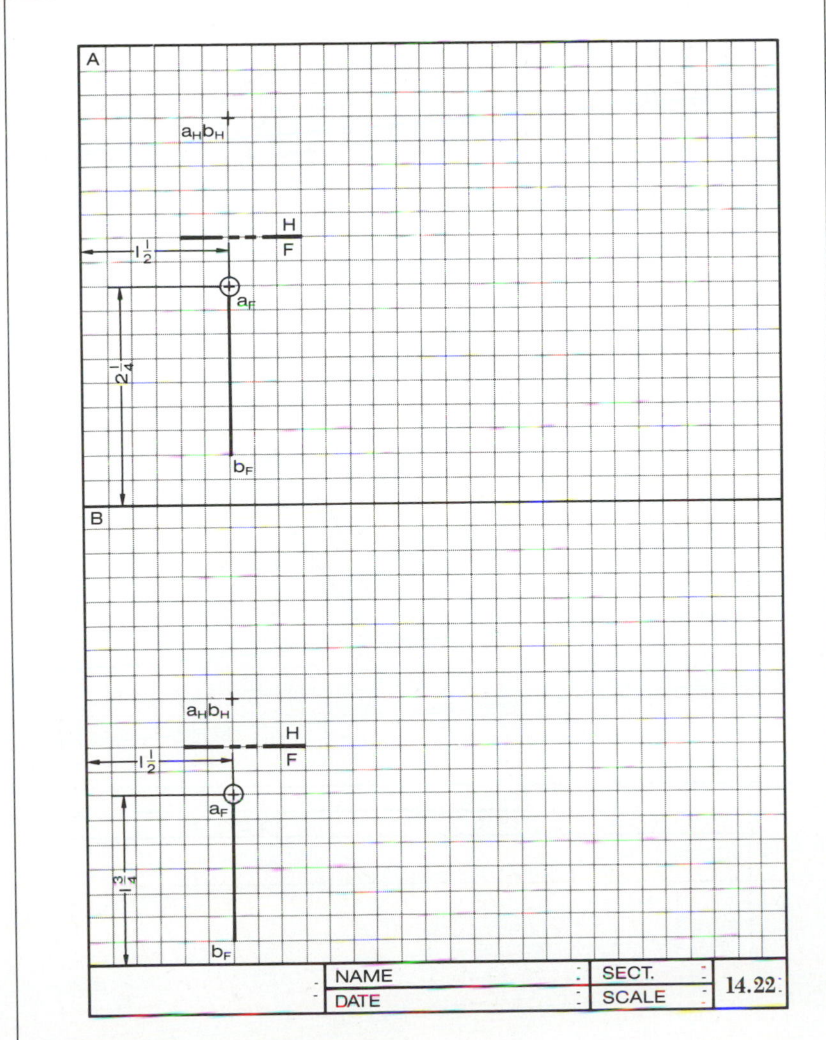

P14.23 Given views H and F of planes abc and mnop, complete the planes in the correct visibility. Solve in the given views.

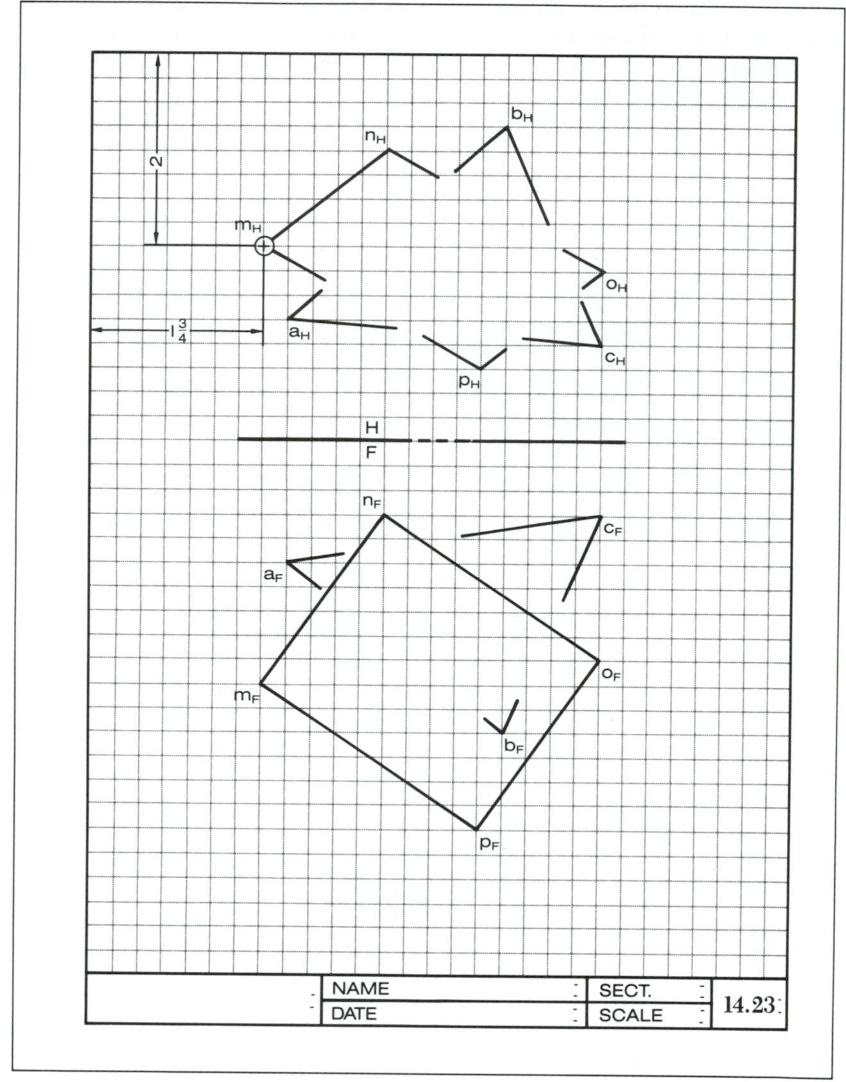

P14.24 Given views H and F of planes abc and mno, complete the planes in the correct visibility. Solve in the given views.

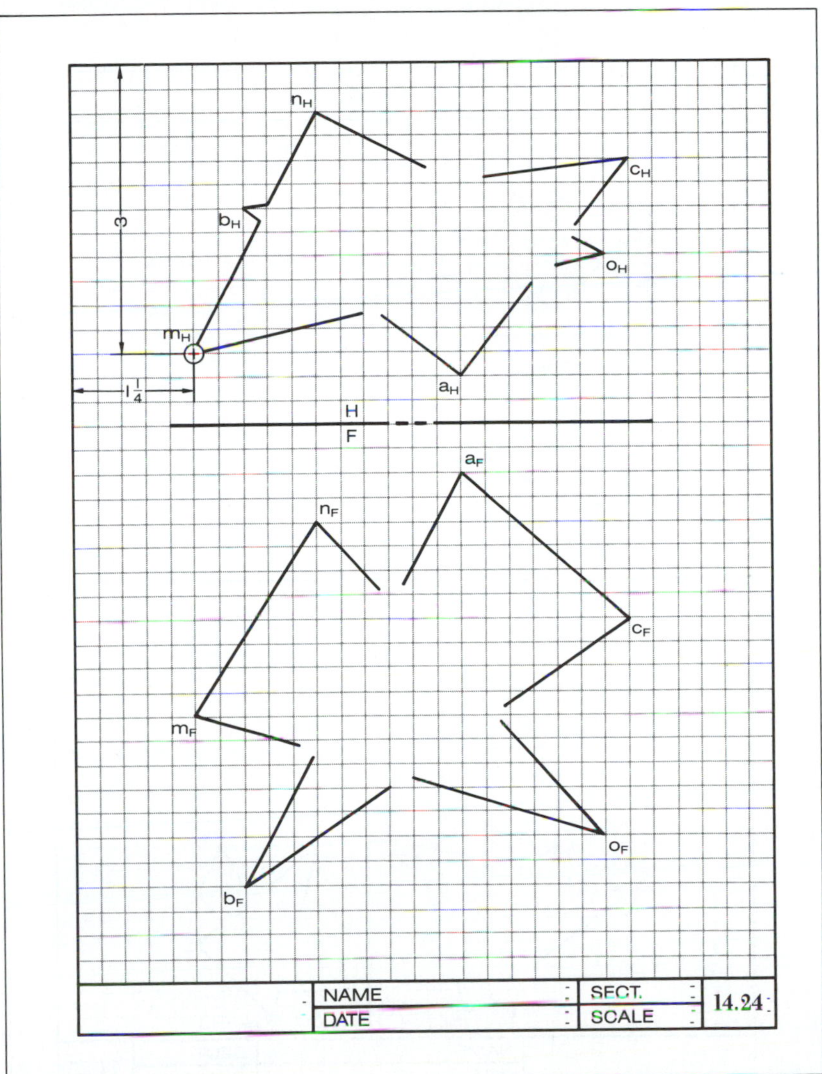

P14.25(a) Given views H and F of planes xyz and abcd, complete the visibility of the planes using the edge view method.

P14.25(b) Given views F and P of planes abc and xyz, complete the visibility of the planes using the edge view method.

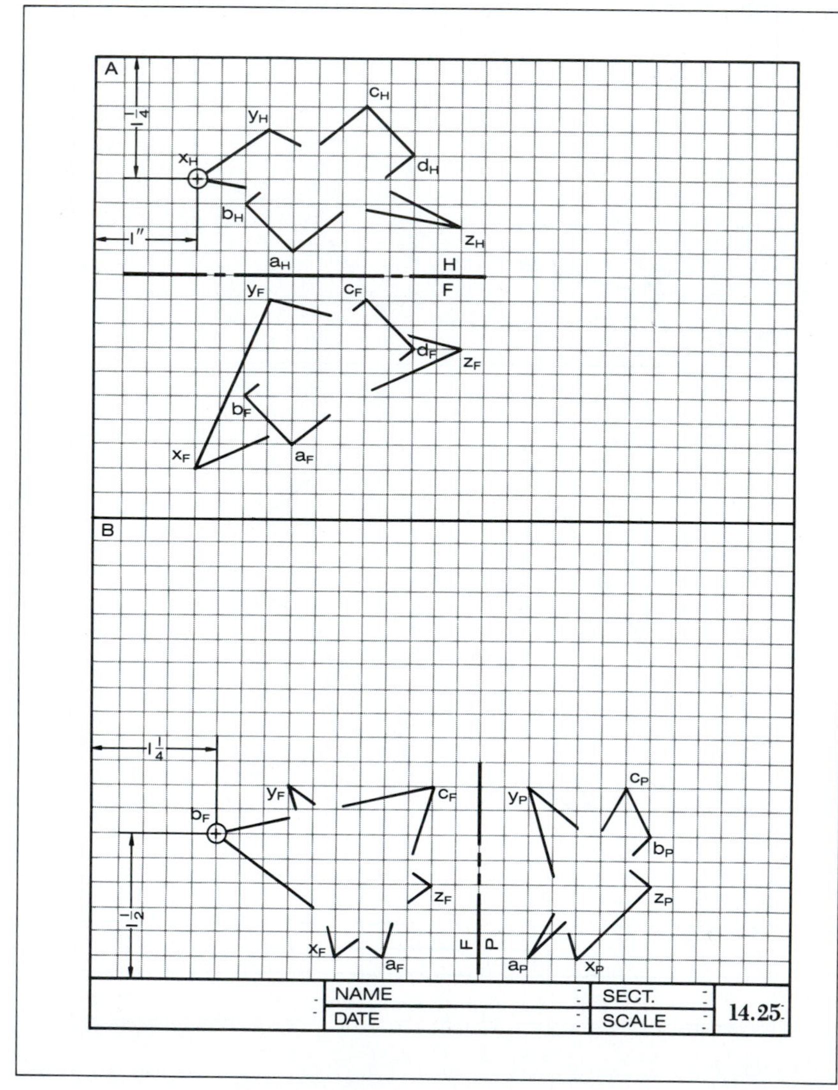

P14.26(a) Given views H and F of planes abc and adc, representing two surfaces that are to be welded together, find the angle between the planes.

P14.26(b) Given views F and P of two planes abc and adc, find the angle between the planes.

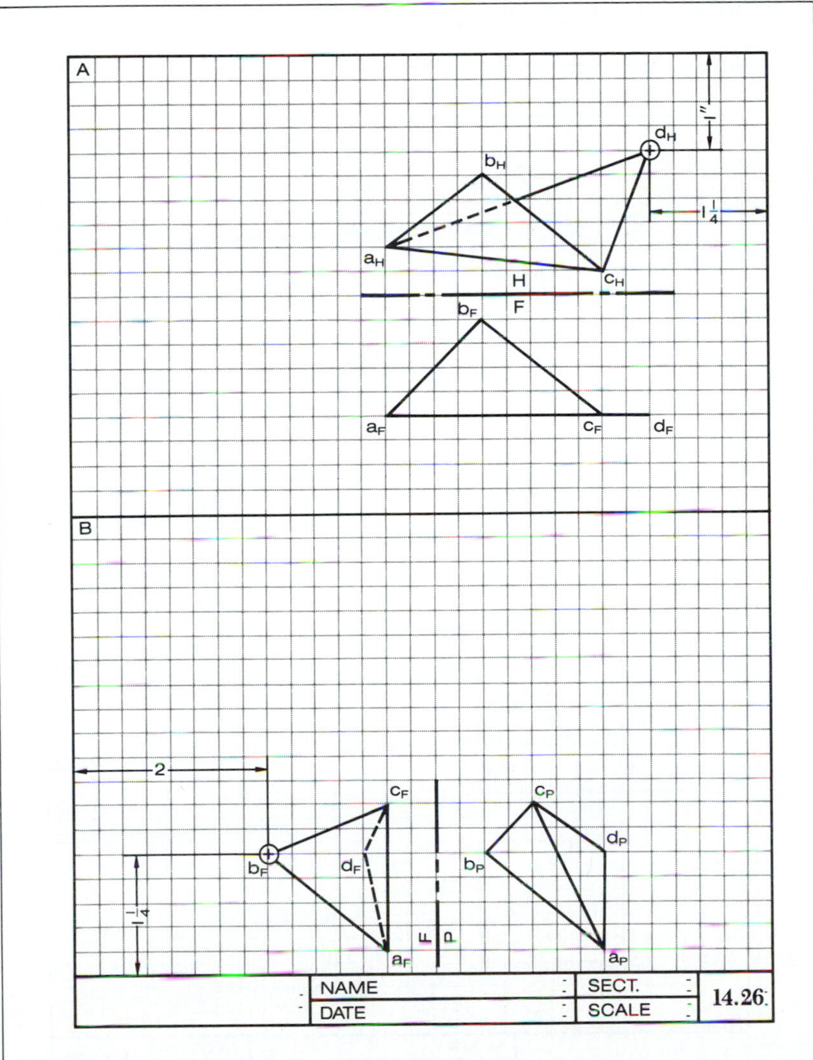

P14.27(a) Given views H and F of planes abcd and abd, find the dihedral angle between the planes.

P14.27(b) Given views H and F of a solid block with an oblique plane 1-2-3 and point x, find the angle between the oblique plane and the right end of the block. A hole is to be drilled with a center at point x perpendicular to plane 1-2-3. Show the center line of the hole and the point (y) where it breaks out of the block in the given views.

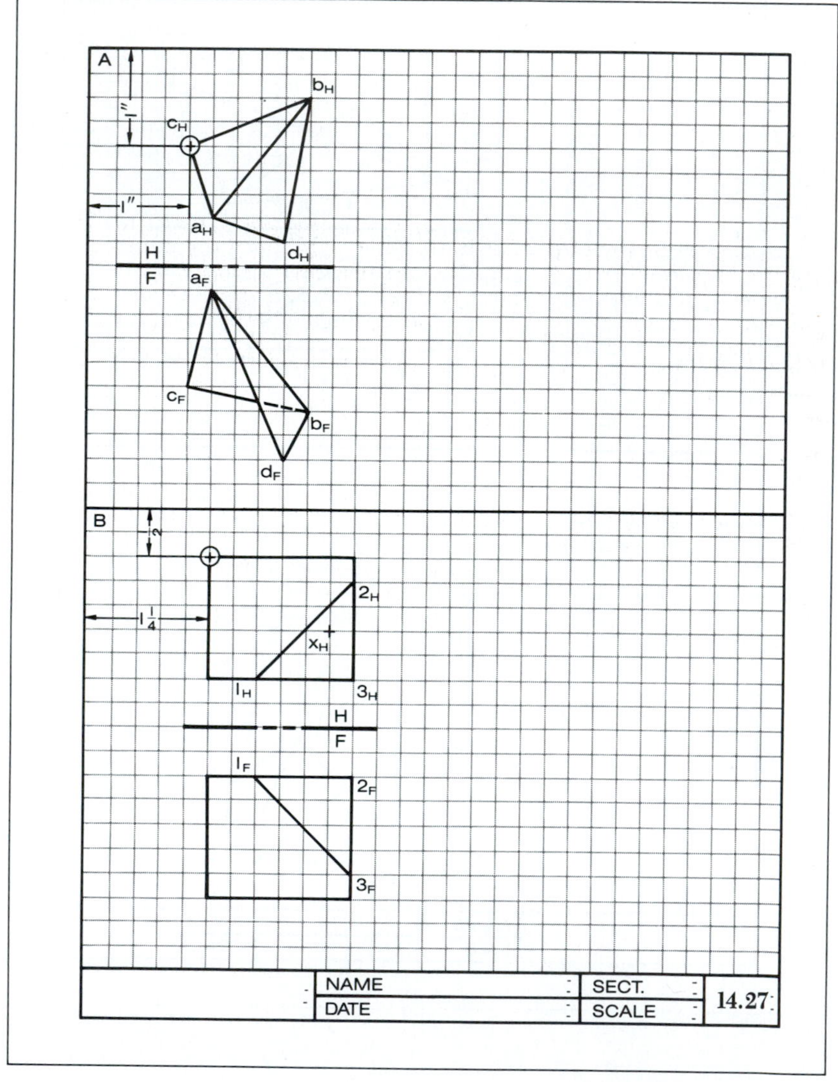

Computer Graphics

OBJECTIVES

After completing this chapter you should have a good basic understanding of:

1. The role that computer-aided design and drafting can play in engineering today.

2. Typical applications and the expected places where computers and computer graphics systems can improve productivity.

3. The hardware components of a computer graphics system, including graphic input and output methods.

4. The software components that make up a computer graphics system.

5. The software structure of typical computer-aided-drawing software packages.

6. The procedures necessary to produce various types of basic engineering drawings on a computer-graphics system.

Graphical tools have evolved since people first began to communicate through graphics. We have progressed from paint on the walls of caves to erasable ink on plastic, each of which required only a person skilled in the use of manual tools. Today there is a relatively new tool that is now within the reach of almost every person: computer graphics. The use of the computer is creating a revolution that will be at least as far reaching as the industrial revolution and perhaps farther reaching. This chapter will provide an introduction to the topic, not an exhaustive presentation of all of the aspects of this new set of tools. The following section will provide a definition of computer graphics and describe very briefly some applications. The system components, hardware and software, will be presented with pictures of some of the hardware. The subsequent sections of the chapter will present the steps to create some simple engineering drawings using computer graphics.

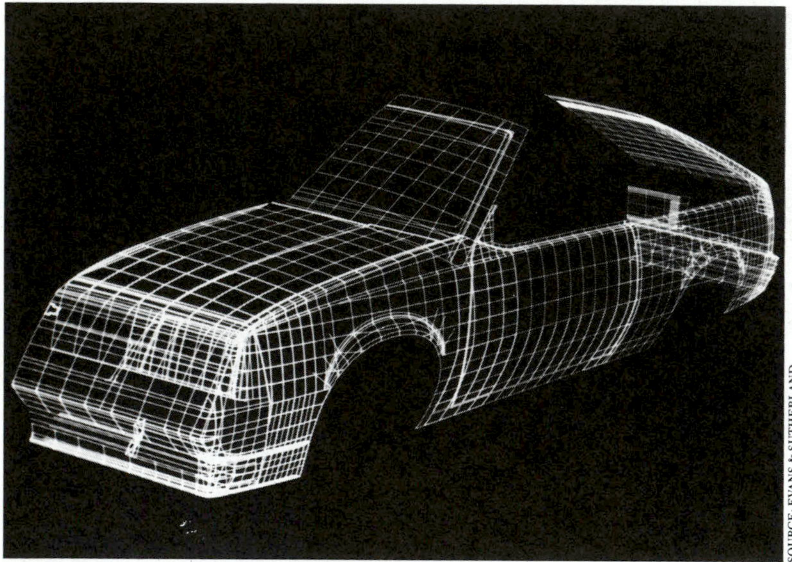

SOURCE: EVANS & SUTHERLAND

FIGURE 15-1 A three-dimensional drawing of a compact car design ***(Section 15.1)***

15.1 COMPUTER GRAPHICS IN ENGINEERING

Computer graphics can be defined as the creation and display of pictures (graphics) on a televisionlike screen, or on a mechanical plotter, with the aid of a computer system. It can best be thought of as a communications tool. Two-dimensional images such as graphs, charts, maps, and plans are an important part of our daily communications. Three-dimensional objects are also important, even though they are displayed as two-dimensional images. (See Figure 15-1, which shows a three-dimensional object in two dimensions on a computer-graphics terminal.) Such objects can be shown in color with the proper lighting and shading so that images appear real. Because the stored information in the computer's memory can include locations in three dimensions (using x-, y-, and z-coordinates), the system user can make changes rapidly once the original information has been entered. Today, creating the first drawing or first set of information may take as long as manual methods. However, new systems are becoming easier to use, and most systems require little knowledge of the inner working of the computer. All systems require a solid understanding of the graphics and graphics conventions presented in this book.

From an engineering point of view, one of the most important uses of computer graphics is the creation of the set of information needed by the computer to display the representation of three-dimensional images. If properly created by the user and the system, this set of information, called the *database*, can be used by the designer/engineer, the people who create the product or process, and the people who check the product or process. In most of today's manufacturing, process plants, or construction, the ideas, plans, and specifications are communicated by words, numbers, and two-dimensional images on paper. The computer allows this communication to be accomplished by each member of the design/production team that has access to the database through a terminal. Today we still do a major part of our communications through the use of paper hardcopy. Members of each group can have the authority to change, add, or delete information from the database. Once this is done the rest of the users have instant access to the new information. This has cut the time from design to production by 30 to 50 percent in automotive, aircraft, and similar industries.

Engineering is a career that is centered around the concept of designing products and bringing them to market or designing plants or buildings and making them ready for use. This concept is illustrated in Figure 15-2(a). As the figure shows, an idea is conceived, a design is created based on the idea, the design is analyzed and the item constructed or manufactured, and, finally, the item is checked for quality. The various disciplines within the field of engineering have available today computer graphics and analysis systems that allow them to do part or all of this process more efficiently.

While Figure 15-2(a) makes it seem that the process is a straightforward, beginning-to-end process, nothing could be further from the truth. Figure 15-2(b) is a more true-to-life representation of the design/engineering process. In this diagram experiences at each stage provide "feedback" that influences previous stages. Thus manufacturing difficulties can cause redesign, and quality problems can cause changes in both design and manufacturing. It is within this framework that some possible applications of computer graphics will be presented.

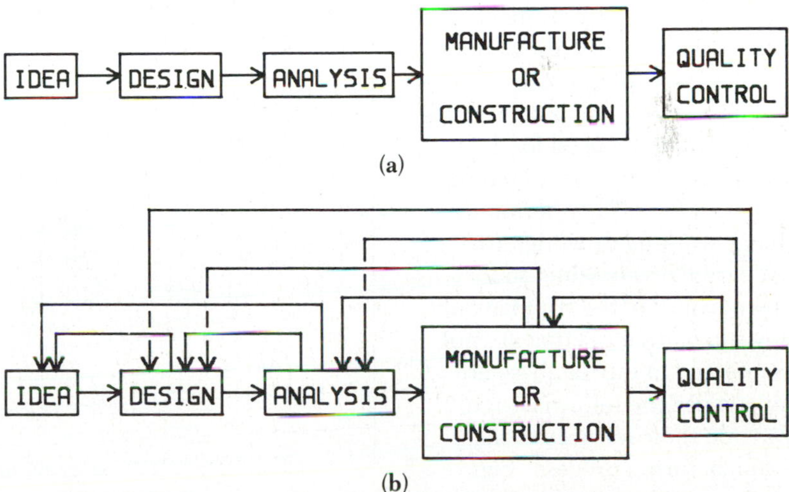

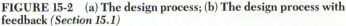

FIGURE 15-2 (a) The design process; (b) The design process with feedback ***(Section 15.1)***

15.2 COMPUTER-AIDED DESIGN/ COMPUTER-AIDED MANUFACTURING (CAD/CAM)

Mechanical Engineering

In typical design situations, as outlined in Figure 15-2, a product is designed by defining the geometry and adding the size of the individual parts, and the mechanism is analyzed to assure that it can withstand the working stresses. The engineers then have the opportunity to try different mechanisms and different combinations of materials so that the design can be optimized. As part of this process, they create the database (dimensions, material properties, and assembly information) to be used for design modifications and for planning the manufacturing operations. Today working drawings are prepared for a product so that any portions of the company or any subcontractors that are not connected to the database are able to use the information to provide their services to the manufacturing process. Figure 15-3 is an example of typical computer-generated drawings of a part.

When the designers are finished with the preliminary work, they call in manufacturing engineers to discuss the design. The manufacturing engineers offer suggestions for design improvements that will make the fabrication processes more economical. Through computer-aided simulation and modeling the design engineers can demonstrate the product design and function for the manufacturing personnel. Once any changes have been incorporated in the product design, the access to the database is made available to the manufacturing engineers.

Using computer-aided techniques, the manufacturing engineers plan the manufacturing and quality control operations. These operations include development of the floor plans for the plant layout, simulation of the material movement, simulation and planning of material handling (including the use of robots), programming of material removal and fabrication machines, placing of the assembly machines, finishing processes (such as painting, plating, or anodizing), quality control, and packaging of the product for shipment. One of the most important aspects of this planning is the placement of the personnel who will be responsible for the manufacturing.

Prototypes generally are built and tested before volume production is begun. Manufacturing engineers and design engineers must look at the prototypes and determine any design and manufacturing changes

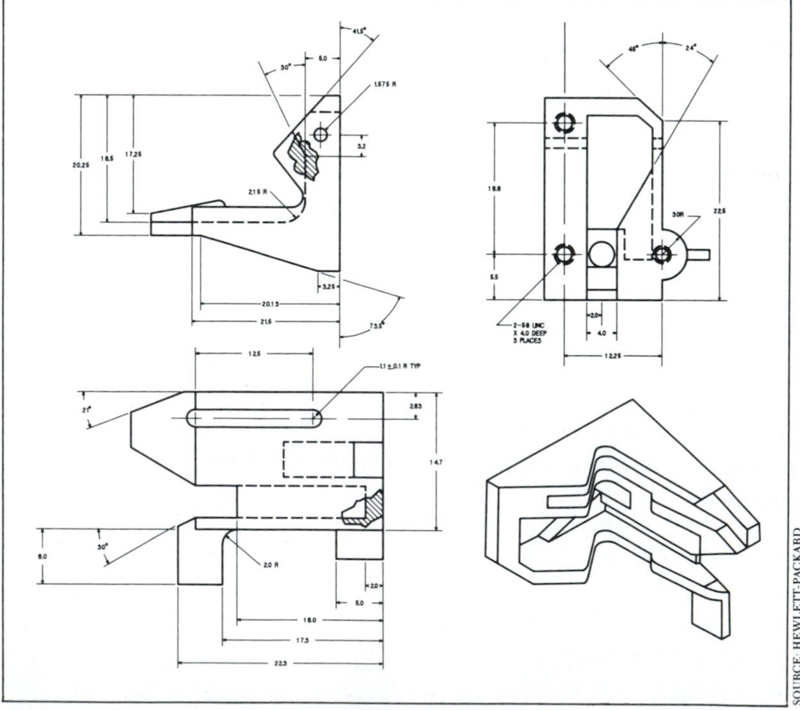

FIGURE 15-3 Detail drawing of a part with an accompanying isometric pictorial ***(Section 15.2)***

that are needed. Using computers, the changes can be made relatively easily. As the engineers change the design they also change the database, and each person working with the product has immediate updates of the information.

Civil Engineering

Another example of the use of computer graphics is in the field of construction. Today many of the civil engineers in charge of designing and constructing highways use computers and computer graphics where they used to use slide rules or calculators and pen-and-paper graphics.

If a road is to be constructed in a new area, the engineers contract to have stereo photographs taken of the area where the road will be located. These photographs are then interpreted by specially trained personnel to establish the elevations of certain grid points. The set of locations on the ground and its matching elevations are fed into a computer, which generates contour maps of the area. Figure 15-4 shows such a computer-generated contour map. The engineers then study the maps and lay out the center line of the proposed road. Next the lane widths, curves, and grades of the street or highway are established using the computer graphics system. Figure 15-5 shows such a street plan in a residential or commercial area.

When the road layout has been finished the engineers use the results of soil tests to determine the design requirements for the roadbed and the slopes at the edges of the road. The combination of these slopes, the road elevations, and the contours of the earth's surface must be used to determine the amount of earth to be moved from one area to another. This can also be done by computer. In addition a computer system can be used to calculate the amounts of fill gravel, concrete, and asphalt that will be needed for the road. The combination of the soil analysis and the ex-

SOURCE: HEWLETT-PACKARD

FIGURE 15-4 A computer-generated contour map *(Section 15.2)*

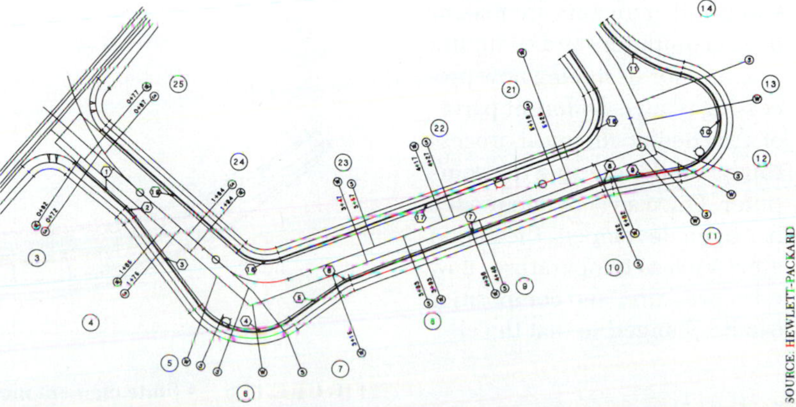

SOURCE: HEWLETT-PACKARD

FIGURE 15-5 Details of a residential street, including utility locations *(Section 15.2)*

pected loads are used to determine the type of compaction required to produce a stable roadbed. All of these items are needed by the contractor to prepare his bid. The computer graphics system can then be used to create the drawings that are needed for the contractor who will do the road construction. While the road is being constructed and after the project is finished the actual elevations and distances can be checked against the original design. This is, in effect, a form of quality control.

Civil engineers and construction engineers often need a system with three-dimensional capability. Such systems allow the relative position of objects and land features to be easily seen through alternative views created by rotation or projection. The three views of the dam shown in Figure 15-6 demonstrate the three-dimensional drawing capability required by civil engineers.

Chemical Engineering

Chemical engineers are making use of computers and computer graphics for designing new processing plants. Different parts of a particular chemical process can be modeled using the computer. Once an accurate model has been developed, the parameters such as temperature, flow rate, pressure, and composition can be changed so that the en-

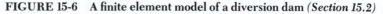

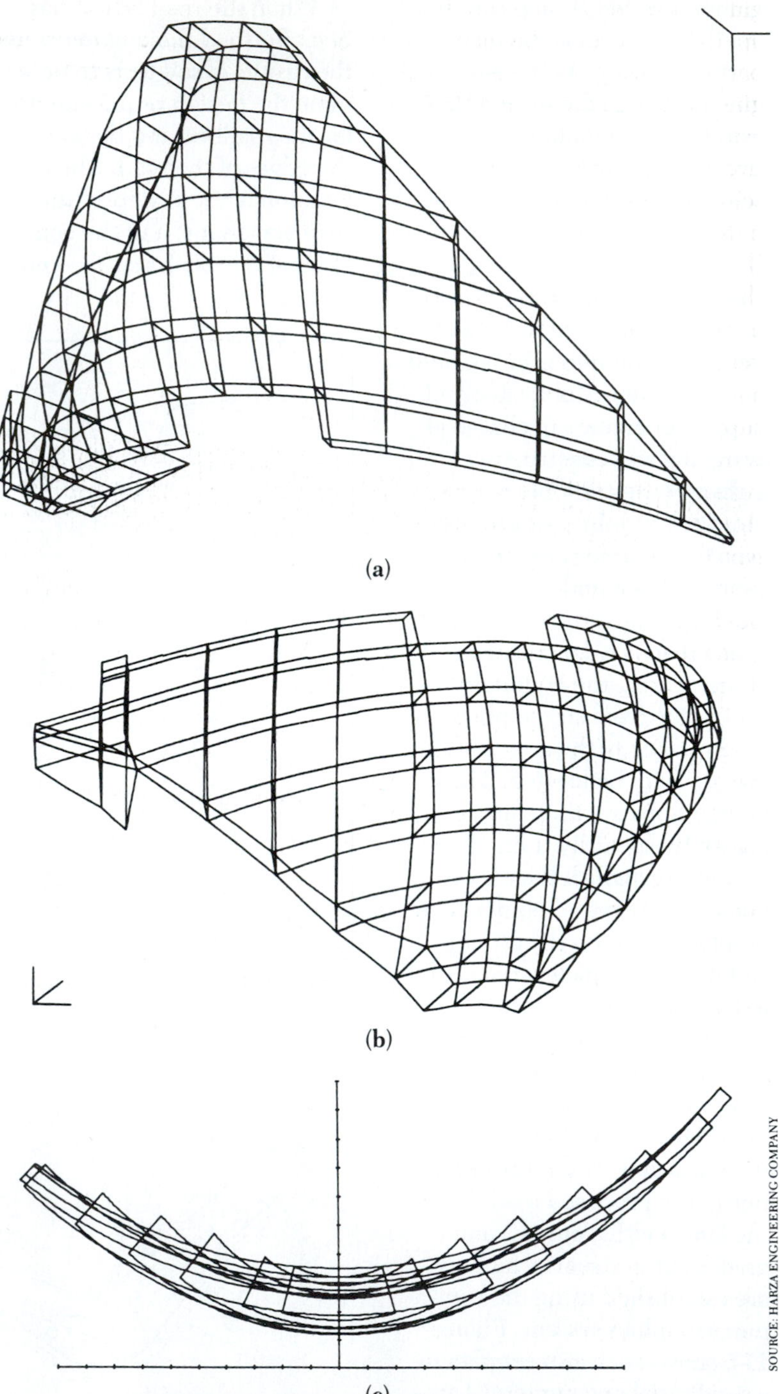

FIGURE 15-6 A finite element model of a diversion dam *(Section 15.2)*

gineer can chose the optimum operating procedure. Once each part has been optimized, the engineer can combine the parts to establish the model for the entire process. This is done as a schematic, and when the computer model works well enough the engineer then must convert the schematic to a physical layout of the different processing vessels, the piping, the controls, and the structures needed to support the processing hardware. This physical layout results in dimensioned drawings that are used by the contractor who does the construction of the plant. Computer programs have been developed that allow the engineer to build a three-dimensional model of a plant. This requires the creation of realistic symbols of the various parts of the plant. The symbols are then placed in the proposed location and such things as clearances for personnel and fire-fighting vehicles can be checked.

Today controls for processing are computerized so that an operator can see a colored schematic of the process or part of the process on a CRT. This schematic can show temperatures, pressures, and flow rates through the use of color. Colors such as red can be used to signify a dangerous condition or shades of reds, oranges, and yellows can represent various temperature ranges or pressure ranges. The entire process can be controlled by computers so that the need for personnel can be reduced.

Structural Engineering

Architectural and engineering firms that design and supervise the construction of new buildings can make extensive use of computers and computer graphics. The architect can use the computer graphics to do a preliminary layout of a floor plan of a building. The client can work with the architect to establish locations of key features in the building. Figure 15-7 shows a computer-generated floor plan for a residence. Computer graphics systems can also be used to create renderings of the building's exterior and interior, showing various colors and textures. This allows the client and architect to experiment with the options and find which combinations are pleasing and cost effective.

Once the graphical model has been created in the computer (the database) the engineer and architect can work together to develop the structure and place the mechanical and electrical components of the building. These include the lighting, the heating and air conditioning, the plumbing, and the elevators or escalators. Many times these features are put on different layers, which can be thought of as pieces of clear plastic that are laid over the floor plan. One layer might contain the electrical wiring and another the plumbing. The architect and engineer must establish that the various components are not in physical conflict so they will need drawings showing all layers, but the electrical contractor is only concerned with

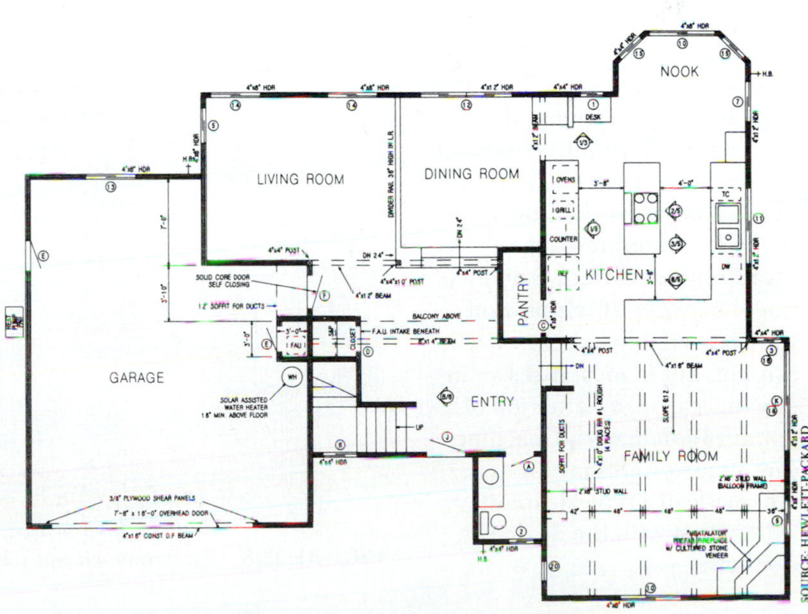

FIGURE 15-7 Architectural floor plan *(Section 15.2)*

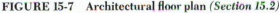

where the lighting fixtures and wires are supposed to be placed and will need a drawing showing only the floor plan and the electrical layer.

Electrical Engineering

Electrical engineers who are involved in digital electronic design use computers and computer graphics systems for the initial design, printed circuit (PC) board layout, and optimization of component placement on the PC board. When circuits are designed, the engineer first establishes the necessary circuit logic. This is done by using the logic symbols for those components to create a graphical plan. The engineer then draws a schematic diagram of the circuit using symbols to represent the individual physical devices or components. Integrated circuits (ICs) can contain many components on one chip. The engineer and PC board designer have to work together to choose the number of ICs required to meet the design constraints, choose the location on the PC board for the ICs, and optimize the PC board routing of the conductive traces between IC chips. Software is available to help with the routing, but in most cases an individual can do a better job of optimization than the machine can. Figure 15-8 shows a schematic of a typical circuit worked out with the device symbols.

CAD/CAM Advantages

As you can see, using a computer has changed the methods for producing new designs, not the basic concepts or the graphics required. The benefits are the speed or ability to look at alternative solutions or to make changes when necessary. Many times the actual design time is shortened, but the real benefits lie in the opportunity to look at alternatives or to make changes rapidly when the situations dictate. However, it is important to remember that, while computer-aided systems can aid the engineering team in each step of the design process, the greatest advantages are realized if the design process is integrated so that information developed early is available for later steps through the database stored in the computer.

15.3 GENERAL COMPUTER GRAPHICS SYSTEM HARDWARE

The hardware in a computer graphics system consists of the physical devices that you can see and touch. Rather than try to cover all of the possible devices that can be classed as computer graphics equipment, this section will concentrate on the devices that would be part of a computer-aided drafting and/or design system.

A Typical Workstation

A typical computer graphics workstation or system would consist of a computer (CPU), a televisionlike screen (display device), mechanisms for putting in pictures, words, and commands

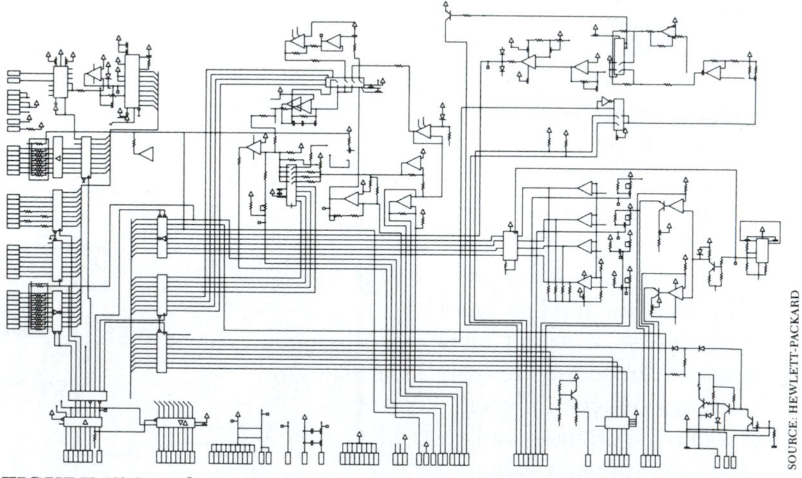

FIGURE 15-8 Electronic-circuit schematic ***(Section 15.2)***

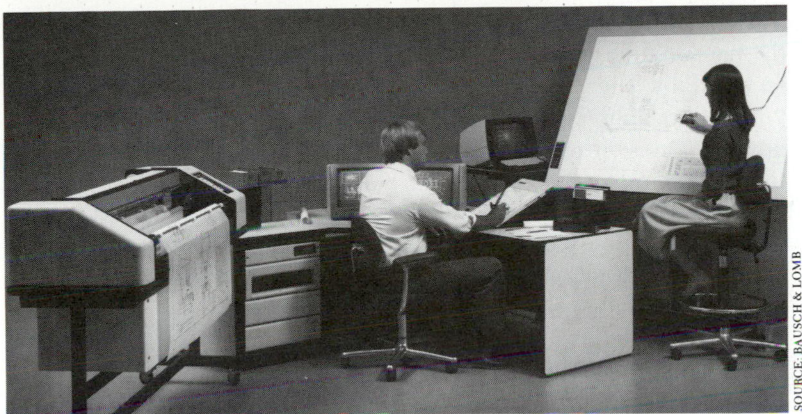

FIGURE 15-9 A computer-aided drafting system ***(Section 15.3)***

FIGURE 15-10 A dual-screen computer-graphics workstation ***(Section 15.3)***

(input devices), mechanisms for transferring finished drawings to paper or plastic film (output devices), and mechanisms for storing the information electronically (storage devices). The following paragraphs will discuss the hardware parts and provide names for these devices in order that you may build a working vocabulary of the most common computer-graphics-system components. Figure 15-9 shows a complete computer graphics system, Figure 15-10 shows a computer-graphics workstation, and Figure 15-11 shows a microcomputer-based system that has all of the components of a basic drafting system.

The term *workstation* is used to signify the hardware that must be present for a drafter, designer, or engineer to do graphical layouts. The workstation either contains a computer or is electronically linked to a computer. There is always a televisionlike screen called a CRT (cathode ray tube) and, in

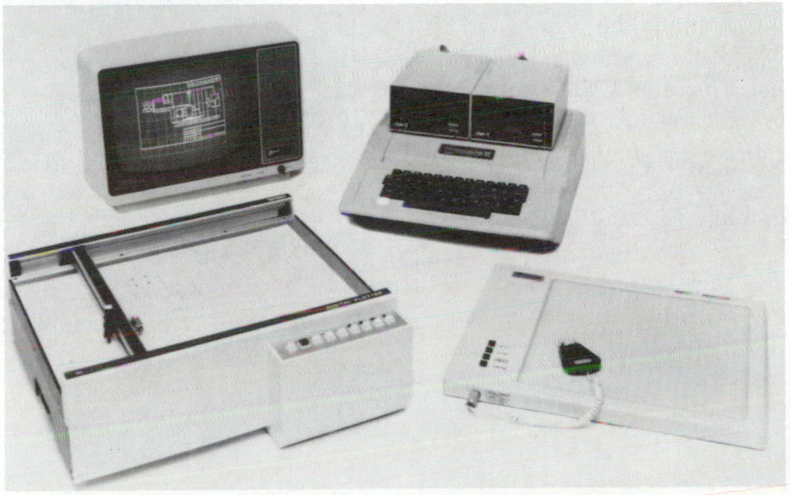

FIGURE 15-11 A microcomputer-based computer-aided drafting system ***(Section 15.3)***

most cases, a keyboard for commands. Many times one of the following devices is used for input of graphical information and commands: digitizer, light pen, joystick, or mouse. The computer itself, which is the heart of a computer graphics system, is sometimes called a CPU or central processing unit. If there is one for each workstation, the CPU is probably called a microcomputer or minicomputer. If a system is designed for multiple users, each workstation may consist of a display and various input devices, while the computer, which supports several workstations, may be located in another room. In such a case the CPU may be a large minicomputer or a mainframe computer. In almost all cases there are microprocessors (a computer on a chip) imbedded in the electronic hardware to perform specific tasks.

DISPLAYS. Many systems have two CRT displays. One is for the communications between the designer and the system and is called an *alphanumeric terminal.* The other is for displaying graphical information and is called a *graphics display.* Figure 15-10 showed a computer graphics workstation that has the alphanumeric terminal on the left and the graphics terminal on the right. There are various types of graphics CRT display devices. One commonly found today is the *raster display,* which is similar to a television set. Raster display pictures are made up of individual lighted dots (picture elements) called *pixels* or *pels.* The pictures displayed range from low to medium-high resolution (picture quality). The quality is determined by the appearance of the picture displayed and, in general, the number of dots that can be turned on to display the picture. If there are a large number of dots in both the vertical and horizontal directions the lines and arcs appear smooth like those drawn with a pencil and the resolution is said to be high. Typical resolutions range from 512 H × 250 V to 1024 H × 800 V and higher. Figure 15-12 shows a high-resolution raster terminal.

Two other types of displays that are more expensive but that provide special features are the *storage CRT* and the *vector refresh CRT.* The storage CRT can provide very high resolutions for picture display (4096 H × 3072 V), and they can produce many

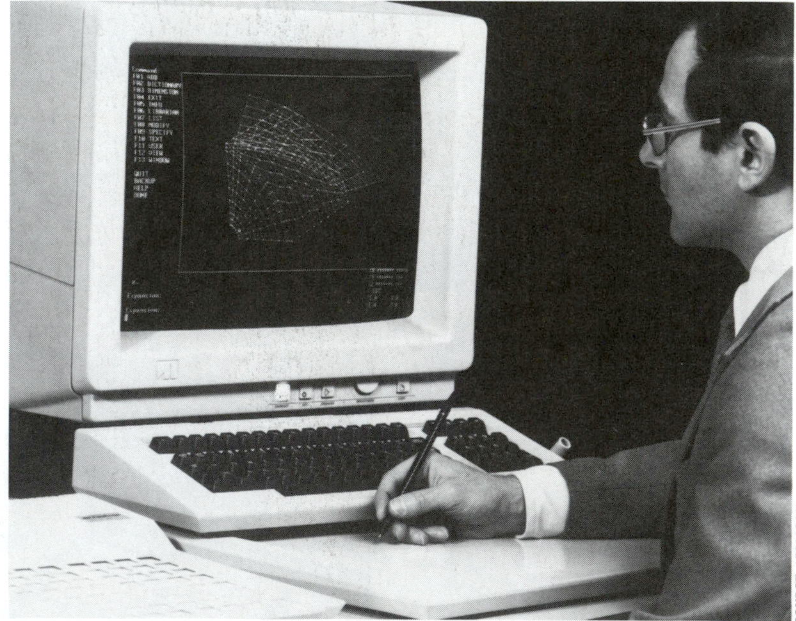

SOURCE: APPLICON

FIGURE 15-12 A high-resolution raster terminal *(Section 15.3)*

different styles and widths of lines on the screen. These lines can be displayed in one of two ways (modes). One is easily visible but can only be erased by erasing the entire picture. The other mode (called *write-through*) provides lines that are somewhat less visible but that can be moved dynamically or erased (see Figure 15-13).

All of the points and lines on a vector refresh display can be moved and erased. The lines are all very visible. An added feature allows lines to be displayed at various intensities. This feature allows creation of more realistic displays by having the lines that are brighter represent the part details that are closer to the viewer. These displays are used where the designers require animation of the picture so they can look at it from different angles to see how the mechanism's components fit or work together. The picture must be continually redrawn by the display's microprocessor or the computer. This requires relatively expensive processing power as compared to the storage or raster displays (see Figure 15-14).

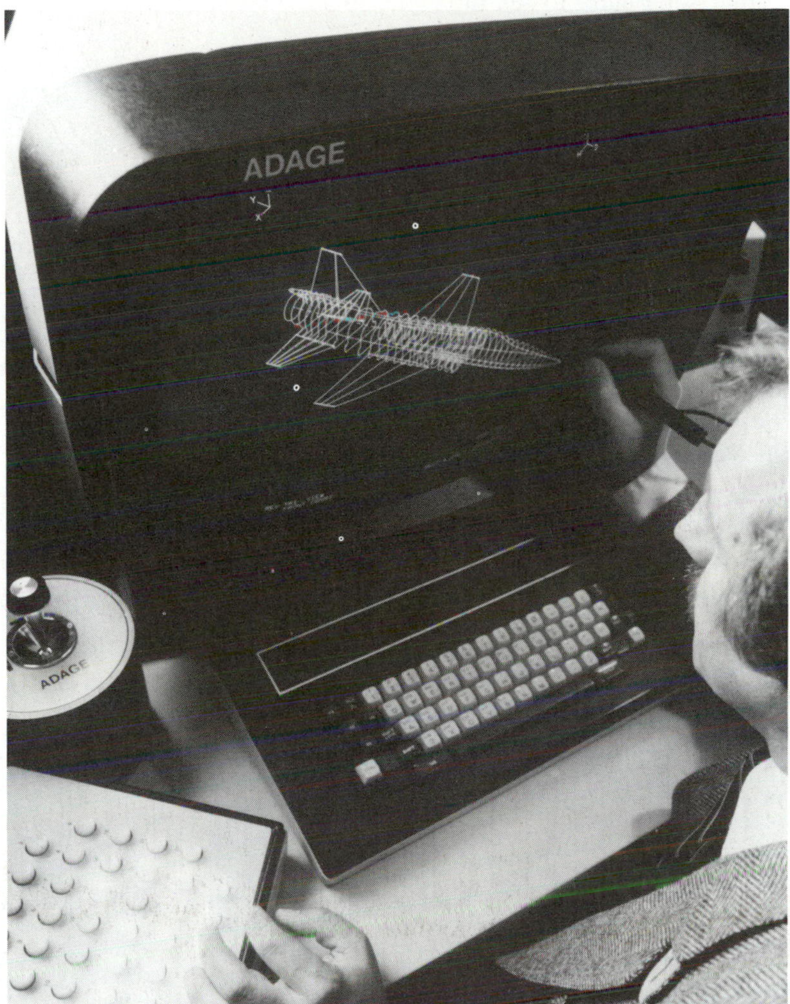

SOURCE: ADAGE, INC.

FIGURE 15-14 A high-resolution, three-dimensional vector refresh display ***(Section 15.3)***

SOURCE: McAUTO

FIGURE 15-13 A high-resolution storage terminal ***(Section 15.3)***

KEYBOARDS. Almost every system has a typewriterlike keyboard to provide communications from the system user to the system (see Figure 15-15). In addition to the alphanumeric keys, some keyboards have function keys that are used to simplify input of commands. The function keys can also allow the user to pick items from a *menu*, which is displayed on the screen or attached to the keyboard. A set of function keys can also control a graphics cursor on the CRT screen. These keys are identified by an arrow indicating the direction of movement of the cursor (up, down, left, or right). Figure 15-16 shows a separate set of function keys that light up when a drawing function is selected.

DIGITIZING TABLETS. The *digitizer* (Figure 15-17) is a device commonly used for graphic input. It usually consists of a hand-

FIGURE 15-15 A keyboard for a computer-graphics display *(Section 15.3)*

FIGURE 15-16 A function key input device with keys that light up when activated *(Section 15.3)*

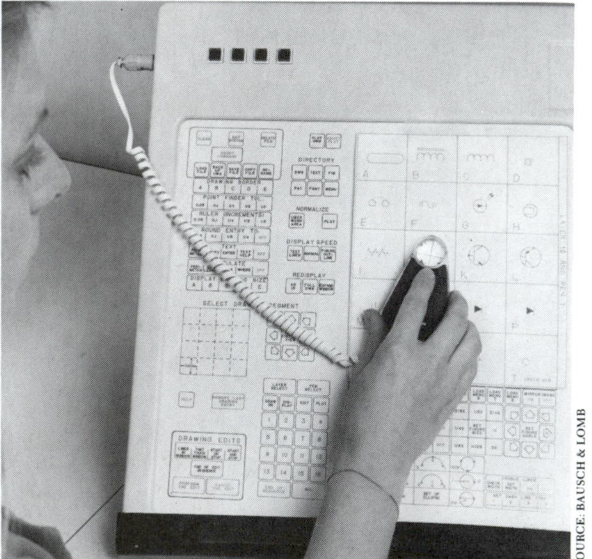

FIGURE 15-17 A common digitizer showing details of the menu and cursor *(Section 15.3)*

held cursor or puck (Figure 15-18) that allows the user to input coordinates that identify the endpoints of lines. Thus the user can "trace" a drawing or sketch that has been previously created and see the lines and drawing features appear on the CRT screen. Digitizers can also be used to control a graphics cursor so that a movement of the digitizer puck provides a corresponding movement of the graphics cursor on the CRT screen. A third use for the digitizer is as the location for a complicated menu, allowing the screen area to be used for graphics display. In this case the coordinates input by the digitizer puck are interpreted as a selection of a menu item, which can be much faster and easier than typing multiple keystrokes to input the selection. Some response from the CRT, such as a message being displayed or the bell being sounded, lets the user know that the proper selection has been made.

LIGHT PENS. A *light pen* is a device that allows the system user to point to a lighted object on the screen and to have the location (screen coordinates) of the object be detected by the computer (see Figure 15-14). Once the screen coordinates have been identified there are a variety of actions that can take place. If the lighted object is a menu location, then the menu item can be selected. If the lighted object is a part of a picture, then that particular object can be deleted, moved, enlarged, or made smaller. It is possible to point to a figure that represents a cursor and to move the cursor around the screen. It is also possible to select one end of a line, "stretch" the line so that it connects two screen locations, and then fix the line in that position. This process, called "rubber-banding," is not limited to light pens but is also available when using cursors controlled by function keys, digitizers, or joysticks. One drawback with today's systems is that light pens have to be held in position touching the vertical screen and the user's arm can get tired. Some systems have the CRT mounted in a horizon-

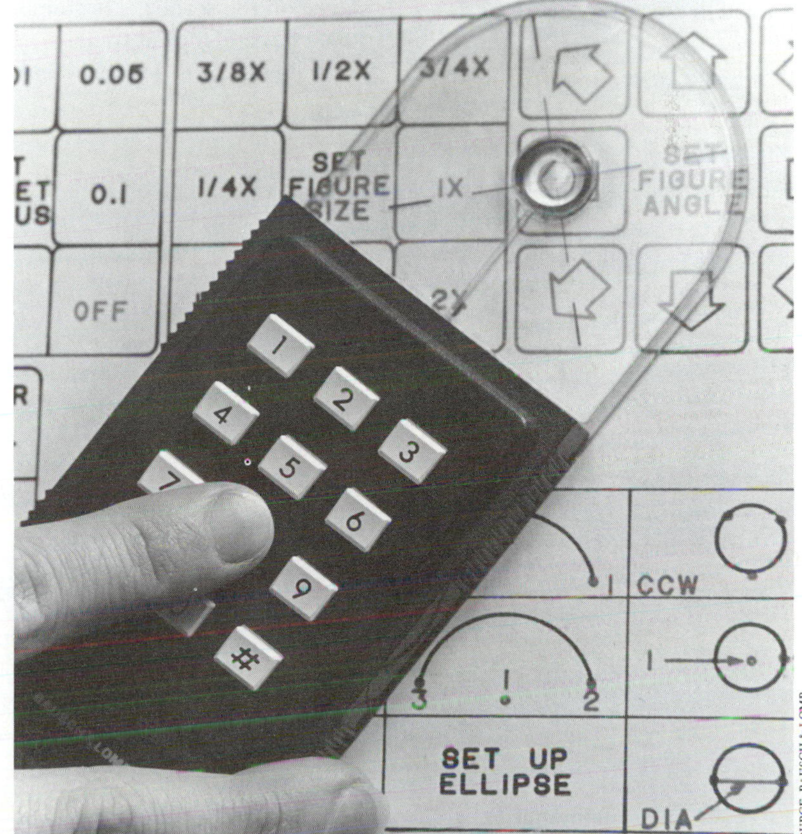

FIGURE 15-18 A digitizer cursor with multiple buttons *(Section 15.3)*

tal position to alleviate this problem. A light pen for a microcomputer is shown in Figure 15-19.

JOYSTICKS. A *joystick* is one of the devices that can be used to control an electronic cursor displayed on the screen. The joystick can be designed to move the cursor in any direction. Some are designed so that the longer the cursor is sent in a particular direction, the faster the cursor moves. One problem the joystick presents is that it is difficult to hold the screen cursor in a single position, which may be necessary to complete a figure. The system software can overcome this limitation by allowing the user to specify the closest grid point. A joystick for a microcomputer is shown in Figure 15-20.

OTHER INPUT DEVICES. A *mouse* is a device that moves along the top of a desk or drawing table. The distance that it moves is proportional to the movement of the cursor on the screen. A mouse can be advantageous when the user is modifying an existing drawing that is displayed on the screen. Drafters do not have to look at their hands when using a mouse, and when they stop moving the mouse it stays in one position. A disadvantage is that it may be difficult to use a mouse to input a sketch previously drawn on paper. This disadvantage is also inherent in the joystick and light pen input devices. An optical mouse is shown in Figure 15-21.

Another input device is the *touchscreen*. This screen is a modification to the CRT display that allows the user to touch part of the screen and have the system respond. This can be used effectively with menus displayed on the screen when the user wants to choose an item. For people who are not familiar with a typewriter keyboard, this system can provide an easy way to interact with the computer. However, touchscreens have a relatively low resolution and

SOURCE: SYMTEK

FIGURE 15-19 A light pen ***(Section 15.3)***

FIGURE 15-20 A joystick for a microcomputer ***(Section 15.3)***

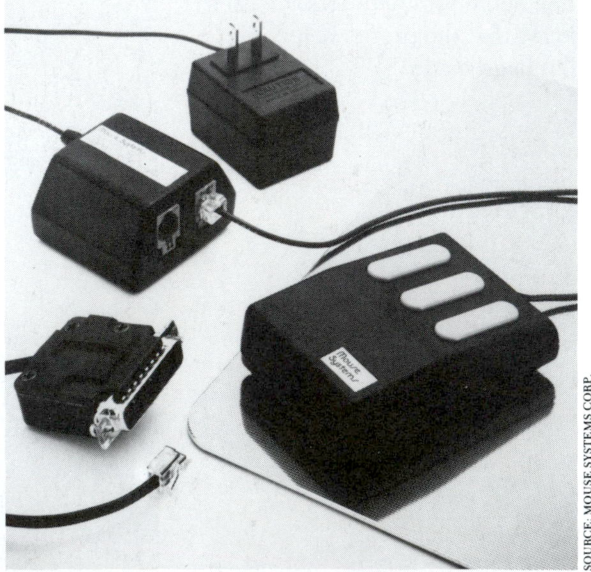

SOURCE: MOUSE SYSTEMS CORP.

FIGURE 15-21 An optical mouse ***(Section 15.3)***

drawing would be difficult without sophisticated software. A touchscreen is shown in Figure 15-22.

Hardcopy Output Devices and File Storage Devices

PEN PLOTTERS. Engineers, drafters, and designers have always dealt with drawings of objects that are to be designed and then manufactured or constructed. Although it is now possible for these individuals to create a drawing on the screen, there are still other workers who must make the parts or supervise the construction projects. Today many members of the technical team must have a paper copy, called a *hardcopy*, of the drawings.

Pen plotters are the devices most commonly used to create drawings because they provide drawings of similar quality to those created by hand. They can put lines on paper at any scale and can do the lettering faster than any human. Plotters can also be used to fill areas on masks (patterns) for printed circuit boards, which is a tedious task for a drafter. Pen plotters come in many sizes to handle the various standard drawing sizes. They also come in three versions: the flatbed plotter, the drum plotter, and the modified drum plotter.

The *flatbed plotter* normally allows the system user to put a single sheet of paper in a fixed position on a plotter's table and the pen does all of the moving. This, of course, is similar to the way a drafter works. Figure 15-23 shows a flatbed plotter.

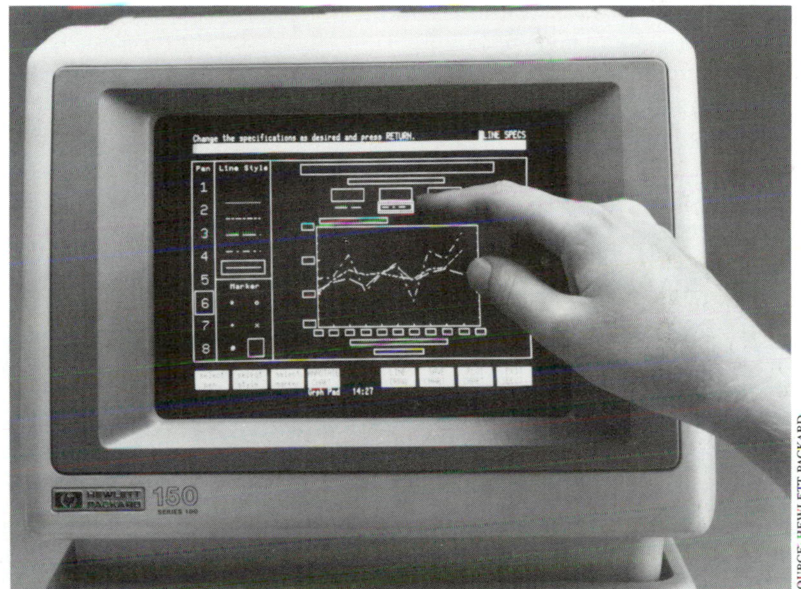

SOURCE: HEWLETT-PACKARD

FIGURE 15-22 **A touchscreen** ***(Section 15.3)***

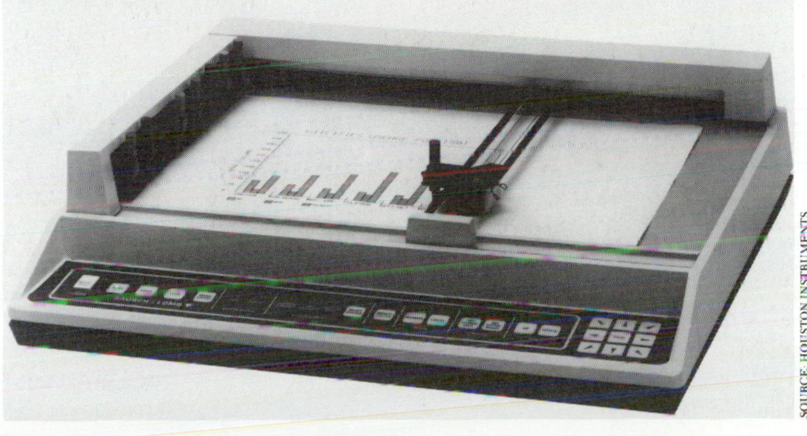
SOURCE: HOUSTON INSTRUMENTS

FIGURE 15-23 **A multiple-pen flatbed plotter** ***(Section 15.3)***

The *drum plotter*, on the other hand, can be set up with a roll of paper many feet long. Both the pen and the paper move. The pen moves on one axis and the paper moves on the other, thus providing access to all points on the paper. The *modified drum plotter* operates in the same way but is limited to a single sheet of standard-size drawing paper. A drum plotter can make many drawings with one setup while the modified drum plotter requires one setup for each drawing. Figures 15-24 and 15-25 show drum plotters with both the schematic diagram for an electrical circuit and the mask for a printed circuit board layout.

Pen plotters can be equipped with many pens. These can provide a choice of colors and/or a choice of line widths. Four pen widths can provide as many as 16 different line widths by making multiple strokes. They can also draw on paper or on plastic film (mylar) and can create transparencies for graphic presentations.

PRINTER-PLOTTERS. *Printer-plotters* are manufactured with a variety of mechanisms. They include impact dot-matrix, ink jet, and electrostatic mechanisms, the first two of which can provide multiple color printouts. The *impact dot-matrix printer* provides color by using a multicolor ribbon. The *ink jet printer* provides color by using different inks and controlling the amount sent to the print head. The *electrostatic printer* can provide only one color at a time. Each of these devices operates on the raster principle. This means that each creates the picture either by making multiple passes with

SOURCE: HEWLETT-PACKARD

FIGURE 15-24 A drum plotter plotting a schematic diagram *(Section 15.3)*

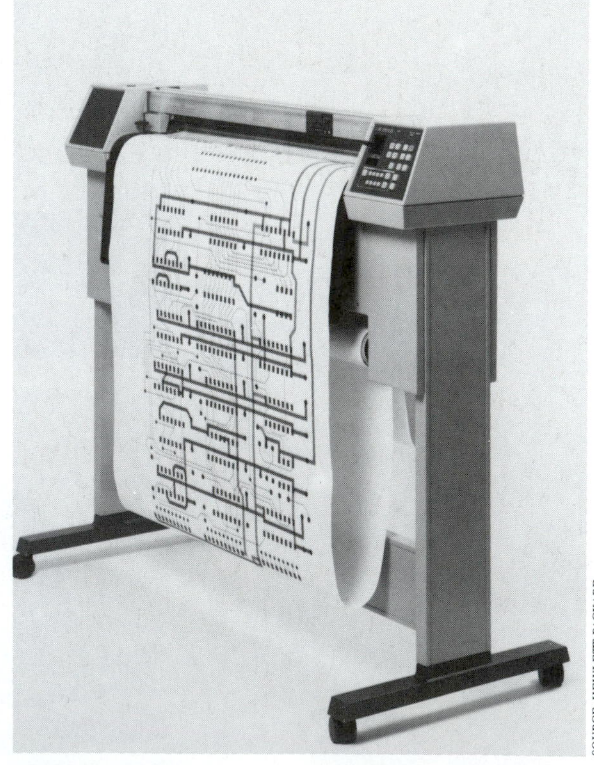

SOURCE: HEWLETT-PACKARD

FIGURE 15-25 A drum plotter plotting a printed circuit mask *(Section 15.3)*

the print head while the paper advances after each pass, or making a single pass with multiple print heads. Figure 15-26 shows a dot matrix printer-plotter.

DISK STORAGE. Disks used for computer storage of information are similar to phonograph records in that they have a spinning disk and a device to retrieve the information. Disks are coated with a magnetic material that can retain information by retaining the magnetic field induced by the write head. As the disk spins, the read head can be moved from the outer edge to the middle of the disk, thus providing random access to the information. This is the principle employed no matter which type of disk is used: *hard disks* and *floppy disks*. The floppy disks are used most often with small microcomputer- or minicomputer-based systems. A typical floppy disk and disk drive are shown in Figures 15-27 and 15-28.

TAPE STORAGE. Magnetic tape, similar to that used in tape cassettes, can store information for file purposes. Although tape drives are slow, they can provide inexpensive storage of information. Generally, tape drives are used on minicomputers and mainframe computers and are not found on microcomputer-based systems.

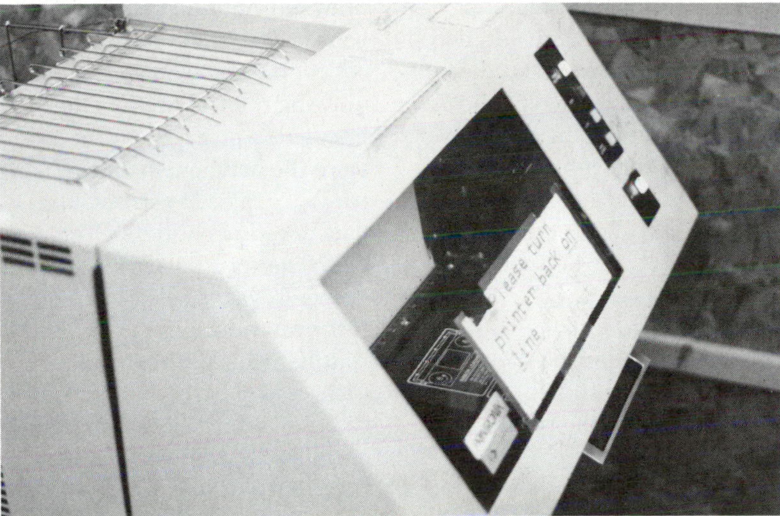

FIGURE 15-26 A dot-matrix printer-plotter ***(Section 15.3)***

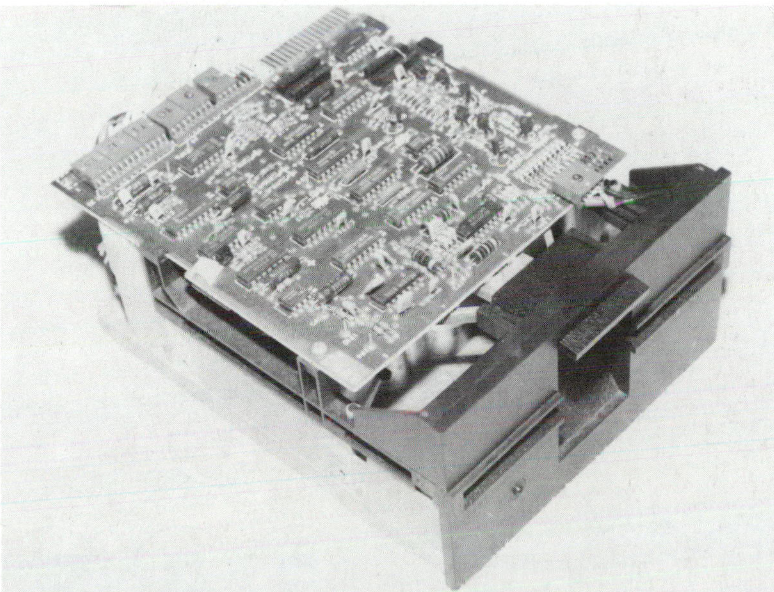

FIGURE 15-27 A Tandon floppy disk drive ***(Section 15.3)***

FIGURE 15-28 A Scotch floppy disk ***(Section 15.3)***

15.4 GRAPHICS SOFTWARE

A large computer graphics program is very complex. It may be composed of 50 or even more than 100 separate modules, each of which has a specific task to perform. These modules may be subroutines, functions, or separate program segments. As a result, special portions of the program are generally created and assigned the task of coordinating and controlling the actions of the many individual segments.

The Menu

Central control of the program is typically accomplished through a *menu*, which is a list of tasks that the program is designed to perform. When the user selects a desired task from the menu, the program activates the necessary routines to do what was requested. Some of the necessary tasks may be to remove the menu from the display screen, display additional messages or display another menu of additional choices that must be made before the task can be accomplished, clear these messages from the screen, and, finally, branch to the necessary task routines or even replace a portion of the program that is in memory with another portion. This last action is known as *overlaying* and is generally done when the program is too large for the amount of memory that is available.

Figure 15-29 shows a diagram of a very simple program that consists only of a menu and two subroutines that can be accessed from it. When you select one of the tasks from the menu, control is passed to that subroutine, which proceeds to execute the task. The last statement that is executed in a subroutine is RETURN. The system keeps track of the order in which subroutines are called, so the return is always to the program segment that called the subroutine. In this simple example, the calling segment is the menu routine.

Figure 15-30 adds a second level of subroutines to the example program shown in Figure 15-29. In the first example the line-drawing or point-drawing routines had to accomplish all functions associated with lines or points. Looking at the line-drawing routine in Figure 15-30, you can see that it only needs to collect the data. It can call a drawing routine to draw the line, a data storage routine to store the data, and a data retrieval routine to retrieve the data to redraw or delete the line. The return path from the drawing, data storage, or data retrieval routine is to the line routine, which in turn will return control to the menu. You can then select that same menu option again or you can select the point routine, which will behave in the same way as the line

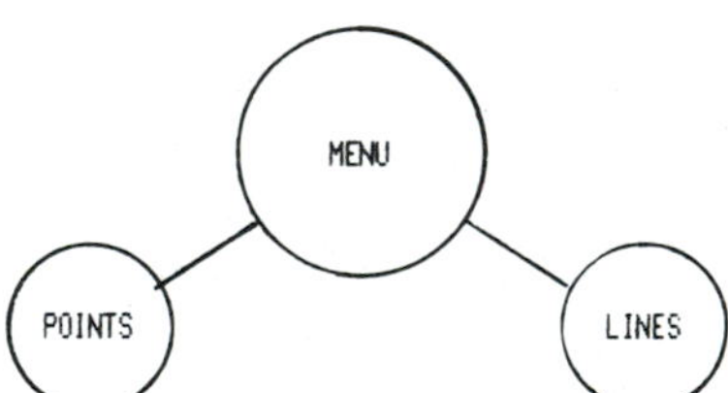

FIGURE 15-29 Interaction of menu and task subroutines *(Section 15.4)*

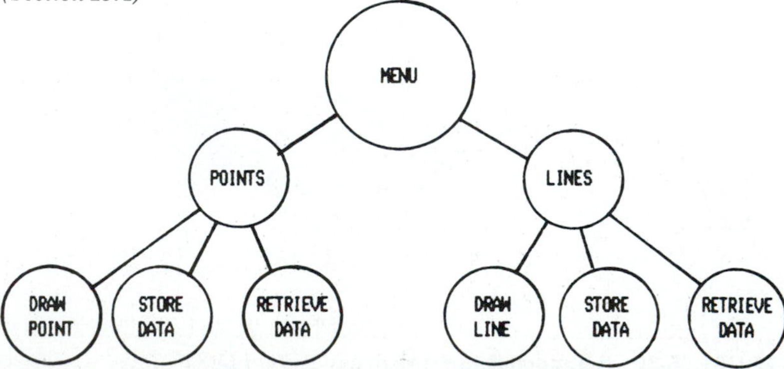

FIGURE 15-30 Adding subroutines *(Section 15.4)*

routine. No matter how many levels of subroutines are added, the return path to the menu always retraces the calling path. Communications channels in a program are from a calling segment to a subroutine it calls and back again. There is only one main program segment. All other segments are subroutines or functions that are called from the main segment or another subroutine. When each subroutine or function has accomplished the task it was to do, it returns control to the segment that called it. As a result, even though control may pass from the main segment through a succession of subroutines, ultimately the control is returned by retracing the same path to the main routine.

Figure 15-31 shows schematically the interactions of the many parts of FIRSTDRAW, an inexpensive computer-aided drawing package intended primarily for instructional use. Note that the diagram somewhat resembles an upside-down tree. The main segment initialization routine is the trunk of the tree and all other routines branch out from it. Not surprisingly, this type of structure is known as a *tree structure*.

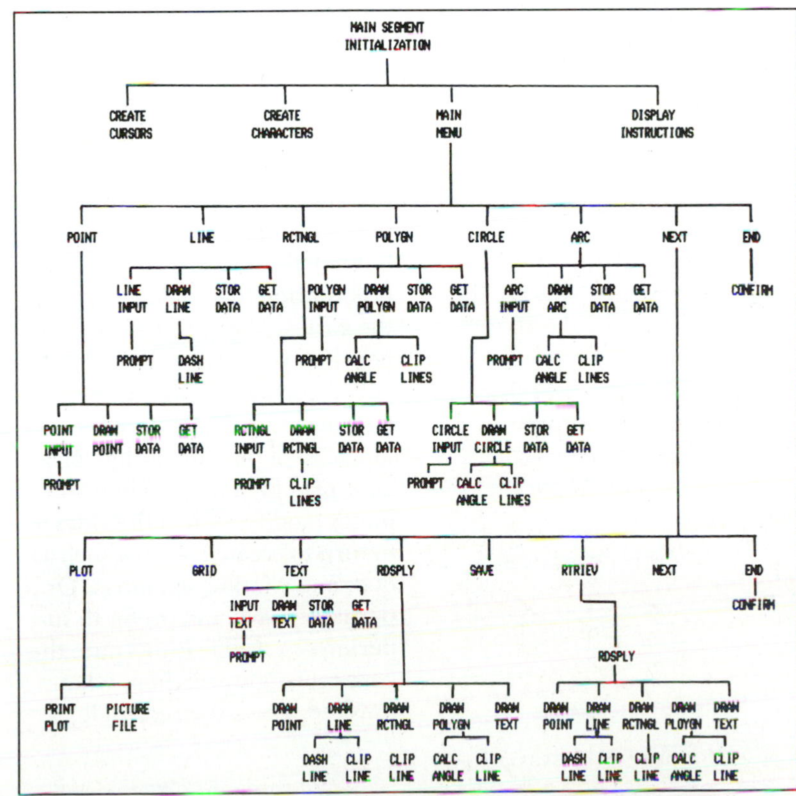

FIGURE 15-31 Interactions of the various parts of FIRSTDRAW *(Section 15.4)*

The menus that appear on the CRT when the program is first run are often called *default menus*. These default menus may also be a series of pictures and words on a digitizing tablet or they may be on a series of labeled function keys. Some computer graphics programs permit you to replace portions of these menus with custom menus or to add additional menu items or even additional menus. An example of this is provision for you to create special shapes, such as electrical circuit element symbols, that can be added to a symbols menu. Such symbols speed up the drafting process, since you only have to create the symbol from scratch one time rather than each time it is used.

Mode Settings

Another form of program control is through the provision for different *modes* of operation. For instance, adding a feature to the drawing is one mode of operation, while deleting a feature is another mode. As you work with different parts of the program creating the various features of the drawing, the program must keep track of the mode selections that are active and make sure that they are not changed until you want them changed.

Most programs assign *default-mode* settings that are in effect when the program starts and other default settings that you may set each time a certain task is selected from the menu. These defaults are generally selected by the programmer as being those most often desired. You can generally change the default modes in one of two ways. Sometimes all of the default-mode settings are stored in a configuration file that is read when the program is started up, and you are given the option of reading that file and changing its contents. The modified file is then resaved with the new default modes. A second common method for changing a mode is pressing certain key sequences or function keys while the program is running. In this case the mode is changed temporarily but the defaults are not changed. The next time the program is run the original default mode is restored. Many times a program provides both the capability of changing the default settings and of changing modes during operation.

Prompts and Other Messages

In order to operate a program successfully, you need to be able to select the desired modes of operation and you must receive feedback as to what modes are active. This feedback is generally provided in the form of prompts and messages that are displayed in an area of the screen reserved for that purpose. If the system has both an alphanumeric and a graphics screen, the prompts and messages are displayed on the alphanumeric screen. Another form of feedback that is employed is audio (sound) to signal a successful operation or, perhaps, a problem (such as when you attempt to do something that the system will not permit). A third common form of feedback is to change the way a word or symbol appears on the screen (such as blinking a symbol or object that has been selected for deletion so that you can verify the selection).

During program operation, you must also be kept advised of the progress being made and what should be done next. This type of information is generally given through *prompts*. You are probably already familiar with the simple prompts that a computer displays on the screen when it is started up. Some examples of these prompts are the A> prompt used by the IBM PC's disk operating system and the OK used by the IBM PC's BASIC language interpreter. Another example is the right bracket (]) used by Apple's DOS 3.3.

Drawing Routines

When you start a computer graphics program and prepare to create a drawing, you will probably find that there is more than one way to create many of the elements that will make up the completed drawing. For instance, a rectangle can be drawn by drawing four separate lines, but the program you are using might also offer a rectangle routine that will draw it as an object rather than as a collection of lines. A square might be drawn as four lines, or as a rectangle with equal length sides, or even as a regular four-sided polygon. Similarly, an arrowhead can be drawn using the lines task routine, but the program may also have a special arrowhead routine that will create arrowheads of a consistant size in any desired orientation, thus relieving the user of the task of figuring out where to draw the lines. It is generally, but not always, best to select the task routine that makes the task the easiest. An exception might be when you wish to later delete only a portion of a figure. For instance, you might draw a rectangle to depict the outline of a floor plan for a room. Then you might modify each of the sides in turn to create features such as door and window openings. Or perhaps you would prefer to superimpose wall features onto the base rectangle and then, when finished, erase the rectangle, leaving the superimposed features intact. In the first case you would probably use the lines routine to draw the rectangle. In the second case you would probably use the rectangle routine. Some computer-aided drawing programs, such as FIRST-DRAW, permit you to create temporary figures and objects that do not become part of the stored drawing data. You can remove these temporary figures easily by simply redisplaying the drawing (in the case of FIRST-DRAW), changing a mode selection, or performing some similar action.

MODE SELECTION. When a task has been selected, it will be activated with certain default modes. For example, LINES may default to ADD and SOLID line style. You will normally be kept informed of the active modes through *status messages.* Some systems will display the status messages continuously, while others will present them only when you request to see them. The latter is generally true of programs that have a large number of possible selections, many of which will be changed infrequently.

ENTERING THE REQUIRED POINTS. Once you have selected the modes you are going to need in order to accomplish a particular task, you are ready to select the points that are needed to define the figure. If you are plotting points, you need only locate where the point is to be placed. To draw a line, on the other hand, you will need to locate both ends. For a circle, you may need to locate the center and then a point on the circumference. In any case, the program should use prompts to inform you as to what to do next. These prompts may be in the form of messages printed on the screen or in the form of audible signals. Some systems show a small picture of the type of figure being drawn and indicate with dots and numbers how to create it (see Figure 15-32).

Drawing Aids

In addition to prompts, a program may make various drawing aids available to you. A commonly provided drawing aid is a pattern of dots that can be displayed on the screen. This pattern is referred to as a *dot grid*, or simply a *grid*. Some grids dis-

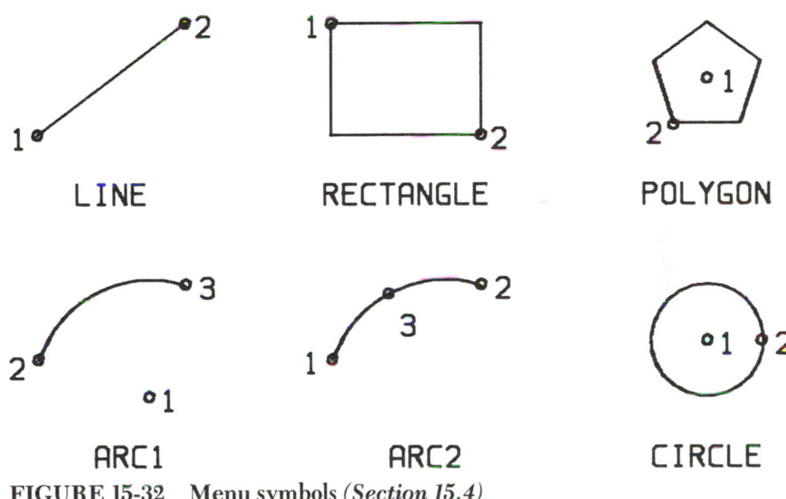

FIGURE 15-32 Menu symbols *(Section 15.4)*

play a full pattern of dots, as shown in Figure 15-33, while others use an open grid pattern like that shown in Figure 15-34. The grid spacing may be available in more than one size and may be available in an isometric or other pattern in addition to the traditional square or rectangular pattern. Some programs permit the grid to be put up at various angles. You may also be able to select grid spacings that correspond to the drawing scale that you have chosen.

CONSTRUCTION LINES. Some systems may provide horizontal and vertical ruler lines and a full-screen cursor rather than a grid (see Figure 15-35). Most provide for the drawing of construction lines that can be removed easily when the drawing is complete. If the program permits drawing on more than one "level," the construction lines may be placed on a different level than the drawing itself.

Storing and Retrieving the Drawing

Some drawings are created in one session, hard copied, and never modified. This, however, is the exception. In most cases you will want to be able to save the drawing so you can retrieve it later to add details, to modify it, or to use it. Drawings stored

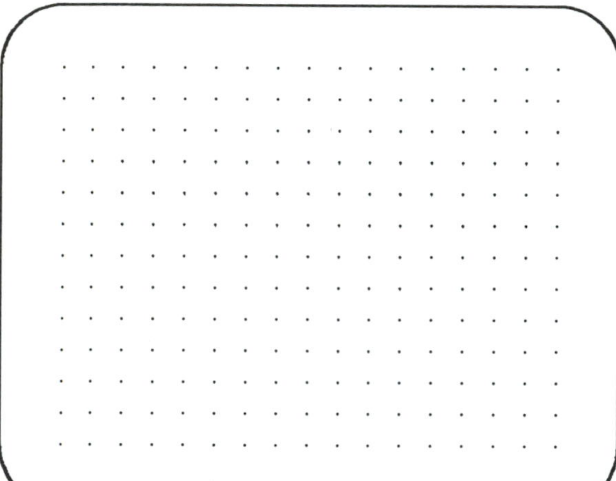

FIGURE 15-33 **Full grid pattern** ***(Section 15.4)***

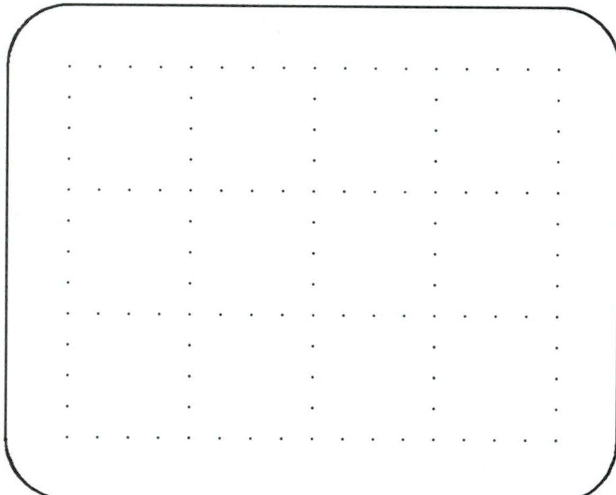

FIGURE 15-34 **Open grid pattern** ***(Section 15.4)***

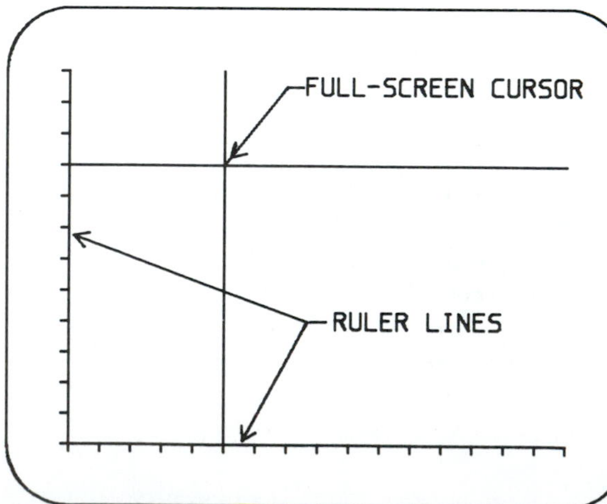

FIGURE 15-35 **Full-screen cursor and ruler lines** ***(Section 15.4)***

on disk take up much less space than do hardcopy drawings, and they are easily retrieved, viewed, hard copied if desired, or modified. Thus, disk storage adds flexibility as well as space savings.

FILE NAMING CONVENTIONS. It is important that you devise a system of file names to permit easy identification and retrieval of your drawings. The file identification includes the name you give it and may also include an extension (such as .DAT) that indicates the type of file (DATA), when the file was created (date and time), and the version number. Some systems keep only the current version. Others keep the current version and a backup version. Still others will keep all versions. Figure 15-36 is an example of file directory information for both an IBM PC and a DEC VAX. Note that the PC directory listing shows the name of the file, an optional extension showing the file type, the amount of space it occupies in bytes, and the date and time of day it was created. The IBM PC operating system will overwrite old file versions with the new versions unless the program has a provision to do otherwise. The VAX directory lists the file name, a mandatory extension showing file type, the version number, the amount of space it occupies in blocks (each block is 512 bytes) and the date and time of day the file was created. The VAX retains all versions of any given file and assigns them version numbers. You can, of course, purge the file of old versions.

Hardcopy

Almost every drafter wants a hardcopy of the drawing at some point. Most like to get a hardcopy at various stages of the drawing process as an aid to checking the drawing and marking changes to be made, and a hardcopy of the completed drawing is usually desired as well. We have already discussed the two types of hardcopy devices in general use: the dot matrix printer (printer-plotter), with dot resolution graphics capability, and the pen plotter.

```
           IBM PC FILE DIRECTORY


 Volume in drive B has no label
 Directory of  B:\

PARTIV          41344   8-20-84  10:15p
PARTIII         20608   8-20-84   8:55p
PARTIV   BAK    40960   8-19-84   4:10p
MYSAMPLE BAS     2048   8-22-84   9:23a
        4 file(s)      54784 bytes free

A>




          DEC VAX FILE DIRECTORY


Directory DEPT$DISK:[EG.MILLER]

DEMO2.FOR;1            31       29-OCT-1983 07:39
EX40.FOR;39            20       13-AUG-1983 10:24
EX44.FOR;2              8       17-AUG-1983 17:45
EX43.FOR;30             4       15-AUG-1983 13:23
BWORK.FOR;2             8       17-AUG-1983 17:46
C7GRAF.FOR;30          32       26-AUG-1983 16:38
C7LCHRT.FOR;27         32       30-AUG-1983 09:19
CH7CPS.FOR;1           20       21-AUG-1983 10:21
DEMO1.FOR;3             5       14-OCT-1983 09:47
EX45.FOR;4              5       27-OCT-1983 16:40
EX45.FOR;3              5        4-OCT-1983 07:49
GRAPH.FOR;3             5       21-FEB-1984 10:58
INSTR.FOR;1             4       21-APR-1984 10:32
TEST.FOR;2              4       16-JUN-1984 14:22
TUTOR1.FOR;1            4       24-AUG-1984 15:51
TUTOR1.FOR;2            4       25-AUG-1984 15:51
TEST.FOR;1              4       15-JUN-1984 10:01

Total of 17 files, 195 blocks.

$
```

FIGURE 15-36 Two examples of file directory information ***(Section 15.4)***

Printer-plotters are very popular, since most systems need a dot matrix printer for normal text output. The primary advantage of printer plotters is the ease with which you can get a hardcopy of the drawing. Disadvantages of printers as graphics hardcopy devices are that they generally print a "screen dump," and are thus limited to the same resolution as the screen, and that the drawing is not produced to scale. Some drafting programs convert the screen image to take advantage of the printer's maximum resolution, but the result is still marginal. As a result of these disadvantages, the printer-plotter is generally used for draft copy rather than final drawings.

Pen plotters also draw in discrete steps. However, the step size is typically about 0.005 inch, which produces a very high quality image. Pen plotters also have the advantage of being able to create a drawing that can be scaled. A disadvantage is cost, both of the plotter and of the materials. Another disadvantage is that plotters may require some setup and thus are often less convenient to use.

Program Interaction

Figure 15-37 shows the first menu for FIRSTDRAW, which contains the following selections: point (PT), line (LN), rectangle (RC), polygon (PG), circle (CR), arc (AC), next menu (NX), end the program (END), and arrowheads (AW). Figure 15-38 shows the second menu, which appears when number key 7 on the first menu is pressed. The selections on this menu are: plotting (PLT), grid (GD), text (TX), redisplay (RD), save picture in disk file (SV), retrieve picture file from disk (RT), next menu (NX), end the program (END), and help (?).

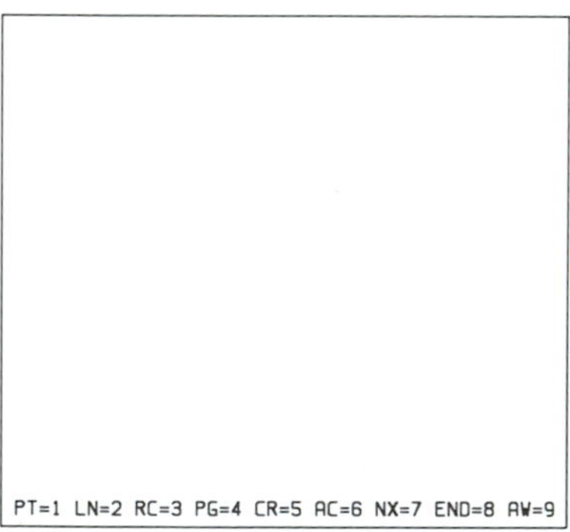

FIGURE 15-37 FIRSTDRAW menu one *(Section 15.4)*

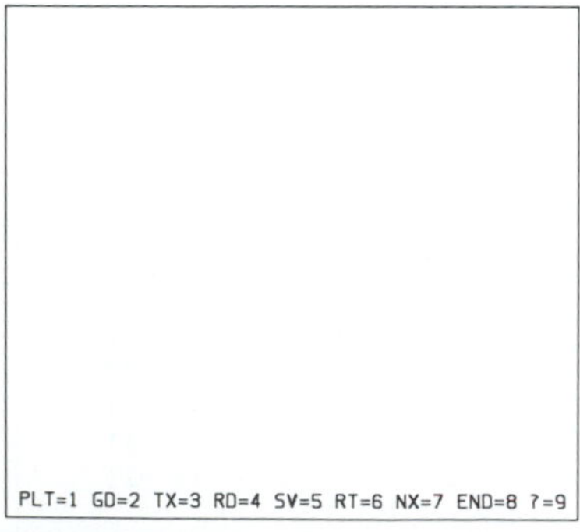

FIGURE 15-38 FIRSTDRAW menu two *(Section 15.4)*

In menu one, when you press key 1, the screen will appear as shown in Figure 15-39(a), with the cursor at one position on the screen. Pressing the space bar and then moving the cursor allows you to see the dot left on the screen, as shown in Figure 15-39(b).

If you return to menu one and select the line task, the screen will look like the one

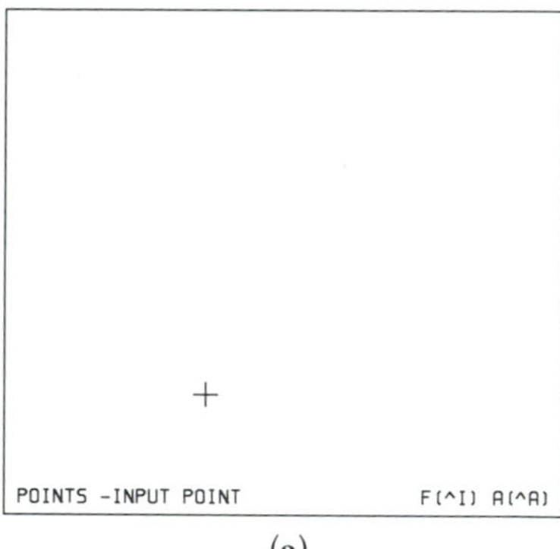

(a)

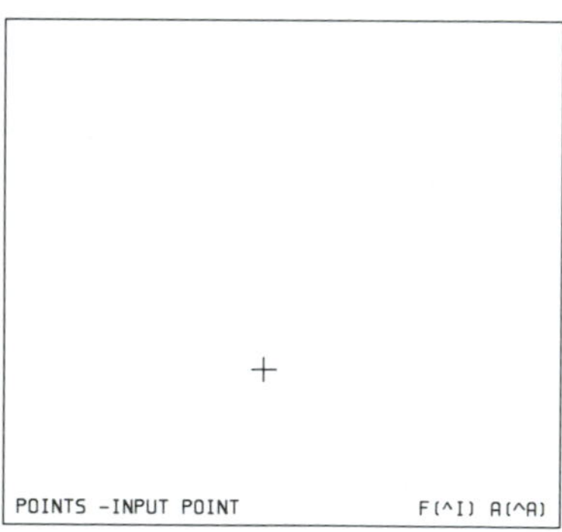

(b)

FIGURE 15-39 FIRSTDRAW screens in point (PT) mode *(Section 15.4)*

shown in Figure 15-40(a). The prompt INPUT START tells you to locate the desired position for one end of the line and press the space bar. Figure 15-40(b) shows the cursor moved to the location of the other end of the line and the dot that remains on the screen to locate the first end. When you press the space bar, the line will appear as shown in Figure 15-40(c).

Selecting circle (CR) from menu one produces the screen shown in Figure 15-41(a). The prompt INPUT CENTER tells you to locate the cursor at the desired center of the circle and press the space bar. Figure 15-41(b) shows the cursor moved away from the center to the desired radius of the circle. Note that the center of the circle is indicated by the dot and the prompt tells you to input one point on the circumference of the circle. When the space bar is pressed again the circle appears. You are then prompted to input another circle, as shown in Figure 15-41(c).

15.5 DRAWING WITH A COMPUTER GRAPHICS SYSTEM

We will now cover the steps necessary to create a simple drawing on a computer graphics system. There are several computer-aided drawing packages on the market. Since they vary widely in the specific actions required to accomplish any given task, we will assume here that you are going to use either EnerGraphics, one of the less expensive commercial packages, or FIRSTDRAW. EnerGraphics is available for IBM PC and PC-

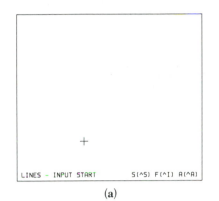

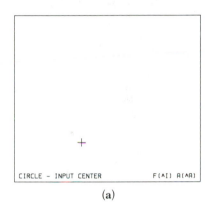

LINES - INPUT START S(^S) F(^I) A(^A)

(a)

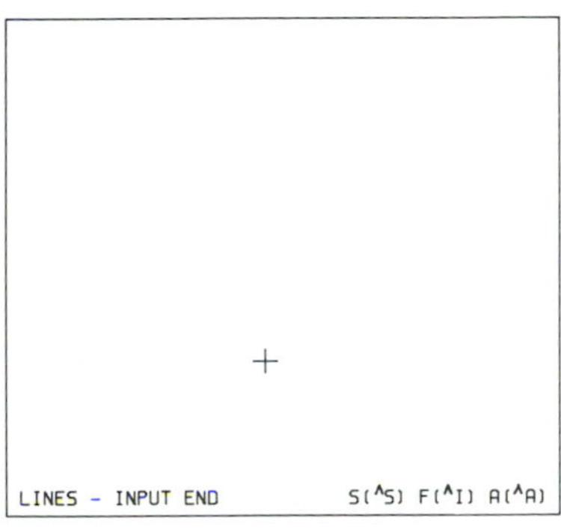

(b)

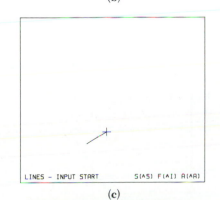

LINES - INPUT START S(^S) F(^I) A(^A)

(c)

FIGURE 15-40 FIRSTDRAW screens in line (LN) mode ***(Section 15.4)***

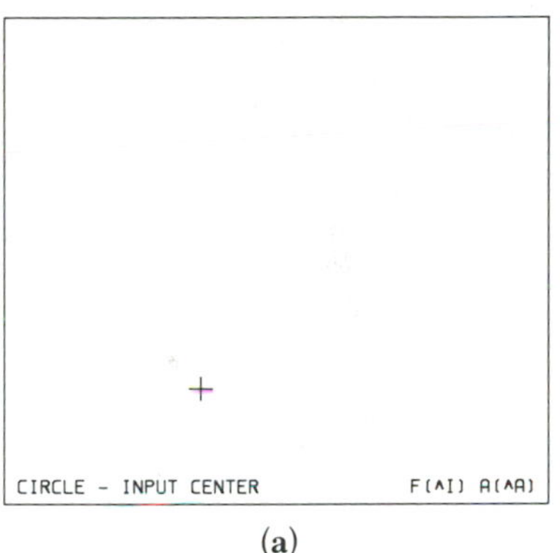

(a)

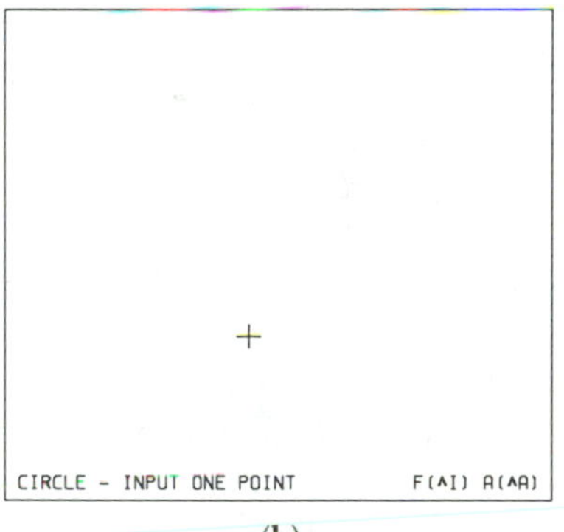

(b)

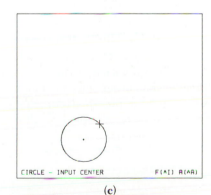

CIRCLE - INPUT CENTER F(^I) A(^A)

(c)

FIGURE 15-41 FIRSTDRAW screens in circle (CR) mode ***(Section 15.4)***

compatible computers. FIRSTDRAW is available for Apple IIe and IBM PC and PC-compatible computers. Our instructions will be based on FIRSTDRAW with suggestions for using EnerGraphics, though you may use another computer-aided drawing package if you wish. The drawing we are going to create is shown in Figure 15-42. As you have learned, the first logical step in making a drawing on paper is to create a rough sketch. Sketching allows you to establish proportion and scale so that your drawing will fit on the paper. Preliminary sketches are equally helpful when creating drawings on a computer graphics system. The time spent creating a sketch will be repaid many times in savings in the actual drawing process. If you need to review sketching techniques, you should refer to Chapters 1 and 2.

Using Grid Paper

Grid paper with eighth-inch (3 mm) or quarter-inch (6 mm) grid spacing is very helpful when creating orthographic views or oblique views. Isometric grid paper with quarter-inch (6 mm) or half-inch (12 mm) grid spacing is good for pictorials. Grids smaller than one-eighth inch for rectangular or one-quarter inch for isometric tend to be difficult to work with on the screen.

Establishing a Scale

Select a scale that will permit you to fit the drawing on the paper without crowding while still making it large enough to be easily read. Full size is generally the preferred scale. Sometimes a scale larger than full is desirable for small objects. More often a smaller scale such as half size or quarter size is required because the object is too large for full scale. The scale should relate to the grid (for instance, one grid space equals one inch).

Chapter 3 discusses the scales commonly used by various disciplines. We will assume here that the maximum vertical and horizontal dimensions of the part shown in Figure 15-42 are each 4 inches and that the part is to be drawn full size. Use grid paper with quarter-inch (6 mm) grid if available.

Blocking In the Views (Outside-In)

Block out an area 4 inches square and then subdivide it as shown in Step 1 of Figure 15-42. Refer to Chapters 1 and 2 if you are not sure how to do this. By blocking in the views you can

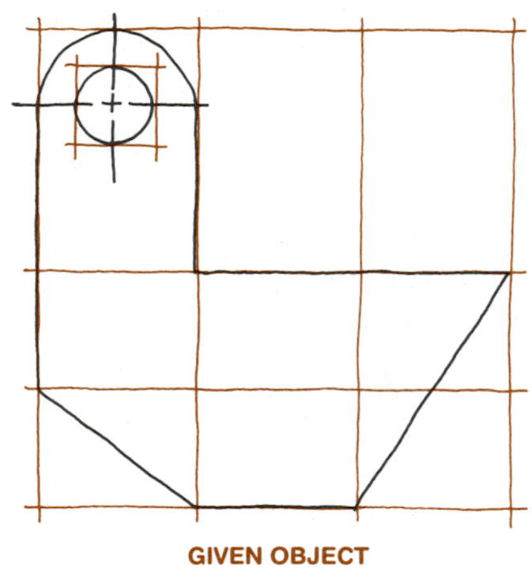

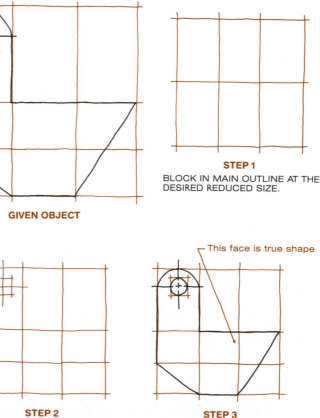

GIVEN OBJECT

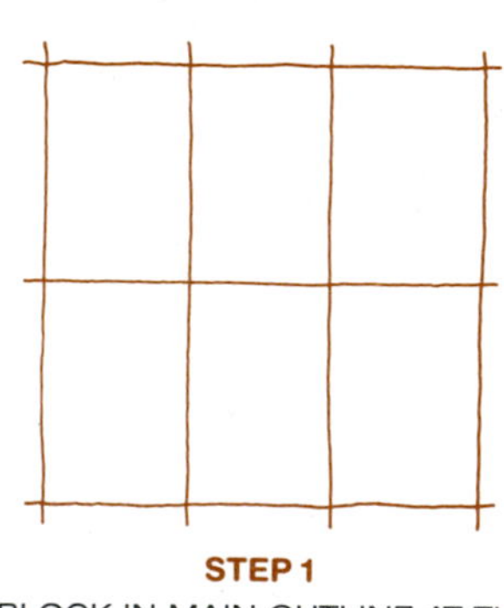

STEP 1
BLOCK IN MAIN OUTLINE AT THE DESIRED REDUCED SIZE.

STEP 2
BLOCK IN DETAILS OF THE OBJECT USING PROPORTIONAL DISTANCES.

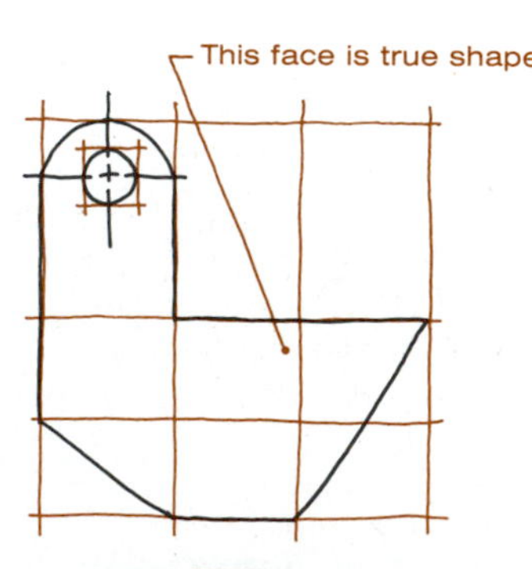

STEP 3
DARKEN FINAL OBJECT.

FIGURE 15-42 Sample figure *(Section 15.5)*

make sure that everything can be positioned correctly on the paper before you spend a lot of time on details. You should use a very light touch and a hard pencil lead so that the outline will not have to be erased later.

Sketching the Details

Now that you have the overall layout of your sketch established, you are ready to begin sketching in the details (see Steps 2 and 3 of Figure 15-42). They should also be drawn very lightly so that an error will be easy to correct. Finally, you are ready to darken in the final features of your sketch. Again, refer to Chapters 1 and 2 if necessary.

Laying Out the Drawing Area on the Screen

CONSIDERATIONS OF SCALE. With your finished sketch before you, you are ready to create the drawing on the computer CRT. Depending on the drafting package you are using, you may need to create the entire drawing in the space available on the screen or you may be able to use more space by "scrolling around" and displaying only a portion of the drawing at one time. Another technique is "zooming in" on features to create them, thus magnifying the portion of the drawing you are working on and displaying that portion on the entire screen at a larger scale.

USE OF MULTIPLE SCREENS. Another technique for creating large drawings on a system is to use multiple screens (views) that are each a portion of a large drawing. The screens are displayed and worked on in turn. Likewise, they may be plotted or print-plotted one at a time and then joined to form the completed drawing. If the plotter is large enough, the entire drawing can be plotted at one time. The object we are drawing can be displayed full size on the CRT in one view.

CREATING A BORDER AND TITLE BLOCK. If your drawing package does not have a built-in border and title block, you should create one and save it in a file on your disk. Then, each time you start a drawing, you can retrieve and display the border and title block and achieve the same effect as using preprinted drawing paper. We will use FIRST-DRAW's border and title block for now, but later in this chapter you will have an opportunity to create a new one.

USE OF DOT GRID AND CONSTRUCTION LINES. When you sketched freehand, you used a grid as a guide. When you drew with instruments, you dispensed with the grid because the tools you needed to locate and create parts of the drawing were available in the form of the T square, triangles, scale, compass, and other instruments. While many drawing packages provide continuous readouts of the location of the cursor relative to an origin, most users find dot grids (or ruler lines) and construction lines to be indispensable tools for creating drawings on the CRT. In fact many drawing packages provide the capability of "snapping" the cursor to grid points. Snapping means that if the cursor is close to the dot grid point it draws to the grid point. This snap-to-grid capability can make the drawing task much easier and is an important reason to establish a scale that conforms to the grid spacing. Of course, the grid will not be part of your final drawing.

In addition to the grid, you should make liberal use of construction lines. Mark the outlines of figures and the centers of circles, arcs, and regular polygons just as you did when creating your sketch. If the grid spacing of the sketch and the CRT drawing compare, it will be easy to locate these lines. The lines will not be part of your final drawing, so they should be kept separate from the rest of the drawing, if possible. Many drawing packages have provision to do this through the use of a temporary-mode selection that does not store the data for the construction lines. Other packages provide multiple drawing levels so that the construction lines can be placed on a different level than the rest of the drawing and then not saved or plotted.

If you are using FIRST-DRAW, go to the second menu and select GRID. When prompted for size, select LARGE, and when prompted for type, select RECTANGULAR. Now go to the first menu and select LINE. Select TMP (temporary) and DASH line mode. Now draw each construction line by moving the cursor to the start of the line and pressing the space bar. Then move the cursor to the end of the line and press the space bar again. The prompt line will keep you advised as to whether you are starting or ending a line. Your grid and construction lines should look like Figure 15-43. Note that all of the construction lines are shown dashed. Also note the similarity to Step 2 of Figure 15-42. If you are using EnerGraphics, draw the construction lines solid, since using dashed lines requires drawing or removing each dash separately and is a time-consuming process. To draw the lines, move the cursor to the start of each line and press F1. Then move the cursor to the end of the line and press F1 again. The system will create the line.

Drawing the Entities

When the grid and construction lines are in place, it is time to begin creating the drawing itself. The drawing will be composed of a number of different elements or entities. The drawing we are creating uses only lines, circles, and arcs, but in many drawings there will be points, straight lines, rectangles, polygons, circles, arcs, arrowheads, irregular shapes that will be constructed from straight lines of various lengths, and, of course, numbers, letters, and words. FIRSTDRAW and most other systems will allow you to create these entities one at a time. To speed up the drawing process, you should enter all of each entity before going back to the menu to select another. For instance, when you select CIRCLE from the menu,

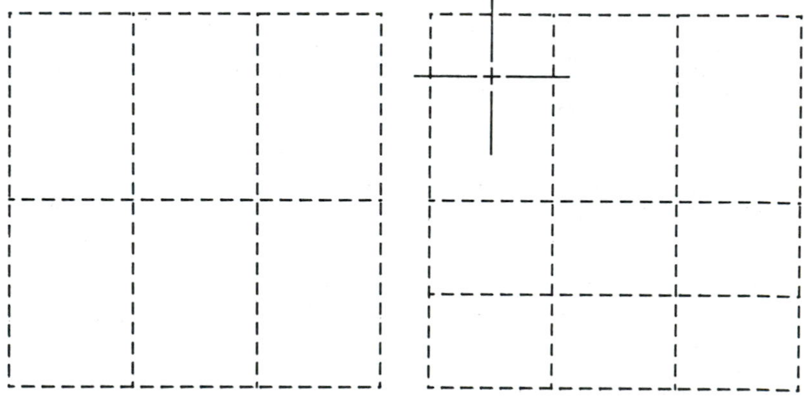

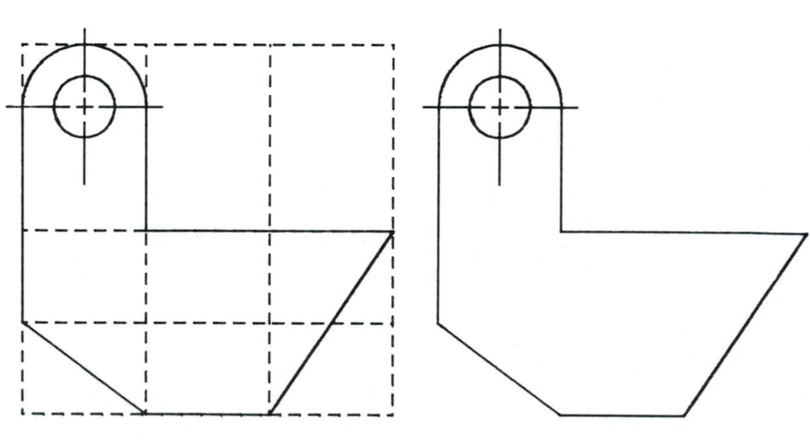

FIGURE 15-43 Drawing the figure ***(Section 15.5)***

draw all of the circles before going back to select something else. Keep in mind that certain drawing elements may need to be created before others. For instance, an arc may need to be drawn before the straight lines connecting it to another drawing feature.

For our drawing we will create the circle first. Select CIRCLE from the FIRSTDRAW menu then mark the center of the circle (Step 2 of Figure 15-43) followed by a point on the circle. It is easiest to select a point on either the horizontal or vertical axis. If you are using EnerGraphics you will define the circle with two points on diagonally opposite sides of the circle. Move the cursor to the point where the circle intersects the horizontal axis and press F1 and F5. Then move the cursor to the opposite side of the circle. As the cursor moves, circles of increasing size will be drawn. When the circle is the correct size, press F1 again to permanently locate the circle there.

We will now draw the arc. Select ARC from the FIRSTDRAW menu, mark the center of the arc, then the beginning point of the arc on the horizontal axis to the left of the center. Finally, mark the end of the arc on the horizontal axis to the right of the center. The system will now draw the arc. If you are using EnerGraphics you will define an arc with three points: the beginning point, the ending point, and an intermediate point on the arc. Move the cursor to the beginning point of the arc and press F1. Move the cursor to the ending point of the arc and press F1 again. A line will be drawn temporarily between the two points and the cursor will be positioned at the center of the arc. Move the cursor upward. Each time the cursor moves, the system will draw an arc through the three points represented by the beginning, the cursor, and the ending point. When the cursor has been moved to where the arc is the correct size, press F1 to permanently locate the arc.

Now draw in the lines with FIRSTDRAW by selecting LINE and then moving the cursor to the start of a line, pressing the space bar, moving the cursor to the end of the line, and pressing the space bar again. If you are using EnerGraphics, move the cursor to the start of a line and press F1. Then move the cursor to the end of the line and press F1 again to draw the line.

When all of the lines are completed, your drawing should look like Step 3 of Figure 15-43. Now go to the second FIRSTDRAW menu and select REDRAW menu and select REDISPLAY. The drawing will be erased and then redrawn without the grid or the construction lines. Check it for completeness. Your completed drawing should look like Step 4 of Figure 15-43. If you are using EnerGraphics, erase the construction lines by moving the cursor to a unique point (one that is not shared with an arc, a circle, or another line you want to keep) and press Del (delete). That line will be erased. Parts of the drawing you want to keep may be erased too, but the data will still be in memory and when you redisplay (redraw) the drawing they will be restored. To redraw the EnerGraphics drawing, press F7 when the menu is displayed.

15.6 MODIFYING AN EXISTING DRAWING

Computer-aided drawing packages really come into their own when it is time to modify the drawing. Drawings done on paper or plastic film are difficult to modify. First, it is necessary to carefully erase the unwanted parts, taking care not to unduly damage the drawing, and then to add the new features. If there are many changes to make it may be necessary to recreate the entire drawing on a clean sheet of paper. By contrast, the drawing done on the computer can be retrieved from the disk file, displayed on the screen, and the unwanted parts removed as easily as they were drawn in the first place. The new features can be added and the modified drawing saved again and, if desired, sent to the plotter to produce a fresh, clean drawing without erasure marks. Many times several variations of a single drawing are needed and can be created easily by doing one basic drawing and then adding the custom features for each variation by retrieving and modifying the basic drawing and saving each version under a different name. If the drawing package supports multiple levels, the basic drawing can be done on one level and the custom features for each variation on separate levels. This technique is particularly effective for building construction drawings. The basic floor plan is drawn and then used as the basis for electrical, plumbing, equipment, and furniture layouts, each of which need to be separate drawings.

15.7 SOME DRAWING EXERCISES USING A SYSTEM

Creating a Border and Title Block

Many drawing software packages will provide a border line and title block but creating one is a good exercise. The border line and title block should allow you to use the maximum amount of screen area. In order to do this select the rectangle option and move the cursor first to the lower left corner of the CRT screen to establish one corner of the rectangle forming the border line. Next move the cursor to the upper right corner of the screen and fix the opposite corner of the rectangle.

The title block provided with FIRSTDRAW occupies the bottom of the screen area. Figure 15-44 shows the construction of a title block that occupies the lower-right corner of the drawing. The block was created with FIRSTDRAW by using the rectangle drawing routine and putting three rectangles together to form the five rectangular spaces. The outer rectangle was the first one drawn, then the lines that form the top of the bottom space and the bottom of the top space were drawn. Finally, the middle rectangle in the bottom space was drawn. This provides five spaces using only six sets of coordinates. Had this been created using individual lines instead of rectangles, twelve sets of coordinates would have been required. A border can be added, as shown in Figure 15-45, using one more rectangle.

If you are using EnerGraphics, you will need to create the title block using lines, since EnerGraphics does not offer a rectangle routine. Select 2D DRAWING from the main menu and then select CREATE 2D GRAPHICS. When the submenu comes up, select BEGIN

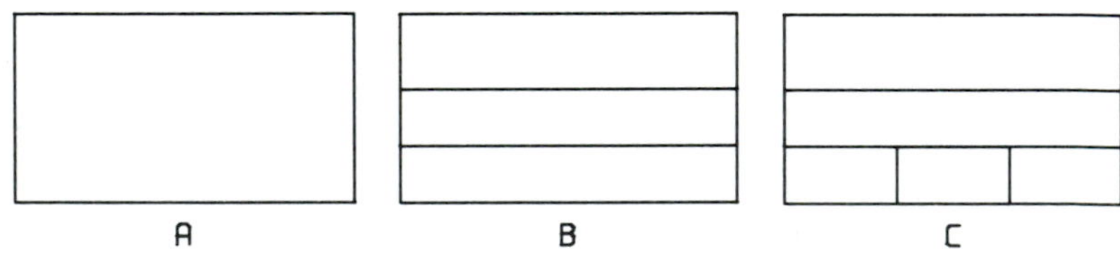

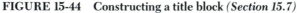

FIGURE 15-44 **Constructing a title block** ***(Section 15.7)***

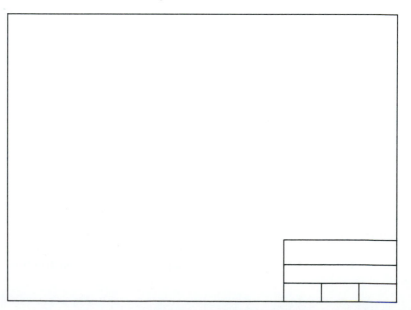

FIGURE 15-45 **Title block and border** ***(Section 15.7)***

EDITING. The screen that is displayed next will have a border and tick marks to indicate grid coordinates. Set the cursor speed at F (fast) and, starting in one corner, press F1. Now move the cursor to the same position in the adjacent corner and press F1 twice more. This will create one line and start the next one. You will not be able to see the line since it will be on top of the line the program uses to define the drawing space and grid coordinates. Proceed around the screen in this fashion until the border is complete. Now, move the cursor to the start of the outline of the title block and proceed to draw the title block in the same way you drew the border.

When you have finished the drawing, save it on your data disk as BRDR2. Plot or print-plot the drawing to verify that the border and title block are complete.

Drawing the Three Principal Views of a Cut Block from the Isometric Drawing

A typical drawing required of engineering graphics students is the orthographic representation of a cut block. For this exercise, you are provided with an isometric drawing and asked to draw the three orthographic views. We will use the block shown in Figure 15-46(a), and we will draw the top, front, and right-side (profile) views.

Go to FIRSTDRAW's second menu. Retrieve BRDR2 and

FIGURE 15-46 Drawing the three principal views of an object ***(Section 15.7)***

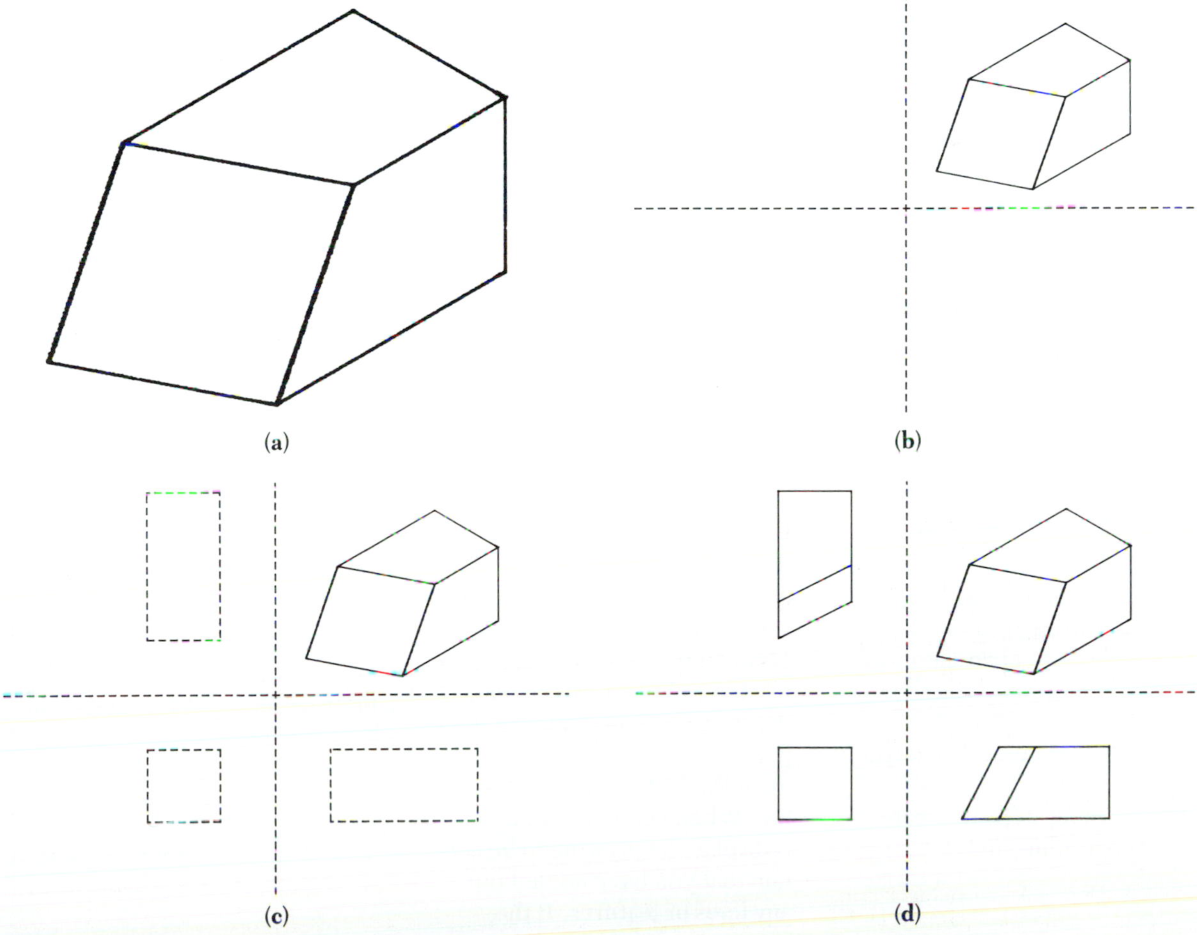

then bring up the rectangular grid. Count the number of vertical grid spaces and horizontal grid spaces available. Make a note of these two numbers. Next measure the overall height, width, and depth of the isometric figure. Establish a scale factor (it may be full size) and convert the dimensions of the isometric figure into screen grid units. Use the rectangle drawing routine in the temporary mode to establish the width and depth of the top view, the width and height of the front view, and the height and depth of the side view. Remember to leave about 12 mm, or ½ inch between the border line and the object on all sides. Leave at least that much space between views as well. If the drawing is to be dimensioned, then allow more space between views and between the views and the border. Refer to the isometric view shown in Figure 15-46(a) and make a list of the end-point coordinates of each line. Draw these lines in each of the views. Redisplay the drawing to eliminate the temporary rectangles that surround each view. Check your work by attempting to sketch the isometric pictorial on a piece of isometric grid paper using your drawing as a reference. If the isometric sketch is complete, then save your drawing. Figures 15-46(b), 15-46(c), and 15-46(d) show an isometric view of a cut block and the accompanying steps to create the orthographic views.

If you are using EnerGraphics, draw all of the lines using F1. Plan to remove the temporary rectangles when the drawing is complete by moving the cursor to each unwanted line and pressing Del. When you have erased all of the unwanted lines, press F7 to redraw (redisplay) the drawing so you can check that none of the needed lines have been permanently erased.

Drawing an Isometric of a Cut Block from the Three Principal Views

The procedure for creating an isometric drawing from three orthographics views is the reverse of the process just described. Again, you are trying to draw the isometric pictorial as big as possible to fill the screen area. Retrieve BRDR2 and display FIRSTDRAW's small isometric grid. Determine the height, width, and depth from the orthographic views. Calculate a scale factor that will allow you to convert the dimensions of the orthographic views to screen isometric grid units. Make notes on the orthographic drawing of the equivalent screen units (three-dimensional coordinates) for ends of lines and key features such as the centers of circles. Draw an isometric box using the maximum height, width, and depth. Use dashed lines in temporary mode for this task. Now fill in the features of the drawing. When you have done this, redisplay the drawing to be certain that you have not left out any lines or features. If the drawing is correct, save and plot the pictorial drawing.

If you are using EnerGraphics, you should use the three-dimensional routines to create your isometric drawing. Rather than drawing the figure using the cursor, as with FIRSTDRAW, you will need to create three data tables for each face. The first table will list the coordinates of each vertex of the face, the second table will list the line connections between points, and the third table will describe the orientation of the face in three-dimensional space.

To create your tables, draw a sketch of each face as it is shown on your orthographic drawing, then draw a set of axes on each view. Label the axes H for horizontal and V for vertical. Sketch the selected face as if you were creating a true-size two-dimensional drawing of it in the plane of the HV axes. Number the points. See Figure 15-47 for an example of one face of the cut block. Remember that cut faces

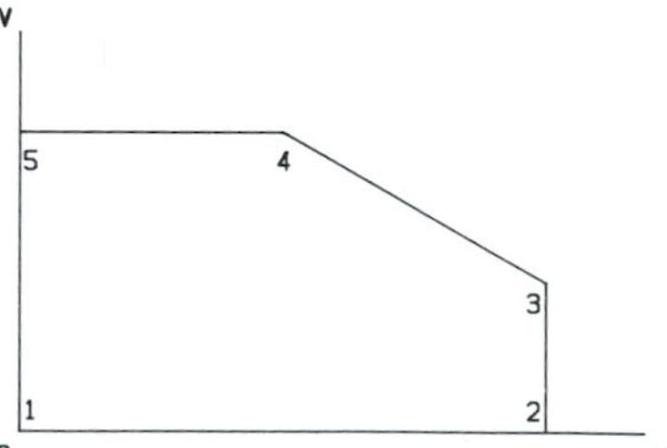

FIGURE 15-47 A face of the cut block *(Section 15.7)*

TABLE 15-1 Values for EnerGraphics Plot

(a) Corner Coordinates			*(b)* Line Connections			*(c)* 3D Face Orientation			
No. of points = 5			*No. of lines = 5*			*Orientation points*			
#	*H*	V	#	*Pt#*	*Pt#*		X	Y	Z
1	0	0	1	1	2	O:	0	4	0
2	4	0	2	2	3	H:	0	0	0
3	4	1.125	3	3	4	A:	0	4	2.25
4	2	2.25	4	4	5				
5	0	2.25	5	5	1				

Note: 0: is origin of HV plot, H: is a point on H axis, and A: is any other point not on HV axis.

do not show true size in any principal view, but the cut face must be drawn true size and shape in your HV sketch. Note that the axes on your sketch are called H and V instead of X and Y. That is because you can place the face in any desired orientation and each point will have x-, y-, and z-coordinates that will depend on the orientation you choose when you assemble the faces to create the object.

Now write the coordinates of each corner relative to the axes you just drew. You will probably need to calculate the coordinates of the cut face. When you have done this, make a table showing the number of each corner and its coordinates, as shown in Table 15-1(a). Make a second table showing the lines that connect the points as in Table 15-1(b). Now you are ready to create the table to orient the face in 3D space, as was done in Table 15-1(c). To do this you need to determine the x-, y-, and z-coordinates of three points on the face. The first one is O (for origin). The second one is H, which means it must be on the H axis of your sketch. The third point, A, can be any other point that is not on the H axis. Figure 15-48 shows the coordinate system EnerGraphics uses. Note that the vertical axis is called Z. You must orient your faces to fit this coordinate system. The cut

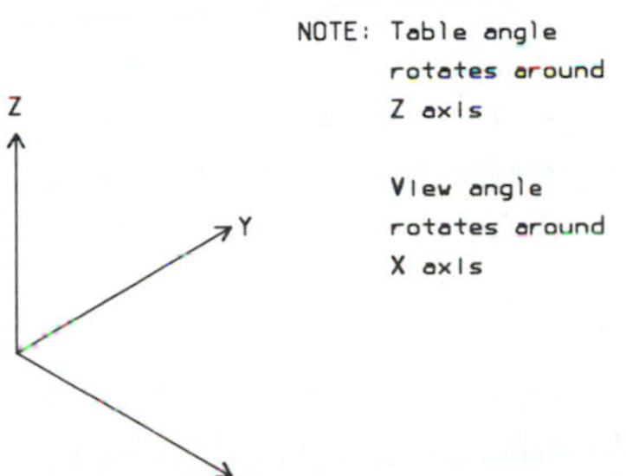
NOTE: Table angle rotates around Z axis
View angle rotates around X axis
Z
Y
X
ENERGRAPHICS COORDINATE SYSTEM

FIGURE 15-48 EnerGraphics coordinate system *(Section 15.7)*

block in our example has seven faces, so if you wish to draw it as a solid and be able to rotate it, you will need to create all seven faces (the back face can be created by duplicating the front face). If you wish to draw it as a solid in normal isometric position and not rotate it, you can do so with four faces, since three faces are hidden. How many faces would you need to create for a wire-frame drawing?

When you have finished the tables, you are ready to input the data into the three-dimensional drawing routine. Enter the tables into the system and describe the orientation of each of the faces separately. The first time you build the tables you may find them confusing. It will be easier if you create one face at a time and display the drawing to make sure the face is properly described and oriented before starting on the next face. Otherwise it will be difficult to determine which face is at fault if the figure is not correct. When you have entered the data to create and orient all of the faces, you will need to set the rotation angles so that the figure will appear to be an isometric. Set the table angle at 60 degrees and the viewing angle at 30 degrees. Now you can display the figure. If it is too large or too small, change the ZOOM (scale) level and redisplay the figure.

If your orthographic drawing is composed only of straight lines, the procedures just outlined are sufficient to create the drawing. If, however, the drawing includes some cylindrical or partial cylindrical features, you will need to employ either a built-in routine that draws ellipses or to use the four-center ellipse method described in Section 4.20. You will also need to use the temporary line mode to mark the center and draw the parallelogram that will surround the ellipse. After you have done this, locate the tangent points where the ellipse will touch the circle. Draw two temporary lines from each obtuse corner of the parallelogram perpendicular to the opposite side. Use the intersection points as centers of the small arcs and the tangent points as the beginning and ending points for the arcs. Now use the obtuse corners as the centers of the large arcs and the tangent points as the starting points and draw the two large arcs. Redisplay the drawing to affirm that you have created an isometric ellipse.

15.8 CREATING AND SAVING A DRAWING

An isometric drawing of a bracket was shown in Figure 1-1, and the dimensions were provided. Use your computer-aided drawing system to create a scale drawing of the bracket. The steps required are as follows:

1. Determine the overall dimensions of the bracket (H, W, D).
2. Establish a scale to be used on the screen. The scale should be in terms of the number of millimeters per screen grid unit or the fraction of an inch per screen unit. Be sure to allow about 25 mm, or 1 inch, between the border and the views and between views.
3. Retrieve one of the border/title blocks from the disk and display it on the screen.
4. Draw in the temporary boxes that would contain the views using the overall height, width, and depth. Use the dashed-line mode if available. Check the spacing to verify that you have enough space for dimensions.
5. Draw the views inside the boxes. Do all of the circles first, then the arcs. If you are using EnerGraphics, you will need to enter the two end points of the arc and then an intermediate point, as in previous examples. The intermediate point may be difficult to locate accurately when drawing the quarter-circle radii. Sometimes it helps to draw a circle, draw the arc on top of the circle, and then erase the circle. Another approach with EnerGraphics is to calculate the coordinates of a point on the arc and then move the cursor to those coordinates (watch the x and y values on the prompt line). Next, draw the straight lines, center lines, dashed or hidden lines, and the rectangles. Dashed lines can be drawn with EnerGraphics using a series of short lines. Rectangles are drawn with EnerGraphics as four connected lines.
6. When you are finished, redisplay the drawing. (If you are using EnerGraphics, delete any unwanted temporary lines first.) Verify that the views are correct by checking them against the isometric drawing.
7. Select TEXT mode and fill in the title block with the appropriate information.
8. Save the drawing in a disk file and plot a copy of the drawing using the plotter or printer-plotter.

A Hex Nut

A drawing of a typical hex nut is shown in Appendix Table A-11, along with the dimensions for different size nuts. Make a drawing of a ½-inch, or 12-mm, diameter hex nut at 1 inch = 0.25 inch, or 4:1 in SI units.

Follow the steps shown previously for the bracket. Only two views are required. Provide the front and right-side views in this case. Use construction lines (temporary lines and arcs) and be sure to redisplay the drawing to get rid of these lines before saving the drawing in a file on the disk. Your finished drawing

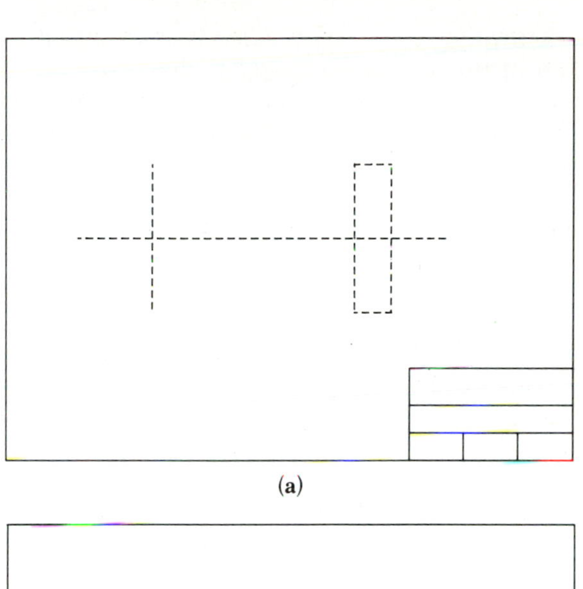

(a)

should be similar to the one developed in Figure 15-49. Figure 15-50 shows the steps to create a drawing of a bolt.

If you are using EnerGraphics, draw the front view of the nut by first drawing a circle. Then calculate the coordinates of the vertices of the nut. Note that you will need only the x- or y-coordinate for each vertex, since a vertical line through the x-coordinate or a horizontal line through the y-coordinate will in-

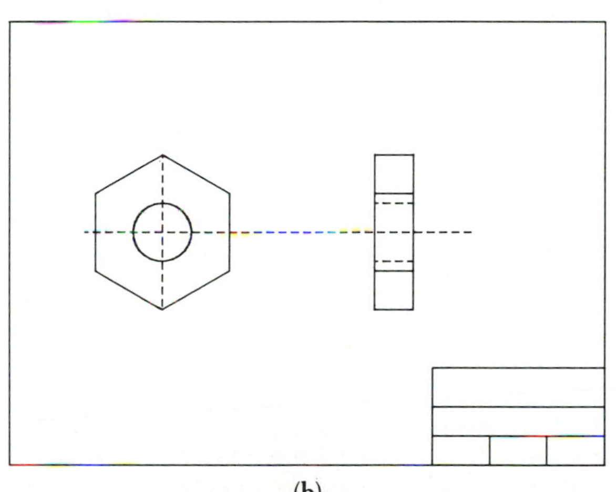

(b)

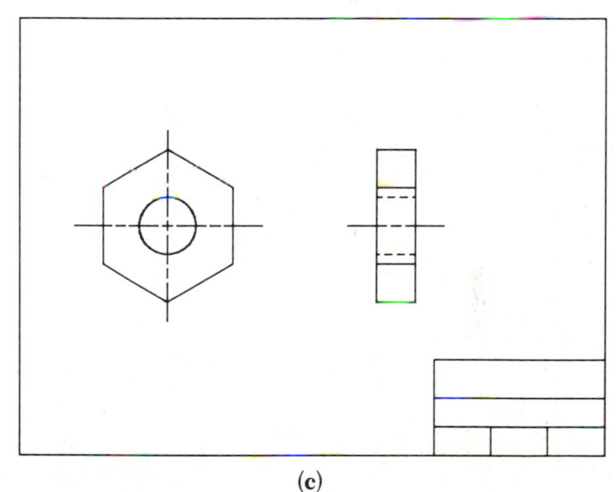

(c)

FIGURE 15-49 Drawing a hex nut *(Section 15.8)*

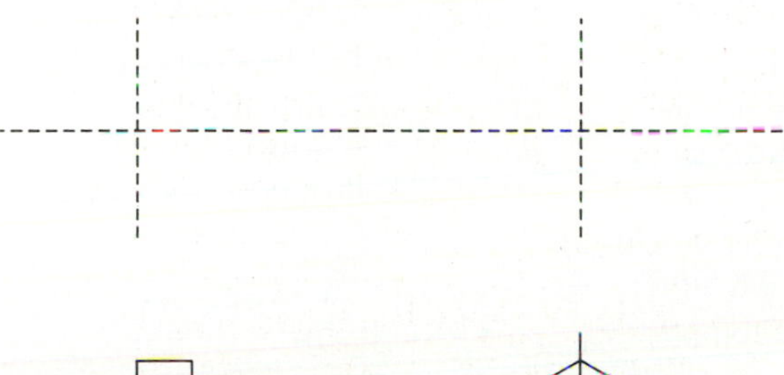

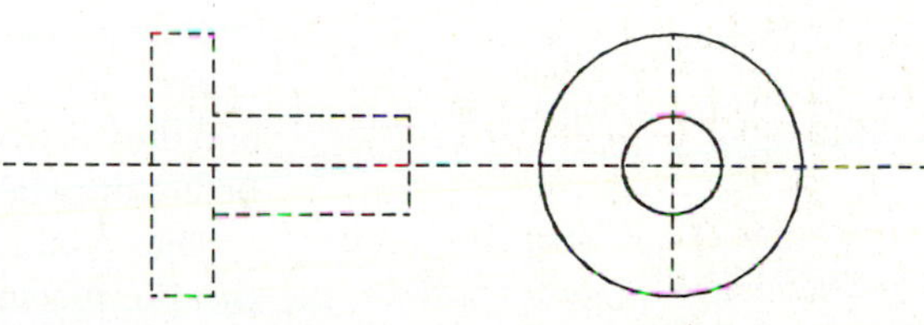

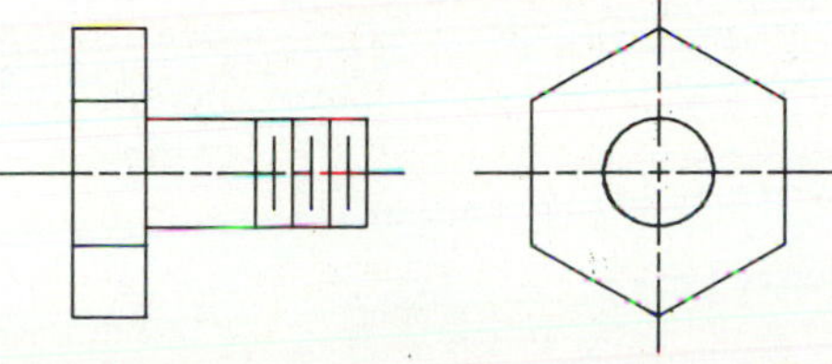

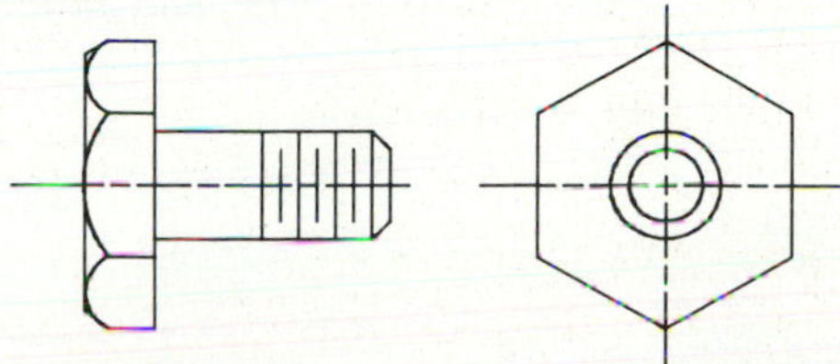

FIGURE 15-50 Drawing a bolt *(Section 15.8)*

tersect the circle at a vertex. Using your calculated coordinates, move the cursor to a vertex and draw a line to the next vertex. Proceed until all six sides are completed and then erase the circle.

A Simple Half-Section

Draw the simple half-section that was shown in Figure 6-14(c). To create the crosshatching, you should draw temporary lines in dash mode at a 45 degree angle and then trace over the appropriate portions with permanent solid lines. Lines at 45 degrees can be created by moving the cursor one step vertically for each horizontal step with the cursor in either medium or fast speed. This technique will work with either FIRSTDRAW or EnerGraphics. It is helpful to have the grid displayed.

Retrieving, Dimensioning, and Resaving Each of the Drawings

As an example of dimensioning using a computer-aided drawing system, we will dimension the bracket that was drawn in the previous section. You may find it helpful to review the dimensioning guidelines covered in Chapter 10 before you proceed.

To begin dimensioning, choose the letter size you prefer and make an estimate of the numbers of dimensions that will need to be "stacked" between views or between the views and the border. The first dimension should be 9 mm, or 0.375 inches, away from the view. The other dimensions must be separated by 6 mm, or 0.25 inches. The total distance that will be required for dimensions must be determined and then be compared to the distance between views. If there is not enough space you may have to move the views further apart. On some systems this is easy to do and on others it requires erasing and redrawing the entire view. It is a good idea to sketch the views and the necessary dimensions before drawing with the system.

Once you have determined that the spacing is correct, begin by drawing the extension lines and dimension lines in temporary mode (if available). Add the numbers and letters in permanent mode, positioning them so that they are properly spaced. (If you are using EnerGraphics, you may be tempted to adjust the text size to fit the amount of space available for each individual dimension. However, this is poor practice. You should select one text size and use it for all of the dimensions.) Next put in the permanent extension and dimension lines, adjusting for any changes in line lengths. Finally, put in the arrowheads on the ends of the dimension lines. Now redisplay the drawing and verify that all of the lines and figures are in the proper position. Plot the drawing.

Similar steps are required for dimensioning any figure. Systems that have a range of character sizes provide you with more flexibility while dimensioning objects but, as noted, it is good practice to choose one workable text size for dimensioning any one drawing and stick with it. You may use larger and smaller text sizes, if you so desire, for the various boxes of the title block or, in some cases, for notes.

PROBLEMS

These questions require you to know something about the computer-graphics system you will be using.

P15.1 Specify the brand and type of computer (mini-, micro-, mainframe) that you will be using.

P15.2 Specify the type of CRT display that is attached to your system. Specify the resolution and whether it is color or monochrome, raster or vector, storage or refresh.

P15.3 What type of cursor appears on your CRT screen (full screen or small plus sign)? What controls that cursor (function keys, joystick, mouse, digitizer, or other)? Does your system have a lightpen?

P15.4 If your system has a digitizer, how large is the digitizer and what resolution does it have? Is it used for graphical input, as a menu device, or both?

P15.5 What other types of graphical output are available for your system (plotters, printer-plotters, screen copy devices)? What size drawing(s) can you produce with your plotter?

P15.6 What is the name for the software package that your system uses? Is your computer-graphics software produced by the hardware manufacturer or by a separate firm?

P15.7 Where is the user menu displayed (screen, digitizer, function keypad, separate card requiring keyboard input)? Does your system have more than one option for menu item selection?

P15.8 Does your system save only current drawing files or does it provide for automatic saving of the previous files? Does it allow you to save the same file under different names? What is the format for drawing file names? What is the drawing file extension (.DAT, .DWG, .PIC, and so on)?

P15.9 Does your software allow you to zoom in on an area? If so, how is the area defined? Can you put a circle or rectangle around the chosen area? Can you choose a magnification ratio (2:1, 3:1)? If so, what are the choices for magnification? Does your software allow you to return to the original size in one step?

P15.10 Can you pan across an entire drawing or is only a small part of the drawing displayed?

P15.11 Can you delete items that you have created? Can you also delete an entire area? If so, how is the area defined?

P15.12 Does your software allow you to put a grid on the screen or a set of ruled lines as axes on the edge of the screen? How many different grid sizes and types (orthographic or pictorial) or drawing sizes are possible on your system? Is it possible to have the drawing "snap" to the grid or ruled lines?

P15.13 Does your system provide a continuous update of the current cursor coordinates? If it does not provide a continuous update, is it possible to query the system for the location?

P15.14 Will your system let you use temporary or construction lines (sometimes called templates or template lines) while creating a drawing?

P15.15 Is it possible on your system to create additional menus by creating additional graphical figures? Are there limits on the number of menus that you can use? Is there a library of menus that have already been created? Is it possible to create your own lettering fonts?

P15.16 Has the dimensioning process been automated in any way? If so, describe how the system helps you dimension an object?

P15.17 For additional exercise, use your computer graphics system to solve appropriate problems at the end of various chapters in this text. Have your instructor approve the problems you select.

Appendixes

TABLE A-1 Screw Threads,[a] American National Standard Unified and American National

Nominal diameter	*Coarse*[b] *NC UNC*		*Fine*[b] *NF UNF*		*Extra fine*[c] *NEF UNEF*		Nominal diameter	*Coarse*[b] *NC UNC*		*Fine*[b] *NF UNF*		*Extra fine*[c] *NEF UNEF*	
	Thds. per inch	*Tap drill*[d]	*Thds. per inch*	*Tap drill*[d]	*Thds. per inch*	*Tap drill*[d]		*Thds. per inch*	*Tap drill*[d]	*Thds. per inch*	*Tap drill*[d]	*Thds. per inch*	*Tap drill*[d]
0 (.060)			80	3/64			1	8	7/8	12	59/64	20	61/64
1 (.073)	64	No. 53	72	No. 53	....	....	1 1/16	....	....	....	....	18	1
2 (.086)	56	No. 50	64	No. 50	....	....	1 1/8	7	63/64	12	1 3/64	18	1 5/64
3 (.099)	48	No. 47	56	No. 45	....	....	1 3/16	....	....	....	....	18	1 9/64
4 (.112)	40	No. 43	48	No. 42	....	....	1 1/4	7	1 7/64	12	1 11/64	18	1 3/16
5 (.125)	40	No. 38	44	No. 37	....	....	1 5/16	....	....	....	....	18	1 17/64
6 (.138)	32	No. 36	40	No. 33	....	....	1 3/8	6	1 7/32	12	1 19/64	18	1 5/16
8 (.164)	32	No. 29	36	No. 29	....	....	1 7/16	....	....	....	....	18	1 3/8
10 (.190)	24	No. 25	32	No. 21	....	....	1 1/2	6	1 11/32	12	1 27/64	18	1 7/16
12 (.216)	24	No. 16	28	No. 14	32	No. 13	1 9/16	....	....	....	....	18	1 1/2
1/4	20	No. 7	28	No. 3	32	7/32	1 5/8	....	....	....	....	18	1 9/16
5/16	18	F	24	I	32	9/32	1 11/16	....	....	....	....	18	1 5/8
3/8	16	5/16	24	Q	32	11/32	1 3/4	5	1 9/16	....	....	....	....
7/16	14	U	20	25/64	28	13/32	2	4 1/2	1 25/32	....	....	....	....
1/2	13	27/64	20	29/64	28	15/32	2 1/4	4 1/2	2 1/32	....	....	....	....
9/16	12	31/64	18	33/64	24	33/64	2 1/2	4	2 1/4	....	....	....	....
5/8	11	17/32	18	37/64	24	37/64	2 3/4	4	2 1/2	....	....	....	....
11/16	....	....	....	....	24	41/64	3	4	2 3/4	....	....	....	....
3/4	10	21/32	16	11/16	20	45/64	3 1/4	4	....	....	....	....	....
13/16	....	....	....	....	20	49/64	3 1/2	4	....	....	....	....	....
7/8	9	49/64	14	13/16	20	53/64	3 3/4	4	....	....	....	....	....
15/16	....	....	....	....	20	57/64	4	4	....	....	....	....	....

Continued on page 554

[a]ANSI B1.1.

[b]Classes 1A, 2A, 3A, 1B, 2B, 3B, 2, and 3.

[c]Classes 2A, 2B, 2, and 3.

[d]For approximate 75% full depth of thread. For decimal sizes of numbered and lettered drills, see Table A-3.

TABLE A-1 (continued)

Nominal diameter	*8-pitch[b] series 8N and 8UN*		*12-pitch[b] series 12N and 12UN*		*16-pitch[b] series 16N and 16UN*		*Nominal diameter*	*8-pitch[b] series 8N and 8UN*		*12-pitch[b] series 12N and 12UN*		*16-pitch[b] series 16N and 16UN*	
	Thds. per inch	*Tap drill[c]*	*Thds. per inch*	*Tap drill[c]*	*Thds. per inch*	*Tap drill[c]*		*Thds. per inch*	*Tap drill[c]*	*Thds. per inch*	*Tap drill[c]*	*Thds. per inch[e]*	*Tap drill[c]*
1/2	....	....	12	27/64	....	....	2 1/16	....	....	....	....	**16**	2
9/16	....	....	12[d]	31/64	....	....	2 1/8	....	....	12	2 3/64	16	2 1/16
5/8	....	....	12	35/64	....	....	2 3/16	....	....	....	....	**16**	2 1/8
11/16	....	....	12	39/64	....	....	2 1/4	8	2 1/8	12	2 11/64	16	2 3/16
3/4	....	....	12	43/64	16[d]	11/16	2 5/16	....	....	....	....	**16**	2 1/4
13/16	....	....	12	47/64	16	3/4	2 3/8	....	....	12	2 19/64	16	2 5/16
7/8	....	....	12	51/64	16	13/16	2 7/16	....	....	....	....	**16**	2 3/8
5/16	....	....	12	55/64	16	7/8	2 1/2	8	2 3/8	12	2 27/64	16	2 7/16
1	8[d]	7/8	12	59/64	16	15/16	2 5/8	....	....	12	2 35/64	16	2 9/16
1 1/16	....	....	12	63/64	16	1	2 3/4	8	2 5/8	12	2 43/64	16	2 11/16
1 1/8	8	1	12[d]	1 3/64	16	1 1/16	2 7/8	....	....	12	....	16	....
1 3/16	....	....	12	1 7/64	16	1 1/8	3	8	2 7/8	12	....	16	....
1 1/4	8	1 1/8	12	1 11/64	16	1 3/16	3 1/8	....	....	12	....	16	....
1 5/16	....	....	12	1 15/64	16	1 1/4	3 1/4	8	....	12	....	16	....
1 3/8	8	1 1/4	12[d]	1 19/64	16	1 5/16	3 3/8	....	....	12	....	16	....
1 7/16	....	....	12	1 23/64	16	1 3/8	3 1/2	8	....	12	....	16	....
1 1/2	8	1 3/8	12[d]	1 27/64	16	1 7/16	3 5/8	....	....	12	....	16	....
1 9/16	....	....	....	....	16	1 1/2	3 3/4	8	....	12	....	16	....
1 5/8	8	1 1/2	12	1 35/64	16	1 9/16	3 7/8	....	....	12	....	16	....
1 11/16	....	....	....	....	16	1 5/8	4	8	....	12	....	16	....
1 3/4	8	1 5/8	12	1 43/64	16[d]	1 11/16	4 1/4	8	....	12	....	16	....
1 13/16	....	....	....	....	16	1 3/4	4 1/2	8	....	12	....	16	....
1 7/8	8	1 3/4	12	1 51/64	16	1 13/16	4 3/4	8	....	12	....	16	....
1 15/16	....	....	....	....	16	1 7/8	5	8	....	12	....	16	....
2	8	1 7/8	12	1 59/64	16[d]	1 15/16	5 1/4	8	....	12	....	16	....

[a]ANSI B1.1.

[b]Classes 2A, 3A, 2B, 3B, 2, and 3.

[c]For approximate 75% full depth of thread.

[d]This is a standard size of the Unified or American National threads of the coarse, fine, or extra fine series.

[e]Boldface type indicates American National threads only.

TABLE A-2 Screw Threads,[a] Metric

Coarse (general purpose)		*Fine*	
Nominal size & thd. pitch	***Tap drill diameter, mm***	***Nominal size & thd. pitch***	***Tap drill diameter, mm***
M1.6 × 0.35	1.25	—	—
M1.8 × 0.35	1.45	—	—
M2 × 0.4	**1.6**	—	—
M2.2 × 0.45	1.75	—	—
M2.5 × 0.45	**2.05**	—	—
M3 × 0.5	**2.50**	—	—
M3.5 × 0.6	**2.90**	—	—
M4 × 0.7	**3.30**	—	—
M4.5 × 0.75	3.75	—	—
M5 × 0.8	**4.20**	—	—
M6.3 × 1	**5.30**	—	—
M7 × 1	6.00	—	—
M8 × 1.25	**6.80**	**M8 × 1**	**7.00**
M9 × 1.25	7.75	—	—
M10 × 1.5	**8.50**	**M10 × 1.25**	**8.75**
M11 × 1.5	9.50	—	—
M12 × 1.75	**10.30**	**M12 × 1.25**	**10.50**
M14 × 2	**12.00**	**M14 × 1.5**	**12.50**
M16 × 2	**14.00**	**M16 × 1.5**	**14.50**
M18 × 2.5	15.50	M18 × 1.5	16.50
M20 × 2.5	**17.50**	**M20 × 1.5**	**18.50**
M22 × 2.5	19.50	M22 × 1.5	20.50
M24 × 3	**21.00**	**M24 × 2**	**22.00**
M27 × 3	24.00	M27 × 2	25.00
M30 × 3.5	**26.50**	**M30 × 2**	**28.00**
M33 × 3.5	29.50	M30 × 2	31.00
M36 × 4	**32.00**	**M36 × 3**	**33.00**
M39 × 4	35.00	M39 × 3	36.00
M42 × 4.5	**37.50**	**M42 × 3**	**39.00**
M45 × 4.5	40.50	M45 × 3	42.00
M48 × 5	**43.00**	**M48 × 3**	**45.00**
M52 × 5	47.00	M52 × 3	49.00
M56 × 5.5	**50.50**	**M56 × 4**	**52.00**
M60 × 5.5	54.50	M60 × 4	56.00
M64 × 6	**58.00**	**M64 × 4**	**60.00**
M68 × 6	62.00	M68 × 4	64.00
M72 × 6	**66.00**	—	—
M80 × 6	**74.00**	—	—
M90 × 6	**84.00**	—	—
M100 × 6	**94.00**	—	—

[a]Metric Fasteners Standard, IFI-500.

NOTE: Preferred sizes for commercial threads and fasteners are shown in **boldface** type.

TABLE A-3 Inch Twist Drill Sizes, Numbered and Lettered[a]

No.	*Size*	*No.*	*Size*	*No.*	*Size*	*Letter*	*Size*
80	0.0135	53	0.0595	26	0.1470	*A*	0.2340
79	0.0145	52	0.0635	25	0.1495	*B*	0.2380
78	0.0160	51	0.0670	24	0.1520	*C*	0.2420
77	0.0180	50	0.0700	23	0.1540	*D*	0.2460
76	0.0200	49	0.0730	22	0.1570	*E*	0.2500
75	0.0210	48	0.0760	21	0.1590	*F*	0.2570
74	0.0225	47	0.0785	20	0.1610	*G*	0.2610
73	0.0240	46	0.0810	19	0.1660	*H*	0.2660
72	0.0250	45	0.0820	18	0.1695	*I*	0.2720
71	0.0260	44	0.0860	17	0.1730	*J*	0.2770
70	0.0280	43	0.0890	16	0.1770	*K*	0.2810
69	0.0292	42	0.0935	15	0.1800	*L*	0.2900
68	0.0310	41	0.0960	14	0.1820	*M*	0.2950
67	0.0320	40	0.0980	13	0.1850	*N*	0.3020
66	0.0330	39	0.0995	12	0.1890	*O*	0.3160
65	0.0350	38	0.1015	11	0.1910	*P*	0.3230
64	0.0360	37	0.1040	10	0.1935	*Q*	0.3320
63	0.0370	36	0.1065	9	0.1960	*R*	0.3390
62	0.0380	35	0.1100	8	0.1990	*S*	0.3480
61	0.0390	34	0.1110	7	0.2010	*T*	0.3580
60	0.0400	33	0.1130	6	0.2040	*U*	0.3680
59	0.0410	32	0.1160	5	0.2055	*V*	0.3770
58	0.0420	31	0.1200	4	0.2090	*W*	0.3860
57	0.0430	30	0.1285	3	0.2130	*X*	0.3970
56	0.0465	29	0.1360	2	0.2210	*Y*	0.4040
55	0.0520	28	0.1405	1	0.2280	*Z*	0.4130
54	0.0550	27	0.1440				

[a]All dimensions are in inches.

NOTE: Drills designated in common fractions are available in diameters 1/64″ to 1 3/4″ in 1/64″ increments, 1 3/4″ to 2 1/4″ in 1/32″ increments, and 2 1/4″ to 3 1/2″ in 1/16″ increments.

TABLE A-4 Metric Twist Drill Sizes[a]

Drill diameter		*Drill diameter*		*Drill diameter*		*Drill diameter*		*Drill diameter*		*Drill diameter*	
mm	*in.*	*mm*	*in.*	*mm*	*in.*	*mm*	*in.*	*mm*	*in.*	*mm*	*in.*
.40	.0157	**1.20**	.0472	**3.20**	.1260	**7.50**	.2953	**19.00**	.7480	**48.00**	1.8898
.42	.0165	**1.25**	.0492	3.30	.1299	7.80	.3071	19.50	.7677	**50.00**	1.9685
.45	.0177	**1.30**	.0512	**3.40**	.1339	**8.00**	.3150	**20.00**	.7874	51.50	2.0276
.48	.0189	1.35	.0531	3.50	.1378	8.20	.3228	20.50	.8071	**53.00**	2.0866
.50	.0197	**1.40**	.0551	**3.60**	.1417	**8.50**	.3346	**21.00**	.8268	54.00	2.1260
.52	.0205	1.45	.0571	3.70	.1457	8.80	.3465	21.50	.8465	**56.00**	2.2047
.55	.0217	**1.50**	.0591	**3.80**	.1496	**9.00**	.3543	**22.00**	.8661	58.00	2.2835
.58	.0228	1.55	.0610	3.90	.1535	9.20	.3622	23.00	.9055	**60.00**	2.3622
.60	.0236	**1.60**	.0630	**4.00**	.1575	**9.50**	.3740	**24.00**	.9449		
.62	.0244	1.65	.0650	4.10	.1614	9.80	.3858	**25.00**	.9843		
.65	.0256	**1.70**	.0669	**4.20**	.1654	**10.00**	.3937	**26.00**	1.0236		
.68	.0268	1.75	.0689	4.40	.1732	10.30	.4055	27.00	1.0630		
.70	.0276	**1.80**	.0709	**4.50**	.1772	**10.50**	.4134	**28.00**	1.1024		
.72	.0283	1.85	.0728	4.60	.1811	10.80	.4252	29.00	1.1417		
.75	.0295	**1.90**	.0748	**4.80**	.1890	**11.00**	.4331	**30.00**	1.1811		
.78	.0307	1.95	.0768	**5.00**	.1969	11.50	.4528	31.00	1.2205		
.80	.0315	**2.00**	.0787	5.20	.0247	**12.00**	.4724	**32.00**	1.2598		
.82	.0323	2.05	.0807	**5.30**	.2087	**12.50**	.4921	33.00	1.2992		
.85	.0335	**2.10**	.0827	5.40	.2126	**13.00**	.5118	**34.00**	1.3386		
.88	.0346	2.15	.0846	**5.60**	.2205	13.50	.5315	35.00	1.3780		
.90	.0354	**2.20**	.0866	5.80	.2283	**14.00**	.5512	**36.00**	1.4173		
.92	.0362	2.30	.0906	**6.00**	.2362	14.50	.5709	37.00	1.4567		
.95	.0374	**2.40**	.0945	6.20	.2441	**15.00**	.5906	**38.00**	1.4961		
.98	.0386	**2.50**	.0984	**6.30**	.2480	15.50	.6102	39.00	1.5354		
1.00	.0394	**2.60**	.1024	6.50	.2559	**16.00**	.6299	**40.00**	1.5748		
1.03	.0406	2.70	.1063	**6.70**	.2638	16.50	.6496	41.00	1.6142		
1.05	.0413	**2.80**	.1102	6.80	.2677	**17.00**	.6693	**42.00**	1.6535		
1.08	.0425	2.90	.1142	6.90	.2717	17.50	.6890	43.50	1.7126		
1.10	.0433	**3.00**	.1181	**7.10**	.2795	**18.00**	.7087	**45.00**	1.7717		
1.15	.0453	3.10	.1220	7.30	.2874	18.50	.7283	46.50	1.8307		

[a]Compiled from manufacturer's catalogs.

NOTE: Preferred sizes are in **boldface** type. Decimal-inch equivalents are for reference only.

TABLE A-5 Screw Threads, Square and Acme

Size	*Threads per inch*	*Size*	*Threads per inch*	*Size*	*Threads per inch*	*Size*	*Threads per inch*
3/8	12	7/8	5	2	2½	3½	1⅓
7/16	10	1	5	2¼	2	3¾	1⅓
½	10	1⅛	4	2½	2	4	1⅓
9/16	8	1¼	4	2¾	2	4¼	1⅓
5/8	8	1½	3	3	1½	4½	1
¾	6	1¾	2½	3¼	1½	over 4½	1

TABLE A-6 Acme Threads, General Purpose

Size	*Threads per inch*	*Size*	*Threads per inch*	*Size*	*Threads per inch*	*Size*	*Threads per inch*
¼	16	¾	6	1½	4	3	2
5/16	14	7/8	6	1¾	4	3½	2
3/8	12	1	5	2	4	4	2
7/16	12	1⅛	5	2¼	3	4½	2
½	10	1¼	5	2½	3	5	2
5/8	8	1⅜	4	2¾	3		

TABLE A-7 Hex Bolts, American National Standard

Nominal size or basic product dia.	E Body dia.	F Width across flats			G Width across corners		H Height			R Radius of fillet		L_T Thread length for bolt lengths 6 in. and shorter	L_T Thread length for bolt lengths Over 6 in.
	Max.	Basic	Max.	Min.	Max.	Min.	Basic	Max.	Min.	Max.	Min.	Basic	Basic
1/4 0.2500	0.260	7/16	0.438	0.425	0.505	0.484	11/64	0.188	0.150	0.03	0.01	0.750	1.000
5/16 0.3125	0.324	1/2	0.500	0.484	0.577	0.552	7/32	0.235	0.195	0.03	0.01	0.875	1.125
3/8 0.3750	0.388	9/16	0.562	0.544	0.650	0.620	1/4	0.268	0.226	0.03	0.01	1.000	1.250
7/16 0.4375	0.452	5/8	0.625	0.603	0.722	0.687	19/64	0.316	0.272	0.03	0.01	1.125	1.375
1/2 0.5000	0.515	3/4	0.750	0.725	0.866	0.826	11/32	0.364	0.302	0.03	0.01	1.250	1.500
5/8 0.6250	0.642	15/16	0.938	0.906	1.083	1.033	27/64	0.444	0.378	0.06	0.02	1.500	1.750
3/4 0.7500	0.768	1 1/8	1.125	1.088	1.299	1.240	1/2	0.524	0.455	0.06	0.02	1.750	2.000
7/8 0.8750	0.895	1 5/16	1.312	1.269	1.516	1.447	37/64	0.604	0.531	0.06	0.02	2.000	2.250
1 1.0000	1.022	1 1/2	1.500	1.450	1.732	1.653	43/64	0.700	0.591	0.09	0.03	2.250	2.500
1 1/8 1.1250	1.149	1 11/16	1.688	1.631	1.949	1.859	3/4	0.780	0.658	0.09	0.03	2.500	2.750
1 1/4 1.2500	1.277	1 7/8	1.875	1.812	2.165	2.066	27/32	0.876	0.749	0.09	0.03	2.750	3.000
1 3/8 1.3750	1.404	2 1/16	2.062	1.994	2.382	2.273	29/32	0.940	0.810	0.09	0.03	3.000	3.250
1 1/2 1.5000	1.531	2 1/4	2.250	2.175	2.598	2.480	1	1.036	0.902	0.09	0.03	3.250	3.500
1 3/4 1.7500	1.785	2 5/8	2.625	2.538	3.031	2.893	1 5/32	1.196	1.054	0.12	0.04	3.750	4.000
2 2.0000	2.039	3	3.000	2.900	3.464	3.306	1 11/32	1.388	1.175	0.12	0.04	4.250	4.500
2 1/4 2.2500	2.305	3 3/8	3.375	3.262	3.897	3.719	1 1/2	1.548	1.327	0.19	0.06	4.750	5.000
2 1/2 2.5000	2.559	3 3/4	3.750	3.625	4.330	4.133	1 21/32	1.708	1.479	0.19	0.06	5.250	5.500
2 3/4 2.7500	2.827	4 1/8	4.125	3.988	4.763	4.546	1 13/16	1.869	1.632	0.19	0.06	5.750	6.000
3 3.0000	3.081	4 1/2	4.500	4.350	5.196	4.959	2	2.060	1.815	0.19	0.06	6.250	6.500
3 1/4 3.2500	3.335	4 7/8	4.875	4.712	5.629	5.372	2 3/16	2.251	1.936	0.19	0.06	6.750	7.000
3 1/2 3.5000	3.589	5 1/4	5.250	5.075	6.062	5.786	2 5/16	2.380	2.057	0.19	0.06	7.250	7.500
3 3/4 3.7500	3.858	5 5/8	5.625	5.437	6.495	6.198	2 1/2	2.572	2.241	0.19	0.06	7.750	8.000
4 4.0000	4.111	6	6.000	5.800	6.928	6.612	2 11/16	2.764	2.424	0.19	0.06	8.250	8.500

NOTE: For additional requirements refer to ANSI B18.2.1.

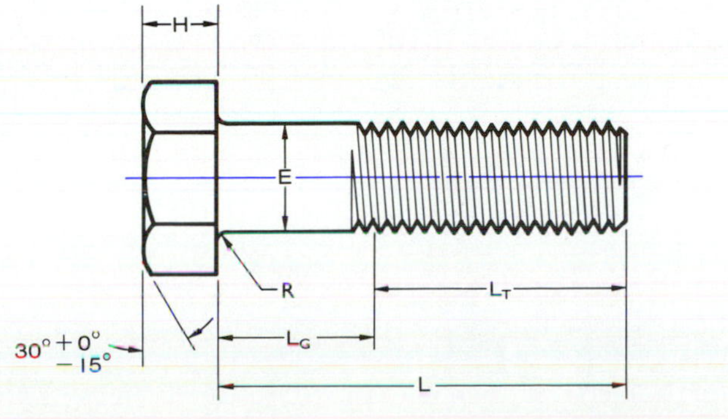

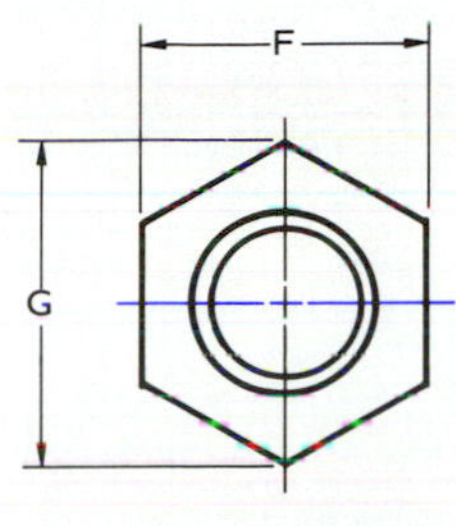

TABLE A-8 Square Bolts, American National Standard

Nominal size or basic product dia.		E	F			G		H			R		L_T	
		Body dia.	*Width across flats*			*Width across corners*		*Height*			*Radius of fillet*		*Thread length for bolt lengths*	
													6 in. and shorter	*Over 6 in.*
		Max.	*Basic*	*Max.*	*Min.*	*Max.*	*Min.*	*Basic*	*Max.*	*Min.*	*Max.*	*Min.*	*Basic*	*Basic*
1/4	0.2500	0.260	3/8	0.375	0.362	0.530	0.498	11/64	0.188	0.156	0.03	0.01	0.750	1.000
5/16	0.3125	0.324	1/2	0.500	0.484	0.707	0.665	13/64	0.220	0.186	0.03	0.01	0.875	1.125
3/8	0.3750	0.388	9/16	0.562	0.544	0.795	0.747	1/4	0.268	0.232	0.03	0.01	1.000	1.250
7/16	0.4375	0.452	5/8	0.625	0.603	0.884	0.828	19/64	0.316	0.278	0.03	0.01	1.125	1.375
1/2	0.5000	0.515	3/4	0.750	0.725	1.061	0.995	21/64	0.348	0.308	0.03	0.01	1.250	1.500
5/8	0.6250	0.642	15/16	0.938	0.906	1.326	1.244	27/64	0.444	0.400	0.06	0.02	1.500	1.750
3/4	0.7500	0.768	1 1/8	1.125	1.088	1.591	1.494	1/2	0.524	0.476	0.06	0.02	1.750	2.000
7/8	0.8750	0.895	1 5/16	1.312	1.269	1.856	1.742	19/32	0.620	0.568	0.06	0.02	2.000	2.250
1	1.0000	1.022	1 1/2	1.500	1.450	2.121	1.991	21/32	0.684	0.628	0.09	0.03	2.250	2.500
1 1/8	1.1250	1.149	1 11/16	1.688	1.631	2.386	2.239	3/4	0.780	0.720	0.09	0.03	2.500	2.750
1 1/4	1.2500	1.277	1 7/8	1.875	1.812	2.652	2.489	27/32	0.876	0.812	0.09	0.03	2.750	3.000
1 3/8	1.3750	1.404	2 1/16	2.062	1.994	2.917	2.738	29/32	0.940	0.872	0.09	0.03	3.000	3.250
1 1/2	1.5000	1.531	2 1/4	2.250	2.175	3.182	2.986	1	1.036	0.964	0.09	0.03	3.250	3.500

NOTE: For additional requirements refer to ANSI B18.2.1.

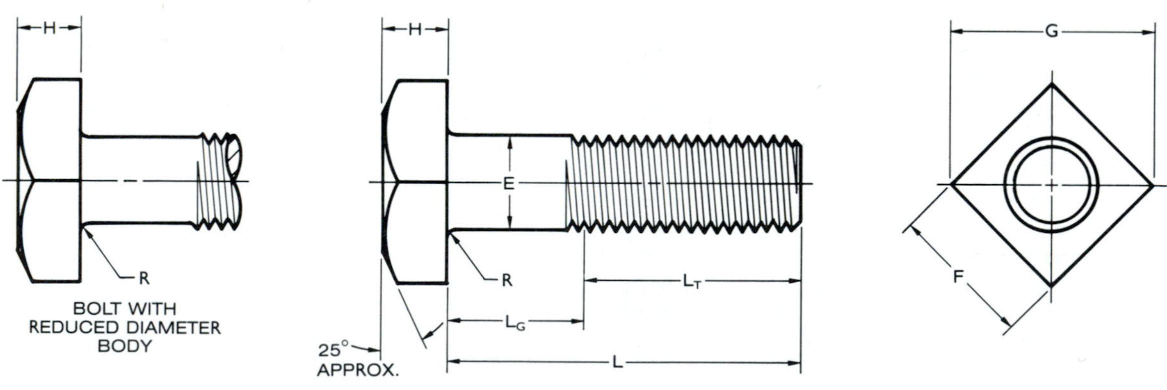

TABLE A-9 Hex Nuts, American National Standard

Nominal size or basic major dia. of thread		*F*			*G*		*H*			*H₁*		
		Width across flats			*Width across corners*		*Thickness hex nuts*			*Thickness hex jam nuts*		
		Basic	*Max.*	*Min.*	*Max.*	*Min.*	*Basic*	*Max.*	*Min.*	*Basic*	*Max.*	*Min.*
1/4	0.2500	7/16	0.438	0.428	0.505	0.488	7/32	0.226	0.212	5/32	0.163	0.150
5/16	0.3125	1/2	0.500	0.489	0.577	0.557	17/64	0.273	0.258	3/16	0.195	0.180
3/8	0.3750	9/16	0.562	0.551	0.650	0.628	21/64	0.337	0.320	7/32	0.227	0.210
7/16	0.4375	11/16	0.688	0.675	0.794	0.768	3/8	0.385	0.365	1/4	0.260	0.240
1/2	0.5000	3/4	0.750	0.736	0.866	0.840	7/16	0.448	0.427	5/16	0.323	0.302
9/16	0.5625	7/8	0.875	0.861	1.010	0.982	31/64	0.496	0.473	5/16	0.324	0.301
5/8	0.6250	15/16	0.938	0.922	1.083	1.051	35/64	0.559	0.535	3/8	0.387	0.363
3/4	0.7500	1 1/8	1.125	1.088	1.299	1.240	41/64	0.665	0.617	27/64	0.446	0.398
7/8	0.8750	1 5/16	1.312	1.269	1.516	1.447	3/4	0.776	0.724	31/64	0.510	0.458
1	1.0000	1 1/2	1.500	1.450	1.732	1.653	55/64	0.887	0.831	35/64	0.575	0.519
1 1/8	1.1250	1 11/16	1.688	1.631	1.949	1.859	31/32	0.999	0.939	39/64	0.639	0.579
1 1/4	1.2500	1 7/8	1.875	1.812	2.165	2.066	1 1/16	1.094	1.030	23/32	0.751	0.687
1 3/8	1.3750	2 1/16	2.062	1.994	2.382	2.273	1 11/64	1.206	1.138	25/32	0.815	0.747
1 1/2	1.5000	2 1/4	2.250	2.175	2.598	2.480	1 9/32	1.317	1.245	27/32	0.880	0.808

NOTE: For additional requirements refer to ANSI B18.2.2.

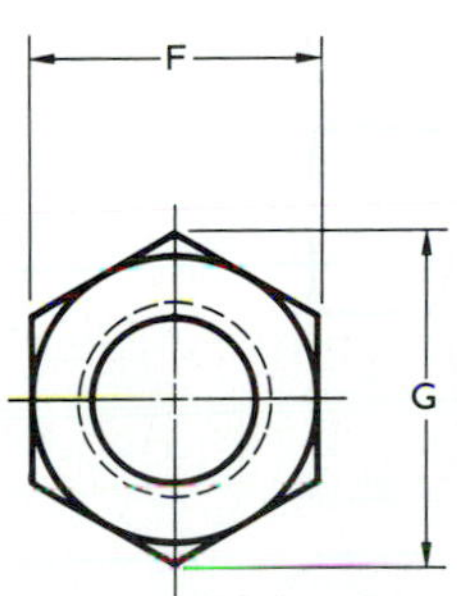

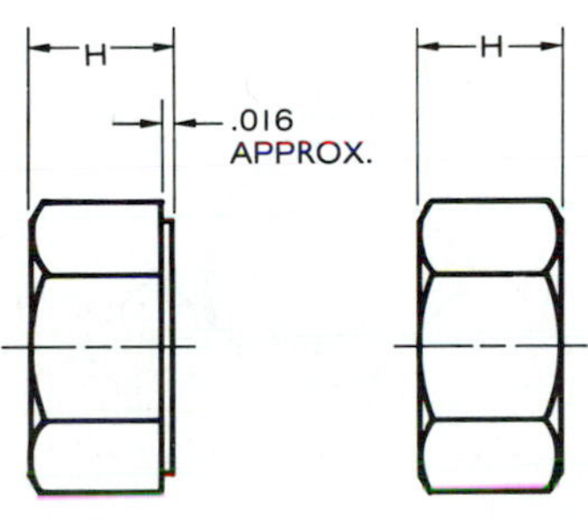

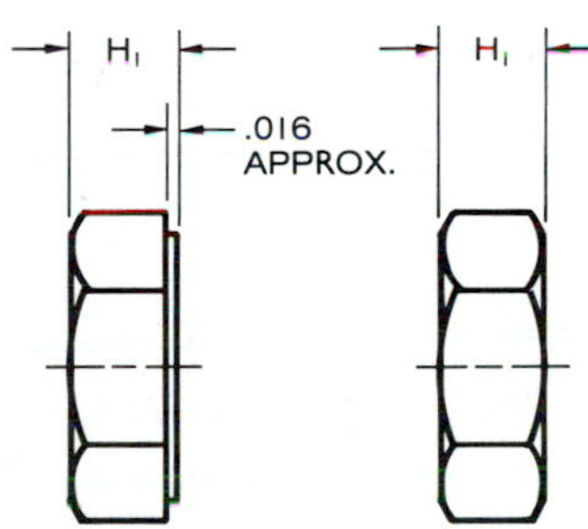

TABLE A-10 Square Nuts, American National Standard

Nominal size or basic major dia. of thread		*F*			*G*		*H*		
		Width across flats			*Width across corners*		*Thickness*		
		Basic	*Max.*	*Min.*	*Max.*	*Min.*	*Basic*	*Max.*	*Min.*
1/4	0.2500	7/16	0.438	0.425	0.619	0.584	7/32	0.235	0.203
5/16	0.3125	9/16	0.562	0.547	0.795	0.751	17/64	0.283	0.249
3/8	0.3750	5/8	0.625	0.606	0.884	0.832	21/64	0.346	0.310
7/16	0.4375	3/4	0.750	0.728	1.061	1.000	3/8	0.394	0.356
1/2	0.5000	13/16	0.812	0.788	1.149	1.082	7/16	0.458	0.418
5/8	0.6250	1	1.000	0.969	1.414	1.330	35/64	0.569	0.525
3/4	0.7500	1 1/8	1.125	1.088	1.591	1.494	21/32	0.680	0.632
7/8	0.8750	1 5/16	1.312	1.269	1.856	1.742	49/64	0.792	0.740
1	1.0000	1 1/2	1.500	1.450	2.121	1.991	7/8	0.903	0.847
1 1/8	1.1250	1 11/16	1.688	1.631	2.386	2.239	1	1.030	0.970
1 1/4	1.2500	1 7/8	1.875	1.812	2.652	2.489	1 3/32	1.126	1.062
1 3/8	1.3750	2 1/16	2.062	1.994	2.917	2.738	1 13/64	1.237	1.169
1 1/2	1.5000	2 1/4	2.250	2.175	3.182	2.986	1 5/16	1.348	1.276

NOTE: For additional requirements refer to ANSI B18.2.2.

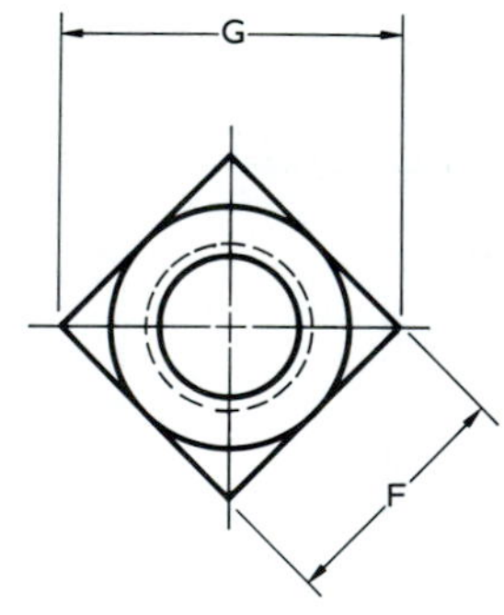

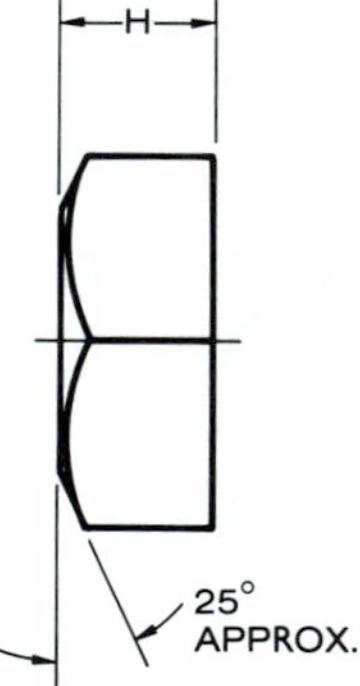

TABLE A-11 Square and Hex Machine Screw Nuts, American National Standard

Nominal size[a] or basic thread diameter		F			G		G1		H	
		Width across flats			Width across corners Square		Hex		Thickness	
		Basic	Max.	Min.	Max.	Min.	Max.	Min.	Max.	Min.
0	0.0600	5/32	0.156	0.150	0.221	0.206	0.180	0.171	0.050	0.043
1	0.0730	5/32	0.156	0.150	0.221	0.206	0.180	0.171	0.050	0.043
2	0.0860	3/16	0.188	0.180	0.265	0.247	0.217	0.205	0.066	0.057
3	0.0990	3/16	0.188	0.180	0.265	0.247	0.217	0.205	0.066	0.057
4	0.1120	1/4	0.250	0.241	0.354	0.331	0.289	0.275	0.098	0.087
5	0.1250	5/16	0.312	0.302	0.442	0.415	0.361	0.344	0.114	0.102
6	0.1380	5/16	0.312	0.302	0.442	0.415	0.361	0.344	0.114	0.102
8	0.1640	11/32	0.344	0.332	0.486	0.456	0.397	0.378	0.130	0.117
10	0.1900	3/8	0.375	0.362	0.530	0.497	0.433	0.413	0.130	0.117
12	0.2160	7/16	0.438	0.423	0.619	0.581	0.505	0.482	0.161	0.148
1/4	0.2500	7/16	0.438	0.423	0.619	0.581	0.505	0.482	0.193	0.178
5/16	0.3125	9/16	0.562	0.545	0.795	0.748	0.650	0.621	0.225	0.208
3/8	0.3750	5/8	0.625	0.607	0.884	0.833	0.722	0.692	0.257	0.239

[a]Where specifying nominal size in decimals, zeros preceding decimal and in the fourth decimal place shall be omitted.

[b]Square machine screw nuts shall have tops and bottoms flat, without chamfer. The bearing surface shall be perpendicular to the axis of the threaded hole pitch cylinder within a tolerance of 4 degrees.

[c]Hexagon machine screw nuts shall have tops flat with chamfered corners. Diameter of top circle shall be equal to the maximum width across flats within a tolerance of minus 15 percent. The bearing surface shall be perpendicular to the axis of the threaded hole pitch cylinder within a tolerance of 4 degrees.

[d]Bottoms of hexagon machine screw nuts are normally flat, but for special purposes may be chamfered, if so specified by purchaser.

NOTE: For additional requirements refer to ANSI B18.6.3.

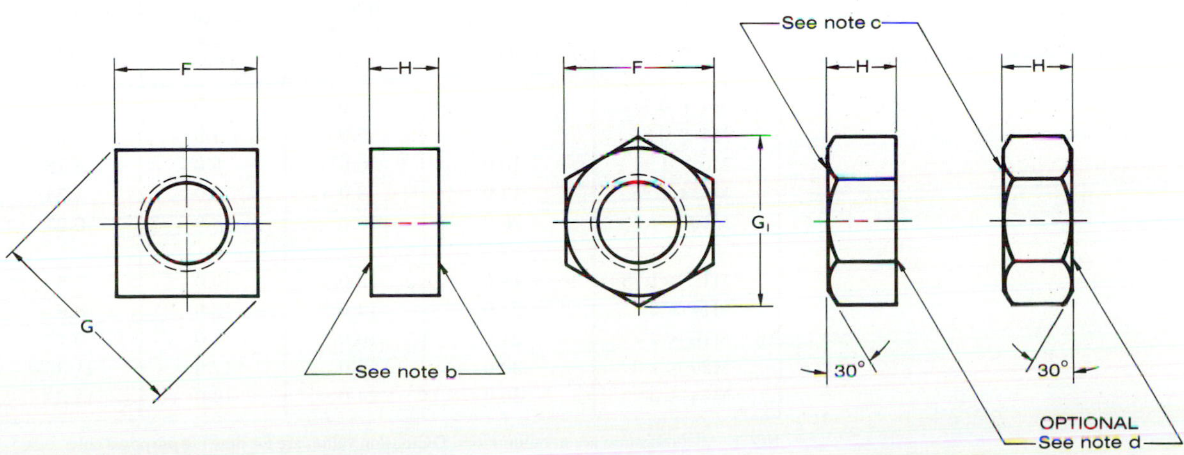

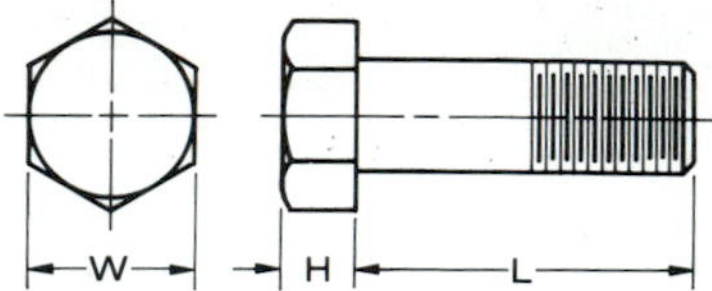

TABLE A-12 Metric Hex Bolts, Hex Cap Screws, and Socket Cap Screws

Nominal size and thread pitch	Hex bolts and hex cap screws		Bolts	Hex cap screws		
	W	A	H	H	B	T
	Width across flats max.	Width across corners max.	Head height max.	Head height max.	Washer face diameter min.	Washer face thickness max.
M5 × 0.8	8.00	9.24	3.98	3.65	7.0	0.5
M6 × 1	10.00	11.55	4.38	4.15	8.9	0.5
M8 × 1.25	13.00	15.01	5.68	5.50	11.6	0.6
M10 × 1.5	16.00	18.49	6.85	6.63	14.6	0.6
M12 × 1.75	18.00	20.78	7.95	7.76	16.6	0.6
M14 × 2	21.00	24.25	9.25	9.09	19.6	0.6
M16 × 2	24.00	27.71	10.75	10.32	22.5	0.8
M20 × 2.5	30.00	34.64	13.40	12.88	27.7	0.8
M24 × 3	36.00	41.57	15.90	15.44	33.2	0.8
M30 × 3.5	46.00	53.12	19.75	19.48	42.7	0.8
M36 × 4	55.00	63.51	23.55	23.38	51.1	0.8
M42 × 4.5	65.00	75.06	27.05	26.97	59.8	1.0
M48 × 5	75.00	86.60	31.07	31.07	69.0	1.0
M56 × 5.5	85.00	98.15	36.20	36.20	78.1	1.0
M64 × 6	95.00	109.70	41.32	41.32	87.2	1.0
M72 × 6	105.00	121.24	46.45	46.45	96.3	1.2

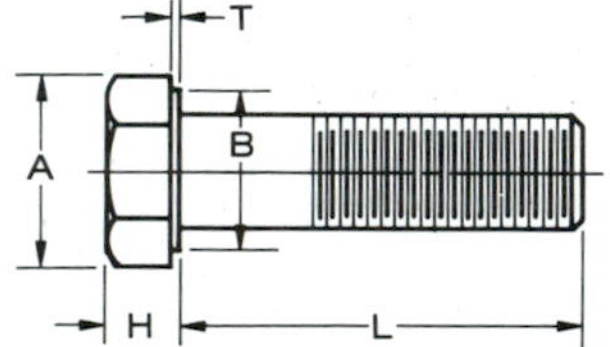

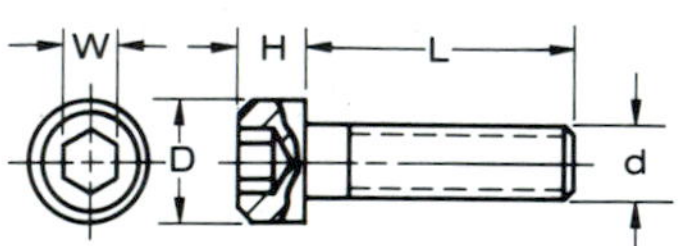

d Nominal screw diameter and thread pitch	Hex Socket Cap Screws			
	Head		Socket	
	D	H	W	S
	Head diameter max.	Head height max.	Hex socket size nom.	Hex socket depth
M4 × 0.7	7.0	4.0	3.0	2.4
M5 × 0.8	8.5	5.0	4.0	3.1
M6 × 1	10.0	6.0	5.0	3.78
M8 × 1.25	13.0	8.0	6.0	4.78
M10 × 1.5	16.0	10.0	8.0	6.25
M12 × 1.75	18.0	12.0	10.0	7.5
M14 × 2	21.0	14.0	12.0	8.6
M16 × 2	24.0	16.0	14.0	9.7
M20 × 2.5	30.0	20.0	17.0	11.8
M24 × 3	36.0	24.0	19.0	14.0

NOTE: All dimensions are in millimeters. Dimension values are for drawing purposes only.

TABLE A-13 Metric Hex Nuts (Style 1 and Style 2)

Nominal size and thread pitch	*W* Width across flats max.	*A* Width across corners max.	*D* Bearing face diameter min.	H_1 Nut thickness Style 1 max.	H_2 Nut thickness Style 2 max.	Washer face thickness max.
M1.6 × 0.35	3.20	3.70	2.3	1.3	—	0.4
M2 × 0.4	4.00	4.62	3.1	1.6	—	0.4
M2.5 × 0.45	5.00	5.77	4.1	2.0	—	0.4
M3 × 0.5	5.50	6.35	4.6	2.4	2.9	0.4
M3.5 × 0.6	6.00	6.93	5.1	2.8	3.3	0.4
M4 × 0.7	7.00	8.08	6.0	3.2	3.8	0.4
M5 × 0.8	8.00	9.24	7.0	4.7	5.1	0.5
M6 × 1	10.00	11.55	8.9	5.2	5.7	0.5
M8 × 1.25	13.00	15.01	11.6	6.8	7.5	0.6
M10 × 1.5	16.00	18.48	14.6	8.4	9.3	0.6
M12 × 1.75	18.00	20.78	16.6	10.8	12.0	0.6
M14 × 2	21.00	24.25	19.4	12.8	14.1	0.6
M16 × 2	24.00	27.71	22.4	14.8	16.4	0.8
M20 × 2.5	30.00	34.64	27.9	18.0	20.3	0.8
M24 × 3	36.00	41.57	32.5	21.5	23.9	0.8
M30 × 3.5	46.00	53.12	42.5	25.6	28.6	0.8
M36 × 4	55.00	63.51	50.8	31.0	34.7	0.8

Type 1: M1.6–M36; *Type 2: M3–M36*

NOTE: All dimensions are in millimeters. Dimension values are for drawing purposes only.

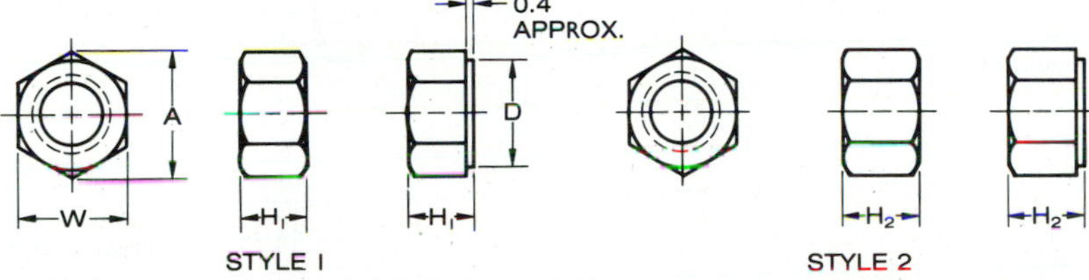

TABLE A-14 Cap Screws, Slotted[a] and Socket Head,[b] American National Standard

Nominal size D	*Flathead[a]*	*Roundhead[a]*		*Fillister head[a]*		*Socket head[b]*		
	A	*B*	*C*	*E*	*F*	*G*	*J*	*S*
0 (.060)	....	....	....	....	....	.096	.05	.054
1 (.073)	....	....	....	....	....	.118	1/16	.066
2 (.086)	....	....	....	....	....	.140	5/64	.077
3 (.099)	....	....	....	....	....	.161	5/64	.089
4 (.112)	....	....	....	....	....	.183	3/32	.101
5 (.125)	....	....	....	....	....	.205	3/32	.112
6 (.138)	....	....	....	....	....	.226	7/64	.124
8 (.164)	....	....	....	....	....	.270	9/64	.148
10 (.190)	....	....	....	....	....	5/16	5/32	.171
1/4	1/2	7/16	.191	3/8	11/64	3/8	3/16	.225
5/16	5/8	9/16	.245	7/16	13/64	15/32	1/4	.281
3/8	3/4	5/8	.273	9/16	1/4	9/16	5/16	.337
7/16	13/16	3/4	21/64	5/8	19/64	21/32	3/8	.394
1/2	7/8	13/16	.355	3/4	21/64	3/4	3/8	.450
9/16	1	15/16	.409	13/16	3/8	....	....	....
5/8	1 1/8	1	7/16	7/8	27/64	15/16	1/2	.562
3/4	1 3/8	1 1/4	35/64	1	1/2	1 1/8	5/8	.675
7/8	1 5/8	....	....	1 1/8	19/32	1 5/16	3/4	.787
1	1 7/8	....	....	1 5/16	21/32	1 1/2	3/4	.900
1 1/8	2 1/16	....	....	....	....	1 11/16	7/8	1.012
1 1/4	2 5/16	....	....	....	....	1 7/8	7/8	1.125
1 3/8	2 9/16	....	....	....	....	2 1/16	1	1.237
1 1/2	2 13/16	....	....	....	....	2 1/4	1	1.350

[a]ANSI B18.6.2.
[b]ANSI B18.3.

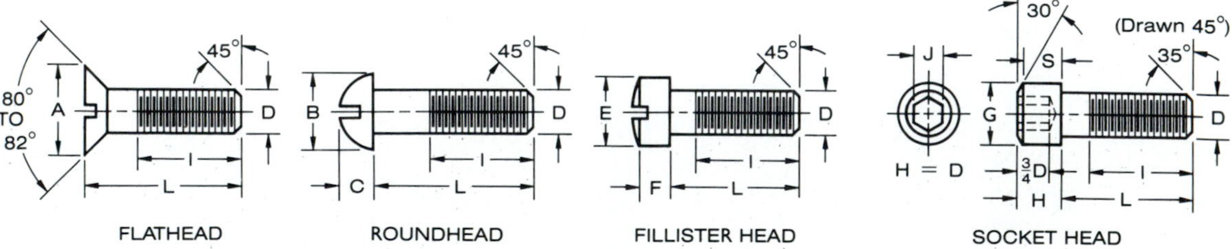

TABLE A-15 Machine Screws[a], American National Standard

Nominal dia.	Roundhead		Flathead	Fillister head			Oval head		Truss head	
	A	H	A	A	H	O	A	C	A	H
0	0.113	0.053	0.119	0.096	0.045	0.059	0.119	0.021		
1	0.138	0.061	0.146	0.118	0.053	0.071	0.146	0.025	0.194	0.053
2	0.162	0.069	0.172	0.140	0.062	0.083	0.172	0.029	0.226	0.061
3	0.187	0.078	0.199	0.161	0.070	0.095	0.199	0.033	0.257	0.069
4	0.211	0.086	0.225	0.183	0.079	0.107	0.225	0.037	0.289	0.078
5	0.236	0.095	0.252	0.205	0.088	0.120	0.252	0.041	0.321	0.086
6	0.260	0.103	0.279	0.226	0.096	0.132	0.279	0.045	0.352	0.094
8	0.309	0.120	0.332	0.270	0.113	0.156	0.332	0.052	0.384	0.102
10	0.359	0.137	0.385	0.313	0.130	0.180	0.385	0.060	0.448	0.118
12	0.408	0.153	0.438	0.357	0.148	0.205	0.438	0.068	0.511	0.134
1/4	0.472	0.175	0.507	0.414	0.170	0.237	0.507	0.079	0.573	0.150
5/16	0.590	0.216	0.635	0.518	0.211	0.295	0.635	0.099	0.698	0.183
3/8	0.708	0.256	0.762	0.622	0.253	0.355	0.762	0.117	0.823	0.215
7/16	0.750	0.328	0.812	0.625	0.265	0.368	0.812	0.122	0.948	0.248
1/2	0.813	0.355	0.875	0.750	0.297	0.412	0.875	0.131	1.073	0.280
9/16	0.938	0.410	1.000	0.812	0.336	0.466	1.000	0.150	1.198	0.312
5/8	1.000	0.438	1.125	0.875	0.375	0.521	1.125	0.169	1.323	0.345
3/4	1.250	0.547	1.375	1.000	0.441	0.612	1.375	0.206	1.573	0.410

[a]ANSI B18.6.3. Dimensions given are maximum values, all in inches.

Continued on page 568

NOTE: Thread length: screws 2 in. long or less, thread entire length; screws over 2 in. long, thread length l = 1¾ in. minimum. Threads are coarse or fine series, class 2. Heads may be slotted or recessed as specified, excepting hexagon form, which is plain or may be slotted if so specified. Slot and recess proportions vary with size of fastener; draw to look well.

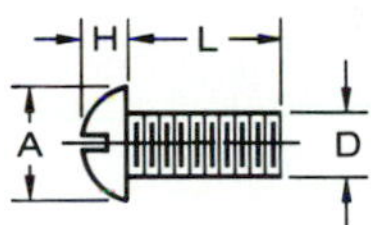

ROUNDHEAD

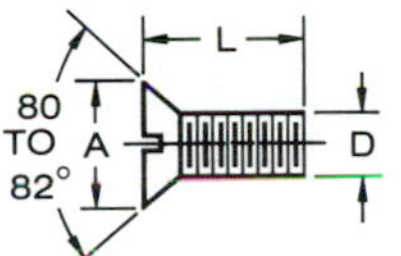

FLATHEAD

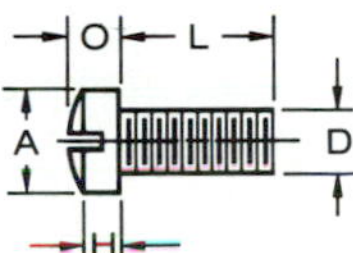

FILLISTER HEAD

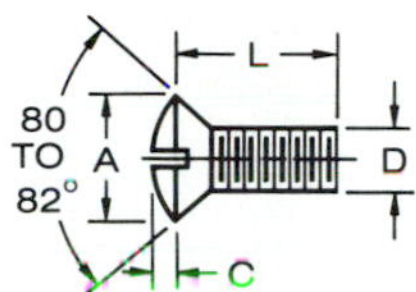

OVAL HEAD

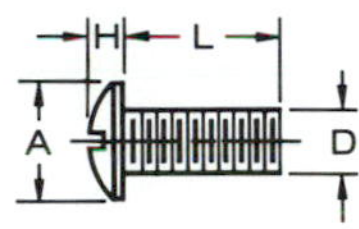

TRUSS HEAD

TABLE A-15 (continued)

Nominal dia.	*Binding head*				*Pan head*			*Hexagon head*		*100° flathead*
	A	***O***	***F***	***U***	***A***	***H***	***O***	***A***	***H***	***A***
2	0.181	0.046	0.018	0.141	0.167	0.053	0.062	0.125	0.050	
3	0.208	0.054	0.022	0.162	0.193	0.060	0.071	0.187	0.055	
4	0.235	0.063	0.025	0.184	0.219	0.068	0.080	0.187	0.060	0.225
5	0.263	0.071	0.029	0.205	0.245	0.075	0.089	0.187	0.070	
6	0.290	0.080	0.032	0.226	0.270	0.082	0.097	0.250	0.080	0.279
8	0.344	0.097	0.039	0.269	0.322	0.096	0.115	0.250	0.110	0.332
10	0.399	0.114	0.045	0.312	0.373	0.110	0.133	0.312	0.120	0.385
12	0.454	0.130	0.052	0.354	0.425	0.125	0.151	0.312	0.155	
¼	0.513	0.153	0.061	0.410	0.492	0.144	0.175	0.375	0.190	0.507
5/16	0.641	0.193	0.077	0.513	0.615	0.178	0.218	0.500	0.230	0.635
⅜	0.769	0.234	0.094	0.615	0.740	0.212	0.261	0.562	0.295	0.762

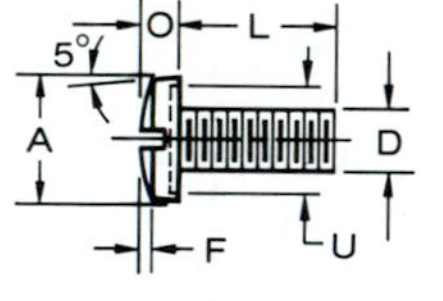

BINDING HEAD

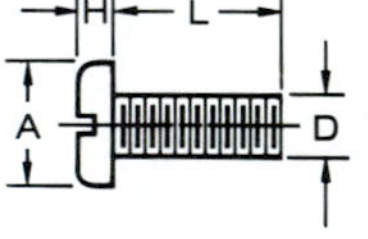

PAN HEAD

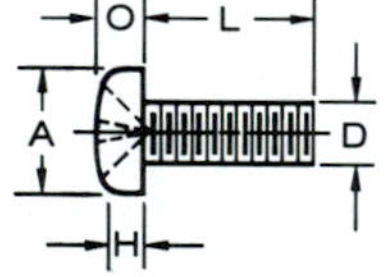

PAN HEAD (RECESSED)

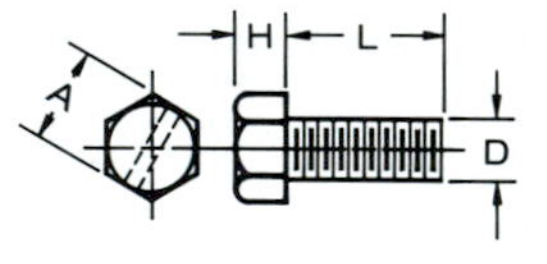

HEXAGON HEAD

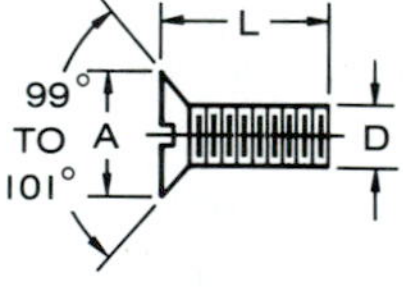

100° FLAT HEAD

TABLE A-16 Metric Machine Screws[a]

Nominal size & thd. pitch	*Max. dia. D, mm*	*Flatheads & oval head*			*Pan heads*		*Hex head*		*Slot width*
		C	*E*	*P*	*Q*	*S*	*T*	*U*	*J*
M2 × 0.4	2.00	3.60	1.20	3.90	1.35	1.60	3.20	1.27	0.7
M2.5 × 0.45	2.50	4.60	1.50	4.90	1.65	1.95	4.00	1.40	0.8
M3 × 0.5	3.00	5.50	1.80	5.80	1.90	2.30	5.00	1.52	1.0
M3.5 × 0.6	3.50	6.44	2.10	6.80	2.25	2.50	5.50	2.36	1.2
M4 × 0.7	4.00	7.44	2.32	7.80	2.55	2.80	7.00	2.79	1.4
M5 × 0.8	5.00	9.44	2.85	9.80	3.10	3.50	8.00	3.05	1.6
M6.3 × 1	6.30	11.87	3.60	12.00	3.90	4.30	10.00	4.83	1.9
M8 × 1.25	8.00	15.17	4.40	15.60	5.00	5.60	13.00	5.84	2.0
M10 × 1.5	10.00	18.98	5.35	19.50	6.20	7.00	15.00	7.49	2.5
M12 × 1.75	12.00	22.88	6.35	23.40	7.50	8.30	18.00	9.50	2.5

[a]Metric Fasteners Standard IFI-513.

NOTES:

Length of thread: On screws 36 mm long or shorter the threads extend to within one thread of the head; on longer screws the thread extends to within two threads of the head.

Points: Machine screws are regularly made with sheared ends, not chamfered.

Threads: Coarse (general-purpose) threads series are given.

Recessed heads: Two styles of cross-recesses are available on all screws except hexagon head.

TABLE A-17 Metric Machine Screw Lengths[a]

Nominal size	*Lengths*																					
	2.5	*3*	*4*	*5*	*6*	*8*	*10*	*13*	*16*	*20*	*25*	*30*	*35*	*40*	*45*	*50*	*55*	*60*	*65*	*70*	*80*	*90*
M2 × 0.4	PH[b]	A[c]	A	A	A	A	A	A	A	A												
M2.5 × 0.45		PH	A	A	A	A	A	A	A	A	A			Min. Thd. Length—28 mm								
M3 × 0.5			PH	A	A	A	A	A	A	A	A	A										
M3.5 × 0.6				PH	A	A	A	A	A	A	A	A	A			Min. Thd. Length—38 mm						
M4 × 0.7				PH	A	A	A	A	A	A	A	A	A	A								
M5 × 0.8					PH	A	A	A	A	A	A	A	A	A	A	A						
M6.3 × 1						A	A	A	A	A	A	A	A	A	A	A	A	A				
M8 × 1.25							A	A	A	A	A	A	A	A	A	A	A	A	A	A	A	
M10 × 1.5								A	A	A	A	A	A	A	A	A	A	A	A	A	A	A
M12 × 1.75									A	A	A	A	A	A	A	A	A	A	A	A	A	A

[a]Metric Fasteners Standard, IFI-513.

[b]PH = recommended lengths for only pan and hex head metric screws.

[c]A < recommended lengths for all metric screw head-styles.

TABLE A-18 Set Screws, Hexagon Socket,[a] Slotted Headless,[b] and Square Head,[c] American National Standard

Dia. *D*	Cup and flat-point dia. *C*	Oval-point radius *R*	Cone-point angle *Y*: 118° for these lengths and shorter	Cone-point angle *Y*: 90° for these lengths and longer	Full and half dog points: Dia. *P*	Full and half dog points: Length, Full *Q*	Full and half dog points: Length, Half *q*	Socket width *J*
5	1/16	3/32	1/8	3/16	0.083	0.06	0.03	1/16
6	0.069	7/64	1/8	3/16	0.092	0.07	0.03	1/16
8	5/64	1/8	3/16	1/4	0.109	0.08	0.04	5/64
10	3/32	9/64	3/16	1/4	0.127	0.09	0.04	3/32
12	7/64	5/32	3/16	1/4	0.144	0.11	0.06	3/32
1/4	1/8	3/16	1/4	5/16	5/32	1/8	1/16	1/8
5/16	11/64	15/64	5/16	3/8	13/64	5/32	5/64	5/32
3/8	13/64	9/32	3/8	7/16	1/4	3/16	3/32	3/16
7/16	15/64	21/64	7/16	1/2	19/64	7/32	7/64	7/32
1/2	9/32	3/8	1/2	9/16	11/32	1/4	1/8	1/4
9/16	5/16	27/64	9/16	5/8	25/64	9/32	9/64	1/4
5/8	23/64	15/32	5/8	3/4	15/32	5/16	5/32	5/16
3/4	7/16	9/16	3/4	7/8	9/16	3/8	3/16	3/8
7/8	33/64	21/32	7/8	1	21/32	7/16	7/32	1/2
1	19/32	3/4	1	1 1/8	3/4	1/2	1/4	9/16

[a]ANSI B18.3. Dimensions are in inches. Threads coarse or fine, class 3A. Length increments: 1/8 in. to 1/2 in. by (1/16 in.); 1/2 in. to 1 in. by (1/8 in.); 1 in. to 2 in. by (1/4 in.); 2 in. to 6 in. by (1/2 in.).

[b]ANSI B18.6.2. Threads coarse or fine, class 2A. Slotted headless screws standardized in sizes No. 0 to 3/4 in. only. Slot proportions vary with diameter. Draw to look well.

[c]ANSI B18.6.2. Threads coarse, fine, or 8-pitch, class 2A. Square head setscrews standardized in sizes No. 10 to 1 1/2 in. only.

NOTE: The full dog point is not available in hex socket.

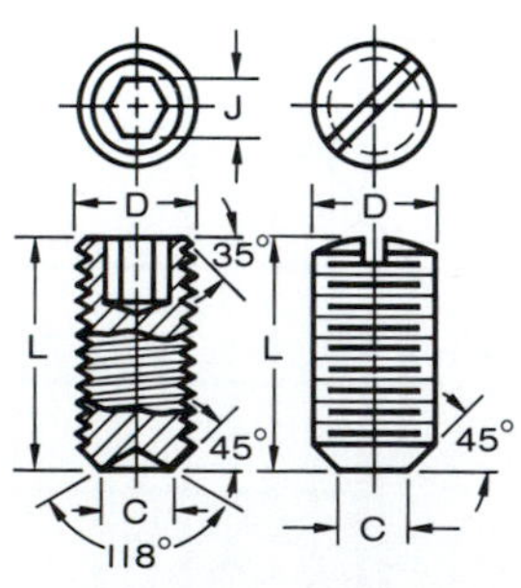

CUP POINT FLAT POINT

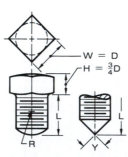

W = D
H = $\frac{3}{4}$D
L
L
R
Y

OVAL POINT CONE POINT

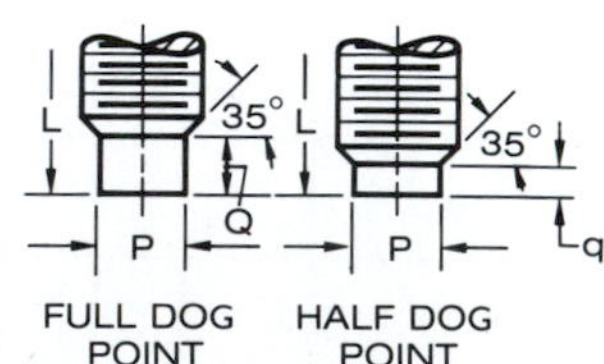

FULL DOG POINT HALF DOG POINT

(ALL SIX POINT TYPES ARE AVAILABLE IN ALL THREE HEAD TYPES)

TABLE A-19 Socket Head Shoulder Screws,[a] American National Standard

Shoulder diameter D			*Head*[b]			*Thread*		*Shoulder lengths*[d]
Nominal	*Max.*	*Min.*	*Dia.* *A*	*Height* *H*	*Hexagon*[c] *J*	*Specification* *E*	*Length* *I*	
¼	0.2480	0.2460	⅜	3⁄16	⅛	10-24NC-3	⅜	¾-2½
5⁄16	0.3105	0.3085	7⁄16	7⁄32	5⁄32	¼-20NC-3	7⁄16	1 -3
⅜	0.3730	0.3710	9⁄16	¼	3⁄16	5⁄16-18NC-3	½	1 -4
½	0.4980	0.4960	¾	5⁄16	¼	⅜-16NC-3	⅝	1¼-5
⅝	0.6230	0.6210	⅞	⅜	5⁄16	½-13NC-3	¾	1½-6
¾	0.7480	0.7460	1	½	⅜	⅝-11NC-3	⅞	1½-8
1	0.9980	0.9960	15⁄16	⅝	½	¾-10NC-3	1	1½-8
1¼	1.2480	1.2460	1¾	¾	⅝	⅞-9NC-3	1⅛	1½-8

[a]ANSI B18.3. Dimensions are in inches.

[b]Head chamfer is 30° to 45°.

[c]Socket depth = ⅔ *H* approx.

[d]Shoulder-length increments: shoulder lengths from ¼ in. to ¾ in., ⅛-in. intervals; shoulder lengths from ¾ in. to 5 in., ¼-in. intervals; shoulder lengths from 5 in. to 8 in., ½-in. intervals.

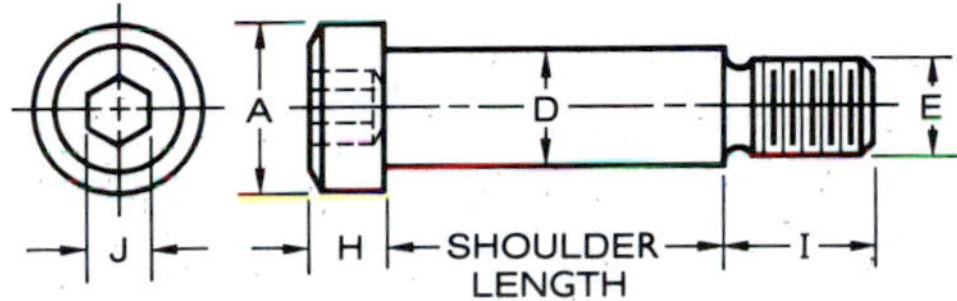

TABLE A-20 Shaft Center Sizes

Shaft diameter D	A	B	C	*Shaft diameter D*	A	B	C
3⁄16 to 7⁄32	5⁄64	3⁄64	1⁄16	1⅛ to 115⁄32	5⁄16	5⁄32	5⁄32
¼ to 11⁄32	3⁄32	3⁄64	1⁄16	1½ to 131⁄32	⅜	3⁄32	5⁄32
⅜ to 17⁄32	⅛	1⁄16	5⁄64	2 to 231⁄32	7⁄16	7⁄32	3⁄16
9⁄16 to 25⁄32	3⁄16	5⁄64	3⁄32	3 to 331⁄32	½	7⁄32	7⁄32
13⁄16 to 13⁄32	¼	3⁄32	3⁄32	4 and over	9⁄16	7⁄32	7⁄32

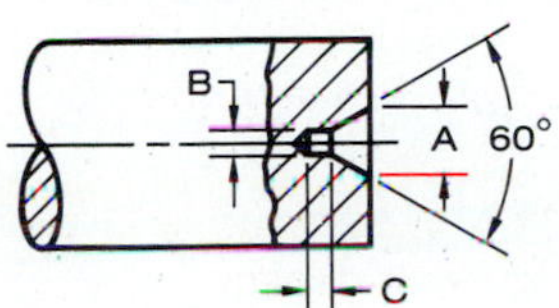

TABLE A-21 Keys—Square, Flat, Plain Taper,[a] and Gib Head

Shaft diameters	*Square stock key*	*Flat stock key*	*Gib head taper stock key*					
			Square			*Flat*		
			Height	*Length*	*Height to chamfer*	*Height*	*Length*	*Height to chamfer*
D	*W = H*	*W × H*	*C*	*F*	*E*	*C*	*F*	*E*
1/2 to 9/16	1/8	1/8 × 3/32	1/4	7/32	5/32	3/16	1/8	1/8
5/8 to 7/8	3/16	3/16 × 1/8	5/16	9/32	7/32	1/4	3/16	5/32
15/16 to 1 1/4	1/4	1/4 × 3/16	7/16	11/32	11/32	5/16	1/4	3/16
1 5/16 to 1 3/8	5/16	5/16 × 1/4	9/16	13/32	13/32	3/8	5/16	1/4
1 7/16 to 1 3/4	3/8	3/8 × 1/4	11/16	15/32	15/32	7/16	3/8	5/16
1 13/16 to 2 1/4	1/2	1/2 × 3/8	7/8	19/32	5/8	5/8	1/2	7/16
2 5/16 to 2 3/4	5/8	5/8 × 7/16	1 1/16	23/32	3/4	3/4	5/8	1/2
2 7/8 to 3 1/4	3/4	3/4 × 1/2	1 1/4	7/8	7/8	7/8	3/4	5/8
3 3/8 to 3 3/4	7/8	7/8 × 5/8	1 1/2	1	1	1 1/16	7/8	3/4
3 7/8 to 4 1/2	1	1 × 3/4	1 3/4	1 3/16	1 3/16	1 1/4	1	13/16
4 3/4 to 5 1/2	1 1/4	1 1/4 × 7/8	2	1 7/16	1 7/16	1 1/2	1 1/4	1
5 3/4 to 6	1 1/2	1 1/2 × 1	2 1/2	1 3/4	1 3/4	1 3/4	1 1/2	1 1/4

[a]Plain taper square and flat keys have the same dimensions as the plain parallel stock keys, with the addition of the taper on top. Gib head taper square and flat keys have the same dimensions as the plain taper keys, with the addition of the gib head.

NOTE: *Stock lengths for plain taper and gib head taper keys*: The minimum stock length equals 4W, and the maximum equals 16W. The increments of increase of length equal 2W.

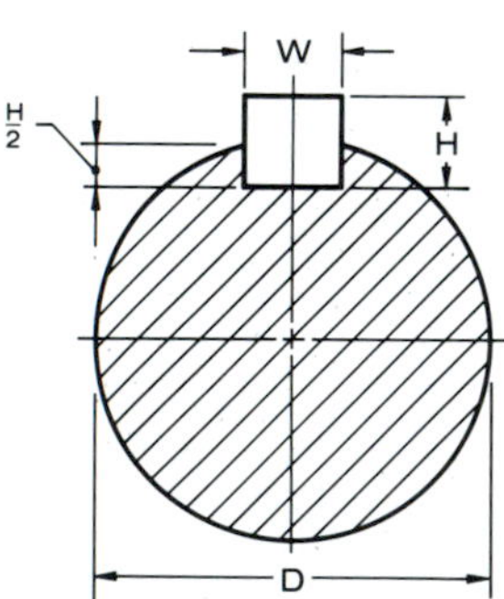

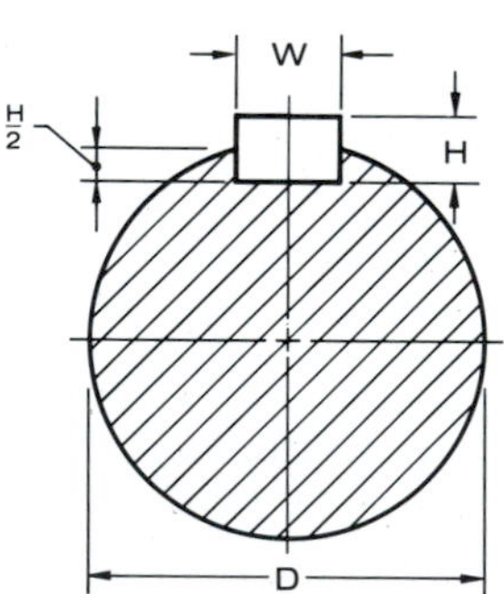

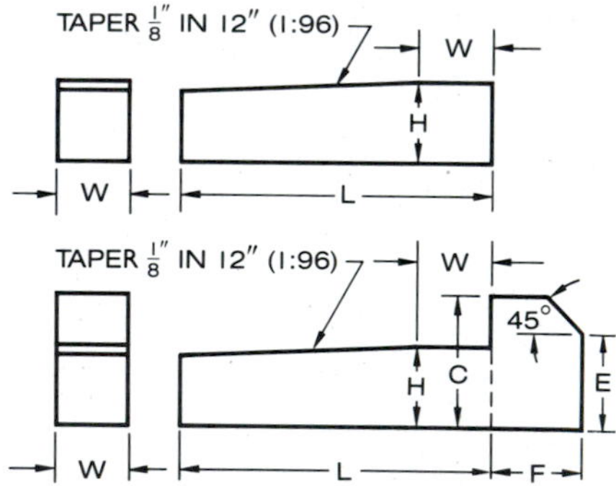

TABLE A-22 Inch Standard Small Rivets,[a] American National Standard

Nominal size or basic shank diameter		*Diameter of shank*	*Pan head*					*Button head*			*Countersunk head*		*Flathead*	
		D	***A***	***H***	R_1	R_2	R_3	***A***	***H***	***R***	***A***	***H***	***A***	***H***
1/16	0.062	0.064	0.118	0.040	0.019	0.052	0.217	0.122	0.052	0.055	0.118	0.027	0.140	0.027
3/32	0.094	0.096	0.173	0.060	0.030	0.080	0.326	0.182	0.077	0.084	0.176	0.040	0.200	0.038
1/8	0.125	0.127	0.225	0.078	0.039	0.106	0.429	0.235	0.100	0.111	0.235	0.053	0.260	0.048
5/32	0.156	0.158	0.279	0.096	0.049	0.133	0.535	0.290	0.124	0.138	0.293	0.066	0.323	0.059
3/16	0.188	0.191	0.334	0.114	0.059	0.159	0.641	0.348	0.147	0.166	0.351	0.079	0.387	0.069
7/32	0.219	0.222	0.391	0.133	0.069	0.186	0.754	0.405	0.172	0.195	0.413	0.094	0.453	0.080
1/4	0.250	0.253	0.444	0.151	0.079	0.213	0.858	0.460	0.196	0.221	0.469	0.106	0.515	0.091
9/32	0.281	0.285	0.499	0.170	0.088	0.239	0.963	0.518	0.220	0.249	0.528	0.119	0.579	0.103
5/16	0.313	0.317	0.552	0.187	0.098	0.266	1.070	0.572	0.243	0.276	0.588	0.133	0.641	0.113
11/32	0.344	0.348	0.608	0.206	0.108	0.292	1.176	0.630	0.267	0.304	0.646	0.146	0.705	0.124
3/8	0.375	0.380	0.663	0.225	0.118	0.319	1.286	0.684	0.291	0.332	0.704	0.159	0.769	0.135
13/32	0.406	0.411	0.719	0.243	0.127	0.345	1.392	0.743	0.316	0.358	0.763	0.172	0.834	0.146
7/16	0.438	0.443	0.772	0.261	0.137	0.372	1.500	0.798	0.339	0.387	0.823	0.186	0.896	0.157

[a]ANSI B18.1.1.

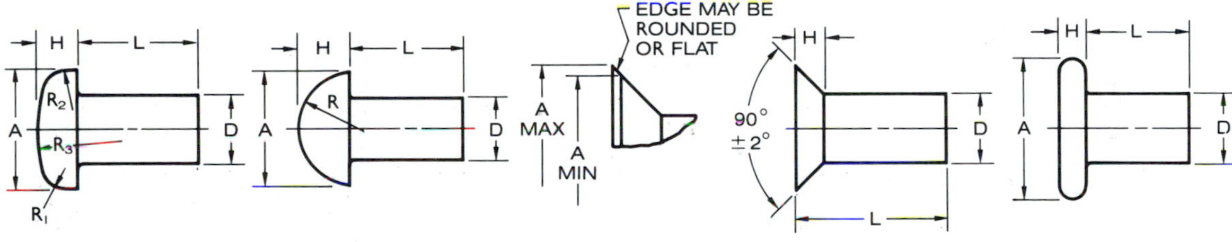

TABLE A-23 Inch Large Rivets,[a] American National Standard

Nominal size	*Button*			*High button*				*Pan*			*Cone*			*Flat*	
	A	*H*	*G*	*A*	*H*	*F*	*G*	*A*	*B*	*H*	*A*	*B*	*H*	*A*	*H*
½	0.875	0.406	0.443	0.781	0.500	0.656	0.094	0.800	0.500	0.381	0.875	0.469	0.469	0.936	0.260
⅝	1.094	0.500	0.553	0.969	0.594	0.750	0.188	1.000	0.625	0.469	1.094	0.586	0.578	1.194	0.339
¾	1.312	0.593	0.664	1.156	0.688	0.844	0.282	1.200	0.750	0.556	1.312	0.703	0.687	1.421	0.400
⅞	1.531	0.687	0.775	1.344	0.781	0.937	0.375	1.400	0.875	0.643	1.531	0.820	0.797	1.647	0.460
1	1.750	0.781	0.885	1.531	0.875	1.031	0.469	1.600	1.000	0.731	1.750	0.938	0.906	1.873	0.520

[a]ANSI B18.1.2.

NOTE: Large rivets are available in length increments of ⅛ inch.

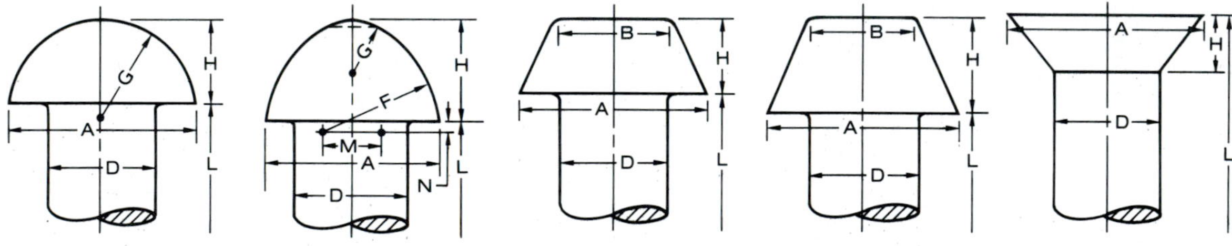

TABLE A-24 Plain Washers,[a] American National Standard

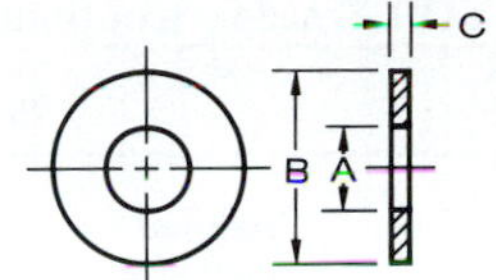

Preferred sizes of type-A plain washers[b]					
Nominal washer size[c]			*Inside diameter*	*Outside diameter*	*Nominal thickness*
			A	*B*	*C*
....	....		0.078	0.188	0.020
....	....		0.094	0.250	0.020
....	....		0.125	0.312	0.032
No. 6	0.138		0.156	0.375	0.049
No. 8	0.164		0.188	0.438	0.049
No. 10	0.190		0.219	0.500	0.049
3/16	0.188		0.250	0.562	0.049
No. 12	0.216		0.250	0.562	0.065
1/4	0.250	N	0.281	0.625	0.065
1/4	0.250	W	0.312	0.734	0.065
5/16	0.312	N	0.344	0.688	0.065
5/16	0.312	W	0.375	0.875	0.083
3/8	0.375	N	0.406	0.812	0.065
3/8	0.375	W	0.438	1.000	0.083
7/16	0.438	N	0.469	0.922	0.065
7/16	0.438	W	0.500	1.250	0.083
1/2	0.500	N	0.531	1.062	0.095
1/2	0.500	W	0.562	1.375	0.109
9/16	0.562	N	0.594	1.156	0.095
9/16	0.562	W	0.625	1.469	0.109
5/8	0.625	N	0.656	1.312	0.095
5/8	0.625	W	0.688	1.750	0.134
3/4	0.750	N	0.812	1.469	0.134
3/4	0.750	W	0.812	2.000	0.148
7/8	0.875	N	0.938	1.750	0.134
7/8	0.875	W	0.938	2.250	0.165
1	1.000	N	1.062	2.000	0.134
1	1.000	W	1.062	2.500	0.165
1 1/8	1.125	N	1.250	2.250	0.134
1 1/8	1.125	W	1.250	2.750	0.165
1 1/4	1.250	N	1.375	2.500	0.165
1 1/4	1.250	W	1.375	3.000	0.165
1 3/8	1.375	N	1.500	2.750	0.165
1 3/8	1.375	W	1.500	3.250	0.180
1 1/2	1.500	N	1.625	3.000	0.165

Continued on page 576

[a]ANSI B27.2. For complete listings, see the standard.

[b]Preferred sizes are for the most part from series previously designated "Standard Plate" and "SAE." Where common sizes existed in the two series, the SAE size is designated "N" (narrow) and the Standard Plate "W" (wide).

[c]Nominal washer sizes are intended for use with comparable nominal screw or bolt sizes.

TABLE A-24 (continued)

Preferred sizes of type-A plain washers[b]					
Nominal washer size[c]			*Inside diameter*	*Outside diameter*	*Nominal thickness*
			A	*B*	*C*
1½	1.500	W	1.625	3.500	0.180
1⅝	1.625		1.750	3.750	0.180
1¾	1.750		1.875	4.000	0.180
1⅞	1.875		2.000	4.250	0.180
2	2.000		2.125	4.500	0.180
2¼	2.250		2.375	4.750	0.220
2½	2.500		2.625	5.000	0.238
2¾	2.750		2.875	5.250	0.259
3	3.000		3.125	5.500	0.284

TABLE A-25 Lock Washers,[a] American National Standard

			Preferred series					
Nominal washer size[b]		*Inside diameter, min.*	*Regular*		*Extra duty*		*Hi-collar*	
			Outside diameter, max.	*Thickness, min.*	*Outside diameter, max.*	*Thickness, min.*	*Outside diameter, max.*	*Thickness, min.*
No. 2	0.086	0.088	0.172	0.020	0.208	0.027	. . . .	. . . .
No. 3	0.099	0.101	0.195	0.025	0.239	0.034	. . . .	. . . .
No. 4	0.112	0.115	0.209	0.025	0.253	0.034	0.173	0.022
No. 5	0.125	0.128	0.236	0.031	0.300	0.045	0.202	0.030
No. 6	0.138	0.141	0.250	0.031	0.314	0.045	0.216	0.030
No. 8	0.164	0.168	0.293	0.040	0.375	0.057	0.267	0.047
No. 10	0.190	0.194	0.334	0.047	0.434	0.068	0.294	0.047
No. 12	0.216	0.221	0.377	0.056	0.497	0.080	. . . .	. . . .
1/4	0.250	0.255	0.489	0.062	0.535	0.084	0.365	0.078
5/16	0.312	0.318	0.586	0.078	0.622	0.108	0.460	0.093
3/8	0.375	0.382	0.683	0.094	0.741	0.123	0.553	0.125
7/16	0.438	0.446	0.779	0.109	0.839	0.143	0.647	0.140
1/2	0.500	0.509	0.873	0.125	0.939	0.162	0.737	0.172
9/16	0.562	0.572	0.971	0.141	1.041	0.182	. . . .	. . . .
5/8	0.625	0.636	1.079	0.156	1.157	0.202	0.923	0.203
11/16	0.688	0.700	1.176	0.172	1.258	0.221	. . . .	. . . .
3/4	0.750	0.763	1.271	0.188	1.361	0.241	1.111	0.218
13/16	0.812	0.826	1.367	0.203	1.463	0.261	. . . .	. . . .
7/8	0.875	0.890	1.464	0.219	1.576	0.285	1.296	0.234
15/16	0.938	0.954	1.560	0.234	1.688	0.308	. . . .	. . . .
1	1.000	1.017	1.661	0.250	1.799	0.330	1.483	0.250
1 1/16	1.062	1.080	1.756	0.266	1.910	0.352	. . . .	. . . .
1 1/8	1.125	1.144	1.853	0.281	2.019	0.375	1.669	0.313
1 3/16	1.188	1.208	1.950	0.297	2.124	0.396	. . . .	. . . .
1 1/4	1.250	1.271	2.045	0.312	2.231	0.417	1.799	0.313
1 5/16	1.312	1.334	2.141	0.328	2.335	0.438	. . . .	. . . .
1 3/8	1.375	1.398	2.239	0.344	2.439	0.458	2.041	0.375
1 7/16	1.438	1.462	2.334	0.359	2.540	0.478	. . . .	. . . .
1 1/2	1.500	1.525	2.430	0.375	2.638	0.496	2.170	0.375

[a]ANSI B27.1.

[b]Nominal washer sizes are intended for use with comparable nominal screw or bolt sizes.

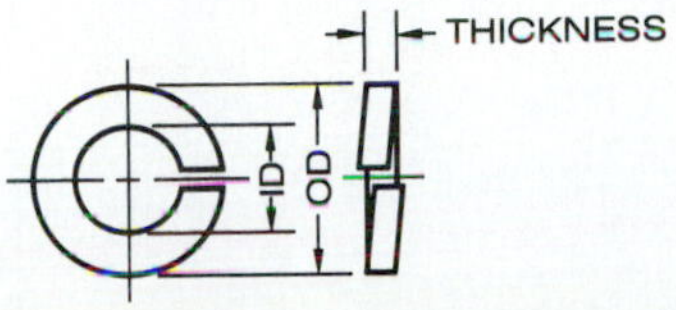

TABLE A-26 Cotter Pins,[a] American National Standard

Diameter, nominal	*Diameter A*		*Outside eye diameter B, min.*	*Hole sizes recom-mended*	*Diameter, nominal*	*Diameter A*		*Outside eye diameter B, min.*	*Hole sizes recom-mended*
	Max.	*Min.*				*Max.*	*Min.*		
0.031	0.032	0.028	1/16	3/64	0.188	0.176	0.172	3/8	13/64
0.047	0.048	0.044	3/32	1/16	0.219	0.207	0.202	7/16	15/64
0.062	0.060	0.056	1/8	5/64	0.250	0.225	0.220	1/2	17/64
0.078	0.076	0.072	5/32	3/32	0.312	0.280	0.275	5/8	5/16
0.094	0.090	0.086	3/16	7/64	0.375	0.335	0.329	3/4	3/8
0.109	0.104	0.100	7/32	1/8	0.438	0.406	0.400	7/8	7/16
0.125	0.120	0.116	1/4	9/64	0.500	0.473	0.467	1	1/2
0.141	0.134	0.130	9/32	5/32	0.625	0.598	0.590	1 1/4	5/8
0.156	0.150	0.146	5/16	11/64	0.750	0.723	0.715	1 1/2	3/4

[a]ANSI B5.20.

NOTE: All dimensions are in inches.

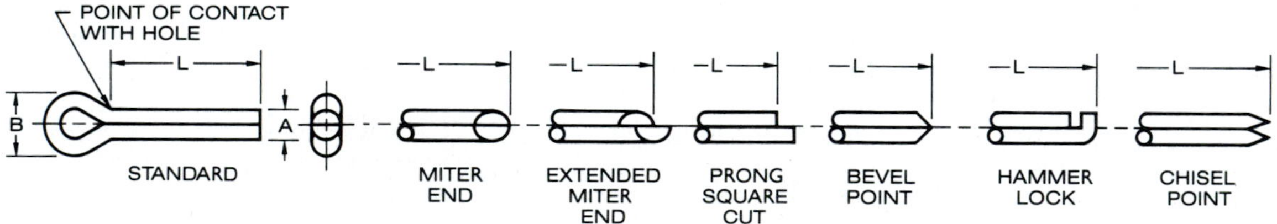

TABLE A-27 Taper Pins,[a] American National Standard

Number	7/0	6/0	5/0	4/0	3/0	2/0	0	1	2	3	4	5	6	7	8
Size (large end)	.0625	.0780	.0940	.1090	.1250	.1410	.1560	.1720	.1930	.2190	.2500	.2890	.3410	.4090	.4920
Shaft diameter (approx.)[b]		7/32	1/4	5/16	3/8	7/16	1/2	9/16	5/8	3/4	13/16	7/8	1	1 1/4	1 1/2
Drill size (before reamer)[b]		.0595	.0785	.0935	.104	.120	.1405	.1495	.166	.189	.213	1/4	9/32	11/32	13/32
Length, L															
.375	X	X													
.500	X	X	X	X	X	X	X								
.625	X	X	X	X	X	X	X								
.750	....	X	X	X	X	X	X	X	X	X					
.875	....	....	....	....	X	X	X	X	X	X					
1.000	....	....	X	X	X	X	X	X	X	X	X	X			
1.250	....	....	....	....	....	X	X	X	X	X	X	X	X		
1.500	....	....	....	....	....	....	X	X	X	X	X	X	X		
1.750	....	....	....	....	....	....	....	X	X	X	X	X	X		
2.000	....	....	....	....	....	....	....	X	X	X	X	X	X	X	X
2.250	....	....	....	....	....	....	....	....	X	X	X	X	X	X	X
2.500	....	....	....	....	....	....	....	....	X	X	X	X	X	X	X
2.750	....	....	....	....	....	....	....	....	....	X	X	X	X	X	X
3.000	....	....	....	....	....	....	....	....	....	X	X	X	X	X	X
3.250	....	....	....	....	....	....	....	....	....	....	....	....	X	X	X
3.500	....	....	....	....	....	....	....	....	....	....	....	....	X	X	X
3.750	....	....	....	....	....	....	....	....	....	....	....	....	X	X	X
4.000	....	....	....	....	....	....	....	....	....	....	....	....	X	X	X
4.250	....	....	....	....	....	....	....	....	....	....	....	....	....	....	X
4.500	....	....	....	....	....	....	....	....	....	....	....	....	....	....	X

[a]ANSI B5.20.

[b]Suggested sizes; not American National Standard.

NOTE: All dimensions are in inches. Standard reamers are available for pins given above the shaded area. To find small diameter of pin, multiply the length by .02083 and subtract the result from the larger diameter.

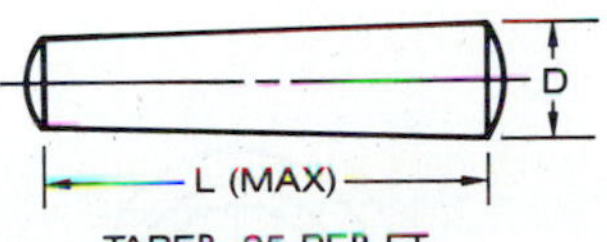

TABLE A-28 Straight Pins,[a] American National Standard

Nominal diameter	*Diameter A*		*Chamfer B*
	Max.	*Min.*	
0.062	0.0625	0.0605	0.015
0.094	0.0937	0.0917	0.015
0.109	0.1094	0.1074	0.015
0.125	0.1250	0.1230	0.015
0.156	0.1562	0.1542	0.015
0.188	0.1875	0.1855	0.015
0.219	0.2187	0.2167	0.015
0.250	0.2500	0.2480	0.015
0.312	0.3125	0.3095	0.030
0.375	0.3750	0.3720	0.030
0.438	0.4375	0.4345	0.030
0.500	0.500	0.4970	0.030

[a]ANSI B5.20.

NOTE: All dimensions are in inches. These pins must be straight and free from burrs or any other defects that will affect their serviceability.

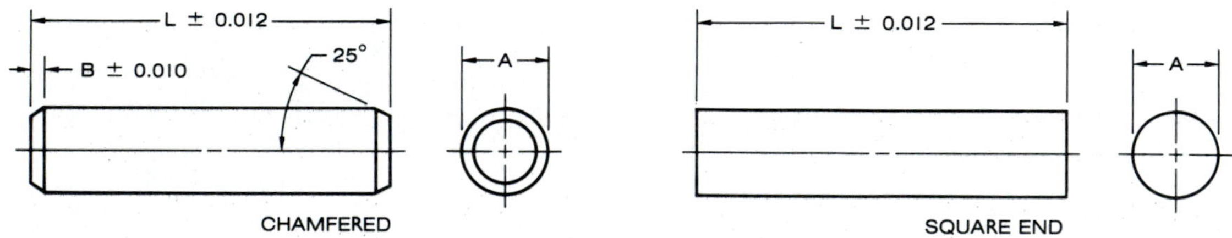

Table A-28

Table A-29

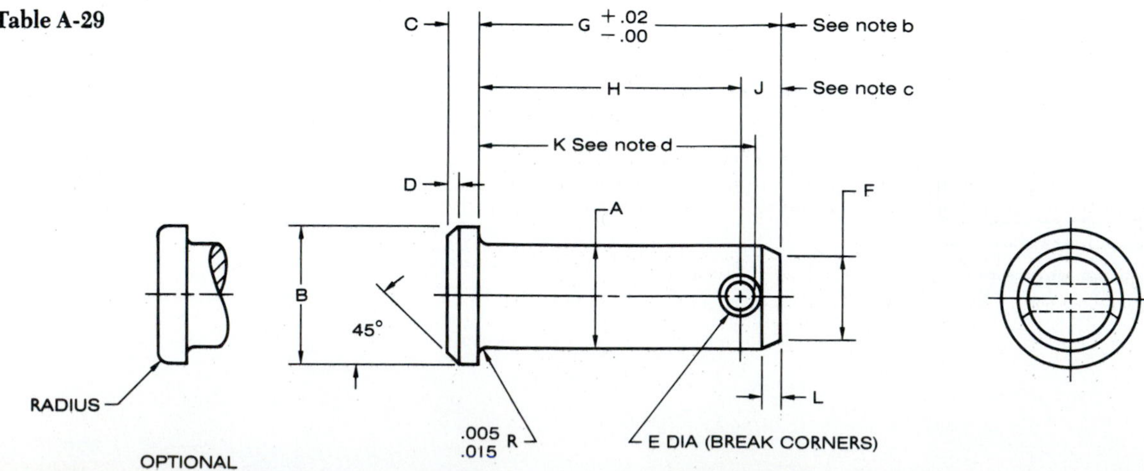

TABLE A-29 Clevis Pins,[a] American National Standard (see art at bottom of page 580)

Nominal size[b] or basic pin diameter		A Shank diameter		B Head diameter		C Head height		D Head chamfer	E Hole diameter		F Point diameter		G[c] Pin length
		Max.	Min.	Max.	Min.	Max.	Min.	±0.01	Max.	Min.	Max.	Min.	Basic
3/16	0.188	0.186	0.181	0.32	0.30	0.07	0.05	0.02	0.088	0.073	0.15	0.14	0.58
1/4	0.250	0.248	0.243	0.38	0.36	0.10	0.08	0.03	0.088	0.073	0.21	0.20	0.77
5/16	0.312	0.311	0.306	0.44	0.42	0.10	0.08	0.03	0.119	0.104	0.26	0.25	0.94
3/8	0.375	0.373	0.368	0.51	0.49	0.13	0.11	0.03	0.119	0.104	0.33	0.32	1.06
7/16	0.438	0.436	0.431	0.57	0.55	0.16	0.14	0.04	0.119	0.104	0.39	0.38	1.19
1/2	0.500	0.496	0.491	0.63	0.61	0.16	0.14	0.04	0.151	0.136	0.44	0.43	1.36
5/8	0.625	0.621	0.616	0.82	0.80	0.21	0.19	0.06	0.151	0.136	0.56	0.55	1.61
3/4	0.750	0.746	0.741	0.94	0.92	0.26	0.24	0.07	0.182	0.167	0.68	0.67	1.91
7/8	0.875	0.871	0.866	1.04	1.02	0.32	0.30	0.09	0.182	0.167	0.80	0.79	2.16
1	1.000	0.996	0.991	1.19	1.17	0.35	0.33	0.10	0.182	0.167	0.93	0.92	2.41

Nominal size[b] or basic pin diameter		H Head to center of hole		J[d] End to center ref.	K[e] Head to edge of hole ref.		L Point length		Recommended cotter pin nominal size	
		Max.	Min.	Basic	Max.	Min.	Max.	Min.		
3/16	0.188	0.504	0.484	0.09	0.548	0.520	0.055	0.035	1/16	0.062
1/4	0.250	0.692	0.672	0.09	0.736	0.708	0.055	0.035	1/16	0.062
5/16	0.312	0.832	0.812	0.12	0.892	0.864	0.071	0.049	3/32	0.093
3/8	0.375	0.958	0.938	0.12	1.018	0.990	0.071	0.049	3/32	0.093
7/16	0.438	1.082	1.062	0.12	1.142	1.114	0.071	0.049	3/32	0.093
1/2	0.500	1.223	1.203	0.15	1.298	1.271	0.089	0.063	1/8	0.125
5/8	0.625	1.473	1.453	0.15	1.548	1.521	0.089	0.063	1/8	0.125
3/4	0.750	1.739	1.719	0.18	1.830	1.802	0.110	0.076	5/32	0.156
7/8	0.875	1.989	1.969	0.18	2.080	2.052	0.110	0.076	5/32	0.156
1	1.000	2.239	2.219	0.18	2.330	2.302	0.110	0.076	5/32	0.156

[a]ANSI B18.8.1.

[b]Where specifying nominal size in decimals, zeros preceding decimal shall be omitted.

[c]Lengths tabulated are intended for use with standard clevises, without spacers. When required, it is recommended that other pin lengths be limited wherever possible to nominal lengths in 0.06 in. increments.

[d]Basic "J" dimension (distance from center line of hole to end of pin) is specified for calculating hole location from underside of head on pins of lengths not tabulated.

[e]Reference dimension provided for convenience in design layout and is not subject to inspection.

TABLE A-30 Wire Gage Standards

No. of wire	*American or Brown & Sharpe for non-ferrous metals*	*Birmingham, or Stubs' iron wire*[a]	*American S. & W. Co.'s (Washburn & Moen) std. steel wire*	*American S. & W. Co.'s music wire*	*Imperial wire*	*Stubs' steel wire*[a]	*Steel Manufacturers' sheet gage*[b]	No. of wire
7–0's	.651354	....	.4900	....	.500	....	.....	7–0's
6–0's	.580049	....	.4615	.004	.464	....	.....	6–0's
5–0's	.516549	.500	.4305	.005	.432	....	.....	5–0's
4–0's	.460	.454	.3938	.006	.400	....	.....	4–0's
000	.40964	.425	.3625	.007	.372	....	.....	000
00	.3648	.380	.3310	.008	.348	....	.....	00
0	.32486	.340	.3065	.009	.324	....	.....	0
1	.2893	.300	.2830	.010	.300	.227	.....	1
2	.25763	.284	.2625	.011	.276	.219	.....	2
3	.22942	.259	.2437	.012	.252	.212	.2391	3
4	.20431	.238	.2253	.013	.232	.207	.2242	4
5	.18194	.220	.2070	.014	.212	.204	.2092	5
6	.16202	.203	.1920	.016	.192	.201	.1943	6
7	.14428	.180	.1770	.018	.176	.199	.1793	7
8	.12849	.165	.1620	.020	.160	.197	.1644	8
9	.11443	.148	.1483	.022	.144	.194	.1495	9
10	.10189	.134	.1350	.024	.128	.191	.1345	10
11	.090742	.120	.1205	.026	.116	.188	.1196	11
12	.080808	.109	.1055	.029	.104	.185	.1046	12
13	.071961	.095	.0915	.031	.092	.182	.0897	13
14	.064084	.083	.0800	.033	.080	.180	.0747	14
15	.057068	.072	.0720	.035	.072	.178	.0763	15
16	.05082	.065	.0625	.037	.064	.175	.0598	16
17	.045257	.058	.0540	.039	.056	.172	.0538	17
18	.040303	.049	.0475	.041	.048	.168	.0478	18

[a]The difference between the Stubs' Iron Wire Gage and the Stubs' Steel Wire Gage should be noted, the first being commonly known as the English Standard Wire, or Birmingham Gage, which designates the Stubs' soft wire sizes, and the second being used in measuring drawn steel wire or drill rods of Stubs' make.

[b]Now used by steel manufacturers in place of old U.S. Standard Gage. Recognized standard in the U.S. for wire and sheet metal of copper and other metals except steel and iron.

NOTE: Compiled from manufacturer's catalogs. Dimensions are in decimal parts of an inch.

TABLE A-30 (continued)

No. of wire	*American or Brown & Sharpe for non-ferrous metals*	*Birming-ham, or Stubs' iron wire*[a]	*American S. & W. Co.'s (Washburn & Moen) std. steel wire*	*American S. & W. Co.'s music wire*	*Imperial wire*	*Stubs' steel wire*[a]	*Steel Manu-facturers' sheet gage*[b]	*No. of wire*
19	.03589	.042	.0410	.043	.040	.164	.0418	19
20	.031961	.035	.0348	.045	.036	.161	.0359	20
21	.028462	.032	.0317	.047	.032	.157	.0329	21
22	.025347	.028	.0286	.049	.028	.155	.0299	22
23	.022571	.025	.0258	.051	.024	.153	.0269	23
24	.0201	.022	.0230	.055	.022	.151	.0239	24
25	.0179	.020	.0204	.059	.020	.148	.0209	25
26	.01594	.018	.0181	.063	.018	.146	.0179	26
27	.014195	.016	.0173	.067	.0164	.143	.0164	27
28	.012641	.014	.0162	.071	.0149	.139	.0149	28
29	.011257	.013	.0150	.075	.0136	.134	.0135	29
30	.010025	.012	.0140	.080	.0124	.127	.0120	30
31	.008928	.010	.0132	.085	.0116	.120	.0105	31
32	.00795	.009	.0128	.090	.0108	.115	.0097	32
33	.00708	.008	.0118	.095	.0100	.112	.0090	33
34	.006304	.007	.0104	....	.0092	.110	.0082	34
35	.005614	.005	.0095	....	.0084	.108	.0075	35
36	.005	.004	.0090	....	.0076	.106	.0067	36
37	.004453	....	.0085	....	.0068	.103	.0064	37
38	.003965	....	.0080	....	.0060	.101	.0060	38
39	.003531	....	.0075	....	.0052	.099	.....	39
40	.003144	....	.0070	....	.0048	.097	.....	40

TABLE A-31 Woodruff Keys,[a] American National Standard

Key no.[b]	*Nominal sizes* A × B	E	F	G	*Maximum sizes* H	D	C	Key no.[b]	*Nominal sizes* A × B	E	F	G	*Maximum sizes* H	D	C
204	1/16 × 1/2	3/64	1/32	5/64	.194	.1718	.203	808	1/4 × 1	1/16	1/8	3/16	.428	.3130	.438
304	3/32 × 1/2	3/64	3/64	3/32	.194	.1561	.203	809	1/4 × 1 1/8	5/64	1/8	13/64	.475	.3590	.484
305	3/32 × 5/8	1/16	3/64	7/64	.240	.2031	.250	810	1/4 × 1 1/4	5/64	1/8	13/64	.537	.4220	.547
404	1/8 × 1/2	3/64	1/16	7/64	.194	.1405	.203	811	1/4 × 1 3/8	3/32	1/8	7/32	.584	.4690	.594
405	1/8 × 5/8	1/16	1/16	1/8	.240	.1875	.250	812	1/4 × 1 1/2	7/64	1/8	15/64	.631	.5160	.641
406	1/8 × 3/4	1/16	1/16	1/8	.303	.2505	.313	1008	5/16 × 1	1/16	5/32	7/32	.428	.2818	.438
505	5/32 × 5/8	1/16	5/64	9/64	.240	.1719	.250	1009	5/16 × 1 1/8	5/64	5/32	15/64	.475	.3278	.484
506	5/32 × 3/4	1/16	5/64	9/64	.303	.2349	.313	1010	5/16 × 1 1/4	5/64	5/32	15/64	.537	.3908	.547
507	5/32 × 7/8	1/16	5/64	9/64	.365	.2969	.375	1011	5/16 × 1 3/8	3/32	5/32	8/32	.584	.4378	.594
606	3/16 × 3/4	1/16	3/32	5/32	.303	.2193	.313	1012	5/16 × 1 1/2	7/64	5/32	17/64	.631	.4848	.641
607	3/16 × 7/8	1/16	3/32	5/32	.365	.2813	.375	1210	3/8 × 1 1/4	5/64	3/16	17/64	.537	.3595	.547
608	3/16 × 1	1/16	3/32	5/32	.428	.3443	.438	1211	3/8 × 1 3/8	3/32	3/16	9/32	.584	.4065	.594
609	3/16 × 1 1/8	5/64	3/32	11/64	.475	.3903	.484	1212	3/8 × 1 1/2	7/64	3/16	19/64	.631	.4535	.641
807	1/4 × 7/8	1/16	1/8	3/16	.365	.2500	.375	....	....	..	..	..	....	....	....

[a]ANSI B17.2.

[b]Key numbers indicate nominal key dimensions. The last two digits give the nominal diameter B in eighths of an inch, and the digits before the last two give the nominal width A in thirty-seconds of an inch.

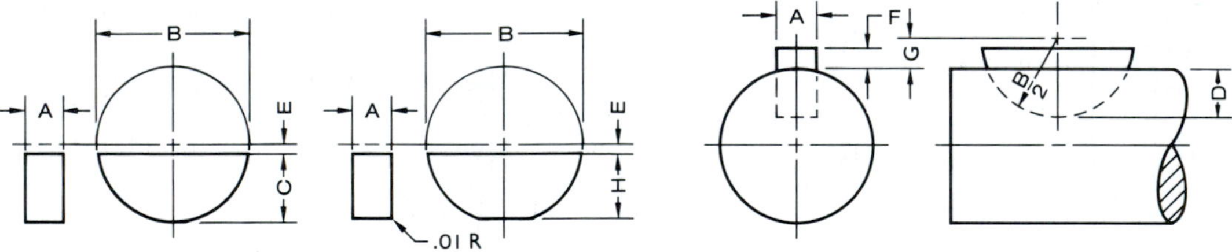

TABLE A-32 Woodruff Key Sizes for Different Shaft Diameters[a]

Shaft diameter	*5/16 to 3/8*	*7/16 to 1/2*	*9/16 to 3/4*	*13/16 to 15/16*	*1 to 1 3/16*	*1 1/4 to 1 7/16*	*1 1/2 to 1 3/4*	*1 13/16 to 2 1/8*	*2 3/16 to 2 1/2*
Key numbers	204	304 305	404 405 406	505 506 507	606 607 608 609	807 808 809	810 811 812	1011 1012	1211 1212

[a]Compiled from industrial drafting manuals.

TABLE A-33 Woodruff Key-Seat Dimensions[a]

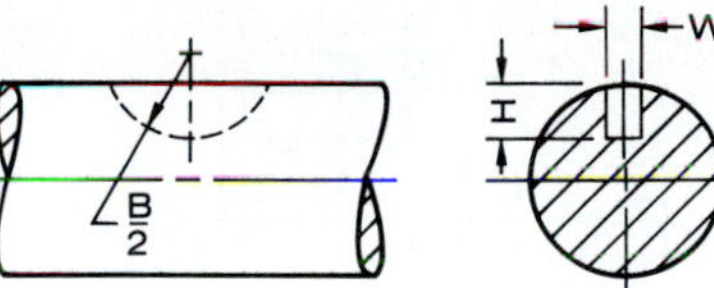

Key no.	Nominal size	Key slot			
		Width W		Depth H	
		Max.	Min.	Max.	Min.
204	1⁄16 × 1⁄2	0.0630	0.0615	0.1718	0.1668
304	3⁄32 × 1⁄2	0.0943	0.0928	0.1561	0.1511
305	3⁄32 × 5⁄8	0.0943	0.0928	0.2031	0.1981
404	1⁄8 × 1⁄2	0.1255	0.1240	0.1405	0.1355
405	1⁄8 × 5⁄8	0.1255	0.1240	0.1875	0.1825
406	1⁄8 × 3⁄4	0.1255	0.1240	0.2505	0.2455
505	5⁄32 × 5⁄8	0.1568	0.1553	0.1719	0.1669
506	5⁄32 × 3⁄4	0.1568	0.1553	0.2349	0.2299
507	5⁄32 × 7⁄8	0.1568	0.1553	0.2969	0.2919
606	3⁄16 × 3⁄4	0.1880	0.1863	0.2193	0.2143
607	3⁄16 × 7⁄8	0.1880	0.1863	0.2813	0.2763
608	3⁄16 × 1	0.1880	0.1863	0.3443	0.3393
609	3⁄16 × 1 1⁄8	0.1880	0.1863	0.3903	0.3853
807	1⁄4 × 7⁄8	0.2505	0.2487	0.2500	0.2450
808	1⁄4 × 1	0.2505	0.2487	0.3130	0.3080
809	1⁄4 × 1 1⁄8	0.2505	0.2487	0.3590	0.3540
810	1⁄4 × 1 1⁄4	0.2505	0.2487	0.4220	0.4170
811	1⁄4 × 1 3⁄8	0.2505	0.2487	0.4690	0.4640
812	1⁄4 × 1 1⁄2	0.2505	0.2487	0.5160	0.5110
1008	5⁄16 × 1	0.3130	0.3111	0.2818	0.2768
1009	5⁄16 × 1 1⁄8	0.3130	0.3111	0.3278	0.3228
1010	5⁄16 × 1 1⁄4	0.3130	0.3111	0.3908	0.3858
1011	5⁄16 × 1 3⁄8	0.3130	0.3111	0.4378	0.4328
1012	5⁄16 × 1 1⁄2	0.3130	0.3111	0.4848	0.4798
1210	3⁄8 × 1 1⁄4	0.3755	0.3735	0.3595	0.3545
1211	3⁄8 × 1 3⁄8	0.3755	0.3735	0.4060	0.4015
1212	3⁄8 × 1 1⁄2	0.3755	0.3735	0.4535	0.4485

[a]Compiled from industrial drafting manuals.

NOTE: Dimensions in inches. Key numbers indicate the nominal key dimensions. The last two digits give the nominal diameter *B* in eighths of an inch, and the digits preceding the last two give the nominal width *A* in thirty-seconds of an inch. Thus 204 indicates a key 2⁄32 by 4⁄8, or 1⁄16 by 1⁄2 in.

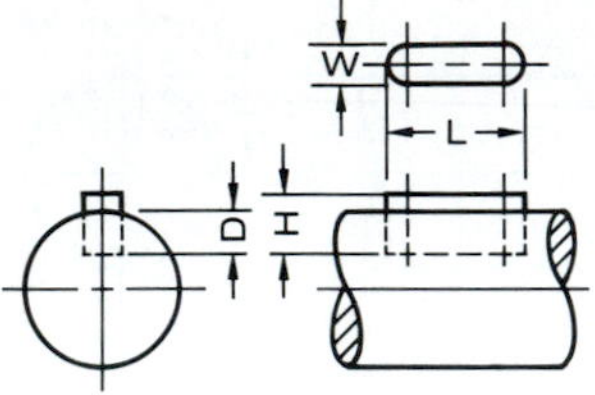

TABLE A-34 Pratt and Whitney Round-End Keys[a]

Key no.	*L*[b]	*W or D*	*H*	*Key no.*	*L*[b]	*W or D*	*H*
1	1/2	1/16	3/32	22	1 3/8	1/4	3/8
2	1/2	3/32	9/64	23	1 1/38	5/16	15/32
3	1/2	1/8	3/16	F	1 3/8	3/8	9/16
4	5/8	3/32	9/64	24	1 1/2	1/4	3/8
5	5/8	1/8	3/16	25	1 1/2	5/16	15/32
6	5/8	5/32	15/64	G	1 1/2	3/8	9/16
7	3/4	1/8	3/16	51	1 3/4	1/4	3/8
8	3/4	5/32	15/64	52	1 3/4	5/16	15/32
9	3/4	3/16	9/32	53	1 3/4	3/8	9/16
10	7/8	5/32	15/64	26	2	3/16	9/32
11	7/8	3/16	9/32	27	2	1/4	3/8
12	7/8	7/32	21/64	28	2	5/16	15/32
A	7/8	1/4	3/8	29	2	3/8	9/16
13	1	3/16	9/32	54	2 1/4	1/4	3/8
14	1	7/32	21/64	55	2 1/4	5/16	15/32
15	1	1/4	3/8	56	2 1/4	3/8	9/16
B	1	5/16	15/32	57	2 1/4	7/16	21/32
16	1 1/8	3/16	9/32	58	2 1/2	5/16	15/32
17	1 1/8	7/32	21/64	59	2 1/2	3/8	9/16
18	1 1/8	1/4	3/8	60	2 1/2	7/16	21/32
C	1 1/8	5/16	15/32	61	2 1/2	1/2	3/4
19	1 1/4	3/16	9/32	30	3	3/8	9/16
20	1 1/4	7/32	21/64	31	3	7/16	21/32
21	1 1/4	1/4	3/8	32	3	1/2	3/4
D	1 1/4	5/16	15/32	33	3	9/16	27/32
E	1 1/4	3/8	9/16	34	3	5/8	15/16

[a]Compiled from industrial drafting manuals. All dimensions are in inches.

[b]The length L may vary from the table, but equals at least 2W.

TABLE A-35 Running and Sliding Fits,[a] American National Standard

Nominal size range,[b] inches		Class RC 1			Class RC 2			Class RC 3			Class RC 4		
		Limits of clearance	Standard limits		Limits of clearance	Standard limits		Limits of clearance	Standard limits		Limits of clearance	Standard limits	
Over	To		Hole H5	Shaft g4		Hole H6	Shaft g5		Hole H7	Shaft f6		Hole H8	Shaft f7
0	0.12	0.1 0.45	+0.2 0	−0.1 −0.25	0.1 0.55	+0.25 0	−0.1 −0.3	0.3 0.95	+0.4 0	−0.3 −0.55	0.3 1.3	+0.6 0	−0.3 −0.7
0.12	0.24	0.15 0.5	+0.2 0	−0.15 −0.3	0.15 0.65	+0.3 0	−0.15 −0.35	0.4 1.2	+0.5 0	−0.4 −0.7	0.4 1.6	+0.7 0	−0.4 −0.9
0.24	0.40	0.2 0.6	+0.25 0	−0.2 −0.35	0.2 0.85	+0.4 0	−0.2 −0.45	0.5 1.5	+0.6 0	−0.5 −0.9	0.5 2.0	+0.9 0	−0.5 −1.1
0.40	0.71	0.25 0.75	+0.3 0	−0.25 −0.45	0.25 0.95	+0.4 0	−0.25 −0.55	0.6 1.7	+0.7 0	−0.6 −1.0	0.6 2.3	+1.0 0	−0.6 −1.3
0.71	1.19	0.3 0.95	+0.4 0	−0.3 −0.55	0.3 1.2	+0.5 0	−0.3 −0.7	0.8 2.1	+0.8 0	−0.8 −1.3	0.8 2.8	+1.2 0	−0.8 −1.6
1.19	1.97	0.4 1.1	+0.4 0	−0.4 −0.7	0.4 1.4	+0.6 0	−0.4 −0.8	1.0 2.6	+1.0 0	−1.0 −1.6	1.0 3.6	+1.6 0	−1.0 −2.0

[a]ANSI B4.1.

[b]For larger sizes, see the standard.

Continued on page 588

NOTES:

RC 1 *Close sliding fits* are intended for the accurate location of parts which must assemble without perceptible play.

RC 2 *Sliding fits* are intended for accurate location, but with greater maximum clearance than class RC 1. Parts made to this fit move and turn easily but are not intended to run freely, and in the larger sizes may seize with small temperature changes.

RC 3 *Precision running fits* are about the closest fits which can be expected to run freely, and are intended for precision work at slow speeds and light journal pressures, but are not suitable where appreciable temperature differences are likely to be encountered.

RC 4 *Close running fits* are intended chiefly for running fits on accurate machinery with moderate surface speeds and journal pressures, where accurate location and minimum play are desired.

RC 5, RC 6 } *Medium running fits* are intended for higher running speeds, or heavy journal pressures, or both.

RC 7 *Free running fits* are intended for use where accuracy is not essential, or where large temperature variations are likely to be encountered, or under both these conditions.

RC 8, RC 9 } *Loose running fits* are intended for use where wide commercial tolerances may be necessary, together with an allowance, on the external member.

Basic hole system. Limits are in thousandths of an inch.
Limits for hole and shaft are applied algebraically to the basic size to obtain the limits of size for the parts.
Data in **boldface** are in accordance with ABC agreements.
Symbols H5, g5, etc., are hole and shaft designations used in ABC System.

TABLE A-35 (continued)

Nominal size range,[b] inches Over – To	Class RC 1 Limits of Clearance	Class RC 1 Standard limits Hole H5	Class RC 1 Standard limits Shaft g4	Class RC 2 Limits of Clearance	Class RC 2 Standard limits Hole H6	Class RC 2 Standard limits Shaft g5	Class RC 3 Limits of Clearance	Class RC 3 Standard limits Hole H7	Class RC 3 Standard limits Shaft f6	Class RC 4 Limits of Clearance	Class RC 4 Standard limits Hole H8	Class RC 4 Standard limits Shaft f7
1.97 – 3.15	0.4 1.2	+0.5 0	−0.4 −0.7	0.4 1.6	+0.7 0	−0.4 −0.9	1.2 3.1	+1.2 0	−1.2 −1.9	1.2 4.2	+1.8 0	−1.2 −2.4
3.15 – 4.73	0.5 1.5	+0.6 0	−0.5 −0.9	0.5 2.0	+0.9 0	−0.5 −1.1	1.4 3.7	+1.4 0	−1.4 −2.3	1.4 5.0	+2.2 0	−1.4 −2.8
4.73 – 7.09	0.6 1.8	+0.7 0	−0.6 −1.1	0.6 2.3	+1.0 0	−0.6 −1.3	1.6 4.2	+1.6 0	−1.6 −2.6	1.6 5.7	+2.5 0	−1.6 −3.2
7.09 – 9.85	0.6 2.0	+0.8 0	−0.6 −1.2	0.6 2.6	+1.2 0	−0.6 −1.4	2.0 5.0	+1.8 0	−2.0 −3.2	2.0 6.6	+2.8 0	−2.0 −3.8
9.85 – 12.41	0.8 2.3	+0.9 0	−0.8 −1.4	0.7 2.8	+1.2 0	−0.7 −1.6	2.5 5.7	+2.0 0	−2.5 −3.7	2.2 7.2	+3.0 0	−2.2 −4.2
12.41 – 15.75	1.0 2.7	+1.0 0	−1.0 −1.7	0.7 3.1	+1.4 0	−0.7 −1.7	3.0 6.6	+2.2 0	−3.0 −4.4	2.5 8.2	+3.5 0	−2.5 −4.7

[b]For larger sizes, see the standard.

TABLE A-35 (continued)

Nominal size range,[b] inches Over – To	Class RC 5 Limits of clearance	Class RC 5 Standard limits Hole H8	Class RC 5 Standard limits Shaft e7	Class RC 6 Limits of clearance	Class RC 6 Standard limits Hole H9	Class RC 6 Standard limits Shaft e8	Class RC 7 Limits of clearance	Class RC 7 Standard limits Hole H9	Class RC 7 Standard limits Shaft d8	Class RC 8 Limits of clearance	Class RC 8 Standard limits Hole H10	Class RC 8 Standard limits Shaft c9	Class RC 9 Limits of clearance	Class RC 9 Standard limits Hole H11	Class RC 9 Standard limits Shaft
0 – 0.12	0.6 1.6	+0.6 −0	−0.6 −1.0	0.6 2.2	+1.0 −0	−0.6 −1.2	1.0 2.6	+1.0 0	− 1.0 − 1.6	2.5 5.1	+1.6 0	− 2.5 − 3.5	4.0 8.1	+ 2.5 0	− 4.0 − 5.6
0.12 – 0.24	0.8 2.0	+0.7 −0	−0.8 −1.3	0.8 2.7	+1.2 −0	−0.8 −1.5	1.2 3.1	+1.2 0	− 1.2 − 1.9	2.8 5.8	+1.8 0	− 2.8 − 4.0	4.5 9.0	+ 3.0 0	− 4.5 − 6.0
0.24 – 0.40	1.0 2.5	+0.9 −0	−1.0 −1.6	1.0 3.3	+1.4 −0	−1.0 −1.9	1.6 3.9	+1.4 0	− 1.6 − 2.5	3.0 6.6	+2.2 0	− 3.0 − 4.4	5.0 10.7	+ 3.5 0	− 5.0 − 7.2
0.40 – 0.71	1.2 2.9	+1.0 −0	−1.2 −1.9	1.2 3.8	+1.6 −0	−1.2 −2.2	2.0 4.6	+1.6 0	− 2.0 − 3.0	3.5 7.9	+2.8 0	− 3.5 − 5.1	6.0 12.8	+ 4.0 − 0	− 6.0 − 8.8
0.71 – 1.19	1.6 3.6	+1.2 −0	−1.6 −2.4	1.6 4.8	+2.0 −0	−1.6 −2.8	2.5 5.7	+2.0 0	− 2.5 − 3.7	4.5 10.0	+3.5 0	− 4.5 − 6.5	7.0 15.5	+ 5.0 0	− 7.0 −10.5
1.19 – 1.97	2.0 4.6	+1.6 −0	−2.0 −3.0	2.0 6.1	+2.5 −0	−2.0 −3.6	3.0 7.1	+2.5 0	− 3.0 − 4.6	5.0 11.5	+4.0 0	− 5.0 − 7.5	8.0 18.0	+ 6.0 0	− 8.0 −12.0
1.97 – 3.15	2.5 5.5	+1.8 −0	−2.5 −3.7	2.5 7.3	+3.0 −0	−2.5 −4.3	4.0 8.8	+3.0 0	− 4.0 − 5.8	6.0 13.5	+4.5 0	− 6.0 − 9.0	9.0 20.5	+ 7.0 0	− 9.0 −13.5
3.15 – 4.73	3.0 6.6	+2.2 −0	−3.0 −4.4	3.0 8.7	+3.5 −0	−3.0 −5.2	5.0 10.7	+3.5 0	− 5.0 − 7.2	7.0 15.5	+5.0 0	− 7.0 −10.5	10.0 24.0	+ 9.0 0	−10.0 −15.0
4.73 – 7.09	3.5 7.6	+2.5 −0	−3.5 −5.1	3.5 10.0	+4.0 −0	−3.5 −6.0	6.0 12.5	+4.0 0	− 6.0 − 8.5	8.0 18.0	+6.0 0	− 8.0 −12.0	12.0 28.0	+10.0 0	−12.0 −18.0
7.09 – 9.85	4.0 8.6	+2.8 −0	−4.0 −5.8	4.0 11.3	+4.5 0	−4.0 −6.8	7.0 14.3	+4.5 0	− 7.0 − 9.8	10.0 21.5	+7.0 0	−10.0 −14.5	15.0 34.0	+12.0 0	−15.0 −22.0
9.85 – 12.41	5.0 10.0	+3.0 −0	−5.0 −7.0	5.0 13.0	+5.0 0	−5.0 −8.0	8.0 16.0	+5.0 0	− 8.0 −11.0	12.0 25.0	+8.0 0	−12.0 −17.0	18.0 38.0	+12.0 0	−18.0 −26.0
12.41 – 15.75	6.0 11.7	+3.5 −0	−6.0 −8.2	6.0 15.5	+6.0 0	−6.0 −9.5	10.0 19.5	+6.0 0	−10.0 13.5	14.0 29.0	+9.0 0	−14.0 −20.0	22.0 45.0	+14.0 0	−22.0 −31.0

[b]For larger sizes, see the standard.

TABLE A-36 Clearance Locational Fits,[a] American National Standard

Nominal size range,[b] inches	Class LC 1			Class LC 2			Class LC 3			Class LC 4			Class LC 5		
	Limits of clearance	*Standard limits*		*Limits of clearance*	*Standard limits*		*Limits of clearance*	*Standard limits*		*Limits of clearance*	*Standard limits*		*Limits of clearance*	*Standard limits*	
Over — *To*		*Hole H6*	*Shaft h5*		*Hole H7*	*Shaft h6*		*Hole H8*	*Shaft h7*		*Hole H10*	*Shaft h9*		*Hole H7*	*Shaft g6*
0 – 0.12	0	+0.25	+0	0	+0.4	+0	0	+0.6	+0	0	+1.6	+0	0.1	+0.4	−0.1
	0.45	−0	−0.2	0.65	−0	−0.25	1	−0	−0.4	2.6	−0	−1.0	0.75	−0	−0.35
0.12 – 0.24	0	+0.3	+0	0	+0.5	+0	0	+0.7	+0	0	+1.8	+0	0.15	+0.5	−0.15
	0.5	−0	−0.2	0.8	−0	−0.3	1.2	−0	−0.5	3.0	−0	−1.2	0.95	−0	−0.45
0.24 – 0.40	0	+0.4	+0	0	+0.6	+0	0	+0.9	+0	0	+2.2	+0	0.2	+0.6	−0.2
	0.65	−0	−0.25	1.0	−0	−0.4	1.5	−0	−0.6	3.6	−0	−1.4	1.2	−0	−0.6
0.40 – 0.71	0	+0.4	+0	0	+0.7	+0	0	+1.0	+0	0	+2.8	+0	0.25	+0.7	−0.25
	0.7	−0	−0.3	1.1	−0	−0.4	1.7	−0	−0.7	4.4	−0	−1.6	1.35	−0	−0.65
0.71 – 1.19	0	+0.5	+0	0	+0.8	+0	0	+1.2	+0	0	+3.5	+0	0.3	+0.8	−0.3
	0.9	−0	−0.4	1.3	−0	−0.5	2	−0	−0.8	5.5	−0	−2.0	1.6	−0	−0.8
1.19 – 1.97	0	+0.6	+0	0	+1.0	+0	0	+1.6	+0	0	+4.0	+0	0.4	+1.0	−0.4
	1.0	−0	−0.4	1.6	−0	−0.6	2.6	−0	−1	6.5	−0	−2.5	2.0	−0	−1.0
1.97 – 3.15	0	+0.7	+0	0	+1.2	+0	0	+1.8	+0	0	+4.5	+0	0.4	+1.2	−0.4
	1.2	−0	−0.5	1.9	−0	−0.7	3	−0	−1.2	7.5	−0	−3	2.3	−0	−1.1
3.15 – 4.73	0	+0.9	+0	0	+1.4	+0	0	+2.2	+0	0	+5.0	+0	0.5	+1.4	−0.5
	1.5	−0	−0.6	2.3	−0	−0.9	3.6	−0	−1.4	8.5	−0	−3.5	2.8	−0	−1.4
4.73 – 7.09	0	+1.0	+0	0	+1.6	+0	0	+2.5	+0	0	+6.0	+0	0.6	+1.6	−0.6
	1.7	−0	−0.7	2.6	−0	−1.0	4.1	−0	−1.6	10	−0	−4	3.2	−0	−1.6

[a]ANSI B4.1.

[b]For larger sizes, see the standard.

NOTES:

LC *Clearance locational fits* are intended for parts which are normally stationary but which can be freely assembled or disassembled. They run from snug fits for parts requiring accuracy of location, through the medium clearance fits for parts such as spigots, to the looser fastener fits where freedom of assembly is of prime importance.

Basic hole system. Limits are in thousandths of an inch.

Limits for hole and shaft are applied algebraically to the basic size to obtain the limits of size for the parts.

Data in **boldface** are in accordance with ABC agreements.

Symbols H6, h5, etc., are hole and shaft designations used in ABC System.

TABLE A-36 (continued)

Nominal size range,[b] inches Over – To	Class LC 1 Limits of Clearance	Class LC 1 Standard limits Hole H6	Class LC 1 Standard limits Shaft h5	Class LC 2 Limits of Clearance	Class LC 2 Standard limits Hole H7	Class LC 2 Standard limits Shaft h6	Class LC 3 Limits of Clearance	Class LC 3 Standard limits Hole H8	Class LC 3 Standard limits Shaft h7	Class LC 4 Limits of Clearance	Class LC 4 Standard limits Hole H10	Class LC 4 Standard limits Shaft h9	Class LC 5 Limits of Clearance	Class LC 5 Standard limits Hole H7	Class LC 5 Standard limits Shaft g6
7.09 – 9.85	0 2.0	+1.2 −0	+0 −0.8	0 3.0	+1.8 −0	+0 −1.2	0 4.6	+2.8 −0	+0 −1.8	0 11.5	+7.0 −0	+0 −4.5	0.6 3.6	+1.8 −0	−0.6 −1.8
9.85 – 12.41	0 2.1	+1.2 −0	+0 −0.9	0 3.2	+2.0 −0	+0 −1.2	0 5	+3.0 −0	+0 −2.0	0 13	+8.0 −0	+0 −5	0.7 3.9	+2.0 −0	−0.7 −1.9
12.41 – 15.75	0 2.4	+1.4 −0	+0 −1.0	0 3.6	+2.2 −0	+0 −1.4	0 5.7	+3.5 −0	+0 −2.2	0 15	+9.0 −0	+0 −6	0.7 4.3	+2.2 −0	−0.7 −2.1

[b]For larger sizes, see the standard.

TABLE A-36 (continued)

Nominal size range,[b] inches Over – To	Class LC 6 Limits of clearance	Class LC 6 Standard limits Hole H9	Class LC 6 Standard limits Shaft f8	Class LC 7 Limits of clearance	Class LC 7 Standard limits Hole H10	Class LC 7 Standard limits Shaft e9	Class LC 8 Limits of clearance	Class LC 8 Standard limits Hole H10	Class LC 8 Standard limits Shaft d9	Class LC 9 Limits of clearance	Class LC 9 Standard limits Hole H11	Class LC 9 Standard limits Shaft c10	Class LC 10 Limits of clearance	Class LC 10 Standard limits Hole H12	Class LC 10 Standard limits Shaft	Class LC 11 Limits of clearance	Class LC 11 Standard limits Hole H13	Class LC 11 Standard limits Shaft
0 – 0.12	0.3 1.9	+1.0 0	−0.3 −0.9	0.6 3.2	+1.6 0	− 0.6 − 1.6	1.0 3.6	+1.6 −0	− 1.0 − 2.0	2.5 6.6	+ 2.5 − 0	− 2.5 − 4.1	4 12	+ 4 − 0	− 4 − 8	5 17	+ 6 − 0	− 5 − 11
0.12 – 0.24	0.4 2.3	+1.2 0	−0.4 −1.1	0.8 3.8	+1.8 0	− 0.8 − 2.0	1.2 4.2	+1.8 −0	− 1.2 − 2.4	2.8 7.6	+ 3.0 − 0	− 2.8 − 4.6	4.5 14.5	+ 5 − 0	− 4.5 − 9.5	6 20	+ 7 − 0	− 6 − 13
0.24 – 0.40	0.5 2.8	+1.4 0	−0.5 −1.4	1.0 4.6	+2.2 0	− 1.0 − 2.4	1.6 5.2	+2.2 −0	− 1.6 − 3.0	3.0 8.7	+ 3.5 − 0	− 3.0 − 5.2	5 17	+ 6 − 0	− 5 − 11	7 25	+ 9 − 0	− 7 − 16
0.40 – 0.71	0.6 3.2	+1.6 0	−0.6 −1.6	1.2 5.6	+2.8 0	− 1.2 − 2.8	2.0 6.4	+2.8 −0	− 2.0 − 3.6	3.5 10.3	+ 4.0 − 0	− 3.5 − 6.3	6 20	+ 7 − 0	− 6 − 13	8 28	+10 − 0	− 8 − 18
0.71 – 1.19	0.8 4.0	+2.0 0	−0.8 −2.0	1.6 7.1	+3.5 0	− 1.6 − 3.6	2.5 8.0	+3.5 −0	− 2.5 − 4.5	4.5 13.0	+ 5.0 − 0	− 4.5 − 8.0	7 23	+ 8 − 0	− 7 − 15	10 34	+12 − 0	− 10 − 22

[b]For larger sizes, see the standard.

Continued on page 592

TABLE A-36 (continued)

Nominal size range,[b] inches	Class LC 6			Class LC 7			Class LC 8			Class LC 9			Class LC 10			Class LC 11		
	Limits of Clearance	*Standard limits*		*Limits of Clearance*	*Standard limits*		*Limits of Clearance*	*Standard limits*		*Limits of Clearance*	*Standard limits*		*Limits of Clearance*	*Standard limits*		*Limits of Clearance*	*Standard limits*	
Over To		*Hole H9*	*Shaft f8*		*Hole H10*	*Shaft e9*		*Hole H10*	*Shaft d9*		*Hole H11*	*Shaft c10*		*Hole H12*	*Shaft*		*Hole H13*	*Shaft*
1.19 – 1.97	1.0 5.1	+2.5 0	−1.0 −2.6	2.0 8.5	+4.0 0	− 2.0 − 4.5	3.0 9.5	+4.0 −0	− 3.0 − 5.5	5 15	+ 6 − 0	− 5 − 9	8 28	+10 − 0	− 8 −18	12 44	+16 − 0	−12 −28
1.97 – 3.15	1.2 6.0	+3.0 0	−1.2 −3.0	2.5 10.0	+4.5 0	− 2.5 − 5.5	4.0 11.5	+4.5 −0	− 4.0 − 7.0	6 17.5	+ 7 − 0	− 6 −10.5	10 34	+12 − 0	−10 −22	14 50	+18 − 0	−14 −32
3.15 – 4.73	1.4 7.1	+3.5 0	−1.4 −3.6	3.0 11.5	+5.0 0	− 3.0 − 6.5	5.0 13.5	+5.0 −0	− 5.0 − 8.5	7 21	+ 9 − 0	− 7 −12	11 39	+14 − 0	−11 −25	16 60	+22 − 0	−16 −38
4.73 – 7.09	1.6 8.1	+4.0 0	−1.6 −4.1	3.5 13.5	+6.0 0	− 3.5 − 7.5	6 16	+6 −0	− 6 −10	8 24	+10 − 0	− 8 −14	12 44	+16 − 0	−12 −28	18 68	+25 − 0	−18 −43
7.09 – 9.85	2.0 9.3	+4.5 0	−2.0 −4.8	4.0 15.5	+7.0 0	− 4.0 − 8.5	7 18.5	+7 −0	− 7 −11.5	10 29	+12 − 0	−10 −17	16 52	+18 − 0	−16 −34	22 78	+28 − 0	−22 −50
9.85 – 12.41	2.2 10.2	+5.0 0	−2.2 −5.2	4.5 17.5	+8.0 0	− 4.5 − 9.5	7 20	+8 −0	− 7 −12	12 32	+12 − 0	−12 −20	20 60	+20 − 0	−20 −40	28 88	+30 − 0	−28 −58
12.41 – 15.75	2.5 12.0	+6.0 0	−2.5 −6.0	5.0 20.0	+9.0 0	− 5 −11	8 23	+9 −0	− 8 −14	14 37	+14 − 0	−14 −23	22 66	+22 − 0	−22 −44	30 100	+35 − 0	−30 −65

[b]For larger sizes, see the standard.

TABLE A-37 Transition Locational Fits,[a] American National Standard

Nominal size range,[b] inches Over – To	Class LT 1			Class LT 2			Class LT 3			Class LT 4			Class LT 5			Class LT 6		
	Fit	Standard limits		Fit	Standard limits		Fit	Standard limits		Fit	Standard limits		Fit	Standard limits		Fit	Standard limits	
		Hole H7	Shaft js6		Hole H8	Shaft js7		Hole H7	Shaft k6		Hole H8	Shaft k7		Hole H7	Shaft n6		Hole H7	Shaft n7
0 – 0.12	−0.10 +0.50	+0.4 −0	+0.10 −0.10	−0.2 +0.8	+0.6 −0	+0.2 −0.2							−0.5 +0.15	+0.4 −0	+0.5 +0.25	−0.65 +0.15	+0.4 −0	+0.65 +0.25
0.12 – 0.24	−0.15 +0.65	+0.5 −0	+0.15 −0.15	−0.25 +0.95	+0.7 −0	+0.25 −0.25							−0.6 +0.2	+0.5 −0	+0.6 +0.3	−0.8 +0.2	+0.5 −0	+0.8 +0.3
0.24 – 0.40	−0.2 +0.8	+0.6 −0	+0.2 −0.2	−0.3 +1.2	+0.9 −0	+0.3 −0.3	−0.5 +0.5	+0.6 −0	+0.5 +0.1	−0.7 +0.8	+0.9 −0	+0.7 +0.1	−0.8 +0.2	+0.6 −0	+0.8 +0.4	−1.0 +0.2	+0.6 −0	+1.0 +0.4
0.40 – 0.71	−0.2 +0.9	+0.7 −0	+0.2 −0.2	−0.35 +1.35	+1.0 −0	+0.35 −0.35	−0.5 +0.6	+0.7 −0	+0.5 +0.1	−0.8 +0.9	+1.0 −0	+0.8 +0.1	−0.9 +0.2	+0.7 −0	+0.9 +0.5	−1.2 +0.2	+0.7 −0	+1.2 +0.5
0.71 – 1.19	−0.25 +1.05	+0.8 −0	+0.25 −0.25	−0.4 +1.6	+1.2 −0	+0.4 −0.4	−0.6 +0.7	+0.8 −0	+0.6 +0.1	−0.9 +1.1	+1.2 −0	+0.9 +0.1	−1.1 +0.2	+0.8 −0	+1.1 +0.6	−1.4 +0.2	+0.8 −0	+1.4 +0.6
1.19 – 1.97	−0.3 +1.3	+1.0 −0	+0.3 −0.3	−0.5 +2.1	+1.6 −0	+0.5 −0.5	−0.7 +0.9	+1.0 −0	+0.7 +0.1	−1.1 +1.5	+1.6 −0	+1.1 +0.1	−1.3 +0.3	+1.0 −0	+1.3 +0.7	−1.7 +0.3	+1.0 −0	+1.7 +0.7
1.97 – 3.15	−0.3 +1.5	+1.2 −0	+0.3 −0.3	−0.6 +2.4	+1.8 −0	+0.6 −0.6	−0.8 +1.1	+1.2 −0	+0.8 +0.1	−1.3 +1.7	+1.8 −0	+1.3 +0.1	−1.5 +0.4	+1.2 −0	+1.5 +0.8	−2.0 +0.4	+1.2 −0	+2.0 +0.8
3.15 – 4.73	−0.4 +1.8	+1.4 −0	+0.4 −0.4	−0.7 +2.9	+2.2 −0	+0.7 −0.7	−1.0 +1.3	+1.4 −0	+1.0 +0.1	−1.5 +2.1	+2.2 −0	+1.5 +0.1	−1.9 +0.4	+1.4 −0	+1.9 +1.0	−2.4 −0.4	+1.4 −0	+2.4 +1.0
4.73 – 7.09	−0.5 +2.1	+1.6 −0	+0.5 −0.5	−0.8 +3.3	+2.5 −0	+0.8 −0.8	−1.1 +1.5	+1.6 −0	+1.1 +0.1	−1.7 +2.4	+2.5 −0	+1.7 +0.1	−2.2 +0.4	+1.6 −0	+2.2 +1.2	−2.8 +0.4	+1.6 −0	+2.8 +1.2

[a] ANSI B4.1.

[b] For larger sizes, see the standard.

Continued on page 594

NOTES:

LT *Transition locational fits* are a compromise between clearance and interference fits, for application where accuracy of location is important, but either a small amount of clearance or interference is permissible.

Basic hole system. Limits are in thousandths of an inch.
Limits for hole and shaft are applied algebraically to the basic size to obtain the limits of size for the mating parts.
Data in **boldface** are in accordance with ABC agreements.
"Fit" represents the maximum interference (minus values) and the maximum clearance (plus values).
Symbols H7, js6, etc., are hole and shaft designations used in ABC System.

TABLE A-37 (continued)

Nominal size range,[b] inches	Class LT 1			Class LT 2			Class LT 3			Class LT 4			Class LT 5			Class LT 6		
	Fit	Standard limits		Fit	Standard limits		Fit	Standard limits		Fit	Standard limits		Fit	Standard limits		Fit	Standard limits	
Over To		Hole H7	Shaft js6		Hole H8	Shaft js7		Hole H7	Shaft k6		Hole H8	Shaft k7		Hole H7	Shaft n6		Hole H7	Shaft n7
7.09 – 9.85	−0.6 +2.4	+1.8 −0	+0.6 −0.6	−0.9 +3.7	+2.8 −0	+0.9 −0.9	−1.4 +1.6	+1.8 −0	+1.4 +0.2	−2.0 +2.6	+2.8 −0	+2.0 +0.2	−2.6 +0.4	+1.8 −0	+2.6 +1.4	−3.2 +0.4	+1.8 −0	+3.2 +1.4
9.85 – 12.41	−0.6 +2.6	+2.0 −0	+0.6 −0.6	−1.0 +4.0	+3.0 −0	+1.0 −1.0	−1.4 +1.8	+2.0 −0	+1.4 +0.2	−2.2 +2.8	+3.0 −0	+2.2 +0.2	−2.6 +0.6	+2.0 −0	+2.6 +1.4	−3.4 +0.6	+2.0 −0	+3.4 +1.4
12.41 – 15.75	−0.7 +2.9	+2.2 −0	+0.7 −0.7	−1.0 +4.5	+3.5 −0	+1.0 −1.0	−1.6 +2.0	+2.2 −0	+1.6 +0.2	−2.4 +3.3	+3.5 −0	+2.4 +0.2	−3.0 +0.6	+2.2 −0	+3.0 +1.6	−3.8 +0.6	+2.2 −0	+3.8 +1.6

[b]For larger sizes, see the standard.

TABLE A-38 Interference Locational Fits,[a] American National Standard

Nominal size range,[b] inches		Class LN 1			Class LN 2			Class LN 3		
		Limits of interference	Standard limits		Limits of interference	Standard limits		Limits of interference	Standard limits	
Over	To		Hole H6	Shaft n5		Hole H7	Shaft p6		Hole H7	Shaft r6
0	0.12	0 0.45	+0.25 −0	+0.45 +0.25	0 0.65	+0.4 −0	+0.65 +0.4	0.1 0.75	+0.4 −0	+0.75 +0.5
0.12	0.24	0 0.5	+0.3 −0	+0.5 +0.3	0 0.8	+0.5 −0	+0.8 +0.5	0.1 0.9	+0.5 −0	+0.9 +0.6
0.24	0.40	0 0.65	+0.4 −0	+0.65 +0.4	0 1.0	+0.6 −0	+1.0 +0.6	0.2 1.2	+0.6 −0	+1.2 +0.8
0.40	0.71	0 0.8	+0.4 −0	+0.8 +0.4	0 1.1	+0.7 −0	+1.1 +0.7	0.3 1.4	+0.7 −0	+1.4 +1.0
0.71	1.19	0 1.0	+0.5 −0	+1.0 +0.5	0 1.3	+0.8 −0	+1.3 +0.8	0.4 1.7	+0.8 −0	+1.7 +1.2
1.19	1.97	0 1.1	+0.6 −0	+1.1 +0.6	0 1.6	+1.0 −0	+1.6 +1.0	0.4 2.0	+1.0 −0	+2.0 +1.4
1.97	3.15	0.1 1.3	+0.7 −0	+1.3 +0.8	0.2 2.1	+1.2 −0	+2.1 +1.4	0.4 2.3	+1.2 −0	+2.3 +1.6
3.15	4.73	0.1 1.6	+0.9 −0	+1.6 +1.0	0.2 2.5	+1.4 −0	+2.5 +1.6	0.6 2.9	+1.4 −0	+2.9 +2.0
4.73	7.09	0.2 1.9	+1.0 −0	+1.9 +1.2	0.2 2.8	+1.6 −0	+2.8 +1.8	0.9 3.5	+1.6 −0	+3.5 +2.5
7.09	9.85	0.2 2.2	+1.2 −0	+2.2 +1.4	0.2 3.2	+1.8 −0	+3.2 +2.0	1.2 4.2	+1.8 −0	+4.2 +3.0
9.85	12.41	0.2 2.3	+1.2 −0	+2.3 +1.4	0.2 3.4	+2.0 −0	+3.4 +2.2	1.5 4.7	+2.0 −0	+4.7 +3.5
12.41	15.75	0.2 2.6	+1.4 −0	+2.6 +1.6	0.3 3.9	+2.2 −0	+3.9 +2.5	2.3 5.9	+2.2 −0	+5.9 +4.5

[a]ANSI B4.1.

[b]For larger sizes, see the standard.

NOTES:

LN *Interference locational fits* are used where accuracy of location is of prime importance, and for parts requiring rigidity and alignment with no special requirements for bore pressure. Such fits are not intended for parts designed to transmit frictional loads from one part to another by virtue of the tightness of fit, as these conditions are covered by force fits.

Limits are in thousandths of an inch.

Limits for hole and shaft are applied algebraically to the basic size to obtain the limits of size for the parts.

Data in **boldface** are in accordance with ABC agreements.

Symbols H7, p6, etc., are hole and shaft designations used in ABC System.

TABLE A-39 Force and Shrink Fits,[a] American National Standard

Nominal size range,[b] inches		Class FN 1			Class FN 2			Class FN 3			Class FN 4			Class FN 5		
		Limits of interference	Standard limits		Limits of interference	Standard limits		Limits of interference	Standard limits		Limits of interference	Standard limits		Limits of interference	Standard limits	
Over	To		Hole H6	Shaft		Hole H7	Shaft s6		Hole H7	Shaft t6		Hole H7	Shaft u6		Hole H8	Shaft x7
0	0.12	0.05	**+0.25**	+0.5	**0.2**	**+0.4**	**+0.85**				0.3	+0.4	+ 0.95	**0.3**	**+0.6**	**+ 1.3**
		0.5	**−0**	+0.3	**0.85**	**−0**	**+0.6**				0.95	−0	+ 0.7	**1.3**	**−0**	**+ 0.9**
0.12	0.24	0.1	**+0.3**	+0.6	**0.2**	**+0.5**	**+1.0**				0.4	+0.5	+ 1.2	**0.5**	**+0.7**	**+ 1.7**
		0.6	**−0**	+0.4	**1.0**	**−0**	**+0.7**				1.2	−0	+ 0.9	**1.7**	**−0**	**+ 1.2**
0.24	0.40	0.1	**+0.4**	+0.75	**0.4**	**+0.6**	**+1.4**				0.6	+0.6	+ 1.6	**0.5**	**+0.9**	**+ 2.0**
		0.75	**−0**	+0.5	**1.4**	**−0**	**+1.0**				1.6	−0	+ 1.2	**2.0**	**−0**	**+ 1.4**
0.40	0.56	0.1	**+0.4**	+0.8	**0.5**	**+0.7**	**+1.6**				0.7	+0.7	+ 1.8	**0.6**	**+1.0**	**+ 2.3**
		0.8	**−0**	+0.5	**1.6**	**−0**	**+1.2**				1.8	−0	+ 1.4	**2.3**	**−0**	**+ 1.6**
0.56	0.71	0.2	**+0.4**	+0.9	**0.5**	**+0.7**	**+1.6**				0.7	+0.7	+ 1.8	**0.8**	**+1.0**	**+ 2.5**
		0.9	**−0**	+0.6	**1.6**	**−0**	**+1.2**				1.8	−0	+ 1.4	**2.5**	**−0**	**+ 1.8**
0.71	0.95	0.2	**+0.5**	+1.1	**0.6**	**+0.8**	**+1.9**				0.8	+0.8	+ 2.1	**1.0**	**+1.2**	**+ 3.0**
		1.1	**−0**	+0.7	**1.9**	**−0**	**+1.4**				2.1	−0	+ 1.6	**3.0**	**−0**	**+ 2.2**
0.95	1.19	0.3	**+0.5**	+1.2	**0.6**	**+0.8**	**+1.9**	**0.8**	**+0.8**	**+ 2.1**	1.0	+0.8	+ 2.3	**1.3**	**+1.2**	**+ 3.3**
		1.2	**−0**	+0.8	**1.9**	**−0**	**+1.4**	**2.1**	**−0**	**+ 1.6**	2.3	−0	+ 1.8	**3.3**	**−0**	**+ 2.5**
1.19	1.58	0.3	**+0.6**	+1.3	**0.8**	**+1.0**	**+2.4**	**1.0**	**+1.0**	**+ 2.6**	1.5	+1.0	+ 3.1	**1.4**	**+1.6**	**+ 4.0**
		1.3	**−0**	+0.9	**2.4**	**−0**	**+1.8**	**2.6**	**−0**	**+ 2.0**	3.1	−0	+ 2.5	**4.0**	**−0**	**+ 3.0**

[a]ANSI B4.1.

[b]For larger sizes, see the standard.

NOTES:

FN 1 *Light drive fits* are those requiring light assembly pressures and produce more or less permanent assemblies. They are suitable for thin sections or long fits or in cast-iron external members.

FN 2 *Medium drive fits* are suitable for ordinary steel parts or for shrink fits on light sections. They are about the tightest fits that can be used with high-grade cast-iron external members.

FN 3 *Heavy drive fits* are suitable for heavier steel parts or for shrink fits in medium sections.

FN 4, FN 5 *Force fits* are suitable for parts which can be highly stressed, or for shrink fits where the heavy pressing forces required are impractical.

Limits are in thousandths of an inch.

Limits for hole and shaft are applied algebraically to the basic size to obtain the limits of size for the parts.

Data in **boldface** are in accordance with ABC agreements.

Symbols H7, s6, etc., are hole and shaft designations used in ABC System.

TABLE A-39 (continued)

Nominal size range,[b] inches		Class FN 1			Class FN 2			Class FN 3			Class FN 4			Class FN 5		
		Limits of interference	Standard limits		Limits of interference	Standard limits		Limits of interference	Standard limits		Limits of interference	Standard limits		Limits of interference	Standard limits	
Over	To		Hole H6	Shaft		Hole H7	Shaft s6		Hole H7	Shaft t6		Hole H7	Shaft u6		Hole H8	Shaft x7
1.58 –	1.97	0.4 1.4	+0.6 −0	+1.4 +1.0	0.8 2.4	+1.0 −0	+2.4 +1.8	1.2 2.8	+1.0 −0	+ 2.8 + 2.2	1.8 3.4	+1.0 −0	+ 3.4 + 2.8	2.4 5.0	+1.6 −0	+ 5.0 + 4.0
1.97 –	2.56	0.6 1.8	+0.7 −0	+1.8 +1.3	0.8 2.7	+1.2 −0	+2.7 +2.0	1.3 3.2	+1.2 −0	+ 3.2 + 2.5	2.3 4.2	+1.2 −0	+ 4.2 + 3.5	3.2 6.2	+1.8 −0	+ 6.2 + 5.0
2.56 –	3.15	0.7 1.9	+0.7 −0	+1.9 +1.4	1.0 2.9	+1.2 −0	+2.9 +2.2	1.8 3.7	+1.2 −0	+ 3.7 + 3.0	2.8 4.7	+1.2 −0	+ 4.7 + 4.0	4.2 7.2	+1.8 −0	+ 7.2 + 6.0
3.15 –	3.94	0.9 2.4	+0.9 −0	+2.4 +1.8	1.4 3.7	+1.4 −0	+3.7 +2.8	2.1 4.4	+1.4 −0	+ 4.4 + 3.5	3.6 5.9	+1.4 −0	+ 5.9 + 5.0	4.8 8.4	+2.2 −0	+ 8.4 + 7.0
3.94 –	4.73	1.1 2.6	+0.9 −0	+2.6 +2.0	1.6 3.9	+1.4 −0	+3.9 +3.0	2.6 4.9	+1.4 −0	+ 4.9 + 4.0	4.6 6.9	+1.4 −0	+ 6.9 + 6.0	5.8 9.4	+2.2 −0	+ 9.4 + 8.0
4.73 –	5.52	1.2 2.9	+1.0 −0	+2.9 +2.2	1.9 4.5	+1.6 −0	+4.5 +3.5	3.4 6.0	+1.6 −0	+ 6.0 + 5.0	5.4 8.0	+1.6 −0	+ 8.0 + 7.0	7.5 11.6	+2.5 −0	+11.6 +10.0
5.52 –	6.30	1.5 3.2	+1.0 −0	+3.2 +2.5	2.4 5.0	+1.6 −0	+5.0 +4.0	3.4 6.0	+1.6 −0	+ 6.0 + 5.0	5.4 8.0	+1.6 −0	+ 8.0 + 7.0	9.5 13.6	+2.5 −0	+13.6 +12.0
6.30 –	7.09	1.8 3.5	+1.0 −0	+3.5 +2.8	2.9 5.5	+1.6 −0	+5.5 +4.5	4.4 7.0	+1.6 −0	+ 7.0 + 6.0	6.4 9.0	+1.6 −0	+ 9.0 + 8.0	9.5 13.6	+2.5 −0	+13.6 +12.0
7.09 –	7.88	1.8 3.8	+1.2 −0	+3.8 +3.0	3.2 6.2	+1.8 −0	+6.2 +5.0	5.2 8.2	+1.8 −0	+ 8.2 + 7.0	7.2 10.2	+1.8 −0	+10.2 + 9.0	11.2 15.8	+2.8 −0	+15.8 +14.0
7.88 –	8.86	2.3 4.3	+1.2 −0	+4.3 +3.5	3.2 6.2	+1.8 −0	+6.2 +5.0	5.2 8.2	+1.8 −0	+ 8.2 + 7.0	8.2 11.2	+1.8 −0	+11.2 +10.0	13.2 17.8	+2.8 −0	+17.8 +16.0
8.86 –	9.85	2.3 4.3	+1.2 −0	+4.3 +3.5	4.2 7.2	+1.8 −0	+7.2 +6.0	6.2 9.2	+1.8 −0	+ 9.2 + 8.0	10.2 13.2	+1.8 −0	+13.2 +12.0	13.2 17.8	+2.8 −0	+17.8 +16.0

[b]For larger sizes, see the standard.

Continued on page 598

TABLE A-39 (continued)

Nominal size range,[b] inches Over – To	Class FN 1			Class FN 2			Class FN 3			Class FN 4			Class FN 5		
		Standard limits			Standard limits			Standard limits			Standard limits			Standard limits	
		Hole H6	Shaft		Hole H7	Shaft s6		Hole H7	Shaft t6		Hole H7	Shaft u6		Hole H8	Shaft x7
9.85 – 11.03	2.8 4.9	+1.2 −0	+4.9 +4.0	4.0 7.2	+2.0 −0	+7.2 +6.0	7.0 10.2	+2.0 −0	+10.2 + 9.0	10.0 13.2	+2.0 −0	+13.2 +12.0	15.0 20.0	+3.0 −0	+20.0 +18.0
11.03 – 12.41	2.8 4.9	+1.2 −0	+4.9 +4.0	5.0 8.2	+2.0 −0	+8.2 +7.0	7.0 10.2	+2.0 −0	+10.2 + 9.0	12.0 15.2	+2.0 −0	+15.2 +14.0	17.0 22.0	+3.0 −0	+22.0 +20.0
12.41 – 13.98	3.1 5.5	+1.4 −0	+5.5 +4.5	5.8 9.4	+2.2 −0	+9.4 +8.0	7.8 11.4	+2.2 −0	+11.4 +10.0	13.8 17.4	+2.2 −0	+17.4 +16.0	18.5 24.2	+3.5 −0	+24.2 +22.0
13.98 – 15.75	3.6 6.1	+1.4 −0	+6.1 +5.0	5.8 9.4	+2.2 −0	+9.4 +8.0	9.8 13.4	+2.2 −0	+13.4 +12.0	15.8 19.4	+2.2 −0	+19.4 +18.0	21.5 27.2	+3.5 −0	+27.2 +25.0

[b]For larger sizes, see the standard.

TABLE A-40 Preferred Hole Basis Clearance Fits—Cylindrical Fits,[a] American National Standard

Basic size		Loose running			Free running			Close running			Sliding			Locational clearance		
		Hole H11	Shaft c11	Fit	Hole H9	Shaft d9	Fit	Hole H8	Shaft f7	Fit	Hole H7	Shaft g6	Fit	Hole H7	Shaft h6	Fit
1	MAX.	1.060	0.940	0.180	1.025	0.980	0.070	1.014	0.994	0.030	1.010	0.998	0.018	1.010	1.000	0.016
	MIN.	1.000	0.880	0.060	1.000	0.955	0.020	1.000	0.984	0.006	1.000	0.992	0.002	1.000	0.994	0.000
1.2	MAX.	1.260	1.140	0.180	1.225	1.180	0.070	1.214	1.194	0.030	1.210	1.198	0.018	1.210	1.200	0.016
	MIN.	1.200	1.080	0.060	1.200	1.155	0.020	1.200	1.184	0.006	1.200	1.192	0.002	1.200	1.194	0.000
1.6	MAX.	1.660	1.540	0.180	1.625	1.580	0.070	1.614	1.594	0.030	1.610	1.598	0.018	1.610	1.600	0.016
	MIN.	1.600	1.480	0.060	1.600	1.555	0.020	1.600	1.584	0.006	1.600	1.592	0.002	1.600	1.594	0.000
2	MAX.	2.060	1.940	0.180	2.025	1.980	0.070	2.014	1.994	0.030	2.010	1.998	0.018	2.010	2.000	0.016
	MIN.	2.000	1.880	0.060	2.000	1.955	0.020	2.000	1.984	0.006	2.000	1.992	0.002	2.000	1.994	0.000

[a]ANSI B4.2.

NOTE: All dimensions are in mm.

TABLE A-40 (continued)

Basic size		Loose running			Free running			Close running			Sliding			Locational clearance		
		Hole H11	*Shaft c11*	*Fit*	*Hole H9*	*Shaft d9*	*Fit*	*Hole H8*	*Shaft f7*	*Fit*	*Hole H7*	*Shaft g6*	*Fit*	*Hole H7*	*Shaft h6*	*Fit*
2.5	MAX.	2.560	2.440	0.180	2.525	2.480	0.070	2.514	2.494	0.030	2.510	2.498	0.018	2.510	2.500	0.016
	MIN.	2.500	2.380	0.060	2.500	2.455	0.020	2.500	2.484	0.006	2.500	2.492	0.002	2.500	2.494	0.000
3	MAX.	3.060	2.940	0.180	3.025	2.980	0.070	3.014	2.994	0.030	3.010	2.998	0.018	3.010	3.000	0.016
	MIN.	3.000	2.880	0.060	3.000	2.955	0.020	3.000	2.984	0.006	3.000	2.992	0.002	3.000	2.994	0.000
4	MAX.	4.075	3.930	0.220	4.030	3.970	0.090	4.018	3.990	0.040	4.012	3.996	0.024	4.012	4.000	0.020
	MIN.	4.000	3.855	0.070	4.000	3.940	0.030	4.000	3.978	0.010	4.000	3.988	0.004	4.000	3.992	0.000
5	MAX.	5.075	4.930	0.220	5.030	4.970	0.090	5.018	4.990	0.040	5.012	4.996	0.024	5.012	5.000	0.020
	MIN.	5.000	4.855	0.070	5.000	4.940	0.030	5.000	4.978	0.010	5.000	4.988	0.004	5.000	4.992	0.000
6	MAX.	6.075	5.930	0.220	6.030	5.970	0.090	6.018	5.990	0.040	6.012	5.996	0.024	6.012	6.000	0.020
	MIN.	6.000	5.855	0.070	6.000	5.940	0.030	6.000	5.978	0.010	6.000	5.988	0.004	6.000	5.992	0.000
8	MAX.	8.090	7.920	0.260	8.036	7.960	0.112	8.022	7.987	0.050	8.015	7.995	0.029	8.015	8.000	0.024
	MIN.	8.000	7.830	0.080	8.000	7.924	0.040	8.000	7.972	0.013	8.000	7.986	0.005	8.000	7.991	0.000
10	MAX.	10.090	9.920	0.260	10.036	9.960	0.112	10.022	9.987	0.050	10.015	9.995	0.029	10.015	10.000	0.024
	MIN.	10.000	9.830	0.080	10.000	9.924	0.040	10.000	9.972	0.013	10.000	9.986	0.005	10.000	9.991	0.000
12	MAX.	12.110	11.905	0.315	12.043	11.950	0.136	12.027	11.984	0.061	12.018	11.994	0.035	12.018	12.000	0.029
	MIN.	12.000	11.795	0.095	12.000	11.907	0.050	12.000	11.966	0.016	12.000	11.983	0.006	12.000	11.989	0.000
16	MAX.	16.110	15.905	0.315	16.043	15.950	0.136	16.027	15.984	0.061	16.018	15.994	0.035	16.018	16.000	0.029
	MIN.	16.000	15.795	0.095	16.000	15.907	0.050	16.000	15.966	0.016	16.000	15.983	0.006	16.000	15.989	0.000
20	MAX.	20.130	19.890	0.370	20.052	19.935	0.169	20.033	19.980	0.074	20.021	19.993	0.041	20.021	20.000	0.034
	MIN.	20.000	19.760	0.110	20.000	19.883	0.065	20.000	19.959	0.020	20.000	19.980	0.007	20.000	19.987	0.000
25	MAX.	25.130	24.890	0.370	25.052	24.935	0.169	25.033	24.980	0.074	25.021	24.993	0.041	25.021	25.000	0.034
	MIN.	25.000	24.760	0.110	25.000	24.883	0.065	25.000	24.959	0.020	25.000	24.980	0.007	25.000	24.987	0.000
30	MAX.	30.130	29.890	0.370	30.052	29.935	0.169	30.033	29.980	0.074	30.021	29.993	0.041	30.021	30.000	0.034
	MIN.	30.000	29.760	0.110	30.000	29.883	0.065	30.000	29.959	0.020	30.000	29.980	0.007	30.000	29.987	0.000

NOTE: All dimensions are in mm.

Continued on page 600

TABLE A-40 (continued)

Basic size		Loose running			Free running			Close running			Sliding			Locational clearance		
		Hole H11	*Shaft c11*	*Fit*	*Hole H9*	*Shaft d9*	*Fit*	*Hole H8*	*Shaft f7*	*Fit*	*Hole H7*	*Shaft g6*	*Fit*	*Hole H7*	*Shaft h6*	*Fit*
40	MAX.	40.160	39.880	0.440	40.062	39.920	0.204	40.039	39.975	0.089	40.025	39.991	0.050	40.025	40.000	0.041
	MIN.	40.000	39.720	0.120	40.000	39.858	0.080	40.000	39.950	0.025	40.000	39.975	0.009	40.000	39.984	0.000
50	MAX.	50.160	49.870	0.450	50.062	49.920	0.204	50.039	49.975	0.089	50.025	49.991	0.050	50.025	50.000	0.041
	MIN.	50.000	49.710	0.130	50.000	49.858	0.080	50.000	49.950	0.025	50.000	49.975	0.009	50.000	49.984	0.000
60	MAX.	60.190	59.860	0.520	60.074	59.900	0.248	60.046	59.970	0.106	60.030	59.990	0.059	60.030	60.000	0.049
	MIN.	60.000	59.670	0.140	60.000	59.826	0.100	60.000	59.940	0.030	60.000	59.971	0.010	60.000	59.981	0.000
80	MAX.	80.190	79.850	0.530	80.074	79.900	0.248	80.046	79.970	0.106	80.030	79.990	0.059	80.030	80.000	0.049
	MIN.	80.000	79.660	0.150	80.000	79.826	0.100	80.000	79.940	0.030	80.000	79.971	0.010	80.000	79.981	0.000
100	MAX.	100.220	99.830	0.610	100.087	99.880	0.294	100.054	99.964	0.125	100.035	99.988	0.069	100.035	100.000	0.057
	MIN.	100.000	99.610	0.170	100.000	99.793	0.120	100.000	99.929	0.036	100.000	99.966	0.012	100.000	99.978	0.000
120	MAX.	120.220	119.820	0.620	120.087	119.880	0.294	120.054	119.964	0.125	120.035	119.988	0.069	120.035	120.000	0.057
	MIN.	120.000	119.600	0.180	120.000	119.793	0.120	120.000	119.929	0.036	120.000	119.966	0.012	120.000	119.978	0.000
160	MAX.	160.250	159.790	0.710	160.100	159.855	0.345	160.063	159.957	0.146	160.040	159.986	0.079	160.040	160.000	0.065
	MIN.	160.000	159.540	0.210	160.000	159.755	0.145	160.000	159.917	0.043	160.000	159.961	0.014	160.000	159.975	0.000
200	MAX.	200.290	199.760	0.820	200.115	199.830	0.400	200.072	199.950	0.168	200.046	199.985	0.090	200.046	200.000	0.075
	MIN.	200.000	199.470	0.240	200.000	199.715	0.170	200.000	199.904	0.050	200.000	199.956	0.015	200.000	199.971	0.000
250	MAX.	250.290	249.720	0.860	250.115	249.830	0.400	250.072	249.950	0.168	250.046	249.985	0.090	250.046	250.000	0.075
	MIN.	250.000	249.430	0.280	250.000	249.715	0.170	250.000	249.904	0.050	250.000	249.956	0.015	250.000	249.971	0.000
300	MAX.	300.320	299.670	0.970	300.130	299.810	0.450	300.081	299.944	0.189	300.052	299.983	0.101	300.052	300.000	0.084
	MIN.	300.000	299.350	0.330	300.000	299.680	0.190	300.000	299.892	0.056	300.000	299.951	0.017	300.000	299.968	0.000
400	MAX.	400.360	399.600	1.120	400.140	399.790	0.490	400.089	399.938	0.208	400.057	399.982	0.111	400.057	400.000	0.093
	MIN.	400.000	399.240	0.400	400.000	399.650	0.210	400.000	399.881	0.062	400.000	399.946	0.018	400.000	399.964	0.000
500	MAX.	500.400	499.520	1.280	500.155	499.770	0.540	500.097	499.932	0.228	500.063	499.980	0.123	500.063	500.000	0.103
	MIN.	500.000	499.120	0.480	500.000	499.615	0.230	500.000	499.869	0.068	500.000	499.940	0.020	500.000	499.960	0.000

NOTE: All dimensions are in mm.

TABLE A-41 Preferred Hole Basis Transition and Interference Fits—Cylindrical Fits,[a] American National Standard

Basic size		*Locational transn.*			*Locational transn.*			*Locational interf.*			*Medium drive*			*Force*		
		Hole H7	*Shaft k6*	*Fit*	*Hole H7*	*Shaft n6*	*Fit*	*Hole H7*	*Shaft p6*	*Fit*	*Hole H7*	*Shaft s6*	*Fit*	*Hole H7*	*Shaft u6*	*Fit*
1	MAX.	1.010	1.006	0.010	1.010	1.010	0.006	1.010	1.012	0.004	1.010	1.020	−0.004	1.010	1.024	−0.008
	MIN.	1.000	1.000	−0.006	1.000	1.004	−0.010	1.000	1.006	−0.012	1.000	1.014	−0.020	1.000	1.018	−0.024
1.2	MAX.	1.210	1.206	0.010	1.210	1.210	0.006	1.210	1.212	0.004	1.210	1.220	−0.004	1.210	1.224	−0.008
	MIN.	1.200	1.200	−0.006	1.200	1.204	−0.010	1.200	1.206	−0.012	1.200	1.214	−0.020	1.200	1.218	−0.024
1.6	MAX.	1.610	1.606	0.010	1.610	1.610	0.006	1.610	1.612	0.004	1.610	1.620	−0.004	1.610	1.624	−0.008
	MIN.	1.600	1.600	−0.006	1.600	1.604	−0.010	1.600	1.606	−0.012	1.600	1.614	−0.020	1.600	1.618	−0.024
2	MAX.	2.010	2.006	0.010	2.010	2.010	0.006	2.010	2.012	0.004	2.010	2.020	−0.004	2.010	2.024	−0.008
	MIN.	2.000	2.000	−0.006	2.000	2.004	−0.010	2.000	2.006	−0.012	2.000	2.014	−0.020	2.000	2.018	−0.024
2.5	MAX.	2.510	2.506	0.010	2.510	2.510	0.006	2.510	2.512	0.004	2.510	2.520	−0.004	2.510	2.524	−0.008
	MIN.	2.500	2.500	−0.006	2.500	2.504	−0.010	2.500	2.506	−0.012	2.500	2.514	−0.020	2.500	2.518	−0.024
3	MAX.	3.010	3.006	0.010	3.010	3.010	0.006	3.010	3.012	0.004	3.010	3.020	−0.004	3.010	3.024	−0.008
	MIN.	3.000	3.000	−0.006	3.000	3.004	−0.010	3.000	3.006	−0.012	3.000	3.014	−0.020	3.000	3.018	−0.024
4	MAX.	4.012	4.009	0.011	4.012	4.016	0.004	4.012	4.020	0.000	4.012	4.027	0.007	4.012	4.031	−0.011
	MIN.	4.000	4.001	−0.009	4.000	4.008	−0.016	4.000	4.012	−0.020	4.000	4.019	−0.027	4.000	4.023	−0.031
5	MAX.	5.012	5.009	0.011	5.012	5.016	0.004	5.012	5.020	0.000	5.012	5.027	−0.007	5.012	5.031	−0.011
	MIN.	5.000	5.001	−0.009	5.000	5.008	−0.016	5.000	5.012	−0.020	5.000	5.019	−0.027	5.000	5.023	−0.031
6	MAX.	6.012	6.009	0.011	6.012	6.016	0.004	6.012	6.020	0.000	6.012	6.027	−0.007	6.012	6.031	−0.011
	MIN.	6.000	6.001	−0.009	6.000	6.008	−0.016	6.000	6.012	−0.020	6.000	6.019	−0.027	6.000	6.023	−0.031
8	MAX.	8.015	8.010	0.014	8.015	8.019	0.005	8.015	8.024	0.000	8.015	8.032	−0.008	8.015	8.037	−0.013
	MIN.	8.000	8.001	−0.010	8.000	8.010	−0.019	8.000	8.015	−0.024	8.000	8.023	−0.032	8.000	8.028	−0.037
10	MAX.	10.015	10.010	0.014	10.015	10.019	0.005	10.015	10.024	0.000	10.015	10.032	−0.008	10.015	10.037	−0.013
	MIN.	10.000	10.001	−0.010	10.000	10.010	−0.019	10.000	10.015	−0.024	10.000	10.023	−0.032	10.000	10.028	−0.037
12	MAX.	12.018	12.012	0.017	12.018	12.023	0.006	12.018	12.029	0.000	12.018	12.039	−0.010	12.018	12.044	−0.015
	MIN.	12.000	12.001	−0.012	12.000	12.012	−0.023	12.000	12.018	−0.029	12.000	12.028	−0.039	12.000	12.033	−0.044

[a]ANSI B4.2.

NOTE: All dimensions are in mm.

Continued on page 602

TABLE A-41 (continued)

Basic size		Locational transn.			Locational transn.			Locational interf.			Medium drive			Force		
		Hole H7	*Shaft k6*	*Fit*	*Hole H7*	*Shaft n6*	*Fit*	*Hole H7*	*Shaft p6*	*Fit*	*Hole H7*	*Shaft s6*	*Fit*	*Hole H7*	*Shaft u6*	*Fit*
16	MAX.	16.018	16.012	0.017	16.018	16.023	0.006	16.018	16.029	0.000	16.018	16.039	−0.010	16.018	16.044	−0.015
	MIN.	16.000	16.001	−0.012	16.000	16.012	−0.023	16.000	16.018	−0.029	16.000	16.028	−0.039	16.000	16.033	−0.044
20	MAX.	20.021	20.015	0.019	20.021	20.028	0.006	20.021	20.035	−0.001	20.021	20.048	−0.014	20.021	20.054	−0.020
	MIN.	20.000	20.002	−0.015	20.000	20.015	−0.028	20.000	20.022	−0.035	20.000	20.035	−0.048	20.000	20.041	−0.054
25	MAX.	25.021	25.015	0.019	25.021	25.028	0.006	25.021	25.035	−0.001	25.021	25.048	−0.014	25.021	25.061	−0.027
	MIN.	25.000	25.002	−0.015	25.000	25.015	−0.028	25.000	25.022	−0.035	25.000	25.035	−0.048	25.000	25.048	−0.061
30	MAX.	30.021	30.015	0.019	30.021	30.028	0.006	30.021	30.035	−0.001	30.021	30.048	−0.014	30.021	30.061	−0.027
	MIN.	30.000	30.002	−0.015	30.000	30.015	−0.028	30.000	30.022	−0.035	30.000	30.035	−0.048	30.000	30.048	−0.061
40	MAX.	40.025	40.018	0.023	40.025	40.033	0.008	40.025	40.042	−0.001	40.025	40.059	−0.018	40.025	40.076	−0.035
	MIN.	40.000	40.002	−0.018	40.000	40.017	−0.033	40.000	40.026	−0.042	40.000	40.043	−0.059	40.000	40.060	−0.076
50	MAX.	50.025	50.018	0.023	50.025	50.033	0.008	50.025	50.042	−0.001	50.025	50.059	−0.018	50.025	50.086	−0.045
	MIN.	50.000	50.002	−0.018	50.000	50.017	−0.033	50.000	50.026	−0.042	50.000	50.043	−0.059	50.000	50.070	−0.086
60	MAX.	60.030	60.021	0.028	60.030	60.039	0.010	60.030	60.051	−0.002	60.030	60.072	−0.023	60.030	60.106	−0.057
	MIN.	60.000	60.002	−0.021	60.000	60.020	−0.039	60.000	60.032	−0.051	60.000	60.053	−0.072	60.000	60.087	−0.016
80	MAX.	80.030	80.021	0.028	80.030	80.039	0.010	80.030	80.051	−0.002	80.030	80.078	−0.029	80.030	80.121	−0.072
	MIN.	80.000	80.002	−0.021	80.000	80.020	−0.039	80.000	80.032	−0.051	80.000	80.059	−0.078	80.000	80.102	−0.121
100	MAX.	100.035	100.025	0.032	100.035	100.045	0.012	100.035	100.059	−0.002	100.035	100.093	−0.036	100.035	100.146	−0.089
	MIN.	100.000	100.003	−0.025	100.000	100.023	−0.045	100.000	100.037	−0.059	100.000	100.071	−0.093	100.000	100.124	−0.146
120	MAX.	120.035	120.025	0.032	120.035	120.045	0.012	120.035	120.059	−0.002	120.035	120.101	−0.044	120.035	120.166	−0.109
	MIN.	120.000	120.003	−0.025	120.000	120.023	−0.045	120.000	120.037	−0.059	120.000	120.079	−0.101	120.000	120.144	−0.166
160	MAX.	160.040	160.028	0.037	160.040	160.052	0.013	160.040	160.068	−0.003	160.040	160.125	−0.060	160.040	160.215	−0.150
	MIN.	160.000	160.003	−0.028	160.000	160.027	−0.052	160.000	160.043	−0.068	160.000	160.100	−0.125	160.000	160.190	−0.215
200	MAX.	200.046	200.033	0.042	200.046	200.060	0.015	200.046	200.079	−0.004	200.046	200.151	−0.076	200.046	200.265	−0.190
	MIN.	200.000	200.004	−0.033	200.000	200.031	−0.060	200.000	200.050	−0.079	200.000	200.122	−0.151	200.000	200.236	−0.265

NOTE: All dimensions are in mm.

TABLE A-41 (continued)

Basic size		Locational transn.			Locational transn.			Locational interf.			Medium drive			Force		
		Hole H7	*Shaft k6*	*Fit*	*Hole H7*	*Shaft n6*	*Fit*	*Hole H7*	*Shaft p6*	*Fit*	*Hole H7*	*Shaft s6*	*Fit*	*Hole H7*	*Shaft u6*	*Fit*
250	MAX.	250.046	250.033	0.042	250.046	250.060	0.015	250.046	250.079	−0.004	250.046	250.169	−0.094	250.046	250.313	−0.238
	MIN.	250.000	250.004	−0.033	250.000	250.031	−0.060	250.000	250.050	−0.079	250.000	250.140	−0.169	250.000	250.284	−0.313
300	MAX.	300.052	300.036	0.048	300.052	300.066	0.018	300.052	300.088	−0.004	300.052	300.202	−0.118	300.052	300.382	−0.298
	MIN.	300.000	300.004	−0.036	300.000	300.034	−0.066	300.000	300.056	−0.088	300.000	300.170	−0.202	300.000	300.350	−0.382
400	MAX.	400.057	400.040	0.053	400.057	400.073	0.020	400.057	400.098	−0.005	400.057	400.244	−0.151	400.057	400.471	−0.378
	MIN.	400.000	400.004	−0.040	400.000	400.037	−0.073	400.000	400.062	−0.098	400.000	400.208	−0.244	400.000	400.435	−0.471
500	MAX.	500.063	500.045	0.058	500.063	500.080	0.023	500.063	500.108	−0.005	500.063	500.292	−0.189	500.063	500.580	−0.477
	MIN.	500.000	500.005	−0.045	500.000	500.040	−0.080	500.000	500.068	−0.108	500.000	500.252	−0.292	500.000	500.540	−0.580

NOTE: All dimensions are in mm.

TABLE A-42 Preferred Shaft Basis Clearance Fits—Cylindrical Fits,[a] American National Standard

Basic size		Loose running			Free running			Close running			Sliding			Locational clearance		
		Hole C11	*Shaft h11*	*Fit*	*Hole D9*	*Shaft h9*	*Fit*	*Hole F8*	*Shaft h7*	*Fit*	*Hole G7*	*Shaft h6*	*Fit*	*Hole H7*	*Shaft h6*	*Fit*
1	MAX.	1.120	1.000	0.180	1.045	1.000	0.070	1.020	1.000	0.030	1.012	1.000	0.018	1.010	1.000	0.016
	MIN.	1.060	0.940	0.060	1.020	0.975	0.020	1.006	0.990	0.006	1.002	0.994	0.002	1.000	0.994	0.000
1.2	MAX.	1.320	1.200	0.180	1.245	1.200	0.070	1.220	1.200	0.030	1.212	1.200	0.018	1.210	1.200	0.016
	MIN.	1.260	1.140	0.060	1.220	1.175	0.020	1.206	1.190	0.006	1.202	1.194	0.002	1.200	1.194	0.000
1.6	MAX.	1.720	1.600	0.180	1.645	1.600	0.070	1.620	1.600	0.030	1.612	1.600	0.018	1.610	1.600	0.016
	MIN.	1.660	1.540	0.060	1.620	1.575	0.020	1.606	1.590	0.006	1.602	1.594	0.002	1.600	1.594	0.000
2	MAX.	2.120	2.000	0.180	2.045	2.000	0.070	2.020	2.000	0.030	2.012	2.000	0.018	2.010	2.000	0.016
	MIN.	2.060	1.940	0.060	2.020	1.975	0.020	2.006	1.990	0.006	2.002	1.994	0.002	2.000	1.994	0.000
2.5	MAX.	2.620	2.500	0.180	2.545	2.500	0.070	2.520	2.500	0.030	2.512	2.500	0.018	2.510	2.500	0.016
	MIN.	2.560	2.440	0.060	2.520	2.475	0.020	2.506	2.490	0.006	2.502	2.494	0.002	2.500	2.494	0.000

[a]ANSI B4.2.

NOTE: All dimensions are in mm.

Continued on page 604

TABLE A-42 (continued)

Basic size		Loose running			Free running			Close running			Sliding			Locational clearance		
		Hole C11	*Shaft h11*	*Fit*	*Hole D9*	*Shaft h9*	*Fit*	*Hole F8*	*Shaft h7*	*Fit*	*Hole G7*	*Shaft h6*	*Fit*	*Hole H7*	*Shaft h6*	*Fit*
3	MAX.	3.120	3.000	0.180	3.045	3.000	0.070	3.020	3.000	0.030	3.012	3.000	0.018	3.010	3.000	0.016
	MIN.	3.060	2.940	0.060	3.020	2.975	0.020	3.006	2.990	0.006	3.002	2.994	0.002	3.000	2.994	0.000
4	MAX.	4.145	4.000	0.220	4.060	4.000	0.090	4.028	4.000	0.040	4.016	4.000	0.024	4.012	3.000	0.020
	MIN.	4.070	3.925	0.070	4.030	3.970	0.030	4.010	3.988	0.010	4.004	3.992	0.004	4.000	3.992	0.000
5	MAX.	5.145	5.000	0.220	5.060	5.000	0.090	5.028	5.000	0.040	5.016	5.000	0.024	5.012	5.000	0.020
	MIN.	5.070	4.925	0.070	5.030	4.970	0.030	5.010	4.988	0.010	5.004	4.992	0.004	5.000	4.992	0.000
6	MAX.	6.145	6.000	0.220	6.060	6.000	0.090	6.028	6.000	0.040	6.016	6.000	0.024	6.012	6.000	0.020
	MIN.	6.070	5.925	0.070	6.030	5.970	0.030	6.010	5.988	0.010	6.004	5.992	0.004	6.000	5.992	0.000
8	MAX.	8.170	8.000	0.260	8.076	8.000	0.112	8.035	8.000	0.050	8.020	8.000	0.029	8.015	8.000	0.024
	MIN.	8.080	7.910	0.080	8.040	7.964	0.040	8.013	7.985	0.013	8.005	7.991	0.005	8.000	7.991	0.000
10	MAX.	10.170	10.000	0.260	10.076	10.000	0.112	10.035	10.000	0.050	10.020	10.000	0.029	10.015	10.000	0.024
	MIN.	10.080	9.910	0.080	10.040	9.964	0.040	10.013	9.985	0.013	10.005	9.991	0.005	10.000	9.991	0.000
12	MAX.	12.205	12.000	0.315	12.093	12.000	0.136	12.043	12.000	0.061	12.024	12.000	0.035	12.018	12.000	0.029
	MIN.	12.095	11.890	0.095	12.050	11.957	0.050	12.016	11.982	0.016	12.006	11.989	0.006	12.000	11.989	0.000
16	MAX.	16.205	16.000	0.315	16.093	16.000	0.136	16.043	16.000	0.061	16.024	16.000	0.035	16.018	16.000	0.029
	MIN.	16.095	15.890	0.095	16.050	15.957	0.050	16.016	15.982	0.016	16.006	15.989	0.006	16.000	15.989	0.000
20	MAX.	20.240	20.000	0.370	20.117	20.000	0.169	20.053	20.000	0.074	20.028	20.000	0.041	20.021	20.000	0.034
	MIN.	20.110	19.870	0.110	20.065	19.948	0.065	20.020	19.979	0.020	20.007	19.987	0.007	20.000	19.987	0.000
25	MAX.	25.240	25.000	0.370	25.117	25.000	0.169	25.053	25.000	0.074	25.028	25.000	0.041	25.021	25.000	0.034
	MIN.	25.110	24.870	0.110	25.065	24.948	0.065	25.020	24.979	0.020	25.007	24.987	0.007	25.000	24.987	0.000
30	MAX.	30.240	30.000	0.370	30.117	30.000	0.169	30.053	30.000	0.074	30.028	30.000	0.041	30.021	30.000	0.034
	MIN.	30.110	29.870	0.110	30.065	29.948	0.065	30.020	29.979	0.020	30.007	29.987	0.007	30.000	29.987	0.000

NOTE: All dimensions are in mm.

TABLE A-42 (continued)

Basic size		Loose running			Free running			Close running			Sliding			Locational clearance		
		Hole C11	*Shaft h11*	*Fit*	*Hole D9*	*Shaft h9*	*Fit*	*Hole F8*	*Shaft h7*	*Fit*	*Hole G7*	*Shaft h6*	*Fit*	*Hole H7*	*Shaft h6*	*Fit*
40	MAX.	40.280	40:000	0.440	40.142	40.000	0.204	40.064	40.000	0.089	40.034	40.000	0.050	40.025	40.000	0.041
	MIN.	40.120	39.840	0.120	40.080	39.938	0.080	40.025	39.975	0.025	40.009	39.984	0.009	40.000	39.984	0.000
50	MAX.	50.290	50.000	0.450	50.142	50.000	0.204	50.064	50.000	0.089	50.034	50.000	0.050	50.025	50.000	0.041
	MIN.	50.130	49.840	0.130	50.080	49.938	0.080	50.025	49.975	0.025	50.009	49.984	0.009	50.000	49.984	0.000
60	MAX.	60.330	60.000	0.520	60.174	60.000	0.248	60.076	60.000	0.106	60.040	60.000	0.059	60.030	60.000	0.049
	MIN.	60.140	59.810	0.140	60.100	59.926	0.100	60.030	59.970	0.030	60.010	59.981	0.010	60.000	59.981	0.000
80	MAX.	80.340	80.000	0.530	80.174	80.000	0.248	80.076	80.000	0.106	80.040	80.000	0.059	80.030	80.000	0.049
	MIN.	80.150	79.810	0.150	80.100	79.926	0.100	80.030	79.970	0.030	80.010	79.981	0.010	80.000	79.981	0.000
100	MAX.	100.390	100.000	0.610	100.207	100.000	0.294	100.090	100.000	0.125	100.047	100.000	0.069	100.035	100.000	0.057
	MIN.	100.170	99.780	0.170	100.120	99.913	0.120	100.036	99.965	0.036	100.012	99.978	0.012	100.000	99.978	0.000
120	MAX.	120.400	120.000	0.620	120.207	120.000	0.294	120.090	120.000	0.125	120.047	120.000	0.069	120.035	120.000	0.057
	MIN.	120.180	119.780	0.180	120.120	119.913	0.120	120.036	119.965	0.036	120.012	119.978	0.012	120.000	119.978	0.000
160	MAX.	160.460	160.000	0.710	160.245	160.000	0.345	160.106	160.000	0.146	160.054	160.000	0.079	160.040	160.000	0.065
	MIN.	160.210	159.750	0.210	160.145	159.900	0.145	160.043	159.960	0.043	160.014	159.975	0.014	160.000	159.975	0.000
200	MAX.	200.530	200.000	0.820	200.285	200.000	0.400	200.122	200.000	0.168	200.061	200.000	0.090	200.046	200.000	0.075
	MIN.	200.240	199.710	0.240	200.170	199.885	0.170	200.050	199.954	0.050	200.015	199.971	0.015	200.000	199.971	0.000
250	MAX.	250.570	250.000	0.860	250.285	250.000	0.400	250.122	250.000	0.168	250.061	250.000	0.090	250.046	250.000	0.075
	MIN.	250.280	249.710	0.280	250.170	249.885	0.170	250.050	249.954	0.050	250.015	249.971	0.015	250.000	249.971	0.000
300	MAX.	300.650	300.000	0.970	300.320	300.000	0.450	300.137	300.000	0.189	300.069	300.000	0.101	300.052	300.000	0.084
	MIN.	300.330	299.680	0.330	300.190	299.870	0.190	300.056	299.948	0.056	300.017	299.968	0.017	300.000	299.968	0.000
400	MAX.	400.760	400.000	1.120	400.350	400.000	0.490	400.151	400.000	0.208	400.075	400.000	0.111	400.057	400.000	0.093
	MIN.	400.400	399.640	0.400	400.210	399.860	0.210	400.062	399.943	0.062	400.018	399.964	0.018	400.000	399.964	0.000
500	MAX.	500.880	500.000	1.280	500.385	500.000	0.540	500.165	500.000	0.228	500.083	500.000	0.123	500.063	500.000	0.103
	MIN.	500.480	499.600	0.480	500.230	499.845	0.230	500.068	499.937	0.068	500.020	499.960	0.020	500.000	499.960	0.000

NOTE: All dimensions are in mm.

Continued on page 606

TABLE A-43 Preferred Shaft Basis Transition and Interference Fits—Cylindrical Fits,[a] American National Standard

Basic size		*Locational transn.*			*Locational transn.*			*Locational interf.*			*Medium drive*			*Force*		
		Hole K7	***Shaft h6***	***Fit***	***Hole N7***	***Shaft h6***	***Fit***	***Hole P7***	***Shaft h6***	***Fit***	***Hole S7***	***Shaft h6***	***Fit***	***Hole U7***	***Shaft h6***	***Fit***
1	MAX.	1.000	1.000	0.006	0.996	1.000	0.002	0.994	1.000	0.000	0.986	1.000	−0.008	0.982	1.000	−0.012
	MIN.	0.990	0.994	−0.010	0.986	0.994	−0.014	0.984	0.994	−0.016	0.976	0.994	−0.024	0.972	0.994	−0.028
1.2	MAX.	1.200	1.200	0.006	1.196	1.200	0.002	1.194	1.200	0.000	1.186	1.200	−0.008	1.182	1.200	−0.012
	MIN.	1.190	1.194	−0.010	1.186	1.194	−0.014	1.184	1.194	−0.016	1.176	1.194	−0.024	1.172	1.194	−0.028
1.6	MAX.	1.600	1.600	0.006	1.596	1.600	0.002	1.594	1.600	0.000	1.586	1.600	−0.008	1.582	1.600	−0.012
	MIN.	1.590	1.594	−0.010	1.586	1.594	−0.014	1.584	1.594	−0.016	1.576	1.594	−0.024	1.572	1.594	−0.028
2	MAX.	2.000	2.000	0.006	1.996	2.000	0.002	1.994	2.000	0.000	1.986	2.000	−0.008	1.982	2.000	−0.012
	MIN.	1.990	1.994	−0.010	1.986	1.994	−0.014	1.984	1.994	−0.016	1.976	1.994	−0.024	1.972	1.994	−0.028
2.5	MAX.	2.500	2.500	0.006	2.496	2.500	0.002	2.494	2.500	0.000	2.486	2.500	−0.008	2.482	2.500	−0.012
	MIN.	2.490	2.494	−0.010	2.486	2.494	−0.014	2.484	2.494	−0.016	2.476	2.494	−0.024	2.472	2.494	−0.028
3	MAX.	3.000	3.000	0.006	2.996	3.000	0.002	2.994	3.000	0.000	2.986	3.000	−0.008	2.982	3.000	−0.012
	MIN.	2.990	2.994	−0.010	2.986	2.994	−0.014	2.984	2.994	−0.016	2.976	2.994	−0.024	2.972	2.994	−0.028
4	MAX.	4.003	4.000	0.011	3.996	4.000	0.004	3.992	4.000	0.000	3.985	4.000	−0.007	3.981	4.000	−0.011
	MIN.	3.991	3.992	−0.009	3.984	3.992	−0.016	3.980	3.992	−0.020	3.973	3.992	−0.027	3.969	3.992	−0.031
5	MAX.	5.003	5.000	0.011	4.996	5.000	0.004	4.992	5.000	0.000	4.985	5.000	−0.007	4.981	5.000	−0.011
	MIN.	4.991	4.992	−0.009	4.984	4.992	−0.016	4.980	4.992	−0.020	4.973	4.992	−0.027	4.969	4.992	−0.031
6	MAX.	6.003	6.000	0.011	5.996	6.000	0.004	5.992	6.000	0.000	5.985	6.000	−0.007	5.981	6.000	−0.011
	MIN.	5.991	5.992	−0.009	5.984	5.992	−0.016	5.980	5.992	−0.020	5.973	5.992	−0.027	5.969	5.992	−0.031
8	MAX.	8.005	8.000	0.014	7.996	8.000	0.005	7.991	8.000	0.000	7.983	8.000	−0.008	7.978	8.000	−0.013
	MIN.	7.990	7.991	−0.010	7.981	7.991	−0.019	7.976	7.991	−0.024	7.968	7.991	−0.032	7.963	7.991	−0.037
10	MAX.	10.005	10.000	0.014	9.996	10.000	0.005	9.991	10.000	0.000	9.983	10.000	−0.008	9.978	10.000	−0.013
	MIN.	9.990	9.991	−0.010	9.981	9.991	−0.019	9.976	9.991	−0.024	9.968	9.991	−0.032	9.963	9.991	−0.037

[a]ANSI B4.2.

NOTE: All dimensions are in mm.

TABLE A-43 (continued)

Basic size		Locational transn.			Locational transn.			Locational interf.			Medium drive			Force		
		Hole K7	*Shaft h6*	*Fit*	*Hole N7*	*Shaft h6*	*Fit*	*Hole P7*	*Shaft h6*	*Fit*	*Hole S7*	*Shaft h6*	*Fit*	*Hole U7*	*Shaft h6*	*Fit*
12	MAX.	12.006	12.000	0.017	11.995	12.000	0.006	11.989	12.000	0.000	11.979	12.000	−0.010	11.974	12.000	−0.015
	MIN.	11.988	11.989	−0.012	11.977	11.989	−0.023	11.971	11.989	−0.029	11.961	11.989	−0.039	11.956	11.989	−0.044
16	MAX.	16.006	16.000	0.017	15.995	16.000	0.006	15.989	16.000	0.000	15.979	16.000	−0.010	15.974	16.000	−0.015
	MIN.	15.988	15.989	−0.012	15.977	15.989	−0.023	15.971	15.989	−0.029	15.961	15.989	−0.039	15.956	15.989	−0.044
20	MAX.	20.006	20.000	0.019	19.993	20.000	0.006	19.986	20.000	−0.001	19.973	20.000	−0.014	19.967	20.000	−0.020
	MIN.	19.985	19.987	−0.015	19.972	19.987	−0.028	19.965	19.987	−0.035	19.952	19.987	−0.048	19.946	19.987	−0.054
25	MAX.	25.006	25.000	0.019	24.993	25.000	0.006	24.986	25.000	−0.001	24.973	25.000	−0.014	24.960	25.000	−0.027
	MIN.	24.985	24.987	−0.015	24.972	24.987	−0.028	24.965	24.987	−0.035	24.952	24.987	−0.048	24.939	24.987	−0.061
30	MAX.	30.006	30.000	0.019	29.993	30.000	0.006	29.986	30.000	−0.001	29.973	30.000	−0.014	29.960	30.000	−0.027
	MIN.	29.985	29.987	−0.015	29.972	29.987	−0.028	29.965	29.987	−0.035	29.952	29.987	−0.048	29.939	29.987	−0.061
40	MAX.	40.007	40.000	0.023	39.992	40.000	0.008	39.983	40.000	−0.001	39.966	40.000	−0.018	39.949	40.000	−0.035
	MIN.	39.982	39.984	−0.018	39.967	39.984	−0.033	39.958	39.984	−0.042	39.941	39.984	−0.059	39.924	39.984	−0.076
50	MAX.	50.007	50.000	0.023	49.992	50.000	0.008	49.983	50.000	−0.001	49.966	50.000	−0.018	49.939	50.000	−0.045
	MIN.	49.982	49.984	−0.018	49.967	49.984	−0.033	49.958	49.984	−0.042	49.941	49.984	−0.059	49.914	49.984	−0.086
60	MAX.	60.009	60.000	0.028	59.991	60.000	0.010	59.979	60.000	−0.002	59.958	60.000	−0.023	59.924	60.000	−0.057
	MIN.	59.979	59.981	−0.021	59.961	59.981	−0.039	59.949	59.981	−0.051	59.928	59.981	−0.072	59.894	59.981	−0.106
80	MAX.	80.009	80.000	0.028	79.991	80.000	0.010	79.979	80.000	−0.002	79.952	80.000	−0.029	79.909	80.000	−0.072
	MIN.	79.979	79.981	−0.021	79.961	79.981	−0.039	79.949	79.981	−0.051	79.922	79.981	−0.078	79.879	79.981	−0.121
100	MAX.	100.010	100.000	0.032	99.990	100.000	0.012	99.976	100.000	−0.002	99.942	100.000	−0.036	99.889	100.000	−0.089
	MIN.	99.975	99.978	−0.025	99.955	99.978	−0.045	99.941	99.978	−0.059	99.907	99.978	−0.093	99.854	99.978	−0.146
120	MAX.	120.010	120.000	0.032	199.990	120.000	0.012	119.976	120.000	−0.002	119.934	120.000	−0.044	199.869	120.000	−0.109
	MIN.	119.975	119.978	−0.025	119.955	119.978	−0.045	119.941	119.978	−0.059	119.899	119.978	−0.101	119.834	119.978	−0.166

NOTE: All dimensions are in mm.

Continued on page 608

TABLE A-43 (continued)

Basic size		*Locational transn.*			*Locational transn.*			*Locational interf.*			*Medium drive*			*Force*		
		Hole K7	*Shaft h6*	*Fit*	*Hole N7*	*Shaft h6*	*Fit*	*Hole P7*	*Shaft h6*	*Fit*	*Hole S7*	*Shaft h6*	*Fit*	*Hole U7*	*Shaft h6*	*Fit*
160	MAX.	160.012	160.000	0.037	159.988	160.000	0.013	159.972	160.000	−0.003	159.915	160.000	−0.060	159.825	160.000	−0.150
	MIN.	159.972	159.975	−0.028	159.948	159.975	−0.052	159.932	159.975	−0.068	159.875	159.975	−0.125	159.785	159.975	−0.215
200	MAX.	200.013	200.000	0.042	199.986	200.000	0.015	199.967	200.000	−0.004	199.895	200.000	−0.076	199.781	200.000	−0.190
	MIN.	199.967	199.971	−0.033	199.940	199.971	−0.060	199.921	199.971	−0.079	199.849	199.971	−0.151	199.735	199.971	−0.265
250	MAX.	250.013	250.000	0.042	249.986	250.000	0.015	249.967	250.000	−0.004	249.877	250.000	−0.094	249.733	250.000	−0.238
	MIN.	249.967	249.971	−0.033	249.940	249.971	−0.060	249.921	249.971	−0.079	249.831	249.971	−0.169	249.687	249.971	−0.313
300	MAX.	300.016	300.000	0.048	299.986	300.000	−0.018	299.964	300.000	−0.004	299.850	300.000	−0.118	299.670	300.000	−0.298
	MIN.	299.964	299.968	−0.036	299.934	299.968	−0.066	299.912	299.968	−0.088	299.798	299.968	−0.202	299.618	299.968	−0.382
400	MAX.	400.017	400.000	0.053	399.984	400.000	0.020	399.959	400.000	−0.005	399.813	400.000	−0.151	399.586	400.000	−0.378
	MIN.	399.960	399.964	−0.040	399.927	399.964	−0.073	399.902	399.964	−0.098	399.756	399.964	−0.244	399.529	399.964	−0.471
500	MAX.	500.018	500.000	0.058	499.983	500.000	0.023	499.955	500.000	−0.005	499.771	500.000	−0.189	499.483	500.000	−0.477
	MIN.	499.955	499.960	−0.045	499.920	499.960	−0.080	499.892	499.960	−0.108	499.708	499.960	−0.292	499.420	499.960	−0.580

NOTE: All dimensions are in mm.

Index

A

Abrasive belt finishing, 277
Accuracy, 326
Acme thread, 288, 588t
Adhesives, 264
Adjacent views, 143
Aligned sections, 227
Allowance, 385
Alloys
  properties of, 256
  kinds of, 252
  unified numbering system for, 254
Aluminum, 255
American National thread, 288
Angle between a line and a plane, 482
Angle of thread, 287
Angles
  acute, 97, 115
  bisecting, 106
  dimensioning, 329
  laying out, 106
  obtuse, 97, 115
  right, 97, 115
  sketching, 10
  straight, 97
Angular measurement, 97
Arcs, 46, 99
  dimensioning, 330
  sketching, 7
  tangent, 114
Arc welding, 264
Arrowheads, 317
Assembly drawings
  exploded pictorial, 426
  production design, 424
  sectioning techniques, 429
  working, 426
Assembly methods, 264
Asymptote, 130
Auxiliary sections, 233
Auxiliary views
  construction of, 192–194
  definition of, 189
  partial, 191
  plotting curves in, 195–196
  primary, 190, 192
  secondary, 198
Axis
  concentricity and, 409–410
  cylinder, 39
  horizontal, 36
  oblique, 36
  origin, 36
  parabola, 128
  transverse (hyperbola), 130
Axis of thread, 287
Axonometric positions, 42
Azimuth, 473

B

Barrel finishing, 276
Basic hole system, 386
Basic size, 382
Basic shaft system, 387
Bearing, 473
Bill of material, 428
Bisecting
  a line, 101
  an angle, 106
Bisector (parabola), 129
Blanking operation, 265
Bolt circle, 344
Bolt lengths, 306
Bolts, 304–306
  computer drawings of, 549–550
  hex, 559t
  metric, 564t
  square, 560t
Border lines, 6, 544
Boring, 271
Breaks, 238
Brittleness, 257
Broaching, 270
Broken-out sections, 234
Buffing, 278
Buttress thread, 288

C

Cabinet sketch, 36
Callouts, 297
Cap screws, 308, 564t, 566t
Casting
  die, 260
  fillets and rounds and, 169–170
  investment, 259–260
  methods, 258–262
  of parts, 169–170
  permanent mold, 260
  plaster mold, 259
  principles, 261t
  sand, 257
  shell molding, 259
Cathode ray tube (CRT), 523
Cavalier sketch, 36
Center line, 8, 144, 317
  symbol, 340
Chamfer, 295, 351
Chamfer angle on bolthead and nut, 305
Chemical engineering, 520
Chord, 99
Circle(s)
  and arcs, 99
  center of, finding, 113
  circumference of, 99
  concentric, 99
  definition of, 99, 122
  eccentric, 99
  in isometric sketches, 46
  in oblique sketches, 39
  sketching, 7
  and tangent lines, 113–114
Civil engineering, 519

Clearance fits
preferred hole basis, 598t
preferred shaft basis, 603t
Clearance locational fits, 590t
Clevis pins, 581t
Compass, 79
Computer
CRT displays, 524
digitizing tablets, 526
disks, 531
file storage, 529, 531
hardware, 522–531
joysticks, 528
keyboards, 526
light pens, 527
output devices, 529–530
plotters, 529–530
tape, 531
workstations, 522–528
See also computer graphics
Computer-aided design (CAD), 518
Computer-aided manufacturing (CAM), 283, 518
Computer graphics
advantages of, 522
and creating and saving a drawing, 548
drawing with, 539–543, 544–550
drawing routines, 534
in engineering, 516–522
modifying drawings with, 543–544
program interaction, 538
software, 532–538
workstation, 522–528
See also Computer
Computer numerical control (CNC), 283
Concentricity, 409
Cone(s)
element, 101, 122
frustum, 100, 101
oblique, 100, 101
right, 100, 101
truncated, 100, 101
Conic sections, 122
Construction methods, geometric, 104–114
Conventional representation, 234–238
Copper, 256
Copper-base alloys, 256
Cores, casting, 259
Corners, 331
Cotter pins, 578t
Crest, thread, 287
Curve fitting, 80
Curves
irregular, 80
ogee, 119
See also Arcs, Circle(s)
Cut block, computer drawing of, 344–548
Cut threads, 292
Cutting plane, 216
Cylinder(s)
dimensioning, 334
element, 101
hollow, 335
intersecting, 168
oblique, 100, 101
right-circular, 47, 101, 108
solid, 334

D

Database, 517
Datum(s)
cylindrical, 418
dimensioning, 361
flat, 416–417
planes, 340
primary, 404, 417
reference letters, 404
secondary, 404, 417
specifying, 415
the three datum plane system, 416
tertiary, 404, 417
Depth, 37, 143
of thread, 287
Descriptive geometry, 466–486
Design
computer-aided (CAD), 518
guidelines for, 252
Detail drawings, 421–423
Deviation, types of, 391
Diameter, 99
major, 286
minor, 286
symbol, 146
Die casting, 260
Die set, 265
Digitizer, 526
Digitizing tablets, 526
Dihedral angle, 484
Dimension figures, 319
Dimension lines, 316, 318
Dimension(s)
out-of-scale, 358
overall, 328
placement, 319
reference, 328
shape, 337
size, 337
Dimensioning
accuracy in, 326
aligned, 319
angles, 329
arcs, 330
baseline, 398
bored holes, 348
for CAD and CAM, 360
chain, 398
chamfers, 351
contour, 339
coordinate, 402
counterbored holes, 348
counterdrilled holes, 346
countersunk holes, 350
curved outlines, 331
cylinders, 334
datum, 331, 361
direct, 398
drilled holes, 346
dual, 323
half-sections, 357
holes, 341, 344, 346–348, 350

irregular curved outlines, 340
keyseats, 354
keyways, 354
knurling, 355
limit, 314
in limited spaces, 327
machining centers, 353
necks, 356
partially-rounded ends, 333
parts with rounded ends, 332
reamed holes, 347
repetitive features, 328
rounded corners, 331
spotfaced holes, 348
square shapes, 353
symmetrical curved outlines, 340
systems, 320
tabular, 359
tapers, 352
tapped holes, 350
thread reliefs, 356
undercuts, 356
unidirectional, 319
Dimetric axonometric position, 42
Dip, 481
Directrix, 128
Disk storage, 531
Dividers, 82
Drafting machine, 60
Drawing
guide lines, 17
horizontal lines, 63
inclined lines, 64
parallel lines, 66
perpendicular lines, 66
pictorial, 49–51
scales, 68–76
vertical lines, 63
See also Computer graphics; Drawings, Geometric construction; Sketching
Drawings
border lines on, 6, 544
checking, 433–434
computer-generated, 13
detail, 421–423
multiview, 140–141
one-view, 146
tabular, 359
three-view, 150–151
two-view, 146–150
working, 315, 421
See also Assembly drawings
Drill sizes, 556t, 557t
Ductility, 257

E

Edge improvement methods, 265
Editing, 545
Elasticity, 257
Electrical engineering, 522
Element
of cone, 101
of cylinder, 101
Elements, limiting, 39
Elevation views, 194
Ellipse, 40–41
concentric circle method, 124
conjugate diameter method, 125
foci method, 123
four-center method, 126
major axis, 123, 125
minor axis, 123, 125
parallelogram method, 125
tangent points of, 40
trammel method, 126
Engineering geometry, 96–127
Erasing shield, 83
Extension lines, 317, 318
External threads, 293
metric designation for, 298

F

Fasteners, threaded, 264–265, 286–310
Feature control frame, 403
Ferrous metals, 253
Fillets, 169, 330
Finished surfaces, 324–326
FIRSTDRAW, 533
Fit(s)
classes of, 289–290
clearance, 385
English system, 388
force, 389
interference, 385
locational, 389
running and sliding, 388
SI metric system, 392
transition, 385, 601t
transition locational, 593t
Flaws, 280
Focus, 128
Fold line, 190
Force and shrink fits, 389, 596t
Forging, 170, 262
grain flow, 263
Forming, 266
Free-state variation, 411
Frontal plane, 476
Full section, 219

G

Gas welding, 264
Geometric characteristic symbol, 85, 403
Geometric construction
constructing a hexagon, 110, 112
constructing a hyperbola, 130–131
constructing a parabola, 128–129
constructing a pentagon, 109, 112
constructing a square, 106
constructing a triangle, 106
constructing an ellipse, 123
constructing an octagon, 111, 112
dividing a line, 101

Geometric construction *(Cont.)*
drawing a circle tangent to a line at a given point, 113
drawing a circle through three points, 113
drawing parallel lines, 104, 105
drawing perpendicular lines, 104, 105
drawing an ogee curve, 119
finding the center of a circle, 113
laying out an angle, 106
of a line tangent through a point outside a circle, 114
of a line tangent to a circle through a point, 114
of tangent arcs, 114
Geometric figures
angles, 97
circles and arcs, 99
polygons, 99
quadrilaterals, 98
solids, 100
triangles, 98
Grade of a line, 475
Grid, 535
Grid paper, 540
Grinding, 274
Guide lines, 17

H

Half-sections, 221, 357
Hammering, 263
Hard copy, 529, 537
Hardness, 257
Hardware, computer graphic system, 522
Height, 37, 143
Hexagons, 99, 100, 109, 110
Heptagon, 99
Hidden lines, 143
Hole basic system, 391
Hole-making operations, 275
Holes
bored, 348
counterbored, 348
counterdrilled, 346
countersunk, 350
drilled, 346
location dimensioning, 341
spotfaced, 348
tapped, 350
reamed, 347
Honing, 278
Horizontal plane, 476
Hyperbola, 122

I

Inclined plane, 476
Inclined surfaces, 154
Interference locational fits, 595t
Internal threads, 293
metric designation for, 300
Intersecting lines, 469
Intersection(s)
cylinders and prisms, 169
of a line and a plane, 482
Instruments, use of, 59–83
compass, 79
dividers, 82
drafting machine, 60
erasing shield, 83
irregular curves, 80
parallel rule straight-edge, 60
protractor, 77
scales, 68
templates, 80
triangles, 62
T square, 60
International tolerance grade, 391
Investment casting, 259
Iron, 253
ISO metric screw threads, 288, 290
Isometric axonometric position, 42
Isometric sketching
arcs and circles, 46
choice of position, 45
circle positions, 48
inclined planes, 44
right circular cylinder, 47
Isosceles triangle, 98

J

Joystick, 528

K

Keys
Pratt & Whitney, 586t
types of, 572t
Woodruff, 584t
Keyseats
dimensioning, 354
Woodruff, 585t
Keyways, dimensioning, 354
Knuckle thread, 288
Knurling, 355

L

Lapping, 278
Lay, 280
Layouts, 423
Lead, 287
Leaders, 318
Least material condition (LMC), 408
Lettering
aids, mechanical, 19
capital, 14
computer-generated, 20
device, 19
engineering style, 13
guide lines for, 16
height, 16
lowercase, 14
practice, 15
single-stroke Gothic, 14
spacing, 18

strokes, 14
styles, 14
template, 19
typewriter, 19
uppercase, 14
Light pen, 527
Limit dimensioning. *See* Dimensioning
Limits, 382
of size, 384
Line(s)
aximuth, 473
bearing, 473
border, 6, 544
center line 8, 144, 317
converging, 35
dimension, 316
direction and position, 467
dividing into equal parts, 7, 101
dividing the space between, 103
drawing horizontal, 103
drawing vertical, 103
extension, 317
fold, 190
freehand break, 191, 230
frontal, 468
grade, 475
guide, 16–17
hidden, 37, 143
hidden in sections, 219
horizontal, 4, 63, 468
inclined, 6, 64
intersecting, 469
isometric, 44
nonisometric, 44
oblique, 468
parallel, 35, 66, 104, 468
perpendicular, 66, 104, 468
phantom, 172
precedence of, 144
principal, 468
profile, 468
projection, 161, 467
receding, 35
skew, 470, 485
slope, 474
techniques, 143
true length, 199, 470
vertical, 5, 63
visible (or object), 3, 143
visible in sections, 218
visibility, 483
Line of intersection, 485
Line of intersection between two planes
cutting plane method, 486
edge view method, 487

M

Machinability, 257
Machinability index, 257
Machine screw lengths, metric, 569t
Machine screws, 309, 567t, 569t
Machine tools, 267
Machining allowance, 258
Machining center, 353
Machining operations, 267
Magnesium, 256
Major diameter, 286
Malleability, 257
Manufacturing processes
primary, 257
secondary, 266
Material condition symbol
least (LMC), 403, 408
maximum (MMC), 403, 405
regardless of feature size (RFS), 403, 407
Materials engineering, 252
Mechanical engineering, 518
Menu, graphics software, 532
Metals, 252
ferrous, 253
nonferrous, 255–256
properties of, 256
unified numbering system for, 254
See also alloys
Metric system, 323
Microfinishing, 278
Milling, 267
Minor diameter, 286
Mouse (computer input device), 528
Multiview drawing, 140

N

Necks, dimensioning, 356
Nickel-base alloys, 256
Nominal size, 382
Nonferrous metals, 255
Normal surfaces, 153
Not to scale (NTS), 359
Notes, 297
Numerical control (NC), 282–283
Numbering systems, steel, 253
Nuts, 304, 306
hex, 561t
hex metric, 565t
machine screw, 304, 309, 563t
square, 562t

O

Oblique planes, 476
Oblique sketching
axes, 36
choice of position, 41
circles, 39
receding line directions, 38
types of, 36
Oblique surfaces, 155
Octagon, 99, 111
Offset sections, 224
Optimizing, 424
Orthographic projection, 140

P

Parabola, 122
Partial views, 236
Part program, 282
Parts list, 428
Pattern, 258
Pencils, 61, 62
Pentagon, 99, 109

Perforating, 265
Permanent mold casting, 260
Perpendicular lines, 469
Phantom lines, 172
Pictorial drawings
  centering, 49
  computer generated, 51
  sectioning in, 50
Pictorial sketches
  axonometric, 35
  and hidden lines, 37
  oblique, 35
  perspective in, 35
Piercing, 265
Pitch, 287
Pitch diameter, 287
Pixel (or pels), 524
Plane(s)
  dip, 481
  edge view, 478
  front, 36, 37, 42
  frontal, 140, 476
  horizontal, 140, 476
  inclined, 44, 476
  oblique, 45, 476
  parallel, 38
  principal, 476
  profile, 140, 476
  right-side, 42
  slope, 480
  strike, 480
  surfaces of, 153, 475
  top, 42
  true length lines in, 478
  true size, 479
  visibility, 486
Plaster mold casting, 259
Plotters
  drum, 530
  flatbed, 529
  pen, 529
Points
  position of, 466
  projection of, 466, 473
Polishing, 278
Polygons, regular, 99
Powder metallurgy, 266
Preferred basic sizes, 393
Preferred fits, 394
Pressing, 263
Pressworking practices, 265
Primary auxiliary views
  front-adjacent, 192
  side-adjacent, 192, 194
  top-adjacent, 192, 193
Printer-plotters
  electrostatic, 530
  impact dot-matrix, 530
  ink jet, 530
Prisms
  square, 100
  truncated rectangular, 100
  truncated triangular, 100
Projection lines, 161
Projection planes, 140
Prompts, 534
Proportion, 11
Prototype, 518
Protractor, 77
Punching, 265
Pyramids, 100

Q

Quadrant, 99
Quadrilaterals, 98

R

Radius, 99
Rectangle, 98
Reference dimension, 328
Regardless of feature size (RFS), 407
Related views, 143
Removed sections, 231
Representation, conventional, 234–238
Resistance welding, 264
Revision block, 432
Revolved sections, 230
Rhomboid, 98
Rhombus, 98
Right triangle, 98
Rise, 128
Rivets, 573t, 574t
Rolled threads, 292
Root, 287
Rotated features, 237
Roughness, 279
Roughness height, 279
Rounding off, 322
Rounds, 169, 330
Running and sliding fits, 587t
Runouts, 169
  circular, 415
  total, 415

S

Sand casting, 258
Sawing, 267
Scale of drawing, 316, 324. *See also* Dimensioning
Scalene triangle, 98
Scales
  architect, 72
  civil engineer, 70
  description of, 72
  fully divided, 69
  mechanical engineer, 71
  metric, 73
  open divided, 70
Screw thread(s)
  Acme, 287, 558t
  American National, 287, 553t
  angle of, 287
  axis of, 287
  buttress, 287
  classes of fit, 289
  crest, 287
  depth of, 287
  designation of, 297
  engagement, 302
  external, 286
  form, 287
  internal, 286
  ISO metric, 287, 555t
  knuckle, 287
  lead, 287
  major diameter, 286

metric, 287, 555t
minor diameter, 286
pitch, 287
pitch diameter, 287
production, 291
representation, 295
right- and left-hand, 290
root, 287
series, 287
single and multiple start, 291
60° sharp V, 287
specifying length of, 300
square, 287, 558t
Unified National, 287, 553t
uses, 286
Whitworth, 287
Screws, cap, 308, 564t, 566t
Screws, machine, 309, 567t, 569t
Screws, shoulder, 571t
Secondary auxiliary views, 198
Secondary machining operations, 344
Sectional views
description, 213
visualizing, 218
Section(s)
aligned, 227
auxiliary, 233
breaks, 238
broken-out, 234
conventional representation of, 235
"cut" surfaces, 214
cutting plane, 216
full, 50, 219
half, 50, 221
hidden lines in, 219
lining, 216, 235
offset, 224
pictorial, 50
removed, 231
revolved, 320
rotated features, 237
spokes in, 236
stepped, 224
visible lines, 218
Sector, 99
Segment, 99
Selective assembly, 384
Semicircle, 99
Series of thread, 287
Set screws, 310, 570t
Shaft basis system, 391
Shaft center sizes, 571t
Shape analysis, 336
Shape description, 139–172
Shape dimensions, 337
Shaping, 271
Sharp 60° V thread, 287
Sheet sizes, 432
Shell molding, 259
Shoulder screws, 571t
Shrinkage allowance, 258
Sintering, 266
SI (metric) system, 73
Size, 382
Size dimensions, 337
Sketching
angles, 10
blocking in, 9
border lines, 6
circles and arcs, 7
inside-out method, 11
isometric, 43
laying out, 11
line quality, 3
materials, 2
oblique, 35
outside-in method, 11
pictorial, 34
proportions, 11
single view, 1–20
straight lines, 4
techniques, 3
Skew lines, 470, 485
Slope, 474, 480
Software, graphics, 532
Solids, 100
Span, 128
Sphere, 101
Square, 98, 99
Square
across corners, 107
across flats, 107
construction, 106
diagonals, 11
thread, 287
Squeezing, 266
Stamping, 265
Start, single and multiple threads, 291
Steel, 253–254
Stepped section, 224
Stock allowance, 325
Straight-edge, parallel rule, 60
Straight pins, 580t
Strength, 256
Strike, 480
Structural engineering, 521
Studs, 307
Superfinishing, 278
Superfluous dimensions, 358
Surface finish control, 279
Surface finishing processes, 276
Surface(s)
adjacent, 157
configuration of, 160
cylindrical, 166
identification of, 157
inclined, 154
normal, 153
oblique, 155
plane, 153–156
texture of, 279–282
Symbol(s)
center line, 340
counterbored hole, 348
countersunk hole, 350
datum feature, 404
depth, 346
diameter, 146
geometric characteristic, 85, 403
material condition, 403
section lining, 216

Symbol(s) *(Cont.)*
spotfaced hole, 349
square shape, 353
surface texture, 280
tolerancing, 392
Symmetry, 221, 410

T

Tabular drawings, 359
Tap drilling, 294
Taper pins, 579t
Tapers, 352
Taps
bottoming, 293
plug, 293, 294
taper, 293, 294
Tap wrench, 293
Templates, 80, 127
Tetrahedron, 100
Thread(s)
chasing, 293
classes, 289
depth of, 287
engagement, 302
external, 292
form, 287
length, 300
internal, 293
ISO (metric) screw, 289
ISO unified (inch) screw, 288
relief, 356
series, 288
Threaded fasteners, 285–310. *See also* Screw threads
Title block, 430–431, 544
Tolerance(s)
accumulation of, 398
angularity, 413
circularity, 411
class, 290
cylindricity, 412
definition of, 382, 391
flatness, 411
of form, 410
geometric. *See* Tolerancing, geometric
grades, 290
limit systems and, 382
parallelism, 413
perpendicularity, 414
positional, 402–405
position symbols, 290
profile, 412
roundness, 411
runout, 415
of size, 382
specification of, 385
straightness, 410
symbols for, 392
for various processes, 276
zone, 391
Tolerancing, 382
geometric, 401–419
limit-dimension, 383
plus-and-minus, 384
positional, 402
Touchscreen device, 528
Transition fits
preferred hole basis, 601t
preferred shaft basis, 606t
Transition locational fits, 593t
Trapezium, 98
Trapezoid, 98
Triangles
adjustable, 78
construction of, 106
drafting, 62
equilateral, 98
isosceles, 98
right, 98
scalene, 98
Trimetric axonometric position, 42
T square, 60
Turning process, 271

U

Undercuts, dimensioning, 356
Unified inch screw thread series
coarse thread, 289
constant-pitch thread, 289
extra-fine thread, 289
fine thread, 289
standard, 288
Unified thread, 287–288

V

Vanish threads, 301
Vertex, 100, 128
Vibratory finishing, 276
Views
adjacent, 143
alternate positions for, 152
auxiliary, 188–211
choice of, 145
elevation, 144
multiple, 146–151
partial, 236
principal, 142–143
related, 143
sectional, 214–234
single, 146
third principal, construction of, 160–165
Visibility
of lines, 483
of planes, 486
Visible lines, 143

W

Washer face, 304
Washers
lock, 577t
plain, 575t
Waviness, 280
Welding, 264
Whitworth thread, 288
Wire gage standards, 582t
Working drawings, 315, 421
Workstations, 59, 522–524

Z

Zinc-base alloys, 256

INCH AND MILLIMETER EQUIVALENCY TABLE

Common fractions					Decimal			Millimeters
4ths	*8ths*	*16ths*	*32nds*	*64ths*	*To 4 places*	*To 3 places*	*To 2 places*	*To 4 places*
				1/64	.0156	.016	.02	0.3969
			1/32		.0312	.031	.03	0.7938
				3/64	.0469	.047	.05	1.1906
		1/16			.0625	.062	.06	1.5875
				5/64	.0781	.078	.08	1.9844
			3/32		.0938	.094	.09	2.3813
				7/64	.1094	.109	.11	2.7781
	1/8				.1250	.125	.12	3.1750
				9/64	.1406	.141	.14	3.5719
			5/32		.1562	.156	.16	3.9688
				11/64	.1719	.172	.17	4.3656
		3/16			.1875	.188	.19	4.7625
				13/64	.2031	.203	.20	5.1594
			7/32		.2188	.219	.22	5.5563
				15/64	.2344	.234	.23	5.9531
1/4					.2500	.250	.25	6.3500
				17/64	.2656	.266	.27	6.7469
			9/32		.2812	.281	.28	7.1438
				19/64	.2969	.297	.30	7.5406
		5/16			.3125	.312	.31	7.9375
				21/64	.3281	.328	.33	8.3344
			11/32		.3438	.344	.34	8.7313
				23/64	.3594	.359	.36	9.1281
	3/8				.3750	.375	.38	9.5250
				25/64	.3906	.391	.39	9.9219
			13/32		.4062	.406	.41	10.3188
				27/64	.4219	.422	.42	10.7156
		7/16			.4375	.438	.44	11.1125
				29/64	.4531	.453	.45	11.5094
			15/32		.4688	.469	.47	11.9063
				31/64	.4844	.484	.48	12.3031
					.5000	.500	.50	12.7000
				33/64	.5156	.516	.52	13.0969
			17/32		.5312	.531	.53	13.4938
				35/64	.5469	.547	.55	13.8906
		9/16			.5625	.562	.56	14.2875
				37/64	.5781	.578	.58	14.6844
			19/32		.5938	.594	.59	15.0813
				39/64	.6094	.609	.61	15.4781
	5/8				.6250	.625	.62	15.8750
				41/64	.6406	.641	.64	16.2719
			21/32		.6562	.656	.66	16.6688
				43/64	.6719	.672	.67	17.0656
		11/16			.6875	.688	.69	17.4625
				45/64	.7031	.703	.70	17.8594
			23/32		.7188	.719	.72	18.2563
				47/64	.7344	.734	.73	18.6531
3/4					.7500	.750	.75	19.0500
				49/64	.7656	.766	.77	19.4469
			25/32		.7812	.781	.78	19.8438
				51/64	.7969	.797	.80	20.2406
		13/16			.8125	.812	.81	20.6375
				53/64	.8281	.828	.83	21.0344
			27/32		.8438	.844	.84	21.4313
				55/64	.8594	.859	.86	21.8281
	7/8				.8750	.875	.88	22.2250
				57/64	.8906	.891	.89	22.6219
			29/32		.9062	.906	.91	23.0188
				59/64	.9219	.922	.92	23.4156
		15/16			.9375	.938	.94	23.8125
				61/64	.9531	.953	.95	24.2094
			31/32		.9688	.969	.97	24.6063
				63/64	.9844	.984	.98	25.0031
					1.0000	1.000	1.00	25.4000